IEE RADAR, SONAR, NAVIGATION AND AVIONICS SERIES 4

Series Editors: Professor E. D. R. Shearman
P. Bradsell

Advanced radar techniques and systems

Other volumes in this series:

Volume 1 **Optimised radar processors** A. Farina (Editor)
Volume 2 **Radar development to 1945** R. W. Burns (Editor)
Volume 3 **Weibull radar clutter** M. Sekine and Y. H. Mao

Advanced radar techniques and systems

Edited by
Gaspare Galati

Peter Peregrinus Ltd. on behalf of the Institution of Electrical Engineers

Published by: Peter Peregrinus Ltd., on behalf of the
Institution of Electrical Engineers, London, United Kingdom

Peter Peregrinus Ltd.,
The Institution of Electrical Engineers,
Michael Faraday House,
Six Hills Way, Stevenage,
Herts. SG1 2AY, United Kingdom

British Library Cataloguing in Publication Data

A CIP catalogue record for this book
is available from the British Library

ISBN 0 86341 172 X

Printed in England by Short Run Press Ltd., Exeter

Contents

Preface

The fascinating field of radar has a large number of facets; perhaps in no other electronic system are so many technologies and techniques encompassed (consider, for instance, antennas, microwave subsystems and components, high power transmitters, RF, IF and video circuits, digital processing of signals, data extraction, data processing and display, structures and thermal/mechanical parts).

In addition to the challenging research and development tasks in the above areas, the system design of a radar and the related analysis activities are interesting and creative activities. As a matter of fact, a set of user requirements can very often be satisfied by a number of different radar system options (e.g. the radiated signal, the antenna pattern and the scan strategy, leading to the management of the energy in time and space; the processing techniques for the detection, location, recognition and tracking of targets in the presence of natural and man-made interference etc.).

From this point of view, radar is nicely 'deregulated' compared with other electronic systems such as telecommunications and navigation whose standards are necessarily strict due to the need for a common signal format to be shared by many users.

Therefore, teaching activities in the radar field are greatly valuable not only for the wide applications of radar techniques in many civilian (mainly for remote sensing of land, sea and atmosphere and for the control of airspace and air traffic), military and, more recently, consumer environments, but also from the education/training point of view.

To mention Merrill I. Skolnik's preface to the second edition of his well known book *Introduction to radar systems*:

> 'Typical electrical engineering courses cover topics related to circuits, components, devices and techniques that might make up an electrical or electronic system; but seldom is the student exposed to the system itself. It is the system application (whether radar, communications, navigation, control, information processing or energy) that is the *raison d'être* for the electrical engineer.'

In the above-outlined frame, a non-profit cultural activity was jointly organised in 1988 by the Tor Vergata University of Rome and the Education and Training Department of the company Selenia (now Alenia): the two-week advanced course on Modern Radar Techniques and Systems (Accademia dei Lincei, Rome, 1988). The selected areas, in which many research and development activities are carried on, included system architectures, phenomenology, array antennas and signal processing.

The distinguished lecturers were Professor Michel H. Carpentier, Thomson-CSF, France, Mr Michael R. B. Dunsmore, RSRE (now DRE), Malvern, UK, Professor Yu-Hai Mao, Tsinghua University, Beijing, and Professor Ramadas P. Shenoy, Defence Research and Development Organisation of the Government of India.

I was pleased to serve as a scientific co-ordinator and, having considered the success of the initiative, to propose the lecturers to collect in a new book an updated and upgraded version of the content of the course with some additional chapters.

The resulting material is organised as follows:

Chapter 1, by myself and Ing Rodolfo Crescimbeni, with the kind co-operation of Professor Luigi Accardi, contains a brief overview of the basic results of detection and estimation theory as well as optimum filtering related to radar applications.

Chapters 2 and 3, by Professor Toshimitsu Musha and Professor Matsuo Sekine, describe the phenomenology of clutter, with special reference to Weibull clutter, autoregressive models and the related constant false alarm rate (CFAR) processors.

Chapter 4, by Professor Michel Carpentier, describes and analyses pulse compression techniques/technologies and their system aspects, with application examples.

Chapter 5, by Professor Michel Carpentier, deals with Doppler techniques (low, medium and high PRF) and their different applications for surveillance radar, multifunction radar and synthetic aperture radar.

Chapter 6, by Professor Yu-Hai Mao, treats and compares the processing techniques for clutter suppression: moving target indicators (MTI), moving target detectors (MTD), adaptive MTI and optimum MTI.

Chapter 7, by Professor Yu-Hai Mao, deals with the rejection of active interference (jamming) in the space domain (antennas) and in the frequency domain, as well as with the simultaneous rejection of clutter and jamming.

Chapter 8, by Professor Yu-Hai Mao, explains the main concepts related to radar target identification.

Chapter 9, by Professor Yu-Hai Mao, summarises the requirements, the architectural options and the trends of radar signal processing.

Chapter 10, by Professor Ramadas Shenoy, is a comprehensive treatment of phased arrays, including: Part 1, array theory, analysis and synthesis techniques; Part 2, the detailed treatment of array elements and mutual coupling effects; Part 3, active arrays—architectures, T/R modules, feed and control, design criteria and low-sidelobe techniques.

Chapter 11, by Michael Dunsmore, refers to bistatic radars, which are described and analysed in terms of operational requirements and advantages, design aspects and constraints, deployment options, current programmes and future trends.

Chapter 12, by myself and Ing Mario Abbati, describes the working principles, the main design options and the capabilities of space-based radar.

Chapter 13, by Professor Michel Carpentier, concludes the book with a glance at the evolution and trends of radar systems and technologies.

The book is primarily intended for radar engineers and researchers, as well as for students of advanced courses (e.g. PhD) on radar. The required background includes the basic knowledge of the nature of radar and its operation, as well as of the elementary aspects of probability and random processes. In addition, a knowledge of antennas and microwaves would be helpful in fully understanding Chapter 10.

Gaspare GALATI

Contributors

Professor G Galati
Department of Electronic Engineering
and Vito Volterra Centre
Tor Vergata University of Rome
via della Ricerca Scientifica
00133 Rome
Italy

Professor T Musha
Brain Functions Laboratory
Kawasaki-shi
Takatsu-ku
Sakado 3-2-1
KSP Bld. East 212
Japan

Professor M Sekine
Department of Applied Electronics
Tokyo Institute of Technology
Nagatsuta Midory-ku
Yokahama 227
Japan

Professor M Carpentier
23 bis Rue du Coteau
F92370 Chaville
France

Dr Y Mao
Specom Technologies Corporation
3673 Enochs Street
Santa Clara
CA 95051
USA

Professor R P Shenoy
115, 6th Main Road
Between 9th and 11th Cross
Malleswaram
Bangalore 560 003
India

Mr M R B Dunsmore
Defence Research Agency
St Andrews Road
Malvern
Worcestershire
WR14 3PS
United Kingdom

Acknowledgements

In addition to the Chapter Authors, whose enthusiastic and exceptionally qualified contributions have made this book possible, many people have contributed to the publication, and I wish to thank them all.

The support of the company Selenia and, in particular, of its former General Manager, Ing Raffaele Esposito, and of its former Director of the Education and Training Department, Ing Francesco Musto, to the advanced course on Modern Radar Techniques and Systems, from which parts of this book originated, is gratefully acknowledged.

Special thanks is addressed to the experts who reviewed parts of my contributions: Professor Luigi Accardi, Dr Sergio Barbarossa and Dr Albert Bridgewater.

Finally, on a more personal level, I wish to thank my wife Rossella and my daughter Claudia for their patience and understanding during the preparation of the manuscripts.

Gaspare GALATI

List of abbreviations

AAM	Air-to-air missile
AAR	Active aperture radar
ABM	Anti-ballistic missile
A/C	Aircraft
ACR	Area coverage rate
ACEC	Adaptive canceller for extended clutter
ADC	Analogue/digital convertor
AESA	Active electronically scanned array
AEW	Airborne early warning
AFS	Automatic frequency selection
AI	Airborne interception
AIC	Akaike information criterion
AM	Amplitude modulation
AMI	Active microwave instrument (ERS–1)
AR	Auto-regressive
ARM	Anti-radiation missile
ARMA	Autoregressive moving average
ARSR	Air route surveillance radar
ASR	Airport surveillance radar
ATC	Air traffic control
AWACS	Airborne warning and control system (USA)
BAC	Bistatic alerting and cueing system (USA)
BEARS	Bistatic experimental array receiving system (UK)
BFN	Beam forming network
BITE	Built-in test equipment
BMEW	Ballistic missiles early warning
BT	British Telecom
BTT	Bistatic technology transition
BYSON	'Son' of Blue Yeoman – a research radar at RSRE
CAA	Civil Aviation Authority (UK)
CAE	Computer aided engineering
CCIR	Comité Consultatif International des Radiocommunications
CCT	Computer compatible tape
CFAR	Constant false alarm rate
CIS	Commonwealth of Independent States (ex USSR)
CMS	Canonical minimum scattering
CNR	Clutter-to-noise (power) ratio
COHO	Coherent oscillator
CONUS	Continental United States
CPI	Coherent processing interval
CRLB	Cramer–Rao lower bound
CW	Continuous wave
DARPA	Defence Advanced Research Projects Agency (USA)
DBF	Digital beam forming

DF	Direction finding
DFB	Direct feedback
DFB	Distributed feedback
DFT	Discrete Fourier transform
DH–DFB	Double heterostructure direct feedback
DPC	Digital pulse compressor
DPCA	Displaced phase centre antenna
DRA	Defence Research Agency (UK)
ECCM	Electronic counter-countermeasures
ECM	Electronic countermeasures
EFIE	Electric field integral equation
EMC	Electromagnetic compatibility
EMCP	Electromagnetically coupled patch
EMF	Electromotive force
EMP	Electromagnetic pulse
EOS	Earth observation satellite
EOSAT	Electronic ocean reconnaissance satellite
ERP	Effective radiated power
ERR	En-route radar
ERS	Earth resources satellite
ESM	Electronic support measures
FAR	False alarm rate
FET	Field effect transistor
FFT	Fast Fourier transform
FIR	Finite impulse response
FO	Fibre optic
FTF	Fast transversal filter
GaAs	Gallium Arsenide
GLONASS	Global Navigation Satellite System (ex-USSR)
GPS	Global Positioning System (USA)
HBT	Heterojunction bipolar transistor
HDDR	High-density digital recorder
HEMFET	High electron mobility FET
HEMT	High electron mobility transistor
HF	High frequency
HOJ	Home on jammer
HRF	High repetition frequency
IC	Integrated circuit
IF	Intermediate frequency
IFF	Identification friend or foe
IIR	Infinite impulse response
IOC	Integrated optoelectronic circuits
I, Q	In-phase, quadrature
I&Q	In-phase and quadrature (ADC)
ISAR	Inverse SAR
JSTARS	Joint surveillance target attack radar system (USA)
KREMS	Kiernan re-entry measurement site (USA)
LEO	Low earth orbit
LEPI	Leningrad Electrophysics Institute (CIS)
LNA	Low noise amplifier
LO	Local oscillator
LPAR	Large phased array radar
LPI	Low probability of intercept
LRF	Low repetition frequency

LSB	Least significant bit
LSE	Least squares estimation
LSI	Large scale integration
MA	Moving average
MAP	Maximum *a posteriori* probability
MBE	Molecular beam epitaxy
MDV	Minimum detectable velocity
MEP	Minimum error probability
MERA	Molecular electronics for radar applications
MESFET	Metal electrodes semiconductor FET
MFIE	Magnetic field integral equation
MIC	Microwave integrated circuit
ML	Maximum likelihood
MLE	Maximum likelihood estimate
MMIC	Monolithic microwave integrated circuit
MMW	Millimeter wave(s)
MOCVD	Metal organic chemical vapour deposit
MOM	Method of moments
MRF	Medium repetition frequency
MTBF	Mean time between failures
MTD	Moving target detector
MRF	Medium repetition frequency
MSI	Medium scale integration
MTI	Moving target indicator
NAVSPASUR	Naval Space Surveillance System (USA)
N–P	Neyman–Pearson
NRL	Naval Research Laboratory (USA)
OTH	Over the horizon
PAA	Phased array antenna
PAE	Power added efficiency
PAL	Phase alternate line
PCA	Printed circuit antenna
PCAA	Printed circuit antenna array
PCF	Pulse compression filter
PD	Probability of detection
PEC	Perfect electric conductor
PFA	Probability of false alarm
PLL	Phase-lock loop
PM	Phase modulation
PMTC	Pacific Missile Test Center (USA)
PN	Pseudo-noise
PPI	Plan position indicator
PRF	Pulse repetition frequency
PRI	Pulse repetition interval
PSK	Phase shift keying
RABIES	Range Ambiguous Bistatic Experimental System (UK)
RAF	Royal Air Force (UK)
RAM	Radar absorbent material
RAM	Random access memory
R.A.M.	Reliability, availability, maintainability
RAR	Real aperture radar
RASSR	Reliable advanced solid state radar
RCS	Radar cross-section
RCVR	Receiver

REB	Re-entry body
REC	Radio Electronic Combat (CIS)
RF	Radio frequency
RIN	Relative intensity noise
RMS	Root mean square
ROC	Receiver operating characteristics
ROM	Read-only memory
RORSAT	Radar Ocean Reconnaissance Satellite
RPM	Revolutions per minute
RSRE	Royal Signal and Radar Establishment (UK)
RWR	Radar warning receiver
SAM	Surface-to-air missile
SAR	Synthetic aperture radar
SAW	Surface acoustic wave
SBR	Space based radar
SIR	Shuttle imaging radar
SIR	Signal-to-interference (power) ratio
SLAR	Side-looking airborne radar
SLC	Sidelobe canceller/cancellation
SLL	Sidelobe level
S/N	Signal-to-noise (power) ratio, also SNR
SNR	Signal-to-noise (power) ratio
SOJ	Systems on silicon
SOS	Stand-off jammer
SSB	Single sideband
SSI	Small scale integration
SSJ	Self-scanning jammer
SSLC	Single sidelobe canceller
SSPA	Solid-state phased array
SSR	Secondary surveillance radar
SS&CS	Site synchronisation and communication system
STALO	Stable local oscillator
STC	Sensitivity time control
TAR	Terminal area radar
TBIRD	Tactical Bistatic Radar Demonstrator (USA)
T/C	Target-to-clutter ratio
TE	Transversal electric
TEGFET	Two-dimensional electron gas FET
TM	Transversal magnetic
TOJ	Track on jamming
T–R	Transmit–receive
TV	Television
TWT	Travelling wave tube
UHF	Ultra high frequency
UKADGE	UK Air Defence Ground Environment
URR	Ultra-reliable radar
UWB	Ultra-wideband (radar)
VCO	Voltage controlled oscillator
VGA	Variable gain amplifier
VHF	Very high frequency
VHSIC	Very high-speed integrated circuits
VLSI	Very large scale integration
VPA	Variable power amplifier
VSWR	Voltage standing wave ratio
WAP	Wavefront array processor

Chapter 1

Basic concepts on detection, estimation and optimum filtering

Prof. Ing. G. Galati and Ing. R. Crescimbeni

Part 1: Decision and estimation theory [1, 2]

In this Chapter we will introduce and develop the basic concepts of decision and estimation, both of fundamental importance in all analyses and activities related to random phenomena. For example, decision theory is used to establish whether a '1,' or a '0' was sent on a digital communication channel, but also to detect a radar target, by examining the signal received by the radar itself, where not only thermal noise, but also unwanted echoes such as the 'clutter' are present and corrupt the target echo.

1.1 Fundamentals of decision theory [1]

All the problems concerning decision and estimation have the structure described in Fig. 1.1, where the following steps can be recognised:

(*a*) An event occurs (in telecommunications language: 'a message is generated')
(*b*) A signal related to the event reaches the observer
(*c*) The signal itself, corrupted by noise, is observed
(*d*) The receiver must 'guess' which event happened.

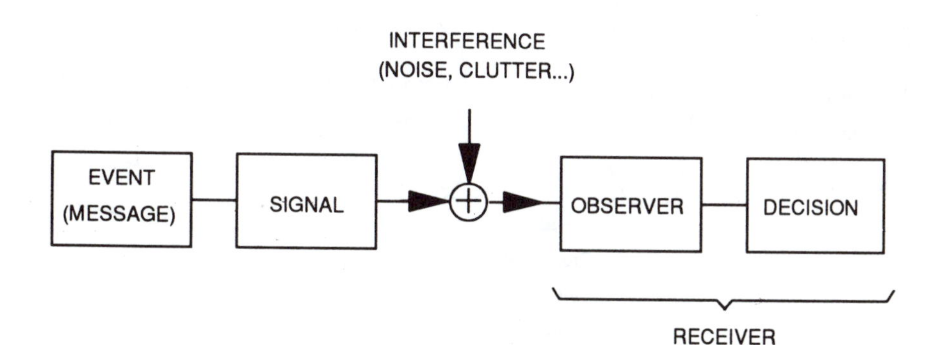

Figure 1.1 *Structure of a decision process*

The decision is generally reached taking account of all the available information: for example, the type of expected signal (waveform), the characteristics of the communication channel and so on.

Before describing the most widely used decision and estimation methods, a brief introduction of the terminology used is necessary.

First, with reference to Fig. 1.2, we define $\boldsymbol{M}$ as the set of the possible events (the events space). $\boldsymbol{M}$ can be a discrete points set, a segment over a line, a set of points over an N-dimensional space, an abstract set etc. A formal important classification of our problems is thus evident:

(i) If $\boldsymbol{M}$ has a finite number K of elements (finite cardinality) we have a *decision* problem: this is the case of binary digital transmission, in which two events ($K=2$) are possible. Formally:

$$\boldsymbol{M}:\{m_k,\ k=0,\ 1,\ \ldots,\ K-1\} \tag{1.1}$$

Such a problem is referred to as a statistical hypothesis testing.

(ii) If the number of elements of $\boldsymbol{M}$ is infinite (infinite cardinality) we speak of *estimation* problems. This is the case in which, for example, we want to reconstruct a continuous-time waveform $m(t)$ from observations corrupted by noise.

Second, we define $\boldsymbol{S}$ as the signals space, that is the space of all the possible waveforms that are sent to the observer. Generally, there is a one-to-one correspondence between the elements of $\boldsymbol{M}$ and the elements of $\boldsymbol{S}$. The set $\boldsymbol{S}$ is introduced in order to functionally separate the information source from the transmission of the information itself. Formally:

$$S\colon \{S_k,\ k=0,\ \ldots,\ K-1\} \tag{1.2}$$

Third, we define $\boldsymbol{Z}$ as the observations space, that is the space of all the possible received signals: it is evident that $\boldsymbol{Z}$ is infinite-dimensionally (while $\boldsymbol{M}$ and $\boldsymbol{S}$ are generally finite-dimensionally where *decision* problems are concerned). This is

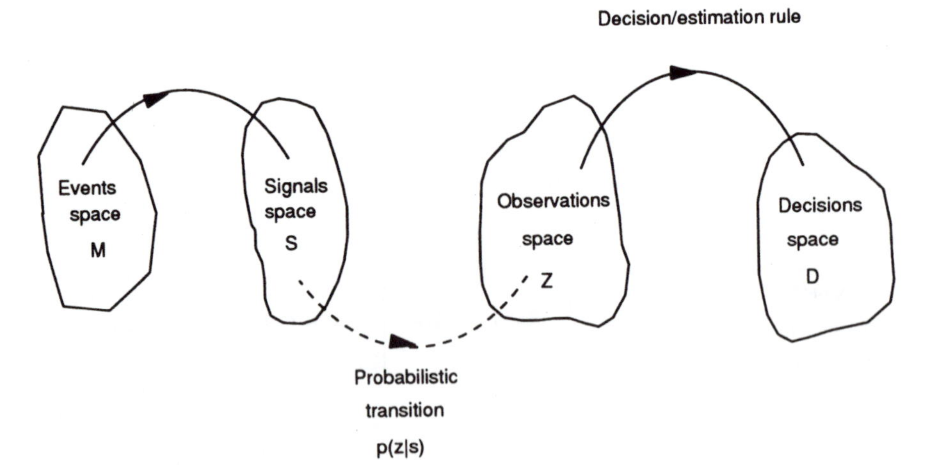

Figure 1.2 *Four steps in a decision process*

due to the continuous nature of the noise that is received along with the signal s_k that conveys the information related to the message m_k. It is extremely important to note that there is a statistical relationship between the elements z of $\boldsymbol{Z}$ and the elements s_k of $\boldsymbol{S}$, represented by the conditional probability distribution functions (PDF) $p(z|s_k)$.

Fourth, we define $\boldsymbol{D}$ as the decision space, that is the space of all the possible results of the decision process. It is clear that $\boldsymbol{D}$ and $\boldsymbol{M}$ have the same number of elements, because the aim of the decision process is to recover the information originated in $\boldsymbol{M}$ and conveyed through $\boldsymbol{S}$.

Finally, there is the decision rule, connecting the observation space to the decision space. Such rules are deterministic (for any observation one and only one decision can be taken). Most of the following paragraphs will deal with the most used rules (criteria) that allow us to associate a decision with every point of $\boldsymbol{Z}$.

In the following, we will only discuss the simpler case of binary decision over single observation; the extension to the case of multiple observation is straightforward, as the case of multiple decision: the reason for this choice is to limit the complexity of the mathematical derivations involved in the decision criteria; further, in signal detection applications generally the binary case only is of interest, while the multiple case is restricted to some applications of communications (multiple level modulations).

In this case of binary decision, $\boldsymbol{M}$ consists of two messages: $\boldsymbol{M}=\{m_0, m_1\}$, and $\boldsymbol{D}$ also has only two elements: $\boldsymbol{D}=\{d_0, d_1\}$; so, m_0 is associated with d_0, and m_1 with d_1. But how should d_0 and d_1 be associated with the observations space? First, the observation space is partitioned† into two regions, $\boldsymbol{Z}_0$ and $\boldsymbol{Z}_1$.

Secondly, a one-to-one correspondence between $\boldsymbol{Z}_i$, $i=0, 1$, and d_i, $i=0, 1$, is established by means of the given decision rules. Therefore, the decision $d(z)$ is

$$d(z)=\begin{cases} d_0 & \text{if } z \in Z_0 \\ d_1 & \text{if } z \in Z_1 \end{cases} \tag{1.3}$$

In Fig. 1.3 a representation of the binary decision procedure is sketched.

1.2 Error probabilities and decision criteria

A binary decision rule is evaluated in terms of the error probabilities pertaining. The probability of a first-kind error, α, is defined as

$$\alpha=\text{Prob}(d=d_1|s=s_0)=P(d_1|m_0) \tag{1.4}$$

i.e. α is the probability of deciding for the event m_1 when the event m_0 actually happened (and hence the signal s_0 was generated).

In a similar way, the probability of a second-kind error, β, is defined as:

$$\beta=\text{Prob}(d=d_0|s=s_1)=P(d_0|m_1) \tag{1.5}$$

† We recall that a partition of a set $\boldsymbol{Z}$ into n subsets $\boldsymbol{Z}_0, \boldsymbol{Z}_1, \ldots \boldsymbol{Z}_{n-1}$ is defined as follows:

$$\begin{cases} Z_0 \cap Z_1 \cap \ldots Z_{n-1}=\emptyset \\ Z_0 \cup Z_1 \cup \ldots Z_{n-1}=Z \end{cases}$$

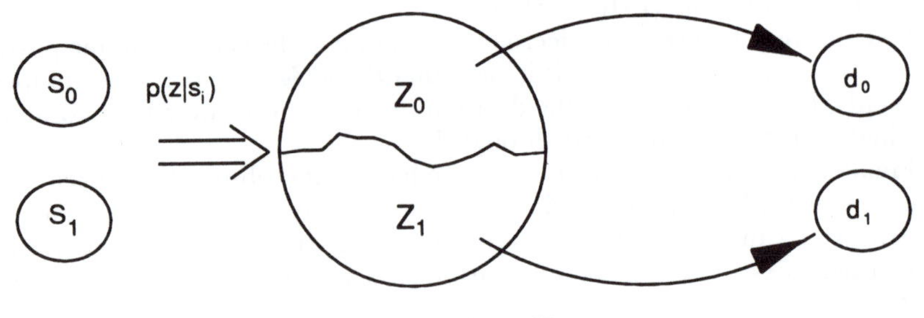

Figure 1.3 *Representation of a binary decision procedure*

In classical mathematical statistics [2] the symbols H_0 and H_1 are used in place of m_0 and m_1; the m_0 event is referred to as the 'null hypothesis' H_0, and corresponds to the absence of statistically meaningful differences between the 'sample', i.e. the observation and the 'population'.

As explained in Section 1.1, the observations space $\mathbf{Z}$ is partitioned in two regions $\mathbf{Z}_0$ and $\mathbf{Z}_1$, and the following decision rule depending on the observation $\boldsymbol{z}$ (vector observations, i.e. observations consisting of sequences of N real or complex numbers) is applied:

$$\text{if } \boldsymbol{z} \in \mathbf{Z}_0,\ d_0 \text{ is decided}$$

$$\text{if } \boldsymbol{z} \in \mathbf{Z}_1,\ d_1 \text{ is decided}$$

Using the conditional PDFs $p(\boldsymbol{z}|m_0)$ and $p(\boldsymbol{z}|m_1)$, the error probabilities, are given by the following relationships:

$$P(d_1|m_0) = \alpha = \int_{Z_1} p(\boldsymbol{z}|m_0) d\boldsymbol{z} \tag{1.6}$$

$$P(d_0|m_1) = \beta = \int_{Z_0} p(\boldsymbol{z}|m_1) d\boldsymbol{z} \tag{1.7}$$

and the correct decision probabilities are expressed as

$$P(d_0|m_0) = \int_{Z_0} p(\boldsymbol{z}|m_0) d\boldsymbol{z} \tag{1.8}$$

$$P(d_1|m_1) = \int_{Z_1} p(\boldsymbol{z}|m_1) d\boldsymbol{z} \tag{1.9}$$

Table 1.1 *Probabilities associated with a binary decision*

		Event	
		m_0	m_1
Decision	d_0	Correct decision Probability: $1-\alpha$	Error of the second kind Probability: β
	d_1	Error of the first kind Probability: α	Correct decision Probability: $1-\beta$

Recalling that the subsets Z_0 and Z_1 are a partition of the observation space Z, from eqns. 1.6 and 1.8 it results:

$$P(d_1|m_0)+P(d_0|m_0)=1 \tag{1.10}$$

and similarly from eqns. 1.7 and 1.9

$$P(d_0|m_1)+P(d_1|m_1)=1 \tag{1.11}$$

The four probabilities associated with a binary decision procedure are shown in Table 1.1.

The probabilities $P_0=P(m_0)$ and $P_1=P(m_1)$ are the '*a priori* probabilities' of the events m_0 and m_i, respectively. Depending on whether the *a priori* probabilities are known or unknown, one speaks of a 'Bayesian' or 'non-Bayesian' decision approach, respectively. Bayesian approaches, described later, require additional information and therefore can achieve better performance with respect to non-Bayesian ones. The problem is in the *a priori* probabilities themselves, because we hardly ever have a good idea of their value (with the relevant exception of digital communications where the possible transmitted symbols are considered equally probable).

In the case of detection systems, such as radar, α is called 'false alarm probability', P_{FA}, and β is called 'probability of a miss'; $1-\beta$ is the 'detection probability', P_D. In classical mathematical statistics, α is called the 'level of significance' and $1-\beta$ is the 'power of the test'.

In the following Sections we shall briefly discuss the most popular decision rules, trying to reduce as much as possible the complete mathematical derivations; an attempt will be made to define the field of applications of each rule, along with its advantages and drawbacks.

The following criteria will be discussed:

- Maximum likelihood
- Neyman–Pearson
- Minimum error probability
- Bayes (minimum risk)

All these criteria lead to a comparison between a function of the observation, namely the likelihood ratio $\Lambda(\boldsymbol{z})$, that we are going to define, with a suitable threshold, whose value is a characteristic of each of them.

1.2.1 *Maximum likelihood (ML) decision criterion*

According to this criterion, the decision corresponding to the event (message) that most likely caused the observation is selected.

It is only required that the conditional PDF of the observations $p(\mathbf{z}|m_i)$, given every possible message, be known.

The decision rule is the following:

$$d(\mathbf{z}) = \begin{cases} d_0 & \text{if} \quad p(\mathbf{z}|m_0) > p(\mathbf{z}|m_1) \\ d_1 & \text{if} \quad p(\mathbf{z}|m_1) < p(\mathbf{z}|m_0) \end{cases} \tag{1.12}$$

Let us define the likelihood ratio as

$$\Lambda(\mathbf{z}) = \frac{p(\mathbf{z}|m_1)}{p(\mathbf{z}|m_0)} \tag{1.13}$$

so that the decision rule can be written in a more compact way as

$$\Lambda(\mathbf{z}) \underset{d_0}{\overset{d_1}{\gtrless}} 1 \tag{1.14}$$

and the decision space $\mathbf{Z}$ is partitioned into the two regions:

$$\begin{cases} Z_0 = \{\mathbf{z}\text{: } \Lambda(\mathbf{z}) < 1\} \\ Z_1 = \{\mathbf{z}\text{: } \Lambda(\mathbf{z}) > 1\} \end{cases} \tag{1.15}$$

The ML criterion is a very simple decision rule, but often, owing to its simplicity, it does not represent practical problems very accurately; conversely, the ML method is a very powerful tool for *estimation* problems.

1.2.1.1 *Example of application of the ML criterion*

Let us suppose we have two observations (signals) z_0 and z_1 that are related to the messages m_0 and m_1 by the following PDFs:

$$p(z|m_0) = \frac{1}{\sqrt{2\pi}} e^{-z^2/2} \tag{1.16}$$

$$p(z|m_1) = \frac{1}{\sigma\sqrt{2\pi}} e^{-(z-\mu)^2/2\sigma^2} \tag{1.17}$$

with $\sigma \geqslant 1$.

The likelihood ratio is

$$\Lambda(z) = \frac{1}{\sigma} e^{(z^2/2)-((z^2-2\mu z+\mu^2)/2\sigma^2)}$$

and the ML test is

$$e^{((\sigma^2-1)z^2+2\mu z-\mu^2)/2\sigma^2} \underset{d_0}{\overset{d_1}{\gtrless}} \sigma$$

i.e.

$$(\sigma^2-1)z^2+2\mu z-(\mu^2+2\sigma^2 \cdot \ln \sigma) \underset{d_0}{\overset{d_1}{\gtrless}} 0 \tag{1.18}$$

Therefore, d_1 is decided when

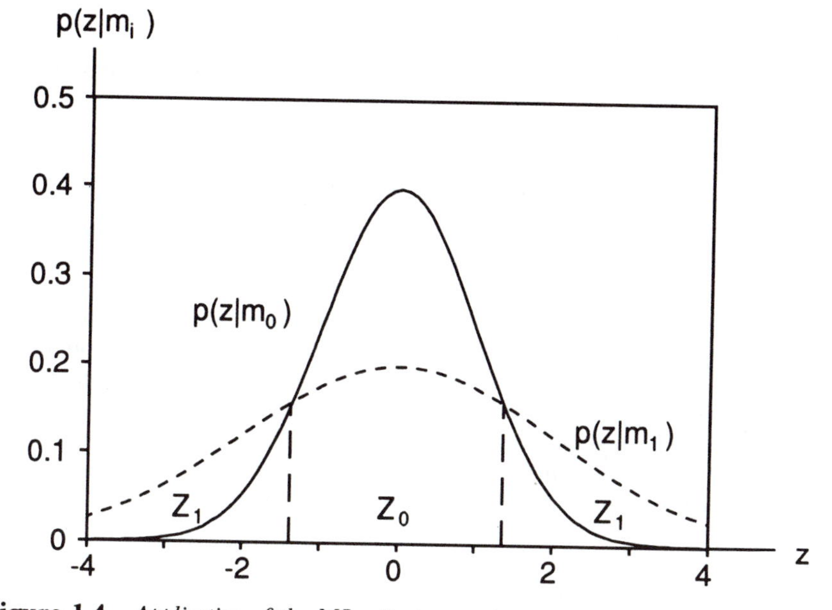

Figure 1.4 *Application of the ML criterion: variance test*

$$z < -\frac{\mu}{\sigma^2-1} - \sqrt{\frac{\mu^2}{(\sigma^2-1)^2} + \frac{\mu^2 + 2\sigma^2 \ln \sigma}{\sigma^2-1}}$$

and when

$$z > -\frac{\mu}{\sigma^2-1} + \sqrt{\frac{\mu^2}{(\sigma^2-1)^2} + \frac{\mu^2 + 2\sigma^2 \ln \sigma}{\sigma^2-1}}$$

When $\sigma^2 = 1$ and $\mu > 0$ the test (eqn. 1.18) becomes

$$z \underset{d_0}{\overset{d_1}{\gtrless}} \frac{\mu}{2}$$

while when $\mu = 0$, $\sigma^2 > 1$ the problem is referred to as *the variance test*, as the two PDFs differ in the variance only; this test is

$$(\sigma^2 - 1)z^2 - 2\sigma^2 \ln \sigma \underset{d_0}{\overset{d_1}{\gtrless}} 0$$

i.e.

$$|z| \underset{d_0}{\overset{d_1}{\gtrless}} \sqrt{\frac{2\sigma^2 \ln \sigma}{\sigma^2 - 1}} \tag{1.19}$$

The regions of decision of the variance test are shown in Fig. 1.4 for the case $\sigma = 2$, in which

$$|z| \underset{d_0}{\overset{d_1}{\gtrless}} 1{\cdot}36$$

1.2.2 *Neyman–Pearson (NP) criterion*

In order to introduce this rule, let us recall the definition of error probabilities α and β, widely used in radar techniques:

$$\alpha = P\{d_1|m_0\} = P_{FA}$$

$$\beta = P\{d_0|m_1\} = 1 - P_D$$

where in detection applications the event m_1 corresponds to the presence of an object to be detected (e.g. a radar target), and the event m_0 indicates the absence of targets.

In order to have a 'good' decision, it is necessary that the 'false alarm probability' $P(d_1|m_0)$ is as low as possible, whilst the 'detection probability' $P(d_1|m_1)$ should be as high as possible. However, this is not possible: in fact, when $P\{d_1|m_1\}$ grows, generally $P\{d_1|m_0\}$ increases too.

A convenient strategy is to fix one of the two probabilities at a given value, while the other is optimised (constrained by the other value). This is precisely the Neyman–Pearson criterion, that can be expressed more formally as follows:

Fix $P\{d_1|m_0\}$ *at a given value* α_0, *and then maximise* $P\{d_1|m_1\}$

i.e. in signal detection terms:

Fix P_{FA} *and then maximise* P_D

This rule corresponds to choosing, among all regions $\mathbf{Z}_1$ where $P\{d_1|m_0\} = \alpha_0$, the one that maximises $P\{d_1|m_1\}$.

The NP criterion is a problem of constrained optimisation: using the Lagrange multiplier λ, and considering the functional

$$\Gamma = P\{d_1|m_1\} - \lambda[P\{d_1|m_0\} - \alpha_0] \tag{1.20}$$

the problem can be expressed as follows:

$$\max_{Z_1} \Gamma \tag{1.21}$$

where the optimum value of $P(d_1|m_1)$ is a function of λ, λ being itself a function of the 'desired' value α_0. Recalling that

$$P\{d_1|m_0\} = \int_{Z_1} p(\mathbf{z}|m_0)d\mathbf{z} \tag{1.22}$$

$$P\{d_1|m_1\} = \int_{Z_1} p(\mathbf{z}|m_1)d\mathbf{z} \tag{1.23}$$

we obtain the following expression for the functional Γ:

$$\begin{aligned}\Gamma &= \int_{Z_1} p(\mathbf{z}|m_1)d\mathbf{z} - \lambda\left[\int_{Z_1} p(\mathbf{z}|m_0)d\mathbf{z} - \alpha_0\right] \\ &= \int_{Z_1} [p(\mathbf{z}|m_1) - \lambda p(\mathbf{z}|m_0)]d\mathbf{z} + \lambda\alpha_0\end{aligned} \tag{1.24}$$

It is readily understood that the above functional is maximised if we choose for $\mathbf{Z}_1$ the region where the integrand is positive; in formulae:

$$\mathbf{Z}_1 = \{\mathbf{z}\colon [p(\mathbf{z}|m_1) - \lambda p(\mathbf{z}|m_0)] > 0\}$$
$$\mathbf{Z}_0 = \{\mathbf{z}\colon [p(\mathbf{z}|m_1) - \lambda p(\mathbf{z}|m_0)] < 0\} \tag{1.25}$$

In terms of likelihood ratio, we can write:

$$\Lambda(\mathbf{z}) \underset{d_0}{\overset{d_1}{\gtrless}} \lambda \tag{1.26}$$

(If $\lambda = 1$ we again obtain the ML criterion.)

The sets $\mathbf{Z}_0$ and $\mathbf{Z}_1$ depend on λ, which in turn depends on the desired value α_0 of $P\{d_1|m_0\}$.

The procedure for exploiting the Neyman–Pearson criterion is the following: one computes $P\{d_1|m_0\}$ as a function of λ and then finds the value (or values) of λ which makes $P\{d_1|m_0\}$ equal to α_0 (see also the examples below).

The choice of α_0 seems somehow arbitrary. Actually, there exist curves called receiver operating characteristics (ROC) that plot $P\{d_1|m_1\}$ versus $P\{d_1|m_0\}$ as a function of one or more parameters; such curves have some interesting properties (see for example Reference 1); in particular, they are always concave downwards. From these curves it is possible to determine an 'optimum' value (the knee) for $P\{d_1|m_0\}$, such that a small decrease of its value causes a fast decrease of $P\{d_1|m_1\}$, whilst any increase has very small effect (the saturation zone, where the rate of change is nearly zero).

In radar applications the choice of α_0 is based upon operational considerations, i.e. the need to keep the false alarm rate within acceptable bounds (e.g. a few false alarms per second). A typical value of α_0 for radar is 10^{-6}.

1.2.2.1 *First example of application of the Neyman–Pearson criterion*

Let us suppose that the statistics of the received signal z are related to the events m_0 and m_1 by the following PDFs:

$$p(z|m_0) = \frac{1}{\sqrt{2\pi}}\, e^{-z^2/2} \tag{1.27}$$

$$p(z|m_1) = \frac{1}{\sqrt{2\pi}}\, e^{-(z-3)^2/2} \tag{1.28}$$

Let us fix the false alarm probability to be: $P_{FA} = P\{d_1|m_0\} = 0.25$. The likelihood ratio is

$$\Lambda(z) = \frac{e^{-(z-3)^2/2}}{e^{-z^2/2}} \tag{1.29}$$

from which we get the following decision rule:

$$e^{(6z-9)/2} \underset{d_0}{\overset{d_1}{\gtrless}} \lambda \Rightarrow z \underset{d_0}{\overset{d_1}{\gtrless}} \frac{\ln\lambda}{3} + 1.5 \tag{1.30}$$

From the desired false alarm probability

$$P_{FA} = P\{d_1|m_0\} = \int_{(\ln\lambda/3)+1.5}^{\infty} p(z|m_0)\,dz = 0.25 \tag{1.31}$$

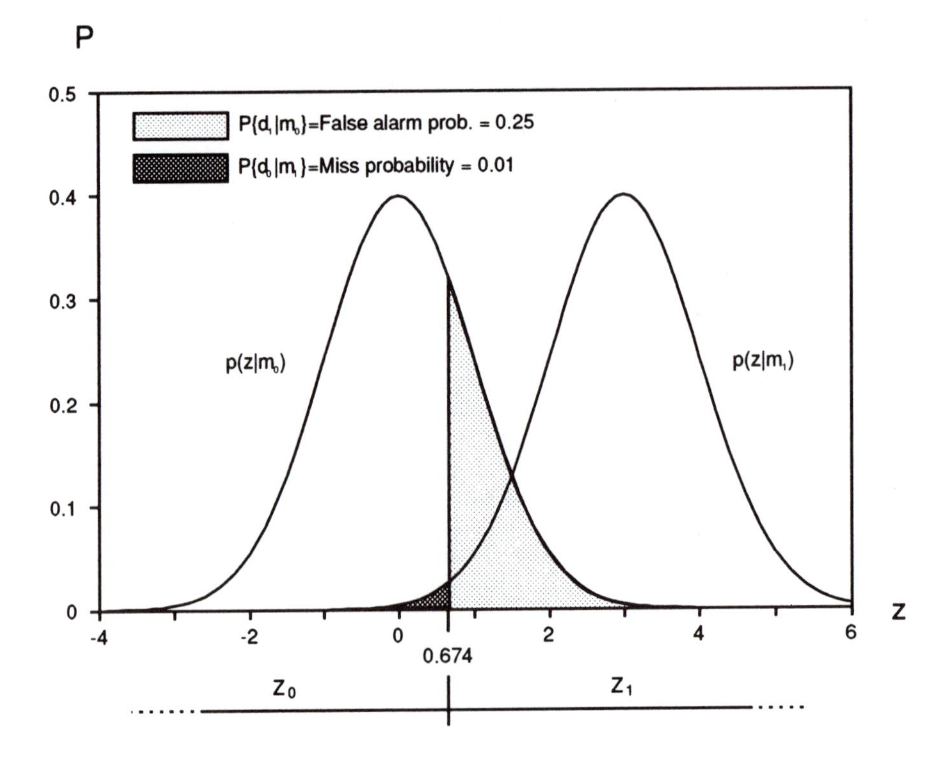

Figure 1.5 *Decision regions in Neyman–Pearson criterion*

the 'threshold' value t_0 is obtained:

$$t_0 = \frac{\ln \lambda}{3} + 1.5 = 0.674 \tag{1.32}$$

The 'probability of a miss' is given by

$$1 - P_D = P\{d_0|m_1\} = \int_{-\infty}^{t_0} p(z|m_1)dz = 0.01 \tag{1.33}$$

It is important to note that the explicit computation of the parameter λ is not necessary.

Fig. 1.5 shows the two regions $\mathbf{Z}_0$ and $\mathbf{Z}_1$ separated by the threshold value.

Fig. 1.6 shows some ROCs for different expected values of z in the hypothesis $\boldsymbol{H}_1$ (in our example we have an expected value, μ, equal to 3), obtained in the following way: the threshold value is calculated for different values of λ, then the corresponding values of P_{FA} and P_D are obtained by the relationships

$$P_{FA} = \int_{(\ln \lambda)/\mu + (\mu/2)}^{\infty} \frac{1}{\sqrt{2\pi}} e^{-z^2/2} \, dz \tag{1.34}$$

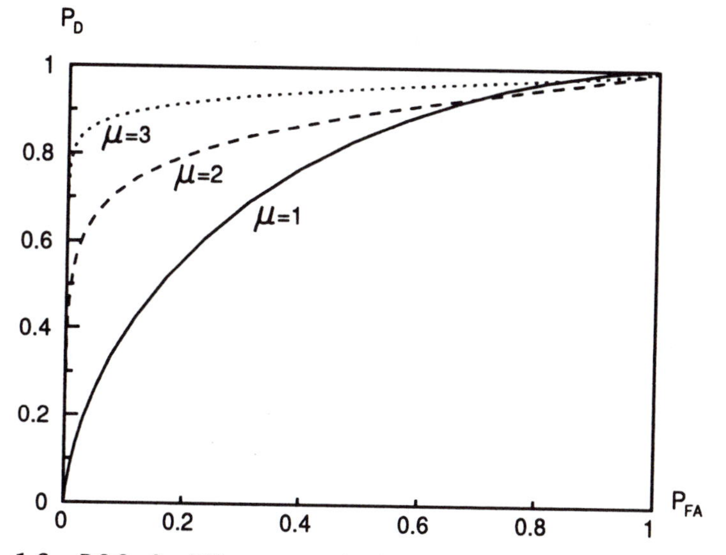

Figure 1.6 *ROCs for different expected values of z*

$$P_D = \int_{(\ln\lambda)/\mu + (\mu/2)}^{\infty} \frac{1}{\sqrt{2\pi}} e^{-(z-\mu)^2/2}\, dz \tag{1.35}$$

where μ is the aforementioned expected value. The function P_D versus P_{FA} is then plotted and the procedure can be repeated for different values of μ.

1.2.2.2 *Second example of application of the Neyman–Pearson criterion*
Let us suppose that z is an envelope-detected radar signal ($z \geqslant 0$) originated by a steady signal (the echo of a 'fixed' target in the radar case) with amplitude 'a' corrupted by additive Gaussian noise with unit variance. For the hypothesis H_0 (absent signal) the related PDF is that of Rayleigh:

$$p(z|m_0) = ze^{\{-z^2/2\}}.\, U(z) \tag{1.36}$$

and in the hypothesis H_1 (signal present) the related PDF is that of Rice (see Appendix 4 for the derivations):

$$p(z|m_1) = ze^{-(z^2+a^2)/2} I_0(za).\, U(z) \tag{1.37}$$

where $U(.)$ is the unit step function. The likelihood ratio is

$$\Lambda(z) = e^{\{-a^2/2\}} I_0(za) \tag{1.38}$$

where $I_0(.)$ is the zero-order modified Bessel function of the first kind; the decision rule is therefore

$$I_0(za) \underset{d_0}{\overset{d_1}{\gtrless}} \lambda e^{(a^2/2)} \tag{1.39}$$

As $I_0(.)$ is a monotonic function, this rule is equivalent to comparing the observation z with a threshold t_0. The false alarm probability is easily evaluated:

$$\alpha = P_{FA} = \int_{t_0}^{\infty} p(z|m_0)dz = e^{-t_0^2/2} \tag{1.40}$$

The threshold value for $\alpha = 0.25$ is

$$t_0 = \sqrt{-2 \ln \alpha} = 1.665 \tag{1.41}$$

The detection probability is known as the Marcum function $Q(a, t_0)$ (see, for example, References 3 and 4).

In Fig. 1.7 a plot of the two PDFs is reported; for the Rice one, the assumed value for the parameter a is 2.5 ($a^2/2$ = signal-to-noise power ratio, SNR). For $a = 2.5$ and $P_{FA} = 0.25$, the detection probability is 0.86.

1.2.2.3 *Third example of application of the Neyman–Pearson criterion*
Let us assume that the received signal z has the following statistics:

$$p(z|m_0) = ze^{\{-z^2/2\}}U(z) \tag{1.42}$$

$$p(z|m_1) = \frac{z}{\gamma^2} e^{\{-z^2/2\gamma^2\}}U(z) \tag{1.43}$$

where $\gamma^2 = 1 + SNR$: this is a 'noise-on-noise' problem with scalar observation.

The likelihood ratio is:

$$\Lambda(z) = \frac{1}{\gamma^2} e^{-z^2(\gamma^{-2}-1)/2} \tag{1.44}$$

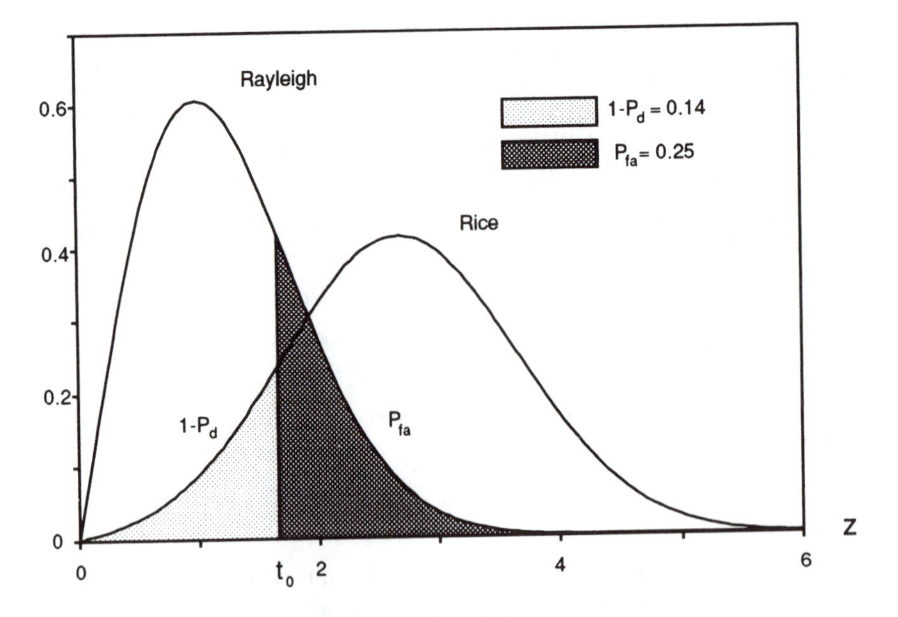

Figure 1.7 *Rayleigh and Rice PDFs (a = 2.5)*

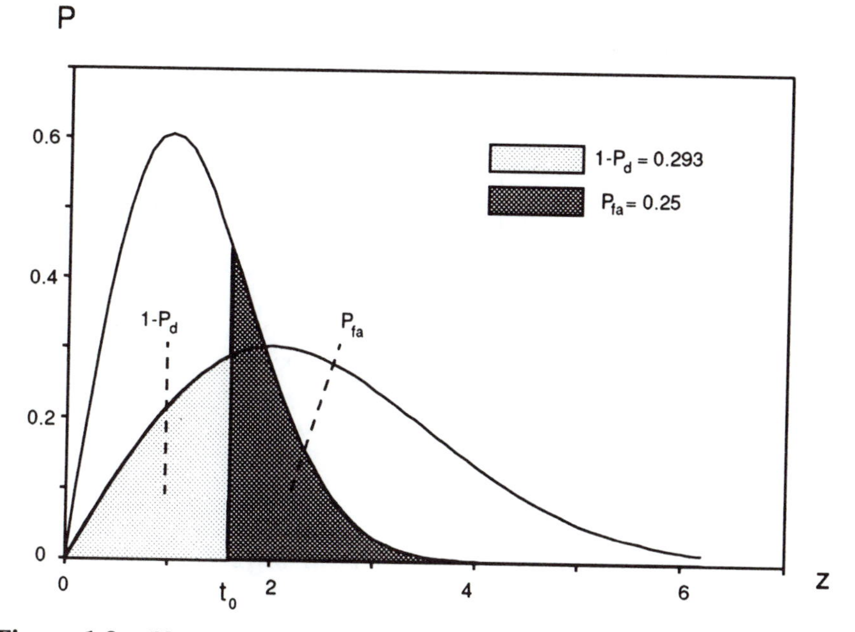

Figure 1.8 *'Noise on noise' problem, scalar observation*

and the threshold can be calculated as in the previous example; fixing $P_{FA}=0.25$, the threshold t_0 is 1.665. With this value, it is possible to calculate the detection probability; the general expression is $P_D=P_{FA}^{1/\gamma^2}$.

If we assume that the parameter γ^2 is equal to four, we obtain a complementary probability of detection $1-P_D=0.293$ and a probability of detection $P_D=0.707$ (see Fig. 1.8).

Assuming now the same SNR as the previous example, that is $\gamma^2=2.5^2/2=3.125$, and maintaining the same $P_{FA}=0.25$, the complementary probability of detection $1-P_D$ becomes 0.342, and the probability of detection P_D decreases to 0.658.

1.2.2.4 *Fourth example of application of the Neyman–Pearson criterion: noise-on-noise problem, vector observation*†

In this example we will use some concepts about pencils of matrices, discussed in some detail in Appendix 2. Moreover, some concepts involving the (complex) Gaussian multivariate are also recalled, and the Appendix 3 is given as a reference. We will use the following notations:

- the operator $E(\cdot)$ denotes the expected value (statistical average);
- the superscrips $*$, T and H denote conjugated, transposed and conjugated transposed, respectively.

Let us suppose we have a Gaussian complex random vector $\boldsymbol{z}$ with dimension N (a zero mean value is assumed for sake of simplicity, but a mean vector $\boldsymbol{a}$ can be

† This section may be omitted from a first reading.

introduced without difficulties), and whose covariance matrix is M (we recall the pertaining definition: $M = E\{z^* z^T\}$).

Two hypotheses are given: the observation z is taken from a population with $M = M_0$, or from a population with $M = M_1$, with M_0 and M_1 being two Hermitian and positive definite matrices. In synthesis,

$$\begin{cases} d_0: & M = M_0 \\ d_1: & M = M_1 \end{cases} \tag{1.45}$$

The probability density function of the received signal is the complex Gaussian multivariate, i.e.

$$p(z|H_0) = \frac{1}{(\pi)^N \det(M_0)} e^{-z^T M_0^{-1} z^*} \tag{1.46}$$

$$p(z|H_1) = \frac{1}{(\pi)^N \det(M_1)} e^{-z^T M_1^{-1} z^*} \tag{1.47}$$

the log-likelihood ratio $l(z) = \ln \Lambda(z)$ is simply

$$l(z) = \ln \frac{\det(M_0)}{\det(M_1)} + [z^T M_0^{-1} z^* - z^T M_1^{-1} z^*] \tag{1.48}$$

and the test, according to the Neyman–Pearson criterion, is

$$l(z) \underset{H_0}{\overset{H_1}{\gtrless}} \ln \lambda \tag{1.49}$$

where

$$l(z) = \ln \frac{\det(M_0)}{\det(M_1)} + z^T Q z^* \tag{1.50}$$

and $Q = M_0^{-1} - M_1^{-1}$ is a Hermitian matrix.

With the position

$$T = \ln(\lambda) - \ln \frac{\det(M_0)}{\det(M_1)} \tag{1.51}$$

the test becomes

$$q = z^T (M_0^{-1} - M_1^{-1}) z^* = z^T Q z^* \underset{d_0}{\overset{d_1}{\gtrless}} T \tag{1.52}$$

i.e. it corresponds to the computation of the Hermitian form $q = z^T Q z^*$ and its comparison with the threshold T.

The Hermitian form q can be reduced to principal axes (see Section 1.8.3.4) as follows:

$$q = \sum_{i=1}^{N} \lambda_Q(i) |t_i|^2 = t^H \Lambda_Q t \tag{1.53}$$

where $\lambda_Q(i)$, $i = 1, 2, \ldots N$ are the eigenvalues of Q^* (and of Q), $\Lambda_Q = \mathrm{diag}[\lambda_Q(i)]$ and the vector t is defined by

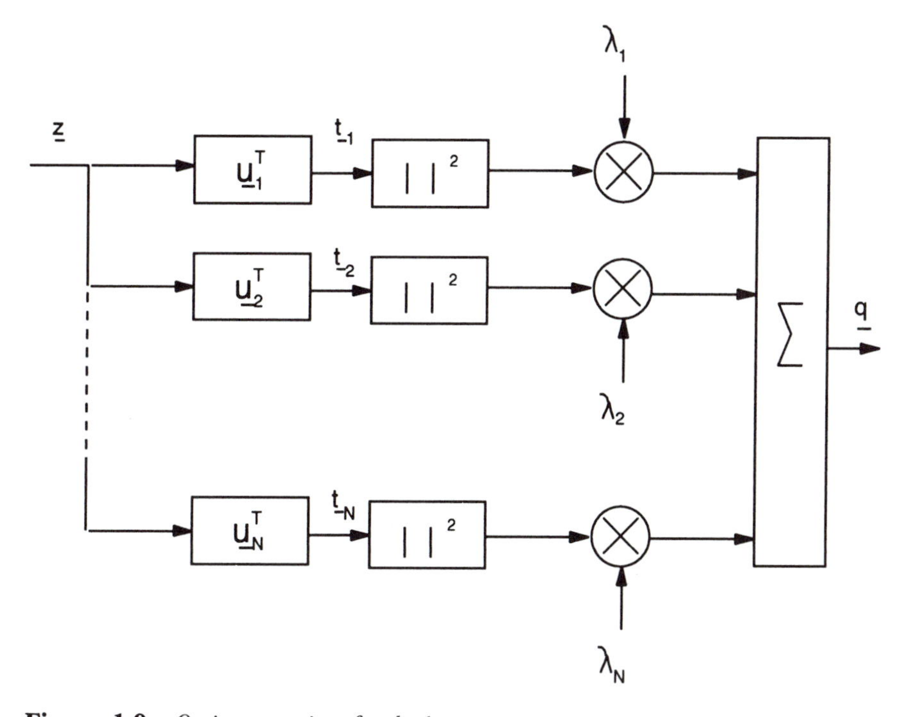

Figure 1.9 *Optimum receiver for the 'noise on noise' problem, vector observation*

$$\boldsymbol{t} = V^H \boldsymbol{z} \tag{1.54}$$

where V is the unitary matrix of the eigenvectors of Q^*; that is

$$Q^* = V\,\Lambda_Q V^H \tag{1.55}$$

[Of course, one could define, instead,

$$\boldsymbol{t} = U^T \boldsymbol{z} \tag{1.56}$$

where U is the unitary matrix of the eigenvectors of Q (note that $U = V^*$).]

The resulting optimum test is

$$q = \sum_{i=1}^{N} \lambda_Q(i)\,|t_i|^2 = \sum_{i=1}^{N} \lambda_Q(i)\,|\boldsymbol{u}_i^T.\boldsymbol{z}|^2 \underset{d_0}{\overset{d_1}{\gtrless}} T \tag{1.57}$$

where $\boldsymbol{u}_i$, $i = 1, 2, \ldots N$, are the eigenvectors of Q. The pertaining block diagram is shown in Fig. 1.9 [39].

An equivalent way to define the optimum test is based upon the concept of simultaneous reduction to principal axes of the two Hermitian forms in the square brackets of expr. 1.48 (see Section 1.8.3.5). Let P be the (generally non-unitary, supposed invertible) matrix of the eigenvectors of the pencil of matrices (M_0^{-*}, M_1^{-*}), and $\tilde{\Lambda} = \mathbf{diag}(\tilde{\lambda}_i)$ the diagonal matrix of the pertaining (real) eigenvalues of the pencil (i.e. the solutions of the system: $M_0^{-*}\boldsymbol{u}_k = \lambda M_1^{-*}\boldsymbol{u}_k$):

$$M_0^{-*}P = M_1^{-*}P\tilde{\Lambda} \tag{1.58}$$

where P satisfies the condition

$$P^T M_1^{-1} P^* = I \tag{1.59}$$

The transformation

$$\boldsymbol{r} = P^{-1}\boldsymbol{z} \tag{1.60}$$

reduces both the Hermitian forms

$$f_0 = \boldsymbol{z}^T M_0^{-1} \boldsymbol{z}^* \tag{1.61}$$

and

$$f_1 = \boldsymbol{z}^T M_1^{-1} \boldsymbol{z}^* \tag{1.62}$$

to weighted sums of squared moduli:

$$f_1 = \sum_{i=1}^{N} |\boldsymbol{r}_i|^2 \tag{1.63}$$

$$f_0 = \sum_{i=1}^{N} \tilde{\lambda}(i)|\boldsymbol{r}_i|^2 \tag{1.64}$$

where $\tilde{\lambda}(i)$, $i = 1, 2, \ldots N$, are the eigenvalues of the pencil. Therefore, the optimum test (eqn. 1.52) becomes

$$\sum_{i=1}^{N} [\tilde{\lambda}(i) - 1]\, |\boldsymbol{r}_i|^2 \underset{d_0}{\overset{d_1}{\gtrless}} T$$

In the particular case in which $M_1 = \mathrm{I}$* the matrix P is unitary and the eigenvalues of the pencil are the reciprocal of those of M_0. We recall that the eigenvalues of the pencil are also the eigenvalues of the matrix $M_1^* M_0^{-*}$, i.e. of $M_1 M_0^{-1}$.

Once the test has been defined, the remaining problem is how to compute the error probabilities α and β.

According to the theory shown in Appendix 3, Section 1.9.5, the error probabilities can be evaluated [33, 34] by the inversion of the characteristic function $\Phi_q(j\omega)$ (expr. 1.755) or of the Laplace transform characteristic function $G_q(s)$ (expr. 1.757). The latter, more straightforward, method can be applied when the eigenvalues of M^*Q^* are greater than zero, i.e. M^*Q^* is positive definite†.

The resulting cumulative density function of the decision variable q is [24, 33, 34]

$$P[q > T] = \sum_{i=1}^{N} A_i\, e^{-T/2\mu_i} \tag{1.65a}$$

* In radar applications, such a case can be applied when the Doppler frequency of the target is uniformly distributed in the Nyquist interval.

† It is easy to show, that when an eigenvalue tends to zero, its contribution to the error probability simply vanishes.

where

$$A_i = \prod_{\substack{k=1 \\ k \neq i}}^{N} \frac{\mu_i}{\mu_i - \mu_k} \tag{1.65b}$$

and μ_i, $i = 1 \ldots N$, are the eigenvalues of the matrix M^*Q^*; they are supposed to be distinct.

The error probabilities α and β can be evaluated by putting $M_z = M_0$ and $M_z = M_1$, respectively, in expr. 1.65; however, this expression has the advantage of allowing the computation of α and β also in mismatched conditions, i.e. when the covariance matrix of the interference, or of the useful signal, differs from the covariance matrix used in the design of the processor (for instance, in the design of the filters and of the multiplicating coefficients shown in Fig. 1.9) [39].

1.2.3 *Minimum error probability (MEP) criterion*

According to the (minimum) error probability criterion, often referred to as the 'ideal observer' criterion, the total error probability P_e (in the usual case of two possible events) is given by

$$P_e = P\{m_0\}P\{d_1|m_0\} + P\{m_1\}P\{d_0|m_1\}$$

where $P\{m_i\}$ are the *a priori* probabilities of occurrence of the events.

Recalling eqn. 1.11 one gets

$$P_e = P\{m_0\}P\{d_1|m_0\} + P\{m_1\} - P\{m_1\}P\{d_1|m_1\} \tag{1.66}$$

and, recalling eqns. 1.6 and 1.9,

$$P_e = \int_{Z_1} [P\{m_0\}p(\mathbf{z}|m_0) - P\{m_1\}p(\mathbf{z}|m_1)]\, d\mathbf{z} + P\{m_1\} \tag{1.67}$$

It is readily shown, again, that the total error probability is minimised by choosing the region Z_1 such as that the integrand is negative, i.e.

$$\mathbf{Z}_1 = \left\{ z\colon \Lambda(z) > \frac{P\{m_0\}}{P\{m_1\}} \right\} \tag{1.68}$$

If the *a priori* probabilities are equal (equally likely message case), the MEP criterion coincides with the ML one (the threshold value for the maximum likelihood ratio becomes 1).

Further, writing eqn 1.68 as

$$\Lambda(z) = \frac{p(\mathbf{z}|m_1)}{p(\mathbf{z}|m_0)} \underset{d_0}{\overset{d_1}{\gtrless}} \frac{P\{m_0\}}{P\{m_1\}} \Rightarrow \frac{p(m_1|\mathbf{z})}{p(m_0|\mathbf{z})} \underset{d_0}{\overset{d_1}{\gtrless}} 1 \tag{1.69}$$

it can be noted that the MEP decision rule is equivalent to the maximum *a posteriori probability* (MAP) criterion, in which comparison is made between the probabilities of the causes, having observed the effects. The MAP criterion can be stated as follows: it is decided for the most likely event given the observation **z**, i.e. for the event with the greatest *a posteriori* probability.

1.2.4 *Bayes (minimum risk) criterion*

In this criterion there are defined 'costs' for each conditioned decision:

$$C_{ij} = \text{cost in deciding } d_i \text{ when } m_j \text{ is true}$$

The 'Bayes risk' B is an 'average cost' for the decision:

$$B = \sum_{i=0}^{N-1} \sum_{j=0}^{N-1} C_{ij} P(m_j) P\{d_i | m_j\} = \sum_{i=0}^{N-1} \sum_{j=0}^{N-1} C_{ij} P(m_j) \int_{Z_i} p(\mathbf{z}|m_j)\, d\mathbf{z} \qquad (1.70)$$

that can be specialised to the binary case ($N=2$) as follows:

$$B = C_{00} P\{m_0\} + C_{01} P\{m_1\} + \int_{Z_1} [(C_{10} - C_{00}) P\{m_0\} p(\mathbf{z}|m_0) - (C_{01} - C_{11}) P\{m_1\} p(\mathbf{z}|m_1) d\mathbf{z} \qquad (1.71)$$

where the identities

$$\int_{Z_i} p(\mathbf{z}|m_i) d\mathbf{z} + \int_{Z_j} p(\mathbf{z}|m_i) d\mathbf{z} = 1 \qquad i, j = 0, 1 \qquad (1.72)$$

have been used.

The minimisation of the average cost B is obtained by selecting the region Z_1 such that the integrand of eqn. 1.71 is negative, resulting in the following expression:

$$\Lambda(\mathbf{z}) = \frac{P(\mathbf{z}|m_1)}{P(\mathbf{z}|m_0)} \underset{d_0}{\overset{d_1}{\gtrless}} \frac{(C_{10} - C_{00}) P\{m_0\}}{(C_{01} - C_{11}) P\{m_1\}} \qquad (1.73)$$

One should not be surprised at the presence of 'direct costs' C_{ii}, that is, costs for a right decision: for example, in the case of a decision as to whether a patient has got illness A or illness B, we would have different costs for the therapy, and there would be different results for taking the right or the wrong decision.

Of course, in most practical cases, the costs of a wrong decisions are the highest, i.e. $C_{01} > C_{11}$ and $C_{10} > C_{00}$; in many cases $C_{00} = C_{11} = 0$.

If one does not have precise information on the costs, $C_{00} = C_{11} = 0$ and $C_{10} = C_{01} = 1$ should be assumed, obtaining back the expression of the MEP (the expression B, eqn. 1.71, becomes equal to that of P_e, eqn. 1.67).

1.2.5 *Applicability of the decision criteria to radar*

As observed in previous sections, in radar applications it is very hard to define the Bayes costs C_{ij}; moreover, it is also practically impossible to define or evaluate the *a priori* probabilities $P(m_0)$ and $P(m_1)$, that is the probabilities that, in a given resolution interval, a target is or is not present. These are the main reasons why Bayes and MEP criteria cannot be used.

On the contrary, for the same reasons, the Neyman–Pearson criterion is particularly well suited to radar applications, owing to its concept of 'P_{FA} threshold' to be fixed *a priori*, whilst P_D is to be maximised.

1.3 Fundamentals of estimation theory

The estimation problems are natural extensions of the decision ones, in the case of an infinite number of possible events (see Section 1.1). For example, besides deciding whether a certain signal was transmitted or not, we also want to know, with a given degree of accuracy, its amplitude and/or its phase. We recall (see Section 1.1) that the basic difference between a decision (hypothesis testing) problem and an estimation problem is the following: in the former, we have two (generally speaking, K) alternative cases for the observed events: a case m_0 representing the null hypothesis and an alternative case m_1; in the latter we have an infinite number of cases corresponding to the values of the parameter to be estimated (or, from another point of view, an unknown parameter which can take infinite different values).

In radar problems, the *detection* of a target is generally followed by the *estimation* of related quantities, such as its azimuthal angle, its range, its Doppler frequency and so on.

As in decision problems, the received signal is corrupted by disturbances of various nature. What we attempt to obtain is a 'reasonable estimate' $\hat{\boldsymbol{\theta}}$ of the unknown parameter(s) $\boldsymbol{\theta}$, starting from statistical information about signal and disturbances (noise, clutter etc.).

As in decision problems, various estimation methods are available, each suitable for a particular application. If the parameter to be estimated (scalar or vector) is a constant, we speak of *parameter (or 'point') estimation*; on the other hand, when varying parameters are to be estimated (the evolution parameters of a dynamic system) we speak of *state estimation* [1].

Among the most common algorithms for parametric estimation we can have:

- Maximum likelihood estimation
- Bayes (several methods)
- Linear minimum-variance methods
- Least-square methods

As in the previous sections, we will try in the following sections to sketch only the fundamental characteristics of each method, without going too deeply into the mathematical details.

Before doing so, we briefly introduce the main properties of the estimators along with an important bound in defining the performance of the estimators: the Cramér–Rao lower bound.

1.3.1 *Parameter estimators and their properties*

1.3.1.1 *Parameter estimation and information in the sample*

Let us suppose that a random variable X depending on an unknown (generally speaking, vector) parameter $\boldsymbol{\theta}$ is observed N times (to estimate $\boldsymbol{\theta}$) resulting in a vector observation $\boldsymbol{x} = [x_1, x_2, \ldots, x_N]^T$.

A (point) estimate is the random (vector) variable:

$$\hat{\boldsymbol{\theta}} = \varphi(\boldsymbol{x}) \tag{1.74}$$

where $\varphi(\cdot)$ is a suitable function; the PDF of $\hat{\boldsymbol{\theta}}$ depends on the unknown parameter θ and is denoted by $p(\hat{\boldsymbol{\theta}}|\boldsymbol{\theta})$.

In *classical estimation theory*, $\boldsymbol{\theta}$ is an unknown (either scalar or vector) parameter, where 'unknown' means that no *a priori* information about it is available. Conversely, in *Bayesian estimation* $\boldsymbol{\theta}$ is the value of a random variable (or random vector) $\boldsymbol{\Theta}$ and the available (*a priori*) information about it is the conditioned density, $f_{x/\boldsymbol{\Theta}}(\boldsymbol{x}|\boldsymbol{\theta})$ or, for short, $p(\boldsymbol{x}|\boldsymbol{\theta})$, called the *likelihood function* when considered as a function of $\boldsymbol{\theta}$.

The Bayesian estimation problems are equivalent to the prediction problems, in which one or more values of a random variable Θ with a known density function $f_\Theta(\theta)$ have to be predicted based on a set $\boldsymbol{x}$ of (measured) values of X, when the conditioned density $f_{x/\theta}(\boldsymbol{x}|\theta)$ is known. Note that in this case Bayes' formula supplies the conditioned density $f_{\Theta/x}(\theta|\boldsymbol{x})$ or, for short, $p(\theta|\boldsymbol{x})$, namely:

$$p(\theta|\boldsymbol{x}) = \frac{p(\boldsymbol{x}|\theta)p(\theta)}{\int_{-\infty}^{\infty} p(\boldsymbol{x}|\theta)p(\theta)\, d\theta} \tag{1.75}$$

When probability is interpreted objectively, i.e. as a *measure of averages* rather than of *belief*, both classical and Bayesian approaches can be followed, depending on the problem to be solved.

As an example, if the azimuthal accuracy of a radar is evaluated by comparison of N measurements x_i, $i = 1, \ldots, N$ with the azimuth x_0 of a reflector (or of a transponder) generating a fixed echo (or reply) in a known position, the underlying probabilistic model is that of the random variable X and a classical estimation problem results.

Conversely, if azimuth measurements are performed on many aircraft targets in a 'live' test, the underlying model is a product space $S_X \cdot S_A$, where S_X is the space of all measurements on a particular aircraft and S_A is the space of all aircraft; the product space, of course, is the space of all measurements during the test. In such a situation, a Bayesian approach is appropriate because the 'true' azimuth of each aircraft can be seen as a random variable with a given probability density function, related to the statistics of the navigation errors.

The discussion in this chapter is compatible with both approaches; when the Bayesian one is used, $p(\boldsymbol{x}|\boldsymbol{\theta})$ denotes the conditioned density function of the observation $\boldsymbol{x}$, otherwise the notation $p(\boldsymbol{x}|\boldsymbol{\theta})$ is equivalent to $p(\boldsymbol{x}; \boldsymbol{\theta})$, i.e. the density function of $\boldsymbol{x}$, which depend on the parameter $\boldsymbol{\theta}$.

Every estimation algorithm is a mapping from the message, or sample, $\boldsymbol{x}$ to the estimated value $\hat{\boldsymbol{\theta}}$, characterised by its (conditioned) density function $p(\hat{\boldsymbol{\theta}}|\boldsymbol{\theta})$. The 'goodness' of an estimation method is a measure of how likely the estimate $\hat{\boldsymbol{\theta}}$ is 'close' to the unknown parameter $\boldsymbol{\theta}$, i.e. how much the pertaining PDF $p(\hat{\boldsymbol{\theta}}|\boldsymbol{\theta})$ is clustered on the value $\boldsymbol{\theta}$.

The design and performance of a given estimation algorithm depend on the PDF of the observation (often referred to as sample) $\boldsymbol{x}$, namely on $p(\boldsymbol{x}|\boldsymbol{\theta})$, where $x : x_k$, $k = 1, \ldots, N$.

In most applications the N elements of the observation are independent, identically distributed (IID) random variables with an assigned PDF $f_{x/\theta}(x|\boldsymbol{\theta})$

or, for short, $f(x|\boldsymbol{\theta})$. Therefore the PDF of the observation, i.e. the likelihood function, $p(\boldsymbol{x}|\boldsymbol{\theta})$ can be written as follows:

$$p(\boldsymbol{x}|\boldsymbol{\theta}) = \prod_{k=1}^{N} f(x_k|\boldsymbol{\theta}) \tag{1.76}$$

Considering for the moment estimation problems of a *scalar* parameter θ we can call, after R. A. Fisher, *information in the sample* the quantity

$$I(\theta) = N \int_{-\infty}^{\infty} \left[\frac{1}{f(x|\theta)} \frac{\partial f(x|\theta)}{\partial \theta} \right]^2 f(x|\theta) \, dx \tag{1.77}$$

or, in shorter form

$$I(\theta) = NE\left[\left(\frac{f'}{f} \right)^2 \right] \tag{1.78}$$

where the prime denotes differentiation with respect to θ and the expected value is taken with respect to $f(x|\theta)$.

This expression is valid for IID elements of the observation but can be immediately generalised as follows:

$$I(\theta) = \sum_{k=1}^{N} E\left[\frac{f_k'}{f_k} \right]^2 \tag{1.79}$$

It is often convenient to express the likelihood function in logarithmic form:

$$L(\boldsymbol{x}|\theta) = \ln p(\boldsymbol{x}|\theta) \tag{1.80}$$

and considering eqn. 1.76

$$L(\boldsymbol{x}|\theta) = \sum_{k=1}^{N} \ln f(x_k|\theta) \tag{1.81}$$

In such a way, eqn. 1.79 can be written as follows:

$$I(\theta) = E[L'(\boldsymbol{x}|\theta)]^2 = \int_{-\infty}^{\infty} [L'(\boldsymbol{x}|\theta)]^2 p(\boldsymbol{x}|\theta) \, d\boldsymbol{x} \tag{1.82}$$

where, as usual, the prime denotes differentiation with respect to θ.

The following properties of the Fisher information follow from the definition (eqn. 1.79): if two different independent samples are observed ($x_1, \ldots, x_N$ and $y_1, \ldots, y_N$), the total information is the sum of the component information:

$$I(\theta) = I_1(\theta) + I_2(\theta) \tag{1.83}$$

moreover

$$I(\theta) \geqslant 0 \tag{1.84}$$

($I(\theta) = 0$ if and only if $p(\boldsymbol{x}|\theta)$ does not depend on θ).

An alternative, often useful, expression for the sample information can be obtained using the following identities:

$$\int_{-\infty}^{\infty} p(\boldsymbol{x}|\theta)\, d\boldsymbol{x} = 1 \tag{1.85}$$

$$\frac{\partial p(\boldsymbol{x}|\theta)}{\partial \theta} = p(\boldsymbol{x}|\theta)\, \frac{\partial \ln p(\boldsymbol{x}|\theta)}{\partial \theta} \tag{1.86}$$

and as a consequence

$$\int_{-\infty}^{\infty} p(\boldsymbol{x}|\theta)\, \frac{\partial \ln p(\boldsymbol{x}|\theta)}{\partial \theta}\, d\boldsymbol{x} = 0 \tag{1.87}$$

Differentiating eqn. 1.87 w.r.t. θ:

$$\int_{-\infty}^{\infty} \frac{\partial p(\boldsymbol{x}|\theta)}{\partial \theta}\, \frac{\partial \ln p(\boldsymbol{x}|\theta)}{\partial \theta}\, d\boldsymbol{x} + \int_{-\infty}^{\infty} p(\boldsymbol{x}|\theta)\, \frac{\partial^2 \ln p(\boldsymbol{x}|\theta)}{\partial \theta^2}\, d\boldsymbol{x} = 0 \tag{1.88}$$

and by substitution of eqn. 1.86 into the first integral of eqn. 1.88:

$$\int_{-\infty}^{\infty} p(\boldsymbol{x}|\theta) \left[\frac{\partial \ln p(\boldsymbol{x}|\theta)}{\partial \theta}\right]^2 d\boldsymbol{x} = -\int_{-\infty}^{\infty} p(\boldsymbol{x}|\theta)\, \frac{\partial^2 \ln p(\boldsymbol{x}|\theta)}{\partial \theta^2}\, d\boldsymbol{x} \tag{1.89}$$

The resulting alternative expression for $I(\theta)$ is

$$I(\theta) = -\int_{-\infty}^{\infty} L''(\boldsymbol{x}|\theta) p(\boldsymbol{x}|\theta)\, d\boldsymbol{x} = -E[L''(\boldsymbol{x}|\theta)] \tag{1.90}$$

where, as usual, the double prime $''$ denotes the second derivative w.r.t. θ.

A comprehensive explanation of the rationale of the definitions in eqns. 1.79 and 1.82, in the frame of the maximum likelihood estimation, is contained in paragraph 36 of Reference 2.

1.3.1.2 *Properties of the estimators and Cramèr-Rao bound*

An important definition to be used in the following is that of *sufficient statistic (sufficient estimator)*. If the PDF of the observation $\boldsymbol{x}$ can be factored in the following form:

$$p(\boldsymbol{x}|\boldsymbol{\theta}) = g(\boldsymbol{t}|\boldsymbol{\theta}) h(\boldsymbol{x}) \tag{1.91}$$

where $\boldsymbol{t} = T(\boldsymbol{x})$, i.e. $\boldsymbol{t}$ does not depend on $\boldsymbol{\theta}$, we say that $\boldsymbol{t}$ is a sufficient statistic, and that $\boldsymbol{t}$ is sufficient for $\boldsymbol{\theta}$. Of course, if $\boldsymbol{t}$ is a sufficient statistic, any function of $\boldsymbol{t}$ is a sufficient statistic.

The rationale for the name 'sufficient' is that in decision theory a sufficient statistic is one that provides enough information about the observation to make a proper decision.

A notable subset of sufficient statistics are the *statistics of the exponential type* with respect to the (scalar) parameter θ, i.e. of the form

$$p(\boldsymbol{x}|\theta) = \exp[A(\theta) q(\boldsymbol{x}) + B(\theta)] \cdot h(\boldsymbol{x}) \tag{1.92}$$

where the functions $A(\theta)$ and $B(\theta)$ depend only on θ and the functions $h(\boldsymbol{x})$ and $q(\boldsymbol{x})$ depend only on $\boldsymbol{x}$.

As an example, it is readily seen that a Gaussian density with mean μ and variance σ^2 is of the exponential type with respect both to the parameter μ and to the parameter σ^2.

For any estimator $\hat{\boldsymbol{\theta}}$ with known PDF $p(\hat{\boldsymbol{\theta}}|\boldsymbol{\theta})$, it is possible to define a conditional mean $E(\hat{\boldsymbol{\theta}}|\boldsymbol{\theta})$ and a conditional variance $\mathrm{var}(\hat{\boldsymbol{\theta}}|\boldsymbol{\theta})$, or $\sigma_{\hat{\theta}}^2$, as follows:

$$E(\hat{\boldsymbol{\theta}}|\boldsymbol{\theta}) = \int \hat{\boldsymbol{\theta}} p(\hat{\boldsymbol{\theta}}|\boldsymbol{\theta})\,\mathrm{d}\hat{\boldsymbol{\theta}}$$

$$\mathrm{var}(\hat{\boldsymbol{\theta}}|\boldsymbol{\theta}) = \sigma_{\hat{\theta}}^2 = \int \|\hat{\boldsymbol{\theta}} - E(\hat{\boldsymbol{\theta}}|\boldsymbol{\theta})\|^2 p(\hat{\boldsymbol{\theta}}|\boldsymbol{\theta})\,\mathrm{d}\hat{\boldsymbol{\theta}}$$

where $\|\boldsymbol{v}\|$ denotes the Euclidean norm of any vector $\boldsymbol{v}$, i.e. the square root of the sum of the squared moduli of its component.

In terms of what is known from the theory of measurements, it is possible to refer to estimator *accuracy* and *precision*, where the *accuracy* is the capability of the estimator to give an estimate $\hat{\boldsymbol{\theta}}$ whose expected value is as close as possible to the value $\boldsymbol{\theta}$, and the *precision* is an indication of the conditional variance.

A sufficient estimator satisfying the relationship

$$E(\hat{\boldsymbol{\theta}}|\boldsymbol{\theta}) = \boldsymbol{\theta} \tag{1.93}$$

for any value of $\boldsymbol{\theta}$ is said to be (conditionally) *unbiased*: this estimator is an accurate one because it satisfies the first requirement. However, the 'goodness' does not depend only on the conditional mean but, obviously, also on the spread of $\hat{\boldsymbol{\theta}}$. Therefore, the 'best' estimator in a class of unbiased estimators is the one that minimises the conditional variance of the estimation.

As a general rule, when the number of observations increases, the complexity of the estimation algorithm becomes higher and higher, and in this case it is important to define the properties of the estimator as a function of the number of observations. Generally, it is much easier to define the asymptotic properties of the estimators which refer to the case of a number of observations that increases without limit.

For example we speak of *asymptotically unbiased* estimators when

$$\lim_{N\to\infty} E\{\hat{\boldsymbol{\theta}}_N|\boldsymbol{\theta}\} = \boldsymbol{\theta} \tag{1.94}$$

where N is the number of observations and $\hat{\boldsymbol{\theta}}_N$ is the estimate of $\boldsymbol{\theta}$ based on N observations.

Another important definition is that of an (unconditionally) *consistent* estimator, i.e. an estimator which converges in probability to the unknown parameter. Mathematically:

$$\lim_{N\to\infty} P\{\|\hat{\boldsymbol{\theta}}_N - \boldsymbol{\theta}\| < \varepsilon\} = 1 \tag{1.95}$$

An important theorem, generally known as the Cramèr–Rao lower bound (CRLB) or Frèchet inequality states that there is a lower bound to the conditional variance of any scalar estimator. Therefore, if we are able to find an unbiased estimator whose variance equals that predicted by the said theorem, we can be sure that the estimator is the best in the class: it is also known as an *efficient* estimator.

The CRLB or Frèchet inequality for a scalar parameter θ with a (conditioned) bias $b(\theta)$ (by definition, $b(\theta) = E(\hat{\theta}|\theta) - \theta$, and $b'(\theta) = \partial b(\theta)/\partial\theta$) can be written as

$$\text{var}(\hat{\theta}|\theta) = \sigma_{\hat{\theta}}^2 \geqslant \frac{[1 + b'(\theta)]^2}{I(\theta)} \tag{1.96a}$$

under the hypotheses of differentiability of the bias term b and of a non-zero information $I(\theta)$, the sample information $I(\theta)$ is as defined in Section 1.3.1.1.

The CRLB is reached by a sufficient estimator $\hat{\theta}$ if, for a non-zero constant k (possibly depending on θ),

$$\frac{\partial \ln p(\boldsymbol{x}|\boldsymbol{\theta})}{\partial \theta} = k[\hat{\theta} - E(\hat{\theta}|\theta)] \tag{1.97a}$$

with a probability of 1.

If the estimator is unbiased, eqns. 1.96*a* and 1.97*a* reduce to

$$\sigma_{\hat{\theta}}^2 \geqslant \frac{1}{I(\theta)} \tag{1.96b}$$

and

$$\frac{\partial \ln p(\boldsymbol{x}|\theta)}{\partial \theta} = k[\hat{\theta} - \theta] \tag{1.97b}$$

The CRLB (eqns. 1.96 and 1.97) can be derived as follows:

(i) The PDF of the observation, $p(\boldsymbol{x}|\theta)$, should be greater than zero and differentiable in a given part D_x of the $\boldsymbol{x}$ domain; this part shall not depend on θ.

(ii) $I(\theta)$ should always be greater than zero.

Let

$$b(\theta) = E(\hat{\theta}|\theta) - \theta \tag{1.98}$$

be the bias of θ.

By definition, we have

$$\int_{D_x} p(\boldsymbol{x}|\theta)\; d\boldsymbol{x} = 1 \tag{1.99}$$

$$\int_{D_x} \hat{\theta} p(\boldsymbol{x}|\theta)\; d\boldsymbol{x} = \theta + b(\theta) \tag{1.100}$$

Using hypothesis (i) above, eqns. 1.99 and 1.100 may be differentiated w.r.t. θ (the derivative being indicated by a prime) giving the following result:

$$\int_{D_x} p'(\boldsymbol{x}|\theta)\; d\boldsymbol{x} = 0 \tag{1.101}$$

$$\int_{D_x} \hat{\theta} p'(\boldsymbol{x}|\theta)\; d\boldsymbol{x} = 1 + b'(\theta) \tag{1.102}$$

Multiplying eqn. 1.101 by $E((\hat{\theta}|\theta)$ and subtracting the result from eqn. 1.102 one gets

$$\int_{D_x} [\hat{\theta} - E(\hat{\theta}|\theta)]p'(\boldsymbol{x}|\theta)\ d\boldsymbol{x} = 1 + b'(\theta) \tag{1.103}$$

or

$$\int_{\theta_x} [\hat{\theta} - E(\hat{\theta}|\theta)] \frac{p'(\boldsymbol{x}|\theta)}{p(\boldsymbol{x}|\theta)} p(\boldsymbol{x}|\theta)\ d\boldsymbol{x} = 1 + b'(\theta) \tag{1.104}$$

i.e.

$$\int_{D_x} a(\boldsymbol{x})b(\boldsymbol{x})\ d\boldsymbol{x} = 1 + b'(\theta) \tag{1.105}$$

where by definition

$$a(\boldsymbol{x}) = [\hat{\theta} - E(\hat{\theta}|\theta)][p(\boldsymbol{x}|\theta)]^{1/2} \tag{1.106}$$

and

$$b(\boldsymbol{x}) = L'(\boldsymbol{x}|\theta) \cdot [p(\boldsymbol{x}|\theta)]^{1/2} \tag{1.107}$$

and, as usual,

$$L'(\boldsymbol{x}|\theta) = \frac{\partial \ln p(\boldsymbol{x}|\theta)}{\partial \theta}$$

Recalling that (Schwartz inequality):

$$\left[\int_{D_x} a(\boldsymbol{x})b(\boldsymbol{x})\ d\boldsymbol{x}\right]^2 \leqslant \left[\int_{D_x} a^2(\boldsymbol{x})\ d\boldsymbol{x}\right]\left[\int_{D_x} b^2(\boldsymbol{x})\ d\boldsymbol{x}\right] \tag{1.108}$$

eqn. 1.105 implies that (see eqns. 1.106 and 1.107)

$$\left\{\int_{D_x} [\hat{\theta} - E(\hat{\theta}|\theta)]^2 p(\boldsymbol{x}|\theta)\ d\boldsymbol{x}\right\} \cdot \left\{\int_{D_x} L'^2(\boldsymbol{x}|\theta)p(\boldsymbol{x}|\theta)\ d\boldsymbol{x}\right\} \geqslant [1 + b'(\theta)]^2 \tag{1.109}$$

As expression 1.109 is identical to expression 1.96, with $I(\theta)$ given by eqn. 1.81,

$$\mathrm{var}(\hat{\theta}|\theta) \geqslant \frac{[1 + b'(\theta)]^2}{I(\theta)} \tag{1.110}$$

and we have demonstrated the CRLB.

Of course, the equality sign in eqn. 1.108 holds if

$$b(\boldsymbol{x}) = ka(\boldsymbol{x}) \tag{1.111}$$

with k being a constant (i.e. not depending on $\boldsymbol{x}$).

Introducing eqns. 1.106 and 1.107 to 1.111 gives

$$k[(\hat{\theta} - E(\hat{\theta}|\theta)] = L'(\boldsymbol{x}|\theta) \tag{1.112}$$

whence, if the estimator is unbiased, eqn. 1.97*b* results.

Integration of eqn. 1.112 w.r.t. θ yields

$$L(\boldsymbol{x}|\theta) = A(\theta) \cdot \hat{\theta} - A(\theta)E(\hat{\theta}|\theta) + A_2(\boldsymbol{x}) \tag{1.113}$$

where $A(\cdot)$ is a function of θ only and $A_2(\cdot)$ is a function of $\boldsymbol{x}$ only.

Remembering that $L(\boldsymbol{x}|\theta)=\ln[p(\boldsymbol{x}|\theta)]$ gives

$$p(\boldsymbol{x}|\theta)=\exp\{A(\theta)q(\boldsymbol{x})+B(\theta)\}h(\boldsymbol{x}) \qquad (1.114)$$

where $q(\boldsymbol{x})=\hat{\theta}$, $B(\theta)=-A(\theta)E(\hat{\theta}|\theta)$ and $h(\boldsymbol{x})=e^{A_2(\boldsymbol{x})}$.

Therefore, if an estimate $\hat{\theta}$ which reaches the CRLB does exist, the likelihood function is of the exponential type.

Moreover, it is readily shown that *if (a) the hypotheses (i) and (ii), under which the CRLB was derived, hold and (b) the likelihood function is of the exponential type, the estimator $\hat{\theta}=q(\boldsymbol{x})$ of eqn. 1.114 has the smallest variance of all the estimates with some bias $b(\theta)$ [2].*

The demonstration in Reference 2 can be outlined as follows: differentiating eqn. 1.113 w.r.t. θ we come back to eqn. 1.112, i.e. we have that $\hat{\theta}-E(\hat{\theta}|\theta)$ and $L'(\boldsymbol{x}|\theta)$ are proportional. Hence eqn. 1.110 is satisfied with the equality sign.

In a similar way it is shown that *when the above conditions (a) and (b) hold and, moreover, the estimator $\hat{\theta}=q(\boldsymbol{x})$ is unbiased, $q(\boldsymbol{x})$ is the minimum variance estimator.* In this case, the information $I(\theta)$ is the derivative w.r.t. θ of the term $A(\theta)$ of eqn. 1.114.

From the preceding expressions, it is possible to introduce a new property of the estimators: an estimator is said to be *asymptotically efficient* if it converges to the CRLB if the sample size N is increasing; mathematically:

$$\lim_{N\to\infty}\,[\mathrm{var}(\hat{\theta}|\theta)I(\theta)]=[1+b'(\theta)]^2 \qquad (1.115)$$

The CRLB can be easily computed after a transformation of the parameter θ into ϕ. Let $\phi=g(\theta)$ be a one-to-one and differentiable function with respect to θ:

$$\text{CRLB for var}\{\hat{\phi}\}=\left[\frac{\mathrm{d}g(\theta)}{d\theta}\right]^2\cdot[\text{CRLB for var}\{\hat{\theta}\}] \qquad (1.116)$$

A limitation of the above treatment is that it only applies to scalar parameters. However, there is a form of CRLB for estimating vectors which defines a bound over the diagonal terms of the covariance matrix.

Let us suppose we have a set of N observations, on the basis of which we wish to estimate the unknown parameters $\theta_1,\ldots,\theta_m$, with $m\leq N$.

It can be shown that the vector form of the CRLB for unbiased estimates is written as†

$$\mathrm{cov}\{\hat{\boldsymbol{\theta}}\}\geq J^{-1} \qquad (1.117)$$

where $\mathrm{cov}(\hat{\theta})$ is the covariance matrix of the estimator, with elements $[\mathrm{cov}(\hat{\boldsymbol{\theta}}]_{i,j}=E\{(\hat{\theta}_i-\theta_i)(\hat{\theta}_j-\theta_j)]\}$ and where J^{-1} is the inverse of the Fisher information matrix J whose elements are given by

$$J_{ij}=E\left\{\frac{\partial}{\partial\theta_i}\ln p(\boldsymbol{x}|\boldsymbol{\theta})\cdot\frac{\partial}{\partial\theta_j}\ln p(\boldsymbol{x}|\boldsymbol{\theta})\right\},\ i,j=1,2,\ldots,m \qquad (1.118)$$

† When A and B are two $(N\times N)$ matrices, $A\geq B$ means that the matrix $A-B$ is semi-definite positive.

The above expressions imply that

$$\mathrm{var}(\theta_j|\boldsymbol{\theta}) \geqslant (J^{-1})_{jj}, j=1, 2, \ldots, m \tag{1.119}$$

In the next section, devoted to the maximum likelihood estimation, an example of application of the CRLB will be given.

1.3.2 *Maximum likelihood estimation*

The rationale of this method is very similar to that of the maximum likelihood decision rule: also in this case knowledge of the conditioned PDF $p(\boldsymbol{x}|\boldsymbol{\theta})$ of the observation, given the parameter, is only required.

We say that, for a given observation $\boldsymbol{x}$, $\hat{\boldsymbol{\theta}}_{ML}$ is a maximum likelihood estimation (MLE) if

$$p(\boldsymbol{x} \mid \hat{\boldsymbol{\theta}}_{ML}) \geqslant p(\boldsymbol{x}|\hat{\boldsymbol{\theta}}) \tag{1.120}$$

for any estimation $\hat{\boldsymbol{\theta}}$, that is $\hat{\boldsymbol{\theta}}_{ML}$ maximises the likelihood function for a given observation $\boldsymbol{x}$; in other words, $\hat{\boldsymbol{\theta}}_{ML}$ is the $\boldsymbol{\theta}$ value that most likely caused the observed value(s) to occur.

Often the log-likelihood $L(\boldsymbol{x}|\boldsymbol{\theta}) = \ln p(\boldsymbol{x}|\boldsymbol{\theta})$ is used to simplify the calculations.

This is a simple and powerful method; it is the estimation method that requires the minimum *a priori* information about the parameter to be estimated. Moreover, it can be shown that the ML estimator is a good approximation of more complicated estimators, when the number of observations becomes greater.

1.3.2.1 *Properties of ML estimators*

As stated previously, it is often useful to describe the behaviour of an estimator when the number of independent observations N approaches infinity (asymptotic behaviour).

It is possible to prove, under general conditions, the following properties:

(*a*) The ML estimate is *consistent*: this means that, under certain regularity conditions, the ML estimate converges in probability to the correct value as N tends to infinity (see above for the rigorous definition of consistency). The required regularity conditions for the consistency related to a scalar parameter θ are:

   (i) The likelihood function $p(\boldsymbol{x}|\theta)$ may be differentiated w.r.t. θ under the integral sign when integrated w.r.t. $\boldsymbol{x}$

   (ii) The portion of the $\boldsymbol{x}$ space on which $p(\boldsymbol{x}|\theta)$ is strictly positive does not depend on θ.

   Most, but not all, the conditioned density functions are regular; for example, the negative exponential density

$$p(\boldsymbol{x}|\theta) = \exp(-x/\theta), \; x \geqslant 0$$
$$p(\boldsymbol{x}|\theta) = 0, \; x < 0 \tag{1.121a}$$

   (with $0 < \theta < +\infty$) is regular, but the density

$$p(\boldsymbol{x}|\theta) = \exp[-(x-\theta)], \; x \geqslant \theta$$
$$p(\boldsymbol{x}|\theta) = 0, \; x < \theta \tag{1.121b}$$

(with $0<\theta<+\infty$) is not regular because it does not satisfy condition (ii).

(*b*) The ML estimate is *asymptotically efficient* (see eqn 1.115 for the pertaining definition), but only under strong regularity conditions; it is even possible to build asymptotically unbiased estimators (biased for N finite) that present an asymptotic variance smaller than that of the ML estimate.

(*c*) The ML estimate is *asymptotically Gaussian* (normal). This property means that the estimated parameter $\hat{\boldsymbol{\theta}}_{ML}$ is an asymptotically Gaussian random variable, with mean θ and variance $1/I(\theta)$; it has been shown that ML estimates are also asymptotically unbiased, so that the mean value of the Gaussian PDF equals θ for $N\rightarrow\infty$†.

(*d*) The ML estimate is *invariant* with respect to any invertible transformation $f(\cdot)$; in other words, if $\hat{\theta}_{ML}$ is the ML estimate of $\boldsymbol{\theta}$, $f(\hat{\theta}_{ML})$ is the ML estimate of $f(\theta)$.

An *efficient* estimator is an unbiased one whose (conditional) variance satisfies the CRLB with the equality sign. An efficient estimator for a given problem may either exist or not; it can be readily shown using

(i) the following property (eqn. 1.112) of an efficient estimator $\hat{\theta}_{\text{eff}}$:

$$L'(\boldsymbol{x}|\theta)=k[\hat{\theta}_{\text{eff}}-\theta] \tag{1.122}$$

(with k being a non-zero constant not depending on $\boldsymbol{x}$)

(ii) the following relationship originating from the definition of the ML estimate, $\hat{\theta}_{ML}$:

$$L'(\boldsymbol{x}|\theta)=0 \quad \text{for } \theta=\hat{\theta}_{ML} \tag{1.123}$$

that $\hat{\theta}_{ML}=\hat{\theta}_{\text{eff}}$, i.e. if the efficient estimator $\hat{\theta}_{\text{eff}}$ exists the ML estimate coincides with it.

If there are many independent observations, and if their PDFs satisfy some regularity conditions, and if a limited number of parameters θ_i, $i=1, \ldots r$, is to be estimated, then the ML method is proved to be 'better' as N is greater. But, if N is not very high and/or the number r of parameters increases with N, no guarantee exists that the ML method is really 'good'.

In some cases, when the ML estimator is biased, it is possible to build other unbiased estimators with minimum variance.

In the following some examples of the ML estimation method are shown.

1.3.2.2 *Example of ML estimation*

Let us consider an observation consisting of a given number N of IID samples z_i, $i=1, 2, \ldots, N$ of a Gaussian variable z, with a mean value μ and a non-zero variance σ^2. In this case the vector of parameters that we want to estimate is

$$\boldsymbol{\theta}=\begin{bmatrix}\mu\\ \sigma^2\end{bmatrix}$$

The conditioned probability density function of the N observations is

$$p(\boldsymbol{z}|\boldsymbol{\theta})=(\sqrt{2\pi}\sigma)^{-N}\exp\left[-\frac{1}{2\sigma^2}\sum_{i=1}^{N}(z_i-\mu)^2\right] \tag{1.124}$$

† This means that one must calculate *before* the limit as $N\rightarrow\infty$ of the PDF of $\hat{\theta}_{ML}$, and only *after* is it possible to calculate the mean and the variance: the contrary should not be correct.

It is possible to maximise this function by equating to zero the derivative of the logarithm of $p(\mathbf{z}|\boldsymbol{\theta})$ with respect to the vector $\boldsymbol{\theta}$, that is with respect to its components, obtaining the following system of two scalar equations:

$$\left.\frac{\partial \ln p(\mathbf{z}|\mu, \sigma^2)}{\partial \mu}\right|_{\mu=\mu_{ML}} = 0 \tag{1.125}$$

$$\left.\frac{\partial \ln p(\mathbf{z}|\mu, \sigma^2)}{\partial \sigma^2}\right|_{\sigma^2=\sigma^2_{ML}} = 0 \tag{1.126}$$

Let L indicate the logarithm of $p(\mathbf{z}|\boldsymbol{\theta})$ (eqn. 1.124):

$$L = -\frac{N}{2}\log(2\pi) - \frac{N}{2}\log\sigma^2 - \frac{1}{2\sigma^2}\sum_{i=1}^{N}(z_i-\mu)^2 \tag{1.127}$$

Equating to zero the derivatives of L with respect to the components of $\boldsymbol{\theta}$, we obtain

$$\frac{\partial L}{\partial \mu} = \frac{1}{\sigma^2}\sum_{i=1}^{N}(z_i-\mu) = 0 \tag{1.128}$$

$$\frac{\partial L}{\partial(\sigma^2)} = -\frac{N}{2\sigma^2} + \frac{1}{2\sigma^4}\sum_{i=1}^{N}(z_i-\mu)^2 = 0 \tag{1.129}$$

Thus the vector $\hat{\boldsymbol{\theta}}_{ML}$ has the following components, whose behaviour is worthy of examination:

$$\hat{\mu}_{ML} = \frac{1}{N}\sum_{i=1}^{N} z_i \tag{1.130}$$

$$\hat{\sigma}^2_{ML} = \frac{1}{N}\sum_{i=1}^{N}(z_i-\hat{\mu}_{ML})^2 \tag{1.131}$$

The mean value estimator $\hat{\mu}_{ML}$ (often called the 'sample mean') is an *unbiased* one, since the expected value of $\hat{\mu}_{ML}$ is μ itself:

$$E[\hat{\mu}_{ML}] = \frac{1}{N}\{E[z_1] + E[z_2] + \ldots + E[z_N]\} = \mu \tag{1.132}$$

while the variance estimator $\hat{\sigma}^2_{ML}$ is *biased,* as is shown below. The expected value of the estimated variance is given by

$$E[\hat{\sigma}^2_{ML}] = \frac{1}{N}\{E[(z_1-\hat{\mu}_{ML})^2] + \ldots + E[(z_N-\hat{\mu}_{ML})^2]\} \tag{1.133}$$

The above expression can be simplified, noting that each term in square brackets can be written as

$$(z_i-\hat{\mu}_{ML})^2 = \{(z_i-\mu) - (\hat{\mu}_{ML}-\mu)\}^2 \tag{1.134}$$

and that the expectation operator applied to this expression gives

$$E(z_i-\hat{\mu}_{ML})^2=\sigma^2-2E\{(z_i-\mu)\cdot(\hat{\mu}_{ML}-\mu)\}+E\{(\hat{\mu}_{ML}-\mu)\}^2 \qquad (1.135)$$

By substitution of expression 1.135 into expression 1.133 and using expression 1.130 one gets

$$E[\hat{\sigma}^2_{ML}]=\sigma^2+\frac{\sigma^2}{N}-\frac{2}{N}E\left\{\left(\frac{1}{N}\sum_{i=1}^{N}z_i-\mu\right)\quad[(z_1-\mu)+\cdots+(z_N-\mu)]\right\} \qquad (1.136)$$

The last term of the above expression equals $-2\sigma^2/N$, owing to the fact that the N random variables z_i are uncorrelated. Finally the expected value of $\hat{\sigma}^2_{ML}$ is given by

$$E[\hat{\sigma}^2_{ML}]=\frac{N-1}{N}\sigma^2 \qquad (1.137)$$

and this shows that this estimator is biased; it is asymptotically (i.e. when the number of observations goes to infinity) unbiased†.

The fact that the estimator of the variance $\hat{\sigma}^2_{ML}$ is biased is due to the contemporary estimation of the mean value μ; if we had assumed that the mean value were known (equal to zero, for example), we would have obtained a different result, as shown below.

The conditioned probability function of the N independent zero-mean Gaussian variables is now given by

$$p(\mathbf{z}|\sigma^2)=\prod_{i=1}^{N}p(z_i|\sigma^2)=(\sqrt{2\pi}\sigma)^{-N}\exp\left[-\frac{1}{2\sigma^2}\sum_{i=1}^{N}z_i^2\right] \qquad (1.138)$$

(the parameter to be estimated is the variance only). Taking again the natural logarithm of $p(\mathbf{z}|\sigma^2)$ and equating to zero the derivative with respect to σ^2 we obtain the following result:

$$\hat{\sigma}^2_{ML}=\frac{1}{N}\sum_{i=1}^{N}z_i^2 \qquad (1.139)$$

and, taking the expected value of $\hat{\sigma}^2_{ML}$, it is easy to verify that now this estimator is *unbiased*.

1.3.2.3 *Example of application of CRLB*

Let us suppose we have N noisy observations of a parameter θ, given by the expression

$$z_i=\theta+n_i \qquad i=1,2,\ldots N \qquad (1.140)$$

where n_i are zero mean independent Gaussian random variables. We can write the conditional density function as

$$p(\mathbf{z}|\theta)=(2\pi\sigma^2)^{-N/2}\exp\left\{-\frac{1}{2\sigma^2}\sum_{i=1}^{N}(z_i-\theta)^2\right\} \qquad (1.141)$$

† It can be shown that the empiric variance $\sigma^2_{emp}=[1/(N-1)]\sum_{1}^{N}(z_i-\hat{\mu}_{ML})^2$ is the unbiased estimator with minimum variance.

In order to find the ML estimate of θ, let us maximise $p(\mathbf{z}|\theta)$ by equating to zero the derivative of its natural logarithm. We obtain

$$\frac{1}{\sigma^2}\sum_{i=1}^{N}(z_i-\theta)=0 \tag{1.142}$$

so that the ML estimate is

$$\hat{\theta}_{ML}=\frac{1}{N}\sum_{i=1}^{N}z_i \tag{1.143}$$

that is, the simple arithmetic mean of the samples (sample mean). The estimate does not depend on the noise variance σ^2.

It is interesting to note two facts:

(*a*) The expected value of $\hat{\theta}_{ML}$ is θ, because the expected value of z_i given θ is θ itself, so that this estimator is unbiased
(*b*) The conditional variance of $\hat{\theta}_{ML}$ is:

$$\mathrm{var}(\hat{\theta}_{ML}|\theta)=E\left\{\left(\frac{1}{N}\sum_{i=1}^{N}(z_i-\theta)\right)^2|\theta\right\}=\frac{\sigma^2}{N} \tag{1.144}$$

so that, as is intuitive, the precision of the estimate is improved as N increases.

Let us now find the CRLB for the stated problem. In our case, the second form of the sample information, i.e. $I(\theta)=-E[L''(\mathbf{x}|\theta)]$ is used. We have

$$E\left[\frac{\partial^2}{\partial\theta^2}\ln p(\mathbf{z}|\theta)\right]=-\frac{N}{\sigma^2} \tag{1.145}$$

So we have obtained the important result that this ML estimator is efficient, in the sense that it satisfies the CRLB with the equality sign.

1.3.2.4 *Comments on the application of the maximum likelihood estimation with one example*†

The examples of the application of ML estimation shown before lead to solutions in closed form, that is $\hat{\boldsymbol{\theta}}_{ML}$ is expressed as a well-defined function of the observations. This is due to the particularly simple relationships of the probabilistic models, with respect to the observations, that we have considered up to now. But in most practical applications the mathematical relationships are much more complicated.

Typical applications of the ML criterion, leading to complex mathematical expressions, are:

(*a*) Estimation of the time delay (determination of the range in radar systems).
(*b*) Estimation of the frequency (e.g. of the Doppler shift)
(*c*) Estimation of the source direction (e.g. of the azimuth angle of a target).

In these cases and in many others, by applying the ML criterion, one obtains a system of non-linear equations that can be resolved numerically with iterative

† This section may be omitted from a first reading.

methods, or can be linearised around some working point: in any case, very rarely can a closed form exact solution be determined.

As an example of the solution of a practical problem we will discuss here the classical problem of the azimuth estimator for a pulsed radar or, equivalently, a secondary surveillance radar (SSR) [32].

Consider an SSR, scanning in azimuth with constant period T_s and illuminating a steady target located at an absolute bearing angle θ_T. On each scan, the target provides N samples of a sine wave of unknown amplitude A, modulated by the antenna pattern and embedded in white Gaussian noise.

The signal envelope is detected, providing N amplitude observations z_i, $i=1, 2, \ldots, N$, each one being associated with the azimuthal direction θ_i at which the antenna boresight is aimed at the relevant time instant. Owing to the lack of synchronism between the scan rate and the pulse repetition frequency (PRF), at each scan the sequence of values θ_i are affected by a random jitter with peak value $\Delta\theta$.

The azimuth estimator processes the observations z_i to provide an estimate $\hat{\theta}$ of the true target azimuth value θ_T. The estimator benefits from *a priori* knowledge of the antenna beam shape, whereas the maximum echo voltage A, normalised with respect to the RMS noise voltage (or, equivalently, the peak signal-to-noise ratio (SNR) $R_0=\frac{1}{2}A^2$) is generally unknown.

The ML estimator for this case is readily obtained under some hypotheses: the beam shape quantisation effect is assumed to be negligible (this means that $\Delta\theta$ is small w.r.t. the antenna beamwidth) and other causes of error are completely neglected.

Let R_i be the signal-to-noise power ratio of the ith sample:

$$R_i=\tfrac{1}{2}A^2g^2(\theta_i-\theta_T) \tag{1.146}$$

where $g(.)$ is the one-way voltage antenna radiation pattern (assumed hereafter to be Gaussian shaped for sake of simplicity):

$$g(\theta_i-\theta_T)=e^{-K[(\theta_i-\theta_T)/\theta_B]^2/2} \tag{1.147}$$

where θ_B is the -3 dB beamwidth and $K=4\ln 2$.

The joint probability density function of the N observations z_i, conditioned to the unknown parameters A and θ_T, i.e. the likelihood function, in the hypothesis of independent noise samples, is written

$$p(z_1\ldots z_N|\theta_T, A)=\prod_{i=1}^{N} p_{z_i}(z_i|\theta_T, A) \tag{1.148}$$

where $p_{z_i}(.)$ is the PDF of the ith amplitude sample.

The ML estimate of θ_T and A is obtained by maximising as usual the logarithm of the likelihood function, indicated as L, with respect to the unknown parameters; that is solving the system

$$\begin{cases} \dfrac{\partial L(z_1\ldots z_N|\theta_T, A)}{\partial\theta_T}=0 \\[2ex] \dfrac{\partial L(z_1\ldots z_N|\theta_T, A)}{\partial A}=0 \end{cases} \tag{1.149}$$

Let us denote $g(\theta_i-\theta_T)$ by g_i and $dg(\theta_i-\theta_T)/d\theta_T$ by g_i'. In the case of SSR replies, the PDF of the amplitudes is the Rice function† (see also Section 1.10.1):

$$p_{z_i}(z)=z\exp\left\{-\frac{z^2+A^2g_i^2}{2}\right\}I_0(Ag_iz)U(z) \qquad (1.150)$$

and the ML estimation procedure (eqn. 1.149) leads to the following equations [32]:

$$-A^2\sum_{i=1}^{N}g_ig_i'+A\sum_{i=1}^{N}\frac{I_1(Az_ig_i)}{I_0(Az_ig_i)}z_ig_i'=0 \qquad (1.151)$$

and

$$-A\sum_{i=1}^{N}g_i^2+\sum_{i=1}^{N}\frac{I_1(Az_ig_i)}{I_0(Az_ig_i)}z_ig_i=0 \qquad (1.152)$$

This solution leads to complicated non-linear processing of the samples z_i and is of no practical use. In the following it is shown how, by the application of reasonable hypotheses, expr. 1.149 can be reduced to linear processing.

Owing to the negligible beam quantisation (that is, $\Delta\theta=\theta_{i+1}-\theta_i\ll\theta_B$), to the even symmetry of the antenna pattern $g(.)$ and to the odd symmetry of its derivative $g'(.)$, the first term in expr. 1.151 is nearly zero; moreover, in case of high values of A (large SNR), the ratio $I_1(.)/I_0(.)$ can be replaced by 1. Hence the approximate estimate $\hat{\theta}$ of the azimuth angle is

$$A\sum_{i=1}^{N}z_ig'(\theta_i-\hat{\theta})=0 \qquad (1.153)$$

Eqn. 1.153 shows that the estimation procedure consists of a convolution of the observed data z_i with the first derivative of the antenna pattern, followed by a zero-crossing test. This procedure does not depend on A, and an estimate of this parameter is not necessary in order to obtain the estimate of θ_T; according to the simplifying assumptions adopted, the non-linear estimation problem has been reduced to a linear one, also decoupling the estimates of θ_T and A.

Incidentally, in the above hypotheses, from eqn. 1.152 the estimate of A is

$$\hat{A}=\frac{\sum_{i=1}^{N}z_ig(\theta_i-\hat{\theta})}{\sum_{i=1}^{N}g^2(\theta_i-\hat{\theta})} \qquad (1.154)$$

In practical applications, the estimation of the azimuth angle can be performed by a discrete-time filter whose impulse response $h_i=h(\theta_i)$, $i=1,\ldots,N$,

† For a primary radar the same PDF applies, provided that $g(.)$ is the two-way antenna pattern. In the following formulas, $I_k(\cdot)$ is the modified Bessel function of the first kind, order k.

depends on the radiation pattern of the antenna. The optimal impulse response $h(.)$ is theoretically infinite, but in practice it is truncated to few beamwidths, when expressed in angular units, in order to save the processor computation time and memory: this obviously introduces a degradation in the estimator performance, along with the noise present in the amplitude data, as evaluated in Reference 32.

The errors introduced by these two sources (i.e. truncation and noise) can be evaluated through the Cramér–Rao lower bound for unbiased estimators. As has been shown in Section 1.3.1, as the number of observations increases, the ML estimate approaches the CRLB; namely

$$\sigma^2(\hat{\theta}) \geqslant \sigma^2_{CR}(\hat{\theta}) = \left\{ E\left[\left(\frac{\partial L}{\partial \theta_T} \right)^2 \right] \right\}^{-1} \tag{1.155}$$

From the approximtion of large values of SNR of the derivative of the likelihood function (eqn. 1.151) we have

$$E\left\{ \left(\frac{\partial L}{\partial \theta_T} \right)^2 \right\} = A^2 \sum_{i=1}^{N} \sum_{j=1}^{N} g_i' g_j' E\{z_i z_j\} \tag{1.156}$$

Exploiting the symmetry properties of the sequences g_i, g_i', $E\{z_i\}$ and $E\{z_i^2\}$, and recalling that z_i and z_j are statistically independent, the CRLB for the estimate of the azimuth can be evaluated in the following form:

$$\sigma^2_{CR}(\hat{\theta}) = A^{-2} \cdot \left[\sum_{i=1}^{N} (g_i')^2 \ \mathrm{var}(z_i) \right]^{-1} \tag{1.157}$$

it should be noted that $\sigma^2_{CR}(\hat{\theta})$ depends on the peak SNR R_0 and on the number of observations. Similarly it is possible to show that the CRLB for the estimate of the signal (target) amplitude A is given by

$$\sigma^2_{CR}(\hat{A}) = \left[\sum_{i=1}^{N} g_i^2 \ \mathrm{var}(z_i) \right]^{-1} \tag{1.158}$$

When the pattern has a Gaussian shape (expr. 1.147), under some conditions (negligible beam quantisation, large observation interval, large SNR), a closed form expression [32] can be given for $\sigma^2_{CR}(\hat{\theta})$:

$$\sigma^2_{CR}(\hat{\theta}) = \frac{1}{\sqrt{(\pi K)}} \frac{\theta_B^2}{N_B R_0} \cong 0.339 \frac{\theta_B^2}{N_B R_0} \tag{1.159}$$

where, as before, $K = 4 \ln 2$, and where $N_B = \theta_B / \Delta\theta$ is the number of replies in the half-power beamwidth.

1.3.3 *Bayes estimation methods*

In estimation theory there are no methods equivalent to Neyman–Pearson and MEP criteria, owing to the fact that we consider continuous variables, for which it is not possible to define probability values.

Indeed, it is possible to define Bayesian approach methods, defining some 'cost functions' $C(\boldsymbol{\theta} | \hat{\boldsymbol{\theta}})$ that assign a cost to each combination of 'estimated-value/actual-value' for the unknown parameter.

Bayesian estimators require knowledge of the *joint* PDF of the observation and of the parameters to be estimated, $p(\boldsymbol{z}, \boldsymbol{\theta})$, while for ML estimation it is sufficient to know the *conditional* PDF $p(\boldsymbol{z}|\boldsymbol{\theta})$†.

In Bayesian estimation the expected value of the cost function, i.e. the Bayes cost, is given by

$$B(\hat{\boldsymbol{\theta}}) = E\{C[\boldsymbol{\theta}, \hat{\boldsymbol{\theta}}(\boldsymbol{z})]\} = \int_{D_z} \int_{D_\theta} C[\boldsymbol{\theta}, \hat{\boldsymbol{\theta}}(z)]p(\boldsymbol{z}, \boldsymbol{\theta})d\boldsymbol{\theta}d\boldsymbol{z} \tag{1.160}$$

For any given cost function $C(\boldsymbol{\theta}|\hat{\boldsymbol{\theta}})$ the estimate $\hat{\theta}_B$ is optimum according to the Bayesian approach if

$$B(\hat{\boldsymbol{\theta}}_B) \leqslant B(\hat{\boldsymbol{\theta}}) \text{ for every } \hat{\boldsymbol{\theta}} \tag{1.161}$$

A large number of Bayes estimators can be defined, depending on the cost function chosen, for instance:

- Quadratic cost function
- Uniform cost function
- Absolute value cost function.

In the following we will sketch their characteristics.

1.3.3.1 *Quadratic cost function (mean square error) estimator*

The mean square error cost function related to a vector of dimension K to be estimated is:

$$C_{MS}(\boldsymbol{\theta}, \hat{\boldsymbol{\theta}}) = ||\boldsymbol{\theta} - \hat{\boldsymbol{\theta}}||^2 = \sum_{i=1}^{K} (\theta_i - \hat{\theta}_i)^2 \tag{1.162}$$

Since $B(\hat{\boldsymbol{\theta}})$ can be written as $E\{B(\hat{\boldsymbol{\theta}}|\boldsymbol{z})\}$, one must minimise the function

$$B_{MS}(\hat{\boldsymbol{\theta}}|\boldsymbol{z}) = \int_{D_\theta} ||\boldsymbol{\theta} - \hat{\boldsymbol{\theta}}||^2 p(\boldsymbol{\theta}|\boldsymbol{z})d\boldsymbol{\theta} \tag{1.163}$$

By equating to zero the derivative with respect to $\hat{\boldsymbol{\theta}}$, we obtain the final expression

$$\hat{\boldsymbol{\theta}}_{MS} = \int_{D_\theta} \boldsymbol{\theta} p(\boldsymbol{\theta}|\boldsymbol{z})d\boldsymbol{\theta} \tag{1.164}$$

that is the expected value of $\boldsymbol{\theta}$ conditioned to the observation z. Finally, using the Bayes rule, the result can be expressed in a more suitable form for computation:

$$\hat{\boldsymbol{\theta}}_{MS} = \frac{\int_{D_\theta} \boldsymbol{\theta} p(\boldsymbol{z}|\boldsymbol{\theta})p(\boldsymbol{\theta})d\boldsymbol{\theta}}{\int_{D_\theta} p(\boldsymbol{z}|\boldsymbol{\theta})p(\boldsymbol{\theta})d\boldsymbol{\theta}} \tag{1.165}$$

where only $p(\boldsymbol{\theta})$ and $p(z|\boldsymbol{\theta})$ must be known.

† It is worth noting that in the Bayesian approach the unknown parameter θ is a random vector.

1.3.3.2 *Uniform cost function estimator*

The uniform cost function for a vector of dimension N to be estimated can be written as

$$C_{UC}(\boldsymbol{\theta}, \hat{\boldsymbol{\theta}}) = \begin{cases} 0 & \text{if } |\theta_k - \hat{\theta}_k| < \varepsilon \quad k = 1, 2, \ldots N \\ 1 & \text{otherwise} \end{cases} \tag{1.166}$$

where the ε value is small. This function gives zero penalty if all the components of the estimation error are less than ε, and penalty one in the opposite case (one component or more greater than ε).

In the same way as for the quadratic cost function, the uniform cost function can be minimised by minimising its conditioned expected value w.r.t. $\boldsymbol{z}$, $B_{UC}(\hat{\boldsymbol{\theta}}|\boldsymbol{z})$:

$$B_{UC}(\hat{\boldsymbol{\theta}}|\boldsymbol{z}) = \int_{|\theta_k - \hat{\theta}_k| > \varepsilon, \forall k} p(\boldsymbol{\theta}|\boldsymbol{z}) d\boldsymbol{\theta} \tag{1.167}$$

since an error occurs whenever one or more than one estimated component has an error greater than ε.

The above integral can be written as

$$B_{UC}(\hat{\boldsymbol{\theta}}|\boldsymbol{z}) = \int_{D_\theta} p(\boldsymbol{\theta}\ |\boldsymbol{z}) d\boldsymbol{\theta} - \int_{\hat{\boldsymbol{\theta}}-\boldsymbol{\varepsilon}}^{\hat{\boldsymbol{\theta}}+\boldsymbol{\varepsilon}} p(\boldsymbol{\theta}|\boldsymbol{z}) d\boldsymbol{\theta} \tag{1.168}$$

and the first integral equals 1. Applying the mean-value theorem to the second integral (for *small* ε) we get

$$B_{UC}(\hat{\boldsymbol{\theta}}|\boldsymbol{z}) = 1 - (2\varepsilon)^N p(\hat{\boldsymbol{\theta}}|\boldsymbol{z}) \tag{1.169}$$

and to minimise the cost function it is necessary to choose $p(\hat{\boldsymbol{\theta}}|\boldsymbol{z})$ as large as possible. This is expressed by

$$p(\hat{\boldsymbol{\theta}}_{UC}|\boldsymbol{z}) \geqslant p(\hat{\boldsymbol{\theta}}|\boldsymbol{z}) \tag{1.170}$$

In practice, this estimator tends to maximise the *a posteriori* probability, and for this reason it is called the MAP estimator; it is also called the conditional mode estimator, because it leads to the maximum value of the PDF $p(\hat{\boldsymbol{\theta}}_{UC}|\boldsymbol{z})$, that is the (conditional) mode.

1.3.4 *Least-squares method and applications*

In the following sections we will try to outline the least-squares method, and its application to estimation problems. Firstly, the generalities of the method are described; then, two kinds of applications of least-squares estimation are presented.

Several concepts involving matrices are recalled and used in the next three sections; the interested reader is invited to refer to Appendix 2, where the main ideas about these topics are summarised.

1.3.4.1 *Generalities of the method*

The general least-squares problem is set out, in a very intuitive way, as follows: let us suppose we have a number M of measurements of a certain phenomenon, from which we want to estimate N parameters linearly dependent on those measurements. We wish to express the linear relationship as follows:

$$\boldsymbol{z} = A\boldsymbol{x} \tag{1.171}$$

where $\boldsymbol{z}$ is an $(M \times 1)$ vector of observed (measured) values, $\boldsymbol{x}$ is an $(N \times 1)$ vector of parameters, and the matrix A represents the linear dependence between observed values and parameters. Generally, the number of observations M is different from and often greater than the number of parameters to be estimated N, so that the system of equations does not have exact solutions. In this case, we are interested in finding a particular solution $\boldsymbol{x}_{LS}$ that minimises a particular norm of the error $\boldsymbol{\varepsilon} = |A\boldsymbol{x} - \boldsymbol{z}|$.

It is readily found that the usual square norm is convenient, for several reasons: firstly, a simple solution form is obtained, whilst other norms (the absolute value, for instance) do not produce general solutions that can be expressed in a simple and compact way; moreover, the square norm intrinsically 'weights' large errors more and the smaller ones less.

The square norm q of the error is given by

$$\begin{aligned} q &= \boldsymbol{\varepsilon}^T \boldsymbol{\varepsilon} = (A\boldsymbol{x} - \boldsymbol{z})^T (A\boldsymbol{x} - \boldsymbol{z}) \\ &= \boldsymbol{x}^T A^T A \boldsymbol{x} + \boldsymbol{z}^T \boldsymbol{z} - \boldsymbol{x}^T A^T \boldsymbol{z} - \boldsymbol{z}^T A \boldsymbol{x} \end{aligned} \tag{1.172}$$

where the superscript T denotes the transposition.

The derivative of q with respect to the vector $\boldsymbol{x}$ (i.e. the vector whose components are the partial derivatives of q w.r.t. the components of $\boldsymbol{x}$) can be readily obtained using the following identities for any square matrix B:

$$\frac{\partial}{\partial \boldsymbol{x}} \boldsymbol{x}^T B \boldsymbol{x} = (B + B^T)\boldsymbol{x}$$

$$\frac{\partial}{\partial \boldsymbol{x}} \boldsymbol{x}^T B \boldsymbol{z} = B\boldsymbol{z}$$

$$\frac{\partial}{\partial \boldsymbol{x}} \boldsymbol{z}^T B \boldsymbol{x} = B^T \boldsymbol{z}$$

This gives

$$\frac{\partial q}{\partial \boldsymbol{x}} = A^T A \boldsymbol{x} + A^T A \boldsymbol{x} - A^T \boldsymbol{z} - A^T \boldsymbol{z} = 2A^T (A\boldsymbol{x} - \boldsymbol{z}) \tag{1.173}$$

which is equal to a null vector $\boldsymbol{0}$ if $\boldsymbol{x}$ is equal to the least squares solution x_{LS}, i.e.

$$A^T (A\boldsymbol{x}_{LS} - \boldsymbol{z}) = \boldsymbol{0}$$

The resulting solution is

$$\boldsymbol{x}_{LS} = (A^T A)^{-1} A^T \boldsymbol{z} \tag{1.174}$$

It may be shown that this extremal minimises q.

The matrix

$$P = (A^T A)^{-1} A^T \boldsymbol{z} \tag{1.175}$$

is often called the pseudoinverse of A; when A is square, from expr. 1.175, it is readily seen that P is the inverse of A.

In the *complex field*, the square norm of the error is

$$q=\boldsymbol{\varepsilon}^H\boldsymbol{\varepsilon}=(A\boldsymbol{x}-\boldsymbol{z})^H(A\boldsymbol{x}-\boldsymbol{z})=\boldsymbol{x}^HA^HA\boldsymbol{x}+\boldsymbol{z}^H\boldsymbol{z}-\boldsymbol{x}^HA^H\boldsymbol{z}-\boldsymbol{z}^HA\boldsymbol{x} \quad (1.176)$$

where the superscript H denotes complex conjugated transposed.

Taking the N partial derivatives with respect to the components of $\boldsymbol{x}^H$ (that is, the gradient of q w.r.t. the vector $\boldsymbol{x}^H$) we obtain (using differentiation rules similar to the real field ones, see for instance [10] and paragraph 1.8.2.7):

$$\nabla_{x^H}(q)=A^H(A\boldsymbol{x}-\boldsymbol{z}) \quad (1.177)$$

that is equal to a null vector $\boldsymbol{0}$ if

$$A^HA\boldsymbol{x}_{LS}=A^H\boldsymbol{z}\Rightarrow\boldsymbol{x}_{LS}=(A^HA)^{-1}A^H\boldsymbol{z}=P\boldsymbol{z} \quad (1.178)$$

where P is the complex pseudoinverse of A. It is readily found that if $M=N$, then $P=A^{-1}$.

In some cases it is possible to know in advance that some measurements must have more importance than others in defining the error: this is accomplished by introducing coefficients (weights) that multiply each term of the error expression. Considering again the real field, these weights are written as an $(M\times 1)$ vector $\boldsymbol{W}$, so that the square norm becomes

$$q=(A\boldsymbol{x}-\boldsymbol{z})^T\boldsymbol{W}\boldsymbol{W}^T(A\boldsymbol{x}-\boldsymbol{z}) \quad (1.179)$$

the $(M\times M)$ matrix $(\boldsymbol{W}\boldsymbol{W}^T)$ is often defined in the literature as the 'weight matrix'; its direct definition is not immediately obvious, but is easy to understand the meaning of each term starting from the weights vector.

If we write the weight matrix $\boldsymbol{W}\boldsymbol{W}^T$ as W', the least-squares solution is

$$\boldsymbol{x}_{LS}=(A^TW'A)^{-1}A^TW'\boldsymbol{z} \quad (1.180)$$

1.3.4.2 *Least-squares estimation*

If N measurements are to be considered affected by some additive noise, the observation $\boldsymbol{z}$ is expressed as

$$\boldsymbol{z}=A\boldsymbol{x}+\boldsymbol{n}$$

where $\boldsymbol{x}$ is the vector to be estimated and $\boldsymbol{n}$ is the noise vector; the solution is readily found by following a 'deterministic' or a 'stochastic' approach. These approaches are dealt with in the following two subsections.

The deterministic approach

Let us examine first an application of the least-squares method to the problem of parameter estimation. This method does not use any stochastic information about the variable to be estimated; indeed it reduces the problem of parameter estimation to a problem of deterministic optimisation.

Again, one starts from the linear observer model (let us indicate this, as before, by $\boldsymbol{\theta}$, the parameter to be estimated):

$$\boldsymbol{z}=A\boldsymbol{\theta}+\boldsymbol{n} \quad (1.181)$$

trying to select an estimate $\hat{\boldsymbol{\theta}}$ *of* $\boldsymbol{\theta}$ so as to minimise the quadratic performance index

$$J(\boldsymbol{\theta})=\tfrac{1}{2}(\boldsymbol{z}-A\hat{\boldsymbol{\theta}})^TW'(\boldsymbol{z}-A\hat{\boldsymbol{\theta}}) \quad (1.182)$$

where the weights matrix W', assumed to be positive definite and symmetric, assigns different costs to each of the errors $(\boldsymbol{z}-A\hat{\boldsymbol{\theta}})$.

Setting to zero the first derivative of the index J, the estimator is determined:

$$\hat{\boldsymbol{\theta}}_{LS}=(A^TW'A)^{-1}A^TW'\boldsymbol{z} \tag{1.183}$$

As mentioned above, the estimator depends only on A (observations matrix) and on W' (weights matrix), and no PDF information is used.

In the complex field, as seen above, the superscript T (transpose) is replaced by the superscript H (Hermitian), so the estimator is given by the expression

$$\hat{\boldsymbol{\theta}}_{LS}=(A^HW'A)^{-1}A^HW'\boldsymbol{z} \tag{1.184}$$

Linear minimum-variance method (stochastic approach)

According to this method one looks for an unbiased linear estimator of the following type:

$$\hat{\boldsymbol{\theta}}_{LS}=\boldsymbol{b}+C\boldsymbol{z} \tag{1.185}$$

where $\hat{\boldsymbol{\theta}}_{LS}$ and $\boldsymbol{b}$ are $[K\times 1]$, C is $[K\times N]$ and $\boldsymbol{z}$ is $[N\times 1]$, while the observation $\boldsymbol{z}$ is assumed to be a linear function of the parameter $\boldsymbol{\theta}$, such as:

$$\boldsymbol{z}=A\boldsymbol{\theta}+\boldsymbol{n} \tag{1.186}$$

where, as usual, $\boldsymbol{n}$ is the noise vector.

The aim is to choose the vector $\boldsymbol{b}$ and the matrix C in order to minimise the error variance, obtaining a *linear minimum-variance* (LMV) *estimator*: it is a *linear sub-optimum* with respect to the (non-linear) conditional mean estimator, eqns. 1.164 and 1.165.

The calculations are rather cumbersome, and the intermediate steps necessary to determine $\boldsymbol{b}$ and C are not given, but we only present the final result.

The linear minimum variance estimate of $\boldsymbol{\theta}$, on the hypothesis that $E\{\boldsymbol{n}\}=\boldsymbol{0}$, and that $\boldsymbol{\theta}$ and $\boldsymbol{n}$ are uncorrelated, is given by

$$\hat{\boldsymbol{\theta}}_{LMV}=\boldsymbol{\mu}_\theta+M_\theta A^T[AM_\theta A^T+M_n]^{-1}(\boldsymbol{z}-A\boldsymbol{\mu}_\theta) \tag{1.187}$$

where

$$\boldsymbol{\mu}_\theta=E\{\boldsymbol{\theta}\} \qquad M_\theta=E[\boldsymbol{\theta}\boldsymbol{\theta}^T] \qquad M_n=E[\boldsymbol{n}\boldsymbol{n}^T] \tag{1.188}$$

($\boldsymbol{\mu}_\theta$ is the (vector) mean, and M_θ and M_n are the covariance matrices of the parameter to be estimated and of the noise, respectively).

The previous result can also be expressed in the form [1]

$$\hat{\boldsymbol{\theta}}_{MLV}=[A^TM_n^{-1}A+M_\theta^{-1}]^{-1}\,(A^TM_n^{-1}\boldsymbol{z}+M_\theta^{-1}\boldsymbol{\mu}_\theta) \tag{1.189}$$

In the first form the matrix to be inverted has the same dimension N of the observation vector $\boldsymbol{z}$; in the second form it has the dimension K of the parameter $\boldsymbol{\theta}$ (very often $K \ll N$). So the second form is often preferred to the first for computation time and precision reasons.

For this minimum-variance linear estimation method the knowledge of the (vector) mean and of the covariance matrix of the variable to be estimated is requested.

In any case, it can be shown that the variance of the LMV estimate is generally less than the LS estimate.

1.4 Applications of decision theory to signal detection

Until now, so far as the detection problem is concerned, we have considered only the particular cases in which just one observation is made, and the decision must follow from this information only.

This is not true of most practical cases, where more scalar or vector observations can either come from different channels, or be received in different time instants. Some cases relevant to radar are treated in the following sections.

The main concepts related to signals may be found in Appendix 1, Section 1.7.

1.4.1 *Detection of a known signal with additive Gaussian noise*

We shall now examine in more detail a decision problem that has practical applications in radar: the detection on the basis of N pulses received from a nonfluctuating target in the 'time-on-target' interval.

The target echo is a narrowband signal, expressed by the relationship

$$s_0(t) = A(t)\cos[2\pi(f_0 + f_D)t + \phi(t)] \tag{1.190}$$

where

f_0 = transmitted frequency

$f_D = \dfrac{2V_R f_0}{c}$ = Doppler shift due to the radial velocity V_R of the target

The observation vector $\boldsymbol{z}$ is the complex envelope of the target echo plus an additional interference $\boldsymbol{n}$ (e.g. receiver noise, clutter etc.) sampled at N integer multiples of the pulse repetition interval T:

$$\boldsymbol{z} = \boldsymbol{s} + \boldsymbol{n} \tag{1.191}$$

where

$$\boldsymbol{s}\colon s_i = s_0(iT), \qquad i = 1, \ldots N \tag{1.192}$$

and $s_0(t)$ is given by eqn. 1.190.

There are two possibilities:

(*a*) The target echo is a deterministic signal; that is, both its amplitude $A(t)$ and its initial phase $\phi(t)$† are known

(*b*) The target echo is partially known (no knowledge of the initial phase is available but the amplitude $A(t)$ is known).

In the following the basics of the treatment of both cases will be derived.

Case (a) Deterministic target: coherent detection

Let $\boldsymbol{z}$ be the vector of the complex envelopes of the N echoes corrupted by the interference:

$$\boldsymbol{z} = \{x_i + jy_i\}, \qquad i = 1, \ldots, N \tag{1.193}$$

while

$$\boldsymbol{s} = \boldsymbol{z} - \boldsymbol{n} \tag{1.194}$$

† In practical applications the only phase term is that due to the Doppler effect and $\Phi(t)$ is a constant: $\Phi(t) = \varphi_0$. Hence, the term 'initial phase'. The Doppler frequency f_D is supposed known.

is the complex vector (completely known, according to our hypotheses) of the N target echoes.

Let us suppose that the additive noise $\boldsymbol{n}$ is stationary, zero-mean Gaussian, with covariance matrix M *a priori* known:

$$M = [M_{ij}] \qquad M_{ij} \equiv E[n_i^* n_j] \tag{1.195}$$

where n_i are the noise samples. M is Hermitian; the transpose equals the complex conjugate: $M^T = M^*$.

It is easy to show that the joint probability distribution function of the noise only is the complex multivariate function (Appendix 3, Section 1.9.2)

$$p(\boldsymbol{z}|H_0) = \frac{1}{(\pi)^n \det[M]} \cdot \exp\left\{ -[\boldsymbol{z}^T M^{-1} \boldsymbol{z}^*] \right\} \tag{1.196}$$

Similarly, the joint PDF of signal plus noise is given by

$$p(\boldsymbol{z}|H_1) = \frac{1}{(\pi)^n \det[M]} \cdot \exp\left\{ -[(\boldsymbol{z}-\boldsymbol{s})^T M^{-1} (\boldsymbol{z}-\boldsymbol{s})^*] \right\} \tag{1.197}$$

The log-likelihood ratio can be written as

$$\begin{aligned} L(\boldsymbol{z}) &= \ln p(\boldsymbol{z}|H_1) - \ln p(\boldsymbol{z}|H_0) \\ &= -(\boldsymbol{z}-\boldsymbol{s})^T M^{-1} (\boldsymbol{z}-\boldsymbol{s})^* + \boldsymbol{z}^T M^{-1} \boldsymbol{z}^* \end{aligned} \tag{1.198}$$

i.e.

$$L(\boldsymbol{z}) = -\boldsymbol{s}^T M^{-1} \boldsymbol{s}^* + \boldsymbol{s}^T M^{-1} \boldsymbol{z}^* + \boldsymbol{z}^T M^{-1} \boldsymbol{s}^*$$

recalling that M^{-1} is Hermitian and therefore that $\boldsymbol{s}^T M^{-1} \boldsymbol{z}^* = (\boldsymbol{z}^T M^{-1} \boldsymbol{s}^*)^*$ we get

$$L(\boldsymbol{z}) = -\boldsymbol{s}^T M^{-1} \boldsymbol{s}^* + 2\mathrm{Re}[\boldsymbol{z}^T M^{-1} \boldsymbol{s}^*] \tag{1.199}$$

Putting the product $M^{-1}\boldsymbol{s}^*$ equal to $\boldsymbol{w}$, we have found the following decision rule for the hypotheses H_0 (target absent) and H_1 (target present):

$$\mathrm{Re}[\boldsymbol{z}^T \boldsymbol{w}] \underset{H_0}{\overset{H_1}{\gtrless}} \frac{1}{2} \boldsymbol{s}^T \boldsymbol{w} + \lambda \tag{1.200}$$

where λ is a constant.

Using the Neyman-Pearson criterion, the previous expression becomes

$$\mathrm{Re}[\boldsymbol{z}^T \boldsymbol{w}] \underset{H_0}{\overset{H_1}{\gtrless}} T, \quad \boldsymbol{w} = M^{-1} s^* \tag{1.201}$$

where the new parameter T depends, as shown in Section 1.2, on the chosen value of the false alarm probability P_{FA}.

The conclusion is that the optimum test is equivalent to a linear processing of the received signals, i.e. to the computation of

$$v = \boldsymbol{w}^T \boldsymbol{z} = \sum_{i=1}^{N} w_i z_i \tag{1.202}$$

and to the comparison of $\mathrm{Re}[v]$ with a threshold, whose value depends on the chosen criterion. (In case of the Neyman–Pearson criterion, the threshold depends on P_{FA}.) The processing is linear up to the comparison with the threshold.

Case (b) Initial phase unknown: non-coherent detection
In most radar and sonar applications, the initial phase φ_0 is not known. In this case, some modifications to the test must be done: the likelihood ratio must be averaged with respect to φ_0, which is assumed to be uniformly distributed in the $(0, 2\pi)$ interval.

The useful signal $\boldsymbol{s}$ can be written as follows:

$$\boldsymbol{s} = \boldsymbol{s}_0 \exp(j\varphi_0) \tag{1.203}$$

where s_0 is the deterministic term, assumed to be completely known, and φ_0 is the random, uniformly distributed phase term.

The likelihood ratio $\Lambda(\boldsymbol{z}; \varphi_0)$ depends on the observation $\boldsymbol{z}$ and on φ_0 and its expression is derived from eqns. 1.199 and 1.203.

The result is

$$\Lambda(\boldsymbol{z}; \varphi_0) = \exp[-\boldsymbol{s}_0^T \boldsymbol{w}_0 + 2\mathrm{Re}(\boldsymbol{z}^T \boldsymbol{w}_0 \, \mathrm{e}^{-j\varphi_0})] \tag{1.204}$$

where, by definition

$$\boldsymbol{w}_0 = M^{-1} \boldsymbol{s}_0^* \tag{1.205}$$

With the position

$$\boldsymbol{z}^T \boldsymbol{w}_0 = |\boldsymbol{z}^T \boldsymbol{w}_0| \exp(j\alpha) \tag{1.206}$$

the last term in eqn. 1.204 becomes

$$2\mathrm{Re}(\boldsymbol{z}^T \boldsymbol{w}_0 \, \mathrm{e}^{-j\varphi_0}) = 2|\boldsymbol{z}^T \boldsymbol{w}_0| \cos(\alpha - \varphi_0)$$

and the likelihood ratio, averaged over the random phase φ_0, is

$$\bar{\Lambda}(z) = E_{\varphi_0}[\Lambda(\boldsymbol{z}; \varphi_0)] = \exp[-\boldsymbol{s}_0^T \boldsymbol{w}_0] \frac{1}{2\pi} \int_0^{2\pi} \exp[2|\boldsymbol{z}^T \boldsymbol{w}_0| \cos(\alpha - \varphi_0)] \, \mathrm{d}\varphi_0 \tag{1.207}$$

where the argument α is a constant w.r.t. the integration variable.

The computation of the integral of the right hand side of eqn. 1.207 leads to a modified Bessel function of the first kind and of zero order $I_0(\cdot)$ [38]; in fact

$$I_0(x) = \frac{1}{2\pi} \int_0^{2\pi} \exp(x \cos \theta) \, \mathrm{d}\theta \tag{1.208}$$

Using eqn. 1.208 the averaged likelihood ratio gives

$$\bar{\Lambda}(\boldsymbol{z}) = \exp[-s_0^T \boldsymbol{w}_0] I_0(2|\boldsymbol{z}^T \boldsymbol{w}_0|) \tag{1.209}$$

and the test is

$$I_0(2|\boldsymbol{z}^T \boldsymbol{w}_0|) \underset{H_0}{\overset{H_1}{\gtrless}} \eta \exp(\boldsymbol{s}_0^T \boldsymbol{w}_0) \tag{1.210}$$

where η is a constant.

As the modified Bessel function $I_0(\cdot)$ is a monotonic function, the test can be simply reduced to

$$|\boldsymbol{z}^T \boldsymbol{w}_0| \underset{H_0}{\overset{H_1}{\gtrless}} T \tag{1.211}$$

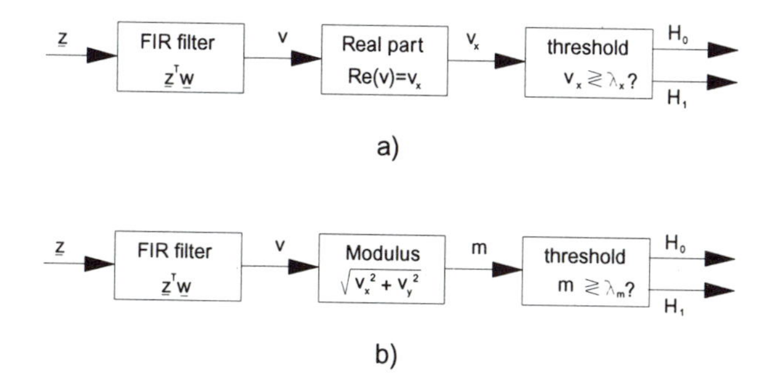

Figure 1.10 *Case of coherent detection (a) and unknown initial phase (b)*

This means that in practical radar and sonar applications, in which the initial phase is uniformly distributed, the optimum detector includes the computation of the inner product of the received sequence $\boldsymbol{z}$ with the weighting vector $\boldsymbol{w}_0$ defined by eqn. 1.205 and the comparison of the modulus of the inner product with a threshold.

Such a test, often referred to as 'envelope detection', is less efficient than the coherent one, expr. 1.201, case (a) above, as less *a priori* information is known about the signal to be detected.

The block diagrams of Fig. 1.10 show the different detection methods schematically: it is seen that the same linear processing is followed by another linear processing [case (a)] or by an envelope detector [case (b)]; in Fig. 1.11 a comparison of detection probabilities for both cases shows that, for $P_{FA}=10^{-6}$, $P_D=80\%$, the signal-to-noise loss due to the lack of knowledge of the initial phase [difference between case (b) and case (a)] is about 0.5 dB, and has a little dependence on P_{FA} and P_D.

The curves shown in Fig. 1.11 are obtained in the following way: for each value of the false alarm probability P_{FA} the corresponding threshold value is determined by inversion of the pertaining cumulative distribution function (for envelope detection it is the integral of the Rayleigh distribution, easily written in closed form; for the coherent case it is the error function, erf(.), the integral of the Gaussian PDF). The probability of detection P_D is then calculated, for a certain number of values of the signal-to-noise ratio, numerically integrating the corresponding PDF, with the threshold value found as the lower limit of integration. Again, for the coherent case, the integrand function is the Gaussian PDF, and for the envelope detection it is the Rician PDF.

1.4.2 *Optimum detection: Neyman–Pearson criterion, vector observations*

Let us consider the case in which the useful signal vector $\boldsymbol{s}$: s_k, $k=1, 2, \ldots N$ can be modelled as a Gaussian stochastic process, with assigned covariance matrix

$$M_s = [M_{s_{ij}}] \qquad M_{s_{ij}} \equiv E[s_i^* s_j] \tag{1.212}$$

and assigned vector of mean values

$$\boldsymbol{\mu}_s = [\mu_{s_i}] \qquad \mu_{s_i} = E\{s_i\},\ i = 1 \ldots n \tag{1.213}$$

whilst the additive, zero-mean Gaussian noise has the covariance matrix given by eqns. 1.195 (indicated hereafter, for the sake of clarity, by M_n).

The calculation of log-likelihood ratio $L(\boldsymbol{z})$ is carried out in the same way as in Section 1.4.1. Assuming $\boldsymbol{z} = \boldsymbol{s} + \boldsymbol{n}$ and dropping the terms that do not depend on the observation $\boldsymbol{z}$ we obtain the following expression for the log-likelihood ratio $L(\boldsymbol{z})$ (neglecting immaterial constants):

$$L(\boldsymbol{z}) = -(\boldsymbol{z} - \boldsymbol{\mu}_s)^T (M_s + M_n)^{-1} (\boldsymbol{z} - \boldsymbol{\mu}_s)^* + \boldsymbol{z}^T M_n^{-1} \boldsymbol{z}^* \tag{1.214}$$

and letting

$$Q = M_n^{-1} - (M_s + M_n)^{-1} \tag{1.215}$$

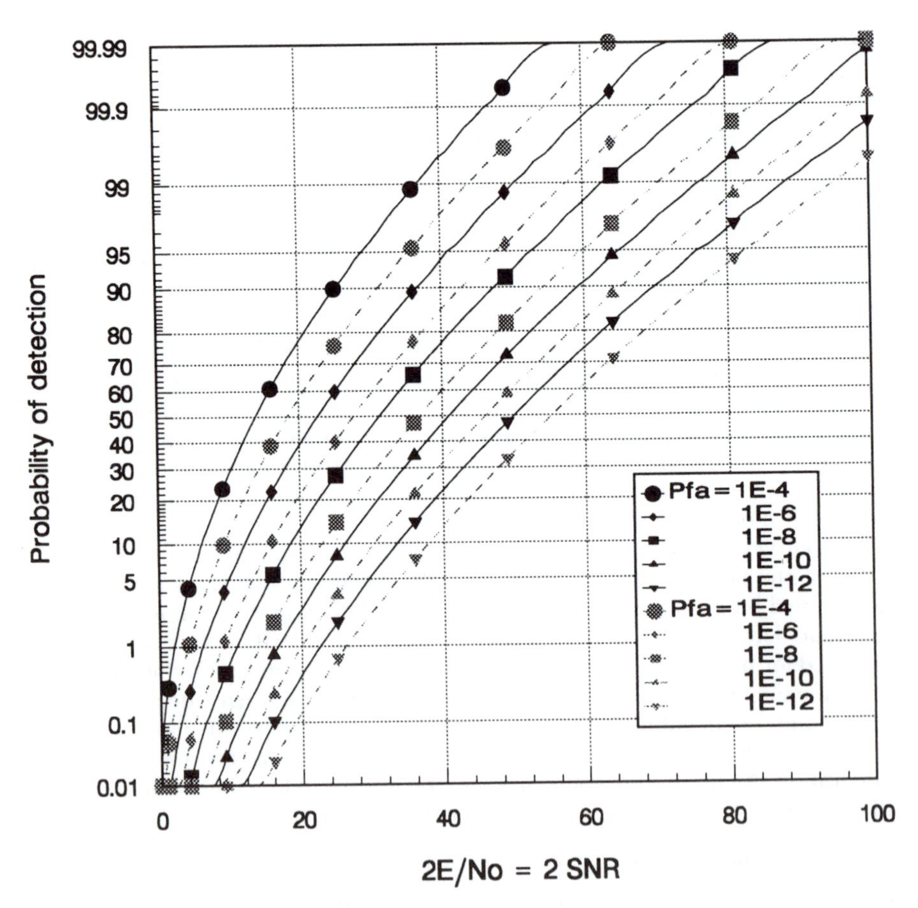

Figure 1.11 *Detection probability for the cases of coherent detection (continuous line) and unknown initial phase (dotted line)*

the following expression is immediately derived:

$$L(z) = z^T Q z^* + \mu_s^T (M_s + M_n)^{-1} z^* + z^T (M_s + M_n)^{-1} \mu_s^* - \mu_s^T (M_s + M_n)^{-1} \mu_s^* \quad (1.216)$$

Remembering that the matrix $(M_s + M_n)^{-1}$ is Hermitian (see Section 1.8.2.2) because M_s and M_n are both Hermitian, we can use the identity $\mu_s^T (M_s + M_n)^{-1} z^* = [z^T (M_s + M_n)^{-1} \mu_s^*]^*$ and eqn. 1.216 becomes:

$$\begin{aligned} L(z) &= z^T Q z^* + [z^T (M_s + M_n)^{-1} \mu_s^*]^* + z^T (M_s + M_n)^{-1} \mu_s^* - \mu_s^T (M_s + M_n)^{-1} \mu_s^* \\ &= z^T Q z^* + 2\mathrm{Re}[z^T (M_s + M_n)^{-1} \mu_s^*] - \mu_s^T (M_s + M_n)^{-1} \mu_s^* \end{aligned} \quad (1.217)$$

Finally, including in the threshold λ' the last term of expr. 1.217 (which does not depend on z) the test becomes

$$z^T Q z^* + 2\ \mathrm{Re}[z^T h] \underset{H_0}{\overset{H_1}{\gtrless}} \lambda' \quad (1.218)$$

where, by definition, $h = (M_s + M_n)^{-1} \mu_s^*$

Again, (see Section 1.4.1.*b*) eqn. 1.218 requires the knowledge of the initial phase φ_0 of the steady component μ_s of the signal. When the initial phase is uniformly distributed in the $(0, 2\pi)$ interval, the averaging of the likelihood ratio with respect to φ_0 leads to the following test (see eqn. 1.207):

$$\exp(z^T Q z^*) I_0(2|z^T h_0|) \underset{H_0}{\overset{H_1}{\gtrless}} \eta \exp(m_s^T h_0) \quad (1.219)$$

where η is a constant equal to $\exp(\lambda')$, m_s is the mean value of the signal without the initial phase term φ_0:

$$m_s = \mu_s \exp(-j\varphi_0) \quad (1.220)$$

and h_0 is the optimum weighting vector for a deterministic signal equal to the mean value m_s

$$h_0 = (M_s + M_n)^{-1} m_s^* \quad (1.221)$$

Taking the logarithm of eqn. 1.219 and introducing, as usual, the threshold T related to the desired P_{FA}, the test can be written as follows:

$$z^T Q z^* + \ln I_0(2|z^T h_0|) \underset{H_0}{\overset{H_1}{\gtrless}} T \quad (1.222)$$

It is now possible to consider the following particular cases:

(*a*) *Deterministic signal*: The deterministic signal is equal to μ_s and the likelihood ratio, computed in paragraph 1.4.1.a, can be obtained putting equal to zero the power of the stochastic component of the useful signal in eqns. 1.214–1.218. This gives

$$\mathrm{Re}[z^T M_n^{-1} s^*] = \mathrm{Re}[z^T w] \underset{H_0}{\overset{H_1}{\gtrless}} \frac{\lambda}{2} \quad (1.223)$$

with $\boldsymbol{w}=M_n^{-1}\,\boldsymbol{\mu}_s^*$ (see eqn. 1.201).

When the initial phase of the target signal is unknown, averaging with respect to it leads to the test

$$|\boldsymbol{z}^T\boldsymbol{w}_0| \underset{H_0}{\overset{H_1}{\gtrless}} T, \quad \boldsymbol{w}_0=M_n^{-1}\boldsymbol{m}_s^* \tag{1.224}$$

where $\boldsymbol{m}_s$ is defined by eqn. 1.220.

(*b*) *Zero-mean stochastic signal*: Equating $\boldsymbol{\mu}_s$ to zero in eqn. 1.218 we obtain the expression:

$$\boldsymbol{z}^T Q\boldsymbol{z}^* \underset{H_0}{\overset{H_1}{\gtrless}} \lambda' \tag{1.225}$$

This problem was discussed in Section 1.2.2.4.

In the case of a deterministic signal [case (*a*)] the optimum processor is linear (eqn. 1.223); on the contrary, in case (*b*) the optimum processor involves a quadratic form (eqn. 1.225), but also, in this case, there exists a sub-optimum linear processor of the form

$$|\boldsymbol{z}^T\boldsymbol{w}| \underset{H_0}{\overset{H_1}{\gtrless}} \lambda \tag{1.226}$$

where $\boldsymbol{w}$ is the eigenvector of the pencil of matrices (see Appendix 2)

$$M_n\boldsymbol{w}=\lambda M_s\boldsymbol{w} \tag{1.227}$$

associated with the minimum (non-zero) eigenvalue of the pencil. Such a suboptimum linear processor is one which maximises the output signal-to-interference ratio, SIR_0, for a fixed input SIR, i.e. the Rayleigh quotient (see Section 1.8.3.6):

$$\mathrm{SIR}_0=\frac{\boldsymbol{w}^H M_s\boldsymbol{w}}{\boldsymbol{w}^H M_n\boldsymbol{w}} \tag{1.228}$$

leading to the eigenproblem of eqn. 1.227; the maximum value of the output SIR is readily shown to be equal to the inverse of the minimum eigenvalue of the pencil.

(*c*) *Non-zero mean stochastic signal*: In the most general case, the useful signal has a fixed-amplitude component $\boldsymbol{\mu}_s$ (of which the deterministic term $\boldsymbol{m}_s$ without the initial phase is relevant) and a Gaussian component with covariance matrix M_s.

The likelihood ratio test for the problem at hand is given by eqn. 1.222, which may be conveniently approximated considering the fact that the term

$$2|\boldsymbol{z}^T\boldsymbol{h}_0|$$

(where $\boldsymbol{h}_0$ is given by eqn. 1.221) is, on average, smaller than unity for small values of the signal-to-interference power ratio.

In such a situation, using only the first term of the power series expansion of the modified Bessel function $I_0(\cdot)$ (see for instance eqn. 10.3.17 of Reference 7) one may obtain

$$\ln I_0(x) \cong x^2/4$$

and therefore

$$\ln I_0(2|\boldsymbol{z}^T\boldsymbol{h}_0|) \cong |\boldsymbol{z}^T\boldsymbol{h}_0|^2 \tag{1.229}$$

and eqn. 1.222 becomes

$$\boldsymbol{z}^T Q\boldsymbol{z}^* + |\boldsymbol{z}^T\boldsymbol{h}_0|^2 \underset{H_0}{\overset{H_1}{\gtrless}} T \tag{1.230}$$

i.e.

$$\boldsymbol{z}^T R\boldsymbol{z}^* \underset{H_0}{\overset{H_1}{\gtrless}} T \tag{1.231}$$

with the matrix R being defined as follows:

$$R = Q + \boldsymbol{h}_0\boldsymbol{h}_0^H \tag{1.232}$$

i.e. recalling eqns. 1.221 and 1.215

$$R = M_n^{-1} - (M_s + M_n)^{-1} + (M_s + M_n)^{-1}\boldsymbol{m}_s^*\boldsymbol{m}_s^T(M_s + M_n)^{-1} \tag{1.233}$$

In deriving the last expression, the Hermitian property for $(M_s + M_n)^{-1}$ has been used.

Eqns. 1.231, 1.232 and 1.233 show that in the small-signal case the likelihood ratio detector for a target with a deterministic (steady) component plus a zero-mean Guassian (fluctuating) component still reduces to the computation of a quadratic form, i.e. to the block diagram of Fig. 1.9.

Such a detector is much easier to be implemented than the one of eqn. 1.222; moreover, it is readily seen that it reduces to the detector for a deterministic signal when the power of the fluctuating signal component approaches zero, i.e. M_s is negligible with respect to M_n, while it reduces to the detector for zero-mean stochastic signals when the deterministic component $\boldsymbol{m}_s$ tends to zero. In other words, the small signal approximation leads to the optimum detector in these two extreme cases, while it leads to a suboptimum solution (but tailored to the most critical situation of small signal-to-noise ratio) in the intermediate cases.

Part 2: Optimum filtering

In this Part some significant aspects of optimum filtering are introduced. The process of filtering consists of passing the signal through a linear system to modify it in a convenient manner. Here, we are interested in a particular kind of filtering in which the optimality criterion to be attained is the maximisation of the signal-to-noise power ratio. In particular, in the case of the so-called *matched filters*, the output peak signal-to-noise power ratio, SNR, is maximised. Both the continuous and the discrete-time cases are discussed.

1.5 Continuous-time optimum filtering

1.5.1 *The matched filter*

Let us consider a signal $s(t)$ with finite energy E:

$$E = \int_{-\infty}^{\infty} |s(t)|^2 dt \tag{1.234}$$

and Fourier transform $S(f)$:

$$S(f) = \int_{-\infty}^{\infty} s(t) e^{-j2\pi ft} dt \tag{1.235}$$

Let us consider the signal $s(t)$ to be the input of a linear system (filter) with transfer function $H(f)$; the instantaneous power of the output signal $s_0(t)$ is given by the expression

$$|s_0(t)|^2 = \left| \int_{-\infty}^{\infty} S(f) H(f) e^{j2\pi ft} df \right|^2 \tag{1.236}$$

In the same way, when a white noise of one-sided spectral power density N_0 is present at the filter input, we can express the mean output noise power as

$$N = \frac{N_0}{2} \int_{-\infty}^{\infty} |H(f)|^2 df \tag{1.237}$$

We are interested in maximising the ratio

$$R = \frac{|s_0(t)|^2_{max}}{N} \tag{1.238}$$

where the numerator is the maximum value of the instantaneous power of the output signal and N is the noise power, as shown above†. From exprs. 1.236–1.238 we obtain

† It is worth noting that the ratio R is the peak signal-to-noise power ratio which is different from the signal-to-noise ratio (S/N) generally considered in communication theory, because the S/N ratio uses the average value of signal power of a received symbol, not the maximum value.

$$R = \frac{\left| \int_{-\infty}^{\infty} S(f)H(f)e^{j2\pi fT} df \right|^2}{N_0/2 \int_{-\infty}^{\infty} |H(f)|^2 df} \tag{1.239}$$

where we indicate by T the time at which the maximum value of $|s_0(t)|^2$ occurs.

We recall here Schwartz's inequality, that states (in one of its several forms):

$$\int P^*P dx \int Q^*Q dx \geqslant \left| \int P^*Q dx \right|^2 \tag{1.240}$$

where P and Q are two complex functions of the real variable x; the equality sign only applies when the two functions are linearly related, that is when $P = kQ$, $k =$ constant.

In our case, we set

$$P^* = S(f)e^{j2\pi fT} \text{ and } Q = H(f) \tag{1.241}$$

From expressions 1.239–1.241, we have:

$$R \leqslant \frac{\int_{-\infty}^{\infty} |H(f)|^2 df \int_{-\infty}^{\infty} |S(f)|^2 df}{N_0/2 \int_{-\infty}^{\infty} |H(f)|^2 df} = \frac{\int_{-\infty}^{\infty} |S(f)|^2 df}{N_0/2} \tag{1.242}$$

remembering (Parseval's theorem) that

$$\int_{-\infty}^{\infty} |S(f)|^2 df = E \text{ (signal energy)} \tag{1.243}$$

it is concluded that

$$R \leqslant \frac{2E}{N_0} \tag{1.244}$$

The equality sign (the maximum value for the ratio R) is reached when $P = kQ$, that is when

$$H(f) = G_a S^*(f) e^{-j2\pi fT} \tag{1.245}$$

where G_a is a constant.

Eqn. 1.245 defines the optimum filter in the frequency domain; it is interesting to characterise it in the time domain. Let us take the inverse Fourier transform of $H(f)$ to obtain the filter impulse response $h(t)$:

$$h(t) = G_a \int_{-\infty}^{\infty} S^*(f) e^{[-j2\pi f(T-t)]} df \tag{1.246}$$

we obtain

$$h(t) = G_a \left[\int_{-\infty}^{\infty} S(f) e^{j2\pi f(T-t)} df \right]^* \tag{1.247}$$

and remembering that $s(t)=\int_{-\infty}^{\infty} S(f)e^{j2\pi ft}df$:

$$h(t)=G_a s^*(T-t) \tag{1.248}$$

Eqn. 1.248 shows that the impulse response of the optimum filter is a time-reversed, conjugated and delayed version of the input signal $s(t)$, that is, it is *matched* to the input signal.

In the more general case, when noise is not white (coloured noise), the derivation of the matched filter can be carried out in a similar way. It can be shown that, if the spectral density of the non-white noise is $N_i(f)$, with the positions

$$P^*=S(f)e^{j2\pi fT}/N_i(f) \text{ and } Q=N_i(f)\cdot H(f) \tag{1.249a}$$

the transfer function of the filter is given by

$$H(f)=\frac{G_a' S^*(f)e^{-j2\pi fT}}{[N_i(f)]^2} \tag{1.249b}$$

where G_a' is a constant.

This general expression for non-white noise can be written as

$$H(f)=\frac{G_a'}{N_i(f)}\left(\frac{S(f)}{N_i(f)}\right)^* e^{-j2\pi fT} \tag{1.250}$$

and the maximum value for the ratio R is:

$$R\leqslant\frac{\left|\int_{-\infty}^{\infty}\left|\frac{S(f)}{N_i(f)}\right|^2 df\right|^2}{\int_{-\infty}^{\infty}\left|\frac{S(f)}{N_i(f)}\right|^2\frac{1}{N_i(f)}df} \tag{1.251}$$

The matched filter for non-white noise can be interpreted as the cascade of two filters. The first one, whose transfer function is $1/N_i(f)$, is the 'whitening' filter, that is one that makes the noise spectrum uniform (white). The second one is matched to the signal filtered by the whitening filter, i.e. to

$$s_w(t)=\int_{-\infty}^{\infty}\frac{S(f)}{N_i(f)}e^{j2\pi ft}df \tag{1.252}$$

1.5.2 *The matched filter as a correlator*

From expr. 1.248, recalling the convolution operator (see Section 1.7.6 expr. 1.490)

$$x(t)*y(t)=\int_{-\infty}^{\infty}x(\tau)y(t-\tau)d\tau \tag{1.253}$$

we find that the matched filter output $s(t)*h(t)$, with $h(t)$ given by eqn. 1.248, is

$$s_0(t)=s(t)*h(t)=\int_{-\infty}^{\infty}h(\tau)s(t-\tau)d\tau$$

$$=G_a\int_{-\infty}^{\infty}s^*(T-\tau)s(t-\tau)d\tau \tag{1.254}$$

i.e. using the variable $\theta = T - \tau$:

$$s_0(t) = G_a \int_{-\infty}^{\infty} s^*(\theta)s(\theta + t - T)d\theta \qquad (1.255)$$

that is, recalling (Section 1.7.1.1, expr. 1.371) the definition of the autocorrelation function $R_s(\cdot)$, we get

$$s_0(t) = R_s(t - T) \qquad (1.256)$$

(to within insignificant gain constant); i.e. the matched filter output is the autocorrelation of the transmitted signal $s(t)$, with a time delay T corresponding, as shown before, to the processing time.

At the time the input is perfectly aligned with the matched filter response, i.e. at $t = T$, the output is

$$s_0 = R_s(0) = \int_{-\infty}^{\infty} |s(\tau)|^2 d\tau = E \qquad (1.257)$$

where E is the signal energy.

If $h(t) = s^*(T - t)$ is the impulse response of the matched filter, and if a white noise with spectral density $N_0/2$ is present at its input, the noise power at its output is

$$N = \frac{N_0}{2} \int_{-\infty}^{\infty} |h(\tau)|^2 d\tau = \frac{N_0}{2} \int_{-\infty}^{\infty} |s(\theta)|^2 d\theta = \frac{N_0}{2} E \qquad (1.258)$$

Therefore, the peak (i.e. for $t = T$) signal-to-noise ratio at the matched filter output is

$$SNR_0 = \frac{s_0^2}{\frac{N_0}{2} E} \qquad (1.259)$$

which, substituting expr. 1.257 for s_0, yields

$$SNR_0 = \frac{2E}{N_0} \qquad (1.260)$$

which can be compared with eqn. 1.244.

This fact leads to an important and interesting property of the matched filter for white noise: the maximum signal-to-noise power ratio is always twice the signal energy divided by the noise spectral density, and it is not dependent on the signal shape. The only problem is that we are not always able to make a matched filter for the given signal shape, because in some cases a non-causal filter could be realised (i.e. one with the impulse response starting before the impulse application).

1.5.3. *Practical applications and examples of matched filtering*

Two important properties of practical interest are given:

(i) From expr. 1.255 for the matched filter output:

$$s_0(t) = \int_{-\infty}^{\infty} s^*(\tau)s[\tau + (t - T)]d\tau \qquad (1.261)$$

considering the *complex envelope* $\tilde{s}(t)$ (see Section 1.7) such that

$$s(t) = \tilde{s}(t)e^{j\omega_c t} \tag{1.262}$$

where ω_c is the angular frequency of an IF or RF carrier, we obtain

$$s_0(t) = e^{j\omega_c(t-T)} \int_{-\infty}^{\infty} \tilde{s}(\tau)\tilde{s}(\tau + [t-T])d\tau$$

$$= e^{j\omega_c t}e^{-j\omega_c T}R_s(t-T) \tag{1.263}$$

this means that the output signal from a matched filtered baseband signal (complex signal, I, Q components) and that of an IF or RF only differ by the carrier $e^{j\omega_c t}$ and by a (irrelevant) phase term $e^{-j\omega_c T}$, corresponding to the processing delay introduced by the filter.

(ii) If we consider, instead of the signal $s(t)$, another signal

$$\bar{s}(t) = s(t)e^{ja} \tag{1.264}$$

where α is a phase term that does not depend on time, one immediately obtains from expr. 1.255 the corresponding output:

$$\bar{s}_0(t) = \int_{-\infty}^{\infty} s^*(\theta)s(\theta + t - T)\ \mathrm{d}\theta \tag{1.265}$$

that is, α does not influence the matched filter output.

1.5.3.1 *Examples of matched filter application*

(a) RF pulse with rectangular envelope

Let us consider a rectangular RF pulse of duration T seconds and unit energy:

$$s(t) = \begin{cases} \sqrt{\dfrac{2}{T}}\cos(2\pi f_c t) & 0 \leq t \leq T \\ 0 & \text{elsewhere} \end{cases} \tag{1.266}$$

where f_c is a (large) integer multiplier of $1/T$. The impulse response of a filter matched to $s(t)$ is then $s(t)$ itself, because $\cos(2\pi f_c t) = \cos[2\pi f_c(T-t)]$.

The output of the matched filter, when excited by the signal $s(t)$, is the self-convolution of $s(t)$ itself, giving

$$s_0(t) = \begin{cases} (t/T)\ \cos(2\pi f_c t) & 0 \leq t \leq T \\ (2 - t/T)\ \cos(2\pi f_c t) & T \leq t \leq 2T \\ 0 & \text{elsewhere} \end{cases} \tag{1.267}$$

This is shown in Fig. 1.12 from which it is possible to see that the maximum output is achieved at $t = T$. It is also evident that, if the pulse were sampled at a rate lower than $2f_c$, bad timing would cause a very high loss, owing to the high steepness of the $s_0(t)$ curve.

The same result could be obtained using property (i): the carrier can be 'cancelled' and the unit-energy signal $s(t)$ would be

$$s(t) = \begin{cases} \sqrt{\dfrac{1}{T}} & 0 \leq t \leq T \\ 0 & \text{elsewhere} \end{cases} \tag{1.268}$$

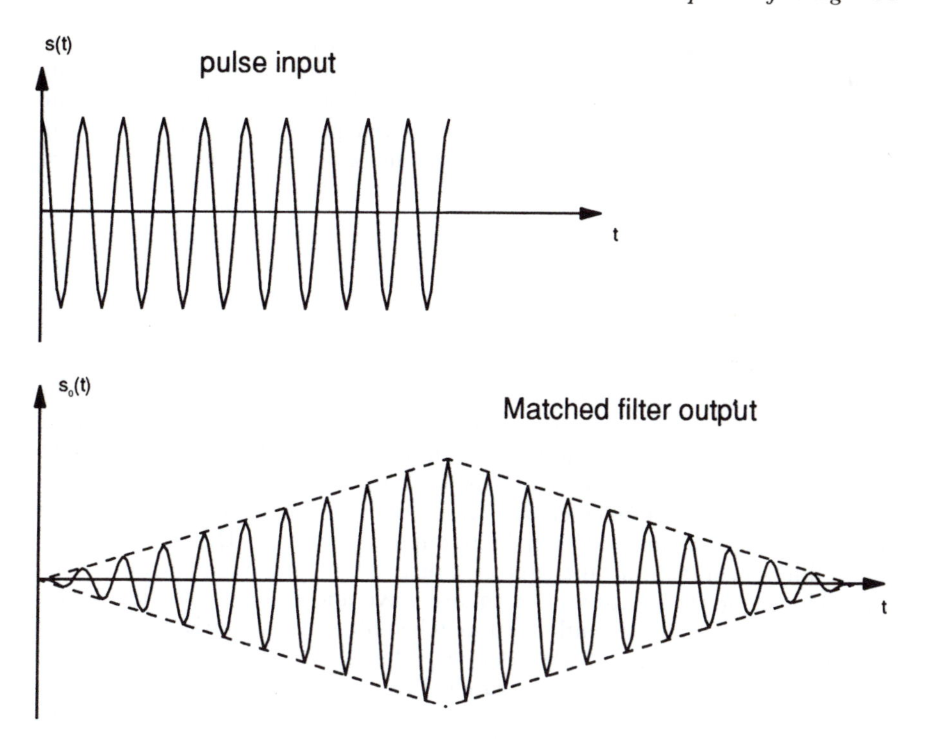

Figure 1.12 *Example of matched filter for RF or IF pulse*

The matched filter impulse response is also the same as $s(t)$ in this case; the output from the matched filter, when excited by $s(t)$, is now

$$s_0(t) = \begin{cases} (t/T) & 0 < t \leq T \\ (2 - t/T) & T < t \leq 2T \\ 0 & \text{elsewhere} \end{cases} \tag{1.269}$$

(*b*) Let us now consider a phase-modulated pulse, with duration $T = kt_1$, such as

$$s(t) = \begin{cases} e^{j\alpha_1}\cos(2\pi f_c t) & 0 < t \leq t_1 \\ e^{j\alpha_2}\cos(2\pi f_c t) & t_1 < t \leq kt_1 \\ 0 & \text{elsewhere} \end{cases} \tag{1.270}$$

with $\alpha_1 - \alpha_2 = \pi$ and $k = 1.5$ (this particular type of signal is known as '3-elements Barker code'); from property (i) we can neglect the carrier, as shown above, and take into account the phase difference only [property (ii)], not the absolute phase values. This leads to a 'baseband' signal described by the following relationships:

$$\tilde{s}(t) = \begin{cases} 1 & 0 < t \leq t_1 \\ -1 & 1 < t \leq kt_1 \\ 0 & \text{elsewhere} \end{cases} \tag{1.271}$$

The output of this pulse through its matched filter can be easily calculated by means of the simple method shown in Fig. 1.13, adding the products of the corresponding pieces, or 'chips', of the signal and of the shifted version of the signal itself, and so determining the representative points in the graph showing the matched filter output; the filter output is linear in all the points comprised between them.

1.6 Discrete-time optimum filtering

It is possible to extend all the considerations of the previous section to the discrete-time case. Digital signal processing is becoming more and more common in every branch of communications and radar techniques, owing to the possibility of an exact and reliable implementation of complicated algorithms.

So it is quite easy to implement a matched filter for every waveform, using the infinite impulse response (IIR) or finite impulse reponse (FIR) techniques for digital filters, whilst in the analogue case just a few types of matched filters are realisable (integrate-and-dump for square wave and resonant circuits for RF pulses are examples of currently used matched filters). A remarkable exception are surface acoustic wave (SAW) devices, allowing straightforward and exact implementation at IF of the filter matched to any waveform of finite duration (the main limitation being the maximum duration of the waveform to be compressed, owing to the limited velocity of the acoustic wave, namely 3 mm/μs, and the need to keep the dimension of the substrate acceptable).

1.6.1 *Introduction and definitions*

Most of the definitions which follow are contained in Sections 1.7 and 1.8, to which the interested reader can refer.

Given a complex random vector $\boldsymbol{x}$: x_i, $i=1, 2, \ldots N$, we can define its covariance matrix as

$$M_x = E\{\boldsymbol{x}^*\boldsymbol{x}^T\} \tag{1.272}$$

and the power of $\boldsymbol{x}$ (its mean square value) as

$$P_x = E\{|\boldsymbol{x}|^2\} \tag{1.273}$$

Given a linear processor $\boldsymbol{w}$: w_i, $i=1, 2, \ldots N$, that is a vector of complex coefficients that multiply the values of x_i in the form $z=\boldsymbol{w}^T\boldsymbol{x}$, the output power is

$$E\{|z|^2\} = E\{\boldsymbol{w}^H\boldsymbol{x}^*(\boldsymbol{x}^T\boldsymbol{w})\} = \boldsymbol{w}^H M_x \boldsymbol{w} \tag{1.274}$$

Let us now consider the signal $\boldsymbol{x}$ as useful signal $\boldsymbol{s}$ plus an additive noise $\boldsymbol{n}$:

$$\boldsymbol{x} = \boldsymbol{s} + \boldsymbol{n} \tag{1.275}$$

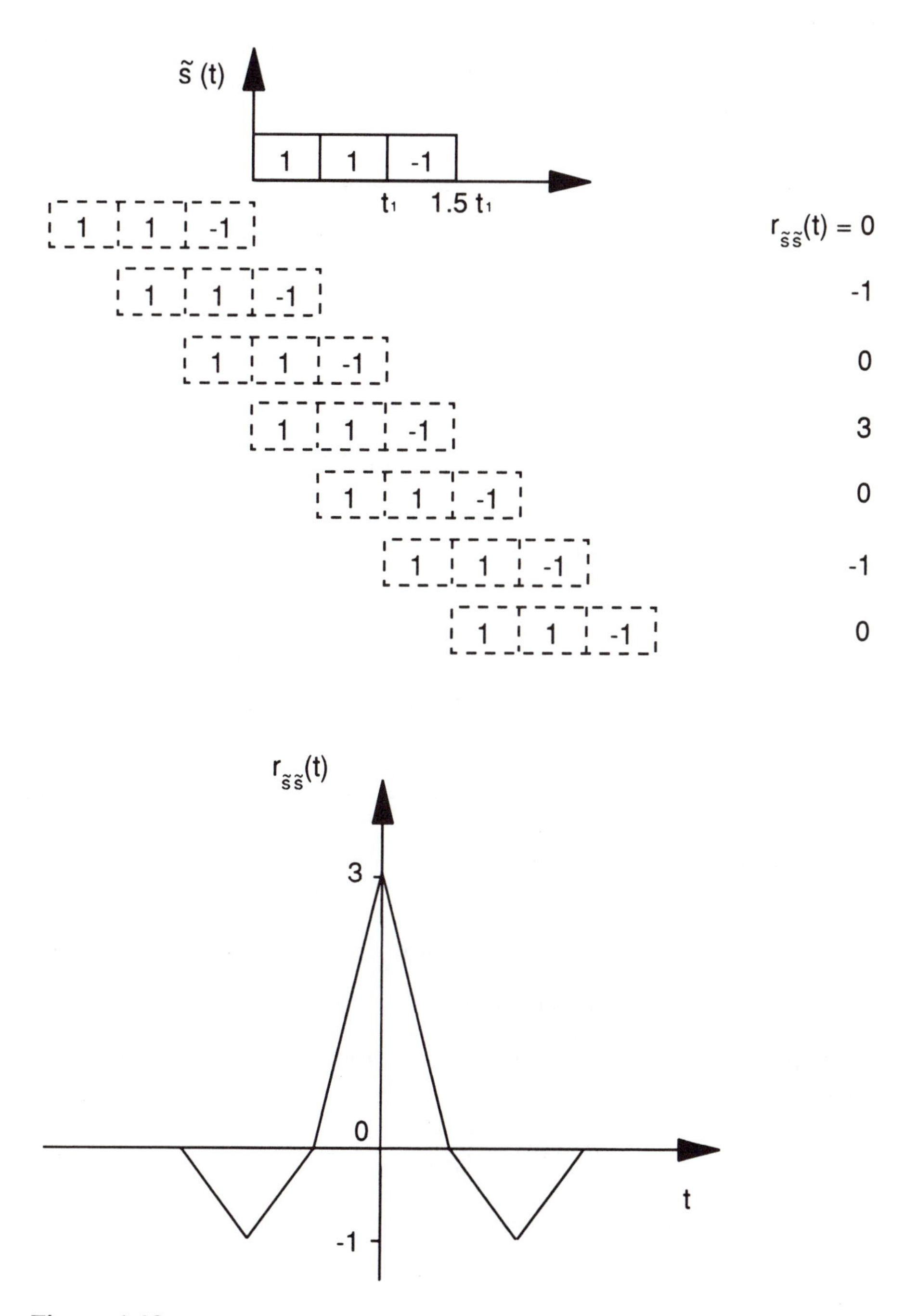

Figure 1.13 *Autocorrelation function of the Barker 3 code*

where $\boldsymbol{s}$ can be:

(i) A *deterministic* signal, of the type $\boldsymbol{s}=\alpha\tilde{\boldsymbol{s}}$, with $\tilde{\boldsymbol{s}}$ completely known, and α a complex random variable, with $E\{|\alpha|^2\}\equiv P_\alpha$, whose argument and modulus are the unknown initial phase and amplitude of the signal, respectively†

(ii) A *random* signal, with covariance matrix M_s.

The processor output is

$$z=\boldsymbol{w}^T\boldsymbol{x}=\boldsymbol{w}^T\boldsymbol{s}+\boldsymbol{w}^T\boldsymbol{n} \tag{1.276}$$

and the signal-to-interference ratio (SIR) at the output of the linear processor $\boldsymbol{w}$ is given by

$$SIR_z(\boldsymbol{w})=\frac{\boldsymbol{w}^H S\boldsymbol{w}}{\boldsymbol{w}^H M_n\boldsymbol{w}} \tag{1.277}$$

where the matrix S has the following meaning:

In case (i)

$$S=E\{\alpha^*\tilde{\boldsymbol{s}}^*\cdot\alpha\tilde{\boldsymbol{s}}^T\}=\tilde{\boldsymbol{s}}^*E\{|\alpha|^2\}\tilde{\boldsymbol{s}}^T=P_\alpha\tilde{S} \tag{1.278}$$

where P_α is the mean square value of α, and $\tilde{S}$ is the dyadic product $\tilde{\boldsymbol{s}}^*\cdot\tilde{\boldsymbol{s}}^T$.

In case (ii) S is the covariance matrix of the signal

$$S=M_s \tag{1.279}$$

1.6.2 *Optimum filtering for a deterministic signal [case (i)]*

In the deterministic signal case the SIR can be expressed as a function of known quantities:

$$SIR_z(w)=\frac{|\boldsymbol{w}^T\tilde{\boldsymbol{s}}|^2}{\boldsymbol{w}^H M_n\boldsymbol{w}}P_\alpha \tag{1.280}$$

the symbols having been defined in Section 1.6.1.

We recall that, for a stationary noise process, the covariance matrix can be written as follows:

$$M_n=P_n\tilde{M}_n \tag{1.281}$$

where P_n is the noise power and $\tilde{M}_n$ is a normalised covariance matrix, i.e. one with unitary diagonal elements.

Expr. 1.280 can therefore be rewritten using the normalised covariance matrix $\tilde{M}_n$ of the noise:

$$SIR_z(\boldsymbol{w})=\frac{|\boldsymbol{w}^T\tilde{\boldsymbol{s}}|^2}{\boldsymbol{w}^H\tilde{M}_n\boldsymbol{w}}\frac{P_\alpha}{P_n}=\eta(\boldsymbol{w})SIR_i \tag{1.282}$$

† It may be convenient to choose α such that the signal $\tilde{s}$ is normalised in power, i.e.

$$\frac{1}{N}\tilde{\boldsymbol{s}}^T\tilde{\boldsymbol{s}}^*=\frac{1}{N}\sum_{i=1}^{N}|\tilde{s}_i|^2=1$$

In such a case, P_α is the (average) signal power.

where $SIR_i = P_a/P_n$ is the input signal-to-interference ratio and $\eta(\boldsymbol{w})$ the improvement of the SIR due to the processor, often referred to as the *improvement factor*.

Since M_n is Hermitian, it can be decomposed (see Section 1.8, exprs. 1.596 and 1.597) as follows:

$$M_n = (U\Lambda^{1/2})(U\Lambda^{1/2})^H = RR^H \tag{1.283}$$

with

$$R = \sum_{i=1}^{N} \lambda_i^{1/2} \boldsymbol{u}^i (\boldsymbol{u}^i)^H \tag{1.284}$$

indicating with λ_i and $\boldsymbol{u}^i$ the eigenvalues and the orthonormal eigenvectors of M_n, respectively.

Note that

$$M_n^{-1} = (R^{-1})^H R^{-1} \tag{1.285}$$

The vector $\boldsymbol{w}$ that maximises the SIR given by expr. 1.280 must be found. The numerator of that expression can be written

$$|\boldsymbol{w}^T \tilde{\boldsymbol{s}}|^2 = |\boldsymbol{w}^T R^* (R^{-1})^* \tilde{\boldsymbol{s}}|^2 = |(R^H \boldsymbol{w})^T (R^{-1} \tilde{\boldsymbol{s}}^*)^*|^2 \tag{1.286}$$

and letting

$$\begin{cases} R^H \boldsymbol{w} = \boldsymbol{a} \\ R^{-1} \tilde{\boldsymbol{s}}^* = \boldsymbol{b} \end{cases} \tag{1.287}$$

we obtain

$$|\boldsymbol{w}^T \tilde{\boldsymbol{s}}|^2 = |\boldsymbol{a}^T \boldsymbol{b}^*|^2 \tag{1.288}$$

From Schwartz's inequality

$$|\boldsymbol{a}^T \boldsymbol{b}^*|^2 \leq |\boldsymbol{a}|^2 \cdot |\boldsymbol{b}|^2 = (\boldsymbol{a}^H \boldsymbol{a})(\boldsymbol{b}^H \boldsymbol{b}) \tag{1.289}$$

and by substitution of eqns. 1.285–1.287 we obtain

$$|\boldsymbol{w}^T \tilde{\boldsymbol{s}}|^2 \leq (\boldsymbol{w}^H R R^H \boldsymbol{w})(\tilde{\boldsymbol{s}}^T R^{-H} R^{-1} \tilde{\boldsymbol{s}}^*) \tag{1.290}$$

which, using exprs. 1.283 and 1.285, becomes

$$|\boldsymbol{w}^T \tilde{\boldsymbol{s}}|^2 \leq \boldsymbol{w}^H M_n \boldsymbol{w} \; \tilde{\boldsymbol{s}}^T M_n^{-1} \tilde{\boldsymbol{s}}^* \tag{1.291}$$

Substituting this result in expr. 1.280, the new expression for the signal-to-interference ratio is obtained:

$$SIR_z(\boldsymbol{w}) \leq \tilde{\boldsymbol{s}}^T M_n^{-1} \tilde{\boldsymbol{s}}^* \cdot P_a \equiv SIR_0 \tag{1.292}$$

therefore, the quantity

$$SIR_0 \equiv \tilde{\boldsymbol{s}}^T M_n^{-1} \tilde{\boldsymbol{s}}^* \cdot P_a \tag{1.293}$$

represents the maximum SIR at the output of the linear processor, and the quantity

$$\eta_0(\boldsymbol{w}) = \tilde{\boldsymbol{s}}^T \tilde{M}_{\tilde{n}}^{-1} \tilde{\boldsymbol{s}}^* \tag{1.294}$$

(where $\tilde{s}$ and $\tilde{M}_{\tilde{n}}$ are both normalised in power) represents the maximum *improvement factor* of the processor.

Defining

$$\boldsymbol{w}_0 = M_n^{-1}\tilde{\boldsymbol{s}}^* \tag{1.295}$$

it is readily obtained from expr. 1.292 that the output SIR is

$$SIR_0 = \tilde{\boldsymbol{s}}^T \boldsymbol{w}_0 \cdot P_a \tag{1.296}$$

moreover, substituting expr. 1.295 in expr. 1.280 one obtains

$$SIR_z(\boldsymbol{w}_0) = \boldsymbol{w}_0^T \tilde{\boldsymbol{s}} P_a \tag{1.297}$$

that is, considering eqn. 1.296:

$$SIR_z(\boldsymbol{w}_0) = SIR_0 \tag{1.298}$$

Therefore it has been shown that the choice

$$\boldsymbol{w} = \boldsymbol{w}_0 = M_n^{-1}\tilde{\boldsymbol{s}}^* \tag{1.299}$$

maximises the SIR_z†.

In the particular case of white noise, expr. 1.299 reduces to

$$\boldsymbol{w}_0 = \boldsymbol{s}^* \tag{1.300}$$

which is the discrete-time form of the matched filter expr. 1.248. Of course, multiplication of $\boldsymbol{w}_0$ by any constant scalar does not modify the optimality of the filter.

1.6.3 *Optimum filtering for random signals*

In this section, two optimisation criteria for the case of unknown signal are shown: the former based on the maximisation of the signal-to-noise ratio, and the latter based on a particular technique that allows one to predict the interference (see References 29–31 for more details about these techniques).

1.6.3.1 *Maximisation of the SIR and noise whitening*

In the case of a random signal (case (ii) in Section 1.6.1), the SIR is expressed by the relationship

$$SIR_z(\boldsymbol{w}) = \frac{\boldsymbol{w}^H M_s \boldsymbol{w}}{\boldsymbol{w}^H M_n \boldsymbol{w}} \tag{1.301}$$

The maximisation of $SIR_z(\boldsymbol{w})$ leads to the generalised eigenproblem (see Section 1.8.3.6 for more details):

† If, instead of expr. 1.272 for the covariance matrix, we adopted the alternative definition

$$C_x = E\{\boldsymbol{x}\boldsymbol{x}^H\} = M_x^*$$

expr. 1.274 would become

$$E\{|z|^2\} = \boldsymbol{w}^T C_x \boldsymbol{w}^*$$

and so on for the subsequent expressions, in which M should be replaced by C^*. In particular, the expression for the optimum filter would be

$$\boldsymbol{w}_0 = (C^{-1})^* \boldsymbol{s}^*$$

$$M_n w = \mu M_s w \tag{1.302}$$

The maximum of SIR_z (let it be SIR_0) is obtained for $w = w_{\mu_0}$, where w_{μ_0} is the eigenvector corresponding to the minimum non-zero eigenvalue μ_0 of the generalised eigenproblem (see Section 1.8.3.6). From exprs. 1.301 and 1.302 one obtains

$$SIR_z(w = w_{\mu_0}) = \frac{1}{\mu_0} \tag{1.303}$$

When the covariance matrices M_s and M_n are both normalised in power, the quantity $\eta = 1/\mu$ is called 'SIR gain' or 'SIR improvement factor', as it represents the ratio between the output SIR and the input SIR. In such conditions, the quantity $\eta_0 = 1/\mu_0$ is the maximum (optimum) *improvement factor*.

We will now show that, when the random signal s tends to a deterministic one, w_{μ_0} tends to the value of w_0 given by expr. 1.295.

Under these circumstances the covariance matrix of the signal tends (to within the multiplicative constant P_a) to the dyadic product $\tilde{s}^* \tilde{s}^T$ (see Section 1.6.1), and expr. 1.302 becomes

$$M_n w = \mu P_a \tilde{s}^* \tilde{s}^T w \tag{1.304}$$

and, with the position $w = w_0 = M_n^{-1} \tilde{s}^*$, we have

$$\tilde{s}^* = \mu \tilde{s}^* \tilde{s}^T M_n^{-1} \tilde{s}^* P_a \tag{1.305}$$

and recalling expr. 1.293,

$$\tilde{s}^* = \mu \tilde{s}^* \cdot (SIR_0) \tag{1.306}$$

When $\mu = \mu_0 = (SIR_0)^{-1}$ expr. 1.306 becomes an identity.

As in the continuous-time case, the optimum filter for coloured interference can be decomposed into a whitening processor followed by a filter matched to the signal after whitening.

The whitening process consists of a multiplication by the matrix $B_n = \Lambda^{-1/2} U^T$ (see Section 1.9.4 eqn. 1.736), where it is readily shown that

$$M_n = (B_n^T B_n^*)^{-1} \tag{1.307}$$

In the case of a deterministic signal s, the optimum filter weights

$$w_0 = M_n^{-1} s^* \tag{1.308}$$

can be rewritten, using expr. 1.307, as

$$w_0 = B_n^T (B_n^* s^*) \tag{1.309}$$

and the optimum filtering of a given vector z can be rewritten as follows:

$$w_0^T z = (B_n^* s^*)^T B_n z \tag{1.310}$$

showing the whitening operation, i.e. $B_n z$, followed by the matched filter, i.e. a weighting vector $B_n^* s^*$, where $B_n s$ is the useful signal after whitening.

In the case of a random signal, the covariance matrices of the noise, M_n, and of the signal, M_s, after whitening become, respectively,

$$M_n' = B_n^* M_n B_n^T = I \tag{1.311}$$

$$M_s' = B_n^* M_s B_n^T \tag{1.312}$$

The SIR can be written, using the previous expr. 1.301, as follows:

$$SIR_z(\boldsymbol{w}) = \frac{\boldsymbol{w}^H B_n^{-*} M_s' B_n^{-T} \boldsymbol{w}}{\boldsymbol{w}^H B_n^{-*} \; B^{-T} \boldsymbol{w}} \tag{1.313}$$

and letting

$$\boldsymbol{v} = B_n^{-T} \boldsymbol{w} \tag{1.314}$$

we have

$$SIR_z(\boldsymbol{v}) = \frac{\boldsymbol{v}^H M_s' \boldsymbol{v}}{\boldsymbol{v}^H \boldsymbol{v}} \tag{1.315}$$

corresponding to the detection problem of a random signal with covariance matrix M_s' in white noise.

Therefore, the optimum filter $\boldsymbol{v}_0$ (i.e. after whitening) corresponds to finding the eigenvector associated with the *maximum* eigenvalue of the matrix

$$M_s' = B_n^* M_s B_n^T \tag{1.316}$$

i.e. to solving the problem:

$$M_s' \boldsymbol{v} = \lambda \boldsymbol{v} \tag{1.317}$$

It can be readily shown that the solution is the same as the 'pencil of matrices' problem:

$$M_s \boldsymbol{w} = \lambda M_n \boldsymbol{w} \tag{1.318}$$

with

$$\boldsymbol{w} = B^T \boldsymbol{v} \tag{1.319}$$

Conversely, one could whiten the signal using the matrix B_s such that

$$M_s^{-1} = B_s^T B_s^* \tag{1.320}$$

and obtaining new covariance matrices (compare with eqns. 1.311–1.312)

$$M_n'' = B_s^* M_n B_s^T \tag{1.321}$$

$$M_s'' = I$$

and a new weight vector

$$\boldsymbol{u} = B_s^{-T} \boldsymbol{w} \tag{1.322}$$

The resulting problem is the maximisation of the quantity

$$SIR_z(\boldsymbol{u}) = \frac{\boldsymbol{u}^H \boldsymbol{u}}{\boldsymbol{u}^H M_n'' \boldsymbol{u}} \tag{1.323}$$

i.e. the search for the minimum eigenvalue and the associated eigenvector of the matrix M_n''. Such a problem has been analysed in Reference 37 to define an optimum *moving target indicator* filter.

1.6.3.2 *Optimisation on the basis of linear prediction*

The optimisation method for interference rejection described in this section is based upon linear prediction techniques. Let x_i, $i = 1, \ldots N$ be the N available

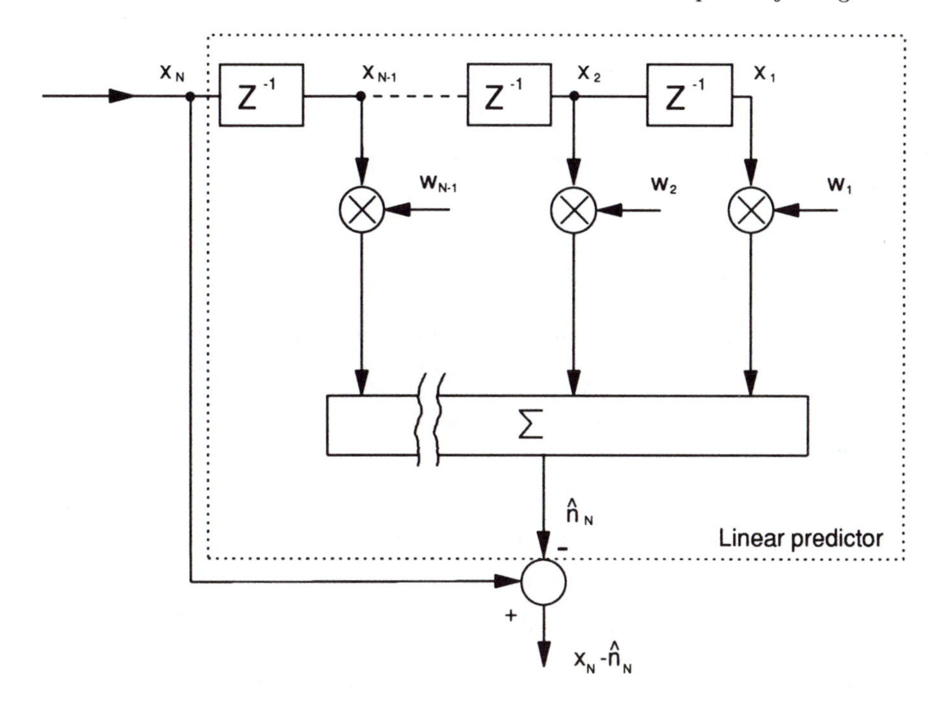

Figure 1.14 *Rejection filter based on the linear prediction of the interference*

samples of the signal plus interference ($x_i = s_i + n_i$). In the hypothesis of interference having much larger power than the useful signal, the rejection can be implemented by estimating the nth sample $\hat{n}_N$ of the interference, and by subtracting it from the received signal $x_N = s_N + n_N$.

According to the scheme of Fig. 1.14, the prediction is obtained by a weighted combination (FIR filtering) of the $N-1$ available data of the vector $\tilde{x} = \tilde{s} + \tilde{n}$, where the $\sim$ sign indicates all the vectors with $N-1$ elements. This is the classical problem of estimating a random variable n_N from a set of data $n_1, n_2, \ldots, n_{N-1}$ (see, for example, Chapter 13 of Reference 13).

The weights vector

$$\tilde{\boldsymbol{w}}^T = [w_1, w_2, \ldots, w_{N-1}] \tag{1.324}$$

must be determined in order to minimise the mean square error

$$e_{ms}^2 = E[|n_N - \hat{n}_N|^2] = E[|n_N - \tilde{\boldsymbol{w}}^T \tilde{\boldsymbol{x}}|^2] \tag{1.325}$$

that is, the interference power at the output of the filter.

The mean square error can be written as follows:

$$\begin{aligned} e_{ms}^2 &= E\{(n_N - \tilde{\boldsymbol{w}}^T \tilde{\boldsymbol{x}})^* (n_N - \tilde{\boldsymbol{w}}^T \tilde{\boldsymbol{x}})^T\} \\ &= E\{|n_N|^2 + \tilde{\boldsymbol{w}}^H \tilde{\boldsymbol{x}}^* \tilde{\boldsymbol{x}}^T \tilde{\boldsymbol{w}} - 2Re(\tilde{\boldsymbol{w}}^H \tilde{\boldsymbol{x}}^* n_N)\} \\ &= P_N + \tilde{\boldsymbol{w}}^H \tilde{M}_x \tilde{\boldsymbol{w}} - 2Re(\tilde{\boldsymbol{w}}^H \boldsymbol{r}_{nx}) \end{aligned} \tag{1.326}$$

where P_N is the power of the Nth interference sample, $\tilde{M}_x$ is the $(N-1)\times(N-1)$ covariance matrix of the input

$$\tilde{M}_x = E(\tilde{\boldsymbol{x}}^* \boldsymbol{x}^T)$$

and $\tilde{\boldsymbol{r}}_{nx}$ is the correlation vector between the input and the Nth intereference sample

$$\tilde{\boldsymbol{r}}_{nx} = E(\tilde{\boldsymbol{x}}^* n_N)$$

The minimum of the mean square error is obtained by setting its gradient with respect to the weight vector equal to zero. As the weight vector is complex, the stationary point, i.e. the minimum, is found (see Section 1.82 of Appendix 2) by setting the gradient equal to zero with respect to either $\tilde{\boldsymbol{w}}^*$ or $\tilde{\boldsymbol{w}}$.

As shown in Section 1.8.2.7, the gradient of $\tilde{\boldsymbol{w}}^H \tilde{M}_x \tilde{\boldsymbol{w}}$ with respect to $\tilde{w}^*$ is $\tilde{M}_x \tilde{\boldsymbol{w}}$ and the gradient of $2Re(\tilde{\boldsymbol{w}}^H \tilde{\boldsymbol{r}}_{nx})$ is simply $\tilde{\boldsymbol{r}}_{nx}$.

Therefore, the gradient is

$$\nabla_{w^*} e^2_{ms} = \tilde{M}_x \tilde{\boldsymbol{w}} - \tilde{\boldsymbol{r}}_{nx}$$

and the weight vector minimising the mean square error is

$$\tilde{\boldsymbol{w}} = \tilde{M}_x^{-1} \tilde{\boldsymbol{r}}_{nx} \tag{1.327}$$

The same result can be obtained by means of the orthogonality principle, stating that the mean square is minimum if $\tilde{\boldsymbol{w}}$ is such that the error is orthogonal to the data (see, for example, page 408 of Reference 13), leading to the equations:

$$E\{(n_N - \tilde{\boldsymbol{w}}^T \tilde{\boldsymbol{x}}) x_i^*\} = 0, \quad i = 1, 2, \ldots, N-1 \tag{1.328}$$

From expr. 1.328 we have:

$$\begin{cases} E\{x_1^* n_N\} - w_1 E\{x_1^* x_1\} - w_2 E\{x_1^* x_2\} \ldots - w_{N-1} E\{x_1^* x_{N-1}\} = 0 \\ E\{x_2^* n_N\} - w_1 E\{x_2^* x_1\} - w_2 E\{x_2^* x_2\} \ldots - w_{N-1} E\{x_2^* x_{N-1}\} = 0 \\ E\{x_{N-1}^* n_N\} - w_1 E\{x_{N-1}^* x_1\} - w_2 E\{x_{N-1}^* x_2\} \ldots - w_{N-1} E\{x_{N-1}^* x_{N-1}\} = 0 \end{cases} \tag{1.329}$$

(Yule–Walker equations)
i.e. in matrix form

$$\tilde{\boldsymbol{r}}_{nx} = \tilde{M}_x \tilde{\boldsymbol{w}} \tag{1.330}$$

which is equal to eqn. 1.327.

When $\tilde{\boldsymbol{x}}$ is considered a 'data' vector with known covariance matrix and n_N is a random quantity to be estimated, eqns. 1.329, called Yule–Walker equations, can be solved to obtain the optimum coefficients $\tilde{\boldsymbol{w}}$. We recall that $\tilde{\boldsymbol{r}}_{nx}$ is the correlation vector between the input 'data' and the quantity to be estimated and $\tilde{M}_x$ is the $(N-1)\times(N-1)$ covariance matrix of the input.

On the other hand, when n_N is the Nth interference sample and in the usual hypothesis of orthogonality between the useful signal and the interference, the vector $\tilde{\boldsymbol{r}}_{nx}$ is equal to the 'reduced' correlation vector of the interference, $\tilde{\boldsymbol{r}}$:

$$\tilde{\boldsymbol{r}}^T = [E\{n_1^* n_N\}, E\{n_2^* n_N\}, \ldots E\{n_{N-1}^* n_N\}] \tag{1.331}$$

From eqns. 1.329 or 1.330 the 'normal equations' are obtained:

$$\tilde{\boldsymbol{r}} - \tilde{M}_x \tilde{\boldsymbol{w}} = \tilde{\boldsymbol{0}} \tag{1.332}$$

where $\tilde{\mathbf{0}}$ is the vector of $(N-1)$ zeroes.

As previously stated, the interference power is assumed to be larger than that of the signal, so that the interference covariance matrix $\tilde{M}_n$ can be substituted for $\tilde{M}_x$; the final expression for the optimum weights is therefore

$$\tilde{\boldsymbol{w}}_0 = \tilde{M}_n^{-1}\tilde{\boldsymbol{r}} \tag{1.333}$$

and the complete weights vector is then (see Fig. 1.14)

$$\boldsymbol{w}_0 = \begin{bmatrix} -\tilde{\boldsymbol{w}}_0 \\ 1 \end{bmatrix} \tag{1.334}$$

Exprs. 1.330 and 1.331 are directly related to the expression of optimum filtering for the deterministic signal (expr. 1.295). In fact, when the following signal is considered:

$$\boldsymbol{s} = \begin{bmatrix} \tilde{\mathbf{0}} \\ 1 \end{bmatrix} = \begin{bmatrix} 0 \\ 0 \\ \cdots \\ 1 \end{bmatrix} \tag{1.335}$$

(i.e. $N-1$ 'zeroes' and a 'one') the optimum filtering equations system, eqn. 1.295,

$$M_n \boldsymbol{w}_0 = \mu \boldsymbol{s}^* \tag{1.336}$$

reduces to the following block-matrix system:

$$\begin{bmatrix} \tilde{M}_n & \tilde{\boldsymbol{r}} \\ \tilde{\boldsymbol{r}}^T & \boldsymbol{P}_N \end{bmatrix} \cdot \begin{bmatrix} -\tilde{\boldsymbol{w}}_0 \\ 1 \end{bmatrix} = \begin{bmatrix} \tilde{\mathbf{0}} \\ \mu \end{bmatrix} \tag{1.337}$$

where $\boldsymbol{P}_N = E\{n_N^* n_N\}$.

The above $N \times N$ system is readily decomposed into a single equation (which determines the constant μ) and an $(N-1) \times (N-1)$ system

$$-\tilde{M}_n \tilde{\boldsymbol{w}}_0 + \tilde{\boldsymbol{r}} = \tilde{\mathbf{0}} \tag{1.338}$$

identical to expr. 1.333.

Of course, other choices of $\boldsymbol{s}$ are possible, e.g. with the 'one' in the central position:

$$\boldsymbol{s}^T = [0 \quad \ldots \quad 0 \quad 1 \quad 0 \quad \ldots \quad 0] \tag{1.339}$$

resulting in better SIR improvements [43].

1.6.3.3 *Comparison of the two methods*

The two techniques (SIR maximisation and linear prediction) can be compared in terms of:

(*a*) Algorithm complexity and subsequent processing load
(*b*) SIR improvement [29, 31].

The first item is the most important in real-time applications.

With linear prediction, the weights are evaluated by solving the normal equations, eqn. 1.332. Generally, the inversion of a $(p \times p)$ matrix requires a

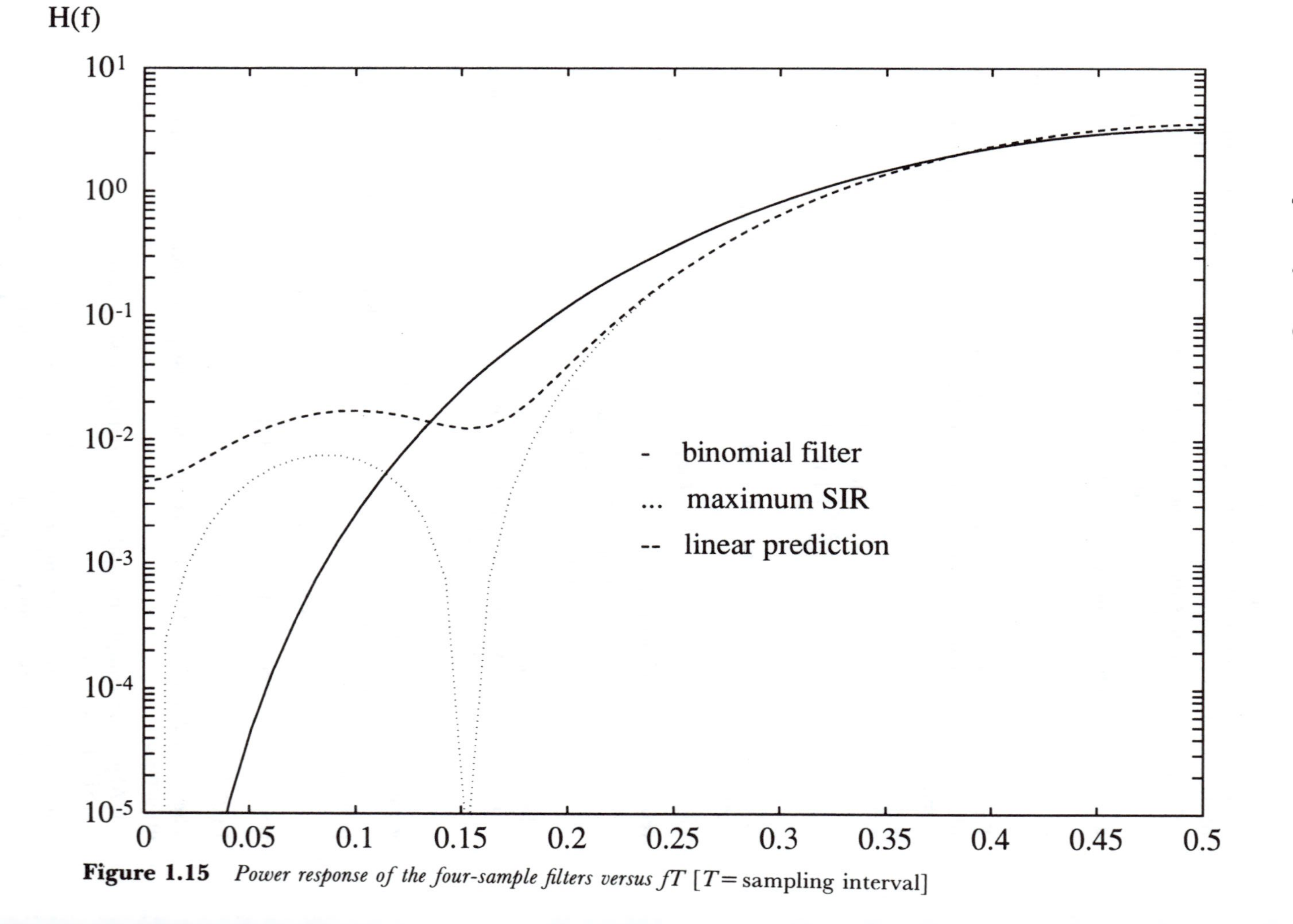

Figure 1.15 *Power response of the four-sample filters versus fT* [T = sampling interval]

number of operations (sums and products) of the order of p^3; such a number reudces to p^2 for Toeplitz matrices (covariance matrices for the stationary case).

On the other hand, the SIR maximisation technique requires asymptotically convergent computations for the search of the eigenvector associated with the minimum eigenvalue; in any case, the number of operations required is dependent on the desired accuracy and is generally larger than the linear prediction case and hardly compatible with real-time applications.

It can be shown [29] that, when the SIR improvement is the parameter to be maximised, linear prediction is a sub-optimum method, and tends to the optimum one when the minimum eigenvector λ_m of the covariance matrix of the interference, M_n, tends to zero (i.e. the SIR improvement becomes larger and larger because the interference spectrum 'shrinks').

A quantitative evaluation of the optimisation techniques can be done [29] in terms of SIR improvement and frequency response; in Fig. 1.15 a comparative case is shown. A Toeplitz (stationary case) interference covariance matrix is assumed to be of the form

$$M = [r_{ij}] \quad r_{ij} = \rho^{(i-j)^2}, \; i, j = 1, 2, \ldots N \tag{1.340}$$

The case of a transversal filter with four samples and a correlation coefficient ρ between two successive samples of the interference equal to 0.8 has been considered; the power spectrum of the interference has been assumed to be Gaussian, while the signal is assumed to be totally unknown.

The frequency response curves are obtained by the set of weights $\boldsymbol{w}_0$ that are given by eqn. 1.336 and by eqns. 1.333 and 1.334, respectively. In the same Figure, for comparison, the curve of the binomial filter† whose weights are

$$w_k = \begin{bmatrix} N-1 \\ k-1 \end{bmatrix}, \; k = 1, 2, \ldots N \tag{1.341}$$

is also shown. It has been shown [29] that if N is not greater than 4 and ρ is not smaller than 0.8, the improvement factor of the optimum (maximum SIR) filter exceeds that of the linear prediction filter by less than 1.5 dB.

† A binomial filter may be thought of as being obtained by cascading $N-1$ two-sample filters with weights $+1$ and -1; the transfer function of a single stage is $h_1(Z) = 1 - Z^{-1}$; therefore that of an N-sample binomial filter is $h_N(Z) = (1 - Z)^{-N}$, whence the binomial expression for the weights.

1.7 Appendix 1: Representation of signals

A *signal* is a physical quantity dependent on an independent variable, which is normally *time*.

Signals carry *information* and are generally corrupted by *interferences* such as thermal noise. For example, the output signal of a radar receiver can be made up of target echoes (useful signals or, for short, signals), unwanted echoes (clutter), active interference and receiver noise.

This appendix contains a short outline of the main topics concerning signals, their characterisation in time and frequency domains, as well as the characterisation of random processes. Special attention is paid to the representation of *narrowband signals and processes*, which are of great importance in radar and telecommunications applications.

1.7.1 Continuous and discrete signals: Representation in the time and frequency domains

1.7.1.1 *Fourier representation*

A continuous-time signal will be indicated by the (real or complex) function $x(t)$, while a discrete-time signal will be indicated by the (real or complex) series x_n (or $x[n]$).

The fact that a signal can be represented by a complex function should not be considered merely as a convention; it is a useful tool when dealing with, for example, the Fourier transform.

A signal $x_T(t)$ is periodic with period T if

$$x_T(t) = x_T(t+T) \quad \forall t \tag{1.342}$$

If $x_T(t)$ is absolutely integrable, i.e.

$$\int_{-T/2}^{T/2} |x_T(t)|\, dt < \infty \tag{1.343}$$

it is possible to define an associated Fourier series

$$x_T(t) = \frac{a_0}{2} + \sum_{k=1}^{\infty} a_k \cos\left(\frac{2\pi kt}{T}\right) + \sum_{k=1}^{\infty} b_k \sin\left(\frac{2\pi kt}{T}\right) \tag{1.344}$$

where, owing to the orthogonal properties of the functions $\cos(2\pi kt/T)$ and $\sin(2\pi kt/T)$ in the interval $[-T/2, T/2]$, i.e.

$$\int_{-T/2}^{+T/2} \phi\left(\frac{2\pi nt}{T}\right) \psi\left(\frac{2\pi mt}{T}\right) dt = \begin{cases} \dfrac{T}{2} & \text{if} \quad n = m \\ 0 & \text{if} \quad n \neq m \end{cases} \tag{1.345}$$

where $\phi(.)$ and $\psi(.)$ denote cos(.) or sin(.), the Fourier coefficients a_k and b_k are given by

$$\begin{cases} a_k = \dfrac{2}{T} \displaystyle\int_{-T/2}^{T/2} x_T(t) \cos\left(\dfrac{2\pi kt}{T}\right) dt \\ b_k = \dfrac{2}{T} \displaystyle\int_{-T/2}^{T/2} x_T(t) \sin\left(\dfrac{2\pi kt}{T}\right) dt \end{cases} \quad k=0, 1, \ldots, \infty \tag{1.346}$$

We shall not discuss the convergence problems of the Fourier series. We only recall that, if $x_T(t)$ has bounded variation in $[0, T)$, then the Fourier series converges to $x_T(t)$ anywhere; $x_T(t)$ itself is continuous, and converges to $\frac{1}{2}[x(t_0^-) + x(t_0^+)]$ on a discontinuity point t_0.

This well known representation of a periodic signal by its 'harmonics' a_k and b_k at positive integer multiples of the basic frequency $F \equiv 1/T$, can be written in a more compact way by means of the complex exponential

$$\exp\left\{\pm j \frac{2\pi k}{T} t\right\} = \cos\left(\frac{2\pi k}{T} t\right) \pm j \sin\left(\frac{2\pi k}{T} t\right) \tag{1.347}$$

and by considering both positive and negative harmonic frequencies, resulting in the 'complex exponential' form of the Fourier series:

$$x_T(t) = \sum_{n=-\infty}^{+\infty} c_n \exp\left\{j \frac{2\pi}{T} nt\right\} \tag{1.348}$$

where

$$c_n = \frac{1}{T} \int_{-T/2}^{T/2} x_T(t) \exp\left\{-j \frac{2\pi}{T} nt\right\} dt \quad \forall n \tag{1.349}$$

It is readily shown that exprs. 1.348 and 1.349 are equivalent to expr. 1.347, and that the relationship between the 'two-sided' Fourier coefficient c_k and the 'one-sided' ones a_k and b_k is

$$c_{-k} + c_k = a_k; \quad j(c_k - c_{-k}) = b_k \tag{1.350}$$

or, in other words

$$\begin{cases} c_k = \frac{1}{2}\,(a_k - jb_k) \\ c_0 = \frac{1}{2}\,a_0 \\ c_{-k} = \frac{1}{2}\,(a_k + jb_k) \end{cases} \tag{1.351}$$

If $x_T(t)$ is real, a_k and b_k are also real and the Fourier coefficients c_k are Hermitian:

$$c_{-k} = c_k^* \tag{1.352}$$

If $x_T(t)$ is real and symmetrical, i.e. $x_T(-t) = x_T(t)$, then $b_k = 0$ and the Fourier coefficients are also real and symmetrical:

$$c_{-k} = c_k; \quad \mathrm{Im}(c_k) = 0 \tag{1.353}$$

As an example, the Fourier series expansion for a train of rectangular (Fig. 1.16*a*) and Gaussian (Fig. 1.16*b*) pulses is reported, limited to the first four

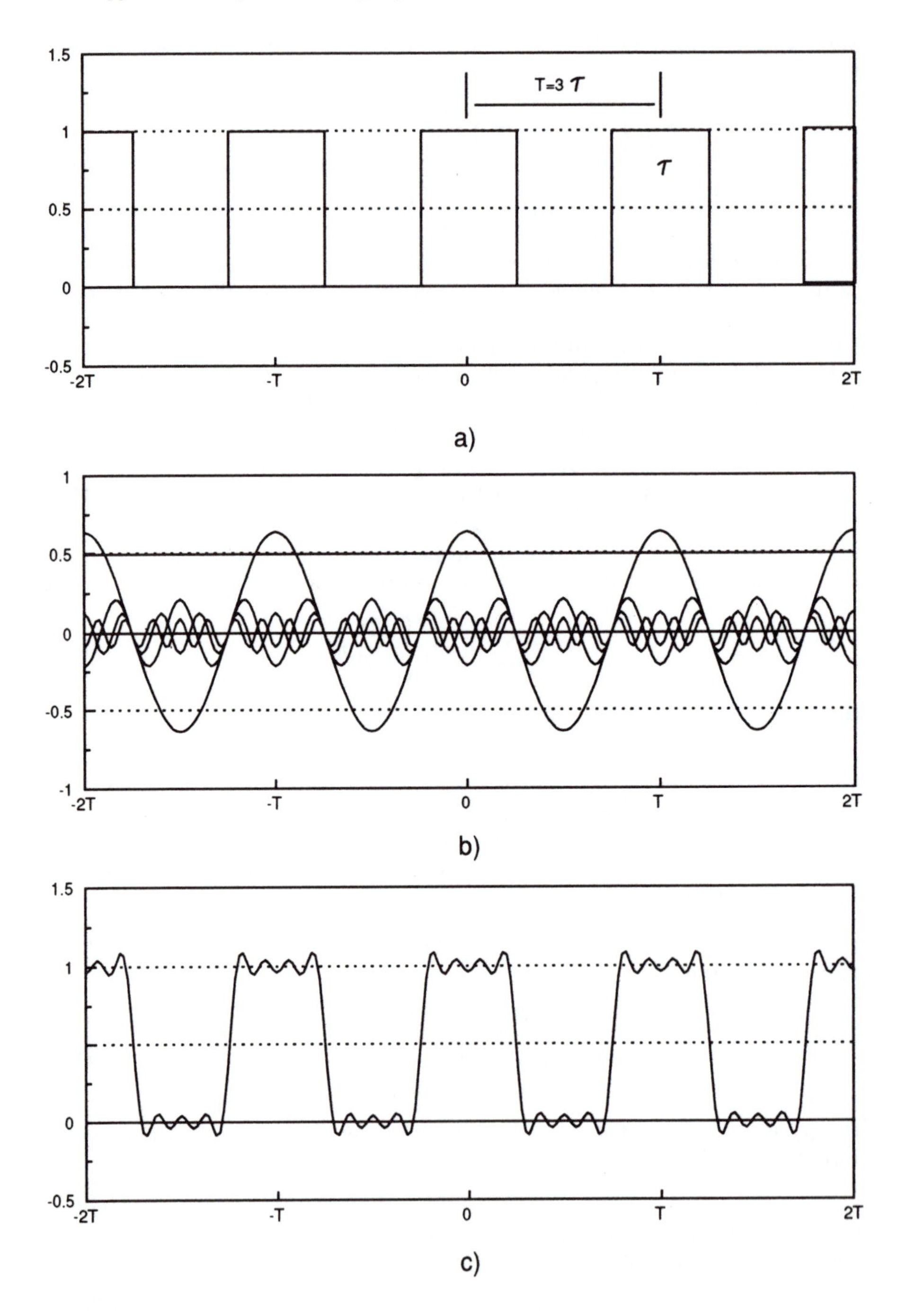

Figure 1.16 *(a) First four harmonics of a square pulses train and partial reconstruction of the signal*

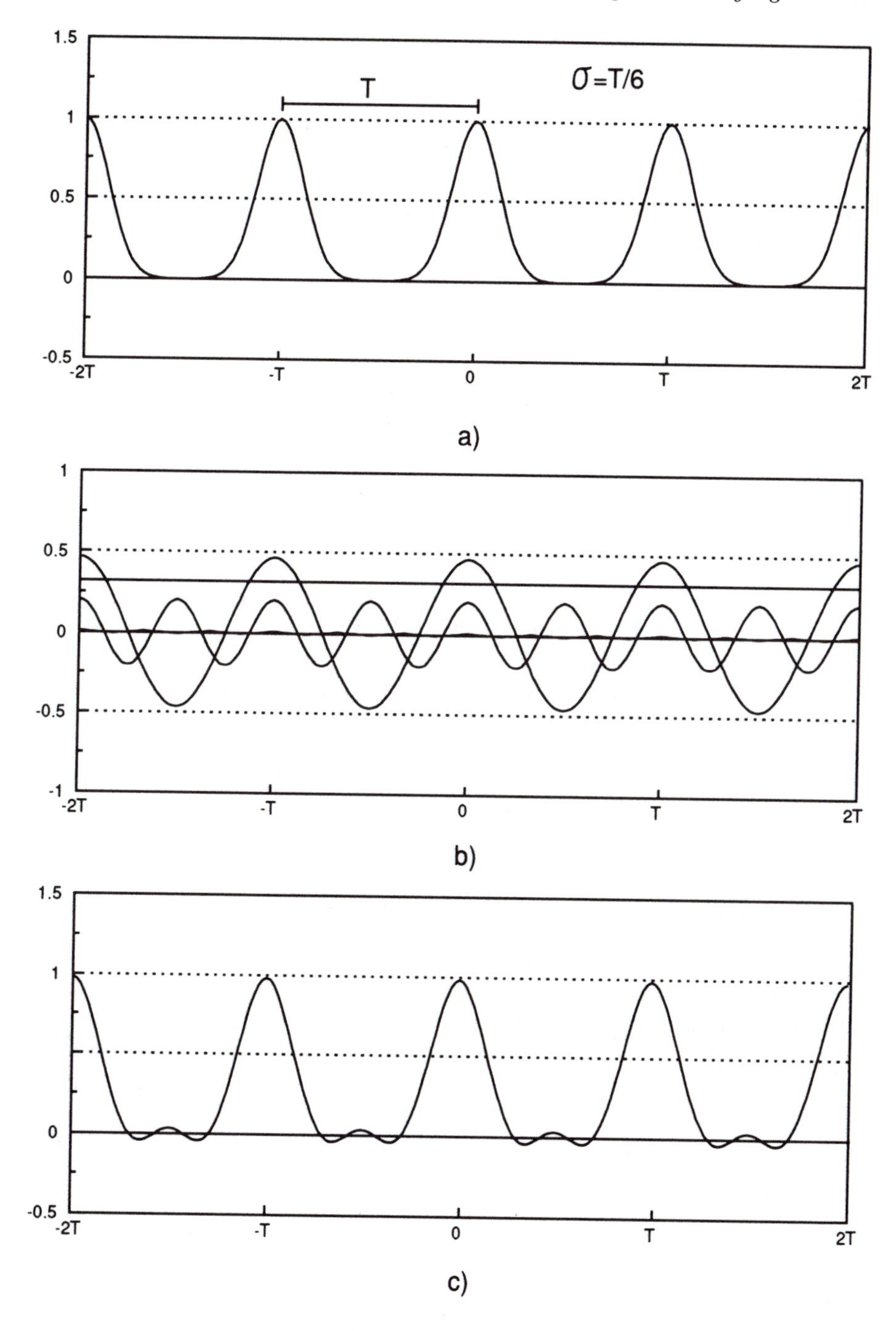

Figure 1.16 *(b) First four harmonics of a Gaussian pulses train and partial reconstruction of the signal*

harmonics: it is possible to see that the partial reconstruction is quite good even with a small number of harmonics.

The sequence

$$|c_n|^2, n=-\infty, \ldots, 0, \ldots, \infty \tag{1.354}$$

of the squared moduli of the Fourier coefficients is the *spectral density* of $x_T(t)$. According to the Parseval theorem [20], its sum equals the energy of the signal in one period:

$$\sum_{n=-\infty}^{+\infty} |c_n|^2 = \frac{1}{T}\int_{-T/2}^{T/2} |x_T(t)|^2 dt \tag{1.355}$$

Another quantity of interest is the average of the product $x_T^*(t)x_T(t+\tau)$, i.e. the *autocorrelation function* $R_{x_T}(t+\tau)$ (for short, ACF) of $x_T(t)$:

$$R_{x_T}(\tau) = \frac{1}{T}\int_{-T/2}^{T/2} x_T^*(t)x_T(t+\tau)dt \tag{1.356}$$

Using the Fourier representation (expr. 1.348), and remembering that

$$\frac{1}{T}\int_{-T/2}^{T/2} \exp\left\{j\frac{2\pi}{T}(n-m)\tau\right\}d\tau = \begin{cases} 0, & n \neq m \\ 1, & n=m \end{cases} \tag{1.357}$$

we have the result that the ACF of $x_T(t)$ has the following Fourier series representation (to be compared with expr. 1.348):

$$R_{x_T}(\tau) = \sum_{n=-\infty}^{\infty} |c_n|^2 \exp\left\{j\frac{2\pi}{T}n\tau\right\} \tag{1.358}$$

The series $|c_n|^2$ is the **power spectral density** of $x_T(t)$.

Let us now consider aperiodic signals. An heuristic way to obtain a Fourier representation for an aperiodic signal $x(t)$ is to consider it as the limit of a periodic signal $x_T(t)$ with period T going to infinity and considering the discrete variable $f=n/T$, that becomes continuous for $T\to\infty$; see Fig. 1.17 for the 'spectra' of some rectangular pulse trains, as their period tends to infinity.

From exprs. 1.348 and 1.349 the Fourier integral representation [13] is obtained:

$$x(t) = \int_{-\infty}^{\infty} X(f)e^{j2\pi ft}df \tag{1.359}$$

$$X(f) = \int_{-\infty}^{\infty} x(t)e^{-j2\pi ft}dt \tag{1.360}$$

where $X(f)$ is related to c_n by

$$c_n = \frac{1}{T}X\left(\frac{n}{T}\right) \tag{1.361}$$

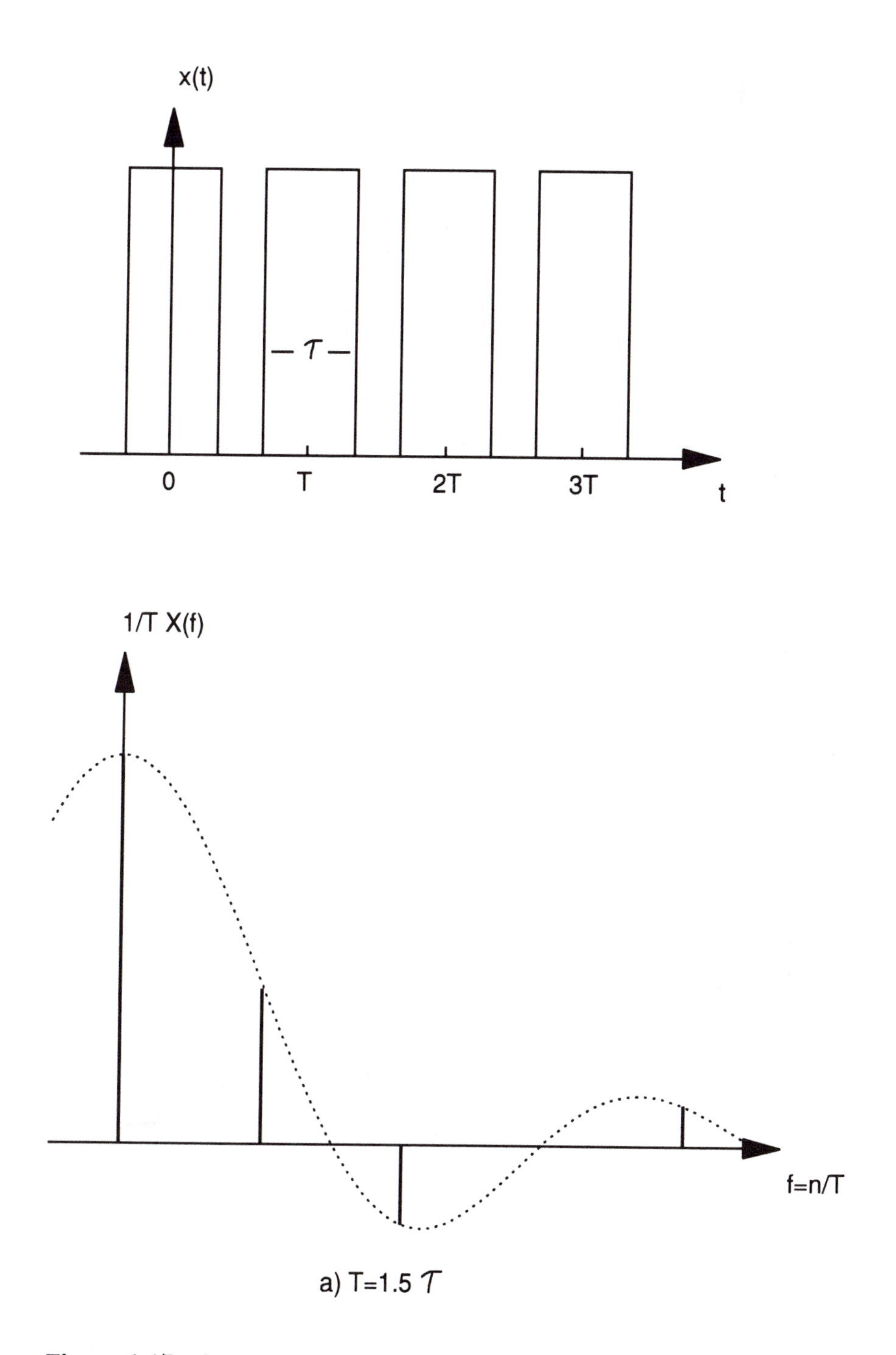

Figure 1.17 *Spectra of periodic pulse trains (variable duty cycle)*

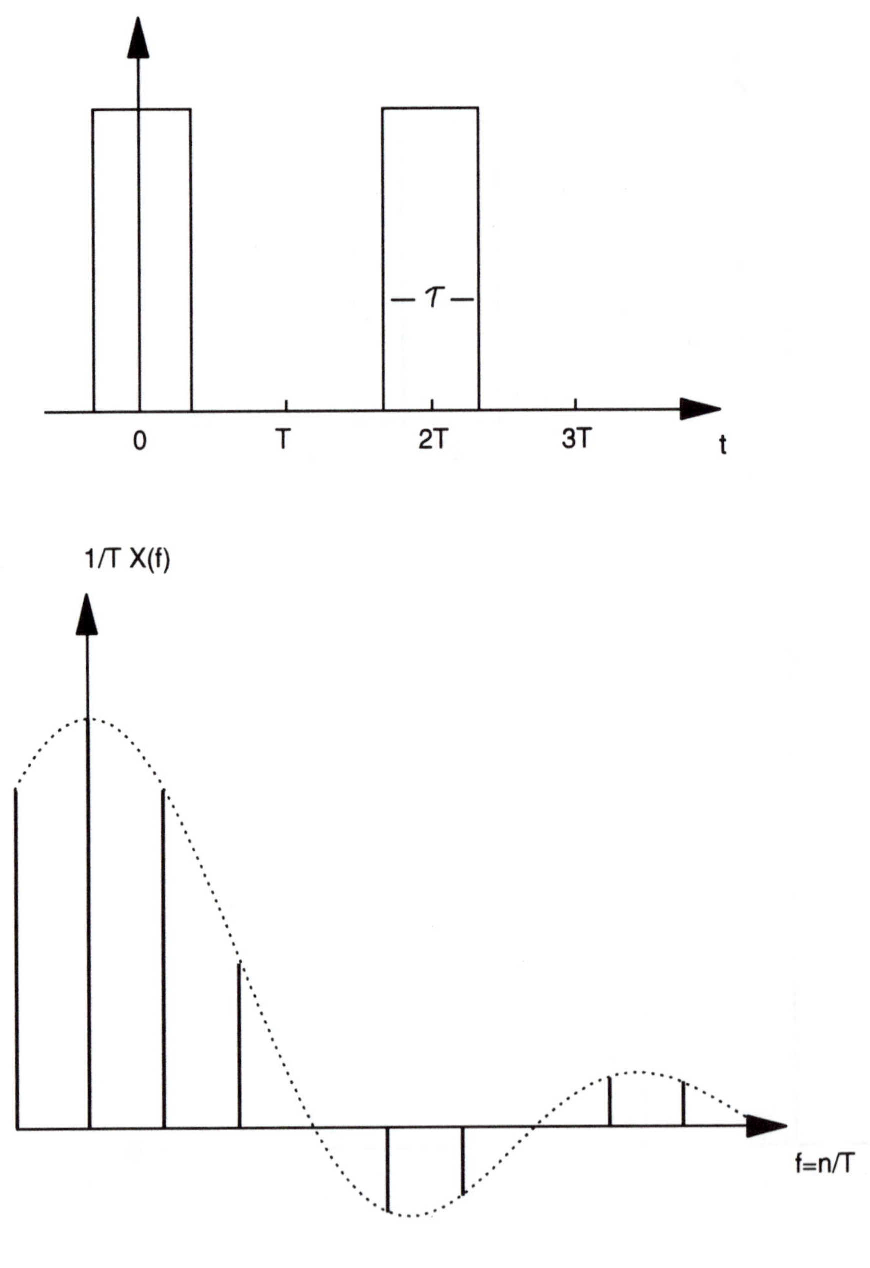

Figure 1.17 *(continued)*

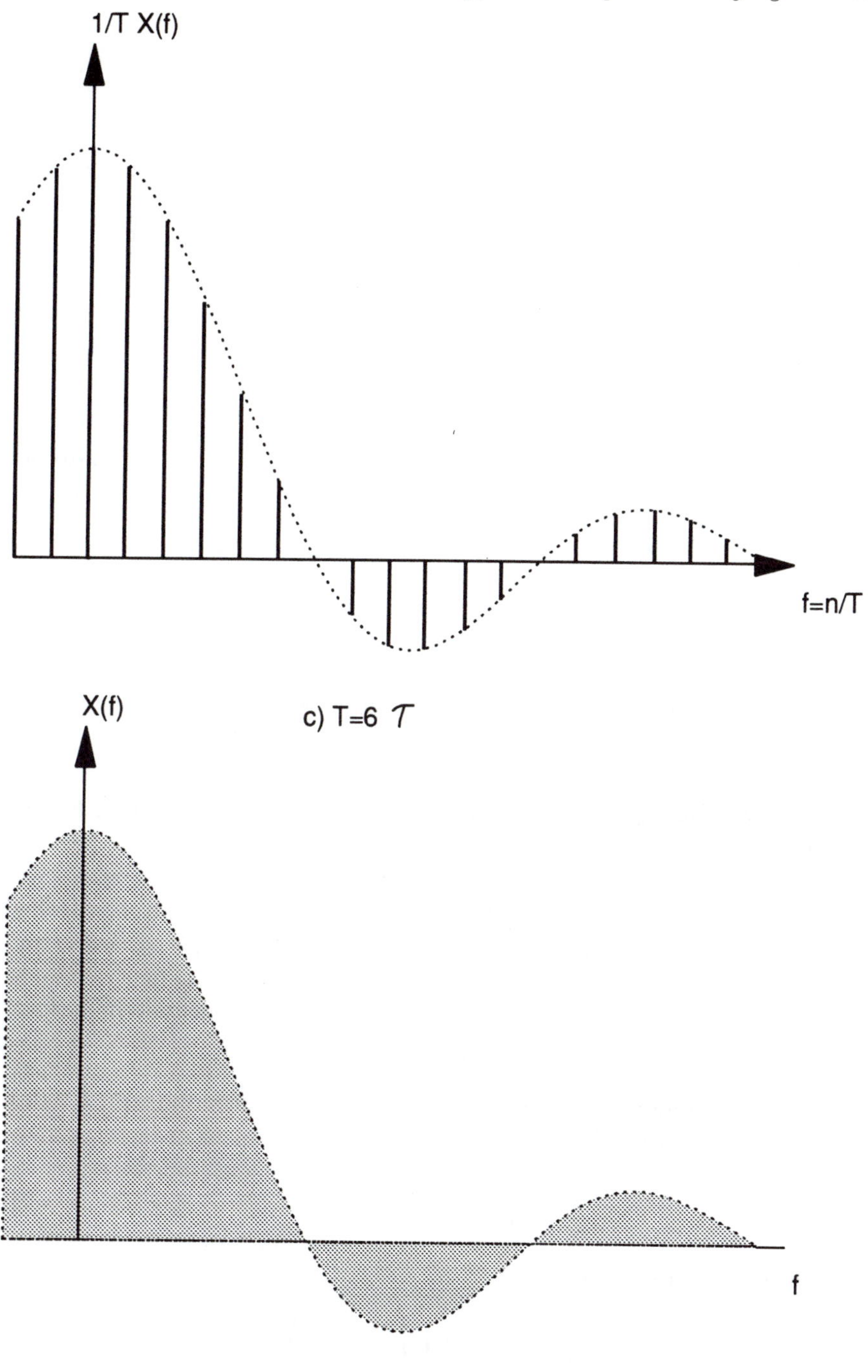

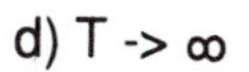

Figure 1.17 *(continued)*

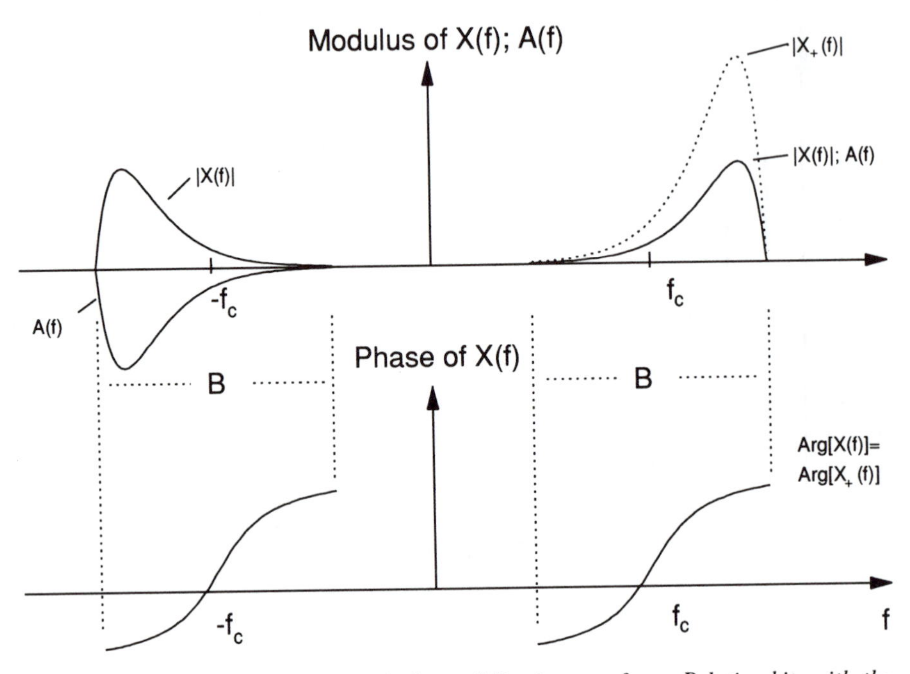

Figure 1.18 *Plot of modulus and phase of Fourier transform: Relationship with the analytic signal*

$X(f)$ is called the *spectrum* of $x(t)$. The functions $x(t)$ and $X(f)$ are a Fourier transform pair:

$$x(t) \overset{F}{\leftrightarrow} X(f) \tag{1.362}$$

Generally, modulus and phase (argument) of $X(f)$ are plotted (see Fig. 1.18). The meaning of the spectrum results from expr. 1.359, showing that $x(t)$ can be considered as the sum of an infinite number of complex exponentials in the form

$$[X(f)df] \exp(j2\pi ft). \tag{1.363}$$

A sufficient condition for the existence of the Fourier transform (expr. 1.360) is that the signal $x(t)$ be absolutely integrable:

$$\int_{-\infty}^{\infty} |x(t)| dt < \infty \tag{1.364}$$

from which, recalling that

$$\int_a^b |x(t)|^2 dt \leqslant \left(\int_a^b |x(t)| dt \right)^2 \tag{1.365}$$

a sufficient condition for the existence of the Fourier representation is that the energy of the signal is finite, i.e.

$$\int_{-\infty}^{\infty} |x(t)|^2 dt < \infty \tag{1.366}$$

Another expression for the Fourier transform is obtained by considering exprs. 1.346, 1.351 and 1.361:

$$X(f) = \tfrac{1}{2}[X_c(f) - jX_s(f)] \tag{1.367}$$

where

$$X_c(f) = 2\int_{-\infty}^{\infty} x(t)\,\cos(2\pi ft)\,dt \tag{1.368}$$

$$X_s(f) = 2\int_{-\infty}^{\infty} x(t)\,\sin(2\pi ft)\,dt \tag{1.369}$$

and $X_c(f)$, $X_s(f)$ are related to a_k and b_k by

$$\begin{cases} \dfrac{1}{T}X_c\left(\dfrac{k}{T}\right) = a_k \\ \dfrac{1}{T}X_s\left(\dfrac{k}{T}\right) = b_k \end{cases} \tag{1.370}$$

If $x(t)$ is real, $X_c(f)$ is twice the real part of its spectrum and $X_s(f)$ is twice its imaginary part.

In a similar manner to the derivation of the Fourier transform, it is possible to consider the limit, for $T \rightarrow \infty$, of the ACF, namely

$$R_x(\tau) = \lim_{T\rightarrow\infty} R_{xT} = \lim_{T\rightarrow\infty} \int_{-T/2}^{T/2} x_T^*(t)x_T(t+\tau)\,dt \tag{1.371}$$

It can be shown that, if the limit exists and if

$$\int_{-\infty}^{\infty} |R_x(\tau)|^2 < \infty \tag{1.372}$$

then $R_x(\tau)$ is the inverse Fourier transform of the squared modulus of $X(f)$:

$$R_x(\tau) \overset{F}{\leftrightarrow} |X(f)|^2 \tag{1.373}$$

The quantity $|X(f)|^2$ is the *power spectral density* of $x(t)$.

1.7.1.2 *Some properties of the Fourier integral*

Three important properties related to the Fourier integral will now be presented, because they will be useful in what follows:

(i) *Translation property*
If $x(t)$ and $X(f)$ are a signal and its Fourier transform, respectively,

$$x(t) \overset{F}{\leftrightarrow} X(f) \tag{1.374}$$

the following relationships hold [the demonstration follows immediately from the definitions (exprs. 1.359 and 1.360) of Fourier transform]:

$$x(t-t_0)\overset{F}{\leftrightarrow}X(f)e^{-j2\pi ft_0} \tag{1.375}$$

$$x(t)e^{+j2\pi f_0 t}\overset{F}{\leftrightarrow}X(f-f_0) \tag{1.376}$$

(ii) *Convolution property*
If $x(t)$ and $y(t)$ are two signals whose Fourier transforms are $X(f)$ and $Y(f)$, respectively,

$$\begin{aligned} x(t)&\overset{F}{\leftrightarrow}X(f)\\ y(t)&\overset{F}{\leftrightarrow}Y(f)\end{aligned} \tag{1.377}$$

and if the convolution between two functions $\alpha(t)$ and $\beta(t)$ is indicated by $\alpha(t)*\beta(t)$ and defined by

$$\alpha(t)*\beta(t)\equiv\int_{-\infty}^{\infty}\alpha(\tau)\beta(t-\tau)d\tau \tag{1.378}$$

the following relationships hold (as above, the demonstration follows immediately from the definition of Fourier transform):

$$x(t)\cdot y(t)\overset{F}{\leftrightarrow}X(f)*Y(f) \tag{1.379}$$

$$x(t)*y(t)\overset{F}{\leftrightarrow}X(f)\cdot Y(f) \tag{1.380}$$

(iii) *Symmetry properties*
In a similar way to the Fourier series, we have:

(a) If $x(t)$ is real, $X(f)=X^*(-f)$
(b) If $x(t)$ is real and symmetrical, $X(f)$ is real and symmetrical.

1.7.1.3 *Linear, fixed parameters systems*

A system with (real or complex) input $x(t)$ and (real or complex) output $y(t)$ is linear if any linear combination of its inputs

$$x(t)=\sum_{i=1}^{M}\alpha_i x_i(t) \tag{1.381}$$

produces the linear combination of the pertaining outputs:

$$y(t)=\sum_{i=1}^{M}\alpha_i y_i(t) \tag{1.382}$$

By 'fixed parameters' it is meant that the input $x(t+\tau)$ produces the output $y(t+\tau)$ for any value of the 'delay' τ.

Using the above properties, it is easily shown that the input

$$x'(t) = \frac{dx}{dt} = \lim_{\Delta t \to 0} \frac{x(t+\Delta t) - x(t)}{\Delta t} \tag{1.383}$$

produces the output

$$y'(t) = \frac{dy}{dt} \tag{1.384}$$

and, more generally, $x^{(m)}(t) = d^m x(t)/dt^m$ produces the output $y^{(m)}(t) = d^m y(t)/dt^m$.

An important class of linear, fixed parameter systems has the input–output relationship expressed by linear differential equations with fixed parameters.

A straightforward way to characterise a linear, fixed parameter system is its response to a unit step, i.e. to an input $u(t)$ such that

$$\begin{cases} u(t) = 0 & t \leqslant 0 \\ u(t) = 1 & t > 0 \end{cases} \tag{1.385}$$

The input $x(t)$ at a given time t can be expressed by a sum of unit steps at previous times $t - \tau_k$ (see Fig. 1.19):

$$x(t) = \sum_{k \in K} [x(t-\tau_k) - x(t-\tau_{k-1})]u(t-\tau_k) \tag{1.386}$$

The pertaining output is

$$y(t) = \sum_{k \in K} [x(t-\tau_k) - x(t-\tau_{k-1})]g(t-\tau_k) \tag{1.387}$$

where $g(.)$ is the step response of the system; for a causal system, $g(t) = 0$ for $t \leqslant 0$. Considering the limits of the above expressions 1.386 and 1.387 of $x(t)$ and

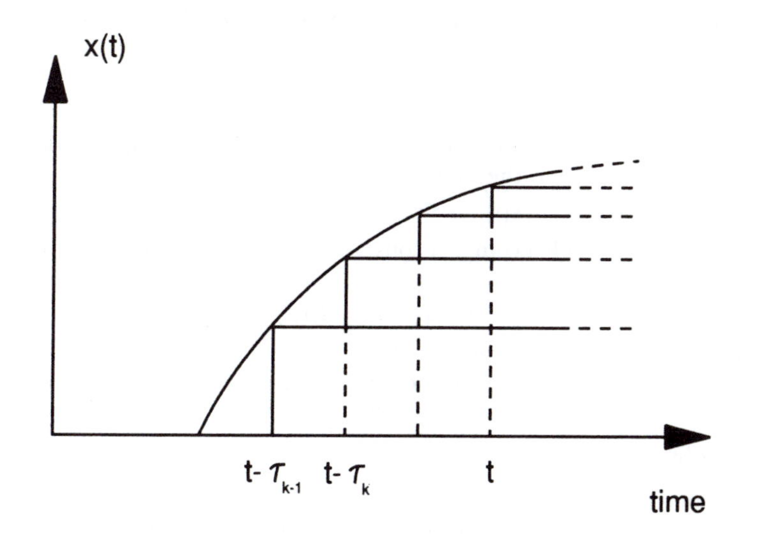

Figure 1.19 *Representation of a function as the sum of unit steps*

$y(t)$ when the intervals $\tau_{k-1}-\tau_k$ (assumed to be of equal durations, for the sake of simplicity) tend to zero, we obtain for a causal system

$$x(t)=\int_{-\infty}^{t} \frac{dx(\tau)}{d\tau} u(t-\tau)d\tau \tag{1.388}$$

and

$$y(t)=\int_{-\infty}^{t} \frac{dx(\tau)}{d\tau} g(t-\tau)d\tau \tag{1.389}$$

Expr. 1.389 (known as the Duhamel's integral) gives the desired input–output relationship as a convolution between the step response and the derivative of the input. A more common expression can be obtained by developing the integral expr. 1.389 by parts:

$$y(t)=|x(\tau)g(t-\tau)|_{-\infty}^{t}-\int_{-\infty}^{t} x(\tau)\frac{dg(t-\tau)}{d\tau} d\tau \tag{1.390}$$

Assuming $x(-\infty)\cdot g(+\infty)=0$, we have:

$$y(t)=\int_{-\infty}^{t} x(\tau)h(t-\tau)d\tau \tag{1.391}$$

where, by definition, the 'impulse response' $h(t)$ is the derivative of the step response. It represents the 'memory' of the past inputs $x(t-\tau_k)$ at time t (see Fig. 1.20). Note that the above derivation did not require the use of the 'impulse' or 'Dirac delta' functions; this use often leads students to forget that any physical system has a **finite dynamic range**.

The Fourier transform of $y(t)$ (expr. 1.391) gives, according to the convolution property (ii), the well known input–output relationship

$$Y(f)=H(f)X(f) \tag{1.392}$$

where the 'frequency response' or 'system function' $H(f)$ is defined by

$$H(f)=\int_{-\infty}^{\infty} h(\tau)e^{-j2\pi f\tau}d\tau \tag{1.393}$$

The meaning of $H(f)$ is clear when one considers that the response of a linear, fixed-parameter signal to the input

$$x(t)=e^{j2\pi ft} \qquad -\infty<t<+\infty \tag{1.394}$$

is proportional (through a complex constant $c_f=a_f e^{j\phi_f}$) to the output (Section 9.1 of Reference 12):

$$y(t)=a_f e^{j\phi_f}e^{j2\pi ft}=H(f)e^{j2\pi ft} \tag{1.395}$$

Therefore, such a system attenuates by a factor $a_f=|H(f)|$ the components of the input spectrum, giving them also a phase rotation ϕ_f. So, the term 'filter' is synonymous with linear, fixed parameter systems.

1.7.1.4 *The analytical signal*

From property (iii) (*a*) of the Fourier integral, we have the result that a real signal can be completely represented by its spectrum of non-negative frequencies only (see again Fig. 1.18), i.e. by

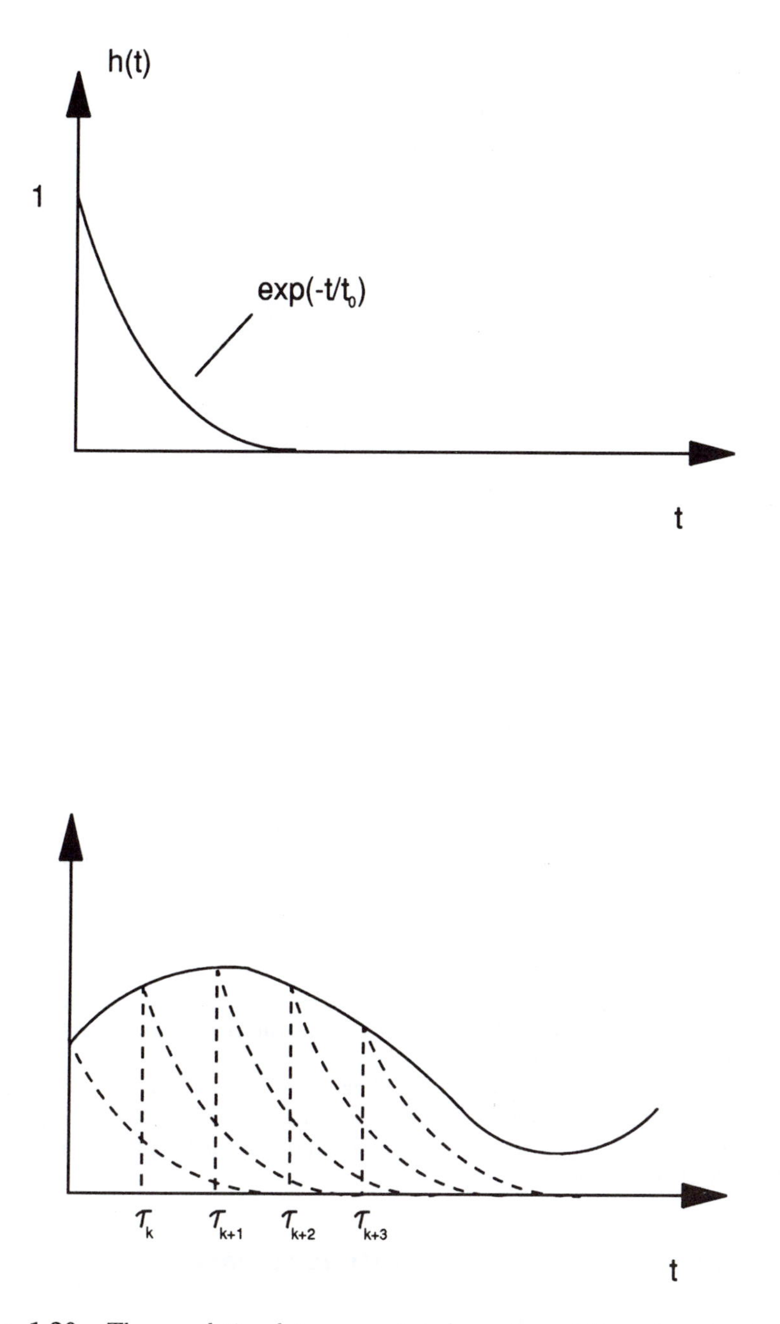

Figure 1.20 *The convolution between an impulse response h(t) and a signal x(t) (baseband waveforms):* $y(t) \cong \Sigma_k x(\tau_k) h(t-\tau_k) \cdot \Delta\tau$

—— $x(t)$
--- $h(t-\tau_k)$

$$\begin{cases} X_+(f)=2X(f) & f\geqslant 0 \\ X_+(f)=0 & f<0 \end{cases} \tag{1.396}$$

Moreover, it is convenient to consider the symmetric part $S(f)$ and the antisymmetric part $A(f)$ of $X_+(f)$ (Fig. 1.18), so that

$$X_+(f)=S(f)+A(f) \tag{1.397}$$

It is immediately seen (remembering expr. 1.396) that

$$\begin{cases} S(f)=X(f) \\ A(f)=X(f)\cdot \text{sign}(f) \end{cases} \tag{1.398}$$

where

$$\text{sign}(f)=\begin{cases} -1 & f<0 \\ +1 & f>0 \end{cases} \tag{1.399}$$

The antisymmetrical part $A(f)$ corresponds to a purely imaginary signal; therefore it is convenient to consider it as j times the output of a real (although not physically realisable) filter, known as the quadrature filter or the *Hilbert filter* (in formulas: $A(f)=jH_H(f)X(f)$). The frequency response of the Hilbert filter is

$$H_H(f)=-j\ \text{sign}(f) \tag{1.400}$$

and the impulse response is real and equals $h(t)=1/\pi t$, $-\infty<t<0$ and $0<t<\infty$ [14]. Note that the modulus of $H_H(f)$ is unity, while its phase is $-90°$ for positive frequencies and $+90°$ for negative frequencies; i.e. the Hilbert filter is equivalent to a perfect phase shifter.

Let us now evaluate the inverse Fourier transform $x_+(t)$ of $X_+(f)$; $x_+(t)$ is often called the *analytic signal* or *analytic part* of $x(t)$. From exprs. 1.397 and 1.398 we have:

$$x_+(t)=x(t)+j\hat{x}(t) \tag{1.401}$$

This means that the real part of the analytic signal is the original signal $x(t)$, and its imaginary part $\hat{x}(t)$ is the output of the Hilbert filter. For example, the Hilbert transform of $\cos(2\pi ft)$ is $\sin(2\pi ft)$, and the pertaining anaytic signal is $\exp(j2\pi ft)$.

The concept of analytic part and Hilbert transform can be used to represent band-limited signals (see Reference 14, pp. 118, 251–257); however, a simpler and more intuitive way, based on complex exponentials for positive frequencies only, can also be used.

1.7.2 Sampling of continuous-time signals

1.7.2.1 *The sampling theorem*

The *bandwidth* B of a signal is the frequency interval containing the significant (from the application point of view) part of its spectrum. *Band-limited* signals, either baseband or narrowband, will be considered in the following. A *baseband* real signal has its bandwidth centred around the zero frequency; a *narrowband* signal has its bandwidth centred around a 'central' frequency f_c, with $f_c \gg B$.

According to an important theorem due to Shannon and Nyquist (see Reference 14, for example), all the information contained in the waveform of a baseband signal is carried by a sequence of samples of the waveform, collected at a rate depending on the bandwidth of the signal itself: more precisely, if the baseband signal has the maximum frequency $B/2$, the sampling frequency shall not be less than twice $B/2$, i.e. of $f_N = B$ (Nyquist frequency).

The meaning of the sampling process is that the waveform can be completely described (or reconstructed) from its samples taken at the said rate.

The sampling process, in any physical system, is realised as the product of an analogue signal $x(t)$ by a periodic signal $c_T(t, \tau)$; that is a train of pulses of period T and duration $\tau \ll T$, and is written as follows:

$$c_T(t, \tau) = \frac{T}{\tau} \sum_{k=-\infty}^{\infty} \mathrm{rect}_\tau(t - kT) \tag{1.402}$$

where

$$\mathrm{rect}_\tau(t) = \begin{cases} 1 & -\dfrac{\tau}{2} < t < \dfrac{\tau}{2} \\ 0 & \text{elsewhere} \end{cases} \tag{1.403}$$

The spectrum of $c_T(t, \tau)$ is shown in Fig. 1.21, and when $\tau/T \rightarrow 0$, it tends to a 'comb' of harmonics with equal amplitude and constant phase.

The ideal sampling of a continuous signal $x(t)$ at a sampling frequency f_s is referred to, in many textbooks, as the result of the multiplication of $x(t)$ by a 'comb of Dirac pulses':

$$\sum_{n=-\infty}^{\infty} \delta\left(t - \frac{n}{f_s}\right) \tag{1.404}$$

We shall avoid this misleading (and mathematically incorrect, unless the distribution theory is used) form, and shall simply say that the result of ideal sampling (thereafter referred to as 'sampling' for short) of $x(t)$ is a discrete signal, i.e. a sequence of (real or complex) numbers x_n (or $x[n]$), such that

$$x[n] \equiv \frac{1}{f_s} x\left(\frac{n}{f_s}\right) \equiv T_s x(nT_s) \quad \forall n \tag{1.405}$$

where T_s is the sampling period.

In such a frame, the Fourier coefficients (expr. 1.346) can be considered as the results of ideally sampling the real and imaginary parts (exprs. 1.368 and 1.369) of the spectrum of a single period of a periodic signal, namely:

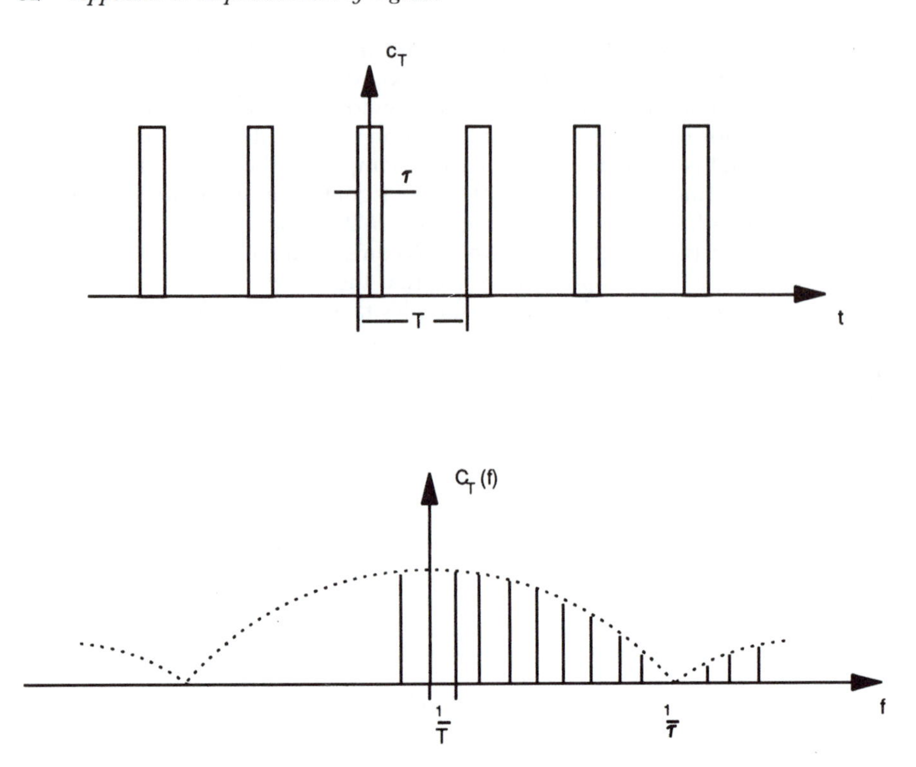

Figure 1.21 *A rectangular pulse train (sampling function) and its spectrum*

$$\begin{cases} a_k = \dfrac{1}{T} X_c\left(\dfrac{k}{T}\right) \\ b_k = \dfrac{1}{T} X_s\left(\dfrac{k}{T}\right) \end{cases} \tag{1.406}$$

By the use of the convolution property (ii), it is easy to show that the spectrum of a sampled signal (baseband) is composed of a number of replicas of the original spectrum, each centred around any integer multiple of the sampling frequency $F = 1/T$ (see Fig. 1.22). The signal can be reconstucted (in principle) in this way by low-pass filtering of the sampled signal, i.e. discarding all the parts of the spectrum outside the original band, which arise from the sampling process.

It is important to note that the original signal must be band-limited; if the signal has spectrum components outside half the Nyquist frequency, a phenomenon known as 'aliasing' appears (see Fig. 1.22 again) and the complete reconstruction of the signal is not possible. For this reason, before the sampling, the signal is generally passed through a low-pass filter to cut off the unwanted frequency components.

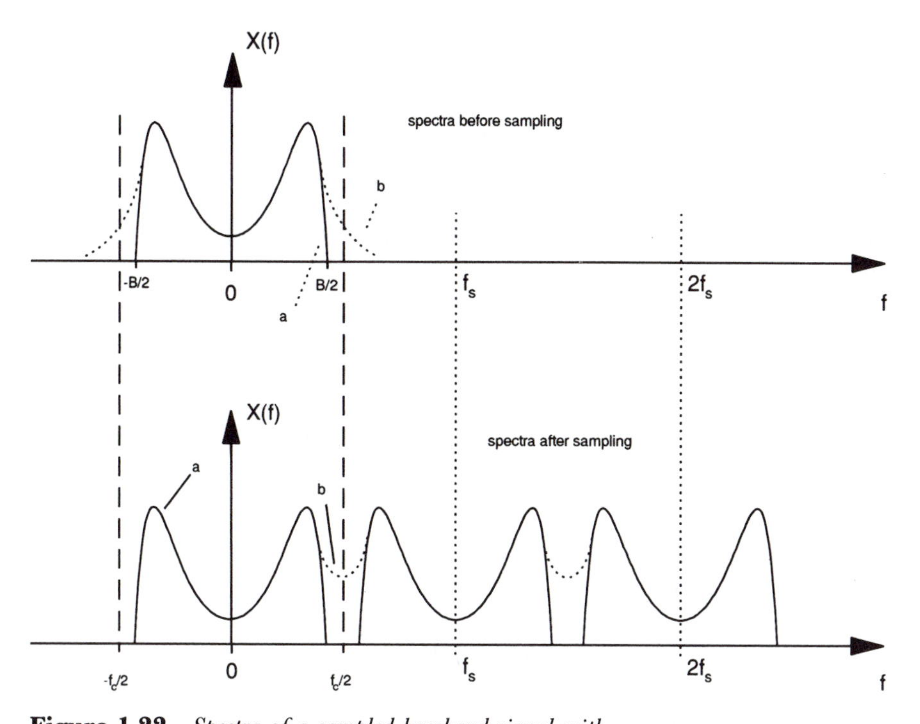

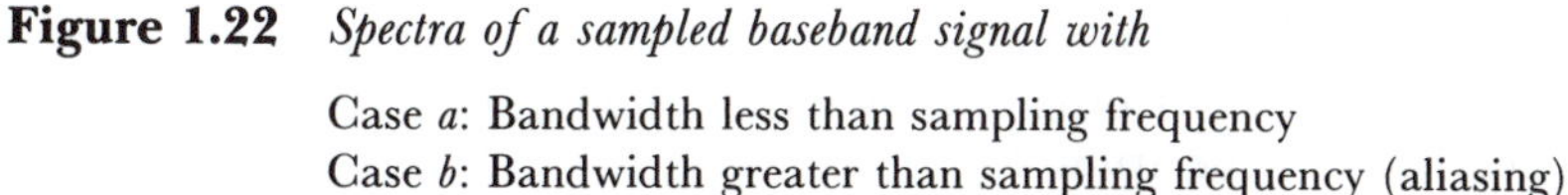

Figure 1.22 *Spectra of a sampled baseband signal with*

Case *a*: Bandwidth less than sampling frequency
Case *b*: Bandwidth greater than sampling frequency (aliasing)

The sampling theorem can be equally applied to the narrowband signals; in this case it is possible to show that the minimum sampling rate required is not necessarily twice the maximum frequency of the signal, but the signal bandwidth itself (see Section 1.7.4).

The spectra of sampled signals and the aliasing error are analysed in Sections 1.7.2.2 and 1.7.3.1.

1.7.2.2 *Spectra of sampled signals*

The Fourier transform of a sampled signal $x_T(t, \tau)$ (see expr. 1.402)

$$x_T(t, \tau) = x(t) \cdot c_T(t, \tau) \tag{1.407}$$

can be obtained by means of the convolution property (expr. 1.379). In the limit case of ideal sampling, the spectrum $X_F(f)$ of the sampled signal can be directly evaluated:

$$X_F(f) = F\{\lim_{\tau\to 0} x_T(t, \tau)\} = \int_{-\infty}^{\infty} \lim_{\tau\to 0} \frac{1}{\tau} \sum_{k=-\infty}^{\infty} Tx(t) \cdot \mathrm{rect}_\tau(t-kT)e^{-j2\pi ft}dt \tag{1.408}$$

Assuming that the integration and summation operators commute and that the limit exists, we have

$$X_F(f) = \sum_{k=-\infty}^{\infty} Tx(kT)e^{-j2\pi Tkf} \tag{1.409}$$

Comparing the Fourier series (expr. 1.348) with expr. 1.409 obtained for $X_F(f)$, it is readily seen that $X_F(f)$ is periodic, with period $F=1/T$, and its Fourier series coefficients x_k are the samples $x[k] \equiv Tx(kT)$, $\forall k$; in other words (see expr. 1.409)

$$X_F(f) = \sum_{k=-\infty}^{\infty} x[k] \exp\left\{-j\frac{2\pi}{F}kf\right\} \tag{1.410}$$

Rigorously, expr. 1.410 is, by definition, the Fourier transform (in other words, the *spectrum*) of the sequence $\{x[k]\}$; a sufficient condition for its existence is

$$\sum_{k=-\infty}^{\infty} |x[k]| < \infty \tag{1.411}$$

from which the condition of finite energy

$$\sum_{k=-\infty}^{\infty} |x[k]|^2 < \infty \tag{1.412}$$

can be derived (but not *vice versa*).

From the Fourier series properties (see expr. 1.349) the $\{x[k]\}$ sequence can be obtained from its spectrum

$$x[k] = \frac{1}{F}\int_{f_0-(F/2)}^{f_0+(F/2)} X_F(f) \exp\left\{j\frac{2\pi}{F}kf\right\}df \tag{1.413}$$

where the term f_0 is an arbitrary constant related to the periodicity of both $X_F(f)$ and the complex exponential $e_f(f) = \exp\{j(2\pi/F)kf\}$.

1.7.3 Relationship between continuous and sampled signals

1.7.3.1 *Relationship between the spectra of continuous and sampled signals*

The relationship between the spectrum $X(f)$ of a signal $x(t)$ and the spectrum $X_F(f)$ of the sampled signal $x[k]$ can be obtained by considering the inverse Fourier transform (expr. 1.359) for $t=kT$ and by substituting the expression of the sampled signal:

$$x[k] = Tx(kT) \tag{1.414}$$

We have:

$$x[k] = T\int_{-\infty}^{\infty} X(f) \exp\{j2\pi fkT\}df \tag{1.415}$$

The integration domain can be divided into contiguous intervals of width $F=1/T$:

$$x[k]=\frac{1}{F}\sum_{i=-\infty}^{\infty}\int_{iF}^{(i+1)F} X(f)\exp\left\{j\frac{2\pi}{F}kf\right\}df \tag{1.416}$$

which, remembering that the complex exponential is periodic, yields

$$x[k]=\frac{1}{F}\int_{0}^{F}\sum_{i=-\infty}^{\infty} X(f+iF)\exp\left\{j\frac{2\pi}{F}kf\right\}df \tag{1.417}$$

By comparison of expr. 1.417 with expr. 1.413, we see that the spectrum of the sampled signal is the sum of an infinite number of shifted replicas of the spectrum of the original signal:

$$X_F(f)=\sum_{i=-\infty}^{\infty} X(f+iF) \tag{1.418}$$

The difference

$$\Delta X(f)=X_F(f)-X(f)=\sum_{\substack{i=-\infty\\ i\neq 0}}^{\infty} X(f+iF) \tag{1.419}$$

is the 'aliasing' error and depends on the values of $X(f)$ outside the interval $[-F/2;\ F/2]$.

If these values are negligible, the aliasing error is also negligible and the sampled signal has a spectrum very close, in the $[-F/2;\ F/2]$ interval, to the spectrum of the original signal.

The above derivation can be used to analyse the effect of sampling the spectrum $X(f)$ with a sampling interval $\Delta f=1/\theta$. From the time–frequency duality of the Fourier transform such 'spectral sampling' is equivalent to the 'periodisation' of the original signal $x(t)$, yielding a periodic signal $x_\theta(t)$:

$$x_\theta(t)=\sum_{i=-\infty}^{\infty} x(t+i\theta) \tag{1.420}$$

As above, if $x(t)$ is negligible outside the $[-\theta/2;\ \theta/2]$ interval, the 'aliased' signal $x_\theta(\tau)$ is very close to $x(t)$ in that interval.

More interesting from the practical point of view is the case in which sampling is applied to $X_F(f)$ with a sampling interval $\Delta f=F/N$, N integer. In this case the sampled signal $x[k]$ is transformed into a periodic sequence

$$x_{NT}[k]=\sum_{i=-\infty}^{\infty} x[k+iN] \tag{1.421}$$

or equivalently

$$x_{NT}[k]\equiv T\sum_{i=-\infty}^{\infty} x(kT+iNT)\quad \forall k \tag{1.422}$$

The periodic sequence with period N so obtained is completely described by N contiguous elements (samples). The same considerations apply to the sequence

$$X_F[n] \equiv X_F(n\Delta f) \quad \forall n \tag{1.423}$$

From exprs. 1.410 and 1.423, recalling that $\Delta f = F/N$, we have

$$X_F[n] = \sum_{k=-\infty}^{\infty} x[k] \exp\left\{-j\frac{2\pi}{N}kn\right\} \tag{1.424}$$

Owing to the periodicity, with period N, of the exponential sequence

$$e_N[k] = \exp\left(-j\frac{2\pi}{N}kn\right) \tag{1.425}$$

it is convenient to divide the summation of expr. 1.424 into summations of N terms each; from expr. 1.421, we have:

$$X_F[n] = \sum_{k=0}^{N-1} x_{NT}[k] \exp\left\{-j\frac{2\pi}{N}kn\right\} \tag{1.426}$$

Exploiting the properties of the sequence $e_N[k]$ (expr. 1.425), it is possible to represent $x_{NT}[k]$ in terms of $X_F[n]$, with the result

$$x_{NT}[k] = \frac{1}{N}\sum_{n=0}^{N-1} X_F[n] \exp\left\{j\frac{2\pi}{N}kn\right\} \tag{1.427}$$

Exprs. 1.426 and 1.427 are the representation of a periodic sequence in terms of discrete Fourier series.

Considering the first period of both time and frequency domain representations, i.e. applying a rectangular window of length N in both domains, the finite sequences

$$x_k = x_{NT}[k], \; k = 0, 1, \ldots, N-1 \tag{1.428}$$

and

$$X_n = X_F[n], \; n = 0, 1, \ldots, N-1 \tag{1.429}$$

are obtained.

Exprs. 1.428 and 1.429 are related to each other by the discrete Fourier transform (DFT) operator

$$x_k \underset{DFT^{-1}}{\overset{DFT}{\rightleftarrows}} X_n$$

In formulas

$$X_n = \sum_{k=0}^{N-1} x_k \exp\left\{-j\frac{2\pi}{N}kn\right\} \tag{1.430}$$

and

$$x_k = \frac{1}{N} \sum_{n=0}^{N-1} X_n \exp\left\{ j \frac{2\pi}{N} kn \right\} \tag{1.431}$$

The DFT can be implemented with a reduced computational burden [fast Fourier transform (FFT) and other algorithms] and is one of the main tools for signal analysis.

The conceptual relationships between continuous, periodic and sampled signals are shown in Fig. 1.23*a*. It is a very general scheme describing the relationships between signals in the time domain (horizontal upper line) and in the frequency domain (horizontal lower line). The single headed arrows show that the process is one-way, indicating that, after sampling (first step), some of the characterisation of the signal is lost, and only in particular cases (band-limited signals) can the process be reversed; the process of sampling a signal in time corresponds to periodisation of its spectrum in frequency, and *vice versa*, while the final step of windowing a signal (or a spectrum) allows us to consider a limited set of coefficients as a complete characterisation of the signal (spectrum) itself.

The vertical arrows are double-headed, and this means that the operations are reversible without any loss of information. For a continuous signal the time domain representation and the spectrum are linked via the Fourier transform, which reduces to a Fourier series for sampled signals or periodic spectra. The discrete Fourier series then applies when the signls (spectra) are made periodic (sampled) and finally the discrete Fourier transform (DFT) applies when both signals and spectra are 'windowed'. For further details on sampling, discrete signals, DFT and FFT, interested readers are addressed to some of the many books on these topics, e.g. References 17–19.

1.7.3.2 *Practical implications*

Let us now consider the particular case in which the signal is time-limited, i.e.

$$\begin{cases} x(t) \neq 0 & 0 \leq t \leq NT \\ x(t) = 0 & \text{elsewhere} \end{cases} \tag{1.432}$$

In this case the 'periodisation' of Fig. 1.23*a* does not imply any computation nor does it modify the shape of the signal $x(t)$ in the $[0, NT]$ interval. Therefore the finite length sequence

$$x_k, \qquad k = 0, 1, \ldots, N-1 \tag{1.433}$$

is directly obtained by the first N samples of the original signal $x(t)$. However, the DFT of x_k *does not* supply N samples of the spectrum, unless the reciprocal of the sampling interval $F = 1/T$ is greater than the bandwidth B of the signal, according to the Nyquist theorem.

However, strictly speaking, a time-limited signal cannot be band-limited too†, and therefore the Nyquist criterion can be met only approximately at a large enough sampling frequency $1/T$, i.e. when the duration NT of the signal is

† See Reference 20 for a complete discussion of the concepts of signal duration and Fourier transform duration, and the lower bound of their product, that can be found by applying the uncertainty principle.

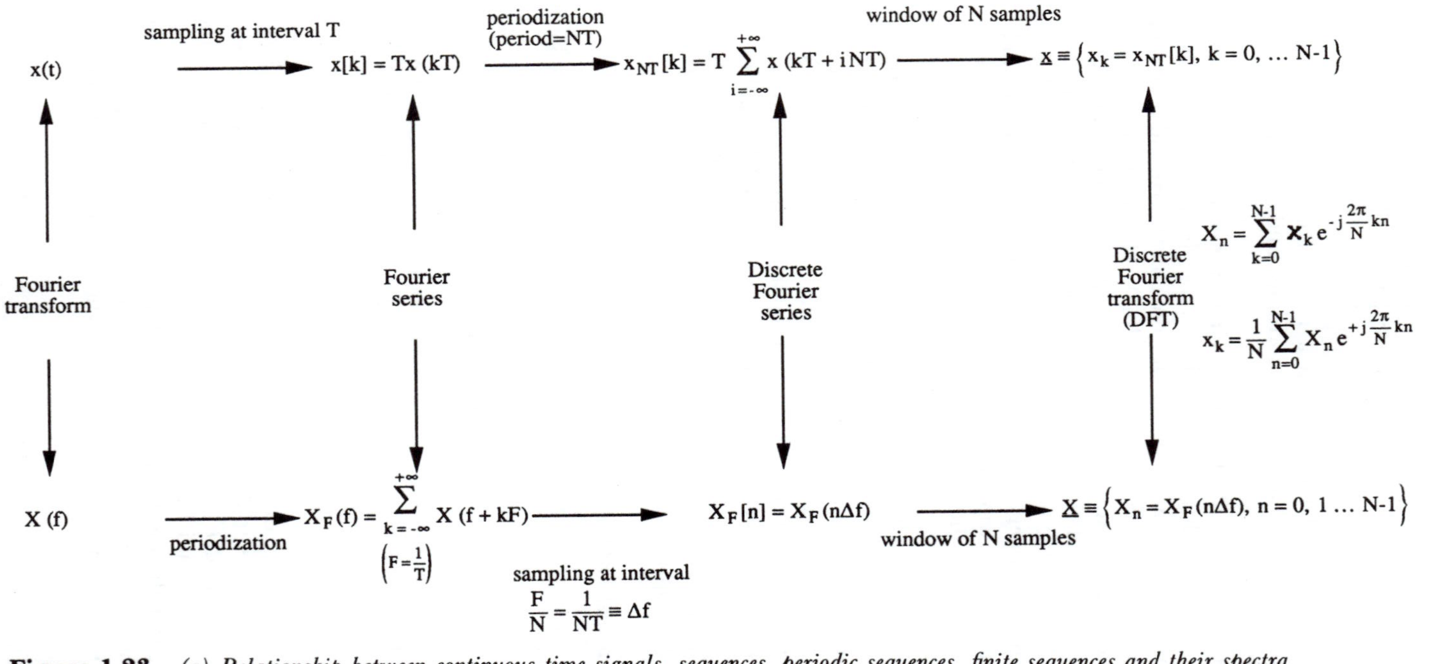

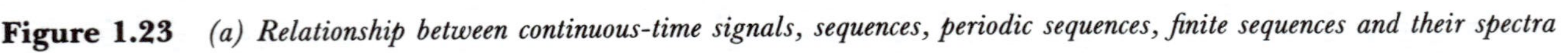
Figure 1.23 *(a) Relationship between continuous-time signals, sequences, periodic sequences, finite sequences and their spectra*

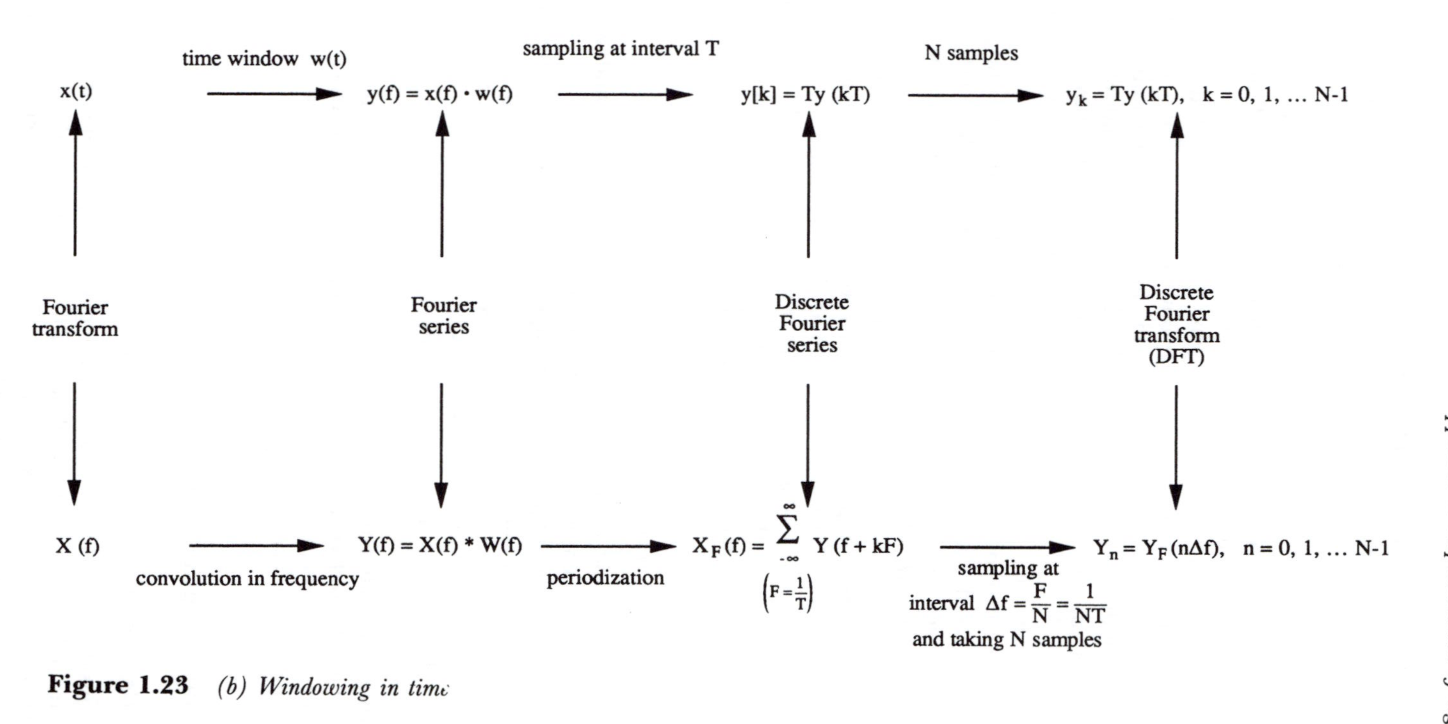

Figure 1.23 *(b) Windowing in time*

fixed by a large enough number of samples. The price to be paid for reducing the spectral aliasing and to make the DFT result closer to a sampling of the spectrum $X(f)$ is, therefore, an increase of the *sampling, storage* and *processing* burden.

To obtain a time-limited signal of duration NT, the multiplication of $x(t)$ by a 'window' $w(t)$ (either rectangular or 'weighted') of duration NT can be used, according to the block diagram of Fig. 1.23*b*.

In the frequency domain, the windowing (see the convolution property above) modifies the spectrum through convolution with the Fourier transform of the window, $W(f)$; the spectral resolution is affected and, generally speaking, is of the order of the reciprocal of the window length. Many window types have been proposed and their characteristics evaluated in the literature (Reference 18, Sections 3.7.9–3.7.17, and Reference 21) in view of trade-offs between the width of the main lobe of $W(f)$ and its sidelobe structure.

In the simple case of a rectangular window, the signal shape in the time domain is preserved, i.e. $x(t) = y(t)$ within the window, but the spectrum $Y(f)$ can be very much different from $X(f)$ owing to the large sidelobe level of $W(f) = [\sin(\pi f NT)]/\pi f NT$.

Dual considerations could occur when a frequency domain window is used, i.e. a filter $H(f)$ whose impulse response $h(t)$ would modify, through convolution, the shape of $x(t)$. Such a filter can be used for anti-aliasing purposes, as explained before; an ideal low-pass filter with bandwidth B equal to the sampling frequency is the exact dual of the rectangular window in time discussed before, and equivalent considerations apply for the time domain effects of the filter.

1.7.3.3 *Relevance to radar signals*

In a pulsed monostatic surveillance radar, both sampling and windowing are strictly tied to the operations of the radar itself.

The pulse repetition frequency is the sampling frequency of the radar signal (for each constant range line, or range bin), while the dwell time (defined as the time interval in which a steady point scatterer is within the main lobe of the antenna) corresponds to the window; the shape of the window is the two-way pattern of the antenna for continuous (mechanical) scanning, while it is rectangular for step (electronic) scanning.

To exploit the different spectra of the useful signals on one hand, and the various kinds of interference on the other hand, processing schemes based on the FFT or, more generally, on digital filtering (e.g. FIR filters) are often used, representing practical applications of the scheme of Fig. 1.23*b*.

1.7.4 Band-limited signals

1.7.4.1 *Baseband and narrowband signals*

Let us consider a *baseband* real signal, i.e. one with a symmetrical spectrum centred around zero frequency and limited to the $(-B/2; +B/2)$ interval.

The *up-conversion*, or double sideband modulation, of the baseband signal $x(t)$ with bandwidth B generates the narrowband signal

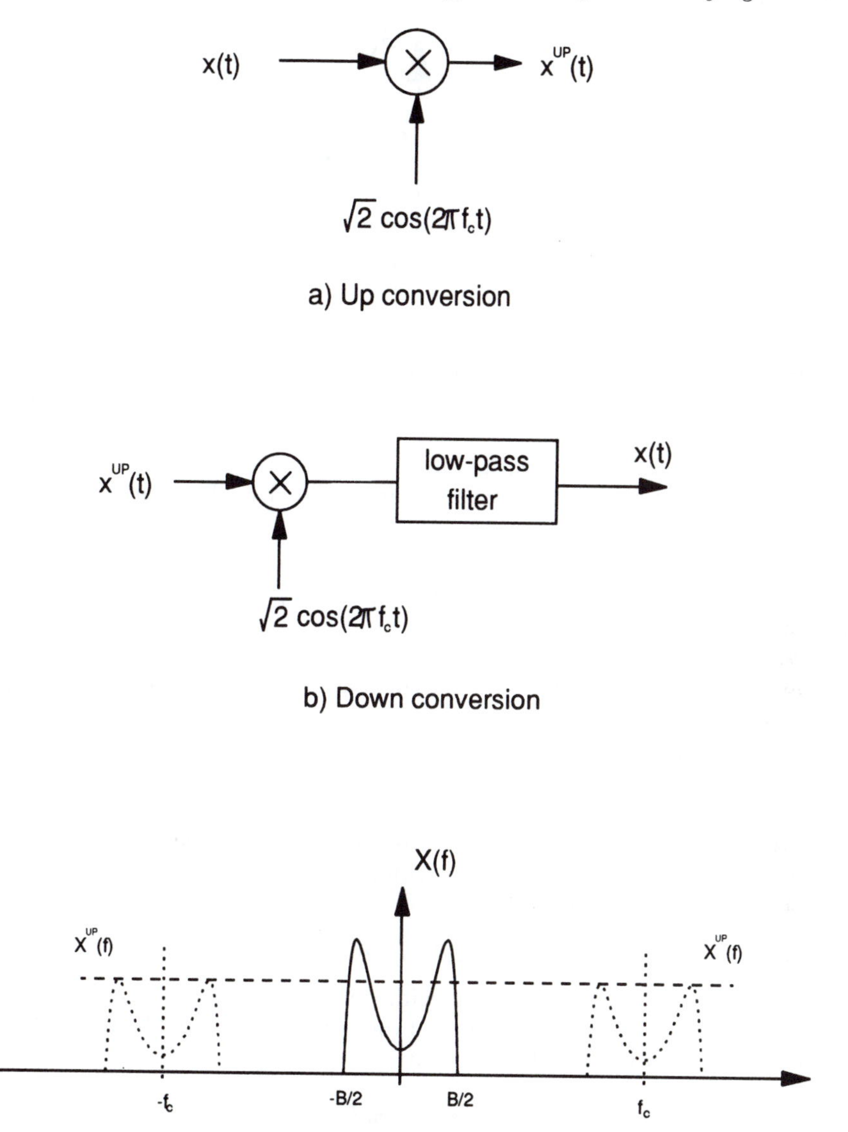

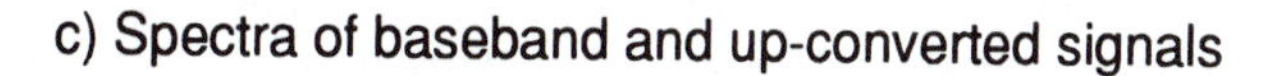

Figure 1.24 *Up and down conversion processes*

$$x^{UP}(t) = \sqrt{2}\; x(t)\; \cos(2\pi f_c t) \tag{1.434}$$

where f_c is the *carrier frequency*, assumed to be much greater than B (see Fig. 1.24*a*). By the use of the Euler formula $\cos(2\pi f_c t) = \frac{1}{2}(e^{j2\pi f_c t} + e^{-j2\pi f_c t})$ and of the

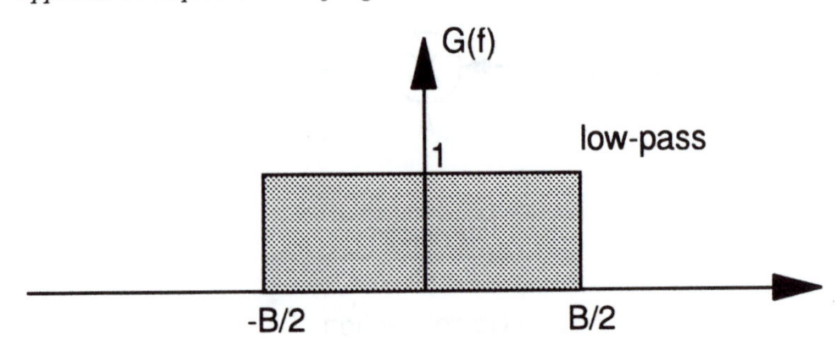

Figure 1.25 *Spectrum of the response of an ideal low-pass filter*

translation property of the Fourier integral (expr. 1.376), the Fourier transform $X^{UP}(f)$ of the up-converted signal is simply related to the Fourier transform $X(f)$ of $x(t)$ as follows:

$$X^{UP}(f) = \frac{\sqrt{2}}{2} [X(f+f_c) + (X(f-f_c)] \tag{1.435}$$

as shown in Fig. 1.24*c*.

As $x^{UP}(t)$ is narrowband, its analytic part $x_+^{UP}(t)$ is the inverse Fourier transform of $2(\sqrt{(2)}/2)\ X(f-f_c), f \geqslant 0$, i.e. a translated replica of the spectrum of $x(t)$. It results in (expr. 1.376)

$$x_+^{UP}(t) = \sqrt{2}x(t)\ \exp\{j2\pi f_c t\} \tag{1.436}$$

The original signal $x(t)$ can be obtained by the up-converted signal by *down-conversion*, i.e. by multiplication with the *reference signal* $\sqrt{2}\ \cos(2\pi f_c t)$ and low-pass filtering (see Fig. 1.24*b*):

$$x(t) = [\sqrt{2}x^{UP}(t)\ \cos(2\pi f_c t)]_{LP} \tag{1.437}$$

where the 'ideal' low-pass filter, with the frequency response shown in Fig. 1.25, rejects the unwanted spectral components generated by the multiplication (beat) and the out-of-band noise.

1.7.4.2 *Reconstruction of the baseband signal: the radar case*

In practical applications things are slightly more complicated than described in the previous section, as the reference signal at the down conversion side can be frequency or phase offset from the one at the up-conversion side. Therefore, one could attempt to reconstruct $x(t)$ by the following processing (to be compared with expr. 1.437):

$$\xi(t) = \{\sqrt{2}x^{UP}(t)\ \cos[2\pi f_c t + \phi(t)]\}_{LP} \tag{1.438}$$

where the phase term $\phi(t)$ includes both the frequency offset, i.e. a linear term, and the phase offset, i.e. a constant term. From exprs. 1.434 and 1.438:

$$\xi(t) = x(t)\ \cos\ \phi(t) \tag{1.439}$$

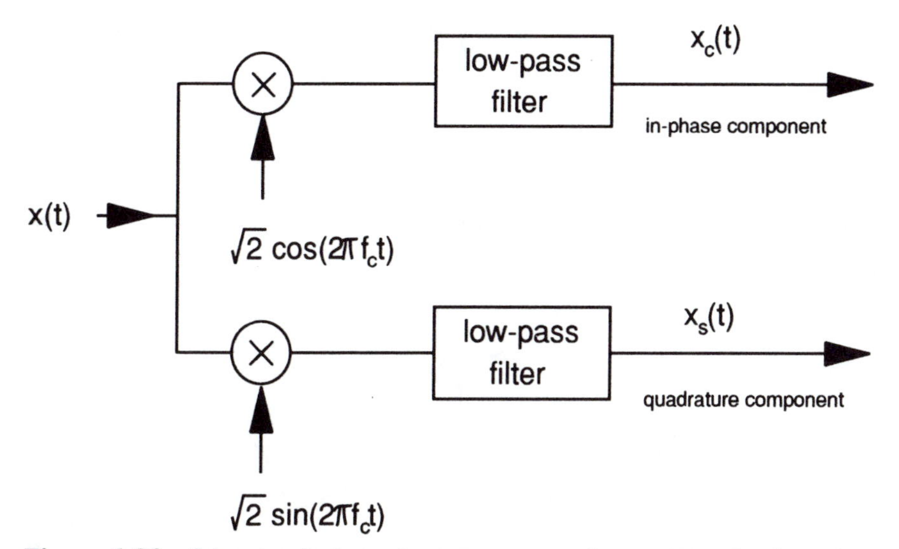

Figure 1.26 *Scheme to obtain quadrature components from a narrow-band signal*

As a consequence, when $\phi(t)$ is an odd multiple of 90°, $\xi(t)$ is equal to zero, and the information in $x^{UP}(t)$ is completely lost.

In monostatic radar applications the phase term is related to the target 'range' R (distance from the phase centre of the target to the phase centre of the radar antenna):

$$\phi(t) = 2\,\frac{2\pi}{\lambda}\,R(t) \tag{1.440}$$

where λ is the operating wavelength and the coefficient 2 is due to the two-way operation. Therefore, a target at a fixed range R will not be detected if

$$R = \frac{2k+1}{8}\,\lambda \qquad k = \text{an integer} \tag{1.441}$$

This is the well-known phenomenon of the 'blind phases' and is avoided by down-converting into two channels, the in-phase (I) channel and the quadrature (Q) channel, as shown in Fig. 1.26. The pertaining receiver is often called the 'phase detector' or 'coherent detector' as it preserves both amplitude and phase information, and is therefore preferred to the single channel scheme of Fig. 1.24*b*.

A second typical situation requiring I and Q representation of baseband signals in radar applications arises when multiple scatterers (e.g. rain and land echoes) are present in the same resolution cell. The spectra of single target (*a*) and multiple scatterers (*b*) echoes are shown in Fig. 1.27; the former is symmetrical around some value $\bar{f}$, generally unknown at the radar side, including the Doppler shift. The latter has no symmetry around any 'carrier frequency', and therefore can only be down-converted by the coherent detector of Fig. 1.26, where, in principle, the choice of f_c is arbitrary. In practice f_c is the transmitted frequency, i.e. the one used to generate the transmitted signal.

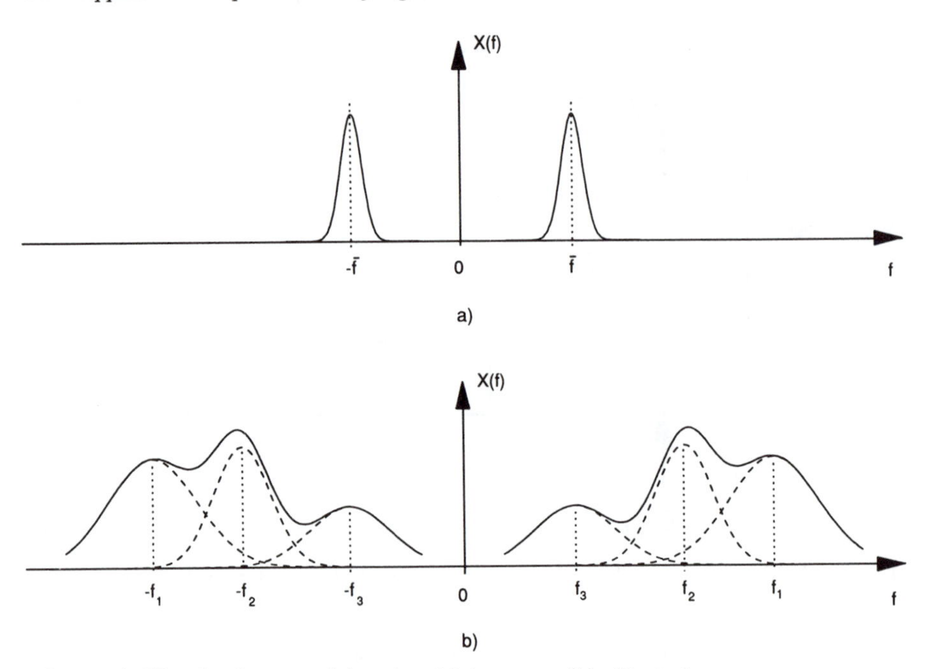

Figure 1.27 *Single target (a) and multiple targets (b): Typical spectra*

1.7.4.3 *Complex representation of band-limited signals*

A narrowband signal $x(t)$ can be written in the following way:

$$x(t) = \sqrt{2}\ A(t)\ \cos\{2\pi f_c t - \phi(t)\} \tag{1.442}$$

where, by definition, the variations of $A(t)$ and $\phi(t)$ are 'slow', i.e. their bandwidth is small with respect to the 'central frequency' f_c. By the use of trigonometrical identities, expr. 1.442 becomes

$$\begin{aligned} x(t) &= \sqrt{2}A(t)\ \cos\ \phi(t)\ \cos(2\pi f_c t) + \sqrt{2}\ A(t)\ \sin\ \phi(t)\ \sin(2\pi f_c t) \\ &= \sqrt{2}\ [x_c(t)\ \cos(2\pi f_c t) + x_s(t)\ \sin(2\pi f_c t)] \end{aligned} \tag{1.443}$$

where

$$\begin{cases} x_c(t) \equiv A(t)\ \cos\ \phi(t) \\ x_s(t) \equiv A(t)\ \sin\ \phi(t) \end{cases} \tag{1.444}$$

are baseband (if f_c is inside the signal bandwidth B) and band-limited signals.

Both the above considerations (arising from practical applications) and the analytical signal theory lead to the representation of a narrow band signal with central frequency f_c by means of two components, one said to be 'in phase' and the other 'in quadrature' according to the following definitions:

$$\begin{cases} x_c(t) = [\sqrt{2}x(t)\ \cos(2\pi f_c t)]_{LP} \\ x_s(t) = [\sqrt{2}x(t)\ \sin(2\pi f_c t)]_{LP} \end{cases} \tag{1.445}$$

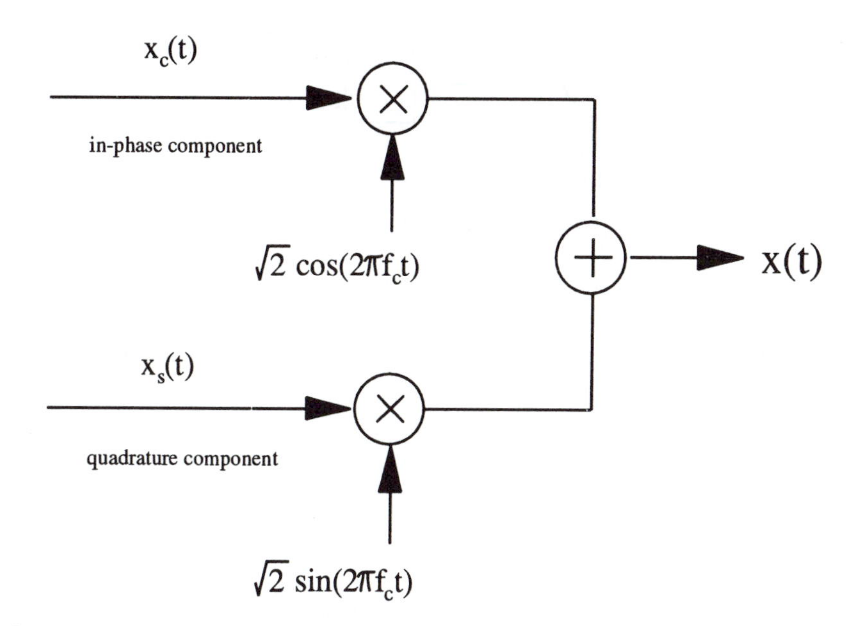

Figure 1.28 *Scheme to obtain the pass-band signal from its quadrature components*

These I and Q components $x_c(t)$ and $x_s(t)$ are obtained as shown in Fig. 1.26.

Given $x_c(t)$ and $x_s(t)$, the original signal $x(t)$ is obtained from the scheme of Fig. 1.28 (from the hypothesis of having a real original signal). It is easy to show that in this case the overall system depicted in Figs. 1.26 and 1.28 is equivalent to a band-pass filter whose bandwidth B is centred around $\pm f_c$ (Fig. 1.29).

As explained before, the choice of f_c is arbitrary, i.e. one is free to choose any value of frequency around which to build the two component signals $x_c(t)$ and $x_s(t)$. It is clear that the choice is generally driven by convenience: for instance, in the case of band-translated signals, it is obvious that the carrier frequency is the best choice for symmetry reasons.

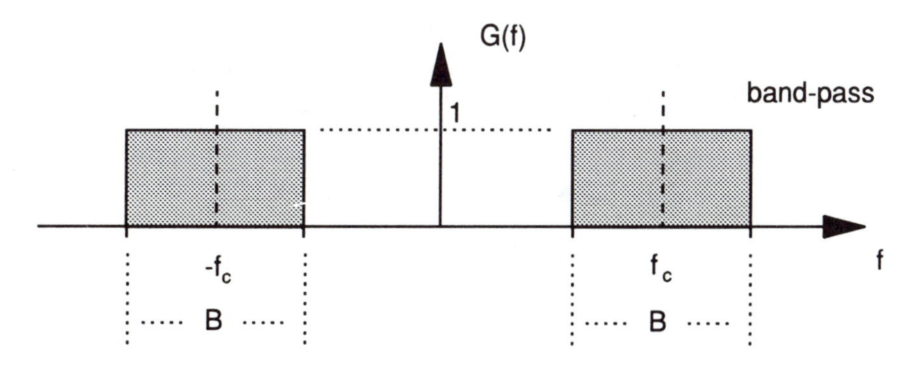

Figure 1.29 *Frequency response of an ideal band-pass filter*

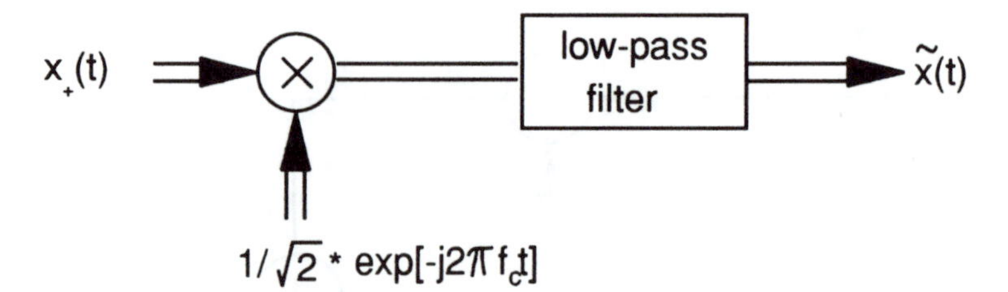

Figure 1.30 *From analytic signal to complex envelope (complex down-conversion)*

It is possible to represent baseband signals by introducing the complex signal, called the *complex envelope* of $x(t)$:

$$\tilde{x}(t) = x_c(t) - jx_s(t) \tag{1.446}$$

or, in an alternative form,

$$\tilde{x}(t) = |\tilde{x}(t)| e^{-j\phi(t)} \tag{1.447}$$

where

$$|\tilde{x}(t)| = \sqrt{x_c^2(t) + x_s^2(t)} = A(t) \tag{1.448}$$

and

$$\phi(t) = \tan^{-1}\left(\frac{x_s(t)}{x_c(t)}\right) \tag{1.449}$$

Remembering that the analytic part of $A\cos(2\pi f_c t - \phi)$ is $A\exp\{j2\pi f_c t - \phi\}$, from expr. 1.442 the analytic part $x_+(t)$ of the narrowband signal $x(t)$ is simply $1/\sqrt{2}$ times the up-converted version of the complex envelope $\tilde{x}(t)$:

$$x_+(t) = \sqrt{2}A(t)\exp\{j2\pi f_c t - \phi(t)\} = \sqrt{2}\tilde{x}(t)\exp\{j2\pi f_c t\} \tag{1.450}$$

Expr. 1.450 is shown in Figs. 1.30 and 1.31, and can be compared with Figs. 1.26 and 1.28, respectively.

We can also write

$$\tilde{x}(t) = [\sqrt{2}x(t)e^{-j2\pi f_c t}]_{LP} \tag{1.451}$$

The reason for the name 'complex envelope' is readily understood, noting that the modulus of $\tilde{x}(t)$ is the envelope of the narrowband signal. The phase

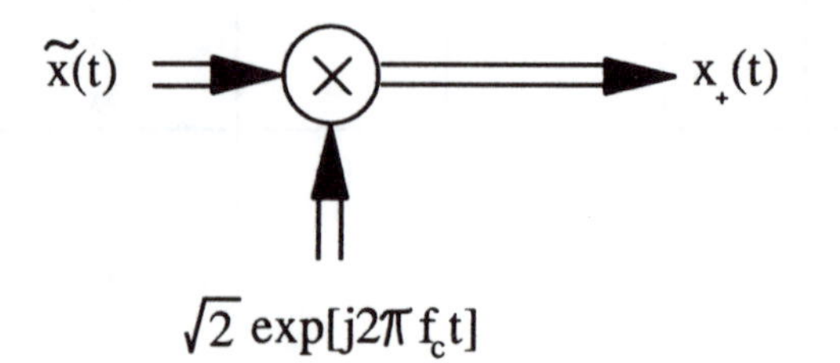

Figure 1.31 *From complex envelope to analytic signal (complex up-conversion)*

$2\pi f_c t - \phi(t)$ is referred to as the instantaneous phase of the signal: it is composed of the carrier angle plus the relative phase due to the narrowband components themselves.

1.7.4.4 *Reconstruction of the components of a band-limited signal*

A narrowband signal $x^{UP}(t)$ (Fig. 1.32*a*) can be represented, as well as by the $x_c(t)$ and $x_s(t)$ components, by a real baseband signal $x^0(t)$, whose spectrum $X^0(f)$ is shown in Fig. 1.32*b*, with a bandwidth twice that of the original bandwidth, or, in an equivalent way, by the correspondent analytic signal $x^0(t) + j\hat{x}^0(t)$, whose spectrum is shown in Fig. 1.32*c*.

Fig. 1.33 shows the conceptual scheme required to obtain the analytic signal $x^0(t) + j\hat{x}^0(t)$ from $x^{UP}(t)$, and in Fig. 1.34 the scheme required to obtain back $x^{UP}(t)$ from the analytic signal is shown (the Hartley modulator, used in communications techniques for single sideband modulation).

The scheme of Fig. 1.33 is of practical interest when one wants to obtain the $x_c(t)$ and $x_s(t)$ components, but, because of problems of matching criticality, it is not advisable to use the two-mixers scheme of Fig. 1.26 (in which the pair of down-converters has to be carefully matched in amplitude and phase). In this case, a suitable procedure to follow is to shift down the analytic signal of the quantity $B/2$, obtaining the complex envelope of the signal, and then to sample it†.

The related signals (see Figs. 1.32 and 1.35) are the following:

- Down-converted (real) signal:

$$x^0(t) = \left[x^{UP}(t)\sqrt{2}\,\cos\left(2\pi\left(f_c - \frac{B}{2}\right)t\right)\right]_{LP} \tag{1.452}$$

- Analytic part of $x^0(t)$:

$$x^0_+(t) = x^0(t) + j\hat{x}^0(t) \tag{1.453}$$

- Complex envelope of $x^{UP}(t)$, expressed by means of the in-phase and quadrature components $x_c(t)$ and $x_s(t)$:

$$\tilde{x}(t) = x_c(t) - jx_s(t) \tag{1.454}$$

The complex envelope can be obtained by frequency shifting of the amount $B/2$ the analytic part of $x^0(t)$:

$$\tilde{x}(t) = x^0_+(t)\ \exp\{-j\pi Bt\}. \tag{1.455}$$

From expr. 1.455 it follows that the samples of $\tilde{x}(t)$ taken at a sampling frequency $f_s(t) = pB$, $p = 1$ or $p = 2$, can be obtained from the samples of $x^0_+(t)$ by multiplying factors such as ± 1, $\pm j$, i.e. without any true multiplication:

$$\tilde{x}\left(\frac{k}{pB}\right) = x^0_+\left(\frac{k}{pB}\right)\ \exp\left\{-j\frac{\pi}{p}k\right\} \qquad k = 0, 1, \ldots \infty, p = 1, 2 \tag{1.456}$$

† With regard to the scheme based on Figs. 1.32 and 1.33 it is worth pointing out that, from the practical realisation point of view, it may be convenient to apply Hilbert filtering directly to the narrowband signal of Fig. 1.32*a* where there is a large signal band around null frequency: in this way a steep transition from $+\pi/2$ to $-\pi/2$ is not required, allowing us so to realise a less critical filter.

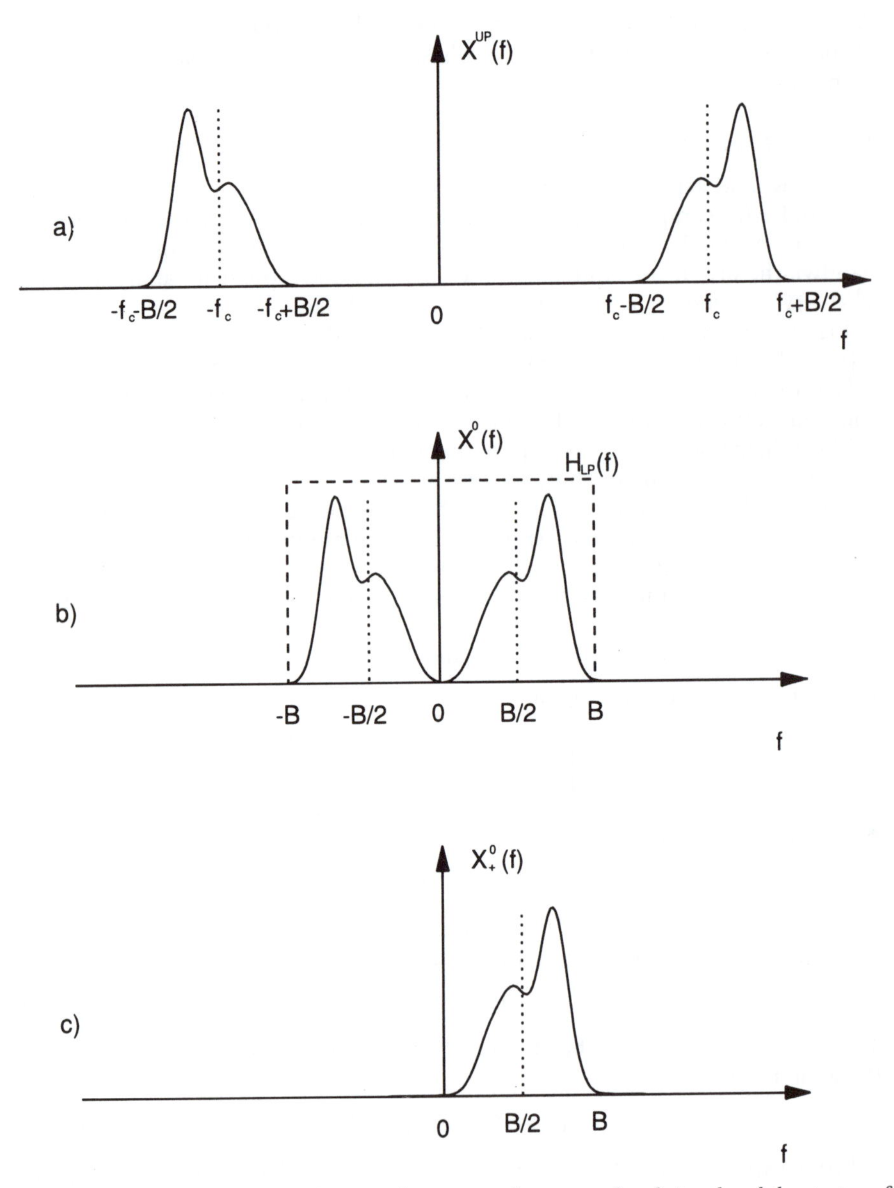

Figure 1.32 *Relationship between the spectra of a narrow-band signal and the spectra of the associated analytic signal*

The spectra of the sampled signal $\tilde{x}(k/pB)$ are shown in Fig. 1.35 for $p=1$ (curve *c*) and $p=2$ (curve *d*).

The main drawback of the above procedure is the need for an analogue Hilbert filter. This can be avoided when it is possible (and convenient) to use the higher sampling frequency $f_s=2B$ (i.e. $p=2$). In this case it is possible to

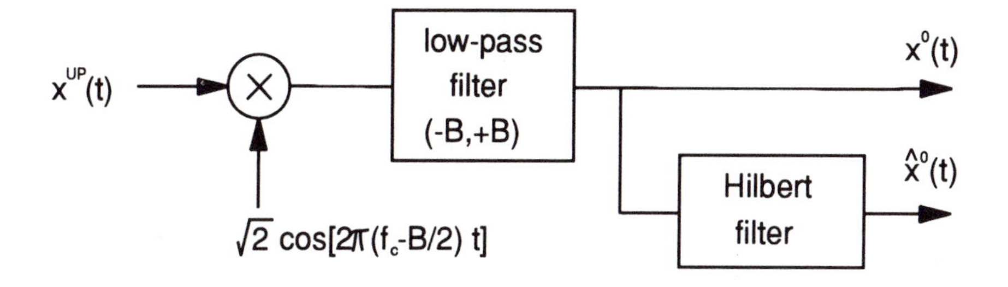

Figure 1.33 *Scheme to obtain the analytic signal from a narrow-band signal*

sample $x^0(t)$ without any spectral aliasing (see Fig. 1.32*b*) obtaining the spectrum of Fig. 1.36*a*, and, through digital Hilbert filtering, the discrete signal $x^0(k/2B) + j\hat{x}^0(k/2B)$, whose spectrum, shown in Fig. 1.36*b*, only differs from that of Fig. 1.35*d* for a frequency shift of $B/2$, i.e. of $f_s/4$, corresponding to a sampling delay of one quarter of the sampling period.

Moreover, the digital Hilbert filter may also be avoided if it is sufficient to get the in-phase and quadrature components at a sampling period $1/B$ and time-shifted with each other by $1/2B$. In such a case it may be convenient to work directly on the real down-converted signal $x^0(t)$. It can be expressed as follows in terms of quadrature components (from now on, we drop the inessential $\sqrt{2}$ coefficient, for sake of clarity):

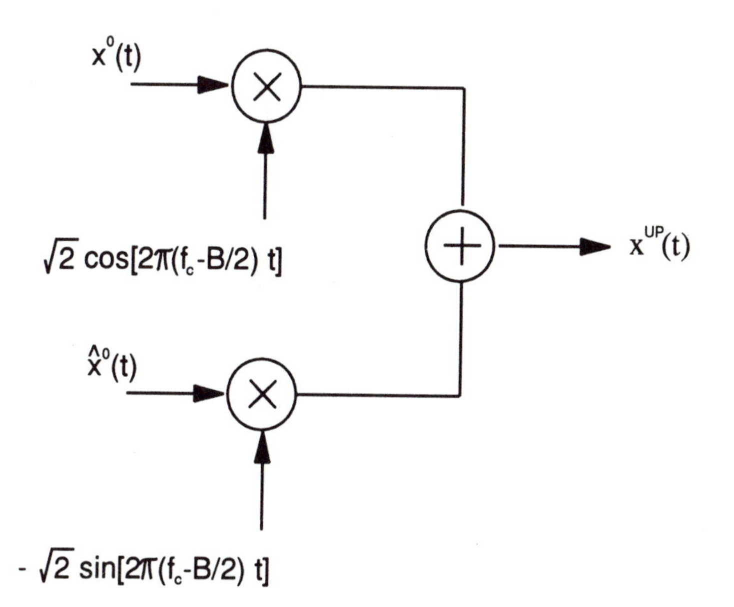

Figure 1.34 *Scheme to obtain SSB band-pass signal from its analytic signal (Hartley modulator)*

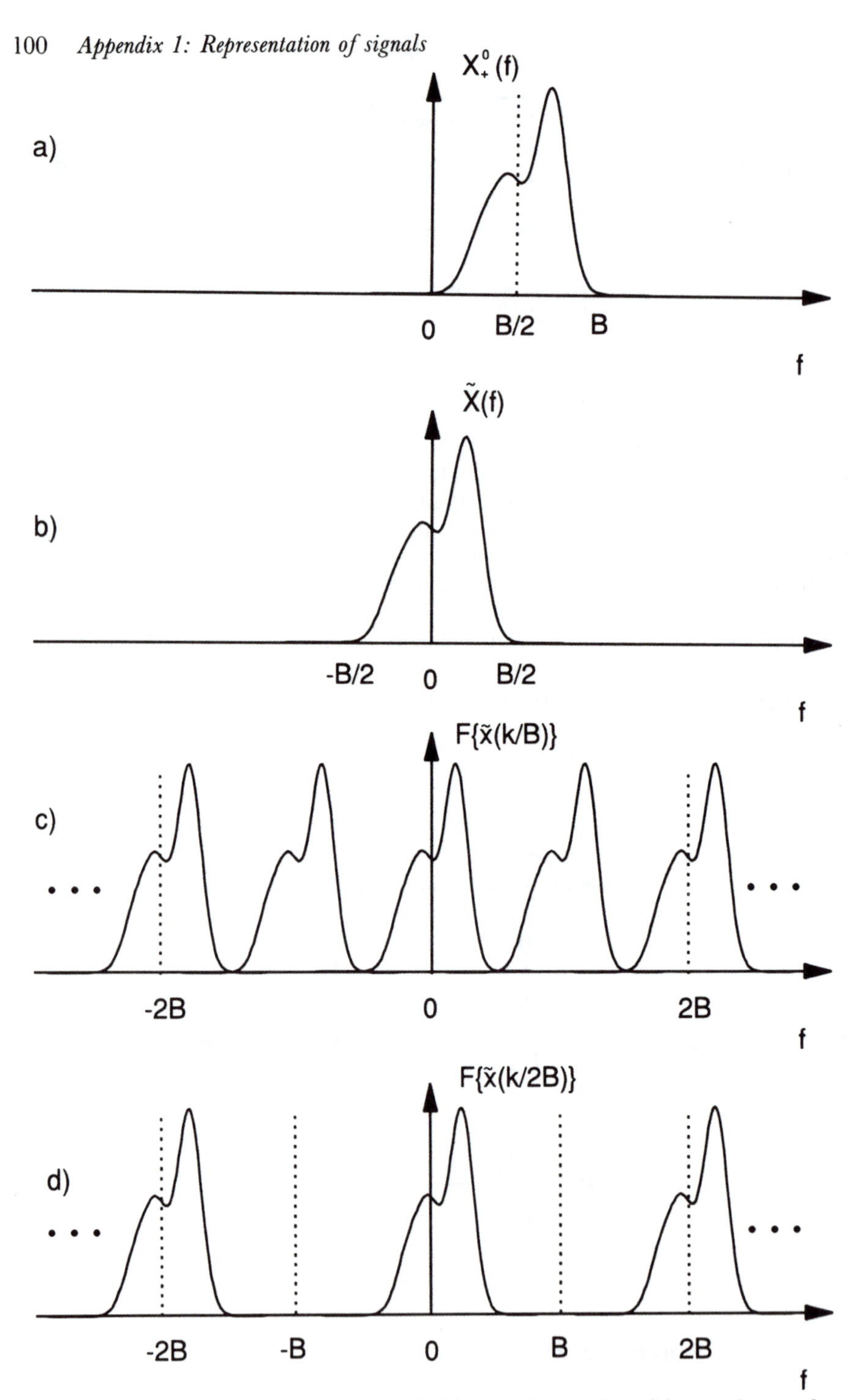

Figure 1.35 *Spectra of analytic signal (a), complex envelope (b), complex envelope sampled at rate B (c) and 2B (d)*

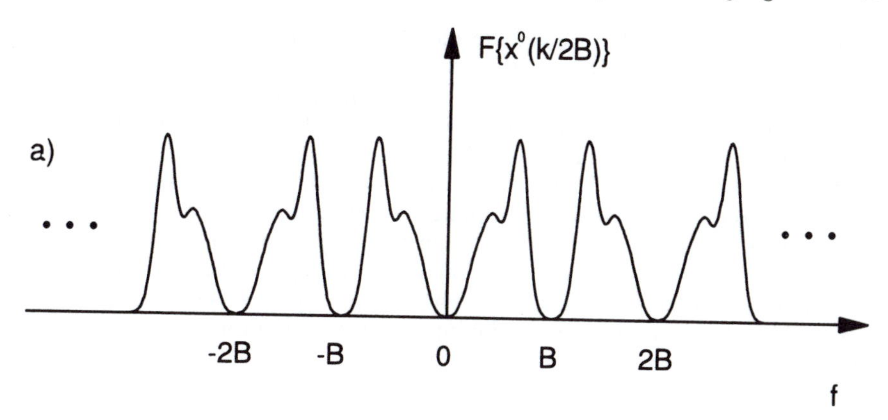

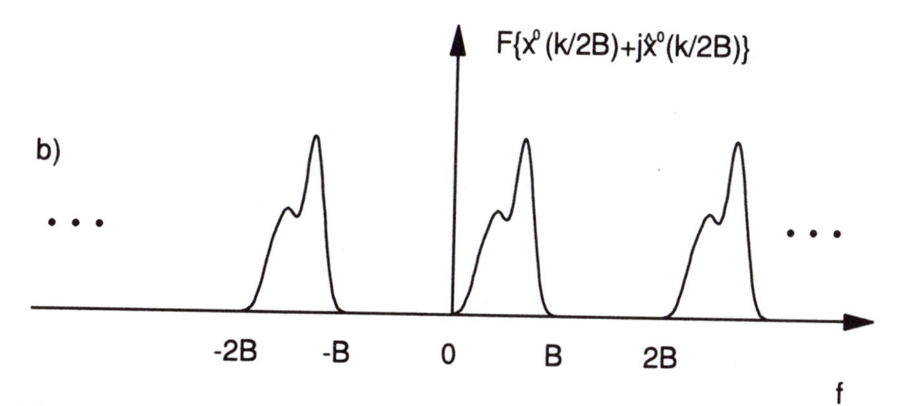

Figure 1.36 *Spectra of a sampled narrow-band signal before (a) and after (b) digital Hilbert filtering*

$$x_0(t) = x_c(t)\cos\left(2\pi \frac{B}{2} t\right) + x_s(t)\sin\left(2\pi \frac{B}{2} t\right) \tag{1.457}$$

Sampling this signal at the rate $f_s = 2B$, we obtain

$$x_0\left(\frac{k}{2B}\right) = x_c\left(\frac{k}{2B}\right)\cos\left(k\frac{\pi}{2}\right) + x_s\left(\frac{k}{2B}\right)\sin\left(k\frac{\pi}{2}\right) \tag{1.458}$$

yielding, for $k = 0, 1, \ldots,$ the sequence

$$x_c(0),\ x_s\left(\frac{1}{2B}\right),\ -x_c\left(\frac{2}{2B}\right),\ -x_s\left(\frac{3}{2B}\right)\ldots \tag{1.459}$$

and again this shows that we get alternately, to within periodic changes of signs, the samples of x_c and x_s. It is worth noting that, operating in this way, no Hilbert filter is required (this device is a very critical one if realised with analogue techniques, although much simpler if digital techniques are used), but the sampling period for x_c and x_s is double, i.e. $1/B$.

Finally, let us consider the case (made possible, in some applications, by modern circuitry) in which sampling is performed directly at IF or RF (remember Fig. 1.21 and the convolution property), with an oscillator locked to the 'carrier' frequency. It is shown that, again, the quadrature components can be obtained, sampling at a frequency that depends only on the signal bandwidth. Let us make the simplifying hypothesis that the central frequency f_c is an odd multiple of half the bandwidth:

$$f_c = (2n+1)B/2 \qquad n \text{ integer} \tag{1.460}$$

The signal can be expressed in the usual form as

$$x^{UP}(t) = x_c(t)\cos\left[2\pi(2n+1)\frac{B}{2}t\right] + x_s(t)\sin\left[2\pi(2n+1)\frac{B}{2}t\right] \tag{1.461}$$

sampling at the frequency $f_s = 2B$ we have

$$x^{UP}\left(\frac{k}{2B}\right) = x_c\left(\frac{k}{2B}\right)\cos\left[(2n+1)k\frac{\pi}{2}\right] + x_s\left(\frac{k}{2B}\right)\sin\left[(2n+1)k\frac{\pi}{2}\right] \tag{1.462}$$

Let us consider the even ($k=2m$) samples of $x^{UP}(t)$; we have, respectively,

$$\cos[(2n+1)m\pi] = (-1)^m \tag{1.463}$$

and

$$\sin[(2n+1)m\pi] = 0 \tag{1.464}$$

so we can write (for the *even* samples)

$$x^{UP}\left(\frac{m}{B}\right) = (-1)^m x_c\left(\frac{m}{B}\right) \tag{1.465}$$

For the odd samples ($k=2m+1$) of $x^{UP}(t)$ we have

$$\cos\left[(2n+1)(2m+1)\frac{\pi}{2}\right] = 0 \tag{1.466}$$

and

$$\sin\left[(2n+1)(2m+1)\frac{\pi}{2}\right] = (-1)^{m+n} \tag{1.467}$$

so we can write (for the *odd* samples)

$$x^{UP}\left(\frac{2m+1}{2B}\right) = (-1)^{m+n+1} x_s\left(\frac{2m+1}{2B}\right) \tag{1.468}$$

So it has been shown that, for any narrowband signal, the samples of x_c are determined by the samples of the signal itself taken at times m/B, $m = 0, \pm 1,$

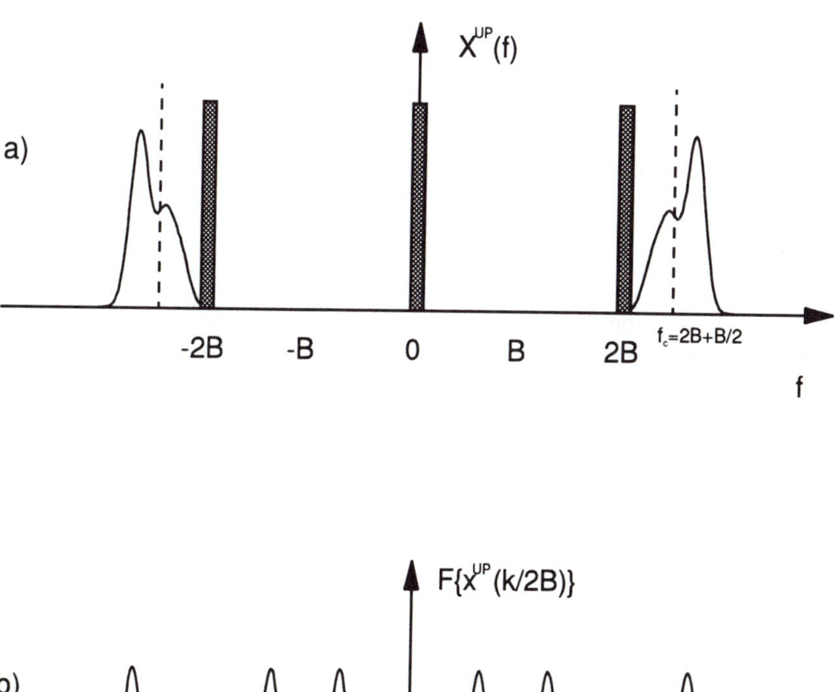

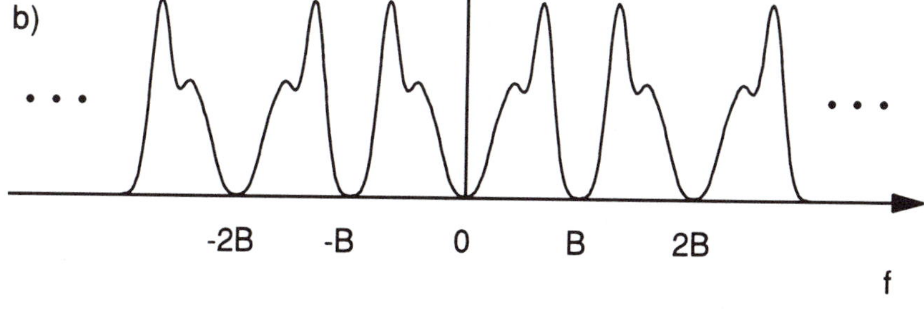

Figure 1.37 *Spectra of narrow-band signal (a) sampled directly at RF (b)*

± 2 . . . etc., while the samples of x_s are determined by samples taken at times $(2m+1)/2B$, $m=0$, ± 1, ± 2 . . . ; the two sample trains are shifted from each other by a quantity $1/2B$.

Fig. 1.37 shows (*a*) the usual spectrum of a narrowband signal, along with (*b*) the spectrum resulting from the sampling of the same signal at a rate $f_s = 1/2B$.

1.7.4.5 *Properties of the complex envelope and definitions*

A certain number of properties and definitions can be introduced for the complex envelope of a signal $x(t)$ with finite energy:

(*a*) If the signal energy is

$$E = \int_{-\infty}^{\infty} x^2(t)\,dt \tag{1.469}$$

for the complex envelope we have

$$\int_{-\infty}^{\infty} |\tilde{x}(t)|^2 dt = E \tag{1.470}$$

(*b*) The mean frequency of the envelope is defined as the first moment of the energy spectrum, according to the following relationship:

$$\bar{f} = \frac{1}{E} \int_{-\infty}^{\infty} f|\tilde{X}(f)|^2 df \tag{1.471}$$

where $\tilde{X}(f)$ is the Fourier tansform of $\tilde{x}(t)$. With a suitable choice of f_c it is always possible to have $\bar{f}=0$.

(*c*) The mean time of the envelope is defined again as the first moment of the squared magnitude

$$\bar{t} = \frac{1}{E} \int_{-\infty}^{\infty} t|\tilde{x}(t)|^2 dt \tag{1.472}$$

Again, with a suitable choice of the origin of the time axis, it is possible to have $\bar{t}=0$.

(*d*) Defining:

$$\overline{f^2} = \frac{1}{E} \int_{-\infty}^{\infty} f^2 |\tilde{X}(f)|^2 df \tag{1.473}$$

we can introduce the mean-square bandwidth (a measure of the frequency spread of the signal) as

$$\sigma_f^2 = \overline{f^2} - (\bar{f})^2 \tag{1.474}$$

In a similar way we define

$$\overline{t^2} = \frac{1}{E} \int_{-\infty}^{\infty} t^2 |\tilde{x}(t)|^2 dt \tag{1.475}$$

and

$$\sigma_t^2 = \overline{t^2} = (\bar{t})^2 \tag{1.476}$$

as the mean-square duration of the signal.

(*e*) The correlation coefficient between two signals $x_1(t)$ and $x_2(t)$ is, by definition,

$$\rho = \frac{1}{\sqrt{E_1 E_2}} \int_{-\infty}^{\infty} x_1(t) x_2(t) dt \tag{1.477}$$

Representing the two signals in terms of complex envelope with respect to the same carrier f_c, the complex correlation is obtained as

$$\tilde{\rho} = \frac{1}{\sqrt{E_1 E_2}} \int_{-\infty}^{\infty} \tilde{x}_1^*(t) \tilde{x}_2(t) dt \tag{1.478}$$

so that $\rho = \mathrm{Re}[\tilde{\rho}]$.

1.7.5 Radar signals and accuracies of measurements

In pulse radar applications, the root mean square bandwidth σ_f and the root mean square duration σ_t of the transmitted signals (the quantities σ_f and σ_t have been defined in the previous section) are related to the minimum achievable RMS errors in time delay (i.e. range) measurements and in Doppler frequency (i.e. radial velocity), respectively.

Considering the case in which the origin of the time and frequency axes are such as to produce $\bar{f}=0$ and $\bar{t}=0$, and defining, according to the common usage,

$$\begin{cases} \alpha = 2\pi\sigma_t \\ \beta = 2\pi\sigma_f \end{cases} \tag{1.479}$$

the above referenced minimum RMS errors [22] result:

$$\text{RMS error in time} = \frac{1}{\beta\sqrt{2E/N_0}} \tag{1.480}$$

$$\text{RMS error in Doppler frequency} = \frac{1}{\alpha\sqrt{2E/N_0}} \tag{1.481}$$

where E is the signal energy and N_0 is the noise (two-sided) power spectral density. From the Schwartz inequality it follows that:

$$\alpha\beta \geqslant \pi \tag{1.482}$$

where the equality sign holds for a Gaussian shaped pulse only.

On the other hand, there is no upper bound for the $\alpha\beta$ product: therefore, the theoretical accuracy both in time and frequency can be increased without limit, by the use of suitable (wideband long duration) waveforms.

From the applications point of view, to correctly use exprs. 1.480 and 1.481, the waveform to be considered is the return from a point scatterer, i.e. a train of pulses (see Section 1.7.3.3 and Fig. 1.38*a*) whose overall duration is, by definition, the dwell time t_D. The radial displacement of the scatterer in the dwell time is normally negligible with respect to the RMS error in time: the pulse train is at (nearly) constant range and the time (range) measurement is performed on each pulse of the train. On the other hand, the Doppler shift is normally negligible with respect to the bandwidth of each pulse of the train, and its measurement is performed on the sequence of I and Q samples of the pulses of the train.

Therefore, the time accuracy (expr. 1.480) is related to the effective bandwidth β of the (single) pulse (Fig. 1.38*b*), while the Doppler accuracy (expr. 1.481) is related to the effective duration α of the sequence of I, Q samples (Fig. 1.38*a*) which is of the order of t_D.

1.7.6 Complex representation for bandpass linear systems

We are now able to develop a complex representation for bandpass linear systems. We recall from Section 1.7.1.3 that a linear time-invariant system is completely characterised by its impulse response $h(t)$, or alternatively by the transfer function $H(f)$:

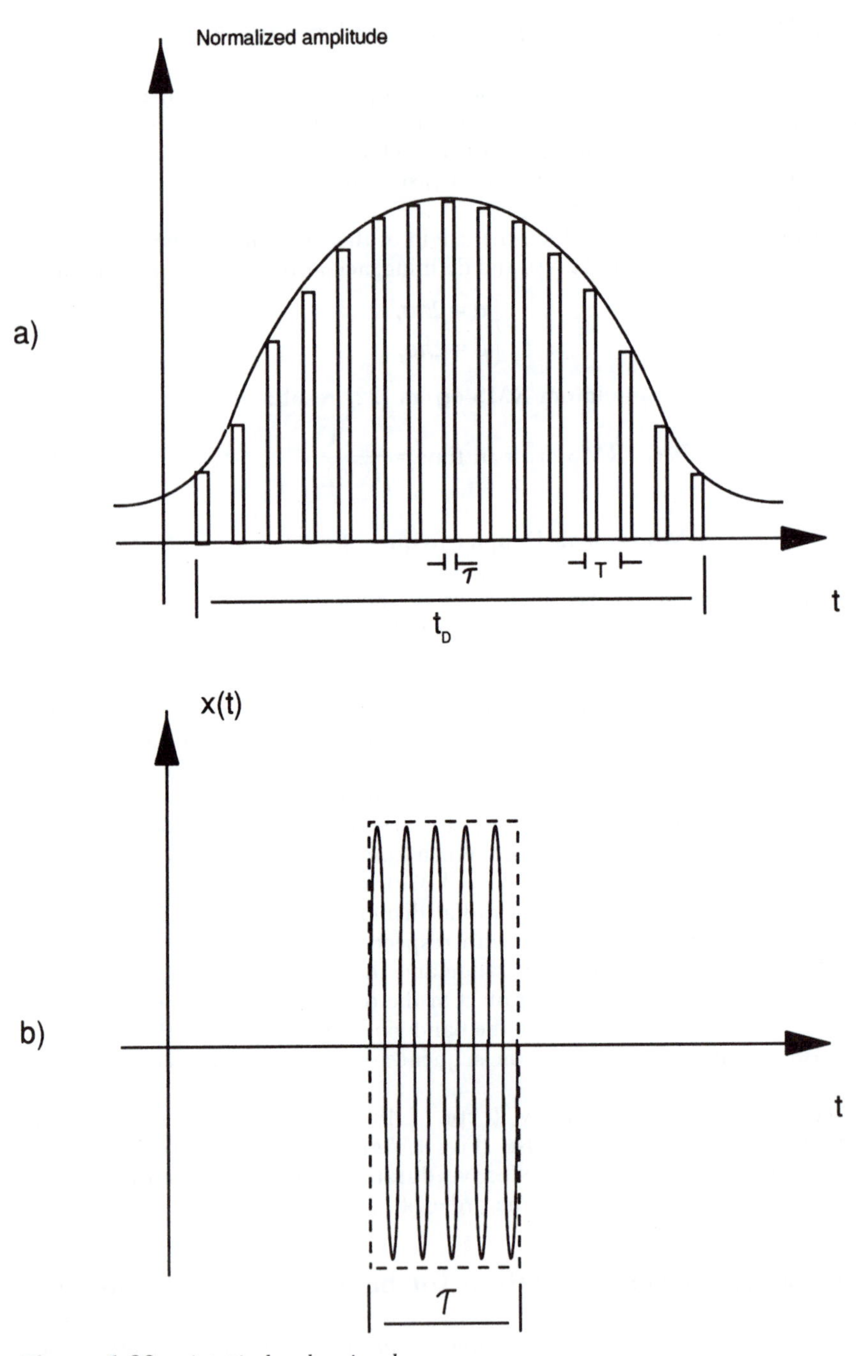

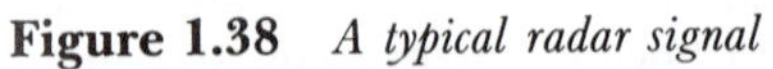

Figure 1.38 *A typical radar signal*

a The train of pulses in the dwell time
b A single pulse

$$H(f) = \int_{-\infty}^{\infty} h(\tau) e^{-j2\pi f\tau} d\tau \tag{1.483}$$

According to the previous results we can define the two low-pass functions

$$h_c(t) = [h(t) \cos(2\pi f_c t)]_{LP} \tag{1.484}$$

$$h_s(t) = [h(t) \sin(2\pi f_c t)]_{LP} \tag{1.485}$$

(different by a factor of $\sqrt{2}$ from expr. 1.445 for convenience reasons only) so that the impulse response can be written in terms of low-pass components as

$$h(t) = 2h_c(t) \cos(2\pi f_c t) + 2h_s(t) \sin(2\pi f_c t) \tag{1.486}$$

Finally, the complex impulse response is written as

$$\tilde{h}(t) = h_c(t) - jh_s(t) \tag{1.487}$$

and we have the representation

$$h(t) = \text{Re}[2\tilde{h}(t) e^{j2\pi f_c t}] \tag{1.488}$$

Let us now assume a narrowband waveform

$$x(t) = \sqrt{2}\ \text{Re}[\tilde{x}(t) e^{j2\pi f_c t}] \tag{1.489}$$

and derive the expression for the output of this signal through the bandpass linear system previously defined (an important requirement for the derivation is that the two carrier frequencies are the same). It is known that the output signal from a linear system is given by the convolution integral

$$y(t) = h(t) * x(t) = \int_{-\infty}^{\infty} h(t-\tau) x(\tau) d\tau \tag{1.490}$$

Developing the convolution integral and defining

$$\tilde{y}(t) = \int_{-\infty}^{\infty} \tilde{h}(t-\tau) \tilde{x}(\tau) d\tau \tag{1.491}$$

we obtain

$$y(t) = \sqrt{2}\ \text{Re}[\tilde{y}(t) e^{j2\pi f_c t}] \tag{1.492}$$

and the important result that the complex envelope of the output from a bandpass system is obtained from the convolution of the complex envelope of the input with the complex envelope of the impulse response.

1.7.7 Complex representation of narrowband random processes

1.7.7.1 *Definition and general properties of random processes*

Without any attempt to be rigorous, a *random process* (or *random signal*) is defined as a family $\boldsymbol{n}_\zeta(t)$ of (real or complex) time functions depending on the outcome ζ of a well defined experiment. In this appendix the index ζ is omitted, and a bold face letter is used to denote a random process.

The (statistical) mean of a random process $\boldsymbol{n}(t)$ is

$$\eta(t) = E\{\boldsymbol{n}(t)\} \tag{1.493}$$

Very often, as in this appendix, zero-mean processes are considered.

The *autocorrelation function* (ACF) of a real random process $\boldsymbol{n}(t)$ is

$$R_n(t_1, t_2) = E\{\boldsymbol{n}(t_1) \cdot \boldsymbol{n}(t_2)\} \tag{1.494}$$

while, for a complex process, it is defined as

$$R_n(t_1, t_2) = E\{\boldsymbol{n}^*(t_1) \cdot \boldsymbol{n}(t_2)\} \tag{1.495}$$

The *average power* of the process is

$$P(t_1) = R_n(t_1, t_1) \tag{1.496}$$

A zero-mean *wide-sense stationary* process is one whose autocorrelation function depends on the time lag $\tau = t_1 - t_2$ only:

$$R_n(t_1, t_2) = R_n(t_1 + h, t_2 + h) \quad \forall h \tag{1.497}$$

so that

$$R_n(t_1, t_2) = R_n(t_2 - t_1) = R_n(\tau) \tag{1.498}$$

There exists a class of stationary processes for which a single function (i.e. any randomly chosen realisation) is sufficient to obtain, with unitary probability, all the statistical information: such processes are said to be *ergodic*; see Reference 13 for more details on this extremely important class of processes.

For an ergodic process, the autocorrelation function can be evaluated from a single realisation:

$$R_n(\tau) = \lim_{T \to \infty} \frac{1}{T} \int_{-T/2}^{T/2} \boldsymbol{n}^*(t)\boldsymbol{n}(t+\tau)dt \tag{1.499}$$

In Section 1.7.1.1 the ACF of a periodic signal was related to the squared moduli of its Fourier coefficients, and the ACF of a finite energy signal was related to the squared modulus of its Fourier transform.

For a periodic, random process $\boldsymbol{n}_T(t)$, i.e. one such that, with a unitary probability

$$\boldsymbol{n}_T(t + T) = \boldsymbol{n}_T(t) \tag{1.500}$$

the Fourier coefficients

$$\boldsymbol{c}_k = \frac{1}{T} \int_{-T/2}^{T/2} \boldsymbol{n}_T(t) e^{-j2\pi kt/T} \tag{1.501}$$

are random variables: it can be shown [1, 12] that they are uncorrelated:

$$E\{\boldsymbol{c}_k^* \boldsymbol{c}_i\} = \begin{cases} 0, & i \neq k \\ |\boldsymbol{c}_k|^2 \neq 0, & i = k \end{cases} \tag{1.502}$$

Using the Fourier series representation of $\boldsymbol{n}_T(t)$

$$\boldsymbol{n}_T(t) = \sum_{-\infty}^{\infty} \boldsymbol{c}_k \exp\left\{ j \frac{2\pi}{T} kt \right\} \tag{1.503}$$

the ACF of $\boldsymbol{n}_T(t)$

$$R_T(\tau) = E\{\boldsymbol{n}_T^*(t)\boldsymbol{n}_T(t+\tau)\} \tag{1.504}$$

can be represented in terms of the Fourier coefficients. Exploiting the fact that they are uncorrelated, it follows that

$$R_T(\tau) = \sum_{-\infty}^{\infty} |\boldsymbol{c}_k|^2 \exp\left\{j \frac{2\pi}{T} k\tau\right\} \tag{1.505}$$

The sequence of $|\boldsymbol{c}_k|^2$ is the power spectrum of $\boldsymbol{n}_T(t)$; it is the Fourier representation of the (periodic) ACF of $\boldsymbol{n}_T(t)$.

For a non-periodic random process $\boldsymbol{n}(t)$, whose realisations, in general, do not have finite energy, a Fourier representation is only possible if time 'segments' $\boldsymbol{n}_T(t)$ of duration T are considered. It is possible to define the Fourier coefficients

$$\boldsymbol{c}_k = \frac{1}{T} \int_{-T/2}^{T/2} \boldsymbol{n}_T(t) \exp\left\{-j \frac{2\pi}{T} kt\right\} \tag{1.506}$$

which, depending on the sample function, are random variables. It can be shown that, unlike in the case of a periodic process, the coefficients $\boldsymbol{c}_k$ are correlated; however, they become uncorrelated when $T \to \infty$:

$$\lim_{T\to\infty} \boldsymbol{c}_k^* \boldsymbol{c}_i = \begin{cases} 0, & i \neq k \\ \dfrac{1}{T} S\left(\dfrac{k}{T}\right), & i = k \end{cases} \tag{1.507}$$

The limiting process, as in Section 1.7.1.1, is done in such a way that, when T tends to infinity, k does the same, leaving the ratio $f_k = k/T$ constant. The spectral lines of the periodic process become closer and closer, and generate the continuous power spectral density (PSD) $S(f)$. Exploiting the property given by expr. 1.507, it is readily shown that the ACF of $\boldsymbol{n}(t)$ is equal to the Fourier transform of its PSD. More exactly, the PSD of $\boldsymbol{n}(t)$ has to be *defined* as the Fourier transform of its ACF:

$$S(f) = \int_{-\infty}^{\infty} R_n(\tau) \exp\{-j2\pi f\tau) d\tau \tag{1.508}$$

This is the well-known Wiener–Kintchine (also called Bochner–Kintchine) relationship.

1.7.7.2 *Complex representation*

The derivation of the low-pass components for a random bandpass process around a given frequency f_c is straightforward, and similar to the case of deterministic signals (see Fig. 1.26), being

$$\boldsymbol{n}(t) = \sqrt{2}\boldsymbol{n}_c(t) \cos(2\pi f_c t) + \sqrt{2}\boldsymbol{n}_s(t) \sin(2\pi f_c t) \tag{1.509}$$

where $\boldsymbol{n}_c(t)$ and $\boldsymbol{n}_s(t)$ are, respectively, (see expr. 1.445):

$$\begin{cases} \boldsymbol{n}_c(t) = [\sqrt{2} \cos(2\pi f_c t)\boldsymbol{n}(t)]_{LP} \\ \boldsymbol{n}_s(t) = [\sqrt{2} \sin(2\pi f_c t)\boldsymbol{n}(t)]_{LP} \end{cases} \tag{1.510}$$

the only difference is that $\boldsymbol{n}_c(t)$ and $\boldsymbol{n}_s(t)$ are both random processes, whose properties will be shown later. The reconstruction method is the same as that of Fig. 1.28.

In complex notation we can write again (see expr. 1.446):

$$\tilde{\boldsymbol{n}}(t) = \boldsymbol{n}_c(t) - j\boldsymbol{n}_s(t) \tag{1.511}$$

or

$$\tilde{\boldsymbol{n}}(t) = [\sqrt{2}\boldsymbol{n}(t)e^{-j2\pi f_c t}]_{LP} \tag{1.512}$$

We detail here without proof [6] some statistical properties of $\tilde{\boldsymbol{n}}(t)$.

The power spectrum of $\tilde{\boldsymbol{n}}(t)$ (the random process is intended to have zero power spectral density outside the interval $\pm[f_c - B/2; f_c + B/2]$) is given by (see Fig. 1.39)

$$S_{\boldsymbol{n}}(f) = 2[S_{\boldsymbol{n}}(f + f_c)]_{LP} \tag{1.513}$$

so that the ACF of $\tilde{\boldsymbol{n}}(t)$ is given by

$$E[\tilde{\boldsymbol{n}}^*(t)\tilde{\boldsymbol{n}}(t+\tau)] = 2\int_{-B/2}^{B/2} S_{\tilde{n}}(f + f_c)e^{j2\pi f_c \tau}\,df \tag{1.514}$$

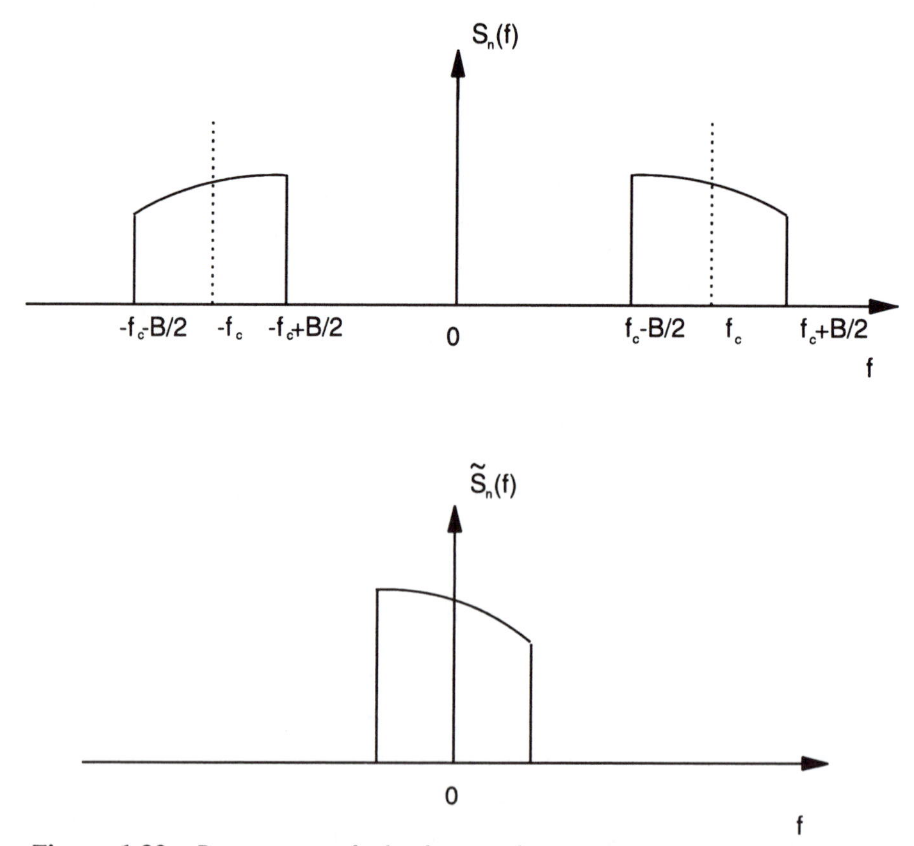

Figure 1.39 *Power spectra of a bandpass random process*

while the following relationship holds:

$$E[\tilde{\boldsymbol{n}}(t_1)\tilde{\boldsymbol{n}}(t_2)] = 0 \quad \forall t_1, t_2 \tag{1.515}$$

The power spectrum of a complex process is even if and only if that of the narrowband process is symmetrical around the carrier frequency f_c.

As far as the low-pass components of $\boldsymbol{n}(t)$ are concerned, it is possible to prove the following relationship between $\boldsymbol{n}_c(t)$ and $\boldsymbol{n}_s(t)$:

$$E[\boldsymbol{n}_c(t)\boldsymbol{n}_c(t+\tau)] = E[\boldsymbol{n}_s(t)\boldsymbol{n}_s(t+\tau)] = \tfrac{1}{2}\,\mathrm{Re}\{E[\tilde{\boldsymbol{n}}^*(t)\tilde{\boldsymbol{n}}(t+\tau)]\} \tag{1.516}$$

and

$$R_{sc}(\tau) = E[\boldsymbol{n}_s(t)\boldsymbol{n}_c(t+\tau)] = -E[\boldsymbol{n}_c(t)\boldsymbol{n}_s(t+\tau)] = E[\boldsymbol{n}_s(t)\boldsymbol{n}_c(t-\tau)]$$
$$= \tfrac{1}{2}\,\mathrm{Im}\{E[\tilde{\boldsymbol{n}}^*(t)\tilde{\boldsymbol{n}}(t+\tau)]\} \tag{1.517}$$

In terms of power spectra we can write for both the in-phase and quadrature components

$$S_c(f) = S_s(f) = \frac{1}{2}\left[\frac{S_{\tilde{n}}(f) + S_{\tilde{n}}(-f)}{2}\right] \tag{1.518}$$

and

$$S_{sc}(f) = \frac{j}{2}\left[\frac{S_{\tilde{n}}(f) - S_{\tilde{n}}(-f)}{2}\right] \tag{1.519}$$

where $S_{sc}(f)$ is the cross spectral density between the two components. From this expression it is possible to see that the quadrature processes are uncorrelated $(R_{sc}(\tau)=0)$ if the spectrum has even symmetry around the central frequency. Moreover, at any instant t^*, $\boldsymbol{n}_c(t^*)$ and $\boldsymbol{n}_s(t^*)$ are always uncorrelated, because (expr. 1.517) $R_{sc}(0)=0$.

1.7.7.3 *White and Gaussian random processes*

It is possible to define *complex white noise* as a process $\boldsymbol{n}(t)$ whose spectrum is constant in the interval $\pm[f_c - B/2; f_c + B/2]$ (see Fig. 1.40). According to exprs. 1.518 and 1.519 we have

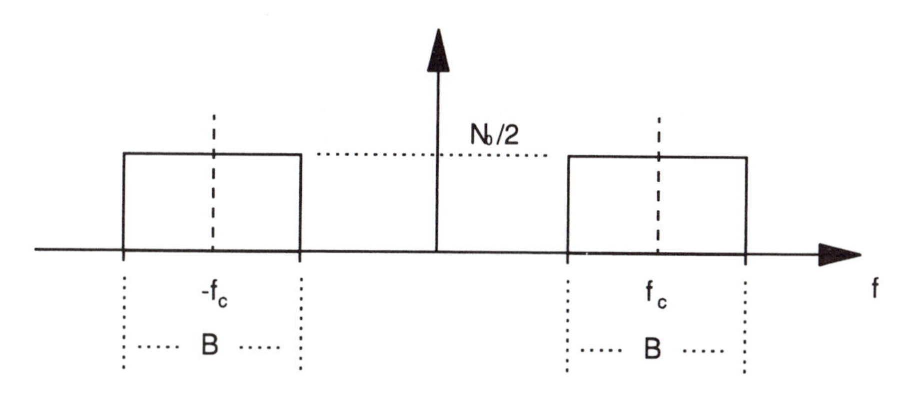

Figure 1.40 *Spectrum of bandpass white noise*

$$S_c(f) = S_s(f) = \begin{cases} \frac{N_0}{2}, & |f| \leq \frac{B}{2} \\ 0, & \text{elsewhere} \end{cases} \tag{1.520}$$

and, owing to the symmetry around the central frequency,

$$S_{sc}(f) = 0 \tag{1.521}$$

The autocorrelation function of this process $\boldsymbol{n}(t)$ is given by

$$R_n(\tau) = N_0 \left(\frac{\sin(\pi B \tau)}{\pi \tau / 2} \right) \tag{1.522}$$

In many cases of interest the processes are *Gaussian* random processes. It is possible to show that, if $\boldsymbol{n}(t)$ is a Gaussian random process, then $\boldsymbol{n}_c(t)$ and $\boldsymbol{n}_s(t)$ are jointly Gaussian, owing to the property of linearity of normal processes.

A more rigorous definition of a *stationary Gaussian random process* with zero mean can be given as the process with Gaussian components that satisfies the relationship

$$E[\tilde{\boldsymbol{n}}(t)\tilde{\boldsymbol{n}}(t+\tau)] = 0 \quad \forall t, \tau \tag{1.523}$$

It is important to stress that a complex process whose real and imaginary parts are both Gaussian is not necessarily Gaussian, because expr. 1.523 must also be satisfied.

Moreover, the statistics of the in-phase and quadrature components of a Gaussian process are not influenced by possible phase rotations: this means that the covariance functions of, say, $\tilde{\boldsymbol{n}}(t)$ and $\tilde{\boldsymbol{n}}(t)e^{j\theta}$ are the same, since

$$E\{\tilde{\boldsymbol{n}}^*(t)e^{-j\theta}\tilde{\boldsymbol{n}}(t+\tau)e^{j\theta}\} = E\{\tilde{\boldsymbol{n}}^*(t)\tilde{\boldsymbol{n}}(t+\tau)\} \tag{1.524}$$

We note that $\boldsymbol{n}_s(t)$ and $\boldsymbol{n}_c(t)$ are independent Gaussian random variables of zero mean and variance σ^2, so that their joint probability density function can be written as

$$p(\boldsymbol{n}_c, \boldsymbol{n}_s) = \frac{1}{2\pi\sigma^2} \exp\left(-\frac{\boldsymbol{n}_c^2 + \boldsymbol{n}_s^2}{2\sigma^2} \right) \tag{1.525}$$

1.7.7.4 *Envelope and phase representation*

We have seen in previous sections the possibility of expressing a narrowband random process $\boldsymbol{n}(t)$ in terms of in-phase and quadrature components. It is also possible to represent the same process in terms of its envelope and phase components as follows:

$$\boldsymbol{n}(\boldsymbol{t}) = \rho(t) \cos[2\pi f_c t + \psi(t)] \tag{1.526}$$

where

$$\rho(t) = [\boldsymbol{n}_c^2(t) + \boldsymbol{n}_s^2(t)]^{1/2} \tag{1.527}$$

and

$$\psi(t) = \tan^{-1}\left[\frac{\boldsymbol{n}_s(t)}{\boldsymbol{n}_c(t)} \right] \tag{1.528}$$

The function $\rho(t)$ is called the *envelope* of $\boldsymbol{n}(t)$ and $\psi(t)$ is the *phase* of $\boldsymbol{n}(t)$.

It is possible to show [12] that the joint probability density function of $\rho(t)$ and $\psi(t)$ is given by the following expression:

$$p(\rho, \psi) = \frac{1}{2\pi} \frac{\rho}{\sigma^2} \exp\left(-\frac{\rho^2}{2\sigma^2}\right) \tag{1.529}$$

It is readily seen that the PDF is independent of the angle $\psi(t)$, and this means that the two random processes are statistically independent. One can thus express the joint PDF as the product of the single PDFs; since the angle $\psi(t)$ is uniformly distributed inside the range $[0, 2\pi]$:

$$p(\psi) = \begin{cases} \dfrac{1}{2\pi}, & 0 \leqslant \psi \leqslant 2\pi \\ 0, & \text{elsewhere} \end{cases} \tag{1.530}$$

the PDF for the envelope $\rho(t)$ is

$$p(\rho) = \begin{cases} \dfrac{\rho}{\sigma^2} \exp\left(-\dfrac{\rho^2}{2\sigma^2}\right), & \rho \geqslant 0 \\ 0, & \text{elsewhere} \end{cases} \tag{1.531}$$

where σ^2 is the variance of the original narrowband process. The function expressed by expr. 1.531 is the well-known Rayleigh function (see Section 1.10.1).

1.8 Appendix 2: Linear algebra in the complex field

In this appendix the fundamental concepts of linear algebra applied to the complex field will be described. Generally, no attempt will be made at an exhaustive or rigorous treatment (in particular, the linear transformations and the matrices representing them will not be distinguished, either formally or conceptually). The aim will be to collect as many possible concepts as are necessary for the comprehension of this appendix and for the sections concerning signal detection and estimation.

For a complete treatment, the interested reader is invited to consult the vast amount of available literature and, in particular, References 9, 15 and 16.

1.8.1 Unitary complex space C^N: Some definitions

The *unitary*† *complex space* C^N is defined as the space of all the N-dimensional complex vectors, hereafter indicated by a bold-face letter, or by an N-tuple of its components, in which the *inner product*

$$(\boldsymbol{x}, \boldsymbol{y}) = \boldsymbol{x}^H \boldsymbol{y} = \sum_{i=1}^{N} x_i^* y_i \tag{1.532}$$

† In the case of real space R^n, the term Euclidean is used.

is defined. We adopt the following notations for the superscripts:

* means *conjugate*
^{H} means *conjugate* (of the) *transpose*, or *adjoint*.

Orthogonal vectors: two vectors $\boldsymbol{x}$ and $\boldsymbol{y}$ are *orthogonal* if their inner product is zero, that is:

$$\boldsymbol{x}^H\boldsymbol{y}=0 \tag{1.533}$$

The zero vector $\boldsymbol{0}$ is orthogonal to any vector in the same space.

The Euclidean norm of a vector is defined as

$$\|\boldsymbol{x}\|=\sqrt{\boldsymbol{x}^H\boldsymbol{x}}=\sqrt{\sum_{i=1}^{N}|x_i|^2} \tag{1.534}$$

An orthonormalised set is a set of L vectors $\boldsymbol{x}^i$, $i=1, 2, \ldots, L$, such that

$$(\boldsymbol{x}^i)^H\boldsymbol{x}^j=\delta_{ij}=\begin{cases}1, & i=j\\ 0, & i\neq j\end{cases} \tag{1.535}$$

Cauchy–Schwartz inequality: Given two vectors belonging to the same space, the following relationship holds:

$$|\boldsymbol{x}^H\boldsymbol{y}|\leqslant\|\boldsymbol{x}\|\cdot\|\boldsymbol{y}\| \tag{1.536}$$

where the equality applies when $\boldsymbol{x}=\alpha\boldsymbol{y}$, α being a (real or complex) scalar.

The angle θ between two vectors is defined† as follows:

$$\cos\theta=\frac{\boldsymbol{x}^H\boldsymbol{y}}{\|\boldsymbol{x}\|\cdot\|\boldsymbol{y}\|} \tag{1.537}$$

1.8.2 Matrices

1.8.2.1 *General*

A table of $N\times M$ complex numbers ordered in N rows and M columns is a *matrix*; any linear transformation of C^N into C^M can be represented by a suitable $N\times M$ matrix, if we have already defined two bases in C^N and C^M. We shall consider square ($N\times N$) non-singular matrices (see Section 1.8.2.3), corresponding to linear transformations of C^N into C^N: in this case the transformation can be considered as a change of basis.

If a vector $\boldsymbol{x}$ has the components $x_1, x_2, \ldots, x_N$ in a given basis‡, the transformed vector $\boldsymbol{y}$ has components

$$\begin{cases} y_1=a_{11}x_1+\cdots+a_{1N}x_N\\ \quad\cdots\\ y_N=a_{N1}x_1+\cdots+a_{NN}x_N\end{cases} \tag{1.538}$$

† An alternative definition is: $\cos\theta=\mathrm{Re}(\boldsymbol{x}^H\boldsymbol{y})/(\|\boldsymbol{x}\|\cdot\|\boldsymbol{y}\|)$. The orthogonality criterion is consistent with this definition.

‡ By definition, a system of *linearly independent* vectors $\boldsymbol{e}_1, \boldsymbol{e}_2, \ldots, \boldsymbol{e}_n$ in a complex space C^N is called a basis for C^N if it is possible to express any vector $\boldsymbol{x}\in C^N$ by means of N coefficients, $\alpha, \alpha_2, \ldots, \alpha_N$, as

$$\boldsymbol{x}=\alpha_1e_1+\alpha_2e_2+\ldots+\alpha_Ne_N$$

in the basis defined by the transformation

$$A = \begin{bmatrix} a_{11} & \dots & a_{1N} \\ a_{21} & \dots & a_{2N} \\ \dots & \dots & \dots \\ a_{N1} & \dots & a_{NN} \end{bmatrix} \tag{1.539}$$

In short, the transformation can be expressed, using the well-known 'row by column' product between matrices (and vectors), as

$$\boldsymbol{y} = A\boldsymbol{x} \tag{1.540}$$

The *inverse transformation* is again *linear* and, if the inverse A^{-1} of A exists, can be written as

$$\boldsymbol{x} = A^{-1}\boldsymbol{y} \tag{1.541}$$

A transformation that leaves unchanged the vector basis is said to be the *identity transformation*, and the related matrix is indicated generally by I, whose structure is simply

$$I = \begin{bmatrix} 1 & 0 & 0 & \dots & 0 \\ 0 & 1 & 0 & \dots & 0 \\ \dots & \dots & \dots & \dots & \dots \\ 0 & 0 & 0 & \dots & 1 \end{bmatrix} \tag{1.542}$$

Two linear transformations of C^N into itself can be applied to a vector, obtaining a third transformation, called the *product* of the two:

$$\boldsymbol{y} = A\boldsymbol{x}; \quad \boldsymbol{z} = B\boldsymbol{y} = B(A\boldsymbol{x}) \Rightarrow \boldsymbol{z} = C\boldsymbol{x} \tag{1.543}$$

where the matrix C is the product of B and A; note that generally the matrix product is not commutative, that is $AB \neq BA$.

The *inverse* A^{-1} of a matrix A (if it exists) is a matrix (see above) which satisfies the relationship

$$AA^{-1} = I \tag{1.544}$$

The operation of *transposition* of a matrix inverts the order of rows and columns, i.e. the element a_{ij} takes the place, in the new matrix, of the element a_{ji}; the transpose of a matrix A is indicated by the symbol A^T.

Finally, we define the *conjugate transpose* of a matrix A, and indicate it by A^H, a matrix in which the element a^*_{ij} takes the place of a_{ji}. The operation of conjugation and transposition are commutative, so we can write

$$A^H = (A^*)^T = (A^T)^* \tag{1.545}$$

1.8.2.2 *Properties of square matrices*

(*a*) If we indicate by the symbol '+' the operations ^{T}, * or ^{H} (transposition, conjugation or both), it is easy to see that

$$(A+B)^+ = A^+ + B^+ \tag{1.546}$$

(*b*) If '+' denotes ^{T}, ^{H} or $^{-1}$ (inversion), we have

$$(AB)^+ = B^+A^+ \tag{1.547}$$

(*c*) The operators $*$, ^{T}, $^{-1}$, ^{H} are commutative; for example,

$$(A^H)^{-1} = (A^{-1})^H \tag{1.548}$$

which allows us to use the compact notations A^{-T}, A^{-*} etc.

(*d*) *The trace* of a matrix is the sum of the elements of its principal diagonal: $\mathrm{tr}(A) = \sum_{i=1}^N a_{ii}$, and the following relationships hold:

$$\mathrm{tr}(A + B) = \mathrm{tr}(A) + \mathrm{tr}(B) \tag{1.549}$$

$$\mathrm{tr}(AB) = \mathrm{tr}(BA) \tag{1.550}$$

(*e*) *Hermitian matrices*: A matrix Q is Hermitian if $Q^H = Q$. An important property is the following: given any matrix B, the matrix $C = B^H B$ is Hermitian. Moreover, if B is invertible, we have for such a C,

$$B^{-H} C B^{-1} = B^{-H} B^H B B^{-1} = I \tag{1.551}$$

(*f*) *Hermitian forms*: A Hermitian form $H(\boldsymbol{x}, \boldsymbol{x})$ is the second-order real homogeneous polynomial

$$H(\boldsymbol{x}, \boldsymbol{x}) = \sum_{i=1}^{N} \sum_{k=1}^{N} h_{ik} x_i^* x_k \tag{1.552}$$

with $h_{ik} = h_{ki}^*$.

If H is the matrix of the coefficients h_{ik}, the Hermitian form (expr. 1.552) can be written as

$$H(\boldsymbol{x}, \boldsymbol{x}) = \boldsymbol{x}^H \cdot H \cdot \boldsymbol{x} = (\boldsymbol{x}, H\boldsymbol{x}). \tag{1.553}$$

The concept of Hermitian form is the extension to the complex case of the well-known concept of *quadratic forms*†.

(*g*) *Positive (semi-) definite Hermitian matrices: B* is positive (semi-) definite if

$$\boldsymbol{x}^H B \boldsymbol{x} \overset{(\geqslant)}{>} 0, \qquad \forall \boldsymbol{x} \neq \boldsymbol{0} \tag{1.554}$$

The definition of negative (semi-) definite matrices is obvious.

(*h*) *Similar matrices*: Two matrices A and B are said to be *similar* if there exists an invertible matrix C such that

$$B = C^{-1} A C \tag{1.555}$$

This definition arises from the following considerations: if a linear relationship exists between two vectors $\boldsymbol{x}'$ and $\boldsymbol{y}'$ in the form

$$\boldsymbol{y}' = A\boldsymbol{x}' \tag{1.556}$$

and if $\boldsymbol{x}'$ and $\boldsymbol{y}'$ are linear functions of two vectors $\boldsymbol{x}$ and $\boldsymbol{y}$ according to the law

† Note that, when writing a quadratic form $Q(\boldsymbol{x}, \boldsymbol{x})$ as $\boldsymbol{x}^T A \boldsymbol{x} = \sum_{i=1}^N \sum_{j=1}^N a_{ij} x_i x_j$, with no loss of generality it can be assumed that $a_{ij} = a_{ji}$, i.e. A symmetrical; conversely, the condition $h_{ij} = h_{ji}^*$ ensures that the Hermitian form is real.

$$\begin{cases} \boldsymbol{x}' = C\boldsymbol{x} \\ \boldsymbol{y}' = C\boldsymbol{y} \end{cases} \tag{1.557}$$

where C is an invertible transformation, the linear relationship between $\boldsymbol{x}$ and $\boldsymbol{y}$ is given by

$$\boldsymbol{y} = B\boldsymbol{x} \tag{1.558}$$

where B is defined by expr. 1.555.

(*i*) *Congruent matrices*: Two real symmetric matrices A and B are said to be *congruent* (or that a congruence relationship exists between them) if there exists a real invertible matrix C such that

$$B = C^T A C \tag{1.559}$$

(*j*) *Diagonalisation*: A Hermitian definite positive matrix A can be diagonalised (transformed into a diagonal matrix) by a non unique *unitary similarity transformation* C as follows:

$$C^H A C = D_A \tag{1.560}$$

where $C^H = C^{-1}$ (Section 1.8.2.4) and

$$D_A = \begin{bmatrix} \alpha_1 & 0 & 0 & \ldots & 0 \\ 0 & \alpha_2 & 0 & \ldots & 0 \\ \cdot & \cdot & \cdot & \cdot & \cdot \\ 0 & 0 & 0 & \ldots & \alpha_N \end{bmatrix} \tag{1.561}$$

is a diagonal matrix with *real* and *positive* elements α_i [9].

(*k*) *Jacobian matrix*: The Jacobian matrix of an N-term function $f(\boldsymbol{x})$ with respect to the N-components vector $\boldsymbol{x}$ is an $N \times N$ matrix whose terms are the first partial derivatives of the components f_i with respect to the elements x_j, that is

$$J = \begin{bmatrix} \frac{\partial f_1}{\partial x_1} & \frac{\partial f_1}{\partial x_2} & \ldots & \frac{\partial f_1}{\partial x_N} \\ \frac{\partial f_2}{\partial x_1} & \frac{\partial f_2}{\partial x_2} & \ldots & \frac{\partial f_2}{\partial x_N} \\ \ldots & \ldots & \ldots & \ldots \\ \frac{\partial f_N}{\partial x_1} & \frac{\partial f_N}{\partial x_2} & \ldots & \frac{\partial f_N}{\partial x_N} \end{bmatrix} \tag{1.562}$$

It is obvious that, when dealing with linear transformations of C^N into C^N, represented by an $N \times N$ matrix A (see expr. 1.538), we are speaking of an N-term linear function, and the partial derivatives are the a_{ij} elements themselves. In this case, the Jacobian matrix is the matrix A itself. In the next section an important application of this matrix will be shown.

(*l*) *Simultaneous diagonalisation of two Hermitian matrices*: A necessary and sufficient condition that two Hermitian matrices A and B can be simultaneously transformed into diagonal form (that is, both reduced to diagonal form by the same transformation T) is that they commute (Reference 25, p. 56): $AB = BA$†.

1.8.2.3 *Determinant of a square matrix* ($N \times N$)

The definition of the determinant of a square matrix is:

$$\det(A) = \sum_{k_1, k_2, \ldots, k_N} (-1)^m . a_{1,k_1} a_{2,k_2} \ldots a_{N,k_N} \tag{1.563}$$

where $(k_1, k_2, \ldots, k_N)$ is a permutation of the first N integers, and the sum is extended to all the permutations, whose number is $N!$. In the previous expression m is the number of inversions in the indexes permutation $k_1, k_2, \ldots, k_N$ with respect to the fundamental permutation $1, 2, \ldots, N$.

Properties of the determinant

(*a*) If a row (column) of a matrix is a linear combination of other rows (columns), $\det(A) = 0$. Particular cases of a determinant being equal to zero are when a row (column) is proportional or equal to another row (column) or identically zero.
(*b*) The determinant changes its sign if two rows (columns) are exchanged with each other.
(*c*) For a *triangular matrix** L the determinant is the product of the elements of the principal diagonal:

$$\det(L) = \prod_{n=1}^{N} l_{nn} \tag{1.564}$$

(*d*) The transposition operation leaves the determinant value unchanged:

$$\det(A) = \det(A^T) \tag{1.565}$$

(*e*) The determinant of a Hermitian matrix is real, owing to the above property.
(*f*) The determinant of a product of matrices is simply

$$\det(A \cdot B) = \det(A) \cdot \det(B) \tag{1.566}$$

(*g*) If A is a non-singular‡ matrix,

$$\det(A^{-1}) = \frac{1}{\det(A)} \tag{1.567}$$

† Here we prove the necessity only. If such a transformation exists, it must be

$$TAT^{-1} = D_A$$

$$TBT^{-1} = D_B$$

where D_A and D_B are diagonal matrices. Then we have

$$AB = (T^{-1} D_A T)(T^{-1} D_B T) = T^{-1} D_A D_B T = T^{-1} D_B D_A T = BA$$

* In a triangular matrix the elements below or above the principal diagonal are zero.
‡ That is, its determinant is $\neq 0$.

(*h*) For any complex constant c:

$$\det(cA) = c^N \cdot \det(A) \tag{1.568}$$

At this point we recall the properties of the *Jacobian determinant*, that is the determinant of the Jacobian matrix (see expr. 1.562). It can be shown that, when considering a (not necessarily linear) transformation of C^N into C^N, the elementary volume $dV = dx_1 \cdot dx_2 \cdot \ldots \cdot dx_N$ transforms into $dV' = |\det(J)| \cdot dx'_1 \cdot dx'_2 \cdot \ldots \cdot dx'_N$, where the primed terms are the components of the vector after the transformation.

In the case of linear transformation, we have already seen that the Jacobian matrix is the transformation matrix itself (generally indicated by A), so that the Jacobian determinant is the determinant of A.

1.8.2.4 *Unitary matrices*

A matrix P is *unitary* if

$$P^H P = I \tag{1.569}$$

i.e. its conjugate transpose equals its inverse: $P^H = P^{-1}$.

In the real field, an *orthogonal* matrix R is one whose transpose equals its inverse: $R^T = R^{-1}$.

Decomposing P into a set of N column vectors,

$$P = [\boldsymbol{p}^1 | \boldsymbol{p}^2 | \ldots | \boldsymbol{p}^N] \tag{1.570}$$

it is readily seen that, for a unitary matrix,

$$(\boldsymbol{p}^i)^H \boldsymbol{p}^j = \delta_{ij} \tag{1.571}$$

that is, the columns are orthonormal.

Norms and angles between vectors are *invariant* with respect to *unitary transformations*. In fact, given two vectors $\boldsymbol{r}$ and $\boldsymbol{s}$, their inner product, after a unitary transformation P, is

$$(P\boldsymbol{r})^H P\boldsymbol{s} = \boldsymbol{r}^H P^H P \boldsymbol{s} = \boldsymbol{r}^H \boldsymbol{s} \tag{1.572}$$

that means

$$(P\boldsymbol{r}, P\boldsymbol{s}) = (\boldsymbol{r}, \boldsymbol{s}) \tag{1.573}$$

The above relationships hold for the norm of a vector, being

$$\|\boldsymbol{x}\|^2 = (\boldsymbol{x}, \boldsymbol{x}) \tag{1.574}$$

The determinant of a unitary matrix P has absolute value 1. In fact, from the definition of a unitary matrix (expr. 1.569), we have

$$\begin{aligned} \det(I) &= \det(P^H P) = \det(P^H) \cdot \det(P) \\ &= \det(P)^* \cdot \det(P) \end{aligned} \tag{1.575}$$

and, as a consequence, $|\det(P)| = \sqrt{\det(I)} = 1$

1.8.2.5 *Hermitian forms after linear (and unitary) transformations*

If $Q(\boldsymbol{x}, \boldsymbol{x}) = \boldsymbol{x}^H Q \boldsymbol{x} = (\boldsymbol{x}, Q\boldsymbol{x})$ is a generic Hermitian form, for any matrix Q, and if we define a linear transformation $\boldsymbol{x} = A\boldsymbol{t}$, for any *invertible* matrix A, the form $Q(\boldsymbol{x}, \boldsymbol{x})$ can be written in terms of $\boldsymbol{t}$ as follows:

$$Q(\boldsymbol{x}, \boldsymbol{x}) = (A\boldsymbol{t})^H QA\boldsymbol{t} = \boldsymbol{t}^H R\boldsymbol{t} \tag{1.576}$$

with

$$R = A^H QA \tag{1.577}$$

If the transformation defined by A is *unitary*, expr. 1.576 becomes (substituting the matrix A with a new unitary matrix P):

$$Q(\boldsymbol{x}, \boldsymbol{x}) = \boldsymbol{t}^H S\boldsymbol{t} \tag{1.578}$$

where now we have

$$S = P^H QP \tag{1.579}$$

and, from the properties of unitary matrices, $S = P^{-1}QP$.

The matrix Q can be reduced in diagonal form by unitary transformations (see Section 1.8.2.2), namely:

$$U^H QU = \Lambda = \text{diag}(\lambda_1, \lambda_2, \ldots, \lambda_N) \tag{1.580}$$

hence the Hermitian form can be written:

$$\begin{aligned} Q(\boldsymbol{x}, \boldsymbol{x}) &= \boldsymbol{x}^H Q\boldsymbol{x} = \boldsymbol{x}^H U\Lambda U^H \boldsymbol{x} \\ &= \boldsymbol{t}^H \Lambda \boldsymbol{t} = (\boldsymbol{t}, \Lambda \boldsymbol{t}) \end{aligned} \tag{1.581}$$

where $\boldsymbol{t} = U^H\boldsymbol{x}$, and t_i, $i = 1, 2, \ldots, N$ are the components of $\boldsymbol{t}$ (Section 1.8.3.4).

Therefore we can write

$$Q(\boldsymbol{x}, \boldsymbol{x}) = \sum_{i=1}^{N}\sum_{j=1}^{N} q_{ij}x_i^* x_j = \sum_{i=1}^{N} \lambda_i |t_j|^2 \tag{1.582}$$

This important result follows: the set of values assumed by $(\boldsymbol{x}, Q\boldsymbol{x})$ on the sphere $(\boldsymbol{x}, \boldsymbol{x}) = 1$ is equal to the set of values assumed by $(\boldsymbol{t}, \Lambda\boldsymbol{t})$ on the sphere $(\boldsymbol{t}, \boldsymbol{t}) = 1$.

1.8.2.6 *A significant integral*

The indefinite integral of the exponential of a quadratic form $(\boldsymbol{x}, A\boldsymbol{x})$

$$I_N = \int_{-\infty}^{\infty} \cdots \int_{-\infty}^{\infty} \exp\{-\boldsymbol{x}^T A\boldsymbol{x}\} d\boldsymbol{x} \tag{1.583}$$

(where $d\boldsymbol{x} = dx_1 dx_2 \ldots, dx_N$) has many applications (Reference 25, pp. 97–99). By using the orthogonal transformation T, reducing A to diagonal form and making the change $\boldsymbol{x} = T\boldsymbol{y}$, it follows that

$$(\boldsymbol{x}, A\boldsymbol{x}) = (\boldsymbol{y}, \Lambda\boldsymbol{y}) = \sum_{i=1}^{N} y_i^2 \lambda_i \tag{1.584}$$

where λ_i, $i = 1, \ldots, N$ are the eigenvalues of A (see Section 1.8.3).

The absolute value of the Jacobian of the transformation is $+1$, being the transformation $\boldsymbol{x} = T\boldsymbol{y}$ unitary. Therefore, remembering that

$$\int_{-\infty}^{\infty} \exp\{-\alpha x^2\} dx = \sqrt{\frac{\pi}{\alpha}} \tag{1.585}$$

from exprs. 1.583 and 1.584 it follows that

$$I_N=\prod_{i=1}^{N}\int_{-\infty}^{\infty}\exp\{-\lambda_i y_i^2\}dy_i=\frac{\pi^{N/2}}{(\lambda_1\lambda_2\ .\ .\ .\ ,\ \lambda_N)^{1/2}} \tag{1.586}$$

Remembering that $\det(A)=\prod_{i=1}^{N}\lambda_i$, the following significant result is obtained:

$$I_N=\frac{\pi^{N/2}}{[\det(A)]^{1/2}} \tag{1.587}$$

This result is extended to the complex case, considering a Hermitian matrix H, a complex vector $\boldsymbol{z}=\boldsymbol{x}+j\boldsymbol{y}$, and the Hermitian form $(\boldsymbol{z}, H\boldsymbol{z})$. The integral is

$$J_N=\int_{-\infty}^{\infty}\ .\ .\ .\int_{-\infty}^{\infty}\exp\{-\boldsymbol{z}^H H\boldsymbol{z}\}d\boldsymbol{z} \tag{1.588}$$

where $d\boldsymbol{z}=dx_1\ .\ .\ .\ ,\ dx_N\ dy_1\ .\ .\ .\ ,\ dy_N$. With some algebra (Reference 25, p. 99) the following result can be obtained:

$$J_N=\frac{\pi^N}{\det(H)} \tag{1.589}$$

1.8.2.7 *Extremal points and gradients in the complex field [35]*

In many applications it is necessary to compute the gradient of a (real or complex) function with respect to a complex vector $\boldsymbol{w}$. Some useful properties are reported here; the interested reader is addressed to Reference 35 for demonstrations and details.

First property
Let $g(z, z^*)$ be a complex function of a complex number z and of its conjugated z^*, analytic w.r.t. z and z^* independently.

Let $f(x, y)$ be the complex function of the real and imaginary parts, x and y, of the argument z $(z=x+jy)$, such that $g(z, z^*)=f(x, y)$. Then the partial derivative (treating z^* as a constant)

$$\frac{\partial g(z, z^*)}{\partial z} \tag{1.590}$$

gives the same result as

$$\frac{1}{2}\left(\frac{\partial f(x, y)}{\partial x}-j\frac{\partial f(x, y)}{\partial y}\right) \tag{1.591}$$

Similarly, the partial derivative

$$\frac{\partial g(z, z^*)}{\partial z^*} \tag{1.592}$$

gives the same result as

$$\frac{1}{2}\left(\frac{\partial f(x, y)}{\partial x}+j\frac{\partial f(x, y)}{\partial y}\right) \tag{1.593}$$

(*a*) For example, if

$$g(z, z^*) = a|z|^2 + bz \tag{1.594}$$

i.e.

$$f(x, y) = a(x^2 + y^2) + b(x + jy) \tag{1.595}$$

we have

$$\frac{\partial g}{\partial z}=\frac{1}{2}\left(\frac{\partial f}{\partial x}-j\frac{\partial f}{\partial y}\right)=a(x-jy)+b=az^*+b \tag{1.596}$$

and

$$\frac{\partial g}{\partial z^*}=\frac{1}{2}\left(\frac{\partial f}{\partial x}+j\frac{\partial f}{\partial y}\right)=a(x+jy)=az \tag{1.597}$$

(*b*) If, given a complex constant $v = a + jb$,

$$g(z, z^*) = 2\ \mathrm{Re}(vz^*) \tag{1.598}$$

we have

$$f(x, y) = 2(ax + by) \tag{1.599}$$

and

$$\frac{\partial g}{\partial z}=\frac{1}{2}(2a-2jb)=v^* \tag{1.600}$$

$$\frac{\partial g}{\partial z^*}=v \tag{1.601}$$

(*c*) If, given a complex constant $v = a + jb$,

$$g(z, z^*) = vz^* \tag{1.602}$$

we have

$$f(x, y) = ax + by + j(bx - ay) \tag{1.603}$$

and

$$\frac{\partial g}{\partial z}=\frac{1}{2}(a+jb-j(b-ja))=0 \tag{1.604}$$

$$\frac{\partial g}{\partial z^*}=\frac{1}{2}(a+jb+j(b-ja))=v \tag{1.605}$$

Second property
Let $p(\boldsymbol{w})$ be a real scalar function of the complex vector $\boldsymbol{w}$ and let us express this function as

$$p(\boldsymbol{w}) = g(\boldsymbol{w}, \boldsymbol{w}^*) \tag{1.606}$$

where g is a real valued function of $\boldsymbol{w}$ and of $\boldsymbol{w}^*$, analytic with respect to each element of $\boldsymbol{w}$ and of $\boldsymbol{w}^*$.

Therefore a necessary and sufficient condition to obtain a stationary point of $p(\boldsymbol{w})$ is that the gradient of $g(\boldsymbol{w}, \boldsymbol{w}^*)$ w.r.t. either $\boldsymbol{w}$ or $\boldsymbol{w}^*$ is equal to zero.

In many cases the gradient $\nabla_{w^*}g$ can be computed using the *first property*. In particular it is readily seen that, for any vector $\boldsymbol{u}$ and any matrix A

$$\begin{cases} \nabla_{w^*}(\boldsymbol{w}^H\boldsymbol{u}) = \boldsymbol{u} \\ \nabla_{w^*}(\boldsymbol{u}^H\boldsymbol{w}) = \boldsymbol{0} \\ \nabla_{w^*}(\boldsymbol{w}^H A\boldsymbol{w}) = A\boldsymbol{w} \end{cases} \tag{1.607}$$

1.8.3 Eigenvalues and eigenvectors

1.8.3.1 *Definitions and some properties*

Eigenvectors (*characteristic vectors*) of a square matrix A are those non-zero vectors $\boldsymbol{u}^i$ which do not change their orientation, but are just scaled by a factor λ_i, i.e. by the corresponding *eigenvalue* (*characteristic value*), when multiplied by the matrix A:

$$A\boldsymbol{u}^i = \lambda_i \boldsymbol{u}^i \qquad (\boldsymbol{u}^i \neq \boldsymbol{0}) \tag{1.608}$$

Expr. 1.608 can be written, for the general case of a vector $\boldsymbol{u}$, as follows:

$$(A - \lambda I)\boldsymbol{u} = \boldsymbol{0} \qquad (\boldsymbol{u} \neq \boldsymbol{0}) \tag{1.609}$$

or, in explicit form,

$$\begin{bmatrix} a_{11}-\lambda & a_{12} & \dots & a_{1N} \\ a_{21} & a_{22}-\lambda & \dots & a_{2N} \\ \dots & \dots & \dots & \dots \\ a_{N1} & a_{N2} & \dots & a_{NN}-\lambda \end{bmatrix} \cdot \begin{bmatrix} u_1 \\ u_2 \\ \dots \\ u_N \end{bmatrix} = \begin{bmatrix} 0 \\ 0 \\ \dots \\ 0 \end{bmatrix} \tag{1.610}$$

so that the eigenvalues λ_i are the solutions ('characteristic' roots or values) of the Nth-degree algebraic equation:

$$\det(A - \lambda I) = 0 \tag{1.611}$$

since the homogeneous equation system of expr. 1.609 (1.610) admits non-trivial solutions only if equality 1.611 is satisfied.

Substituting in expr. 1.610 each eigenvalue found, the corresponding eigenvector is derived. Each eigenvector and associated eigenvalue form an *eigenpair*.

Some properties of the eigenvectors (eigenvalues) are reported here without proof:

(*a*) Eigenvectors associated with distinct eigenvalues of a Hermitian matrix are orthogonal (see Section 1.8.3.3).

(*b*) If the $N \times N$ Hermitian matrix A has N distinct eigenvalues, its eigenvectors form a basis in C^N.

(*c*) The matrix U whose columns are the distinct eigenvectors of a Hermitian matrix A (see Section 1.8.2.4) is unitary; the unitary transformation

$$U^{-1}AU = U^H AU = \Lambda_A \tag{1.612}$$

diagonalises A; Λ_A is the diagonal matrix whose elements are the eigenvalues of A. In other words,

$$A = U\Lambda_A U^{-1} \tag{1.613}$$

1.8.3.2 *Normalisation of eigenvectors and similarity transformations*

We showed that an eigenvector defines a particular direction that does not change under a linear transformation. Therefore, it is possible to normalise the eigenvectors in various ways, in particular according to the relationship

$$\|\boldsymbol{u}^i\| = (\boldsymbol{u}^i)^H \cdot \boldsymbol{u}^i = 1 \tag{1.614}$$

However, it should not be overlooked that such a normalisation is still arbitrary to the extent of a complex multiplier of modulus one: in other words, if $\boldsymbol{u}^i$ is a normalised (with unity norm) eigenvector, then $e^{j\psi}\boldsymbol{u}^i$, with ψ being any real constant, is another normalised eigenvector.

Eqn. 1.613 can be written

$$AU = U\Lambda_A \tag{1.615}$$

where Λ_A is the diagonal matrix of the eigenvalues of A and U is the matrix whose columns are the corresponding eigenvectors.

When the eigenvalues of A are all distinct, it can be shown that any set of N eigenvectors of A corresponding to the N eigenvalues λ_i gives a non-singular matrix U, so that eqns. 1.612 and 1.613 apply.

The eigenvectors $\boldsymbol{y}^i$ and the eigenvalues μ^i of a matrix S similar to $A(S = B^{-1}AB$, where B is any invertible matrix) are obtained by solving the usual problem

$$S\boldsymbol{y}^i = \mu_i \boldsymbol{y}^i \tag{1.616}$$

that is

$$B^{-1}AB\boldsymbol{y}^i = \mu_i \boldsymbol{y}^i \tag{1.617}$$

$$AB\boldsymbol{y}^i = \mu_i B\boldsymbol{y}^i \tag{1.618}$$

Comparing eqn. 1.618 with the eigenvector equation for A:

$$A\boldsymbol{u}^i = \lambda_i \boldsymbol{u}^i \tag{1.619}$$

it follows that $\boldsymbol{u}^i = B\boldsymbol{y}^i$ and $\mu_i = \lambda_i$.

Therefore, the similar matrices A and S have the same eigenvalues; the eigenvectors $\boldsymbol{y}^i$ of S and the eigenvectors $\boldsymbol{u}^i$ of A are related by the transformation through the matrix B. In other words, a similarity transform through B maintains the eigenvalues and multiplies the eignevectors by B^{-1}.

1.8.3.3 *Eigenvalues and eigenvectors of the adjoint of a matrix*

The eigenvalues of the adjoint (or conjugate transpose) A^H of a matrix A are obtained by solving the problem

$$\det(A^H - \lambda I) = 0 \tag{1.620}$$

or, recalling expr. 1.565: $\det(A^* - \lambda I) = 0$,

$$\det(A - \lambda^* I) = 0 \tag{1.621}$$

therefore, they are the complex conjugate of the eigenvalues of A. If A is Hermitian (self-adjoint) its eigenvalues are real, because they are complex conjugates of themselves.

It can be shown [9] that a necessary and sufficient condition for A to be positive (semi-)definite is that all the eigenvalues (characteristic roots) of A are positive (non-negative).

The eigenvectors $\boldsymbol{v}^i$ of the adjoint of any matrix A are the complex conjugate of the *left-hand eigenvectors* $\boldsymbol{w}^i$ of A, defined by

$$(\boldsymbol{w}^i)^T A = \lambda_i (\boldsymbol{w}^i)^T \tag{1.622}$$

i.e. $\boldsymbol{v}^i = (\boldsymbol{w}^i)^*$. To show this, let us consider the form of expr. 1.615

$$AU = U\Lambda \tag{1.623}$$

of the eigenvectors equation for the adjoint A^H:

$$A^H V = V\Lambda^* \tag{1.624}$$

where V is the matrix whose columns are the eigenvectors $\boldsymbol{v}^i$ of A^H and Λ^* is the diagonal matrix of its eigenvalues. By taking the adjoint of each term of expr. 1.624 one gets

$$V^H A = \Lambda V^H \tag{1.625}$$

Defining a matrix W whose rows are the left-hand eigenvectors $\boldsymbol{w}^i$ of A, it is readily shown that expr. 1.622 can be written as

$$WA = \Lambda W \tag{1.626}$$

Comparing expr. 1.626 with 1.625, it follows, as expected, that $W = V^H$.

Moreover, the eigenvectors $\boldsymbol{u}^i$ of A are orthogonal to the eigenvectors $\boldsymbol{v}^i$ of its adjoint:

$$(\boldsymbol{u}^i)^H \boldsymbol{v}^k = \delta_{ik} \tag{1.627}$$

In particular, if A is self-adjoint, its eigenvectors form an orthogonal set (that can be normalised).

1.8.3.4 *Reduction of a Hermitian form to principal axes*

Any Hermitian form

$$f = \boldsymbol{x}^H H \boldsymbol{x} \tag{1.628}$$

H being a Hermitian matrix, can be reduced by the unitary transformation

$$\boldsymbol{x} = U\boldsymbol{\xi} \tag{1.629}$$

(where U is the matrix of the eigenvectors of H) to the canonical form

$$f = \boldsymbol{\xi}^H \Lambda_H \boldsymbol{\xi} = \sum_{i=1}^{N} \lambda_i |\xi_i|^2 \tag{1.630}$$

where λ_i, $i = 1, 2, \ldots, n$ are the eigenvalues of H and $\Lambda_H = \mathrm{diag}(\lambda_1, \lambda_2, \ldots, \lambda_N)$. In fact, from expr. 1.628 and 1.629 it follows that

$$f=\xi^H U^H H U \xi \qquad (1.631)$$

and, recalling that (expr. 1.615) $U\Lambda_H = HU$ and that $U^H U = I$,

$$f=\xi^H U^H U \Lambda_H \xi = \xi^H \Lambda_H \xi \qquad (1.632)$$

1.8.3.5 *The pencil of matrices and the simultaneous reduction of two Hermitian forms to principal axes*

In many applications it is necessary to generalise the eigenproblem (see Section 1.8.3.1) by searching the 'principal' vectors $\boldsymbol{u}^k$ and the 'principal' values λ_k such that, for two Hermitian matrices A and B (for the *pencil* of matrices A, B),

$$A\boldsymbol{u}^k = \lambda_k B \boldsymbol{u}^k \qquad (1.633)$$

When the pencil is 'regular', i.e. B is positive definite, the equation

$$\det(A - \lambda B) = 0 \qquad (1.634)$$

(see also Section 1.8.3.3) has N real roots $\lambda_1, \lambda_2, \ldots, \lambda_N$. Each one of these roots λ_k defines a 'principal vector' $\boldsymbol{u}^k$, that is $\boldsymbol{u}^k$ is the solution of eqn . 1.633 when the pertaining value of λ_k is substituted in it. These N principal vectors $\boldsymbol{u}^1, \boldsymbol{u}^2, \ldots, \boldsymbol{u}^N$ can be arranged in a matrix

$$U = [\boldsymbol{u}^1 | \boldsymbol{u}^2 | \ldots | \boldsymbol{u}^N] \qquad (1.635)$$

It can be shown (Reference 9, chap. X, pp. 310–315 and 337–338) that the principal vectors of the pencil can be chosen so that

$$(\boldsymbol{u}^i)^H B \boldsymbol{u}^k = \delta_{ik} \quad i, k = 1, 2, \ldots, N \qquad (1.636)$$

or, in matrix form,

$$U^H B U = I \qquad (1.637)$$

Note that, unless $B = I$, U is *not* unitary: $U^{-1} = U^H B$, $U^{-1} \neq U^H$.

From expr. 1.637 and from Section 1.8.3.4 it follows that the transformation

$$\boldsymbol{x} = U\xi \qquad (1.638)$$

reduces the Hermitian form $f_B = \boldsymbol{x}^H B \boldsymbol{x}$ to a sum of squared moduli

$$f_B = \sum_{i=1}^{N} |\xi_i|^2 \qquad (1.639)$$

Let us consider now the form $f_A = \boldsymbol{x}^H A \boldsymbol{x}$. Expr. 1.633 in matrix form becomes

$$AU = BU\Lambda \qquad (1.640)$$

i.e.

$$U^H A U = U^H B U \Lambda \qquad (1.641)$$

and, using expr. 1.637,

$$U^H A U = \Lambda \qquad (1.642)$$

Therefore, the transformation $\boldsymbol{x} = U\xi$ also reduces the Hermitian form f_A to the diagonal form:

$$f_A = \xi^H \Lambda \xi = \sum_{i=1}^{N} \lambda_i |\xi_i|^2 \tag{1.643}$$

In the particular case $B=I$, the characteristic equation of the pencil coincides with the characteristic equation of A, and the principal vectors coincide with the characteristic vectors (eigenvectors) of A, that are orthonormal, according to expr. 1.636.

Multiplying both sides of expr. 1.633 by B^{-1} it follows that the generalised eigenproblem is equivalent to the eigenproblem for the (generally not Hermitian, but similar to the diagonal, real matrix Λ: $U^{-1}FU=\Lambda$) matrix

$$F = B^{-1}A \tag{1.644}$$

From the above discussion it follows that the eigenvalues of F are real, and its eigenvectors satisfy expr. 1.636.

1.8.3.6 *Extremal properties of the eigenvalues of the pencil A, B*

There are two principles (see, for example, Reference 28, chap. 6), known as 'min–max and max–min principles', that, at least from the theoretical point of view, allow us to find the eigenvalues of a matrix (in increasing or decreasing order).

In this, an important role is played by the *Rayleigh ratio* of a matrix A, which is expressed as

$$R(\boldsymbol{x}) = \frac{\boldsymbol{x}^H A x}{\boldsymbol{x}^H \boldsymbol{x}} \tag{1.645}$$

It can be shown that the Rayleigh ratio is bounded by the minimum and maximum eigenvalues, that is:

$$\lambda_m \leqslant R(\boldsymbol{x}) \leqslant \lambda_M \tag{1.646}$$

Under the constraint that $\boldsymbol{x}$ is orthogonal to the eigenvector $\boldsymbol{x}^1, \boldsymbol{x}^2, \ldots, \boldsymbol{x}^{j-1}$ (the corresponding eigenvalues are set in increasing order), the Rayleigh ratio $R(\boldsymbol{x})$ is minimised by the nearest eigenvector $\boldsymbol{x}^j$, and the corresponding value of $R(\boldsymbol{x})$ is exactly λ_j.

Using this property it is theoretically possible to find the eigenvalues in increasing order. In fact, the first eigenpair $(\boldsymbol{x}^1, \lambda_1)$ is immediately found by minimising the ratio $R(\boldsymbol{x})$ without considering any orthogonality condition. The second eigenpair $(\boldsymbol{x}^2, \lambda_2)$ can be found simply by determining the minimum of $R(\boldsymbol{x})$, subjected to the orthogonality conditions with respect to $\boldsymbol{x}^1$; after that, the third eigenpair $(\boldsymbol{x}^3, \lambda_3)$ can be found in the same way, by determining the minimum of $R(\boldsymbol{x})$, subjected to the simultaneous orthogonality condition with respect to $\boldsymbol{x}^1$ and $\boldsymbol{x}^2$, and so on for the remaining eigenpairs.

But it is also possible to give an extremal principle for the second eigenpair (and, as a consequence, for all those remaining) even if the first one is not known.

Let us suppose that, with $(\boldsymbol{x}^1, \lambda_1)$ unknown, we impose the limitation that all the vectors $\boldsymbol{x}$ we are testing to find the minimum of $R(\boldsymbol{x})$ are orthogonal to an arbitrary vector $\boldsymbol{z}$. It is obvious that, if $\boldsymbol{z}=\boldsymbol{x}^1$, the minimum is again exactly λ_2; if $\boldsymbol{z}$ is not equal to $\boldsymbol{x}^1$, it can still be shown that $R(\boldsymbol{x})$ is *not greater* than λ_2, that is:

$$\min_{x^H \cdot z=0} R(\boldsymbol{x}) \leqslant \lambda_2 \tag{1.647}$$

This leads to the 'max–min principle': maximising expr. 1.647 with respect to $\boldsymbol{z}$, the second eigenvalue is found:

$$\lambda_2 = \max_{z} \, [\min_{x^H \cdot z=0} R(\boldsymbol{x})] \tag{1.648}$$

This principle can be easily generalised in order to find the $(j+1)$th eigenvalue, taking into account j constraints, according to the following expression:

$$\lambda_{j+1} = \max_{z_1, z_2, \ldots, z_j} \, [\min_{x^H \cdot z_1=0, \ldots, x^H \cdot z_j=0} R(\boldsymbol{x})] \tag{1.649}$$

In a complementary way the 'min–max principle' can be stated: in this case, the first step is maximising the Rayleigh ratio, and then the identification of the minimum is performed.

The Rayleigh ratio is to be maximised over an *arbitrary* bidimensional subspace S_2, and it can be shown that, in this case,

$$\max_{x \in S_2} R(\boldsymbol{x}) \geqslant \lambda_2. \tag{1.650}$$

If $R(\boldsymbol{x})$ is minimised over *all* the possible bidimensional subspaces S_2, the minimum possible value is λ_2:

$$\lambda_2 = \min_{S_2} \, [\max_{x \in S_2} R(\boldsymbol{x})] \tag{1.651}$$

Also the min–max principle, expressed by expr. 1.651, can be generalised from the case of bidimensional spaces to the case of j-dimensional spaces:

$$\lambda_{j+1} = \min_{S_j} \, [\max_{x \in S_j} R(\boldsymbol{x})] \tag{1.652}$$

The above mentioned principles can be extended to the case of a pencil of matrices (A, B). The Rayleigh ratio is now written

$$R(\boldsymbol{x}) = \frac{\boldsymbol{x}^H A \boldsymbol{x}}{\boldsymbol{x}^H B \boldsymbol{x}} \tag{1.653}$$

where A and B are intended to be Hermitian and B is also definite positive. We can derive the following properties:

(*a*) The roots of the characteristic equation

$$\det(A - \lambda B) = 0 \tag{1.654}$$

$$\lambda_M \geqslant \cdots \geqslant \lambda_i \geqslant \lambda_j \geqslant \cdots > \lambda_m$$

with A and B Hermitian, B positive definite, are all real and positive.

(*b*) The values of λ_M and λ_m are the bounds of the Rayleigh ratio $R(\boldsymbol{x})$:

$$\lambda_M \leqslant R(\boldsymbol{x}) \leqslant \lambda_m \tag{1.655}$$

the equality signs being attained when

$$A\boldsymbol{x} = \lambda_M B\boldsymbol{x} \tag{1.656}$$

or, respectively,

$$Ax = \lambda_m Bx \tag{1.657}$$

In other words, the Rayleigh ratio has a maximum, λ_{max}, corresponding to the maximum principal value of the pencil (A, B), and to the related eigenvector (principal vector); the Rayleigh ratio has a minimum, λ_{min}, corresponding to the minimum principal value of the pencil (A, B) and the related eigenvector.

1.9 Appendix 3: The Gaussian multivariate

1.9.1 The Gaussian multivariate in the real field

A (real) random variable is called Gaussian (or 'normal') if its probability density function (PDF) is

$$p(x) = \frac{1}{\sqrt{2\pi\sigma^2}} \exp\left[-\frac{(x-\bar{x})^2}{2\sigma^2} \right] \tag{1.658}$$

where $\bar{x}$ and σ^2 are the mean value and the variance, respectively.

The shape of this PDF is the well-known 'bell'; the importance of the normal distribution in most fields involving probability theory and statistics is so great that no attempt is made here to discuss it.

We recall here [13] that the *characteristic function* $\Phi_x(\omega)$ of a random variable x is the expected value of $e^{j\omega x}$, that is, if x has the PDF $p(.)$:

$$\Phi_x(\omega) = \int_{-\infty}^{\infty} p(x) e^{j\omega x} dx = E[e^{j\omega x}] \tag{1.659}$$

For a Guassian PDF we have

$$\Phi_x(\omega) = \exp\left[-\frac{\sigma^2\omega^2}{2} + j\omega\bar{x} \right] \tag{1.660}$$

If the mean value $\bar{x}$ is zero, the characteristic function is $\sqrt{(2\pi/\sigma)}$ times a Gaussian function with variance equal to $1/\sigma^2$

$$\Phi_x(\omega) = \exp\left[-\frac{\sigma^2\omega^2}{2} \right] \tag{1.661}$$

Considering now the general case of a real random vector $\boldsymbol{x}$, i.e. an N-tuple of real random variables, we can define a Gaussian probability measure on the N-dimensional linear vector space R^N as one which admits a density function (with respect to the Lebesgue measure), this function being the exponential of a quadratic form.

The Gaussian multivariate is first considered in the particular case in which its N variates y_i are independent, with mean values $\bar{y}_i$ and variances

$$\sigma_i^2, \; i = 1, 2, \ldots, N \tag{1.662}$$

In this case, the multivariate PDF $p_Y(\boldsymbol{y})$ is simply the product of N Gaussian PDFs:

$$p_Y(\mathbf{y}) = \frac{1}{(2\pi)^{N/2} \cdot (\sigma_1 \sigma_2 \ldots \sigma_N)} \exp\left[-\frac{1}{2} \sum_{i=1}^{N} \frac{(y_i - \bar{y}_i)^2}{\sigma_i^2} \right] \tag{1.663}$$

and can be written as the exponential of a quadratic form:

$$p_Y(\mathbf{y}) = \exp[-\tfrac{1}{2}(\mathbf{y}-\bar{\mathbf{y}})^T \Lambda^{-1} (\mathbf{y}-\bar{\mathbf{y}}) - c] \tag{1.664}$$

where $\bar{\mathbf{y}}$ is the (vector) mean and

$$\Lambda = \text{diag}(\sigma_1^2, \sigma_2^2 \ldots \sigma_N^2) \tag{1.665}$$

is the (diagonal) covariance matrix.

Finally, the constant c is defined as follows:

$$\begin{aligned} \exp(c) &= (2\pi)^{N/2}(\sigma_1 \cdot \sigma_2 \ldots \sigma_N) \\ &= [(2\pi)^N \det(\Lambda)]^{1/2} \end{aligned} \tag{1.666}$$

Let us consider the following transformation (see paragraph 1.8.2.5 and 1.8.2.6 of Appendix 2):

$$(\mathbf{x} - \bar{\mathbf{x}}) = U(\mathbf{y} - \bar{\mathbf{y}}) \tag{1.667}$$

with U being invertible and *orthogonal* (*unitary* when considered in the complex field):

$$U^{-1} = U^T \tag{1.668}$$

It is well known (see also paragraph 1.8.2.2 of Appendix 2) that for any invertible transformation $\mathbf{y} = f(\mathbf{x})$ the PDF of $\mathbf{x}$ is

$$p_X(\mathbf{x}) = p_Y(\mathbf{y})|_{\mathbf{y}=f(\mathbf{x})} \cdot |\det(J)| \tag{1.669}$$

where J is the Jacobian matrix of the transformation $\mathbf{y} = f(\mathbf{x})$ (see paragraph 1.8.2.2 and 1.8.2.3 of Appendix 2); in the present case, U being orthogonal, $J = U^T$ and $|\det(J)| = 1$ (see also paragraph 1.8.2.4).

From eqns. 1.664 and 1.669 with $J = U^T$, the PDF of the vector $\mathbf{x}$ becomes

$$p_X(\mathbf{x}) = \frac{1}{[(2\pi)^N \det \Lambda]^{1/2}} \exp[-\tfrac{1}{2}(\mathbf{x}-\bar{\mathbf{x}})^T U \Lambda^{-1} U^T (\mathbf{x}-\bar{\mathbf{x}})] \tag{1.670}$$

The matrix $U\Lambda^{-1}U^T$ is readily recognised as the inverse M_x^{-1} of the covariance matrix M_x of $\mathbf{x}$:

$$M_x = E[(\mathbf{x}-\bar{\mathbf{x}})(\mathbf{x}-\bar{\mathbf{x}})^T] \tag{1.671}$$

and, using eqn. 1.667,

$$M_x = UE[(\mathbf{y}-\bar{\mathbf{y}})(\mathbf{y}-\bar{\mathbf{y}})^T]U^T = U\Lambda U^T \tag{1.672}$$

Moreover, $\det(M_x) = \det(\Lambda)$.

Therefore the expression (eqn. 1.670) of the (real) Gaussian multivariate reduces to

$$\boxed{p(\mathbf{x}) = \frac{1}{(2\pi)^{N/2}(\det M_x)^{1/2}} \exp[-\tfrac{1}{2}(\mathbf{x}-\bar{\mathbf{x}})^T M_x^{-1} (\mathbf{x}-\bar{\mathbf{x}})]} \tag{1.673}$$

The matrix Λ in eqns. 1.665 and 1.670 is readily recognised as the diagonal matrix of the eigenvalues of M_x and the matrix U (eqns. 1.667 and 1.670) is the orthogonal matrix whose columns are the eigenvalues of M_x. The inverse transformation of eqn. 1.667:

$$(\boldsymbol{y}-\bar{\boldsymbol{y}}) = U^T(\boldsymbol{x}-\bar{\boldsymbol{x}}) \tag{1.674}$$

reduces the covariance matrix M_x to diagonal form, i.e. transforms $\boldsymbol{x}$ into a vector $\boldsymbol{y}$ of independent variates (this concept is also discussed in Section 1.8.3.4).

If any linear transformation is applied to the Gaussian random vector $\boldsymbol{x}$, the result $\boldsymbol{z}=A\boldsymbol{x}+\boldsymbol{b}$ is still a Gaussian random vector with covariance matrix AMA^T and mean vector $\bar{\boldsymbol{z}}=A\bar{\boldsymbol{x}}+\boldsymbol{b}$, where A is an $N\times N$ matrix and $\boldsymbol{b}$ an $N\times 1$ vector. In particular, a linear combination of the elements of the Gaussian random vector $\boldsymbol{x}$ is a Gaussian random variable.

This property can be used to define the Gaussian random vector $\boldsymbol{x}$ as one such that any linear combination v of its components:

$$v=\sum_{i=1}^{N} a_i x_i = \boldsymbol{a}^T\boldsymbol{x} \tag{1.675}$$

is a Gaussian random variable for any choice of the coefficients (the vector $\boldsymbol{a}$).

Let us recall the concept of the characteristic function of an N-element random vector $\boldsymbol{x}$: such a function is defined as the expected value of $\exp(j\boldsymbol{\omega}^T\boldsymbol{x})$, where $\boldsymbol{\omega}^T\equiv[\omega_1\omega_2\ldots\omega_N]$, (or equivalently as the N-dimensional Fourier transform of the PDF of $\boldsymbol{x}$):

$$\Phi_x(\boldsymbol{\omega}) = E[\exp(j\boldsymbol{\omega}^T\boldsymbol{x})] \tag{1.676}$$

It is easily shown (Reference 13, page 186) that if $\boldsymbol{x}$ satisfies the above definition, i.e. the linear combination v defined above is Gaussian for every $\boldsymbol{a}$, the characteristic function of $\boldsymbol{x}$ (supposed zero-mean) is

$$\Phi_x(\boldsymbol{\omega}) = \exp[-\tfrac{1}{2}\boldsymbol{\omega}^T M_x \boldsymbol{\omega}] \tag{1.677}$$

where M_x is the covariance matrix of $\boldsymbol{x}$.

From the Fourier inversion of eqn. 1.677, the zero-mean Gaussian multivariate PDF gives

$$p(\boldsymbol{x}) = \frac{\exp[-\tfrac{1}{2}\boldsymbol{x}^T M_x^{-1}\boldsymbol{x}]}{(2\pi)^{N/2}(\det M_x)^{1/2}} \tag{1.678}$$

i.e. the same result as eqn. 1.673.

1.9.1.1 *The conditional Gaussian multivariate*

For a given (real) Gaussian multivariate $p(\boldsymbol{x})$ (eqn. 1.673), with $\boldsymbol{x}$ being N-dimensional (i.e. $\boldsymbol{x}\in R^N$), we can consider an M-dimensional subspace S_1, with $M<N$, and find the conditional density with respect to S_1.

Let S_2 be the ($N-M$ dimensional) subspace orthogonal to S_1, and let the $M\times N$ matrix P_1 and the $(N-M)\times N$ matrix P_2 be the orthogonal projectors on

S_1 and S_2, respectively. Every $\boldsymbol{x} \in R^N$ is identified by the ordered pair $(\boldsymbol{x}_1, \boldsymbol{x}_2)$, with $\boldsymbol{x}_1 \in R^M$ and $\boldsymbol{x}_2 \in R^{N-M}$, namely:

$$\boldsymbol{x}_1 = P_1 \boldsymbol{x} \tag{1.679a}$$

$$\boldsymbol{x}_2 = P_2 \boldsymbol{x} \tag{1.679b}$$

and, concerning the mean values:

$$\bar{\boldsymbol{x}}_1 = P_1 \bar{\boldsymbol{x}} \tag{1.680a}$$

$$\bar{\boldsymbol{x}}_2 = P_2 \bar{\boldsymbol{x}} \tag{1.680b}$$

The quadratic form of eqn. 1.673, namely:

$$q = q(\boldsymbol{x}) = (\boldsymbol{x} - \bar{\boldsymbol{x}})^T A (\boldsymbol{x} - \bar{\boldsymbol{x}}) \tag{1.681}$$

(with $A \equiv M_x^{-1}$) can be expressed in terms of $\boldsymbol{x}_1$, $\boldsymbol{x}_2$ and of the four matrices:

$$A_{ij} = P_i A P_j^T, \; ij = 1, 2 \tag{1.682}$$

Using block matrices the quadratic form is written as follows:

$$q = \begin{bmatrix} \boldsymbol{x}_1 - \bar{\boldsymbol{x}}_1 \\ \boldsymbol{x}_2 - \bar{\boldsymbol{x}}_2 \end{bmatrix}^T \begin{bmatrix} A_{11} & A_{12} \\ A_{21} & A_{22} \end{bmatrix} \begin{bmatrix} \boldsymbol{x}_1 - \bar{\boldsymbol{x}}_1 \\ \boldsymbol{x}_2 - \bar{\boldsymbol{x}}_2 \end{bmatrix} \tag{1.683}$$

It results:

$$q = \sum_{i=1}^{2} \sum_{j=1}^{2} (\boldsymbol{x}_i - \bar{\boldsymbol{x}}_1)^T A_{ij} (\boldsymbol{x}_j - \bar{\boldsymbol{x}}_j) \tag{1.684}$$

The joint density function of the pair $(\boldsymbol{x}_1, \boldsymbol{x}_2)$ is therefore

$$p(\boldsymbol{x}_1, \boldsymbol{x}_2) = \frac{[\det(A)]^{1/2}}{(2\pi)^{N/2}} \exp\{-\tfrac{1}{2} q\} \tag{1.685}$$

where q is given by eqn. 1.684.

The (marginal) PDF of $\boldsymbol{x}_1$ is:

$$p(\boldsymbol{x}_1) = \int_{R^{N-M}} p(\boldsymbol{x}_1, \boldsymbol{x}_2) \, d\boldsymbol{x}_2 \tag{1.686}$$

which, using eqns. 1.684 and 1.685, becomes

$$p(\boldsymbol{x}_1) = \frac{[\det A]^{1/2}}{(2\pi)^{N/2}} \exp\{-\tfrac{1}{2}(\boldsymbol{x}_1 - \bar{\boldsymbol{x}}_1)^T A_{11} (\boldsymbol{x}_1 - \bar{\boldsymbol{x}}_1)\} \cdot \int_{R^{N-M}} \exp[-\tfrac{1}{2} q_1(\boldsymbol{x}_2)] \, d\boldsymbol{x}_2 \tag{1.687}$$

where

$$\begin{aligned} q_1(\boldsymbol{x}_2) &= (\boldsymbol{x}_2 - \bar{\boldsymbol{x}}_2)^T A_{22} (\boldsymbol{x}_2 - \bar{\boldsymbol{x}}_2) + (\boldsymbol{x}_1 - \bar{\boldsymbol{x}}_1)^T A_{12} (\boldsymbol{x}_2 - \bar{\boldsymbol{x}}_2) + (\boldsymbol{x}_2 - \bar{\boldsymbol{x}}_2)^T A_{21} (\boldsymbol{x}_1 - \bar{\boldsymbol{x}}_1) \\ &= (\boldsymbol{x}_2 - \bar{\boldsymbol{x}}_2)^T A_{22} (\boldsymbol{x}_2 - \bar{\boldsymbol{x}}_2) + 2(\boldsymbol{x}_2 - \bar{\boldsymbol{x}}_2)^T A_{21} (\boldsymbol{x}_1 - \bar{\boldsymbol{x}}_1) \end{aligned} \tag{1.688}$$

Using the notations $\mathbf{X}_2 \equiv \boldsymbol{x}_2 - \bar{\boldsymbol{x}}_2$, $\mathbf{X}_1 \equiv \boldsymbol{x}_1 - \bar{\boldsymbol{x}}_1$ and $q_1(\boldsymbol{x}_2) \equiv q_1$ gives

$$q_1 = \mathbf{X}_2^T A_{22} \mathbf{X}_2 + 2\mathbf{X}_2^T A_{21} \mathbf{X}_1 \tag{1.689}$$

and with the position

$$\boldsymbol{m} = -A_{22}^{-1} A_{21} \boldsymbol{X}_1 \tag{1.690}$$

eqn. 1.689 becomes

$$q_1 = \boldsymbol{X}_2^T A_{22} \boldsymbol{X}_2 - 2\boldsymbol{X}_2^T A_{22} \boldsymbol{m} \tag{1.691}$$

Recalling that

$$(\boldsymbol{X}_2 - \boldsymbol{m})^T A_{22} (\boldsymbol{X}_2 - \boldsymbol{m}) = \boldsymbol{X}_2^T A_{22} \boldsymbol{X}_2 - 2\boldsymbol{X}_2^T A_{22} \boldsymbol{m} + \boldsymbol{m}^T A_{22} \boldsymbol{m} \tag{1.692}$$

we get

$$q_1 = (\boldsymbol{X}_2 - \boldsymbol{m})^T A_{22} (\boldsymbol{X}_2 - \boldsymbol{m}) - \boldsymbol{m}^T A_{22} \boldsymbol{m} \tag{1.693}$$

Using eqn. 1.693, the integral of eqn. 1.687 is easily solved as follows:

$$\int_{R^{N-M}} \exp[-\tfrac{1}{2} q_1(\boldsymbol{x}_2)]\, d\boldsymbol{x}_2 = \int_{R^{N-M}} \exp[-\tfrac{1}{2}(\boldsymbol{X}_2 - \boldsymbol{m})^T A_{22} (\boldsymbol{X}_2 - \boldsymbol{m})]\, d\boldsymbol{x}_2 \times \exp[\tfrac{1}{2} \boldsymbol{m}^T A_{22} \boldsymbol{m}] \tag{1.694}$$

The integral in the right hand side of eqn. 1.694 (Section 1.8.2.6) is equal to

$$\frac{(2\pi)^{(N-M)/2}}{[\det(A_{22})]^{1/2}}$$

and therefore eqn. 1.687 for $p(\boldsymbol{x}_1)$ reduces to:

$$p(\boldsymbol{x}_1) = \left[\frac{\det(A)}{\det(A_{22})}\right]^{1/2} \cdot \frac{1}{(2\pi)^{M/2}} \cdot \exp[-\tfrac{1}{2} \boldsymbol{X}_1^T A_{11} \boldsymbol{X}_1 + \tfrac{1}{2} \boldsymbol{m}^T A_{22} \boldsymbol{m}] \tag{1.695}$$

The conditional PDF $p(x_2|x_1)$ is readily obtained dividing $p(\boldsymbol{x}_1, \boldsymbol{x}_2)$ (eqn. 1.685), by $p(\boldsymbol{x}_1)$, (eqn. 1.695). It results

$$p(\boldsymbol{x}_2|\boldsymbol{x}_1) = \frac{[\det(A_{22})]^{1/2}}{(2\pi)^{(N-M)/2}} \cdot \exp\{-\tfrac{1}{2}[\boldsymbol{X}_2^T A_{22} \boldsymbol{X}_2 + 2\boldsymbol{X}_2^T A_{21} \boldsymbol{X}_1 + \boldsymbol{m}^T A_{22} \boldsymbol{m}]\} \tag{1.696}$$

where, recalling eqn. 1.689, the quadratic form q_2 in the square brackets of eqn. 1.696 is equal to

$$q_2 = q_1 + \boldsymbol{m}^T A_{22} \boldsymbol{m} \tag{1.697}$$

which, using eqn. 1.693 for q_1, becomes:

$$q_2 = (\boldsymbol{X}_2 - \boldsymbol{m})^T A_{22} (\boldsymbol{X}_2 - \boldsymbol{m}) \tag{1.698}$$

i.e. substituting eqn. 1.690 for $\boldsymbol{m}$:

$$q_2 = [\boldsymbol{x}_2 - \bar{\boldsymbol{x}}_2 + A_{22}^{-1} A_{21} (\boldsymbol{x}_1 - \bar{\boldsymbol{x}}_1)]^T A_{22} [\boldsymbol{x}_2 - \bar{\boldsymbol{x}}_2 + A_{22}^{-1} A_{21} (\boldsymbol{x}_1 - \bar{\boldsymbol{x}}_1)] \tag{1.699}$$

Therefore, the conditional PDF $p(\boldsymbol{x}_2|\boldsymbol{x}_1)$ is a Gaussian function with inverse covariance matrix A_{22} and mean value equal to

$$\bar{\boldsymbol{x}}_{2|1} = \bar{\boldsymbol{x}}_2 - A_{22}^{-1} A_{21} (\boldsymbol{x}_1 - \bar{\boldsymbol{x}}_1) \tag{1.700}$$

The expression of the inverse matrix A_{22}^{-1} and of the matrix $A_{21} (= A_{12}^T)$ is easily obtained from

$$\begin{bmatrix} A_{11} & A_{12} \\ A_{21} & A_{22} \end{bmatrix} \begin{bmatrix} M_{11} & M_{12} \\ M_{21} & M_{22} \end{bmatrix} = I = \begin{bmatrix} 1_M & 0 \\ 0 & 1_{N-M} \end{bmatrix} \tag{1.701}$$

(where the diagonal matrices of dimension M and $N-M$ are denoted 1_M and 1_{N-M}, respectively), yielding:

$$A_{21}M_{11} + A_{22}M_{21} = 0 \tag{1.702}$$

$$A_{21}M_{12} + A_{22}M_{22} = I \tag{1.703}$$

From eqn. 1.702, it follows, premultiplying by A_{22}^{-1} and postmultiplying by M_{11}^{-1}:

$$A_{22}^{-1}A_{21} = -M_{21}M_{11}^{-1} \tag{1.704}$$

Solving by substitution the eqns 1.702 and 1.703 we obtain

$$A_{22}^{-1} = M_{22} - M_{21}M_{11}^{-1}M_{12} \tag{1.705}$$

$$A_{21} = -(M_{22} - M_{21}M_{11}^{-1}M_{12})^{-1}M_{21}M_{11}^{-1} \tag{1.706}$$

Thus, the conditional PDF $p(\boldsymbol{x}_2|\boldsymbol{x}_1)$ for each $\boldsymbol{x}_2 \in \boldsymbol{R}^{N-M}$, $\boldsymbol{x}_1 \in \boldsymbol{R}^M$ is Gaussian with mean value (eqns. 1.700 and 1.704)

$$\bar{\boldsymbol{x}}_{2|1} = \bar{\boldsymbol{x}}_2 + M_{21}M_{11}^{-1}(\boldsymbol{x}_1 - \bar{\boldsymbol{x}}_1) \tag{1.707}$$

and covariance matrix (eqn. 1.705):

$$M_{2|1} = M_{22} - M_{21}M_{11}^{-1}M_{12}. \tag{1.708}$$

When $\boldsymbol{x}_1$ and $\boldsymbol{x}_2$ are scalar, i.e. $N=2$, $M=1$, the conditional PDF is the well-known expression

$$p(x_1, x_2) = \frac{1}{2\pi\sigma_1\sigma_2\sqrt{1-\rho^2}} \cdot \exp\left\{-\frac{1}{2}q\right\} \tag{1.709}$$

with

$$q = \frac{1}{1-\rho^2}\left[\frac{(x_1-\bar{x}_1)^2}{\sigma_1^2} - 2\rho\frac{(x_1-\bar{x}_1)(x_2-\bar{x}_2)}{\sigma_1\sigma_2} + \frac{(x_2-\bar{x}_2)^2}{\sigma_2^2}\right] \tag{1.710}$$

while the marginal PDF is

$$p(x_1) = \frac{1}{\sigma_1\sqrt{2\pi}} \exp\left\{-\frac{(x_1-\bar{x}_1)^2}{2\sigma_1^2}\right\} \tag{1.711}$$

The conditional PDF of x_2 is

$$p(x_2|x_1) = \frac{p(x_2, x_1)}{p(x_1)} = \frac{1}{\sigma_2\sqrt{1-\rho^2}\sqrt{2\pi}} \cdot \exp\left\{-\frac{1}{2}q_2\right\} \tag{1.712}$$

with

$$q_2 = \frac{1}{1-\rho^2}\left[\frac{(x_1-\bar{x}_1)^2}{\sigma_1^2} - 2\rho\frac{(x_1-\bar{x}_1)(x_2-\bar{x}_2)}{\sigma_1\sigma_2} + \frac{(x_2-\bar{x}_2)^2}{\sigma_2^2} - \frac{(x_1-\bar{x}_1)^2}{\sigma_1^2}(1-\rho^2)\right] \tag{1.713}$$

i.e.

$$q_2 = \frac{1}{1-\rho^2}\left[-\rho\frac{x_1-\bar{x}_1}{\sigma_1}+\frac{x_2-\bar{x}_2}{\sigma_2}\right]^2 \tag{1.714}$$

The resulting expression for the conditional density function is

$$p(x_2|x_1) = \frac{1}{\sqrt{(2\pi)}\sigma_2\sqrt{(1-\rho^2)}}\exp\left\{-\frac{\left[(x_2-\bar{x}_2)-\rho\frac{\sigma_2}{\sigma_1}(x_1-\bar{x}_1)\right]^2}{2\sigma_2^2(1-\rho^2)}\right\} \tag{1.715}$$

i.e. a Gaussian PDF with mean value

$$m = \bar{x}_2+\rho\frac{\sigma_2}{\sigma_1}(x_1-\bar{x}_1) \tag{1.716}$$

(to be compared with eqn. 1.707) and variance

$$\sigma^2 = \sigma_2^2(1-\rho^2) \tag{1.717}$$

(to be compared with eqn. 1.708).

1.9.2 The complex Gaussian multivariate [10, 11]

It is well known that a complex random variable z is, by definition, a pair (x, y) of (real) random variables with an assigned joint (bivariate) PDF.

In a similar way, an N-element complex random vector $\boldsymbol{z}$ (i.e. an N-tuple of complex random variables $z_i = x_i + jy_i$, $i = 1 \ldots N$) can be represented by the real vector $\boldsymbol{z}_R$ defined as follows:

$$\boldsymbol{z}_R^T = [x_1 y_1\, x_2 y_2 \ldots x_N y_N] \tag{1.718}$$

An $N \times N$ complex square matrix A with elements

$$a_{ik} = \alpha_{ik} + j\beta_{ik}, \qquad i, k = 1, 2 \ldots N \tag{1.719}$$

can be represented by a $2N \times 2N$ real matrix A_R made up by $N \times N$ blocks:

$$A_R = \begin{bmatrix} B_{11} \cdots B_{1N} \\ \vdots \quad\quad \vdots \\ B_{N1} \cdots B_{NN} \end{bmatrix} \tag{1.720}$$

Each block B_{ik} has the following structure:

$$B_{ik} = \begin{bmatrix} \alpha_{ik} & -\beta_{ik} \\ \beta_{ik} & \alpha_{ik} \end{bmatrix} \tag{1.721}$$

The reason that the blocks of A_R are defined by eqn. 1.721 is understood when matrix-by-vector multiplication is performed: the complex vector $A\boldsymbol{z}$ corresponds to the real vector $A_R\boldsymbol{z}_R$ where A_R and $\boldsymbol{z}_R$ are defined by eqns. 1.718 and 1.720. As a matter of fact, the ith element $(A\boldsymbol{z})_i$ of $A\boldsymbol{z}$ is

$$(A\boldsymbol{z})_i = \sum_{k=1}^{N} a_{ik}z_k = \sum_{k=1}^{N} \alpha_{ik}x_k - \beta_{ik}y_k + j(\beta_{ik}x_k + \alpha_{ik}y_k) \tag{1.722}$$

and, according to the block-matrix multiplication rules, the ith pair of terms $((A_R z_R))_i$ of the product $A_R z_R$ is

$$((A_R z_R))_i = \sum_{k=1}^{N} B_{ik} \begin{bmatrix} x_k \\ y_k \end{bmatrix} \tag{1.723}$$

If B_{ik} is given by eqn. 1.721, it is readily seen that the ith pair of elements of the vector $A_R z_R$ (eqn. 1.723) are the real and imaginary part of Az (eqn. 1.722), respectively. The above result may be extended [11] to matrix-by-matrix products.

If the only real matrices being considered have the form of eqns. 1.720 and 1.721, it may be stated [11] that an isomorphism (i.e. a one-to-one, invertible, ordered correspondence compatible with additions and multiplications) exists between the multiplication of real, $2N \times 2N$, matrices and the multiplication of complex, $N \times N$ matrices. This includes, as a special case, matrix by vector multiplications, recalling that real vectors of the form of eqn. 1.718 are isomorphic to complex vectors with elements $x_k + jy_k$, $k = 1, \ldots, N$.

The following properties hold:

(*a*) If the real matrix A_R is symmetric, the corresponding matrix A is Hermitian and conversely. This property is readily shown considering that if A_R is symmetric, according to eqns. 1.720 and 1.721 we have, for every pair (i, k)

$$\begin{cases} \alpha_{ik} = \alpha_{ki} \\ \beta_{ik} = -\beta_{ki} \end{cases} \tag{1.724}$$

and, considering eqn. 1.719, the matrix A is Hermitian.

(*b*) The quadratic forms defined by a symmetric matrix A_R are equal to the Hermitian forms defined by its isomorphic Hermitian matrix A, i.e. [11]

$$z_R^T A_R z_R = z^H A z \tag{1.725}$$

(*c*) If A_R is symmetric positive definite, A is Hermitian positive definite

(*d*) $\det(A_R) = [\det(A)]^2$. (1.726)

From the above properties, the expression of the complex Gaussian multivariate for the case of narrow band processes is easily derived.

Let us consider again the N-elements, zero-mean, complex random vector z and its representation z_R given by eqn. 1.718. The covariance matrix M_R of z_R is defined as follows:

$$M_R = E(z_R z_R^T) \tag{1.727}$$

and can be written (similarly to eqns. 1.720 and 1.721) as follows:

$$M_R = \begin{bmatrix} M_{11} \ldots M_{1N} \\ \vdots \qquad \vdots \\ M_{N1} \ldots M_{NN} \end{bmatrix} \tag{1.728}$$

where

$$M_{ik} = \begin{bmatrix} E(x_i x_k) & E(x_i y_k) \\ E(y_i x_k) & E(y_i y_k) \end{bmatrix} \tag{1.729}$$

The covariance matrix of the complex vector $\boldsymbol{z}$ is defined as follows:

$$M = E(\boldsymbol{z}^*\boldsymbol{z}^T) \tag{1.730}$$

with elements

$$m_{ik} = E(x_i x_k) + E(y_i y_k) + j[E(x_i y_k) - E(y_i x_k)] \tag{1.731}$$

(Note that in some text books and papers alternative definitions such as $M = 1/2E(\boldsymbol{z}^*\boldsymbol{z}^T)$, $M = E(\boldsymbol{z}\boldsymbol{z}^H)$, $M = 1/2E(\boldsymbol{z}\boldsymbol{z}^H)$ are used; with these the following derivations change in obvious ways.)

In general, the blocks M_{ik} do not have the symmetrical structure (eqn. 1.721) required for the isomorphism between M and M_R, and the whole statistical information of the (zero-mean) vector is contained in M_R, *not* in M.

However, in many applications the elements of the random vector $\boldsymbol{z}$ are the samples of the complex envelope of a zero-mean, band-limited Gaussian process (i.e. x_i and y_i, $i = 1, \ldots N$, are the samples of the I and Q baseband components of the process). In such a case (see Sections 8-5 of Reference 12 and Section 1.7.4 here) we have (for $i = 1, 2 \ldots N$; $k = 1, 2 \ldots N$)

$$E(x_i^2) = E(y_i^2) = \sigma_i^2 \tag{1.732a}$$

$$E(x_i x_k) = E(y_i y_k) = r_{ik}\sigma_i\sigma_k, \; i \neq k \tag{1.732b}$$

$$E(x_i y_i) = 0 \tag{1.732c}$$

$$E(x_i y_k) = -E(y_i x_k) = -s_{ik}\sigma_i\sigma_k, \; i \neq k \tag{1.732d}$$

Of course, $r_{ik} = r_{ki}$ and $s_{ik} = -s_{ki}$.

If x_i and y_i, $i = 1, 2, \ldots N$, are the I and Q samples of a narrow-band stationary Gaussian process with power spectral density $S(f)$ and central frequency f_0, then [Section 8-5 of Reference 12]

$$\sigma_i^2 = \sigma^2 = 2\int_0^\infty S(f)\, df \tag{1.733a}$$

$$r_{ik}\sigma_i\sigma_k = \rho_c(\tau)\cdot\sigma^2 = 2\int_0^\infty S(f)\cos[2\pi(f - f_0)\tau]\, df \tag{1.733b}$$

$$s_{ik}\sigma_i\sigma_k = -\rho_s(\tau)\cdot\sigma^2 = -2\int_0^\infty S(f)\sin[2\pi(f - f_0)\tau]\, df \tag{1.733c}$$

where ρ_c and ρ_s are the correlation coefficients of the I and Q components and $\tau = (k - i)T_s$, T_s being the sampling period. In general, r_{ik} are the correlation coefficients of the I (and the Q) samples while s_{ik} are the cross-correlation coefficients between the I and the Q samples; for zero lag ($i = k$) $r_{ik} = 1$ and $s_{ik} = 0$.

Therefore, for band-limited Gaussian processes, the covariance matrix M_R is symmetric and has the block structure of eqn. 1.728. From eqns. 1.729 and 1.710, the blocks have the following form:

$$M_{ii} = \sigma_i^2 \begin{bmatrix} 1 & 0 \\ 0 & 1 \end{bmatrix} \tag{1.734a}$$

$$M_{ik}=\sigma_i\sigma_k\begin{bmatrix} r_{ik} & -s_{ik} \\ s_{ik} & r_{ik}\end{bmatrix}, \ i\neq k \tag{1.734b}$$

For example, when $N=2$ and the process is stationary with $\sigma^2=1$, the covariance matrix is

$$M_R=\begin{bmatrix} 1 & 0 & r & -s \\ 0 & 1 & s & r \\ r & s & 1 & 0 \\ -s & r & 0 & 1\end{bmatrix} \tag{1.735}$$

where $r=r_{12}=r_{21}=E(x_1x_2)=E(y_1y_2)$ and $s=s_{12}=-s_{21}=-E(x_1y_2)=E(x_2y_1)$. In the particular case of (band-limited and) stationary processes, M_R is both symmetric and centrosymmetric, as in the example given by eqn. 1.735.

For any band-limited process, i.e. when eqn. 1.732 applies, the complex covariance matrix M (see eqns. 1.730 and 1.731) has the following elements m_{ik}, $i, k=1 \ldots N$:

$$m_{ii}=2\sigma_i^2 \tag{1.736a}$$

$$m_{ik}=2\sigma_i\sigma_k(r_{ik}-js_{ik}), \ i\neq k \tag{1.736b}$$

Comparing eqns. 1.734 and 1.736 it results that the matrices M_R and $\frac{1}{2}M^*$ are isomorphic†.

At this point, it is easy to derive the expression of the zero-mean complex Gaussian N-dimensional multivariate from the $2N$-dimensional multivariate. The latter has the well-known form (eqn. 1.673):

$$p_R(\mathbf{z}_R)=\frac{1}{(2\pi)^N(\det M_R)^{1/2}}\exp\left(-\frac{\mathbf{z}_R^TM_R^{-1}\mathbf{z}_R}{2}\right) \tag{1.737}$$

We recall that M_R is symmetric, M is Hermitian and both are positive definite; the inverse matrices M_R^{-1} and M^{-1} have the same properties.

Due to the above-discussed property (*b*) of the isomorphism, we have

$$\mathbf{z}_R^TM_R^{-1}\mathbf{z}_R=2\mathbf{z}^HM^{-*}\mathbf{z} \tag{1.738}$$

therefore, recalling that M^{-1} is Hermitian, the argument of the exponential can be written as follows:

$$\tfrac{1}{2}\,\mathbf{z}_R^TM_R^{-1}\mathbf{z}_R=\mathbf{z}^TM^{-1}\mathbf{z}^* \tag{1.739}$$

Moreover, due to the property (*d*),

$$\det(M_R)=\left(\det\left(\frac{1}{2}M\right)\right)^2=\frac{1}{2^{2N}}(\det M)^2 \tag{1.740}$$

Substituting eqns. 1.738 and 1.740 into eqn. 1.737, the expression of the N-dimensional complex Gaussian multivariate* is obtained:

† Of course, using the definition $M=1/2E(\mathbf{z}\mathbf{z}^H)$, the matrices M_R and M would be isomorphic.

* Using the definition: $M=1/2E(\mathbf{z}^*\mathbf{z}^T)$ one should get

$$p(z)=\frac{1}{(2\pi)^N\det M}\exp[-\tfrac{1}{2}\mathbf{z}^TM^{-1}\mathbf{z}^*]$$

$$p(\mathbf{z}) = \frac{1}{\pi^N \det M} \exp(-\mathbf{z}^T M^{-1} \mathbf{z}^*) \tag{1.741}$$

The conditional complex multivariate is easily derived either in the same way as the derivation in the real field (eqns. 1.679–1.708) or by exploitation of the isomorphism.

It results that the conditional PDF, $p(\mathbf{z}_2|\mathbf{z}_1)$, for each $\mathbf{z}_2 \in C^{N-M}$, $\mathbf{z}_1 \in C^M$ is Gaussian with a mean value of

$$\bar{\mathbf{z}}_{2|1} = \bar{\mathbf{z}}_2 + M_{21}^* M_{11}^{-*} (\mathbf{z}_1 - \bar{\mathbf{z}}_1) \tag{1.742}$$

and the covariance matrix

$$M_{2|1} = M_{22} - M_{21} M_{11}^{-1} M_{12} \tag{1.743}$$

i.e.

$$p(\mathbf{z}_2|\mathbf{z}_1) = \frac{1}{\pi^N \det M_{2|1}} \exp[-(\mathbf{z}_2 - \bar{\mathbf{z}}_{2|1})^T M_{2|1}^{-1} (\mathbf{z}_2 - \bar{\mathbf{z}}_{2|1})^*] \tag{1.744}$$

1.9.3 Moments of the Gaussian complex multivariate [27]

We report in this section some important properties concerning the moments of the Gaussian complex multivariate.

Let $\mathbf{z}(\boldsymbol{t})$ be a Gaussian, zero-mean, stationary complex process. Taking two samples of $\mathbf{z}(\boldsymbol{t})$ at instants t_1 and t_2 it follows that

$$E\{\mathbf{z}(t_1)\mathbf{z}^*(t_2)\} = R(t_1 - t_2) \tag{1.745}$$

$$E\{\mathbf{z}(t_1)\mathbf{z}(t_2)\} = 0 \tag{1.746}$$

where $R(.)$ is the covariance function of the process. It can be shown moreover that, starting from these properties, expr. 1.741 of the complex multivariate follows.

Let us assume we have N samples of z taken at times t_i, $i = 1, 2, \ldots, N$. The following properties can be proved:

$$E\{z_{m_1}^* \ldots z_{m_r}^* z_{n_1} \ldots z_{n_s}\} = 0 \quad r \neq s \tag{1.747}$$

$$E\{z_{m_1}^* \ldots z_{m_r}^* z_{n_1} \ldots z_{n_s}\}$$
$$= \sum_{\pi} (E\{z_{m_{\pi(1)}}^* z_{n_1}\}) \cdot (E\{z_{m_{\pi(2)}}^* z_{n_2}\}) \ldots (E\{z_{m_{\pi(r)}}^* z_{n_s}\}) \quad r = s \tag{1.748}$$

where π is a permutation of the set of integers $[1, 2, \ldots, r]$.

For example, according to expr. 1.747,

$$E\{z_1^* z_2 z_3\} = 0 \tag{1.749}$$

since $r = 1$ and $s = 2$ are different; conversely, according to expr. 1.748 we obtain, for $r = s = 2$,

$$E\{z_1^* z_2^* z_3 z_4\} = E\{z_1^* z_3\} E\{z_2^* z_4\} + E\{z_2^* z_3\} E\{z_1^* z_4\}. \tag{1.750}$$

1.9.4 Reduction of the Gaussian multivariate to independent variates; whitening and colouring transformations

In Section 1.8.2.2 it has been shown that any Hermitian matrix can be reduced in diagonal form in many ways. This applies also to the covariance matrix M of a zero-mean complex Gaussian vector $\mathbf{z}$; the transformation

$$\mathbf{y} = U^T \mathbf{z} \tag{1.751}$$

where U is the matrix formed by the eigenvectors of M, is one means to do that; the matrix U is Hermitian, that is $U^H = U^{-1}$. Note that if the components of the transformed vector $\mathbf{y}$ are required to have unit variances, another transformation has to be used, i.e.

$$\mathbf{y} = D^{-1} U^T \mathbf{z} \tag{1.752}$$

where D is any solution of

$$\Lambda = D^* D^T \tag{1.753}$$

and where Λ is the diagonal matrix of eigenvalues of M. The commonly used solution of expr. 1.753 is

$$D = \Lambda^{1/2} \tag{1.754}$$

i.e. the square root† of Λ.

Using the former transformation (eqn. 1.751), and recalling that $M^{-1} = U\Lambda^{-1}U^H$ and that $U^T = U^{-*}$, it follows that the Hermitian form in the exponential of the complex multivariate (expr. 1.741) becomes

$$\mathbf{z}^T M^{-1} \mathbf{z}^* = (U^{-T}\mathbf{y})^T M^{-1} (U^{-T}\mathbf{y})^* = \mathbf{y}^T U^{-1} M^{-1} U \mathbf{y}^* = \mathbf{y}^T \Lambda^{-1} \mathbf{y}^* = \sum_{i=1}^{N} \frac{|y_i|^2}{\lambda_i} \tag{1.755}$$

Therefore, after the transformation, and recalling that $\det(M) = \Pi_{i=1}^N \lambda_i$ (see Sections 1.8.2.3, 1.8.2.4 and 1.8.2.5) exprs. 1.741 and 1.751 yield

$$p(\mathbf{y}) = \prod_{i=1}^{N} \frac{1}{2\pi\lambda_i} \exp\left\{ -\frac{|y_i|^2}{2\lambda_i} \right\} = \frac{1}{(2\pi)^N \prod_{i=1}^{N} \lambda_i} \exp\left\{ -\frac{1}{2} \sum_{i=1}^{N} \frac{|y_i|^2}{\lambda_i} \right\} \tag{1.756}$$

i.e. the extension to the complex case of the expr. 1.663.

Of course, using the transformation of expr. 1.752, the resulting PDF of $\boldsymbol{t}$ is again in the form of expr. 1.756, with $\lambda_i = 1$, $i = 1, 2, \ldots, N$.

Let us now consider a complex random vector $\mathbf{z}$: its covariance matrix M_z can be written:

$$M_z = E\{\mathbf{z}^* \mathbf{z}^T\} = U_z \Lambda_z U_z^H \tag{1.757}$$

where Λ_z is the diagonal matrix of the eigenvalues of M_z, and U_z is the square matrix whose columns are the eigenvectors of M_z.

† The square root of a diagonal matrix is simply another diagonal matrix whose elements are the square root of the former ones.

The transformation

$$\boxed{t=\Lambda_z^{-1/2}U_z^T z} \tag{1.758}$$

provides a vector $\boldsymbol{t}$ whose covariance matrix is the identity matrix $\boldsymbol{I}$, that is $\boldsymbol{t}$ is a *white* random process. In fact, the covariance matrix of $\boldsymbol{t}$ is

$$M_t=E\{\boldsymbol{t}^*\boldsymbol{t}^T\}=\Lambda^{-1/2}U_z^H E\{\boldsymbol{z}^*\boldsymbol{z}^T\}U_z\Lambda^{-1/2} \tag{1.759}$$

and, recalling expr. 1.757, and that $U^H=U^{-1}$, we have

$$M_z=\Lambda_z^{-1/2}U_z^H U_z\Lambda_z U_z^H U_z\Lambda_z^{-1/2}=I \tag{1.760}$$

The transformation in expr. 1.758 is called 'whitening process'. If the vector $\boldsymbol{z}$ is Gaussian, the components of $\boldsymbol{t}$ are also independent (see expr. 1.755) and have equal power.

Conversely, applying the inverse transformation

$$\boxed{z=U_z^*\Lambda_z^{1/2}t} \tag{1.761}$$

to the white vector $\boldsymbol{t}$ ($M_t=I$), one obtains a vector $\boldsymbol{z}$ whose covariance matrix is

$$M_z=U_z\Lambda_z U_z^H \tag{1.762}$$

this last operation is called the 'colouring process'.

It is important to note that the whitening process does not require the vector $\boldsymbol{z}$ to be Gaussian (this only implies the independence of the components of the transformed vector).

These properties are widely used to generate, by means of a computer, random numbers with an assigned covariance matrix.

1.9.5 Characteristic function of a Hermitian form [24]

Let us consider the characteristic function $\Phi_q(j\omega)\equiv E[j\omega q]$ of the Hermitian form

$$q=\boldsymbol{z}^H Q\boldsymbol{z} \tag{1.763}$$

where $\boldsymbol{z}$ is a complex Gaussian random vector with zero mean and covariance matrix M_z, and Q is a Hermitian matrix.

The problem (which is relevant in many applications) of evaluating the PDF $p(q)$ of the Hermitian form can be solved by Fourier inversion of the characteristic function of q:

$$p(q)=\frac{1}{2\pi}\int_{-\infty}^{\infty}\Phi_q(j\omega)e^{-j\omega q}d\omega \tag{1.764}$$

The evaluation of $\Phi_q(j\omega)$ [24] can be obtained by a reduction of q to principal axes [9] (see Appendix 2, Section 1.8.3.4). In other words, q is rewritten in terms of the independent zero-mean Gaussian variables t_i, i.e. of the vector $\boldsymbol{t}$, by the use of the *non-unitary transformation* (see exprs. 1.758 and 1.759):

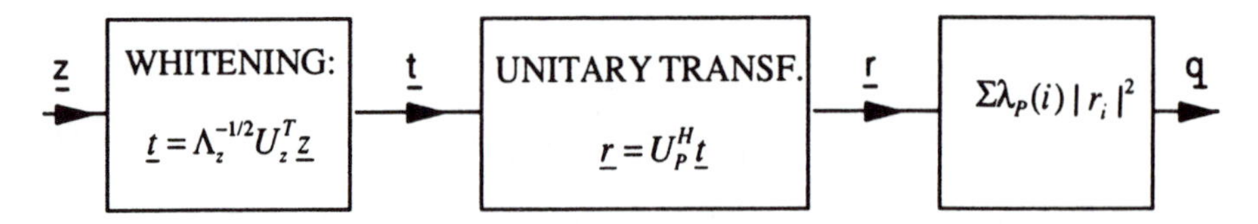

Figure 1.41 *Overall transformation of a complex Gaussian vector*

$$\begin{cases} \boldsymbol{t}=\Lambda_z^{-1/2}U_z^T\boldsymbol{z} \\ \boldsymbol{z}=U_z^*\Lambda^{1/2}\boldsymbol{t} \end{cases} \quad (1.765a, b)$$

(where Λ_z and U_z are the matrices containing the eigenvalues and eigenvectors of M_z, respectively, and where the components of $\boldsymbol{t}$ are independent with *unitary power* ($M_t=I$)) followed by the unitary transformation:

$$\begin{cases} \boldsymbol{r}=U_P^H\boldsymbol{t} \\ \boldsymbol{t}=U_P\boldsymbol{r} \end{cases} \quad (1.766a, b)$$

where U_P is the unitary matrix of the (normalised) eigenvectors of the Hermitian matrix:

$$P=\Lambda_z^{1/2}U_z^T Q U_z^* \Lambda_z^{1/2} \quad (1.767)$$

The resulting expression of q is readily obtained by substitution of expr. 1.765*b* in expr. 1.763, with P given by expr. 1.767.

$$q=\boldsymbol{t}^H P\boldsymbol{t} \quad (1.768)$$

that is, using expr. 1.766*b*:

$$q=\boldsymbol{r}^H U_P^H P U_P \boldsymbol{r} \quad (1.769)$$

i.e.

$$q=\boldsymbol{r}^H\Lambda_P\boldsymbol{r}=\sum_{i=1}^{N}\lambda_P(i)|r_i|^2 \quad (1.770)$$

where $\lambda_P(i)$ are the eigenvalues of P. Generally they are different from those of Q, because P is not similar to Q (Appendix 2, expr. 1.555) due to the term $\Lambda_z^{1/2}$ in expr. 1.767.

The covariance matrix of $\boldsymbol{r}$ is (see expr. 1.766)

$$M_r=U_P^T E\{\boldsymbol{t}^*\boldsymbol{t}^T\}U_P^*=U_P^T M_t U_P^*=U_P^T I U_P^*=I \quad (1.771)$$

i.e. the components of $\boldsymbol{r}$ are independent and with unitary variance.

The overall transformation (see exprs. 1.765 and 1.766)

$$\boldsymbol{r}=U_P^H\Lambda_z^{-1/2}U_z^T\boldsymbol{z}=R\boldsymbol{z} \quad (1.772)$$

corresponds to the block diagram of Fig. 1.41, where the computation of the Hermitian form, according to expr. 1.770, is also shown†.

† The matrix R in expr. 1.772 can be written as follows:

$$R=\sum_{i=1}^{N}\lambda_z^{-1/2}(i)\boldsymbol{v}(i)\boldsymbol{u}_z^T(i)$$

where $\boldsymbol{v}(i)$ is the ith column of the unitary matrix U_P^H (note that the set $\boldsymbol{v}(i)$, $i=1, 2, \ldots, N$ is orthogonal to the set $u_p(i)$, $i=1, 2, \ldots, N$).

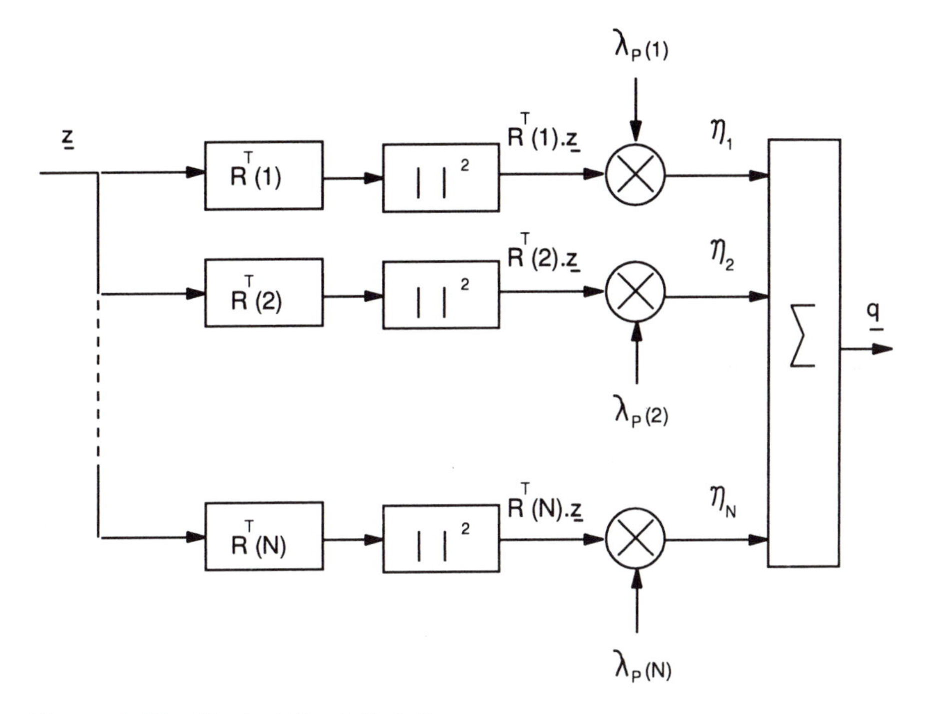

Figure 1.42 *'Bank of filters' block diagram*

The Hermitian form q (exprs. 1.770, 1.772) can be written as a sum of squared moduli of dot products between the ith row $\boldsymbol{R}^T(i)$ of R and the vector $\boldsymbol{z}$:

$$q = \sum_{i=1}^{N} \lambda_P(i) |\boldsymbol{R}^T(i) \cdot \boldsymbol{z}|^2 \tag{1.773}$$

leading to the block diagram of Fig. 1.42, showing a 'bank' of N filters (i.e. convolutors of the input signal $\boldsymbol{z}$ with $\boldsymbol{R}^T(i)$, $i = 1, 2, \ldots, N$), each followed by a multiplier, a squared modulus and final adder. The N squared moduli y_i to be added are independent random variables†.

Let us now consider the statistics of q in terms of the variates y_i defined by (see expr. 1.770):

† Note that the value of this diagram is conceptual only, as it requires a knowledge of the covariance matrix of the observation $\boldsymbol{z}$ (of course, depending on the hypothesis to be tested); a practical way to write q as a weighted sum of squared moduli is by the transformation:

$$\boldsymbol{s} = U_Q^H \boldsymbol{z}$$
$$\boldsymbol{z} = U_Q \boldsymbol{s}$$

resulting in (see also Fig. 1.9 in Section 1.7):

$$q = \boldsymbol{z}^H Q \boldsymbol{z} = \boldsymbol{s}^H U_Q^H Q U_Q \boldsymbol{s} = \boldsymbol{s}^H \Lambda_Q \boldsymbol{s} = \sum_{i=1}^{N} \lambda_Q(i) |s_i|^2$$

where, however, the components s_i of $\boldsymbol{s}$ are not independent.

$$y_i = \lambda_P^{1/2}(i) r_i, \quad i = 1, 2, \ldots, N \tag{1.774}$$

The PDF of y_i is (see Section 1.9.4):

$$p_i(y_i) = \frac{1}{2\pi\lambda_P(i)} \exp\left\{-\frac{|y_i|^2}{2\lambda_P(i)}\right\} \tag{1.775}$$

By using the polar representation of y_i:

$$y_i = \rho_i \exp\{j\phi_i\}, \quad \rho > 0 \tag{1.776}$$

expr. 1.775 can be written

$$p_i(\rho_i, \phi_i) = \frac{\rho_i}{2\pi\lambda_P(i)} \exp\left\{-\frac{\rho_i^2}{2\lambda_P(i)}\right\} \tag{1.777}$$

Recalling that the phase of a narrowband Gaussian process is uniformly distributed in $(0, 2\pi]$ and independent of the modulus (see Section 1.7.7), the PDF of ρ_i becomes

$$p(\rho_i) = \frac{\rho_i}{\lambda_P(i)} \exp\left\{-\frac{\rho_i^2}{2\lambda_P(i)}\right\}, \quad \rho_i > 0 \tag{1.778}$$

and that of $\eta_i = \rho_i^2$ follows:

$$p(\eta_i) = \frac{1}{2\lambda_P(i)} \exp\left\{-\frac{\eta_i}{2\lambda_P(i)}\right\}, \quad \eta_i \geqslant 0. \tag{1.779}$$

The characteristic function of q is readily obtained considering that q is a weighted sum of the exponentially-distributed independent random variables η_i:

$$\Phi_q(j\omega) = \prod_{i=1}^{N} E\{e^{j\omega\lambda_P(i)\eta_i}\} = \prod_{i=1}^{N} \frac{1}{1 - 2j\omega\lambda_P(i)} \tag{1.780}$$

When Q is positive semidefinite, it is convenient to define a Laplace-transform characteristic function

$$G_q(s) = \int_0^\infty e^{-sq} p(q)\, dq \tag{1.781}$$

which, in the present instance, yields

$$G_q(s) = \prod_{i=1}^{N} \frac{1}{1 + 2s\lambda_P(i)} \tag{1.782}$$

With regard to the eigenvalues $\lambda_P(i)$ of P, it can be shown that they are also the eigenvalues of $M_z^* Q$†. Therefore, exprs. 1.780 and 1.782 can be written [24]

$$\Phi_q(j\omega) = \frac{1}{\det(I - 2j\omega M_z^* Q)} \tag{1.783}$$

$$G_q(s) = \frac{1}{\det(I + 2sM_z^* Q)} \tag{1.784}$$

1.10 Appendix 4: The Ricean distribution and the Q function

1.10.1 Introduction—the Ricean distribution

This appendix recalls the basic statistical properties of the envelope of a sine wave with additive Gaussian noise. The distributions of the envelope of a Gaussian process are useful in evaluating the radar detection probability of non-fluctuating targets in radar and the bit error probability in incoherent digital communication systems.

Let us consider a sine wave with a carrier frequency f_c and an additive Gaussian noise whose spectrum is allocated in a narrow band around f_c. The case of noise alone is first described.

The real and the imaginary parts, X_N and Y_N, of the complex envelope of the narrow band Gaussian noise [12] are independent and equally distributed; their joint density function is therefore

$$f_{X_N Y_N}(x, y)\mathrm{d}x\ \mathrm{d}y = \frac{1}{2\pi\sigma^2} \exp\left\{-\frac{x^2 + y^2}{2\sigma^2}\right\} \mathrm{d}x\ \mathrm{d}y \tag{1.785}$$

where σ^2 is the variance (power) of the noise.

The joint PDF of the envelope $R = \sqrt{X_N^2 + Y_N^2}$ and of the argument $\Theta = \mathrm{tg}^{-1}(Y_N/X_N)$ is readily obtained recalling that the Jacobian of the transformation from X_N, Y_N to R, Θ is equal to R. Therefore

$$f_{R\Theta}(r, \theta) = r f_{X_N Y_N}(r \cos\theta, r \sin\theta) \tag{1.786}$$

From eqns. 1.785 and 1.786

$$f_{R\Theta}(r, \theta)\ \mathrm{d}r\ \mathrm{d}\theta = \frac{r}{2\pi\sigma^2} \exp\left\{-\frac{r^2}{2\sigma^2}\right\} \mathrm{d}r\ \mathrm{d}\theta \tag{1.787}$$

By integration of eqn. 1.787 w.r.t. the envelope r the marginal PDF of the argument is obtained:

$$f_\Theta(\theta) = \frac{1}{2\pi} \int_0^\infty -d\left[\exp\left(-\frac{r^2}{2\sigma^2}\right)\right] = \frac{1}{2\pi} \tag{1.788}$$

† We recall that two similar matrices (see Appendix 2, Section 1.8. 3.2) have the same eigenvalues. In this case P and $M_z^* Q$ are similar, since

$$(U_z^* \Lambda_z^{1/2}) P (U_z^* \Lambda_z^{1/2})^{-1} = U_z^* \Lambda_z U_z^T Q = M_z^* Q$$

Therefore, the argument is uniformly distributed and independent of the envelope.

The PDF of the envelope results from expressions 1.787 and 1.788, namely

$$f_R(r) = \frac{r}{\sigma^2} \exp\left\{-\frac{r^2}{2\sigma^2}\right\} U(r) \tag{1.789}$$

where $U(\cdot)$ is the unit step; $f_R(r)$ is the well known Rayleigh density function.

From eqn. 1.789 the PDF of the instantaneous power $W = R^2$ is readily obtained:

$$f_w(w) = \frac{1}{2\sigma^2} \exp\left\{-\frac{w}{2\sigma^2}\right\} U(w) \tag{1.790}$$

It is an exponential PDF with a mean value equal to $2\sigma^2$.

Let us now consider a deterministic signal $s(t)$

$$s(t) = A(t) \cos(2\pi f_c t + \phi(t)) \tag{1.791}$$

whose complex envelope, dropping the time dependance for the sake of convenience, can be simply written as follows:

$$X_s + jY_s = A \cos \phi + jA \sin \phi \tag{1.792}$$

Therefore, the complex envelope of the received signal (deterministic signal plus noise) has the following components:

$$\begin{cases} X = X_N + X_s \\ Y = Y_N + Y_s \end{cases} \tag{1.793}$$

where X_N and Y_N are distributed according to eqn. 1.785. The resulting PDF of X and Y is

$$f_{XY}(x, y) = \frac{1}{2\pi\sigma^2} \exp\left(-\frac{(x - X_s)^2 + (y - Y_s)^2}{2\sigma^2}\right) \tag{1.794}$$

Again, with a transformation from rectangular to polar co-ordinates the joint PDF of the envelope R and of the argument Θ is obtained. From eqns. 1.785, 1.786, 1.792 and 1.794 the following result is readily obtained:

$$f_{R\Theta}(r, \theta) = \frac{r}{2\pi\sigma^2} \exp\left(-\frac{Z^2}{2\sigma^2}\right) U(r) \tag{1.795}$$

where

$$Z^2 = r^2 + A^2 - 2Ar \cos \theta \cos \phi - 2Ar \sin \theta \sin \phi \tag{1.796}$$

i.e.

$$Z^2 = r^2 + A^2 - 2Ar \cos(\theta - \phi) \tag{1.797}$$

From eqns. 1.795 and 1.797 the joint PDF is:

$$f_{R\Theta}(r, \theta) = \frac{r}{2\pi\sigma^2} \exp\left(-\frac{r^2 + A^2}{2\sigma^2}\right) \exp\left[\frac{Ar}{\sigma^2} \cos(\theta - \phi)\right] U(r) \tag{1.798}$$

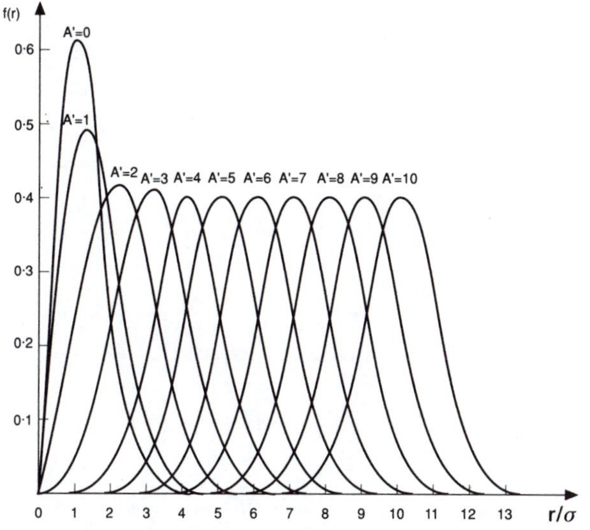

Figure 1.43 *Ricean probability density function $f(r)$ for various values of the normalized peak amplitude $A' = A/\sigma$ of the useful signal.*

Note that the envelope and the argument are not independent.

Integrating the joint PDF with respect to θ and using the identity

$$\frac{1}{2\pi}\int_0^{2\pi} \exp(x \cos \alpha)\, d\alpha = I_0(x) \tag{1.799}$$

the PDF of the envelope (Ricean distribution, after S. O. Rice) is given by

$$f_R(r) = \frac{r}{\sigma^2} \exp\left(-\frac{r^2 + A^2}{2\sigma^2}\right) I_0\left(\frac{A\,r}{\sigma^2}\right) U(r) \tag{1.800}$$

where I_0 is the modified Bessel function of the first kind, order zero, whose series expansion is

$$I_0(x) = \sum_{k=0}^{\infty} \frac{(x^2/2)^k}{(k!)^2} \tag{1.801}$$

The Ricean PDF (which reduces to the Rayleigh one for $A = 0$) is plotted in Fig. 1.43 for various values of the normalised envelope $A' = A/\sigma$ (the square of A' is twice the signal-to-noise ratio, SNR).

For large values of A' and of $v = r/\sigma$, the following approximation holds:

$$f_R(r) \cong \frac{1}{\sigma\sqrt{2\pi}} \exp\left[-\frac{(r-A)^2}{2\sigma^2}\right] \tag{1.802}$$

i.e. the envelope is approximately Gaussian.

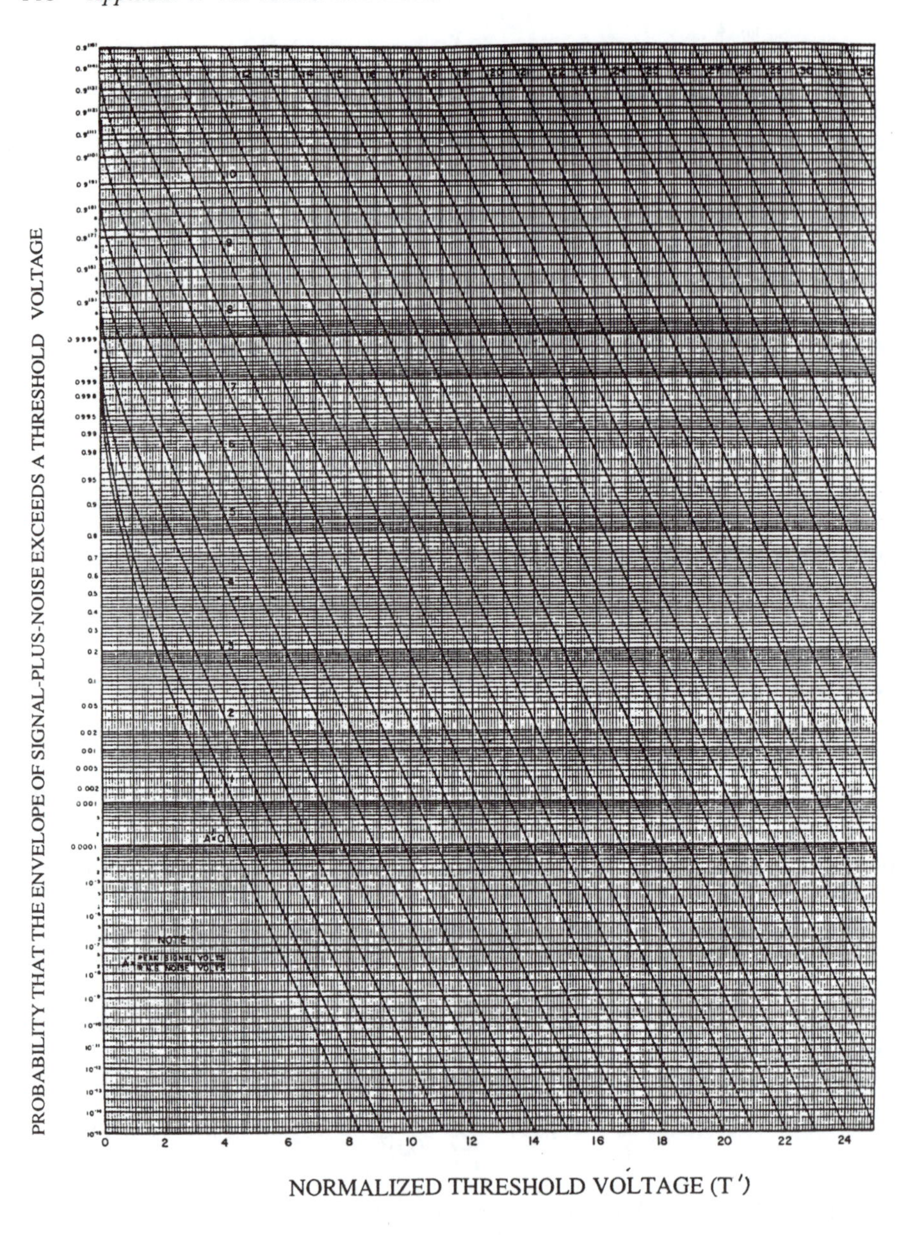

Figure 1.44 *Probability that the Ricean variate exceeds the normalized threshold T' for various values of the normalized useful signal amplitude A'*

Eqn. 1.802 is readily obtained neglecting the noise component which is in quadrature with respect to the useful signal; as a matter of fact, the effect of that component on the envelope decreases as the SNR increases.

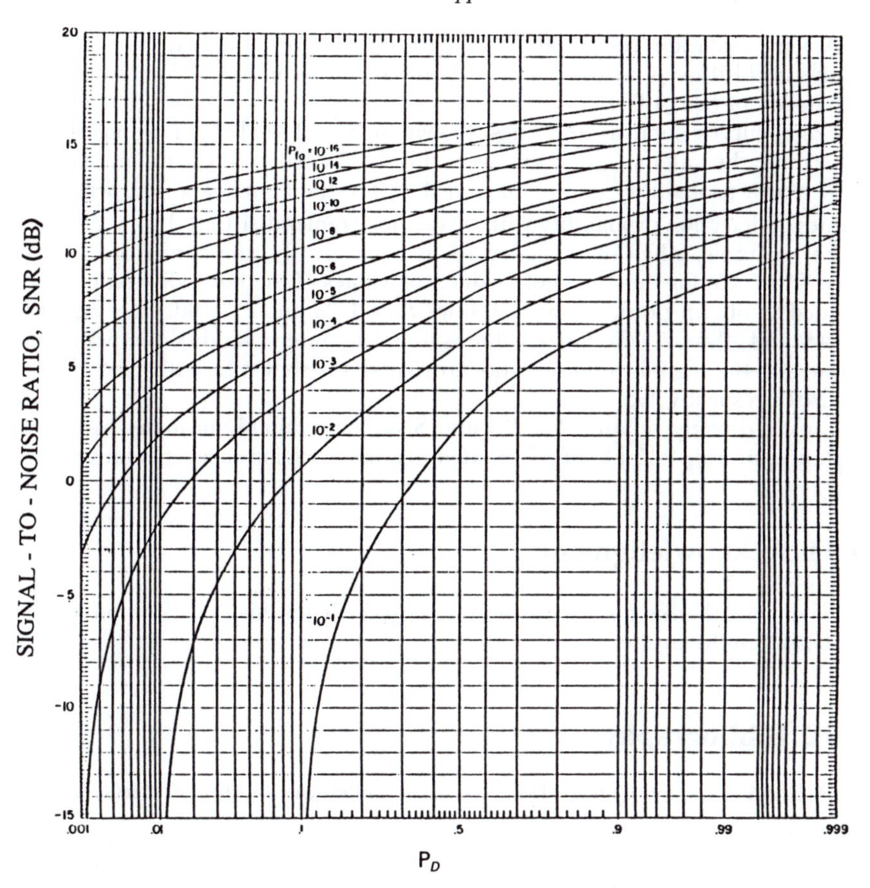

Figure 1.45 *Detection probability for a non fluctuating signal*

Using the normalised quantities A' and v, the Ricean PDF can be written as follows:

$$f_V(v) = v \exp\left(-\frac{v^2 + A'^2}{2}\right) I_0(A'v)\, U(v) \tag{1.803}$$

1.10.2 The Q function

The probability that the Ricean variate exceeds a threshold T is obtained integrating the Ricean PDF from T to infinity; using the normalised threshold $T' = T/\sigma$ one gets

$$\text{Prob}(r > T) = Q(A', T') = \int_{T'}^{\infty} v \exp\left(-\frac{v^2 + A'^2}{2}\right) I_0(A'v)\, \mathrm{d}v \tag{1.804}$$

No closed form exists for eqn. 1.804; it is called the Q function or the Marcum function [3, 4] and often denoted $Q(\alpha, \beta)$ where $\alpha = A' = A/\sigma$ and $\beta = T/\sigma$.

A very accurate computation of $Q(\alpha, \beta)$ can be obtained by computer programs [40] using algorithms based on term-to-term integration of the series expansion of eqn. 1.801.

From curves such as the ones shown in Fig. 1.44 [41] it is possible to determine the value of the normalised threshold needed to obtain an assigned false alarm probability (curve marked $A'=0$ in the figure) and the pertaining values of detection probability with A' varying (remember that $A'=\sqrt{2\cdot \mathrm{SNR}}$).

For example, when the assigned P_{FA} is equal to 10^{-6} the curves of Fig. 1.44 give $T'=5.25$; for this threshold value, the values of the normalised amplitude needed to obtain detection probabilities $P_D=80\%$ and $P_D=90\%$ are equal to $A'=6$ and $A'=6.5$ (the latter value is found by interpolation); the pertaining values of the signal-to-noise ratio are equal to SNR = 12.5 dB and SNR = 13.2 dB, respectively.

Curves directly relating P_D, P_{FA} and SNR are available in many books; for instance, in Fig. 1.45 [42] P_D is plotted against SNR for various values of P_{FA}.

1.11 Acknowledgment

A sincere thanks goes to Prof. Luigi Accardi for his contribution on the Gaussian multivariate.

1.12 Bibliography

1 MELSA, J.L., and COHN, D.L.: 'Decision and estimation theory' (McGraw–Hill, 1978)
2 VAN DER WAERDEN, B.L.: 'Mathematical statistics' (Springer–Verlag, 1969)
3 MARCUM, J.I.: 'A statistical theory of target detection by pulsed radar.' Rand Research Memorandum RM–754, December 1947
4 MARCUM, J.I.: 'Tables of Q function'. US Air Force Project Rand Research Memorandum M–339, January 1950
5 LEVINE, B.: 'Fondaments théorique de la radiotechnique statistique' (MIR, 1979)
6 VAN TREES, H.L.: 'Detection, estimation and modulation theory: Vol. 1, 2, 3' (John Wiley and Sons, 1968)
7 DI FRANCO, J.V., and RUBIN, W.L.: 'Radar detection' (Prentice Hall, Englewood Cliffs, 1968)
8 SCHWARTZ, M.: 'Information transmission, modulation and noise' (McGraw–Hill, 1989)
9 GANTMACHER, F.R.: 'The theory of matrices' (Chelsea Publishing Company, 1977)
10 MONZINGO, R.A., and MILLER, T.W.: 'Introduction to adaptive arrays' (John Wiley and Sons, 1980)
11 GOODMAN, N.R.: 'Statistical analysis based on a certain multivariate complex Gaussian distribution', *Ann. Math. Stat.*, 1963, **34**, pp. 152–177
12 DAVENPORT, W.B., Jr., and ROOT, W.L.: 'An introduction to the theory of random signals and noise' (McGraw–Hill, 1958)
13 PAPOULIS, A.: 'Probability, random variables and stochastic processes' (McGraw–Hill, 1984)
14 PAPOULIS, A.: 'Signal analysis' (McGraw–Hill, 1986)
15 WILKINSON, J.H.: 'The algebraic eigenvalue problem' (Oxford University Press, London, 1965)

16 GRENANDER, U., and SZEGÖ, G.: 'Toeplitz forms and their applications' (Chelsea Publishing Co., 1984)
17 BELLANGER, M.: 'Digital processing of signals: Theory and practice' (John Wiley and Sons, 1984)
18 KUNT, M.: 'Traitement numerique des signaux' *in* 'Traité d'electricité, Vol. XX' (Georgi, 1980)
19 OPPENHEIMER, A.V., and SCHAFER, R.W.: 'Digital signal processing' (Prentice–Hall, 1975)
20 PAPOULIS, A.: 'The Fourier integral and its applications' (McGraw–Hill, 1962)
21 HARRIS, F.J.: 'On the use of windows for the harmonic analysis with the discrete Fourier transform', *Proc. IEEE,* 1978, **66**, (1), pp. 51–83
22 WOODWARD, P.M.: 'Probability and information theory, with applications to radar' (McGraw–Hill, 1953)
23 LEVANON, N.: 'Radar principles' (Wiley and Sons, 1988)
24 TURIN, G.L.: 'The characteristic function of Hermitian quadratic forms in complex normal variables', *Biometrika,* 1960, **47**, pp. 199–201
25 BELLMAN, R.: 'Introduction to matrix analysis' (McGraw–Hill, 1970)
26 ACCARDI, L.: 'Calcolo delle probabilità' (Università di Tor Vergata, Rome, 1991)
27 REED, I.S.: 'On a moment theorem for complex Gaussian processes', *IRE Trans.*, 1962, **IT–8**, pp 194–195
28 STRANG, G.: 'Linear algebra and its applications' (Academic Press, 1976)
29 CHIUPPESI, F., GALATI, G., and LOMBARDI, P.: 'Optimization of rejection filters' *IEE Proc.* Pt. F, 1980, **127**, (5)
30 D'ADDIO, E., and GALATI, G.: 'Adaptivity and design criteria of a latest-generation MTD processor' *IEE Proc.,* Pt. F, 1985, **132**, (1), pp. 58–65
31 BARBAROSSA, S., D'ADDIO, E., and GALATI, G.: 'Comparison of optimum and linear prediction techniques for clutter cancellation' *IEE Proc.*, Pt. F, 1987, **134**, (3), pp. 278–282
32 GALATI, G., and STUDER, F.A.: 'Maximum likelihood azimuth estimation applied to SSR/IFF systems' *IEEE Trans.* 1990, **AES–26** (1), pp. 28–43
33 BLONDY, P.: 'Methode de calcul numerique de probabilité de detection de signaux fluctuants' *in* 'New devices, techniques and systems in radar', AGARD CP 197, The Hague, 1976, pp. 31.1–31.9
34 BARRET, M.J.: 'Error probability for optimal and suboptimal quadratic receivers in rapid Rayleigh fading channels' *IEEE J.* 1987, **SAC–5** (2), pp. 302–304
35 BRANDWOOD, D.H.: 'A complex gradient operator and its application in adaptive array theory' IEE Proc. 130 Parts F and H, Feb. 1983, pp. 11–16
36 ALOISIO, V. DI VITO, A., and GALATI, G.: 'Optimisation of Doppler filters for fluctuating targets', *IEICE Trans.*, 1992, **E75-B**, (10), pp. 1090–1104
37 HSIAO, J.K.: 'On the optimisation of MTI clutter rejection', *IEEE Trans.*, 1974, **AES-10** (5), pp. 622–629
38 ANDREWS, L.C.: 'Special functions for engineers and applied mathematicians' (MacMillan Publishing Company, 1985)
39 ALOISIO, V., DI VITO, A., and GALATI, G.: 'An optimal detector for moderately fluctuating targets'. IEE Int. Conf. Radar '92, Brighton, 12–13 October 1992, pp. 110–113
40 SHNIDMAN, D.A.: 'Evaluation of the Q function'. IEEE., 1974, **COM-22**, (3), pp. 342–346
41 GREENFIELD, S.E. and MARINO, P.F.: 'Statistical comparison of digital detection techniques: Technical Report 101' (Sylvania Electronic Systems, East 189 B Street, Needham Heights 94, Massachusetts, 1961 (revised))
42 BLAKE, L.V.: 'A guide to basic pulse-radar maximum-range calculation' (Naval Research Laboratory Report 5868, 1962)
43 KRETSCHMER, F. F. and FENG LING, C. L.: 'Effects of the main tap position in adaptive clutter processing'. IEEE 1985 Int. Radar Conf., Arlington, 6–9 May 1985, pp. 303–307

Chapter 2

Models of clutter

Prof. T. Musha and Prof. M. Sekine

2.1 Statistical properties of clutter

Statistical properties of clutter amplitudes have been investigated by means of radars with a rotating as well as a stationary antenna which is directed in a desired direction. These properties are summarised in this Chapter based on our observations. We used an L-band, long-range, air-route-surveillance radar (ARSR) of frequency 1.3 GHz, beam width 1.2°, pulse width 3.0 μs, pulse repetition rate 350 Hz, antenna scan rate 6.0 r.p.m. when it is rotating, elevation angle 2.0°, and transmitted power 2 MW. Return signals were recorded on magnetic tape as video signals of inphase (I) and quadrature (Q) components. The amplitude of the return signal x is expressed as $x=(I^2+Q^2)^{1/2}$. Sample intervals between adjacent range bins and range sweeps are 0.13 n.m.i. and 0.1044°, respectively. Before numerical analysis, signals were converted into 10 bit numbers, and hence values of I and Q range from -512 to $+511$, and the signal-to-noise ratio of the IF amplifier was 30 dB; horizontal polarisation was used.

2.1.1 *Ground clutter*

The observation areas are illustrated in Fig. 2.1, where the ARSR for the Narita airport is located at Yamada which is marked with a cross.

2.1.1.1 *Power spectral density and amplitude distribution of ground clutter measured with a stationary antenna*

The area ranging between 26 and 31.6 nm which is indicated by 1 in Fig. 2.1 was covered with a stationary antenna. The power spectral density (PSD) of the echo amplitude fluctuations is shown in Fig. 2.2 where it is proportional to $1/f^2$ (f is frequency); when random fluctuations are successively added and the total displacement is sampled at equal sampling time, the sequence of displacements has this type of PSD. The sampling time is 2.8 ms which is equal to the pulse repetition time. This is easily understood in the following way. The time rate of the displacement $x(t)$ is subject to random variation:

$$dx(t)/dt=r(t) \tag{2.1}$$

where $r(t)$ is a random variable with white PSD. The absolute value of the Fourier transform of x, $|X(f)|$, is inversely proportional to f; hence the PSD of x is proportional to $1/f^2$. Therefore, the clutter amplitude is interpreted as a sum

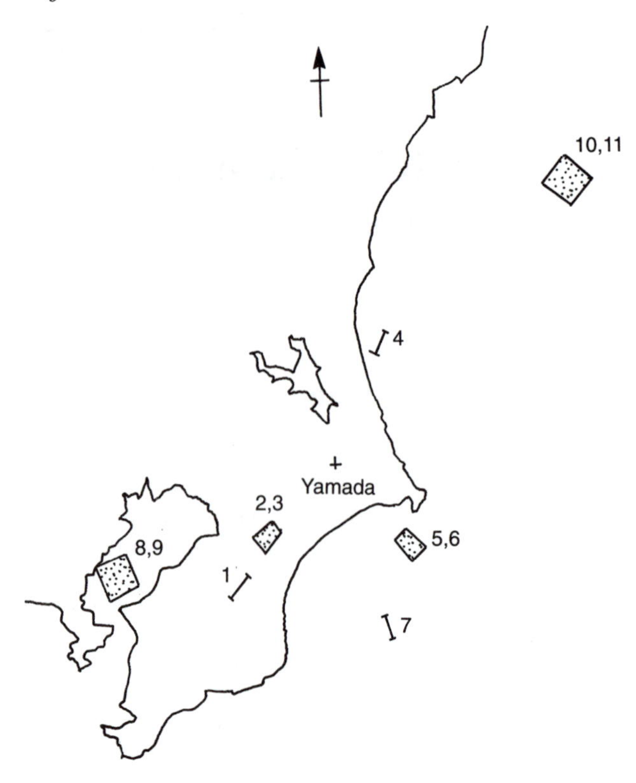

Figure 2.1 *Area in the vicinity of Tokyo*

of random variables. The amplitude distribution after having passed a log-amplifier is Gaussian as shown in Fig. 2.4, namely it is log-normal or sometimes Weibull [1, 2, 3].

2.1.1.2 *Power spectral density measured with a rotating antenna*

The PSD of return signals from the area ranging from 16 to 21.6 nm spreading over 256 range sweeps in area 2 in Fig. 2.1 is shown in Fig. 2.3. The horizontal scale is the angular wave number K in $\deg^{-1}$ (reciprocal of wave number in degrees). For low wave numbers (larger wavelengths), the PSD is proportional to K^{-2} which suggests that for wavelengths large compared with the beam width, echoes are the sum of random signals just as we observed in Fig. 2.4. For short wavelengths (smaller than 1.5°), however, the PSD becomes proportional to $K^{-3.5}$. This critical angle is close to the beam width (1.2°) which gives a spatial resolution; wavelengths shorter than 1.2° fall within a beam width, and hence fluctuations are strongly correlated to each other.

2.1.1.3 *PSD in the scan-to-scan cancellation*

All the echo signals of two successive scans coming from area 3 were stored on magnetic tape, and amplitudes in the second scan were subtracted from the

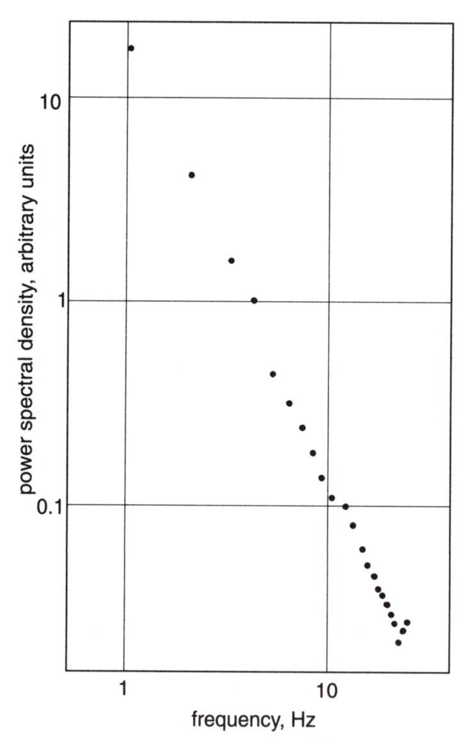

Figure 2.2 *Power spectral density of clutter amplitude in area 1*

corresponding echoes in the first scan. The PSD along the range sweep is plotted in Fig. 2.5. It is approximately proportional to $1/K$. The resolution of the antenna has no influence on the PSD for wavelengths shorter than the beam width, unlike the case of a single scan. In the scan-to-scan cancellation, the common clutter pattern inherent in the strucure of the reflectors is eliminated, and the fluctuation components across 10 s (the antenna rotation period) are picked up. The $1/K$ dependence of the PSD reflects a weaker mutual correlation than that of $1/K^2$ dependence. There are no theoretical bases for the mechanism of $1/f$ or $1/K$ PSD. The amplitude distribution is plotted in Fig. 2.6, in which the scan-to-scan amplitude x_s is defined as

$$x_s = \{(I_1 - I_2)^2 + (Q_1 - Q_2)^2\}^{1/2} \quad (2.1)$$

Since the horizontal scale is linear, it is seen that the scan-to-scan amplitude distribution of the ground clutter obeys a log-normal distribution.

2.1.2 *Sea clutter*

2.1.2.1 *Rotating antenna*

The wind speed was 7 ~ 14 knots and the sea wave of the observed area was recognised on the PPI. Range bins between 20 and 25.6 nm in area 4 were examined; echoes were observed with a stationary antenna. The PSD of echoes did not show any definite results.

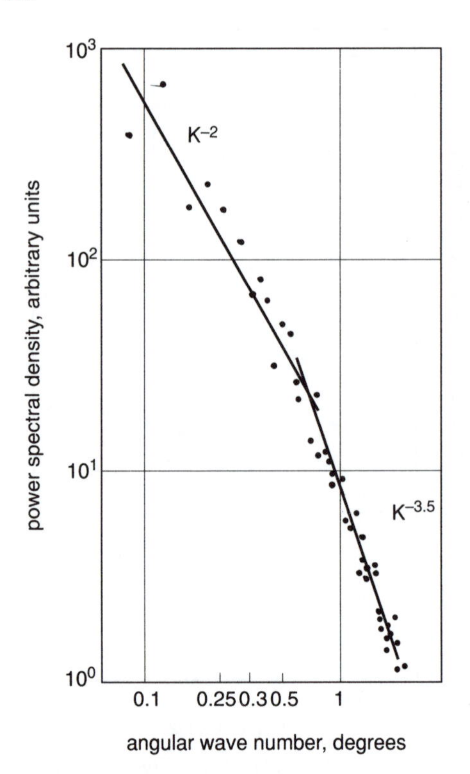

Figure 2.3 *PSD along the range sweep of area 1*

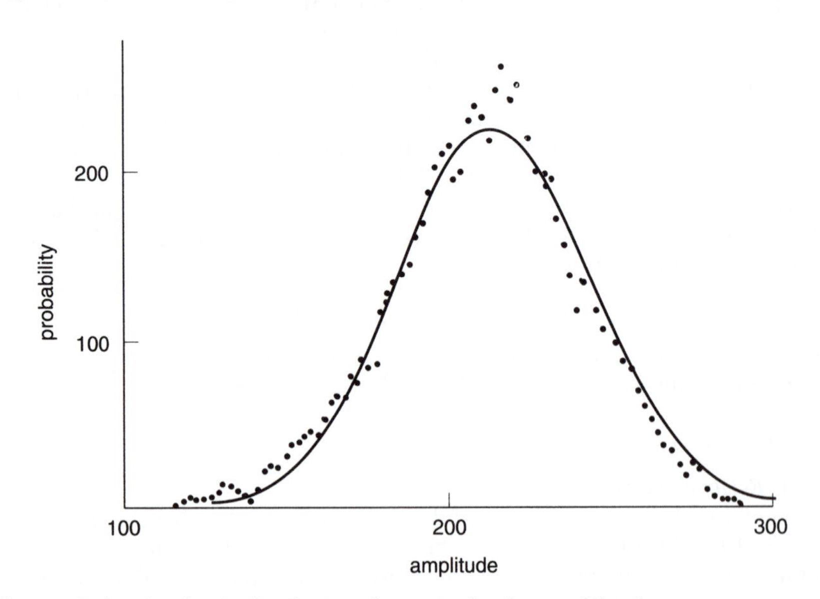

Figure 2.4 *Amplitude distribution of area 1 after log-amplification*

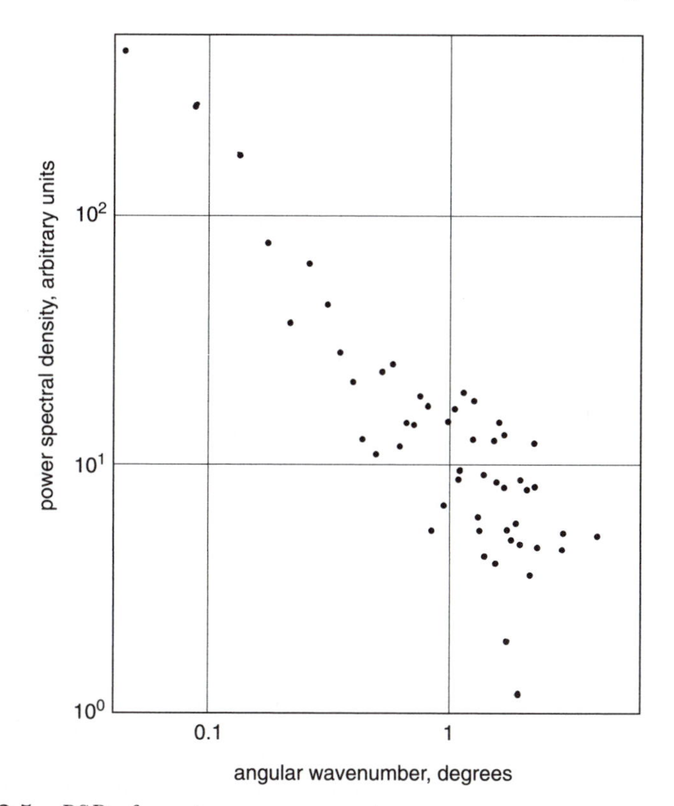

Figure 2.5 *PSD of scan-to-scan range sweep*

The PSD of 256 range sweeps between 130° and 155° (area 5) was estimated by averaging over 45 range bins between 16 and 21.6 nm; the result is plotted in Fig. 2.7. This is proportional to $1/K^2$, which is essentially that of random walks. The amplitude distribution is shown in Fig. 2.8; this is well approximated by a Rayleigh distribution for small amplitudes and by a log-normal distribution for large amplitudes. The small amplitude part is enhanced by amplifier noise.

The scan-to-scan PSD is proportional to $1/K$ and is almost identical to that in Fig. 2.5, and the amplitude distribution obeys a log-normal distribution as shown in Fig. 2.9. The $1/K$ spectrum is peculiar to the ground and sea clutter after scan-to-scan processes.

2.1.3 *Weather clutter*

The weather clutter was examined under conditions of storm and wind with wind speeds of 4 ~ 14 knots.

2.1.3.1 *Rotating antenna*

A range interval between 25 and 30.6 nm to 164° from the north (area 7) was examined. No definite form of the PSD was obtained, and the amplitude distribution was Gaussian around $x = 200$.

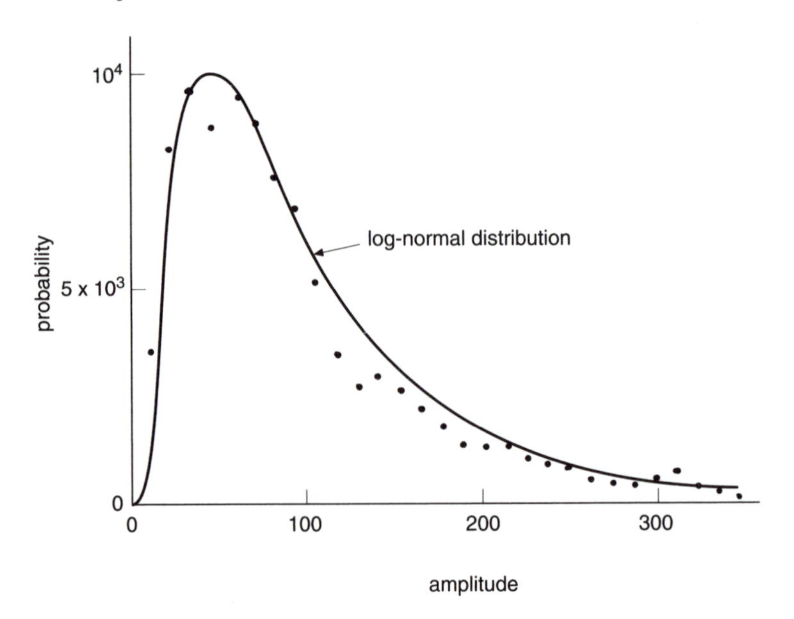

Figure 2.6 *Scan-to-scan amplitude distribution*

Weather clutter was analysed in a range interval between 42 and 47.6 nm with an azimuthal interval between 250.7 and 261.1° above the Tokyo Bay (areas 8 and 9). We have obtained PSD of $1/K$ with and without scan-to-scan cancellation; in these cases the amplitude distributions were of Weibull type with shape parameter $c = 1.8$. For a range interval between 60 and 65.6 nm and an azimuthal interval between 47.0 and 57.4° (areas 10 and 11), we also found $1/K$ PSD with and without scan-to-scan cancellation. The amplitude distribution was of the Rayleigh type; or, more generally, it should be classified as a Weibull distribution [4, 5, 6, 7, 8].

2.1.4 *Atmospheric fluctuations*

The PSD of echo signals reflected by a distant object was derived. The distant object we examined was Mt. Fuji about 97 nm west of the ARSR. No clouds were observed, and the antenna was directed towards Mt. Fuji. The PSD is shown in Fig. 2.10. This is Lorentzian where the turnover frequency is 5 Hz. Turbulence in a viscous fluid takes place when the Reynolds number R exceeds 1000; $R = L^2/\nu\tau$ where L is a length of interest (turbulence size *et al.*), ν is a kinematic viscosity, and τ is a time constant associated with turbulence [9]. For air at temperature 20°C, we have $\nu = 0.15$ cm^2/s. Let L be equal to half the wavelength, 11.5 cm, and we obtain $1/\tau = 1000 \times 0.15/11.5^2 = 1.13$/s. It is probable that the PSD of Fig. 2.10 reflects air turbulence at high altitude because the top of Mt. Fuji as a reflector of the radar beam in the present case was covered with snow and there were no plants which may cause fluctuations in the reflection coefficient.

The present results are summarised in the following way. The return signals after scan-to-scan cancellation for any type of clutter obey the $1/K$ or $1/f$

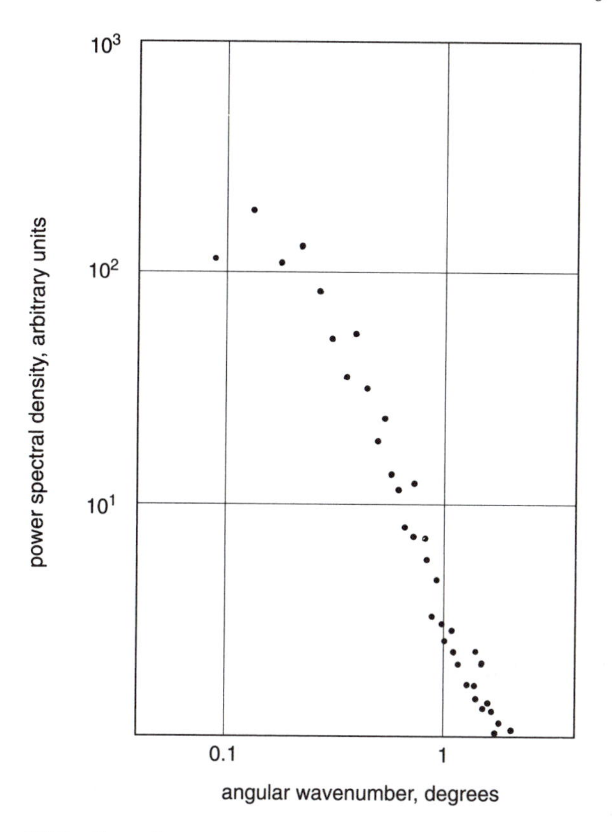

Figure 2.7 *PSD of sea clutter*

spectrum. The physical background for this spectrum is not clear. Furthermore, return signal fluctuations are partly attributable to air turbulence. Our observations are listed in Table 2.1.

Recently, various types of clutter have been observed with various spatial resolutions, wavebands, sea states and grazing angles. For example, ground clutter obeys the Weibull distribution [1, 2, 3, 10] and sea-clutter amplitudes obey the log-normal [11, 12], contaminated-normal [11], Weibull [13, 14, 15], log-Weibull [16] and k distributions [16]. Recently, it has been observed that sea ice amplitudes obey the Weibull distribution [17]. Thus, recently, there has been a continuing great interest in Weibull clutter.

2.2 Weibull distribution

2.2.1 *Original derivation of the Weibull distribution*

Let us follow the original idea of W. Weibull who proposed what we now call the Weibull distribution. Let us assume a chain consisting of n rings as shown in Fig. 2.11, and a force x is applied along the chain. It is assumed that the ring is broken when a force larger than the threshold value X is applied to it, where X is a stochastic variable with a probability function $P(X)$. The probability $F(x)$ that the chain is broken when a force x is applied is given by

$$F(x) = P(X \leqq x) = 1 - P(X > x) \tag{2.2}$$

Therefore, the probability that the chain consisting of n rings is not broken with a force x is equal to

$$\begin{aligned}\{P(X > x)\}^n &= \{1 - F(x)\}^n \\ &= \exp[-n\phi(x)]\end{aligned} \tag{2.3}$$

where

$$1 - F(x) = \exp[-\phi(x)] \tag{2.4}$$

This is the probability that the chain is not broken by a force x, and hence it is a monotonically decreasing function of x or $\phi(x)$ is a positive, monotonically increasing function of x above x_u and is zero below this value. The simplest candidate for this function is given by

$$\phi(x) = ((x - x_u)/b)^c \tag{2.5}$$

from which we have

$$F(x) = 1 - \exp[-((x - x_u)/b)^c] \tag{2.6}$$

for $x > x_u$. The probability $f(x)dx$ that the chain is just broken by a force $x \sim x + dx$ is equal to a derivative of $F(x)$ with respect to x:

$$f(x) = (c/b)\,((x - x_u)/b)^{c-1} \exp[-((x - x_u)/b)^c] \tag{2.7}$$

This is the Weibull distribution.

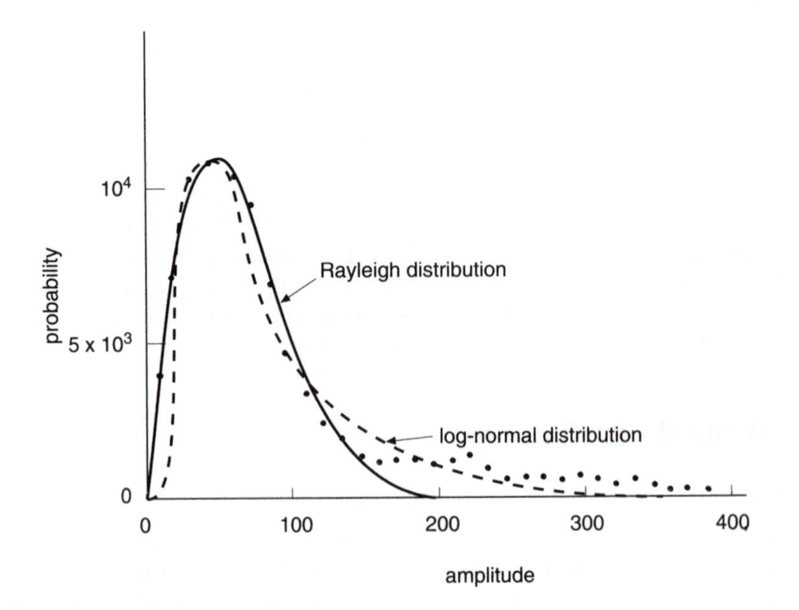

Figure 2.8 *Amplitude distribution of sea clutter*

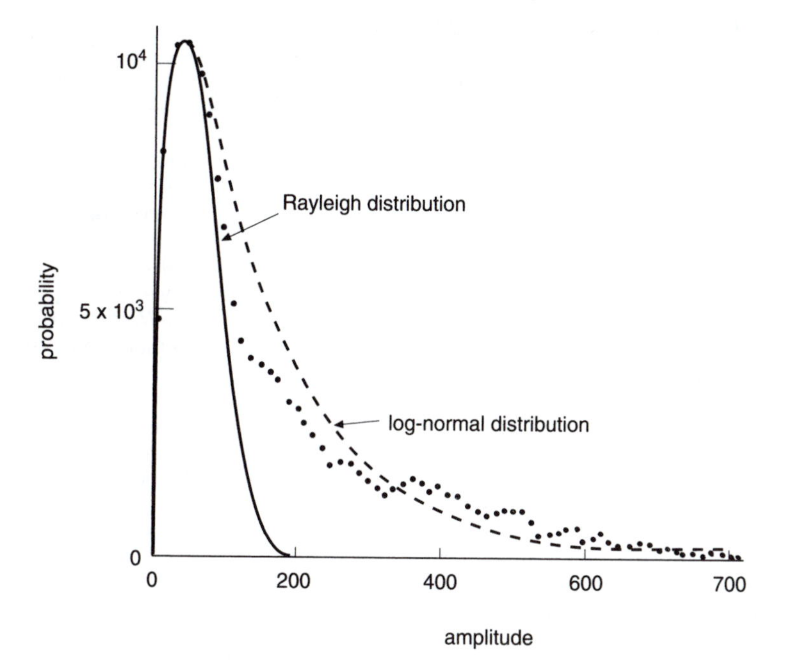

Figure 2.9 *Scan-to-scan amplitude distribution of sea clutter*

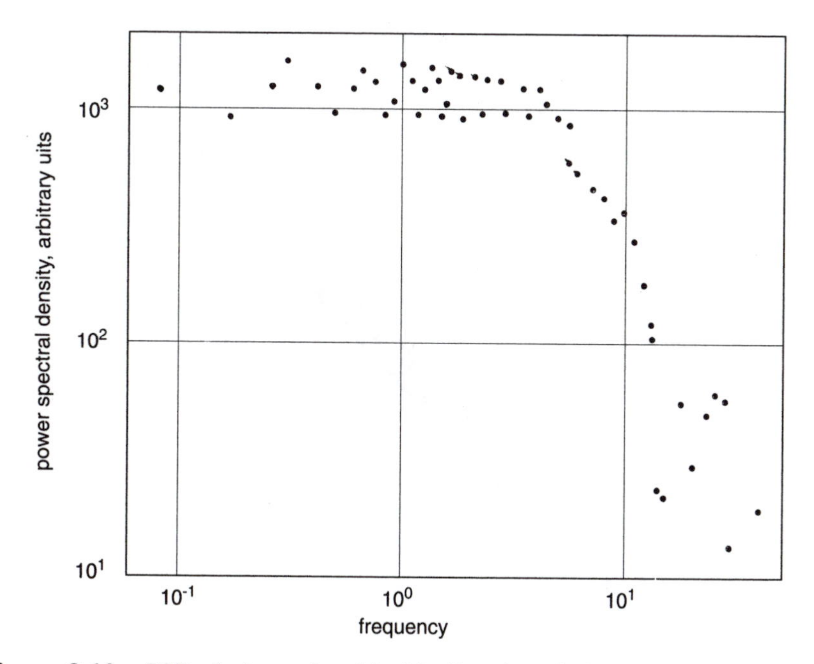

Figure 2.10 *PSD of echoes reflected by Mt. Fuji through clear sky*

Table 2.1 *Summary of the observed spectra*

	Power spectrum	Distribution
Ground clutter		
stationary antenna	$1/f^2$	Gaussian
rotating antenna	$1/K^2 \times 1/K^{3.5}$	Weibull
scan-to-scan	$1/K$	log-normal
Sea clutter		
stationary antenna	?	Weibull [15]
rotating antenna	$1/K^2$	Rayleigh + log-normal, log-Weibull [16] Weibull (sea ice) [17]
scan-to-scan	$1/K$	log-normal
Weather clutter		
stationary antenna	?	Gaussian
rotating antenna	$1/K$	Weibull [4, 5, 6, 7, 8]
scan-to-scan	$1/K$	Weibull Reyleigh

Eqn. (2.6) is rewritten as

$$f(x)dx = \exp[-((x-x_u)/b)^c]d\{(x-x_u)/b\}^c$$
$$= \exp[-y^c]dy^c \tag{2.8}$$

if one puts

$$y = (x-x_u)/b \geqq 0 \tag{2.9}$$

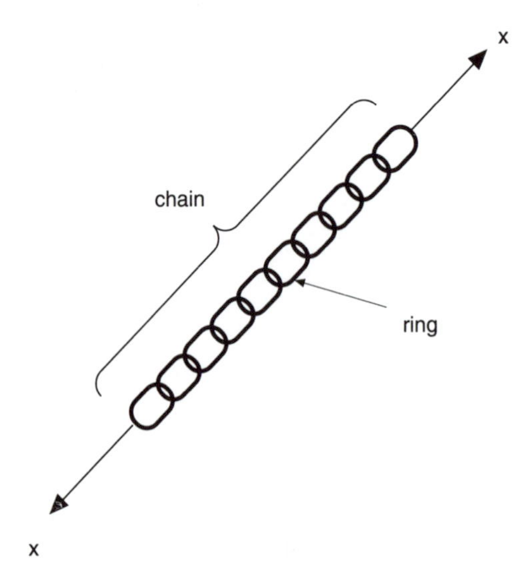

Figure 2.11 *A chain of rings*

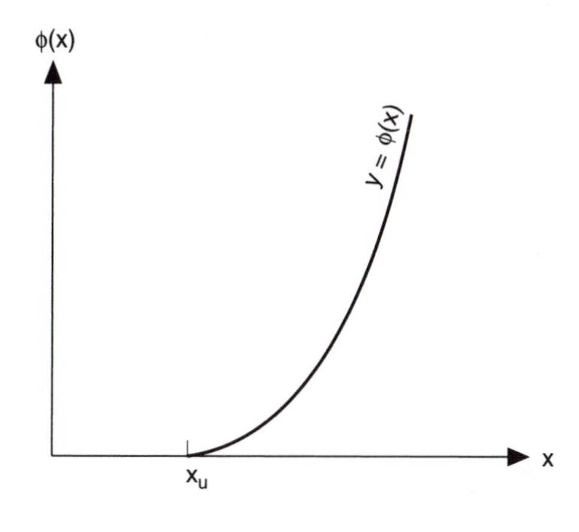

Figure 2.12 *Behaviour of $\phi(x)$*

It is reasonable that b is called the scale parameter, c the shape parameter, and x_0 the location parameter. It is clear that the Weibull distribution is applicable to describing various patterns of the lifetime distribution of products which are composed of many elements.

2.2.2 *Derivation of the Weibull distribution from the extreme value theory*

From the statistical theory of extreme values, a Weibull distribution is also derived. According to Gumbel [18], we consider the stability postulate

$$F^n(x) = F(a_n x + b_n) \tag{2.10}$$

This is a linear transformation which does not change the form of the distribution. As the asymptotic distribution, we consider the distribution of the largest value in samples of size n which are taken from the same population. The two parameters a_n and b_n are functions of n. Fisher and Tippett [19] derived the following three asymptotes:

$$F(x) = \exp[-\exp[-\alpha(x-u)]], \qquad \alpha > 0 \tag{2.11}$$

$$F(x) = \exp[-(v/x)^k], \qquad x>0,\ v>0,\ k>0 \tag{2.12}$$

$$F(x) = \exp[-(x/v)^k], \qquad x<0,\ v<0,\ k>0 \tag{2.13}$$

Gnedenko [20] proved that only three asymptotes of eqns. 2.11, 2.12 and 2.13 exist under the condition of the stability postulate of eqn. 2.10. Eqn. 2.13 is especially important in the derivation of the Weibull distribution. This equation was derived from the assumption that there exists $F(0) = 1$; that is the variate is non-positive and the distribution $F^n(x)$ satisfies $F^n(x) = F(a_n x)$. Here $F^n(x)$ is the distribution of the largest value, where $x = \max(x_1, x_2, \ldots, x_n)$ in n samples $(x_1, x_2, \ldots, x_n)$ from the same population.

Now change the sign of x, that is, $y = -x$, and consider n samples $(y_1, y_2, \ldots, y_n) = (-x_1, -x_2, \ldots, -x_n)$ out of the population obeying the distribution $F(y)$. Then the smallest value is written as

$$\begin{aligned} y &= \min(y_1, y_2, \ldots, y_n) \\ &= \min(-x_1, -x_2, \ldots, -x_n) \\ &= -\max(x_1, x_2, \ldots, x_n) \end{aligned} \tag{2.14}$$

If the distribution of the smallest values, $[1 - F(y)]^n$, satisfies

$$F^n(x) = (1 - F(y))^n = F(a_n x) = 1 - F(a_n y) \tag{2.15}$$

then $F(y)$ is finally written as

$$F(y) = 1 - F(-y) = 1 - \exp[-(y/v)^k] \tag{2.16}$$

where $b = -v > 0, y > 0, k > 0$. This is a Weibull distribution.

2.2.3 *Properties of the Weibull distribution*

When $c = 1$, we have $\phi(x) = \text{constant} \times \exp[-x/b]$, and a lifetime distribution of the chain becomes an exponential function of x. This is the case for a Poisson process in which the life of the product ends completely randomly; in other words, failure occurs randomly in time. When $c < 1$, on the other hand, failure occurs more frequently in the early stage of a new product; after initial failures are repaired or removed, the product becomes more stable. When $c > 1$, failure occurs more frequently as time goes on; this situation represents the fatigue effect.

The Weibull distribution becomes a Rayleigh distribution when $c = 2$. This is identical to the χ^2-distribution for two degrees of freedom. The Weibull distributions for various values of the shape parameter c are plotted in Fig. 2.13.

The Weibull distribution when $x_u = 0$ becomes

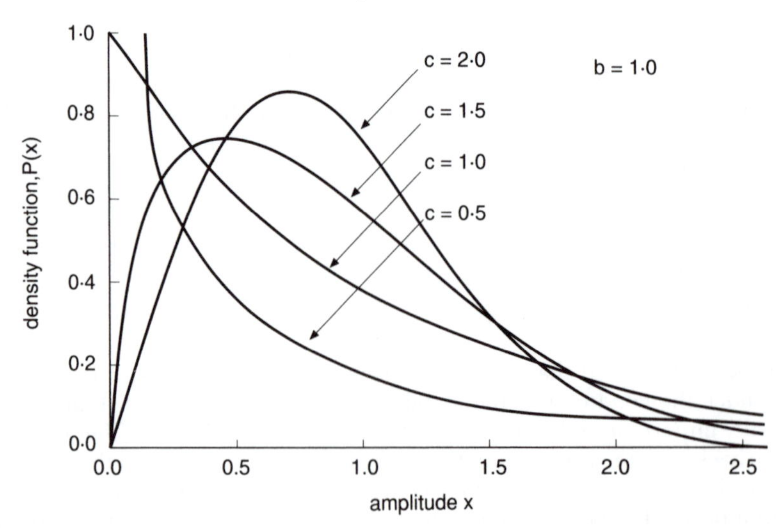

Figure 2.13 *Weibull distributions for various values of the shape parameter c*

$$f(x)dx = (c/b)(x/b)^{c-1}\exp[-(x/b)^c]dx \tag{2.17}$$

The mean value of x, $\langle x \rangle$, is given by

$$\langle x \rangle = \int_0^\infty xf(x)dx = (c/b)\int_0^\infty x(x/b)^{c-1}\exp[-(x/b)^c]dx$$

$$= b\int_0^\infty \xi^{1/c}\exp[-\xi]d\xi = b\Gamma(1/c+1) \tag{2.18}$$

where $\Gamma(x)$ is a gamma function defined as $\Gamma(x) = \int t^{x-1}\exp[-t]dt$. The variance is equal to $\langle x^2 \rangle - \langle x \rangle^2$, and $\langle x^2 \rangle$ is evaluated as

$$\langle x^2 \rangle = \int_0^\infty x^2(c/b)(x/b)^{c-1}\exp[-x/b)^c]dx$$

$$= b^2\int_0^\infty \xi^{2/c}\exp[-\xi]d\xi = b^2\Gamma(2/c+1) \tag{2.19}$$

from which

$$\langle x^2 \rangle - \langle x \rangle^2 = b^2\{\Gamma(2/c+1) - \Gamma(1/c+1)^2\} \tag{2.20}$$

Suppose x obeys the Weibull distribution given by eqn. 2.20, and let a new variable X be introduced which is defined as

$$X = \log x \tag{2.21}$$

The probability density function $p(X)$ is related to $f(x)$ as

$$p(X)dX = f(x)dx$$

$$= c\exp[X']\exp[-\exp[X']]dX \tag{2.22}$$

where

$$X' = c\{X - \log b\} \tag{2.23}$$

The probability densities of X' for various values of the shape parameter c are plotted in Fig. 2.14.

These distributions look like distorted Gaussian distributions, or log-normal distributions for x. The mean and variance of X are given by

$$\langle X \rangle = \log b - \gamma/c \tag{2.24}$$

$$\langle X^2 \rangle - \langle X \rangle^2 = (1/6)(\pi/c)^2 \tag{2.25}$$

where $\gamma = 0.577316\ldots$ is Euler's constant.

2.2.4 *Weibull plot*

The probability function $P(x)$ for the Weibull distribution is given by

$$P(x) = \int_0^x p(r)dr = 1 - \exp[-(x/b)^c] \tag{2.26}$$

Therefore,

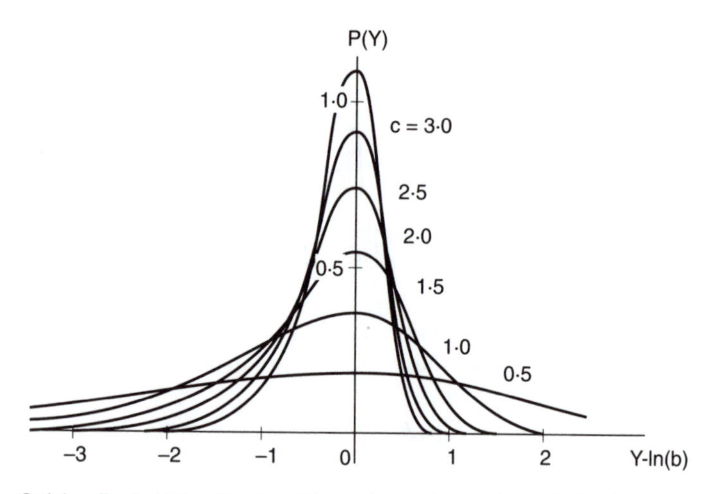

Figure 2.14 *Probability density of log x for various values of the shape parameter c*

$$\exp[-(x/b)^c] = 1 - \int_0^x p(r)dr$$

and then

$$(x/b)^c = -\log\{1 - \int_0^x p(r)dr\}$$

Again taking its logarithm gives

$$c\{\log x - \log b\} = cX - c\log b$$

$$= \log\left\{-\log\left\{1 - \int_0^x p(r)dr\right\}\right\}$$

We define

$$Y = \log\left\{-\log\left\{1 - \int_0^x p(r)dr\right\}\right\} \tag{2.27}$$

$$X = \log x \tag{2.28}$$

and we observe a linear relation between X and Y:

$$Y = cX - \log b \tag{2.29}$$

When the original distribution is a Weibull distribution, plots of Y against X will lie on a straight line, and b and c are read from this plot.

2.2.5 *Weibull distribution and log-normal distribution*

The intensity of an echo signal amplitude depends on distance r of a reflector from the radar and the two-way attenuation is given by $h(r)$ as a function of r. This trend is compensated by sensitivity time control (STC). Suppose reflectors are distributed as $f(r)$ per unit area, and reflectors in a small area da at r cause echo signals whose amplitude z at the receiver has a probability density function $g(z; h(r))$ in log z. Then the probability density function $G(z)$ of echo signals in log scale coming from an area with constant width extending from r_1 to r_2 in range is given as

$$G(z) \propto \int_{r_1}^{r_2} f(r)g(z; h(r))dr \tag{2.30}$$

According to our observation of sea ice, $g(z; h(r))$ is well approximated by a log-normal function, and hence $g(z; h(r))$ can be rewritten as

$$g(z; h(z)) = ((2\pi)^{1/2}\sigma) \exp[-(\log z - \log h(r))^2/(2\sigma^2)] \tag{2.31}$$

Unless the radar beam is scattered by air turbulence or dust, we can approximate

$$h(z) = \text{constant}/r^4 \tag{2.32}$$

When reflectors are uniformly distributed, $f(z)$ remains constant, and eqn. 2.30 is rewritten as

$$G(z) \int_{r_1}^{r_2} ((2\pi)^{1/2}\sigma) \exp[-(\log z - 4\log(1/r) - \text{constant})^2/(2\sigma^2)]dr$$

$$= \int_{r_1}^{r_2} ((2\pi)^{1/2}\sigma) \exp[-(\log z - m)^2/(2\sigma^2)]dm \tag{2.33}$$

Since

$$dm = 4dr/r \tag{2.34}$$

log-normal distributions are added with a weight of $4/r$. The results are plotted in Fig. 2.15 for three values of $r_2/r_1 = 3/5$, 2 and 3, where (*a*) refers to the weighting function which is proportional to $1/r$, (*b*) to the total probability density function in logarithmic scale, and (*c*) to a Weibull plot of curve (*b*). When the data area covers a relatively large area in range, the Weibull plot of clutter amplitudes becomes a straight line except for small amplitudes; the small amplitude part is often mixed up with amplifier noise, and moreover a so-called log-amplifier does not have a logarithmic property at smaller inputs. In addition to these, small amplitudes do not contribute to the false alarm and a small-amplitude tail of the clutter amplitude distribution is not important. Fig. 2.16 shows a Weibull plot of clutter amplitudes at 9410 MHz which were observed with sea ice in Hokkaido over the area ranging from 460 m to 1996 m from the radar antenna.

It should be noted that, in the frequency domain after applying the digital Fourier transform to the clutter data, amplitude distribution very often obeys a Weibull distribution and the CFAR technique is used in the frequency domain, as discussed in Reference 3.

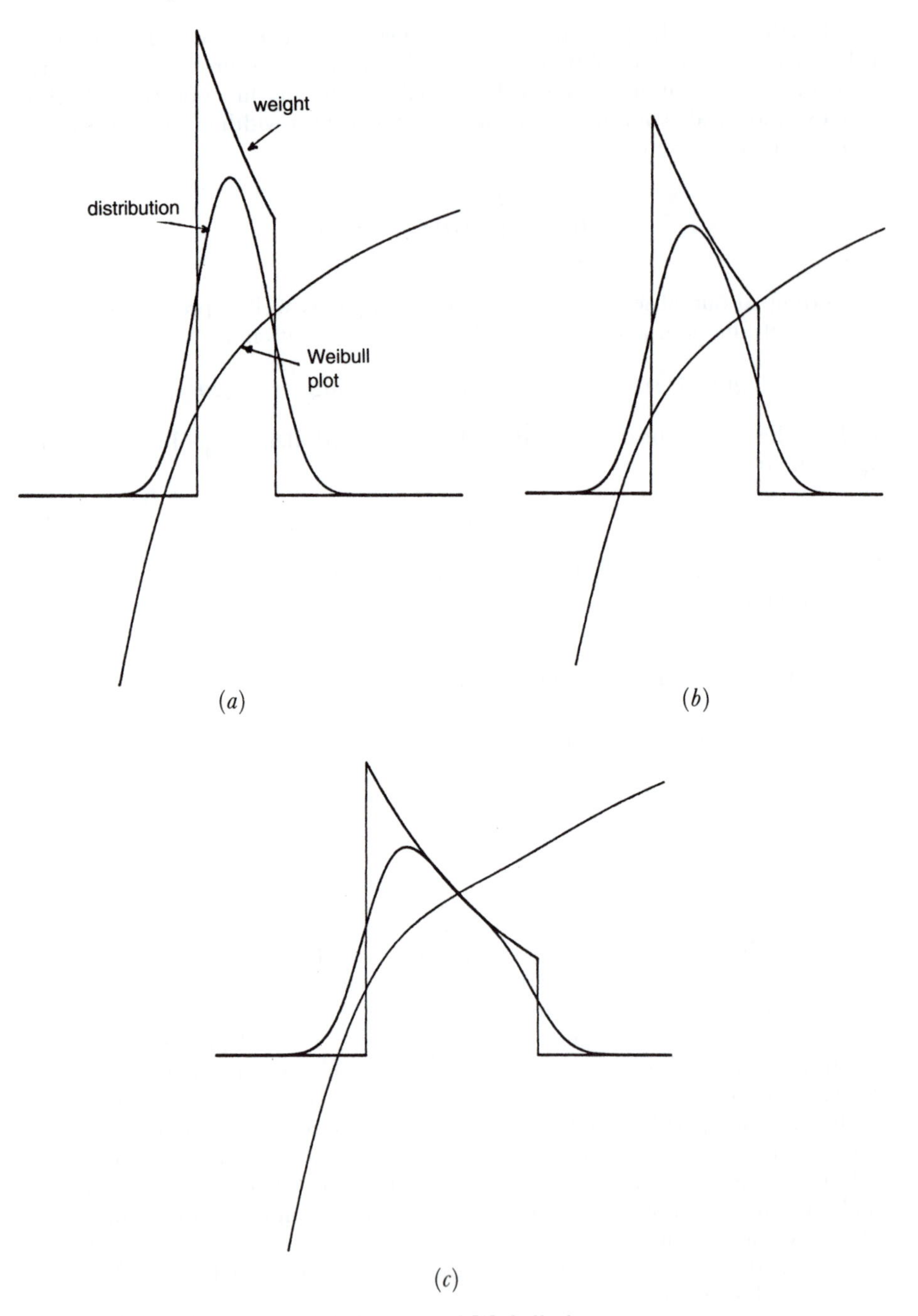

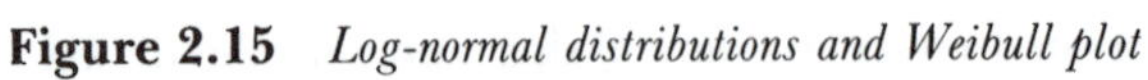

Figure 2.15 *Log-normal distributions and Weibull plot*

a $r_2/r_1=5/3$
b $r_2/r_1=2$
c $r_2/r_1=3$

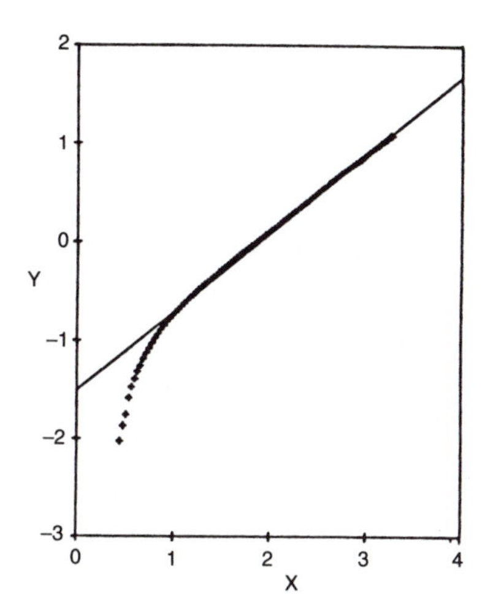

Figure 2.16 *Weibull plot of observed clutter amplitudes of sea ice*
$r_2/r_1 = 1996/460$

2.3 Application of the linear prediction method to clutter reduction

2.3.1 *Autoregressive linear prediction*

Observed clutter amplitudes are more or less correlated to one another; this is caused by the spatial correlation of the reflectors and a finite radar beamwidth. Suppose a radar pulse with duration τ is emitted and reflected back by a point target (see Fig. 2.17). Loci of the head and tail of a radar pulse are indicated by two parallel lines. It is seen that two objects which are separated by a distance smaller than $c\tau/2$ cannot be separated at the receiver position (c = speed of light in free space). Therefore, the spatial resolution of this radar system is equal to $c\tau/2$. If the sampling time of the return signals is larger than τ, a return signal of a certain point target is missed. In our observations the sampling time is always equal to $\tau/2$. When the spatial correlation of the reflectors is less than this resolution, a correlation between the observed return signals along the range direction will be found best between the adjacent bins. A marine radar in the X band, with which we often made observations, has $\tau = 80$ ns; the resolution is 12 m. The structure of the reflectors within this length is smeared out, and a spatial structure to a larger scale gives rise to correlation in the observed clutter amplitude.

The antenna, which rotates at 28 r.p.m. for instance, makes a full rotation in 60/28 s or the radar beam scans over $360 \times 28/60 = 168°$ per second; with a pulse repetition rate of 1680 Hz, the sampling angle $\delta\theta$ along the azimuth direction is $\delta\theta = 0.1°$. When the beamwidth is 1.2°, a point target makes a contribution to return signals of over 12 hits in the sweep. The spatial structure of the reflectors is added to this natural correlation and mathematically involved with it. In this fashion correlations in the observed clutter intensities

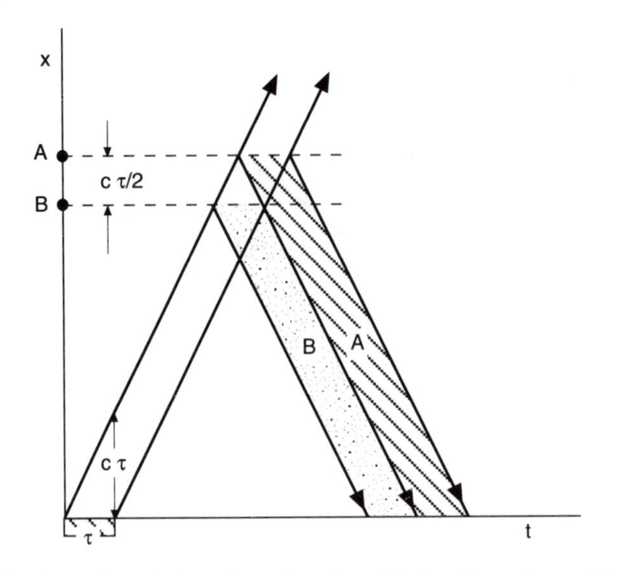

Figure 2.17 *Spatial resolution of a pulse echo with duration τ is equal to $c\tau/2$*

have different characteristics in the range and sweep directions. Since radar returns have a certain amount of correlation as described above, some part of the return signal intensity at a given pixel of the PPI can be predicted from the data in the vicinity of that point. The predicted part is subtracted from the observed ones and we are left with unpredicted errors which are smaller than the original return signal levels. If we have a target, a return signal from it does not obey the statistical nature of the background returns, and it is enhanced after elimination of the correlated parts of the background returns. This technique can increase the target-to-clutter ratio.

Prediction of the signal in terms of the preceding data is done by means of the AR model. Suppose we have observed a signal $x(t)$ as a function of time t. For the purpose of signal processing, the analogue signal is sampled at discrete time points $n\tau$ where n is an integer, and let $x(t)$ at time point $n\tau$ be denoted as x_n. The mean value is subtracted and the prediction technique is applied; afterwards the mean value is added. We now estimate x_n as a weighted average of the preceding p terms $x_{n-1}, x_{n-2}, \ldots, x_{n-p}$ with weighting coefficients $a_1, a_2, \ldots, a_p$. The estimated value $\hat{x}_n$ is given as

$$\hat{x}_n = a_1 x_{n-1} + a_2 x_{n-2} + \ldots + a_p x_{n-p} \tag{2.35}$$

There is a remainder ε_n between the observed and predicted ones. If the correlation is fully used for the prediction, ε_n is no longer correlated with x_n. It would be reasonable to assume that the remainder is random and Gaussian. Therefore, we have

$$x_n = a_1 x_{n-1} + a_2 x_{n-2} + \ldots + a_p x_{n-p} + \varepsilon_n \tag{2.36}$$

Within a homogeneous region, a common set of the prediction coefficients is used. In this case p terms are used for the prediction and hence it is called a prediction of order p. When the clutter is not homogeneous, PARCOR is used as will be described in Section 2.3.5.

The coefficients are determined so that the squared error $\langle \varepsilon_n^2 \rangle$, which is an average over n samples, is minimised. Since

$$\langle \varepsilon_n^2 \rangle = \langle \{x_n - (a_1 x_{n-1} + a_2 x_{n-2} + \cdots + a_{n-p})\}^2 \rangle \tag{2.37}$$

where $\langle \cdot \rangle$ means average over n samples, a condition for minimising the mean squared error is found by equating derivatives of the above equation with respect to a_j ($j = 1, \ldots, p$) to zero:

$$\begin{aligned} d\langle \varepsilon_n^2 \rangle / da_j &= (d/da_j)\langle \{x_n - (a_1 x_{n-1} + a_2 x_{n-2} + \cdots + a_{n-p})\}^2 \rangle \\ &= 2\langle x_{n-j}\{x_n - (a_1 x_{n-1} + a_2 x_{n-2} + \cdots + a_{n-p})\}\rangle \\ &= 2\{C_{-j} - a_1 C_{j-1} - a_2 C_{j-2} - \cdots - a_{n-p} C_{j-p}\} \\ &= 0 \end{aligned} \tag{2.38}$$

where

$$C_j = C_{-j} = \langle x_m x_{m-j} \rangle \tag{2.39}$$

We thus obtain

$$\begin{array}{cccccc} a_1 C_0 + & a_2 C_1 + & a_3 C_2 & + \cdots + & a_p C_{p-1} = & C_1 \\ a_1 C_1 + & a_2 C_0 + & a_3 C_1 & + \cdots + & a_p C_{p-2} = & C_2 \\ a_1 C_2 + & a_2 C_1 + & a_3 C_0 & + \cdots + & a_p C_{p-3} = & C_3 \\ \cdots & \cdots & \cdots & \cdots & \cdots & \cdots \\ a_1 C_{p-1} + & a_2 C_{p-2} + & a_3 C_{p-3} & + \cdots + & a_p C_0 = & C_p \end{array} \tag{2.40}$$

By solving these equations a set of coefficients a_j is obtained. The clutter intensity of a homogeneous reflector decreases in the range distance r as r^{-4}, and therefore this attenuation should be corrected before applying the prediction technique.

2.3.2 *Akaike information criterion (AIC)*

Does a larger p always result in a more accurate prediction? This is not the case and a criterion for the optimum order of the prediction was given by Akaike [21]; this is called Akaike information criterion which is abbreviated as AIC. The AIC for a given model is defined as

$$\begin{aligned} \text{AIC} = -2\{&[\text{maximum logarithmic likelihood}] \\ &- [\text{number of parameters included in the model}]\} \end{aligned} \tag{2.41}$$

The 'logarithmic likelihood' is explained as follows. Given a model of probability function $f(x;\, a)$ for a stochastic variable x; this includes parameters a (symbol a stands for many parameters). The value of x is discrete as $x_1, x_2, \ldots,$ and it is assumed that x_j was observed n_j times among n observations. Then the mean logarithmic likelihood is defined by

$$[\text{mean logarithmic likelihood}] = \sum_{j=1}^{N} (n_j/n) \log[f(x_j;\, a)] \tag{2.42}$$

According to our experience the maximum logarithmic likelihood is not an unbiased estimate of the mean logarithmic likelihood and is always larger than that by the amount of the number of parameters involved; therefore the AIC is defined by eqn. 2.41. The model which yields the smallest AIC is regarded as the best one.

Since the prediction error often obeys a normal distribution, the probability function for ε is given by

$$f(x_m;\,\sigma) = (2\pi\sigma^2)^{1/2}\,\exp\left[-\left(x_m - \sum_{j}^{p} a_j x_{m-j}\right)^2 \Big/ 2\sigma^2\right] \tag{2.43}$$

The likelihood L of the model after n trials is given by

$$L = (2\pi\sigma^2)^{n/2}\,\exp\left[-\sum_{m=1}^{N}\left(x_m - \sum_{j=1}^{p} a_j x_{m-j}\right)^2 \Big/ 2\sigma^2\right] \tag{2.44}$$

The logarithmic likelihood l is defined as

$$l = -(n/2)\,\log(2\pi\sigma^2) - (2\sigma^2)^{-1}\sum_{m=1}\left(x_m - \sum_{j=1} a_j x_{m-j}\right)^2 \tag{2.45}$$

This is maximised under the condition of eqn. 2.40, and σ^*, which maximises l, is given by

$$\begin{aligned}\sigma^{*2} &= (1/n)\sum_{m=1}^{N}\left(x_m - \sum_{j=1}^{p} a_j^* x_{m-j}\right)^2 \\ &= (1/n)\{C(0) - \sum_{j} a_j^* C(j)\}\end{aligned} \tag{2.46}$$

where a_j^* is the coefficient satisfying eqn. 2.40. We then find

$$\begin{aligned}\text{AIC} &= -2l + 2(p+1) \\ &= n(\log(2\pi) + 1) + n\,\log(\sigma^{*2}) + 2(p+1)\end{aligned} \tag{2.47}$$

In this case the parameters to be determined are a_j ($j = 1, 2, \ldots, p$) and σ, and hence the total number of parameters is $(p+1)$. The first term on the right hand side is constant and the second and third terms should be evaluated in determining the optimum number of prediction coefficients. This technique is also very useful in determining the optimal probability distribution model among several models for fitting observed data. Some examples are shown in the following Sections.

2.3.3 *Determination of the optimum probability function for the observed data*

Radar echoes were recorded in February 1987, at the port of Monbetsu in Hokkaido with a marine radar (KODEN MD–3221), which has a frequency of 9410 MHz, a peak transmitted power of 25 kW, horizontal polarisation, the beam width 1.2°, pulse width 80 ns, pulse repetition frequency 1680 Hz, and antenna scan rate of 28 r.p.m.; the input power is logarithmically transformed

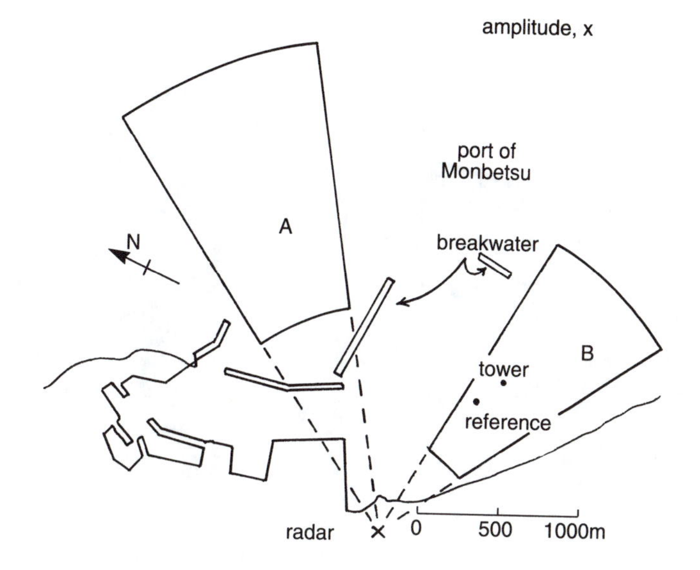

Figure 2.18 *Areas of radar search in Monbetsu and Tokyo*

to widen the dynamic range. The sampling interval along the azimuth is about 0.1°. The map is illustrated in Fig. 2.18. Area A covers open sea surfaces and young sea ice, and area B contains an iron tower and a radar-wave reflector. The return signals were sampled every 40 ns ($\tau = 40$ ns) and converted into 8 bit digital numbers. The echo intensity distribution after STC is plotted in Fig. 2.19 in four levels. Most of the port of Monbetsu was covered with sea ice. The total range of 1536 m under observation was divided into eight regions of 200 m range each; three regions were selected which did not contain open sea surfaces; the intensity distributions are shown in Fig. 2.20. These distributions can be approximated by a log-normal distribution or a Weibull distribution. Which one better approximates the observed data is determined by AIC. By maximising the logarithmic likelihood, the optimal values of the parameters involved in these models are determined, and then the AIC is evaluated. Values of AIC of the optimum log-normal and Weibull distributions are listed in Table 2.2.

Because of the large number of data the absolute value of the AIC is very large, but it is found that the log-normal distribution better approximates the observed distributions in this case.

2.3.4 *Examples of clutter reduction by linear prediction*

The electromagnetic power density is attenuated as r^{-2} from the radial distance r to a reflector, and it is also attenuated as r^{-2} during its path to the receiver; hence the overall attenuation is proportional to r^{-4}. Therefore, the clutter intensity is reduced as r^{-4}. This tendency must be compensated before applying a clutter reduction technique. This is called sensitivity time control (STC). Since real time signal processing is required, the STC must be performed quickly, and this is performed approximately by digital processing as shown in

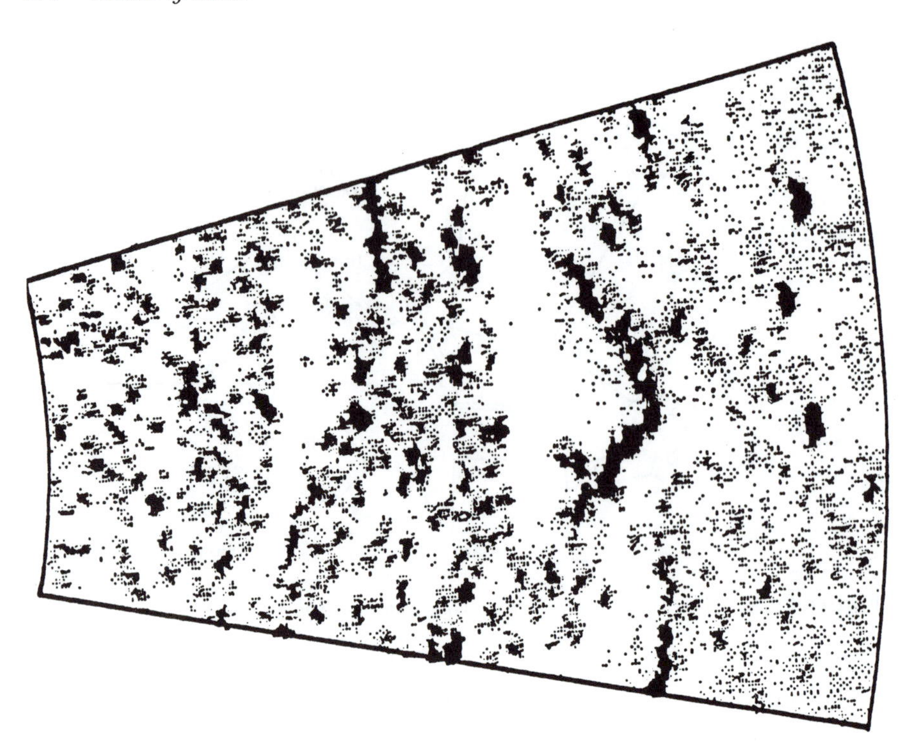

Figure 2.19 *Echo map of the port of Monbetsu after STC*

Fig. 2.21, where operator z^{-1} shifts a step τ back in time and $\lambda < 1$. The first stage is a difference operation:

$$d_n = x_n - x_{n-1} \tag{2.48}$$

and in the last stage we find

$$\begin{aligned} y_n &= d_n + \lambda y_{n-1} \\ &= d_n + \lambda d_{n-1} + \lambda^2 d_{n-2} + \cdots \\ &= x_n - M_n \end{aligned} \tag{2.49}$$

where

$$M_n = (\lambda x_{n-1} + \lambda^2 x_{n-2} + \cdots)/(\lambda + \lambda^2 + \cdots) \tag{2.50}$$

M_n is a weighted mean of the previous xs. The effective number of terms N used in evaluating the mean is controlled through parameter λ. Since the weight $\lambda < 1$, the data in the remote past are not taken into account in evaluating the mean and hence the local mean level of the return signals is subtracted to give y_n. The effective number N is evaluated as

$$N = \lambda/(1-\lambda) \tag{2.51}$$

when $\lambda = 0.9$, $N = 9$. Since signal x is a logarithmic transformation of the return signal, subtraction of the local mean is equivalent to automatic gain control, and this operation is equivalent to STC. AIC given by eqn. 2.47 is rewritten as

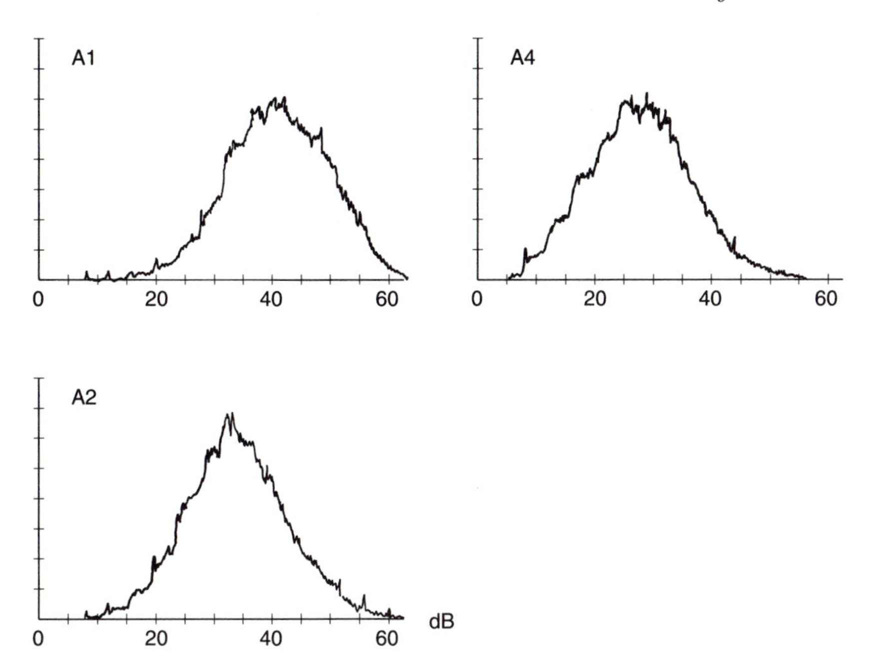

Figure 2.20 *Intensity distributions after STC in three regions selected in area A which were covered with sea ice*

$$\mathrm{AIC} = N\{\log(2\pi) + 1\} + N\log(\sigma^2) + 2(p + 1) \tag{2.52}$$

The first term can be neglected because it is not involved in minimising AIC. This formula was applied to areas A and C in Fig. 2.19 where we used 128 successive data, and p was taken as 1 to 10. Values of AIC are listed in Table 2.3(A) referring to area A. The optimum order of prediction is underlined, and in most cases the order is smaller than 3. The case when the optimum order is unity means that the spatial correlation of the sea ice is smaller than the resolution.

In applying this technique, the order of prediction must be optimised to perform the optimised operation. However, every time the order of the prediction is changed the prediction coefficients need to be recalculated according to eqn. 2.40. In realising the adaptive operation for deriving the prediction error, this situation is not practical and the cumbersome process must be omitted while the same effect on clutter reduction is maintained. This

Table 2.2 *Values of AIC for the log-normal and Weibull distributions*

	Area A1	Area A2	Area A4
Log-normal	2 021 559	2 033 668	2 035 419
Weibull	2 042 557	2 069 732	2 074 054

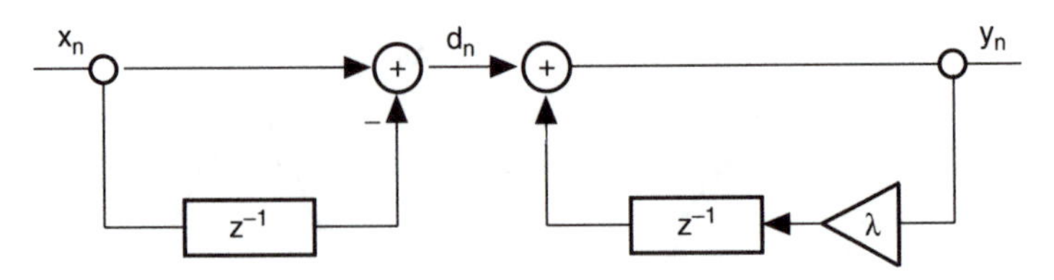

Figure 2.21 *Digital STC circuit*

requirement is satisfied by the PARCOR technique in which the order of prediction is adaptively determined to produce the prediction error without recalculating the prediction coefficents from the beginning.

An example of improvement of the target-to-clutter ratio was described in Reference 22. Data were collected at Yamada using an ARSR with frequency 1.3 GHz, beamwidth 1.2° and horizontal polarisation. The weather clutter was

Table 2.3(A) *Values of AIC for area A with sea ice*

	Sweep no.							
n	0	1	2	3	4	5	6	7
1	469.8	507.2	497.3	510.6	500.0	602.1	499.5	514.0
2	469.3	504.4	498.9	508.7	500.1	603.8	498.2	513.6
3	468.2	506.3	497.5	511.1	499.7	603.7	496.8	514.7
4	470.3	507.9	499.5	508.3	501.0	605.3	499.3	516.2
5	472.0	508.4	498.1	510.4	502.0	607.0	501.4	518.3
6	473.6	510.5	499.8	512.1	504.3	608.6	503.5	516.6
7	475.2	512.5	500.4	514.1	506.1	607.6	496.7	518.9
8	477.3	514.1	502.4	515.1	501.8	608.6	498.7	512.7
9	479.1	515.3	501.0	516.5	503.8	607.8	494.5	514.4
10	478.5	514.9	500.1	518.8	502.6	609.6	496.4	505.9

	Sweep no.							
n	8	9	10	11	12	13	14	15
1	477.8	524.2	484.8	483.2	475.6	496.6	503.1	538.4
2	479.1	516.0	482.6	477.6	477.6	495.8	502.8	539.5
3	479.2	518.1	481.9	479.1	478.5	497.8	504.6	540.7
4	481.3	520.1	483.8	477.7	478.7	494.2	503.4	537.6
5	483.7	522.5	485.7	479.8	480.2	495.0	504.8	536.7
6	481.9	524.4	487.7	478.5	481.8	496.8	506.3	538.8
7	483.4	526.2	489.8	477.3	483.7	479.8	506.6	540.0
8	485.0	525.2	490.8	479.3	485.5	499.1	507.9	535.0
9	480.6	527.1	486.9	480.1	487.4	501.1	510.3	537.0
10	482.5	524.3	487.2	481.2	487.2	500.3	506.3	535.8

Table 2.3(B) *Values of AIC for area C with sea ice*

	Sweep no.							
n	0	1	2	3	4	5	6	7
1	544.7	534.5	541.5	533.5	531.7	520.6	520.4	564.2
2	537.2	529.6	540.2	535.9	528.7	522.5	517.2	561.8
3	537.4	532.0	540.4	525.2	528.1	523.2	516.4	563.1
4	539.2	532.5	541.5	526.4	527.6	525.2	518.3	564.7
5	541.8	535.5	544.7	529.1	530.0	527.4	520.2	567.6
6	536.7	537.1	546.5	530.4	531.9	527.0	522.7	567.2
7	534.8	537.2	537.2	530.4	531.9	527.0	522.7	567.2
8	536.8	538.7	538.5	531.9	533.9	527.7	524.2	569.6
9	538.4	539.7	539.5	533.9	535.0	528.8	522.7	569.8
10	540.3	541.7	541.6	535.7	534.5	530.6	523.3	568.1

	Sweep no.							
n	8	9	10	11	12	13	14	15
1	556.3	566.8	573.6	579.0	567.6	561.6	563.7	556.7
2	557.1	563.0	561.7	567.0	558.7	545.9	550.9	554.4
3	557.1	564.2	562.8	565.8	560.5	547.3	552.9	556.1
4	559.1	566.2	564.8	567.4	562.5	549.3	554.8	557.4
5	560.3	568.7	568.1	569.6	564.8	551.7	556.7	559.4
6	560.4	567.5	569.3	570.8	565.3	553.1	558.3	557.7
7	561.8	567.7	569.7	571.3	567.4	555.2	557.0	555.1
8	563.7	570.1	570.9	572.7	566.4	555.6	558.5	556.8
9	563.0	571.4	566.6	526.9	567.1	556.1	559.6	557.4
10	562.6	570.3	566.1	564.9	567.0	557.7	555.7	559.3

examined by adding a phantom target to it, and it is found that the improvement factor (IF) defined by

$$\mathrm{IF}=\frac{\text{input power from clutter}}{\text{output power from clutter}}\times\frac{\text{output power from target}}{\text{input power from target}}$$

was 22.22 dB. The order of the digital filter is 8.

2.3.5 *Principle of partial autocorrelation (PARCOR) filtering*

The PARCOR method was originally worked out by the Japanese group of Itakura *et al.* [23] for speech signal analysis. With this technique, the prediction error is calculated for any order of prediction iteratively. We have combined the PARCOR method with AIC to adaptively decide the prediction error which minimises the correlation (Fig. 2.22).

The prediction process described above is called 'forward' prediction because data in the past are used for the prediction. Contrary to this, the 'backward' prediction is possible, in which a signal value is estimated from the following ones. We consider a sequence of observed data $\ldots, x_{n-p}, x_{n-p+1}, \ldots, x_{n-2}, x_{n-1},$

$x_n, \ldots$, and x_n is estimated from the preceding $p-1$ data, $x_{n-p+1}, x_{n-p+2}, \ldots, x_{n-2}, x_{n-1}$; this is the forward prediction and a deviation of the predicted value from the real value is called the forward prediction error. The forward prediction error of x_n is denoted as $\varepsilon_{f,n}^{(p-1)}$.

In the same fashion x_{n-p} is estimated from the following $p-1$ data, $x_{n-p+1}, x_{n-p+2}, \ldots, x_{n-1}$ and let $\varepsilon_{b,n}^{(p-1)}$ be the backward prediction error of x_{n-p} estimated from the same samples. When $p=1$, one has $\varepsilon_{f,n}^{(1)}=x_n$ and $\varepsilon_{b,n}^{(1)}=x_{n-1}$. The PARCOR coefficient k_p is defined as

$$k_p=\frac{\langle \varepsilon_{f,n}^{(p-1)}\varepsilon_{b,n}^{(p-1)}\rangle}{\langle\{\varepsilon_f^{(p-1)}\}^2\rangle\langle\{\varepsilon_b^{(p-1)}\}^2\rangle^{1/2}}. \tag{2.53}$$

The mean $\langle\cdot\rangle$ is taken over all possible n. We obtain the following recurrence formula for the prediction errors:

$$\varepsilon_{f,n}^{(p)}=\varepsilon_{f,n}^{(p-1)}-k_p\varepsilon_{b,n}^{(p-1)} \tag{2.54}$$

$$\varepsilon_{b,n}^{(p)}=\varepsilon_{b,n-1}^{(p-1)}-k_p\varepsilon_{f,n-1}^{(p-1)} \tag{2.55}$$

These processes are performed with the lattice-like digital filter which is shown in Fig. 2.23. The PARCOR coefficient is determined such that the mean of $\varepsilon_{f,n}^{(p)}$ and $\varepsilon_{b,n}^{(p)}$ over the preceding data are minimised. However, when the clutter is not uniform or non-stationary, the mean over too many data would yield a wrong result. To avoid this difficulty, a factor c must be multiplied and

$$R_f^{(p)}=\sum_m \gamma^m(\varepsilon_{f,n-m}^{(p)})^2 \tag{2.56}$$

or

$$R_b^{(p)}=\sum_m \gamma^m(\varepsilon_{b,n-m}^{(p)})^2 \tag{2.57}$$

is minimised where $c<1$. The forward and the backward prediction error has the same statistical nature and the PARCOR coefficients determined by minimising eqns 2.56 and 2.57 are identical to each other.

2.3.6 *Application of the PARCOR coefficients to clutter reduction*

The forward prediction error $\varepsilon_{f,n}^{(p)}$ of x_n corresponds to the prediction error of the AR linear prediction model. The optimum order p is determined in terms of AIC in which it is implicitly assumed that the prediction error is a Gaussian

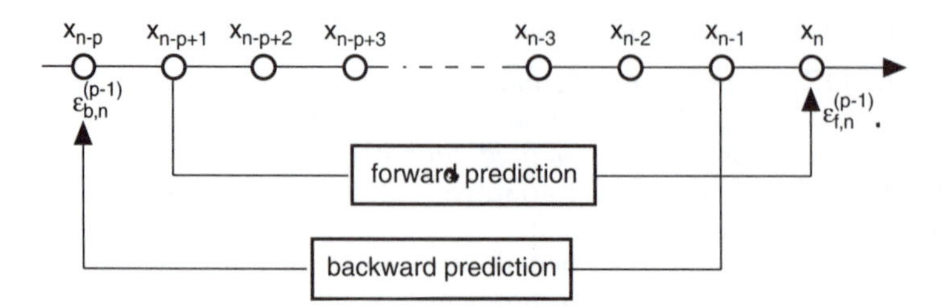

Figure 2.22 *Digital filter for evaluating the (p − 1)th forward and backward prediction errors*

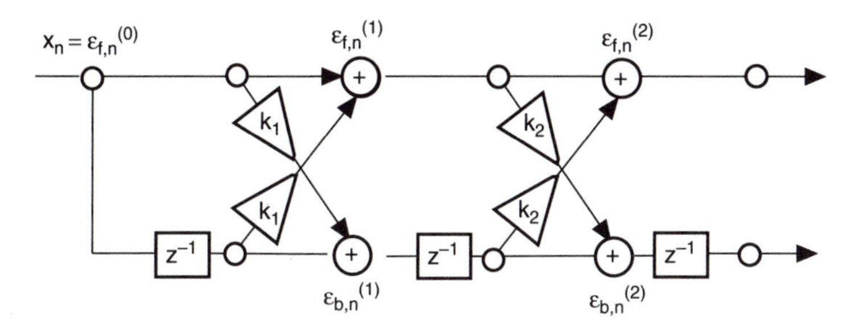

Figure 2.23 *PARCOR filter*

process with a mean which is zero. AIC is calculated in terms of $R_{f,\min}^{(p)}$ which is the minimised $R_f^{(p)}$ and the effective number of terms $1/(1-c)$ as

$$\mathrm{AIC} = (1-c)^{-1}\log(R_{f,\min}^{(p)}) + 2p \qquad (2.58)$$

At every time step AIC is calculated for various values of p and the error with the optimum p is obtained. The total system is shown in Fig. 2.24.

When statistical characteristics of return signals change abruptly, the prediction error becomes relatively large. As shown in Table 2.3 the optimum order of prediction is not so large in the range direction; therefore, only the edge of a target is detected. This shortcoming is corrected in the following four steps. The STC output signal is differentiated and the PARCOR filter is applied, and finally the signal is integrated. The digital circuit and the waveforms in the four steps are illustrated in Figs. 2.25 and 2.26, respectively.

In the final stage, a factor α (=0.999) is introduced to stabilise the integration against a drift of the data after the previous processes. The original PPI map of area B after STC is illustrated in Fig. 2.27 and PPI maps of the same area processed by systems I and II are illustrated in Figs. 2.28 and 2.29. The iron tower and the radar-wave reflector are seen clearly among sea ice. The false alarm rate was obtained by setting a threshold level at a certain amount below the mean level of the radar-wave reflector. Blocks of sea ice gave a strong reflection and the false alarm rate was 20%, which was reduced to 1% after STC and 0.3% after the PARCOR filter.

The CFAR is applied to the final result to further reduce the background level. The digital Weibull/CFAR circuit is applicable to any distribution in the sense that the local mean level is subtracted after logarithmic transformation; this is equivalent to automatic gain control.

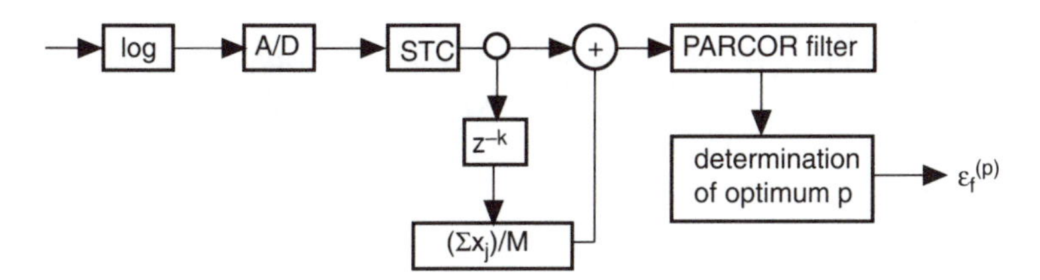

Figure 2.24 *PARCOR clutter reduction system I*

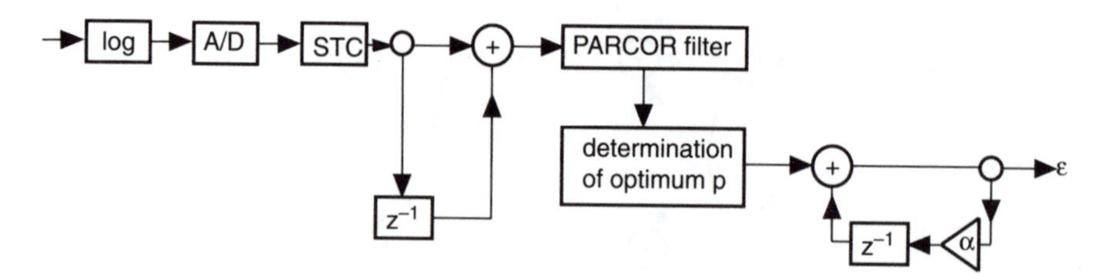

Figure 2.25 *PARCOR clutter reduction system II*

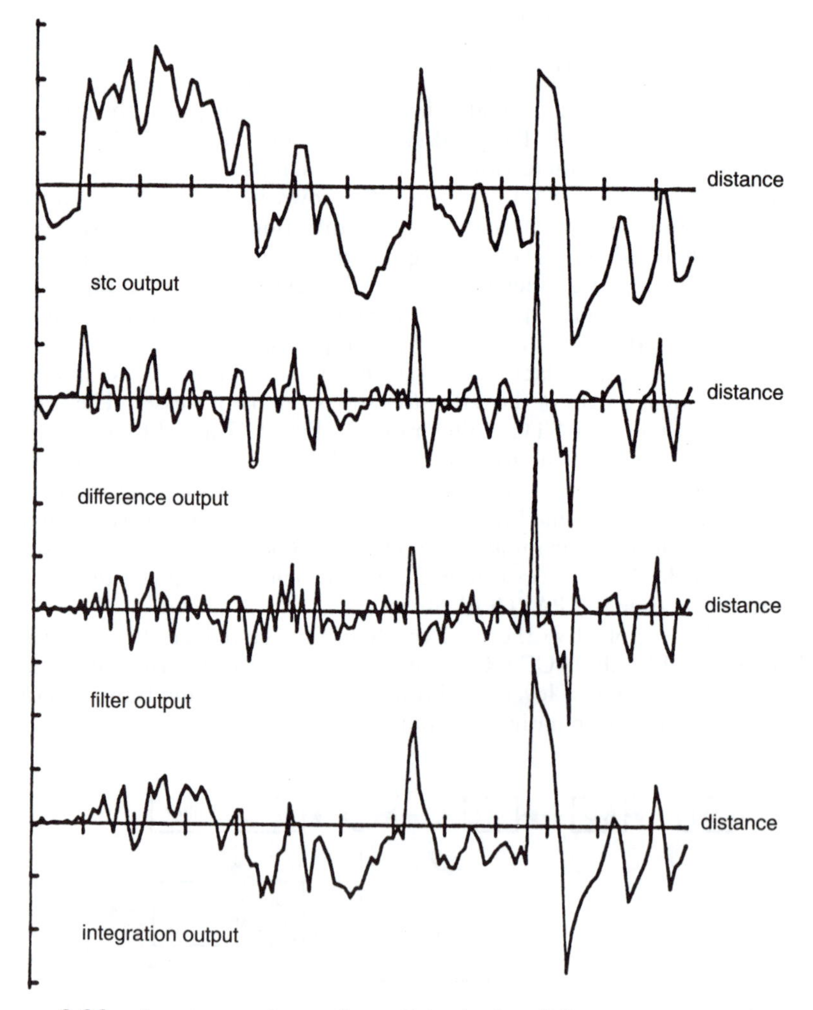

Figure 2.26 *A-scope waveforms of area B in the four different stages*

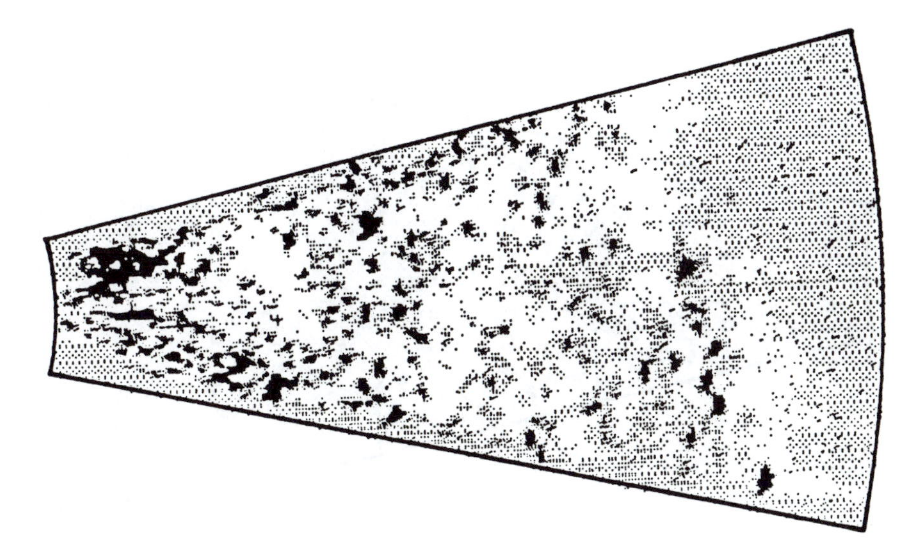

Figure 2.27 *PPI map of area B after STC*

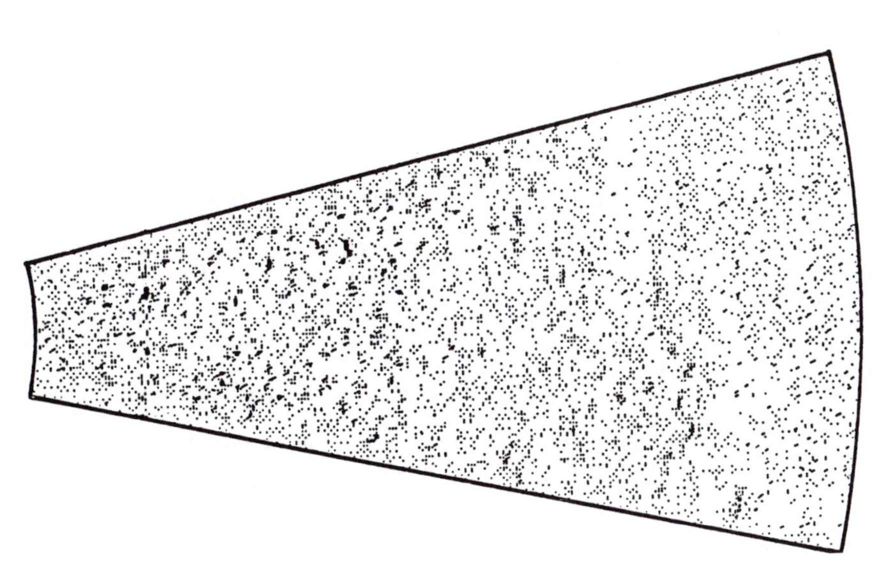

Figure 2.28 *PPI map of area B after process I*

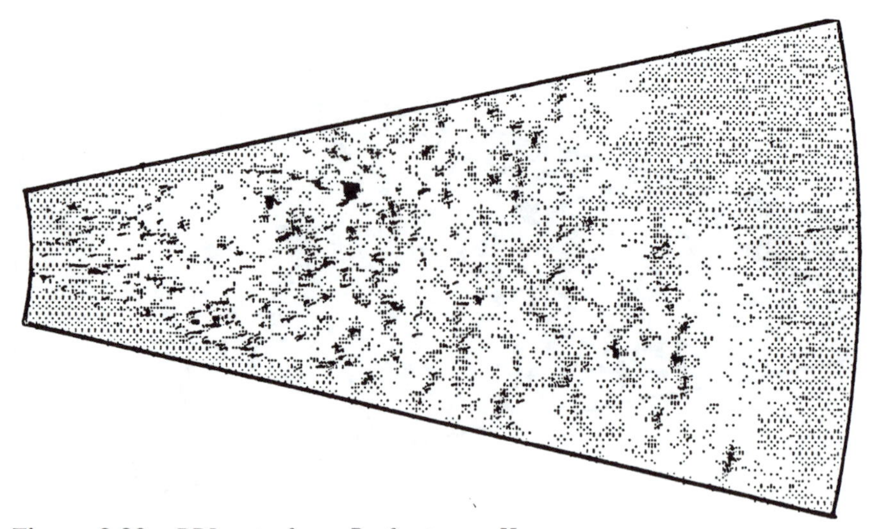

Figure 2.29 *PPI map of area B after process II*

2.4 References

1 SEKINE, M., OHTANI, S., MUSHA, T., IRABU, T., KIUCHI, E., HAGISAWA, T., and TOMITA, Y.: 'Weibull-distributed ground clutter', *IEEE Trans.*, 1981, **AES–17**, pp. 596–598
2 SEKINE, M., OHTANI, S., MUSHA, T., IRABU, T., KIUCHI, E., HAGISAWA, T., and TOMITA, Y.: 'MTI processing and Weibull-distributed ground clutter', *IEEE Trans.*, 1982, **AES–18**, pp. 729–730
3 SEKINE, M., MUSHA, T., TOMITA, Y., HAGISAWA, T., and KIUCHI, E.: 'Weibull-distributed ground clutter in the frequency domain', *Trans. IECE Japan*, 1985, **E68**, pp. 365–370
4 SEKINE, M., MUSHA, T., TOMITA, Y., HAGISAWA, T., IRABU, T., and KIUCHI, E.: 'On Weibull-distributed weather clutter', *IEEE Trans.*, 1979, **AES–15**, pp. 824–830
5 SEKINE, M., MUSHA, T., TOMITA, Y., HAGISAWA, T., IRABU, T., and KIUCHI, E.: 'Suppression of Weibull-distributed weather clutter'. Proc. IEEE International Radar Conf., 1980, pp. 294–298
6 SEKINE, M., MUSHA, T., IRABU, T., KIUCHI, E., HAGISAWA, T., and TOMITA, Y.: 'Non-Rayleigh weather clutter', *IEE Proc. F.*, 1980, **127**, pp. 471–474
7 SEKINE, M., OHTANI, S., MUSHA, T., IRABU, T., HAGISAWA, T., and TOMITA, Y.: 'Suppression of round and weather clutter', *IEE Proc. F.*, 1981, **128**, pp. 175–178
8 SEKINE, M., MUSHA, T., TOMITA, Y., HAGISAWA, T., IRABU, T., and KIUCHI, E.; 'Weibull-distributed weather clutter in the frequency domain', *IEE Proc. F.*, 1984, **131**, pp. 549–552
9 LANDAU, L.D., LANDAU, E., and LIFSHITZ, E.M.: 'Fluid Mechanics'. (Pergamon Press, 1959) chap. 2
10 BOOTH, R.: 'The Weibull distribution applied to the ground clutter backscatter coefficient'. US Army Missile Command, Technical Report RE–TR–69–15, ASTIA Doc. AD 691109, 1969
11 TRUNK, G.V., and GEORGE, S.F.: 'Detection of targets in non Gaussian sea clutter', *IEEE Trans.*, 1970, **AES–6**, pp. 620–628
12 SCHLEHER, D.C.: 'Radar detection in log-normal clutter'. Proc. IEEE International Radar Conference, 1975, pp. 262–267

13 CLARDE, J., and PETER, R.S.: 'The effect of pulse length changes on Weibull clutter'. Royal Radar Establishment Memorandum 3033, 1976
14 SCHLEHER, D.C.: 'Radar detection in Weibull clutter', *IEEE Trans.*, 1976, **AES–12**, pp. 736–743
15 SEKINE, M., MUSHA, T., TOMITA, Y., HAGISAWA, T., IRABU, T., and KIUCHI, E.: 'Weibull-distributed sea clutter', *IEE Proc. F.*, 1983, **130**, p. 476
16 SEKINE, M., MUSHA, T., TOMITA, Y., HAGISAWA, T., IRABU, T., and KIUCHI, E.: 'Log-Weibull distributed sea clutter', *IEE Proc. F.*, 1980, **127**, pp. 225–228
17 OGAWA, H., SEKINE, M., MUSHA, T., AOTA, M., OHI, M., and FUKUSHI, H.: 'Weibull-distributed radar clutter reflected from sea ice', *Trans. IEICE Japan*, 1987, **E70**, pp. 116–120
18 GUMBEL, E.J.: 'Statistics of extremes' (Columbia Univ. Press, 1958)
19 FISHER, R.A., and TIPPET, L.H.C.: 'Limiting forms of the frequency distribution of the largest or smallest member of a sample', *Proc. Cambridge Phil. Soc.*, 1928, **24**, p. 180
20 GNEDENKO, B.V.: 'On the role of the maximal summand in the summation of independent random variables', *Ukarain. Mat. J.*, 1953, **5**, p. 291
21 AKAIKE, H.: 'Information, theory and an extention of the maximum likelihood principle' *in* PETROV, B.N., and CSAKI, F. (Eds.): 2nd International Symposium on Information Theory, Akademiai Kiado, Budapest, 1973, pp. 267–281
22 OGATA, M., SEKINE, M., and MUSHA, T.: 'Clutter suppression using the linear prediction method' *in* MUSHA, T., SUZUKI, T., and OGURA, H. (Eds.): Proceedings of the 1984 International Symposium on noise and clutter rejection in radars and imaging sensors, 1984, pp. 297–302
23 ITAKURA, F., SAITO, S., KOIKE, T., SAWABE, H., and NISHIKAWA, M.: 'An audio response unit based on partial correlation', *IEEE Trans.*, 1972, **COM–20–4**, pp. 729–797

Chapter 3

CFAR techniques in clutter

Prof. T. Musha and Prof. M. Sekine

3.1 Linear and logarithmic CFAR techniques

Constant false alarm rate (CFAR) techniques have been developed by several people [1, 2, 3, 4]. This technique keeps the receiver output level constant against Rayleigh-distributed clutter and effectively suppresses the clutter to the receiver noise level. Depending on whether the dynamic range is small or large, linear and logarithmic CFAR circuits are used. In Reference 4, operations of two CFAR circuits with square-law detectors are discussed and the detection probability and the CFAR loss are evaluated for a finite number of samples. It is concluded that the linear CFAR receiver gives higher detection probability and lower CFAR loss as compared with the logarithmic CFAR receiver. We will calculate the false alarm probability, detection probability and CFAR loss for a finite number of samples by means of the linear detector. Indeed, the square-law detector is suitable for small target-to-clutter ratios, while for large target-to-clutter ratios the linear detector is more appropriate [5].

3.2 Linear and logarithmic CFAR receivers

Block diagrams of the linear and logarithmic CFAR receiver are shown in Figs. 3.1 and 3.2.

The difference between linear and logarithmic CFAR receivers is in taking arithmetic or geometric averages.

3.3 Linear CFAR (Lin CFAR)

First we consider the linear CFAR circuit. If the clutter amplitude x obeys the following Rayleigh distribution:

$$p(x, \sigma) = (2x/\sigma^2)\ \exp(-x^2/\sigma^2) \qquad (3.1)$$

we obtain the mean value $\langle x \rangle$ and the mean-squared value $\langle x^2 \rangle$ as

$$\langle x \rangle = (\pi^{1/2}/2)\sigma \qquad (3.2)$$

$$\langle x \rangle = \sigma^2 \qquad (3.3)$$

Hence the variance of x is calculated as

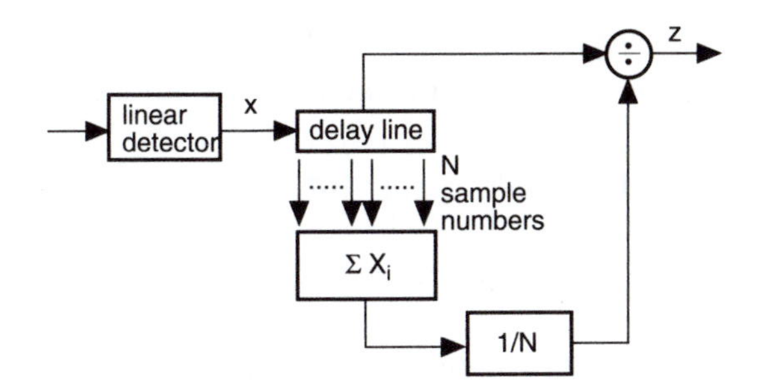

Figure 3.1 *Linear CFAR circuit*

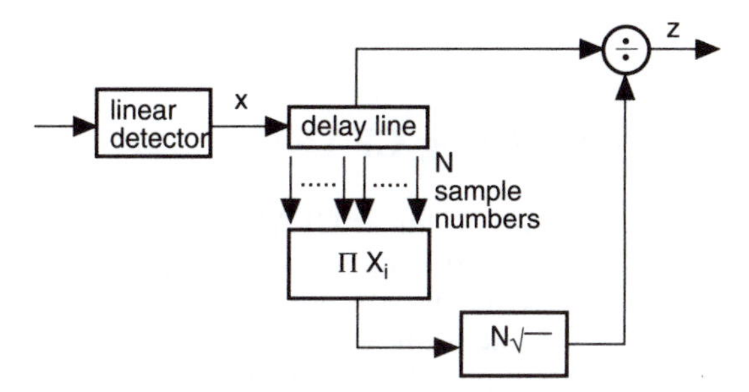

Figure 3.2 *Logarithmic CFAR circuit*

$$E(x) = \langle x^2 \rangle - \langle x \rangle^2 = \sigma^2 - \pi\sigma^2/4 = 0.2146\sigma^2 \tag{3.4}$$

If the clutter return x passes through the linear CFAR receiver as shown in Fig. 3.1, the output z is calculated as

$$z = x/\langle x \rangle = (2x)/\sigma\pi^{1/2} \tag{3.5}$$

Thus the probability density function of the output z is written as

$$P(z) = (\pi/2)z \exp(-\pi/4z^2) \tag{3.6}$$

from which we obtain

$$\langle z \rangle = 1 \tag{3.7}$$

$$\langle z^2 \rangle = 4/\pi \tag{3.8}$$

$$E(z) = \langle z^2 \rangle - \langle z \rangle^2 = 4/\pi - 1 = 0.2732$$

Thus the variance remains constant independently of σ, and CFAR is realised.

The false alarm probability p_{fa} is the probability that a clutter signal above the threshold T is misjudged as a target signal, and this probability is given by

$$p_{fa} = \int_T^\infty p(z)dz = \exp[(-\pi/4)T^2] \tag{3.9}$$

Suppose the amplitude of an echo signal from a target is equal to constant value s. Then eqn. 3.5 is written as

$$z_s = 2(x+s)/(\sigma\pi^{1/2}) \tag{3.10}$$

and the detection probability p_d for the target at the threshold T is given by

$$p_d = \int_T^\infty p(z_s)dz_s = \exp[-\{(\pi^{1/2}/2)T - s/\sigma\}]^2 \tag{3.11}$$

The detection probability p_d depends on target-to-clutter ratio T/C which is defined as

$$T/C = 10\ \log(s^2/\langle x^2\rangle) = 10\ \log(s^2/\sigma^2) \tag{3.12}$$

From eqns. 3.9, 3.11 and 3.12, p_d is rewritten in terms of p_{fa} and T/C as

$$p_d = \exp[-\{(-\ln p_{fa})^{1/2} - 10(T/C)/20\}^2]. \tag{3.13}$$

3.4 Logarithmic CFAR (log CFAR)

Log CFAR is more practical than linear CFAR because of its larger dynamic range [4]. Log CFAR is obtained simply by passing input signals through a logarithmic amplifier and removing the local mean value. This log CFAR circuit corresponding to Fig. 3.2 is shown in Fig. 3.3.

Suppose the input signal amplitude x is converted into y after passing a log-amplifier. Then we have

$$y = k\ \ln(lx), \tag{3.14}$$

where k and l are amplifier constants. The mean value of y is given by

$$\langle y\rangle = k\ \ln(b\sigma) - (k/2)\gamma \tag{3.15}$$

where γ is Euler's constant which is equal to 0.5772. . . . The mean square value of y is given by

$$\langle y^2\rangle = k^2\ \ln^2(b\sigma) - k^2\ \ln(b\sigma)\gamma + (k^2/4)\,[\pi^{1/2}/6 + \gamma^2]$$

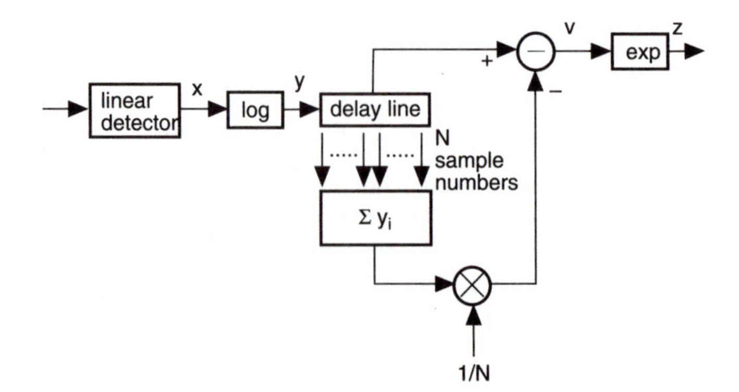

Figure 3.3 *Logarithmic CFAR circuit corresponding to Fig. 3.2.*

From these equations, the variance of y is calculated as

$$E(y) = \langle y^2 \rangle - \langle y \rangle^2 = (k^2/4)(\pi^{1/2}/6) \tag{3.16}$$

Thus the variance of y does not depend on σ but on the amplifier constant k only, and the constant false alarm rate is realised. This is the ordinary log CFAR process.

Subtracting the mean value $\langle y \rangle$ from y, we define

$$v = y - \langle y \rangle \tag{3.17}$$

the output signal z of an anti-logarithmic amplifier is represented by

$$z = m \exp(nv) \tag{3.18}$$

where m and n are constants of the amplifier. Substituting eqns. 3.14 and 3.15 into eqn. 3.17, and then eqn. 3.17 into eqn. 3.18, we obtain

$$z = c(x/\sigma)^{kn} \exp\{(kn/2)\gamma\} \tag{3.19}$$

If the constants are determined such that $kn = 1$, eqn. 3.19 becomes

$$z = (cx/2) \exp(\gamma/2). \tag{3.20}$$

The probability density function of z is derived as

$$p(z) = (2z/c^{1/2} \exp(\gamma)) \exp[-z^{1/2}/c^2 \exp(\gamma)] \tag{3.21}$$

The mean value and the mean squared value of z are given by

$$\langle z \rangle = (c\pi^{1/2})/2 \exp(\gamma/2) \tag{3.22}$$

$$\langle z^2 \rangle = c^2 \exp(\gamma) \tag{3.23}$$

The variance of z is then written as

$$E(z) = \langle z^2 \rangle - \langle z \rangle^2 = c^2 \exp(\gamma)(1 - \pi/4) \tag{3.24}$$

When the threshold level is set at T, the false alarm rate p_{fa} is given by

$$p_{fa} = p(z)dz = \exp[-T^2/c^2 \exp(\gamma)] \tag{3.25}$$

We now put

$$T = KE(z) = K(E(z))^{1/2} \tag{3.26}$$

where K is a constant, and eventually we have

$$p_{fa} = \exp[-K^2(1 - \pi/4)] \tag{3.27}$$

For $K = 8.023$, one has $p_{fa} = 10^{-6}$.

With a non-fluctuating target signal s, eqn. 3.20 is written as

$$z_s = c(x + s)/\sigma \exp(\gamma/2) \tag{3.28}$$

and the detection probability p_d is given by

$$p_d = p(z_s)dz_s = \exp[-\{\pi/c \exp(\gamma/2) - s/\sigma\}^2] \tag{3.29}$$

From Eqns. 3.12 and 3.25, eqn. 3.29 is written as

$$p_d = \exp\{-[(-\ln p_{fa})^{1/2} - 10(r/c)/20]^2\} \tag{3.30}$$

which is the same as eqn. 3.13.

3.5 CFAR loss

In applying these linear and logarithmic CFAR methods to practical problems, we have considered the effect of the number of samples N being finite. The true algebraic or geometrical mean value of the signal cannot be estimated from N samples. The threshold level T_N decided by a finite number of samples is inevitably higher than the asymptotic threshold level T_∞. The CFAR loss depends on the value of $T_N - T_\infty$. Calculation of the CFAR loss for a finite number of samples N was made for a Rayleigh distribution. A Rayleigh-distributed variable x was generated by

$$x = \sigma(-\ln\xi)^{1/2} \tag{3.31}$$

where ξ is a uniformly distributed random number over the interval (0,1). As many random numbers as 2.4×10^7 were used in calculating the low value of the false alarm rate, for example, $p_{fa} = 10^{-6}$. The number of samples N is selected as 2, 4, 8, 16, 32 and 64, and we put $\sigma = 10$.

The false alarm rate is plotted in Figs. 3.4 and 3.5 for the log and lin CFARs; broken lines refer to the case $N = \infty$.

The threshold level which is required to yield $p_{fa} = 10^{-1}$, 10^{-2}, 10^{-3}, 10^{-4}, 10^{-5} and 10^{-6} is given in Table 3.1; log CFAR generally requires a higher threshold level than linear CFAR under a given false alarm rate [6, 7].

Now we define v as the threshold/mean-square-noise ratio, that is

$$v = T/\langle z^2 \rangle \tag{3.32}$$

From eqns. 3.9 and 3.25, p_{fa} is given by

$$p_{fa} = \exp(-v^2) \tag{3.33}$$

Here we have examined cases $N = 4$, 8, 16, 32 and ∞ for lin and log CFAR circuits. The false alarm probability for $N = \infty$ is given by eqn. 3.33. The results of Monte Carlo simulation are shown in Fig. 3.6, where the threshold/mean-square-noise ratio of a log CFAR circuit is higher than that of a lin CFAR circuit for a given false alarm probability.

The detection probability p_d was calculated as a function of T/C, and the results are shown in Figs. 3.7–3.12 for various false alarm rates.

CFAR losses with $p_d = 0.5$ and $p_d = 0.9$ are shown in Figs. 3.13 and 3.14.

3.6 Log CFAR technique for Weibull clutter

The Weibull distribution is again given by

$$P_c(x) = \begin{cases} (c/b)\,(x/b)^{c-1} \exp[-(x/b)^c] & x > 0,\ b > 0,\ c > 0 \\ 0 & \text{otherwise} \end{cases} \tag{3.34}$$

where b and c are scale and shape parameters, respectively. For $c = 2$ we find the Rayleigh distribution given by eqn. 3.1.

Our observation of sea clutter [8] showed $c = 1.585$ using an antenna of an L-band air route surveillance radar (ARSR) with a frequency of 1.33 GHz, a beam width of 1.23°, horizontal linear polarisation, pulse width of 3.0 μs, pulse repetition frequency of 348 Hz and transmitted peak power of 2 MW. Radar echoes were taken from sea state 3 in a range interval of 23–28.6 nm at a fixed azimuth angle of 154°. The grazing angles were 0.5° at 28.6 nm and 0.72° at 23 nm. To determine the value of the shape parameter of Weibull-distributed sea clutter for the purpose of CFAR, a sample area of 23 range bins was selected corresponding to 5.6 nm using a fixed antenna at 154°, where the number of data points was 120 060.

Eqn. 3.34 is also rewritten as

$$Y = cX - c \ln b \tag{3.35}$$

where

$$Y = \ln[-\ln\{1 - \int_0^x P_c(x)\}]$$

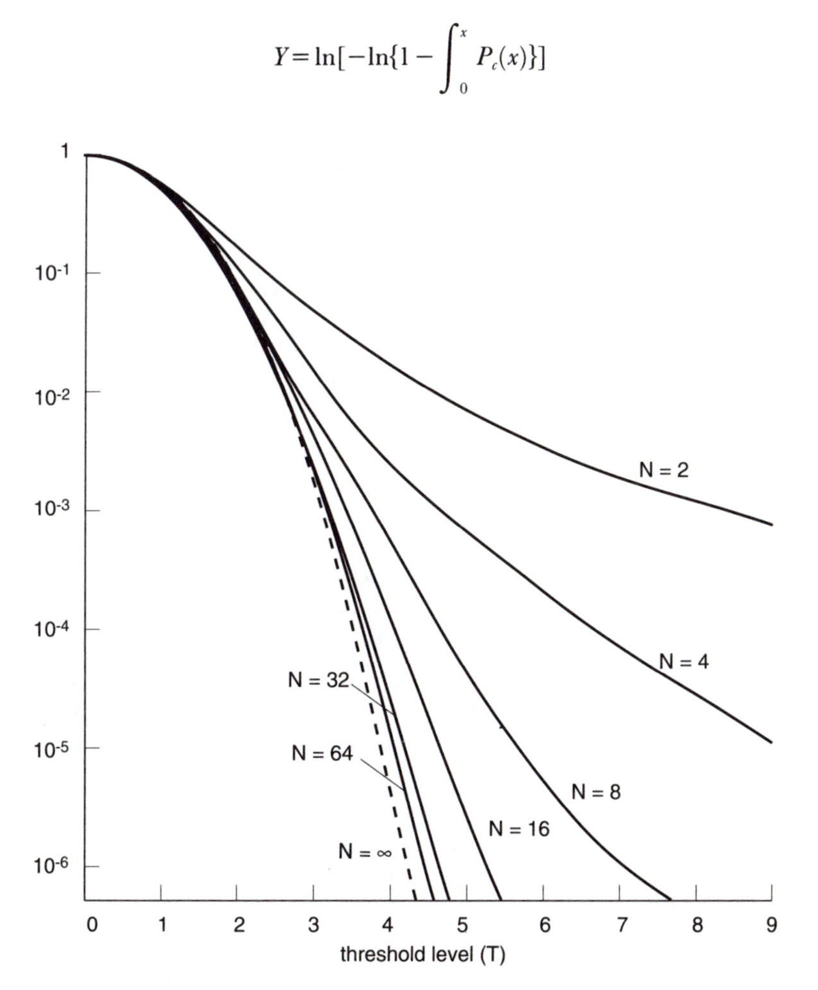

Figure 3.4 *False alarm rate with lin CFAR*

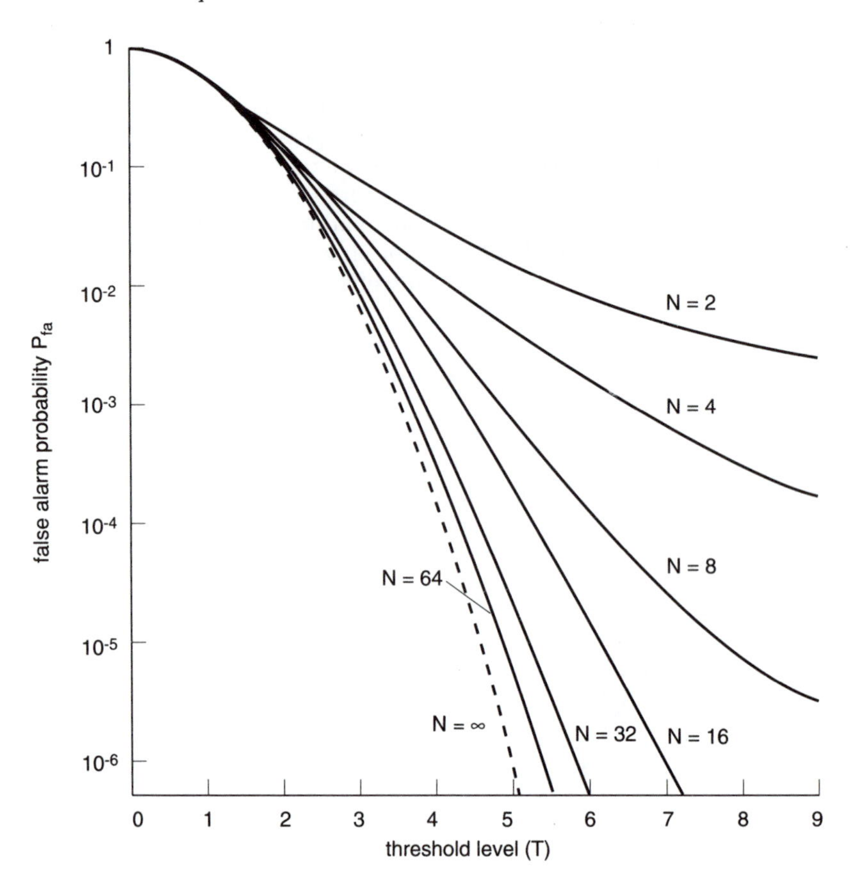

Figure 3.5 *False alarm rate with log CFAR*

Table 3.1 *Threshold level value versus false alarm rate for various numbers of samples*

	$N=$	2	4	8	16	32	64	∞
(10^{-1})	log CFAR	2.534	2.287	2.158	2.092	2.058	2.041	2.025
	lin CFAR	2.223	1.942	1.822	1.765	1.736	1.725	1.712
(10^{-2})	log CFAR	5.600	4.069	3.428	3.136	2.999	2.931	2.864
	lin CFAR	4.548	3.257	2.796	2.600	2.509	2.465	2.421
(10^{-3})	log CFAR	11.101	6.335	4.736	4.077	3.783	3.642	3.508
	lin CFAR	8.372	4.774	3.726	3.316	3.135	3.049	2.966
(10^{-4})	log CFAR	21.108	9.390	6.164	4.997	4.508	4.267	4.050
	lin CFAR	15.096	6.672	4.687	3.986	3.689	3.553	3.424
(10^{-5})	log CFAR	40.533	13.670	7.908	5.950	5.204	4.848	4.529
	lin CFAR	26.879	9.162	5.723	4.649	4.208	4.001	3.829
(10^{-6})	log CFAR	79.000	19.830	10.328	7.009	5.486	5.375	4.960
	lin CFAR	49.898	12.403	7.100	5.283	4.680	4.458	4.194

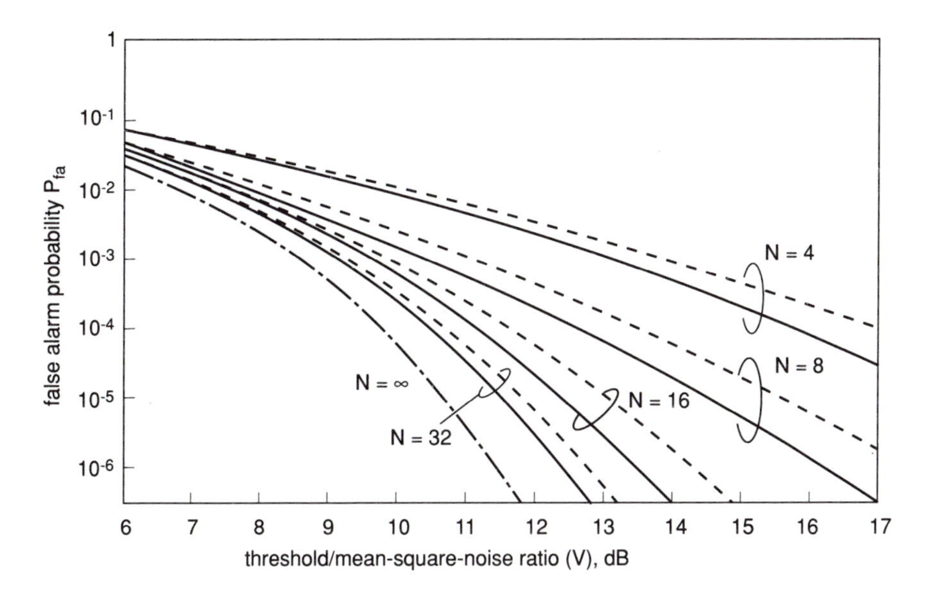

Figure 3.6 *False alarm probability versus threshold/mean-square-noise ratio for lin and log CFAR circuits*

—— lin CFAR
– – – log CFAR

and

$$X = \ln x \tag{3.36}$$

From eqn. 3.35, the value of the shape parameter c is easily estimated from a plot of Y against X. This is shown in Fig. 3.15.

The statistical scatter was estimated by using 120 060 samples. The departure of the empirical distribution from the Weibull distribution was estimated by calculating the root mean square error (RMSE) of the linear fit. The RMSE is the deviation of the data points from the straight line which is obtained by the least-squares method. In Fig. 3.15, one unit in X is 8.7 dB.

If sea clutter returns obey a Weibull distribution, perfect suppression of the clutter cannot be attained by using the conventional CFAR system, since the output clutter level is dependent on the variance and hence on the shape parameter of the original Weibull distribution. To suppress such Weibull-distributed clutter, a new CFAR method should be considered [9, 10, 11].

To see how the false alarm rate increases as c deviates from 2, suppose that the Weibull-distributed signals x in eqn. 3.34 pass through the conventional log CFAR circuit as shown in Fig. 3.3. After passing a log-amplifier of eqn. 3.14, the mean value of y is given by

$$\langle y \rangle = k \ln(b\sigma) - (k/c)\gamma \tag{3.37}$$

The mean square of y is given by

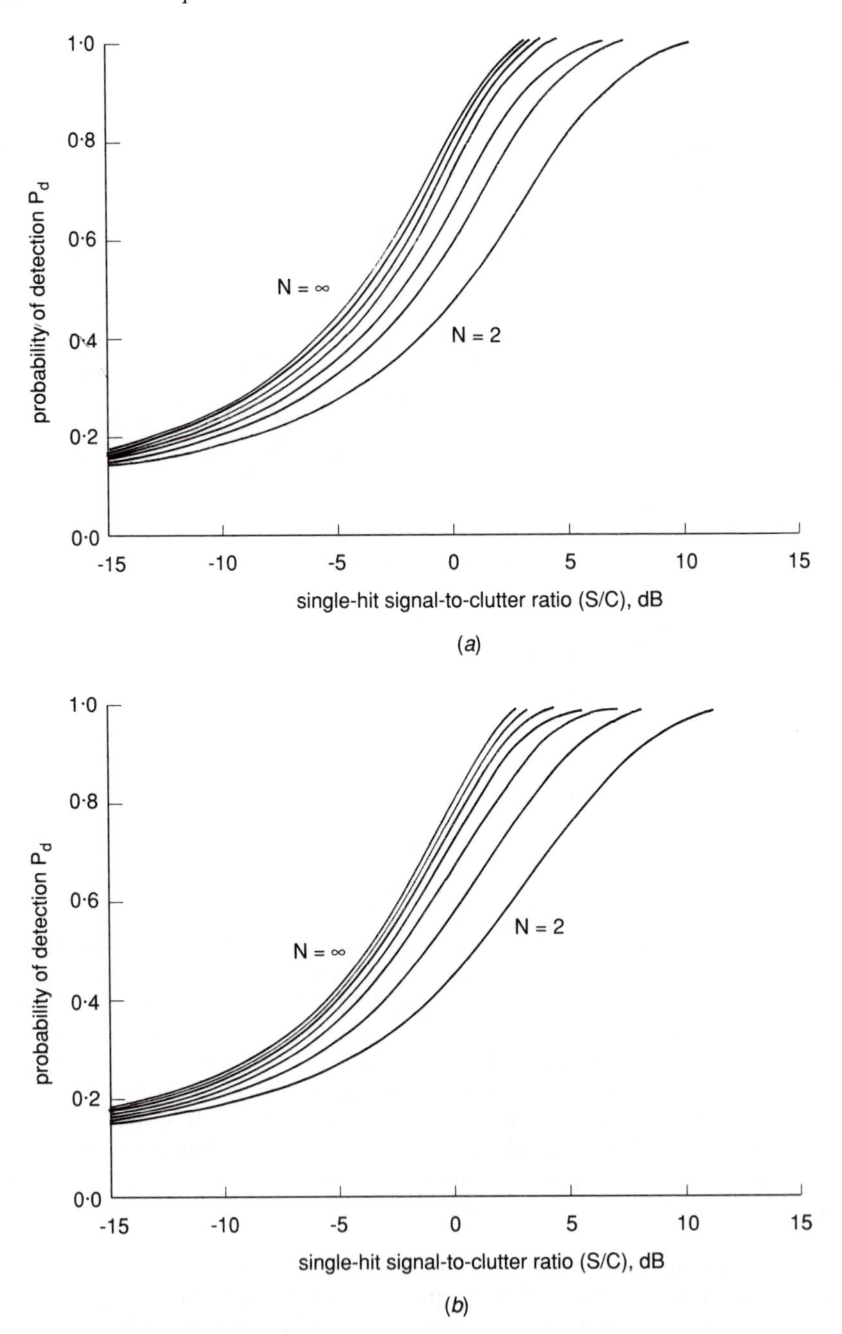

Figure 3.7 *Detection probability of a target as a function of T/C with lin and log CFARs for $p_{fa} = 10^{-1}$*

a lin CFAR
b log CFAR

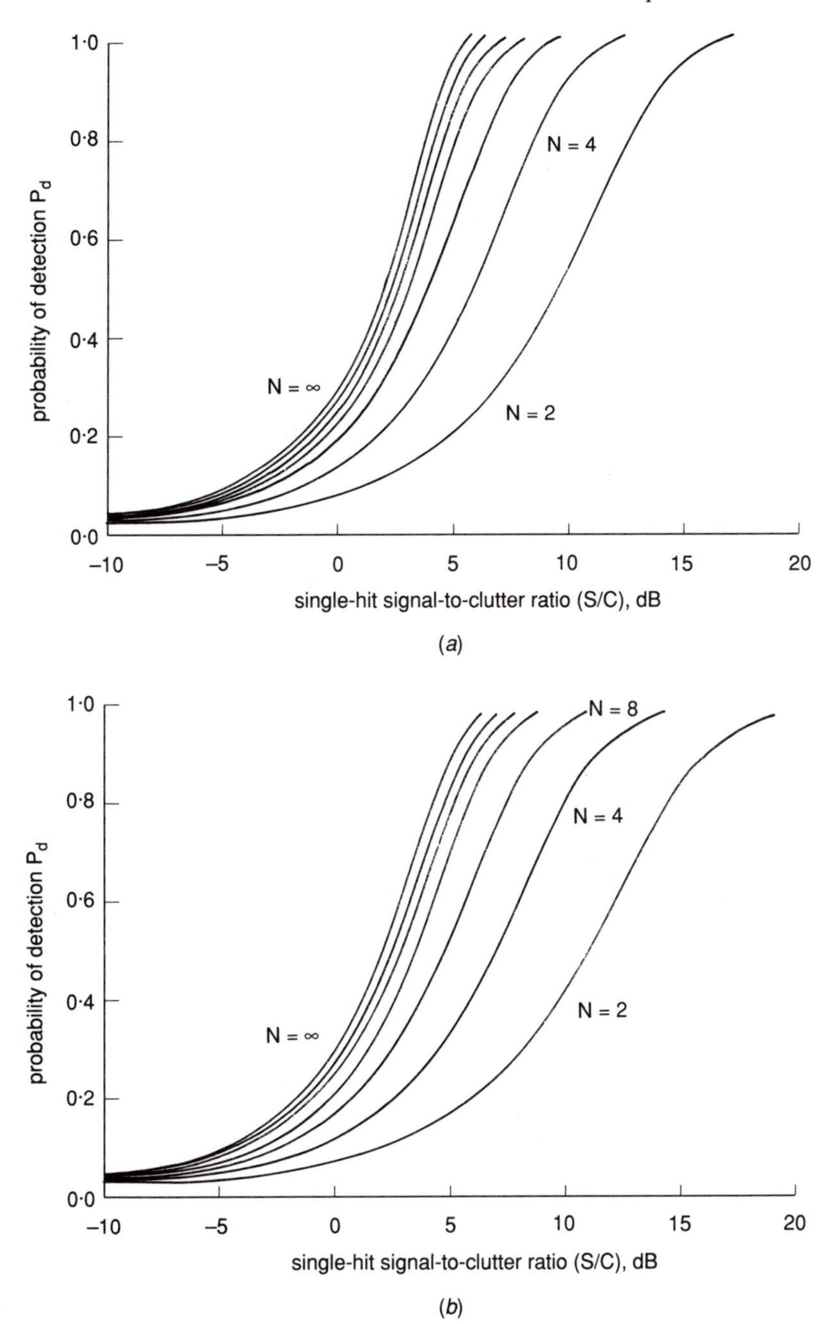

Figure 3.8 *Detection probability of a target as a function of T/C with lin and log CFARs for $p_{fa} = 10^{-2}$*

a lin CFAR
b log CFAR

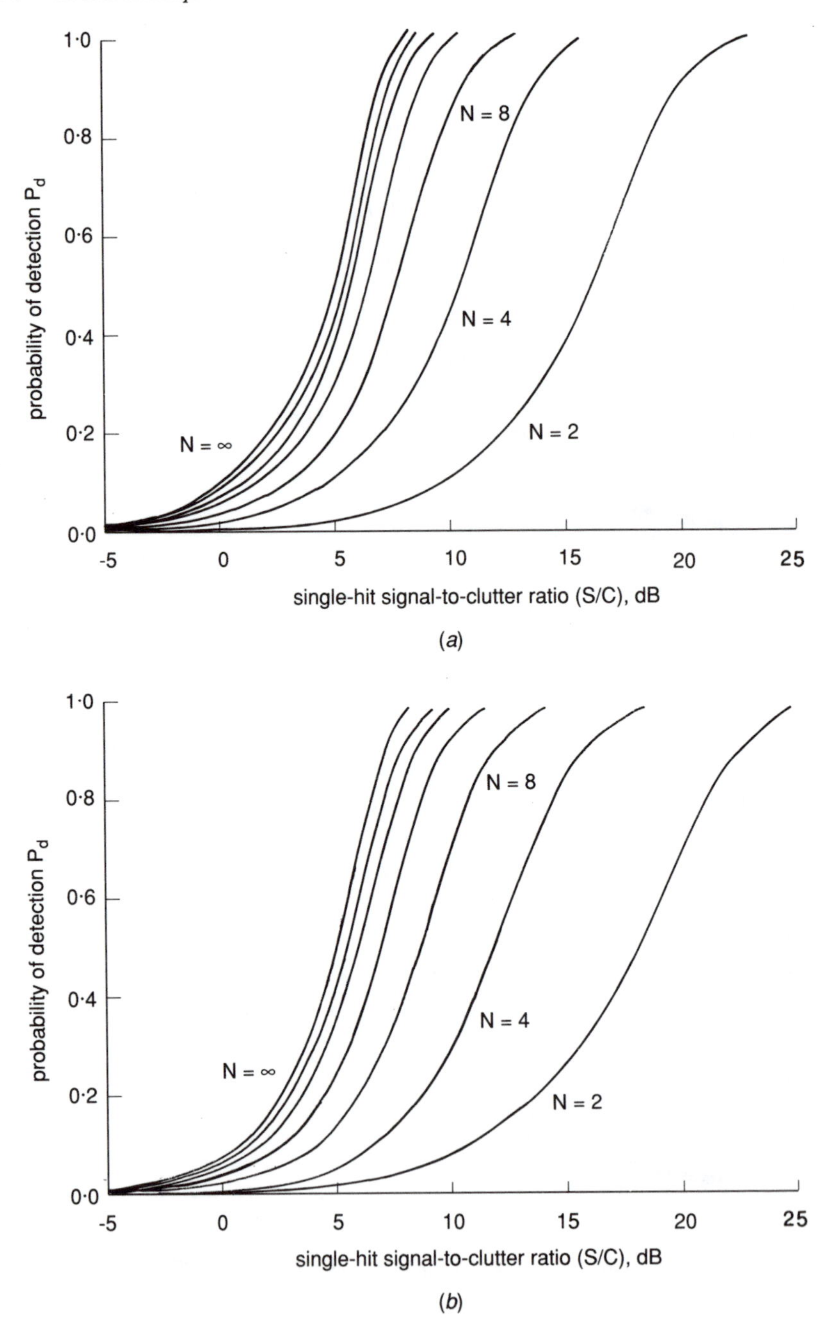

Figure 3.9 *Detection probability of a target as a function of T/C with lin and log CFARs for $p_{fa} = 10^{-3}$*

a lin CFAR
b log CFAR

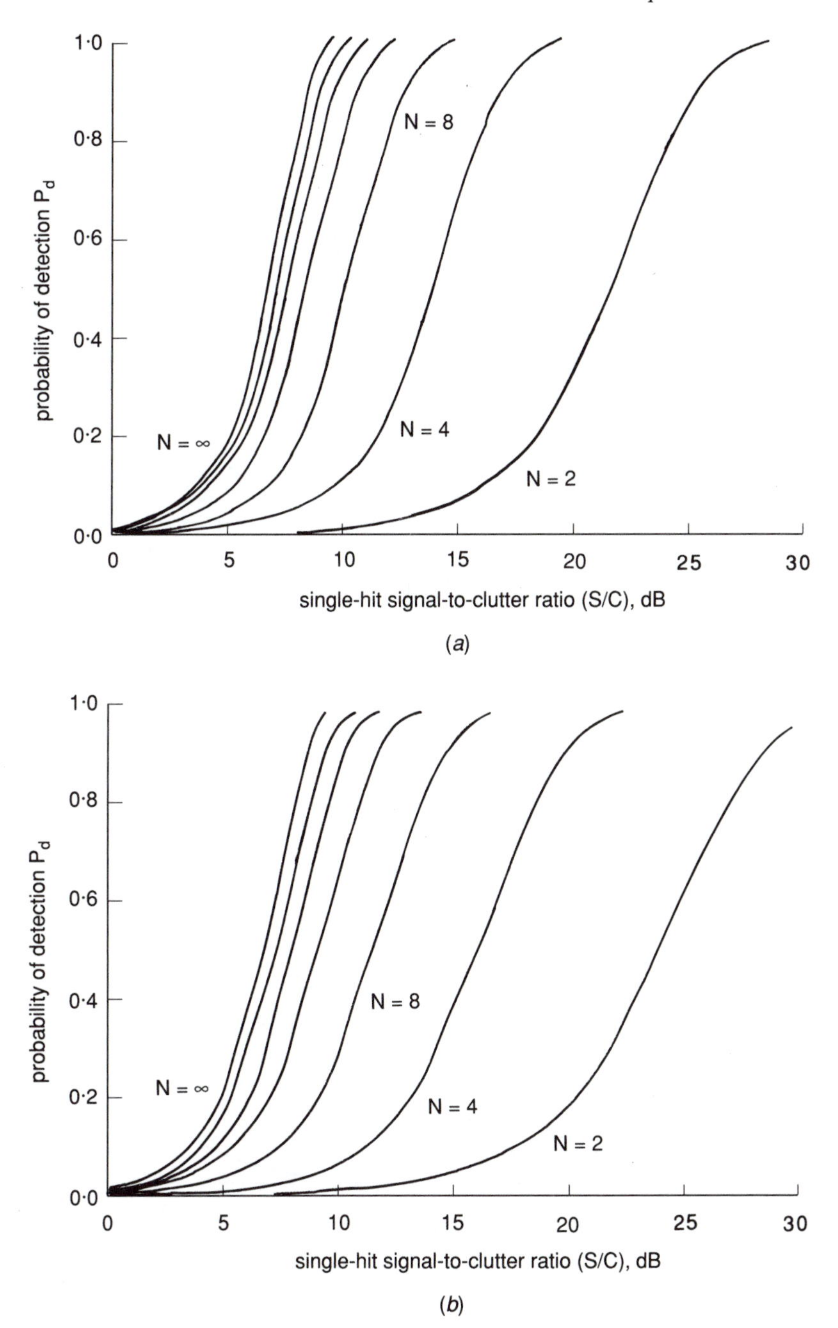

Figure 3.10 *Detection probability of a target as a function of T/C with lin and log CFARs for $p_{fa} = 10^{-4}$*

a lin CFAR
b log CFAR

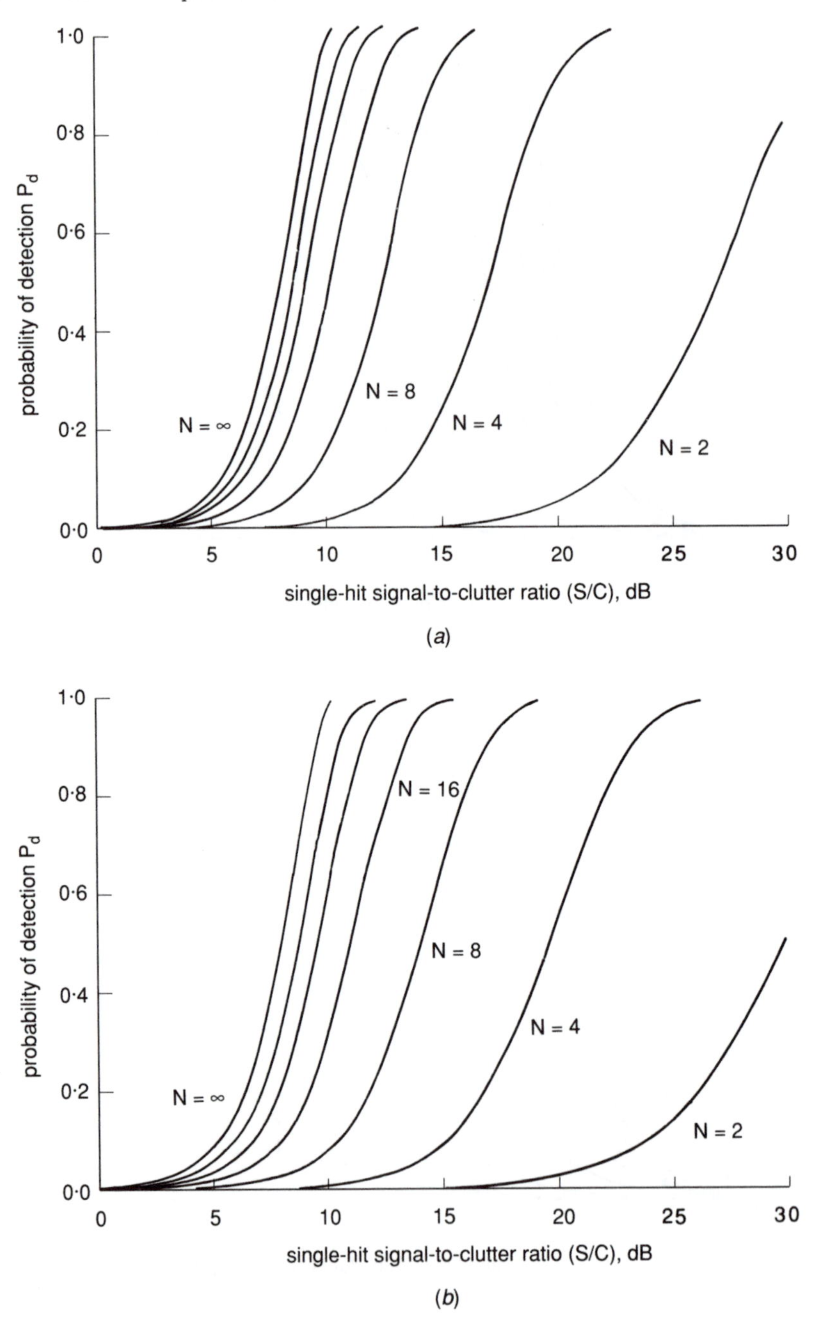

Figure 3.11 *Detection probability of a target as a function of T/C with lin and log CFARs for $p_{fa} = 10^{-5}$*

a lin CFAR
b log CFAR

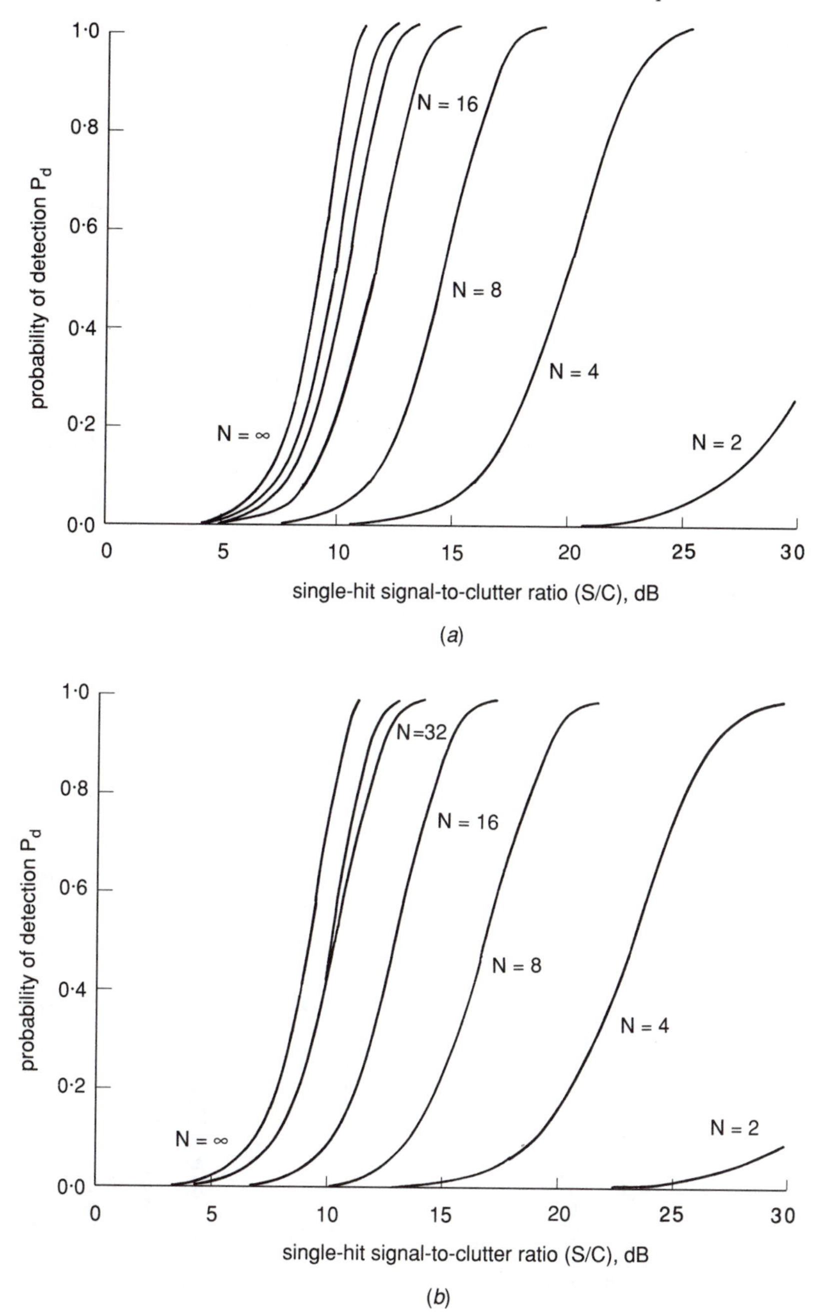

Figure 3.12 *Detection probability of a target as a function of T/C with lin and log CFARs for $p_{fa} = 10^{-6}$*

a lin CFAR
b log CFAR

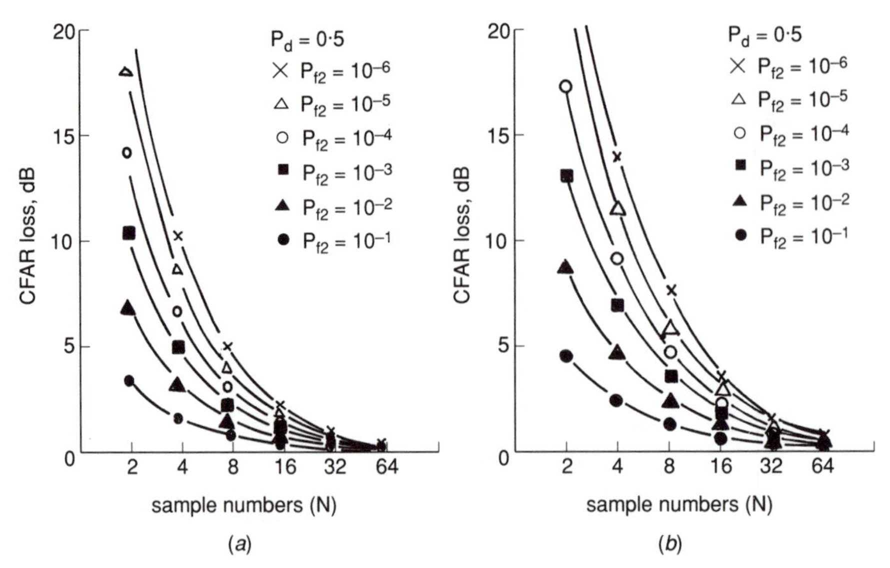

Figure 3.13 *CFAR loss with $p_d = 0.5$ as a function of the number of samples N with lin and log CFARs for various false alarm rates*

a lin CFAR
b log CFAR

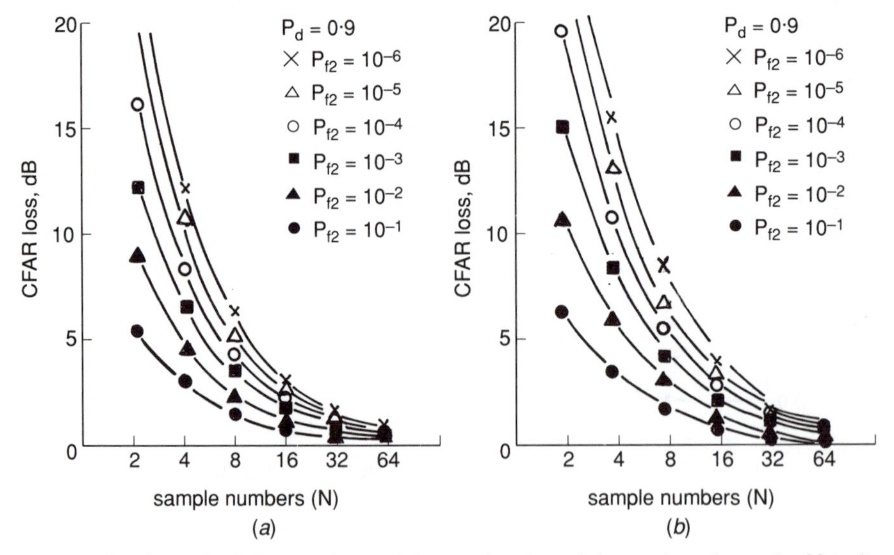

Figure 3.14 *CFAR loss with $p_d = 0.9$ as a function of the number of samples N in lin and log CFARs for various false alarm rates*

a lin CFAR
b log CFAR

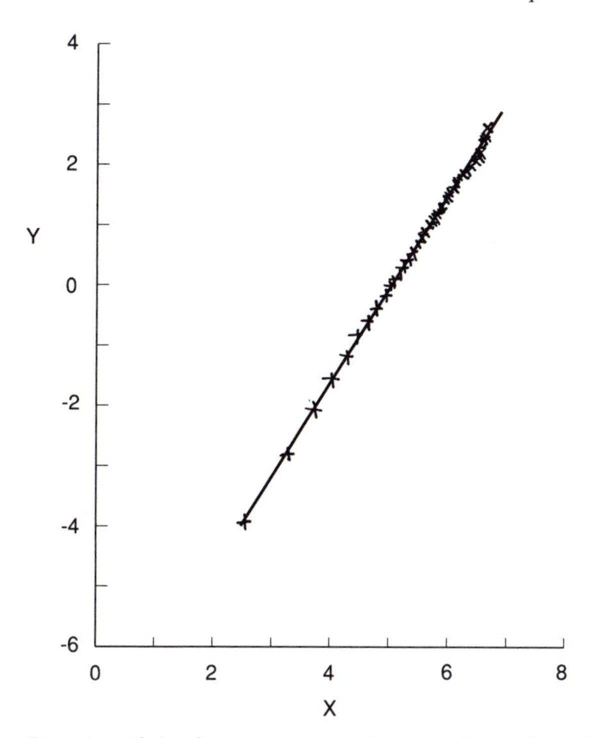

Figure 3.15 *Detection of the shape parameter c from sea clutter data of sea state 3*
$c = 1.585$, RMSE = 0.043

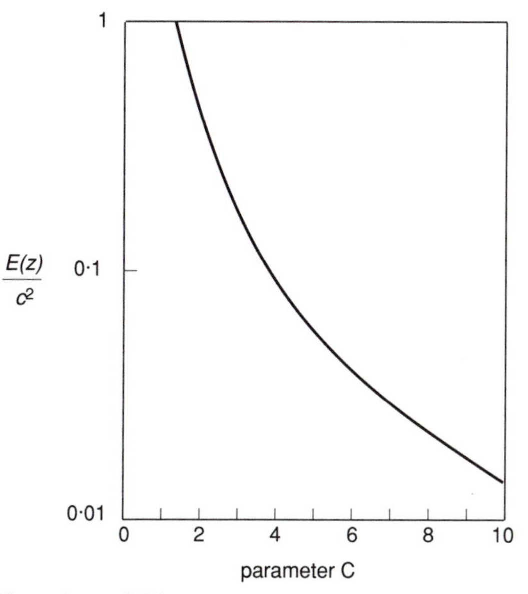

Figure 3.16 *Plots of eqn. 3.34*

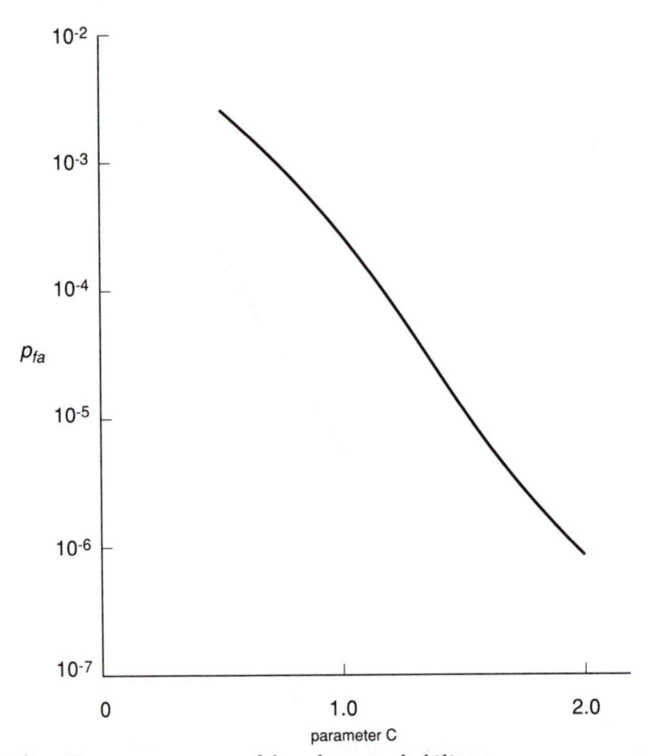

Figure 3.17 *Parameter c versus false alarm probability p_{fa}*

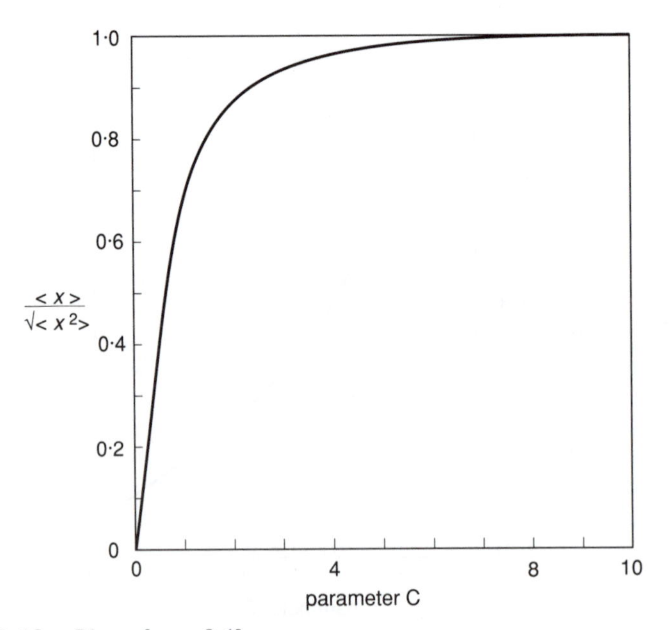

Figure 3.18 *Plots of eqn. 3.49*

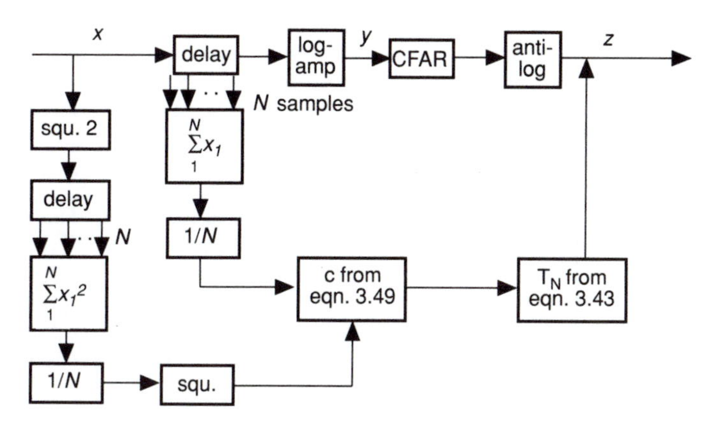

Figure 3.19 *Log CFAR method for Weibull clutter (type 1)*

$$\langle y^2 \rangle = k^2 \ln^2(b\sigma) - (2k^2/c) \ln(b\sigma)\gamma + (k^2/c^2)(\pi^2/6 + \gamma^2) \tag{3.38}$$

from which the variance of y is

$$E(y) = \langle y^2 \rangle - \langle y \rangle^2 = (k^2/c^2)(\pi^2/6) \tag{3.39}$$

The variance of y is dependent on c and the circuit parameter k only but not on the initial variance of the signal. In the case of Rayleigh signals with $c=2$, we have eqn. 3.16. After subtracting the mean value and then passing through an antilog-amplifier of eqn. 3.18, we obtain

$$z = (mx/\sigma) \exp(\gamma/c) \tag{3.40}$$

Here we used the relation $kn=1$. Hence

$$\langle z \rangle = m \exp(\gamma/c)\Gamma(1/c+1) \tag{3.41}$$

$$\langle z^2 \rangle = m^2 \exp(2\gamma/c)\Gamma(2/c+1) \tag{3.42}$$

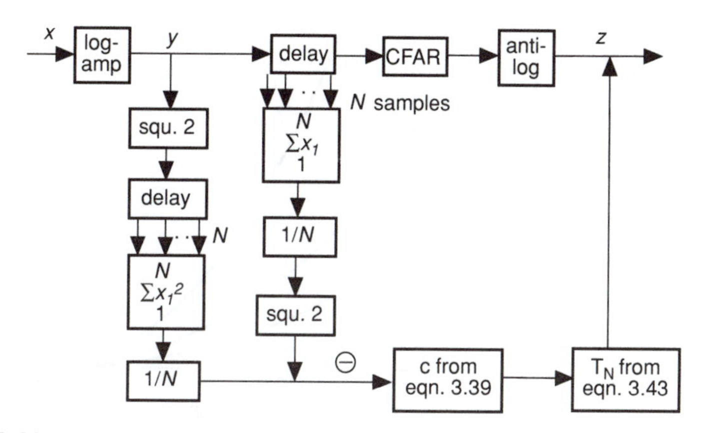

Figure 3.20 *Log CFAR method for Weibull clutter (type 2)*

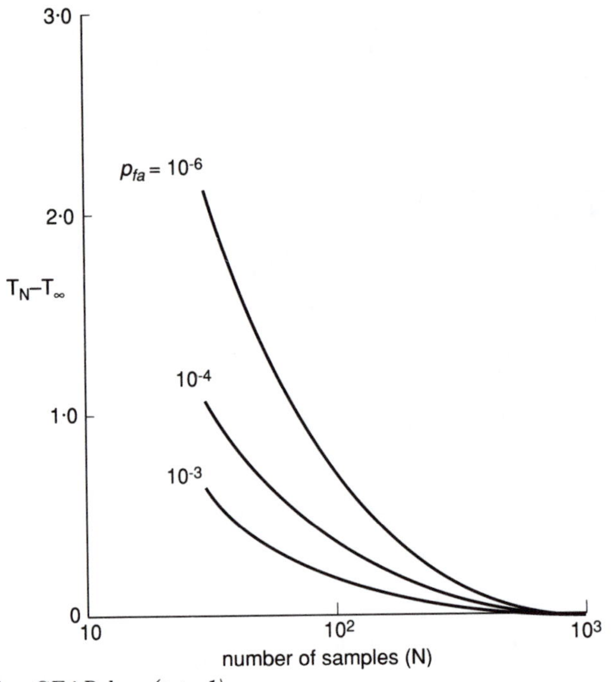

Figure 3.21 *CFAR loss (type 1)*

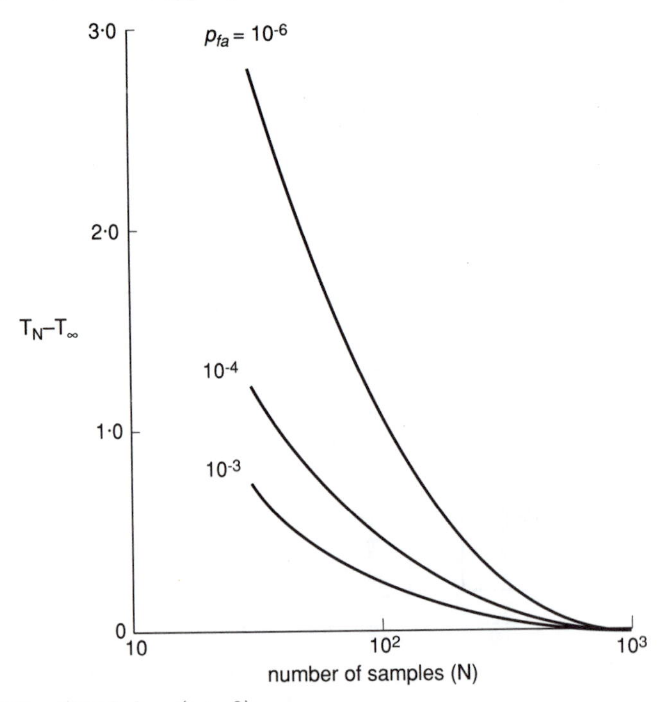

Figure 3.22 *CFAR loss (type 2)*

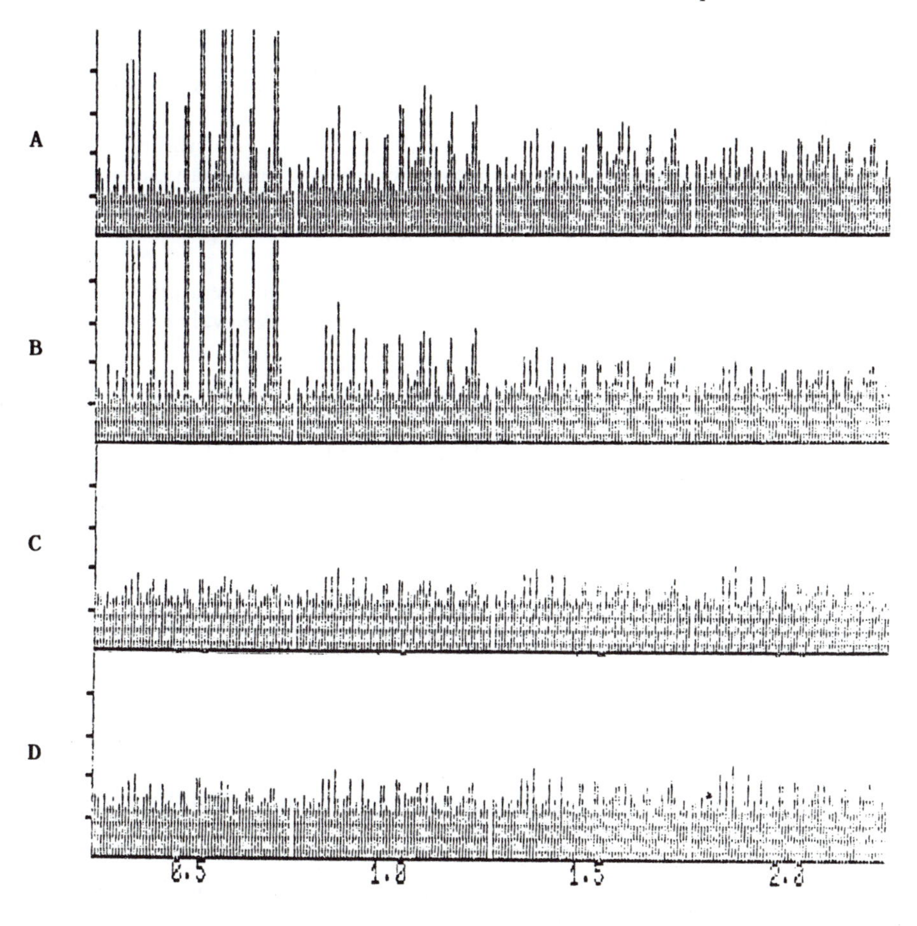

Figure 3.23 *Log CFAR methods for Weibull clutter: sample number N = 16*

A: Weibull distribution generated from a random number of eqn. 3.50
B: Conventional log CFAR method of Fig. 3.3
C: Type 1 of Fig. 3.19
D: Type 2 of Fig. 3.20

$$E(z) = \langle z^2 \rangle - \langle z \rangle^2 = m^2 \exp(2\gamma/c)\,[\Gamma(2/c+1) - \Gamma^2(1/c+1)] \qquad (3.43)$$

where $\Gamma(.)$ is the gamma function. A plot of eqn. 3.34 is given in Fig. 3.16.

The false alarm probability is given by

$$p_{fa} = \int_T^{\infty} p(z)dz = \exp[-(T/m \exp(\gamma/c)]^c \qquad (3.44)$$

Putting $T = KE(z)$, we obtain

$$p_{fa}=\exp[-\{K(\Gamma(2/c+1)-\Gamma^2(1/c+1))\}^c] \tag{3.45}$$

For $c=2$ and $K=8.023$, eqn. 3.27 gives $p_{fa}=10^{-6}$. Fig. 3.17 shows the dependence of p_{fa} on c for $K=8.023$.

For observed sea clutter where $c=1.585$, the false alarm probability when log CFAR is applied becomes larger than those where $c=2$, and many false targets appear on a plan position indicator (PPI). This common difficulty is overcome by the methods which will be described below.

First method: The value of y is determined from the data at the input to the log-amplifier, from which the threshold value is obtained. We evaluate

$$\langle x\rangle=b\Gamma(1/c+1) \tag{3.46}$$

$$\langle x^2\rangle=b^2\Gamma(2/c+1) \tag{3.47}$$

$$E(x)=\langle x^2\rangle-\langle x\rangle^2=b^2[\Gamma(2/c+1)-\Gamma^2(1/c+1)] \tag{3.48}$$

The ratio

$$\frac{\langle x\rangle}{\sqrt{\langle x^2\rangle}}=\frac{\Gamma(1/c+1)}{\sqrt{\Gamma(2/c+1)}} \tag{3.49}$$

does not include b and depends on c only. The function on the right-hand side is plotted in Fig. 3.18.

Once the values of $\langle x\rangle$ and $\langle x^2\rangle$ are obtained we can find the value of c. Then the variance of z is predicted from eqn. 3.34 and Fig. 3.16. Therefore, CFAR will be realised by adjusting the threshold in this way. The block diagram for this method is shown in Fig. 3.19.

This method, however, does not permit a large dynamic range. This difficulty is overcome by the next method.

Second method: The value of c is determined after the log-amplifier by using eqn. 3.39, and then eqn. 3.43 and the results of Fig. 3.16 are used. The circuit structure is shown in Fig. 3.20.

The statistical scatter of the c-value due to a finite number of samples was not taken into account in the above discussion and the results refer to an infinite number of samples. The threshold difference T_N-T_∞ depends on the CFAR

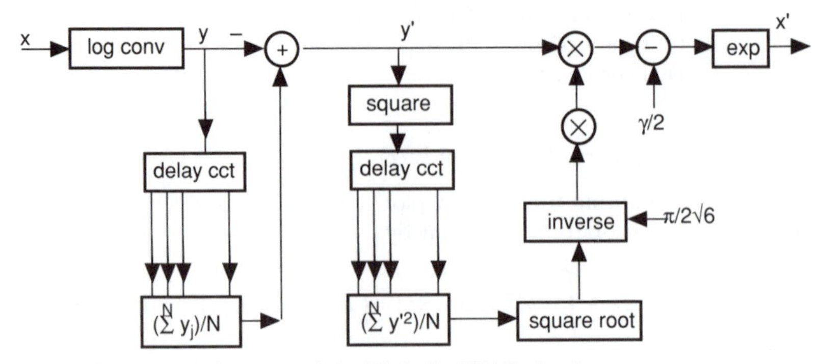

Figure 3.24 *Block diagram of the Weibull CFAR circuit*

Table 3.2 *Standard deviations σ_b and σ_c of estimated values of b and c in Monte Carlo simulation*

		$N=16$			
c=		0.5	1.0	1.5	2.0
$b=1$	$\sigma_b=$	0.577	0.265	0.176	0.132
	$\sigma_c=$	0.144	0.288	0.433	0.577
$b=10$	$\sigma_b=$	5.77	2.65	1.76	1.32
	$\sigma_c=$	0.144	0.288	0.433	0.577
		$N=32$			
c=		0.5	1.0	1.5	2.0
$b=1$	$\sigma_b=$	0.275	0.134	0.0895	0.0671
	$\sigma_c=$	0.0658	0.132	0.197	0.263
$b=10$	$\sigma_b=$	2.75	1.34	0.895	0.671
	$\sigma_c=$	0.0658	0.132	0.197	0.263

loss for $c=1.585$, in which the Weibull variable x was generated from a random number

$$x=b(-\ln \xi)^{1/c} \tag{3.50}$$

which reduces to eqn. 3.31 in the case of $c=2$.

The false alarm probability p_{fa} and the threshold T_N were evaluated as functions of N, from which the relation between T_N-T_∞ and N was derived. The results for the first and the second method are plotted in Figs. 3.21 and 3.22. In the first method, the dynamic range is small but the CFAR loss also becomes small. In the second method, on the other hand, the dynamic range becomes relatively large but the CFAR loss is also large. For a Rayleigh distribution where $c=2$ as shown in Figs. 3.4 and 3.5, the lin CFAR loss and the log CFAR loss are smaller than those for $c=1.585$.

Computer simulation was carried out based on Figs. 3.3, 3.19 and 3.20 for $N=16$ and $c=0.5$, 1.0, 1.5 and 2.0. The results are shown in Fig. 3.23.

As shown in Fig. 3.23, if Weibull-distributed variables pass through the conventional log CFAR circuit of Fig. 3.3, the false alarm probability in B increases with the decrease of the shape parameter value c. This tendency was already explained from the result in Fig. 3.17. From the adaptive CFAR circuit of Figs. 3.19 and 3.20, on the other hand, it is possible to maintain CFAR for various types of clutter that obey Weibull distributions. This is shown in C and D in Fig. 3.23.

Some authors have proposed new methods to maintain CFAR in Weibull clutter; for example, Goldstein [12] proposed the log t detector, which requires logarithmic detection of the mean and standard deviation of the logarithm of the input clutter samples. Based on the use of an additional parallel adaptive detector with a lower threshold, Cole and Chen [13] considered a double adaptive CFAR technique. In addition, Clerk and Peters [14] proposed a similar CFAR detector using a low-level register operating in Weibull clutter.

Recently Farina, Russo, Scannapieco and Barbarossa [15] have investigated similar problems on Weibull clutter. The input Weibull clutter returns are transformed into Gaussian variables and then added to an adaptive linear prediction filter. Hansen [16] has proposed a Weibull CFAR detector that takes into account the nonlinear transformation from the Weibull to the exponential probability density function. In this case, it is necessary to determine the values of the shape and scale parameters of respective Weibull distribution by using a

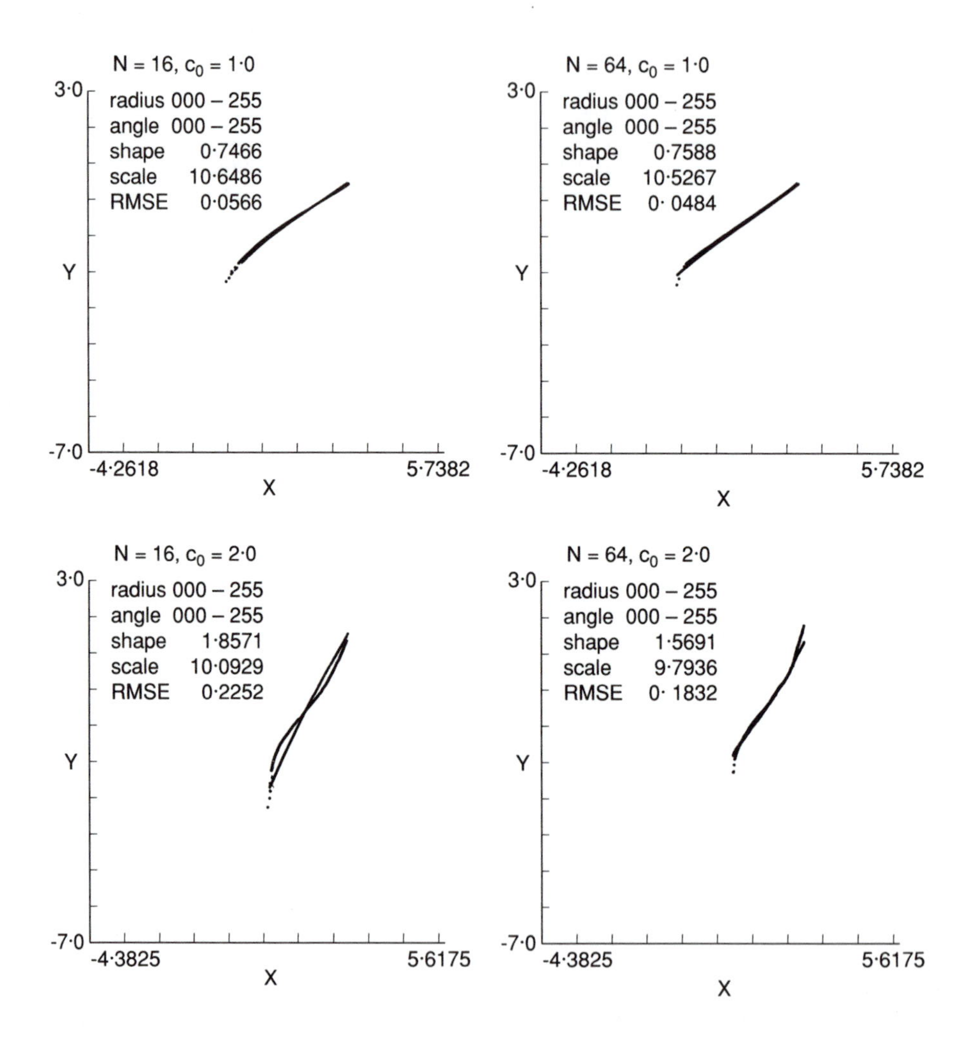

Figure 3.25 *Weibull plots of computer-generated data after Weibull CFAR*

$b = 1$
$c = 0.5$
$b_0 = 10$

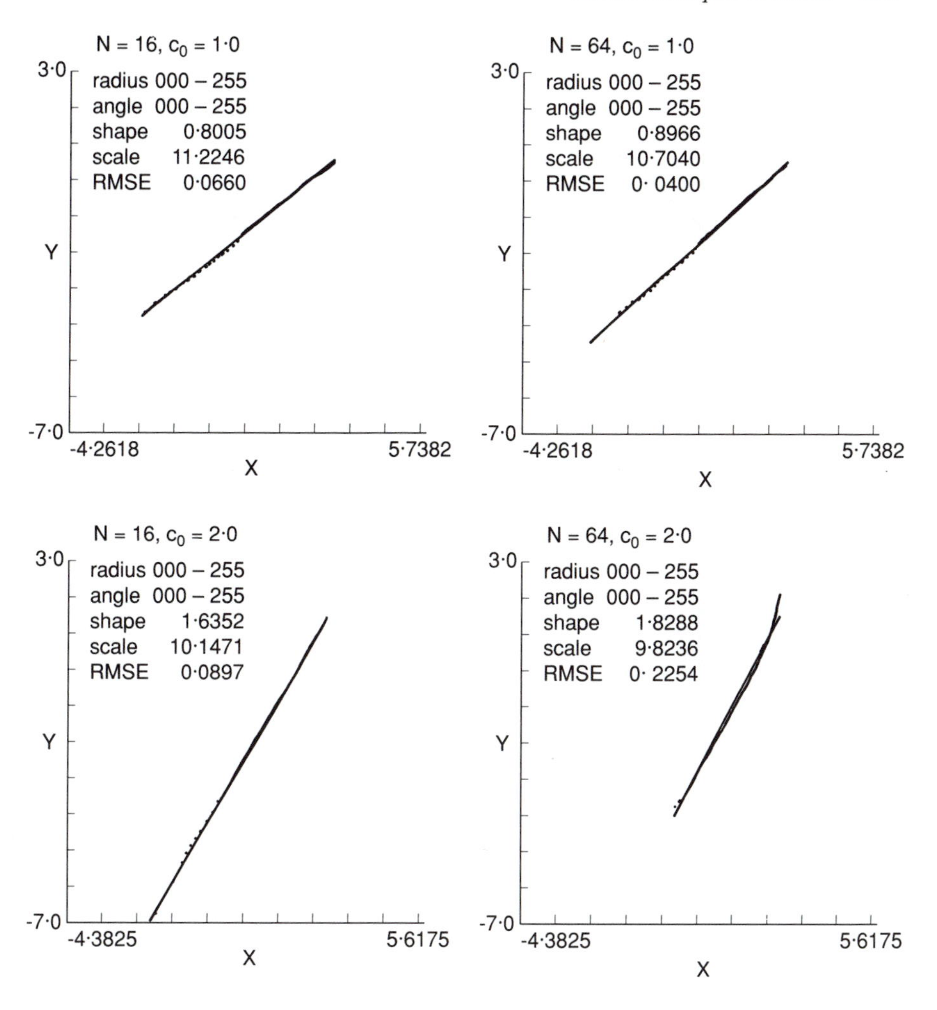

Figure 3.25 *continued*

finite number of data samples passed through a log-amplifier. A Weibull CFAR, which will be described in Section 3.7, is quite similar to Hansen's method.

3.7 Weibull CFAR

The false alarm rate is evaluated when values of the parameters involved in the assumed distribution function and the threshold level are obtained. Since the Weibull distribution is a very flexible distribution, this model is often used for

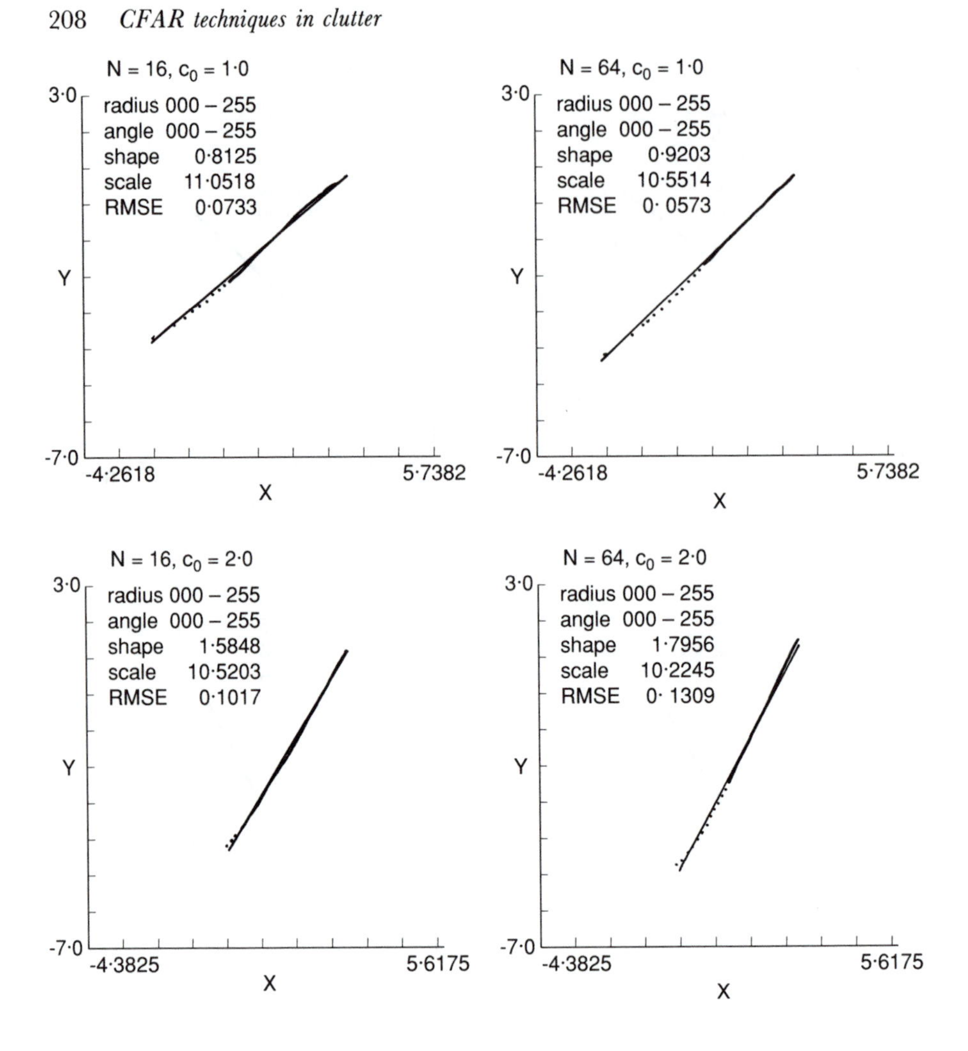

Figure 3.25 *continued*

realising CFAR. The Weibull distribution is a very flexible one, and, apart from the physical meaning, the clutter amplitude is widely expressed in terms of this distribution. As was shown in eqn. 3.39 the variance of log x depends on the shape parameter c, and the constant false alarm is not established after log CFAR.

Eqn. 3.1 is rewritten as

$$p(x, \sigma)dx = \exp[-x^2/2\sigma^2]d(x^2/2\sigma^2)$$

or

$$p(x, \sigma)dx = \exp[-\xi^2]d\xi^2 \tag{3.51}$$

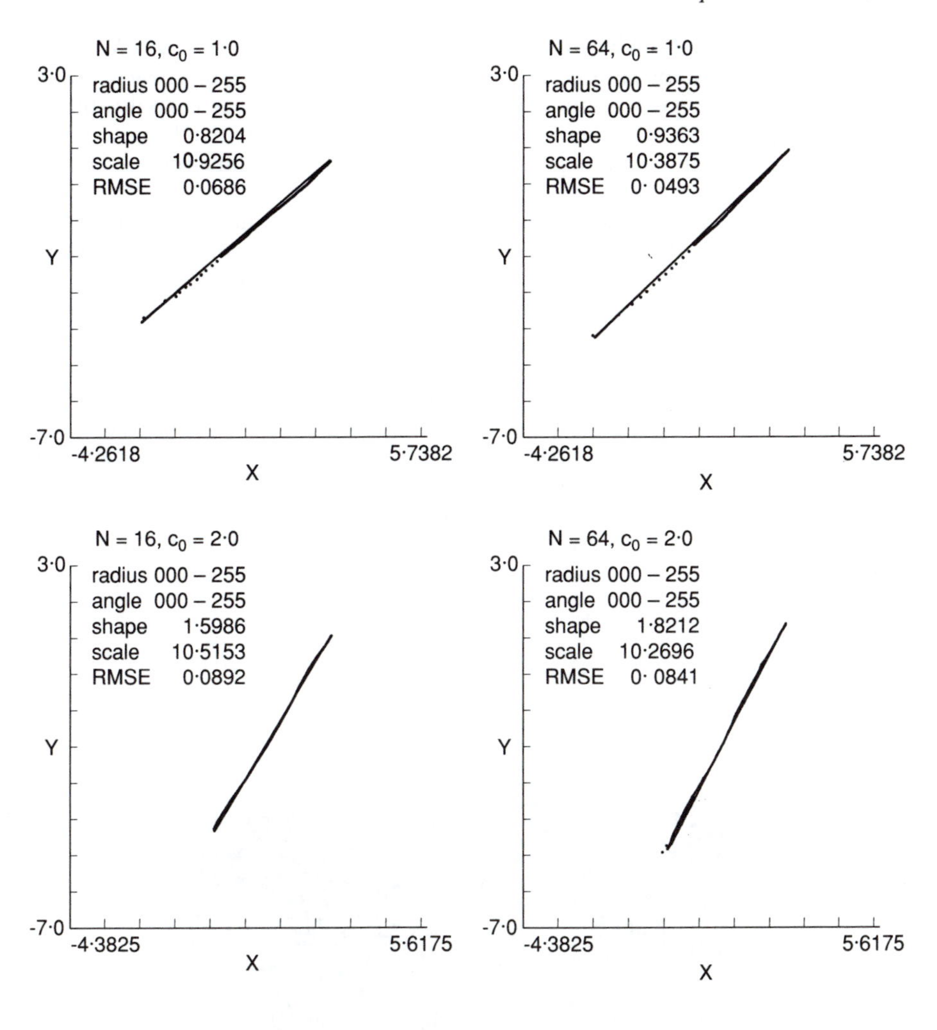

Figure 3.25 *continued*

Table 3.3 *Weibull parameters of weather clutter after Weibull CFAR with b_0 and c_0*

	$b_0=10,\ c_0=1.0$				$b_0=10,\ c_0=2.0$		
$N=$	16	32	64	$N=$	16	32	64
$b=$	10.86	10.54	10.34	$b=$	10.09	9.84	9.74
$c=$	0.79	0.84	0.87	$c=$	1.77	2.00	2.11

where $\xi = x/(2^{1/2}\sigma)$. Looking at eqn. 2.7, we find that if x, which obeys a Weibull distribution with shape parameter c, is transformed as $x \to x^{2/c}$, this new variable will obey a Rayleigh distribution and the log CFAR technique is applicable to it. For this purpose, the shape parameter must be evaluated from a finite number, N of samples. This is performed according to a digital circuit as shown below.

The input signal x is passed through a log-amplifier, and let its output be denoted as y; the output signals are averaged over samples and the average is subtracted from individual y to yield y'; y' is squared and the variance is evaluated over N samples, which is inversely proportional to the shape parameter c. By taking its inverse and multiplying by $\pi/26, y'$ is normalised to unit variance.

It is not necessary to transform the observed distribution into a Rayleigh distribution to realise CFAR; more generally, it is sufficient to transform the observed distribution into a given distribution (not necessary to a Rayleigh distribution). When the observed data obey a Weibull distribution, the parameter values should be estimated and these data are transformed into a Weibull distribution with given shape and scale parameter values to realise CFAR.

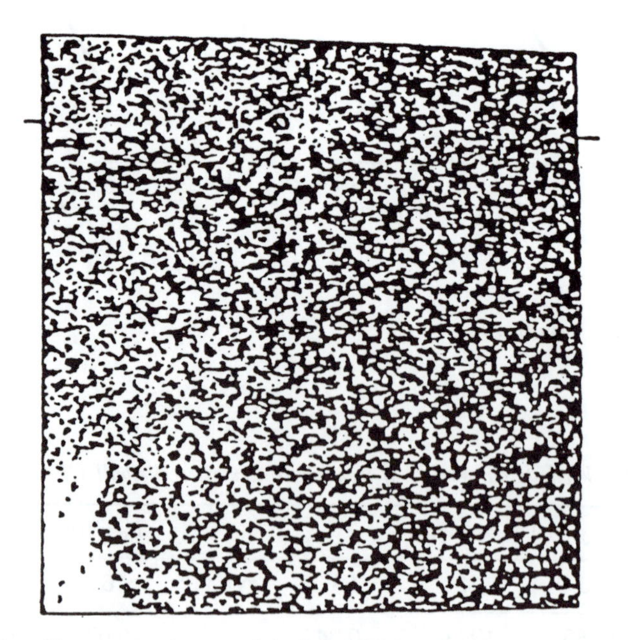

Figure 3.26 *Clutter map of a part of the Bay of Tokyo including a small island and part of a peninsula; the map is shown in Fig. 3.27*

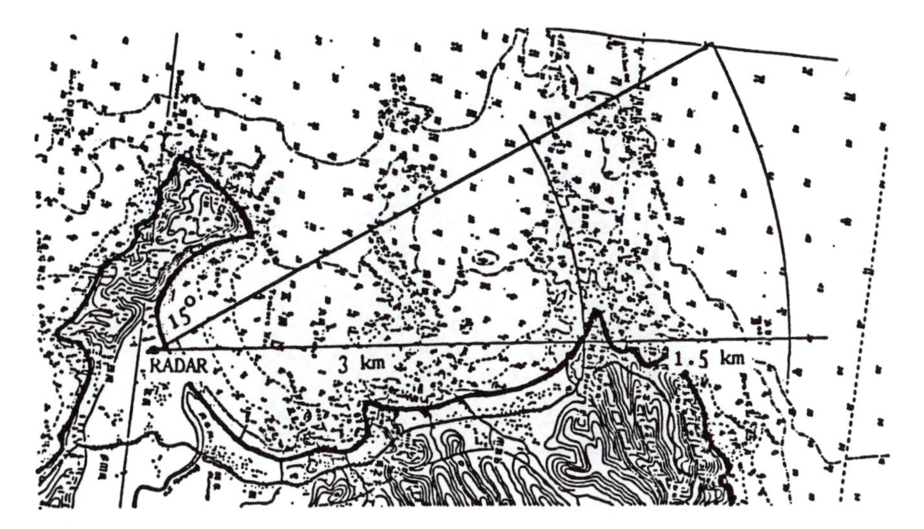

Figure 3.27 *Map of the area of the clutter map in Fig. 3.26*

3.8 Error in estimation of the parameter from a finite number of samples

In the digital circuit given in Fig. 3.24, values of the Weibull parameters are estimated from N samples, and these estimated values inevitably scatter about the true values. If very many samples are used in the estimation, however, not only the computation time becomes longer but also variations in the clutter are smeared, which sometimes bring about worse results than those obtained by using a smaller number of samples. Since this deviation cannot be estimated by an analytical method, Monte Carlo computer simulation was tried for this purpose [17].

A random number ξ, which takes a value between 0 and 1, is generated, and it is transformed into x through

$$x = b\{-\log \xi\}^{1/c} \tag{3.52}$$

The new variable x obeys a Weibull distribution with scale parameter b and shape parameter c, because

$$d\xi = c(x/b)^{c-1} \exp[-(x/c)^c]dx$$

Values of b and c are evaluated from N xs. Estimations of parameter values and variances were tried over 10^6 sample data for $n = 16$ and 64 for $b = 1$, 10, $c = 0.5$, 1.0, 1.5, 2.0; standard deviations σ_b and σ_c for b and c are listed in Table 3.2.

Table 3.4 *Values of T/C*

$N=$	16	32	64
$T/C=$	−0.17	1.39	3.49

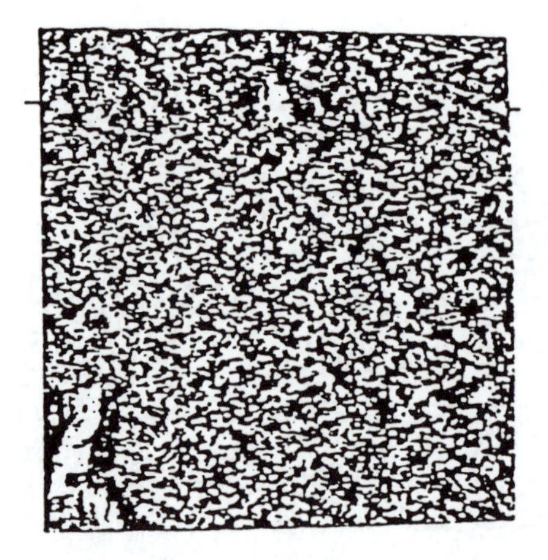

Figure 3.28 *Clutter map for values in Table 3.5*

The standard deviation of the estimated shape parameter c is independent of the assumed b and proportional to the true value of c; $\sigma_c/c=0.288$ and 0.132 for $N=16$ and 32, respectively. On the other hand, we find $\sigma_b=0.265b/c$ and $0.134b/c$ for $N=16$ and 32, respectively.

3.9 Computer simulation of Weibull CFAR

Accuracy of the practical Weibull CFAR was examined by means of computer simulation. As many as 65 536 data were generated, which obey a Weibull distribution with $b=1$ and $c=0.5$, 1.0, 1.5 or 2.0. Values of the parameters were estimated for every set of data where $N=16$ and 64, and nonlinear transformation of the data were carried out such that they would obey a Weibull

Table 3.5 *Values of T/C for three window sizes*

$N=$	16	32	64
$T/C=$	2.90	3.95	4.56
	$N_w=16$		
$N=$	16	32	64
$T/C=$	4.35	4.74	4.84
	$N_w=32$		
$N=$	16	32	64
$T/C=$	3.95	4.27	4.68
	$N_w=64$		

distribution with $b_0 = 10$ and $c_0 = 1.0$ or 2.0. The validity of the transformation was checked by means of the Weibull plot.

The results of computer simulation are plotted in Fig. 3.25. Better Weibull fitting was obtained when $c_0 = 1.0$ rather than $c_0 = 2$ (Rayleigh distribution).

3.10 Weather clutter

A marine radar with frequency 9410 MHz, peak power 6 kW, beam width 2.4°, pulse width 80 ns, pulse repetition rate 1250 Hz and antenna rotation 24 r.p.m. was directed into rain clouds above the Bay of Tokyo. The weather clutter was contaminated by sea clutter, and they were A/D converted into 8 bit data at a sampling time of 40 ns with an A/D converter which was designed and assembled by ourselves. The area which was covered consisted of 256 × 256 pixels.

The data were converted into those which obey a Weibull distribution with $b_0 = 10$ and $c_0 = 1.0$ or 2.0 for $N = 16$, 32 and 64, and the resulting numbers are listed in Table 3.3. The obtained scale parameter is close to 10.

3.11 Weather clutter with targets

When a target is included, how does the Weibull CFAR work to enhance the target? Fig. 3.26 is a clutter map of an area of the Bay of Tokyo as shown in Fig. 3.27; the area ranges from 3 km to 4.5 km from the radar site, and it was rainy.

The clutter map was made without STC with an X-band radar described in Section 3.9. The target-to-clutter ratio, T/C, which is defined as 10 log [(mean power of the target)/(mean clutter power)], of an island in the area is 3.10 dB. The Weibull CFAR was applied with $b_0 = 10$ and $c_0 = 1.0$ for $N = 16$, 32 and 64. Values of T/C are listed in Table 3.4.

The clutter map consists of 256 × 256 sample points, and the island on the clutter map contains about 20 × 20 sample points. In the present case, the island image was suppressed for $N = 16$ and 32, and the T/C is better than the other two for $N = 64$. This is because a target is generally suppressed by its own Weibull parameters when the sampled range for estimating the Weibull parameters is smaller than, or comparable to, the target size. This situation is peculiar to the X-band radar which has a high spatial resolution compared with the L-band air route surveillance radar. This difficulty can be improved by adopting a new technique, 'window Weibull CFAR', which is explained below.

3.12 Window Weibull CFAR

A window with a size N_w in the number of the sample points is taken around a point at which a Weibull CFAR is applied, and the Weibull parameters are evaluated outside the window over N sample points. The optimum window size is comparable to the target. Then the Weibull parameters of the target are excluded from the CFAR process at the target and the target will not be suppressed by itself. The values of T/C for window sizes $N_w = 16$, 32 and 64 are listed in Table 3.5 together with a clutter map (Fig. 3.28) for N = 16 and $N_w = 16$ where $b_0 = 10$ and $c_0 = 1.0$.

As expected T/C is maximised for $N = 32$, and this ratio becomes smaller for $N = 64$.

3.13 Smoothing

The clutter still looks fuzzy, and T/C can be made larger by means of displacement averaging over 16 sample points. Values of T/C are listed in Table 3.5 together with a clutter map for $N_w = 32$ and $N = 64$ which has the largest T/C.

3.14 References

1 CRONEY, J.: 'Clutter on radar displays', *Wireless Engr.*, 1956, **33**, pp. 83–96

2 FINN, H/M., and JOHNSON, R.S.: 'Adaptive detection mode with threshold control as function of spatially sampled clutter-level estimates', *RCA Rev.*, 1968, **29**, pp. 414–464

3 NITZBERG, R.: 'Analysis of the arithmetic mean CFAR normalizer for fluctuating targets', *IEEE Trans.* 1978, **AES-14**, pp. 44–47

4 HANSEN, V.G., and WARD, H.R.: 'Detection performance of the cell averaging LOG/CFAR receiver', *IEEE Trans.*, 1972, **AES–8**, pp. 648–652

5 SKOLNIK, M.I.: 'Introduction to radar systems' (McGraw–Hill, Kogakusha, Ltd., 1962)

6 TATSUKAWA, S., SEKINE, M., and MUSHA, T.: 'False-alarm probability of conventional and logarithmic CFAR receivers', *Trans. IECE Japan*, 1984, **E67**, pp. 563–568

7 TATSUKAWA, S., SEKINE, M., and MUSHA, T.: 'Simulation in radar clutter suppression', *Simulation*, 1984, **4**, pp. 101–107

8 SEKINE, M., MUSHA, T., TOMITA, Y., HAGISAWA, T., IRABU, T., and KIUCHI, E.: 'Weibull-distributed sea clutter', *IEE Proc. F*, 1983, **130**, p. 476

9 SEKINE, M., MUSHA, T., TOMITA, Y., and IRABU, T.: 'Suppression of Weibull-distributed clutters using a cell-averaging LOG/CFAR receiver', *IEEE Trans.* 1978, **AES–14**, pp. 823–826

10 SEKINE, M., MUSHA, T., TOMITA, Y., and IRABU, T.: 'Suppression of Weibull-distributed clutter', *Electron. Commun. Japan*, 1979, **K62–B**, pp. 45–49

11 SEKINE, M., MUSHA, T., TOMITA, Y., HAGISAWA, T., IRABU, T., and KIUCHI, E.: 'Suppression of Weibull-distributed weather clutter'. IEEE International Radar Conference, 1980, pp. 294–298

12 GOLDSTEIN, G.B.: 'False-alarm regulation in log-normal and Weibull clutter', *IEEE Trans.*, 1973, **AES–9**, pp. 84–92

13 COLE, L.G., and CHEN, P.W.: 'Constant false alarm detector for a pulse radar in a maritime environment'. Proc. IEEE, NAECON, 1987, pp. 1101–1113

14 CLARK, J., and PETER, R.S.: 'Constant false alarm detector adaptive to clutter statistics'. RSRE Memo 3150, 1978

15 FARINA, A., RUSSO, A., SCANNAPIECO, F., and BARBAROSSA, S.: 'Theory of radar detection in coherent Weibull clutter', *IEE Proc. F*, 1987, **134**, pp. 174–190

16 HANSEN, V.G.: 'Constant false alarm rate processing in search radars in radar-present and future'. IEE Conf. Publ. 105, 1973, pp. 325–332

17 TATSUKAWA, S.: 'Weibull CFAR at the X band radar', MS Thesis, Tokyo Institute of Technology, 1986

Chapter 4

Pulse compression and equivalent technologies

Prof. M.H. Carpentier

4.1 Recall of basic principles of modern radars

Utilisation of matched filters or correlation receivers ('ideal' receivers) is the basis of the so-called modern radars in comparison with the so-called conventional pulse radars.

4.1.1 *Basic assumptions*

The basic assumptions which support the conclusion that utilisation of ideal receivers is optimum are reviewed first.

Firstly, the noise accompanying the 'useful' return from a target is Gaussian (i.e. it has a Gaussian amplitude distribution). Gaussian means that the corresponding entropy is maximum or the noise has maximum disorder, or that the noise is the least appropriate for detection. This means that, if the actual noise is not Gaussian, using an ideal receiver would not always be optimum, this also means that actual performance could then be better than expected from a theory which is strictly only valid when the noise is Gaussian.

Secondly, the noise accompanying the useful return is assumed to be white in the frequency band of the received useful signal. When the noise is not white, a whitening filter has to be used in cascade with a filter matched to the useful signal as modified by the whitening filter (or a correlation with a reference signal identical to a useful signal as modified by the whitening filter). The result is that the noise at the output of the cascade of a whitening filter, plus a filter matched to the useful signal as modified by the whitening filter, is not white, even if, in many cases, to simplify the practical realisation, the receivers are eventually using a cascade of a whitening filter plus a normal matched filter (a filter adapted to the useful signal not modified by the whitening filter). Moreover it has to be recalled that, in most cases, the really ideal reception of a coloured noise cannot be practically implemented.

Thirdly, the target is assumed to be a point target; this means a target whose radial dimension is zero, and the ideal receiver is that proved to be optimum with regard to the detection and the localisation of that point target. The consequence is that, if the target is definitely not small compared with the range resolution of the radar, the 'ideal receiver' could not be ideal with regard to the localisation of the target and, for instance, better results would be obtained by using a filter matched not to the transmitted signal, but to the signal as modified (in phase, amplitude and duration) by the target. The consequence is also that improvement of range resolution could be obtained in the radar by sacrificing detection performance, which is not sufficiently remembered.

Fourthly, the theory of the ideal receiver assumes that there is *a priori* no particular place or zone where the presence of a target is significantly more likely than elsewhere. If this is not the case, ideal receivers should have to take that into account, which is practically achieved in a tracking radar receiving information from an acquisition radar, but is not always achieved in most practical implementations.

From these assumptions it is derived that the ideal receiver has either (i) to use filters matched to all possible Doppler shifts, which means as many matched filters as the ratio of the maximum shift in Doppler divided by $1/T$, T being the maximum possible duration of the radar measurement; or (ii) to use correlation with replicas of the transmitted signal shifted by all possible Doppler shifts (with the same comment) and by all possible range positions of the target (in fact, it is sufficient to use range positions separated only by $1/\Delta f$, if Δf is the spectral width of the transmitted signal).

Under the above assumptions the output of an ideal receiver is composed of a useful signal (which exists only when there is a useful signal at the input of the radar receiver), accompanied by a Gaussian parasitic signal, and the ratio between the square of the peak of the useful term and the mean value of the square of the parasitic term is

$$R = \frac{\text{Energy of the useful (received) signal } E \times 2}{\text{Spectral density of the noise } b^*}$$

When the received signal is on a carrier (e.g. IF) the peak of the useful term is twice its mean power around the maximum, so it is said that the signal-to-noise ratio at the output of the ideal receiver is

$$\frac{R}{2} = \frac{E}{b}$$

If the useful signal is not amplitude modulated during its duration T, $E = PT$, P being the useful power. On the other hand, if we call N the (mean) power of the noise at the input of the ideal receiver, assumed to be band-limited to Δf, it is clear that $N = b\Delta f$ and

$$\frac{R}{2} = \frac{PT\Delta f}{N} = \frac{P}{N} T\Delta f$$

which indicates that the signal-to-noise ratio at the output of an ideal receiver is equal to the signal-to-noise ratio at the input (in the minimum Δf band of frequencies), but multiplied by $T\Delta f$. And that is true whatever the phase/frequency modulation of the useful signal, provided it is not amplitude modulated.

The first action of the ideal receiver is to improve the signal-to-noise ratio by the product $T\Delta f$ of the time duration of the useful signal (if not amplitude modulated) by its bandwidth (in 'conventional' pulse radars, $T\Delta f \# 1$).

* Power of the (white) noise divided by the bandwidth Δf of the positive frequencies that it occupies.

With regard to the shape of the useful term at the output of an ideal receiver, this is the shape of the autocorrelation function of the useful signal at the input, which means that it is a symmetrical signal, the centre being the maximum of maxima and in practice composed of a central signal of short duration surrounded by secondary signals (sidelobes), the total duration of the signal plus sidelobes being equal to twice the duration of the useful signal T. The 3 dB duration of the central signal being of the order of $1/\Delta f$, the ratio of the shortening of the useful signal is around $T\Delta f$.

So the pulse compression ratio is strictly defined as being the ratio between signal to noise at the output and at the input (for a signal not amplitude modulated) and approximately as the ratio between the duration of the useful input signal and the duration of the central signal of the useful output signal.

4.1.2 *Practical simplifications*

In most cases, cost-effectiveness considerations simplify the equipment in two ways.

Firstly, when using matched filters, if the maximum possible Doppler shift is small compared with the spectral width Δf of the useful signal, cost-effectiveness considerations could lead to the use of only one initial expensive filter achieving one part of the matching, followed by a batch of cheaper filters completing the work, even if that leads to a small reduction in performance (but it is not always possible to do so without heavy degradation, e.g. range discrimination). Secondly, instead of using the maximum available time T for the ideal reception ('coherent integration'), money could be saved (volume of hardware would be reduced) by using, for coherent integration only, a duration several times smaller than the maximum available, and finishing off with non-coherent integration. Loss in sensitivity could be balanced by a reduction of the price (see Section 5.1.5.1). These two points will be better understood using an example.

Let us consider a panoramic radar with an antenna of 9.54 m in span, rotating at 6 r.p.m. The transmitted signal is a succession of pulses with 10 μs duration, repeating at 1 kHz. Each pulse is linearly frequency modulated between 9990 MHz and 10 010 MHz ($\Delta f = 20$ MHz). The antenna is considered to be uniformly illuminated. In that case a useful signal is not identical to the transmitted signal because of the amplitude modulation (on transmission and reception) by the rotating antenna beam, as shown in Fig. 4.1. The amplitude of the corresponding Fourier transform of a useful signal is constituted by successive triangles separated by 1 kHz (PRF) and covering quasi-uniformly a total bandwidth of 20 MHz (see Fig. 4.2): the total spectrum contains 20 000 triangles, phase varying from one triangle to the next according to a parabolic law because of the linear frequency modulation within each pulse.

If we compare the Fourier transform of another useful signal at the same range but shifted up in frequency (due to Doppler) by 1 kHz, it is composed of 19 999 similar triangles plus one on the right and minus one on the left. The phases of the triangles are very similar: rigorously the phases are the same as for a useful signal (not frequency shifted) delayed by 0.5 ns (7.5 cm). The conclusion is that, if the maximum Doppler shift is 10 kHz (150 m/s or 300 kts), and if we use the same ideal receiver for a frequency shift of 10 kHz as

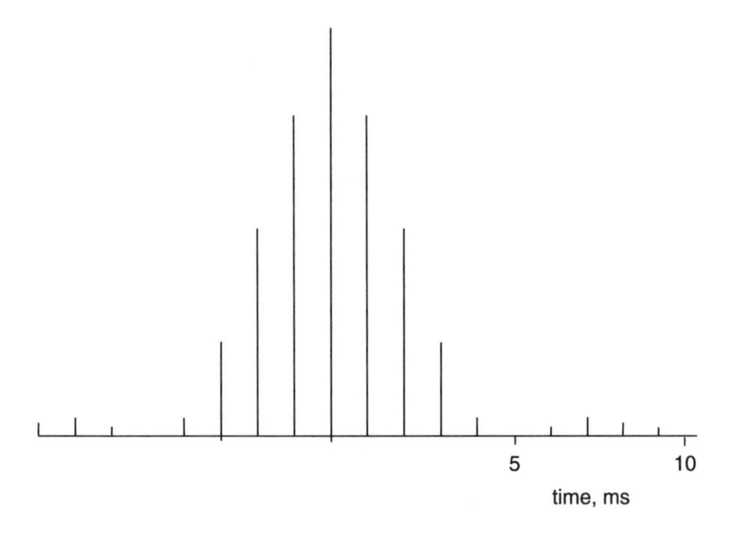

Figure 4.1 *Signal received by a panoramic radar*

for no frequency shift, the maximum loss will be 10 triangles out of 20 000 (strictly negligible) and the relevant error in range measurement will be less than 1 m (negligible). So we shall take into account only Doppler shifts below 1 kHz (and the Doppler shift will be measured modulo 1 kHz).

In practice the ideal receiver will consist of:

- One filter matched to a 10 μs frequency (not amplitude) modulated pulse, i.e. a filter with a rectangular band pass of 20 MHz and a parabolic phase characteristic (opposite to the useful signal law)
- Followed by a batch of a few filters (more than six) in parallel, each having (no action on the phase but) an amplitude transmission like that in Fig.4. 2 (i.e. an impulse response as in Fig. 4.1), similar filters being equally spaced.

In practice, only the first filter will be called a 'pulse compression filter'; the other filters will be called either 'coherent integration filters' or 'Doppler filters' or 'velocity filters'.

If we do not need to directly measure the radial speed of the targets, we could still simplify the receiver by only making the coherent integration during the

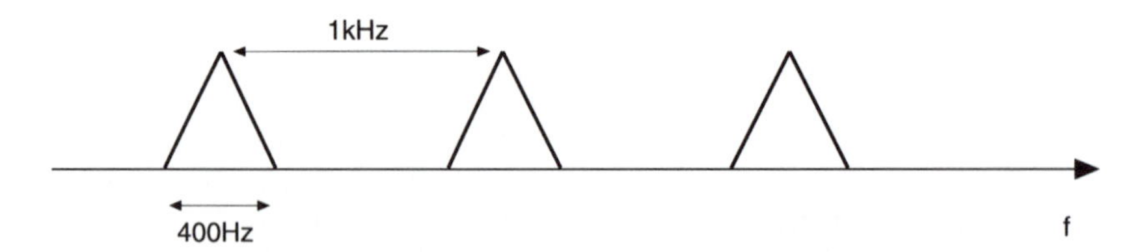

Figure 4.2 *Frequency spectrum associated with the signal received by a panoramic radar*

10 μs of an elementary pulse duration, and using – directly after the pulse compression filter – detection followed by only one low-pass filter (cut-off frequency at about 200 Hz) to achieve non-coherent integration. Loss in sensitivity will be a few decibels (2 to 3), but the money saving could be significant.

4.2 Technology of pulse compression: Various means to realise compression filters

4.2.1 *General*

According to Section 4.1.2, pulse compression is generally devoted to arrangements in which the transmitted signal is frequency or phase modulated (generally not amplitude modulated for convenience reasons, because it is generally not economical to use amplitude modulation) and where, on reception, use is made of a matched filter.

If $\Phi(f)$ is the Fourier transform of the transmitted signal, the receiving matched filter achieving pulse compression has a transmittance equal to the conjugate of $\Phi(f)$, written $\Phi^*(f)$. In most cases, pulse compression is achieved at low frequency, i.e. at intermediate frequency or possibly around zero frequency.

This means that $\Phi(f)$ is the Fourier transform of the useful signal at IF.

In that case the Fourier transform $\Phi(f)$ of a useful signal crossing the pulse compression device is then multiplied by $\Phi^*(f)$, providing at the output a signal whose Fourier transform is $\Phi(f)\ \Phi^*(f) = |\Phi^2(f)|$. That means a symmetrical signal without frequency or phase modulation which is the autocorrelation function of the useful signal.

The essential quality of a pulse compression device is to totally remove the frequency (or phase) modulation of a useful signal crossing it. The phase characteristic of the matched filter is obtained if that condition is fulfilled. (If the useful signal is not frequency modulated this quality is obtained from a filter not altering the phases, which is why the receiver of a conventional radar provides not far from ideal reception.) On the other hand, if a Dirac impulse enters a pulse compression network, the corresponding output has a Fourier transform equal to $\Phi^*(f)$, and then it is identical to a useful signal time-reversed. (This explains why it is not strictly possible to build a matched filter, but only the equivalent of a cascade of a delay exceeding the duration T of a useful signal, followed by a matched filter.)

4.2.2 *Chirp system*

Historically pulse compression has first made use of useful signals, not amplitude but linearly frequency modulated between $f_c - \Delta f/2$ and $f_c + \Delta f/2$ (or vice versa) over the duration T_p of the pulse (see Fig. 4.3). In that case, it could be said – to a first approximation – that any frequency lasts for the same time with the same power, which means that the spectrum of the useful signal ($|\phi(f)^2|$) is constant between $f_c - \Delta f/2$ and $f_c + \Delta f/2$ and then $|\Phi(f)|^2$. In a similar manner it could be considered that the spectrum of the useful signal is zero outside that bandwidth Δf (see Fig. 4.4).

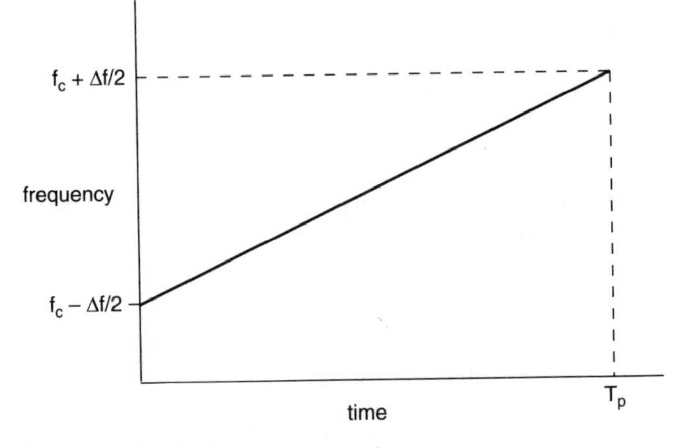

Figure 4.3 *Chirp signal: frequency versus time*

So amplitude matching will be achieved by using a rectangular filter limiting the bandwidth to Δf (zero gain outside $f_c - \Delta f/2$ and $f_c + \Delta f/2$, and constant gain between those two frequencies).

Phase matching will be achieved by using a purely dispersive device removing the frequency modulation of a useful signal, which is achieved by using a dispersive device whose (group) propagation time decreases from $T_0 + T_p$ to T_0 for frequency varying between $f_c - \Delta f/2$ to $f_c + \Delta f/2$ (see Fig. 4.5).

The complete matched filter will be made up of the cascade of the rectangular filter achieving amplitude matching and of the dispersive device achieving phase matching (see Fig. 4.6).

Then if we consider a useful signal crossing that matched filter, the amplitude of the Fourier transform will not be changed and will remain rectangular, while its phase modulation (quadratic law for the phase:

$$\varphi = \frac{\pi}{\Delta f} T_p (f - f_c)^2)$$

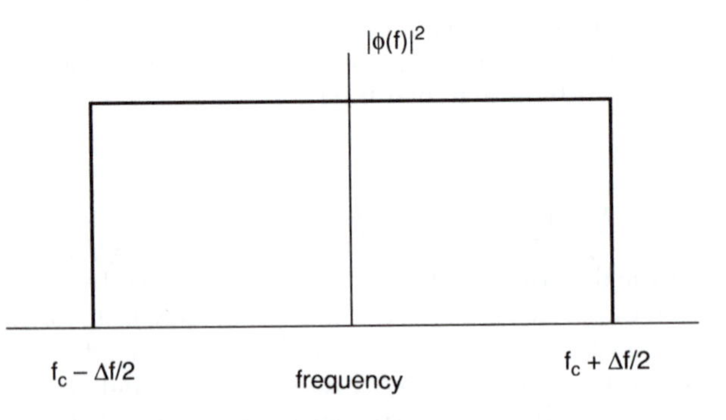

Figure 4.4 *Chirp signal: Simplified frequency spectrum*

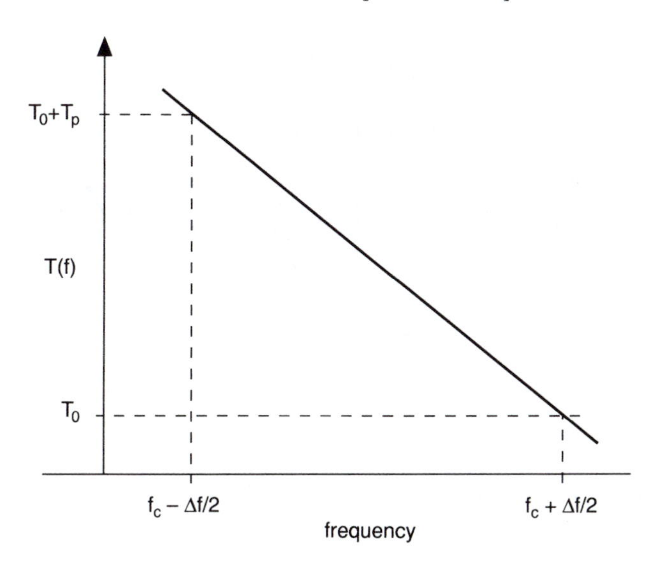

Figure 4.5 *Chirp signal: Group delay versus frequency of the associated matched filter*

is totally removed, so that the output of the matched filter has a real rectangular Fourier transform. This means that a useful signal after crossing the matched filter has an envelope according to

$$\frac{\sin \pi\Delta f(t - t_{0i})}{\pi\Delta f(t - t_{0i})}$$

Such a signal is represented in Fig. 4.7: the central signal is surrounded by sidelobes, the main lobes being at 13.5 dB below the central signal.

In fact, referring to 'instantaneous' frequency (or 'stationary phase'), the real shape of $|\Phi(f)|$ is only rectangular if $T_p\Delta f$ is very large: this explains why we find the infinite envelope of the useful output signal. The real shape of $|\Phi(f)|$ is only close to rectangular, as represented in Fig. 4.8, for $T_p\Delta f = 10$ and $T_p\Delta f = 60$. It follows that a rectangular filter does not provide perfect amplitude matching, but introduces a loss in sensitivity (1 dB for $T_p\Delta f = 10$ and is strictly negligible for $T_p\Delta f$ exceeding 100). It also follows that the real shape of the envelope of the useful signal at the output is only similar to what is shown in Fig. 4.7; Fig. 4.9 shows the real shape of the envelope for $T_p\Delta f = 20$.

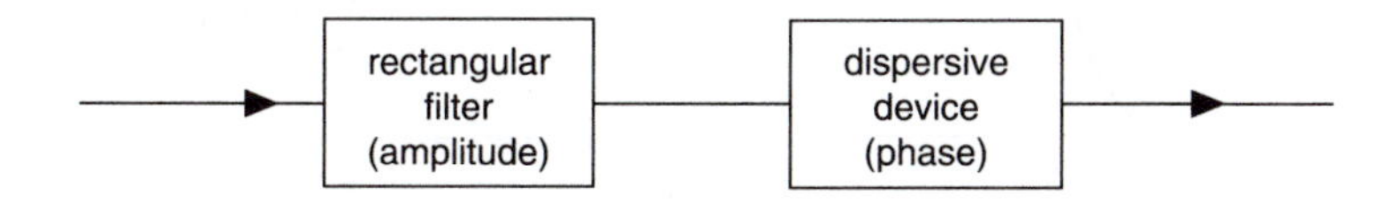

Figure 4.6 *Chirp signal: Block diagram of the matched filter*

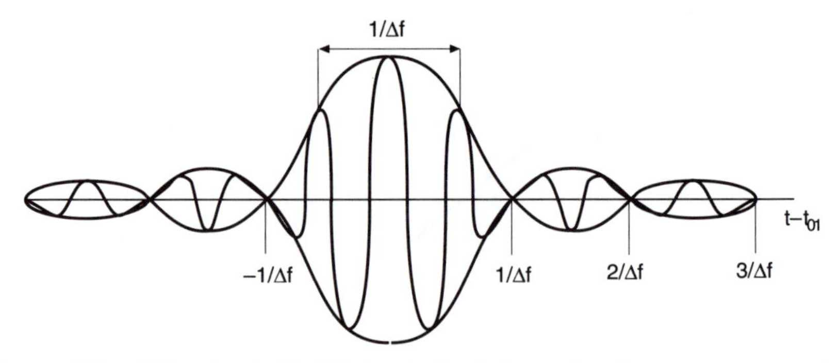

Figure 4.7 *Chirp signal: Useful signal after pulse compression*

4.2.3 *Doppler effect in chirp systems*

In the case of a Doppler shift f_D of the received signal

$$(f_D = \frac{2V_R}{\lambda},\ V_R$$

being the radial velocity of the target and λ the wavelength corresponding to the central frequency of the transmitted signal), this produces three consequences.

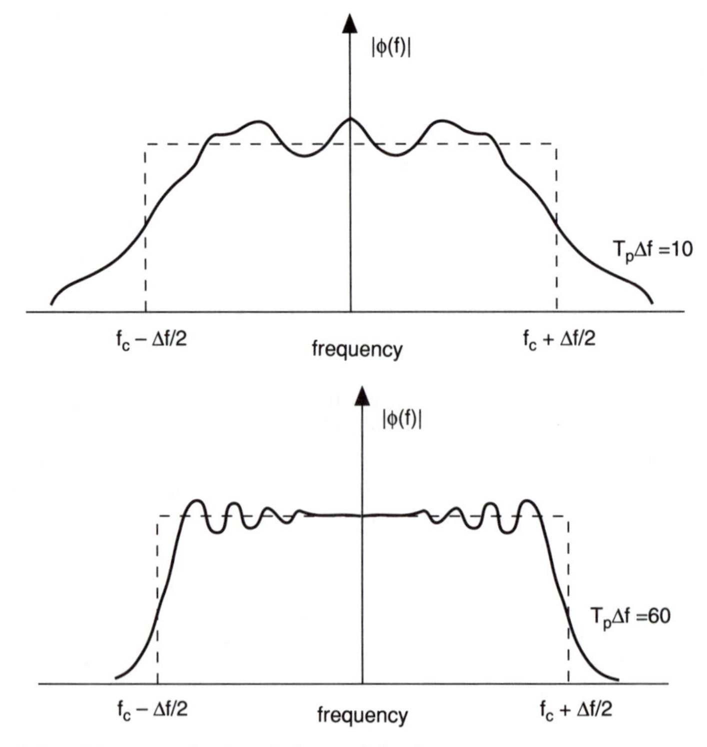

Figure 4.8 *Chirp signal: Actual shape of the frequency spectrum*

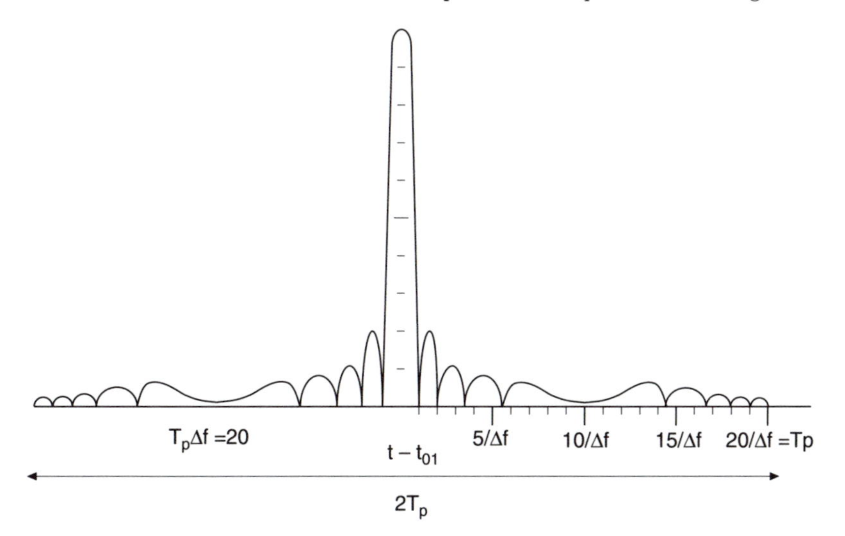

Figure 4.9 *Chirp signal: Actual shape of the amplitude of the useful signal after pulse compression*

Firstly, after crossing the dispersive device, all frequencies contained in the useful signal remain at the output with the same phase, giving an output signal which is not frequency/phase modulated but shifted in time by

$$\frac{T_p f_D}{\Delta f} = T_p f_D \frac{1}{\Delta f}$$

which is $T_p f_d$ times the 'duration' of the output compressed signal.

Secondly, the width of the spectrum of the output signal is multiplied by

$$\frac{\Delta f - |f_D|}{\Delta f}$$

which reduces by the same ratio the power of the useful compressed signal.

Thirdly, the reduction of the width of the spectrum of the compressed signal gives an increase in its duration of

$$\frac{\Delta f}{\Delta f - |f_D|}$$

As a result, the power of the compressed signal is multiplied by

$$\left|\frac{\Delta f - |f_D|}{\Delta f}\right|^2$$

and the amplitude of the compressed signal is multiplied by

$$\frac{\Delta f - |f_D|}{\Delta f}$$

(which is strictly valid only for very large values of $T_p\Delta f$). Fig. 4.10 shows the exact result (when $T_p\Delta f = 20$) for $f_D = -1/T_p$ and $f_D = -10/T_p$.

Numerical applications

Firstly $\lambda = 0.75$ m $T_p = 300$ µs $f = \Delta 200$ kHz

$$\frac{1}{\Delta f} = 5 \text{ µs (750 m)}$$

$$V_R = 300 \text{ m/s } f_D = 800 \text{ Hz } T_p f_D = 0.24$$

$$T_p f_D \frac{I}{\Delta f} = 1.2 \text{ µs (180 m)}$$

$$\frac{\Delta f - |f_D|}{\Delta f} = 0.996 \text{ loss of } 0.03 \text{ dB}$$

$$V_R = 5000 \text{ m/s } f_D = 13\ 300 \text{ Hz } T_p f_D = 4$$

$$T_p f_D \frac{1}{\Delta f} = 20 \text{ µs (3000 m)}$$

$$\frac{\Delta f - |f_D|}{\Delta f} = 0.933 \text{ loss of } 0.6 \text{ dB}$$

Secondly $\lambda = 0.1$ m $T_p = 4$ µs $\Delta f = 10$ MHz

$$\frac{1}{\Delta f} = 0.1 \text{ µs (15 m)}$$

$$V_R = 300 \text{ m/s } f_D = 6000 \text{ Hz } T_p f_D = 0.024$$

$$T_p f_D \frac{1}{\Delta f} = 0.0024 \text{ µs (0.36 m)}$$

$$\frac{\Delta f - |f_D|}{\Delta f} = 0.9994 \text{ loss of } 0.005 \text{ dB}$$

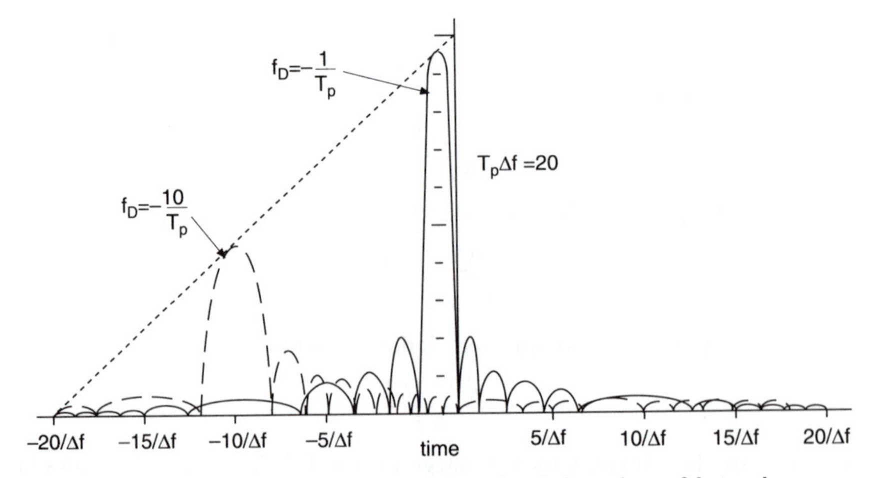

Figure 4.10 *Chirp signal: Effect of the Doppler shift on the useful signal*

Thirdly $\lambda = 0.03$ m $T_p = 20$ μs $\Delta f = 10$ MHz

$$\frac{1}{\Delta f} = 0.1 \ \mu s \ (15 \ m)$$

$$V_R = 1000 \ m/s \ f_D = 66\,700 \ Hz \ T_p f_D = 1.33$$

$$T_p F_D \frac{1}{\Delta f} = 0.13 \ \mu s \ (20 \ m)$$

$$\frac{\Delta f - |f_D|}{\Delta f} = 0.9933 \text{ loss of } 0.06 \text{ dB}$$

Fourthly $\lambda = 0.05$ m $T_p = 100$ μs $\Delta f = 500$ MHz

$$\frac{1}{\Delta f} = 0.002 \ \mu s \ (0.3 \ m)$$

$$V_R = 5000 \ m/s \ f_D = 200\,000 \ Hz \ T_p f_D = 20$$

$$T_p f_D \frac{1}{\Delta f} = 0.04 \ \mu s \ (6 \ m)$$

$$\frac{\Delta f - |f_D|}{\Delta f} = 0.9996 \text{ loss of } 0.003 \text{ dB}$$

4.2.4 *Practical implementation of dispersive filters*

4.2.4.1 *Lumped constants filters*

Historically, dispersive filters were first realised (USA, France) at the end of 1950s by using a cascade of filters as represented in Figs. 4.11 and 4.12 (in fact, the equivalent bridged-T filters) with the associated curves giving (group) propagation time T_g versus frequency f, providing quasi-linear zones (one in Fig. 4.11 and two in Fig. 4.12).

The difference between the required straight line and the actual one introduces imperfections in the pulse compression which will be dealt with later on; the result being that, even by trying to compensate for reception imperfections by action on transmission (see description below of the Carpentier–Adamsbaum patent), the volume of such devices (number N of components) increases rapidly with the pulse compression ratio (in case of the Carpentier–Adamsbaum patent, N various devices like $(T_p\Delta f)^{5/3}$, this gives a multiplication of N by 16 when $T_p\Delta f$ is multiplied by 5).

4.2.4.2 *Volume acoustic waves*

Acoustic delay lines have been used to take this into account by using the dispersive propagation of acoustic vibrations along a metallic film (see Fig. 4.13). The curves are similar to that of Fig. 4.12. Reducing thickness of the film changes the horizontal scale (frequencies). The necessity of obtaining a controlled thickness of the film in practice limited Δf to below a few megahertz. The total length of the film was proportional to the pulse duration T_p (roughly 5000 m/s or 5 m/ms). Such devices are still in use on operational radars with large values for T_p (around 1 ms) and relatively reduced values for Δf, but they are now replaced on new equipment by digital devices.

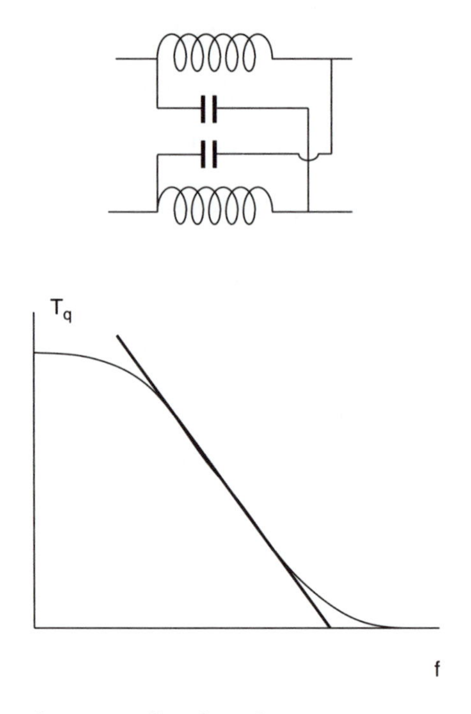

Figure 4.11 *Lumped constants filter for pulse compression*

4.2.4.3. *Love surface acoustic waves*
Increasing the bandwidth Δf was then obtained by replacing volume acoustic waves by Love surface waves propagating along substrates covered by a thin film of another material (MAERFELD, C.: Thomson–CSF, late 1960s). Roughly speaking, the velocity of the acoustic waves varies between the velocity in the substrate at low frequencies and the velocity in the thin film at high frequencies. Values of Δf up to 100 MHz might be obtained, together with relatively small values of T_p (a few tens of microseconds).

4.2.4.4 *Rayleigh surface acoustic waves*
In 1967 Dieulesaint and Hartman from Thomson–CSF used the possibility of propagating surface waves along the surface of a piezoelectric substrate, as represented in Fig. 4.14. Initially such devices used an input comb and an output comb engraved on the substrate in such a way that the high frequencies were propagated during a short time T_1 and the low frequencies during a longer time T_2. Repositioning of the teeth along the output comb allowed for control of the dispersive curve T_g versus frequency. (Other arrangements are now used in which the input and output are on the same side of the devices, but they are basically of the same nature.) Accuracy of the positioning of the teeth directly affects the accuracy of the phase versus frequency curve. For instance, if the central frequency of the acoustic wave which is used for the processing is 60 MHz, then, assuming an acoustic velocity of about 5000 m/s, this will give a wavelength of 83 microns. If we want a phase error below 20 mrad (which will

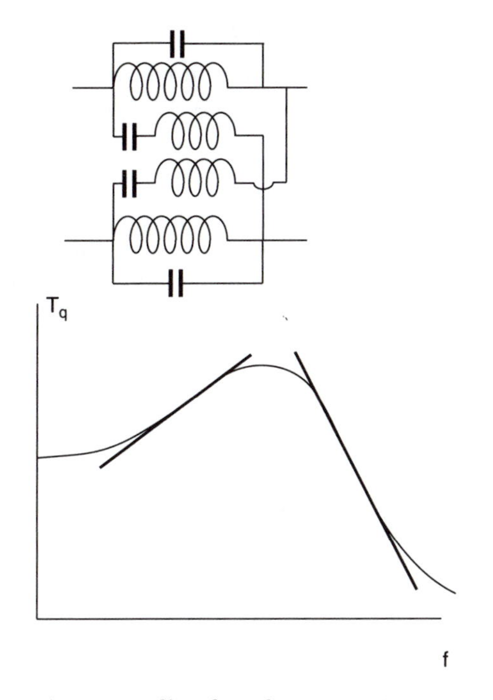

Figure 4.12 *Lumped constants filter for pulse compression*

give parasitic sidelobes around −40 dB, see below), the positioning of the teeth has to be made with an accuracy of around $(20 \times 10^{-3}/2\pi) \times 83 = 0.25$ micron. This was difficult with optical photoengraving but is easily achieved now using electron beams.

Another way to understand how the device represented in Fig. 4.14 works is to say that the impulse response of such a system is as represented in Fig. 4.15 (signal with a frequency modulation, frequency decreasing regularly from the beginning to the end); see Section 4.2.1

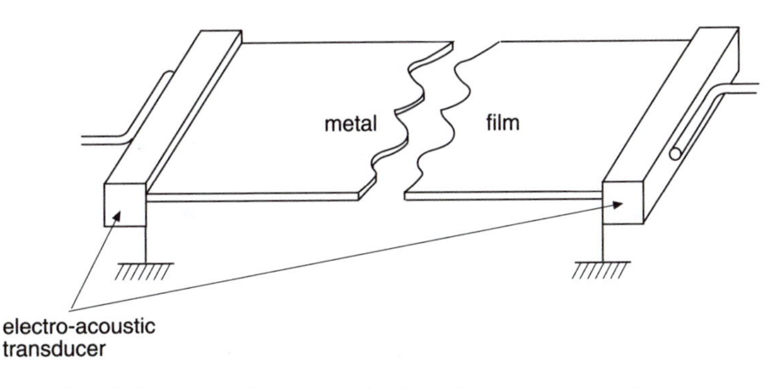

Figure 4.13 *Volume acoustic waves delay line for pulse compression*

4.2.5 *Generation of the transmitted signal*

In order to produce the signal to be transmitted, which, in the case of a chirp signal, is linearly frequency modulated, there are basically two methods: active generation and passive generation.

4.2.5.1 *Active generation*

This consists basically of two methods: the first uses various frequency generators with a strict phase relation between them and the frequency generators are used consecutively in order to approximate the ideal signal by successive pieces of pure frequencies (during which the phase of that frequency remain constant) in such a way that there is no discontinuity when changing frequency (see Fig. 4.16).

If there are N steps, the maximum phase error introduced by approximating the straight line by stairs is given by

$$\frac{\pi}{N^2} T_p \Delta f$$

If it is necessary not to exceed a phase error of about 20 mrad, this means that

$$\frac{\pi}{N^2} T_p \Delta f < 0.02$$

$$N^2 > 50\pi T_p \Delta f$$

Numerical applications:

$$T_p \Delta f = 30 \qquad N > 70$$

$$T_p \Delta f = 100 \qquad N > 125$$

$$T_p \Delta f = 1000 \qquad N > 400$$

The second method consists in (digitally) constructing the sampling of the required signal (centred on a low intermediate frequency, possibly on zero frequency, in which case it is necessary to construct the sampling of both components in phase and in quadrature) at a high enough sampling frequency.

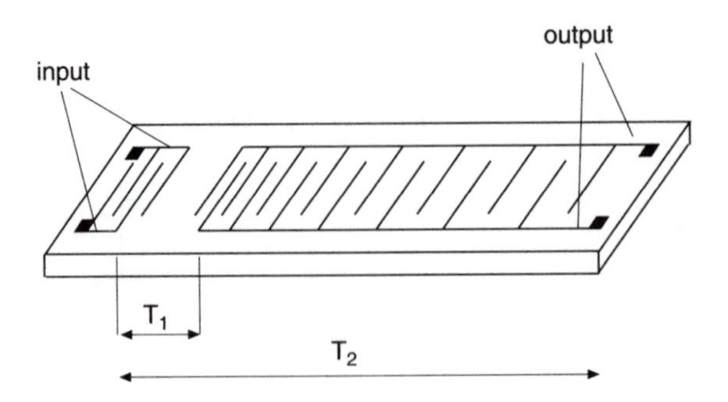

Figure 4.14 *Surface acoustic waves (SAW) delay line for pulse compression*

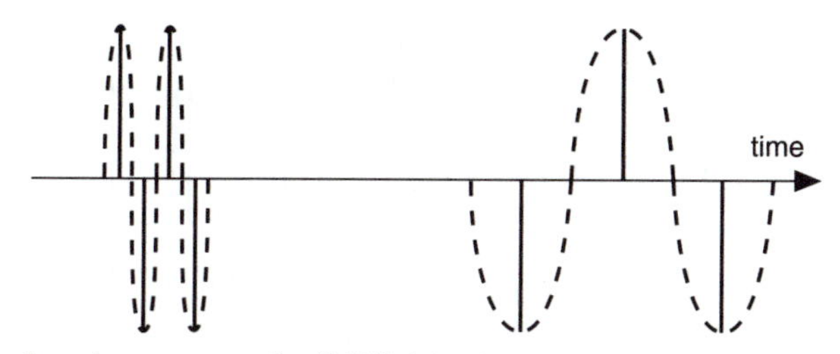

Figure 4.15 *Impulse response of a SAW delay line*

The sampling is then low-pass filtered to provide (in IF) the required signal. Fig. 4.17 gives an example of such a sampling made on an IF signal (the intermediate frequency being slightly larger than $\Delta f/2$).

4.2.5.2 *Passive generation*

This consists in using a dispersive filter also on transmission, with a characteristic 'propagation time versus frequency' conjugate to that used on reception, i.e. with a positive slope when using on reception the filter described in Fig. 4.5. In that case, generation (in IF) of the transmitted signal is obtained using a very short pulse (very short compared with $1/\Delta f$) crossing the dispersive filter. The very short pulse could be either modulated by the IF (only in amplitude, with no phase modulation) or it could be only a very very short video pulse (in which case it has to be short compared with the sum of the intermediate frequency plus $\Delta f/2$). After crossing the dispersive filter, the very short (not phase modulated) signal becomes a long pulse, which is frequency modulated. It is then properly range gated to keep only the good portion of the frequency modulation, and generally hard-clipped, in order to remove parasitic amplitude modulation. Obviously dispersive filters used on transmission and reception have to be conjugate, which means that the propagation time on transmission plus the propagation time on reception has to be constant with frequency.

Fig. 4.18 gives an example of two conjugate curves which could be used. A particular example of such a situation is shown in Fig. 4.19 where curves of

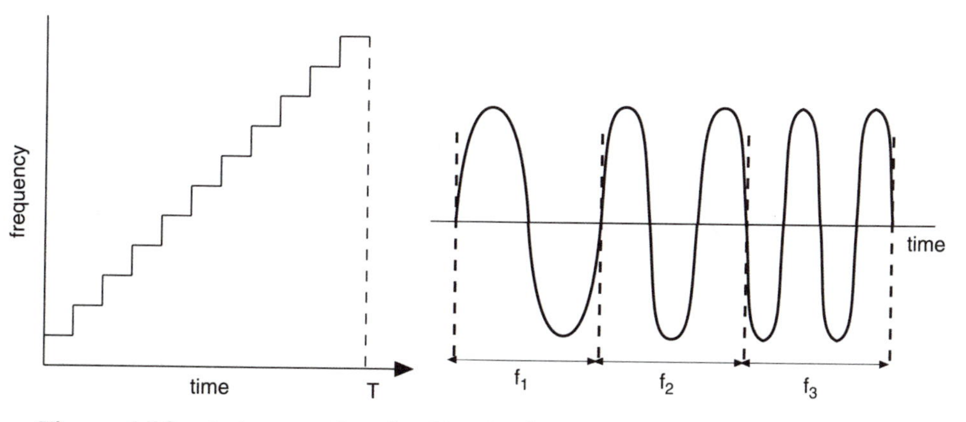

Figure 4.16 *Active generation of a chirp signal*

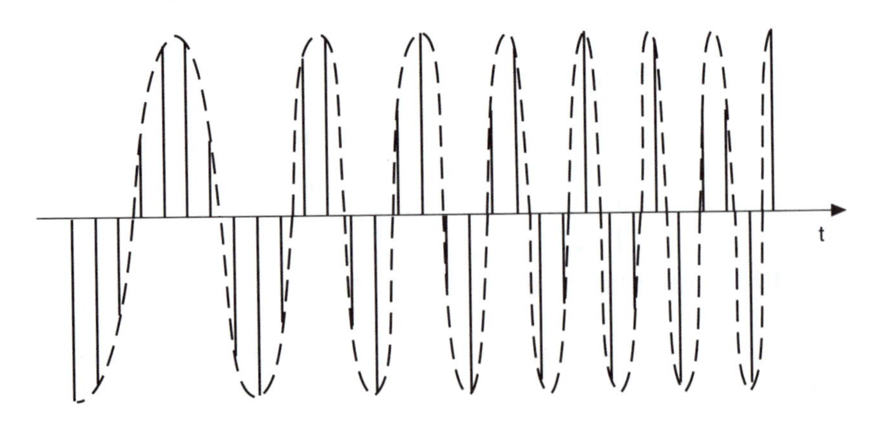

Figure 4.17 *Sampling of a chirp signal*

propagation time versus frequency both show an inflection point. Such a situation is met when 'intrinsic weighting' is used.

It has to be noted that reversing the curve on the left-hand side of Fig. 4.19 by symmetry around the frequency of the inflection point effectively gives the curve of the right-hand side of Fig. 4.19, which explains why it is possible to use the *same* dispersive delay line on transmission *and* on reception, provided such a reversal is done. This is obtained by mixing with a frequency above the frequencies of the spectrum and maintaining the lowest frequencies:

$$-(f_0+\varepsilon)+2f_0 \rightarrow f_0-\varepsilon$$

This was patented in 1959 by A. Adamsbaum and M. H. Carpentier.

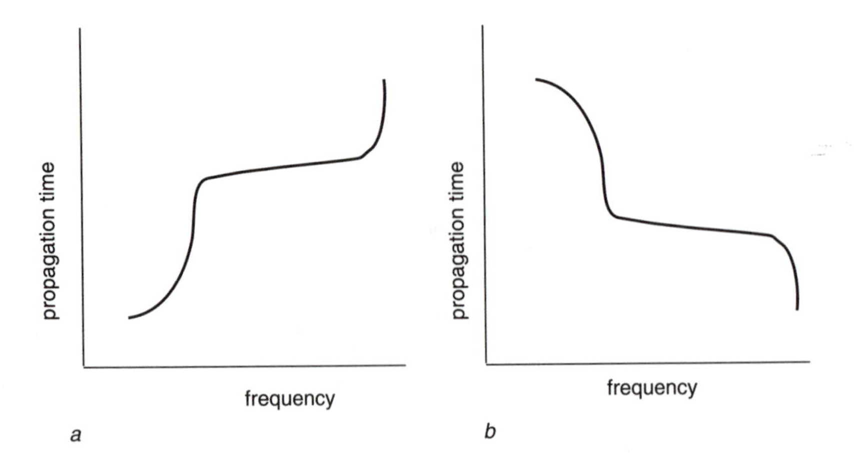

Figure 4.18 *Conjugate curves used in transmission and reception to achieve matching*

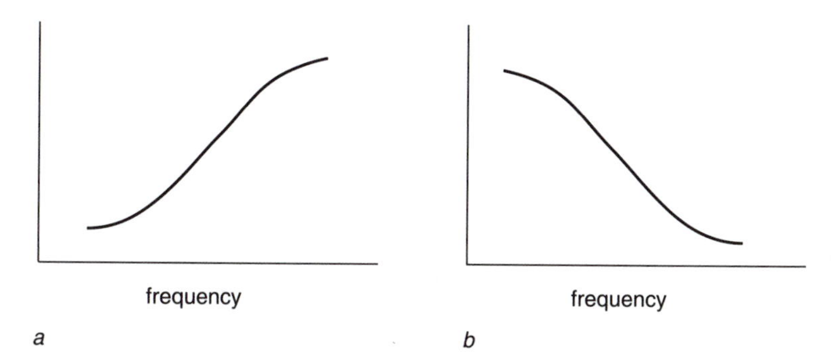

Figure 4.19 *Conjugate curves used in transmission and reception to achieve matching*

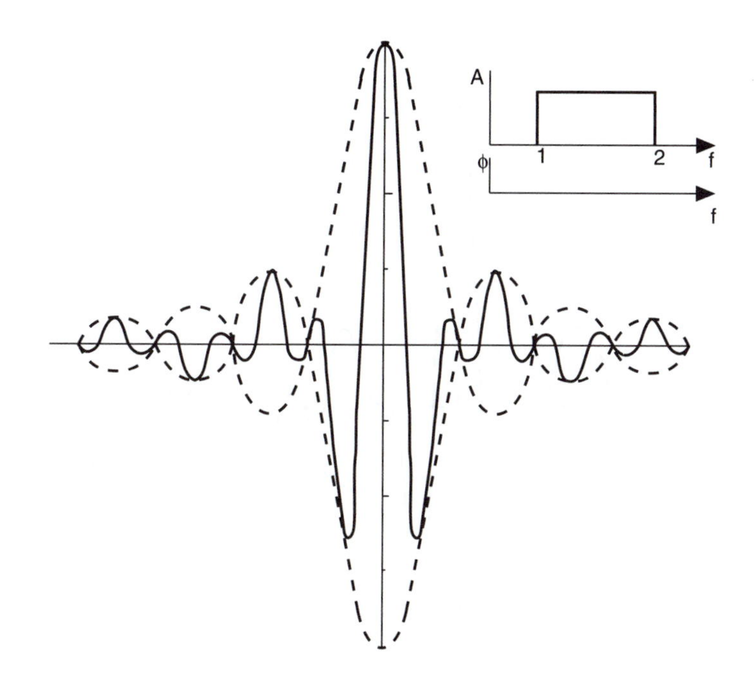

Figure 4.20 *Useful signal after perfect pulse compression*

4.2.6 *Consequences of imperfections in matched filtering*

4.2.6.1 *General*

Imperfections in matched filtering could be considered as differences in the amplitude of the matched filter or in the phase of the matched filter, compared with the ideal.

To give an idea of the influence of imperfections, Fig. 4.20 shows the shape of a chirp signal after a perfect pulse compression, and Fig. 21*a* shows the same shape after using a matched filter with no phase error but with an amplitude

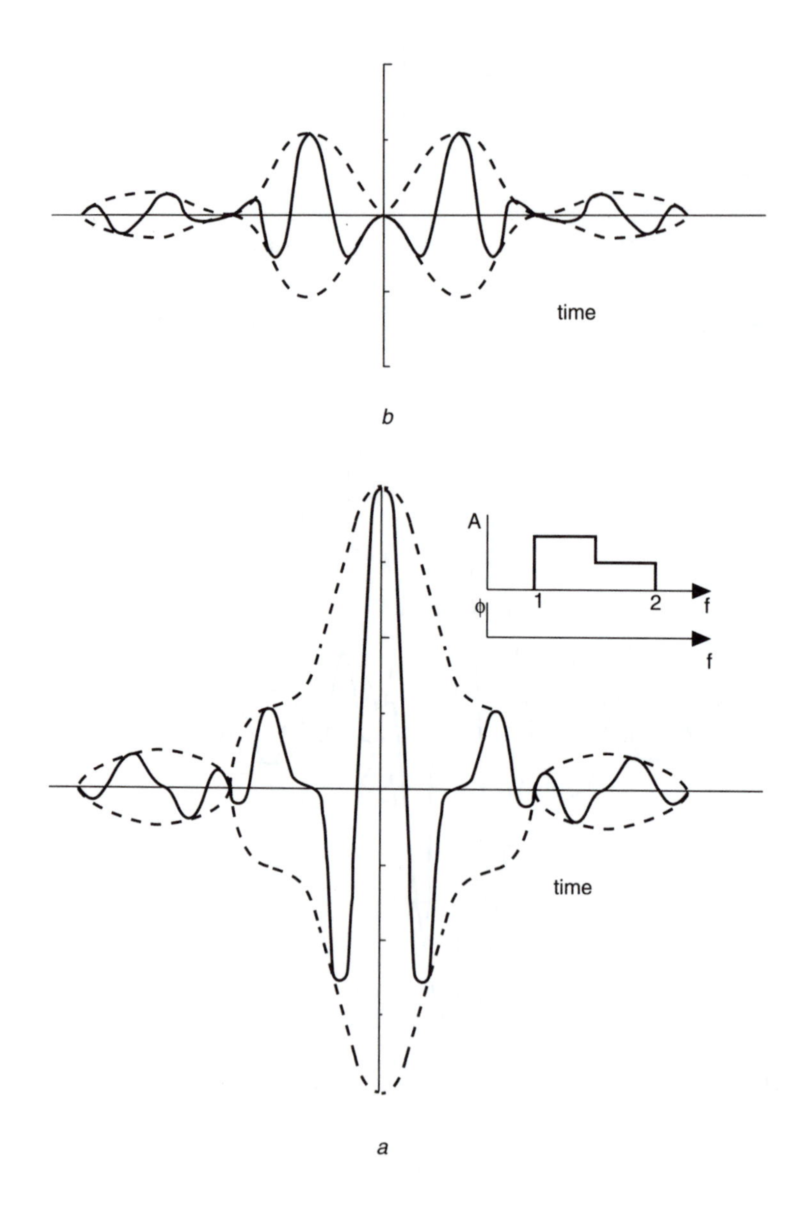

Figure 4.21 *Amplitude mismatching*

a Modified compressed pulse
b The relevant error

transfer function (6 dB attenuation for the right-hand part of the spectrum) as indicated. Fig. 4.21*b* represents the relevant error signal (difference between the perfection and the actual one). The effect of a phase error is indicated in Fig. 4.22*a* where the 'compressed signal' is represented after a phase error (90° error for the right-hand part of the spectrum). Fig. 4.22*b* represents the relative error signal. Figs. 4.21*a* and 4.22*a* are self-explanatory.

4.2.6.2 *'Paired-echoes method'*

This Section will simply present the paired-echoes method of analysing the defects introduced by errors in matched filtering in the case of a chirp signal. For more information, see Reference 1. It could be useful to know that paired-echoes method was also described in 1939 in Reference 2.

It is assumed that filtering on reception is band-pass filtering which cancels frequencies outside

$$f_1 - \frac{\Delta f}{2}, f_1 + \frac{\Delta f}{2}$$

It is assumed that the useful signal has a rectangular spectrum (chirp signal with a very large $T_p\Delta f$), requiring a rectangular spectrum for ideal reception.

So the ideal compressed signal should have a Fourier transform

$$a(f) = K \text{ for } f_1 - \Delta f/2 < f < f_1 + \Delta f/2$$

and $a(f) = 0$ for other frequencies with $\phi(f) = 0$ for any f. But it actually has a Fourier transform

$$a(f)[1 + \Delta_a(f)][\exp(-j\Delta_\phi(f)]$$

in which $\Delta_a(f)$ is the amplitude error, $\Delta_\phi(f)$ is the phase error.

Assuming, of course, that $\Delta_a(f) \ll 1$ and $\Delta_\phi(f) \ll 1$ radian, the compressed signal is as follows:

$$\int a(f) \exp(2\pi jft)df$$

$$+\int a(f)\, \Delta_a(f) \exp(2\pi jft)df$$

$$-j \int a(f)\Delta_\phi(f) \exp(2\pi jft)df$$

or

$$K \int_{f_1-\Delta f/2}^{f_1+\Delta f/2} \exp(2\pi jft)df$$

$$+K \int_{f_1-\Delta f/2}^{f_1+\Delta f/2} \Delta_a(f) \exp(2\pi jft)df$$

$$-jK \int_{f_1-\Delta f/2}^{f_1+\Delta f/2} \Delta_\phi(f) \exp(2\pi jft)df$$

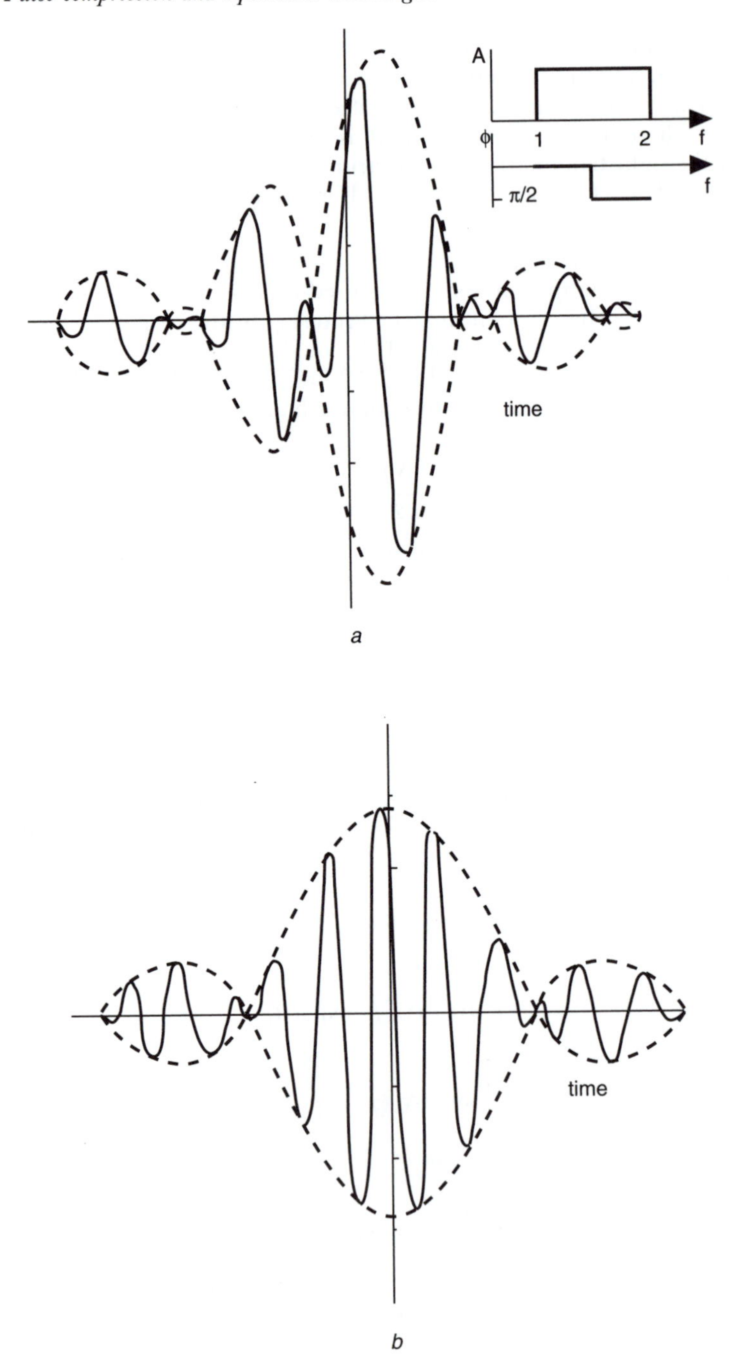

Figure 4.22 *Phase mismatching*

a Modified compressed pulse
b The relevant error

As presented, the compressed signal is the sum of

- 'useful signal' (first term)
- 'amplitude error signal' (second term)
- 'phase error signal' (third term)

Both error signals are very similar (except for $-j$, which is a phase-shift of the IF carrier) and will be dealt with similarly.

$\Delta_a(f)$ (and also $\Delta_\phi(f)$) could be considered as different from zero only between $f_1-\Delta f/2$ and $f_1+\Delta f/2$ and they could then be presented as follows:

$$\begin{aligned}\Delta_a(f) = a_0 & \\ & + a_1 \cos(2\pi \frac{f}{\Delta f} + \alpha_1) \\ & + a_2 \cos(2\pi \frac{f}{\Delta f} 2 + \alpha_2) \\ & + \cdots\cdots\cdots \\ & + a_n \cos(2\pi \frac{f}{\Delta f} n + \alpha_n) \\ & \cdots\cdots\cdots\end{aligned}$$

Thus describing the amplitude error signal as being the sum of the terms

$$Ka_n \int_{f_1-\Delta f/2}^{f_1+\Delta f/2} \cos(2\pi \frac{nf}{\Delta f} + \alpha_n) \exp(2\pi j f t) df$$

Such a term is in fact the sum of two 'echoes' (a 'pair' of echoes), respectively:

$$K\Delta f \frac{a_n}{2} \frac{\sin \pi(t\Delta f + n)}{\pi(t\Delta f + n)} \sin 2\pi\left(f_0 t + \alpha_n + \frac{nf_0}{\Delta f}\right)$$

and

$$-K\Delta f \frac{a_n}{2} \frac{\sin \pi(-t\Delta f + n)}{\pi(-t\Delta f + n)} \sin 2\pi\left(f_0 t - \alpha_n - \frac{nf_0}{\Delta f}\right)$$

The first echo of the pair has the same shape as the useful one (except for a phase shift of the IF carrier), and is situated at $t = n/\Delta f$, with a relative amplitude (versus the maximum of the theoretical useful signal) equal to $a_n/2$. It is added to the useful theoretical signal in phase, quadrature or opposition, depending on α_n and $nf_0/\Delta f$.

The second echo of the pair also has the same shape as the theoretical useful signal, but is situated at $t = -n/\Delta f$.

So the parasitic error signal coming from the amplitude error could be considered as a sum of paired echoes, a periodic (versus f) amplitude error with relative amplitude a_n producing two sidelobes on both sides of the centre with $a_n/2$ as relative amplitude. This permits us to recall the following orders of magnitude (strictly valid only for sine wave error):

1% relative amplitude error produces sidelobes at −46 dB
3% relative amplitude error produces sidelobes at −36 dB
10% relative amplitude error produces sidelobes at −26 dB

Similarly the following orders of magnitude for consequences of phase errors are found

10 mrad phase error produces sidelobes at −46 dB
30 mrad phase error produces sidelobes at −36 dB
0.1 mrad phase error produces sidelobes at −26 dB

4.2.7 *Parallel methods for pulse compression*

Another method to achieve pulse compression by using some kind of passive generation is described in Fig. 4.23: the left-hand part is the IF generation of the long-phase-modulated signal, and the right-hand is the relevant pulse-compression device. We could assume that the input is a small pulse of 0.1 μs duration with a 30 MHz frequency carrier (no phase modulation within that short pulse). The filters A, B, . . . , G are adjacent band-pass filters introducing no phase modification, with a bandwidth of 1.4 MHz centred, respectively, on 25.8, 27.2, 28.6, 30, 31.4, 32.8 and 34.2 MHz. At the output of those seven filters we obtain seven pulses of 0.7 μs duration (no phase modulation) which are simultaneous (their middles are obtained at exactly the same time). Then introduction of delays of 0, 0.7, 1.4, 2.1, 2.8, 3.5 and 4.2 μs provides finally after addition a total signal of 4.9 μs duration with a frequency modulation similar to that represented in Fig. 4.16. Crossing the device represented on the right-hand side of Fig. 4.23 reproduces at the output the initial short pulse of 0.1 μs, since finally all the frequencies will have been delayed by 4.2 μs seconds from left to

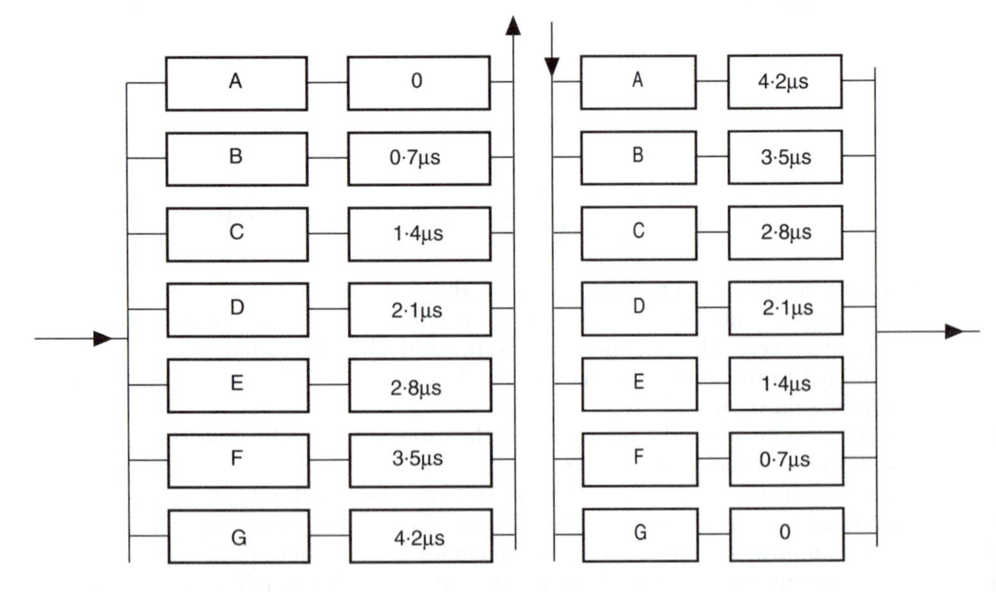

Figure 4.23 *Parallel method for pulse compression: General block diagram*

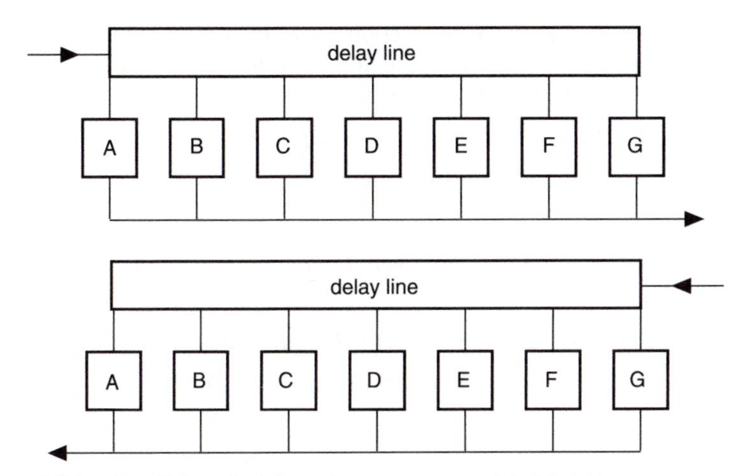

Figure 4.24 *Parallel method for pulse compression: Modified block diagram*

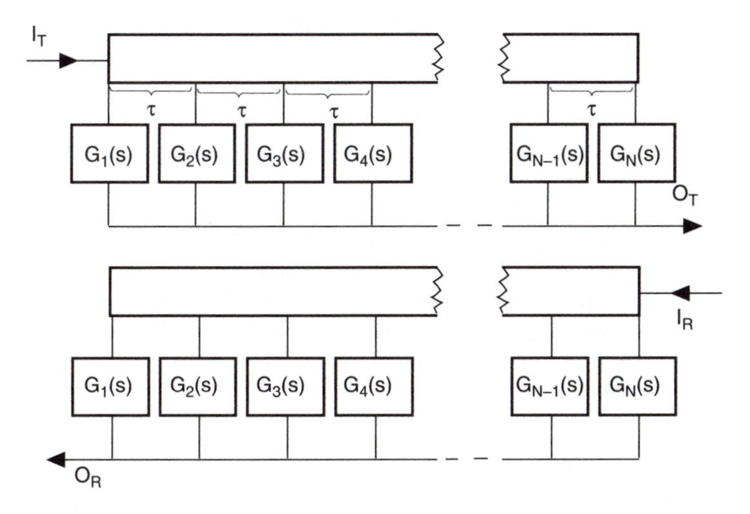

Figure 4.25 *Parallel method for pulse compression: Generalisation*

right. The device permits one to obtain a pulse compression ratio of 49 (7×7). Such a 'parallel' system could be represented as in Fig. 4.24 where the delay line is of 4.2 μs with outputs every 0.7 μs (this is obtained by changing the respective order of filters and delay lines on transmission and on reception). It is then clear that parallel pulse compression is a particular example of the general family of systems using a transverse filter on transmission and the same in reception but reversed in time (the order of successive filters is time-reversed), which is a direct consequence of what has been said at the end of Section 4.2.1.

In general, if we consider the device represented in Fig. 4.25 using a delay line with outputs every τ and filters with transfer functions, respectively, of $G_1(s)$

$G_2(s) \ldots G_{N-1}(s)\ G_N(s)$, the introduction in I_T of a very small pulse (Dirac impulse) will provide in O_T a long (at least the total delay of the delay line) signal whose Fourier transform will be

$$\begin{aligned} G_1(2\pi jf) + G_2(2\pi jf)\exp(-2\pi jf\tau) + \cdots \\ + G_{N-1}(2\pi jf)\exp(-2\pi j(N-2)f\tau) \\ + G_N(2\pi jf)\exp(-2\pi j(N-1)f\tau) = \phi(f) \end{aligned}$$

Using for reception the device between I_R and O_R represents a device whose transmittance is

$$\begin{aligned} T(f) = G_N(2\pi jf) + G_{N-1}(2\pi jf)\exp(-2\pi jf\tau) + \cdots \\ + G_2(2\pi jf)\exp(-2\pi j(N-2)f\tau) \\ + G_1(2\pi jf)\exp(-2\pi j(N-1)f\tau) \\ \equiv \exp(-2\pi j(N-1)f\tau)[G_1(2\pi jf) \\ + G_2(2\pi jf)\exp(+2\pi jf\tau) + \cdots \\ + G_{N-1}(2\pi jf)\exp(+2\pi j(N-2)f\tau) \\ + G_N(2\pi jf)\exp(+2\pi j(N-1)f\tau)] \end{aligned}$$

which is equal to

$$\phi^*(f)\exp(-2\pi j(N-1)f\tau)$$

if

$$\phi_k(2\pi jf) \equiv \phi_k(-2\pi jf) \quad \forall k$$

that is to say if all $\phi_k(s)$ are real functions, which implies that the relevant filters are either pure amplifiers or pass-band filters not introducing phase modification, or cascades of both, which was the case for the device of Fig. 4.24. In order to compress the signal represented at Fig. 4.17, it is possible to use a delay line, with every output of the delay line feeding an amplifier, the gains of the successive amplifiers being respectively:

1	−0.85	0.37	0.31	−0.88	0.94
−0.31	−0.62	0.99	−0.31 etc.		

Clearly the device will be realised by using digital technology, so the gains of the amplifiers will be successively

$$\begin{aligned} & 2^0 \\ & -(2^{-1}+2^{-2}+2^{-4}+2^{-5}+2^{-7}) \\ & 2^{-2}+2^{-3} \\ & 2^{-2}+2^{-4} \\ & \ldots \end{aligned}$$

This method now generally replaces the device described in Section 4.2.4.2.

During the 1960s various optical computers were used on radar prototypes employing the basic philosophy of transverse filters in order to achieve parallel pulse compression. They were using as the delay line an ultrasonic medium transparent to light in devices using coherent light, or they were using as the delay line the mechanical rotation of a drum (whose transparency represented the gains of the successive amplifiers). But not many operational systems have in practice been equipped with such devices for several reasons, of which one was the lack of dynamic range.

4.2.8 *Use of Fourier transformers*

At the same time, optical computers have also been used, by which the electrical received signal was transformed into a variation of the amplitude of coherent light and its Fourier transform was achieved by optical means (using only lens). This Fourier transform was then optically multiplied by $\phi^*(f)$, $\phi^*(f)$ being physically represented by the variation of the transparency of a hologram containing $\phi^*(f)$ or equivalent means. Then a new optical Fourier transform was produced, giving the required result.

Although that system has been used, for instance in synthetic aperture radars where it was possible to process the signal not in real time, it has generally been abandoned, because the progress in digital circuitry has caused them to be replaced by digital systems such as those described in Section 4.2.7.

However, the same evolution of digital circuitry will permit us to use a similar digital method. The Fourier transform of the received signal (after demodulation, and analogue-to-digital encoding) is computed by using an algorithm such as fast Fourier transform (FFT). Then this Fourier transform is digitally multiplied by the conjugate of the Fourier transform of the transmitted signal (also down frequency-shifted, of course), and the result of that multiplication is again digital Fourier transformed by FFT or equivalently.

4.3 Effect of pulse compression on clutter

The effect of pulse compression on clutter depends on its nature.

If there are a very very large number of pieces of clutter in the resolution cell of the radar (roughly defined in case of a ground radar by the product of the horizontal beamwidth, the distance and the range resolution, and correctly modified in the case of volume clutter such as rain or chaff) the pulse compression, because it represents the division of the range resolution by the pulse compression ratio, represents the division of the number of pieces of clutter and then the power of the clutter by the same ratio. But this is a rare situation. In most cases the number of pieces of clutter is not sufficient to consider that the effect of pulse compression is only to reduce the power of the clutter by the pulse compression ratio.

Let us consider the simple situation in which the clutter is composed of a few very powerful pieces of clutter, for instance only a very small number of pieces of clutter but with an equivalent echoing area of $\sigma = 10\ 000\ \text{m}^2$ each, in such a way that the pieces of clutter are at random position with a probability of 0.05 of being present in a resolution cell (as obtained after a pulse compression of 100). That means that, without pulse compression (so with a range resolution multiplied by 100), the probability of having at least one piece of clutter in the resolution cell would have been equal to $1-(1-0.05)^{100}$ which is 0.994, practically 1, which means that a useful target would have nearly always been mixed with a clutter above 10 000 m^2; while, with the pulse compression, the probability that a useful target would be mixed with the clutter is nearly negligible. As a consequence, without pulse compression Doppler filtering of excellent quality will be required, while in the case of pulse compression – if the required detection probability is 0.9 – it will be necessary only to use a system which cancels the clutter as well as the useful targets mixed with it (Doppler filtering with a fairly poor 'sub clutter visibility' will be convenient).

Furthermore, under our assumptions, the probability that without pulse compression there is no clutter at all is 0.0059

a clutter of $\sigma = 10\ 000\ m^2 = 0.0312$

a clutter of $2\sigma = 0.0812$

a clutter of $3\sigma = 0.1396$

a clutter of $4\sigma = 0.1781$

a clutter of $5\sigma = 0.1800$

a clutter of $6\sigma = 0.1500$

a clutter of $7\sigma = 0.1060$

a clutter of $8\sigma = 0.0649$

a clutter of $9\sigma = 0.0349$

a clutter of $10\sigma = 0.0167$

The curves of Fig. 4.26 give the relevant distributions of the clutter power with and without pulse compression (probability that the actual clutter Σ is above σ, versus σ). The mean value of the clutter power is indicated by O: the pulse compression by 100 divides the mean (average) value of the clutter power by 100 but in this case that information is meaningless.

Fig. 4.27 shows real graphs for actual clutter distributions in the S band, in relatively flat country without rocky mountains and with a low density of small houses. In the graphs, O indicates the average value of clutter power: three graphs are represented, corresponding to, respectively, range resolutions of 50 μs, 5 μs and 0.5 μs. It is clear from the Figure that:

- The average clutter power is divided by 10 when the range resolution is improved by 10
- Even in that case, which is very different from the theoretical case given just before, the clutter distribution is very different for a small resolution cell than a large resolution cell.
- If a detection probability of 75% is accepted, the eventual benefit obtained by reducing the range resolution from 50 μs down to 5 μs is about 17 dB and not 10 dB, and the benefit obtained by reducing the range resolution from 5 μs down to 0.5 μs is about 24 dB.

In practice clutter is generally something of a mixture of 'point clutter' and 'scattered clutter'. The effect of improving range resolution is:

- To reduce the power of the scattered clutter by the pulse compression ratio

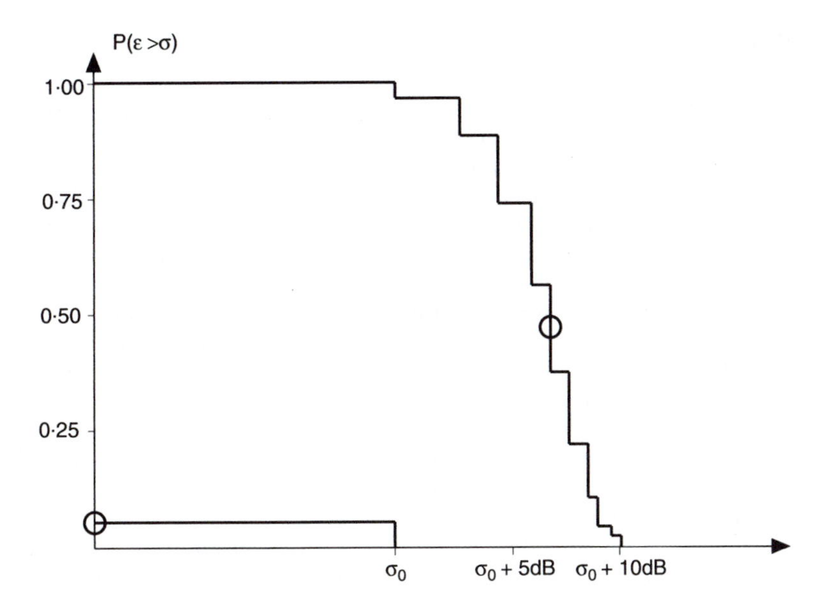

Figure 4.26 *Modification of the clutter distribution coming from pulse compression: Theoretical*

- To reduce the percentage of the zones where a useful target is mixed with a piece of 'point clutter'.

In most of the practical cases, the result is an improvement of detection in clutter conditions above what could be expected in the case of scattered clutter only; this is valid also in case of rain clutter and chaff clutter.

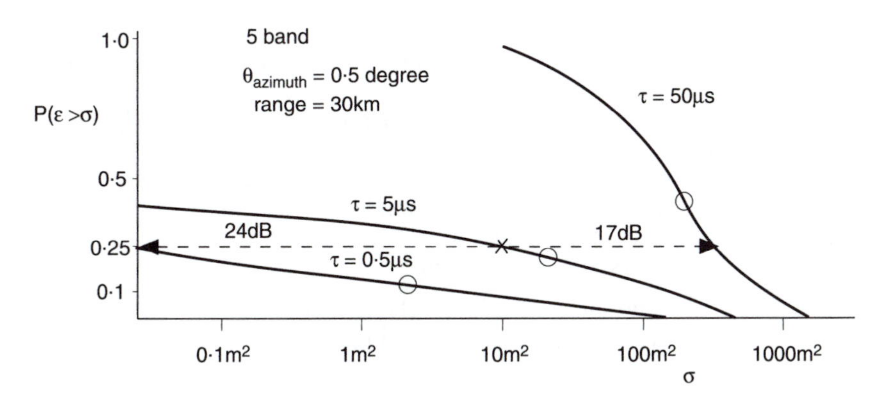

Figure 4.27 *Modification of the clutter distribution coming from pulse compression: Real example*

4.4 Polyphase codes

Up to the present, the transmitted signals which have been considered are basically signals without amplitude modulation during their duration, within which frequency was varying against time.

Transmission of signals with an amplitude modulation within their duration can *a priori* also be considered. Unless it is total amplitude modulation (providing, in fact, a series of pulses not amplitude modulated) such modulation is presently not generally acceptable, since the efficiency of the transmitter unit would be very low. However, in the future, the use of active arrays with a very large number of transmitter–receiver modules will allow for transmission of amplitude modulated signals obtained by the addition of many different elementary signals not amplitude modulated. That is not taken into consideration in the next discussion.

But, on the other hand, even without active arrays, transmitted signals could be used where the 'instantaneous' frequency remains constant, while the phase varies in steps against time.

For instance, the signal is divided into pulses of equal duration τ, the phase being constant within every elementary pulse but hopping from pulse to pulse (see Fig. 4.28). This is called 'using polyphase codes'.

Particular polyphase codes are codes in which the phase could take only two values (basically zero and π) called 'binary phase-coded signals', and among those most utilised are the 'pseudo-randomly binary phase-coded signals'.

The pseudo-randomly binary phase-coded signals most utilised are those using:

- Barker's codes
- Maximum length codes (in France, they are called 'Codes de Galois' since a fairly complete method of studying them consists in using the so-called 'Corps de Galois' or, to use the first historical name, the 'Champs de Galois' (Galois' fields)).

4.4.1 *Barker's codes*

Barker's codes result from the multiplication of a sine wave sin $(2\pi f_0 t)$ by a function $V(t)$ equal to $+1$ or -1, changes possibly only occurring every τ seconds.

Barker's codes have a finite duration and provide transmitted signals $S(t)$ with finite duration.

The autocorrelation function of a Barker's code binary phase-modulated signal is composed of a central peak surrounded by secondary peaks, all with the same amplitude (see Fig. 4.29).

The Barker sequences are the following:

Length 2	+ + (no interest: it is the classical radar)
	− + (not really more interesting)
Length 3	+ + −
Length 4	+ + − +
Length 5	+ + + − +
Length 7	+ + + − − + −

Length 11 + + + − − − + − − + −
Length 13 + + + + + − − + + − + − +

It is easy to prove that no Barker sequence is possible with an even length (above 4). Nobody found an odd Barker sequence above 13. In practice, that means that pulse compression using a Barker sequence could not provide a pulse compression ratio above 13 (which is unfortunate!) As a consequence only sequences with 11 or 13 length are effectively used (the relative height of the central peak compared with the sidelobes peaks is equal to the length of the code: i.e. 13 for the last sequence, or 22.3 dB).

It is clear that parallel methods are very convenient for generating such sequences and achieving the relevant pulse compression.

4.4.2 *Galois codes*

Let us consider Fig. 4.30 where a shift register is completed by a feedback loop via a 'modulo 2 adder' (in which $0+0=0$ $1+0=1$ $0+1=1$ $1+1=0$).

Assuming that initially the shift register only contains 1, the following sequence will get one out of the fix (at the clock frequency):

1 1 1 1 0 0 0 1 0 0 1 1 0 1 0 1 1 1 1 0 0 0 1 . . .

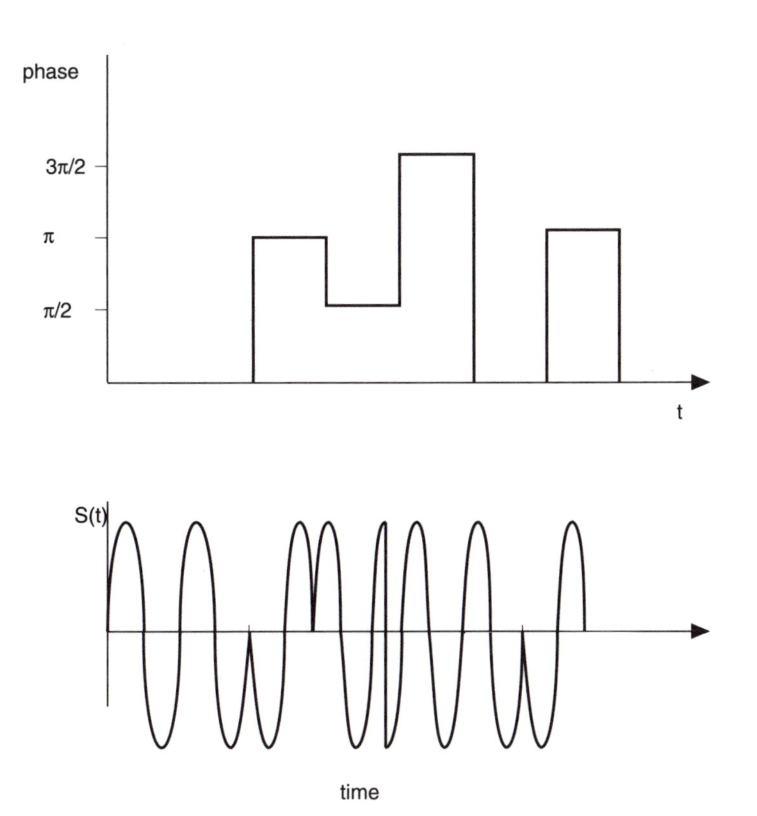

Figure 4.28 *Polyphase code*

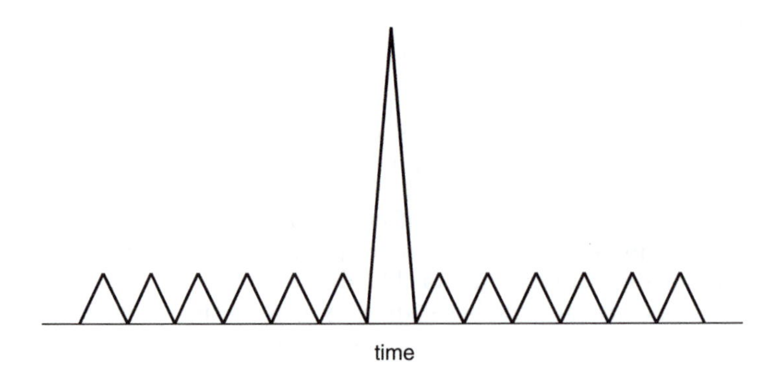

Figure 4.29 *Use of Barker's Code: Shape of the useful compressed signal*

which is a periodic signal whose period is 15 times the clock period. If we used $V(t) = +1$ for an output of 1 and $V(t) = -1$ for an output of 0, the signal $S(t) = V(t)\ \sin(2\pi f_0 t)$ will also be periodic with the same period $(15 = 2^4 - 1)$ times the clock period.

It is clear that during a period the register has contained all possible binary numbers (except 0000), which explains the term 'maximum length codes'. Generally, it is possible, with an N-cell shift register used with the relevant feedback loop, to generate sequences of $(2^N - 1)$, 1 and 0, which repeats regularly. It could be proved that the autocorrelation function of the relevant $V(t)$ (and of $S(t)$ as a consequence) which is generated by that means has the following characteristics (if τ is the clock period):

- Base duration of the central peak, 2τ
- Obvious periodicity of $(2^N - 1)\tau$
- All sidelobes with the same amplitude
- Ratio of the central peak amplitude and the maximum sidelobes amplitude, $2^N - 1$ (e.g. if $N = 10$ the ratio will be 1000 or 60 dB)

Fig. 4.31 represents the autocorrelation function $V(t)$ for $N = 4$. Maximum length codes are often used in continuous wave radars.

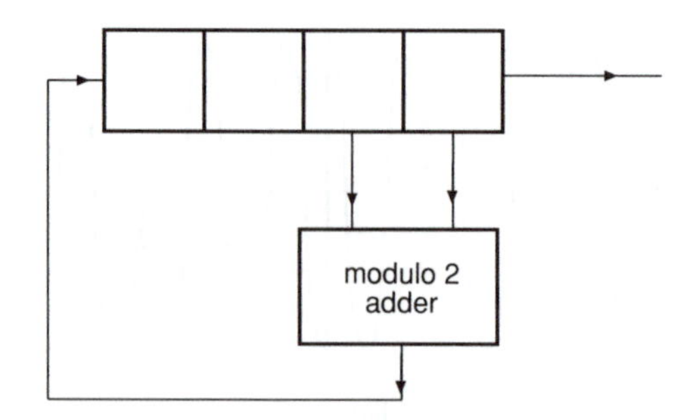

Figure 4.30 *Construction of a Galois code*

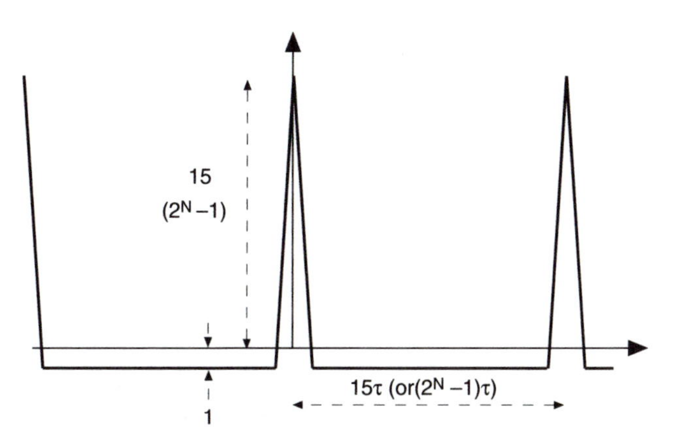

Figure 4.31 *Autocorrelation function of a signal phase-modulated by a Galois code*

4.4.3 *Other polyphase codes*

Many other polyphase codes could obviously be used. For example, polyphase code described in Reference 3 could be used.

An N code corresponds to possible phase values differing by $2\pi/N$. Its length is N^2 which corresponds to values of the phase equal to $2\pi p/N$, p being obtained as follows:

$$\overbrace{0\,0\,0\ldots0}\;\overbrace{0\,1\,2\,3\ldots(N-1)}\;\overbrace{0\,2\,4\ldots2(N-1)}$$
$$\overbrace{\ldots0\quad N-1\quad 2N-1\ldots(N-1)^2}$$

For instance for $N=4$, the sequence is as follows:

$$0\quad 0\quad 0\quad 0\quad \pi/2\quad \pi\quad 3\pi/2\quad 0\quad \pi\quad 0\quad \pi\quad 0\quad 3\pi/2\quad \pi\quad \pi/2$$

Such a code has autocorrelation functions with a central peak (2τ width at the base level) with relatively low sidelobes (see Fig. 4.32 for $N=4$).

For $N=4$ maximum sidelobe amplitude is 1 for a peak amplitude of 16 (N^2). That, which is also true for $N=3$, is no more true for $N \geqslant 5$, the actual results being given in Table 4.1

For N very large the ratio between the amplitude of the maximum peak and that of the worst sidelobe tends to $2\pi N$.

4.4.4 *Two remarks about polyphase codes*

First, it is very important to realise that (see Section 4.2.3), while in the case of a chirp signal a relatively small Doppler shift f_D (small compared with Δf, but significant compared with $1/T_p$) introduces only a shift in position, a reasonable loss in sensitivity and not too drastic changes in the shape of the sidelobes, in the case of a polyphase coded signal a small Doppler shift (about $f_D = 1/T_p$) generally leads to a quasi-complete loss of the useful signal.

In that case, in order to know more about the effect of a Doppler shift, it is necessary to achieve complete computation of the response of a filter matched to the signal, for signals shifted by all possible Doppler shifts (in practice,

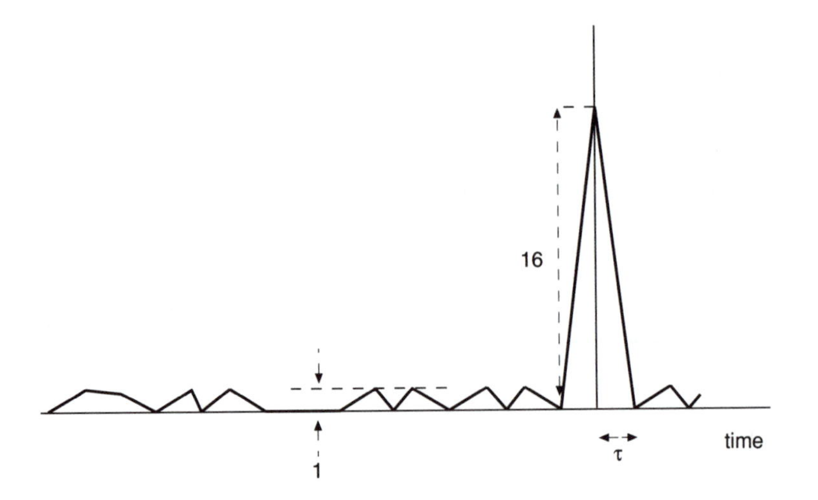

Figure 4.32 *Autocorrelation function of a signal phase-modulated by a Franck's code*

computation has to be made for successive Doppler shifts separated by $1/2T_p$, and not $1/T_p$ as is theoretically necessary).

Secondly, generation of polyphase codes could sometimes be conveniently obtained directly in microwaves by introducing a microwave phase shifter between the transmitter (which could be a simple microwave – CW or otherwise – oscillator) and the antenna.

4.5 Sidelobe reduction

4.5.1 *Weighting filters*

It has been explained in Section 4.2.2 that after pulse compression the chirp signal becomes a short signal surrounded by parasitic sidelobes, the two closest sidelobes having a level of 13.3 dB below the main signal. This means that, if a parasitic target is very close to a useful one with an equivalent echoing area

Table 4.1

N	pulse compression ratio N^2	relative level of the worst sidelobes	$20 \log_{10}\left(\frac{1}{2\pi N}\right)$
3	9	−19 dB	−26 dB
4	16	−24 dB	−28 dB
5	25	−26 dB	−30 dB
6	36	−28 dB	−32 dB
7	49	−30 dB	−33 dB
8	64	−32 dB	−34 dB

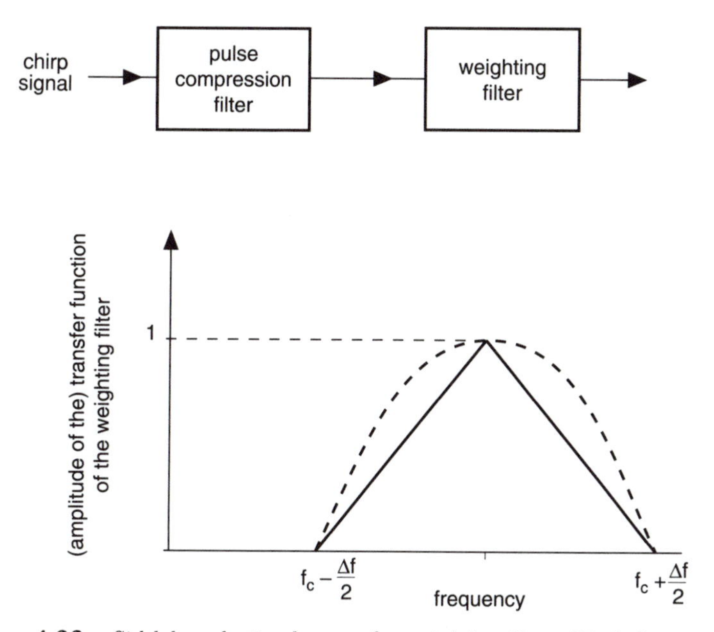

Figure 4.33 *Sidelobe reduction by use of a weighting filter: Block-diagram*

more than 13 dB above the useful one, it will mask it; that is generally not acceptable.

To avoid this, a 'weighting' filter could be introduced in cascade with the normal matched filter, in order to reduce the amplitude of the Fourier transform of the useful signal for frequencies far from the central frequency (compared with the amplitude of that central frequency).

To take a simple example, if the useful chirp signal, after pulse compression (and then with a rectangular spectrum) crosses a filter with a triangular transfer function (as indicated in Fig. 4.33) and no significant phase shift:

- The noise power will be multiplied by 1/3 (easy to compute)
- The shape of the useful signal will be modified as indicated in Fig. 4.34 (new signal in continuous line, compared with the signal without weighting filter, in dotted line) with the maximum amplitude divided by 2 (power multiplied by 1/4)

As a consequence:

- The signal-to-noise ratio has been multiplied by 0.75 (loss of 1.25 dB in sensitivity)
- The 3 dB pulse duration is multiplied by about 1.5
- The level of the first sidelobe varies from −13.3 dB down to −26.5 dB.

This simple example could be generalised: reducing the amplitude of the frequencies far from the centre will permit one to significantly reduce the level of parasitic sidelobes at the price of a loss in sensitivity and a widening of the useful signal.

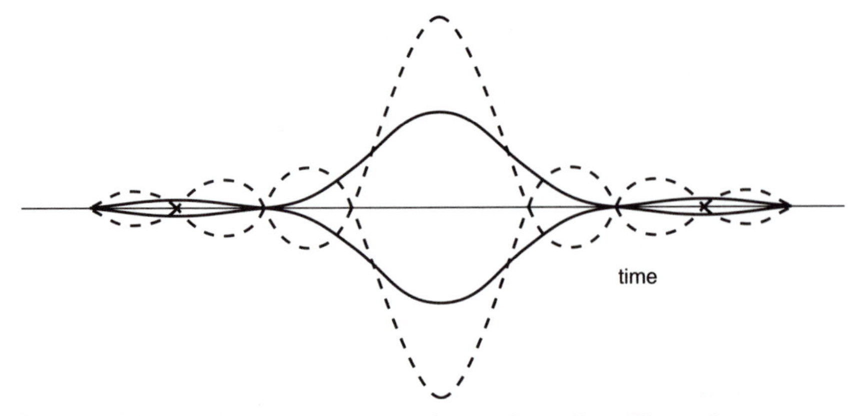

Figure 4.34 *Sidelobe reduction by use of a weighting filter: The result*

Practical shaping curves are not like the theoretical example of Figs. 4.33 and 4.34. Using a transfer function with a cosine shape (see dotted curve in Fig. 4.33) will provide sidelobe level (the worst) at −23 dB, associated with a loss of 0.9 dB and a multiplication of the pulse width by about 1.35. Using a transfer function with a (cosine square) shape will provide sidelobe level (the worst) at −32 dB, associated with a loss of 1.8 dB and a multiplication of the (3 dB) pulse width by about 1.63.

People familiar with antenna are familiar with that situation since, in a similar manner, an antenna pattern is the Fourier transform of the antenna illumination (maximum antenna efficiency for rectangular illumination but with significant sidelobes, and reduction of the sidelobes associated with reduction of efficacy and beam widening, by reduction of illumination on the edges of the antenna etc.).

Polyphase coded signals can have by nature relatively, even extremely, good sidelobes as indicated in Section 4.4.3, but it has to be recalled that, in practice, they occupy a frequency band comparatively much higher than the inverse of the useful part of the compressed signal.

4.5.2 *Intrinsic weighting*

Coming back to chirp signals, a loss of 0.9 dB and *a fortiori* a loss of 1.8 dB could be considered as unacceptable (losing 0.9 dB when transmitting 20 kW mean power means losing nearly 4 kW!). In that case it is generally required to have low sidelobes on compressed signal without losing sensitivity.

That could be obtained if we transmit a signal (which is not amplitude modulated – in order to keep good efficiency in transmission) whose frequency modulation is not linear.

For instance, if we use passive generation with (on transmission) a delay line, the propagation time of which is represented by the continuous line of Fig. 4.35, the signal obtained will have the central frequencies of more duration than the extremer ones; so the spectrum of the signal will be more important around the central frequency f_c than about $f_c - \Delta f/2$ or $f_c + \Delta f/2$.

Specifically, the curve represented is that of

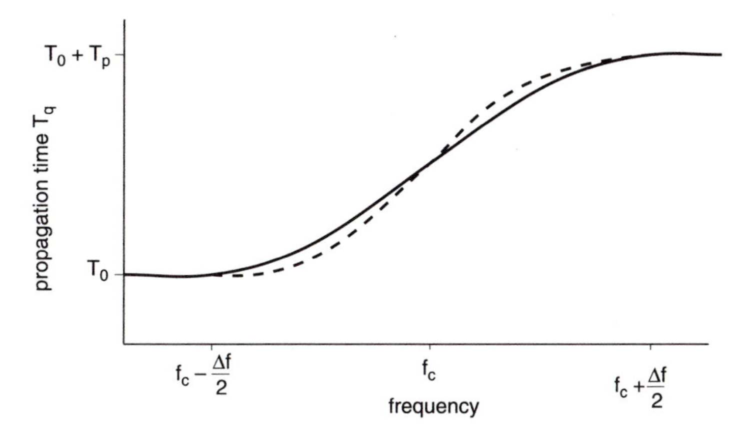

Figure 4.35 *Intrinsic weighting characteristic curves on transmission*

$$T_g = T_0 + \frac{T_p}{2}\left(1 + \sin\left(\frac{\pi}{\Delta f}(f - f_c)\right)\right)$$

which means that

$$\frac{dT_g}{df} = \frac{\pi T_p}{2\Delta f}\cos\left(\frac{\pi}{\Delta f}(f - f_c)\right)$$

This implies that the shape of the spectrum (which is $|\phi(f)|^2$ if $\phi(f)$ is the Fourier transform of the signal) varies according to

$$\left|\cos\left(\frac{\pi}{\Delta f}(f - f_c)\right)\right|$$

As a consequence $|\phi(f)|$ varies as

$$\left|\cos\left(\frac{\pi}{\Delta f}(f - f_c)\right)\right|^{1/2}$$

If then:

- Firstly we use on reception a filter with a propagation time T_g versus f conjugate of the Fig. 4.35 curve (see Fig. 4.36)
- Secondly, amplitude filtering with a shape like

$$\left|\cos\frac{\pi}{\Delta f}(f - f_c)\right|^{1/2}$$

we shall obtain pulse compression

- With sidelobe level at −23 dB (for the worst) (since obviously $\sqrt{\cos} \times \sqrt{\cos} = \cos$)
- But without loss in sensitivity.

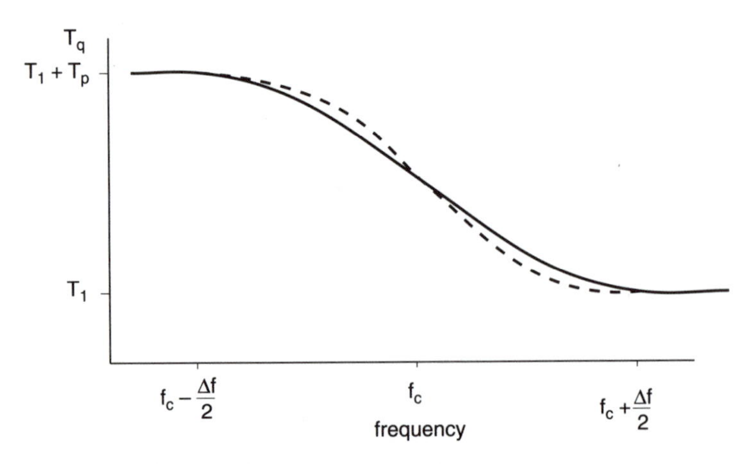

Figure 4.36 *Intrinsic weighting characteristic curves on reception*

If we use on transmission a delay line, the propagation time of which varies as

$$T_g = T_0 + \frac{T_p}{\Delta f}\left(f - f_c + \frac{\Delta f}{2}\right) + \frac{T_p}{2\pi}\sin\left(\frac{2\pi}{\Delta f}(f - f_c)\right)$$

in that case, since

$$\frac{dT_g}{df} = \frac{T_p}{\Delta f} + \frac{T_p}{\Delta f}\cos\left(\frac{2\pi}{\Delta f}(f - f_c)\right)$$

$$\frac{dT_g}{df} = \frac{2T_p}{\Delta f}\cos^2\left(\frac{\pi}{\Delta f}(f - f_c)\right)$$

$|\phi(f)|^2$ will vary as

$$\cos^2\left(\frac{\pi}{\Delta f}(f - f_c)\right)$$

$|\phi(f)|$ will vary as

$$\cos\left(\frac{\pi}{\Delta f}(f - f_c)\right)$$

If we then use on reception the conjugate filter corresponding to the dotted line of Fig. 4.36 and amplitude filtering with a shape like

$$\cos\left(\frac{\pi}{\Delta f}(f - f_c)\right)$$

we shall obtain pulse compression:

- With sidelobe level at −32 dB (for the worst) (since $\cos \times \cos = \cos^2$)

- But again without loss in sensitivity.

Systems of that nature are called by the author 'quasi-chirp', and are used very often.

4.5.3 *Use of non-matched filters*

The matched filter is the ideal solution when the problem is to detect *one* target and to locate it.

When the problem of resolving targets is also very important, detection capability could be partly sacrificed. This clearly was the case described in Section 4.5.1 where a loss in sensitivity was accepted when using a weighting filter, partly losing the information contained in the side frequencies of the spectrum to obtain a reduction of the sidelobes. Many other examples could be given of this nature.

Another example will be presented to indicate that other possibilities could be used. In Section 4.4.2 the signals binary phase-modulated according to a maximum length binary sequence have been presented. In such a signal, the compressed signal is composed of a central useful triangle accompanied by sidelobes with a constant level equal to the useful one divided by $(2^N - 1)$, very close to 2^N if N is significant, which means $-6\ N$ dB.

Remembering that the transmitted signal is based on $V(t)\ \sin(2\pi f_c t)$, where $V(t)$ is a sequence of $(2^N - 1)$ one or minus one, repeated with a periodicity of $(2^N - 1)$ clock times, if, on reception we realise the transverse filter equivalent to a correlation by

$$W(t) = 0.5(V(t) + 1)$$

instead of

$$V(t)$$

we shall obtain

$$\int_{T=(2^N-1)\tau} V(t)\, W(t-\theta)\, dt$$

instead of

$$\int_T V(t)\, V(t-\theta)\, dt$$

with

$$\int_T V(t)\, W(t-\theta)\, dt = 0.5 \int_T V(t)\, dt + 0.5 \int_T V(t)\, V(t-\theta)\, dt$$

$$= 0.5 - 0.5 = 0 \quad \text{for} \quad \tau < \theta < (2^N - 2)\tau$$

The sidelobes have completely disappeared and it is easy to show that the signal-to-noise ratio has been multiplied by

$$2^{N-1}/(2^N - 1) = 0.5 + \frac{0.5}{2^N - 1}$$

very close to 0.5 when N is large.

Total suppression of the sidelobes has then been obtained, accompanied by a loss of 3 dB in sensitivity.

4.6 Compatibility with constant false alarm reception

Hard clipping of the received signal before pulse compression provides constant false alarm reception if the pulse compression ratio is large enough.

In effect, in the absence of a useful signal, the power of the noise after hard clipping and before pulse compression is fixed (fixed by the level of hard limitation). Hence, after pulse compression (after crossing a well defined linear and stationary filter), the power of the noise is again constant. On the other hand, action of pulse compression on noise is equivalent to the addition of $T_p\Delta f$ independent samples of noise (at least, and much more if the bandwidth before hard limiting is larger than the bandwidth Δf), providing – according to the central-limit theorem – a noise with a *Gaussian* distribution. As a consequence hard limiting before pulse compression provides, in the absence of a useful signal, at the output of the pulse compression filter, a Gaussian noise whose power (variance) is constant and known.

It is explained in detail in Section 4.6 of Reference 4 that this hard clipping:

- Introduces a loss in sensitivity of only about 1 dB in the case of a transmitted signal only frequency (or phase) modulated, if $T_p\Delta f$ is large enough
- A higher loss in the case of a transmitted signal also amplitude modulated
- More difficulty in the case of a small target very close to a large one.

4.7 Typical examples of applications of pulse compression

4.7.1 *Very long range radars*

The first family of pulse compression radars comprises the case when it is necessary to transmit high energy in order to have long range. In that case powerful long pulses are transmitted and pulse compression is used to still maintain reasonably good range measurement. An example of parameters corresponding to that family is as follows:

Peak power $= P_c = 50$ MW
Pulse length $= T_p = 2$ ms
Transmitted energy per pulse $= 100\,000$ J
Central transmitted frequency $= f_0 = 100$ MHz $\lambda = 3$ m
Spectral width $= \Delta f = 100$ kHz

$$\left(\frac{1}{\Delta f} = 10\ \mu\text{s} \rightarrow 1500\ \text{m}\right)$$

Pulse compression ratio $= T_p\Delta f = 200$
For a radial velocity V_R of 5000 m/s, corresponding to a Doppler shift of 3333 Hz

$2(T_p\Delta f)\,V_R = 2000$ km/s $\ll 300\ 000$ km/s (velocity of light)

If a chirp signal is used (linear frequency modulation), position change coming from radial velocity is

$$T_p f_d \frac{1}{\Delta f} = 6667 \frac{1}{\Delta f} = 10 \text{ km}$$

This means that the radial range directly measured is that of the radial range of the target

$$T_p \frac{f_D}{\Delta f} \text{ later, 2 s later}$$

4.7.2 *Long range conventional radars with good range measurement*

The second family of pulse compression radars are basically conventional radars in which pulse compression has been introduced to improve the range resolution. An example of parameters corresponding to that family is as follows:

Peak power $= P_c = 30$ MW
Pulse length $= T_p = 5$ μs
Transmitted energy per pulse $= 150$ J
Central transmitted frequency $= f_0 = 3$ GHz $\lambda = 0.1$ m
Spectral width $= \Delta f = 5$ MHz

$$\left(\frac{1}{\Delta f} = 0.2\ \mu\text{s} \rightarrow 30 \text{ m}\right)$$

Pulse compression ratio $= T_p\Delta f = 25$
For a radial velocity of $V_R = 600$ m/s, corresponding to Doppler shift of 12 kHz
$2(T_p\Delta f)\,V_R = 30$ km/s $\ll 300\ 000$ km/s
If chirp signal is used, positive change coming from radial velocity is

$$T_p f_D \frac{1}{\Delta f} = 0.06 \times 30 = 1.8 \text{ m}$$

$$\left(\text{range of target } \frac{T_p f_D}{\Delta f} \text{ later (or earlier), i.e. 3 ms}\right)$$

4.7.3 *Modern radars with relatively long range and good range measurement*

The corresponding family is that of radars with reduced peak power, and thus reduced detectability (by the enemy!), reduced high voltage, reduced volume for the transmitter, allowing for easier mobility. An example of the parameters is as follows:

Peak power $= P_c = 20$ KW
Pulse length $= T_p = 50$ μs
Transmitted energy per pulse $= 1$ J
Central transmitted frequency $= f_0 = 6$ GHz $\lambda = 0.05$ m

Spectral width $= \Delta f = 10$ MHz

$$\left(\frac{I}{\Delta f} = 0.1\ \mu\text{s} \rightarrow 15\ \text{m}\right)$$

Pulse compression ratio $= T_p \Delta f = 500$
For a radial velocity of $V_R = 400$ m/s, corresponding to a Doppler shift f_D of 16 000 Hz

$$2(T_p \Delta f)\, V_R = 400\ \text{km/s} \ll 300\ 000\ \text{km/s}$$

If chirp signal is used, position change coming from radial velocity is

$$T_p f_D \frac{1}{\Delta f} = 0.8 \frac{I}{\Delta f} \Rightarrow 12\ \text{m}$$

$$\frac{T_p f_0}{\Delta f} = 0.03\ \text{s}$$

The range of such a radar could be typically around 100 km (PRF of about 1 kHz) and problems arise for short range detection (at ranges below 15 km), when part of the received pulse is obtained while the receiver is shut because transmission is not finished: It gives a degradation of range measurement on close targets.

The solution could be to transmit, at the end of the chirp signal, a short signal (not frequency modulated) (on a different carrier frequency) of 0.1 μs (2 mJ) which is sufficient for short range detection since

$$\left(\frac{15}{100}\right)^4 = 0.0005$$

4.7.4 *Very accurate radar for instrumentation or equivalent*

In this case a very wide spectrum is used to ensure very good range resolution. For example, Δf will be of the order of 500 MHz ($1/\Delta f = 0.002$ μs corresponding to 0.3 m range resolution). Use of a pulse length of 100 μs would give a pulse compression ratio of 50 000. Against a target at $V_R = 3000$ m/s it is found that $2(T_p \Delta f)\, V_R = 300\ 000$ km/s, the velocity of the light. This means that simultaneously

- During the duration of the transmitted pulse the target has moved by 0.3 m, which is the range resolution
- And/or the Doppler shifts for the extreme frequencies differ by

$$\frac{2V_R}{c} \Delta f = 10\ \text{kHz} = \frac{1}{T_p}$$

The Doppler effect can no longer be considered only as a frequency *shift*. The use of several different pulse compression filters matched to various possible radial velocities is then necessary.

4.8 References

1 COOK, C.E., and BERNFELD, M.: 'Radar signals' (Academic Press, New York, 1967) chap. 12
2 MACCOLL, L.A., and WHEELER, H.A.: *Proc. IRE*, 1939, **27**, pp. 359–385
3 FRANCK, R.L.: 'Polyphase codes with good non-periodic correlation properties' *IEEE Trans.*, 1969, **IT–9**, pp. 43–45
4 CARPENTIER, M.H.: 'Principles of modern radar systems' (Artech House, 1988)
5 CARPENTIER, M.H.: 'Present and future evolution of radars', *Microwave J.*, June 1985, pp 32–67

Chapter 5

Pulse Doppler radars

Prof. M.H. Carpentier

5.1 Analysis of a simple pulse Doppler radar: Fixed station, low PRF (no range ambiguity)

5.1.1 *Introduction*

The radar under consideration is working in S band ($\lambda = 0.1$ m) with an antenna installed at 15 m above the ground, which implies that the distance of the horizon is on average about 15 to 16 km. The antenna span is 2 m and it rotates at 60 r.p.m. The horizontal antenna beamwidth is about 3.7° (at 3 dB) or 65 mrad. The repetition frequency is 7.5 kHz.

The maximum range of the radar being 15 km, the range ambiguity of 20 km is mainly above the maximum range; except in the case of very powerful targets above 20 km, there is no risk of range confusion [no range ambiguity or low repetition frequency (LRF)].

5.1.2 *Nature of clutter*

Assuming that the length of the transmitted pulses is 4 μs, let us examine the nature of the clutter mixed with a useful target at, for instance, 5000 m from the radar.

The resolution cell at that range is $0.065 \times 5000 \times 600 = 195\ 000\ \text{m}^2$, which is very similar to the resolution cell of the 5 μs radar corresponding to curve of Fig. 4.27 of Chapter 4:

$$\frac{0.5}{57.3} \times 30\ 000 \times 750 = 195\ 000\ \text{m}^2$$

This indicates that on a similar clutter (called 'Normandy country clutter' for simplification), the average value of the radar cross section (RCS) of the clutter entering the main beam at 5000 m would be about 20 m^2, but there is a probability of 0.1 that it exceeds 150 m^2 and a probability of 0.2 that it exceeds its mean value of 20 m^2. (In the case of rocky mountain clutter, the level of clutter would be much higher, with a significant probability of clutter above 10 000 m^2, much reduced by reducing the range resolution.)

Clutter entering the radar via the antenna sidelobes depends on the level of sidelobes. Assuming that this level is about −30 dB compared with the main gain, clutter entering via the antenna sidelobes corresponds to a resolution cell of $2\pi \times 5000 \times 600$ which is 10 times the cell of the 50 μs radar of Fig. 4.27 of Chapter 4, which gives a mean value of 2000 m^2 and a clutter which does not

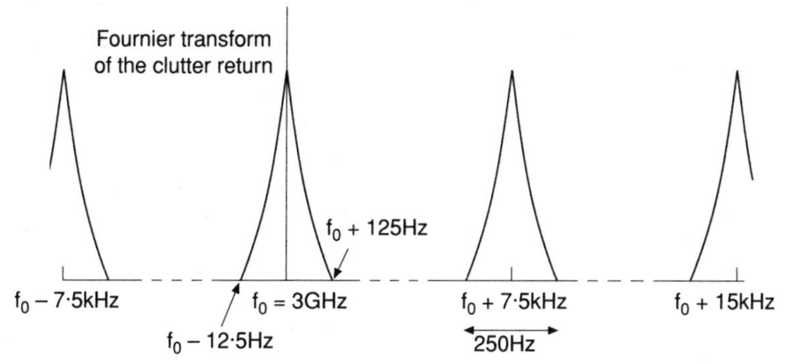

Figure 5.1 *Fixed low repetition frequency (LRF) radar equipment: Frequency spectrum of the clutter section*

vary much from its mean value. We can simply say that the clutter RCS entering via the antenna sidelobes is more or less constant and equal to 2000 m^2. Reduction by 60 dB (twice the relative sidelobe level) finally gives 0.002 m^2.

The signal returned from the clutter, assumed to have zero (radial) velocity, would not present any Doppler shift in the case of a non-rotating antenna (this is almost the case with a tracking radar). In that case it would be a periodic signal like the transmitted one, composed of spectral lines separated by the PRF of 7.5 kHz.

In the case of a rotating antenna, the fixed echo returns present some Doppler shift, maximum Doppler shift corresponding to the relative velocity of a fixed echo compared with the edge of the antenna. That means that, since the edge of the antenna has a speed of 6.3 m/s, the spectrum of a fixed echo return remains between $\pm(6.3 \times 2)/\lambda = 125$ Hz (modulo PRF); see Fig. 5.1, where f_0 is the central frequency of the transmission. That is valid for the clutter entering via the main beam as well as for the clutter entering via the sidelobes.

It could be useful to represent the clutter spectrum (only between f_0 and $(f_0 + 7.5)$ kHz) using vertical logarithmic scales as in Fig. 5.2. For simplicity, only the *mean* RCS of the clutter is taken into account in considering the clutter in the main beam. This simplification will be used later, but it has to be remembered that it is not enough to represent the actual clutter completely. The continuous line represents the clutter in the main beam, and the dotted line represents the clutter via the antenna sidelobes. Both are for a radar without pulse compression, with 4 μs pulse length, as against 'Normandy country clutter'.

5.1.3 *The 'useful' target*

The useful target (here assumed to have an RCS of 1 m^2, represented as fixed in Fig. 5.2 but obviously also fluctuating) will return the transmitted signal, after amplitude modulation by the antenna rotation, shifted by the Doppler shift known as $2V_R/\lambda$ if V_R is the radial speed of the target.

It is well known that sampling by PRF implies an ambiguous radial velocity measurement (stroboscopic effect), since the Doppler shift is measured modulo

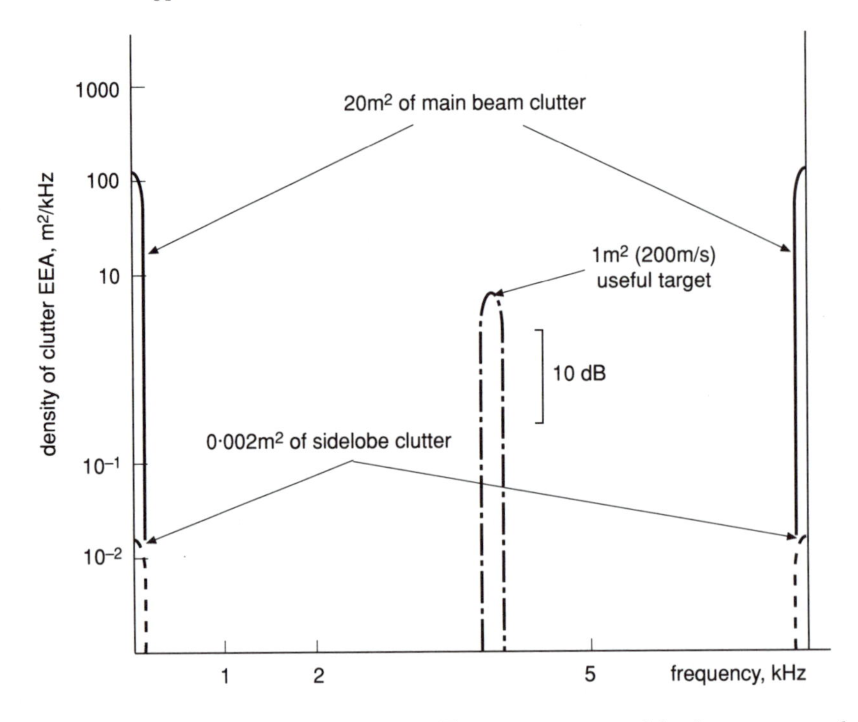

Figure 5.2 *Fixed LRF radar equipment: Frequency spectrum of the clutter return and of a useful target*

the PRF. Here it is measured modulo 7.5 kHz. This implies that the radial velocity could only be measured (by measuring Doppler shift) with an ambiguity of

$$\frac{7500 \times \lambda}{2} = 375 \text{ m/s}$$

It also implies that two useful targets having the same radial velocity modulo 375 m/s will give the same spectral response, and that a target approaching at 300 m/s will give the same response as a target leaving at 75 m/s. Immediate measurement of the radial velocity without any ambiguity could only be possible in our case on targets with maximum velocity of 187 m/s. (Using 0.25 m wavelength instead of 0.1 m, with the same PRF would have multiplied that figure by 2.5, 187 m/s becoming 468 m/s.)

Fig. 5.2 shows in the mixed line the spectral response of a useful target of 1 m^2 RCS at 200 m/s (or at −175 m/s, or at 575 m/s).

5.1.4 *Basic block diagram of the radar*

Fig. 5.3 represents the basic block diagram of such a radar. A stable local oscillator (STALO) (1) provides a very stable frequency (e.g. at 2970 MHz), and an oscillator (2) provides a very stable frequency at 30 MHz (intermediate

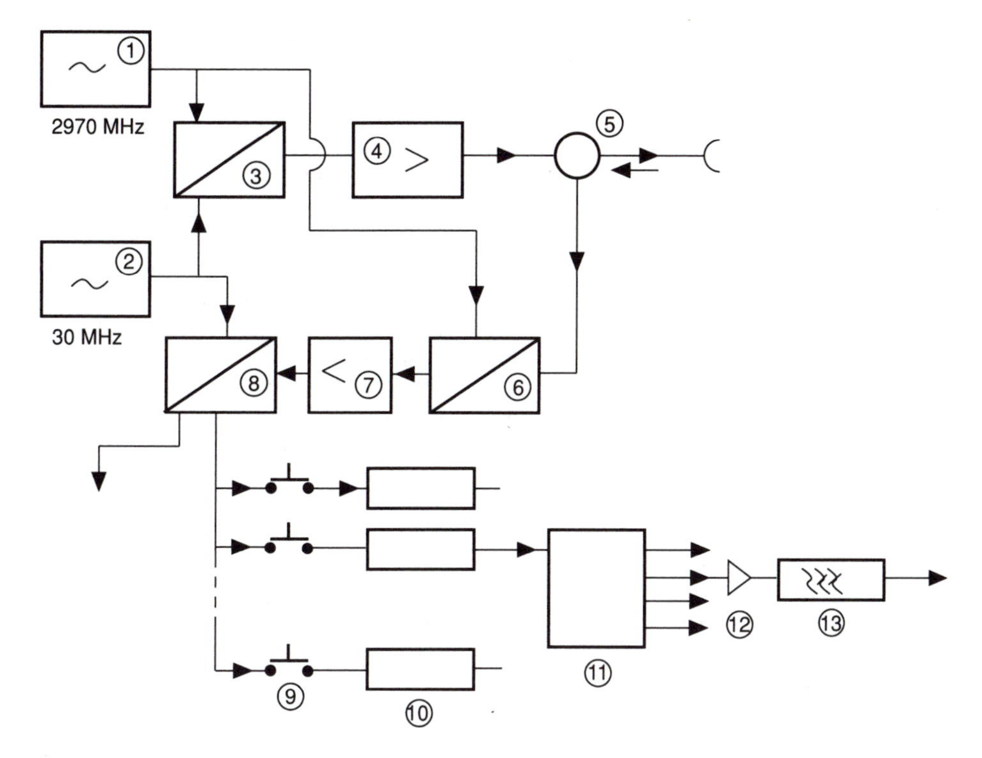

Figure 5.3 *Basic block diagram of a pulse Doppler radar*

frequency IF). Mixing in (3) provides $f_0 = 3000$ MHz which is amplified and pulse modulated in (4).

In the case of pulse compression, a filter for passive generation could be introduced, for instance, somewhere between (2) and (3). The amplifying transmitter could be replaced by an auto-oscillator triggered by signal coming from (3). (5) represents the duplexer system.

On reception, after mixing with the local oscillator signal in (6) (possibly after RF amplification), the received signal is amplified and approximately filtered in (7) (filtering will maintain, for instance, 1 MHz bandwidth in the case of using 4 μs pulses, not phase-modulated, in order to reduce the dynamic range).

The signal provided by (7) is demodulated in (8) in phase (I) and in quadrature (9) (or in an equivalent single sideband demodulator when the following signal processing is analogue), in order not to have 'blind phase' problems which occurred in the past when only a single demodulation was used (blind phase occurs when the Doppler sine wave which is finally obtained is zero around the time when the maximum antenna gain is in the direction of the useful target). Fig. 5.3 represents only one of the two outputs of the I and Q demodulations.

The output(s) of (8) are then 'range gated', which means that the first wire is connected only during the period from 4 to 8 μs following the beginning of transmission and then corresponds to possible returns from targets at 600 m; the second wire is connected only during the next 4 μs; to take care of targets at 1200 m, etc.

Filters (10), all identical, are used to reject the clutter returns. Their ideal transmittance is given in Fig. 5.4. These filters are followed by as many Doppler filter benches as there are range gates (the maximum number of range gates is given here by 20 km, corresponding to range ambiguity divided by pulse length of 0.6 km, minus one, which gives 32). Every Doppler filter bench will be composed of adjacent filters all identical except for their central frequency. The dwelling time of the antenna on a given target is (at −6 dB) about 10 ms, leading to the use of Doppler filters every 100 Hz, with 3 dB width around 100 Hz (see representation of their transmittance in Fig. 5.5).

Every output of any Doppler filter (in *toto* in this case a maximum of $32 \times 72 = 2304$ Doppler filters) is followed by detection.

How such a radar works is easy to understand: the transmitted signal is represented by

$$S(t) = V(t| \sin 2\pi f_0 t$$

where $V(t)$ is equal to 1 during transmission and 0 during the rest of the time. The main difference between a pulse Doppler radar and a conventional one is the fact that the carrier frequency has the same phase all the time. An ideal 'correlation radar' has to achieve the computation of

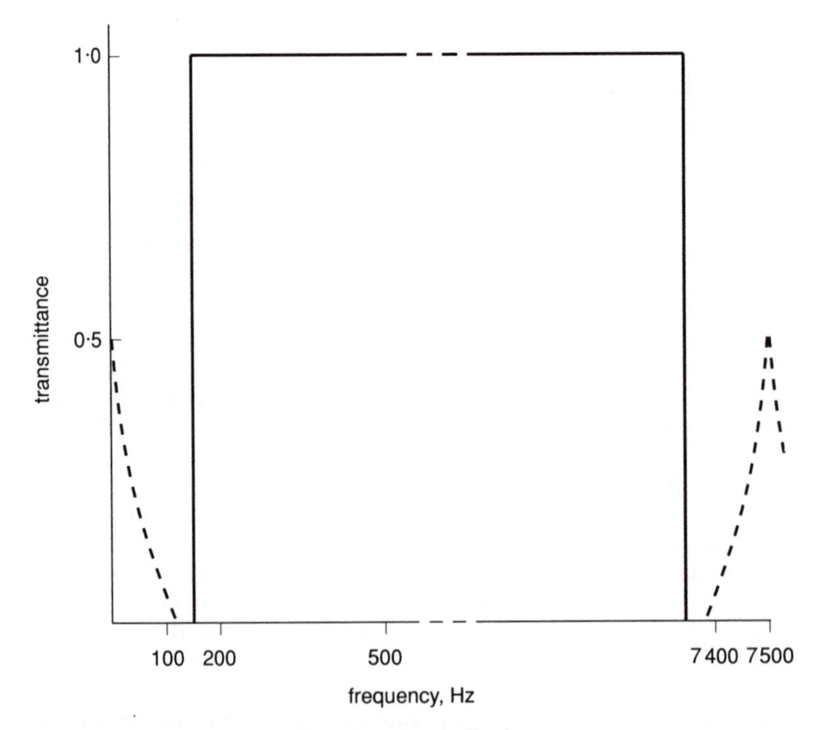

Figure 5.4 *Clutter rejecting filter (anticlutter filter)*

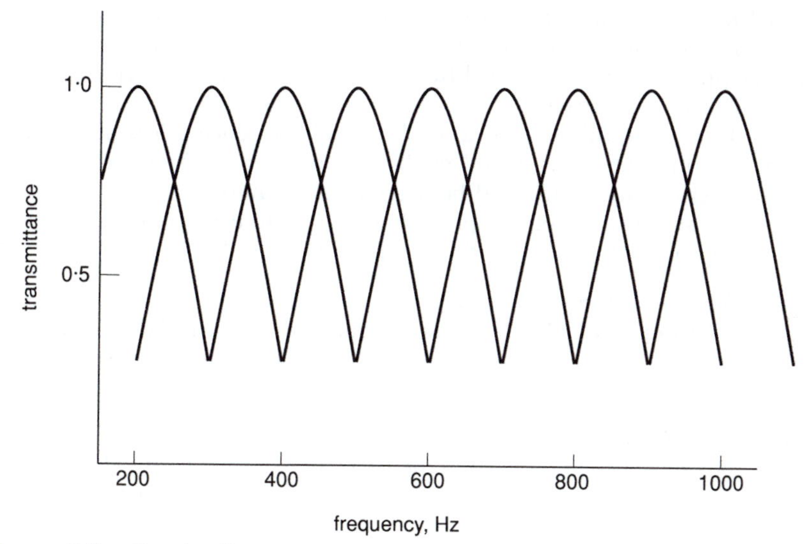

Figure 5.5 *Doppler filters*

$$C(t_0) = \int_T y(t)S(t - t_0)dt$$

for all values of t_0 every 4 μs (range resolution), the integral being carried out during the time $T = 10$ ms (if $y(t)$ is the received signal).

Since

$$C(t_0) = \int_T y(t) \sin 2\pi f_0(t - t_0)V(t - t_0)dt$$

and if we replace $\sin 2\pi f_0(t - t_0)$ by $\sin 2\pi f_0(t)$ (accepting that range is not measured with an accuracy better than half a wavelength), it clearly appears that $C(t_0)$ is obtained for any interesting value of t_0 from

- Demodulation of the received signal $y(t)$ by the carrier (here carried out in two steps)
- Followed by multiplication by all interesting $V(t - t_0)$ (here carried out by 'range gating')
- Followed by integration during T (which is here realised by the Doppler filters for all possible Doppler shifts).

5.1.5 *Practical considerations*

5.1.5.1 *Reduction of the number of Doppler filters*

It is well known that being not far from an optimum solution can often in practice be a better solution, because money saving can be important while the relevant losses would be acceptable.

In the present case, if 200 Hz bandwidth Doppler filters are used instead of 100 Hz (36 Doppler filters per range gate instead of 72), detection (12) being followed by ('not coherent') integration during 10 ms, Fig. 5.6 (derived from Fig. 7.2 of Reference 1) indicates the relevant loss (0.5 dB) in sensitivity. The cost of Doppler filtering is divided by 2, which represents the saving in cost for the 0.5 dB loss in sensitivity.

Using 14 Doppler filters instead of 72 would introduce a loss of 1.5 dB (as indicated in Fig. 5.6), but would enable a reduction by 5 of the cost of the Doppler filtering; and that could be a good compromise.

Complete suppression of Doppler filtering (radar using only MTI) would result in an approximate loss of 6 dB in sensitivity.

5.1.5.2 *PRF wobbulation*

Targets at 375 m/s radial velocity are eliminated in the radar ('blind velocities'). If that is unacceptable several PRFs could be used consecutively or alternately to cope with that problem.

5.1.5.3 *Practical implementation*

Many radars have been produced in the past where elements (9), (10), (11) and (12) were analogue devices, which represent a physically large number of Doppler filters.

Using digital technology has completely changed the nature of the physical implementation. The (32) range-gates (9) and rejecting (anti-clutter) filters (10) are now realised in the same physical digital mobile target indicator (MTI).

For instance, Fig. 5.7 gives a practical example of a simple realisation with a transfer function (in z^{-1}) given by

$$T(z^{-1}) = (1 - z^{-1})\,\frac{1 - z^{-1}\sqrt{3} + z^{-2}}{1 - 0.5z^{-1} + 0.5z^{-2}} = \frac{V_0}{V_i}$$

The relevant curve is given in Fig. 5.8 versus f/f_r, f_r being the pulse repetition frequency.

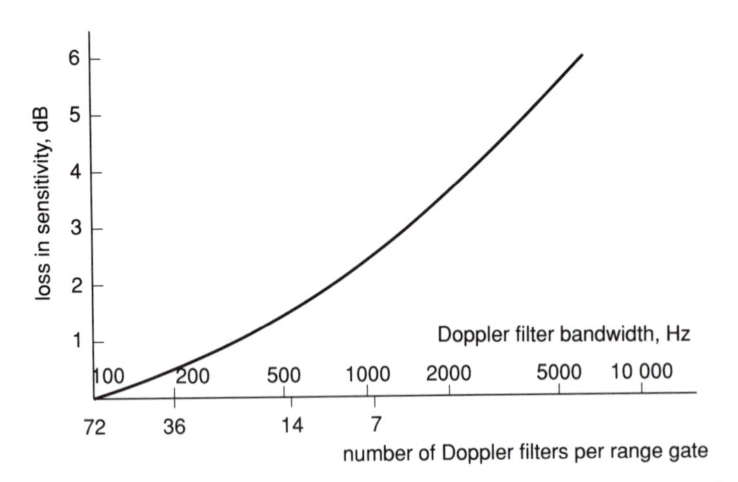

Figure 5.6 *Loss in sensitivity associated with a reduction of the number of Doppler filters*

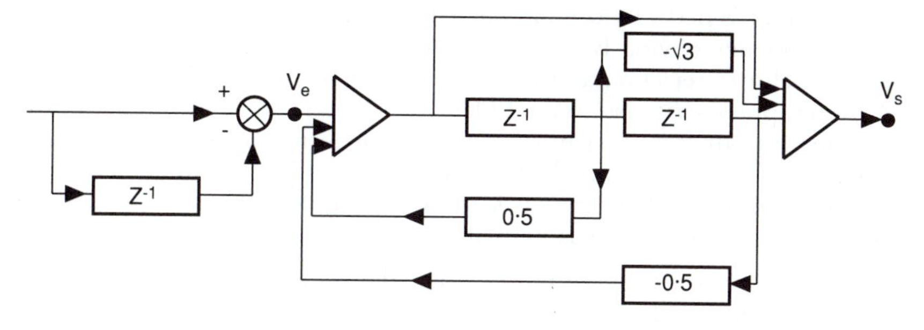

Figure 5.7 *Digital mobile target indicator (MTI)*

But computation of V_o from V_i could be obtained by other means. For instance, the following algorithms could be used which require less accuracy for computation and the introduction of two intermediate parameters u and w:

$$u(1-z^{-1})+w(1.5-0.5z^{-1})=V_i$$

$$w(1-z^{-1})=uz^{-1}$$

$$V_o=u(1-z^{-1})^2+w(2-\sqrt{3})(1-z^{-1})$$

With regard to the Doppler filters bench, at present it is generally obtained by using a fast Fourier transform (FFT).

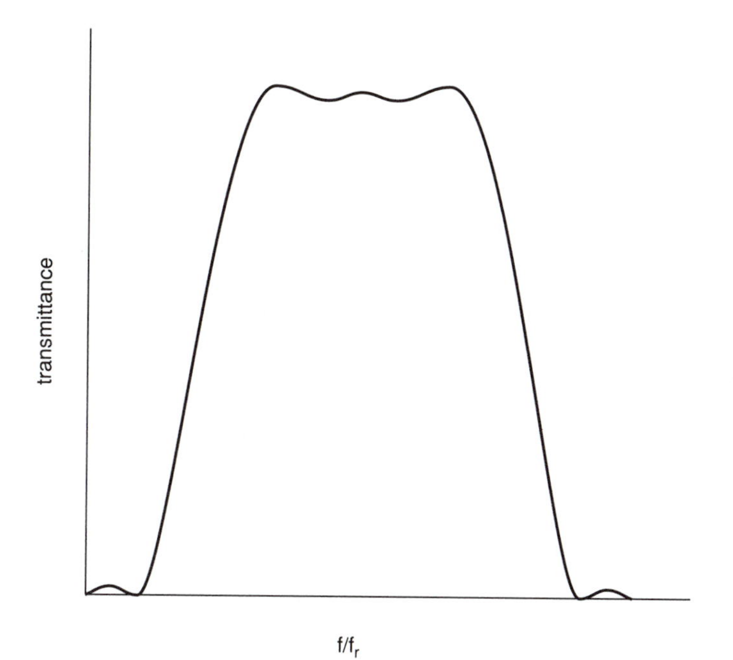

Figure 5.8 *Transfer function of a digital MTI*

Such a radar is perfectly compatible with frequency agilty from time to time, but the frequency obviously has to remain (remarkably) constant during the time when clutter rejection and Doppler filtering are achieved.

Pulse compression is also compatible with such a radar, a pulse compression filter being introduced either in (7) (in analogue form) or somewhere between (10) and (12) in digital form. In that case range gating has to be achieved at least every compressed pulse (clock pulse higher than Δf). The anti-clutter performance of such a radar is limited:

- By the nature of the curve representing the transmittance of the rejecting filter (anti-clutter filter), which depends on its complexity
- By the accuracy of the analogue/digital converter (rejecting the clutter by more than 48 dB requires an A/D conversion of at least $48/6=8$ bits accuracy, plus 1 bit for the sign +1 or 2 bits to take into account the noise instability).
- By the phase instability of the carrier frequency during transmission (phase instability of less than 1 mrad during transmission is needed if clutter rejection better than 60 dB is required, 10 mrad for 40 dB, etc.)
- By the instability of the frequencies provided by the various oscillators used.

See, for instance, Reference 2.

5.2 Similar radars with range ambiguity (medium PRF)

Let us now consider a very similar radar but with a PRF of 15 kHz instead of 7.5 kHz, and with the same maximum range of about 15 km. The velocity ambiguity will be 750 m/s: a target approaching at 375 m/s will give the same response as one leaving at the same speed.

With regard to the clutter situation, if we consider a useful target at 11 km (1 m^2 of RCS) it will be mixed not only with the clutter at the same range, but also with the clutter at 1 km, since the range ambiguity is now 10 km.

The resolution cell at 1 km (for the 4 μs, no pulse compression mode) is about 40 000 m^2, giving on average an RCS of Normandy country clutter of about 4 m^2; but since the range of the target is 11 times larger, the situation is the same as if that clutter were at the same place as the useful target but with an RCS of $4\times(11)^4=60\ 000\ m^2$. Clutter really close to the target is insignificant, and targets at 12 km will be mixed with a clutter equivalent to 10 000 m^2.

Detection of targets at 11 km would (in this context of light clutter) require clutter rejection of at least 70 dB. (After clutter reduction, the mean value of the clutter has to be much less than the useful target in order that the false alarm probability on clutter return be very low.)

5.3 Shipborne surveillance pulse Doppler radar

Let us consider the situation represented in Fig. 5.9 where a ship close to the shore is attacked by a missile flying at a distance of 10 km from the radar, at 250 m/s above the ground on the axis of the ship, the velocity of the ship is 10 m/s. The radar installed on the ship is a C band radar ($\lambda=0.055$ m)

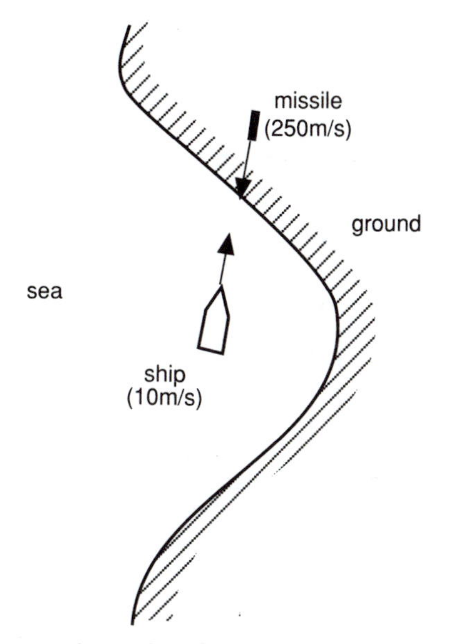

Figure 5.9 *Surveillance radar on board a ship*

transmitting, with a PRF of 4 kHz, 7.5 μs pulses associated with a pulse compression (ratio of 37.5 for a short pulse of 0.2 μs). The antenna rotates at 45 r.p.m. with a span of 3 m. The range ambiguity of such a radar is 37.5 km, and the velocity ambiguity is 110 m/s. The RCS of the attacking missile is assumed to be 0.1 m^2. The terrain (ground) clutter is assumed to be not very heavy (sea clutter is negligible compared with it).

Fig. 5.10 represents the structure of the clutter return: about 8 m^2 of clutter RCS entering in the main beam, with an average Doppler shift corresponding to 10 m/s (365 Hz) and occupying 170 Hz (full-line curve), but the clutter entering the antenna sidelobes is spread over ± 450 Hz (dotted-line curve) on both sides of zero, because it comes from ground whose speed is between $+10$ m/s and -10 m/s owing to the motion of the radar versus ground.

The dotted-line curve represents sidelobe clutter in the case when the average sidelobe level is -27 dB, while the mixed-line curve represents the same clutter when average sidelobe level is -18 dB. (Using 2 μs range resolution instead of 0.2 μs would increase that level by 10 dB.)

'Useful' target at absolute velocity 250 m/s (relative 260 m/s) is present as if its relative velocity were 40 m/s (1450 Hz).

An anti-clutter (rejecting) filter will be used in such a radar, as in Section 5.4, but in order that the rejection be around the frequency of the clutter in the mean beam, the mixer (8) must be fed by the intermediate frequency shifted by 365 Hz, and generally – depending on the velocity of the ship and the angular positions of the antenna – by the Doppler shift of the fixed echoes in the main beam (knowledge of the speed of the ship is then necessary).

The phenomenon of blind velocities (in our case around 10 m/s, 120 m/s, 230 m/s . . . of *relative* velocity) is then introduced and must be taken into account by changing the radar PRF from time to time.

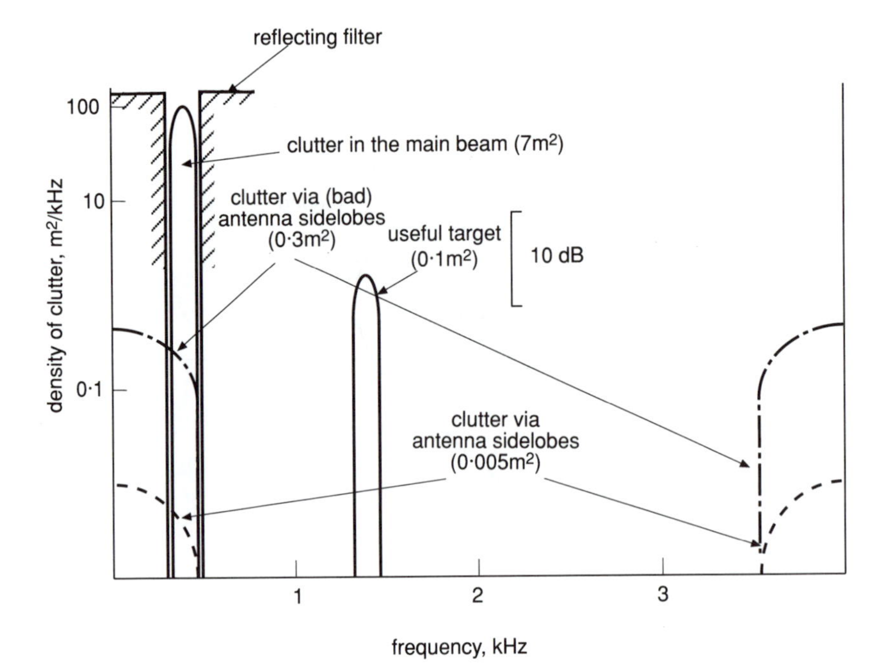

Figure 5.10 *Surveillance radar on board a ship: Associated frequency spectrum of the clutter return*

Doppler filters could follow the rejecting filter as in Section 5.4. If their width is around 200 Hz, that means that the maximum clutter return via the antenna sidelobes will be around 0.002 m^2 (which is small compared with the RCS of the target – 17 dB below – and acceptable). It has to be noted that if the antenna sidelobe level is -18 dB instead of -27 dB, the clutter return via the antenna sidelobes is increased to 0.12 m^2, which is not acceptable (with such bad sidelobes and 2 μs range resolution, it goes up to 1.2 m^2). In that case, rejecting filters should have to reject all frequencies between ± 450 Hz modulo 4 kHz, with more problems with blind velocities (nearly 25% of the velocities would be blind).

This shows the importance of good quality for the antenna sidelobes in this case. The same consideration is still more valid when the radar is on a rapid flying 'platform'.

5.4 Airborne surveillance radar

5.4.1 *General description: LRF mode*

Let us consider an 'airborne early warning' radar (AEW) installed on board a jet aircraft flying at 222 m/s at an altitude $z = 10\ 000$ m above ground.

S band is used in the radar ($\lambda = 0.1$ m). The antenna has a span of 9 m and a height of 1.5 m. It is assumed that it rotates at 6 r.p.m. The horizontal (3 dB)

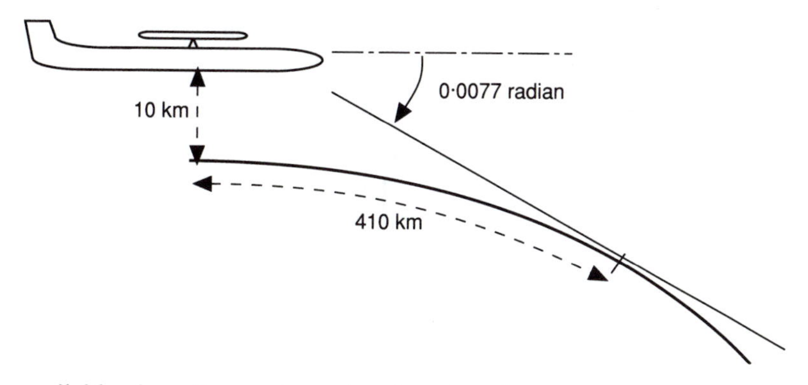

Figure 5.11 *Surveillance radar on board an aircraft*

beamwidth is around 14 mrad and the vertical (elevation) beamwidth is around 80 mrad. It is possible to tilt the beam at negative elevations.

The range resolution (obtained by pulse compression) is assumed to be 0.2 μs. At the altitude of the aircraft, the radar horizon is at about 410 km, which indicates that there is 'normally' no ground return above that range. No clutter rejection is then required for detection of targets at a distance above 410 km (see Fig. 5.11), and the radar could be used as a conventional radar with a low PRF [low repetition frequency (LRF) mode] such as 100 Hz (pulse length on transmission could be about 50 μs, associated with a pulse compression ratio of 250). In that mode the tilt of the antenna will be slightly negative.

5.4.2 *Medium repetition frequency (MRF) mode*

Let us now consider the detection of an aircraft (1 m^2 RCS) flying at low altitude at 100 km on the axis of the AEW aircraft. The antenna is assumed to be tilted down in order that the maximum gain be in direction of the target (see Fig. 5.12).

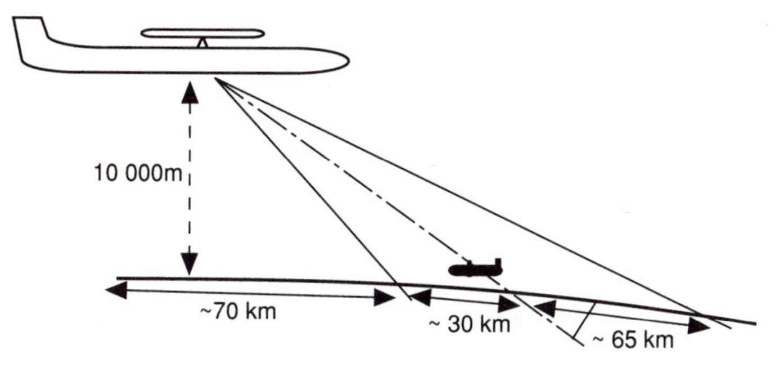

Figure 5.12 *Airborne surveillance radar: Medium repetition frequency (MRF) mode*

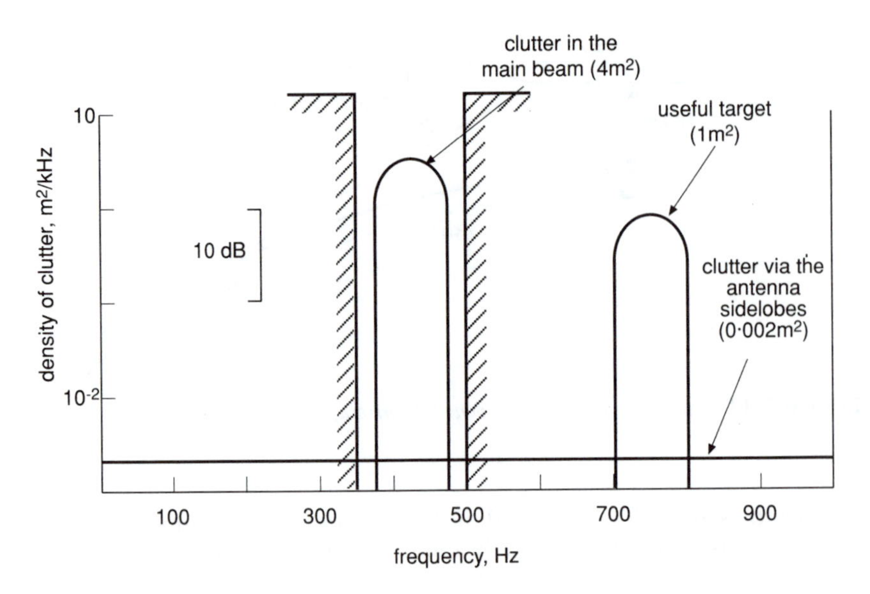

Figure 5.13 *Airborne surveillance radar in MRF mode: Frequency spectrum of the clutter return and of the useful target: Target within the range ambiguity*

The clutter accompanying the 'useful' target is first that entering via the main beam, corresponding to a resolution cell of $0.014 \times 100\ 000 \times 30 = 42\ 000\ \mathrm{m}^2$ (seen at a low incidence angle of a few degrees).

In those conditions mean clutter RCS is about $4\ \mathrm{m}^2$ (in the case of 'Normandy ground clutter', this is not heavy clutter). It is represented in Fig. 5.13, assuming a PRF of 1 kHz (range ambiguity, 150 km; velocity ambiguity, 50 m/s). Clutter entering via the antenna sidelobes is coming from the clutter at the same range of 100 km, as well as from the clutter at 250 km and at 400 km, which can be neglected. An average antenna sidelobe level of −30 dB gives, for the relevant clutter in the same conditions, a figure of about $0.002\ \mathrm{m}^2$.

Fig. 5.13 represents the return of a useful target ($1\ \mathrm{m}^2$ RCS) flying at a ground velocity of 215.5 m/s (222 m/s + 215.5 m/s = 437.5 m/s = 37.5 m/s modulo 50 m/s).

Using a rejecting filter as shown (with position depending on the speed of the AEW plane, on the antenna tilt, and on the horizontal position of the rotating antenna) will enable elimination of the clutter entering via the main beam, together with the associated problems of blind velocities.

Doppler filtering would give a further reduction of the clutter entering via the antenna sidelobes (by 7 to 8 dB if using 150 Hz bandpass for Doppler filters).

If we now consider a target at 160 km, it will be mixed with the clutter existing just below the plane seen with an angle of incidence of 90°, and which for that reason could, in some cases, be very important. The importance of that clutter is reinforced by the fact that it is at a range 16 times shorter.

On average, a figure of 500 to $1000\ \mathrm{m}^2$ could be expected for the equivalent RCS of that clutter (taking into account the disadvantage arising from the

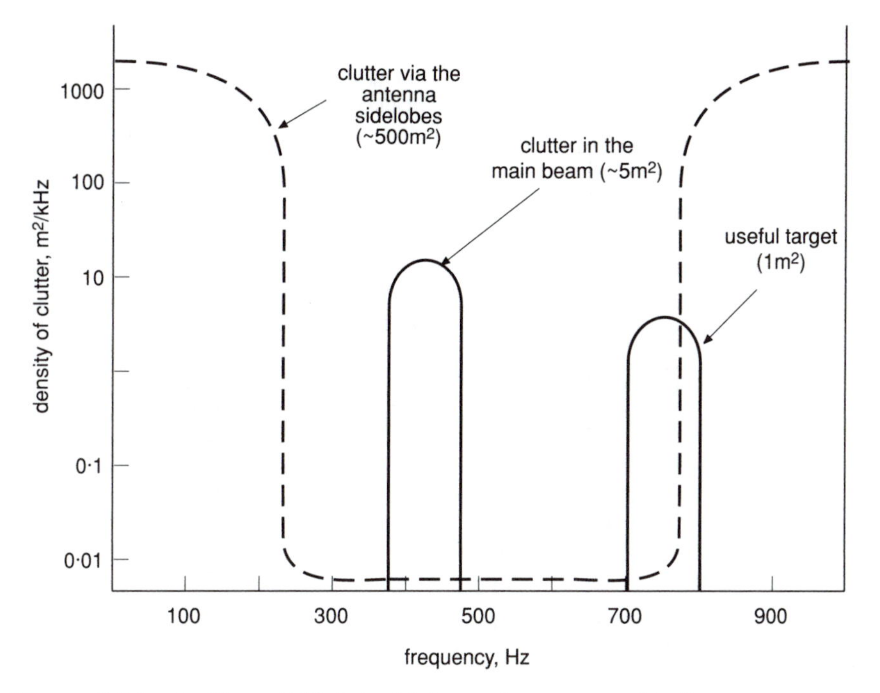

Figure 5.14 *As Figure 5.13, but with target above range ambiguity*

differences in distances and antenna sidelobe level of −30 dB in direction of the ground), but much higher figures could be encountered.

Fig. 5.14 represents the situation corresponding to that context, which is a very bad situation, leaving no real possibility for good detection of a plane at 160 km. 1 kHz PRF is only possible for use at ranges below 150 km.

5.4.3 *High repetition frequency (HRF) mode*

Let us now examine another mode in which a much higher PRF is used, e.g. 30 000 Hz. Range resolution will remain at 0.2 μs, and pulses used on transmission will have a length of a few microseconds with a relatively reduced pulse compression ratio; for instance pulses of about 4 μs could be used, in which case we shall have no detection during 600 m every 5 km; i.e. blind zones between 5000 and 5600 m, 10 000 and 10 600 m, etc.

To overcome the problem, several (two or three) different PRFs of the same order will be used consecutively or alternately. Using several PRFs will also enable one to solve the range ambiguity inherent to such a mode (see Section 5.6).

We will also consider, as an example, the detection of an aircraft flying at low altitude above a 'Normandy type' ground clutter.

Clutter entering the main beam is coming not only from 100 km, but also, since the range ambiguity is of the order of 5000 m, from 95 km, 90 km . . . down to about 70 km and from 105 km, 110 km etc. . . . , up to about 465 km; for instance the total clutter will be about 65 m^2 (see Fig. 5.15) spread over

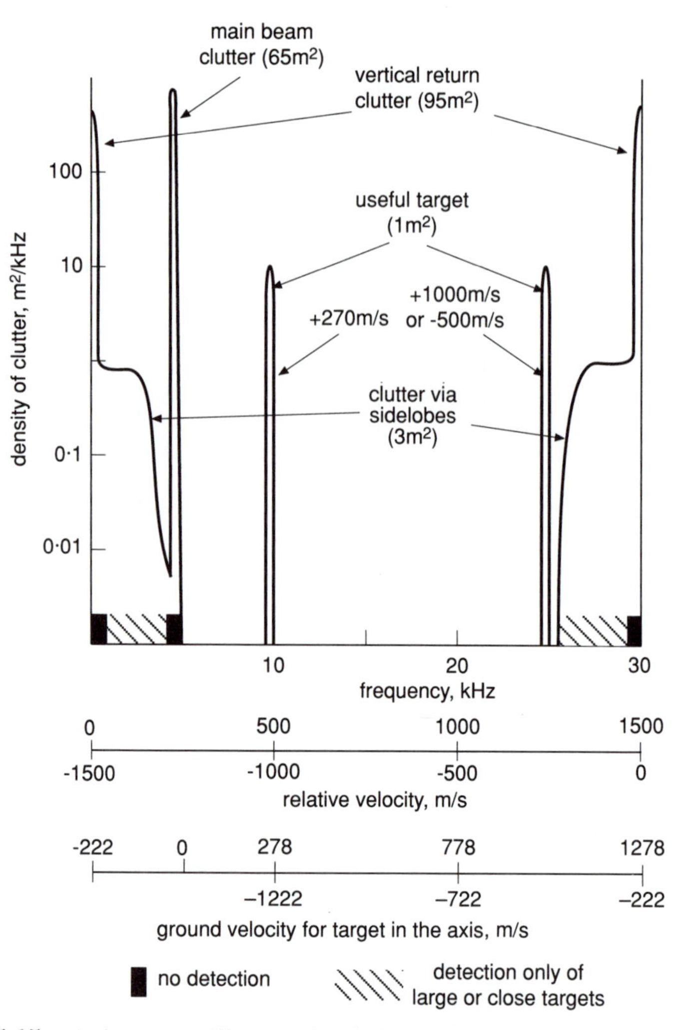

Figure 5.15 *Airborne surveillance radar in high repetition frequency (HRF) mode: Associated frequency spectrum for clutter return*

frequencies between 4340 and 4490 Hz (the antenna is also assumed to rotate at 6 r.p.m.).

Clutter entering via the antenna sidelobes (−30 dB relative average gain as in Section 5.4.2) is composed of the clutter at 15, 20 and 25 km etc. . . . from the radar, equivalent (for the detection of a target at 100 km) to about 3 m^2 RCS spread over the frequencies between −4440 Hz and +4440 Hz (ground speed of the plane =222 m/s) are represented in Fig. 5.15.

But heavy (and possibly very heavy) clutter will also come from the clutter immediately below the radar (vertical return), assumed in Fig. 5.15 to be around 95 m^2, but spread over only about ±250 Hz. This is because the radar altitude is 10 000 m, the range ambiguity 5000 m, and the useful target range

100 km. At a slightly different radar altitude, this clutter would, of course, be present with useful target at other ranges.

Using a narrow rejection filter around zero frequency will permit rejection of that clutter.

Another filter (whose position will depend on the position of the antenna beam, radar speed, etc., here around 4400 Hz) will reject the clutter entering the main beam.

Detection of targets with a relative velocity between 230 m/s and 1270 m/s modulo 1500 m/s depends only on the quality of clutter rejection (see Section 5.1). Fig. 5.15 represents the situation with a 1 m^2 useful target flying on the axis of the radar towards the radar with a (positive) ground speed of 270 m/s (or a negative ground speed of -1230 m/s!), and with a similar useful target flying under the same condition at about 1000 m/s (or a negative ground speed of -500 m/s). Detection of similar targets flying at a relative velocity between ± 210 m/s (ground velocity between -432 m/s and -12 m/s) will only be possible for relatively large targets. Even using 64 Doppler filters (about 450 Hz bandwidth) will give in our example average clutter returns between 0.1 m^2 and 0.5 m^2 in those zones, allowing for detection of targets either much larger than 1 m^2 or at shorter distances from the radar. A similar situation is found for all airborne radars in a HRF mode.

5.5 Combat aircraft radar

Let us examine some situations met in a combat aircraft radar (only in the tracking mode, and only some situations chosen as relatively typical).

The radar is an X band radar ($\lambda = 0.03$ m) using a circular antenna of 56 cm diameter (3 dB beamwidth of 70 mrad). The far-out average sidelobe relative level will be assumed to be about -40 dB.

Several modes will be examined, all with the same range resolution of 1 μs (150 m):

Low repetition frequency (LRF)
PRF = 2 kHz
Range ambiguity = 75 km
Velocity ambiguity = 30 m/s

Medium repetition frequency (MRF)
PRF = 12 kHz
Range ambiguity = 12.5 km
Velocity ambiguity = 180 m/s

High repetition frequency (HRF)
PRF = 200 kHz
Range ambiguity = 750 m
Velocity ambiguity = 3000 m/s

In all cases the speed of the aircraft will be assumed to be 400 m/s (corresponding Doppler shift of 26 667 Hz).

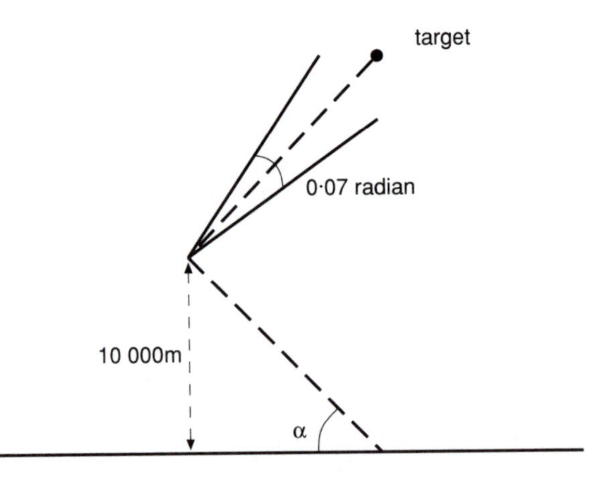

Figure 5.16 *Combat aircraft radar: Look-up mode*

5.5.1 *LRF mode*

5.5.1.1 *Look-up mode*

We will consider the radar at an altitude $Z = 10\ 000$ m; with the radar beam looking upwards and detecting a target at 14 000 m (see Fig. 5.16).

The clutter enters the radar only via the sidelobes, coming from the ground at 14 000 m, and it comes from a surface of

$$2\pi \times 14\ 000 \times 150/\cos\alpha = 19 \times 10^6\ \text{m}^2$$

seen with an incidence angle α (here $\alpha = 45°$).

The mean RCS of the clutter is then (−40 dB relative sidelobe level) $19 \times 10^6 \sigma_0 \times 10^{-8}$ where σ_0 is the mean reflectivity (at X band) for an incidence angle α (for instance $\sigma_0 = 0.2$ for $\alpha = 45°$).

The clutter entering via the antenna sidelobes is, in those conditions, equal to about 0.04 m^2, spread over ±18 900 Hz, which, taking into account a Doppler ambiguity of 2000 Hz will give a result practically constant between 0 and 2000 Hz (see Fig. 5.17). Detection at the same distance, but with a range resolution of 0.2 μs instead of 1 μs, would have given 0.008 m^2 of clutter. A sidelobe level of −35 dB instead of −40 dB (1 μs for range resolution) would have given 0.4 m^2 instead of 0.04 m^2. Detection at 30 km instead of 17 km would have given a similar result (0.02 m^2 instead of 0.04 m^2).

Detection of a 1 m^2 useful target does not present any particular problem without a rejection filter, the use of Doppler filters making detection easier (clutter within the 250 Hz filter being 0.04 m^2/8 = 0.005 m^2).

5.5.1.2 *Look-down mode*

With the radar in the same situation, let us now examine the case of the radar beam looking down (elevation angle of −45°) to detect a target (1 m^2) at 14 140 m flying at low altitude.

(a) *If the target is at a very low level in the axis of the plane* (equipped with the radar), the resolution cell in the main beam has a surface of

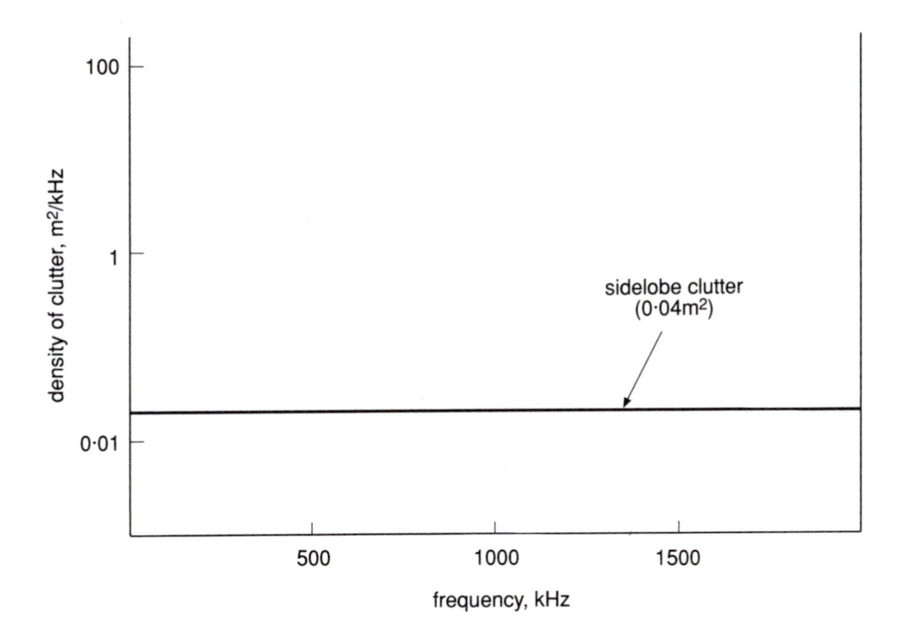

Figure 5.17 *Combat aircraft radar: in LRF look-up mode: Clutter spectrum*

$$\frac{z}{\sin \alpha} \times 0.07 \times \frac{150}{\cos \alpha}$$

z, altitude $=10\ 000$
α, angle of incidence $=45°$
i.e. in our case 210 000 m^2.

With a $\sigma_0 = 0.2$ (for $\alpha = 45°$), it gives about 40 000 m^2 RCS for the clutter entering the radar via the main beam. The spectrum of the clutter is between Doppler frequencies corresponding to the radial velocity of A and the Doppler velocity of B (see Fig. 5.18).

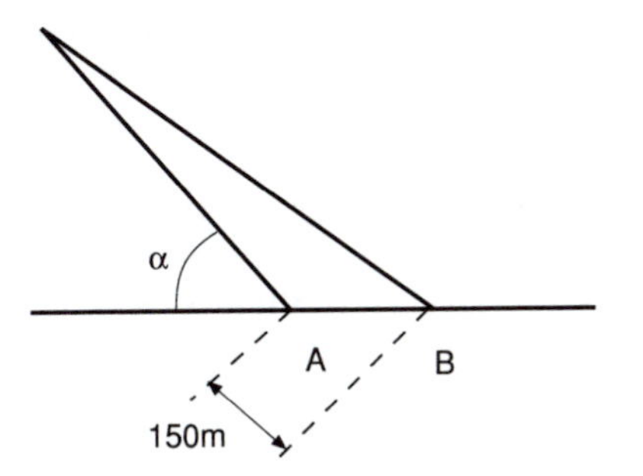

Figure 5.18 *Combat aircraft radar: Look-down mode*

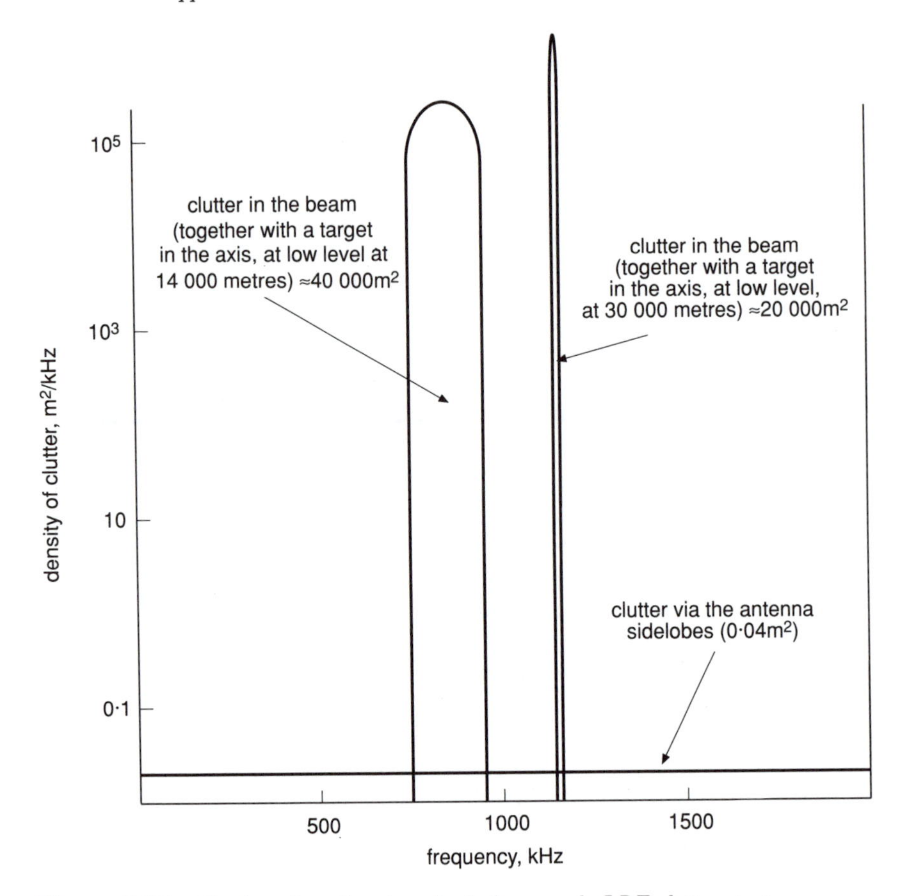

Figure 5.19 *Combat aircraft radar: Look-down mode LRF clutter spectrum*

$$\Delta f_D = \frac{V \sin^3\alpha \times 150}{z \cos \alpha} \frac{2}{\lambda} = 200 \text{ Hz}$$

(with $V = 400$ m/s $z = 10\ 000$ $\lambda = 0.03$ m).

But if no special means are used, the angular position of the target (and its associated clutter) is known only with an accuracy of ± 0.035 rad and the clutter central frequency is known with an accuracy of ± 650 Hz, which obliges us to insert any rejection filter to cover 1300 Hz, leaving practically nothing for the detection of useful targets (see Fig. 5.19).

If the useful target flies at 400 m/s ground speed towards the radar, its relative velocity will be about 283 m/s + 400 m/s, giving a Doppler shift of 45 523 Hz or 1523 Hz. Its spectral position will be the same as a car coming towards the radar with a ground velocity of 18.16 m/s (65 km/h).

For a useful target similarly flying in the axis of the plane but at 30 000 km, clutter level would be similar (maybe 20 000 m^2 instead of 40 000 m^2) but with a spectrum width reduced to 16 Hz around 25 140 Hz (central frequency *a priori* known within ± 300 Hz, which is better but not very good).

(*b*) *If the target is at very low level just below the radar* (radar beam vertical), it will be mixed with the vertical return of the clutter which could be very strong.

An order of magnitude of that return is around 5×10^6 m^2 covering a total spectrum of 4500 Hz, which means that, with the Doppler ambiguity, it will cover all the frequencies, preventing the radar from any detection immediately below it at the ground level.

(*c*) *If the target is at very low level on the side of the radar*, the resolution cell in the main beam is similar to that of Section 5.5.1.2(*a*), giving possibly about 40 000 m^2 RCS of clutter at 14 140 m, spread over

$$\pm \frac{2}{\lambda} V\, 0.935 \approx \pm 1000 \text{ Hz (independent of } \alpha)$$

No detection is possible in that case.

5.5.2 *MRF mode (PRF* = 12 *kHz, range ambiguity* = 12.5 *km, velocity ambiguity* = 180 *m/s)*

5.5.2.1 *Look-up mode*

Let us consider again the radar at an altitude of 10 000 m with the radar beam looking up to detect a target at 14 140 m (see Fig. 5.16).

The clutter enters the radar only via the antenna sidelobes, coming from the ground at 14 140 m (RCS of 0.04 m^2 spread over 18 900 Hz), but also from the ground at 26 390 m (RCS of about 0.02 m^2, but equivalent at 14 140 m to $0.02 \times (14.14/26.39)^4 \approx 0.002$ m^2 negligible); see Fig. 5.20 continuous line.

But if we now consider a target at 30 000 m (always up), clutter entering via the antenna sidelobes is coming from the ground at 30 000 m (roughly 0.02 m^2 RCS) but also from the ground at 17 500 m (about 0.02 m^2 but equivalent to $0.02 \times (30/17.5)^4 \approx 0.2$ m^2 (dotted line of Fig. 5.20)).

5.5.2.2 *Look-down mode*

Let us now also examine, with the radar remaining in the same situation, the case of the radar beam looking down with an elevation angle of $-45°$ to detect a target at 14 140 m, flying at a very low altitude. The situation for clutter entering via the main beam is very similar to what was found in Section 5.5.1.2, as represented in Fig. 5.21, where clutter accompanying a useful target on the axis is represented by the associated rejection filter (which moves depending on the position of the beam, the velocity of the radar etc.), and the clutter accompanying a target on the side, by the associated rejection filter (which always exists, and protects the receiver from leakage from the transmitter).

Obviously, phenomena of blind speeds are found in the MRF mode, as well as problems with range ambiguity itself in measuring target range (see later).

5.5.3 *HRF mode (PRF* = 200 *kHz, range ambiguity* = 750 *m, velocity ambiguity* = 3000 *m/s)*

5.5.3.1 *Look-up mode*

Here again let us consider the radar at an altitude of 10 000 m with the radar beam looking up in direction of a target at 14 140 m (see Fig. 5.16).

The clutter enters the radar only via the antenna sidelobes, coming from the ground at 14 140 m (RCS of 0.04 m^2 spread over $\pm$18 900 Hz), but also from the ground at 10 390 m (RCS of 0.04 m^2 about spread over $\pm$724 Hz, but equivalent to $0.04 \times (14\ 140/10\ 390)^4 = 0.13\ m^2$) from the ground at 11 140 m, at 11 890 m, etc. The final result is represented in Fig. 5.22, total clutter return being equivalent to about 0.5 m^2. Similar results are represented in Fig. 5.22 when the target under consideration is (in look-up mode) at 30 000 m and at 35 000 m.

To complete representation of the situation of the clutter entering the radar via the antenna sidelobes, Fig. 5.23 gives

- Clutter for a radar at $z = 1000$ m looking up to a target at 14 140 m
- Clutter for a radar at $z = 1000$ m looking up to a target at 30 000 m
- Clutter for a radar at $z = 750$ m looking up to a target at 30 000 m

with an important vertical return since 30 000 m = 750 m modulo 750).

5.5.3.2 *Look-down mode*
(*a*) *The first example* is the detection of a low flying aircraft on the axis of the radar aircraft, the altitude of the aircraft being $Z = 10\ 000$ m, the distance of the target from the radar being 14 140 m at an elevation of $-45°$ (distance from the vertical of the radar 10 000 m, as in Sections 5.5.1.2 and 5.5.2.2). The situation for the clutter entering the main beam is not very different from the situation found in Section 5.5.1.2 or 5.5.2.2, and is represented in Fig. 5.24. A rejection filter about 18 900 Hz will cancel the clutter in the beam and permit detection of the useful target for any relative speed between 400 m/s and 2600 m/s modulo 3000 m/s, under the condition that clutter rejection is good enough (see end of Section 5.2.5.3). If the target is flying towards the radar, this field of visibility corresponds to ground speeds between 0 m/s and 2400 m/s. Detection of targets of relative velocity between ±400 m/s (target flying in same direction as the plane with velocity between 0 and 800 m/s) will depend on the RCS of

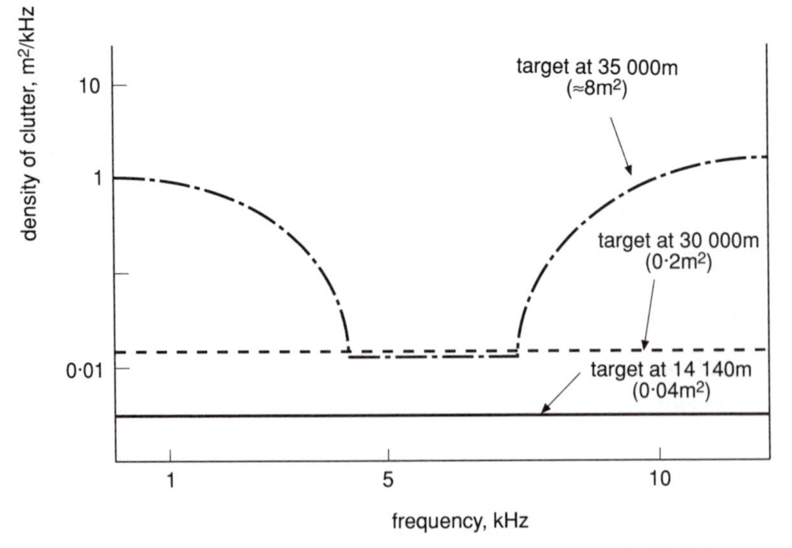

Figure 5.20 *Combat aircraft radar: Medium repetition frequency (MRF) look-up mode clutter return*

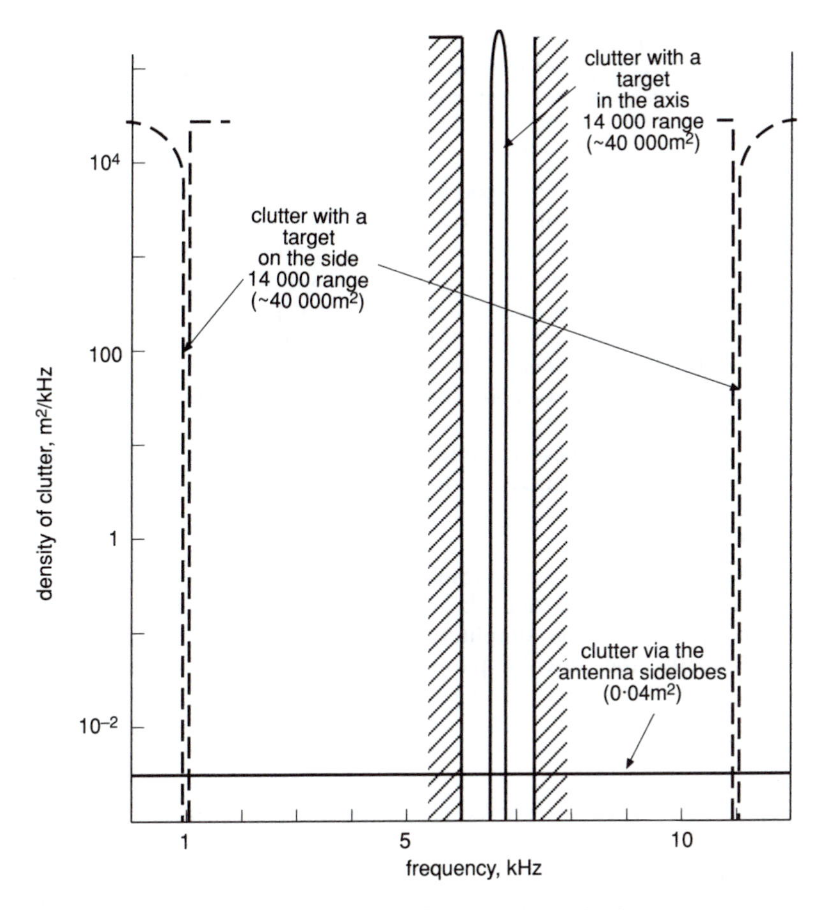

Figure 5.21 *Combat aircraft radar: MRF look-down mode clutter return*

the target (mixed with antenna sidelobe clutter) and could be improved by using Doppler filtering (except of course for relative speed corresponding to the rejected zones around 280 m/s of relative velocity).

(*b*) *The second example* will also be with the radar at $Z = 10\,000$ m against a target at 30 000 m range in front of the radar aircraft (vertical distance of 28 280 m).

In that case clutter in the main beam mixed with the useful target comes not only from 30 000 m but also from 27 750 m or 28 500 m up to about 33 000 m^2. Under the same conditions, its RCS will be about 150 000 m^2, occupying a frequency band of about 600 to 800 Hz around 25 100 Hz (associated with curve B of Fig. 5.22).

(*c*) *The third example* is with the radar aircraft at $Z = 1000$ m altitude against always a low flying target at 30 000 m relative range. Clutter entering the main beam obviously comes from clutter close to the target (at 30 000 m), but also from clutter at 29 250 or 28 500 m down to about 15 000 m, as well as from the clutter at 30 750 m or 31 500 m to very far distant. The final result is a total amount of clutter possibly reaching up to 2×10^6 m^2 or more (always in same conditions as above) spread over frequencies between about 26 600 Hz up to 26 667 Hz. Better quality of clutter rejection is obviously needed.

5.5.4 *Comparison between three modes*

The LRF mode is clearly very good for the detection of targets above the radar (as well as for ground mapping). It is very limited in looking-down modes.

HRF is clearly very good for the detection of targets coming in a direction towards the radar, achieving relative speed measurement with no ambiguity at all on most of the targets. It is not so good (except at short ranges) against targets travelling in the same direction as the radar.

The M.R.F. mode is reasonably good whatever the relative speed of the target, and often represents a good compromise when only one mode is possible (for financial reasons).

Fig. 5.25 illustrates these considerations, showing what kind of detection zones could be achieved (in X band) in typical situations.

5.5.5 *Important remarks*

In any case improving the range resolution of the radar is very efficient against clutter entering either the main beam or the antenna sidelobes. As a bonus improved range resolution simplifies solving the range ambiguities in MRF modes. Reducing the antenna parasitic radiation is very efficient in many situations.

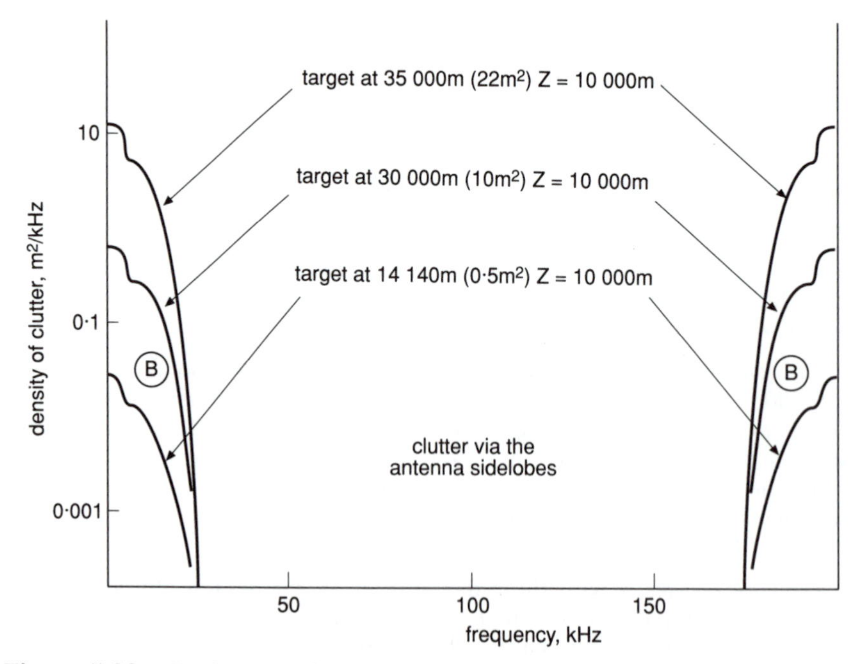

Figure 5.22 *Combat aircraft radar: High repetition frequency (HRF) look-up mode clutter return*

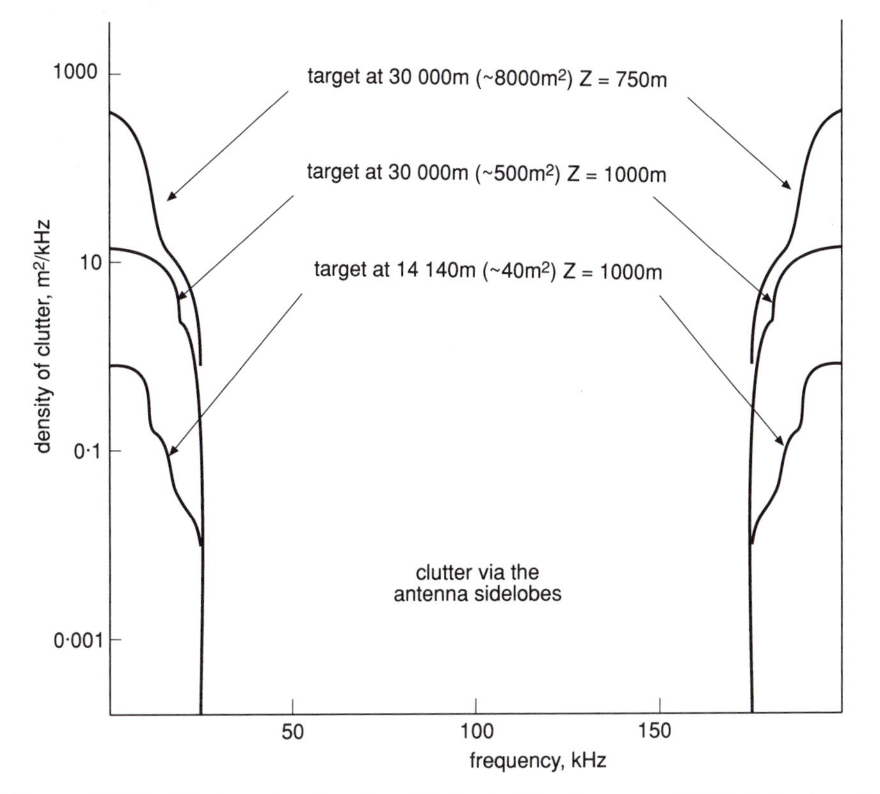

Figure 5.23 *Combat aircraft radar: High repetition frequency (HRF) look-up mode clutter return*

5.6 Solving range ambiguities in MRF and HRF modes

Basically two main means are used to solve range ambiguity in MRF and HRF modes: using several PRF, or several carrier frequencies.

When the resolution range is small compared with the pulse repetition period, using two different PRFs could solve the range ambiguity. For this, a practical example is better than a theoretical explanation.

Let us assume that two pulse repetition periods are used: namely 83 μs (12.0482 kHz) and 89 μs (11.2360 kHz) (83 and 89 are prime numbers). If the range of a target is 831 μs, we have

$$R = 831\ \mu s = 10 \times 83\ \mu s + 1\ \mu s = 9 \times 89\ \mu s + 30$$

What is measured is $R_1 = 1$ and $R_2 = 30$. Knowing that 83.74 = 1 modulo 89, R is computed by

$$R = (R_2 - R_1) \times 74 \times 83 + R_1 \text{ modulo } (83 \times 89)$$

$$R = 29 \times 6142 + 1 \text{ modulo } 7387\ \mu s$$

$$R = 831\ \mu s \text{ modulo } 7387\ \mu s$$

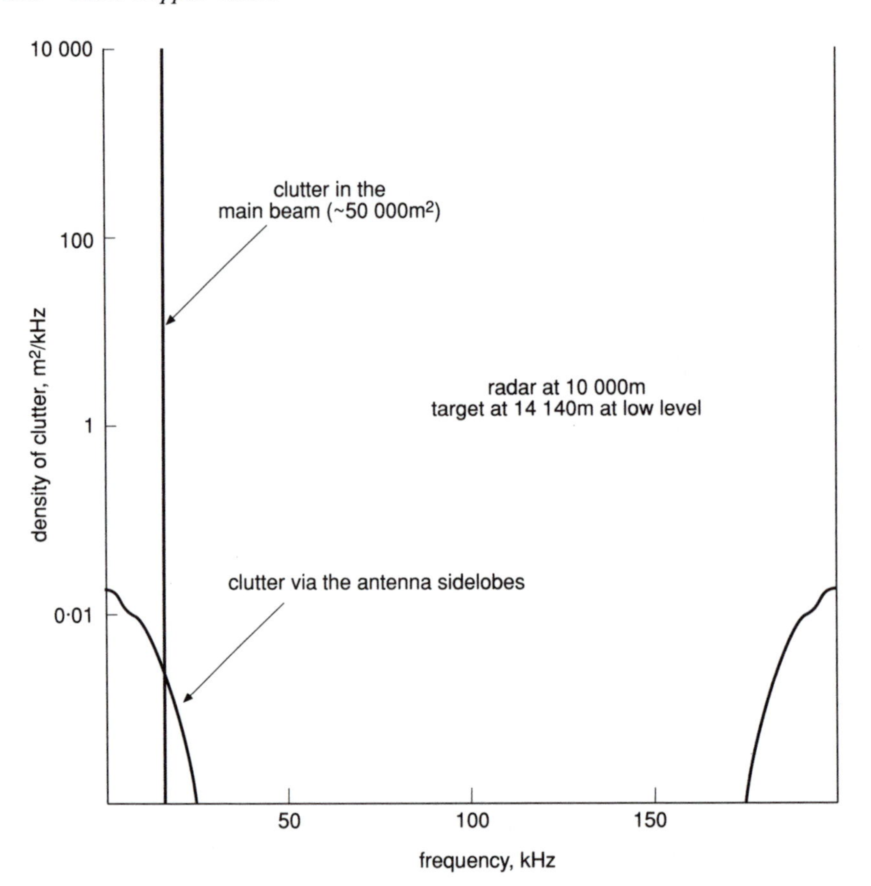

Figure 5.24 *Combat aircraft radar: HRF look-down mode clutter return*

When the resolution range is not small compared with the pulse repetition period, it is possible to use two different carrier frequencies. For instance, one carrier frequency will be used on odd pulses and the other on even pulses (difference between both frequencies being relatively small). The phase difference between the two frequencies obtained in a return from a given target will be proportional to the target range and will permit us to measure it.

5.7 Synthetic aperture radars

Let us consider (see Fig. 5.26) a plane flying along a straight line, equipped with an antenna parallel to its plane of symmetry (on the side of the plane: hence the name of 'side-looking radar'), flying at a ground velocity V.

If we are able to record all the information received during time T, during which the antenna has moved by VT, and if we use it correctly, it is clear that we could obtain an angular resolution corresponding to that of an antenna with a

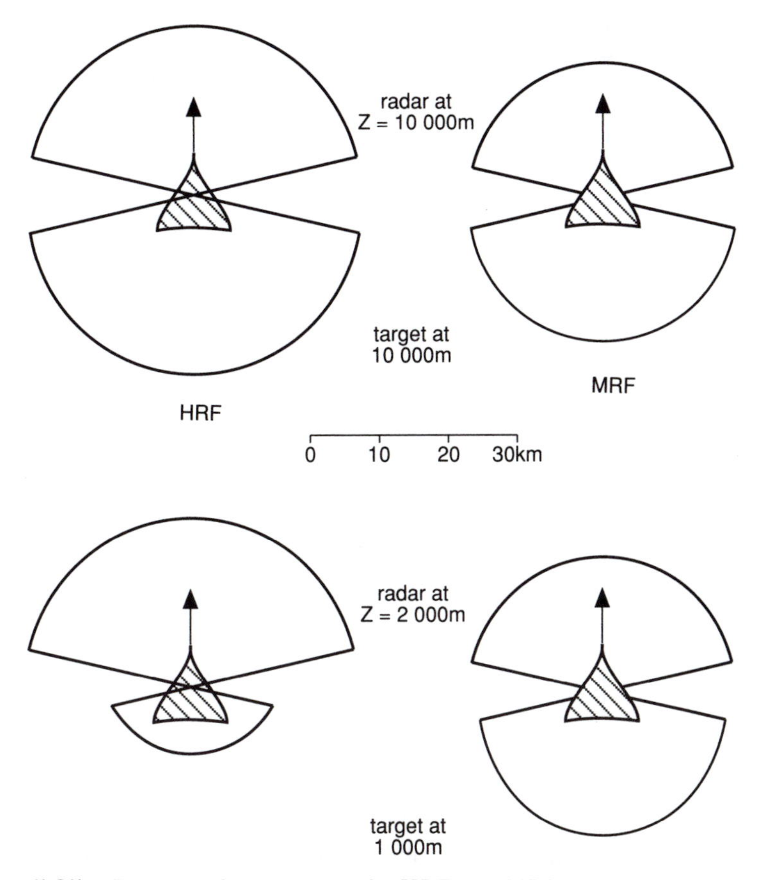

Figure 5.25 *Respective detection zones for HRF and MRF combat aircraft radars*

span equal to VT, around λ/VT, giving, in the direction of the plane motion at a distance D from the plane, a range resolution of $\lambda D/VT$ (if λ is the wavelength).

As an indication, if $\lambda = 0.1$ m, D = 10 000 m, $V = 250$ m/s, $T = 1$ s, $\lambda D/VT = 4$ m.

This is the basic principle of synthetic aperture radars (SAR) generally used as side-looking radars (SLA) and sometimes, for that reason, called SLAR. The principle was probably developed for the first time in 1953 by Michigan University.

Of course, to be relevant, the range resolution in the direction perpendicular to the plane motion has to be similar, which implies fairly good range resolution and the use of a conventional pulse compression associated with a wide transmitted bandwidth (of 50 MHz for instance, which would give a range resolution of 3 m).

The side-looking radar will first be a conventional pulse Doppler radar with very good range accuracy, as represented in Fig. 5.3 (using pulse compression).

But at the output of every range gate [after (9)] the situation will be changed since the purpose of an SAR is normally to take a picture of the ground on

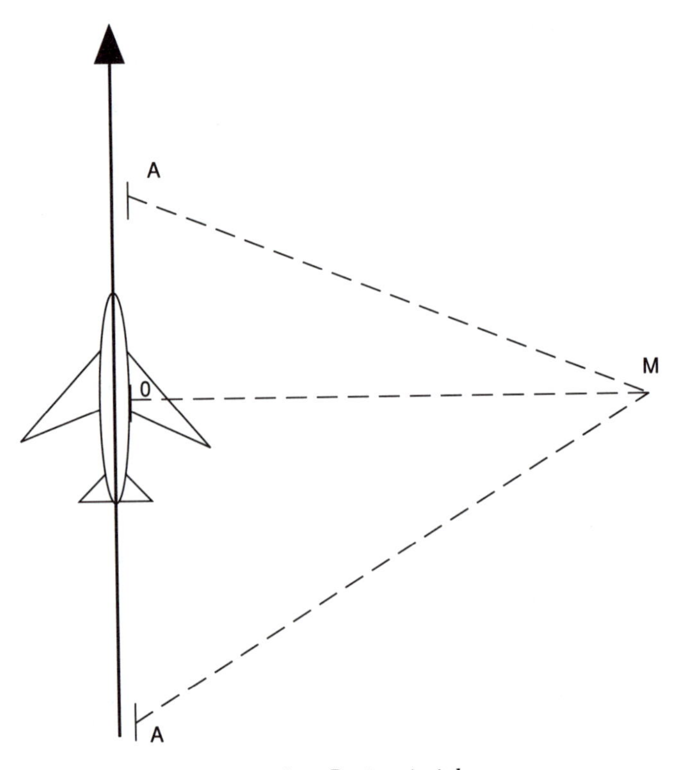

Figure 5.26 *Synthetic aperture radar: Basic principle*

(both) side(s) of the plane. So the following filters replacing (10) and (12) have to be matched to the detection of fixed echoes (fixed compared with the ground).

Coming back to Fig. 5.26, where $OM = D$, it is clear that

$$OA = Vt$$

if V is the ground velocity of the plane

$$AM = D\sqrt{1+\frac{V^2t^2}{D^2}} \approx D + \frac{V^2t^2}{2D}$$

So the phase difference between the transmitted signal and the signal coming from a fixed point M varies as

$$\frac{2\pi}{\lambda} \times 2 \times \frac{V^2t^2}{2D} \quad \text{modulo a fixed phase (hereafter forbidden)}$$

As a consequence, the signal coming from M at the output of (9) varies as the sampling of

$$\cos\left(\frac{2\pi}{\lambda}\frac{V^2t^2}{D}\right)$$

by the pulse repetition frequency.

$$\cos\left(\frac{2\pi}{\lambda}\times\frac{V^2t^2}{D}\right)$$

is a signal linearly frequency modulated since the (instantaneous) frequency is given by $2V^2t/D\lambda$ ($t=0$ when A is at 0).

Pulse compression filtering adapted to chirp modulation (followed by a low-pass filter) will time-compress the signal and provide the relevant convenient signal processing.

With the same parameters as above, it is found that

$$\frac{2V^2T}{D\lambda}=125 \text{ Hz} \quad T=1 \text{ s}$$

If the antenna is nearly omnidirectional, which means its gain is constant for all directions (not too far from the perpendicular to the plane of symmetry of the aircraft), which implies a very small antenna, the signal out of (9) is not amplitude modulated.

If the total time of measurement is T (between $-T/2$ and $+T/2$), the spectrum width of the signal is $2V^2T/D\lambda$, which provides, after pulse compression (after removing the frequency modulation), a signal with a 3 dB duration around $D\lambda/2V^2T$, corresponding to a length in the direction of motion of the aircraft of

$$\frac{\lambda D}{2V^2T}V=\frac{D}{2}\frac{\lambda}{VT}$$

The resolution along the direction of aircraft motion is $(D/2)$ (λ/VT).

If $t=1$ s with the above parameters, the signal has a spectral width of 125 Hz, which provides a compressed pulse of 0.008 s and a signal of $0.008\times250=2$ m.

If we consider that the antenna is directive and that, for instance, it is an antenna with a span of E, uniformly illuminated; in that case the signal out of (9) varies as

$$\cos\left(\frac{2\pi}{\lambda}\frac{V^2t^2}{D}\right)\left[\frac{\sin\dfrac{\pi EVt}{D\lambda}}{\dfrac{\pi EVt}{D\lambda}}\right]^2$$

It is amplitude modulated.

In that case after removal of the frequency modulation by a chirp type pulse compression filter, even if T is very large, the compressed signal cannot be very short.

If T is large (let us say infinite), the compressed signal will have the shape represented in fig. 5.27. The corresponding time resolution could be considered to be $E/2V$, as is normal, which gives $E/2$ as the limit for the metric resolution along the direction of aircraft motion, independent of the wavelength.

In practice, the pulse compression in range (large bandwidth and short time) is generally achieved by classical means, and presently mainly with SAW

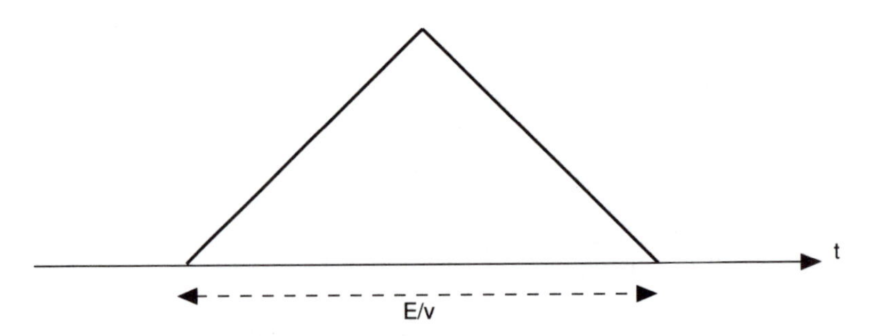

Figure 5.27 *Signal at the output of the synthetic aperture radar*

devices when bandwidths of 50 MHz or more are used (range resolution in the perpendicular better than 3 m).

The pulse compression in lateral direction to achieve 'synthetic aperture processing' depends on completely different parameters (small bandwidth and long time). The original equipments used optical processing, but the present ones use digital processing.

Obviously all the former considerations rely on the assumption that the motion of the radar is strictly along a straight line. Since this is not the case, deviations from the straight line have to be detected and measured (e.g. by use of accelerometers) and the relevant correction has to be introduced in the form of a phase shift between the local oscillator (2) and the mixer (9), or equivalently.

5.8 References

1 CARPENTIER, M.H.: *'Principles of modern radar systems'* (Artech House, London and Boston, 1988)
2 CARPENTIER, M.H.: Present and future evolution of radars, *Microwave J.* June 1985, pp. 48–52
3 KEY, E.L., FOWLE, E.N., and HAGGARTY, R.D.: 'A method of designing signals of largetime–bandwidth products. Paper presented at the IRE National Convention, 22 March 1961.
4 WOODWARD, P.M.: 'Probability and information theory with application to radar' (Pergamon Press, Oxford, 1953)

Chapter 6

MTI, MTD and adaptive clutter cancellation

Prof. Y. H. Mao

6.1 Introduction

Clutter is very harmful in radars whether military or civil radar, since it always appears accompanying the useful target signal. Therefore, how to reject clutter is one of the most important problems facing the radar designer.

There are many ways to reject clutter. We can summarise them as follows:

(i) Preventing the clutter energy from entering the radar antenna:
  (a) Installing the radars in high mountains
  (b) Tilting the radar antenna to higher elevation angles
  (c) Surrounding the radar antenna with a 'clutter shelter fence'

All these methods can be applied to existing radars.

(ii) Enhancing the signal-to-clutter ratio by shaping the beam pattern of the radar antenna: dual beam antennas are often used in modern airport surveillance radar (ASR). The high receiving beam is used for increasing the signal strength of aircraft nearby.

(iii) Enhancing the signal-to-clutter ratio by adopting the polarisation technique: Circular polarisation can reduce the RCS of the raindrop by 15 ~ 30 dB for most microwave radars, but will reduce the RCS of aircraft by only 5 ~ 7 dB. Therefore, more than 10 dB gain can be obtained. A similar technique for increasing the target-to-precipitation echo ratio is to use crossed linear polarisation (RCS is then σ_{HV} or σ_{VH}). While rain echoes are again reduced by 15 ~ 25 dB, the reduction in target RCS is generally greater than 7 dB.

(iv) Reducing the clutter energy by decreasing the size of resolution cell of the radar. This includes: narrowing the beamwidth (it is often limited by the antenna size), narrowing the pulsewidth or adopting pulse compression. This method is often used in shipborne radar to reject sea clutter.

(v) Preventing the receiver from saturation: STC is widely used for this purpose. Utilizing microwave *p–i–n* diodes and gain programmable IF amplifiers can realise digital controlled STC. Logarithmic amplifier (IF or video) only can be used in non-coherent radars owing to the nonlinearity.

(vi) Suppressing the clutter in time domain with CFAR detector or adaptive threshold or clutter map: However, these methods only can obtain super-clutter visibility.

(vii) Suppressing the clutter in frequency domain with MTI or MTD techniques: With these methods, we can obtain sub-clutter visibility.

In this chapter, we will discuss the last method in detail.

6.2 MTI techniques

The MTI technique is one of the oldest methods to reject the radar clutter, and was developed in the Second World War. The first radar equipped with moving target indicator is the CPS–5D. It uses a mercury delay line to delay the echo signal for one period, then subtracts it from the current sweep. Its improvement factor cannot exceed 18 dB. After more than 40 years development, especially in the progress of digital technology, the improvement factor of most advanced MTI radar can now achieve 60 dB. However, there are still many problems to solve.

There are two kinds of MTI radar: the full coherent radar, and coherent on receive MTI radar. The difference between these two types of MTI radar is mainly in the transmitter. The full coherent radar has a master oscillator, e.g. frequency synthesiser, to generate the required transmitting waveform and a local oscillator waveform. So it can maintain coherence between echoes of adjacent sweeps owing to the high stability of the master oscillator. However, a high power amplifier must be used in the transmitter to amplify the small signal from the master oscillator to a sufficient power level.

On the other hand, the coherent on receive MTI radar uses a power oscillator, such as magnetron for the transmitter. A coherent oscillator (COHO) is used to 'remember' the phase of the transmitting waveform. The clutter suppression performance of this type of MTI radar is worse than that of the full coherent radar, owing to the instability of the power oscillator. However, it is much cheaper. Recently, a digital correction unit has been used to correct this instability, so that the improvement factor of the coherent on receive MTI system can achieve as high as 40 to 50 dB.

There are two types of MTI processor: the single channel MTI and the quadrature dual channel MTI. The latter can avoid blind phase and obtain 3 dB more gain than the single channel. Therefore, most of the advanced MTI radars adopt the quadrature dual channel MTI processor with the trade off of more complex hardware. The block diagram of this dual channel MTI is shown in Fig. 6.1.

There are three basic types of MTI filter configurations: nonrecursive, recursive, and canonical. The simplified diagrams are shown in Fig. 6.2.

In practice several basic configurations are used in cascade or in parallel rather than the all-inclusive canonical configuration. Sometimes, we call the filter without feedback the transversal filter or FIR filter, owing to its finite impulse response, and the filter with feedback the IIR filter, owing to its infinite impulse response. The IIR filter can obtain the same transition band as the FIR filter but with a lower number of order, which means less hardware. However, the long transient response makes it susceptible to pulsed interference. As the cost of hardware has decreased considerably in the last decade, we more often now use the FIR filter as the MTI processor.

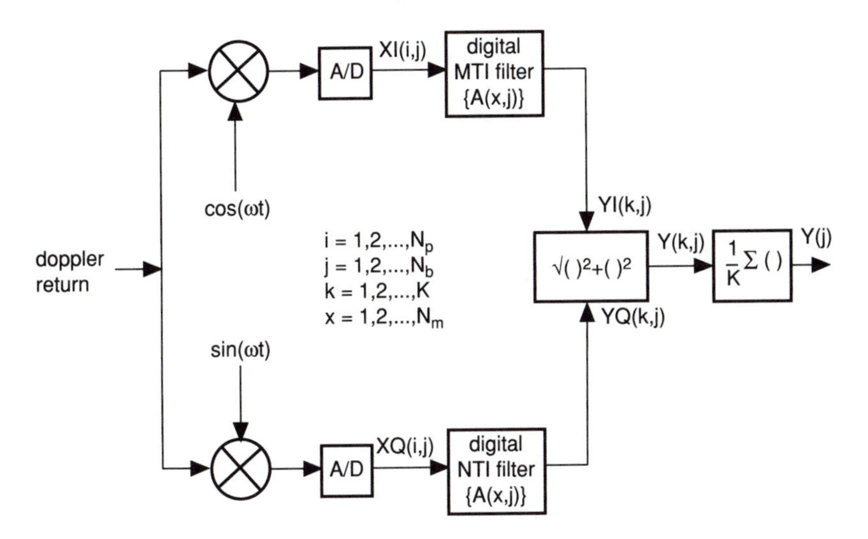

Figure 6.1 *Block diagram of quadrature dual channel MTI*

There are two kinds of FIR MTI filter: the 'fixed window' MTI filter and the 'moving window' MTI filter. If the filter is 'fixed window', only one output is permitted for each block of $N_p = N_m$ pulses, where N_p is the number of pulses within the block, and N_m is the number of taps of the filter; i.e. $K = 1$. However, if the filter is 'moving window', then $N_p > N_m$, and $K = N_p - N_m + 1$ outputs are available.

There are two modes of operation of the filter: uniform sampling mode and nonuniform sampling mode. Sometimes, we call them uniform PRF MTI filter and staggered PRF MTI filter. The latter is used for elimination of blind speeds.

In this Section, we shall describe the recent progress in MTI technology.

6.2.1 *Improvement factor and optimum MTI filter*

It is well known that the improvement factor of an MTI system is defined as the signal-to-clutter ratio at the output of this system compared with that at the input, where the signal is understood to be that averaged uniformly over all radial velocities, that is

$$I = \frac{\bar{S}_o / C_o}{S_i / C_i} \tag{6.1}$$

which can be expressed as

$$I = \frac{\bar{S}_o}{S_i} CA \tag{6.2}$$

where CA is the clutter attenuation, which is defined as the ratio of input clutter power C_i to output clutter power C_o

$$CA = \frac{C_i}{C_o} \tag{6.3}$$

The average improvement factor also can be defined as

$$\bar{I} = \bar{G}\frac{C_i}{C_o} = \bar{G} \times CA \tag{6.4}$$

Many radar manufacturers list this improvement factor as an important specification of their radars. However, it is evident that this specification is a clutter dependent one. In particular, it depends on the spectra of the clutter. Similarly, when a radar is tested in the field to measure the sub-clutter visibility: how to choose a fixed clutter is an important problem. The results are quite different for a fixed clutter, but with some fluctuation compared with a true fixed one. Furthermore, the results for a fixed antenna are also different from those for a radar with a rotating antenna. Therefore, this specification not only depends on the MTI system itself but also on the clutter characteristics and the interaction of radar and clutter.

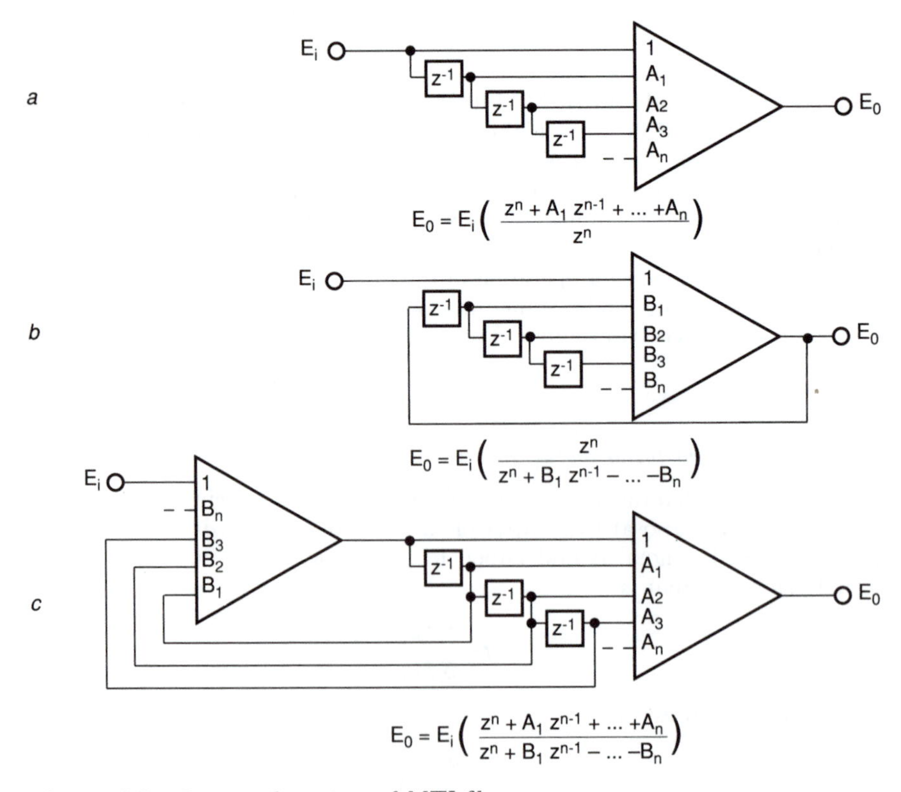

Figure 6.2 *Basic configurations of MTI filters:*

a Nonrecursive;
b Recursive;
c Canonical

The simple MTI system is constructed with a cascade of single delay line cancellers with or without feedback. Shrader [1] showed the degradation of improvement factor of single, dual and triple cancellers due to the spread of clutter power spectrum and scan modulation.

It has been shown that the optimum MTI filter can be constructed by an FIR filter with optimum weights [2–7]. The determination of optimum weights depends on the definition of 'optimum'. In general, 'optimum' means that the output signal-to-clutter ratio, or the improvement factor, is a maximum. If we denote

$$\boldsymbol{A} = E(\boldsymbol{c}^*\boldsymbol{c}^T) = \text{clutter covariance matrix}$$

$$\boldsymbol{S} = E(\boldsymbol{s}^*\boldsymbol{s}^T) = \text{signal covariance matrix}$$

where

$$\boldsymbol{w} = (N \times 1) \text{ vector of filter weights}$$

$$\boldsymbol{s} = (N \times 1) \text{ vector of signal samples}$$

$$\boldsymbol{c} = (N \times 1) \text{ vector of clutter samples}$$

and the asterisk indicates conjugation and T indicates transposition.

The signal-to-clutter power ratio at the output of the filter is:

$$\gamma^2 = \frac{E[|\boldsymbol{w}^T\boldsymbol{s}|^2]}{E[|\boldsymbol{w}^T\boldsymbol{c}|^2]} = \frac{(\boldsymbol{w}, \boldsymbol{S}\boldsymbol{w})}{(\boldsymbol{w}, \boldsymbol{A}\boldsymbol{w})} \tag{6.5}$$

and is equal to the ratio of two quadratic forms.

Then the optimisation of the MTI filter is to maximise this ratio. Under this decision criterion, the calculation of the optimum weights also depends on the knowledge of signal and clutter. If, and only if, the signal and the clutter covariance matrix can be known *a priori*, we can design an optimum MTI filter.

In general, the signal can be known *a priori*, since it is transmitted by the radar. However, the reflected signal contains the Doppler frequency of the target; this Doppler frequency is unknown. We can treat this problem in two ways. In the first method we assume the Doppler frequencies of the signal are uniformly random distributed between $(0, F_p)$ where F_p is the sampling frequency or the pulse repetition frequency of the radar. Then we can design the optimum MTI filter. In another method we assume that the Doppler frequencies of the signal are a number of equally spaced discrete frequencies. This number is often taken as the number of pulse samples. Then we can design the optimum MTD system with a series of Doppler filters.

However, the clutter covariance matrix is still unknown. If we assume some types of clutter model, such as Gaussian power spectral density, then we can design an optimum MTI filter. But this optimum MTI filter is only in the sense of specifying the clutter model. If we can estimate the clutter covariance matrix in real time, then the adaptive MTI system can be realised.

We will discuss the optimum MTI and MTD system first, and then the adaptive MTI and MTD system later.

If the clutter covariance matrix can be known in advance, and the signal is completely known (deterministic signal), then the optimum MTI filter is an FIR filter with the following optimum weights [2]:

$$w_i = k\boldsymbol{A}^{-1}\boldsymbol{s}^* \tag{6.6}$$

where k is an arbitrary constant, $\boldsymbol{A}^{-1}$ is the inverse of the clutter covariance matrix and $\boldsymbol{s}$ is the signal vector reflected from the target.

$$s_n = s_0 \exp[j(2\pi f t_n + \phi)] \tag{6.7}$$

where f is the Doppler frequency of the target.

Since the Doppler frequency cannot be known in advance, a filter bank of many different Doppler frequency channels is required.

If only a single filter channel MTI system is required, the optimal weights of this filter can be obtained by optimising the signal-to-clutter ratio in the average Doppler frequency band. The output signal power of this MTI system is

$$P_s = 1/2 \left| \sum_i w_i s_i \right|^2 \tag{6.8}$$

By use of eqn. 6.7, this can be written as

$$P_s = (s_0^2/2) \sum_i \sum_j w_i w_j \cos[2\pi f(t_i - t_j)] \tag{6.9}$$

This function is periodic with a period of

$$F_p = l_p/(t_i - t_j) \tag{6.10}$$

where the l_p are integers for all i and j, T_i and T_j must be rational numbers. This F_p is commonly referred to as the blind Doppler frequency. When the target Doppler frequency is unknown, one may assume that it has uniform distribution. In this case, the expected signal output is then

$$P_s = (s_0^2/2) \sum_i |w_i|^2 \tag{6.11}$$

The average output clutter power is

$$P_c = \sum_i \sum_j w_i w_j^* a_{ij} \tag{6.12}$$

where a_{ij} are the elements of the clutter covariance matrix $\boldsymbol{A}$. Receiver thermal noise can be included as a real constant added to the diagonal elements of this covariance matrix.

The MTI improvement factor I or ratio of output to input SCR is then

$$I = \left(\sum_i w_i w_j^* \right) \Big/ \left(\sum_i \sum_j w_i w_j^* a_{ij} \right) \tag{6.13}$$

where the elements of the covariance matrix are normalised, so that $a_{ij} = 1$. The quantities a_{ij} are the Fourier transforms of the clutter power density function. If we assume that the clutter power density function is a Gaussian function

$$G(f) = [1/\sqrt{2\pi}\sigma_c] \exp(-f^2/2\sigma_c^2) \tag{6.14}$$

where f is the Doppler frequency and σ_c is the standard deviation or the spectral bandwidth of the clutter. For ground clutter, we can assume that the mean

Doppler frequency is zero. Accordingly, the elements of the clutter covariance matrix $\boldsymbol{A}$ are real and have the form

$$a_{ij} = \exp[-2\pi^2\sigma_c^2(t_i - t_j)^2] \tag{6.15}$$

For a constant PRF MTI filter, this becomes

$$a_{ij} = \exp[-2\pi^2\sigma_c^2(i-j)^2 T_p^2] \tag{6.16}$$

One may notice that this matrix is symmetrical; that is

$$a_{ij} = a_{ji} \tag{6.17}$$

The diagonal elements have the following form:

$$a_{ij} = 1 + \eta_0 \tag{6.18}$$

where η_0 is the normalised receiver noise level, it is usually very small and can be omitted. This diagonal element has the maximum value; hence, one may normalise this matrix such that

$$0 < a_{ij} < a_{ij} = 1; \quad i, j = 1, 2, \ldots, N \tag{6.19}$$

where N is the number of coherent pulses.

Since the covariance matrix is real and symmetrical, the optimal filter weights w_i are components of the eigenvector of this matrix, these w_i are real.

The clutter output P_c is actually a quadratic form. Representing the filter weights $w_1, w_2, \ldots, w_N$ by an N-dimensional vector $\boldsymbol{w}$, this quadratic form can be compactly written as

$$Q(\boldsymbol{w}) = (\boldsymbol{w}, \boldsymbol{A}\boldsymbol{w}) \tag{6.20}$$

where the parentheses represent the inner product of the vector $\boldsymbol{w}$ and the vector $\boldsymbol{A}\boldsymbol{w}$.

The improvement factor thus becomes

$$I = (\boldsymbol{w}, \boldsymbol{w})/(\boldsymbol{w}, \boldsymbol{A}\boldsymbol{w}) \tag{6.21}$$

The reciprocal of this improvement factor is actually the clutter output, or

$$P_c = (\boldsymbol{w}, \boldsymbol{A}\boldsymbol{w}) \tag{6.22}$$

when $\boldsymbol{w}$ is normalised.

The optimal MTI filter means that the improvement factor of this filter is a maximum or the clutter output is a minimum. The problem then becomes to find a vector $\boldsymbol{w}$ such that the quadratic form $Q(\boldsymbol{w})$ is a minimum.

It has been shown [8, 9] that this quadratic form $Q(\boldsymbol{w})$ can be represented as

$$Q(\boldsymbol{w}) = \sum_i d_i^2 \lambda_i \tag{6.23}$$

where

$$w = \sum_{i} d_i w^i \tag{6.24}$$

and λ_i are eigenvalues and w^i are the associated eigenvectors. It is evident that the minimum $Q(w)$ occurs when w takes the eigenvector which is associated with the minimum eigenvalue λ_{min}, since

$$\lambda_1 \geqslant \lambda_2 \geqslant \cdots \geqslant \lambda_{min}$$

Then

$$Q(w) = \lambda_{min} \tag{6.25}$$

Thus the quadratic form has a minimum value which is equal to the smallest eigenvalue λ_{min} of matrix A, and the eigenvector associated with this eigenvalue is the required minimisation vector, i.e. the optimisation weights vector.

If one expands the determinant $|A - I|$, one finds the following valuable relations between the eigenvalues and the elements of matrix A.

$$\sum_{i=1}^{N} \lambda_i = \sum_{i=1}^{N} a_{ii} \tag{6.26}$$

$$\prod_{i=1}^{N} \lambda_i = \det A \tag{6.27}$$

We pointed out earlier that the diagonal elements of matrix A are unity; hence

$$\sum_{i=1}^{N} \lambda_i = N \tag{6.28}$$

Since the quadratic form $Q(w)$ is positive definite, we have the following properties:

$$\det A > 0 \tag{6.29}$$

$$\det A < \prod_{i=1}^{N} a_{ij} \tag{6.30}$$

Hence

$$0 < \prod_{i=1}^{N} \lambda_i < 1 \tag{6.31}$$

From eqns 6.28 and 6.31, one can deduce that

$$0 <= \lambda_{min} <= 1 \tag{6.32}$$

Here λ_{min} is the smallest eigenvalue, which then, in turn, represents the optimal clutter output of an MTI filter. If the clutter spectrum standard deviation $\sigma_c \rightarrow \infty$ or the correlation time $t_i - t_j$ becomes infinite, the off-diagonal elements in matrix $\boldsymbol{A}$ approach zero. In this case,

$$\lambda_1 = \lambda_2 = \cdots = \lambda_N = 1 \tag{6.33}$$

and any normalised vector can satisfy the equations.

On the other hand, if σ_c or the correlation time $t_i - t_j$ approach zero, then every element of matrix $\boldsymbol{A}$ approaches unity. In this case, the determinant $\boldsymbol{A}$ is zero. According to eqn. 6.27, the smallest eigenvalue is zero. Under this condition, the eigen equations are identical and have the following form:

$$w_1 + w_2 + \cdots + w_N = 0 \tag{6.34}$$

There are infinite numbers of solutions. One solution is to set these w_i according to the binomial distribution. This is well known. One sees that the binomial weighting MTI is an optimal MTI only when the standard deviation of the clutter spectrum is zero.

Hsiao [9] derived the λ_{min} of a constant PRF MTI filter, when the clutter has a Gaussian spectral density function of eqn. 6.14. It is equal to

$$\lambda_{min} = [2/\sqrt{2\pi\sigma}] \int_0^{\infty} \exp\{-[f^2/(2\sigma^2)]\} \sum_i \sum_j w_i w_j \cos 2\pi f(i-j) df \tag{6.35}$$

where σ is the normalised standard deviation of the clutter spectral density:

$$\sigma = \sigma_c T \tag{6.36}$$

and w_i are the optimal filter weights for a given σ.

Then the improvement factor can be calculated for a given σ. It is also a function of N, the number of coherent pulses. Fig. 6.3 gives the results of computed improvement factor I for Gaussian clutter spectral density as a function of σ and N [9].

It can be seen from this Figure that the improvement factor degrades seriously with the increase of clutter spectral bandwidth, even with the assumption of a Gaussian shape spectrum. Since the 'tail' of the Gaussian shape spectrum is shorter than that of other real spectra, such as a 'cubic' spectrum, the improvement factor of the optimum MTI filter in real clutter will be worse than this figure. The standard deviation of the clutter spectrum is normalised to the PRF of the radar, so the degradation is more serious in low PRF and higher microwave radars.

There are some assumptions in this optimisation procedure. The first is that this optimisation is derived under the assumption of uniformly distributed Doppler frequency of the target. The second is that it is derived under the assumption of zero mean Gaussian clutter spectral density.

Though the angle θ between the course of the target and the radar beam can be assumed to be uniformly distributed between $\pm\pi$, cos θ is not uniformly distributed between $+1$ and -1, and the minimum velocity of the target is much greater than zero. Therefore, the Doppler frequency of the target is not uniformly distributed between 0 and F_p. Though there are blind speeds

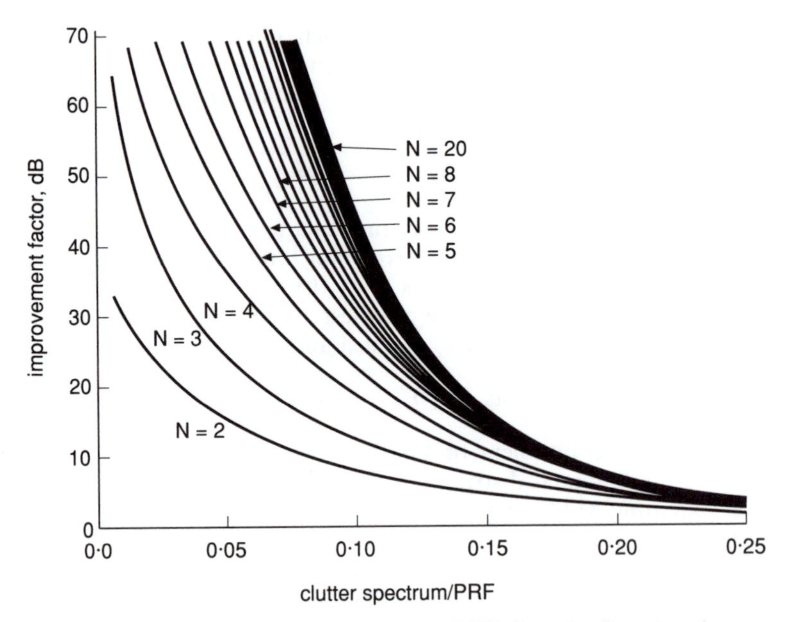

Figure 6.3 *Improvement factor of an optimum MTI filter for Gaussian clutter spectral density [9] © IEEE, 1974*

problems these blind speeds can be avoided by using staggered PRF. This will be discussed later.

Arora *et al.* [10] proposed a method of optimisation which considered the expected Doppler frequency of the target. The procedure involves minimisation of the output clutter power subject to the constraint that the average output signal power over the expected target Doppler range is to be kept constant. Another constraint in the minimisation problem is also incorporated, according to which the transfer function of the system has a null at the dominant clutter Doppler frequency f_c, which may often be zero. This means that the clutter has two components: one is an impulsive component which is located in zero frequency, the other is a distributed component with a Gaussian shape of spectrum and a non-zero mean Doppler shift f_c.

Four special cases are considered:

(i) The target visibility region consists of the entire interval between zero and the first blind speed and there is no linear constraint. This is the case discussed above.
(ii) The target is assumed to lie, with uniform probability, anywhere in the subinterval of the Doppler frequency range $(0, F_p)$, say in $(f_a, F_p - f_a)$, where $f_a < F_p/2$.
(iii) As in case (i), but with the additional constraint that the transfer function be forced to zero at frequency f_c, the mean value of clutter spectrum.
(iv) As in case (ii) but the same linear constraint is included. In this way, the distributed component of clutter is minimised at the filter output while the clutter component at a fixed mean relative velocity is completely suppressed.

With the two constraints specified above, the problem is mathematically formulated as a modified eigenvalue problem. We will omit the detail of mathematical derivation and give the results of computation directly.

Fig. 6.4*a* shows the improvement factors I_2 and I_4 for cases (ii) and (iv) versus normalised standard deviation σ for different numbers of pulses. Fig. 6.4*b* shows the comparison of the improvement factors of optimum filters of cases (i) and (ii) for $N=3$. Fig. 6.5 shows the improvement factor I_3 for case (iii). The transfer functions of these optimum filters for case (iii) are shown in Fig. 6.6.

It can be seen from Fig. 6.4 that the improvement factor of case (ii) is somewhat better than that of case (i). But the improvement factor of case (iii) (Fig. 6.5) is worse than that of case (i). This is due to the fact that the mean of the spread clutter is not equal to zero. It can be seen from the transfer function of case (iii) that the notch is centred at the clutter mean Doppler frequency f_c.

All of the previously mentioned optimisation methods are concerned with the maximisation of the improvement factor of a transversal, or non-recursive filter, of a given order. The solution of this optimisation problem requires an explicit knowledge of the clutter characteristics, e.g. its power spectrum or covariance matrix. Just as we mentioned before, all of these optimisation methods are based on the assumption of Gaussian spectral density function. In many practical situations, however, the assumptions regarding this knowledge are not warranted, due to the non-stationary nature of clutter. Furthermore, it has been shown in earlier study that the performance of the higher order filters is quite sensitive to variations in the assumed clutter model.

Prasad [11] suggested a class of MTI filters which do not require the knowledge of the actual shape of the clutter power spectral density function, but only its bandwidth. This design may be called a worst-case design, in the sense that it tries to maximise the improvement factor corresponding to the most unfavourable clutter characteristics satisfying the bandwidth constraint.

Consider the input to a coherent MTI filter is a discrete time sequence $\{x_1, x_2, \ldots, x_N\}$ of length N, assumed to be sampled at the radar pulse repetition period

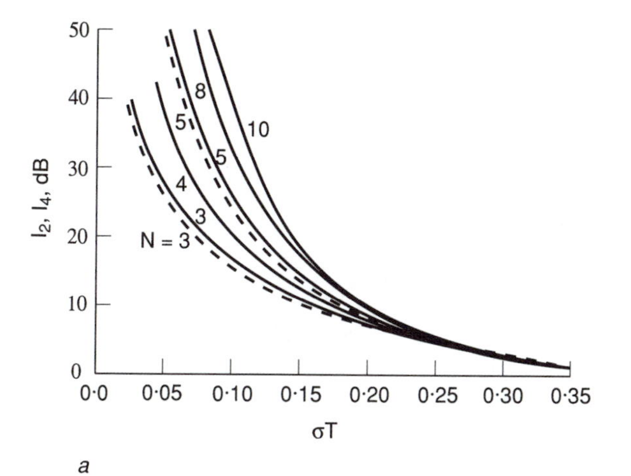

Figure 6.4a *Improvement factor I_2 (solid line) and I_4 (dash line) for $f_a = 0.1$ [10]*

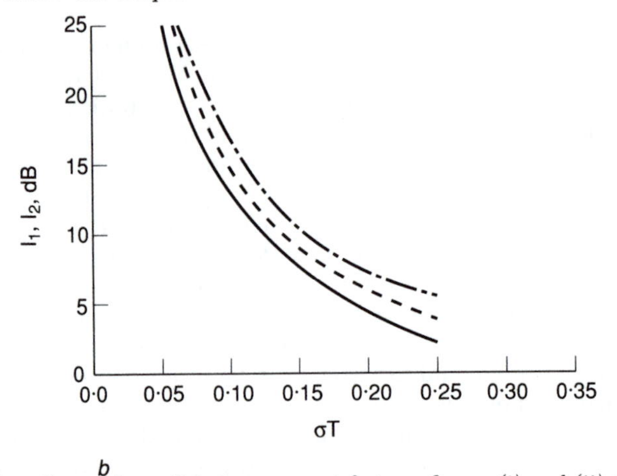

Figure 6.4b *Comparison of the improvement factors of cases (i) and (ii) for N=3 [10]*

—— case (i), $f_a=0$
– – case (ii), $f_a=0.15$
- · · - case (ii), $f_a=0.25$

T_p, which is normalised to unity. In analogy to the Gabor (RMS) bandwidth of the discrete-time signal sequence is defined by

$$BW=1/2\pi\left|\sum_{i=1}^{n-1}(x_i-x_{i+1})^2/\sum_{i=1}^{n}x_i^2\right|^{1/2} \tag{6.37}$$

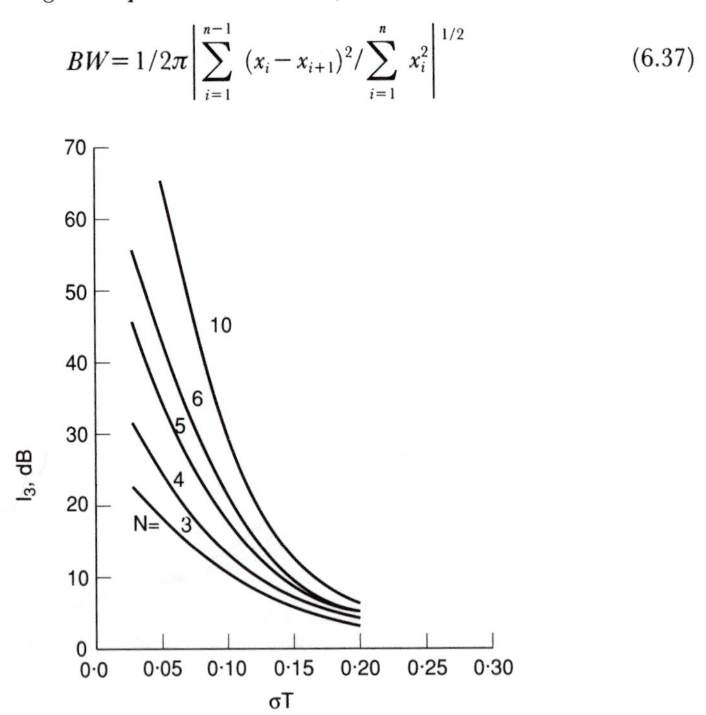

Figure 6.5 *Improvement factor I_3 for case (iii). [10] Gaussian clutter: $f_c T_p = 1/12$*

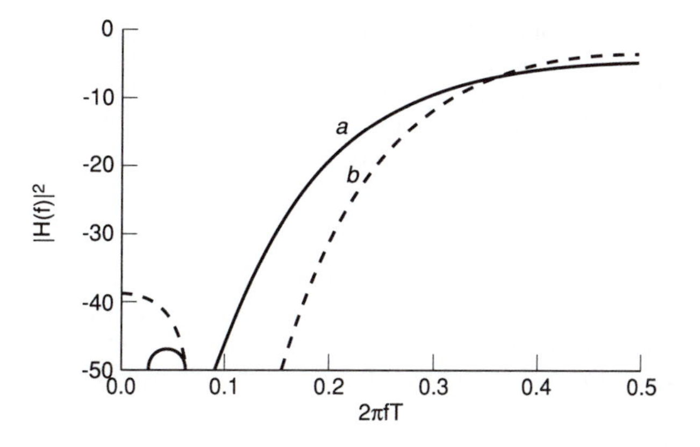

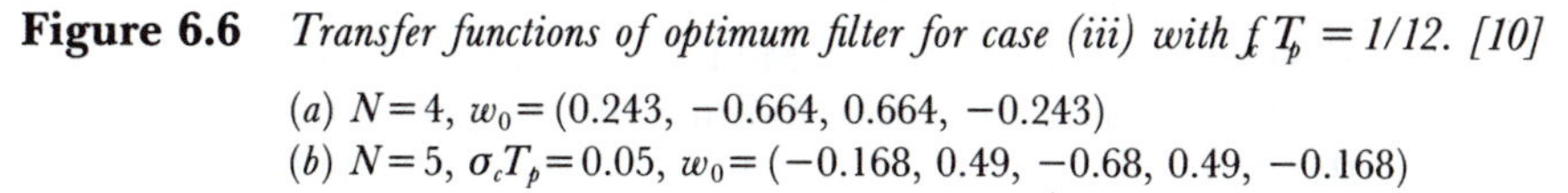

Figure 6.6 *Transfer functions of optimum filter for case (iii) with $f_c T_p = 1/12$. [10]*

(a) $N=4$, $w_0=(0.243, -0.664, 0.664, -0.243)$
(b) $N=5$, $\sigma_c T_p=0.05$, $w_0=(-0.168, 0.49, -0.68, 0.49, -0.168)$

Consider next the class B of essentially bandlimited signals with RMS bandwidth of B Hz or less, defined by

$$B=\{x=(x_1, x_2, \ldots, x_N)^T \in R^N: \Sigma(x_i - x_{i+1})^2/\Sigma x_i^2 < 4\pi^2 B^2$$

which can also be written in the form

$$B=\{x \in R^N: x^T G x/\|x\|^2 \leqslant 4\pi^2 B^2\} \tag{6.38}$$

where

$$G=\begin{bmatrix} 1 & -1 & 0 & 0 & \ldots & 0 & 0 & 0 \\ -1 & 2 & -1 & 0 & \ldots & 0 & 0 & 0 \\ 0 & -1 & 2 & -1 & \ldots & 0 & 0 & 0 \\ . & . & . & . & \ldots & . & . & . \\ 0 & 0 & 0 & 0 & \ldots & 2 & -1 & 0 \\ 0 & 0 & 0 & 0 & \ldots & -1 & 1 & 2 \\ 0 & 0 & 0 & 0 & \ldots & 0 & -1 & 1 \end{bmatrix} \tag{6.39}$$

It can be shown [11] that the improvement factor for the worst case input clutter is

$$I(B)=\begin{cases} 10 \log_{10} \dfrac{\lambda_{m+1}-\lambda_1}{4\pi B^2-\lambda_1}, & \lambda_1 \leqslant 4\pi^2 B^2 \leqslant \lambda_{m+1} \\ 0 & \lambda_{m+1} < 4\pi^2 B^2 \end{cases} \tag{6.40}$$

where λ is the eigenvalue.

Some numerical results have been obtained. First, for the case of $N=3$ and for all clutter power spectra, the optimum weight vector w_o as obtained either from the min–max approach, or from the best average improvement factor approach, surprisingly turns out to be the same, i.e. $w_o = (-0.40825, 0.81650, -0.40825)$. This, however, is explained easily in view of the two constraints of the problem, namely, unity norm and a zero at the origin, thus leaving only a single degree of freedom in the optimisation problem. This leads one to the above unique solution for $n=3$. It is interesting to compare the performance of this filter under various conditions. Table 6.1 [11] gives the optimum average improvement factors for Gaussian, raised cosine and Butterworth (1st- and 2nd- orders) clutter spectra, as well as the corresponding worst-case improvement factor. The latter is seen to be a lower bound on all the rest.

A comparison of the improvement factor of this worst-case optimum MTI filter with that of the conventional optimum filter is shown in Fig. 6.7 for $N=4$.

6.2.2 *Optimisation of frequency response*

All the previously described optimum MTI filters are based on the maximisation of the average improvement factor within the sampling frequency $(0, F_p)$. However, the frequency response of the filter is also very important from the view-point of detection of target signal. Let us discuss this problem in detail.

The frequency response function of the transversal MTI filter is given by the Fourier transform of the weight vector

$$H(\omega) = \sum_{i=1}^{N} w_i \exp[-j(i-1)2\pi f_d T_p], \; 0 \leqslant f_d \leqslant 1/T_p \tag{6.41}$$

It is convenient to introduce the change of variables, $\theta = 2\pi f_d T_p$, so that eqn. 6.41 becomes

$$H(\omega) = \sum_{i=1}^{N} w_i \exp[-j(i-1)\theta], \; 0 \leqslant \theta \leqslant 2\pi \tag{6.42}$$

Now we can calculate the frequency responses of the optimum MTI filters for different numbers of pulses N. The results are shown in Fig. 6.8.

Each of these filters was computed assuming a large narrow band clutter signal centred at zero Doppler frequency. As the number of pulses N is increased, the improvement factor increases. From the Figure we can see that

Table 6.1 *Performance of the 3rd-order filter with optimum weights, against various clutter spectra, each of RMS BW=0.08*

Clutter type	Average improvement factor (dB)
Raised cosine	21
Gaussian	17.5
Butterworth (2nd-order)	15.2
Butterworth (1st-order)	11.3
Worst case	10.0

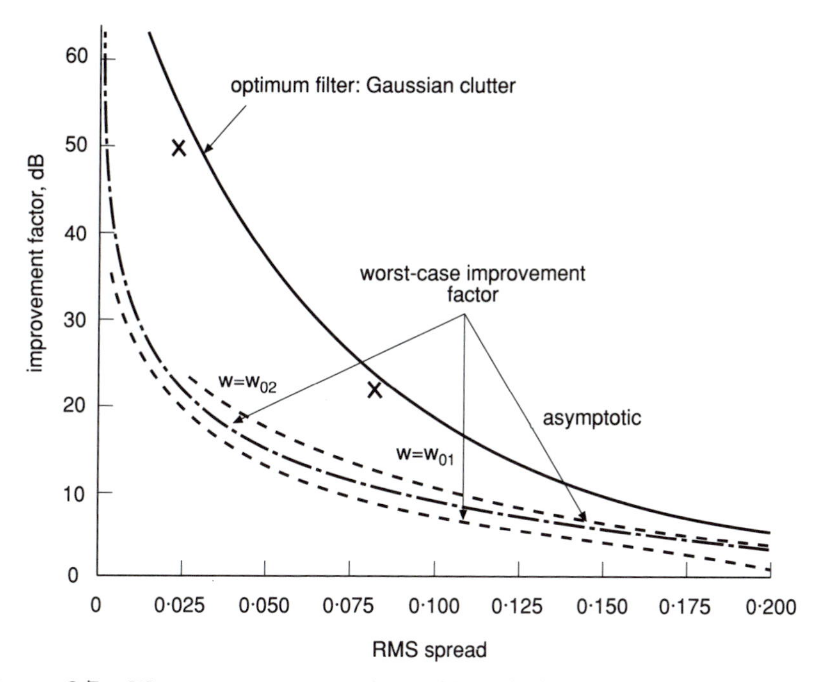

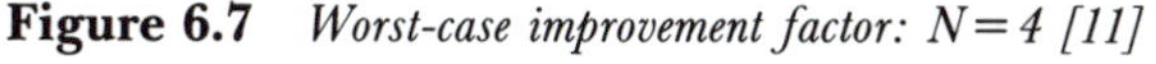

Figure 6.7 *Worst-case improvement factor: N=4 [11]*

$w_{01} = (-0.5, 0.5, 0.5, -0.5)$, $B_1 = 0.225$
$w_{02} = (-0.27060, 0.65328, -0.65328, 0.27060)$, $B_2 = 0.294$
× average performance of worst-case optimum filter with $w = w_{02}$ against Gaussian clutter

this increase is due to the broadening of the notch in the region of heavy interference and the increased gain in the region of lowest interference. This, however, is a disadvantage of maximum improvement factor filters. The broad notch and narrow signal passband can produce large target 'blind speed' regions. Therefore, a broader passband is often desired, even at the expense of some reduction in the improvement factor, especially when the bandwidth of the clutter spectrum is narrow and the number of pulses is large. So it is necessary to develop the design procedure of a sub-optimal MTI filter with the desired frequency response.

Schleher and Schulkind [12] developed a method for optimising a nonrecursive MTI filter based on the principle of quadratic programming. With this method the desired velocity response can be specified.

The quadratic programming algorithm represents a systematic procedure for the minimisation of a quadratic form (objective function) subject to linear constraints. It is guaranteed to converge in a finite number of steps. Furthermore, if overly restrictive constraints are imposed, thereby preventing minimisation of the quadratic objective function, the algorithm will identify this problem and allow the constraints to be relaxed for the purpose of obtaining a

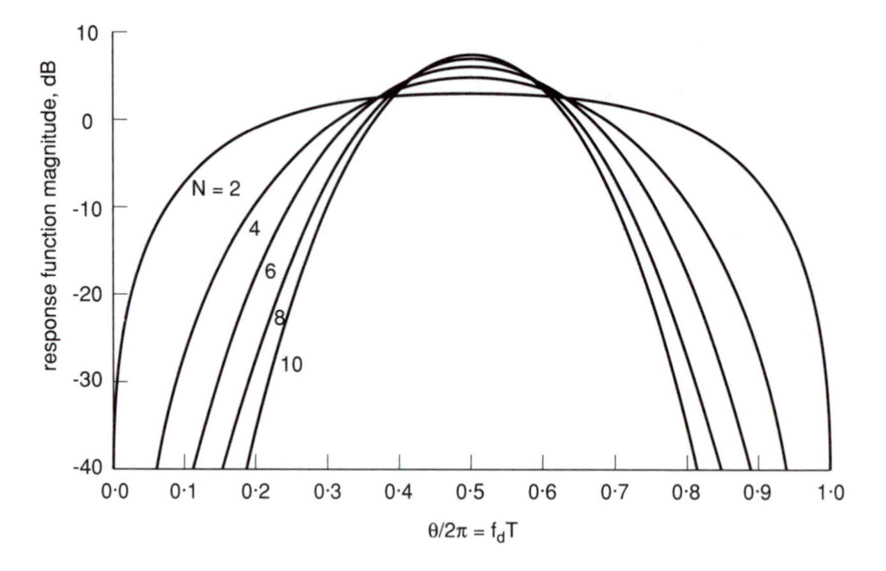

Figure 6.8 *Frequency responses of the optimum MTI filters*

solution. In the design of MTI filters, this property can be used to determine the best velocity characteristic a given number of delay section can achieve, or alternatively, the number of delay sections required to achieve a specified velocity response.

In order to specify the MTI's velocity characteristic in the form of a linear constraint, it is necessary to restrict the class of MTI filters to those with linear phase. This results in the filter weights being symmetric about their central point when an odd number of pulses (even number of delays) are processed, and anti-symmetric about their central point when an even number of pulses (odd number of delays) are processed. Therefore, the filter transfer function for n even ($n \geqslant 2$) is given by

$$H(j\omega) = \exp\left[-i\left(\frac{n-1}{2}\,\omega T_p + \frac{\pi}{2}\right)\right]\left[\sum_{i=0}^{(n/2)-1} (-1)^{i+1} 2|w_i| \sin\left(\frac{n-1}{2} - i\right)\,\omega T\right] \tag{6.43}$$

and for n odd ($n \geqslant 3$)

$$H(j\omega) = \exp\left[-j\left(\frac{n-1}{2}\,\omega T_p\right)\right]\left[(-1)^{(n-1)/2}|w_{(n-1)/2}| + \sum_{i=0}^{(n-3)/2} (-1)^i 2|w_i| \cos\left(\frac{n-1}{2} - i\right)\omega T\right] \tag{6.44}$$

The quadratic programming problem for the optimisation of a nonrecursive MTI can be summarised as follows:

$$\min P_c = \sum_i \sum_j w_i w_j a_{ij} \tag{6.45}$$

subject to

$$\Sigma w_i = 0 \tag{6.46}$$

and m constraints evaluated throughout the MTI's velocity response of the form

$$|H(j\omega)| = \sum_{i=0}^{(n/2)-1} (-1)^{i+1} 2|w_i| \sin\left(\frac{n-1}{2} - i\right) \omega T_p; \; n \text{ even.} \tag{6.47}$$

or

$$|H(j\omega)| = (-1)^{(n-1)/2}|w_{(n-1)/2}| + \sum_{i=0}^{(n-3)/2} (-1)^i 2|w_i| \cos\left(\frac{n-1}{2} - i\right)\omega T_p; \; n \text{ odd} \tag{6.48}$$

In addition, the quadratic programming algorithm requires that the variables (weights) be equal to, or greater than, zero ($w_i \geqslant 0$). This is easily handled for the MTI case by using the absolute value of the weights in the algorithm and including the signs in the constraints:

$$\begin{aligned} w_i &= |w_i| = x_i; \; w_i \text{ positive} \\ w_i &= -|w_i| = -x_i; \; w_i \text{ negative} \end{aligned} \tag{6.49}$$

The quadratic programming problem specified by eqns. 6.45, 6.46, 6.48 and 6.49 can be solved by the application of an algorithm originally developed by Dantzig [13]. This algorithm is a straightforward Simplex method for linear programming.

As an example, assume a requirement exists for a non-recursive MTI filter design with a clutter attenuation of at least 30 dB; the clutter spectrum is Gaussian shaped with $\sigma = 0.025$, and the velocity characteristics are to have a 3 dB point $f_{co}/\sigma = 10$, i.e. $f_{co} = 0.25$. A four pulse processor ($n = 4$) is to be investigated.

Applying eqn. 6.48 at the PRF/2 point ($\omega = \pi/T_p$) and at f_{co} ($\omega = \pi/2T_p$) results in

$$\begin{aligned} H(\omega = \pi/T_p) &= 2w_0 + 2w_1 = 1 \\ H(\omega = \pi/2T_p) &= \sqrt{2}(w_1 - w_0) \geqslant b \end{aligned} \tag{6.50}$$

Applying eqn. 6.45 using

$$a_{ij} = \exp[-(i^2\Omega^2/2)] \tag{6.51}$$

where $\Omega = 2\pi\sigma$ results in

$$P_c = 0.2108 w_0^2 + 0.02452 w_1^2 - 0.14356 w_0 w_1 \tag{6.52}$$

The sum of the weights is zero because n is even.

The quadratic programming problem is then

$$\max\,[0.05255 - P_c] = 0.28196 x_1 - 0.37826 x_1^2$$

Subject to

$$x_1 - x_2 + y = 0.5 \text{ or } 0.4240$$
$$x_1, x_2 \geqslant 0 \tag{6.53}$$

where x_2 is a slack variable and y is an artificial variable unrestricted in sign. The use of an artificial variable is required to find an initial basic feasible solution. The parameters of the quadratic problem are

$$C = \begin{bmatrix} 0.75652 & 0 \\ 0 & 0 \end{bmatrix} \quad p = \begin{bmatrix} 0.28196 \\ 0 \end{bmatrix}$$
$$b = 0.5 \text{ or } 0.4240 \qquad A = [1 - 1] \tag{6.54}$$

where the alternate values for b permit either the velocity constraint ($b = 0.5$) or the *CA* constraint ($b = 0.4240$) to be achieved.

The computational results are shown in Figs. 6.9–6.12. It can be seen from these figures that the velocity responses obtained by this method are better than that of the classical one.

Houts and Burlage [14] suggested a method to maximize the usable bandwidth of MTI filters. They used the Chebyshev high pass filter as an MTI filter. It can come close to the ideal highpass filter but with a trade off in reducing the improvement factor. Three types of MTI filters are compared: three-pulse canceller with binomial coefficients (TPC), five-tap covariance (COV) design, and 15-tap Chebyshev (CHEV) filter. The amplitude response is shown in Fig. 6.13, and the improvement factor versus standard deviation σ_c is shown in Fig. 6.14.

The frequency response of the normalised TPC is

$$H(f) = 1.633 \sin^2(\pi f T) \tag{6.55}$$

The normalised bandwidth for 0 dB gain is defined by

$$0.286 < BW < 0.714 \tag{6.56}$$

which is less than 45% of the total bandwidth.

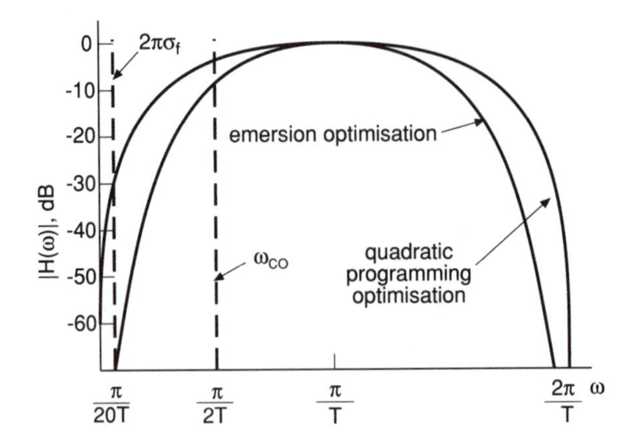

Figure 6.9 *Velocity response for N = 4 [12]*

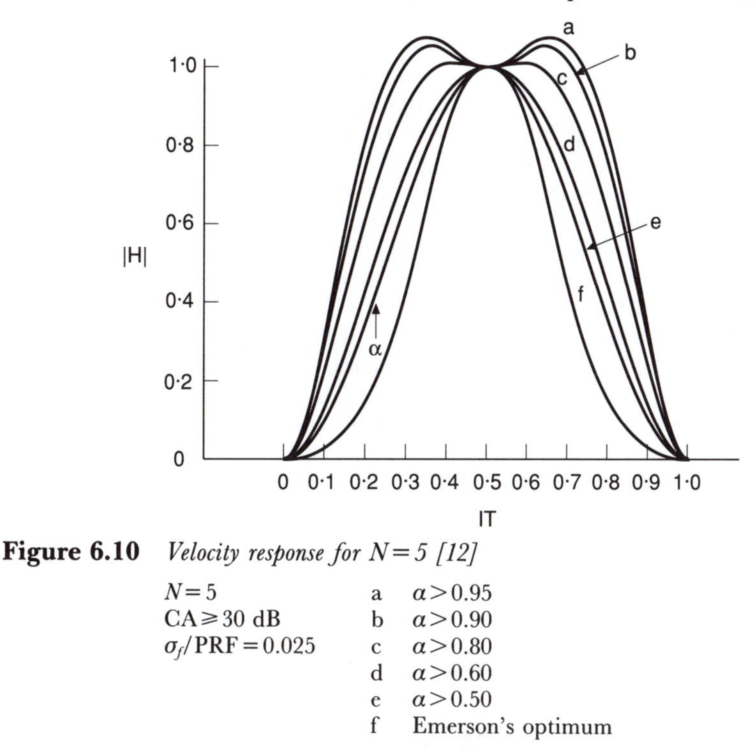

Figure 6.10 *Velocity response for $N=5$ [12]*

$N=5$	a	$\alpha>0.95$
CA$\geqslant$30 dB	b	$\alpha>0.90$
σ_f/PRF$=0.025$	c	$\alpha>0.80$
	d	$\alpha>0.60$
	e	$\alpha>0.50$
	f	Emerson's optimum

It is noted from inspection of Fig. 6.13 that the five-tap covariance (COV) design is a definite improvement over the TPC. Although it would appear that even better response could be obtained with more weights, it has been found that solutions for such COV design specifications still utilise no more than five weights when $\sigma<0.001$. It is evident that the best one is the 15-tap Chebyshev filter. However, the trade off between I and BW is clearly demonstrated in these two Figures. If the overall system improvement factor is limited to 40 dB, and there are enough pulses to be processed, the latter is the best choice. However, if the permitted ripple within the passband is larger, such as 6–7 dB, better results can be obtained.

We shall omit the design detail of the Chebyshev filter, since it can be found in a text book of digital filter design. Some examples of filter design will be shown here.

It will be assumed for the ground clutter example that a signal-to-clutter improvement I of 50 dB is needed. The radar system has PRF$=5$ kHz and transmits 50 pulses per beam dwell, 48 of which can be used in the MTI processor.

According to the specifications of the radar, the signal processor must provide a gain $\Delta S=6$ dB. This means that the integration gain G_i of the fixed window MTI processor must be

$$G_i \geqslant 6\text{ dB} - HDB(f_m)$$

where f_m is the minimum frequency in the passband (Fig. 6.15). Since the minimum passband value $HDB(f_m) \simeq -R_p/2$, R_p is the ripple of the passband, it follows that $G_i > 6 + R_p/2$. The value of N and corresponding R_p values are listed in Table 6.2. The values chosen to attenuate the clutter to acceptable levels are $f_s = 0.004$ and $A_s = 50$ dB for N even, 55 dB for odd values.

Once the ground filter parameters have been chosen, the filter weights and passband lower edge f_p are selected by Chebyshev design programs. The resulting frequency responses for the four filters are shown in Fig. 6.16 and the variation of I as a function of the clutter standard deviation σ in Fig. 6.17 [14].

It is evident that the nine-tap filter has the widest BW, providing some additional 4.5% of f_T over the five-tap design; however, it is necessary to consider the effect on bandwidth of the lower G_i, as indicated in Table 6.2.

The usable bandwidth of the Chebyshev filter is affected by two factors: number of filter weights N and tolerated passband ripple R_p. It can be seen from the relationship between the lower edge of passband f_p and the values of N and R_p (Figs. 6.18, 6.19). A smaller value of f_p means a larger usable bandwidth. The

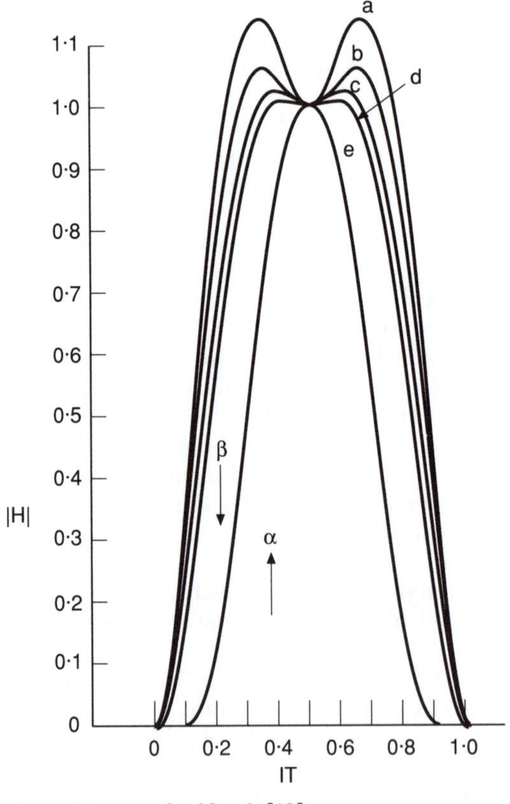

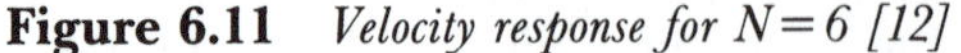

Figure 6.11 *Velocity response for $N=6$ [12]*

$N=6$	a	$\alpha > 1.10$ $\beta < 0.65$
CA $\geq$ 30 dB	b	$\alpha > 1.05$ $\beta < 0.55$
σ_f/PRF $= 0.025$	c	$\alpha > 1.02$ $\beta < 0.54$
	d	$\alpha > 1.00$ $\beta < 0.50$
	e	Emerson's optimum

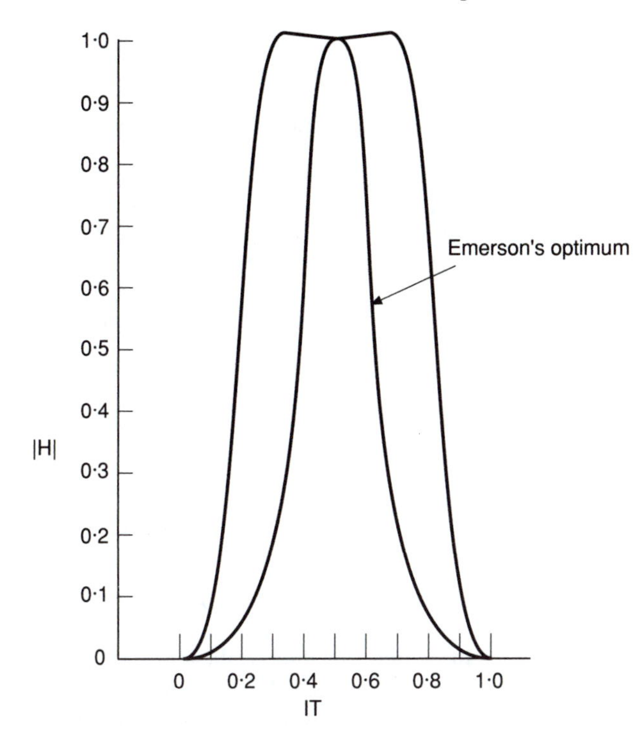

Figure 6.12 *Velocity response for $N=10$ [12]*

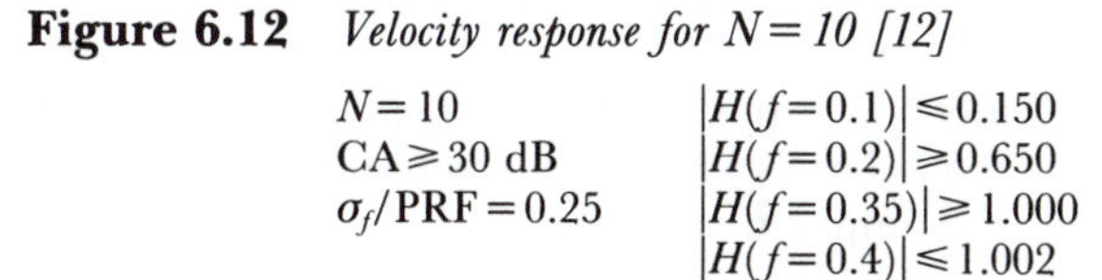

$N=10$ | $|H(f=0.1)|\leqslant 0.150$
CA $\geqslant$ 30 dB | $|H(f=0.2)|\geqslant 0.650$
$\sigma_f/\mathrm{PRF}=0.25$ | $|H(f=0.35)|\geqslant 1.000$
| $|H(f=0.4)|\leqslant 1.002$

usable bandwidth increases with the increasing of the number of weights and passband ripple.

Unfortunately, increasing the number of weights for a fixed window MTI processor has a detrimental effect on the integration gain G_i. A similar problem exists with increasing the passband ripple. Typically, better frequency responses can be obtained with an odd number of weights for narrow stopbands; however, an even value is quite effective for eliminating clutter which essentially

Table 6.2 *Initial R_p specifications consistent with $\Delta S=6$ dB [14]* © *IEEE, 1977*

N	G_i (dB)	R_p (dB)	HDB (f_m)	f_m	BW
5	9.5	7.0	−3.5	0.126	0.748
6	9.0	6.0	−3.0	0.161	0.678
8	7.8	3.6	−1.8	0.137	0.726
9	7.0	2.0	−1.0	0.103	0.793

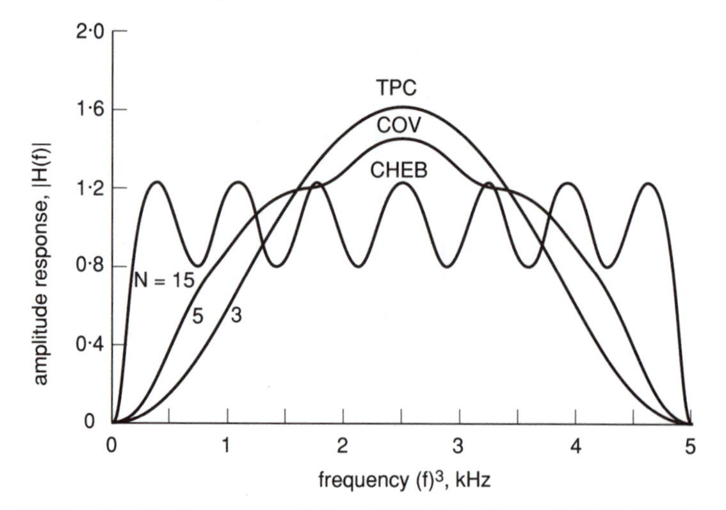

Figure 6.13 *Amplitude response of three MTI filter design [14] © IEEE, 1977*

has only a DC component. These trade offs will become more apparent in the following examples.

The effective manner in which the Chebyshev design procedure accomplished the trade off between I and BW for ground clutter filter is further illustrated in Table 6.3 where the values of BW and I, assuming $\sigma=0.001$ (5 Hz), are tabulated for the four CHEB designs just considered plus two covariance designs and the classic TPC.

It is apparent that even the five-tap CHEB effectively trades 30 bB unnecessary I for an additional 9.6% bandwidth when compared with the five-tap COV design, or 19.3% when compared with the TPC.

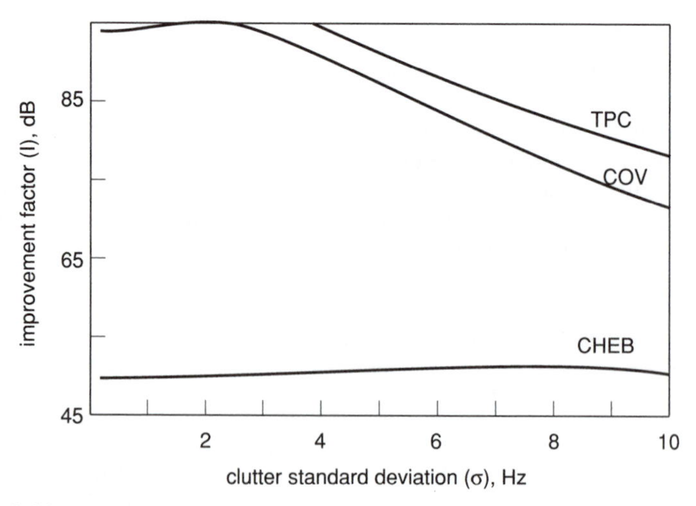

Figure 6.14 *Improvement factors versus σ_c (Hz) of the clutter (PRF = 5 kHz) [14] © IEEE, 1977*

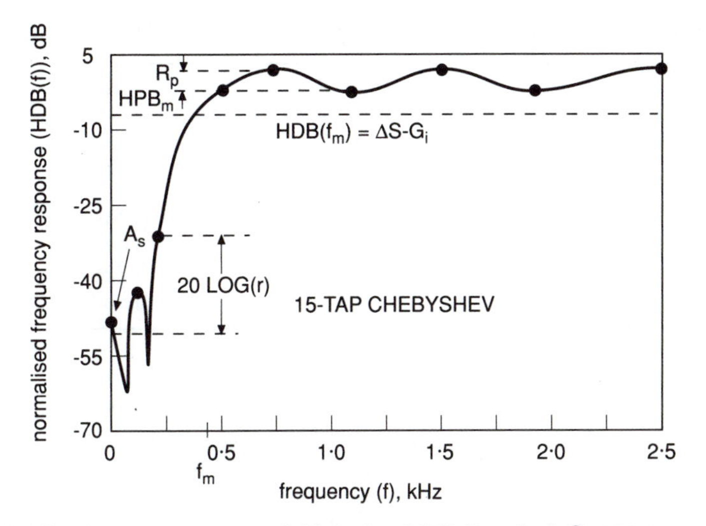

Figure 6.15 *Design parameters of Chebyshev MTI filter [14] © IEEE, 1977*
$f_S = 250$ Hz, $f_p = 502$ Hz, $A_S = 50$ dB, $R_p = 5$ dB, $r = 0.1$

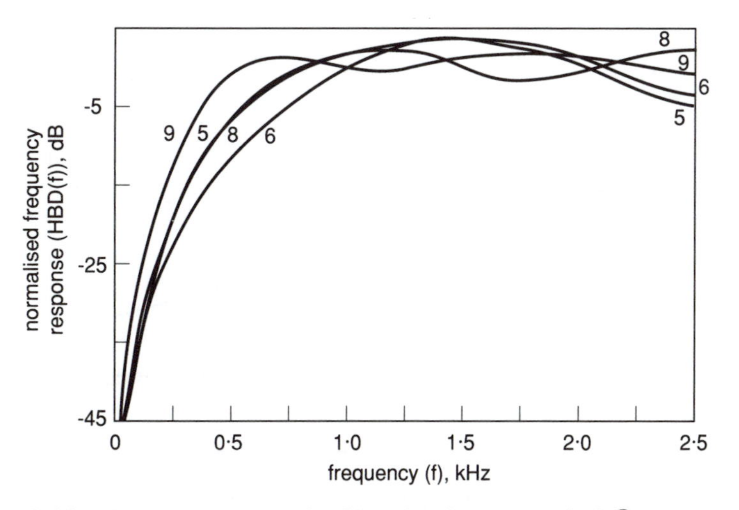

Figure 6.16 *Frequency responses for Chebyshev filter design [14] © IEEE, 1977*
$f_S = 20$ Hz

N	R_p	A_S	f_p
5	7.0	55	624
6	6.0	50	835
8	3.6	50	737
9	2.0	55	557

Table 6.3 *Comparison of design methods for ground clutter [14] © IEEE, 1977*

Design	*N*	*HDB* (f_m)	*BW*	*I*
TPC	3	−6.0	0.626	91
COV	4	−4.8	0.676	87
COV	5	−3.5	0.682	85
CHEB	5	−3.5	0.748	55
CHEB	6	−3.0	0.678	58
CHEB	8	−1.8	0.726	59
CHEB	9	−1.0	0.808	70

Fig. 6.20 shows the frequency responses of the Chebyshev filter for wideband clutter, such as weather clutter, designs. The clutter standard deviation is equal to 0.02 (100 Hz). It is assumed that $I = 20$ dB is sufficient for suppressing this type of clutter. Fig. 6.21 shows the improvement factor versus clutter standard deviation for different designs. Table 6.4 shows the comparison of the CHEB and COV designs.

This comparison is based on the same number of weights. Though the COV method has a higher than needed improvement factor, the usable bandwidths

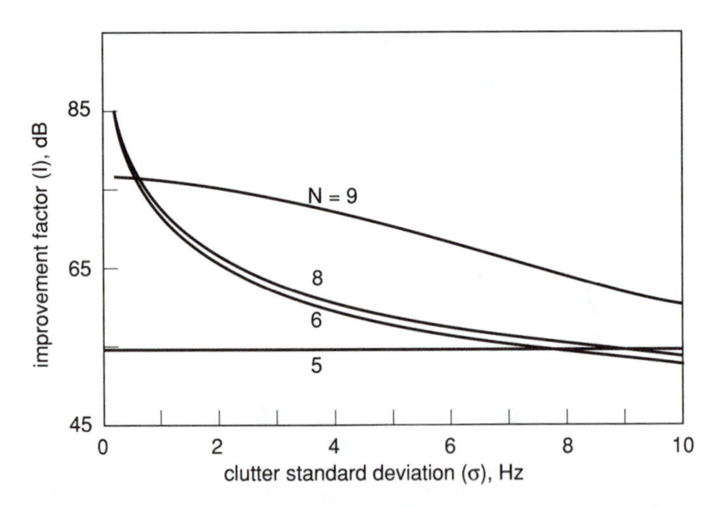

Figure 6.17 *Improvement factors versus clutter standard deviation [14] © IEEE, 1977*

N	R_p	A_S	f_p
5	7.0	55	624
6	6.0	50	835
8	3.6	50	737
9	2.0	55	557

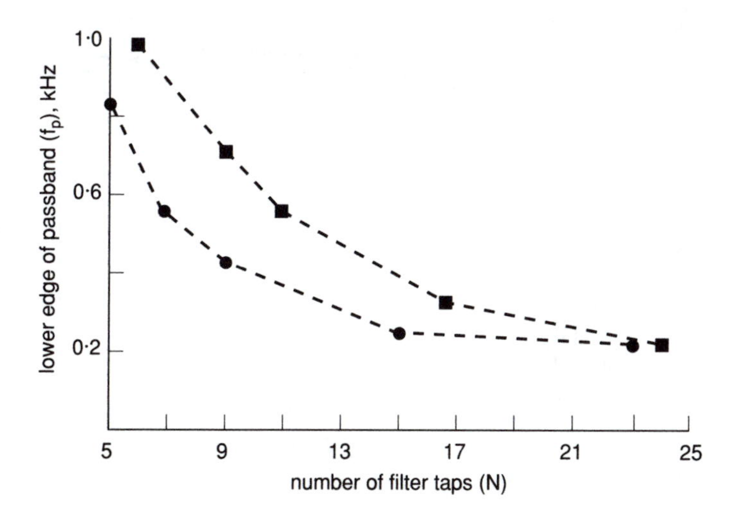

Figure 6.18 *Effect of varying number of weights on usable bandwidth [14] © IEEE, 1977*

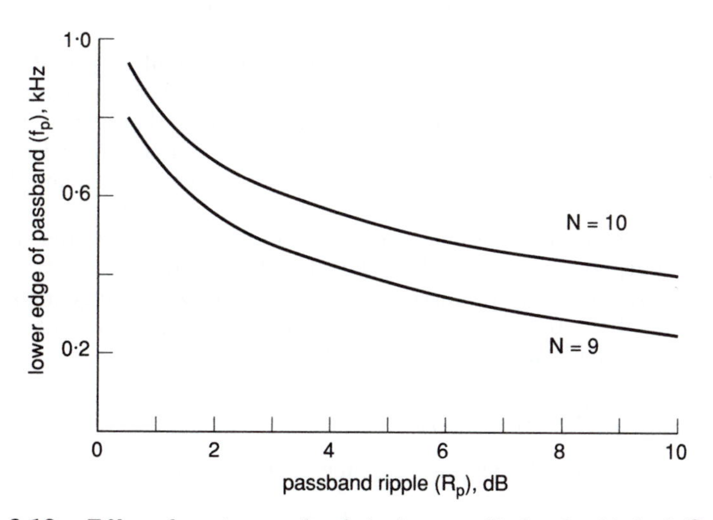

Figure 6.19 *Effect of varying passband ripple on usable bandwidth [14] © IEEE, 1977*

$f_S = 20$ Hz
$A_S = 50$ dB for $N = 10$
$A_S = 55$ dB for $N = 9$

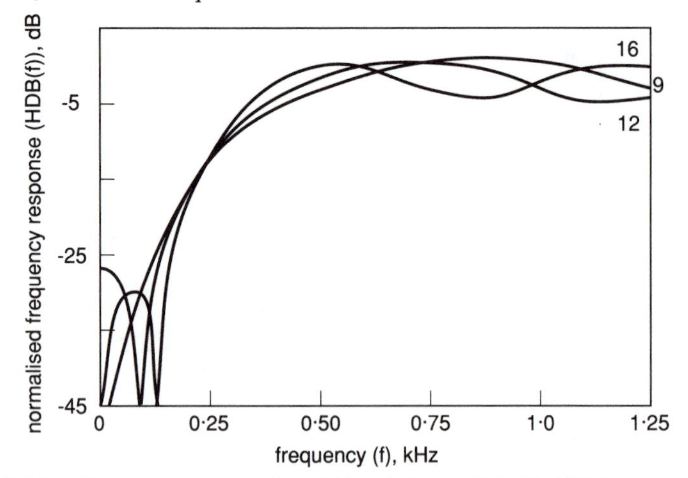

Figure 6.20 *Frequency response for wideband clutter [14] © IEEE, 1977*

N	f_S	A_S	r	f_p	R_p
9	250	30	0.1	393	7.0
12	250	30	0.1	397	5.5
16	250	30	0.1	363	4.0

are 71%, 63% and 40% narrower than that of CHEB designs for $N=9$, 12 and 16, respectively.

The Dolph–Chebyshev MTI filter is especially useful in the short range combat surveillance radar [15]. Where the standard deviation σ of bush clutter is very wide, and the velocity of the target is very slow, so that it embeds into the clutter spectrum, very sharp transition of the passband is required to achieve high improvement factor. Since the PRF of the radar is very high and the speed of antenna rotating is very low, many pulses within the dwell time can be obtained. Therefore, it is possible to use high order, such as 64-order, FIR filter to obtain sharp transition. In Reference 15, the design of the 64-order FIR complex filters has been optimised by using the Remez exchange algorithm so as to obtain about 0.01 transition band, and 27 dB improvement factor for $f_d=0.02$ slow moving target has been achieved.

6.2.3 *Staggered PRF MTI system*

It is well known that the Doppler frequency of a moving target is

$$f_d = 2v/\lambda \tag{6.57}$$

where v is the velocity of the target and λ is the wavelength of the radar. When this Doppler frequency is equal to the PRF or its multiples, 'blind speed' occurs. It is evident that there are two ways to avoid the blind speed problem: changing the PRF or changing the wavelength of the radar. It is very difficult to change

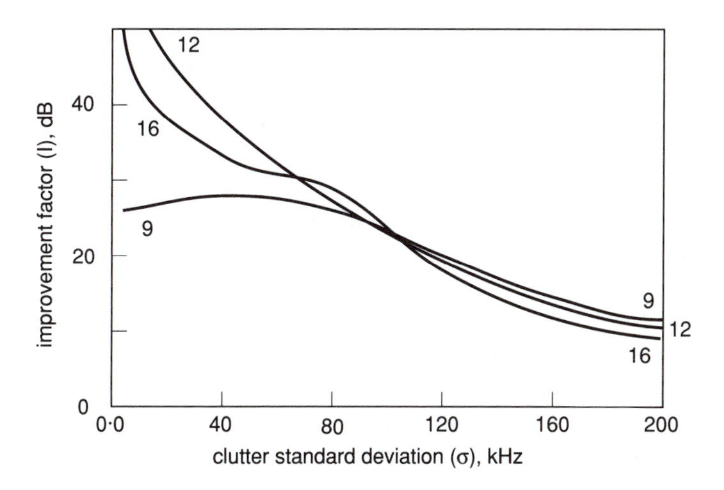

Figure 6.21 *Improvement factor versus clutter standard deviation [14] © IEEE, 1977*

N	R_p	f_p
9	7.0	393
12	5.5	397
16	4.0	363

$f_S = 250$ Hz
$A_S = 30$ dB
$r = 0.1$

the carrier frequency on a pulse-to-pulse basis for an MTI radar, since it destroys the coherence between echoes. Even though it is not impossible, it is very complex to realise, and we will discuss it later. The easier way to avoid the blind speed problem is to change the PRF of the radar on a pulse-to-pulse basis.

In earlier MTI systems, an analogue delay line is often used. Changing the PRF can be realised by adding or subtracting a short delay line with an electronic switch. With this method only very simple PRF staggering can be realised. The frequency response is not so flat, and the loss of improvement

Table 6.4 *Comparison of MTI design methods for wideband clutter [14] © IEEE, 1977*

		Chebyshev			Covariance		
N	HDB (f_m)	f_m	BW	I	f_m	BW	I
9	−4.0	0.079	0.842	23.5	0.255	0.490	94.2
12	−3.0	0.079	0.842	23.7	0.183	0.514	79.1
16	−1.8	0.073	0.854	23.5	0.196	0.609	78.7

factor is considerable. With the development of digital technique, it is now possible to design a more flexible staggered PRF MTI system.

There are three problems concerned with the design of a staggered PRF MTI system:

(i) How to choose the ratio of staggering to obtain the maximum flat frequency response?
(ii) What is the effect of staggering on the improvement factor of the MTI system?
(iii) How can we reduce the effect of the staggering and maximise the improvement factor in the staggered PRF condition?

There are some constraints for these problems.

(i) The minimum interpulse period must be greater than the time interval determined by the maximum detection range.
(ii) The first blind speed after staggering must be greater than the maximum speed of the target.
(iii) The first notch is often the deepest, but it must be smaller than some given value.

According to the first constraint, one may assume that the interpulse durations are varying between a lower and upper bound:

$$T_1 < T_s < T_N, \text{ or } T_N = T_1(1+\alpha) \tag{6.58}$$

where α represents the fractional variation of the interpulse duration. If the periods of staggered waveforms have the relationship $T_1/n_1 = T_2/n_2 = \cdots = T_N/n_N$, where $n_1, n_2, \ldots, n_N$ are integers, and if v_B is the first blind speed of a nonstaggered waveform with a constant period equal to the average period $T_{av} = (T_1 + T_2 + \cdots + T_N)/N$, then the first blind speed v_1 is

$$v_1/v_B = (n_1 + n_2 + \cdots + n_N)/N \tag{6.59}$$

The frequency response of an FIR filter for a uniform PRF MTI system is

$$H(j\omega) = \sum_{n=0}^{N-1} w_n \exp(-j\omega nT) \tag{6.60}$$

where T is the pulse repetition interval (PRI), w_n is the weights of the filter, and N is the number of order of the filter. The frequency response of a transversal filter for a staggered PRF MTI system is

$$H_s(j\omega) = \sum_{n=0}^{N-1} w_n(\exp -j\omega T_n) \tag{6.61}$$

where T_n is the transmitting time of nth pulse. If these T_n are given, we can calculate the frequency response of the filter in staggered mode. However, it is very difficult to synthesise the filter with the tolerated depth of notches in the whole passband. The value of T_n not only affects the frequency response of the MTI system, but also affects the improvement factor of the MTI system.

The average improvement factor of an MTI system is

$$\bar{I} = \bar{G}\,\frac{C_i}{C_o} \tag{6.4}$$

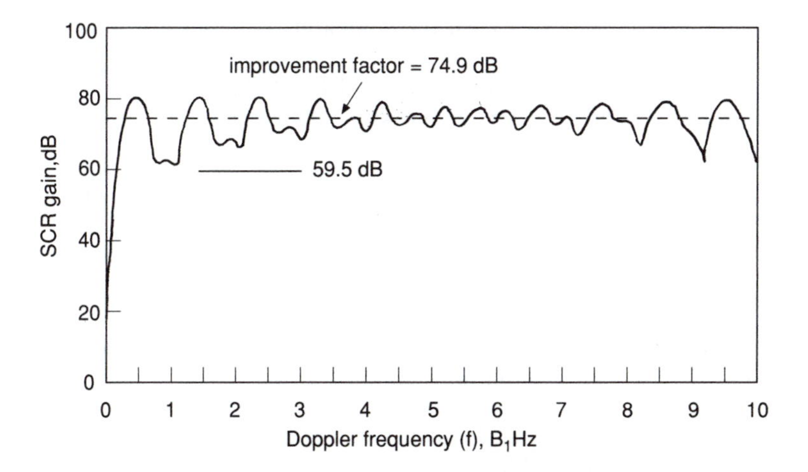

Figure 6.22 *SCR gain versus $f_d(B_1Hz)$ [16] © IEEE, 1973*

Optimum PRF ratio: 1.111, 1.000, 1.091, 1.058
Optimum weights: 1.000, −4.258, 6.242, −4.031, 1.048
$N=5$
$B_u=10B_1$
$\sigma=0.02B_1$

where $\bar{G}$ is the average gain of the MTI filter. Approximately, it is not affected by the staggering. C_i is also not affected by staggering. C_o is mainly determined by the shape (frequency response) of the filter in the stopband (null). As the PRF is staggered, the null of the response becomes narrower compared with that of uniform PRF. This distorted null makes the improvement factor for staggered PRF I_s smaller than that for uniform PRF.

There are several ways to solve these problems:

(i) *Prinsen [16]:* has developed a computer program that combines random search and the gradient method to produce points that combine a high value of improvement factor with good blind speeds performance. At random, a starting point is generated for the gradient algorithm to start. As soon as an optimum has been found, another starting point is generated at random. Each time the new optimum is compared with the best optimum thus far. This process is stopped as soon as the computing time exceeds a preset value.

According to the gradient algorithm, the gradient must be evaluated to find a direction in which a one-dimensional search is done. The gradient is approximated by differences. At the point that results from the one-dimensional search the gradient is computed again and the procedure is repeated. This process is continued until the improvement obtained by the one-dimensional search is less than a preset value.

Prinsen [16] used the golden section ratio, which is equal to 1.61, to determine the starting interval of an optimisation procedure. Some numerical results were obtained. Two examples for $N=5$ and $N=6$ are shown in Figs 6.22 and 6.23. It is assumed that the upper of the velocity region B_u is equal to ten times the first blind speed B_1, and $\sigma=0.02\ B_1$.

The effects of staggered PRF on the stopband responses are shown in Fig. 6.24 for $N=5$ and $\sigma=0.02$.

Fig. 6.25 shows the improvement factor of Fig. 6.22 as a function of clutter spectral width σ. Fig. 6.26 shows the improvement factor of Fig. 6.22 as a function of stagger rate.

(ii) *Roy and Lowenschuss [17]:* developed an interference pattern approach to select the staggering rate of simple MTI filters. Thomas and Abram [18] modified the interference pattern to provide a near optimum design.

The transmission times of a staggered PRF radar vary in a cyclic manner, so that if n periods T_i are employed, the transmission times t_i will be as shown in Fig. 6.27.

The output sequence of an m-order digital filter with n-period stagger can be representd by n sequences uniformly sampled at the rate $(1/T_s)$ of the overall stagger cycle. The output sample at time t_r is computed from the weighted sum of the past input samples, as indicated diagrammatically in Fig. 6.27. The frequency response of the filter may be obtained by considering the output sequences resulting from a sinusoidal input $s(t)=\exp(j\omega t)$. For a sinusoidal input, each sequence $C_r(kT_s+t_r)$, $(k=0, 1, 2, \ldots)$ may be written as

$$C_r(kT_s+t_r)=\{w_0+w_1\exp[-j\omega(T_r-1)]+w_2\exp[-j\omega(T_{r-1}+T_{r-2})]+w_3\exp[-j\omega(T_{r-1}+T_{r-2}+T_{r-3})]+\ldots\}\exp[j\omega(kT_s+t_r)] \quad (6.62)$$

Each $C_r(kT_s+t_r)$ may thus be interpreted as samples of a sinusoid with complex magnitude $C_r(j\omega)$, where

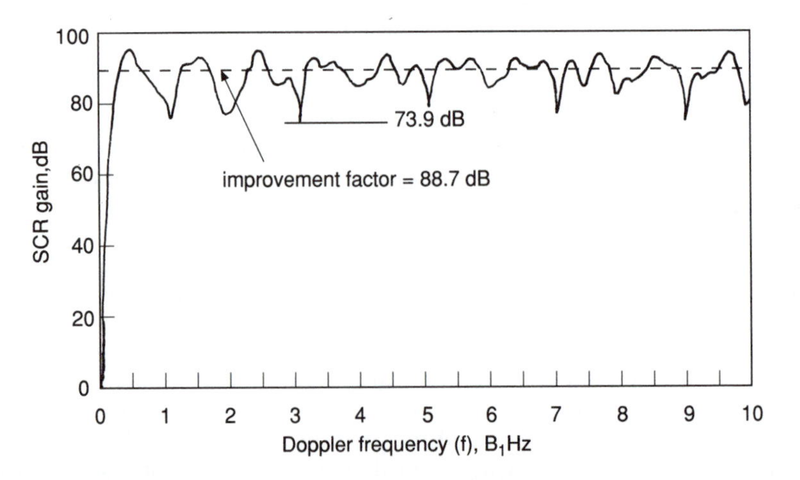

Figure 6.23 *SCR gain versus $f_d(B_1Hz)$ [16] © IEEE, 1973*

Optimum PRF ratio: 1.006, 1.489, 1.000, 1.084, 1.152
Optimum weights: 1.000, −3.533, 8.670, −10.327, 5.213, −1.022
$N=6$
$B_u=10B_1$
$\sigma=0.02B_1$

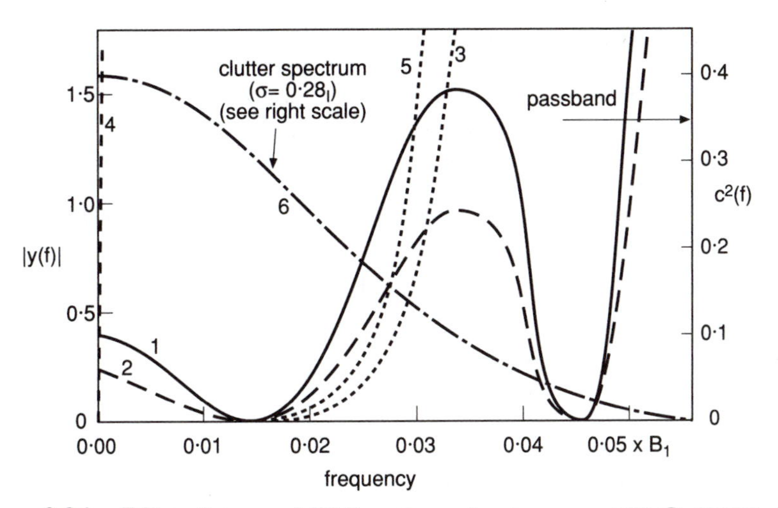

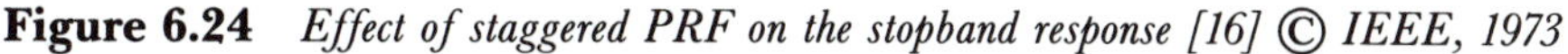

Figure 6.24 *Effect of staggered PRF on the stopband response [16] © IEEE, 1973*

Curve 1: Optimum filter, staggered PRF, I = 74.9 dB (Fig. 6.22)
Curve 2: Optimum filter, uniform PRF, I = 76.8 dB
Curve 3: Binomial filter, uniform PRF, I = 70.5 dB
Curve 4: Binomial filter, staggered PRF, I = 46.3 dB
Curve 5: Vandermonde filter, staggered PRF, I = 68.6 dB
Curve 6: Clutter power spectrum

$$C_r(j\omega) = \sum_{i=0}^{m} w_i \exp[-j\omega(t_r - t_{r-i})]$$

The MTI filter gain at a frequency ω, defined as the mean-square value of its output samples for a sinusoidal input, is found by summing the mean-square values of each envelope $C_r(j\omega)\exp(j\omega t)$. Then

$$|G(\omega)|^2 = \sum_{r=1}^{n} |C_r(j\omega)|^2 \qquad (6.63)$$

If ΔT is now defined as the highest common factor of each T_i, then $|G(\omega)|^2$ has its first true blind speed at $\omega_0 = 2\pi/\Delta T$.

For a 1st-order filter, eqn. 6.63 becomes

$$|G(\omega)|^2 = \sum_{r=1}^{n} |1 - \exp(-j\omega T_r)|^2 = 2\sum_{r=1}^{n} (1 - \cos \omega T_r) \qquad (6.64)$$

If each T_r is part of an arithmetic progression T_{min}, $T_{min} + \Delta T$, $T_{min} + 2\Delta T, \ldots, T_{max}$, then the summation of eqn. (6.64) may be written as

$$|G(\omega)|^2 = n\left[1 - \frac{\sin(n\omega\Delta T/2)}{n \sin(\omega\Delta T/2)} \cos\frac{\omega}{2}(T_{min} + T_{max})\right] \qquad (6.65)$$

The second part of this expression may be recognised as an envelope $\sin(n\omega\Delta T/2)/n\ \sin(\omega\Delta T/2)$, modulating a 'carrier' of $\cos(\omega T_{av})$, as shown in Fig. 6.28 for $n=5$. This is called frequency interference pattern.

Such an envelope gives readily predictable ripple levels, and Table 6.5 gives the depth of the first (deepest) null for various combinations of n and expansion ratios (defined as $T_{av}/\Delta T$).

For higher-order filters, eqn. 6.63 may be expanded into the general form

$$|G(\omega)| = A_0 + A_1 \sum_{i=1}^{n} \cos\ \omega T_i + A_2 \sum_{i=1}^{n} \cos\ \omega(T_i + T_{i-1}) + \cdots$$

$$+ A_m \sum_{i=1}^{n} \cos\ \omega(T_i + T_{i+1} + \cdots + T_{i-m+1}) \tag{6.66}$$

where m is the filter order.

The normalised coefficients $A_0, A_1, \ldots, A_m$ are given in Table 6.6 for $m=1$ to 4.

It is seen that additional weighted interference patterns may be formed with the periods taken two-at-a-time, three-at-a-time, etc., and nulls in the passband will result if these higher order interference patterns have true blind speeds within the passband. For a 2nd-order filter, this may be avoided by rearranging the stagger periods so that the sums of successive periods differ by ΔT (e.g. 15 ΔT, 18 ΔT, 16 ΔT, 19 ΔT, 17 ΔT, 20 ΔT), but for a 3rd-order filter, only

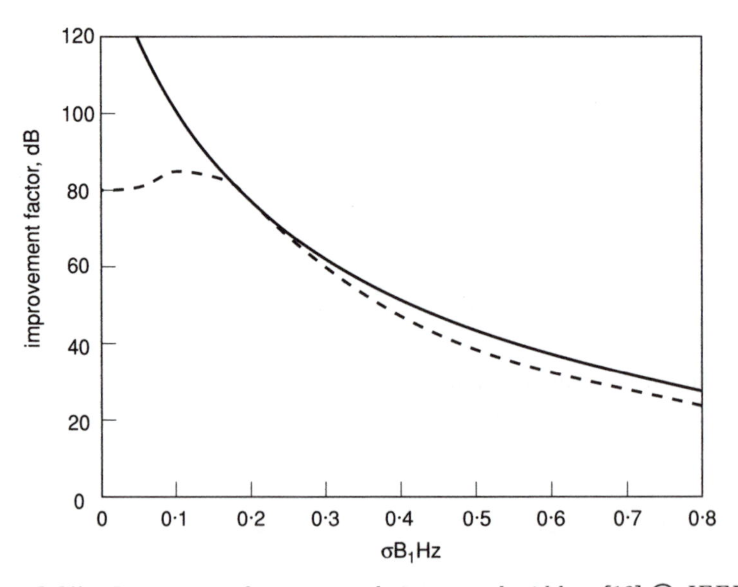

Figure 6.25 *Improvement factor versus clutter spectral width σ [16] © IEEE, 1973*

—— coefficients matched to σ interpulse periods matched to $\sigma=0.2B_1$

– – – both coefficients and interpulse periods matched to $\sigma=0.2B_1$

$N=5$

$B_u=10B_1$

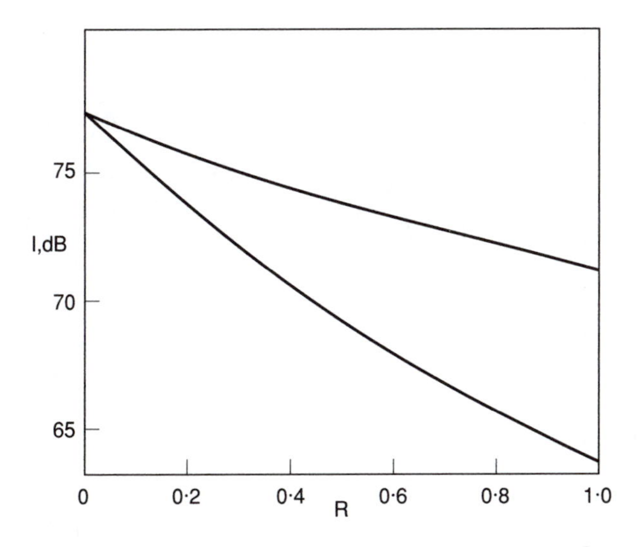

Figure 6.26 *Improvement factor as a function of stagger rate [16] © IEEE, 1973*
$R=|(T_1, T_2, \ldots T_{N-1})-(I, I \ldots I)|$
$N=5$
$B_u=10B_1$
$\sigma=0.02B_1$

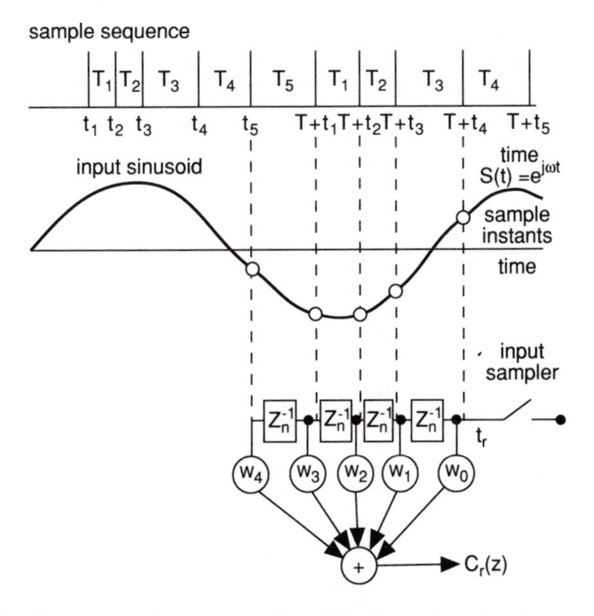

Figure 6.27 *Diagrammatic representation of the formation of each output sequence $C_r(z)$ occurring at time t_r [18]*

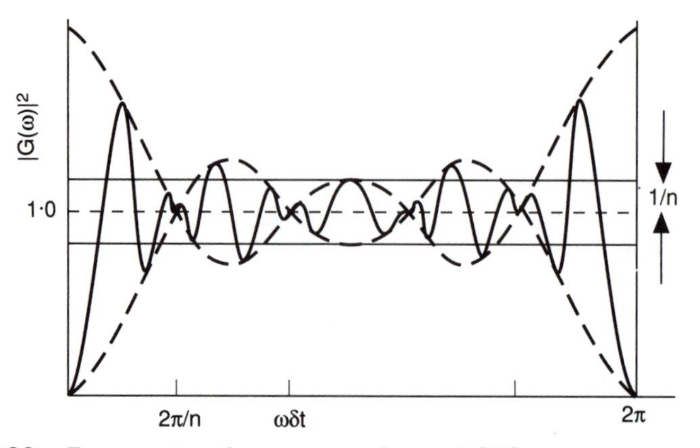

Figure 6.28 *Frequency interference pattern for n=5 [18]*

the sequence length is available as a control on the 3rd-order pattern. A further practical recommendation for such a stagger arrangement is that the alternation of long and short periods maintains an almost constant transmitter duty cycle and avoids low-frequency power-supply ripple.

The ratio of T_{max}/T_{min} required for an acceptable passband ripple may be achieved, for high expansion ratios, by selecting a stagger pattern with a highest common factor of 2 ΔT. Accordingly, the basic response has a true null at $\pi/\Delta T$. Now, if any period T_i is changed to T_i', then, for a 1st-order filter, a term $(\cos \omega T_i - \cos \omega T_i') = 2 \sin \omega[(T_i' - T_i)/2] \sin \omega[(T_i' + T_i)/2]$ is effectively added to the basic response. If $(T_i' - T_i) = \Delta T$, then this envelope extends to $\omega_0 = 2\pi/\Delta T$, the required first blind speed, and the null at $\pi/\Delta T$ is partially filled by a positive peak of the added term. In addition, by careful choice of the stagger period to be changed, it is possible to improve the first null of the original response. Such a technique may be illustrated most clearly by an example.

Table 6.5 *Approximate depth of first null (dB) [18]*

	Expansion ratio $T_{av}/\Delta T$								
n	8	10	12	14	16	18	20	25	30
4	4.6	6.3	7.9	9.2	10.3	11.3	12.2	14.1	15.7
5	2.9	4.5	6.0	7.2	8.3	9.3	10.2	12.1	13.7
6	1.6	3.1	4.5	5.7	6.8	7.8	8.6	10.5	12.1
7	0.7	2.0	3.3	4.5	5.5	6.5	7.3	9.2	10.7
8	0	1.2	2.4	3.5	4.5	5.4	6.2	8.1	9.6
9	—	0.5	1.6	2.6	3.6	4.5	5.3	7.1	8.6
10	—	0	1.0	1.9	2.8	3.7	4.5	6.2	7.7
11	—	—	0.5	1.3	2.2	3.0	3.8	5.5	6.9
12	—	—	0	0.8	1.6	2.4	3.1	4.8	6.2
15	—	—	—	—	0.3	1.0	1.6	3.1	4.5

Table 6.6 *Normalised coefficients for filters of order 1 to 4 [18]*

m	A_0	A_1	A_2	A_3	A_4
1	1	−1	—	—	—
2	1	−1.33	+0.33	—	—
3	1	−1.5	+0.60	−0.1	—
4	1	−1.6	+0.80	+0.23	+0.03

The sequence (12 ΔT, 16 ΔT, 14 ΔT, 18 ΔT) has the response shown in Fig. 6.29*a*. The effect of changing 14 ΔT to 13 ΔT or 15 ΔT is shown in Fig. 6.29*b* and 6.29*c*. Both fill the null at $\pi/\Delta T$ equally but a change to 13 ΔT gives a slight improvement to the first null of the original pattern.

A similar approach may be successfully applied to five and six period stagger sequences with differences of 2 ΔT or 3 ΔT.

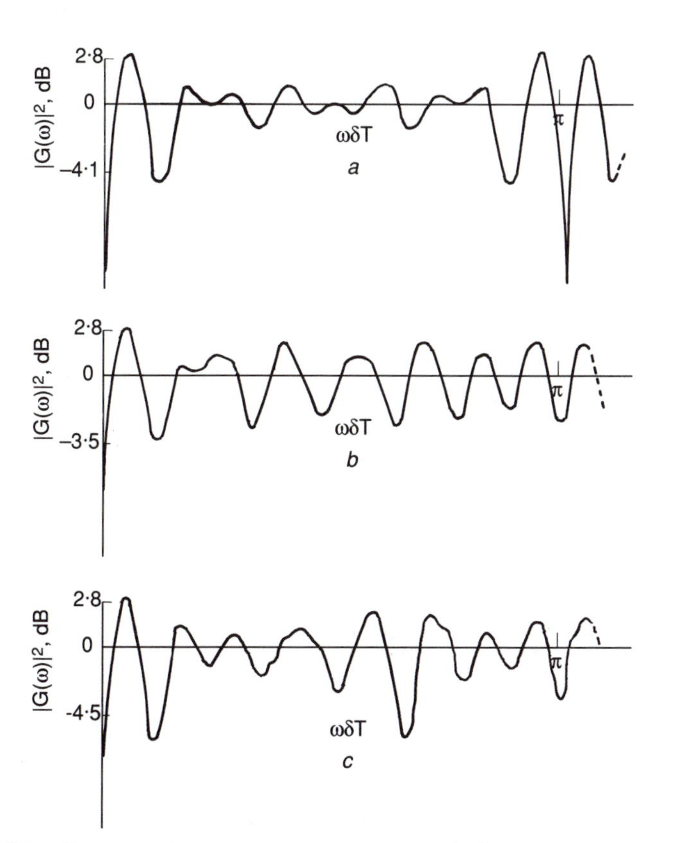

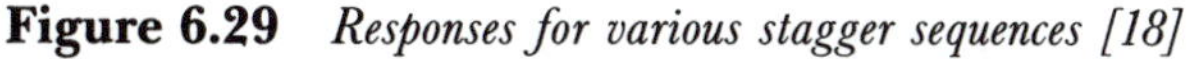

Figure 6.29 *Responses for various stagger sequences [18]*

(*a*) $12\Delta T$, $16\Delta T$, $14\Delta T$, $18\Delta T$
(*b*) $12\Delta T$, $16\Delta T$, $13\Delta T$, $18\Delta T$
(*c*) $12\Delta T$, $16\Delta T$, $15\Delta T$, $18\Delta T$

A disadvantage of this method of stagger selection is its *ad hoc* nature; there is no guarantee that the responses so obtained are in any way optimum. However, by its systematic nature, careful application of the method ensures a satisfactory response and major nulls may be readily avoided.

Although stagger patterns based on multiples of ΔT have a true null at $\omega_0 = 2\pi/\Delta T$, it seems reasonable to expect that the optimum stagger pattern need not be restricted to such integral periods. However, the previously described methods cannot apply to this case. By defining a suitable performance index, it should be possible to apply standard computer optimisation techniques. The difficulties with such optimisation methods are that, for the complicated expression describing the MTI response, there exist numerous local minima, giving no guarantee of a global optimum.

(iii) *Time varying weights:* It has been shown that the velocity response of an MTI filter can be improved by staggered PRF. However, the improvement factor will be degraded by this staggering. This can be compensated by the time varying weights. The use of time varying weights has no appreciable effect on the MTI velocity response curve. Whether the added complexity of utilising time varying weights is desirable depends on whether the stagger limitation is predominant. For a 2nd-order filter, the stagger limitation is often comparable to the basic canceller capability without staggering. For a 3rd-order filter, the stagger limitation usually predominates. When wide notch is desired to reject weather clutter, the staggering of PRF will narrow the notch. In this case, time varying weights are necessary.

The problem is how to vary the weights to obtain optimum improvement factor I.

To avoid reduction of I, we must design a new set of weights w'_n for staggered MTI. The new frequency response is

$$H_v(\omega) = \sum_{n=0}^{N-1} w'_n \exp(-j\omega T_n) \tag{6.67}$$

which has the properties

$$P'_{co} = \int_{-\infty}^{\infty} C(\omega)|H_v(\omega)|^2 d\omega \simeq P_{co} = \int_{-\infty}^{\infty} C(\omega)|H_0(\omega)|^2 d\omega \tag{6.68}$$

where $C(\omega)$ is the spectrum of the clutter, and P_{co} and P'_{co} are the output clutter power before and after readjusting the weights, respectively. In general, the clutter spectrum width σ is smaller than the PRF; thus eqn. 6.68 becomes

$$P'_{co} \simeq \int_{\omega_0-\Delta\omega}^{\omega_0+\Delta\omega} C(\omega)|H_v(\omega)|^2 d\omega \simeq \int_{\omega_0-\Delta\omega}^{\omega_0+\Delta\omega} C(\omega)|H_0(\omega)|^2 d\omega \tag{6.69}$$

where ω_0 is the centre frequency of the clutter, and $\Delta\omega = 2\pi$ (PRF/2). Therefore, the weights readjusting method is such a procedure which leads to

$$|H_0(\omega)|^2 \simeq |H_v(\omega)|^2, \; |\omega - \omega_0| \leqq \Delta\omega \tag{6.70}$$

Prinsen [19] proposed an algorithm, which is termed a Vandermonde filter. The transfer function of an FIR filter for a nonuniformly sampled signal is (see also eqn. 6.61)

$$H_s(f)=\sum_{n=0}^{N-1} w_n \exp(j2\pi/t_n) \tag{6.71}$$

where $t_0, t_1, \ldots, t_{N-1}$ are the sample moments with $t_0=0$. The Taylor series expansion of $H_s(f)$ around $f=0$ is

$$H_s(f)=H_s(0)+\frac{H_s'(0)}{1!}f+\frac{H_s''(0)}{2!}f^2+\ldots \tag{6.72}$$

with

$$H_s^{(k)}(0)=(j2\pi)^k\sum_{n=0}^{N-1} t_n^k w_n,\ k=0, 1, 2, \ldots \tag{6.73}$$

Now we develop a class of filters with maximally flat stopband by requiring that the first M coefficients $H_s^{(k)}$ (0), $k=0, 1, \ldots, M-1$ in eqn. 6.72 be zero.

Then the remaining series has a leading term which is proportional to f^M. If M is maximally flat around $f=0$, each coefficient, being set to zero, yields a homogeneous linear equation in $\{w_n\}$. As there are N variables w_n, we can form a system of maximally $N-1$ linearly independent equations with a non-trivial solution, so $M=N-1$. This yields (using eqn. 6.73)

$$\sum_{n=0}^{N-1} t_n^k w_n=0,\ k=0, 1, \ldots, N-2 \tag{6.74}$$

Since $t_0=0$, only the first equation of the system in eqn. 6.74 has a term containing w_0. Then we have

$$\begin{aligned} w_1+w_2+\cdots+w_{N-1}&=-w_0\\ t_1w_1+t_2w_2+\cdots+t_{N-1}w_{N-1}&=0\\ \cdots\cdots\cdots\cdots\cdots\cdots&\cdots\\ t_1^{N-2}w_1+t_2^{N-2}w_2+\cdots+t_{N-1}^{N-2}w_{N-1}&=0 \end{aligned} \tag{6.75}$$

or

$$Tw=-w_0u_1 \tag{6.76}$$

with

$$w'=(w_1, w_2, \ldots, w_{N-1})$$
$$u_1'=(1, 0, \ldots, 0)$$

and

$$T=\begin{bmatrix} 1, & 1, & \ldots, & 1\\ t_1, & t_2, & \ldots, & t_{N-1}\\ \cdots & \cdots & \cdots & \cdots\\ t_1^{N-2}, & t_2^{N-2}, & \ldots, & t_{N-1}^{N-2} \end{bmatrix} \tag{6.77}$$

The system 6.76 can be solved by inverting the matrix T if T^{-1} exists:

$$w=-w_0T^{-1}u_1 \tag{6.78}$$

Table 6.7 *6-pulse MTI in weather clutter*

	Ground clutter	Weather clutter	
Clutter width σ_c	4 Hz	13 Hz	13 Hz
Clutter centre f_0	0 Hz	±50 Hz	±100 Hz
IF for uniform PRF	64.6 dB	45.2 dB	26.5 dB
IF for staggered PRF	54.6 dB	34.5 dB	21.5 dB
IF for time varying w_n	64.7 dB	44.6 dB	25.8 dB

The matrix is a Vandermonde matrix and T^{-1} exists since $t_i \neq t_j$ for $i \neq j$. One example of a Vandermonde filter is shown in Fig. 6.24.

This method of readjusting the weights of the filter is suitable for the case of only one null in the frequency response. In Reference 20, an example is given of a 3-pulse MTI system. The improvement factor for uniform PRF is equal to 41.0 dB. After taking 23:18 stagger, the improvement factor is reduced to 36.1 dB. If the weights of the filter are readjusted, the improvement factor will be increased to 41.1 dB.

In the case of multi-nulls, which means suppression of multi-clutter, the optimisation approach can be used for design of the time varying weights [20]. However, the optimisation approach is a more general design method at the cost of more computational time.

Set the destination function

$$f(W) = \Sigma[|H_v(\omega_k)|^2 - |H_0(\omega_k)|^2]^2 S_k \tag{6.79}$$

where ω_k is a set of frequency values, s_k is the weights of $f(W)$, $s_k = 1/|H_0(\omega_k)|^4$, $(|H_0(\omega_k)| > 10^{-4})$; or $s_k = 10^{-8}$, $(|H_0(\omega_k)| < 10^{-4})$.

By an optimisation approach, e.g. DPF Newton's method, to minimize $f(W)$, we can obtain the time varying weights, which satisfy eqn. 6.70. It can be shown that the analytical expression of the gradient $\nabla f(W)$ and Hessiann matrix $\nabla^2 f(W)$ exist, so it is easy to program and converge very fast.

An example of this method is shown in Reference 20. In this example, a 6-pulse MTI worked in the environment of weather clutter (clutter width $\sigma_c = 13$ Hz, clutter centre f_0 varied from -100 Hz to $+100$ Hz) and ground clutter (clutter width $\sigma_c = 4$ Hz). The PRF of the radar is equal to 657 Hz, and the stagger ratios are 14, 16, 17, 19, 16. The results are shown in Table 6.7.

Hsiao [9] pointed out that the optimum filter, which is based on the eigenvector associated with the minimùm eigenvalue, can also be applied to the stagger PRF case. The only difference between a stagger PRF MTI and a constant PRF MTI is the difference between the elements of the covariance matrix. He also pointed out that, if the interpulse duration of a stagger PRF MTI is varying between a lower and upper bound randomly,

$$T_1 < T_s < T_2, \; T_2 = T_1(1+\alpha) \tag{6.80}$$

where α represents the fractional variation of the interpulse duration, then

$$\lambda_{min}(T_1) \leqq \lambda_{min}(T_s) \leqq \lambda_{min}(T_2) \qquad (6.81)$$

Therefore, the improvement factor of a staggered PRF MTI will lie within the improvement factor of two constant PRF MTI with lower and upper bound interpulse periods. An example of a triple canceller is shown in Fig. 6.30 [9].

Yan [20] extended the eigenvector method to the multiple clutter situation. Consider a coherent pulse radar which transmits a train of N pulses in a given direction at the instants T_n. Let the clutter spectrum be

$$C(f) = C_1(f) + C_2(f) + \cdots + C_m(f) \qquad (6.82)$$

where $C_n(f)$ is a type of clutter power spectrum, such as land, weather, chaff etc. Therefore, the clutter covariance matrix is $\boldsymbol{C} = [C_{ij}]$, where

$$C_{ij} = \frac{1}{2\pi} \int_{-\infty}^{\infty} C(f) \exp[-j2\pi(T_i - T_j)f]df \qquad (6.83)$$

Let the power spectrum of the target signal be denoted by $S(f)$, and the signal covariance matrix $\boldsymbol{S} = [S_{ij}]$, where

$$S_{ij} = \frac{1}{2\pi} \int_{-\infty}^{\infty} S(f) \exp[-j2\pi(T_i - T_j)f]df \qquad (6.84)$$

It can be shown that the improvement factor of an FIR MTI filter is given by

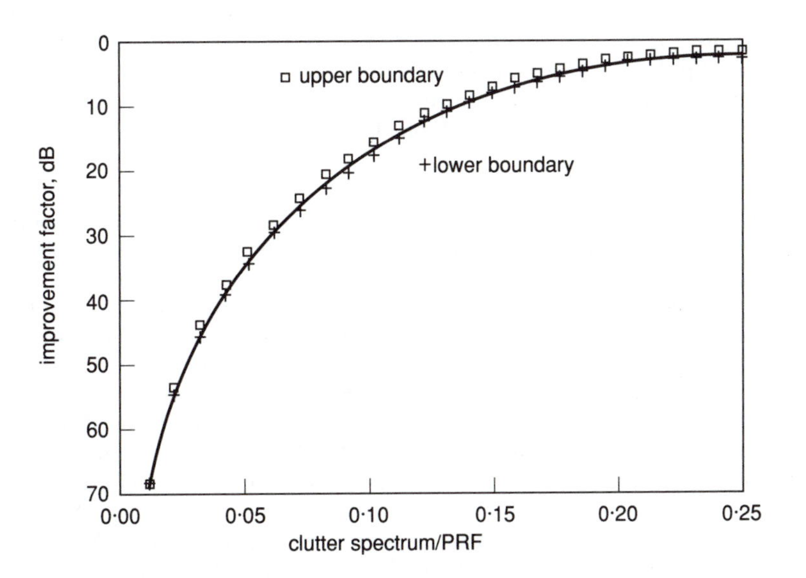

Figure 6.30 *Improvement factor of a staggered PRF MTI with α = 10% [9] © IEEE, 1974*

$$I = \frac{\boldsymbol{W}^T \boldsymbol{S} \boldsymbol{W}^*}{\boldsymbol{W}^T \boldsymbol{C} \boldsymbol{W}^*} \tag{6.85}$$

The problem of design optimised $\boldsymbol{W}$ becomes to find λ_{min} and the associated eigenvector $\boldsymbol{W}$ from the equation $\lambda \boldsymbol{C}\boldsymbol{W}^* = \boldsymbol{S}\boldsymbol{W}^*$. Since $\boldsymbol{C}$ and $\boldsymbol{S}$ are Hermitian of a positive definite matrix, so time varying weights vector $\boldsymbol{W}$ exists.

(*iv*) *Block staggered PRF*: If the number of pulses is large enough during the dwell time of the beamwidth, block-to-block stagger can be used instead of pulse-to-pulse stagger. Block stagger implies that the total number of pulse N_t transmitted during the dwell time are subdivided into N_b blocks of N_p pulses each. Each block has its own PRF. The composite velocity response consisting of the N_b individual block responses has the first true blind velocity when

$$v_b = M(j) v_b(j) \quad j = 1, 2, \ldots, N_b \tag{6.86}$$

holds for all N_b blocks and $M(j)$ is a unique integer for each block.

In an ECM environment, frequency agility is often used to avoid jamming. In this case, however, frequency agility will destroy the coherence between echoes of adjacent sweeps. If clutter rejection capability is still desired, block-to-block frequency jumping can be employed as a compromise.

Blind speed is also a function of carrier frequency of the radar as mentioned before, so the block-to-block frequency agility can also eliminate the blind speed within the required passband. However, the operating frequency of frequency agility radar is often selected within the weakest point of the spectrum of the jamming activity. Therefore, sometimes we combine block stagger with block frequency agility to eliminate the blind speeds.

Houts [21] studied this problem thoroughly. A 3-pulse canceller is used. $N_t = 9$, so there are three blocks with interpulse spacing of 200, 225 and 250 μs, respectively. The improvement factor I of block stagger is about 20 dB higher than that of pulse-to-pulse stagger, and the velocity response is acceptable (Fig. 6.31). When 5% frequency agility between block, i.e., $f_t = 5.00$, 5.25 and 5.50 GHz, respectively, is used with block stagger, a 30 dB null now appears in the passband.

(*v*) *Discussion*: Hansen [22] made a comparison of improvement factor between optimum MTI and MTI with binomial coefficients. The results are shown in Fig. 6.32.

It can be seen from this Figure that the improvement factor of an optimum MTI filter, whose coefficients are varying with the clutter spectral width σ, has few decibels better than an MTI filter with fixed binomial coefficients. This conclusion is true only for the constant PRF case. However, most advanced radar systems use staggered PRF to eliminate the blind speed to obtain continuous tracks. In the case of staggered PRF, the improvement factor of an MTI with binomial coefficients degrades more seriously than that of the optimum MTI.

Galati [23] made a comparison between these two types of MTI system under the condition of staggered PRF. The results are shown in Table 6.8. The MTI system is a triple canceller. Two types of staggering are used: sequence A with ±11% PRT variation; sequence B with ±14% variation. Both sequences

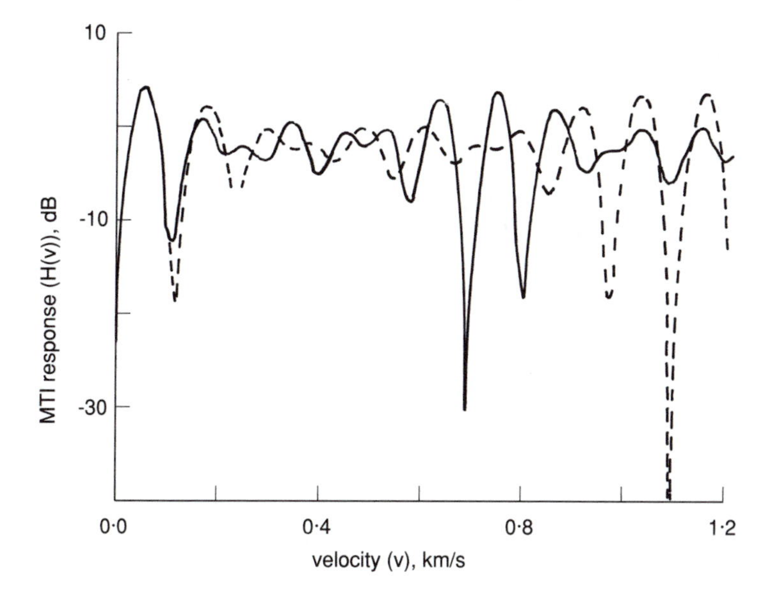

Figure 6.31 *Velocity response of block stagger with and without frequency agility [21]*

--- 0% agility
—— 5% agility

have six periods with the largest possible variation from one period to the successive one. The results of two sub-optimal time varying weights are also listed in this table.

It can be seen from this Table that, in the case of staggered PRF, the improvement factor of an optimum MTI is 16–30 dB better than that of an MTI with fixed binomial coefficients.

Until now, there have been many ways to select the stagger ratio and to determine the weighting coefficients. It is very difficult to say which is the best. Houts [21] calculated the improvement factors and frequency responses with different methods for the same radar. The radar is a phased-array system operating at C band ($f_t = 5.5$ GHz) with a specified maximum unambiguous range of 30 km ($\Delta T_m \geqq 0.2$ ms), and a requirement to detect all targets moving at speeds of up to Mach 3 ($v_b \geqq 1$ km/s). The first blind speed is located at 1.09 km/s, corresponding to interpulse spacings which differ by multiples of 25 μs. The relative improvements are listed in Table 6.9 for a variety of σ values.

The velocity responses are compared in Table 6.10. The response at the worst passband null (N_w) is tabulated along with the average signal gain $\bar{S}$, and the percentages of points >0 dB (P_+) and <−4 dB (P_-).

It can be seen from these Tables that, from the viewpoint of improvement factor, excluding the block stagger, the best one is Prinsen's method. However, it is impossible for the improvement factor of Capon's method to be worse than that of the binomial filter, since it is optimised for every value of σ. The best velocity response is the H&B (MW) method.

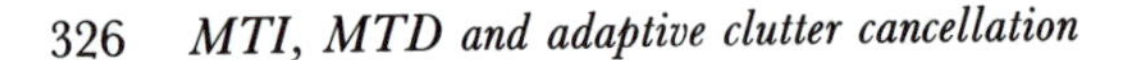

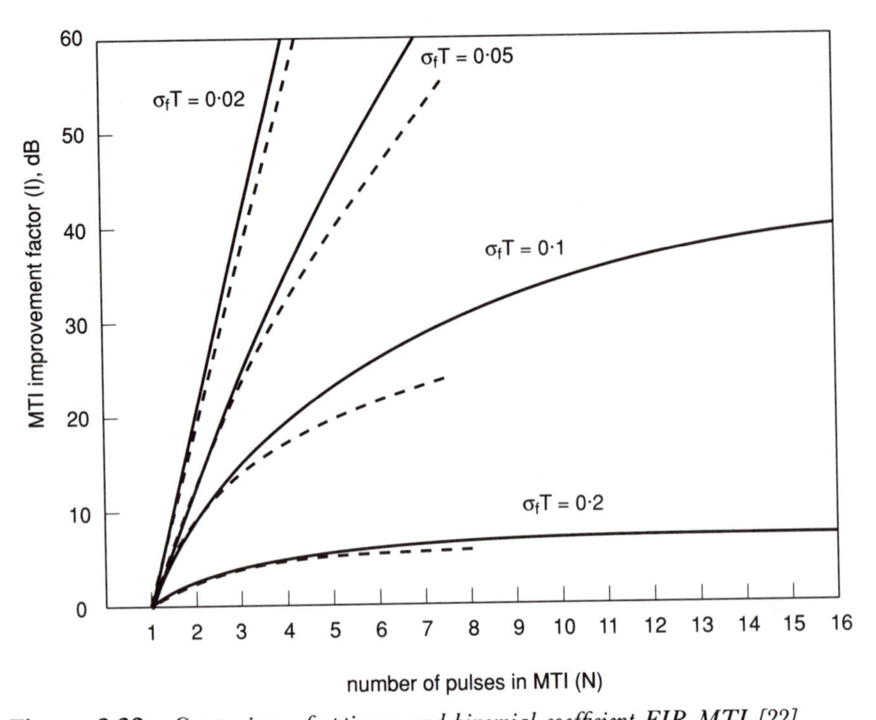

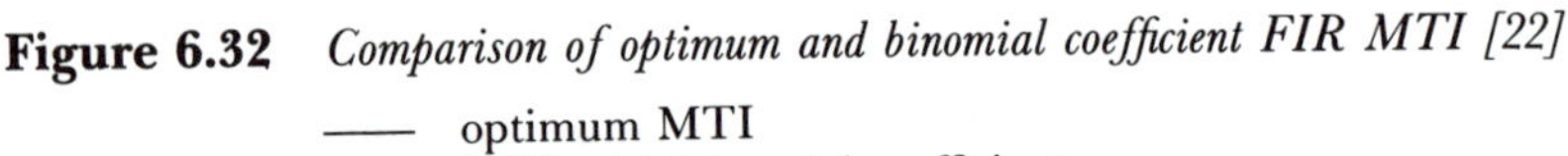

Figure 6.32 *Comparison of optimum and binomial coefficient FIR MTI [22]*

—— optimum MTI
--- MTI with binomial coefficients

6.2.4 *Cancellation of bimodal clutter*

It is very often in a real clutter environment that more than one clutter exists in the same range interval and the same azimuth interval. These clutters have different Doppler frequencies. They appear in the power spectral density as many peaks. The most frequent case is bimodal clutter, such as ground clutter and weather clutter, or sea clutter and weather clutter, etc. We will discuss the simplest, in which one clutter is fixed ground clutter with zero Doppler

Table 6.8 *Improvement factor comparison [23]*

	Uniform PRF			Stagger A ±11%			Stagger B ±14%		
σ	0.023	0.016	0.007	0.023	0.016	0.007	0.023	0.016	0.007
Bino	52.3	61.2	76.1	40.8	43.9	51.0	37.9	40.9	47.9
Opti	56.2	64.8	76.1	56.2	65.1	77.8	56.3	65.1	78.4
SubO	—	—	—	52.0	60.5	76.7	52.0	60.7	78.7
NeaO	56.2	61.0	65.0	55.3	60.0	64.7	55.4	59.9	64.6

SubO–Sub–Optimum [1], NeaO–Near–Optimum, $w_1 \neq 1$, two digits

Table 6.9 *Improvement comparison for various clutter σ [21]*

$\sigma = \sigma_c T_A = 2\sigma_v T_A/\lambda$	0.002	0.004	0.008	0.020
1 Block (0%)	77.2	65.2	53.1	37.3
2 Block (5%)	76.5	64.4	52.4	36.5
3 Pulse (MW)	51.6	45.5	39.3	30.1
4 Pulse (FW)	50.3	44.3	38.1	29.2
5 Binomial	51.8	45.8	39.8	32.3
6 Capon [2]	50.2	44.1	37.9	29.0
7 Prinsen [16]	60.9	54.9	48.9	41.1
8 Ewell & Bush [24]	37.5	36.6	34.1	26.3
9 Houts & Burlage [14]	53.1	48.8	40.2	25.6
10 H & B (MW)	51.7	46.9	38.9	25.1

frequency, and the other is moving clutter, such as weather, chaff, sea or birds clutter. In this case, we can use an ordinary MTI filter to cancel the ground clutter, and then cancel the moving clutter after compensating for its mean Doppler frequency. The key problem of this system is how to estimate the mean Doppler frequency of the clutter and how to compensate for it.

The earlier moving clutter MTI system is the clutter-lock MTI. It measures the mean Doppler frequency of the moving clutter by measuring the phase difference between two adjacent samples of the same range bin at first, and then the signal output from an integrator is used to control a voltage controlled oscillator (VCO) type COHO as a phase-locked loop. The block diagram is shown in Fig. 6.33.

A good example of this clutter-lock MTI is the time-averaged-clutter coherent airborne radar (TACCAR) system [1], which was developed by MIT Lincoln Laboratory to solve the airborne MTI problem. The clutter-lock MTI system can also be implemented by digital technique [25, 26], but the compensation of the mean Doppler frequency of the clutter is still in the analogue COHO channel.

Table 6.10 *Velocity response comparisons [21]*

Example	N_w (dB)	$\bar{S}$ (dB)	P_+ (%)	P_{-4} (%)
1 Block (0%)	−18.0	−1.32	21.5	24.0
2 Block (5%)	−30.2	−1.07	25.0	24.5
3 Pulse (MW)	− 6.6	−0.40	37.0	12.5
4 Pulse (FW)	−11.3	−0.35	36.0	7.5
5 Binomial	−24.6	−0.035	39.0	23.0
6 Capon	−21.1	−0.043	43.0	16.5
7 Prinsen	−15.4	−0.043	38.0	31.0
8 Ewell & Bush	−13.6	−0.001	46.5	20.0
9 Houts & Burlage	−16.1	−0.039	49.0	17.0
10 H & B (MW)	− 6.0	−0.289	38.5	6.0

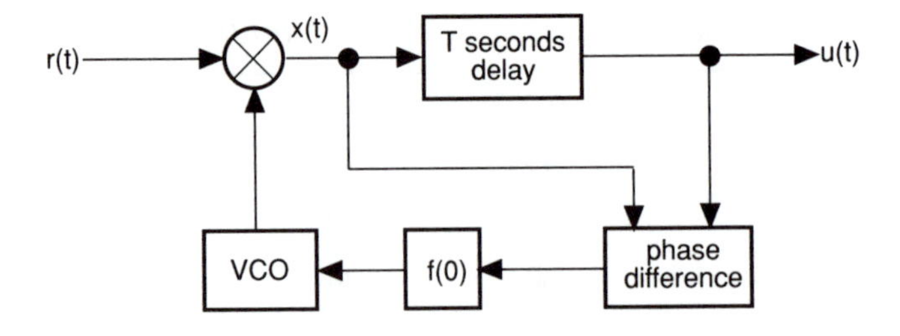

Figure 6.33 *Block diagram of a clutter-lock loop*

Although the clutter-lock MTI system can centre the clutter rejection notch at the average Doppler frequency of the clutter, it cannot simultaneously cancel both ground clutter and moving clutter. The reason is that, if there is more than one clutter with different Doppler shifts within a same range bin, it cannot estimate the 'mean' Doppler shift. Furthermore, since the compensation loop includes the COHO, the MTI for fixed clutter cannot be inserted before the clutter-lock MTI.

In some advanced radar systems, several sets of filter weights are used, each set corresponding to a specified mean Doppler frequency. Then, after several tens of antenna rotation, two antenna rotations are used to test which set of weights is suitable to reject the moving clutter in every azimuth direction.

Houts and Holt [27, 28] presented the design procedure of bandstop digital filters for rejecting moving cluter for this type of radar. The bandstop filter is designed to minimise the transition region for a fixed filter length, specified stopband width, stopband attenuation and passband ripple.

It is evident that this type of adaptive MTI is not a true adaptive one. Since the wind speed may be changed within several tens of seconds, i.e. several antenna rotations, it cannot adapt fast enough for these variations. Furthermore, the Doppler shifts may be different even in different range cells.

The modern bimodal clutter canceller separates the canceller for fixed clutter and the canceller for moving clutter into two independent cascaded parts. The compensation of the mean Doppler shift of the moving clutter is located at the output of the canceller for fixed clutter. This technique is particularly useful in long range radars where unambiguous range must be preserved and pulse repetition frequencies cannot be increased to the point where the velocity spread is a reasonable percentage of Doppler ambiguity, so that a wider null is sufficient to cancel them.

There are two types of moving clutter cancellers: the closed-loop adaptive canceller and the open-loop adaptive canceller. The closed-loop canceller is simpler than the open-loop canceller. However, the adaptation time is longer than that of the open-loop canceller. A block diagram of the closed-loop canceller is shown in Fig. 6.34*b* [29].

The principle of this closed-loop adaptive canceller was analysed by Spafford [6]. It works like the closed-loop sidelobe canceller (Fig. 6.34*a*). Multiple feedback loops are employed. The auxiliary input signals come from a delay structure, so the output signal may be summed with the results from the previous pulse repetition frequency. The narrow band filter in each circuit

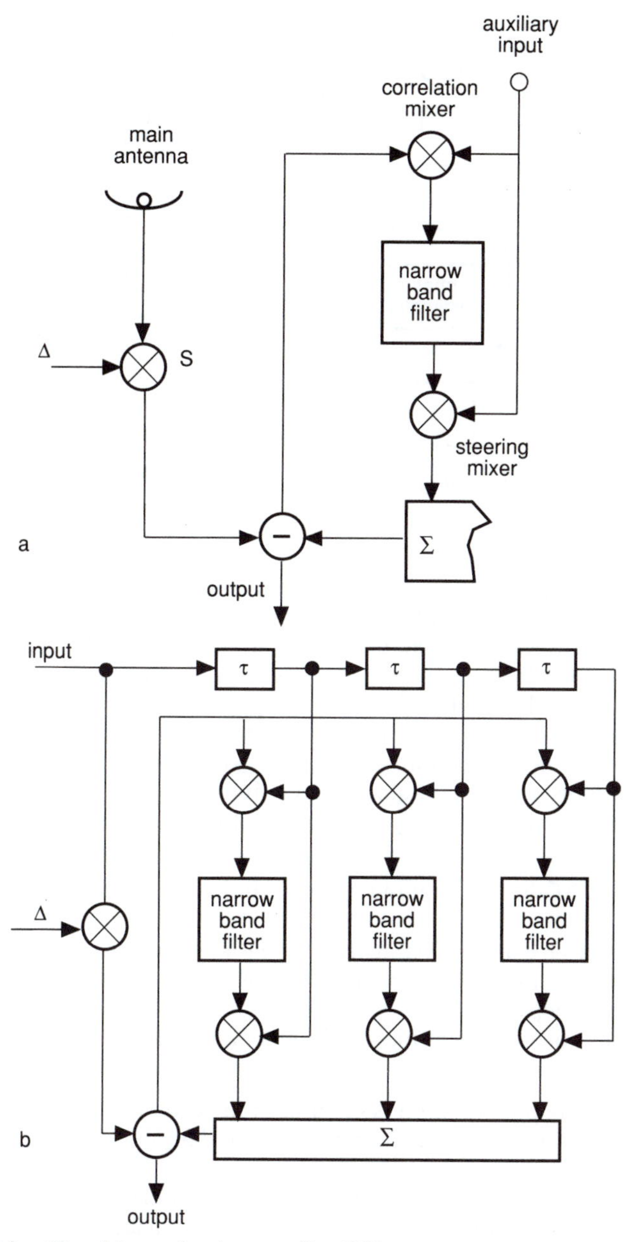

Figure 6.34 *Closed-loop adaptive canceller [29]*

operates so as to take an approximate time average of its input. Its frequency response has a notch at the mean Doppler of the moving clutter.

The system must be constrained so it will not respond to returns from 'point' targets, since they are generally desired targets. The clutter returns usually extend over many range cells, while a target is usually no longer than one or two

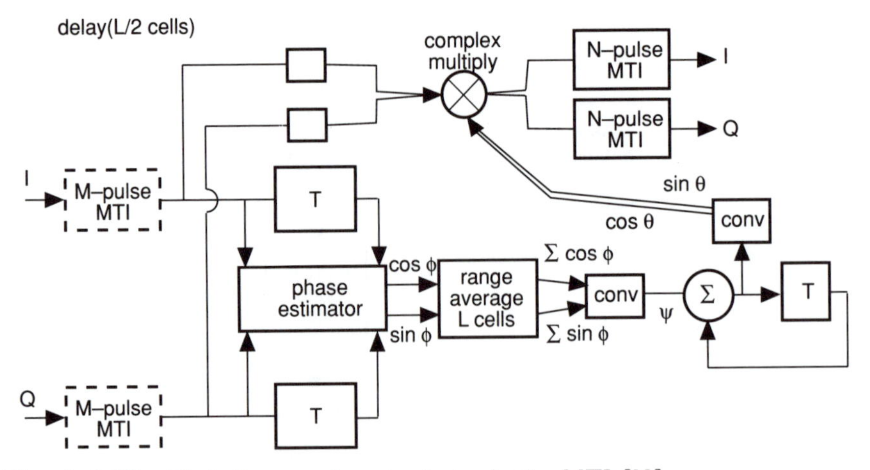

Figure 6.35 *Block diagram of an open-loop adaptive MTI [22]*

range cells. The targets are excluded from influencing the loop weights by controlling the bandwidth and feedback times on the individual loops. Too slow a response time prevents the adaptive loops from following the changing clutter conditions and too fast a response time prevents the optimum weight variance from being achieved, and the adaptive canceller performance suffers. Therefore, a trade off is always performed for the response time of the loops against the accuracy with which the optimum weights can be derived and the following clutter variations, while maintaining target integrity.

There are some shortcomings of the closed-loop adaptive canceller. These are:

(i) Special circuitry is required to avoid instability when the clutter power varies
(ii) Poor clutter cancellation at the leading edge of extended clutter; this is due to the high time constant (of the order of 10 range cells) of the adaptive loops
(iii) Degradation of clutter Doppler frequency estimation when a strong target signal is present.

Recently, open-loop adaptation is often used instead of closed-loop adaptation owing to its high speed of adaptation. A block diagram is shown in Fig. 6.35 [22].

Quadrature channels are used in this system. An ordinary *M*-pulse MTI is used to reject fixed clutter. In each channel, one delay line of *T* periods is inserted to obtain two samples from adjacent sweeps. The pulse-to-pulse phase shift, due to Doppler frequency of the moving clutter, is estimated using an appropriate (but accurate) algorithm, and the quadrature components of the estimated phase shift are then averaged over a range window to reduce the variance of the estimate. The averaged output is next converted back to a phase angle and is added modulo 2π to the phase shift, which was applied to the input data on the previous repetition period. The resultant required phase shift θ_i, as

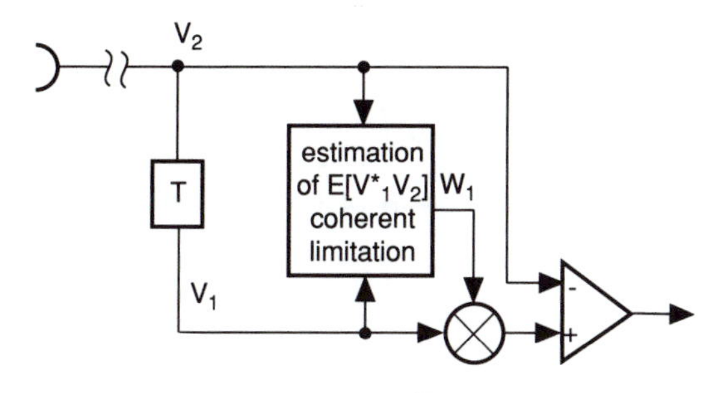

Figure 6.36 *Block diagram of ACEC [33] © IEEE, 1978*

determined separately in each range cell, can be applied to the new data through a complex multiplication as shown in this Figure. The output after the phase correction will have no pulse-to-pulse phase shift due to Doppler and can be cancelled in a conventional M-pulse MTI.

It is important to note that the averaging in range must be carried out on the quadrature component of the phase shift estimates. Any attempt to average the phase angles directly would cause large errors when the pulse-to-pulse phase shift is in the vicinity of 180°. Also storage of the previous sweep phase shift, which is added to each new estimate, ensures that the system can operate continuously in azimuth for any number of sweeps. The memory never needs to be reset.

The performance of an open-loop adaptive canceller largely depends on the accuracy of the phase estimator. The clutter mean Doppler interpulse phase shift, $\phi_D = 2\pi f_D T$, has to be estimated in real time, being unknown and time and space dependent. A typical method is direct estimation of the complex quantity $\exp(j\phi_D)$ instead of the Doppler phase shift ϕ_D itself [30, 31, 32]. Then the estimated Doppler frequency $f_D = \phi_D / 2\pi T$. This approach completely overcomes the phase ambiguity problem in weight computation and directly takes into account the related amplitude and phase processes of the clutter process to be estimated.

In the case of a single delay line canceller, the complex weights that maximise the improvement factor can be shown to be [9]

$$w_1 = a \exp(j2\pi f_D T)$$

$$w_2 = -a \qquad (6.87)$$

where a is an arbitrary constant that can be assumed equal to 1.

The complex adaptive weight, going to the complex multipliers, located after each delay line, provides the shift of the MTI frequency response, and in particular of its stopband central frequency, from zero to f_D. An alternative way is for the estimated complex phase shift to be multiplied with the complex input data to shift their Doppler frequency to zero, and then to be cancelled by a fixed MTI, as in Fig. 6.35.

If we denote the input samples of two adjacent sweeps v_1 and v_2, we have

$$w_1 = \exp(j\hat{\phi}_D) = \exp(\{j \arg(\hat{E}[v_1^* v_2])\} \qquad (6.88)$$

An adaptive canceller for extended clutter (ACEC) was proposed by Galati [33]. The block diagram of ACEC is shown in Fig. 6.36.

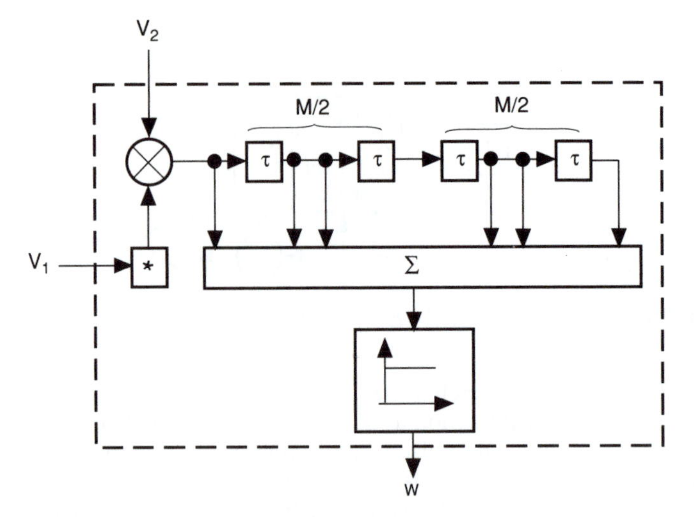

Figure 6.37 *Weights estimator (or M samples) of ACEC [33] © IEEE, 1978*

The complex input video samples relative to the same range cell of two consecutive sweeps v_1 and v_2, after complex conjugation of v_1, go to a complex multiplier which yields the complex signal $v_2 v_1^*$. The estimation of $E[v_2 v_1^*]$ in ACEC is as in the cell average CFAR detector. The complex signal $v_2 v_1^*$ enters a delay line, consisting of M taps each of length τ (range cell duration). So $M+1$ samples of $v_2 v_1^*$, relative to as many contiguous range cells, become available (Fig. 6.37).

In this estimator, the mean value $E[v_1^* v_2]$ in eqn. 6.88 is estimated by cell averaging. This means that

$$\hat{E}[v_1^* v_2] = \sum^{M} v_1^* v_2 \tag{6.89}$$

All such samples, except the central one, are summed to give an estimate of $ME[v_1^* v_2]$. The summed signal is coherent limited, originating in the adaptive weight. The delay of $(M/2)\tau$ introduced by the estimator is compensated for by an identical delay introduced along the path of the signal to be cancelled. So, when sample v_1 comes to the complex multiplier, the corresponding sample (relative to the same range cell) of $v_1^* v_2$ does not affect the weight estimation. This fact allows the sub-clutter visibility of a point target.

But when the target is not in the middle of the delay line it may affect the estimation of the clutter's Doppler frequency. However, if M is large enough (greater than or equal to 4) and the CNR is not less than 2 dB, the clutter Doppler frequency estimation is still good even when near the target. Typical improvement factor degradation due to the target is 3 dB (single canceller). Good clutter cancellation implies cancellation in the leading and trailing edge, i.e. in its first and last range cells. One can see from Fig. 6.37 that, since $M/2$ samples of clutter contribute in these situations to clutter Doppler frequency estimation, if CNR is not low the canceller performance is still good.

The performance of this single adaptive canceller is a function of correlation coefficient and clutter-to-noise ratio. The true mean value of $E[v_1^* v_2]$ is

$$E[v_{c1}^* v_{c2}] = \rho(t)\alpha^2 \exp(j2\pi f_D T) \tag{6.90}$$

where $\rho(T)$ is the correlation coefficient of v_{c1} and v_{c2} apart from the term $\exp(j2\pi f_D T)$, and α^2 is the clutter mean power. If the spectrum of clutter is of Gaussian type, namely

$$S(f) = S_0 \exp[-f^2/(2\sigma_f^2)] \tag{6.91}$$

where σ_f^2 is the second moment of the spectrum. By taking the Fourier transform of eqn. 6.91, the normalised autocorrelation function, which is also a Gaussian function, is obtained:

$$\rho(\tau) = \exp[-\tau^2/(2\sigma_t^2)] \tag{6.92}$$

where σ_t^2 is the second moment of the correlation function, which is related to σ_f by

$$\sigma_t = 1/(2\pi\sigma_f) \tag{6.93}$$

In our case,

$$\rho(T) = \exp[-2(\pi\sigma_c T_{AV})] \tag{6.94}$$

where σ_c is the spectral width of the clutter, and T_{AV} is the average pulse repetition period. Fig. 6.38*a* and 6.38*b* [33] show the improvement factor of ACEC (single canceller) with two and four samples for weights estimation, respectively.

In the case of a nonrecursive canceller of length L, $L > 2$ (e.g. $L = 3$ for the double canceller), a set of nonadaptive weights, $b_1, b_2, \ldots, b_L$, is used to shape the filter response as desired. The canceller output (with no adaptation) at the kth sweep is

$$z_k = \sum_{l=1}^{L} b_l v_{l+k-L}, \quad k = L, L+1, \ldots \tag{6.95}$$

The adaptation algorithm of ACEC modifies the weights so that the output is

$$z_k = \sum_{l=1}^{L} b_l \exp(j\hat{\phi}_{kl}) v_{l+k-L} \tag{6.96}$$

where

$$\hat{\phi}_{kL} \equiv 0, \quad \hat{\phi}_{k1} = \left[\sum_{i=k+1-L}^{k-1} \arg\left(\sum^{M} v_i^* v_{i+1}\right)\right] \bmod 2\pi \tag{6.97}$$

The performance of a double canceller ACEC system is shown in Fig. 6.39 [34]. These simulation results are obtained under ±11% staggered PRF condition. Improvement factors are given as a function of correlation coefficient with the number of range samples and CNR as parameters.

When land clutter and moving clutter exist in the same range cell, a fixed MTI must precede the adaptive canceller. However, radar operation with staggered PRF poses some problems. The main problems are due to the amplitude and phase modulation, from sweep to sweep, of the moving clutter due to the fixed MTI and to the stagger. A suitable compensation for the amplitude modulation is then needed which can be implemented by a real weight in cascade with the complex weight (phasor) used for Doppler compensation. As shown in Fig. 6.40 [34] for weather clutter with $\rho = 0.99$, an improvement of several decibels can be achieved.

The phase modulation also has to be compensated, as it could affect the improvement factor even more than the amplitude modulation. The phase

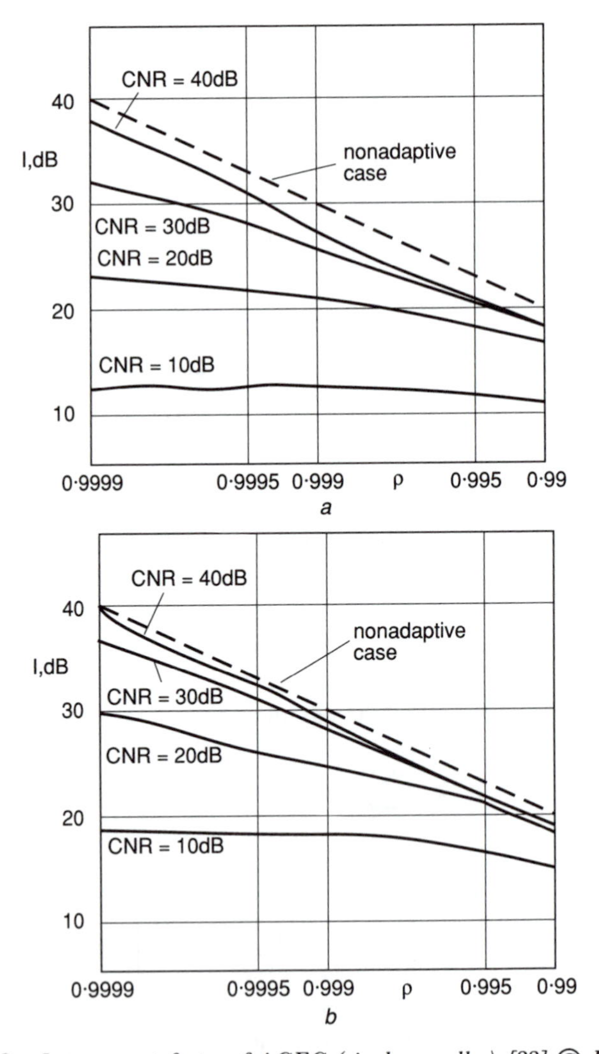

Figure 6.38 *Improvement factor of ACEC (single canceller) [33] © IEEE, 1978*

a Two samples for weight estimation
b Four samples for weight estimation

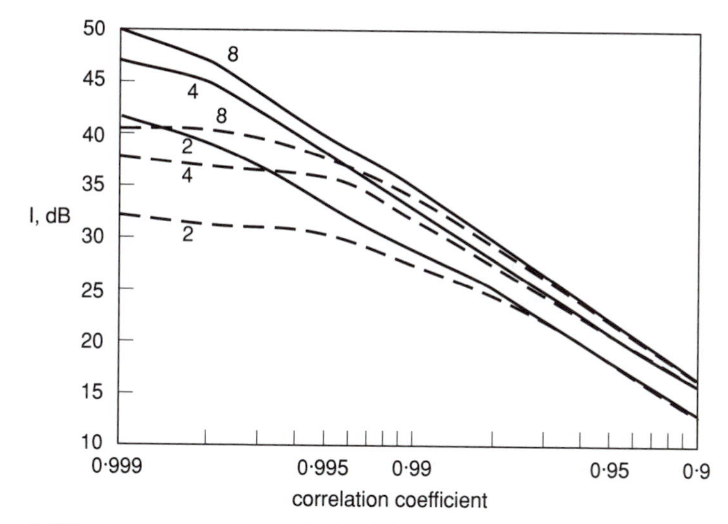

Figure 6.39 *Improvement factor (dB) of the ACEC double canceller versus the clutter correlation coefficient for 2, 4 and 8 range cells [34] © IEEE, 1985*
—— CNR = 40 dB
--- CNR = 30 dB

compensation is automatically performed in the ACEC approach, since the real-time estimation accounts for the stagger.

There is another type of adaptive canceller, which belongs to the 'deferred' type in contrast to the 'real-time' one. The estimated Doppler frequency is averaged in a range-azimuth cell, which may be somewhat greater than the resolution cell, and a Doppler map is formed. The averaged Doppler data are used in the next antenna scan to steer the null of the MTI filter. This method assumes that moving clutter is stationary and homogeneous in the cell and the aircraft signals do not contribute to the average. However, field experience shows that heavy rain has a non-uniform spatial distribution, and, in addition, the average Doppler frequency (and the power) may vary significantly, in an azimuth extent of a beamwidth, owing to the masking effect of nearby land clutter. In addition, while the real-time adaptivity 'naturally' takes into account the stagger of the PRF, the deferred one does not. A performance comparison between the ACEC (real-time adaptation) and Doppler map (deferred adaptation) is shown in Fig. 6.41 [34], in which the Doppler map adaptation is based on the current PRT.

It can be seen from this Figure that the ACEC is not degraded by the phase modulation, whereas the Doppler map approach suffers several dB loss in a wide range of Doppler frequencies. It is noteworthy that, when only rain clutter is present, the Doppler map approach is equivalent to the real-time technique, provided that the phase correction is made proportional to the current PRT value. Conversely, in the bimodal clutter case, compensation with average PRT would give better results (see Table 6.11 for $\rho = 0.95$).

This Table is obtained under the condition of $\pm 14\%$ stagger, and the MTI system is constructed with a triple binomial fixed MTI cascaded with a triple adaptive MTI, where MAP (T_{AV}) means compensation with the average PRT, and MAP (T_i) means adaptivity on the current PRT. These can be expressed by the following equations:

$$\phi_{iAV} = 2\pi \hat{f}_D T_{AV} (\text{mod } 2\pi), \quad i = 1, 2, \ldots \tag{6.98}$$

$$\phi_{ic} = 2\pi \sum_{k=k_0}^{i} \widehat{f_D T_k} (\text{mod } 2\pi), \quad i = 1, 2, \ldots \tag{6.99}$$

It is evident from this Table that, when the Doppler frequency is low, the improvement factor of Doppler map is higher than that of ACEC (real time). As the Doppler frequency of the moving clutter becomes higher, the performance of Doppler map, which is adaptivity on the current PRT (eqn. 6.99), degrades rapidly.

Finally, the time variation of the weights to improve the land clutter cancellation is found to be very detrimental with respect to the cancellation of moving clutter (as shown in Fig. 41, curve *e*), regardless of the estimation technique, and cannot be compensated by reasonable means. Therefore, simultaneous cancellation of land and rain/chaff clutter calls for fixed MTI weights and for real-time adaptivity.

It is also very interesting to study the overall velocity response of the cascaded MTI filters, for different stagger sequences, different values of the mean Doppler frequency of the moving clutter and correlation coefficients and

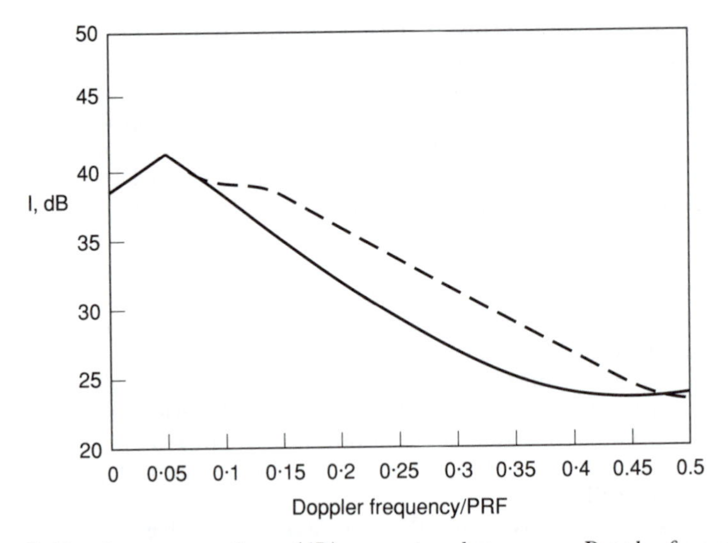

Figure 6.40 *Improvement factor (dB) on moving clutter versus Doppler frequency with and without amplitude equalisation (double canceller) [34] © IEEE, 1985*

—— without amplitude equalisation
--- with amplitude equalisation
$\rho = 0.99$
Stagger 'A'
CNR = 40 dB

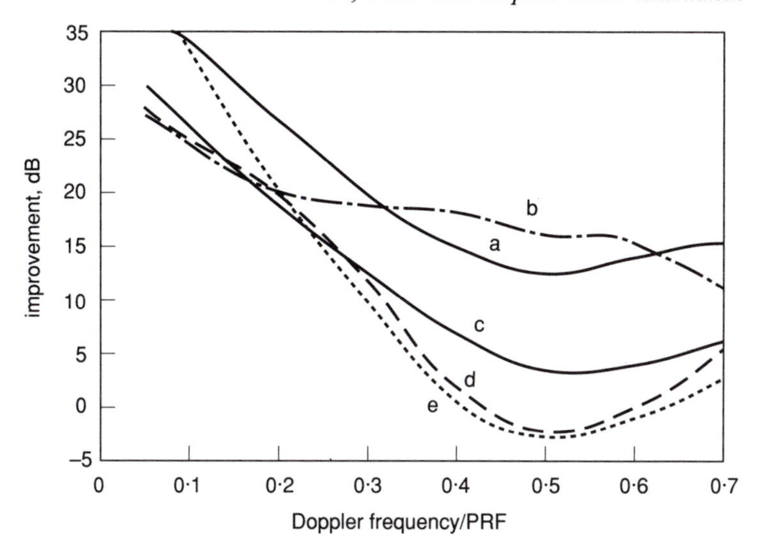

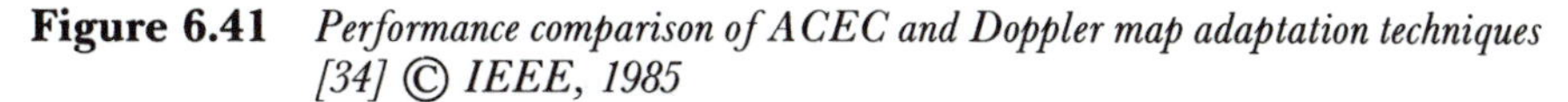

Figure 6.41 *Performance comparison of ACEC and Doppler map adaptation techniques [34] © IEEE, 1985*

a STG A double MTI real time
b STG B triple MTI real time
c STG A double MTI map
d STG B triple MTI map
e STG B triple MTI TVW map
RHO = 0.95

different numbers of processed pulses in both fixed and adaptive MTI. As an example, the velocity responses of the fixed MTI plus adaptive MTI cancellation systems are shown in Fig. 6.42*a*, and *b*.

From these Figures it turns out that, by carefully selecting the filters orders as well as the stagger sequence, it is possible to achieve satisfactory performance in both clutter suppression and velocity response. It is noteworthy that the depth of notch at the Doppler frequency of the moving clutter, in this case $f_D = 0.2$, in the velocity response of Fig. 6.42*a* is deeper than that of Fig. 6.42*b*. This is due to the fact that time varying weights are used and causes the reduced IF.

There is another type of configuration, which is termed the 'parallel radar signal processing scheme' in contrast to the serial (cascaded) one. This means that the three channels of processing (logarithmic channel, fixed MTI channel, and fixed plus adaptive MTI channel) are connected in parallel. The three processing channels, each with its own adaptive threshold quantiser, are combined by a logical OR of their 1 bit (threshold crossing) outputs to feed a single moving window integrator. The block diagrams of the 'parallel' and 'serial' signal processing schemes are shown in Figs. 6.43*a* and 6.43*b*.

Galati [35] compared these two types of configuration. The main disadvantage of the 'parallel' one is the collapsing losses due to the combination of three channels. This loss is slightly greater than 1 dB in the case of a 16- or 18-pulse moving window. In addition, there is also the CFAR losses (even in the absence of clutter) due to the TTI. This loss is about 1 dB for linear video and 1.3 dB for logarithmic video.

However, ACEC also suffers from sensitivity loss in clutter-free environments. Fig. 6.44 [33] shows this loss for the single and double canceller (assuming binary integration). The best way is to use a clutter sensor to switch them [36].

It is interesting to compare the performance of this type of adaptive canceller with that of the optimum MTI.

Hsiao [37] proved that a multiple stage MTI filter is equivalent to a single stage filter with the same number of delays; and the design of each stage of a multiple-stage filter separately, each for a specific clutter type, does not yield an optimum result. Some computation results are given. It is assumed that two clutters with a Gaussian power spectral density function exist within the same range bin. The standard deviation σ of the first one (ground clutter) is equal to 0.01, and the second (volume or weather) clutter has a value of 0.1. The optimal MTI improvement factor is then computed with the clutter cross-section ratio as a variable. It is assumed that the first clutter has a larger radar cross-section. The results are shown in Fig. 6.45*a*. In this Figure it is assumed that the mean Doppler frequency of the second clutter is equal to zero. Fig. 6.45*b* shows the effect of the mean Doppler frequency of the second clutter. In this Figure it is assumed that the RCS ratio of the two clutters is equal to 0 dB (shown as GWCR (ground weather clutter ratio) = 0 on the plot).

If a two-stage MTI system is optimised separately, the first one is a 3-pulse canceller, while the number of pulses in the second stage is assumed to vary from 2 to 10; then the result is as shown in Fig. 6.46. It is evident that the improvement factor of Fig. 6.46 is worse than those of Fig. 6.45*a* and *b*.

Table 6.11 *Improvement factor (dB) comparison on moving clutter [34] © IEEE, 1985*

Estimation correction ($f_c \cdot T_{AV}$)	Real time	MAP (T_{AV})	MAP (T_i)
0.05	26.93	28.21	28.19
0.10	23.35	24.57	24.50
0.15	21.04	22.09	21.80
0.20	19.70	20.54	19.42
0.25	19.15	19.65	16.23
0.30	19.16	19.18	11.74
0.35	19.25	18.75	6.67
0.40	18.50	17.76	1.95
0.45	16.66	15.95	−1.44
0.50	15.45	14.62	−2.82
0.55	15.86	15.12	−2.30
0.60	15.39	14.95	−0.46
0.65	13.18	13.31	2.16
0.70	10.66	11.25	4.98
0.75	8.43	9.38	6.88

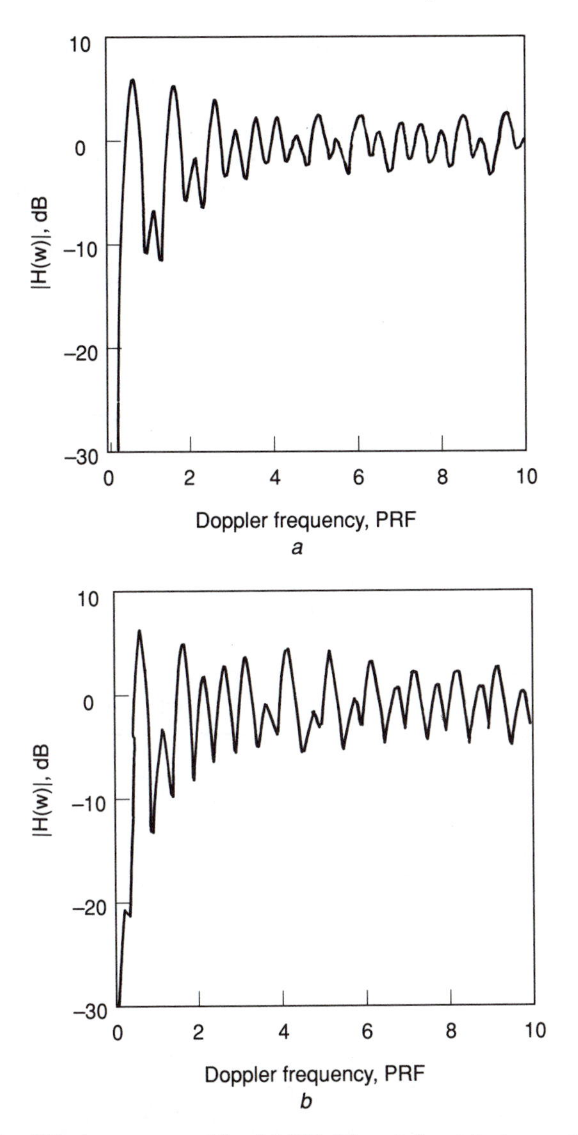

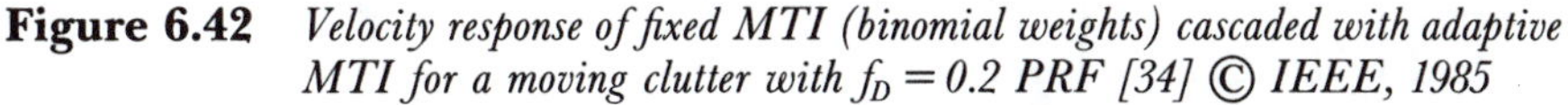

Figure 6.42 *Velocity response of fixed MTI (binomial weights) cascaded with adaptive MTI for a moving clutter with $f_D = 0.2$ PRF [34] © IEEE, 1985*

a Double fixed canceller plus double adaptive canceller with ± 11% stagger

1*b* Triple fixed canceller plus triple adaptive canceller with ± 14% stagger

Hansen [22] also showed the performance comparison between a fixed canceller cascaded with an adaptive canceller (Fig. 6.35) and an optimal MTI. For this, a bimodal clutter model was considered, consisting of a Gaussian land clutter spectrum located at zero Doppler having a spectum width of $\sigma_c T = 0.01$ and chaff (or rain) located at a Dopppler of 1/4 of the radar PRF ($f_D T = 0.25$) with a spectrum width of $\sigma_c T = 0.05$. The power ratio between land clutter and the chaff return is the parameter. The result is shown in Fig. 6.47 [22].

In Fig. 6.47 the improvement factor of the optimum four-pulse FIR MTI canceller is shown in full line as a function of the land clutter-to-chaff power ratio. For comparison, the broken line curve of Fig. 6.47 shows the performance of an adaptive MTI according to Fig. 6.35, which places two zeroes on the land clutter (broken line, fixed MTI section) and one zero adaptively centred on the chaff spectrum. In this case, $Z_3 = 0, j1$ as shown in the Figure. It is seen that, for land clutter-to-chaff power ratios from 10 dB to 55 dB, the simple adaptive MTI performs within 6 dB of the optimum. When chaff returns dominate, the improvement factor of the simple adaptive MTI is much worse than that of the optimum one. This is due to the zeroes of the optimum MTI filter moving along the unit circle adaptively (Fig. 6.48).

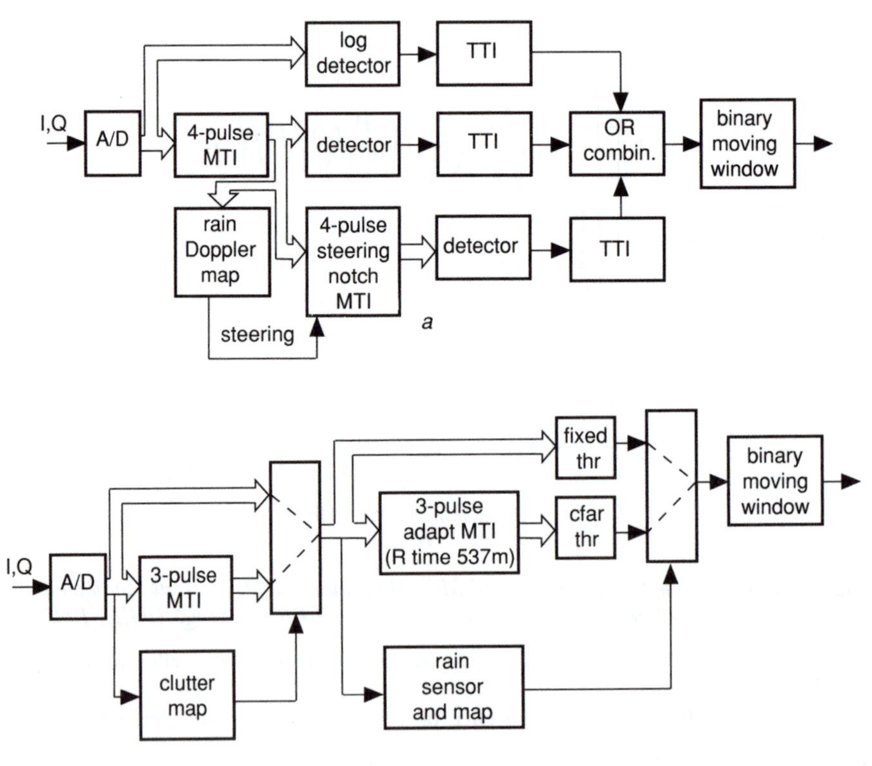

Figure 6.43 *Block diagrams of 'parallel' (a), and 'serial' (b) processing schemes [34] © IEEE, 1984*

TTI = temporal threshold integrator

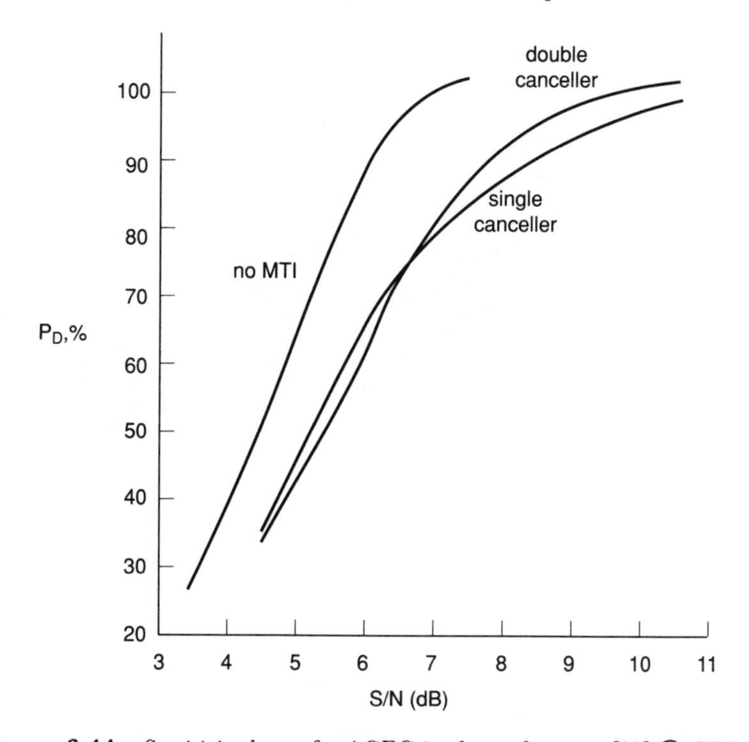

Figure 6.44 *Sensitivity losses for ACEC in clutter-free area [33] © IEEE, 1978*

N = number of pulses in the moving window = 10
M = digital threshold = $M_{opt} = 6$

T = analogue normalised threshold $= \dfrac{V_T}{d_n} = 2.4477$

$P_{fa} = 2.0 \times 10^{-6}$

6.3 Moving target detector

It is well known that the detection of a radar signal is essentially the detection of a signal with unknown time of arrival and unknown Doppler frequency shift. In noncoherent radar, we detect the signal in time domain; the problem of unknown time of arrival can be solved by multiplexing in range, or using a multiple range bin. In coherent radar, or coherent on receive radar, we can reserve the Doppler information of the target signal, and detect them in frequency domain. Similarly, the multiple Doppler filter bank is used to solve the unknown Doppler frequency problem. This type of detector is termed 'moving target detector'. Since the time of arrival of the target signal is unknown, the Doppler filter banks have to connect to each range bin as shown in Fig. 6.49.

Although the principle of MTD was formed in the early stage of development of radar signal detection theory, its realisation is as it was at the beginning of the 1970s. It relies on the development of digital signal processing on one hand, and

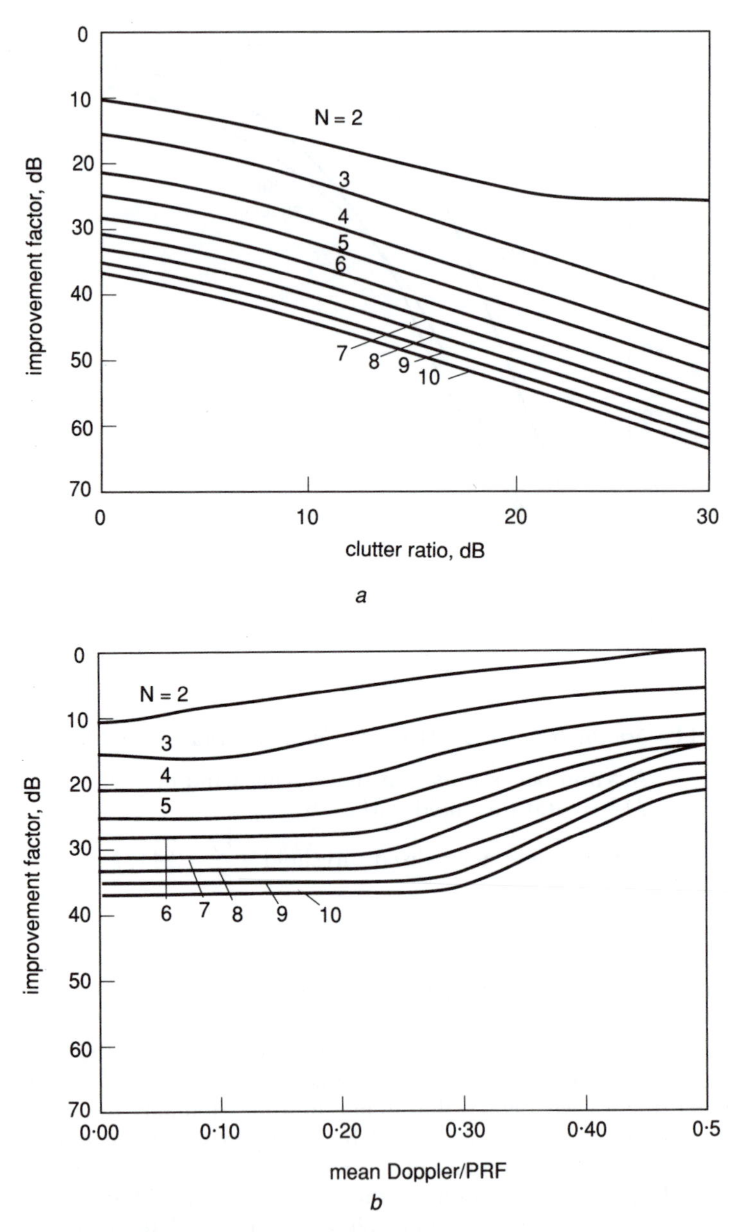

Figure 6.45 *Optimum MTI improvement factor for two-clutter case [37] © IEEE, 1977*

(*a*) FMEAN2/PRF = 0.00
(*b*) GWCR = 0 dB
SIGMA1/PRF = 0.01
SIGMA2/PRF = 0.10
GWCA = 0

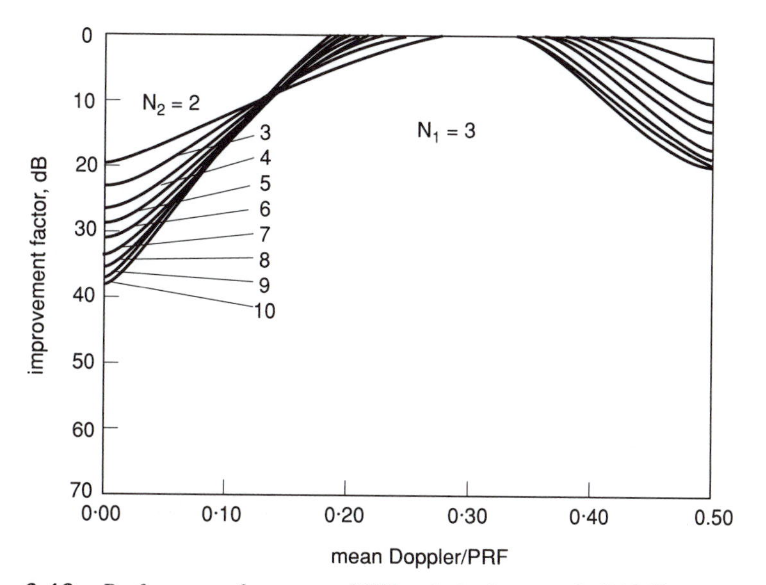

Figure 6.46 *Performance of two-stage MTI optimised separately [37] © IEEE, 1977*
GWCR = 0 dB
SIGMA1/PRF = 0.01
SIGMA2/PRF = 0.10
GWCA = 0

on the development of LSI and VLSI devices on the other. The first experimental MTD system was developed at MIT Lincoln Laboratory [38, 39]. The block diagram is shown in Fig. 6.50. It was implemented with hardwired digital technology and with SSI and MSI devices.

The distinguishing features of the MTD system are as follows.

(i) Doppler filter bank is used instead of comb filter canceller.

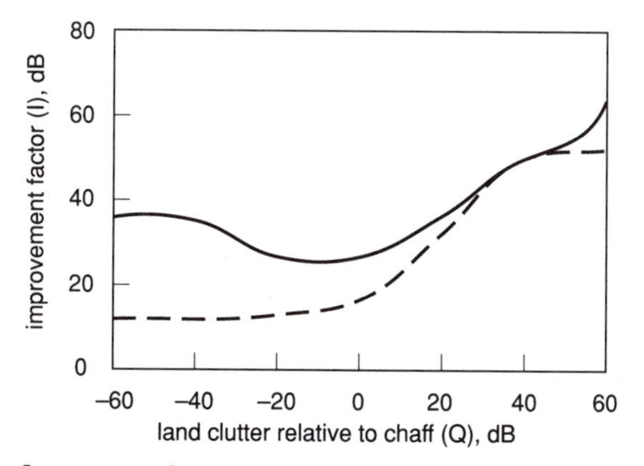

Figure 6.47 *Improvement factor comparison between a four-pulse optimum MTI system and adaptive MTI with fixed zeroes [22]*

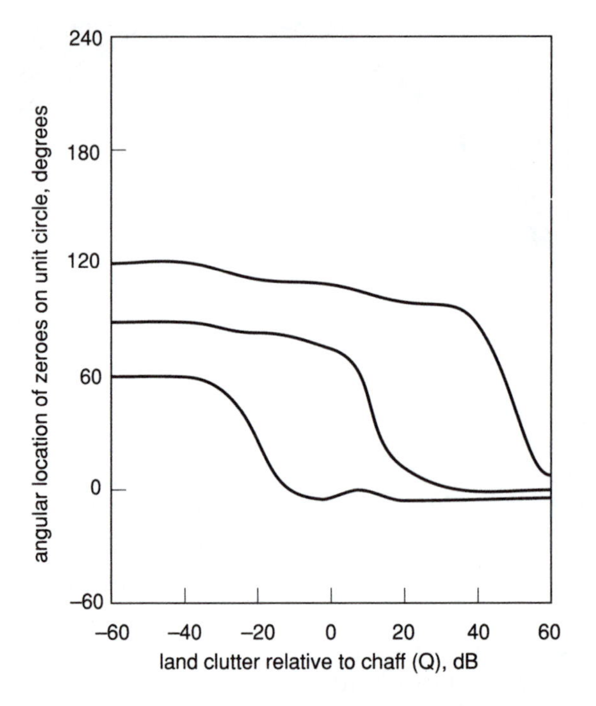

Figure 6.48 *Optimum zero locations of four-pulse FIR MTI against bimodal mixture of land clutter and chaff [22]*

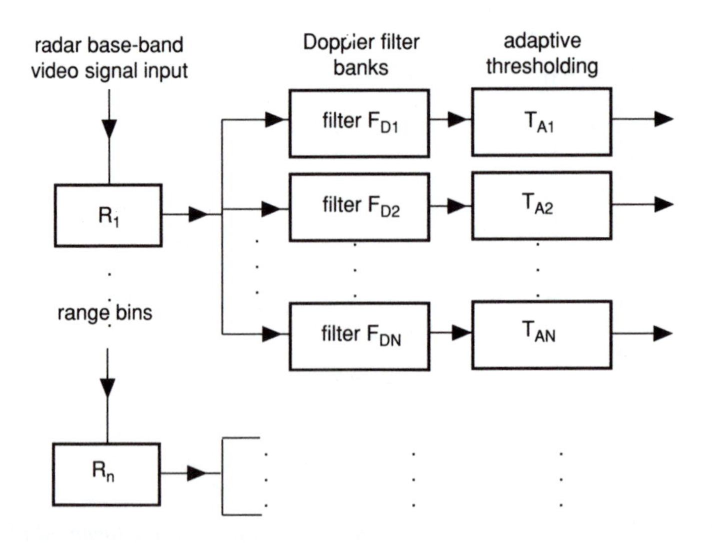

Figure 6.49 *Block diagram of moving target detector*

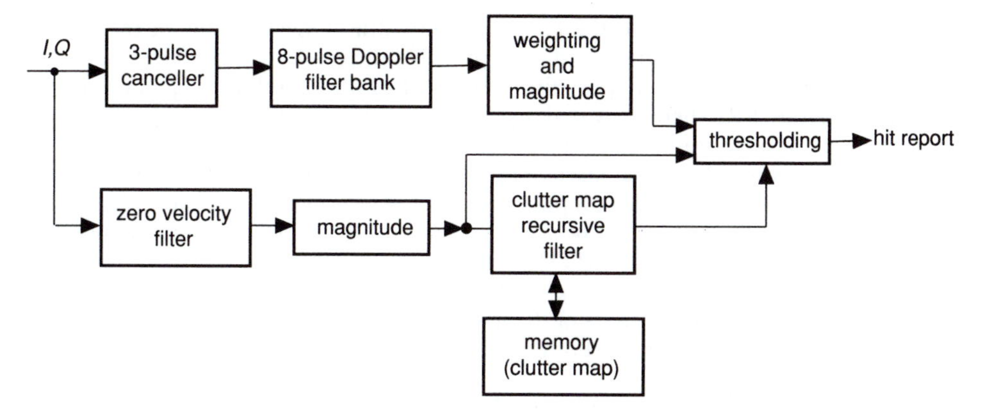

Figure 6.50 *Block diagram of first generation MTD system*

(ii) Block processing is used instead of moving processing.
(iii) 2- or 3-pulse canceller precedes the Doppler filter bank to reject strong fixed clutter and to reduce the required dynamic range of the Doppler processor.
(iv) Adaptive threshold in frequency domain is used to reject moving clutter.
(v) Clutter map is utilised to detect tangential flying target.

The first two features imply that the MTD system contains inherently coherent integration in contrast to ordinary video integration, such as moving window integration. However, it does not mean that the MTI system can follow with non-coherent integration only. Since the MTI system is a moving processing system, the number of output pulses is equal to that of the input in the continuous sliding case, and equal to the number of input pulses minus the order of the MTI filter plus 1 in the block processing case. Therefore, either a non-coherent or coherent integrator can follow the MTI system. On the other hand, if the number of pulses within the beamwidth is large enough, and if coherent integration of all these pulses is not necessary, non-coherent integration can also be used after the MTD system.

In the MTD system only one output is available for each block owing to the block processing. Although the MTD system can also use moving processing instead of block processing, the trade off in hardware complexity is considerable and few advantages can be obtained.

Pulse-to-pulse stagger is not suitable for the MTD system owing to the block processing. Block-to-block stagger is used to eliminate blind speed. Therefore, at least two PRFs must be used for the pulses within the beamwidth. This means that at least two blocks must be within one beamwidth scanning. Each block contains enough pulses with constant inter-pulse period, also called a coherent processing interval (CPI). In MTD–I (see Fig. 6.50), one CPI contains ten pulses, two for the 3-pulse canceller and eight for the 8-point FFT. The total number of pulses within the beamwidth is then equal to 20.

From the viewpoint of matched filter theory, the optimum number of pulses used for coherent integration is equal to the number of pulses within the beamwidth. This is due to the fact that the width of spectral line of the echo pulse is approximately equal to the reciprocal of the dwelling time. The width of Doppler filter response is approximately equal to the sampling frequency (it is equal to the PRF here) divided by the number of pulses to be processed. Therefore, it is desirable that the number of pulses used for coherent integration is much greater than that used for cancellation. Thus the MTD system is suitable for radars with enough pulses within the beamwidth, e.g. more than 18. Although 3-pulse MTD has been constructed [40], the performance is much worse than that of 10-pulse MTD systems.

In the MTD system ground clutter is rejected by the 3-pulse canceller and Doppler filtering. The outputs of the first and last Doppler filters are discarded. The improvement factor of the MTD system is about 20 dB higher than the ordinary MTI system.

Weather clutter is rejected by adaptive-thresholding the output of each Doppler filter. This adaptive threshold is determined by the estimated mean of the moving clutter, which is obtained from the summation of the samples of adjacent range bins just like the cell average CFAR detector in time domain. This is based on the assumption that the Doppler component of moving clutter is homogeneously distributed between same Doppler filter of adjacent range bins.

The radar cross-section of a tangential flying target is usually much greater than that of heading or tailing flying targets. Therefore, the echo signal from a tangential flying large aircraft is possible, exceeding some weak ground clutter. The clutter map formed from the average clutter magnitude of several scans is used as the threshold of a zero Doppler frequency channel to obtain super-clutter visibility.

In MTD–I FFT was used to perform the transform from time domain to frequency domain and to form the equivalent Doppler filter bank. There are two shortcomings to this method. The first is that it can form equally spaced filters only, and the responses of all the filters are the same. The second is that the number of points of FFT must be integer power of 2. But the number of pulses available depends on the number of pulses within the beamwidth divided by two. It may not be equal to the integer power of 2.

The second generation of MTD (MTD–II) [41] uses eight FIR filters for the Doppler filter bank instead of FFT. The shortcoming of FFT can be avoided. A two-pulse canceller is used instead of a three-pulse canceller. The total number of pulses within a CPI is equal to nine.

In MTD–II a parallel microprogrammed processor (PMP) [42] is used instead of the hardwired processor in MTD–I. MSI and LSI devices are used to reduce the volume and power consumption of the processor. Semiconductor memory replaces the core memory.

There are two disadvantages of block processing. The first is that it is difficult to estimate azimuth angle accurately, since there are only two outputs within one beamwidth. The other is that the block processing cannot be compatible with the pulse-to-pulse frequency agility. It can only be compatible with block-to-block frequency agility.

In the third generation of MTD [43, 44] the beamshape match processing is used to solve the first problem. Very high azimuth accuracy and azimuth resolution have been achieved. Different numbers of pulses in consecutive CPIs with different PRF are used to obtain nearly-equal-bandwidth Doppler filters in these CPIs.

We shall discuss the design and performances of the MTD system in detail.

6.3.1 *Improvement factor of an MTD system with ideal filters*

It is well known that the optimum processor for unknown or random target Doppler shift is the maximum likelihood ratio processor. It can be implemented as a Doppler filter bank, where each filter is separately optimized against the specified clutter model at the particular Doppler frequency. In practice the optimisation of the individual filters is made difficult because the processor must be designed to operate against a range of clutter scenarios with varying power level, mean Doppler shift, spectral spreading and possibly spectral shape. Therefore, the maximum likelihood theory can only be used as a reference to determine how well a given practical filter design meets the performance requirements for all variation in the clutter model.

The performance of the maximum likelihood processor is usually presented as signal-to-interference improvement, I_{SIR}, as a function of target Doppler. Interference here refers to the sum of clutter and thermal noise. A simplified measure of performance can be obtained by calculating the average value of I_{SIR} across the unambiguous Doppler interval. This average SIR improvement will be denoted as $\overline{I_{SIR}}$. For the practical Doppler filter bank the average is calculated by using the maximum value of I_{SIR} at each target Doppler frequency. If a single filter is used by the coherent processor it is found that $\overline{I_{SIR}}$ is equal to the improvement factor in the MTI system.

For a CPI consisting of N pulses the signal return is represented as a complex vector:

$$\boldsymbol{s}(f_D)^T = \sqrt{P_{si}}\,\exp(j\phi)\{\exp(j\theta), \exp(j2\theta), \ldots, \exp(jN\theta)\} \tag{6.100}$$

where P_{si} is the input signal power per pulse, ϕ is a random signal phase, and $\theta = 2\pi f_D T$ is the pulse-to-pulse phase shift due to the target Doppler shift. The clutter return is

$$\boldsymbol{c}^T = \{c_1, c_2, \ldots, c_N\} \tag{6.101}$$

where $E\{|c_i|^2\} = P_{ci}$ is the clutter power. The accompanying thermal noise is

$$n^T = \{n_1, n_2, \ldots, n_N\} \tag{6.102}$$

where $E\{|\boldsymbol{n}_i|^2\} = P_{Ni}$ represents the thermal noise power. Finally the total input is

$$\boldsymbol{x} = \boldsymbol{s}(f_D) + \boldsymbol{c} + \boldsymbol{n} \tag{6.103}$$

and the total power per pulse is

$$P_x = E\{|\boldsymbol{x}_i|^2\} = P_s + P_c + P_n \tag{6.104}$$

The input signal-to-interference ratio (SIR) is

$$(SIR)_{IN} = \frac{P_s}{P_c + P_n} \tag{6.105}$$

The Doppler filter has the complex weights

$$\boldsymbol{w}^T = \{w_1, w_2, \ldots, w_N\} \tag{6.106}$$

and the filter output therefore is

$$y = x^T w^* \tag{6.107}$$

The corresponding output power is

$$P_y = E\{|y|^2\} = E\{(x^T w^*)^{T*}(x^T w^*)\} = w^T E\{x^* x^T\} w^* \tag{6.108}$$

The expected value in eqn. 6.108 represents the covariance matrix of the input, which can be written as

$$M_x = P_{si} M_s(f_D) + P_{ci} M_c + P_{ni} M_n \tag{6.109}$$

assuming statistical independence of signal, clutter and thermal noise. By definition,

$$Ms(f_D) = \frac{1}{P_s} (s(f_D)^* s(f_D)^T\} \tag{6.110}$$

$$M_c = \frac{1}{P_c} E\{c^* c^T\} \tag{6.111}$$

The ij-th element of M_c can be determined from the correlation function $\rho_c(\tau)$ of the clutter return using the argument $(ij)T$.

$$M_n = \frac{1}{P_n} E\{n^* n^T\} = I \text{ (identity matrix)} \tag{6.112}$$

The noise gain of the Doppler filter is

$$G_n = \frac{P_{no}}{P_{ni}} = \frac{w^T P_n I w^*}{P_{ni}} = w^T w^* = \sum_{i=1}^{N} |w_i|^2 \tag{6.113}$$

The output SIR ratio is

$$(SIR)_{OUT} = \frac{P_{SO}}{P_{co} + P_{no}} = \frac{P_{si} w^T M_s(f_D) w^*}{P_c w^T M_c w^* + P_{ni} G_n} \tag{6.114}$$

and the SIR improvement is therefore

$$I_{SIR}(f_D) = \frac{(SIR)_{OUT}}{(SIR)_{IN}}$$

$$= \frac{w^T M_s(f_D) w^*}{w^T w^*} \frac{w^T w^* (P_{ci} + P_{ni})}{w^T (P_{ci} M_c + P_{ni} I) w^*} \tag{6.115}$$

This expression has been arranged to identify I_{SIR} as the product of the coherent integration gain G_c (against thermal noise) and the normalised clutter cancellation C. The latter is equivalent to the definition of the MTI improvement factor.

The optimum weight vector corresponding to maximum SIR improvement can be determined as follows [6]:

$$w_{opt}(f_D) = (P_c M_c + P_n I)^{-1} s(f_D)^* \tag{6.116}$$

The average SIR improvement for the optimum processor is then determined by

$$\overline{I_{SIR}} = \frac{1}{PRF} \int_0^{PRF} I_{SIR}(f_D) df_D \tag{6.117}$$

where $I_{SIR}(f_D)$ is given by eqn 6.115 and the filter weights are a function of f_D as shown in eqn. 6.117. For a given clutter model, the value of $\overline{I_{SIR}}$ can be calculated by numerical integration.

Hansen [45] calculated $\overline{I_{SIR}}$ for a Gaussian clutter Doppler spectrum of the form

$$S_c(f_D) = k \exp\left[-\left(\frac{f_D^2}{2\sigma_c^2} \right) \right] \tag{6.118}$$

where σ_c is the standard deviation of the width of the clutter spectral lines. It was further assumed that $P_{ni} \ll P_{ci}$, so that $\overline{I_{SIR}}$ is not limited by thermal noise. The results obtained under the assumption of $f_D = 0$ for several values of N are shown in Fig. 6.51 [45].

These results coincided with the results obtained by Andrews [46]. Fig. 6.52 shows the results obtained by Andrews. The improvement factors for FIR filters with fixed binomial weights are also shown in this Figure as dashed curves. It is noteworthy that the improvement factors of the fixed binomial weights FIR filter are only a few decibels worse than those of the optimum filters (see also Fig. 6.32).

When the number of pulses in a CPI is large the optimum Doppler processor can be configured as the combination of a whitening filter and a coherent integrator. The transfer function of the whitening filter can be written as

$$H_w(f_D) = [1/U_c(f_D)]^{1/2} \tag{6.119}$$

where $U_c(f_D)$ is the clutter spectrum at the filter input. Across one unambiguous PRF interval we can write

$$H_w(f_D) = [1/(\text{rep}_{1/T} S_c(f_D))]^{1/2} \tag{6.120}$$

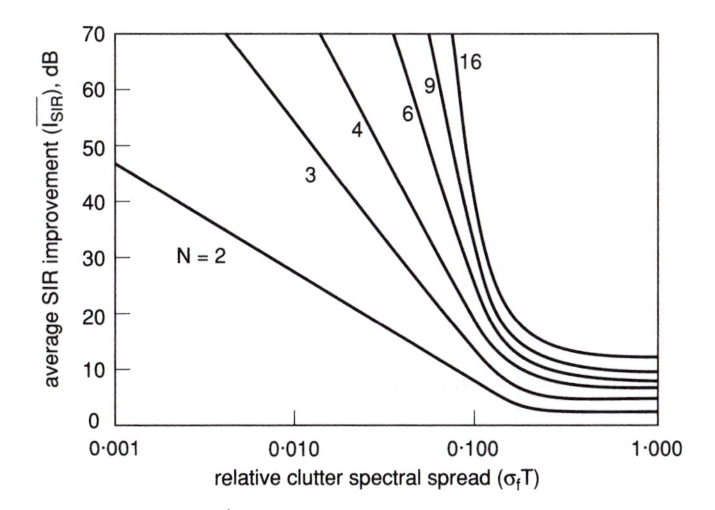

Figure 6.51 *Optimum average SIR improvement for Gaussian clutter spectrum with different standard deviation [45]*

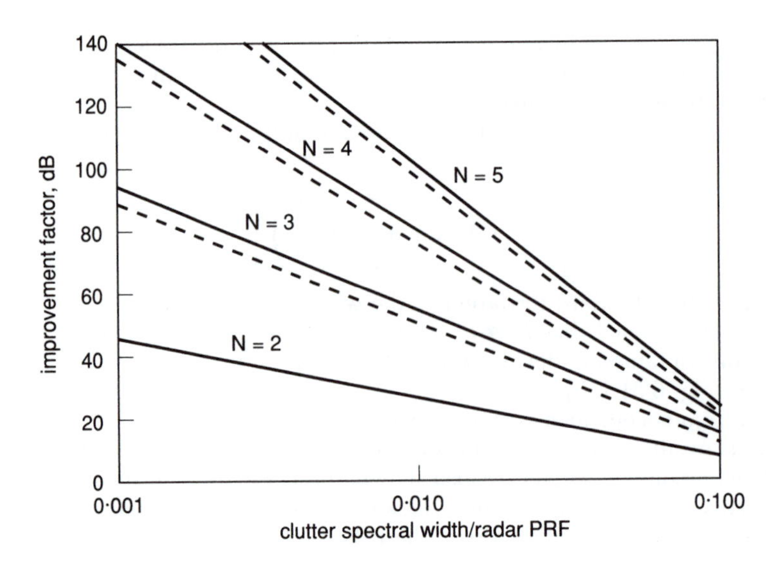

Figure 6.52 *Improvement factor of optimum FIR MTI filter [46]*

where $S_c(f_D)$ is the spectrum of the Doppler spread of the clutter returns. The function $\text{rep}_{1/T}$ is defined as

$$\text{rep}_{1/T} S_c(f_D) = \sum_{j=-\infty}^{\infty} S_c(f_D + j/T) \tag{6.121}$$

The average normalised clutter cancellation due to the whitening filter is

$$C_{AV} = \frac{P_{ci}}{P_{co}} \frac{P_{no}}{P_{ni}} \tag{6.122}$$

We find

$$\begin{aligned} P_{ci} = &= \int_0^{PRF} \text{rep}_{1/T} S_c(f)\, df \\ &= \int_{-\infty}^{\infty} S_c(f)\, df \end{aligned} \tag{6.123}$$

$$P_{c0} = \int_0^{PRF} \text{rep}_{1/T} S_c(f) \frac{1}{\text{rep}_{1/T} S_c(f)}\, df = \text{PRF} \tag{6.124}$$

$$P_{ni} = 1 \text{ (by definition)} \tag{6.125}$$

$$P_{no} = \int_0^{PRF} \frac{1}{\text{rep}_{1/T} S_c(f)}\, df \tag{6.126}$$

Thus

$$C_{AV} = T^2 \int_{-\infty}^{\infty} S_c(f)df \int_0^{PRF} \frac{1}{\text{rep}_{1/T} S_c(f)} df \tag{6.127}$$

The coherent integrator which follows the whitening filter provides a coherent gain which in the ideal case is

$$G_c = 10 \log_{10} N \tag{6.128}$$

at every signal Doppler frequency, and the average value of $\overline{I_{SIR}}$ is therefore

$$\overline{I_{SIR}} = G_c C_{AV} \tag{6.129}$$

Assuming

$$S_c(f) = \frac{1}{\sqrt{2\pi}\sigma_f} \exp\left(\frac{-f^2}{2\sigma_f^2}\right) \tag{6.130}$$

so that for $P_{ci} = 1$ we have

$$C_{AV} = T^2 \int_0^{PRF} \frac{1}{\text{rep}_{1/T} S_c(f)} df \tag{6.131}$$

This expression was evaluated numerically for several values of $\sigma_f T$ and the results are given in Table 6.12 [45].

For large N we can find the asymptotic value of the average I_{SIR} as follows:

$$(\overline{I_{SIR}})_{opt} = C_{AV}(\text{dB}) + 10 \log_{10} N \tag{6.132}$$

In Fig. 6.53 this asymptotic result is shown together with the value shown in Fig. 6.51. Good agreement between the exact calculation and the asymptotic result can be seen.

6.3.2 *Improvement factor of MTD system with practical filters*

In fact, a limited number of filters are available owing to the limited number of pulses available and the complexity of hardware realisation. The improvement factor of an MTD system depends not only on the number of filters but also on the weighting function of the filter. Furthermore, different filters have different improvement factors owing to the different relative frequency location between the filter and the clutter. It is evident that the Doppler filter farthest from the clutter has the best improvement factor. Fig. 6.54 [47] shows the improvement factor for each filter of an 8-pulse Doppler filter bank with uniform weighting. The average improvement factor for all filters is indicated by the dotted curve.

Chebyshev filter designs are attractive for Doppler filter banks owing to their uniform Doppler sidelobes and ease of design. The weights of a Chebyshev filter

Table 6.12 *Upper bound on average clutter cancellation [45]*

$\sigma_f T$	0.07	0.08	0.10	0.12	0.14	0.20
C_{AV} (dB)	85.2	61.0	33.5	19.4	11.6	2.8

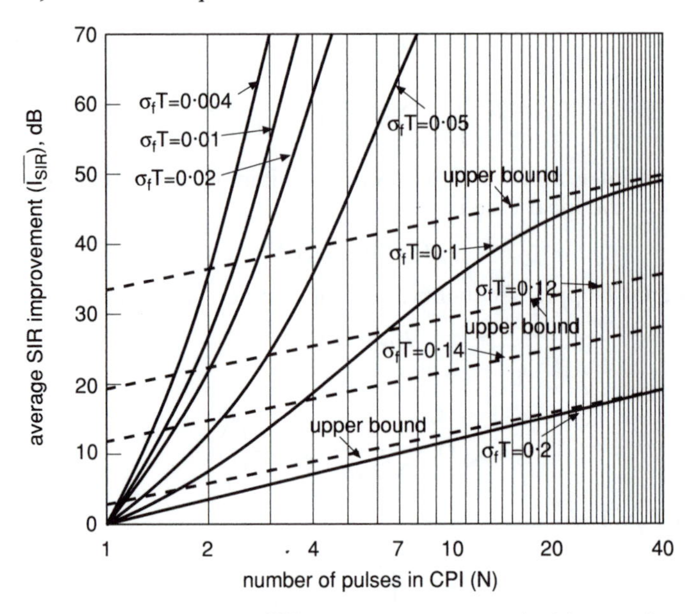

Figure 6.53 *Optimum average SIR improvement compared with upper bound [45]*

for arbitrary N and sidelobe level (SLL) can be determined as in Reference 48. The improvement factor of a 10-pulse Doppler filter bank with 25 dB sidelobe level and 40 dB sidelobe level Chebyshev weighting is shown in Fig. 6.55 [49].

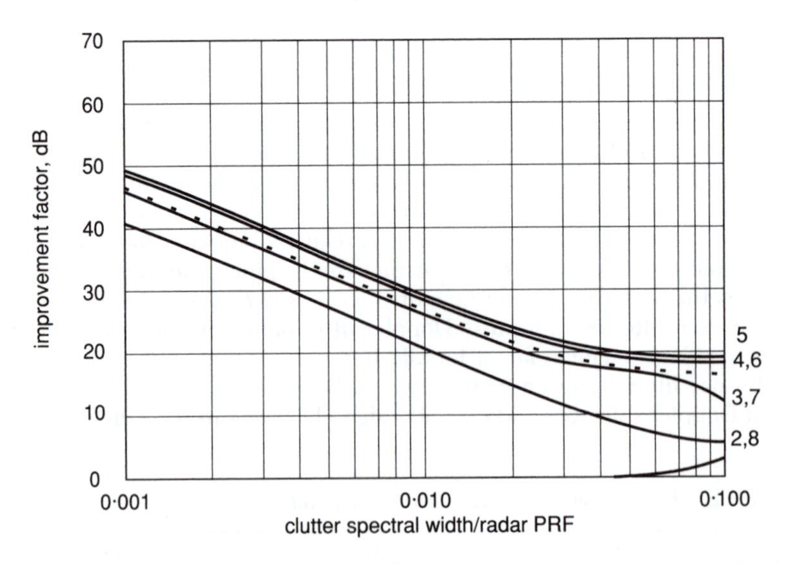

Figure 6.54 *Improvement factor for each filter of an 8-pulse Doppler filter bank with uniform weighting [47]*

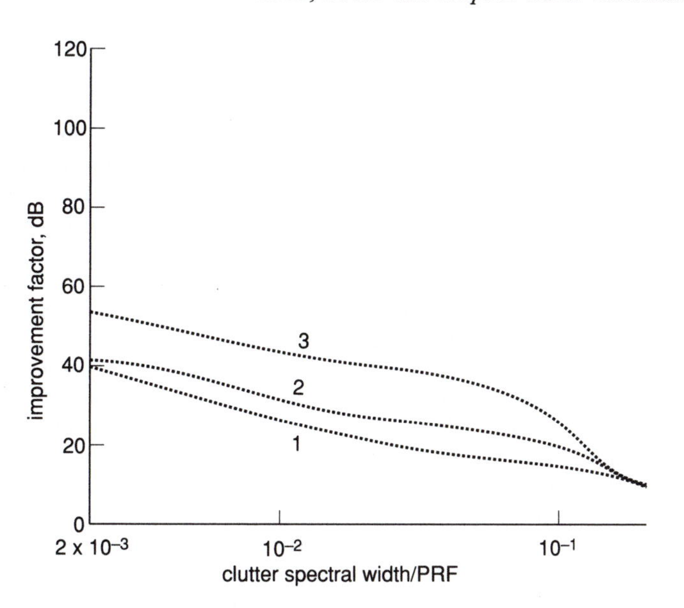

Figure 6.55 *Improvement factor of Chebyshev weighted 10-pulse Doppler filter bank versus clutter spectral width [49]*

1 uniform WT
2 Chebyshev WT (SLL = −25 dB)
3 Chebyshev WT (SLL = −40 dB)

It can be seen from Fig. 6.55 that the improvement factor of a Doppler filter bank with lower sidelobe level is better than that of a Doppler filter bank with higher sidelobe level. This is due to the fact that the lower sidelobe level prevents clutter leakage from the sidelobe. Therefore, lower sidelobe is desired. Fig. 6.56 [45] shows the response of a Doppler filter bank with −68 dB sidelobe Chebyshev weighting for $N=9$ pulses.

These response curves are normalised to the noise gain of the filters and therefore represent the coherent filter gain of the individual filters as given by the first factor of eqn. 6.115. By determining the normalised clutter cancellation for each filter, as given by the second factor of eqn. 6.115, the SIR improvement versus Doppler frequency can be calculated for any specified clutter model. Examples of the results of such calculations are shown in Fig. 6.57*a* and *b* [45] for a Chebyshev filter bank using $N=9$ and assuming a Gaussian clutter spectrum with zero average Doppler shift and relative spectral spread of $\sigma_f T=0.05$ and 0.1, respectively. Only the maximum envelope for the nine filters is shown and a clutter-to-noise ratio of 100 dB was assumed in these graphs and the following calculations in order to prevent thermal noise from limiting the performance in these graphs. The average $\overline{I_{SIR}}$ is obtained by averaging this response curve (in power) across the Doppler interval. The corresponding values of $\overline{I_{SIR}}$ are shown as broken lines in these Figures.

If we repeat these calculations for a number of discrete values of $\sigma_f T$, the curves of $\overline{I_{SIR}}$ versus $\sigma_f T$ can be obtained. The resulting curves of $\overline{I_{SIR}}$ are shown in Fig. 6.58 *a*, *b* and *c* [45] for each of the three Chebyshev filter designs of $N=6$, 9 and 16, respectively. Also shown by broken lines in each of these graphs are the optimum results from Fig. 6.51 for comparison. The average $\overline{I_{SIR}}$ obtained with a given Doppler filter bank depends to some extent on the mean clutter Doppler shift. The best performance is obtained when the clutter spectrum is centred on the rejection region of one of the Doppler filters. The worst case is obtained when the clutter spectrum is offset by 1/2 filter separation. Curves of $\overline{I_{SIR}}$ for each of these cases are therefore shown.

It can be seen from these Figures that, when the width of clutter spectrum is wider than some value, the best performance of the 68 dB Chebyshev Doppler filter bank is very close to that of the optimum filter. This value depends on the number of filters. It is equal to 0.04, 0.06 and 0.07 for $N=6$, 9, and 16, respectively. When the clutter spectrum width is narrower than this value, the average SIR improvement obviously is worse than the optimum value, and is close to a constant. This effect can be explained by the fact that, when the width of clutter spectrum is narrower than this value, almost all the clutter spectrum is within the rejection region of the Doppler filter. Then the SIR improvement is limited by the sidelobe level of the filter. If higher improvement is desired for narrow clutter, a deeper null is required. It can also be explained that the Doppler filter bank is equivalent to the match filter for a sinusoidal signal embedded in white noise, but not coloured noise. The optimum filter for sinusoidal signal embedded in coloured noise is a whitening filter cascaded with a Doppler filter bank. This can be approximated by a 2- or 3-pulse canceller cascaded by a Doppler filter bank. Fig. 6.59 *a* and *b* shows the improvement factors for a 3-pulse (double canceller) MTI cascaded with an 8-pulse Doppler filter bank with uniform weighting and 25 dB Chebyshev weighting, respectively [47].

The average improvement factor for all filters is indicated by the dotted curves. It can be seen from this Figure that the improvement factor for a narrow clutter spectrum is much better than that of the Doppler filter bank without a 2-

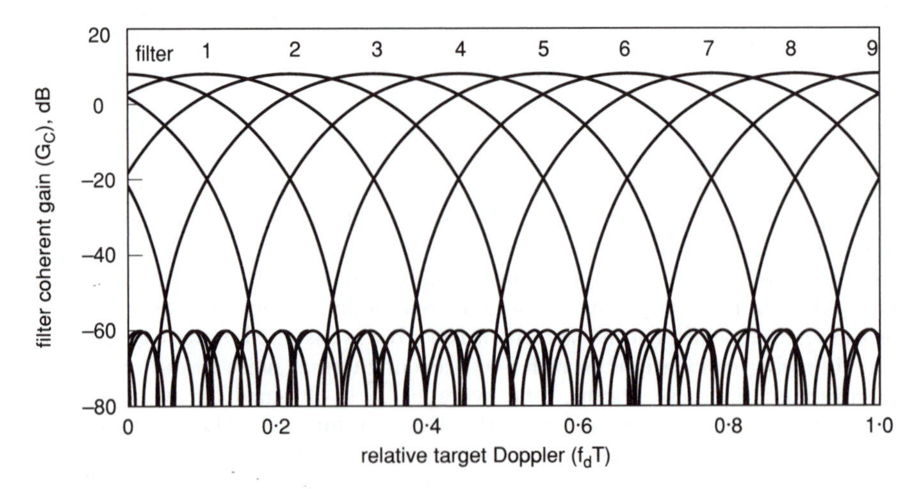

Figure 6.56 *Chebyshev Doppler filter bank with 68 dB sidelobe for N = 9 pulses [45]*

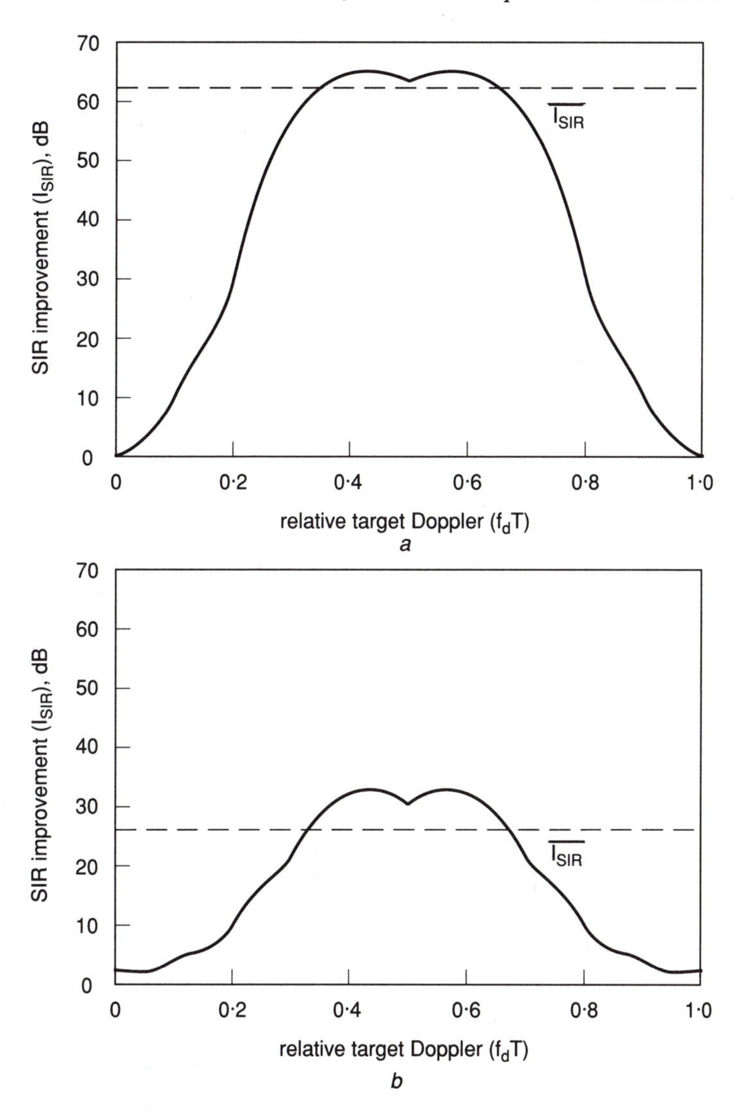

Figure 6.57 *SIR improvement of Chebyshev Doppler filter bank for $N=9$ pulses in Gaussian spectrum clutter with $\sigma_f T=0.05$ and 0.1 [45]*

a CNR = 100 dB
$\sigma_f T=0.05$
zero Doppler

b CNR = 100 dB
$\sigma_f T=0.1$
zero Doppler

or 3-pulse canceller (see Fig. 6.54). The difference in average SIR improvement between uniform weighting and 25 dB Chebyshev weighting is smaller than expected. This is due to most of the clutter spectrum being rejected by the null of the 3-pulse canceller.

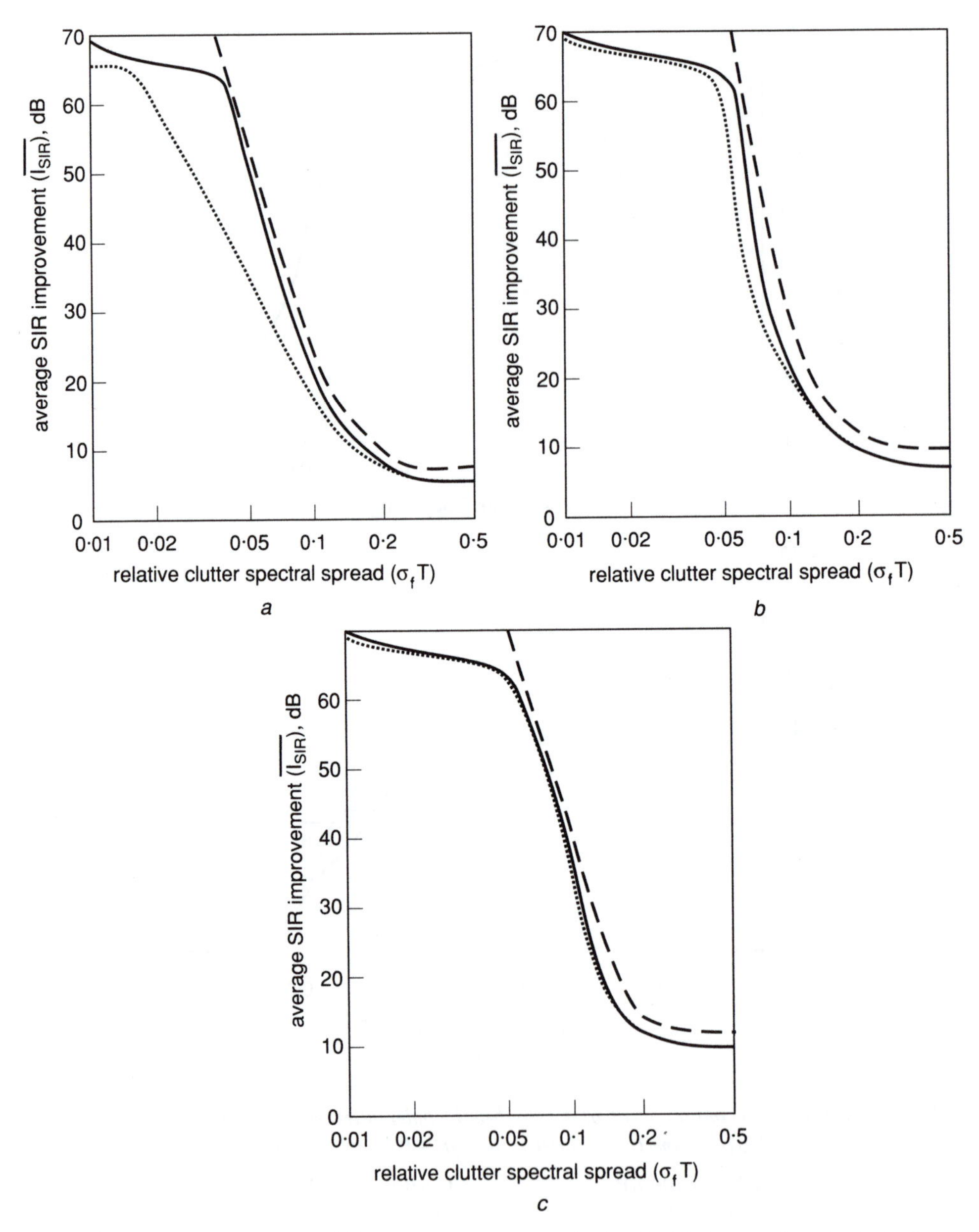

Figure 6.58 *Average SIR improvement curves for 68 dB Chebyshev filter bank [45]*

(*a*) $N=6$
(*b*) $N=9$
(*c*) $N=16$
--- optimum
—— clutter spectrum centred on bandstop region of filter
····· clutter spectrum offset by half filter separation

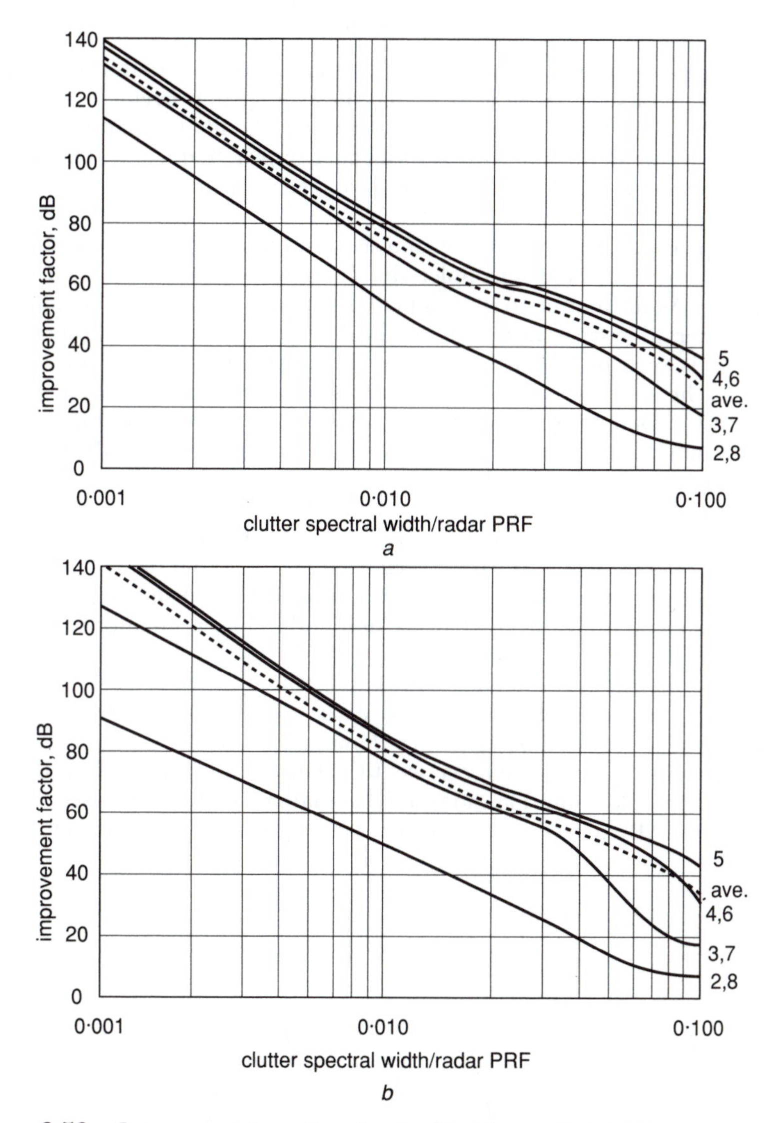

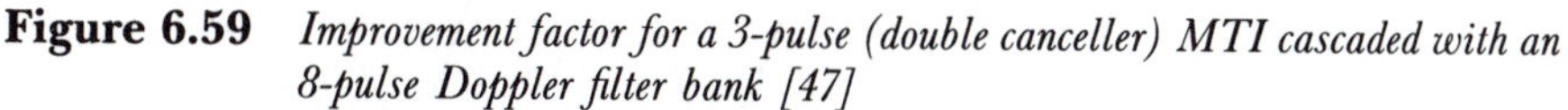

Figure 6.59 *Improvement factor for a 3-pulse (double canceller) MTI cascaded with an 8-pulse Doppler filter bank [47]*

a Uniform amplitude weights
b 25 dB Chebyshev weights

Since the total system improvement factor is limited by the instability of the system, it is very difficult to obtain higher than 60 dB improvement. Therefore, the 68 dB Chebyshev filter bank without canceller is good enough for most cases.

6.3.3 *Moving clutter rejection and CFAR detection of MTD*

It is interesting to study the moving clutter rejection capability of MTD systems. It can be seen from Fig. 6.58 that the average improvement factor for a clutter with mean Doppler frequency equal to 1/2 filter separation is only several dB worse than that with zero mean Doppler frequency. The average SIR improvement versus mean Doppler frequency of the clutter is plotted in Fig. 6.60 [45].

The shape is just like a sine wave, and the amplitude is equal to the difference between the best case (zero Doppler shift) and worst case (offset by 1/2 filter separation). As the number of pulses per CPI (and therefore also the number of Doppler filters) increases from 6 to 16, the difference between the best case and worst case is seen to become negligible. This effect can be explained by the change in the ratio between the Doppler separation between the individual filters and the Doppler rejection region, as listed in Table 6.13 [45].

If one would like to suppress the variation to the order of 5 dB, the number of Doppler filters would have to be increased to approximately 12 parallel filters.

Although the average SIR improvement for moving clutter is good enough in most cases (see Fig. 6.58), the SIR improvement of individual filter is not as good as the average value. It can be seen from Fig. 6.54 that the bottom curve is the improvement factor of the first filter. It means that, if the Doppler shift of the target is equal to the mean Doppler shift of the clutter, no improvement can be obtained.

Another problem in the moving clutter case is that, when there is clutter only in one filter, the output of this filter may cause a false alarm. Therefore, CFAR detection is also required in frequency domain.

Since moving clutter, such as weather clutter and chaff, has Rayleigh distribution for low resolution radar, cell averaging CFAR can be used. The cell averaging CFAR in freqeuncy domain is somewhat different from the cell

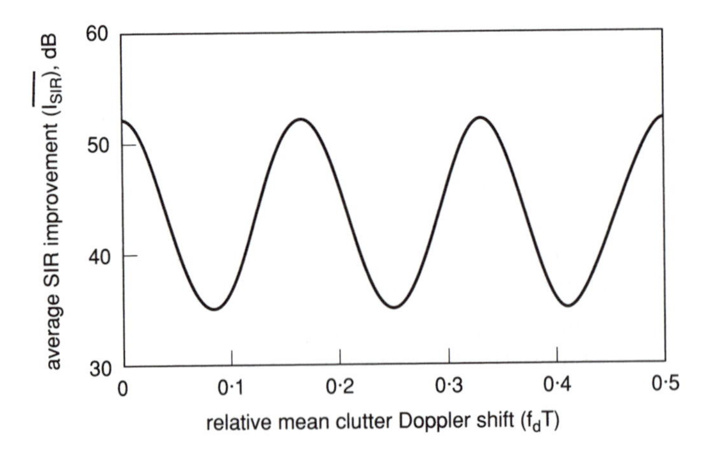

Figure 6.60 *Average SIR improvement for N = 6 as a function of mean Doppler shift of clutter [15]*

$\sigma_f T = 0.05$

Table 6.13 *Ratio of filter separation to Doppler rejection region [45]*

Number of filters, N	6	9	16
Filter separation	0.167	0.111	0.0625
Doppler rejection region	0.22	0.414	0.65
Ratio	0.76	0.268	0.096

averaging CFAR in time domain. Since there are many channels in frequency domain, each channel must have its own CFAR system. This means that the reference samples are taken from adjacent range cells but the same Doppler filter. An adaptive threshold is established based on the estimate of the mean or standard deviation of the moving clutter. Similarly, super-clutter visibility in frequency domain can be obtained. This means that, if the amplitude of the target signal is greater than the adaptive threshold, which is proportional to the mean value of the clutter within the same channel, this signal can be detected.

In the third generation MTD [43, 44], in order to control the false alarms on the edges of clutter patches, the cell averaging CFAR is implemented by averaging the leading and trailing half separately and using the greater of the two averages to establish the adaptive threshold. To improve the resolution capability of the range CFAR, the averages are computed after eliminating the greatest return and its two neighbours from the sum in the CFAR window. This prevents an aircraft present within the CFAR window from incorrectly raising the CFAR threshold.

Another way to realise the CFAR capability is to normalise the output of each channel with the standard deviation of its clutter separately [50]. The block diagram of this system is shown in Fig. 6.61 [50].

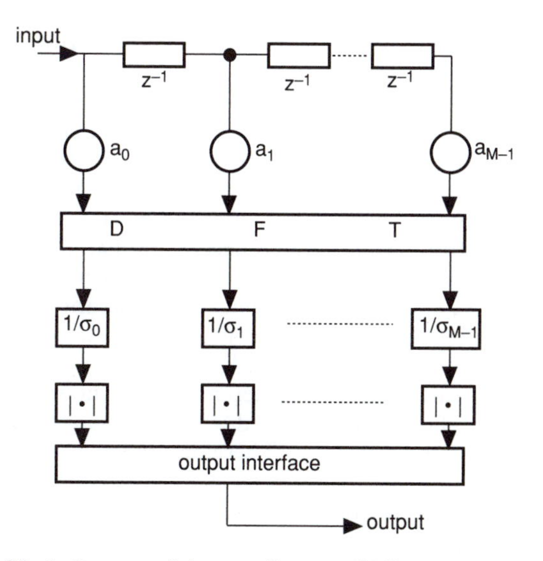

Figure 6.61 *Block diagram of the overall system [50]*

Table 6.14 *Improvement factor of MTD for moving clutter*

Type of interface	Weighting function	Clutter-to-noise	
		30 dB	20 dB
Greatest of	$W(n)=0.5[1-\cos(2\pi n/M)]$	31 dB	21.5 dB
	$W(n)=1-0.5\cos(2\pi n/M)$	22.7 dB	19.5 dB
Sum of	$W(n)=0.5[1-\cos(2\pi n/M)]$	30.9 dB	22.9 dB
	$W(n)=1-0.5\cos(2\pi n/M)$	21.8 dB	20.2 dB

There is an interface unit at the output. It is used to combine the output of multiple filters to a single output. There are two ways to combine the output of multiple channels: the 'greatest of' (GO), and the 'sum of' (SO). The first one selects the greatest output of the filters, and the latter just sums the outputs from all the filters. The output of this interface circuit is then compared with a threshold to decide whether or not a target exists.

Bao [50] calculated the equivalent improvement factor of MTD for moving clutter in both these types of interface circuit. The Doppler filter bank is constructed with an 8-point FFT with weighting. (The improvement factor of this type of Doppler filter bank for fixed clutter has been calculated by Ziemer and Ziegler [51].) The mean Doppler shift of the moving clutter is equal to 0.125, i.e. the frequency of the maximum of the second filter. The shape of the clutter spectrum is assumed to be a Gaussian function with standard deviation of 0.0586. The results are listed in Table 6.14.

Since in most cases the clutter-to-noise ratio is higher than 30 dB, it seems that the 'greatest of' scheme is better than the 'sum of' scheme.

CFAR processing of a real MTD system is much more complex than the interface scheme described above. Fig. 6.62 shows the CFAR detector of GE 592 radar [52].

The requirement of CFAR processing is to maintain a low constant alarm rate in all combinations of noise and clutter without seriously degrading target detectablility. Fig. 6.62 indicates that a signal must pass five tests to be declared a positive detection.

The first test determines whether the signal exceeds the system noise level by an amount appropriate to maintain the desired probability of false alarm in noise. It uses an estimate of the system noise made prior to transmission. In this way, only system noise, not clutter or targets, influences the estimate. The noise estimate is multiplied by eight computer-controlled constants ($K_0, K_1, \ldots, K_7$) to obtain the desired false-alarm rate (FAR) in each Doppler filter. The differences in the constants reflect the differences in the noise gain through each Doppler filter. In the short range interval, the sensitivity time control (STC) varies the noise as a function of range. Therefore, the eight constants must also vary with range.

The second test uses a short-range normaliser to control the FAR in sea and weather clutter. Background estimates, shown as $\overline{C_0}, \overline{C_1}, \ldots, \overline{C_7}$ in Fig. 6.62 are derived by averaging over a limited number of range cells and the two

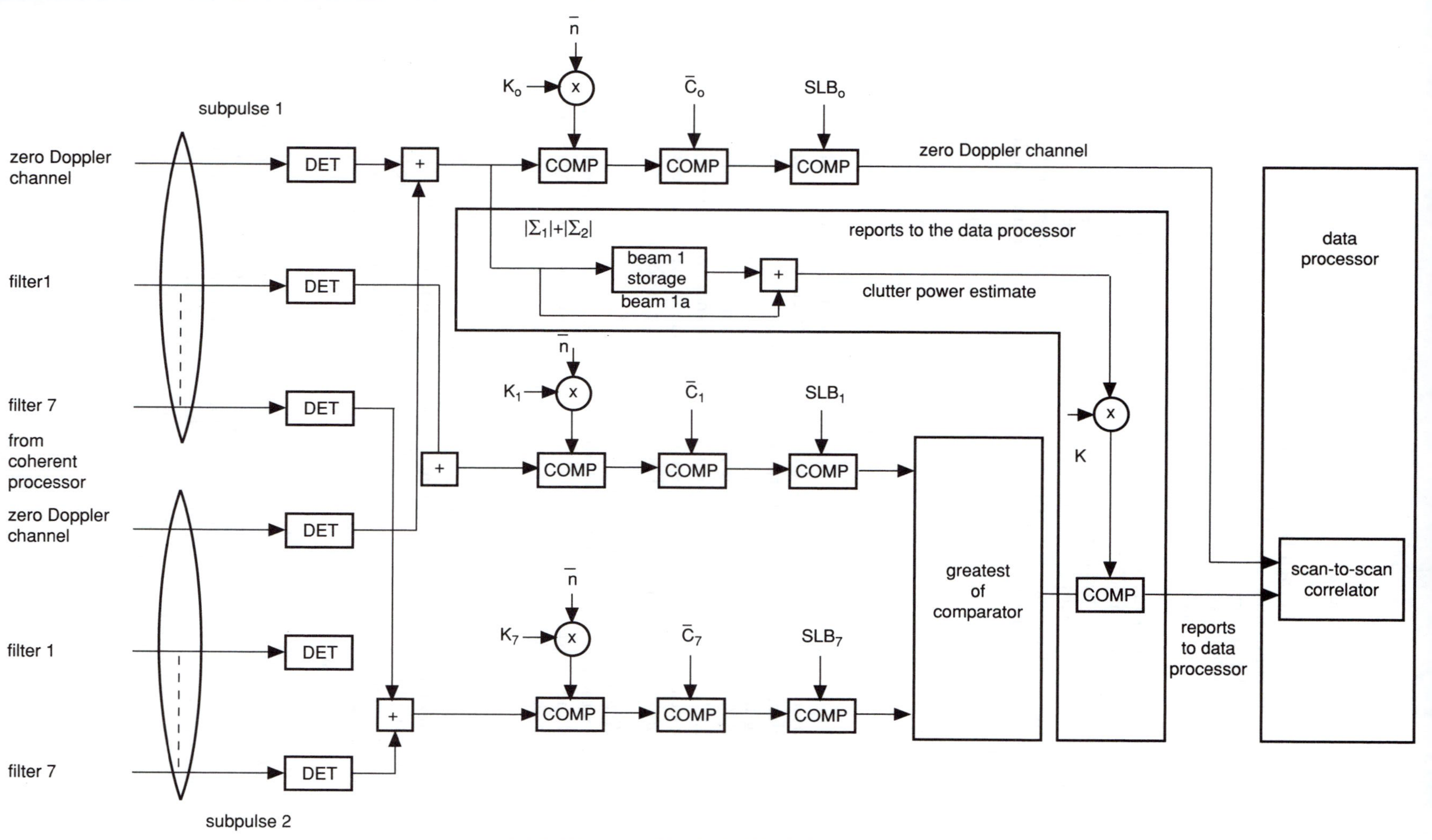

Figure 6.62 *CFAR detector of GE 592 radar [52] © IEEE, 1984*

diversity channels in each Doppler channel. The short-range normaliser is very effective in controlling the FAR in sea clutter and weather clutter.

The third test compares the output of the sum beam to the output of the sidelobe blanker (SLB) beam. This test is useful in rejecting large aircraft returns, clutter and repeater jammers that are detected in the antenna sidelobes.

Range-extended normalisers are not as effective in controlling the FAR in heterogeneous terrain clutter. Therefore, a fourth test is used to set the FAR in terrain clutter. This test is applied to the seven high-Doppler signals which have passed the first two tests. It involves the comparison of the filter outputs with an estimate of the average clutter power on a range-cell-by-range-cell basis. The clutter power estimate is derived by averaging four independent samples of the clutter. These are the outputs of two zero Doppler filters (one per diversity channel) on both 10-pulse transmissions used in the lowest short-range beam position.

This test relates to using a clutter map with power estimates derived by averaging over four scans. The implementation uses frequency rather than time to decorrelate the clutter samples. It has the significant advantage of not requiring multiscan storage and processing for every range-azimuth cell.

A fifth test that is used is an azimuth-scan-to-azimuth-scan correlator. The correlator is implemented in the computer and rejects returns on the basis of ground speed. The primary purpose of the correlator is to eliminate slow moving targets such as birds. A simulation of a four-scan correlator indicated that 86% of the birds flying at 20 to 80 knots would be eliminated while less than 0.2% of aircraft flying between 80 and 3000 knots would be rejected.

Desodt and Larvor [53] suggested four methods of CFAR processing for a shipborne pulse Doppler radar. These methods are as follows:

(i) *Cell averaging CFAR* (*method* 1)
This method is suitable where clutter echoes are weak: far from the radar, behind the horizon line, or in the Doppler filters at half the ambiguous speed. In these areas, it provides as little processing losses as possible.

(ii) *The 'greatest of' CFAR* (*method* 2)
This method is applied in areas where noisy echoes are stronger than in the previous case, and their distribution is presumed not to be homogeneous. So we have chosen the number of reference cells in one side to be smaller than that of the total of two sides in method 1. Of course, that will produce more losses than the first method. One advantage of this method is its robustness in front of clutter edges: the normalisation by the largest average value will avoid detection in such cases.

(iii) *Area averaging CFAR* (*method* 3)
A map describing statistical values over sets of cells is updated at every antenna revolution, by a first order recursive filter. So, at any time, and for any block of range cells, a mean value μ and a standard deviation σ are available. The normalisation operation means that the current value of a cell under test subtracts the mean value μ and divides by σ. This method is designed to cancel sea clutter. It is assumed that the clutter is stationary distributed along the block of cells.

(iv) *CFAR* (*method* 4)
The principle of this method is similar to the previous method. The single difference lies in the non-stationarity of the clutter we expect to meet within one block of range bins, for instance ground clutter echoes. Fortunately, these echoes are assumed to have no velocity, which reduces the application domain of this method. The computation of this method is carried out in the same way as method 3. The statistical characteristics μ and σ do not have the same definition.

In operational work, for each range bin, all of these four methods are computed, but only one is chosen for the output. An external sensor informs the radar how far the coastline is from the ship the radar is settled on. Method 3 is chosen for an offshore bin and method 4 for an inland one. The first criterion used for choosing which method will give the definitive value of a cell is an estimation of the clutter power level around the cell. If the estimated value is small enough, method 1 is chosen. Otherwise the result will be the smallest result between method 2 and the other computed method.

Results from experiments show that no false alarm is due to clutter even in severe conditions, while targets can still be detected. For a fine clutter estimation, four methods were used, each of them being well matched to a specific clutter type (white noise, atmospheric clutter, sea clutter, ground clutter). As often as possible, the estimator providing minimum losses is chosen. In a clutter free area 100% Doppler band is achieved, with reduced losses.

6.3.4 *Accuracy and resolution of MTD system*

As mentioned before, the block processing of MTD results in only two outputs within a beamwidth dwelling time. This will cause inaccuracy in azimuth estimation.

In practice, a large aircraft may produce 100 primitive reports per scan depending on cross-section, range, elevation etc. A primitive report is a single detection output from the MTD Doppler filter bank in a given CPI, range cell and filter combination. A typical value is 35 primitive reports per aircraft. Therefore it is possible to develop an algorithm to estimate the centroid of these primitive reports in order to improve the accuracy and resolution of the position (range and azimuth) estimation.

The third generation MTD developed by Westinghouse Electric for the ASR–9 radar utilises a new algorithm for this purpose [54].

Each primitive report provided for correlation and interpolation (C&I) contains its range, azimuth, Doppler filter number and the detection amplitude encoded in log format. A single target can create a multiplicity of primitives extending in Doppler, range and azimuth, as illustrated in Fig. 6.63.

Clustering is the process which groups together primitives created by a target and associates them with an active target report in the report file. Primitives which enter the C&I process are checked to determine if they are associated with an existing entry in the report file. If primitives fail to associate with an existing report a new entry in the report file is created using adjacent primitives. The primitive having the greatest amplitude is located (filter 4 in Fig. 6.63) and its range is used as the range centre of the new report entry (R_c). A five cell window centred on R_c is used to group the remaining primitives to form this

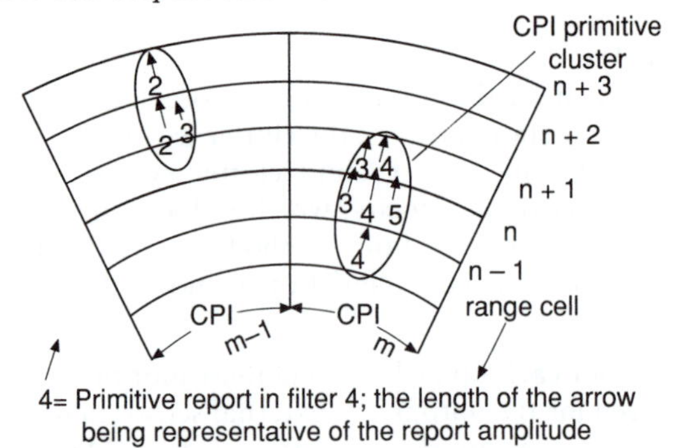

Figure 6.63 *Clustering of primitives from a single CPI [54] © IEEE, 1986*

cluster. Primitives at ranges more than 1/8 n.m. (2 range cells) from R_c are prohibited from associating with a target report.

The association of primitive data with an active target report is accomplished using the report range R_c. Those primitives with ranges within two range cells of R_c are tentatively grouped with that active target report.

6.3.4.1 *Range resolution improvement*

The ASR–9 radar is required to solve two targets spaced in range by 1/8 n.m. or greater. Since large targets can cause amplitude smears with extents up to 5/16 n.m. (5 range cells), an ambiguity exists as to whether two targets or a single large target is present. The method used for resolving two targets is a simplified form of pulse shape matching. If the primitive data extend more than a single range cell beyond R_c, a test is performed to determine if one or two targets should be declared.

A two target case (see Fig. 6.64) having pulse centres 1/8 n.m. apart (e.g. R_c and $R_c \pm 1/8$ n.m.) creates a composite pulse shape which does not have the

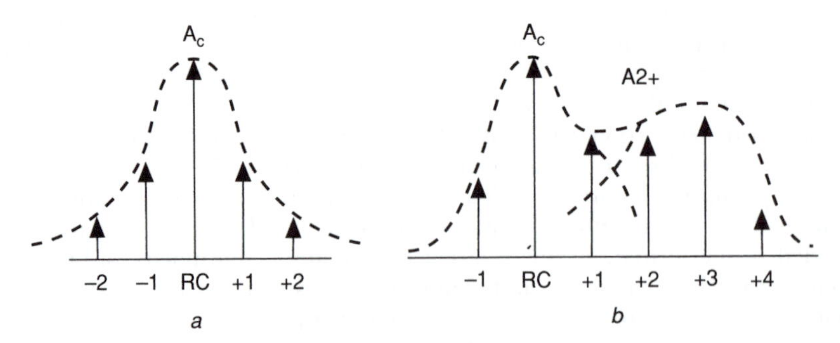

Figure 6.64 *Range resolution example [54] © IEEE, 1986*

a $\dfrac{A_c}{A_c \pm 2} \geqslant P_v$: single target

b $\dfrac{A_c}{A_c \pm 2} < P_v$: two targets

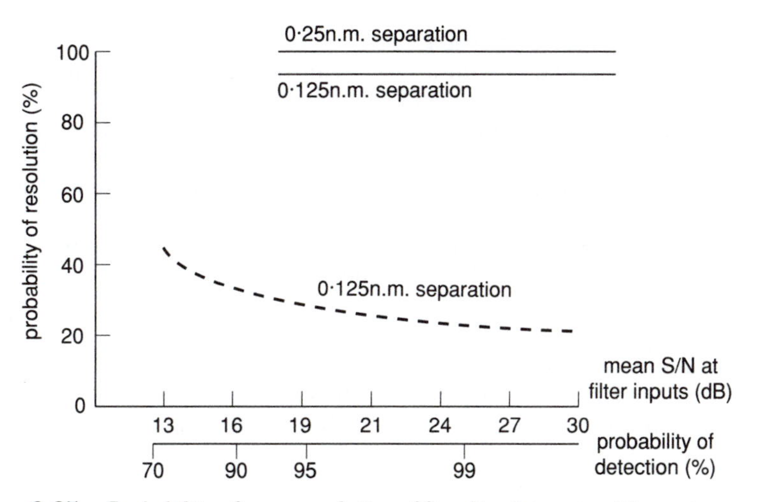

Figure 6.65 *Probability of range resolution of Swerling 1 targets with equal mean cross-section and Doppler [54] © IEEE, 1986*

—— range resolution test
– – – no range resolution test

expected amplitude drop off away from R_c; that is, the ratio of amplitudes of the primitives at R_c and 1/8 n.m. away is less than that expected for a single target. This condition results in a two target declaration. The primitives at $R_c \pm$ 1/8 n.m. range cells for a two target case are used to correlate with another report or to initiate a new report. If the amplitude ratio is within acceptable limits, a single target is declared and the data at $R_c \pm 1/8$ n.m. are discarded.

This range resolution test is constrained to the testing of primitive data collected within a single CPI to avoid fluctuation effects in the amplitudes between CPIs. The range resolution test increases the probability of resolving two targets separated by 1/8 n.m. as shown in Fig. 6.65 [54].

6.3.4.2 *Range accuracy improvement*

The centroid processes employed for the ASR–9 require that some of the primitive data for each CPI be preserved during the dwell on the target. The peak nonzero velocity filter and the peak zero velocity filter amplitudes from correlating primitive data are extracted and placed in the report file for primitives at R_c as well as for the largest adjacent reporting cell. A maximum of four amplitudes are retained from the input primitive data on each CPI for target extents up to a maximum run length of 10 CPIs (5 beamwidths). The remainder of the primitive data associated with the target report are discarded.

A range straddle test is performed to determine if the range centre of the group should be selected at R_c or on the boundary of R_c and its adjacent cell. If the ratio of the amplitudes of the two cells is within preset limits, a straddle condition is declared. This straddle condition biases the range estimate by 1/2 range cell (1/32 n.m.) in the direction of the adjacent cell to provide an improved range accuracy of 0.019 n.m. RMS as determined by computer simulation.

6.3.4.3 *Azimuth accuracy improvement*

The report data associated with a target can be obtained only once per CPI owing to the block processing of MTD. The azimuth extent of a CPI is equal to 1/2 beamwidth for most MTD systems. Therefore, an interpolation algorithm must be used to improve the azimuth accuracy. Five interpolation algorithms are used in ASR–9 radar. They are: single CPI, beamsplit, centre of mass, interpolation and beamshape match algorithms.

Except for the beamshape match algorithm, each of the algorithms used for interpolation can be expressed in the following form:

$$\text{Azimuth} = \theta_1 + (\theta_2 - \theta_1)F(M_2 - M_1) \tag{6.133}$$

where θ_1 and θ_2 are the azimuths of the target report data to be used for interpolation and F is the interpolation function which depends only on the difference of the log amplitude (M_1 and M_2) of the primitives at θ_1 and θ_2, respectively.

(i) *Single CPI*

When only a single CPI of data is present in the report, the azimuth position is set equal to the azimuth at which the single CPI of data was measured. In this case the function F equals 0, so the azimuth is simply θ_1.

$$\text{Azimuth} = \theta_1 \tag{6.134}$$

(ii) *Beamsplit*

This algorithm is used when the entire set of data in the report has been affected by an exception condition. In this case θ_1 and θ_2 are the azimuths of the first and last CPI associated with the target report. When receiver saturation is sensed in the selected data set, a variation of the algorithm is used by selecting θ_1 and θ_2 to be the first and last CPI having the saturation flag set. In both cases the value of the interpolating function F is 1/2.

$$\text{Azimuth} = \theta_1 + (\theta_2 - \theta_1)1/2 \tag{6.135}$$

(iii) *Centre of mass*

The centre of mass algorithm is an interpolation algorithm which is used when only a single CPI of data is available on each PRF. The classic form of this equation is as follows:

$$\text{Azimuth} = \frac{\theta_1 A_1 + \theta_2 A_2}{A_1 + A_2} = \theta_1 + (\theta_2 - \theta_1)\left[\frac{1}{(A_1/A_2)+1}\right] \tag{6.136}$$

where θ_1 and θ_2 are the azimuths associated with the two CPIs and A_1 and A_2 are the linear amplitudes associated with θ_1 and θ_2, respectively.

(iv) *Interpolation*

When detections occur at adjacent azimuths on the same PRF, but not enough data exist to invoke beamshape matching, the interpolation algorithm is employed. It interpolates between two angular positions based on the ratio of the amplitudes at the two azimuths by using a difference curve derived from the antenna pattern. Again, the difference in the two log amplitudes is used as an index to a look-up table:

$$\text{Azimuth} = (\theta_1 + \theta_2)/2 + K(\log A_1 - \log A_2) \tag{6.137}$$

where θ_1, θ_2, A_1, A_2 are as before and K is a multiplier that translates the difference in the amplitude into an angle offset from the centre of the two adjacent azimuths. The parameter K approximates the slope of the antenna pattern log difference curve between the two adjacent azimuths, and is exact for a Gaussian antenna pattern.

(v) *Beamshape match*

Since the actual antenna azimuth pattern varies as a function of elevation angle and frequency in addition to variation in the actual antenna azimuth pattern due to aging and production tolerance, the simple interpolation algorithm will produce error due to ignoring these factors. The variation with elevation is a significant factor since the actual azimuth beamwidth varies by approximately 10% over the specified elevation coverage. A threshold must be selected that results in a small split probability for elevations of interest when only a single target is present.

Since the PRF is alternated between CPIs, samples from the same PRF in the ASR–9 are spaced by 1.4° which is approximately the antenna beamwidth. The matching is performed at 0.088° intervals corresponding to the granularity of the stored reference table and the granularity of the reported primitive azimuths. The reference table consists of three azimuth beamwidths of data requiring the storage of 49 samples. The spacing between the first and last amplitude samples of the central three CPIs is approximately 32 ACPs, thus requiring about 17 iterations to check all possible positions of the target. The offset of the stored pattern providing the minimum mismatch is used to determine the single target position.

An algorithm was developed to minimise the mismatch error. If the minimum error is greater than a preset value a two target condition is declared present. Otherwise, a single target case is declared. If a single target outcome results, the azimuth position of the target is determined as the azimuth associated with the peak of the reference pattern where the best match occurs. The angle off boresight (AOB) relative to the azimuth of the three measured samples is

$$\text{AOB} = \text{Boresight} - \text{Offset} \tag{6.138}$$

Where the boresight of the stored antenna pattern is at the centre of the table and offset is the position of the best match which is the iteration count of the best match, the target azimuth is computed to be

$$\text{Azimuth} = \theta_1 + \text{AOB} \tag{6.139}$$

where θ_1 is the azimuth of the first of the three measured data samples used in the matching process.

The azimuth estimation accuracy of four of these five algorithms is shown in Table 6.15.

Table 6.15 *Azimuth centroid quality and accuracy [54] © IEEE, 1986*

Condition	Report quality	Algorithm	RMS accuracy (beamwidth)	No. of primitives
Single CPI Report	0	Single CPI	0.164	1–3
2 CPI report 1 on each PRF	1	Centre of mass	0.093	2–10
2 CPI report, 1 at least 1 PRF	2	Interpolation	0.043	6–35
⩾3 CPI report on at least 1 PRF	3	Beamshape match	0.036	>30

The centroid algorithm selection process operates on the data received from the association process to select one of four algorithms. The algorithm selection is a function of the signal-to-noise of the echo as shown in Fig. 6.66.

6.3.4.4 *Azimuth resolution improvement*

The beamshape match algorithm provides a means of resolving target returns overlapping in azimuth that are not resolvable by the normal hit/miss processing. Resolution is accomplished by comparing the echo amplitudes with the expected modulation created by the antenna scanning across a single target. Typical amplitude returns for one and two target cases are depicted in Fig. 6.67.

The matching process uses the central three CPIs of data at same PRF. For an even number of CPIs the central three are formed by using the central two and the adjacent CPI with the largest amplitude. These three amplitude values are sequentially compared with a stored replica of the theoretical antenna pattern offset, varying amounts in azimuth to find the point where the best fit occurs. If the error associated with the best fit is sufficiently large, two targets

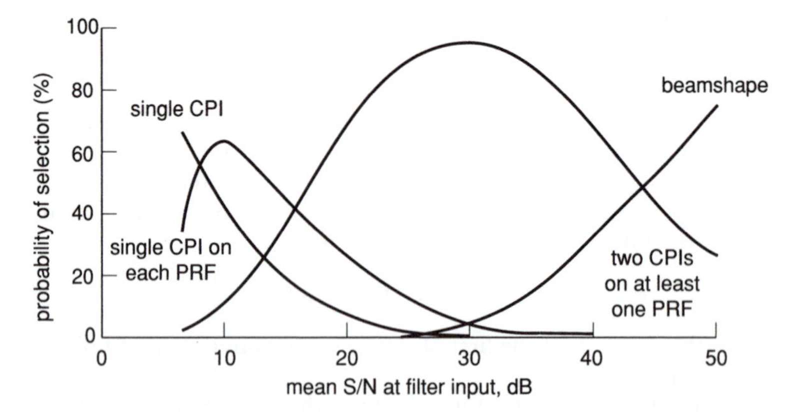

Figure 6.66 *Algorithm selection as a function of S/N [54] © IEEE, 1986*

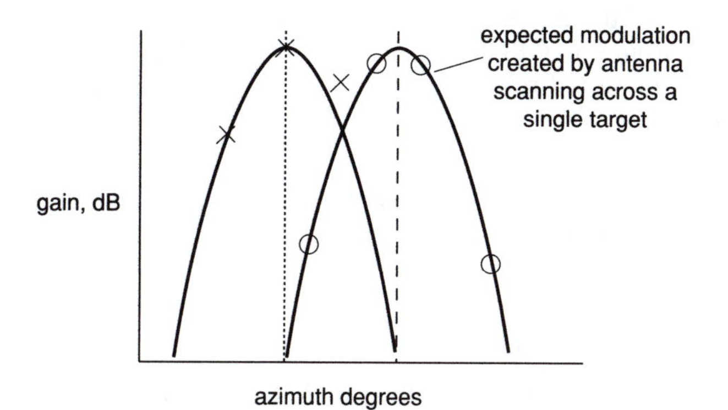

Figure 6.67 *Amplitude patterns for one and two target cases [54] © IEEE, 1986*

○ = typical amplitude pattern for a large single target
× = typical amplitude pattern for two targets in phase, closely spaced in azimuth (data normalised)
····· target 1 location
– – – target 2 location

are declared. Fig. 6.68 shows the resolution of a Swerling I target detectable on only one of two PRFs with equal mean S/N separated only in azimuth.

The ability of the beamshape algorithm to resolve targets closely spaced in azimuth is limited primarily by four factors: (i) noise; (ii) granularity of the

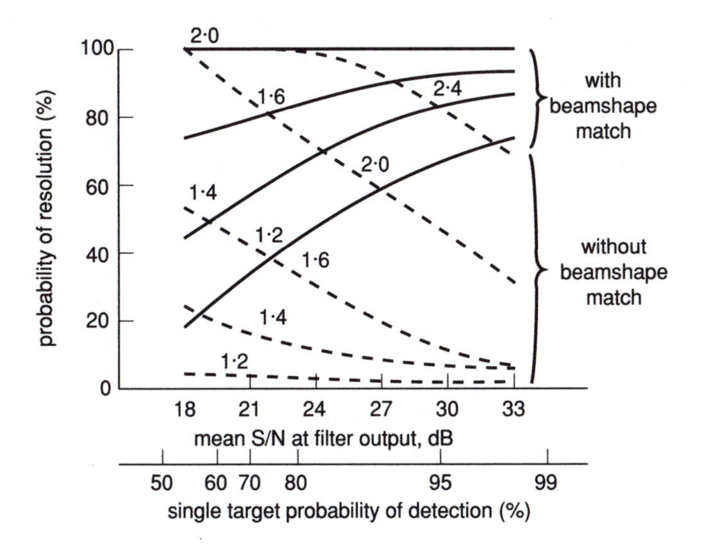

Figure 6.68 *Resolution of Swerling I targets with equal mean S/N [54] © IEEE, 1986*

stored replica of the antenna pattern; (iii) fluctuation of the apparent radar cross-section from one CPI to the next; and (iv) variations of actual antenna pattern from the stored reference pattern.

6.4 Performance comparison between MTI, MTD and optimum MTI

In modern surveillance radar systems, three types of MTI system are used to reject fixed and moving clutter:

(i) Moving window fixed coefficient MTI system cascaded with non-coherent moving window integrator-detector
(ii) Fixed MTI cascaded with adaptive MTI and followed with moving window detector
(iii) Fixed MTI cascaded with coherent integrator (MTD).

We discussed the performance of these three systems in the previous Section. It is interesting to compare their performance in the same conditions. The most important condition is the number of pulses processed in these systems; the available number of pulses of a radar system mainly depends on the tactical requirement, and therefore the technical specifications of the radar. These are: maximum non-ambiguity detection range (PRF), angular resolution and accuracy (beam-width), the data rate (antenna revolution rate). Another important factor is the ECCM capability of the radar, namely the employment of frequency agility. We can divide the number of pulses N into four categories:

(i) $N > 64$: This happens in the case of short range high PRF radar. The MTD system compatible with frequency agility can be considered.
(ii) $N > 32$: This happens in the case of medium range medium PRF radar. An MTI system compatible with frequency agility can be employed. We will discuss this type of MTI system later.
(iii) $N > 16$: This happens in the case of the ASR type of radar. The MTD system is the best selection.
(iv) $N > 8$: This happens in the case of long range surveillance radar. It is very difficult to say which is the best in this case.

Hansen [55] compared the performance of MTI, MTD and optimum processor against clutter in the same condition. The total number of hits per target was assumed to be $N = 18$ in all cases considered. The number of pulses within a CPI was chosen as 6, with the exception of the 3-pulse MTI where three pulses are used. The choice of six pulses in a CPI is a compromise between Doppler resolution and the requirement to use multiple PRFs for blind speed elimination and multiple transmitter frequencies during the time on target for the ECCM requirement. The highest PRF was assumed to be 1000 Hz and the CPI-to-CPI stagger ratios were chosen as 12:13:15. Only fixed frequency radar operation and a Swerling case 0 non-fluctuating target were considered in this comparison.

The clutter models are assumed to have Gaussian shape of spectrum with a standard deviation σ_f and the interpulse period T corresponds to the highest

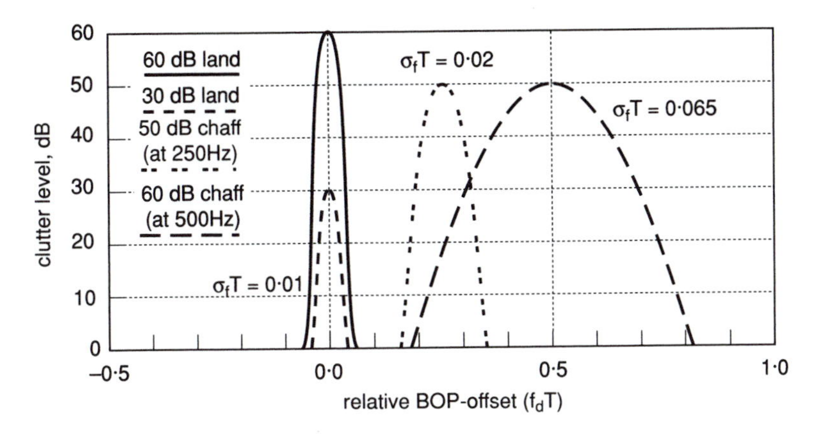

Figure 6.69 *Spectrum of clutter models assumed for performance comparison [55] © IEEE, 1980*

PRF used. Along with thermal noise at 0 dB reference level the following types of clutter were considered:

(i) 60 dB land clutter at zero Doppler with $\sigma_f T = 0.01$
(ii) 30 dB land clutter at zero Doppler with $\sigma_f T = 0.01$
(iii) 50 dB weather clutter (or chaff) at relative Doppler of 0, 0.25, 0.5, with $\sigma_f T = 0.065$
(iv) 50 dB weather clutter (or chaff) at relative Doppler of 0, 0.25, 0.5 with $\sigma_f T = 0.02$
(v) 60 dB land clutter and 50 dB weather clutter ($\sigma_f T = 0.065$) simultaneously
(vi) 30 dB land clutter and 50 dB weather clutter ($\sigma_f T = 0.065$) simultaneously.

Graphic illustrations of the above clutter spectra are shown in Fig. 6.69 for a single Doppler period.

Five types of processor were examined:

(i) 3-pulse adaptive MTI [30]: In this case both zeroes are adaptively centred on the clutter spectrum.
(ii) 6-pulse adaptive MTI: The magnitudes of the weights are assumed to be binomial. Two zeroes are always at zero Doppler to cancel land clutter while the three remaining zeroes are adaptively located to cancel the moving clutter spectrum. A practical implementation of the 6-pulse adaptive MTI uses a cascade of five 2-pulse canceller sections as shown in Fig. 6.70. The responses of this processor for all five zeroes at zero Doppler, for three zeroes at a relative Doppler of 0.3 and for 3 zeroes at a relative Doppler of 0.5, are shown in Fig. 6.71.
(iii) Doppler filter bank consisting of 5 FIR filters: The weights are the same for all clutter models, and the responses of the individual filters are shown in Fig. 6.72. The sidelobe levels are about −33 dB for first filter, −43 dB for second and third filters.

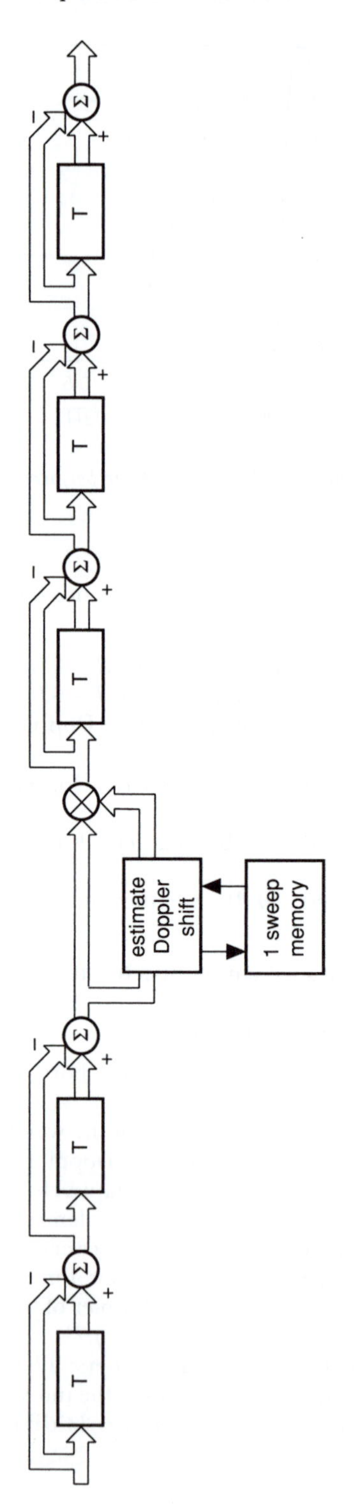

Figure 6.70 *Practical implementation of 6-pulse adaptive MTI [55]*

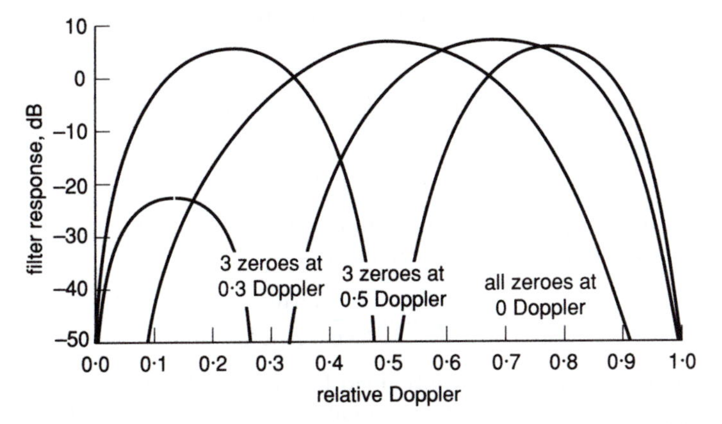

Figure 6.71 *Responses of 6-pulse adaptive MTI for different locations of three moving zeroes [55] © IEEE, 1980*

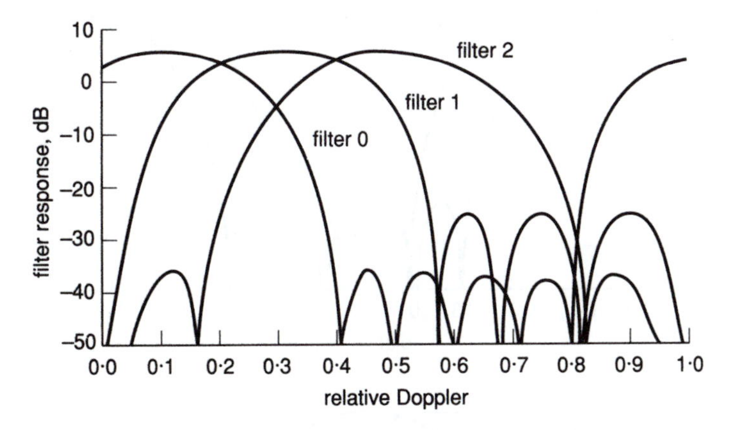

Figure 6.72 *Filter responses for five FIR filter Doppler processors [55] © IEEE, 1980*

Filters 3 and 4 are symmetric to filters 1 and 0

(iv) Doppler filter bank consisting of 6 FIR filters: As with the 5-filter bank, the weights are fixed and the individual filter responses are shown in Fig. 6.73. The sidelobe level is about −40 dB for all filters.

(v) The optimum processor: For a specified clutter model, this processor maximizes the output SIR at each distinct target Doppler frequency. Thus, since target Doppler takes on a continuum of values, a large number of sets of complex weights would be required. To determine the required filter weights is theoretically possible but could become difficult to implement in an actual system.

The optimum weights of this optimum filter are calculated from

$$w_{opt} = R_I^{-1} \cdot s^* \tag{6.140}$$

where R_I^{-1} is the inverse of the interference matrix R_I and s^* is the complex conjugate of the signal vector.

$$R_I = E\{Y \cdot Y^{*T}\} \tag{6.141}$$

where Y^T is the complex matrix of the interference entering the K-tap FIR filter.

The corresponding SIR improvement is

$$I_{SIR} = |w_{opt}|^2 \frac{Tr\{R_I\}}{w_{opt}^T R_I w_{opt}^*} \tag{6.142}$$

SIR improvement curves were obtained for each of the five types of coherent processors against the various clutter models. An examination of this total of 64 response plots shows that typically the 6 and 5 FIR filter processors are somewhat inferior to the optimum as the target Doppler approaches the location of the clutter spectrum. As an example, the SIR improvement against

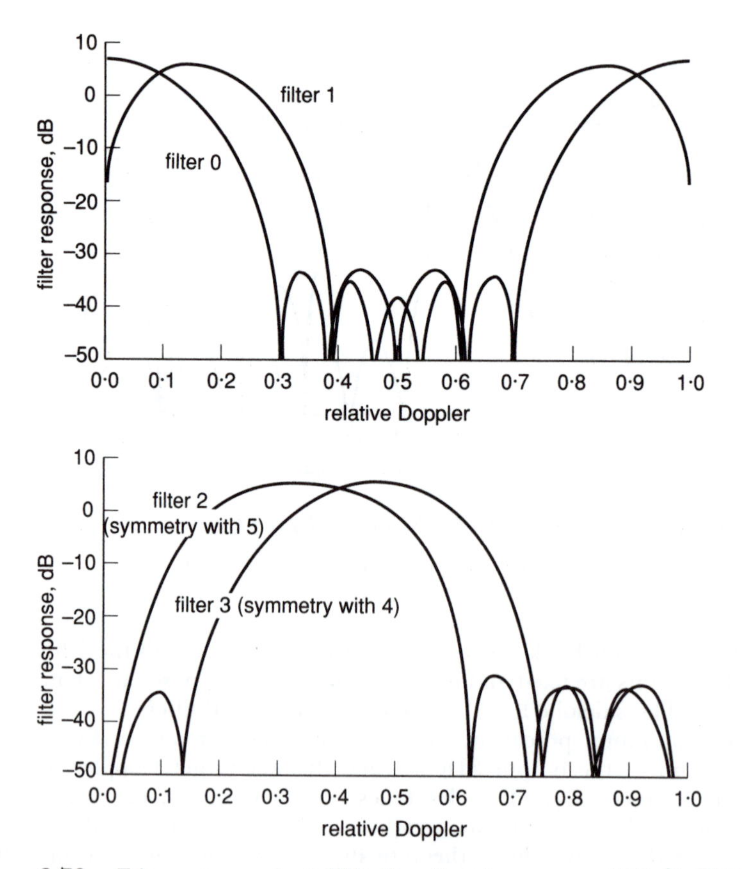

Figure 6.73 *Filter responses for 6 FIR filter Doppler processor [55] © IEEE, 1986*

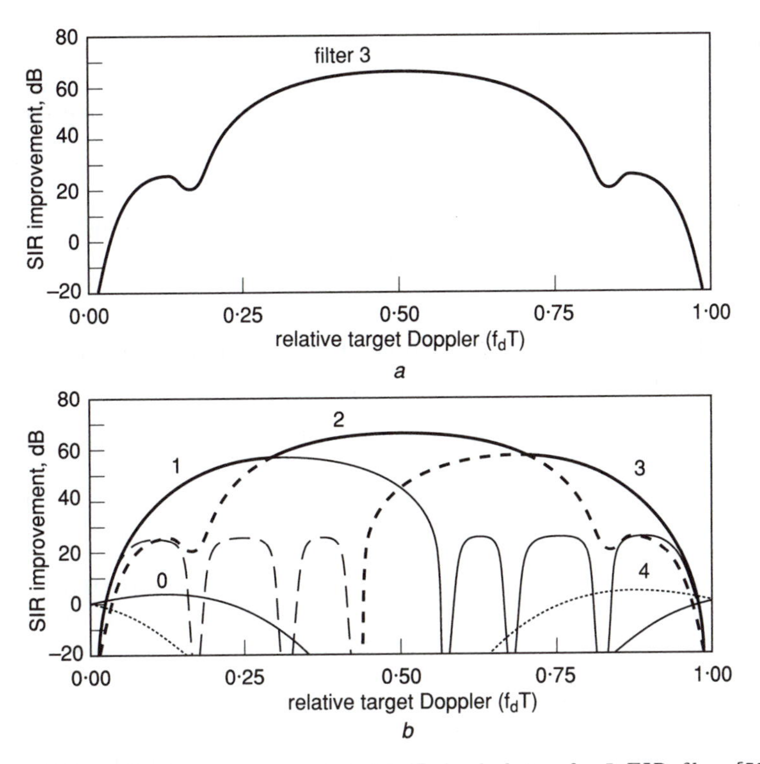

Figure 6.74 *SIR improvement against 60 dB land clutter for 5 FIR filter [55] © IEEE, 1986*

(*a*) I_{SIR} of filter 2,
(*b*) Overall SIR improvement

60 dB land clutter for filter 2 of 5 FIR filter processor is shown in Fig. 6.74*a*, and the overall SIR improvement is shown in Fig. 6.74*b*. The comparison of this filter with the optimum filter is shown in Fig. 6.75.

The performance of a fixed parameter FIR filter against wide spectrum chaff is shown in Fig. 6.76*a*, *b* and *c* for 0.0, 0.25 and 0.5 relative Doppler, respectively. The SIR improvement of optimum processor is also shown in these Figures for comparison.

The performance of the MTI processor is substantially below the optimum in the same condition. The only case where an adaptive 6-pulse MTI appears to be superior to the 6 and 5 FIR filter processors, as based on these SIR improvement plots, is in the case of combined 60 dB land clutter at zero Doppler and 50 dB chaff at a relative Doppler of $f_D T = 0.5$. This bimodal clutter situation represents such a difficult situation for any 6-pulse FIR filter that significant clutter rejection requires the filter zero to be located exactly at the clutter spectra. In this case higher order FIR filters can obtain better results.

D'Addio *et al.* [56] compared the SIR improvement of the MTI and/or adaptive MTI plus coherent integrator (CI) with that of the optimum processor. The coherent integrator is realised with DFT with or without weighting. It is assumed that the coherent integrator is to be tuned to the effective target Doppler frequency, neglecting the shaping losses which in practice may affect the performance of a CI. When two clutters are present and a fixed MTI is used to filter the ground clutter, the effects of ground clutter residues on the tuning process of the adaptive MTI notch are neglected.

The clutter is assumed to have a Gaussian shape spectrum and Gaussian probability density. The spread of the clutter is measured with the correlation coefficient ρ. In the Gaussian assumption case, ρ is defined as

$$\rho = \exp[-2(\pi\sigma_c T)^2] \tag{6.143}$$

where σ_c denotes the standard deviation of the clutter spectrum. Therefore,

$$\sigma_c T = 1/\pi[(-\ln \rho)/2]^{1/2} \tag{6.144}$$

where T is the interpulse period of the radar, which is assumed to be a constant.

The improvement factors of a 3-pulse binomial MTI cascaded with 8-pulse coherent integrator, i.e. 10-pulse MTD, in 40 dB ground clutter are shown in Fig. 6.77. The performance of a 10-pulse optimum processor is also shown in this Figure for comparison.

Three different clutter spreads are assumed in this Figure: $\rho = 0.8$, 0.9, 0.99. This is equivalent to $\sigma_c T = 0.106$, 0.073 and 0.0226, respectively. The performance of MTD is inferior to that of the optimum processor as the clutter spread is increased and the spacing between target and clutter is increased.

For a larger CNR (60 dB) (Fig. 6.78), the performance degradation of the MTD with respect to the optimum processor is more evident.

In the case of bimodal clutter, three types of processors are studied. Fig. 6.79 shows the IF pertinent to a processor consisting of the cascade of a binomial MTI plus an adaptive MTI for a signal with random frequency. The two MTI filters are assumed to have the same number of pulses indicated in the abscissa.

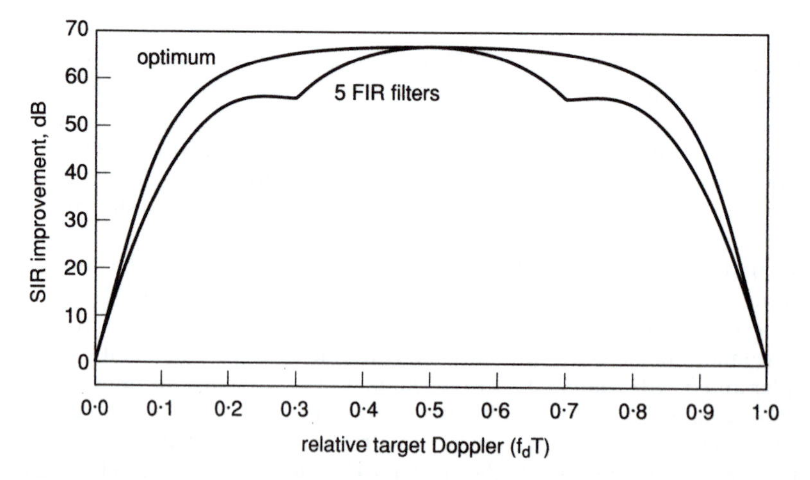

Figure 6.75 *Comparison of SIR improvement of 5 FIR filter and the optimum filter against 60 dB land clutter [55] © IEEE, 1986*

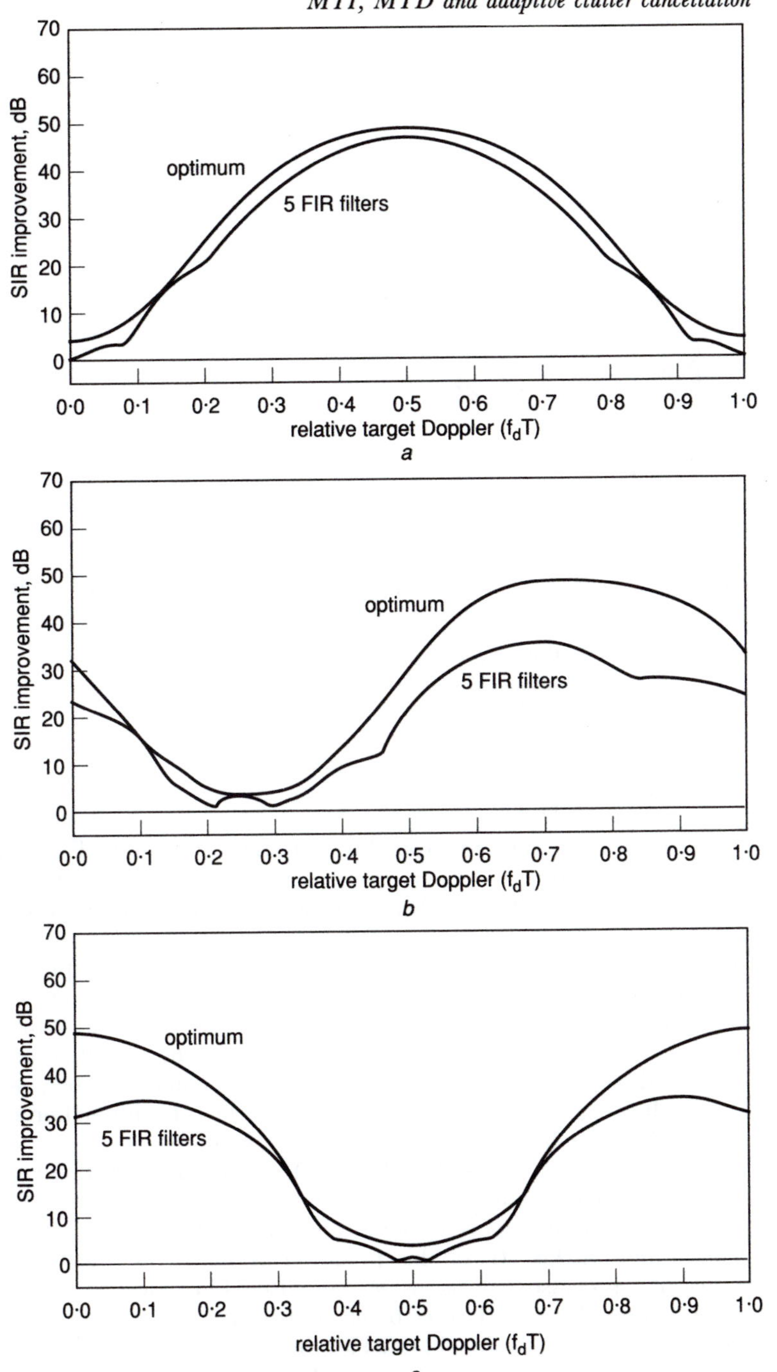

Figure 6.76 *SIR improvement of 5 FIR filter against 50 dB chaff ($\sigma_f T = 0.065$) at relative Doppler of (a) 0.0, (b) 0.25 and (c) 0.5, respectively [55] © IEEE, 1986*

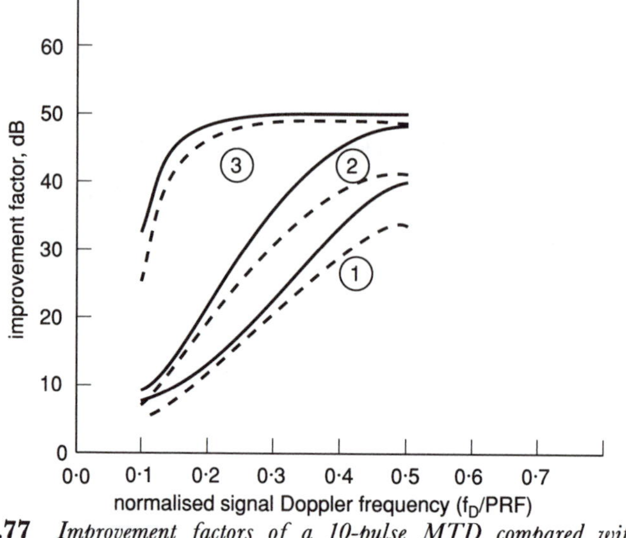

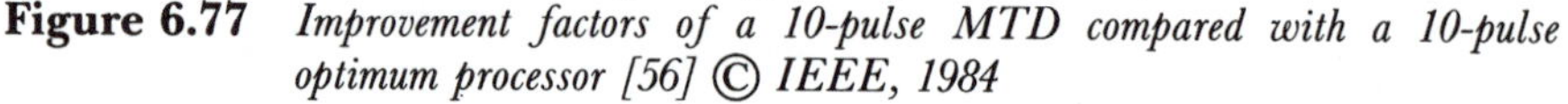

Figure 6.77 *Improvement factors of a 10-pulse MTD compared with a 10-pulse optimum processor [56] © IEEE, 1984*

—— optimum processor
– – – binomial MTI + C1
CNR = 40 dB
$f_c = 0$
1 $\rho = 0.80$
2 $\rho = 0.90$
3 $\rho = 0.99$

The clutter spectra are centred around $f_{c1} = 0$ and $f_{c2} = 0.25$ PRF and have different CNR and ρ values. The performance of an optimum processor is also shown for comparison. The number of orders of the optimum processor $N_{opt} = N_{MTI} + N_{AMTI} - 1$.

Another type of processor is a binomial MTI ($N_{MTI} = 3$) plus adaptive MTI ($N_{AMTI} = 2$) plus coherent integrator ($N_{CI} = N_{OPT} - 3$). The improvement factors of the optimum processor are also calculated for comparison. While the parameters of the first clutter are kept constant, three different situations are considered for the second clutter. The target signal is assumed to be $f_D = 0.25$ PRF. The results are shown in Fig. 6.80. The loss of the conventional processor amounts to several decibels in many operational conditions.

The third processor is the A–MTD. The A–MTD provides three sets of weights each designed for a different level of ground clutter and with very low sidelobes; the most suitable set is automatically selected by means of a measure of the clutter power. The IF of this processor is shown in Fig. 6.81. The performance of a processor consisting of a 3-pulse MTI cascaded with a six-point Hamming-weighted DFT filter bank, is also shown.

Most of the above curves were obtained for the situation where a fixed PRF is used during the time-on-target. When a waveform employing a CPI–CPI PRF stagger is used to eliminate blind speeds an averaging effect on the SIR improvement curves results for high Doppler targets, and this has a significant effect on the detection performance.

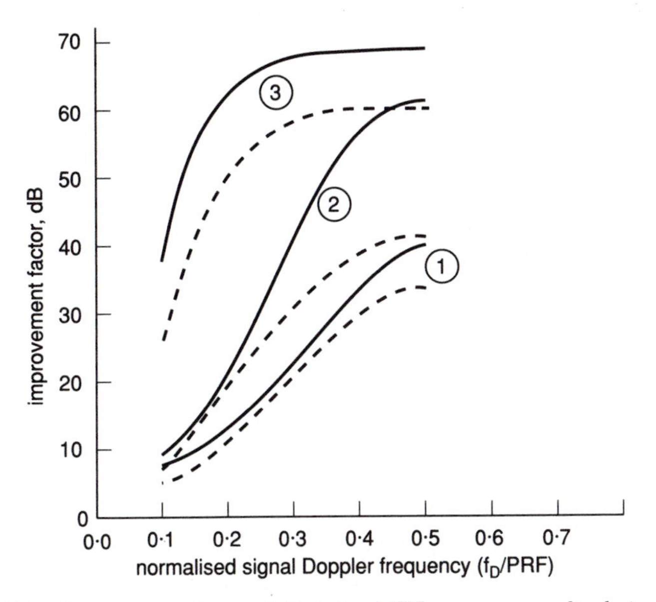

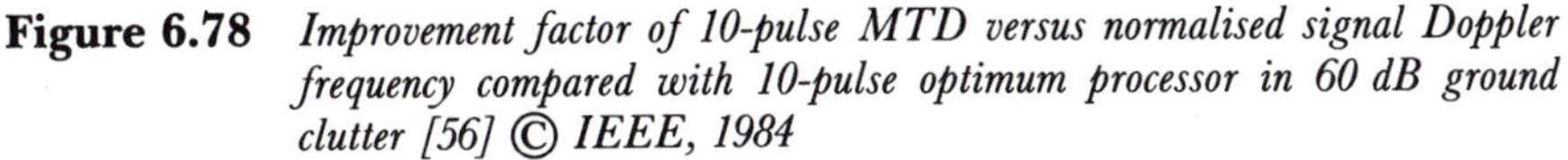

Figure 6.78 *Improvement factor of 10-pulse MTD versus normalised signal Doppler frequency compared with 10-pulse optimum processor in 60 dB ground clutter [56] © IEEE, 1984*

—— optimum processor
--- binomial MTI + C1
CNR = 60 dB
$f_c = 0$
1 $\rho = 0.80$
2 $\rho = 0.90$
3 $\rho = 0.99$

For example, when the PRF periods are staggered in the ratios of 12:13:15, this averaging is illustrated in Fig. 6.82 for the SIR improvement curve of Fig. 6.74*b* [55]. This stagger design was selected to eliminate blind speeds for target Doppler shifts up to $10/T$ [57].

To compare processor performances directly in terms of average SIR improvement curves as shown in Fig. 6.82 would be quite difficult. Following the approach presented in Reference 57 the probability of detection was therefore determined assuming the target Doppler shift to be uniformly distributed between zero and $f_D = 10/T$. Assuming a nonfluctuating target (Swerling case 0) the probability of detection at any given target Doppler and signal-to-interference ratio is obtained from the average SIR improvement curve together with standard results for the probability of three pulses noncoherently integrated. These three pulses are obtained from three CPIs with different PRFs. This approach is slightly inaccurate because PRF stagger will cause a high Doppler target to appear in different FIR filters on successive CPIs, resulting in a combining or collapsing loss in the noncoherent integration process. By averaging across all target Dopplers for a given input SIR, an average probability of detection curves, for the optimum and the 5 FIR filter

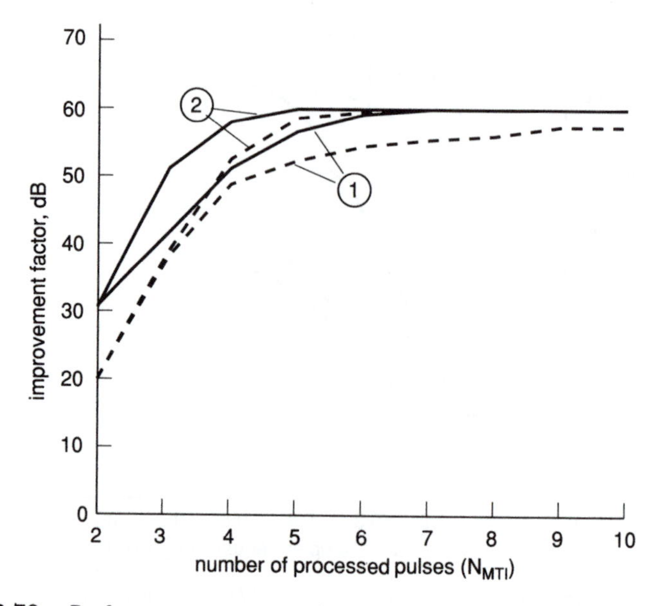

Figure 6.79 *Performance comparison between adaptive MTI (binomial MTI plus adaptive MTI) and optimum processor [56] © IEEE, 1984*

—— optimum processor
--- binomial MTI + AMTI
$\rho_1 = 0.99$
$CNR_1 = 60$ dB
$CNR_2 = 30$ dB
$f_{c1} = 0$
$f_{c2} = 0.25$ PRF
1 $\rho_2 = 0.90$
2 $\rho_2 = 0.99$

processors in the case of 60 dB land clutter and given $P_f = 10^{-6}$, are shown in Fig. 6.83.

From these results the SIR required to obtain a certain average value of P_D is directly determined and different processors can be compared quantitatively. Thus, from Fig. 6.83, it is seen that the 5 FIR filter processor is about 10 dB below the optimum for $P_D = 0.9$ and $P_f = 10^{-6}$ against 60 dB land clutter.

It is also interesting to compare all these processors in their performance of detection in noise only. Many authors have studied the MTI loss with noncoherent and coherent integrators [58–64]. It is more interesting to compare the loss of different processors under the same condition.

Scheleher [65] compared the detection performance of MTI cascaded with noncoherent and coherent integrators, with the optimum processor in noise as well as in clutter and noise. The block diagrams for these three types of processor are shown in Figs. 6.84*a*, *b* and *c* [65].

The improvement factor for the transversal filter depicted in Fig. 6.84*a* is given by

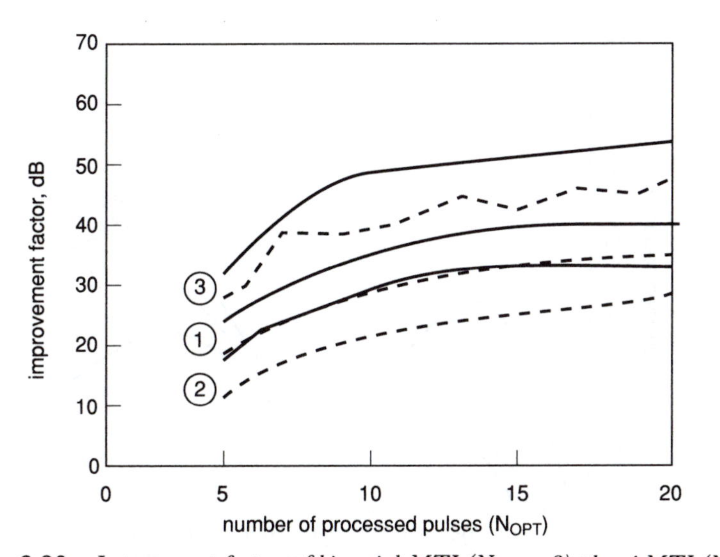

Figure 6.80 *Improvement factors of binomial MTI ($N_{MTI}=3$) plus AMTI ($N_{AMTI}=2$) plus coherent integrator ($N_{CI}=N_{OPT}-3$) versus the total number N_{OPT} of processed pulses [56] © IEEE, 1984*

—— optimum processor
--- binomial MTI + AMTI + C1
$\rho_1=0.99$
$CNR_1=40$ dB
$f_{c1}=0$
$f_{c2}=0.5$ PRF
1 $\rho_2=0.90$ $CNR_2=30$ dB
2 $\rho_2=0.90$ $CNR_2=40$ dB
3 $\rho_2=0.99$ $CNR_2=30$ dB

$$I=\frac{\boldsymbol{w}^T\boldsymbol{M}_s\boldsymbol{w}^*}{\boldsymbol{w}^T\boldsymbol{M}_x\boldsymbol{w}^*} \tag{6.145}$$

where $\boldsymbol{M}_s$ and $\boldsymbol{M}_x$ are the signal and interference covariance matrices and $\boldsymbol{w}^*$ is the complex conjugate weight factor. For Gaussian distributed interference the optimum weights are obtained by maximising eqn. 6.145 and are given by

$$\boldsymbol{w}_0=\boldsymbol{M}_x^{-1}\boldsymbol{s}^* \tag{6.146}$$

where $\boldsymbol{M}_x^{-1}$ is the inverse of the interference covariance matrix and s^* is the complex conjugate of the signal vector.

The improvement factor for the optimum weights can be found from:

$$I_o=\text{trace}[\boldsymbol{M}_x^{-1}\boldsymbol{M}_s] \tag{6.147}$$

The probability of detection is then equal to [66]

$$P_D=Q\{[2P_s/P_n(1+F/I_o)]^{1/2},\ [2\ln(1/P_{fa})]^{1/2}\} \tag{6.148}$$

where F is the clutter-to-noise power ratio, P_s and P_n are the signal and noise powers, P_{fa} is the probability of false alarm and $Q\{\cdot,\cdot\}$ is Marcum's Q function.

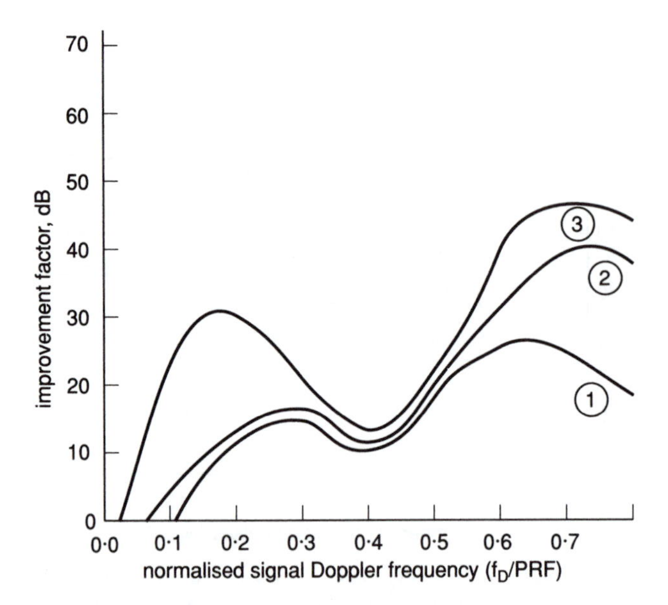

Figure 6.81 *Improvement factor of an 8-pulse A-MTD processor versus target Doppler frequency in two clutter [56] © IEEE, 1984*

$CNR_1 = 50$ dB $\rho_1 = 0.99$ $f_{c1} = 0$
$CNR_2 = 40$ dB $\rho_2 = 0.95$ $f_{c2} = 0.4$
1 MTI + DFT − BANK
2 AMTD BANK
3 optimum if achievable

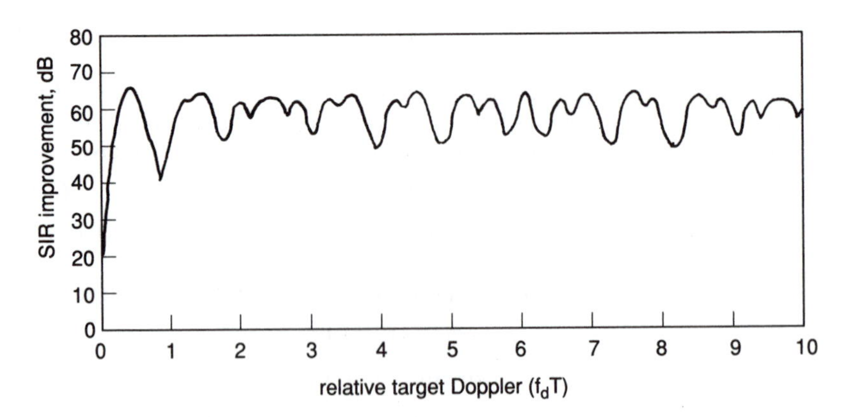

Figure 6.82 *Averaging effect on SIR improvement due to CPI-to-CPI PRF stagger in 60 dB land clutter [55] © IEEE, 1980*

The improvement factor for the cascaded MTI and FFT coherent integrator can be found from

$$I = \frac{|H(f)|^2 \int_0^{f_r} S_i(f)df}{\int_0^{f_r} |H(f)|^2 S_i(f)df} \tag{6.149}$$

where $S_i(f)$ is the normalised clutter-plus-noise power spectral density, f_r is the radar's PRF and

$$H(f) = \sin^{m-1}(\pi x) \frac{\sin[\pi(x-k)]}{n \sin[\pi(x-k)/n]} \tag{6.150}$$

The probability of detection for the cascaded binomial MTI and FFT coherent integrator can be found by substituting eqn. 6.149 into eqn. 6.148.

The characteristic function of the output of the MTI followed by incoherent integration can be obtained by decorrelating the MTI output using an orthogonal tansformation, and is given by [61]

$$\phi(u) = \exp\left[\sum_{i=1}^{n} ug_k^2/(1-2u\lambda_k)\right] \prod_{k=1}^{n} 1/(1-2u\lambda_k) \tag{6.151}$$

where λ_k are the eigenvalues of the covariance matrix at the output of the MTI and g_k are the projections of the signal vector onto the kth eigenvector. An expression for the probability of false alarm can be developed from eqn. 6.151 with $g_k = 0$ as

$$P_{fa} = \sum_{k=1}^{n} \left\{ \prod_{l=1}^{n} \frac{\lambda_k}{\lambda_k - \lambda_l} \right\} \exp\left(-\frac{V_T}{2\lambda_k}\right) \tag{6.152}$$

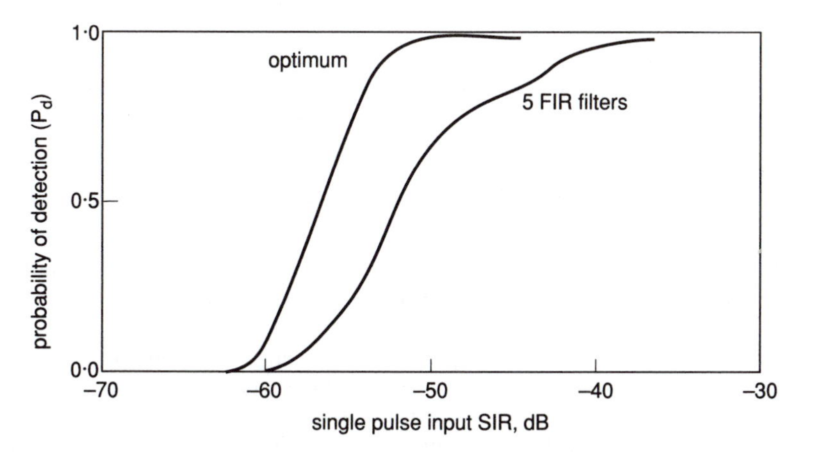

Figure 6.83 *Detectability curves for optimum and 5 FIR filter processors against 60 dB land clutter and given $P_f = 10^{-6}$ [55] © IEEE, 1980*

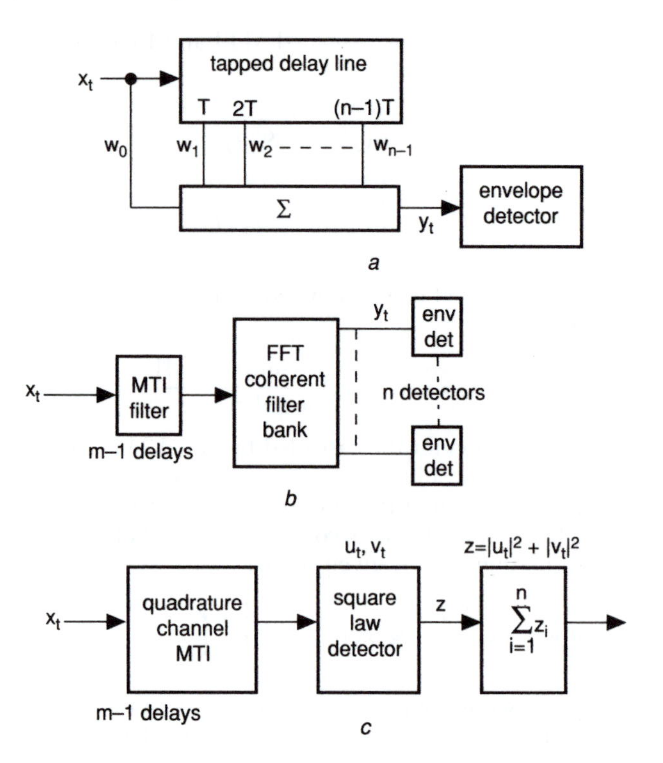

Figure 6.84 *(a) Optimum transversal filter; (b) Cascaded MTI and coherent integrator; (c) MTI and incoherent integrator [65]*

n points
k = filter number
$x = f/f_r$

from which the thresholds V_T can be determined. Eqn. 6.151 can be used to develop a cumulant generating function from which the cumulants can be determined as

$$\chi_l = \sum_{k=1}^{n} (2\lambda_k)^l (l-1)! (lg_k^2 + 1) \tag{6.153}$$

Eqn. 6.153 can be used in a Gram–Charlier series to determine the probability of detection. A three term series with Edgeworth grouping was found to provide good accuracy.

The detection performance comparisons between the various processors are based on the same number of pulses in a radar dwell and various clutter-to-noise power ratios. A nine-pulse optimum transversal filter and a two-pulse MTI followed by an 8-pulse coherent or incoherent integrator are considered. Gaussian clutter spectrum is assumed. Detection curves are generally evaluated at $f_D = f_r/4$. This Doppler frequency provides a response which is equivalent to

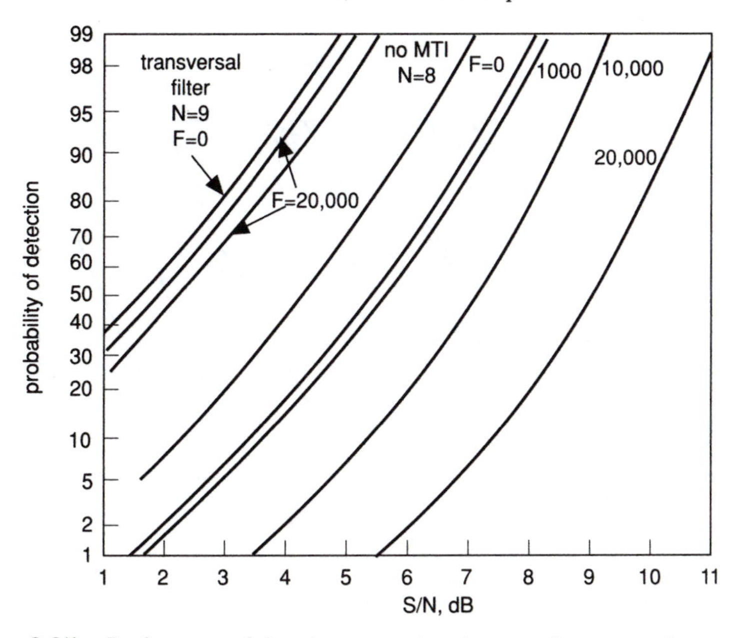

Figure 6.85 *Performance of detection comparison between three types of processors for $\sigma_c T = 0.01$ [65]*

$P_{FA} = 10^{-6}$
$F = P_c / P_N$
$\sigma_c T = 0.01$
$I = 43$ dB
$N = 8$
$M = 2$

the average response of a single canceller MTI when all velocities are equally likely. P_{fa} is assumed equal to 10^{-6} in all cases.

Fig. 6.85 shows the detection performance of the various processors for a narrow clutter spectrum ($\sigma_c T = 0.01$). The single canceller MTI improvement factor for this spectrum is 43 dB.

The detection loss of the optimum transversal filter is of the order of 0.2 dB between no clutter ($F = 0$) and a clutter-to-noise power ratio $F = 20\,000$ (43 dB). The detection performance of the MTI coherent integrator (MTD) in heavy clutter ($F = 20\,000$) is only 0.4 dB worse than that of the optimum, indicating that this processor was effective in removing clutter. The detection losses for a single canceller quadrature MTI with noncoherent integrator are equal to 3.5 dB with noise alone and 6.6 dB when the clutter residual power equals the noise power. The detection performance for an 8-pulse coherent integrator without MTI is also shown for comparison.

Fig. 6.86 depicts the detection performance of the various processors for a medium-width clutter spectrum ($\sigma_c T = 0.05$) which provides a single canceller MTI improvement factor of 29 dB.

The detection loss of the optimum transversal filter is of the order of 0.4 dB between no clutter and a clutter-to-noise power ratio of $F = 1000$ (30 dB). The

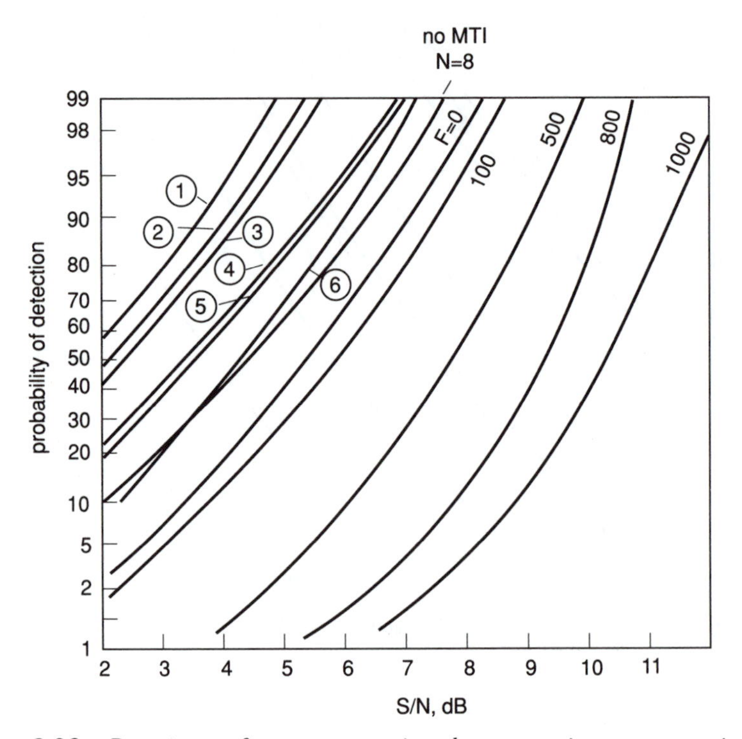

Figure 6.86 *Detection performance comparison between various processors in clutter with $\sigma_c T = 0.05$ [65]*

1 8-pulse transversal filter, $F = 0$

$F = 1000$:
2 9-pulse transversal filter
3 MTI 8-pulse FFT max—no weighting
4 MTI 8-pulse FFT max—Hamming weighting
5 MTI 8-pulse FFT crossover—Hamming weighting
6 MTI 8-pulse FFT crossover—no weighting

$P_{FA} = 10^{-6}$
$F = P_c/P_N$
$\sigma_c T = 0.05$
$I = 29$ dB
$N = 8$
$M = 2$

peak detection performance of the MTD for $F = 1000$ is within the order of 0.3 dB of an optimum processor. This indicates good performance in clutter. However, at the crossover between adjacent Doppler filters, the composite performance (considering detection probabilities in both filters) is down almost 2 dB from the peak. Using a Hamming weighting of the coherent integrator response considerably smooths the overall Doppler response, but the overall performance for $F = 1000$ is uniformly down by the order of 1.4 dB from that of the peak without weighting. The detection loss of the MTI noncoherent

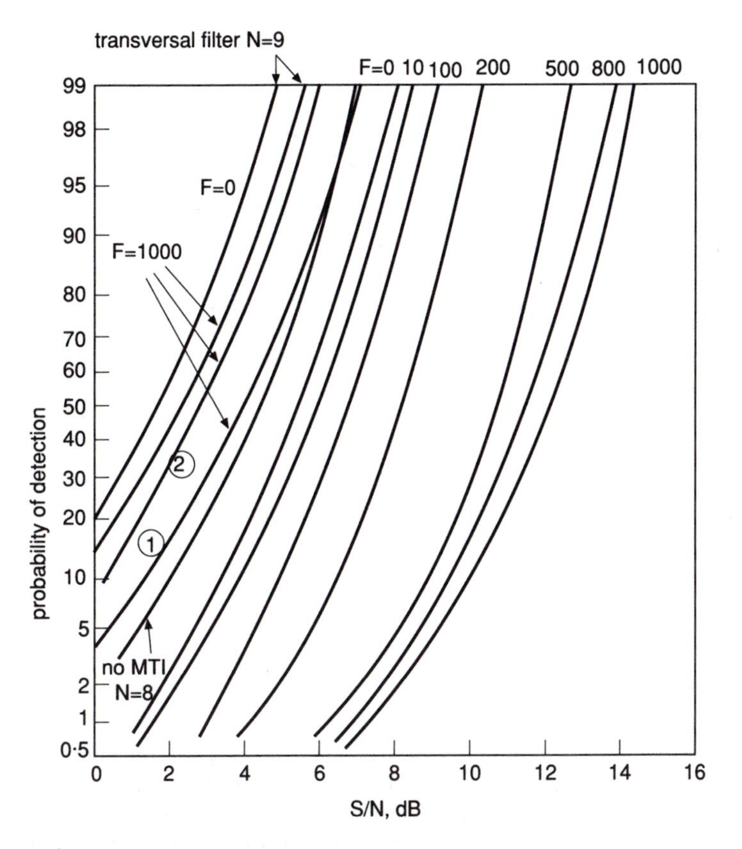

Figure 6.87 *Performance of detection comparison between various processors in clutter with $\sigma_c T = 0.1$ [65]*

1 single canceller MTI 8-pulse FFT—no weighting
2 double canceller MTI 8-pulse FFT—no weighting
$P_{FA} = 10^{-6}$
$F = P_c / P_N$
$\sigma_c T = 0.1$
$I = 23$ dB
$N = 8$
$M = 2$

integrator compared with the optimum is again 3.4 dB for noise alone and 7.6 dB when the residual clutter power is equal to the noise power.

Fig. 6.87 depicts the detection performance of the various processors for a wide clutter spectrum ($\sigma_c T = 0.1$). The single canceller MTI improvement factor is 23 dB. Performance of the optimum transversal filter is again good over strong clutter ($F = 1000$) with a reduction of the order of 0.8 dB with respect to noise-only performance ($F = 0$). The performance of the single-canceller/coherent integrator for $F = 1000$ is poor, down to the order of 1.5 dB with respect to a 9-pulse optimum processor. This indicates that a better MTI filter

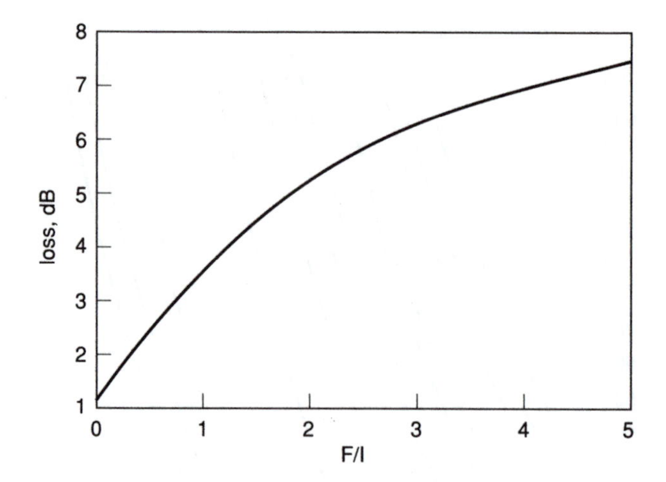

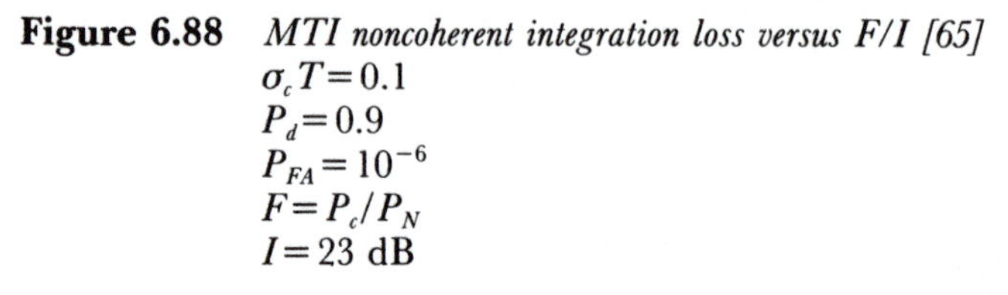

Figure 6.88 *MTI noncoherent integration loss versus F/I [65]*
$\sigma_c T = 0.1$
$P_d = 0.9$
$P_{FA} = 10^{-6}$
$F = P_c / P_N$
$I = 23$ dB

is required. Using a double canceller instead of a single canceller reduces the detection loss for $F = 1000$ to within 0.5 dB of a 9-pulse optimum processor in the same clutter. However, in double canceller condition, the total pulse processed is equal to 10 instead of 9. Detection loss for the MTI noncoherent integrator ranges from 3.5 dB for noise alone to 10.1 dB for $F = 1000$.

The MTI noncoherent integration loss of single canceller quadrature MTI with an 8-pulse video-integrator, compared with the 8-pulse coherent integrator without MTI, is 1.1 dB for noise alone to 7.5 dB for $F = 1000$. It is a function of F/I, which is plotted in Fig. 6.88. The curve indicates that residual clutter power must be made significantly less if good detection performance is to be realised using an MTI noncoherent integrator.

6.5 Adaptive MTI based on spectral estimation principles

All the previously analysed performances of MTI or MTD systems are based on the assumption of Gaussian clutter spectrum. However, this is not true not only in the case of land clutter but also in the cases of weather clutter and chaff. The latter are often considered as homogeneously distributed and therefore Gaussian spectra.

The power spectral density of land clutter covered with bush can be approximated by [67]

$$S_c(f) = 1/[1 + (f/f_c)^3] \tag{6.154}$$

where f_c is the 3 dB point of the clutter spectrum. It has a larger high frequency tail than the Gaussian spectrum.

Janssen and van der Spek [68] measured large amounts of weather clutter data and analysed the spectra of these data. From the measured data they found that, in about one fourth of the spectra, the deviation from Gaussian is considerable; e.g., one or both edges may be too steep or too slight, the peak may be off-centre, there may be more than one peak.

Estes *et al.* [69] reported that the same phenomena can be observed in chaff clutter. It seems that the main reasons for a deviation from the Gaussian shape are a nonlinear relation between height and wind speed and the occurrence of irregular patterns of turbulence.

In fact, the clutter echo is a very complex time sequence. It is a non-stationary time varying sequence. To fit the clutter spectra with any fixed model is impossible and unreasonable. For a time varying sequence, only a short term spectrum can be approximated. It is better to study the clutter echo as a time sequence directly.

It is well known that three types of models are fitted to the stationary random sequences: autoregressive (AR), moving average (MA), and autoregressive moving average (ARMA). The autoregressive model is also termed an all-pole model, the moving-average model an all zero model, and the autoregressive moving average a pole and zero model.

Haykin [70, 71] showed that most of the clutter can be modelled with low order AR sequence. The expression of an AR sequence is

$$x_n = \sum_{i=1}^{M} w_i x_{n-i} + \varepsilon(n) \tag{6.155}$$

where $\varepsilon(n)$ is a white noise sequence with zero mean and σ_n^2 variance, w_i are the AR coefficients, and M is the order of the sequence. The spectrum of this AR sequence can be expressed as

$$S(f) = \frac{P_M T}{\left| 1 - \sum_{i=1}^{M} w_i \cdot \exp(-j2\pi i f T) \right|^2} \tag{6.156}$$

where P_M is the output power and T is the Nyquist sampling period of the signal.

It has been proved that the autoregressive spectral analysis is equivalent to the maximum entropy spectral analysis and linear prediction [72]. The maximum entropy method (MEM) proposed by Burg in 1967 [73] does not assume any value for stochastic processes outside the observation period, but it tends to make the process remain most random and most undetermined, and thus with the largest entropy. This means that the MEM under the constraint of the correlation function of the estimated spectrum within the observation interval equal to the given value

$$r_m = \int_{-\pi}^{\pi} \hat{S}(f) \exp(j2\pi m f T)\, df \tag{6.157}$$

maximises the entropy

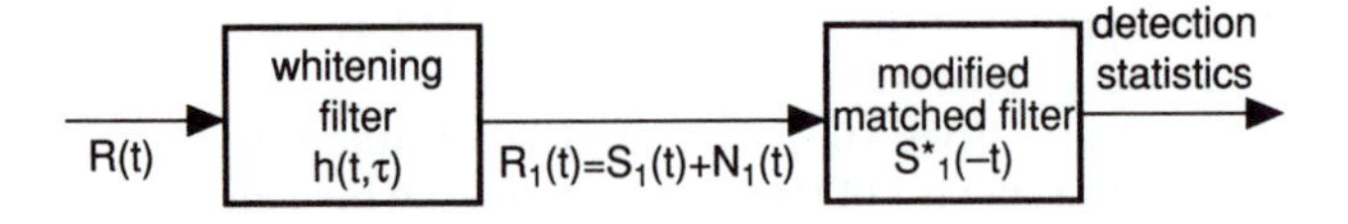

Figure 6.89 *Optimum processor for clutter environment*

$$E=\int_{-\pi}^{\pi} \ln \hat{S}(f)\,df \qquad (6.158)$$

Because this spectral estimation implies an infinite extrapolation of the autocorrelation coefficient, it does increase the spectral resolution considerably, especially in the case of a short data sequence.

Since the data sequence available from a resolution cell of a radar is very short, e.g. 10–20 for microwave surveillance radar, the MEM is suitable for radar data spectral analysis.

From the viewpoint of optimum detection of signal in coloured noise, the optimum processor can be constructed with a whitening filter cascaded with a modified matched filter as in Fig. 6.89.

It is well known that, if the input sequence is an AR process, the whitening filter is a predictive error filter. The output of a predictive error filter is

$$e_n=x_n-\hat{x}_n=x_n-\sum_{i=1}^{M} w_i x_{n-1} \qquad (6.159)$$

where e_n is the error of prediction, $\hat{x}_n$ is the predicted value of the current value x_n of the input sequence. It is evident that, if the input sequence is an AR sequence, eqn. 6.159 is equivalent to eqn. 6.155.

It can be seen from eqn. 6.159 that a whitening filter can be realised with an FIR filter with the weights of $[1, -w_1, -w_2, \ldots, -w_M]$. This means that the whitening filter for the AR type of clutter is very similar to the optimum MTI filter, which was discussed in the previous Section. The only difference is the optimum weights.

6.5.1 *Comparison between different algorithms*

The most important problem in MEM spectral estimation is how to find out the weighting vector $\boldsymbol{w}$ with an effective algorithm.

A good algorithm must combine good performance of estimation and simplicity of computation. The performance of an algorithm includes the accuracy of the estimation, the resolution of peaks, speed of convergence, stability, etc. These performance figures depend not only on the algorithm itself, but also on the criteria of order decision, the number of data points etc. For simplicity, in our comparison we assume that the order of estimation is fixed and the number of data points is equal to three times the order.

Many authors study the performance of estimation of different algorithms, but very few of them study the computational complexity of these algorithms. However, in real-time applications of spectral estimation, the most important problem is the simplicity of computation. We will discuss the computational complexity of 12 different algorithms.

Since Burg presented the MEM spectral estimation algorithm in 1967, more than ten algorithms have been suggested. All of these algorithms can estimate the power spectral density with different performance and different computational complexity. Although all of these algorithms can estimate the weighting vector, they are based on different principles. We can classify these algorithms from several different viewpoints.

(*a*) Processing format of input data:
  (i) Block processing
  (ii) Moving processing
(*b*) Architecture of the filter
  (i) Transversal filter (FIR)
  (ii) Lattice filter
(*c*) Form of the autocorrelation matrix $\boldsymbol{R}$
  (i) Toeplitz matrix (autocorrelation method)
  (ii) Toeplitz and Hankel matrix
  (iii) Centro-symmetry matrix (covariance method)
(*d*) Criteria of output error
  (i) Least mean square (LMS)
  (ii) Least square with constraints
  (iii) Exact least square
(*e*) Type of algorithm
  (i) Recursive algorithm
  (ii) Direct matrix inverse (DMI)

The algorithms which will be compared here include the following 12:

(i) Levinson's algorithm [74]
(ii) Burg's algorithm [73]
(iii) Marple's algorithm [75]
(iv) LUD (lower upper triangular matrix decomposition) algorithm [76]
(v) CR–LS (conjugate reverse–least-squares) algorithm [77]
(vi) LP–LS (linear-phase–least-squares) algorithm [78]
(vii) Modified LUD algorithm [79]
(viii) Widrow–Hoff's LMS algorithm [80]
(ix) Least mean square Kalman filter [81]
(x) LSL (least squares ladder) algorithm [82]
(xi) RLS (recursive least square) algorithm [83]
(xii) FTF (fast transversal filter) algorithm [84]

The first seven algorithms belong to block processing, and the last five algorithms belong to moving processing.

Levinson's algorithm solves the Yule–Walker equations with the recursive formula

$$W_{k,k} = -\left[R_x(k) + \sum_{i=1}^{k-1} W_{k-1,1} R_x(k-1)\right] / P_{k-1} \tag{6.160}$$

$$W_{k,i} = W_{k-1,i} + W_{k,k} W^*_{k-l,k-i} \tag{6.161}$$

$$P_k = (1 - |W_{k,k}|^2) P_{k-l} \tag{6.162}$$

The initial condition is

$$W_{1,1} = -R_x(1)/R_x(0) \tag{6.163}$$

$$P_1 = [1 - |W_{1,1}|^2] R_x(0) \tag{6.164}$$

The criterion for the output error of this algorithm is the least mean square criterion. The main advantage of this algorithm is the simplicity of computation. However, the performance of estimation of this algorithm is the worst.

Burg's algorithm begins from the least square error criterion, adopts a forward and backward prediction concept, and finds the reflection coefficients via minimising the prediction error of forward and backward prediction. The constraint of Burg's algorithm is that the coefficients must satisfy Levinson's recursive equations.

Forward predictive error:

$$f_{M,k} = \sum_{i=0}^{M} W_{M,i} X_{k+M-i} \tag{6.165}$$

Backward predictive error:

$$b_{M,k} = \sum_{i=0}^{M} W^*_{M,i} X_{k+i} \tag{6.166}$$

Sum of error power:

$$e_M = \sum_{k=1}^{N-M} |f_{M,k}|^2 + \sum_{k=1}^{N-M} |b_{M,k}|^2 \tag{6.167}$$

Substituting eqn. 6.163 into eqn. 6.165 and eqn. 6.166, then substituting the results into eqn. 6.167, we can obtain $W_{M,M}$ for the minimum e_M:

$$W_{M,M} = -\frac{2 \sum_{k=1}^{N-M} f_{M-1,k+1} b^*_{M-1,k}}{\sum_{k=1}^{N-M} [|f_{M-1,k+1}|^2 + |b_{M-1,k}|^2]} \tag{6.168}$$

In the calculation of the denominator of eqn. 6.168, the following recursive formula can be used to decrease the number of computations:

$$DEN_M = [1 - |W_{M-1,M-1}|^2]\, DEN_{M-1} - |b_{M-1,N-M+1}|^2 - |f_{M-1,1}|^2 \tag{6.169}$$

Since additional constraint is added in Burg's algorithm, it suffers from deviation of estimation and split of peak.

Ulrych, Clayton and Nuttal suggested a least squares autoregressive spectral estimation algorithm without the constraint present in the Burg's algorithm. This algorithm directly derives the weighting coefficients by the squared sum of forward and backward predictive errors. This means that we can substitute eqn 6.165 and eqn. 6.166 into eqn. (6.167), and then derive the minimum with respect to $W_{M,1}$ to $W_{M,M}$. The following equations are obtained:

$$\begin{bmatrix} R_x(0,0) & \ldots & R_x(0,M) \\ R_x(1,0) & \ldots & \\ \vdots & & \\ R_x(M,0) & \ldots & R_x(M,M) \end{bmatrix} \begin{bmatrix} 1 \\ W_{M,1} \\ \vdots \\ W_{M,M} \end{bmatrix} = \begin{bmatrix} P_M \\ 0 \\ \vdots \\ 0 \end{bmatrix} \tag{6.170}$$

$$P_M = \sum_{j=0}^{M} W_{M,j} R_x(0,j) \tag{6.171}$$

$$R_x(i,j) = \sum_{k=0}^{N-M} (x^*_{M+k-i} x_{M+k-j} + x_{k+i} x^*_{k+j}) \tag{6.172}$$

These equations are the basic equation of Marple's algorithm. This algorithm overcomes the disadvantages of Burg's algorithm. The shortcoming of this algorithm is that the resolution will be degraded in low signal-to-noise ratio and it is difficult to decide the order of the sequence; it also has relatively large computational complexity.

The lower upper triangular matrix decomposition (LUD) algorithm is based on the same equations (eqns. 6.170, 6.171, 6.172) of Marple's algorithm, but does not use the recursive equation to solve for $\boldsymbol{w}$, adopting instead the direct matrix inverse (DMI) method. The LUD algorithm utilises the Hermitian and centro-symmetric property of the matrix $\boldsymbol{R}$; specifically:
The conjugate property:

$$R_x(i,j) = R^*_x(j,i) \tag{6.173}$$

The centro-symmetry:

$$R_x(i,j) = R^*_x(M-i, M-j) \tag{6.174}$$

The recursive relation:

$$R_x(i+1, j+1) = R_x(i,j) + x^*_{M-i} x_{M-j} - x^*_{N-i} x_{N-j} - x_{i+1} x^*_{j+1} + x_{N-M+i+1} x^*_{N-M+j+1} \tag{6.175}$$

The performance of estimation in the LUD algorithm is the same as Marple's algorithm, but the computational complexity has been reduced considerably.

The conjugate reverse least squares (CR–LS) algorithm further utilises the centro-symmetry properties of the covariance matrix to construct a new symmetric real matrix:

$$\tilde{R} = \mathrm{Re}[\boldsymbol{R}] + J\,\mathrm{Im}[\boldsymbol{R}] \tag{6.176}$$

$$\boldsymbol{w} = [(\boldsymbol{a} + J\boldsymbol{b}) + j(J\boldsymbol{a} - \boldsymbol{b})]/(a_0 + b_M) \tag{6.177}$$

where the vectors $\boldsymbol{a}$ and $\boldsymbol{b}$ satisfy the following equations:

$$\boldsymbol{a} = \tilde{R}^{-1} \cdot \begin{vmatrix} 1 \\ 0 \\ \vdots \\ 0 \end{vmatrix}, \qquad \boldsymbol{b} = \tilde{R}^{-1} \cdot \begin{vmatrix} 0 \\ \vdots \\ 0 \\ 1 \end{vmatrix} \tag{6.178}$$

This algorithm reduces by one fourth the time of multiplication of the LUD algorithm, but the time of addition will be increased.

If the location of the main tap of the predictive error filter is changed to the centre, this filter becomes a smoothing filter. If the constraint of linear phase is added to this filter, we get the linear phase least squares (LP–LS) algorithm. This algorithm can simplify the complex matrix of the LUD algorithm to a real matrix of the same order. In addition, the coefficients of this filter are symmetrical. This means that

$$W_0 = W_M,\ W_1 = W_{M-1},\ \ldots,\ W_{M/2}. \quad \text{(If M is even)} \tag{6.179}$$

The time of multiplication can be further reduced.

Widrow–Hoff's LMS algorithm is the simplest moving processing method. They use the steepest descent method to solve for the weighting coefficients

$$\boldsymbol{W}_{k+1} = \boldsymbol{W}_k + 2\mu \boldsymbol{x}_k \varepsilon_k \tag{6.180}$$

$$\varepsilon_k = x_{k+1} - \boldsymbol{W}_k^T \boldsymbol{x}_k \tag{6.181}$$

where μ is a convergence factor. There is only vector multiplication in this algorithm without any division. However, the convergence speed is very slow, and there is a contradiction between the convergence speed and the accuracy of estimation. It is very difficult to choose the value of μ, since it is related to the input data.

If we treat the weighting coefficients as state variables, the Kalman recursive equations can be derived:

Weighting vector:

$$\boldsymbol{W}_{k+1} = \boldsymbol{W}_k + \boldsymbol{K}_{k+1}[X_{k+1} - \boldsymbol{X}_{k+1}\boldsymbol{W}_K] \tag{6.182}$$

Covariance matrix of error:

$$\boldsymbol{P}_{k+1} = [\boldsymbol{I} - \boldsymbol{K}_{k+1}\boldsymbol{X}_{k+1}]\boldsymbol{P}_k \tag{6.183}$$

Gain vector:

$$\boldsymbol{K}_{k+1} = \{1/[G^2 + \boldsymbol{X}_{k+1}\boldsymbol{P}_k\boldsymbol{X}_{k+1}^*]\}(\boldsymbol{P}_k\boldsymbol{X}_{k+1}^*) \tag{6.184}$$

This algorithm also belongs to the moving processing group. The speed of convergence is faster than the Widrow's algorithm, but the computational complexity is much greater than the latter. Both algorithms employ LMS criteria.

Morf and Lee proposed an exact least squares lattice filter (LSL). This algorithm discards the constraints of other algorithms and has good convergence property and fast tracking performance. The computational formulas are as follows:

$$\Delta_{i+1,\,T} = \lambda \Delta_{i+1,\,T-1} + \frac{b_{i,\,T-1} f_{i},\,\hat{T}}{1 - \gamma_{i-1,\,T-1}} \tag{6.185}$$

$$\gamma_{i,\,T} = \gamma_{i-1,\,T} + |b_{i,\,T}|^2 / \sigma_{i,\,T}^b \tag{6.186}$$

$$f_{i+1,\,T} = f_{i,\,T} - K_{i+1,\,T}^b b_{i,\,T-1} \tag{6.187}$$

$$b_{i+1,\,T} = b_{i,\,T-1} - K_{i+1,\,T}^f f_{i,\,T} \tag{6.188}$$

$$K_{i+1,\,T}^f = \Delta_{i+1,\,T} / \sigma_{i,\,T}^f \tag{6.189}$$

$$K_{i+1,\,T}^b = \Delta_{i+1,\,T} / \sigma_{i,\,T-1}^b \tag{6.190}$$

$$\sigma_{i+1,\,T}^f = \lambda \sigma_{i+1,\,T-1}^f + |f_{i+1,\,T}|^2 / (1 - \gamma_{i,\,T-1}) \tag{6.191}$$

$$\sigma^b_{i+1,\,T}=\lambda\sigma^b_{i+1,\,T-1}+|b_{i+1,\,T}|^2/(1-\gamma_{i,\,T}) \tag{6.192}$$

where $\Delta_{i+1,\,T}$ are the partial autocorrelation coefficients, γ is a likelihood variable, $f_{i,\,T}$ and $b_{i,\,T}$ are forward and backward prediction errors, $K^f_{i,\,T}$ and $K^b_{i,\,T}$ are forward and backward reflection coefficients, $\sigma^f_{i,\,T}$ and $\sigma^b_{i,\,T}$ are forward and backward predictive error powers. The performance of this algorithm is very good, but the computational complexity is too large.

The convergence property and tracking performance of the recursive least squares (RLS) transversal filter are just as good as the LSL algorithm, but with less computational complexity. This algorithm solves for the coefficients with least squares sum of output error without any constraint:

Output error:

$$\boldsymbol{e}(n)=\boldsymbol{X}(n)-\boldsymbol{X}(n)\boldsymbol{W} \tag{6.193}$$

Squared sum:

$$v(n)=\sum_{s=0}^{n}\lambda^{n-s}e^2(s) \tag{6.194}$$

Weighting coefficients:

$$\boldsymbol{W}(n)=[\boldsymbol{X}^T(n)\boldsymbol{X}(n)]^{-1}\boldsymbol{X}^T(n)\boldsymbol{X}(n) \tag{6.195}$$

where λ is the forgetting factor. In fact, recursive equations are used:

$$\boldsymbol{W}_T=\boldsymbol{W}_{T-1}+[X_T-\boldsymbol{W}^T_{T-1}X(T)]\,C_T \tag{6.196}$$

$$\boldsymbol{C}_T=-\boldsymbol{X}^*(T)\,R_T^{-1} \tag{6.197}$$

$$\boldsymbol{R}_T^{-1}=\frac{1}{\lambda}\left[R^{-1}_{T-1}-\frac{R^{-1}_{T-1}X(T)X^*(T)R^{-1}_{T-1}}{\lambda+X^T(T)R^{-1}_{T-1}X(T)}\right] \tag{6.198}$$

where $X(T)$ is the input sequence (at time T), G is the gain vector, R_T^{-1} is the inverse of the correlation matrix. These algorithms are moving processing algorithms. When the number of input data points exceeds the order of the filter, the estimated $\boldsymbol{w}$ has no bias, and the variance of estimation reaches the Cramer–Rao bound.

In real-time application the most important performance of an algorithm is the computational complexity. Therefore, we will compare this performance of all these algorithms. The computational complexity of different algorithms to be compared includes: the number of complex additions, the number of complex multiplications, the number of real divisions and the storage capacity for complex data. All these depend on the number of input data points N and on the order of the filter M. The available number of data points depends on the application, and the order of the filter depends on the required performance. However, for most algorithms, the number of data points must be at least two times greater than the order of the filter. We assume that $N=3M$ in this comparison. Therefore, for a given order of the filter, the required number of data points can be determined. However, sometimes the available number of data points is limited, and then the order of the filter is also limited to between $M\simeq N/2$ and $N/3$.

In block processing, the calculation of coefficients is once for every N points block, while, in moving processing, the coefficients are updated for every set of

Table 6.16 *Comparison of computational complexity of different algorithms*

Name of algorithm	No. of complex addition	No. of complex multiplication	No. of real division	Storage capacity
Levinson	$NM+M^2+N/2-3M$	$NM+M^2+N/2-M/2$	M	$2M+3$
Burg	$3NM-M^2-2N-M$	$3NM-M^2-N+3M$	M	$3N+M+2$
Marple	$NM+8M^2+N+7M-8$	$NM+9M^2+2N+25M-3$	$5M+3$	$N+4M+15$
LUD	$M^3/6+MN+M^2+N-8M/3-1$	$M^3/6+MN+M^2/2+N-5M/3$	M^2+3M	$M^2/2+N+M/2$
CR–LS	$M^3/6+MN+M^2/2+N+M/2-1$	$M^3/18+MN+M^2/2+N+M/18$	$M^2/2+7M/2+2$	$M^2/2+N+M/2$
LP–LS	$M^3/12+MN+M^2/2+N-M$	$M^3/24+MN-3M^2/4+N-M/2$	$M^2/2+M/2$	$M^2/4+N+M/4$
MLUD (N>2M)	$M^3/12+MN+3N-7M/4$	$M^3/12+MN-M^2/4-N/2+3M/4$	$M^2/2+7M/2+2$	$M^2/2+N+M/2$
Widrow	$2NM$	$2NM$		$2M+1$
Kalman	$(3M^2+1)N$	$(3M^2+3M)N$	N	M^2+3M
LSL	$7MN$	$6MN$	$6MN$	$6M$
RLS	$(5M^2/2+4M)N$	$(5M^2/2+4M)N$	N	M^2+3M
FTF	$7MN$	$(7M+11)N$	$2N$	$7M+8$

input data, even though the variation of the coefficients is very small. In our comparison, we assume that the number of data points N is the same for both types of algorithm. This means that the moving processing algorithms will calculate the coefficients N times while the block processing algorithms only calculate the coefficients once for every N points block.

The results of our comparison are listed in Table 6.16 and Table 6.17.

Table 6.17 *Quantitative comparison of computational complexity of different algorithms*

	No. of complex addition		No. of complex multiplication		No. of real division	
Name of algorithm	$M=5$, $N=15$	$M=10$, $N=30$	$M=5$, $N=15$	$M=10$, $N=30$	$M=5$, $N=15$	$M=10$, $N=30$
Levinson	92.5	385	105	410	5	10
Burg	165	730	200	800	5	10
Marple	317	1192	452	1507	28	53
LUD	109	519	115	530	40	130
CR-LS	125	551	110	436	32	87
LP-LS	108	453	74	292	15	55
MLUD	122	456	75	351	32	87
Widrow	150	600	150	600		
Kalman	1140	9030	1350	9900	15	30
LSL	525	2100	450	1800	450	1800
RLS	1238	8700	1238	8700	15	30
FTF	525	2100	690	2430	30	60

It can be seen from these results that Levinson's algorithm has the least number of complex additions, but the number of complex multiplications is not the least, and its performance of estimation is the worst. The computation of the LP–LS algorithm is also very simple. However, it implies linear phase constraint; this will affect the distribution of zeroes, so the performance of estimation will be degraded. Therefore, the MLUD is the best algorithm for real-time implementation of MEM spectral estimation.

This algorithm is a modification of the LUD algorithm. It still adopts the direct matrix inverse method. However, some procedures are taken: (*a*) The elements of the matrix $\boldsymbol{R}_x$ are computed by the combination of some basic items other than by eqn. 6.172; (*b*) in some cases ($N>2M$), the Winograd's principle is used to reduce the multiplication time of the basic items; (*c*) the centro-symmetry property of the matrix is applied to simplify an original complex matrix into a new symmetric real matrix just as in the CR–LS algorithm; (*d*) the square root method is applied to decompose the matrix other than Cholesky's decomposition, so as to eliminate root operation and to reduce the time of division.

The performance of this algorithm is the same as LUD and Marple's algorithm. It overcomes peak splitting in the Burg's algorithm.

6.5.2 *Performance comparison between the optimum MTI and the adaptive MTI*

It is very interesting to compare the performance of adaptive MTI based on spectral estimation with that of the optimum MTI which has been discussed in the previous Section.

It is well known that the optimality criterion of these two MTIs are different. The optimum MTI maximises the improvement factor. This leads to an eigenvector method. The adaptive MTI minimises the residual clutter power. This leads to a linear prediction method. Both methods involve an FIR filter implementation, but with different weights.

If the Gaussian clutter spectrum is assumed, and the Doppler frequency of the signal is a random variable, uniformly distributed in the (0, 1/T) interval, the improvement factor of an FIR filter with weight vector $\boldsymbol{w}$ is

$$I_{SIR}=\frac{\boldsymbol{w}^H\cdot\boldsymbol{w}}{\boldsymbol{w}^H\cdot\boldsymbol{R}\cdot\boldsymbol{w}} \tag{6.199}$$

where $\boldsymbol{R}$ is the normalised covariance matrix of clutter.

The maximisation of the improvement factor can be obtained by minimising the quadratic form $\boldsymbol{w}^H\cdot\boldsymbol{R}\cdot\boldsymbol{w}$ with the constraint $\boldsymbol{w}^H\cdot\boldsymbol{w}=$constant [9].

As a consequence, the optimum vector $\boldsymbol{w}_0$ is given by the eigenvector $\boldsymbol{E}_{min}$ associated with the minimum eigenvalue of $\boldsymbol{R}$:

$$\boldsymbol{w}_0=\boldsymbol{E}_{min} \tag{6.200}$$

On the other hand, the weight vectors of the predictive error filter $\boldsymbol{w}_p$ are a solution of the equation

$$\boldsymbol{R}\cdot\boldsymbol{w}_p=\sigma^2 e_i \tag{6.201}$$

Table 6.18 *Expression of the improvement factor for the two types of optimisation [86]*

No of sample	Optimum MTI maximization of I_{SIR}	Linear prediction minimization of residual
2	$(1-\rho)^{-1}$	$(1+\rho^2)/(1-\rho^2)$
3	$\dfrac{2}{2+\rho^4[1-(1+8\rho^{-6})^{1/2}]}$	$\dfrac{(1+\rho^2+3\rho^4+\rho^6)}{\{(\rho^2-1)(\rho^4-1)\}}$
4	$\dfrac{2}{2-\rho+\rho^9-\sqrt{(\rho-\rho^9)^2+4(\rho-\rho^4)^2}}$	$\dfrac{1+3\rho^4+3\rho^6+2\rho^8+\rho^{10}}{(1-\rho^2)^2(1-\rho^6)}$

where σ^2 is the power of the process and $e_i = (0\ldots\ 0\ 1\ 0\ldots 0)^T$ is a unity vector which has all components equal to 0 except the ith one, which is equal to 1. The position of the '1' indicates the position of the predicted sample.

Barbarossa *et al.* [85] derived the relationship between these two sets of weights and showed that the weight vector $\boldsymbol{w}_p$, derived from linear prediction, tends to the optimum vector, given by the eigenvector associated with the minimum eigenvalue of the clutter covariance matrix, only when the minimum eigenvalue is much smaller than the others.

Chiuppesi *et al.* [86] compared the performance of adaptive MTI based on linear prediction with that of the optimum MTI. The expression of improvement factor for these two types of optimum MTI filter were obtained. It was expressed as a function of correlation coefficients ρ of the clutter. These expressions for different numbers of samples are shown in Table 6.18. The relationship between ρ and standard deviation σ_c of Gaussian clutter spectrum has been shown in eqn. 6.143.

Table 6.19 shows the improvement factors of these two types of optimisation calculated from these formulas for four samples.

It can be seen from this Table that the difference in I_{SIR} for the two types of optimisation is very small. In the case of a small value of ρ, the optimum MTI is somewhat better than the predictive error filter. The power frequency responses of these two types of filter are shown in Fig. 6.90*a*, *b*, *c*, *d* and *e* for $\rho = 0.99$, 0.95, 0.9, 0.8 and 0.6, respectively.

Table 6.19 *Comparison of improvement factors of the 4-sample filter for the two types of optimisation [86]*

ρ	Optimum MTI, maximisation of I_{SIR}, I_{SIR} (dB)	Linear prediction, minimisation of ε, I_{SIR} (dB)
0.995	65.21	65.18
0.99	56.17	56.11
0.95	35.12	34.78
0.9	26.03	25.33
0.8	17.03	15.58
0.6	8.55	6.04

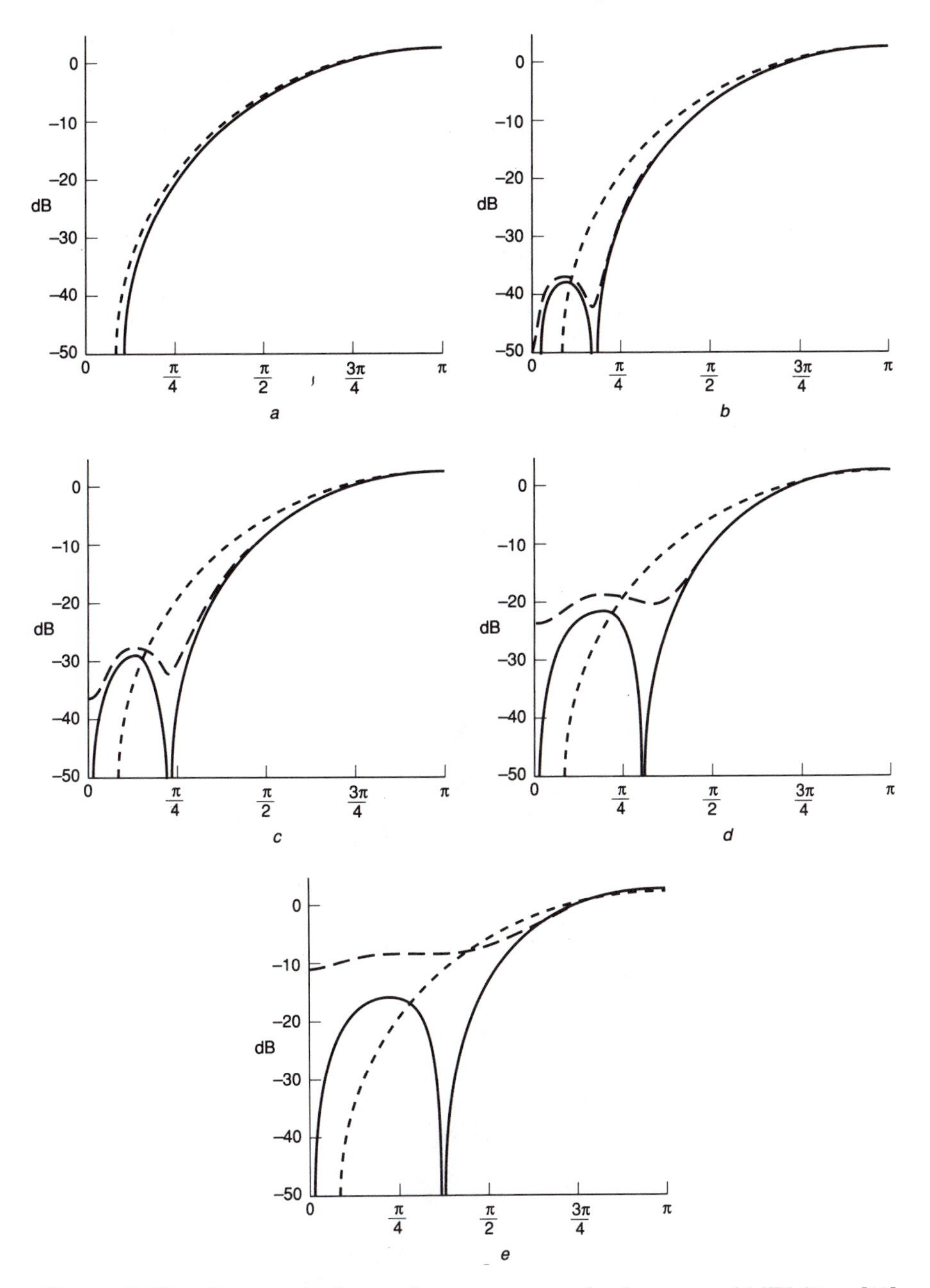

Figure 6.90 *Comparison of power frequency response for three types of MTI filters [86]*

(*a*) $\rho = 0.99$, (*b*) $\rho = 0.95$, (*c*) $\rho = 0.9$, (*d*) $\rho = 0.8$, (*e*) $\rho = 0.6$

—— maximisation of η

--- binomial weights

—– linear prediction

It can be seen from these Figures that the frequency responses of linear prediction filters are very close to that of the optimum MTI in highly correlated clutter, but become flatter in the case of a small value of ρ.

Bucciarelli *et al.* [87] calculated the improvement factor versus $\sigma_c T$ of the predictive error filter as a whitening filter. The results are shown in Fig. 6.91 *a* and *b*.

Fig. 6.91*a* is the improvement factor versus spectral width of linear prediction with side position for the predicted sample, while Fig. 6.91*b* is the improvement factor versus the spectral width of linear prediction with central position for the estimated sample. In fact, the latter is not the predictive error filter, and is termed a smoothing filter. However, the performance of the latter is better than that of the former. If we compare these curves with that of the optimum MTI (see Fig. 6.3), we find that the performance of the latter is just the same as the optimum filter.

It is interesting to study the performance of linear prediction in the case of two clutters, i.e. one ground clutter and one moving clutter. It is assumed that the spectral width of the ground clutter is equal to 0.02, and the moving clutter is 20 dB smaller than the ground clutter. The mean Doppler frequency of the

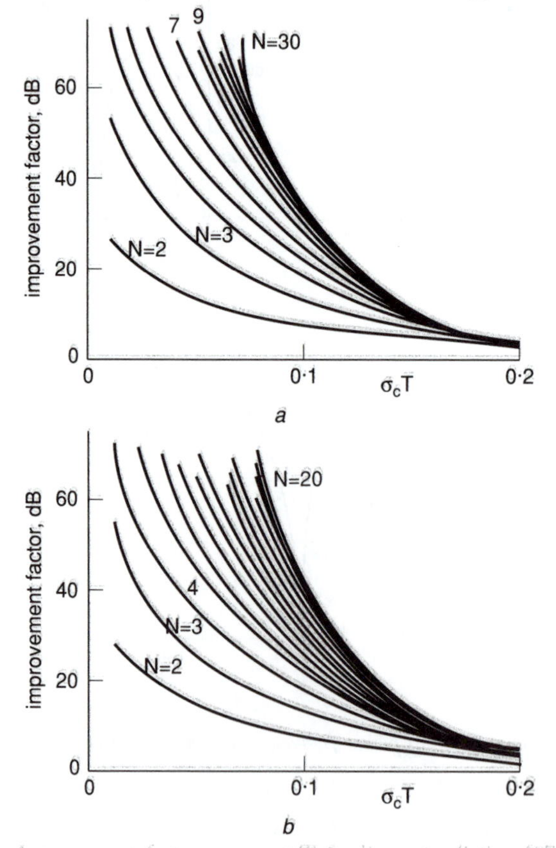

Figure 6.91 *Improvement factors versus $\sigma_c T$ for linear prediction [87]*

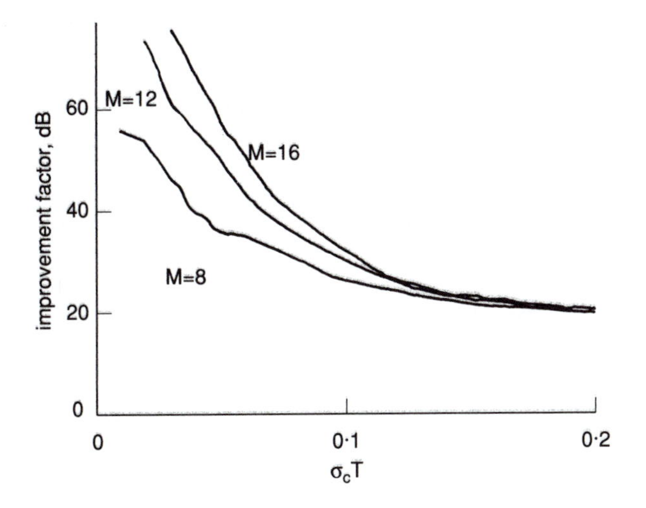

Figure 6.92 *Overall improvement factor in the case of two clutters versus spectral width of the moving clutter [87]*

moving clutter is equal to $1/2T$ (one half the sampling frequency); this appears to be an upper limit. The overall improvement factors versus the spectral width of the moving clutter are shown in Fig. 6.92 [87] and the partial improvement factors for ground clutter and moving clutter are shown in Fig. 6.93*a* and *b* [87], respectively.

The partial improvement factor is obtained by substituting the covariance matrix of one clutter in eqn. 6.199.

In these Figures, curves are plotted with different M as a parameter. M is the order of the filter which is chosen as equal to $N/2$. It can be seen from Fig. 6.93 that, if the spectral width of moving clutter is equal to 0.05, the improvement factor for $M=8$ linear prediction filter for ground clutter is about 37 dB; for moving clutter it is about 18 dB. Better results can be obtained for $M=12$ and 16.

Barbarossa [85] studied the case of two clutters by means of the location of zeroes in the complex plane. The zeroes are the solution of

$$\sum_{k=0}^{M} W_k Z^{M-k} = 0 \tag{6.202}$$

The optimum MTI filter has all zeroes with unity modulus, but phases depending on the clutter spectra. In particular, when only a clutter source is present, all zeroes are located around the phase corresponding to the mean Doppler frequency of the clutter, with a spread depending on the spectral bandwidth of the clutter.

When two clutter sources are present, the partition of zeroes depends on the clutter power ratio Q and on the correlation coefficients ρ_1 and ρ_2 of the clutters.

The zeroes of the linear prediction filter have similar behaviour to the optimum MTI filter, but their modulus could be equal to or less than one.

The behaviour of zeroes of these two types of filters were studied quantitatively [85]. It is assumed that the ground clutter has a narrow spectrum, i.e. $\rho = 0.995$, and the moving clutter has a wide spectrum, i.e. $\rho = 0.85$, and a mean Doppler frequency equal to 0.5. The phase plot of the two types of filter against power ratio Q for $M = 4$ are shown in Fig. 6.94*a* and *b*, respectively.

The behaviour of the zeroes of the two filters are similar, but there are some regions in which two zeroes of the predictive filter have the same phase, in particular for 5 dB $< Q < 17$ dB and 34 dB$< Q < 37$ dB. This fact does not apply to the optimum MTI filter.

The polar co-ordinates representation of Fig. 6.95 shows the migration of zeroes in the complex plane for $Q > 0$ and $M = 4$.

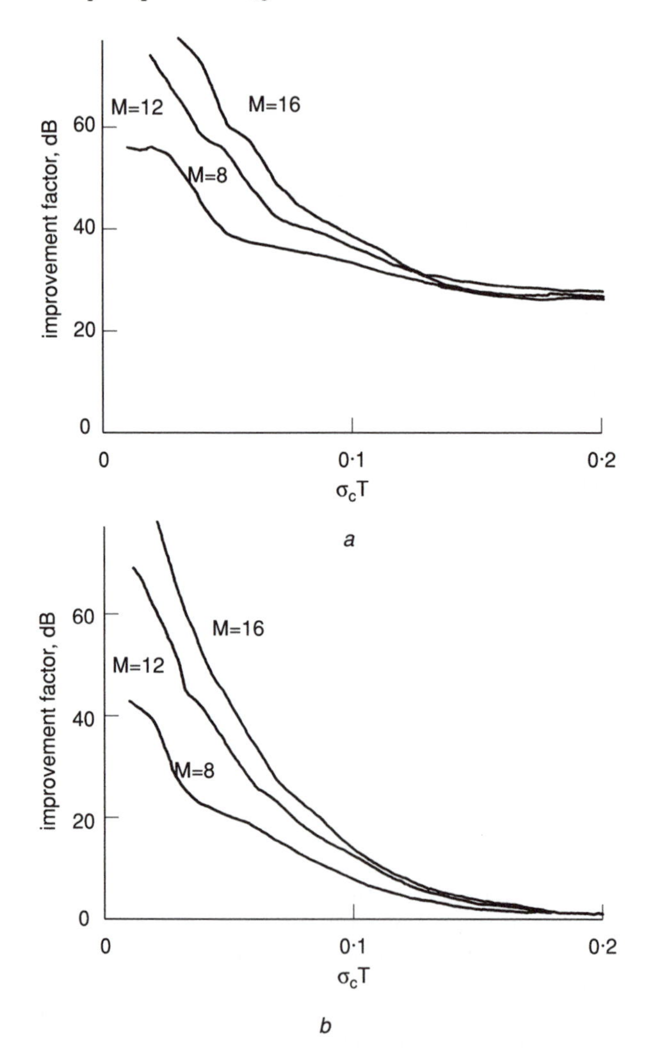

Figure 6.93 *Partial improvement factors for ground clutter (a) and moving clutter (b) [87]*

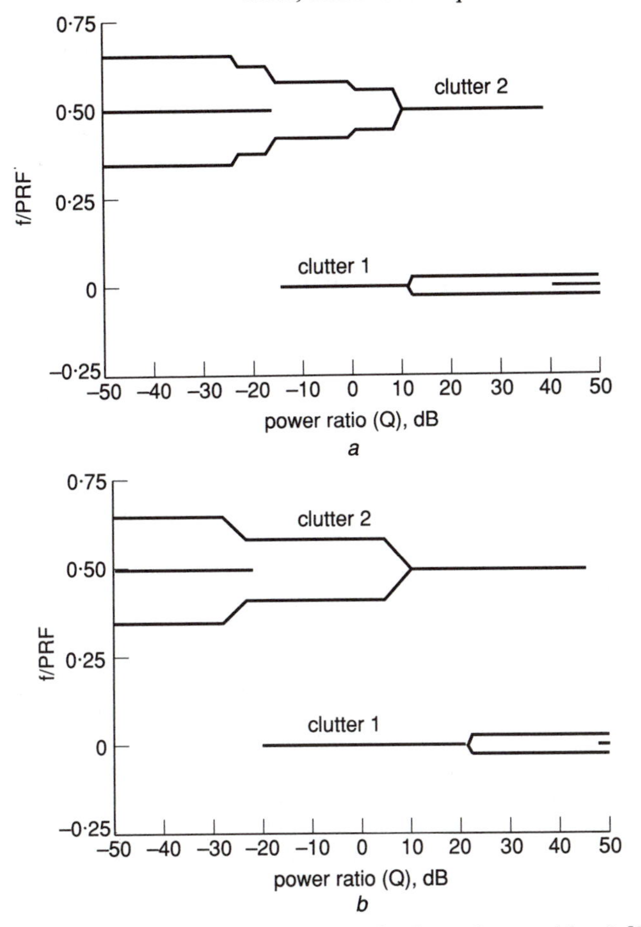

Figure 6.94 *Phase plot against power ratio Q of two clutters. $M=4$ [85]*

a Optimum MTI filter
b Linear predictive filter

In Fig. 6.96*a* the input clutter spectrum (solid line) and the response (dash line) of the optimum MTI filter are shown for $M=4$, $Q=20$ dB and the above-described clutter environment. The effect of receiver noise is neglected. The response of the linear prediction filter is shown in Fig. 6.96*b* for comparison.

The examination of these Figures shows that the predictive error filter (PEF) has a smoother frequency response, owing to its zeroes being inside the unity circle as opposed to the eigenvector (optimum MTI) approach, whose zeroes lie on the unit circle for every stationary disturbance.

The overall improvement factor of the two filtering techniques considered is shown plotted against the power ratio of two clutter Q in Fig. 6.97 in the absence of thermal noise.

The shapes of these curves are very similar to that of the curve in Fig. 6.47, which was derived by Hansen for the open-loop adaptive MTI (Fig. 6.35).

Their performance looks even worse than that of the latter. This is due to the fact that different clutter models are assumed. In the former case, the spectral width of ground clutter is assumed to be 0.016, while the latter is 0.01; the spectral width of moving clutter is 0.09 versus 0.05 and the mean Doppler shift is 0.5 versus 0.25, respectively.

The improvement factor of the eigenvector filter exceeds the one of the PEF by at most 2.5 dB. The IF difference of two filters increases when the thermal noise is present. In Fig. 6.98 the IF of both filters with a single clutter source with a clutter-to-noise ratio CNR of 20 dB are plotted against Q. The above difference is significant for IF values greater than CNR; i.e. when the clutter power at the filter is required to be smaller than the noise power.

6.5.3 *Performance of adaptive MTI realised with different algorithms*

Many authors studied the adaptive MTI based on MEM spectral estimation with different algorithms. The performance of the adaptive MTI mainly depends on the algorithms used. Therefore, we have to pay attention to the relationship between the performance and algorithm.

6.5.3.1 *Performance of adaptive MTI realised with Burg algorithm*

Most of the earlier adaptive MTI based on MEM principles employed Burg's algorithm. A typical example is the adaptive processor established at RADC. The block diagram of this processor is shown in Fig. 6.99.

Nitzberg [88] analysed the performance of this type of adaptive MTI. He assumed that the real clutter has a Gaussian shape spectrum. Therefore, there is a mismatch loss if maximum entropy spectrum is assumed. The other loss is due to the inaccuracy in estimation of the AR coefficients. It is termed 'estimation loss'. The processing loss of different numbers of order relative to the optimum processing for a moving clutter with CNR = 50 dB, mean Doppler shift = 0.28, and spectral width = 0.12745, 0.0881, 0.05 are shown in Table 6.20.

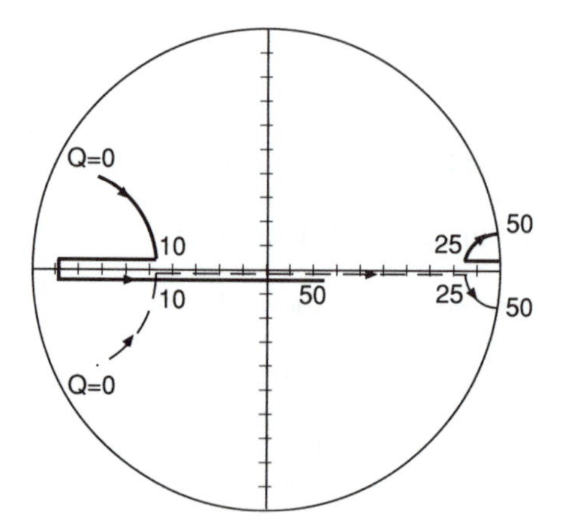

Figure 6.95 *Zeroes migration, in complex plane, of linear prediction filter against Q for $Q > 0$ and $M = 4$ [85]*

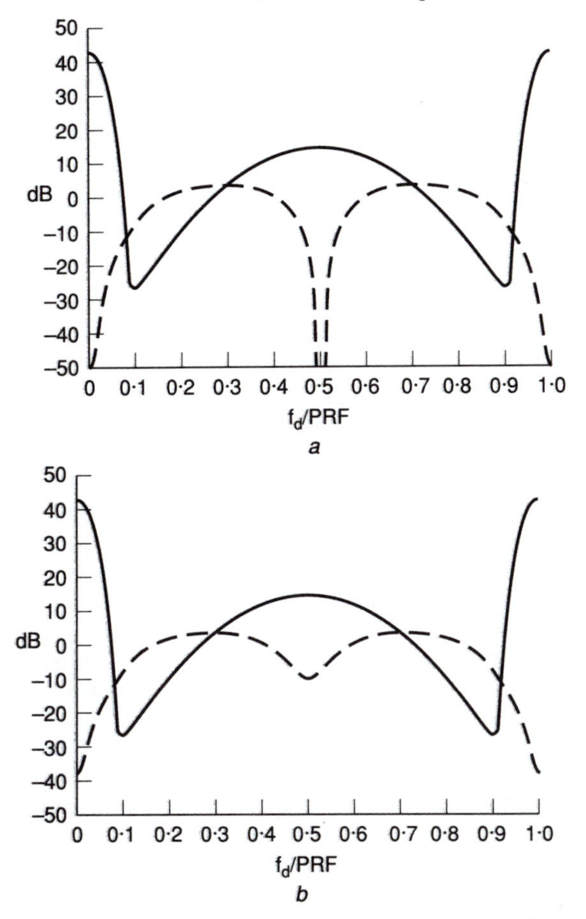

Figure 6.96 *Frequency response (dash line) of the optimum MTI filter (a) and the linear prediction filter (b) [85]*

The solid line in these graphs is the input clutter spectrum

It can be seen from this Table that the mismatch loss decreases with increasing order of the filter, but the estimation loss increases with increasing order of the filter. The total loss depends on the spectral width of the clutter. It decreases with increasing order in wider spectrum, but increases with increasing order in narrower spectrum.

However, this conclusion is based on the assumption that the real clutter has a Gaussian spectrum. If the real clutter has the AR spectrum, the situation will be changed.

6.5.3.2 *Performance of adaptive MTI realised with LUD algorithms*

Mao, *et al.* [89, 90] studied the adaptive MTI with the LUD algorithm. This algorithm calculates the AR coefficients by means of direct matrix inverse, but the number of complex multiplications is much less than N^3 as in the classical algorithm of DMI. The fixed order FIR filter is used. Although this filter operates as a moving window MTI filter, the AR coefficients are calculated from a block of N points of data. Therefore, it works as an open-loop adaptive filter. The problem of convergence time no longer exists. Since no recursive algorithm is used, it is suitable for pipeline architecture.

Computer simulation was performed for this algorithm, and three cases of clutter environment were studied:

(i) *Ground clutter only with $\sigma_c T = 0.0035$, and CNR = 60 dB:* The improvement factors are shown in Table 6.21. The results calculated by Burg's algorithm are also shown in this Table for comparison.

It can be seen from this Table that the improvement factor calculated with the LUD algorithm is better than, or equal to, that of Burg's algorithm in lower order, but worse than that of the latter in higher order of filter.

(ii) *Two moving clutters, i.e. one rain and one chaff clutter:* It is assumed that both clutters have Gaussian shape spectrum. The mean Doppler frequencies are 0.125 and 0.65, the standard deviations of the spectral width are 0.01 and 0.015, and the intensities are equal to 36 and 40 dB, respectively. The results of the computation are shown in Table 6.22.

It can be seen from this Table that there is an optimum number of order ($M_{opt} = 4$) for the LUD algorithm in the two moving clutters case, while the improvement factors calculated by Burg's algorithm increase with the increasing number of order. In this case, an ordinary four-order MTI shows no improvement at all; even the adaptive MTI based on the estimation of the Doppler frequency shift cannot work at all. However, the improvement factors of the adaptive MTI based on MEM spectral estimation are quite satisfied.

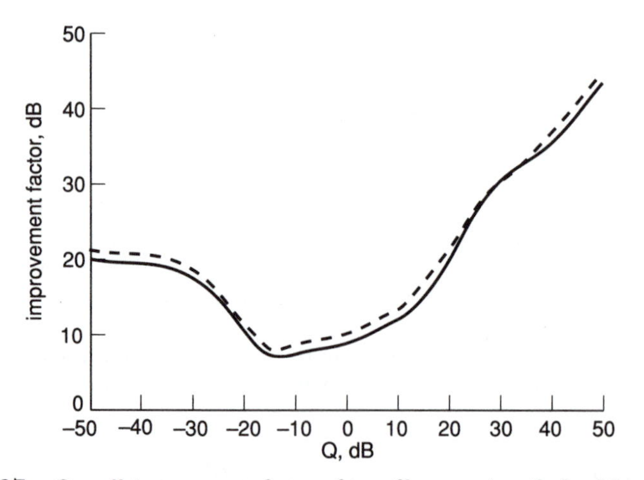

Figure 6.97 *Overall improvement factor of two filters against Q for $M = 4$ [85]*

Dash line: Optimum MTI filter
Solid line: Linear prediction filter

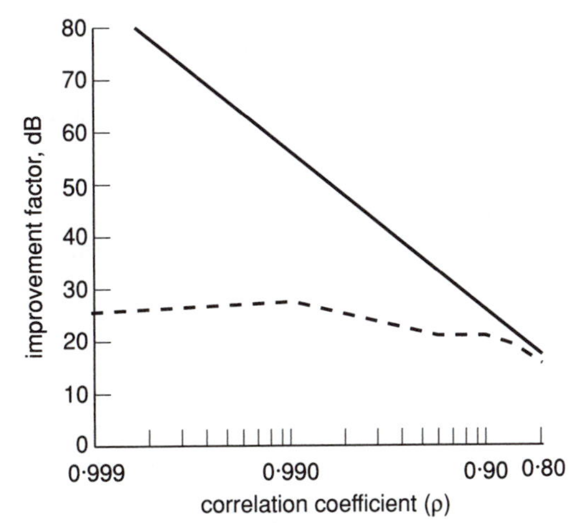

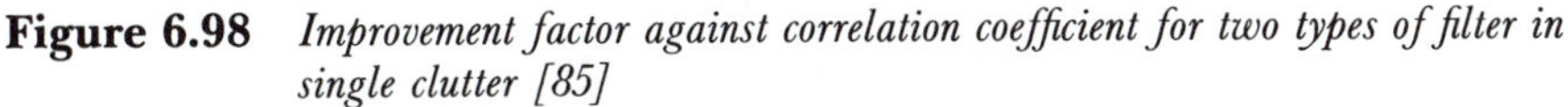

Figure 6.98 *Improvement factor against correlation coefficient for two types of filter in single clutter [85]*

Solid line: Optimum MTI filter;
Dash line: Linear prediction filter
CNR = 20 dB, $M = 4$

(iii) *Three clutter case, i.e. one ground clutter, one weather clutter and one chaff clutter:* The intensities of these three clutters are 60 dB, 36 dB and 40 dB; the mean Doppler frequencies are 0, 0.125 and 0.65; the standard deviations of the spectral width are 0.007, 0.01 and 0.015, respectively. The results of computation are shown in Table 6.23.

It can be seen that, for the adaptive MTI with the filter order equal to the optimum value of 4, the improvement factor can achieve 40 dB, even when

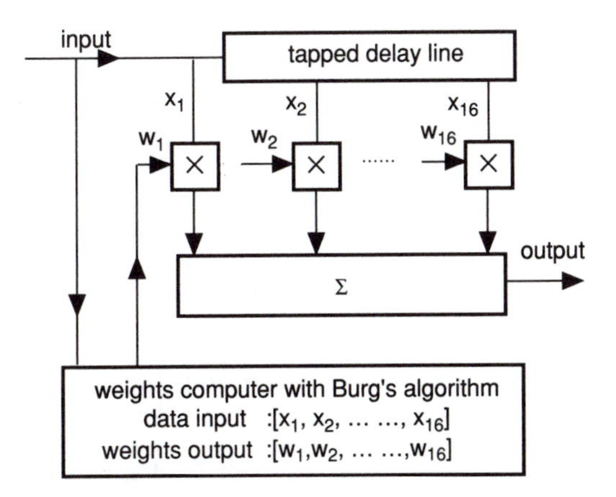

Figure 6.99 *Block diagram of adaptive processor established at RADC [88] © IEEE, 1981*

there exist three kinds of clutters at the same time; but for an ordinary fixed coefficients MTI, it is only equal to 22 dB. It is mainly obtained from the rejection of fixed clutter. In the case of three clutters, the adaptive MTI based on mean Doppler estimation cannot work, just as in the case of two moving clutters.

6.5.3.3 *Performance of adaptive MTI realised with Gram–Schmidt algorithm*
It is well known that the optimum weights of an optimum MTI filter are given by

$$\boldsymbol{w}_{opt}=\boldsymbol{M}^{-1}\boldsymbol{s}^* \tag{6.146}$$

where $\boldsymbol{M}^{-1}$ is the inverse of the interference covariance matrix and $\boldsymbol{s}^*$ is the complex conjugate of the signal vector. Owing to the positive-definite Hermitian property of $\boldsymbol{M}$, the following relationships hold:

$$\boldsymbol{\Lambda}=\boldsymbol{U}^H\boldsymbol{M}\boldsymbol{U} \tag{6.203}$$

$$\left.\begin{aligned}\boldsymbol{M}^{-1}&=\boldsymbol{U}\boldsymbol{\Lambda}^{-1}\boldsymbol{U}^H=\boldsymbol{D}^T\boldsymbol{D}^*\\ \boldsymbol{D}&=\boldsymbol{\Lambda}^{-1/2}\boldsymbol{U}^T\end{aligned}\right\} \tag{6.204}$$

where $\boldsymbol{\Lambda}$ denotes a diagonal matrix formed from real positive eigenvalues of $\boldsymbol{M}$, and $\boldsymbol{U}$ is the matrix of corresponding eigenvectors, having the property $\boldsymbol{U}^H\boldsymbol{U}=\boldsymbol{I}$. The output of the optimum filter $\boldsymbol{y}$ can be obtained from input $\boldsymbol{z}$ by

$$\boldsymbol{y}=\boldsymbol{z}^T\boldsymbol{w}_{opt}=\boldsymbol{z}^T\boldsymbol{M}^{-1}\boldsymbol{s}^*=(\boldsymbol{D}\boldsymbol{z})^T(\boldsymbol{D}\boldsymbol{s})^* \tag{6.205}$$

If we can perform the transformation $\boldsymbol{D}$ for vector $\boldsymbol{z}$

$$\boldsymbol{z}'=\boldsymbol{D}\boldsymbol{z} \tag{6.206}$$

the matrix inverse can be avoided. This can be realised with the Gram–Schmidt orthonormalisation procedure, which is well known in functional analysis theory and is used to obtain an orthonormal basis in a vector space. When applied to stochastic processes, the Gram–Schmidt algorithm gives rise to the following equations:

$$z_1'=z_1/[E\{|z_1|^2\}]^{1/2} \tag{6.207}$$

Table 6.20 *Processing losses (dB) relative to optimum processing [88]*

$\sigma_c T$	0.12745			0.0881			0.05		
No. of order	Mis-match loss	Estimated loss	Total loss	Mis-match loss	Estimated loss	Total loss	Mis-match loss	Estimated loss	Total loss
4	8.4	1.41	9.81	1.80	2.64	4.42	0.80	0.99	1.79
6	4.1	4.40	8.50	1.00	2.04	3.04	0.30	1.97	2.27
8	2.5	4.66	7.16	0.60	2.37	2.97	0.20	2.72	2.92

Table 6.21 *Improvement factors (dB) in ground clutter [89] © IEEE, 1984*

M	2		3		4		5	
N	LUD	Burg	LUD	Burg	LUD	Burg	LUD	Burg
6	75.9	70.9	74.3	74.3	72.5	75.7	73.5	77.0
8	74.6	70.3	73.0	73.6	71.2	74.4	71.1	74.8
12	74.1	70.0	71.9	73.2	70.4	73.8	69.7	74.7
16	73.2	69.5	70.9	72.9	69.3	73.3	68.6	74.0
30	71.6	68.9	68.7	72.0	66.9	72.5	65.8	73.2

Table 6.22 *Improvement factors (dB) in two moving clutters [89] © IEEE, 1984*

M	2		3		4		5	
N	LUD	Burg	LUD	Burg	LUD	Burg	LUD	Burg
6	22.0	15.7	27.6	18.0	37.5	29.2	16.9	33.6
8	21.4	15.5	26.6	17.8	36.3	28.7	14.3	30.7
12	20.6	15.3	25.5	17.4	35.0	28.3	13.5	30.2
16	20.2	15.2	24.7	17.3	34.3	28.2	13.1	29.9
30	19.6	15.1	23.5	16.8	33.0	27.9	12.3	29.4

Table 6.23 *Improvement factors (dB) in three clutters [89] © IEEE, 1984*

M	2		3		4		5	
N	LUD	Burg	LUD	Burg	LUD	Burg	LUD	Burg
6	27.8	27.5	38.7	36.4	39.6	38.9	27.3	38.1
8	27.7	27.2	38.9	36.0	39.9	38.4	24.8	36.6
12	27.0	26.6	37.3	35.0	39.9	36.7	23.7	34.8
16	26.8	26.4	37.0	34.7	37.2	36.3	23.4	34.3
30	26.0	25.7	35.6	33.8	35.3	35.2	22.8	33.3

$$z_i' = \frac{z_i - \sum_{m=1}^{i-1} E\{z_i z_m'^*\} z_m'}{\left[E\left\{\left|z_i - \sum_{m=1}^{i-1} E\{z_i z_m'^*\} z_m'\right|^2\right\}\right]^{1/2}} \quad (i=2, 3, \ldots, N) \qquad (6.208)$$

Eqn. 6.207 means that z_1' is merely a replica of z_1 with unity power. Eqn. 6.208 means that for any successive sample z_i, the first step of the processing requires the estimation of the correlation between z_i and all previous z'_m already calculated; therefore the quantity $E\{z_i z'_m\}z'_m$ represents the component of z_i which is correlated with z'_m. Hence the new sample z'_i is evaluated by progressively cancelling from z_i these correlated components. As a final step, the residual is normalised in order to have unity power. As a consequence, the output $\boldsymbol{z}'$ has an identity covariance matrix; hence the Gram–Schmidt procedure is equivalent to multiplying the incoming signal $\boldsymbol{z}$ by $\boldsymbol{D}$.

Farina [91] suggested a Gram–Schmidt processor scheme for clutter rejection, which is shown in Fig. 6.100.

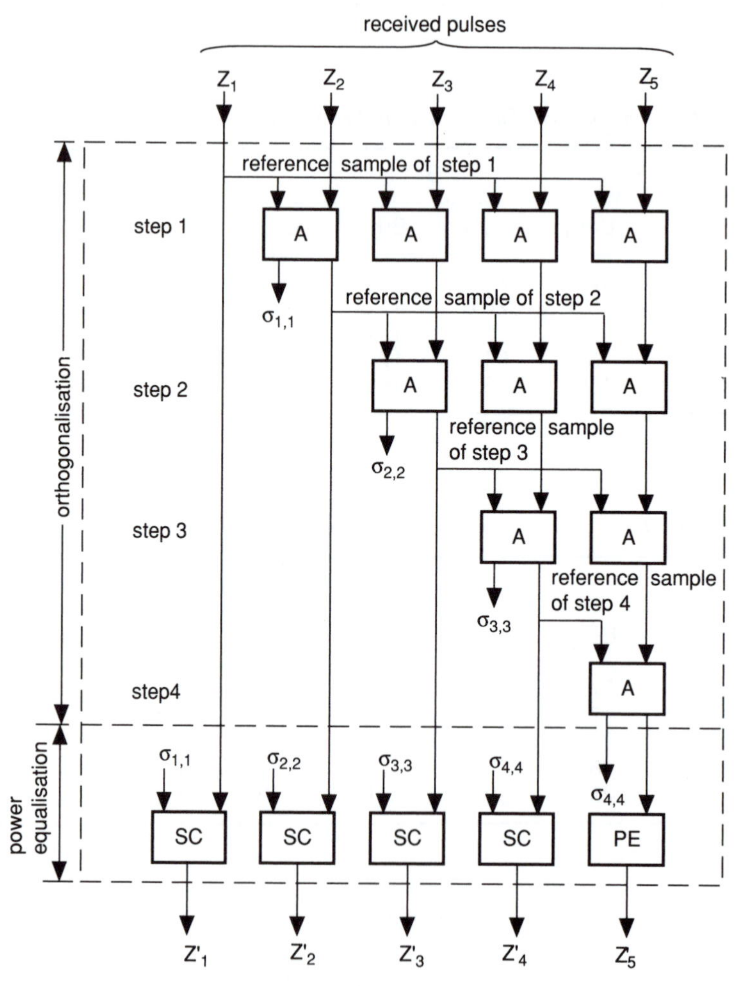

Figure 6.100 *Schematic of the Gram–Schmidt processor [91]*

SC = scaling of signal
PE = power equalizer
$\sigma_{i,i}$ = RMS value of signal at the *i*th level

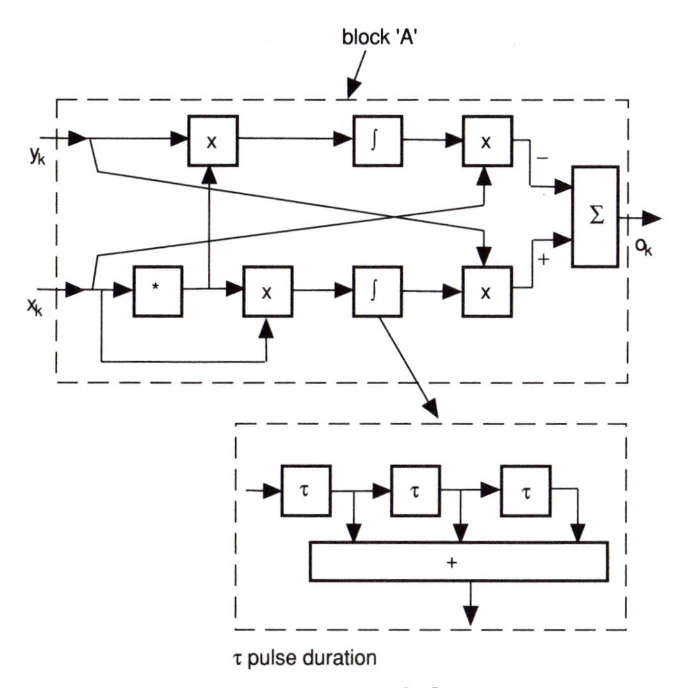

Figure 6.101 *Detailed schematic of block A [91]*

The working principle of the processor is to perform the orthogonalisation and power equalisation separately. The former is obtained through $(N-1)$ steps; at each step, one sample is taken as a reference and all the successive samples are decorrelated from it. This is obtained by means of a set of equal blocks A (Fig. 6.101), each producing an output orthogonal to the reference input of the corresponding step.

In the steady-state condition, the output of A is

$$o_k = y_k E\{x_k x_k^*\} - x_k E\{y_k x_k^*\} \tag{6.209}$$

The expectations are estimated as time averages of independent samples taken from consecutive range cells. It is easy to verify that the following condition holds:

$$E\{o_k x_k^*\} = 0 \tag{6.210}$$

corresponding to the required orthogonalisation of the output sample with the reference input.

Kretschmer [92] also proposed an adaptive MTI processor, which is based on the Gram–Schmidt algorithm. It was developed primarily for the sidelobe canceller.

A two-pulse adaptive MTI making use of the Gram–Schmidt technique is shown in Fig. 6.102*a*. The adaptive processor AP computes the error or residual signal

$$e = x_1 - wx_2 \tag{6.211}$$

In order for w to provide the minimum error, it must satisfy the orthogonalisation principle

$$E[ex_2^*] = E[x_1 x_2^*] - wE[x_2 x_2^*] = 0 \qquad (6.212)$$

$$w = E[x_1 x_2^*]/E[|x_2|^2] \qquad (6.213)$$

A representation of this process is shown in Fig. 6.102*b* for the two-dimensional signals s_1 and s_2. The signal s_2 is weighted by the real weight w to minimise the error e which is seen to be orthogonal or uncorrelated with the signal s_2.

The adaptive processor AP can be realised with a digital processor which is shown in Fig. 6.103.

A 4-pulse adaptive MTI can be constructed with six APs as shown in Fig. 6.104. The MTI clutter improvement factor I_c can be calculated by

$$I_c = \boldsymbol{w}^T \boldsymbol{w}^* / \boldsymbol{w}^T \boldsymbol{M}_c \boldsymbol{w}^* \qquad (6.214)$$

where $\boldsymbol{M}_c$ is the clutter covariance matrix.

The improvement factor of 3- and 5-pulse adaptive MTI were calculated by two independent methods using a double precision matrix-inverse routine and the Gram–Schmidt procedure for various clutter-to-noise ratios C/N. It has been found that a significant improvement in clutter cancellation is obtained by using the centre tap as the main tap. The results of computation are shown in Fig. 6.105*a* and *b* for 3-pulse adaptive MTI and 5-pulse MTI, respectively.

The improvement of clutter plus noise I_{c+n} can be calculated by

$$I_{c+n} = \boldsymbol{w}^T \boldsymbol{w}^* / \boldsymbol{w}^T \boldsymbol{M}_{c+n} \boldsymbol{w}^* \qquad (6.215)$$

where $\boldsymbol{M}_{c+n}$ is the covariance matrix of clutter pulse noise.

The results of computation of I_{c+n} for 3-pulse adaptive MTI and 5-pulse adaptive MTI are shown in Fig. 6.106*a* and *b*, respectively. It can be seen from this Figure that I_{c+n} is not very sensitive to the tap position.

The transfer function for 3-pulse adaptive MTI was calculated for a Gaussian input-clutter spectrum with mean Doppler frequency equal to 0.4, as the normalised standard deviation $\sigma_c T$ is varied (Fig. 6.107).

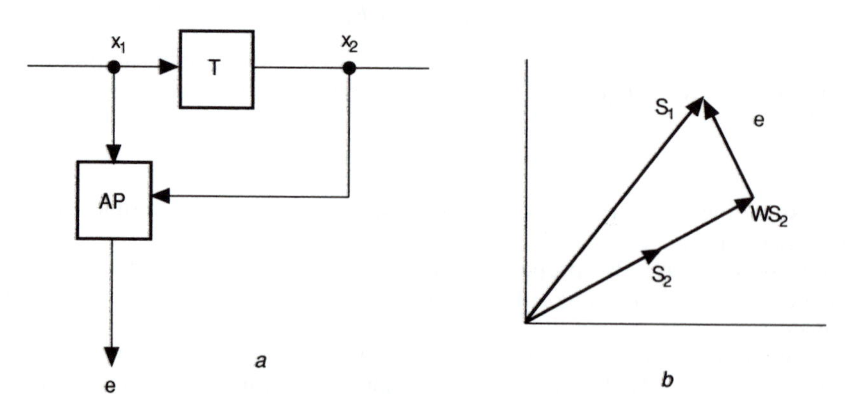

Figure 6.102 *a Two-pulse adaptive MTI*
b Error signal minimisation © IEEE, 1985

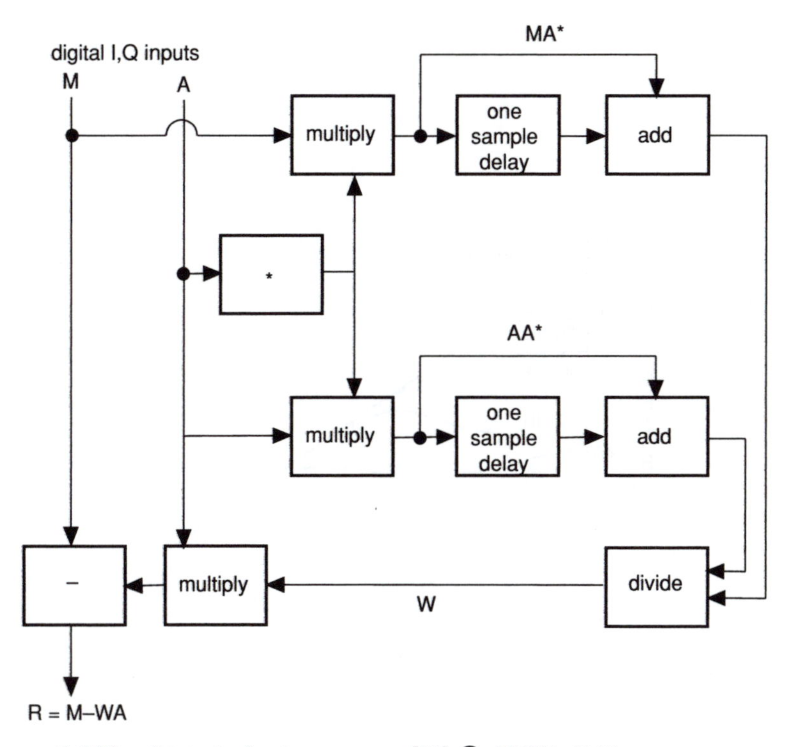

Figure 6.103 *Digital adaptive processor [92] © IEEE, 1985*

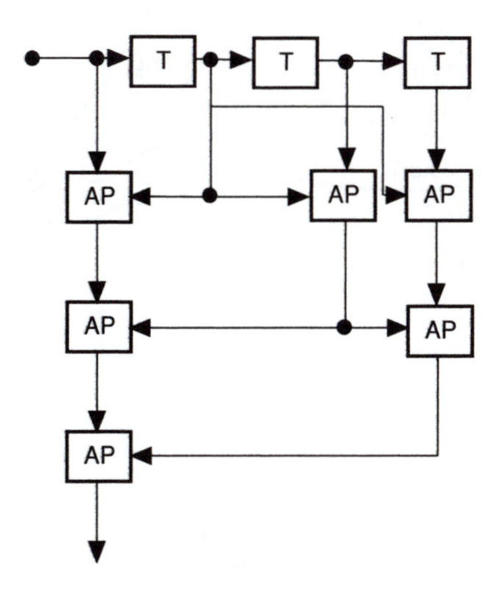

Figure 6.104 *Block diagram of Gram–Schmidt 4-pulse adaptive MTI [92] © IEEE, 1985*

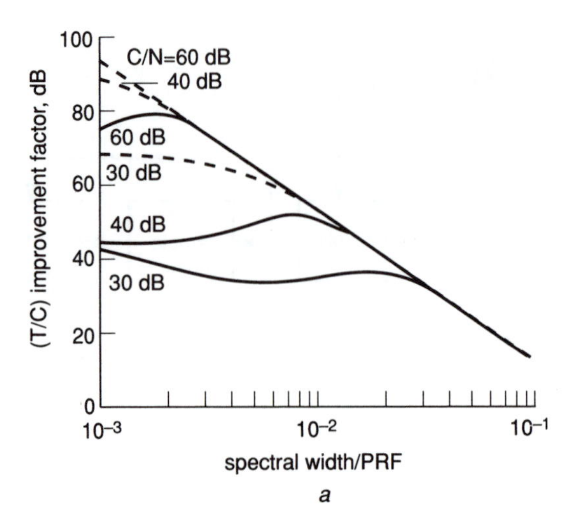

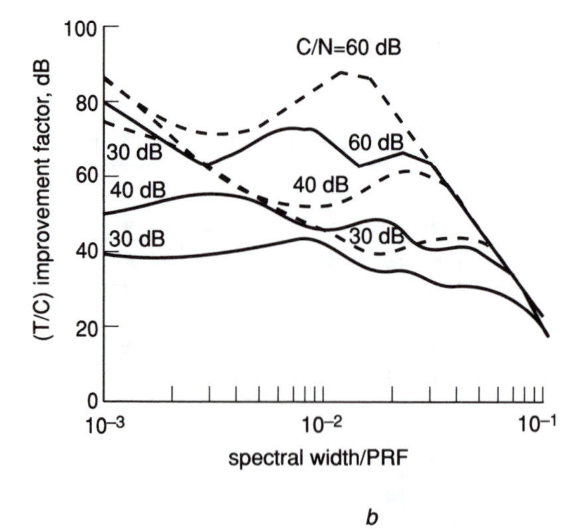

Figure 6.105 *a* I_c *for 3-pulse adaptive MTI*
b I_c *for 5-pulse adaptive MTI [92] © IEEE, 1985*

—— first tap
--- centre tap

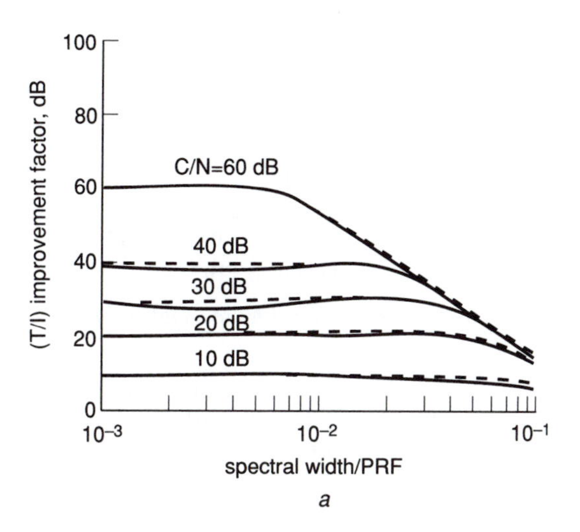

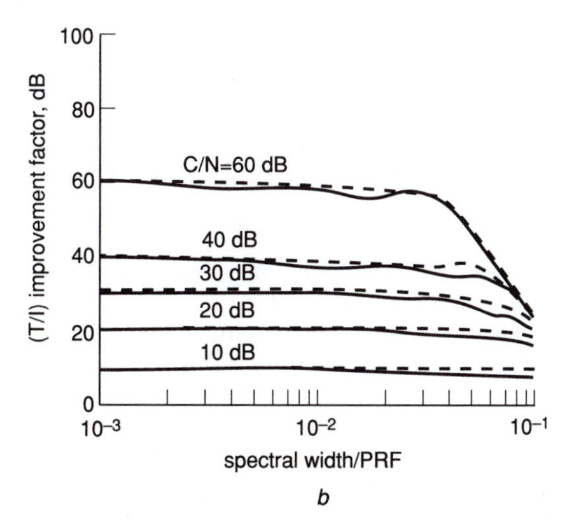

Figure 6.106 *(a) I_{c+n} for 3-pulse adaptive MTI*
(b) I_{c+n} for 5-pulse adaptive MTI [92] © IEEE, 1985

—— first tap
– – – centre tap

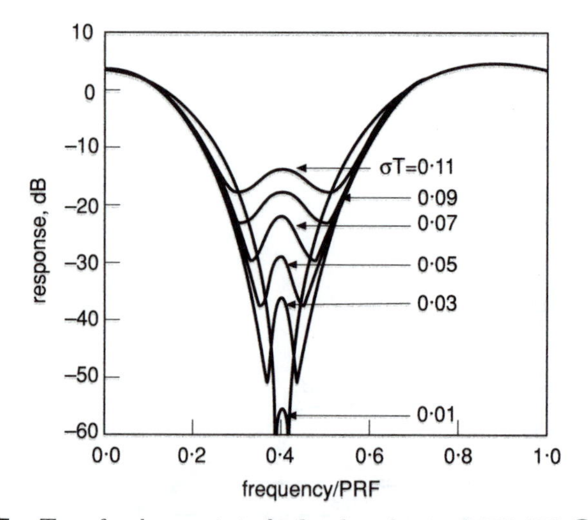

Figure 6.107 *Transfer characteristic for 3-pulse adaptive MTI [92] © IEEE, 1985*

Kretschmer [93] also compared the performance of an adaptive MTI cascaded with a Doppler filter bank (DFB) with that of DFB only.

The improvement factor for the ith channel I_i is then given by

$$I_i = \frac{\boldsymbol{w}'^T_i \boldsymbol{M}_s \boldsymbol{w}'^*_i}{\boldsymbol{w}'^T_i \boldsymbol{M}_x \boldsymbol{w}'^*_i} \tag{6.216}$$

where $\boldsymbol{w}'_i$ is the overall weighting vector of the ith channel, $\boldsymbol{M}_s$ is the covariance matrix of the signal and $\boldsymbol{M}_x$ is the normalised covariance matrix of clutter defined by

$$\boldsymbol{M}_x = E[\boldsymbol{x}\boldsymbol{x}^{*T}] \tag{6.217}$$

For the case of two simultaneous stationary clutter spectra having clutter powers C_1 and C_2, the (m, n)th element of the matrix $\boldsymbol{M}_x$ is written as,

$$M_x(m, n) = [C_1\rho_1(m, n) + C_2\rho_2(m, n)]/(C_1 + C_2) \tag{6.218}$$

where $\rho_i(m, n)$ denotes the normalised complex correlation function of the ith clutter source evaluated at time $(m-n)T$.

The improvement factors of different channels of a 2-pulse adaptive MTI plus DFB with uniform weighting is shown in Fig. 6.108 as a function of clutter spectrum width.

The average improvement factor for a 2- and 3-pulse adaptive MTI preceding a DFB (solid curve) are compared with a DFB without an MTI (dotted curve) in Fig. 6.109*a* and *b*.

It is assumed that the total number of pulses $N = 10$. So in the 2-pulse AMTI case, the number of Doppler filters is equal to 9; in the 3-pulse AMTI case, the number of Doppler filters is equal to 8. The number of filters in DFB to be compared is equal to 10. It is seen that, for clutter spectral widths less than approximately 0.2, the adaptive MTI plus DFB is significantly advantageous in terms of improving signal-to-clutter ratio.

The bimodal case was also considered. It is assumed that the power of ground clutter C_g equals 1000 and the rain clutter power C_r equals 100. Also, the ground clutter $\sigma_g T = 0.0025$ and the rain spectral width $\sigma_r T$ is 0.025. The result of using 3-pulse AMTI plus DFB is compared with using the DFB alone in Fig. 6.110*a*, where the ground clutter is centred at zero frequency, and in Fig. 6.110*b* where the normalised ground clutter frequency is offset by 0.0625. The abscissa in these plots denotes the mean Doppler frequency of the rain spectrum.

For bimodal clutter the AMTI plus DFB is superior to the fixed binomial MTI plus DFB (MTD), as shown in Fig. 6.111*a* for $\mu_g T = 0$, and Fig. 6.111*b* for $\mu_g T = 0.0625$.

6.5.3.4 *Performance of adaptive MTI realised with lattice filter*

Most of the previously mentioned adaptive MTIs are realised with the FIR filter. Some authors employed lattice filters in adaptive MTI. There are several advantages of lattice filters over FIR filters:

(i) The operating criterion for the lattice filter minimises the error output of each filter stage independently. This characteristic makes it possible to cascade any number of stages. However, once the number of stages is fixed, this advantage no longer exists.
(ii) Good convergence properties relative to the FIR filter with Widrow's LMS algorithm. However, the convergence problem is non-existent in an open-loop adaptive FIR filter.
(iii) Guaranteed stability. The stability problem only exists in the closed-loop adaptive algorithm.
(iv) Insensitivity to quantisation noise.

The structure of the lattice filter is shown in Fig. 6.112.

Basically, it is the combination of forward prediction and backward prediction. In Burg's algorithm, the AR coefficients can be calculated from the

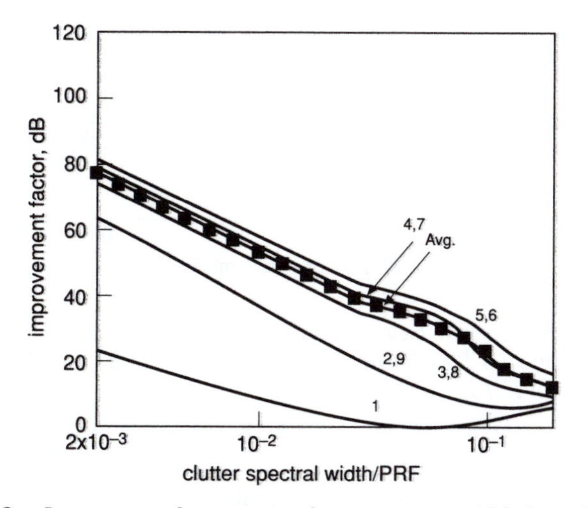

Figure 6.108 *Improvement factor versus clutter spectrum width for a 2-pulse adaptive MTI plus DFB with uniform weighting [93] © IEEE, 1984*

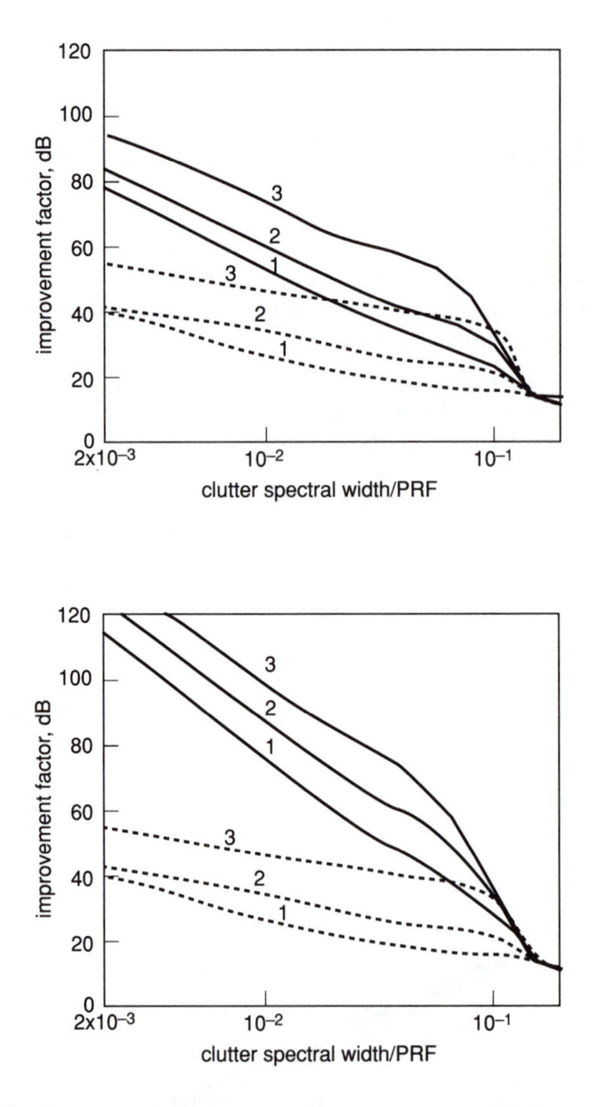

Figure 6.109 *Improvement factors versus clutter spectrum width for an adaptive MTI plus DFB (solid curves) compared with DFB (dotted curves) [93] © IEEE, 1984*

(*a*) 2-pulse AMTI + DFB/DFB, (*b*) 3-pulse AMTI + DFB/DFB

1 uniform WT
2 Chebyshev WT (SLL = 25 lb)
3 Chebyshev WT (SLL = 40 lb)

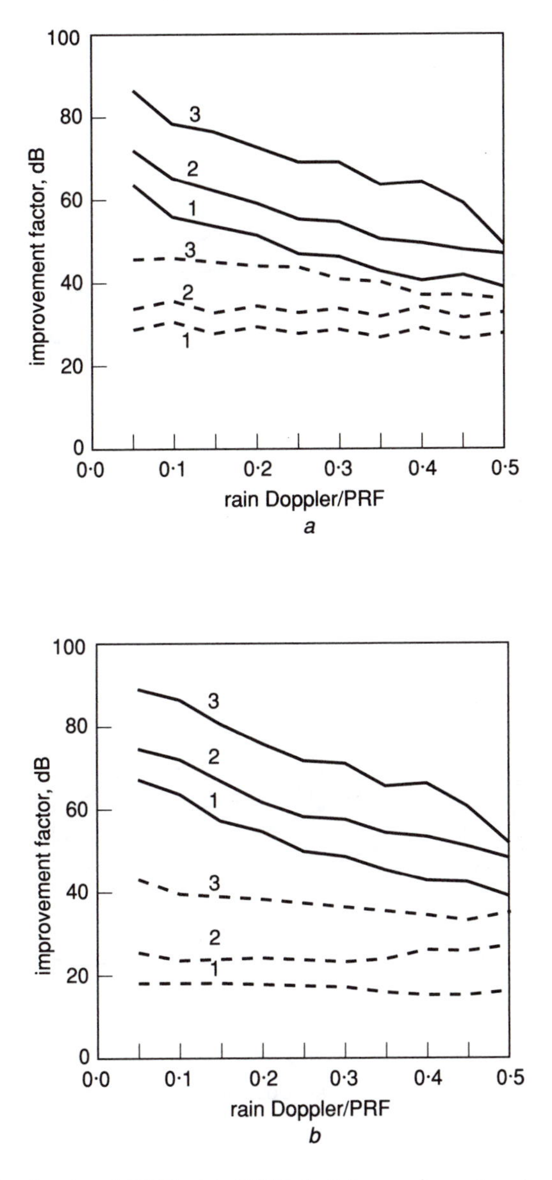

Figure 6.110 *3-pulse AMTI plus DFB (solid curves) compared with DFB (dotted curves) [93] © IEEE, 1984*

(*a*) $\mu_g T = 0$, (*b*) $\mu_g T = 0.0625$

1 uniform WT
2 Chebyshev WT (SLL = 25 lb)
3 Chebyshev WT (SLL = 40 lb)

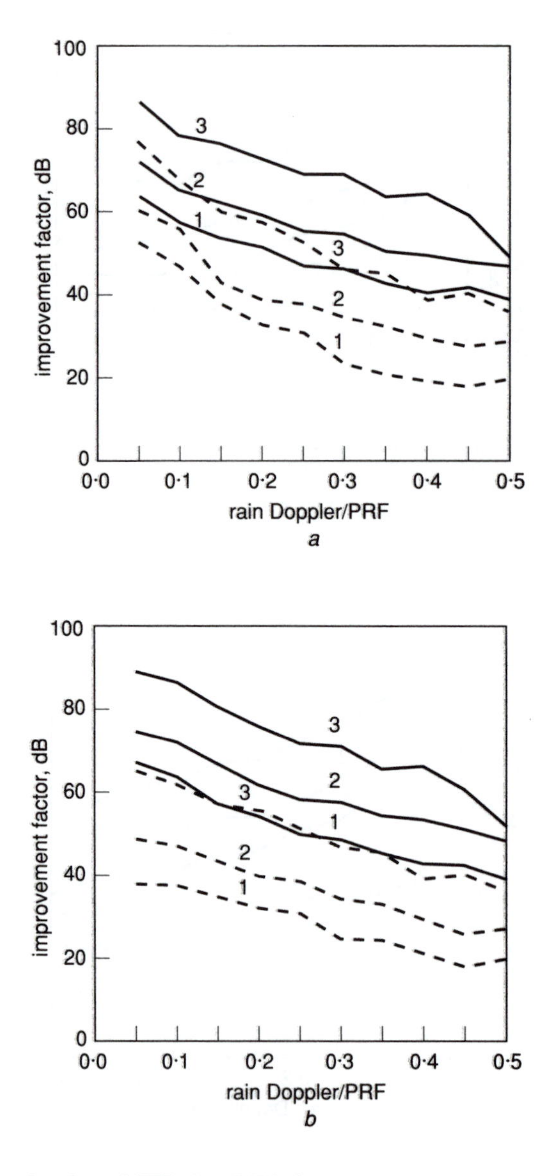

Figure 6.111 *3-pulse AMTI plus DFB (solid curves) compared with MTD (dotted curves) [93] © IEEE, 1984*

(*a*) $\mu_g T = 0$, (*b*) $\mu_g T = 0.0625$

1 uniform WT
2 Chebyshev WT (SLL = 25 lb)
3 Chebyshev WT (SLL = 40 lb)

forward and backward prediction error (see eqn. 6.168), which are related to the input data by eqns. 6.165 and 6.166. The recursive equations for forward and backward prediction errors are related to the reflection coefficients $\rho_m(n)$ by

$$f_m(n) = f_{m-1}(n) + \rho_m(n) b_{m-1}(n-1) \tag{6.219}$$

$$b_m(n) = b_{m-1}(n-1) + \rho_m^*(n) f_{m-1}(n) \tag{6.220}$$

for $n \geqq 1$ and $0 \leqq m \leqq M$, where M is the order of the filter. $\rho_m(n)$ is the reflection coefficient for stage m (also called a partial correlation coefficient) and $\rho_m^*(n)$ is its complex conjugate. The input data at time n are placed in both $f_o(n)$ and $b_0(n)$.

A number of different algorithms have been proposed for calculation of the reflection coefficient. Comparisons of these algorithms using simulated radar data have shown that Burg's harmonic-mean algorithm gives the best results in this application [94]. The optimum values of the reflection coefficients are given by this algorithm as

$$\rho_m(n) = \frac{-2E[f_{m-1}(n) b_{m-1}^*(n-1)]}{E[|f_{m-1}(n)|^2 + |b_{m-1}(n-1)|^2]} \tag{6.221}$$

In some applications where the data are completely stationary, the reflection coefficients may be calculated by block processing, where the expectations of eqn. 6.221 are simply replaced by summations from $m+1$ to N, where N is the size of the data block. In this manner the reflection coefficients are calculated only once for each filter stage and static block of data. In other applications, however, it is desirable to use moving processing, where a new value of the reflection coefficient is calculated with the arrival of each new data sample. This has the advantages of: (*a*) achieving better tracking of non-stationary signals, and (*b*) filter outputs being available immediately and continuously. The shortcomings of moving processing are: (*a*) increased computational complexity, (*b*) reduction of accuracy for stationary signals, compared with the block processing. Rather than totally re-computing the reflection coefficients with each new data sample, they may be computed recursively. Starting with an initial (arbitrary) value, the reflection coefficients are updated by correction terms, which are calculated to provide the greatest reduction of prediction

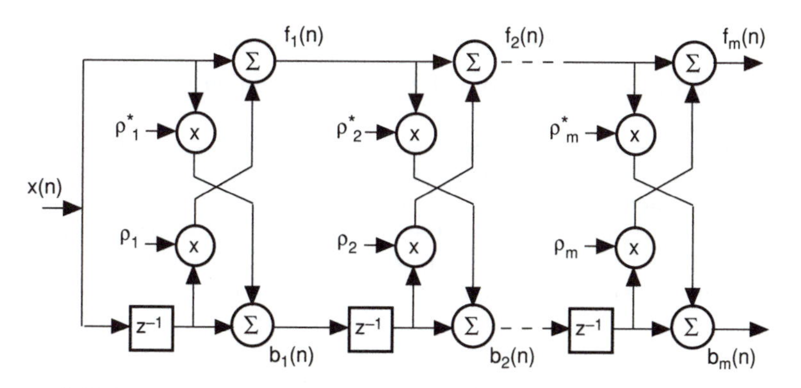

Figure 6.112 *Lattice-structure prediction-error filter*

errors per correction unit. Also it is generally desirable to apply a weighting factor to the prediction-error terms used in these calculations. This allows the filter to essentially 'forget' the effects of samples beyond a certain distance or time (which may not represent current signal statistics), including any start-up effects. It should be noted that weighting the prediction errors is not in any way equivalent to weighting the input samples, which would result in a stationary signal of reduced resolution.

The moving processing of lattice filter can be realised by many methods. Gibson [95] suggested two methods of recursive calculation of the reflection coefficients. They exhibit the controlled convergence properties required for use as a radar clutter suppression filter.

One method is called the standard gradient method. This method consists of replacing the expectations of the various terms used for the reflection coefficient calculation with corresponding averages in which an exponential weighting has been added to the individual values. The reflection coefficients thus estimated are given by

$$\hat{\rho}_{m+1}(n)=\frac{-2\sum_{i=1}^{n}[\mu^{(n-1)}f_m(i)b_m^*(i-1)]}{\sum_{i=1}^{n}\mu^{(n-1)}[|f_m(i)|^2+|b_m(i-1)|^2]} \tag{6.222}$$

where μ is a weighting constant, normally in the range $0<\mu\leqq 1$. This is equivalent to weighting the forward and delayed backward prediction errors by $\mu^{(n-1)/2}$. The exponential form of the weighting function is chosen because it can be applied as part of a simple recursive implementation. Thus, with each new sample, a new value for $\rho_{m+1}(n)$ is calculated from

$$\hat{\rho}_{m+1}(n)=\frac{v_{m+1}(n)}{y_{m+1}(n)} \tag{6.223}$$

where

$$v_{m+1}(n)=\mu v_{m+1}(n-1)-2f_m(n)b_m^*(n-1) \tag{6.224}$$

and

$$y_{m+1}(n)=\mu y_{m+1}(n-1)+|f_m(n)|^2+|b_m(n-1)|^2 \tag{6.225}$$

To apply the algorithm, we start with the initial conditions $v_{m+1}(0)=y_{m+1}(0)=0$ for all m.

The other method of recursively calculating the reflection coefficients is called the simple gradient method, owing to the use of an approximation which results in a simpler implementation than the more common standard gradient method. This method starts by considering the new coefficient as being the sum of the old coefficient and a correction term, which is the difference between the new and old values of the coefficient. The simple gradient algorithm may be written as

$$\hat{\rho}_{m+1}(n)=[1-\gamma_m(n)]\hat{\rho}_{m+1}(n-1)-\frac{2\gamma_m(n)f_m(n)b_m^*(n-1)}{|f_m(n)|^2+|b_m(n-1)|^2} \tag{6.226}$$

where

$$\gamma_m(n) = \frac{|f_m(n)|^2 + |b_m(n-1)|^2}{\sum_{i=1}^{n} [|f_m(i)|^2 + |b_m(i-1)|^2]} \quad (6.227)$$

We see that for the steady-state (constant power) case, $\gamma_m(n) \simeq 1/n$, where n is the number of data samples processed. If, however, $\gamma_m(n)$ is held constant in the calculation, then it may be replaced by using the weighting factor ω as defined in the formula

$$\omega = 1 - \gamma = 1 - 1/n' \quad (6.228)$$

where n' is the theoretical data adaptive length of the filtering action. The resulting constant ω has a value in the range $0 \leqq \omega \leqq 1$, with smaller values giving quicker adaptation. Rewriting eqn. 6.226 with ω gives

$$\hat{\rho}_{m+1}(n) = \omega \hat{\rho}_{m+1}(n-1) + \alpha_m(n) f_m(n) b_m^*(n-1) \quad (6.229)$$

where the adaptive step size $\alpha_m(n)$ equals

$$\alpha_m(n) = \frac{-2(1-\omega)}{|f_m(n)|^2 + |b_m(n-1)|^2} \quad (6.230)$$

An implicit condition on the recursive relationship of eqn. 6.229 is that the power of the prediction error $f_m(n)$ or $b_m(n)$ is not a time-varying function. This condition results in a somewhat poorer response for nonstationary inputs when compared with the standard gradient method. However, this shortcoming is partially offset by the desirable convergence property and the reduced storage and computation time requirements.

Gibson [95] measured the performance of this lattice filter with actual radar data. Fig. 6.113*a*, *b* and *c* show the average improvement factor versus filter order for ground, rain and snow clutter, based on real-life data. In these Figures, the averaging was performed with respect to five (actual) aircraft velocities. In each part of the Figure, three curves are included, two for the adaptive lattice filter with the adaptive constant $\lambda = 0.2$; one for the standard gradient filter (dashed line), one for the simple gradient filter (solid line). The performance of conventional MTI (dotted line) is also shown for comparison.

The results are worse than those obtained from computer simulation [96]. However, the performance of the lattice and MTI filters were approximately the same for ground clutter, while for precipitation clutter the lattice was 4–6 dB better.

Yu [97] proposed an adaptive MTI which is realised by a two-pulse canceller cascaded with a 2-order lattice filter. The two-pulse canceller is used to cancel fixed clutter, while the lattice filter is used to cancel moving clutter. This structure can reduce the number of computations required for the higher order lattice filter. Exact least square ladder (LSL) algorithm is employed to calculate the reflect coefficients of the lattice filter. It is well known that this algorithm has the best performance, but larger computational complexity.

Computer simulation has been performed for this adaptive MTI. A trimodal Gaussian clutter model, i.e. one ground clutter plus two moving clutters, was

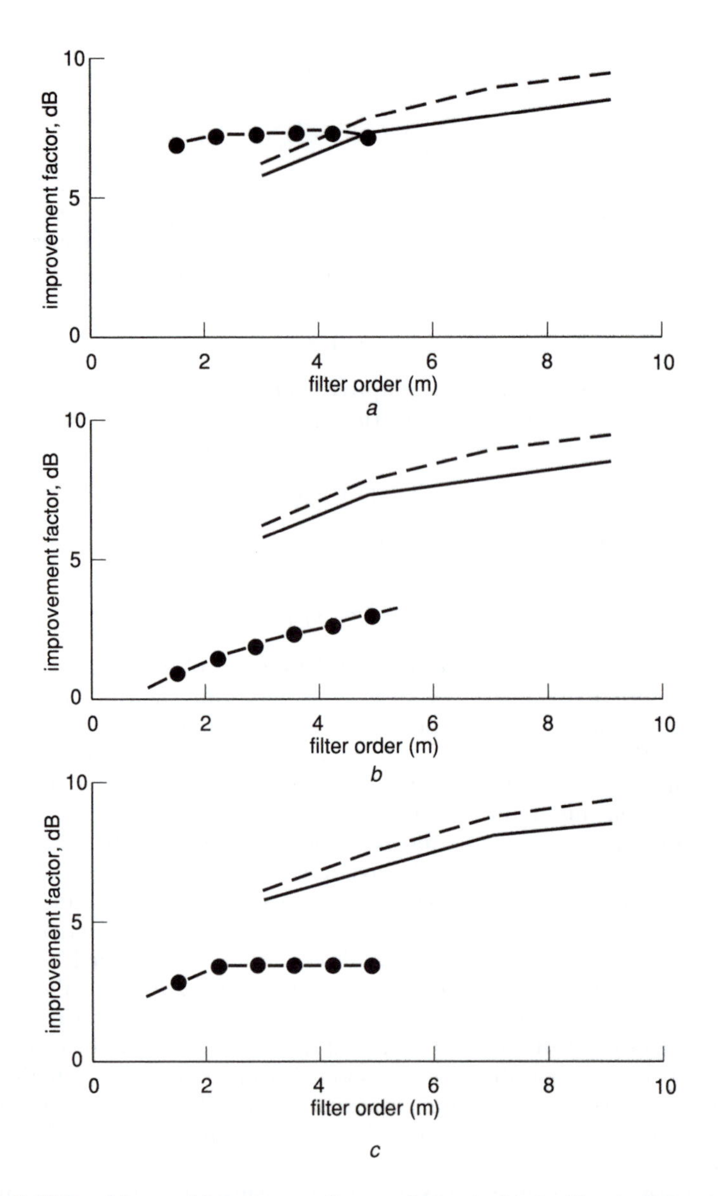

Figure 6.113 *Measured improvement factors of adaptive lattice filter in different clutters [95]*

a Improvement factors for ground clutter
b Improvement factors for rain clutter
c Improvement factors for snow clutter
—— simple gradient filter
--- standard gradient filter
-●- MTI filter

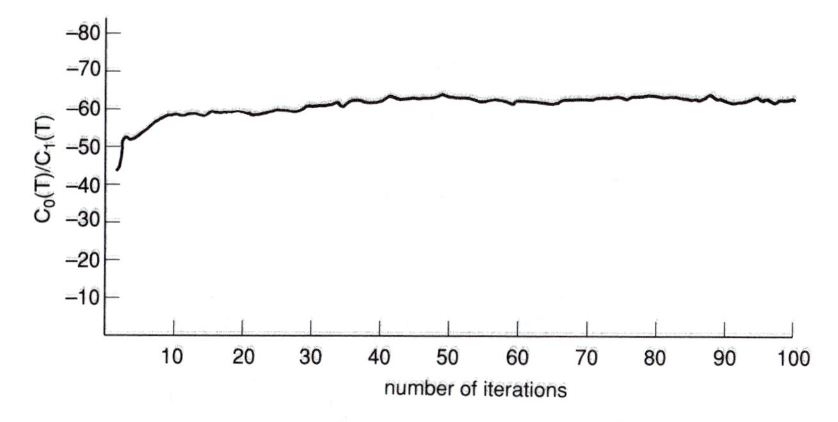

Figure 6.114 *Tracking property of the LSL algorithm [97] © IEEE, 1984*

assumed. The CNRs of these are 76 dB, 37 dB and 40 dB; the spectral widths are 0.007, 0.01 and 0.01; and the mean Doppler frequencies are 0, 0.135 and 0.35, respectively.

The clutter attenuation $C_o(T)/C_i(T)$ versus number of iterations is shown in Fig. 6.114. It can be seen from this Figure that the tracking property of this algorithm is good enough. So it can track the rapid variation of clutter statistics.

To avoid the whitening of target signal, a modified algorithm was proposed [97]. With this modified algorithm, the change in input statistics due to the target signal can produce a peak output, which is used to prevent the whitening of the target signal. The detection performance of this scheme is shown in Fig. 6.115 for target Doppler equal to 0.23 and 0.8. The dashed line is the average detection performance.

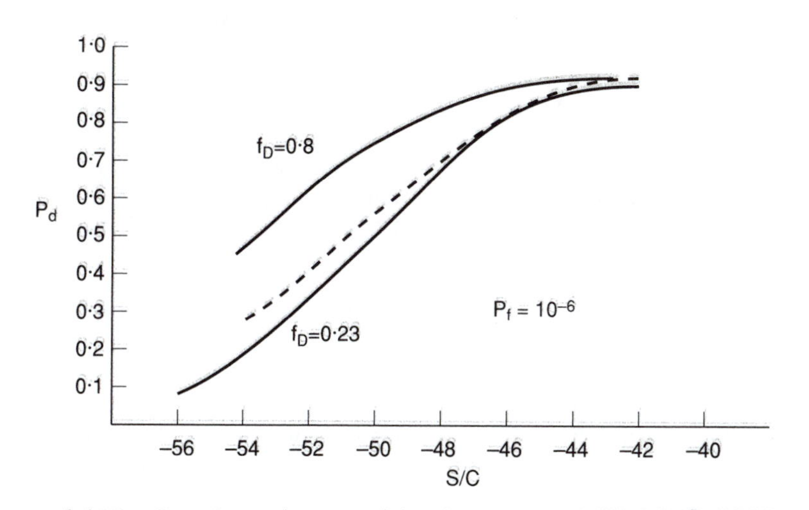

Figure 6.115 *Detection performance of the adaptive lattice MTI [97] © IEEE, 1984*

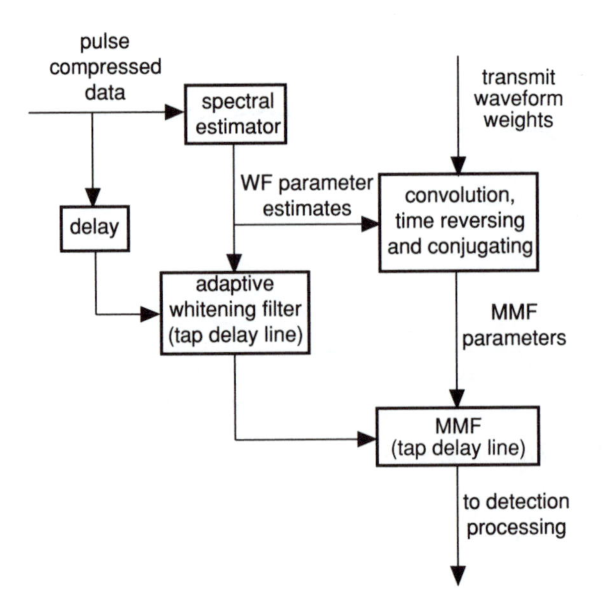

Figure 6.116 *Adaptive filter structure to process coherent burst of N weighted pulses [98] © IEEE, 1979*

6.5.3.5 *Performance of adaptive MTI realised with Kalman filtering*
Kalman filtering is suitable for processing non-stationary sequences of data windows of finite length in contrast to the Wiener filter requiring a window of infinite length of samples from stationary process. The theory provides the optimum filter (which is linear) in a recursive form (well suited for computer implementation) rather than closed form for the impulse response, as in the case of Wiener filtering. Its application requires signal modelling by means of dynamic state equation and the assumption that the probability density of the process is Gaussian.

Bowyer *et al.* [98] proposed an adaptive processor which employs Kalman filtering to estimate the AR coefficients of the clutter. The processor structure for processing the N-pulse burst data is shown in Fig. 6.116.

Let the transmit weights of an N-pulse burst waveform be $\{a_1, a_2, \ldots, a_N\}$, which may, in general, be complex. Then the sampled return from a desired target of unit strength, zero range and Doppler velocity ω_d rad/s will be

$$r(k) = a_k \exp(j\omega_d kT), \quad k = 1, 2, \ldots, N \tag{6.231}$$

The matched filter to match this return will be a tapped delay line filter with weights $\{b_k\}$, where

$$b_k = a^*_{N-k+1} \exp(-j\omega_d kT), \quad k = 1, 2, \ldots, N \tag{6.232}$$

Without loss of generality we can set $\omega_d = 0$, in which case the exponential in eqn. 6.232 can be set to unity. Let the spectral parameter estimates corresponding to the interference be $\alpha_1, \alpha_2, \ldots, \alpha_M$ at any given time sample. Then the impulse response of the whitening filter is given by $\{1, -\alpha_1, -\alpha_2, \ldots, -\alpha_M\}$

and can be implemented as a tapped delay line filter. The return signal from the desired target is passed through the whitening filter along with the interference. In this process the target signal is modified and corresponds to the sequence $\{c_k\}$, $k=1, 2, \ldots, M+N-1$ given by convolving $\{a_k\}$ with $\{1, -\alpha_1, \ldots, -\alpha_M\}$. In order to maximise S/N at the output of the processor, the filter following the whitening filter must be matched to this modified signal return $\{c_k\}$. Thus the modified match filter (MMF) has an impulse response $\{d_k\}$, and is given by

$$d_k = c^*_{M+N-k}, \quad k=1, 2, \ldots, M+N-1 \tag{6.233}$$

The MMF can also be implemented as a tapped delay line filter with weights d_k. The output of the MMF can now be processed further for detection processing.

The performance of this adaptive processor was evaluated. Clutter data were simulated by means of two closely spaced discrete frequency scatterers with a transmit waveform of 16 uniformly weighted coherent pulses. The clutter-to-noise ratio was 37 dB. The clutter spectrum is shown in Fig. 6.117. The number of spectral parameters used in this case was four. The AR parameter of this clutter was estimated with the Kalman filter algorithm from eqns. 6.182 to 6.184. Then the whitening filter was constructed with these coefficients. The frequency response of this whitening filter is also shown in this Figure as a dashed line. Note that the null of the adaptive filter is down -100 dB from the zero frequency gain.

The S/C and S/I ($I=C+N$) of this adaptive filter were also evaluated and compared with the optimum MTI filter. The clutter was assumed as a stationary, first order Gaussian Markov process generated by $C_{k+1} = \Theta C_k + \varepsilon_k$, where C_k is the interference clutter sample, $|\Theta| < 1$, and ε_k is a white Gaussian sequence. Assuming zero mean for the processes involved, the covariance matrix of the clutter for N samples is given by the $N \times N$ Toeplitz matrix

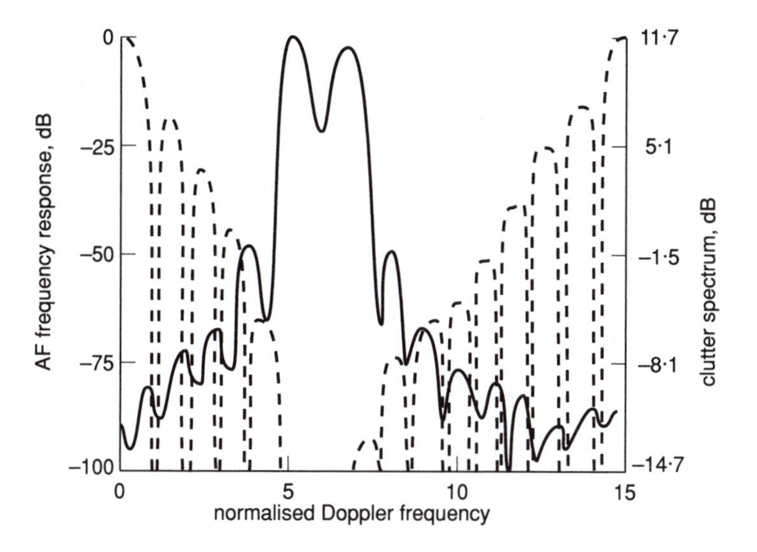

Figure 6.117 *Clutter spectrum and adaptive filter frequency response [98] © IEEE, 1979*

$$\phi_c = \begin{bmatrix} 1 & \Theta & \Theta^2 & \ldots & \Theta^{N-1} \\ \Theta & 1 & \Theta & \ldots & \Theta^{N-2} \\ \cdot & \cdot & \cdot & \cdot & \cdot \\ \Theta^{N-1} & \Theta^{N-2} & \cdot & \cdot & 1 \end{bmatrix} \quad (6.234)$$

Let the vector $\boldsymbol{w}_{opt}$ denote the weights of the optimum FIR filter. Then S/C and S/I are given by

$$S/C = |\boldsymbol{w}^*_{opt}\boldsymbol{s}|^2 / |\boldsymbol{w}^*_{opt}\phi_c\boldsymbol{w}_{opt}| \quad (6.235)$$

and

$$S/I = |\boldsymbol{w}^*_{opt}\boldsymbol{s}|^2 / |\boldsymbol{w}^*_{opt}\phi_I\boldsymbol{w}_{opt}| \quad (6.236)$$

where ϕ_I refers to the interference covariance matrix, the sum of the clutter and thermal noise covariance matrices ϕ_c and ϕ_n. For the case of an adaptive filter, we can obtain S/C and S/I with $\boldsymbol{w}_{opt}$ replaced by the appropriate adaptive filter weights $\boldsymbol{w}_{AF}$.

The performance of the autoregressive adaptive filter (ARAF) was evaluated for a clutter source which consisted of the Markov process just described, with different values for the correlation coefficient θ. This performance was evaluated in terms of S/C and S/I at each iteration of the adaptive process, and is shown in Figs. 6.118 and 6.119, respectively. The performance of adaptive filter realised with Widrow's LMS algorithm (eqns. 6.180 and 6.181) are also shown for comparison. The assumed signal, in this case, is a 10-pulse Taylor-weighted burst. Each point on these curves is an average of 100 Monte Carlo runs of the adaptive process. It can be seen from these Figures that the convergent speed of the Kalman algorithm is very fast. It takes only five times of iterations to reach the value within 5% of the final value. The final values of S/C or S/I are very close to the optimum. While the convergence of LMS algorithm is very slow, the final value may be the same.

As a specific example of the performance of the adaptive filter on real radar data, radar returns from a ballistic missile test at the Kwajalein Missile Range were recorded and used off-line for the adaptive filter processing. In this example the transmitted waveform was a coherent burst of 16 pulses with 50 dB Chebyshev weighting. The recorded data consisted of tank break-up clutter; that is, no targets were present.

Table 6.24 presents the number of false detections reported by the conventional matched filter and the adaptive filter processing of the data from a number of altitude regions. The matched filter detection processing was designed to yield $P_D = 0.99$ and $P_F \simeq 10^{-6}$ in white noise environment for a Swerling I target with SNR = 34 dB. In this Table, the detection probability was maintained at 99% level for a Swerling I target with SNR = 34 dB. The results demonstrate the ability of the adaptive filter to reduce the false detection by as much as an order of magnitude. This reduction in false detection is obtained as a consequence of the clutter suppression properties of the adaptive filter.

The hardware implementation of this adaptive filter was studied. For practical application in a real-time processor, fixed point arithmetic with a small number of bits (yet performing adequately) is desired.

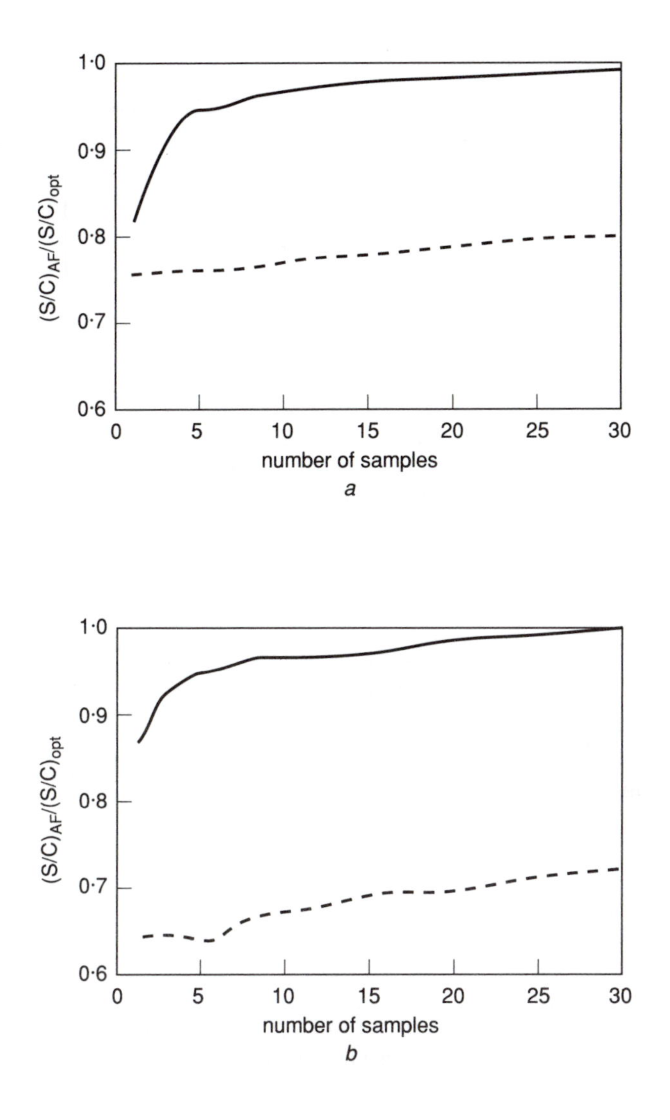

Figure 6.118 *S/C performance of ARAF and LMSAF relative to optimum filter [98] © IEEE, 1979*

(*a*) $\theta = 0.7$
(*b*) $\theta = 0.9$
—— ARAF
--- LMSAF

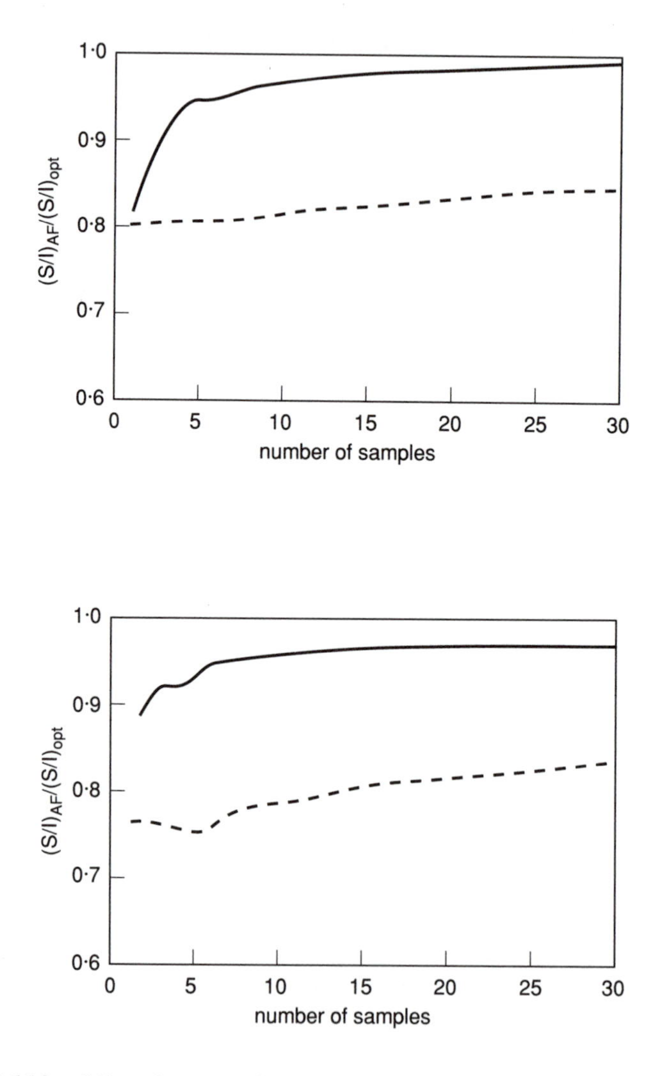

Figure 6.119 *S/I performance of ARAF and LMSAF relative to optimum filter [98] © IEEE, 1979*

(*a*) $\theta = 0.7$
(*b*) $\theta = 0.9$
—— ARAF
--- LMSAF

Table 6.24 *Number of false detections in TBU environment [98]* © *IEEE, 1979*

	Altitude region				
Filter	1	2	3	4	5
Matched	16	16	22	20	20
Adaptive	0	1	4	14	2

In particular, the spectral estimator was analysed for 8-, 10- and 12-bit internal accuracy implementations with an 8-bit complex signal input from a simulated A/D convertor. Four test signals were generated: (i) single frequency signal, (ii) two frequency signal, (iii) single frequency Markov signal (Doppler spread), (iv) two frequency Markov signal (Doppler spread).

The results of computer simulation showed that the 8-bit implementation is inadequate for most cases; the 12-bit implementation is almost as good as floating point implementation; and the 10-bit implementation is 2–3 dB worse than the latter.

6.5.4 *Some problems related to the adaptive MTI*

Enormous work has been done in the past decade on the adaptive MTI based on spectral estimation. Many results of computer simulation have been obtained. It is very difficult to evaluate these results, since they were obtained with different clutter models, different algorithms and different architectures of filter. However, it is worthwhile to discuss some fundamental concepts concerning the adaptive MTI based on spectral estimation.

6.5.4.1 *Clutter models: Gaussian model versus AR model*

Most authors study the performance of the adaptive MTI with the assumption that the clutter model is a stationary random process with Gaussian shape spectrum. Although in some cases of weather clutter and chaff clutter this model is very close to the real situation, it is not true for most land clutter and sea clutter. In general, clutter echoes are a non-stationary random process with arbitrary shape of spectrum.

Haykin [70] showed that radar clutter can be modelled as an AR process very well. If it is true for most cases, then the adaptive MTI filter based on linear prediction is the real optimum MTI filter rather than the optimum MTI proposed by Hsiao [9], since the latter is based on the Gaussian assumption.

Many authors have compared the performance of the prediction-error filter (PEF) with that of the 'optimum MTI filter' [85, 86, 99]. Unfortunately, most adopt the Gaussian clutter model. Therefore, they concluded that the PEF is worse than the optimum MTI filter by about 1–2 dB. If an AR model is employed, the 'optimum' filter belongs to the PEF, since it can whiten the coloured AR process with least mean square error. It is interesting to compare these two types of adaptive filter with the AR clutter model.

Some authors [100, 101] employed the ARMA model to simulate clutter plus noise. In the case of small CNR, say 20–30 dB, the ARMA model is more accurate than the AR model. However, the algorithms for estimating the

ARMA coefficients are more cumbersome than those for the AR coefficients. In practice, the active region of MTI is smaller than the maximum detection range for most surveillance radars. The receiver noise can be neglected within this region.

Although the AR model is also a model for stationary random process, it can model the time-varying process if the AR coefficients are estimated with short data sequence. In this case, however, algorithms with good convergence property are needed.

6.5.4.2 *The best algorithm: DMI versus recursive*

It is well known that direct matrix inversion is needed to solve the Wiener–Hopf equation. In spite of the computational complexity, DMI is the best algorithm for linear prediction. A recursive estimation of $\boldsymbol{M}^{-1}$, directly from the input sample, has been proposed [102]. Farina [103] employed this algorithm to adaptive MTI filters and compared its performance with that of an optimum filter based on real DMI. Although it is possible to attain a steady state loss within 2–3 dB or less, the transient time is greater than 10 iterations.

However, the solution of the Wiener–Hopf equation is the optimum only in the sense of least mean square error. The least mean square criterion is suitable to an infinite length of sequence. In the case of a short data sequence the least square criterion can offer better performance.

Although Burg's algorithm employs the least square criterion, it has the constraint of the Levinson recursive equation. The estimation performance of Marple's algorithm is better than that of the Burg's algorithm, since this constraint is removed. However, the recursive algorithm is employed in Marple's algorithm. It is not suitable for real-time application.

The LUD algorithm is a direct matrix inversion version of Marple's algorithm. It has to be noted that this matrix is not the simple clutter covariance matrix (see eqn. 6.170). It is derived from the forward and backward prediction error with least square criterion. Therefore, the performance of the LUD algorithm is better than that of the DMI algorithm for the Wiener–Hopf equation.

The MLUD further simplifies the LUD algorithm. The computational complexity can be compared with that of the simplest LMS algorithm in the low order case. It is the best algorithm for real-time implementation.

6.5.4.3 *Architecture of the filter: FIR versus lattice filter*

Some authors emphasise the fast convergence property of the lattice filter. It is true compared with the FIR filter with the LMS (steepest descent) algorithm. The Kalman filter can also be realised with the FIR filter. It can be seen from Figs. 6.118 and 6.119 that the convergence property is good enough for most applications. In the case of open-loop adaptation, such as the adaptive MTI realised with Gram–Schmidt algorithm or MLUD algorithm, no convergence problem exists at all.

In hardware realisation, the FIR filter is much simpler than the lattice filter. In the FIR filter, only one complex multiplication is needed for one section, whereas in the case of the lattice filter, two complex multiplications are needed for one section.

Therefore, the open-loop FIR filter is preferred.

6.5.4.4 *Format of processing: Block processing versus moving processing*
There are two different areas of processing: one for estimating AR coefficients (or reflection coefficients) and another for whitening input data.

Computational complexity is the most important problem in the estimation of coefficients. Block processing estimates the weight for each new block of length N, whereas moving processing updates the weight for each new data point. This is the reason that the number of computations is proportional to N for the moving processing algorithm (see Table 6.15). It is evident that moving processing is more complex than block processing. Therefore, block processing is preferred for weight estimation.

The shortcoming of block processing is that its tracking performance is worse than that of moving processing. However, too fast tracking is not desired, since it will whiten not only the clutter but also the signal. The problem of signal whitening will be discussed later. If the size of block is not too long, the tracking performance of block processing is still satisfactory for most practical applications.

There are also two types of whitening filter: block processing and moving processing. They are known as fixed window MTI and moving window MTI in MTI techniques. Just as we mentioned before, if the filter is of the fixed window type, only one output is permitted for each block of $N_b = M$, where N_b is the number of pulses within the block and M is the order of the filter. However, if the filter is of the moving window type, $N = N_b - M + 1$ outputs are available. If the input is not divided in blocks, continuous outputs are available. It is evident that moving processing is more desirable, since it can offer more pulses for further coherent integration.

Therefore, the final format is a moving window FIR filter whose weights are updated for each block. However, the weights cannot be updated within one pulse period, since a transient output will be caused. If the input data are divided in blocks, only $N_b - M + 1$ pulses can be obtained for each block. One way to solve this problem is to update the weights one by one, i.e. only one weight is updated for every pulse.

6.5.4.5 *The whitening of target signal*
This is the most troublesome problem in adaptive MTI based on the spectral estimation principle. Since the received echo signal is a mixture of clutter, target signal and noise, we cannot separate these components before further processing. This means that we cannot obtain a 'pure' clutter covariance matrix from which to calculate the optimum weights for the clutter whitening filter. If the data used for estimating the weights of whitening filter contain the target signal, signal whitening cannot be avoided.

Many ways to solve this problem have been suggested. We will introduce them in detail.

6.5.4.5.1 *The utilisation of the correlation of clutter in range direction*: This method is based on the assumption that the clutter is correlated in the range direction. This means that the clutter spectral parameters are the same between adjacent range cells. Zhang [104] suggested an adaptive clutter suppression scheme which predicts the current value of the cell under processing with the data from two adjacent cells (Fig. 6.120). With this method, the average improvement

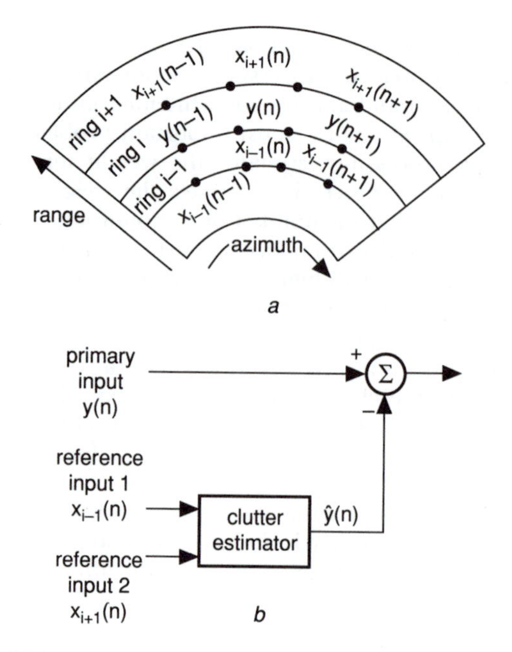

Figure 6.120 *PEF with the data from adjacent range cells as reference input [104]*

factors are enhanced about 4.3 dB in ground clutter, and 5.4 dB in weather clutter, over the previous one [95].

A modified scheme was also suggested [104] This is obtained by applying the output of the above scheme to a one step LPF. The LPF is included to reduce the effect of residual clutter remaining at the output of the above scheme. The average improvement factor of this scheme is superior by 10.2 dB in ground clutter, and 9.9 dB in weather clutter, over the old one [95].

6.5.4.5.2 *The utilisation of correlation of clutter in the azimuth direction*: This method is based on the assumption that the clutter is correlated in the azimuth direction. This means that the clutter of the same range cell but adjacent azimuth cell have the same spectral parameters. Mao [89] suggested that the weights of the adaptive filter are calculated with the previous block of N_b data points to prevent the whitening of target signal. Once a target is detected, the updating of weights stops. The old weights are still used for the next block. In this way, the weights estimated from the data including a target signal are discarded to prevent under-whitening of the clutter in the next block. This is illustrated in Fig. 6.121.

6.5.4.5.3 *Utilisation of correlation of clutters in both range and azimuth direction*: The Z–H schemes were further improved by Watanabe *et al.* [105], with modified multi-segment MEM. This scheme is based on the assumption that the clutters are correlated both in range and azimuth direction.

The reference data sequences and structure of this new scheme are shown in Fig. 6.122*a*, *b* and *c*, respectively.

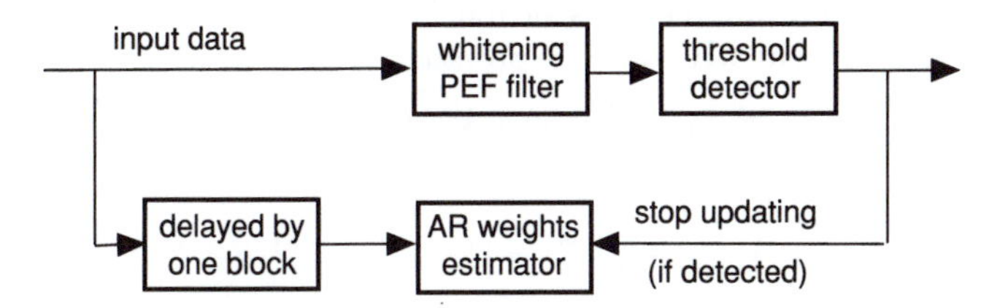

Figure 6.121 *Schematic diagram of delayed estimation [89] © IEEE, 1984*

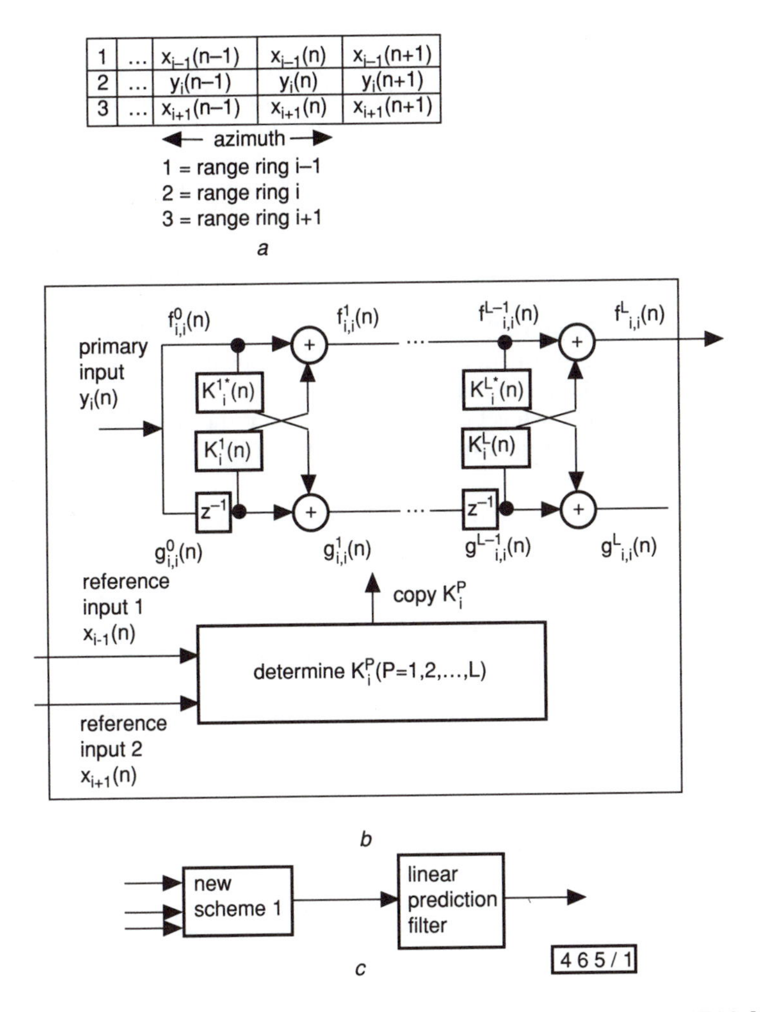

Figure 6.122 *Clutter rejection schemes based on modified multisegment MEM [105]*

The input and output relations of the Pth stage ($P=1, 2, \ldots, L$) for the modified multisegment MEM are expressed as

$$f_{i,j}^{P}(n)=f_{i,j}^{P-1}(n)+K_i^P(n)g_{i,j}^{P-1}(n-1) \tag{6.237}$$

and

$$g_{i,j}^{P}(n)=g_{i,j}^{P-1}(n-1)+K_i^P(n)f_{i,j}^{P-1}(n) \tag{6.238}$$

with

$$f_{i,j}^{0}(n)=q_{i,j}^{0}(n)=x_j(n)+\delta_{i,j}T(n) \tag{6.239}$$

It has been confirmed by examination with real radar data that a superior performance in clutter suppression can be obtained with this algorithm. Table 6.25 shows the average improvement factors using real ground, sea and weather clutter data.

6.5.4.5.4 *Utilisation of correlation of clutters in successive scans*: All the previously mentioned methods, included the adaptation delayed technique suggested by Kirimoto *et al.* [106], are based on the assumption that the clutters of adjacent resolution cells are highly correlated. However, this assumption is not always true owing to the point clutter and the edges of extended clutters. To overcome this shortcoming, a modified scheme was proposed by Watanabe [107]. This modified scheme is based on the assumption that the clutter components of the same resolution cell of several successive scans are highly correlated.

The structure of the modified schemes, which are called MS1 and MS2, are similar to that of S1 and S2 of Reference 104. Fig. 6.123 shows how the data sequences are taken from successive scans. The structure of MS1 and MS2 is shown in Fig. 6.124*a* and *b*, respectively.

This modified scheme has been tested with real radar data. The improvement factors, which are averaged over all possible target Doppler frequencies, are summarised as follows:

Scheme 1: 16.9 dB Modified scheme 1: 20.1 dB
Scheme 2: 19.0 dB Modified scheme 2: 22.9 dB

Further improvements in the performance of the modified schemes can be obtained by increasing the number of reference inputs. On the other hand, the performances of S1 and S2 cannot be improved by an increase in the number of reference inputs.

A new scheme, which is called AR map scheme, was proposed by Mao [90]. This scheme is also based on the assumption that the clutters of the same resolution cell of successive scans are highly correlated, but the echo from a

Table 6.25 *Average improvement factors in decibels [105]*

Type of schemes	S1	S2	MS1	MS2
Ground clutter	15.1	19.6	26.3	29.8
Sea clutter	6.7	9.4	11.3	14.1
Weather clutter	9.3	11.5	14.8	18.0

0	...	y(n–1)	yn	y(n+1)	...
1	...	$x_{j-1,i}(n-1)$	$x_{j-1,i}(n)$	$x_{j-1,i}(n+1)$	...
2	...	$x_{j-2,i}(n-1)$	$x_{j-2,i}(n)$	$x_{j-2,i}(n+1)$	...
m	...	$x_{j-m,i}(n-1)$	$x_{j-m,i}(n)$	$x_{j-m,i}(n+1)$	...

← azimuth →

0 = scan j, range ring i
1 = scan j–1, range ring i
2 = scan j–2, range ring i
m = scan j–m, range ring i

Figure 6.123 *Reference data sequence format [107]*

moving target will exist in any cell on only one scan. This means that the target will move out from the current resolution cell in the next scan. This is true for most practical cases.

An AR map for the clutter is constructed for every resolution cell. Two means can be used to calculate the AR coefficients of the map. One way is to calculate the AR coefficients of each cell in every scan. Then take the moving average of AR coefficients from several scans, say 6 to 8 scans. The mean value of the AR coefficient can represent the spectral parameter of the clutter. The effect of target data to the mean value of AR coefficients can be neglected, if the number of scans is great enough. Since most resolution cells do not contain the target signal, perfect estimation of the clutter spectra can be achieved. Another way to construct the AR map is that the AR coefficient is calculated from the moving

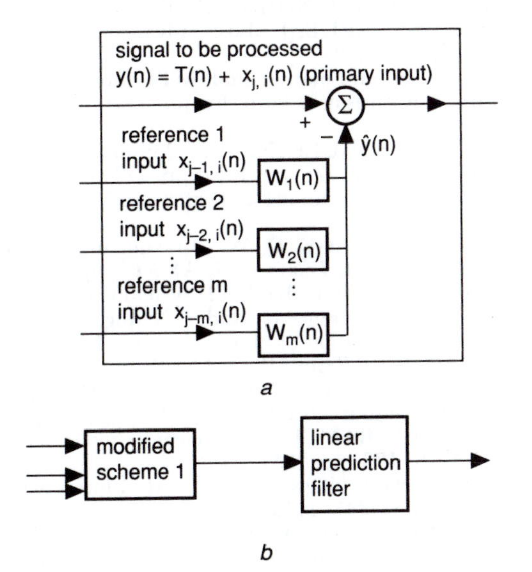

Figure 6.124 *Modified Z–H clutter suppression scheme [107]*

average of the input data. Since the mean values of input data are varying for every scan, calculation of the AR coefficients must also be performed for every scan. Larger memory capacity is needed for this method, since the number of data points is greater than the number of AR coefficients.

In all these methods, the dividing of data points to the resolution cell must synchronise with the rotation of the antenna. This is required to ensure that the calculation data come from the same resolution cell.

With this method not only can the problem of signal whitening be avoided but also the point clutters and clutter edges can be cancelled perfectly.

6.6 References

1 SKOLNIK, M. I. (Ed): 'Radar handbook' (McGraw-Hill Book Co. NY, 1970) pp. 17–15—17–19

2 CAPON, J.: 'Optimum weighting functions for the detection of sample signal in noise', *IEEE Trans.*, 1964, **IT-10**, pp. 152–159

3 RIHACZEK, A. W.: 'Optimum filters for signal detection in clutter', *IEEE Trans.* 1965, **AES-1**, pp. 297–299

4 RUMMLER, W. D.: 'Clutter suppression by complex weighting of coherent pulse trains', *IEEE Trans.*, 1966, **AES-2**, pp. 689–699

5 DELONG, D. F. and HOFSTETTER, E. M.: 'On the design of optimum radar waveform for clutter rejection', *IEEE Trans.*, 1967, **IT-13**, pp. 734–743

6 SPAFFORD, L. J.: 'Optimum radar signal processing in clutter', *IEEE Trans.*, 1968, **IT-14**, pp. 734–743

7 MITCHELL, R. L., and RIHACZEK, A. W.: 'Clutter suppression properties of weighted pulse trains', *IEEE Trans.*, 1968, **AES-4**

8 BRENNAN, L. E., and REED, I. S.: 'Optimum processing of unequally spaced radar pulse trains for clutter rejection', *IEEE Trans.*, 1968, **AES-4**, pp. 474–477

9 HSIAO, J. K.: 'On the optimization of MTI clutter rejection' *IEEE Trans.*, 1974, **AES-10**, pp. 622–629

10 ARORA, K., PRASAD, S., and INDIRESAN, P. V.: 'Class of optimum MTI filters for clutter rejection' *in* CLARKE, J. (Ed): 'Advances in radar techniques' (Peter Peregrinus, 1985) pp. 365–368

11 PRASAD, S.: 'Class of min-max digital MTI filters for clutter rejection' *in* CLARKE, J. (Ed): 'Advances in radar techniques (Petu Peregrinus, 1985) pp. 369–375

12 SCHLEHER, D. C., and SCHULKIND, D.: 'Optimization of non-recursive digital MTI', IEE Int. Conf. Radar, 77, Oct. 1977, pp. 173–176

13 DANTZIG, G.: 'Linear programming and extensions' (Princeton University Press, 1963)

14 HOUTS, R. C., and BURLAGE, D. W.: 'Maximizing the usable bandwidth of MTI signal processors', *IEEE Trans.*, 1977, **AES-13**, pp. 48–55

15 PENG, Y. N., *et al.*: 'A programmable MTD system with high performance'. Proc. CIE 1986 International Conference on Radar, Najing, China, pp. 360–365

16 PRINSEN, P. J. A.: 'Elimination of blind velocities of MTI radar by modulating the interpulse period', *IEEE Trans.*, 1973, **AES-9**, pp. 714–724

17 ROY, R., and LOWENSCHUSS, O.: 'Design of MTI detection filters with nonuniform interpulse periods', *IEEE Trans.*, 1970, **CT-17**, pp. 604–612

18 THOMAS, H. W., and ABRAM, T. M.: 'Stagger period selection for moving target radar', *Proc. IEE*, 1976, **123**, pp. 195–199

19 PRINSEN, P. J. A.: 'A class of high pass digital MTI filters with nonuniform PRF', *Proc. IEEE*, 1973, **61**, pp. 1147–1148c

20 YAN, M. S., and MAO, Y. H.: 'Design of staggered MTI weighting in multi-clutters'. CIE 1986 International Conference on Radar. Nov. 1986

21 HOUTS, R. C.: 'Velocity response characteristics of MTI radars using pulse or block stagger'. IEE Radar '77, pp. 77–81

22 HANSEN, V. G.: 'Topics in radar signal processing', *Microwave J.*, March 1984, pp. 24–46
23 GALATI, G., *et al.*: 'On the cancellation of bimodal clutter in doppler-ambiguous radar'. IEEE 1985 International Radar Conference, pp. 204–209
24 EWELL, G. W., and BUSH, A. M.: 'Constrained improvement MTI radar processor', *IEEE Trans.*, 1975, **AES-11**, pp. 768–779
25 NATHANSON, F. E.: 'Radar design principles' (McGraw-Hill Book Co., NY, 1969) sec. 14.5
26 SHORT, R. D.: 'An adaptive MTI for weather clutter suppression', *IEEE Trans.*, 1982, **AES-18**, pp. 552–562
27 HOUTS, R. C., and HOLT, B. P.: 'Design of bandstop digital filters for rejecting weather or chaff clutter in MTI radars', Proc. Southeastcon. '77, pp. 136–139
28 HOLT, B. P., and HOUTS, R. C.: 'Design considerations in bandstop digital filters for suppressing wideband clutter in MTI radars'. Proc. Southeastcon. '78, pp. 46–50
29 STONER, D. C.: 'Adaptive MTI and signal processing techniques for long range chaff and weather'. IEE Radar '77, pp. 195–198
30 HANSEN, V. G., *et al.*: 'Adaptive digital MTI signal processing'. EASCON '73
31 VOLES: 'The losses due to the error in estimating the velocity in clutter locking MTI systems', *IEEE Trans.*, 1973, **AES-9**, pp. 950–952
32 BENVENUTI, P., and GUARGUAGLINI, P. F.: 'Improvement factor evaluation of open-loop adaptive digital MTI', *Proc. IEEE* 1975, Int. Radar Conf., pp. 52–56
33 GALATI, G., and LOMBARDI, P.: 'Design and evaluation of an adaptive MTI filter', *IEEE Trans.*, 1978, **AES-14**, pp. 899–905
34 GALATI, G., *et al.*: 'On the cancellation of bimodal clutter in Doppler-ambiguous radar'. Proc. 1985 IEEE Int. Radar Conf., pp. 204–209
35 GALATI, G., and NOTTE, S. A.: 'Clutter suppression in air traffic control radars'. Proc. 1984 Int. Symp. on Noise and Clutter Rejection in Radars and Imaging Sensors, pp. 467–472
36 GALATI, G., and ORSINI, M.: 'Clutter sensors in radar signal processing', Proc. Int. Conf. on Digital Signal Processing, Florence, 1984, pp. 593–599
37 HSIAO, J. K.: 'Analysis of a cascaded MTI system in a multiple-clutter environment'. Proc. Southeastcon '77, pp. 339–342
38 MUEHE, C. E.: 'Digital signal processor for airtraffic control radars'. NEREM 1974 Conf. Rec., pp. 73–82
39 O'DONNEL, R. M., *et al.*: 'Advanced signal processing for airport surveillance radar'. EASCON '74, pp. 71–71F
40 McCRAE, V. B., and WARD, H. R.: 'A three-pulse moving target detector'. IEEE 1980 Int. Radar Conf., pp. 206–210
41 O'DONNEL, R. M., and MUEHE, C. E.: 'Automated tracking for aircraft surveillance system', *IEEE Trans.*, 1979, **AES-15**, pp. 508–517
42 MUEHE, C. E., *et al.*: 'The parallel microprogrammed processor (PMP)'. IEE Radar '77, pp. 97–100
43 TAYLOR, J. W., and BRUNINS, G.: 'Design of a new airport surveillance radar (ASR-9)', *Proc. IEEE,* 1985, **73**, pp. 284–289
44 COLE, E. L., *et al.*: 'Novel accuracy and resolution algorithms for the third generation MTD'. IEEE 1986 National Radar Conference, pp. 41–47
45 HANSEN, V. G.: 'Optimum pulse Doppler search radar processing and practical approximation'. IEE Radar '82, pp. 138–143
46 ANDREWS, G. A., Jr.: 'Optimization of radar Doppler filters to maximize moving target detection'. NAECON '74 Record, pp. 279–283
47 ANDREWS, G. A., Jr.: 'Performance of cascaded MTI and coherent integration filters in a clutter environment', NRL Report 7533, 27 March, 1973
48 WARD, H. R.: 'Properties of Dolph-Chebyshev weighting function', *IEEE Trans.*, 1973, **AES-9**, pp. 785–786
49 KRETSCHMER, F. F., LEWIS, B. L., and LIN, F. L. C.: 'Adaptive MTI and Doppler filter bank clutter processing'. IEEE 1984 National Radar Conf., pp. 69–73

50 BAO, Z., PENG, X. Y., and ZHANG, S. H.: 'Study of weather clutter rejection with moving target detection (MTD) processor'. IEE Radar '82, pp. 56–60
51 ZIEMER, R. E., and ZIEGLER, J. A.: 'MTI improvement factors for weighted DFTs', *IEEE Trans.*, 1980, **AES-16**, pp. 393–397
52 PERRY, J. L.: 'Modern radar clutter suppression techniques. A comparison of theoretical and measured results', Intl. Conf. Radar. Paris, 1984, pp. 485–491
53 DESODT, G., and LARVOR, J. P.: 'Medium range radar processing: Experimental results'. IEE Int. Conf. Radar '87, pp. 103–107
54 COLE, E. L., *et al.*: 'Novel accuracy and resolution algorithms for the third generation MTD'. IEEE 1986 National Radar Conf., pp. 41–47
55 HANSEN, V. G., and MICHELSON, D.: 'A comparison of the performance against clutter of optimum, pulsed Doppler and MTI processor'. IEEE 1980 Int. Radar Conf., pp. 211–218
56 D'ADDIO, E., FARINA, A., and STUDER, F. A.: 'Performance comparison of optimum and conventional MTI and Doppler processors', *IEEE Trans.*, 1984, **AES-20**, pp. 707–714
57 HANSEN, V. G.: 'Clutter suppression in search radar'. IEEE Conf. on Decision and Control, New Orleans, Dec. 1977
58 DILLARD, G.: 'Signal-to-noise ratio loss in an MTI cascaded with coherent integration filters'. IEEE 1975 Int. Radar Conf., pp. 117–119
59 TRUNK, G. V.: 'MTI noise integration loss', *Proc. IEEE*, 1977, **65**, pp. 1620–1621c
60 MOORE, J. D., and LAWRENCE, N. B.: 'Signal-to-noise losses associated with hardware-efficient designs of moving target indicators', *IEEE Trans.*, 1980, **ASSP-28**, pp. 35–40
61 DILLARD, G. M., and RICHARD, J. T.: 'Performance of an MTI followed by incoherent integration for nonfluctuating signals'. IEEE 1980 Int. Radar Conf., pp. 194–199
62 SCHLEHER, D. C.: 'MTI detection losses in clutter', *Electron. Lett.*, 1981, **17**, pp. 82–83
63 MULLER, B.: 'MTI loss with coherent integration of weighted pulses', *IEEE Trans.*, 1981, **17**, pp. 549–552
64 WEISS, M., and GERTNER, I.: 'Loss in single-channel MTI with post-detection integration', *IEEE Trans.*, 1982, **AES-18**, pp. 205–207
65 SCHLEHER, D. C.: 'Performance comparison of MTI and coherent Doppler processor'. IEE Radar '82, pp. 154–158
66 SCHLEHER, D. C.: 'MTI detection performance in Rayleigh and Log-normal clutter'. IEEE 1980 Int. Radar Conf., pp. 299–304
67 FISHBEIN, W., *et al.*: 'Clutter attenuation analysis', USA ECOM, Tech. Report, ECOM-2808, Fort Monmouth, NJ, March 1967
68 JANSSEN, L. H., and VAN DER SPEK, G. A.: 'The shape of Doppler spectra from precipitation', *IEEE Trans.*, 1985, **AES-21**, pp. 208–219
69 ESTES, W. J., *et al.*: 'Spectral characteristics of radar echoes from aircraft dispensed chaff', *IEEE Trans.*, 1985, **AES-21**, pp. 8–20
70 HAYKIN, S., *et al.*: 'Maximum-entropy spectral analysis of radar clutter', *Proc. IEEE*, 1982, **70**, pp. 953–962
71 STEHWIEN, W., and HAYKIN, S.: 'Statistical classification of radar clutter'. IEEE 1986 National Radar Conf., pp. 101–106
72 VAN DEN BOS, A.: 'Alternative interpretation of maximum entropy analysis', *IEEE Trans.*, 1971, **IT-17**, pp. 493–494c
73 BURG, J. P.: 'Maximum entropy spectral analysis'. 37th Ann. International Meeting of Society of Explorations Geophysicists, Oklahoma City, Oct. 1967
74 LEVINSON, N.: 'The Wiener (root mean square) error criterion in filter design and prediction', *J. Math. Phys.*, 1947, **25**, pp. 261–278
75 MARPLE, L.: 'A new autoregressive spectrum analysis algorithm', *IEEE Trans.*, 1980, **ASSP-28**, pp. 441–454
76 LIN, D. M., and MAO, Y. H.: 'A fast algorithm of maximum entropy method for spectral estimation suitable for short data sequence', *Scientia Sinica, Series A.*, 1984, **XXVII**, pp. 196–212

77 DENG, X. D.: 'A fast algorithm of conjugate reverse symmetry equation and its application to AR spectral estimation'. 1st Conf. on Signal Processing, Beijing, China, 1984
78 ZHOU, X. Q.: 'Parameter estimation of AR model and its application in adaptive filtering'. Master Thesis, National University of Defense Technology, Changsha, China, 1985
79 LI, W. G., and MAO, Y. H.: 'The comparison of computational complexity of some MEM algorithms and a new fast algorithm of MEM'. To be published
80 WIDROW, B., and HOFF, M. E.: 'Adaptive switching circuits'. IRE WESCON Conv. Record, pt. 4, 1960, pp. 96–104
81 BAYER, D. E., *et al.*: 'Adaptive clutter filtering using autoregressive spectral estimation', *IEEE Trans.*, 1979, **AES-15**, pp. 538–546
82 LEE, D. T., MORF, M., and FRIEDLANDER, B.: 'Recursive least square ladder estimation algorithm', *IEEE Trans.*, 1981, **ASSP-29**, pp. 627–641
83 PORAT, B., *et al.*: 'Square-root covariance ladder algorithms', *IEEE Trans.*, 1982, **AC-27**, pp. 813–829
84 CIOFFI, J. M., and KAILATH, T.: 'Fast, recursive-least-squares transversal filters for adaptive filtering', *IEEE Trans.*, 1984, **ASSP-32**, pp. 304–337
85 BARBAROSSA, S., D'ADDIO, E., and GALATI, G.: 'Comparison of optimum and linear prediction techniques for clutter cancellation', *IEE Proc.*, 1987, **134**, Pt. F, pp. 277–282
86 CHIUPPESI, F., GALATI, G., and LOMBARDI, P.: 'Optimization of rejection filters', *IEE Proc.*, 1980, **127**, Pt. F, pp. 354–360
87 BUCCIARELLI, T., MARTINELLI, G., and PICARDI, G.: 'Clutter cancellation in search radars by spectral AR techniques', *Alta Frequenza*, 1983, **LII**, pp. 389–393
88 NITZBERG, R.: 'Some design details of the application of modern spectral estimation techniques to adaptive processing'. 1st ASSP Workshop on Spectral Estimation, 1981
89 MAO, Y. H., LIN, D. M., and LI, W. G.: 'An adaptive MTI based on maximum entropy spectrum estimation principle'. Proc. 1984 Int. Conf. on Radar, Paris, pp. 103–108
90 MAO, Y. H., and XIE, X.: 'The detection performance of an adaptive MTI based on AR map'. Proc. IEE Int. Conf. Radar-87, pp. 438–442
91 FARINA, A., and STUDER, F. A.: 'Application of Gram-Schmidt algorithm to optimum radar signal processing', *IEE Proc.*, 1984, **131**, Pt. F, pp. 139–145
92 KRETSCHMER, F. F., Jr., and LIN, F. L. C.: 'Effects of the main tap position in adaptive clutter processing'. IEEE 1985 Int. Radar Conf., pp. 303–307
93 KRETSCHMER, F. F., Jr., *et al.*: 'Adaptive MTI and Doppler filter bank clutter processing'. IEEE 1984 National Radar Conf., pp. 69–73
94 GIBSON, C. and HAYKIN, C.: 'A comparison of algorithms for the calculation of adaptive lattice filters'. Proc. 1980 IEE Int. Conf. on ASSP, Vol. 80, pp. 978–983
95 GIBSON, C., and HAYKIN, C.: 'Radar performance studies of adaptive lattice clutter suppression filters', *IEE Proc.*, 1983, **130**, Pt. F, pp. 357–367
96 GIBSON, C., HAYKIN, C., and KESLER, S.: 'Maximum entropy (adaptive) filtering applied to radar clutter', Proc. 1979 IEEE Int. Conf. on ASSP, Vol. 79, pp. 166–169
97 YU, X. L., LU, D. J., and MAO, Y. H.: 'Performance studies of exact least squares ladder algorithm for automatic signal processing in radar system'. Proc. 1984 Int. Symp. on noise and clutter rejection in radars and imaging sensors, pp. 268–273
98 BOWYER, D. E., RAJASEKARAN, P. K., and GEBHART, W. W.: 'Adaptive clutter filtering using spectral estimation', *IEEE Trans.*, 1979, **AES-15**, pp. 538–546
99 CACOPARDI, S., PICARDI, G., and PRESTIFILIPPO, E.: 'Limits of the clutter cancellation techniques by AR models'. Proc. 1984 Int. Symp. on noise and clutter rejection in radars and imaging sensors, pp. 303–308
100 MEHTA, M. S., and RAO, B. V.: 'Application of Kalman filtering to adaptive clutter filtering'. Int. Conf. on Radar, Paris, 1984, pp. 472–478
101 MARTINELLI, G., BURRASCANO, P., and ORLANDI, G.: 'Simultaneous cancellation of two clutters by the extended lattice predictor'. Proc. 1984 Int. Symp. on noise and clutter rejection in radars and imaging sensors, pp. 274–278

102 MONZINGO, R. A., and MILLER, T. W.: 'Introduction to adaptive arrays' (John Wiley, 1980)
103 FARINA, A., STUDER, F. A., and TURCO, E.: 'Adaptive methods to implement the optimum radar signal processor'. Proc. Int. Radar Symp. India, 1983, pp. 42–47.
104 ZHANG, Q. T., and HAYKIN, S.: 'Radar clutter suppression schemes', *Electron. Lett.* 1984, **20**, pp. 1007–1008
105 WATANABE, H., *et al.*: 'New radar, clutter suppression schemes using modified multisegment MEM', *Electron Lett.*, 1986, **22**, pp. 369–370
106 KIRIMOTO, T., *et al.*: 'Adaptive moving target indicators to preserve target signals using adaptation delay technique'. IEE Int. Conf. RADAR '87, pp. 443–447
107 WATANABE, H., *et al.*: 'Modification of the Zhang and Haykin radar clutter suppression schemes', *Electron. Lett.*, 1986, **22**, pp. 92–93

Chapter 7

Rejection of active interference

Prof. Y. H. Mao

7.1 Introduction

It is well known that the most effective measure of ECM is active noise jamming. Therefore, we will discuss the rejection of active interference as a main topic of radar ECCM. The principle methods of active interference rejection are as follows [1]:

(i) *Spatial selectivity*; Utilises direction or entrance path to select desired from undesired signal.
(ii) *Frequency selectivity*; Discriminates against wideband off-frequency and impulse signals.
(iii) *Amplitude selectivity*; Uses amplitude characteristics to discriminate against, for example, high energy or impulsive interference
(iv) *Time selectivity*; Uses time of arrival of signal or time separation between successive returns to distinguish between interference and target return.
(v) *Signal selectivity*; Employs signal statistics, shape, codes etc.

Over 200 radar electronic ECCM techniques have been identified [2]. These include a number of variants of basic techniques and special combinations of techniques. The most important ECCMs are as follows [3]:

(*a*) *Transmitter*
- Frequency agility/PRF agility
- Pulse compression
- Higher frequency (MM/EO) (T)
- Higher average power
- Coherent transmitter

(*b*) *Antenna*
- Lower sidelobes
- Sidelobe canceller/blanker (S)
- Monopulse (T)
- Higher gain (narrower beam)

(*c*) *Receiver*
- CFAR (S)
- Dicke–fix (wideband–limiter–narrowband) (S)
- MTI/Doppler filtering
- Leading edge tracking (T)
- Clean design
- Large dynamic range

(*d*) *System*

Passive strobe (S)

Passive track (TOJ/HOJ) (T)

(S): Search radars, (T): Tracking radars

The strategy of active interference rejection is to prevent the interference entering the radar detector as far as possible. Since there are many nonlinear components in the radar receiver, such as mixer, envelope detector etc., the weak signal will be suppressed by the strong interference signal in these nonlinear components. Therefore, the most effective measures to reject interference are spatial selectivity and frequency selectivity, since spatial selectivity rejects the interference at the antenna, and frequency selectivity rejects the interference before the mixer.

Two types of jamming are often used: standoff jamming (SOJ) and self-screening jamming (SSJ). Standoff jamming interferes with the radar receiver mainly through the sidelobe of the radar antenna, while self-screening jamming interferes with the radar receiver mainly through the main lobe of the antenna.

It is evident that the best way to reject standoff jamming is by spatial selectivity, and the best way to reject self-screen jamming is by frequency selectivity.

Therefore we will discuss these two methods in this Chapter in detail.

7.2 Spatial selectivity

There are two means for spatial selectivity; one is the direct method, and the other is the indirect method. The direct method is to reduce the real sidelobe of the antenna directly. This is known as the low sidelobe antenna technique. The indirect method is to reduce the equivalent sidelobe of the antenna. This is known as adaptive sidelobe canceller. A classical method, which is called sidelobe blanking, also belongs to the indirect method. However, we will discuss the first two techniques in this Section, since they are more effective than the latter.

Crossover range, i.e. the range at which $J/S=1$, is of interest, where J is the power received from the jammer and S is the power received from the target.

In the case of standoff jamming, the crossover range can be given as follows.

$$R_T^4/R_J^2=(B_J/B_R)(P_T/P_J)(G_T/G_J)(G_R/G_{SL})(L_JL_R/L_T)(\sigma/4\pi) \tag{7.1}$$

where

R_T = range from the target
R_J = range from the jammer
B_J = bandwidth of the jammer
B_R = bandwidth of the radar
P_T = power from the target
P_J = power from the jammer
G_T = gain of the transmitting antenna
G_J = gain of the jammer's antenna
G_R = gain of the receiving antenna
G_{SL} = gain of the sidelobe in the direction of the jammer
L_J = loss of the jammer

Table 7.1 *Typical aperture taper and sidelobe level [4]*

Aperture taper	Derivative in which impulses or large peaks appear	Largest sidelobe (dB)
Rectangle	1	−13.4
Circle	1~2	−17.5
Parabola	2	−22.0
Cosine	2	−23.5
Triangle	2	−26.8
Raised cosine	3	−32.0

L_R = loss of the receiver
L_T = loss of the transmitter

From the radar designer's viewpoint, most of the parameters in this equation are determined by other factors and cannot be changed. The only one which has the greatest potential for change is the sidelobe level of the antenna, G_{SL}.

It can be seen from eqn 7.1 that, if R_J = 100 km and the sidelobe level equals −20 dB, the maximum detectable target range is equal to 20 km. Then if the sidelobe level is reduced to −40 dB, the maximum detectable range will be increased to 63 km. If the sidelobe level is reduced to −60 dB, the maximum detectable range will be increased to 200 km. This means that ultra-low sidelobe level can reject standoff jamming effectively.

Low sidelobe requirements apply to both the transmitting and the receiving antenna patterns. Reduction of receiving antenna sidelobes is useful not only against standoff jammers but also against false target deception ECM. Reduction of transmitting antenna sidelobes is useful against ARMs and hostile ESM receivers. Accordingly, attention should be devoted to reduction of sidelobe levels of both the receiving and the transmitting antenna patterns in a search radar.

There are two kinds of radar antenna: array antenna, and reflector antenna. We will discuss the latter only, since it is the most commonly used antenna.

7.2.1 *Low sidelobe antenna technique*

It is well known that the sidelobe level of a reflector type antenna mainly depends on the aperture amplitude illumination or aperture taper. A simple rule of thumb is that: if impulse or large peaks first appear in the nth derivative of the aperture excitation, the sidelobe level is of the order of $-10n$ dB. Some typical aperture tapers and their largest sidelobe levels are shown in Table 7.1.

All these aperture tapers are of very simple form. Better results can be obtained with more complex tapers.

In the case of circular apertures, an illumination function of the form $(1-r^2)^m$ is often used, where r is the direction, and m is any integer from 0 to 3. However, for many antenna applications, illuminations that go to zero at the aperture edge are not practical or desirable. If expressed as a $(1-r^2)^m$ illumination on a pedestal b, more versatile results and appropriate far field patterns can be calculated. Several families of b and m are given in Table 7.2 [4].

It is evident from this Table that there is a contradiction between the sidelobe level and gain. The lower the sidelobe, the smaller the gain becomes.

The use of aperture illuminations of the form $(1-r^2)^m$ results in a far-field pattern given by

$$\Lambda_\nu(u)=\sum_{k=0}^{\infty}\frac{1}{k!}\cdot\frac{\nu!}{(\nu+k)!}\left(\frac{iu}{2}\right)^{2k}=\Lambda_{m+1}(u) \tag{7.2}$$

for which the sidelobe envelope decay rate is controlled by ν. However, it is possible to specify the first sidelobe level other than that which occurs naturally for the given ν. A two parameter family of space factors was originated by Bickmore and Spellmire for line sources, but is extendable to circular apertures as well. The family is given by

$$F(u)=\Lambda\nu[\sqrt{u^2-A^2})] \tag{7.3}$$

If ν is chosen as $-1/2$, the far field is $\Lambda_{-1/2}[(u^2-A^2)^{1/2}]=\cos(u^2-A^2)$, and the associated linear-aperture illumination is the envelope of Dolph–Chebyshev array distributions with a large number of elements. This pattern has been designated as the ideal space factor because it is just the continuous equivalent to the discrete Dolph–Chebyshev optimum array design. Optimum is used here in the sense that the maximum gain is obtained for a given sidelobe level.

However, the aperture illumination needed to produce a far field of the form $\cos(u^2-A^2)^{1/2}$ requires a very large amplitude at the aperture endpoints, and hence is impractical for large apertures. The zeroes of the far field, u_n, occur when

$$u_n=\pm[A^2+(n-1/2)^2\pi^2]^{1/2} \tag{7.4}$$

Taylor showed that an approximation to the ideal space factor could be obtained over the first $\bar{n}$ zeroes of the pattern by introducing a stretch-out parameter $\sigma(\sigma\geqq 1)$ such that

$$\begin{aligned} u_n&=\pm\sigma[A^2+(n-1/2)^2\pi^2]^{1/2}, && 1\leqq n<\bar{n} \\ u_n&=\pm n\pi &&, \bar{n}\leqq n<\infty \end{aligned} \tag{7.5}$$

This means that the first $\bar{n}-1$ sidelobes are of nearly equal amplitude. The stretch-out parameter is given by

Table 7.2 *Far-field pattern parameters for $b+(1-r^2)^m$ circular-aperture illuminations [4]*

	Sidelobe level			G/G_0		
m	b=0	b=1/4	b=1/2	b=0	b=1/4	b=1/2
0	17.6	. . .	. . .	1.00	. . .	. . .
1	24.6	23.7	22.0	0.75	0.87	0.92
2	30.7	32.3	26.5	0.55	0.81	0.88
3	36.1	32.3	30.8	0.45	0.79	0.87

Table 7.3 *$\Lambda_{1/2}[(u^2-A^2)^{1/2}]$ design parameters [5] Reproduced with permission of McGraw-Hill*

Sidelobe level (dB)	β_0 (rad)	G/G_0	A
13.2	0.885	1.000	0
15	0.923	0.993	1.118
20	1.024	0.933	2.320
25	1.116	0.863	3.214
30	1.200	0.801	4.009
35	1.278	0.751	4.755
40	1.351	0.709	5.471

$$\sigma=\frac{\bar{n}\pi}{[A^2+(n-1/2)^2\pi^2]^{1/2}}=\frac{\bar{n}}{[(A/\pi)^2+(n-1/2)^2]^{1/2}} \tag{7.6}$$

The 3 dB beamwidth is now approximately

$$\theta_B=2\sin^{-1}[\sigma\beta_0\lambda/21] \tag{7.7}$$

A table of typical values of sidelobe and beamwidth relationships of the $\Lambda_{-1/2}[(u^2-A^2)^{1/2}]$ far-field function and approximations to it for several values of $\bar{n}$ is given in Reference 5. Narrow-beamwidth antennas with approximate illuminations of Taylor can be realised without significant gain reduction by choosing the appropriate value of $\bar{n}$.

Circular-aperture illuminations which have modified $\Lambda_{-1/2}\ [(u^2-A^2)^{1/2}]$ patterns have been determined by Taylor [6] using tables provided by Hansen [7].

Similar control of the first sidelobe level is obtained for other values of ν. If $\nu=1/2$, the so-called modified $(\sin u)/u$ pattern results; that is, the sidelobe envelope is that of the $(\sin u)/u$ function, but the first sidelobe is something other than 13.2 dB with respect to the main beam. For this case the first sidelobe voltage amplitude is

$$\eta=\frac{4.603\sinh A}{A} \tag{7.8}$$

and the 3 dB beamwidth is determined from

$$\frac{\sinh A}{\sqrt{2A}}=\frac{\sinh(u^2-A^2)^{1/2}}{(u^2-A^2)^{1/2}} \tag{7.9}$$

The relationship between first sidelobe level, 3 dB beamwidth and A is listed in Table 7.3 [5].

Two parameter circular aperture distributions are usually chosen because of their easy integrability. However, there is no way other than 'numerical brute force' methods to find the pair of parameters that, for example, give maximum aperture efficiency for a given sidelobe ratio.

A one parameter circular aperture distribution has been constructed by Hansen [8]. It follows the procedure of Taylor for the line source. The uniform circular aperture has the pattern

$$F(u) = \frac{2J_1(\pi u)}{\pi u} \tag{7.10}$$

where $u = (D/\lambda)\sin\theta$, and D is the aperture diameter. Calling the single parameter H, the close-in zeroes are shifted to give the pattern

$$F(u) = \frac{2J_1[\pi(u^2 - H^2)^{1/2}]}{\pi(u^2 - H^2)^{1/2}}, \; u \geqq H \tag{7.11}$$

and

$$F(u) = \frac{2I_1[\pi(H^2 - u^2)^{1/2}]}{\pi(H^2 - u^2)^{1/2}}, \; u \leqq H \tag{7.12}$$

J_1 and I_1 are the usual Bessel and modified Bessel functions of the first kind and order one. Analogue to the $\sin \pi u/(\pi u)$ case, the pattern form for $u \geqq H$ provides a sidelobe structure much like that of $2J_1(\pi u)/\pi u$, and part of the main beam, while the pattern form for $u \leqq H$ provides a higher main beam peak than unity, thereby raising the sidelobe ratio. Fig. 7.1 shows a typical pattern, for a 10 λ aperture, normalised to 0 dB at $\theta = 0°$.

The transition from I_1 to J_1 is marked by dashed lines ($u = H$), and from this point to the first sidelobe is 17.57 dB. From this transition point to the beam peak is 12.43 dB, giving a 30 dB SLR. Thus the sidelobe ratio is

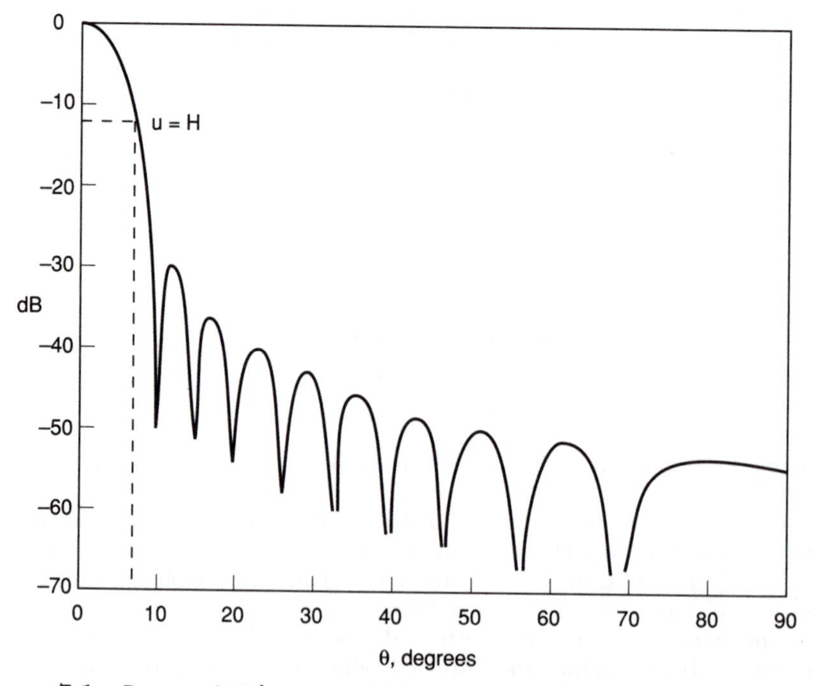

Figure 7.1 *Pattern of 10 λ aperture, SLR = 30 dB [8] © IEEE, 1976*

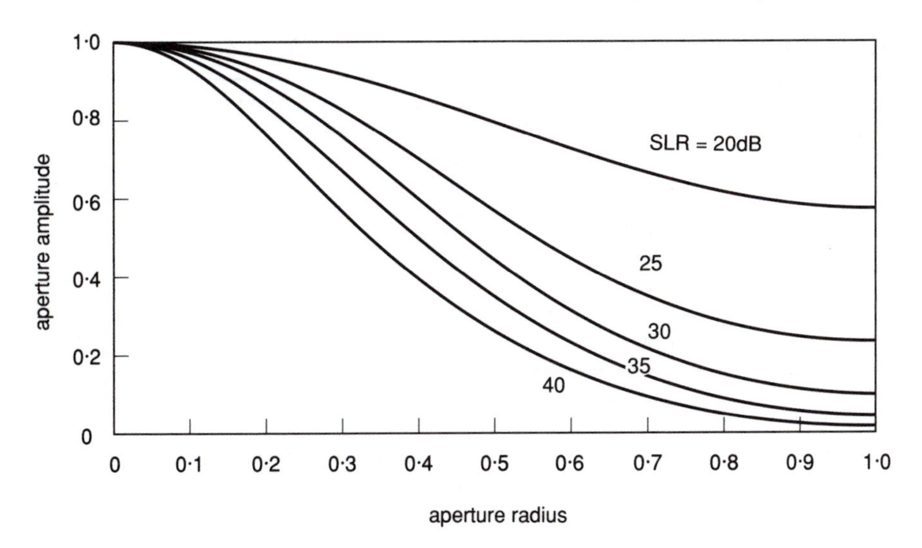

Figure 7.2 *One parameter circular distribution with sidelobe ratio as parameter [8] © IEEE, 1976*

$$\mathrm{SLR} = 17.57\ \mathrm{dB} + 20\log\frac{2I_1(\pi H)}{\pi H} \tag{7.13}$$

This one parameter circular aperture distribution is

$$g(p) = I_0[H(\pi^2 - p^2)^{1/2}] \tag{7.14}$$

Fig. 7.2 shows these radial distributions for sidelobe ratios of 20, 25, 30, 35, and 40 dB.

The beamwidth can be calculated from

$$\beta_0 = 1.0290\lambda/D \tag{7.15}$$

A useful design value is the directivity normalised to uniform aperture directivity, commonly called the aperture illumination (or taper) efficiency. This efficiency is given by

$$\eta = \frac{4I_1^2(\pi H)}{\pi^2 H^2[I_0^2(\pi H) - I_1^2(\pi H)]} \tag{7.16}$$

Table 7.4 shows the relationships between H and performance parameters of this one parameter circular aperture taper.

Offset reflector antennas [9] give excellent performance with respect to both aperture efficiency and sidelobe levels because the blocking effects of the primary feed, subreflector and its support can be eliminated.

Fante [10] presented an offset-fed parabolic cylinder antenna which has a sidelobe level of −50 dB or less over a 15% frequency band. The geometry of the parabolic cylinder (satisfying the equation $z = x^2/4F$) with a line-source feed at its focus is shown in Fig. 7.3.

Table 7.4 *One parameter distribution values [8] © IEEE, 1976*

SLR (dB)	Beamwidth	Efficiency	Edge taper	*H*
17.57	1.0000	1.0000	0 dB	0
20	1.0483	0.9786	4.49	0.4872
25	1.1408	0.8711	12.35	0.8899
30	1.2252	0.7595	19.29	1.1977
35	1.3025	0.6683	25.78	1.4708
40	1.3741	0.5964	31.98	1.7254
45	1.4409	0.5390	38.00	1.9681
50	1.5038	0.4923	43.89	2.2026

It is assumed that the feed is vertically polarised, has a symmetric radiation pattern, and is pointed at the centre of the reflector. The envelope of the relative sidelobe level of the radiation pattern at θ is

$$S(\theta) \simeq \frac{2}{\pi^2}\left(\frac{\lambda}{F}\right)^2 \left|\frac{F_v(\psi_1)}{F_v(\psi_c)}\right|^2 \left(\frac{1}{\theta\psi_B}\right)^2 \tag{7.17}$$

where $S(\theta) \equiv |H_s(\theta)/H_s(\theta)|^2$, λ is the wavelength, $F_v(\psi)$ is the field pattern of the vertically polarised signal, ψ_B is the total angle subtended at the feed by the reflector, and $\psi_c = (\psi_0 + \psi_1)/2$. Eqn. 7.17 is valid provided $\theta \ll 1$, but $2kF \sin\theta \gg 1$. As an example of the utility of Eqn 7.17 let us suppose that we have a parabolic cylinder with $\psi_0 = 0$, $\psi_1 = \pi/2$, $kF = 188.5$, and an edge taper of

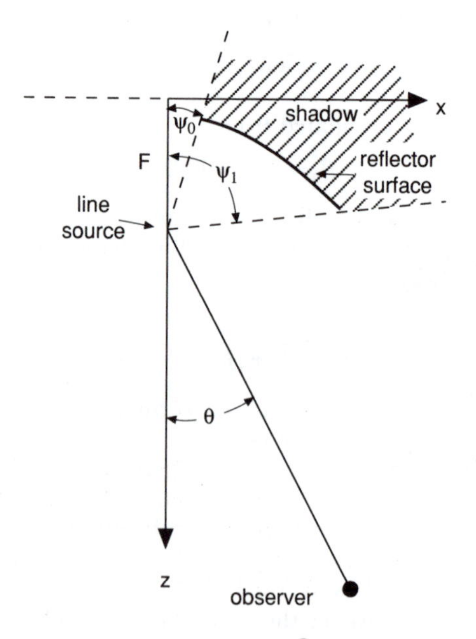

Figure 7.3 *Parabolic cylinder geometry [10] © IEEE, 1980*

−36 dB. The sidelobe level predicted at $\theta = 5.7°$ by eqn 7.17 is −56.6 dB, whereas the exact computer evaluation gives −55.7 dB.

Eqn. 7.17 is the desired result assuming that the reflector surface exactly satisfies the equation $z = x^2/4F$. Unfortunately, because of random errors, the reflector surface will never be exactly a parabolic cylinder, and the phase errors produced by manufacturing imperfections will also contribute to the sidelobe level. For example, suppose that the aperture area is 225 ft^2, its efficiency is 0.42, the operating frequency is 3.34 GHz, and we desire that there be a 99% chance that any sidelobe beyond $\theta = 7°$ will be below $S_0 = 10^{-5}$ (−50 dB). It can be calculated that the required surface tolerance is $\delta \leqq 4.87 \times 10^{-3}$ in.

The design of a feed is an important problem in low sidelobe antenna design, since it has to satisfy the following requirements simultaneously: (i) illuminating the reflector with a relatively high efficiency, (ii) providing the required edge illumination, and (iii) limiting the spillover of the feed radiation beyond the reflector edges.

As an example, a parabolic cylinder antenna, with a 2° (3 dB) beamwidth and −50 dB sidelobes, which can operate from 3.1 ~ 3.6 GHz, was considered. The parameters of the reflector are $\psi_0 = 5°$, $\psi_1 = 80°$, and $F = 8.802$ ft. This leads to a reflector diameter D of 14 ft. This reflector has a gain of 14 dB, so that, in order to get sidelobes smaller than −50 dB, we require a feed antenna such that (i) the edge illumination on the reflector is −36 dB relative to the illumination at the centre, and (ii) the spillover of the feed pattern beyond the edges of the reflector must be smaller than −36 dB. This can be achieved by a feed with a planar aperture of width $b = 10.895$ in, and with the aperture distribution $\cos(\pi\xi/b) + \alpha\cos(3\pi\xi/b)$. It can be shown that the far-field sidelobe level produced by the scattering of the secondary radiation by this feed is at most −61.57 dB, so that blockage is clearly of no concern. If $\alpha = 0.14$, the theoretical radiation pattern is shown in Fig. 7.4 for $f = 3.35$ GHz.

Some measured results of an X-band model are shown in Fig. 7.5*a* and *b* for $f = 9.8$ GHz and $f = 11.23$ GHz, respectively.

7.2.2 *Adaptive sidelobe interference canceller*

The advantages of ultra-low sidelobe antenna are as follows:

(i) Suppressing interference signals in all sidelobe directions for an infinite number of jammers
(ii) Interference signals are rejected before they enter the radar receiver

The shortcomings of this type of antenna are as follows:

(i) Lower antenna gain results. This means that the radar maximum detection range will be reduced, e.g. half the normal one, unless the antenna is made larger.
(ii) It cannot be added to an existing radar system. Even though the antenna can be changed to a new one of low sidelobe design, it will affect the overall specifications of the radar.

These shortcomings can be overcome by using an adaptive sidelobe interference canceller.

Adaptive sidelobe interference cancellation has been under consideration for radar systems for more than 20 years. More than 200 papers have been

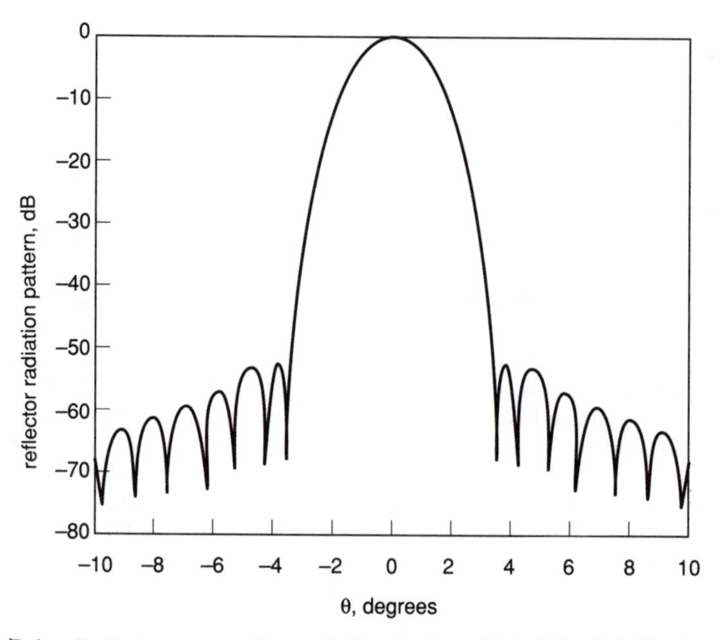

Figure 7.4 *Radiation pattern of a parabolic cylinder at 3.35 GHz [10] © IEEE, 1980*

presented [11]. Two types of antenna were under consideration: array antenna and reflector antenna. We will concentrate on the latter as before.

All the sidelobe cancellers (SLC) have a similar structure, in which some auxiliary antennas are placed near the radar antenna. The signal received through these aerials is multiplied by appropriate weights and then summed to obtain an estimate of the jamming signal received through the antenna sidelobes. Cancellation is performed by subtracting the jamming estimate from the receiver output. The effect is of steering a null in the direction of the interference which will not significantly affect the wanted signal unless it is in the same direction. More than one null can be realised by using several auxiliary antennas and several adaptive loops.

There are two types of adaptive canceller systems: the analogue closed-loop adaptive system and the digital open-loop adaptive system. We will concentrate on the digital open-loop adaptive canceller, since it acts much faster than the former.

7.2.2.1 *The Applebaum adaptive system*

In 1976, Applebaum [12] presented a sidelobe cancellation system which is shown in Fig. 7.6 It consists of a main, high gain antenna whose output is designated as channel 'o' and K auxiliary antennas. The auxiliary antenna gains are designed to approximate to the average sidelobe level of the main antenna gain pattern. The desired target signal received by the auxiliaries is negligible compared with the target signal in the main channel. The purpose of the auxiliaries is to provide independent replicas of jamming signals in the sidelobes of the main pattern for cancellation.

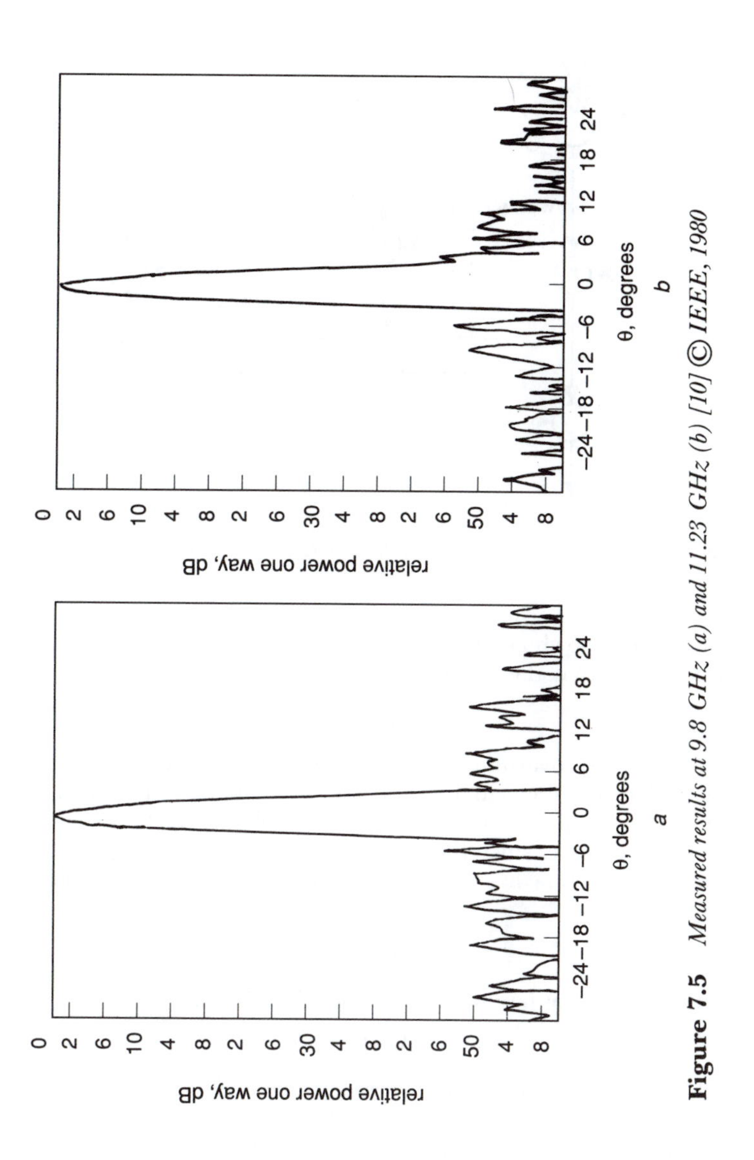

Figure 7.5 *Measured results at 9.8 GHz (a) and 11.23 GHz (b) [10] © IEEE, 1980*

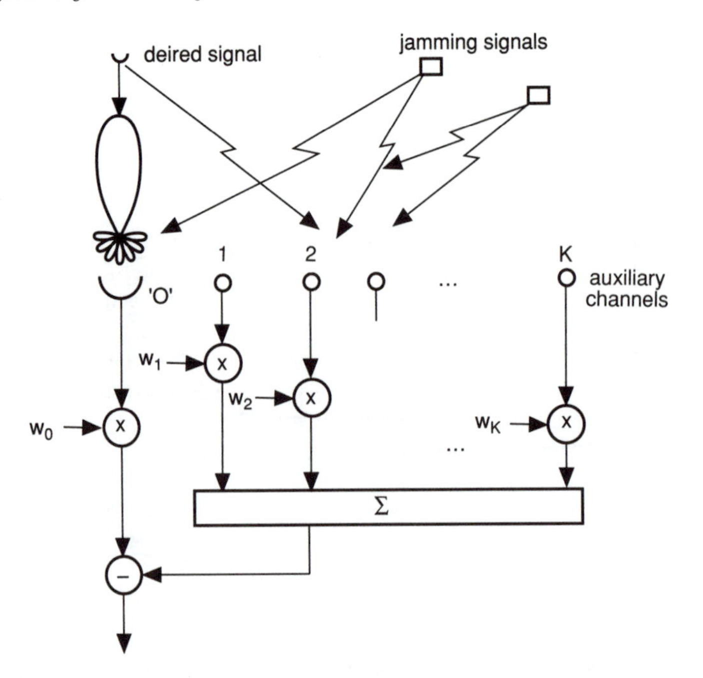

Figure 7.6 *Applebaum sidelobe cancellation system [12] © IEEE, 1976*

The problem is to choose the weights w_k, so that the output SIR is maximised. It can be shown that the optimum weights are determined by

$$\boldsymbol{MW} = \mu S^* \tag{7.18}$$

where $\boldsymbol{M} = [\mu_{kl}]$ = covariance matrix of the interference outputs, $\boldsymbol{W}$ is the weights vector, $\boldsymbol{S}$ is the signal vector, and μ is an arbitrary constant.

To find an appropriate control law for the weights, the first step is to select a T vector. In this case, since the signal collected by the auxiliaries is negligible and the main antenna has a carefully designed pattern, we choose the $K+1$ column vector

$$T = [1, 0, \ldots, 0]^T \tag{7.19}$$

The optimum control law would then be

$$M'W' = \mu T \tag{7.20}$$

where M' is the $K+1$ by $K+1$ covariance matrix of all the channels and W' is the $K+1$ column vector of all the weights.

The control loops for an adaptive array can be implemented using the same circuitry as shown in Fig. 7.7.

The weights w_k are derived by correlating u_k with the sum signal Σ, subtracting the correlation from the desired vector component t_k^*, and then using a high gain amplifier.

For each w_k we have

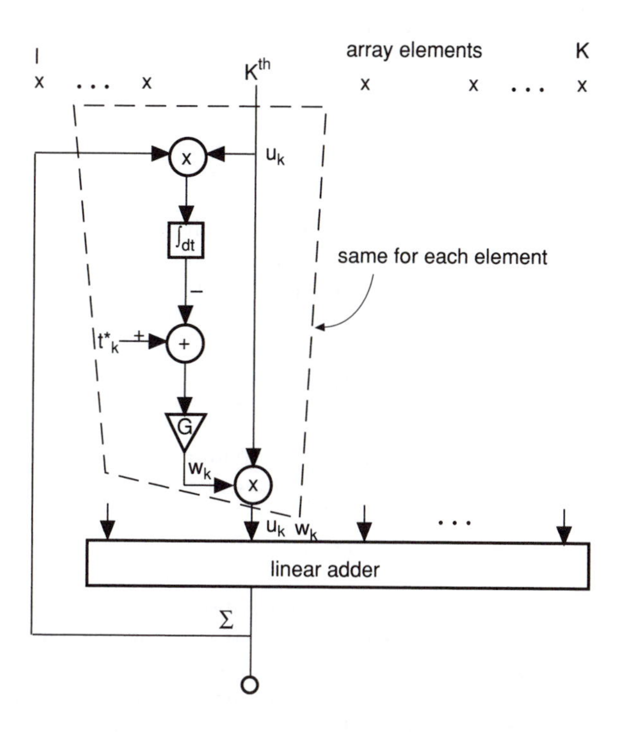

Figure 7.7 *Block diagram of control loop for adaptive array [12] © IEEE, 1976*

$$w_k = G\left\{ t_k^* - u_k^* \sum_{l=1}^{k} w_l u_l \right\} \tag{7.21}$$

or

$$\sum_{l=1}^{k} w_l \left(u_k^* u_l + \frac{\delta_{lk}}{G} \right) = t_k^*, \text{ for } k = 1, \ldots, K \tag{7.22}$$

where

$$\delta_{lk} = \begin{cases} 1, & \text{when } k = 1 \\ 0, & \text{all other } k \end{cases}$$

Recalling that $u_k^* u_l$ is an element of the covariance matrix M, the K equations represented by eqn. 7.22 may be written as

$$\left\{ M + \frac{\delta_{lk}}{G} \right\} W = T^* \tag{7.23}$$

This differs from the optimum control law by the addition of a term inversely proportional to gain. This term introduces an error analogous to the servo error of a type 0 servo. Its effect can be negligible with sufficiently high gain.

The dynamic behaviour of the control loop is determined by the 'integrator' within the loop. In practice this is a high Q, single pole circuit. The differential equations describing the dynamic behaviour of the loops can be shown to be

$$\tau \frac{dw_k}{dt} + w_k = G\left\{ t_k^* - u_k^* \sum_{l=1}^{k} w_l u_l \right\} \tag{7.24}$$

where τ is the time constant of the 'integrator' circuits.

The loops are designed so that the weights w_k will vary slowly compared with the bandwidth of the signals u_l. Hence the weights will be uncorrelated with u_l. Thus if we apply the expectation operator to both sides of eqn. 7.24 we get

$$\tau \frac{d\overline{w_k}}{dt} + \overline{w_k} = G\left\{ t_k^* - \sum \overline{w_l u_k^* u_l} \right\} \tag{7.25}$$

This equation determines the dynamic behaviour of the expected value of w_k. The K equations obtained from eqn. 7.25 may be represented in matrix form as

$$\tau \frac{d\overline{W}}{dt} = -(GM + l_k)\overline{W} + GT^* \tag{7.26}$$

It can be shown easily that matrix equations of this form are stable if the matrix $GM + l_k$ has only positive eigenvalue. Since M is a positive definite Hermitian matrix, $GM + l_k$ is also a positive definite Hermitian matrix. It therefore has only positive eigenvalues. Thus, this implementation is stable under all conditions for assumptions made. In practice, second order effects may make the loop unstable.

Bucciarelli, *et al.* [13], evaluated the performance of the optimum sidelobe canceller.

The optimum weights control law can be obtained by resorting to the linear prediction theory. Let us indicate by the N-dimensional vector x the set of the complex signals received through the auxiliary antennas and by x_M the signal received through the main antenna. The target signal, if present, is contained only in x_M while the vector $\boldsymbol{x}$ contains only jamming due to the low gain of the auxiliary antennas.

Let us indicate by $\boldsymbol{M}$ (N, N) the covariance matrix of the process $\boldsymbol{x}$ and by $\boldsymbol{R}$ the N-dimensional vector containing the covariance of the signal x_M with the auxiliary signals x_i $(i = 1, 2, \ldots, N)$:

$$\boldsymbol{M}(i, j) = E\{x_i x_j^*\},\ M(i, i) = \sigma^2 \tag{7.27}$$

$$\boldsymbol{R}(i, l) = E\{x_i x_M^*\} \tag{7.28}$$

Under the assumption of an interference having much more power than the useful signal, the rejection of the interference in the main channel is achieved by subtracting from x_M the estimation of the jamming signal. The estimation is performed through linear prediction of the jamming x_M on the basis of the auxiliary signal x_i. The optimum vector $\boldsymbol{w}(N, l)$ containing the weights w_i of the linear combination of the data x_i is determined minimising the mean square prediction error. It can be shown that the optimum weights can be obtained by

$$\boldsymbol{w} = \boldsymbol{M}^{-1}\boldsymbol{R} \tag{7.29}$$

The prediction error is

$$v(t)=x_M-\boldsymbol{w}^T\boldsymbol{x} \tag{7.30}$$

therefore the power of jamming residue is

$$\begin{aligned}E\{|v(t)|^2\}&=E\{|x_M-\boldsymbol{w}^T\boldsymbol{x}|^2\}\\&=E\{(x_M-\boldsymbol{w}^T\boldsymbol{x})x_M\}=\sigma^2-\boldsymbol{w}^T\boldsymbol{R}\end{aligned} \tag{7.31}$$

the second equality follows from the orthogonality of the data x_i to the prediction error. The jamming cancellation g, defined as the ratio of input jamming power σ^2 to output residual power $E\{|v(t)|^2\}$, is

$$g=\sigma^2/(\sigma^2-\boldsymbol{w}^T\boldsymbol{R}) \tag{7.32}$$

It is assumed that the processes x_M and x_i ($i=1, 2, \ldots, N$) have the following autocorrelation of Gaussian shape:

$$\begin{aligned}E\{x_M(t)x_M^*(t+\tau)\}&=E\{x_i(t)x_i^*(t+\tau)\}\\&=\sigma^2\exp[-(\tau/\tau_c)^2]\end{aligned} \tag{7.33}$$

where the correlation time τ_c depends on the receiving channel bandwidth. If perfect amplitude and phase equalisation is assumed, between the main and auxiliary receiving channels, x_i and x_M are delayed versions of the same process. The covariance matrix $\boldsymbol{M}$ and the cross-correlation vector $\boldsymbol{R}$ can be evaluated by means of eqn. 7.33 and taking into account the geometrical displacement of the auxiliary antenna around the main antenna. As an example, consider the auxiliary and main antennas set in a line and regularly spaced at a constant distance d between them. The matrix $\boldsymbol{M}$ and the vector $\boldsymbol{R}$ relevant to the jammer, having an incidence angle θ with respect to the array, are as follows:

$$\boldsymbol{M}(i,j)=\rho^{(i-j)^2}\sigma^2 \tag{7.34}$$

$$\boldsymbol{R}(i,l)=\rho^{i^2}\sigma^2 \tag{7.35}$$

where the coefficient ρ is

$$\rho=\exp[-(\Delta t/\tau_c)^2] \tag{7.36}$$

Δt being the time difference with which the jamming signal is received by two contiguous antenna. Indicating by c the velocity of light, the following equation holds:

$$\Delta t=d\sin\theta/c \tag{7.37}$$

We can calculate the optimum weights under these assumptions. Fig. 7.8 shows the jammer cancellation g versus the number of auxiliary channels and with a different value of parameter ρ. Two sets of curves are drawn: those with solid lines refer to jamming only as a disturbance source while those with dashed lines refer to jamming plus thermal noise having a variance σ_n^2.

It can be seen from this Figure that, when the receiver noise cannot be neglected, there is an optimum number of auxiliary antennas for each value of ρ. The optimum number of auxiliary aerials is less than 3 for $\rho>0.9$. If the effect of receiver noise can be neglected, the jammer cancellation increases with the increase of ρ and the number of auxiliary antennas. Since ρ is a function of incidence angle, the jammer cancellation is also a function of incidence angle of the interference.

To show the mutual influence of the steady-state and transient behaviour, digital simulation of the Applebaum system shown in Fig. 7.9 has been performed in order to evaluate the steady-state jammer cancellation and the number of samples needed for the adaptation of all the loops.

The digital integrator to correlated x_1 and $v(t)$ is performed through a single pole filter with a constant:

$$\beta \simeq 1 - T_s/\tau \tag{7.38}$$

where T_s is the sampling interval of the signals and τ is the time constant of the integrator.

This method, which is attractive for its simple implementation, has slow convergence of the adaptation weights. The problem of slow convergence arises whenever there is a wide spread in the eigenvalues of the input signal correlation matrix.

The steady-state and transient responses can be calculated as a function of two parameters: α and ρ. The bandwidth ratio α is defined as

$$\alpha = B_L/2B_c = \tau_c/2\tau_L \tag{7.39}$$

where B_L is the bandwidth of a single closed loop, B_c is the bandwidth of the channel, and τ_c is the correlation time. The parameter τ_L

$$\tau_L = \tau/(1 + G\sigma^2) \tag{7.40}$$

is the closed loop time constant of a canceller having a single aerial, and can be assumed to be the mean value of each identical loop in a canceller with multiple auxiliary antennas.

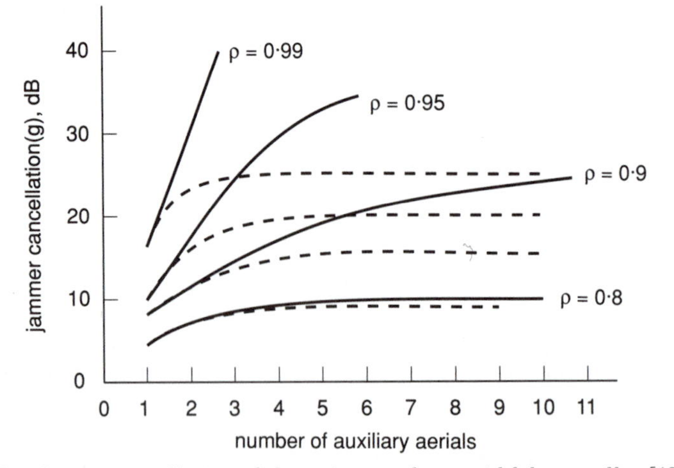

Figure 7.8 *Jammer cancellation of the optimum coherent sidelobe canceller [13]*

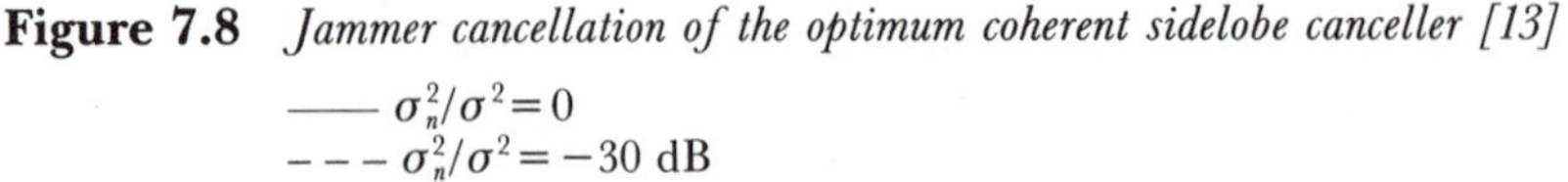

—— $\sigma_n^2/\sigma^2 = 0$
– – – $\sigma_n^2/\sigma^2 = -30$ dB

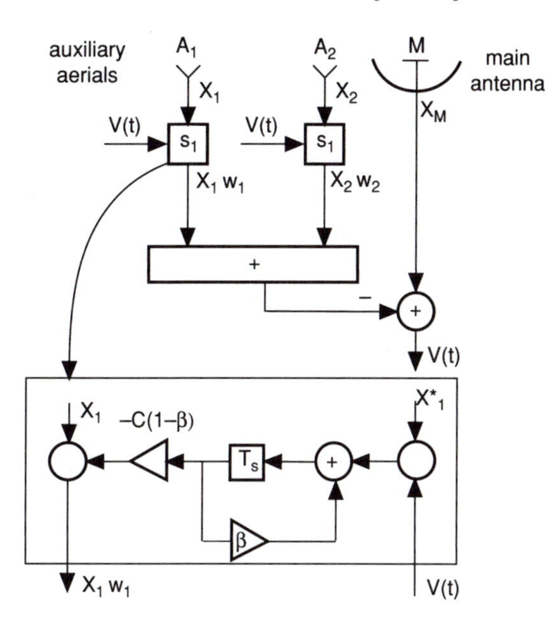

Figure 7.9 *The Applebaum adaptive sidelobe canceller*

The steady-state and transient performance of the Applebaum system are shown in Fig. 7.10*a* and *b*, respectively. It can be seen from these Figures that the α cannot be larger than 0.1 for good steady-state performance, but the transient time increases with decrease of α.

Applebaum's configuration is suitable for an array antenna. An experimental analogue 8-element adaptive array based on Applebaum's algorithm has been realised by Baldwin *et al.* [14]. The array operates at X-band and comprises a linear arrangement of eight dipoles spaced approximately 0.6 λ apart. The signal processor is an analogue implementation of a simple gradient descent algorithm operating at an IF of 60 MHz. Radar waveforms are usually well suited to this adaptive implementation in that target returns are of sufficiently low power and duration that the control circuitry does not respond to them. The schematic diagram of this adaptive array is shown in Fig. 7.11.

Two choices of correlation low-pass filter time constant are available: $\tau_0 = 100$ or 1 ms (integration bandwidths of 10 Hz and 1 kHz) for the slow and fast modes, respectively.

The maximum sidelobe of this 8-element array in the quiescent state is about −13 dB. Interference sources were adjusted to give power 40 dB above single element noise (5 MHz bandwidth) at the IF processor interface. Results with CW sources were only a few dB better than those obtained with noise modulation, indicating that the array was working well over the 5 MHz band. Fig. 7.12*a* and *b* show the adaptive pattern of this array for CW source and three noise sources, respectively.

It can be seen from these Figures that this adaptive array can form −43 to −48 dB nulls in the direction of interference sources. Table 7.5 shows the

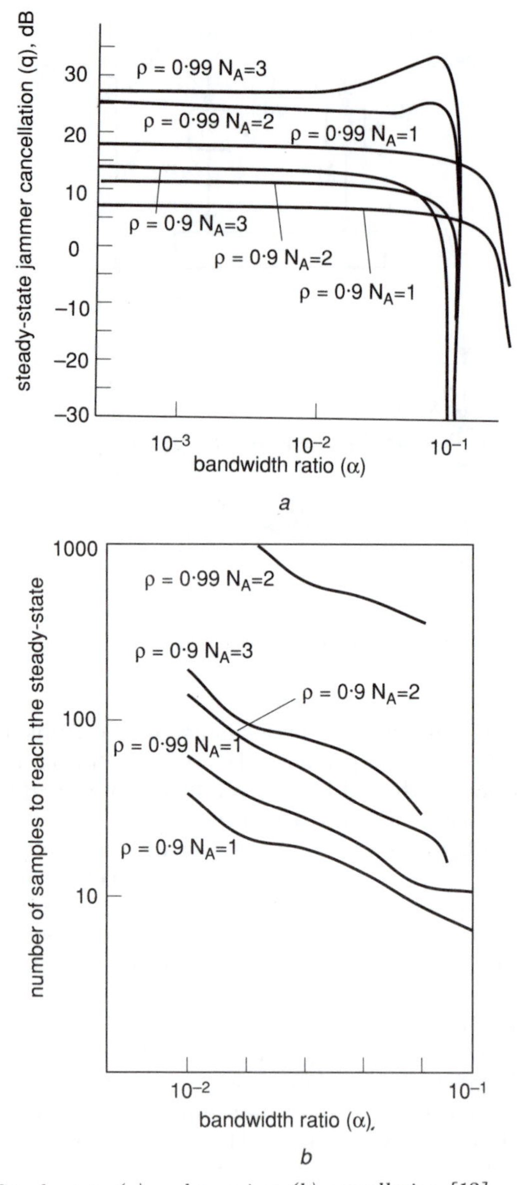

Figure 7.10 *Steady-state (a) and transient (b) cancellation [13]*

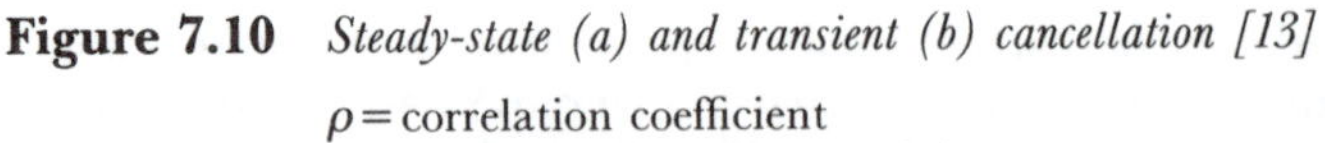
ρ = correlation coefficient
N_A = number of auxiliary aerials

experimental results of SIR improvement for this adaptive array in the fast mode of operation.

It is more interesting to add the adaptive sidelobe canceller to an existing radar with reflector antenna. The principle of this kind of adaptive sidelobe

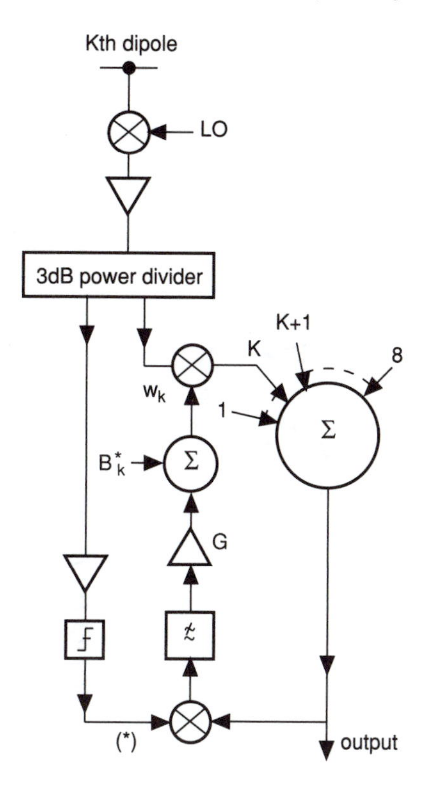

Figure 7.11 *Schematic diagram of an 8-element adaptive array [14] © IEEE, 1980*

interference canceller is very similar to the adaptive array. The difference is that an independent feedback loop is added to each auxiliary antenna. Therefore, the number of steerable nulls is equal to the number of auxiliary channels.

There are many configurations which can realise the adaptive sidelobe canceller. Fielding *et al.* [15] presented three types of configuration: (i) RF/IF canceller, (ii) all-IF canceller, and (iii) post-detector adaptive canceller. Fig. 7.13 shows a typical radar sidelobe canceller which is a modification of an all-IF canceller. The all-IF solution is rejected primarily because of problems in achieving a suitable narrow band filter of the order of 100 Hz.

The performance of cancellation depends on the path matching. Cancellation takes place at a summing point somewhere in the receiver. If it is to be effective all signals emanating from the same interference source must arrive at that summing point having undergone the same delay and not having followed paths which treat different parts of the signal differently. In particular, the frequency responses of the various paths over which the interference has travelled must be similar. Various factors can contribute to poor path matching. They are: (i) propagation paths, (ii) sidelobe frequency sensitivities, (iii) antenna separation, and (iv) paths internal to the radar system. The quality of matching which is required is given in Fig. 7.14.ı+XX

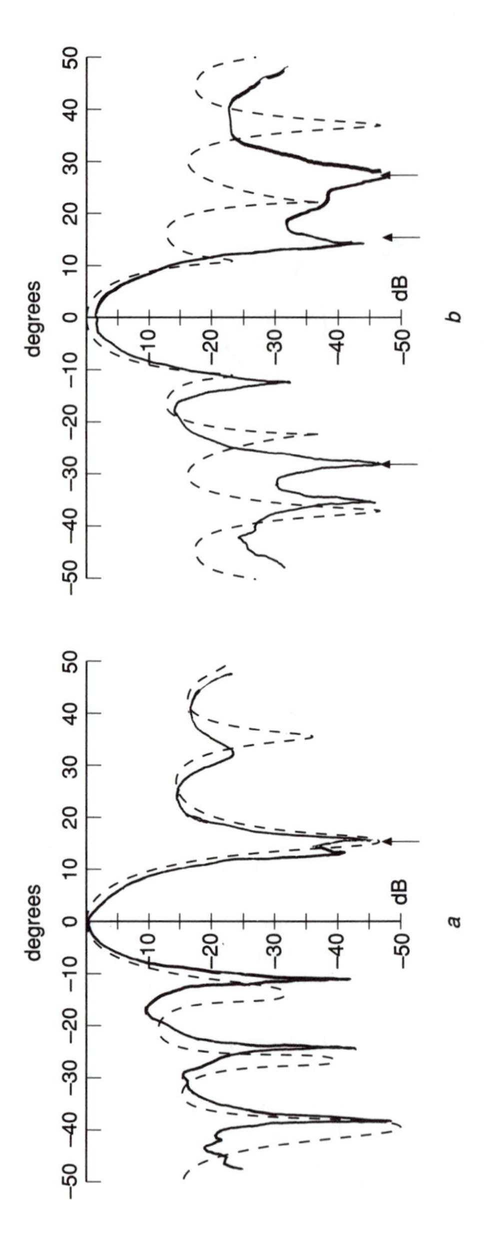

Figure 7.12 *Adapted pattern of 8-element adaptive array [14] © IEEE, 1980*

a CW source at 16°
- - - experiment
—— theory
b Three noise sources
- - - adapted I
—— quiescent I

Table 7.5 *SIR improvements for different interference sources [14] © IEEE, 1980*

	CW source at 16°	2 noise sources	3 noise sources	4 noise sources	6 noise sources
SIR gain	34 dB	29 dB	27 dB	24 dB	18.5 dB

It can be seen for instance that a differential amplitude mismatch of 0.2 dB combined with a differential phase mismatch of 1.2° limits cancellation to 30 dB.

The problem of good cancellation path matching increases rapidly with the bandwidth over which cancellation is required. The cancellation performance

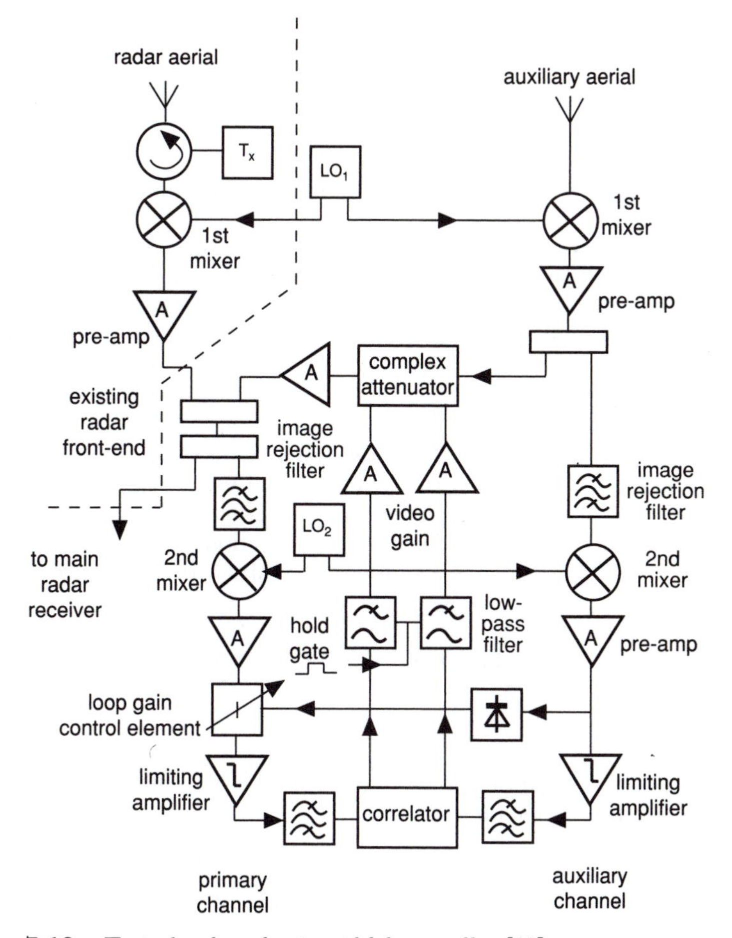

Figure 7.13 *Typical radar adaptive sidelobe canceller [15]*

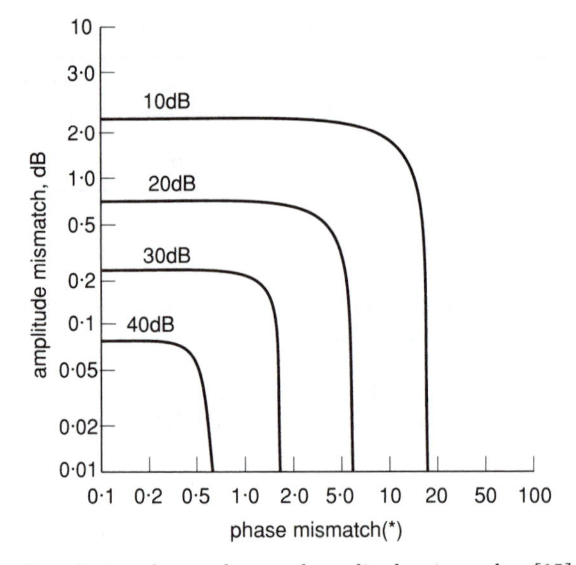

Figure 7.14 *Cancellation due to phase and amplitude mismatches [15]*

for CW, narrow band noise, and wideband noise are shown in Fig. 7.15*a*, *b* and *c*, respectively.

Farina [16] analysed the performance of the sidelobe canceller theoretically. The analysed analogue block diagram of the single sidelobe canceller (SSLC) is shown in Fig. 7.16.

The 'ideal' cancellation, defined as the ratio of input jamming power to output residual power, is given by

$$g_{ID} = (1 - \rho^2)^{-1} \tag{7.41}$$

where ρ is the correlation coefficient between $V_M(t)$ and $V_A(t)$. To achieve the ideal performance, $W(t)$ should reach the optimal value $W_{opt} = -\rho$ in deterministic sense. The plot of g_{ID} against ρ, depicted in Fig. 7.17, shows the role played by ρ on the maximum achievable cancellation.

The actual value of $W(\infty)$ is a random variable, with a variance

$$\sigma_w^2 = (1 + \rho^2) F(\alpha) \tag{7.42}$$

where α is the ratio of the loop bandwidth B_L to twice the channel bandwidth B_c:

$$\alpha = B_L / 2B_c = \tau_c / 2\tau_L \tag{7.43}$$

where τ_c is the correlation time of the signals in each receiving channel, and τ_L is the closed-loop time constant.

$$F(\alpha) = (\pi/2)^{1/2} \alpha \exp(\alpha^2/2)[1/2 - |\mathrm{erf}(\alpha)|] \tag{7.44}$$

In the steady state condition, the cancellation ratio is

$$g_{SSLC} = [(1 - \rho^2) + (1 + \rho^2) F(\alpha)]^{-1} \tag{7.45}$$

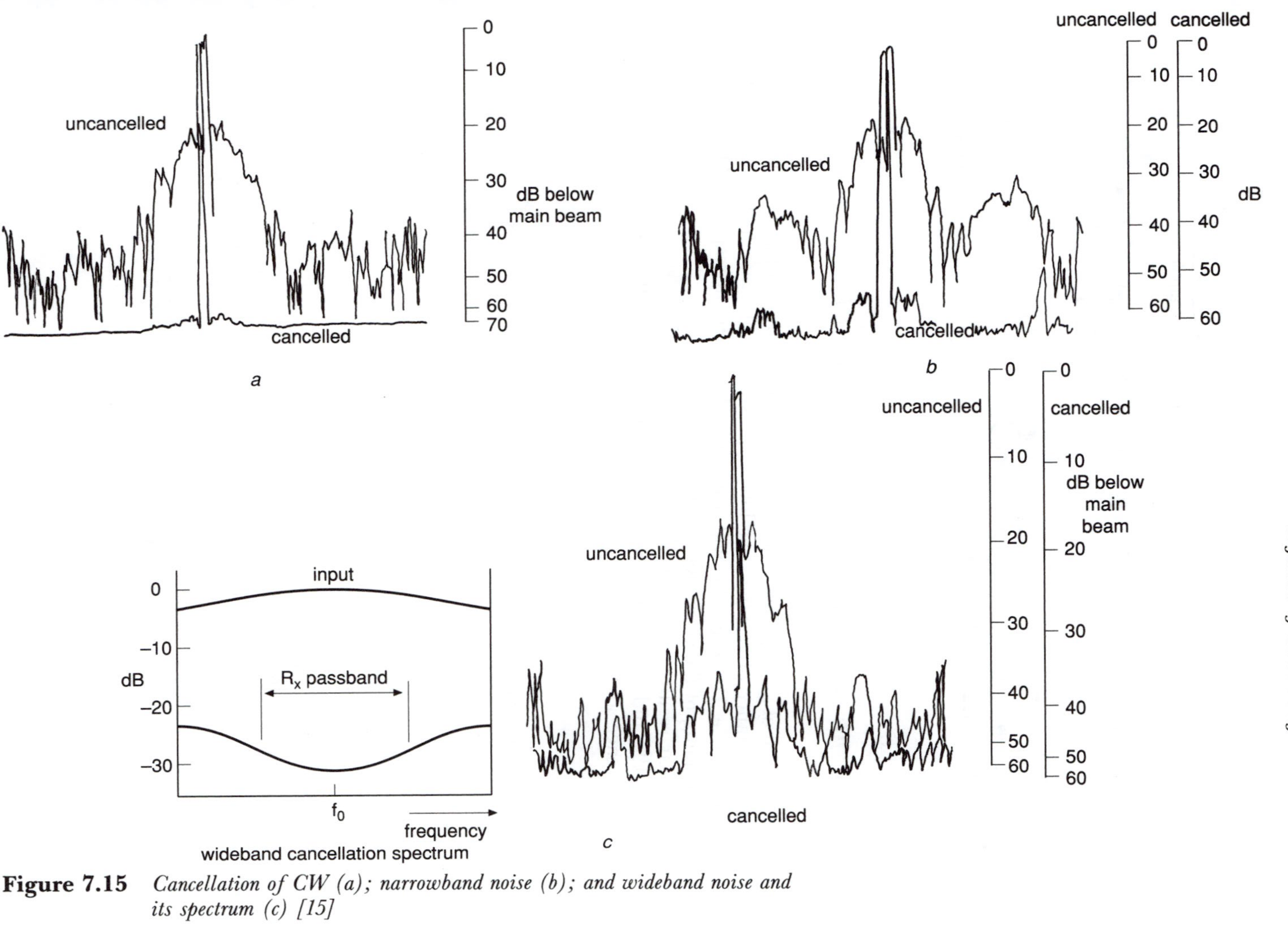

Figure 7.15 *Cancellation of CW (a); narrowband noise (b); and wideband noise and its spectrum (c) [15]*

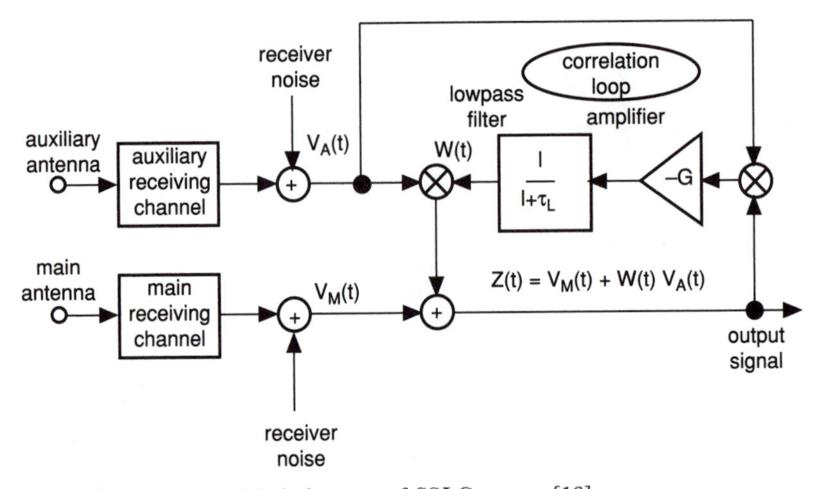

Figure 7.16 *Analogue block diagram of SSLC system [16]*

The plot of g_{SSLC} against α, for some value of ρ, shown in Fig. 7.18, illustrates the importance of keeping α as low as possible to obtain a high cancellation performance. However, this will result in a slower circuit response. This is the shortcoming of the closed-loop adaptive system.

Similar results can be obtained for the double sidelobe canceller (DSLC). Fig. 7.19*a* and *b* show the plots of g_{ID} and g_{DSLC} of DSLS in the case of one interference source.

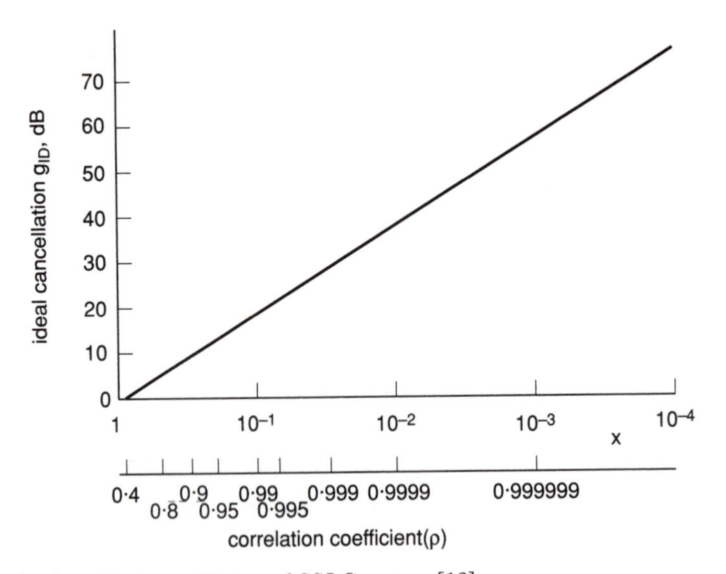

Figure 7.17 *Ideal cancellation of SSLC system [16]*

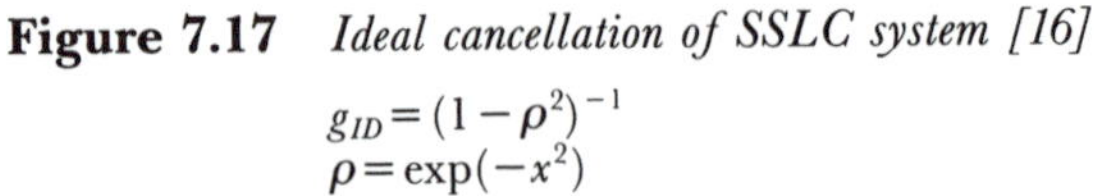

$g_{ID} = (1-\rho^2)^{-1}$
$\rho = \exp(-x^2)$

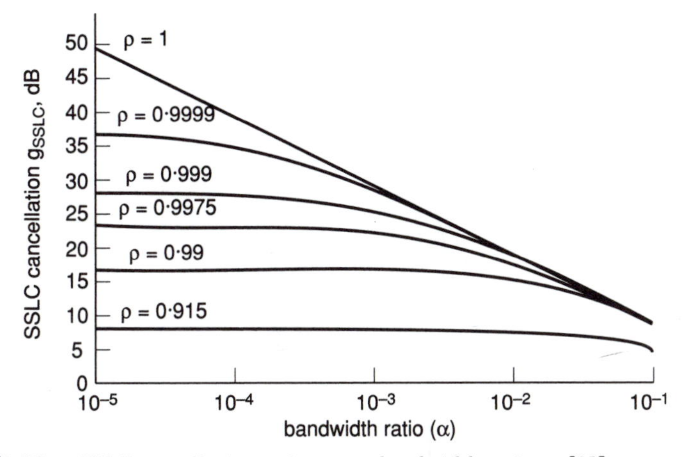

Figure 7.18 *SSLC cancellation ratio versus bandwidth ratio α [16]*

The comparison between SSLC and DSLC is shown in Fig. 7.20. The points on the curve mean equal performance $g_{SSLC}=g_{DSLC}$. Region A_{DSLC} means $g_{DSLC}>g_{SSLC}$, and region A_{SSLC} means $g_{SSLC}>g_{DSLC}$.

7.2.2.2 *The Gram–Schmidt adaptive sidelobe interference canceller*

It has been shown that the main shortcoming of the Applebaum system is its slow convergence property. A way to overcome the problem of slow convergence is to reduce the eigenvalue spread of the input signal correlation matrix through the introduction of an appropriate transformation which resolves the input signals into their eigenvector components. Then by equalising in power the resolved signals, the eigenvalue spread may be considerably reduced. An orthogonal signal set can be obtained by a transformation based on the Gram–Schmidt orthogonalisation procedure, introduced by Giraudon [17] and White [18]. To understand the co-ordinate transformation based on the Gram-Schmidt orthogonalisation procedure, consider the case of two auxiliary aerials and a main antenna, as shown in Fig. 7.21, which receive the signals x_1, x_2 and x_M, respectively.

Generalisation of this scheme to more aerials is straightforward; transformations S_{ij} each having two input signals, and giving one output uncorrelated with the left side input, are used in the canceller. Three blocks of this type need to orthogonalise the input signals, thus obtaining the new signal set x_1, y_1 including the useful output $v(t)$.

Bucciarelli *et al.* [13], suggested that the implementation of the decorrelating blocks S_{ij} be realised with the same block as the Applebaum system, which is shown in Fig. 7.9. Assuming that each block S_{ij} have the same parameters G(amplifier gain) and τ (integrator time constant), each S_{ij} has a time constant

$$\tau_{Li}=\tau/(1+G\sigma_i^2) \tag{7.46}$$

where σ_i^2 is the power of the auxiliary signals (left side input) in the block S_{ij}. The two block S_{11} and S_{12} have the same time constant, and S_{22} also, owing to the power equaliser inserted after Y_1 to restore the input jamming power σ^2.

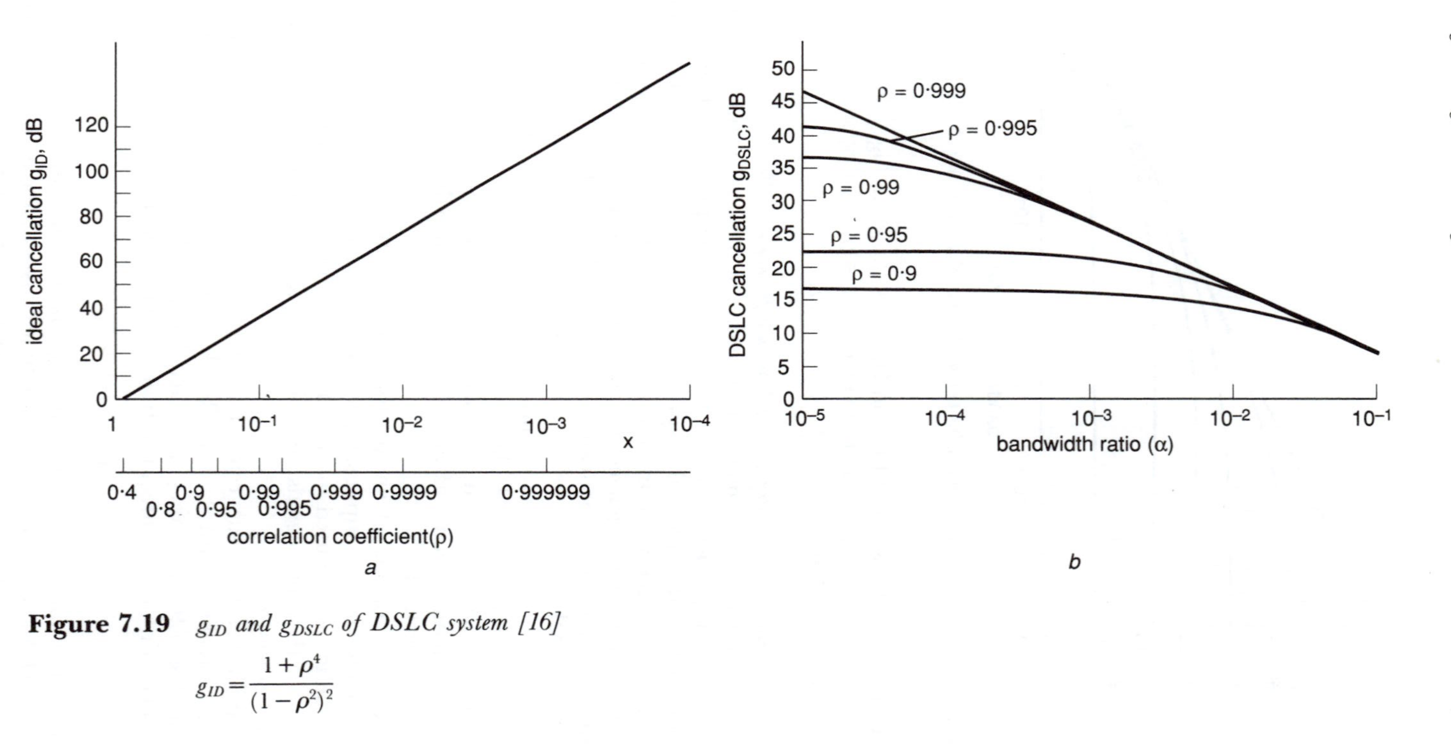

Figure 7.19 *g_{ID} and g_{DSLC} of DSLC system [16]*

$$g_{ID} = \frac{1+\rho^4}{(1-\rho^2)^2}$$

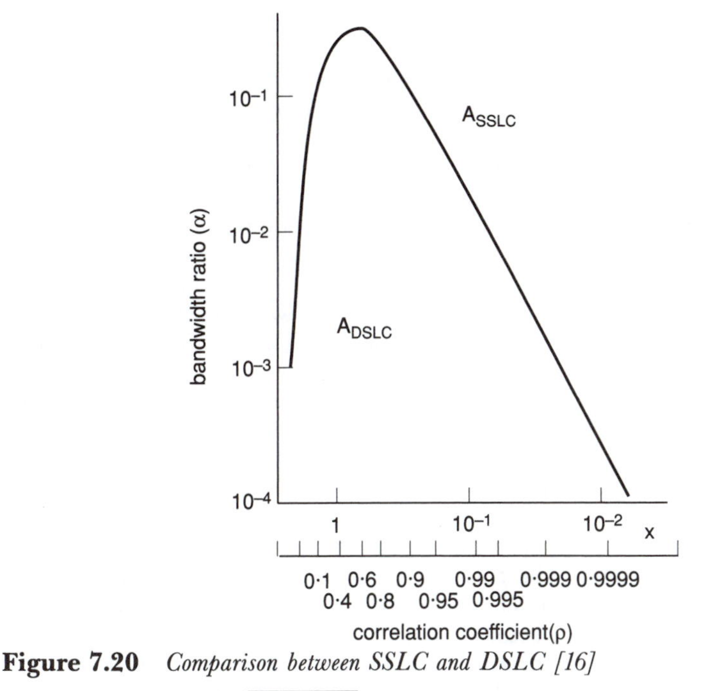

Figure 7.20 *Comparison between SSLC and DSLC [16]*

$$\alpha = 2\sqrt{\frac{2\rho^2(1-\rho^2)}{\pi(1-\rho^4)}}$$

$$\rho = \exp(-x^2)$$

It has been shown [13] that the Gram–Schmidt system with that configuration has the same steady state cancellation as that of the Applebaum system. However, the transient response of the former is better than that of the latter. The overall transient time of the Gram–Schmidt system is equal to the loop time constant τ_L times the number of loops in the main channel. In the case of the Applebaum implementation, it is noted that the transient time duration depends on ρ and increases indefinitely when the correlation coefficient goes to unity. Fig. 7.22 shows the number of samples needed to reach the steady-state cancellation in the Gram–Schmidt case.

These results refer to a number of auxiliary aerials up to nine and a correlation coefficient $\rho = 0.9$. Comparison can be made with the curves of Fig. 7.10*b* for a number N_A of auxiliary aerials up to three.

In fact, the convergence rate of this Gram–Schmidt canceller is faster than that of the Applebaum canceller, but it is still not fast enough for practical requirements. For example, if the bandwidth ratio α is equal to 10^{-2} and $N_A = 2$, the number of samples to reach the steady-state is about 60.

The reason is that, although the global structure of this Gram-Schmidt system is the transversal architecture without feedback, the block S_{ij} is the same as the Applebaum canceller with feedback. The loop time constant τ_L is inversely proportional to the closed-loop bandwidth. To reduce the effect of noise, the loop bandwidth cannot be too large. Therefore, open-loop adaptation has to replace the closed-loop adaptation.

Kretschmer and Lewis [19] proposed an open-loop adaptive processor based on the Gram–Schmidt algorithm. The principle is the same as the Gram–Schmidt adaptive MTI which we have introduced in Section 7.2.2.2.

Some results of computer simulation were shown in Reference 19. In the simulations, various numbers of pulses were averaged (N_{AVG}) for W_{opt} and average residue powers were computed. The simulation results are expressed in terms of the sampled residue power normalised to the average input correlated noise power level I_m in the main channel. Thus the 0 dB level in the simulation results corresponds to I_m and the residue sample power shown in the ensuing plots is in terms of the number of decibels below I_m. The first 200 input samples in the simulations contain only correlated interference and thermal noise, while desired signals were injected at preselected points throughout the next 300 points. Results are presented in Fig. 7.23*a*, *b* and *c* for an input thermal noise level N 50 dB below I_m and for N_{AVG} equal to 2, 4 and 8, respectively.

In these Figures, a desired signal S_m was injected in the main channel at a level 20 dB below I_m beginning at the sample number 200 and recurring every 10 samples thereafter. Prior to sample number 200, correlated and thermal noise alone are presented. The correlated interference levels I_m and I_a in the main channel and the auxiliary channel were the same and the desired signal S_a in the auxiliary channel was zero.

Several general observations can be made from these Figures. First, it is observed that the adaptive digital processor cancels the correlated interference down to the thermal noise level almost at the first few samples. This means that

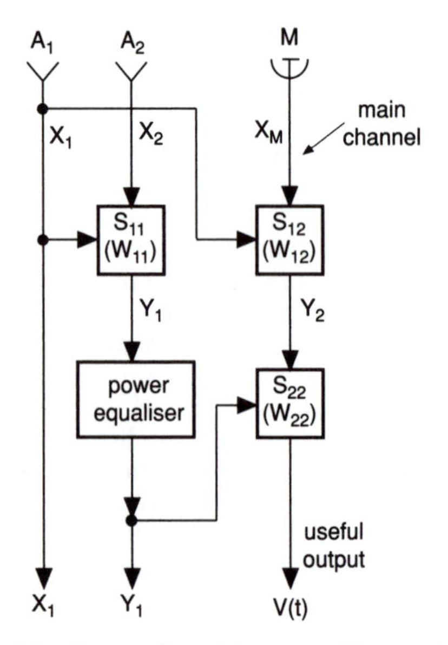

Figure 7.21 *Gram–Schmidt canceller with two auxiliary aerials*

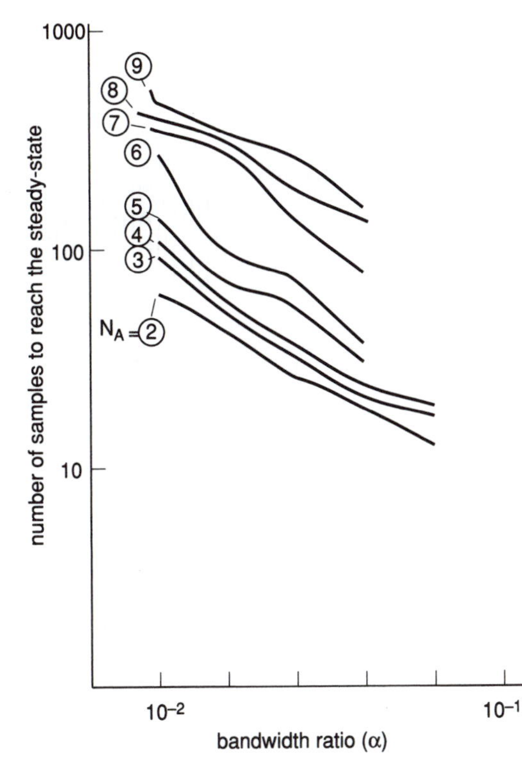

Figure 7.22 *Number of samples needed to reach the steady-state in the Gram–Schmidt canceller [13]*

N_A = auxiliary aerials
correlation coefficient $\rho = 0.9$

no transient time is needed for the adaptation. This is the main advantage of the open-loop processor over the closed-loop processor. Secondly, perfect cancellation of the correlated interference can be obtained. This is because of the effective infinite gain associated with the processor. Since the thermal noise is uncorrelated between these channels, the thermal noise level represents the floor of the residue except where the desired signal (also an uncorrelated signal) appears. Another general observation is that, as N_{AVG} increases, the desired signal is less likely to be reduced by the processor. If no cancellation of the desired signals took place, the signal points would be at the −20 dB level in the plots. It is seen in these plots that the amount of signal cancellation decreases as N_{AVG} increases.

The normalised residue power was averaged over the first region excluding the transient part and over the second region at the points where the desired signals were injected. These averages, expressed in decibels, are denoted by C and CS, respectively. The negative of C also corresponds to the cancellation ratio defined as

$$CR = |\overline{V_m}|^2 / |\overline{V_r}|^2 \tag{7.47}$$

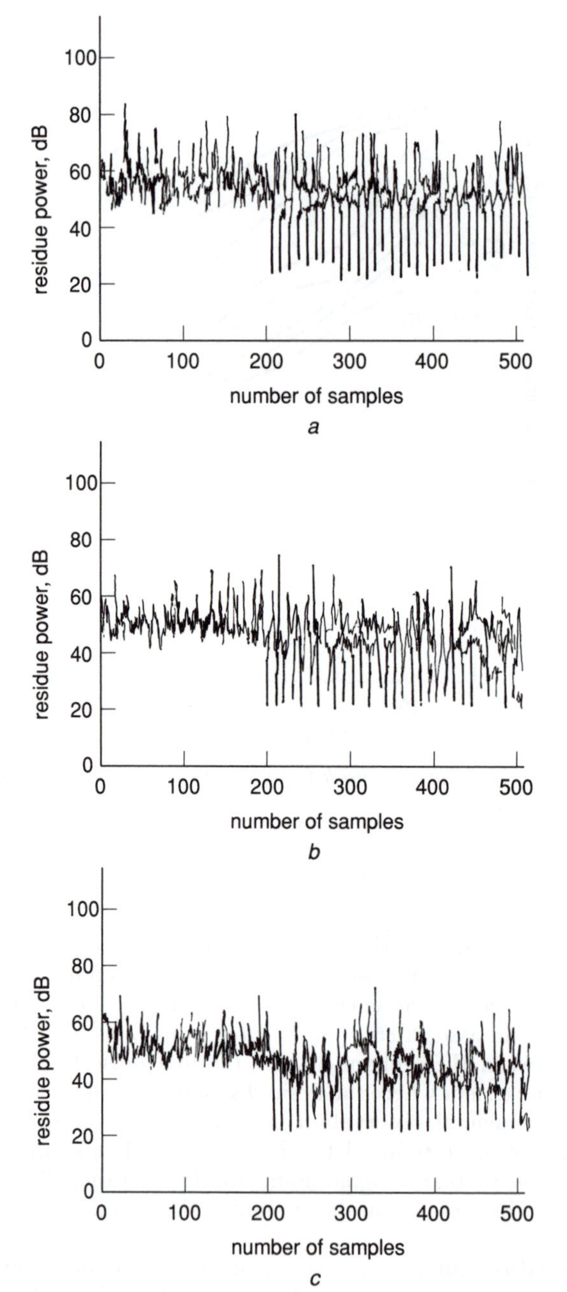

Figure 7.23 *Residue power versus number of samples ($N = -50$ dB, $S_m = -20$ dB) [19] © IEEE, 1978*

a $N_{AVG} = 2$
b $N_{AVG} = 4$
c $N_{AVG} = 8$

Table 7.6 *Performance results using W_{opt} for N_{AVG} = 2, 4 and 8 [19] © IEEE, 1978*

N_{AVG}	2	4	8
C	−50.5	−48.2	−47.5
CS	−23.7	−21.8	−20.9
Q	46.8	46.4	46.6

A figure of merit, similar to the improvement factor used in MTI, was also calculated and is defined herein as the ratio of the average signal-to-interference ratio at the adaptive processor output and input. This improvement factor Q may be expressed as

$$Q = (S/I)_{out}/(S/I)_{in} = -C + CS - [(S/I)_{in}] \text{ dB} \tag{7.48}$$

The interference consists of correlated noise and uncorrelated thermal noise. The latter can be neglected in most cases. The results of computer simulation are listed in Table 7.6.

It is observed that the interference residue increases as N_{AVG} increases but that the target cancellation decreases so that the improvement factor Q remains fairly constant.

It has been found for the case of M auxiliary inputs that good results are obtained by cascading M open-loop processors. Each processor uses one of the auxiliary signal inputs and uses the residue of the previous processor as its main input. This processing is repeated by cascading another M processors and reusing the original M auxiliary signal. Several iterations are generally adequate for good cancellation. In low bandwidth applications, the number of processors can be substantially reduced by commutation and storage of intermediate results.

The Gram–Schmidt algorithm is also one type of direct or sample matrix inversion (DMI or SMI) algorithm. Bristow [20] realised an open-loop canceller with this algorithm. It was implemented with digital circuitry. The block diagram is shown in Fig. 7.24.

CW cancellation performance measurements were carried out using a CW source in the 30 ± 3 MHz frequency band applied to the main and auxiliary IF input ports via an in-phase power splitter, and attenuators and phase shifters as appropriate. The maximum input level to avoid saturation in the A/D converters is −15 dB. Fig. 7.25 shows the variation of cancelled output for equal powers applied to the two inputs. The residual output is approximately constant, the limit to cancellation being due to spurious noise output and quantisation and rounding errors in the digital weight forming. At the maximum input level a cancellation ratio of 33 dB is evident. As the signals get smaller, the accuracy with which the weight can be calculated decreases, and less cancellation is achieved.

Variation of cancellation with input phase angle for equal signal levels at the two inputs is shown in Fig. 7.26. This measurement corresponds to variation of the I and Q weights through all amplitude settings and shows that better than 30 dB cancellation is maintained.

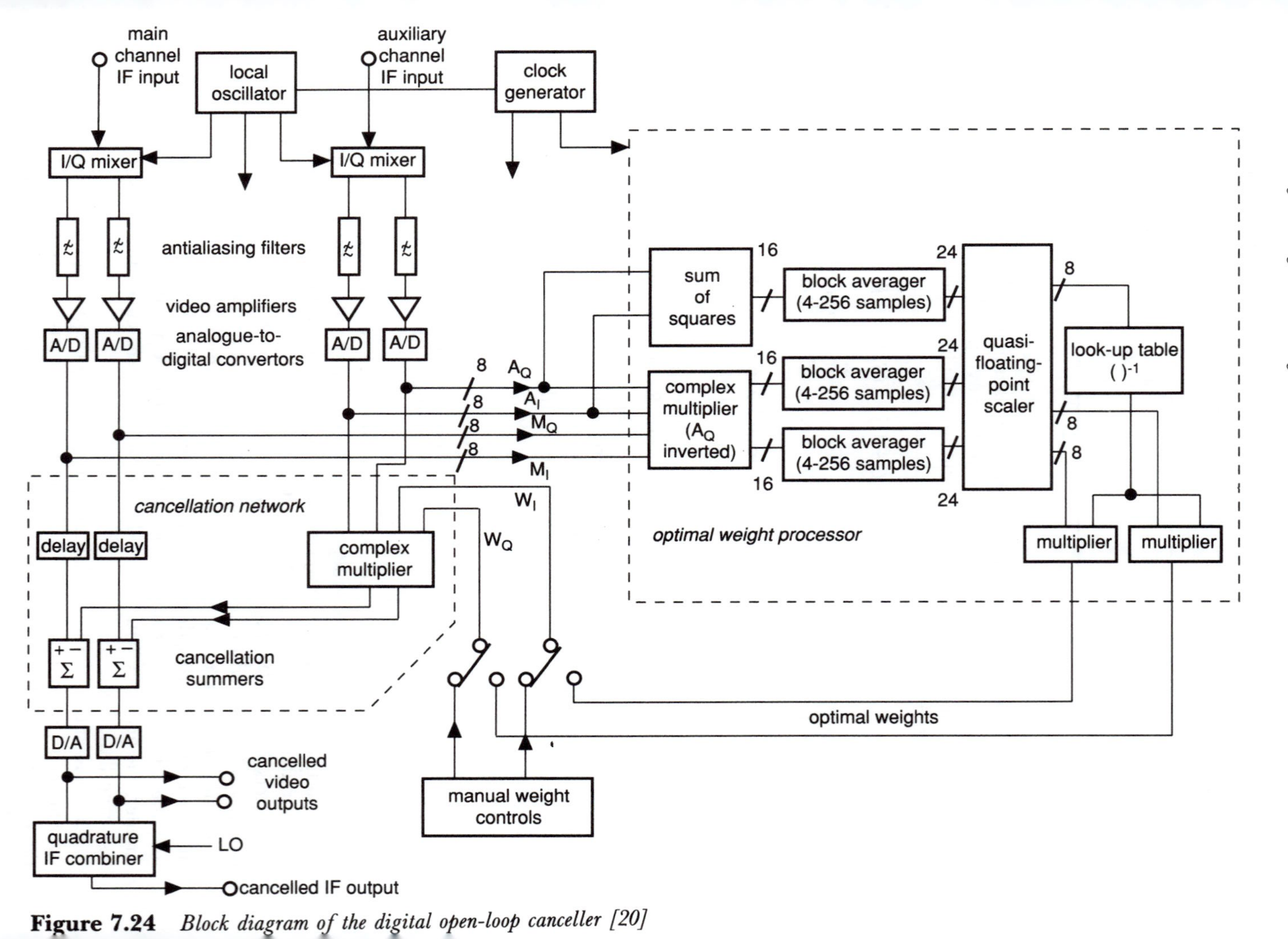

Figure 7.24 *Block diagram of the digital open-loop canceller [20]*

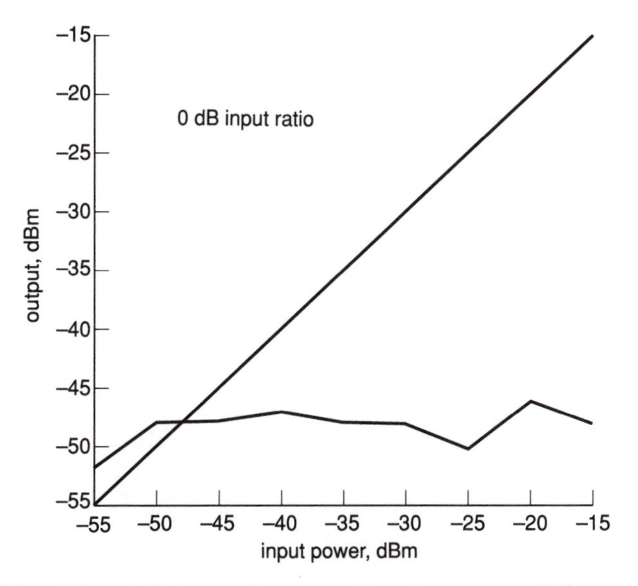

Figure 7.25 *CW cancellation performance against input power [20]*

The broadband performance of the equipment was measured by applying wideband FM noise to the inputs and observing the uncancelled and cancelled spectra (see Fig. 7.27). The CW spike at the centre of these spectra is leakage from the local oscillator used for up conversion.

When the averaging process is performed on a large number of samples (see Fig. 7.27*b*) the cancellation ratio averaged over the 6 MHz passband is about

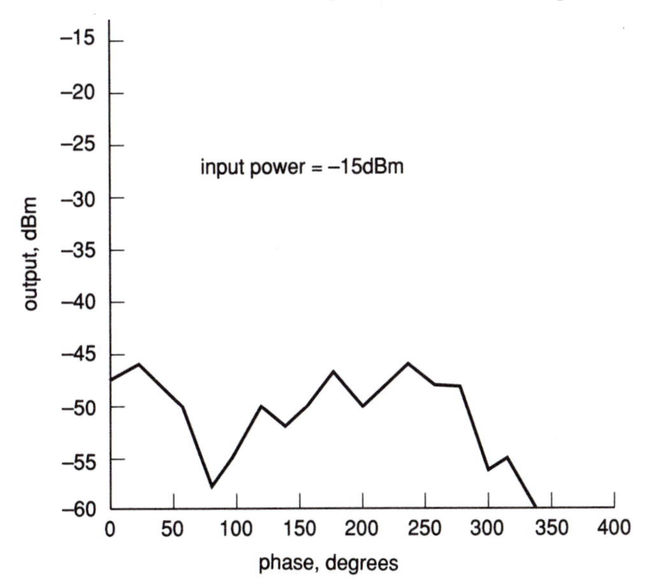

Figure 7.26 *CW cancellation performance against phase [20]*

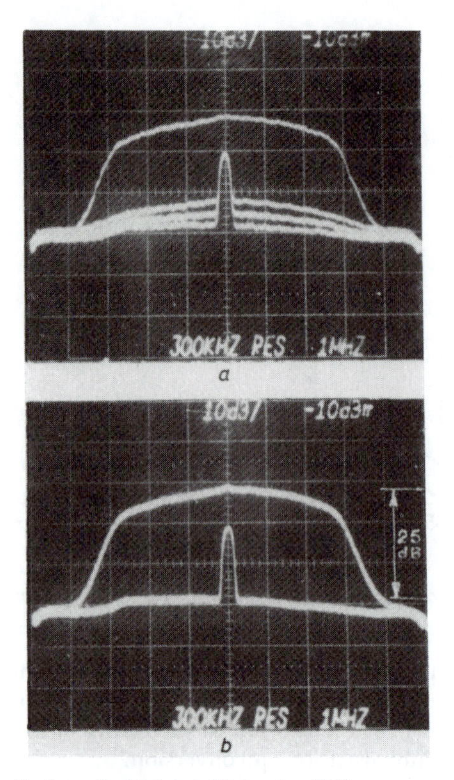

Figure 7.27 *Cancellation of wideband FM noise [20]*

a 3, 7 and 30 sample average
b 100 sample average

25 dB. Fig. 7.27*a* shows progressive loss of performance as the number of samples averaged is reduced. The reason is that the noise on correlation signal will give noisy weights which will result in a higher residue level. The nonlinearity prior to digitisation and computation delay will also cause degradation of cancellation.

For the block-type averaging system used in this equipment the time taken to cancel a pulsed source is determined by the number of samples averaged (and the sample interval) and is independent of signal level. This is the major advantage of the open-loop algorithm over the closed loop, whose response time is very dependent on signal gains and phases.

Fig. 7.28*a* shows the response for a 4-sample average. The 1.8 μs delay between the input step and the cancelled output is made up of 1 μs computation time and 0.8 μs (4 × 200 ns) delay in the block average.

For a 256-sample average (see Fig. 7.28*b*) the delay between input and cancelled output is approximately 50 μs.

Stone [21] also presented experimental results of an open-loop adaptive sidelobe canceller for a rotating antenna. The experimental radar employed in the system has a high PRF so that under some circumstances it is impossible to obtain a set of samples uncontaminated by clutter by sampling just before a transmission. Preprocessing (two-pulse MTI canceller) is therefore provided to minimise the effects of clutter.

The complex weight W is determined by use of the relationship

$$W = \sum_{i=1}^{N} M_i A_i^* / \sum_{i=1}^{N} A_i A_i^* \tag{7.49}$$

where M_i is the ith sample from the main antenna channel, A_i is the ith sample from the auxiliary antenna channel and A_i^* is the complex conjugate of the ith sample of the auxiliary channel.

Bench tests with a noise signal equally split between main and auxiliary channels showed best attenuations of the order of 40 dB. Similar tests with the linear radar receivers interposed between source and SLC resulted in attenuation of the main-channel noise by about 30 dB. The loss of improvement is presumably mainly due to dispersion effects within the receivers. The effect of introducing delays into one of the channels in the bench test described above is a reduction in the improvement achieved of the order of 10 dB per 50 ns.

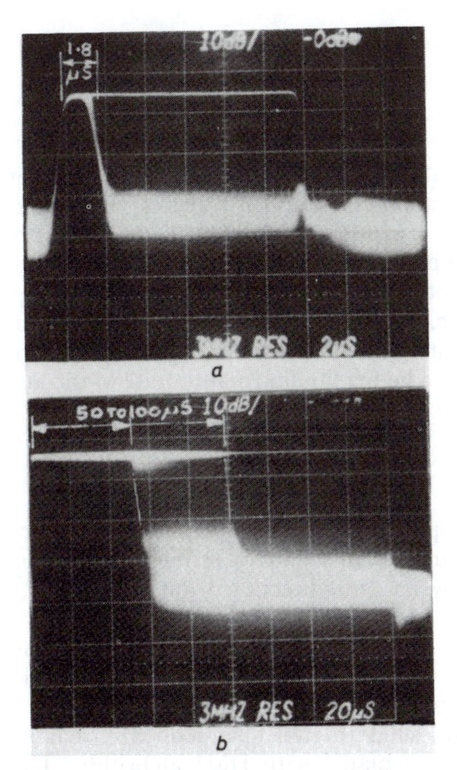

Figure 7.28 *Response time for pulse modulated source [20]*

a 4-sample average
b 256-sample average

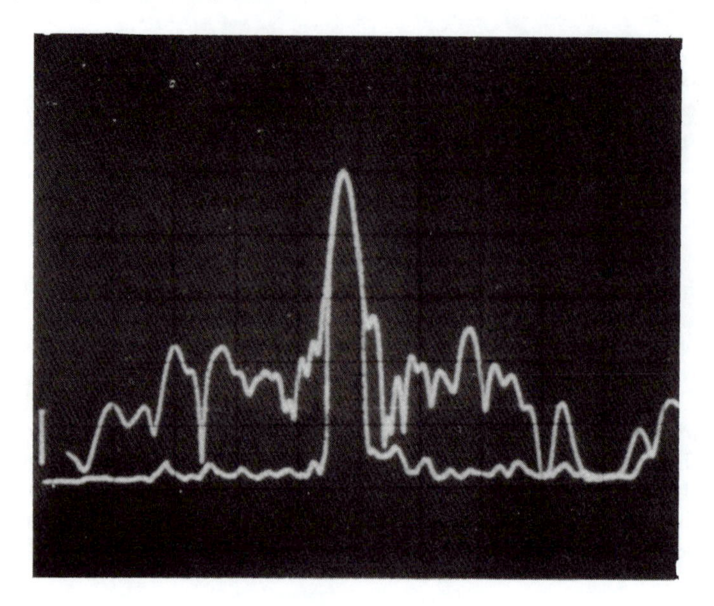

Figure 7.29 *180° of antenna response about the noise source direction, with and without SLC (10 dB per division) [21]*

Fig. 7.29 shows the 180° of the rotating antenna azimuth response about the noise source direction. The upper trace is the uncancelled version. This experiment showed a reduction in the response of the principal sidelobes from about 23 dB to approximately 44 dB below main beam.

The rate of adaptation is an important specification especially for the SLC of high rotation rate, high PRF, and frequency agility radar. Lu *et al.* [22] discussed the design and realisation of a digital open-loop sidelobe canceller for these applications. The maximum rotation rate of the antenna for cancellation loss less than 1 dB is

$$\Omega_{max} \leqq T[(N-1)G'_m]^{-1}(P_J/P_n)^{-1/2} \tag{7.50}$$

where T is the sample interval, N is the number of samples, G'_m is the maximum slope of the primary antenna pattern in the sidelobe range, P_J is the received jamming power by an omnidirectional antenna and P_n is the channel noise power.

Fig. 7.30 shows computer simulation results of cancellation ratio versus adaptation rate. If 0.5 dB cancellation loss is allowed, the adaptation rate should be less than 5 μs.

An experimental single digital open-loop SLC has been constructed. The optimal weight is calculated with DMI algorithm. The cancellation performance has been measured by applying CW, AM and broadband interference to the system input. 18 dB of cancellation ratio of 5–10 μs adaptation transient time have been attained in all cases.

7.2.2.3 *Adaptive sidelobe cancellation by linear prediction*
Gabriel [23] showed that adaptive sidelobe cancellation can be treated as a linear prediction problem. In Reference 23, the emphasis was on the application of spectral analysis technique to super-resolution target detection.

Hung [24] analysed the adaptive sidelobe cancellation with a linear prediction algorithm as an alternative to the sample matrix inversion (SMI) method.

It is assumed that the adaptive sidelobe cancellation system consists of a high-gain main antenna and K auxiliary antennas which have approximately the same gains. In the convention used, a snapshot of the array antennas is denoted by a $(K+1)$-dimensional column vector

$$\tilde{\boldsymbol{u}} = \begin{bmatrix} b \\ \boldsymbol{u} \end{bmatrix} \tag{7.51}$$

where b is the main antenna output, and $\boldsymbol{u} = (u_1, u_2, \ldots, u_K)^T$ is the output of the auxiliary antennas. The weight vector for the snapshots is denoted by

$$\tilde{\boldsymbol{w}} = \begin{bmatrix} b \\ -\boldsymbol{w} \end{bmatrix} \tag{7.52}$$

where $\boldsymbol{w} = (w_1, w_2, \ldots, w_K)^T$ is the weight vector for the auxiliary antenna output. The sidelobe cancellation system output is defined as the residue

$$\varepsilon = \tilde{\boldsymbol{w}}^{\#}\tilde{\boldsymbol{u}} = b - \boldsymbol{w}^{\#}\boldsymbol{u} \tag{7.53}$$

The superscript denotes the conjugation-transposition operation.

Each snapshot of the array antenna has three contributions: the signal S, the interference I, and the white noise N:

$$\tilde{\boldsymbol{u}} = \tilde{\boldsymbol{u}}_s + \tilde{\boldsymbol{u}}_I + \tilde{\boldsymbol{u}}_N \tag{7.54}$$

The optimum weight vector is given by the Wiener solution

$$\boldsymbol{w}_0 = R_{uu}^{-1}\boldsymbol{r}_{ub} \tag{7.55}$$

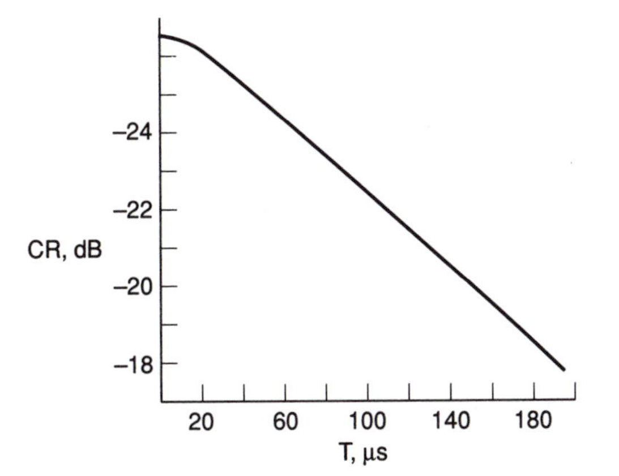

Figure 7.30 *Cancellation ratio versus adaptation rate [22]*

where R_{uu} is the covariance matrix for the outputs of the auxiliary antennas when the signal is absent, and r_{ub} is the cross-correlation vector between the main antenna and the auxiliary outputs.

Let $\tilde{\boldsymbol{u}}_1, \tilde{\boldsymbol{u}}_2, \ldots, \boldsymbol{u}_M$ be a set of M snapshots taken in the absence of the signal. The SMI method estimates $\boldsymbol{w}_0$ using the equation

$$\hat{\boldsymbol{w}}_0 = \hat{R}_{uu}^{-1} \hat{\boldsymbol{r}}_{ub} \tag{7.56}$$

where

$$\hat{R}_{uu} = UU^{\#} \tag{7.57}$$

is the sample covariance matrix, $U = (\boldsymbol{u}_1, \boldsymbol{u}_2, \ldots, \boldsymbol{u}_M)$ is the matrix of auxiliary antenna outputs,

$$\hat{\boldsymbol{r}}_{ub} = U\boldsymbol{b}^* \tag{7.58}$$

is the sample correlation vector, and $\boldsymbol{b} = (b_1, b_2, \ldots, b_M)^T$ is a vector for the main antenna outputs.

The above equations in the SMI method suggest that there is a more general approach to construct an adaptive weight vector for the auxiliary antennas, i.e. identify this vector as a solution of the equation

$$U^{\#}\boldsymbol{w} = \boldsymbol{b}^* \tag{7.59}$$

or, equivalently, the system of linear equations

$$\boldsymbol{u}_m^{\#}\boldsymbol{w} = b_m^*, \; m = 1, 2, \ldots, M \tag{7.60}$$

The solution of eqns 7.56 and 7.59 are identical if rank $(U) = K$.

In the special case of only one auxiliary antenna (i.e. $K = 1$), $\boldsymbol{w}$ has only one component and can be simply written as w_1. If in addition a large number of snapshots is available (i.e. $M > 1$), w_1 is approximately given by the ratio of two expected values,

$$w_1 = -\frac{w_0 E\{u_1 b^*\}}{E\{|u_1|^2\}} \tag{7.61}$$

where

$$w_0 = -1 \tag{7.62}$$

is the weighting coefficient for the main antenna output. This expression for w_1 is essentially the expression for a two-element array with one adaptive loop in the Applebaum coherent sidelobe cancellation method.

The adapted weight vector for the auxiliary antennas can be constructed by the two-step procedure described below:

Step 1: Take a prespecified number, M, of snapshots with the desired signal absent.

Step 2: Identify the adapted weight vector identified as a solution of eqn. 7.60.

A choice of $M = 2K$ is sufficient in most applications. If $\hat{R}_{uu} = UU^{\#}$ is non-singular, so that the solutions of eqns. 7.56 and 7.60 are identical, the expected value of the output signal-to-noise ratio in $|\varepsilon|^2$ is within 3 dB of the optimal value when this value of M is used. The solution of eqn. 7.60 is not unique if rank $(U) \neq K$.

Computer simulation was carried out with this algorithm. It is assumed that the antenna system consists of a linear array with $K' = (21 + K)$ isotropic elements spaced at a distance of half a wavelength, i.e. $d = 1/2\ \lambda$. The output of the first 21 elements are weighted with a Hamming window and summed to produce the main antenna output. The $(k + 21)$th element is identified as the kth auxiliary antenna. The antenna pattern is defined as

$$G(\theta) = -\frac{|s_0(\theta) - \boldsymbol{w}^{\#}\boldsymbol{s}(\theta)|^2}{\sum |\alpha_k|^2 + |\boldsymbol{w}|^2} \tag{7.63}$$

where $s_0(\theta)$ and $s(\theta)$ are the outputs of the main and auxiliary antennas due to a unit amplitude input signal at bearing θ; α_k is the Hamming window for the main antenna. The weight vector is calculated from eqns. 7.56–7.60 with the QR decomposition method.

The simulations were carried out with $K = 3$. The powers of three CW interference sources were $(P_1, P_2, P_3) = (1.0, 0.3162777, 0.1)$, and the bearing angles were $(\theta_1, \theta_2, \theta_3) = (-30°, 17°, 65°)$. The white noise was assigned as -50, -45 and -40 dB below those interference sources. The results of simulation are shown in Fig. 7.31*a* and *b* before and after adaptation. It can be seen that three -55 dB nulls are formed after adaptation.

7.3 Frequency selectivity

All the previously described spatial selectivity anti-jamming methods are suitable to reject standoff jamming. In the case of self-screening jamming (SSJ), any ECCM method based on the spatial selectivity principle is of no avail, since the target and jammer are located at the same position. In this case, the most effective way to reject active interference is frequency selectivity. Frequency selectivity is effective to reject denial ECM but is helpless to reject deception ECM. Denial ECM techniques are generally 'brute force' ECM schemes for most types of radar, while deception ECM techniques are effective mainly for tracking radar.

Emission characteristics for denial jammers can be any of several, e.g. simple CW, pulse or noise. Some form of noise is generally preferable for denial jammers since noise is more nearly universal in its effectiveness against most types of radar emission.

In the early stage of EW, the jammer is generally tuned to the radar carrier frequency to utilise the limited jamming power effectively. Radar changes its frequency to avoid the jamming. In this stage, the main problem was the tuning rate. Many measures have been developed to increase the tuning rate of both the radar and jammer side.

Frequency agility is the top speed of tuning for the radar side. The definition of frequency agility radar is that: if a radar can change its carrier frequency over the full operating frequency band from pulse to pulse, this radar is a frequency agility radar [25]. It is evident that this is the limit of the tuning rate for a radar.

There is also a limit of tuning rate of the jammer. The fastest tuning rate of a jammer is that the jammer can acquire the radar frequency and tune on this

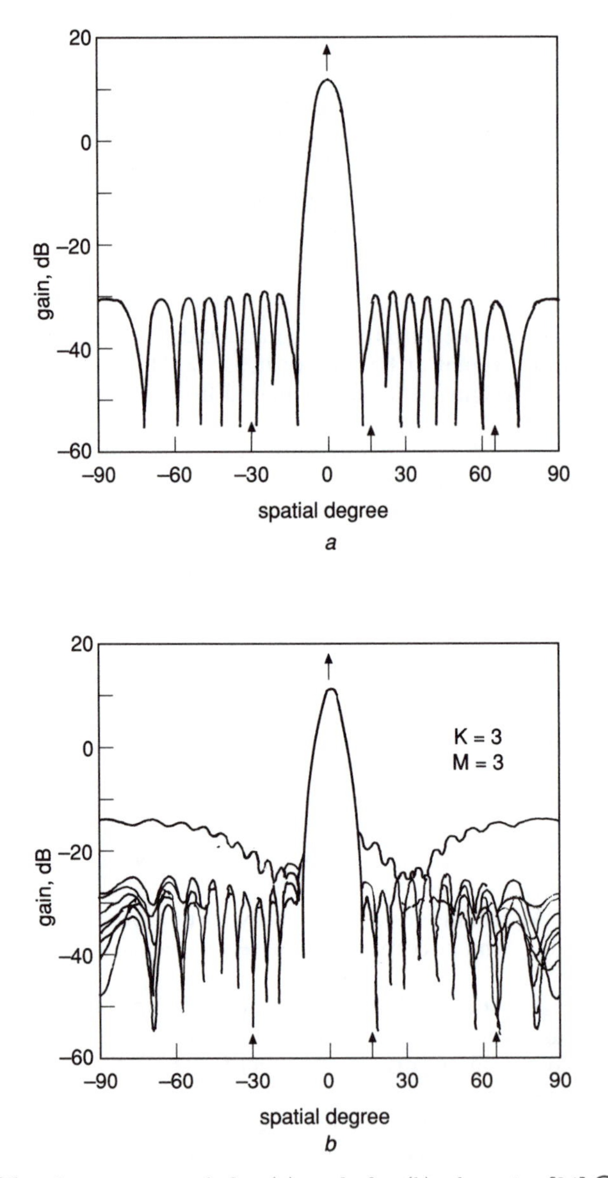

Figure 7.31 *Antenna pattern before (a), and after (b) adaptation [24] © IEEE, 1985*

frequency within the radar pulsewidth. This is realised with electronic tuning by a voltage controllable oscillator (VCO). However, a narrow-band noise jammer cannot jam the frequency agility radar even with this fastest tuning rate. The reason is that, before the jammer receives a radar transmission pulse, it has no way of knowing the frequency of the radar pulse. Thus, there is no way of tuning the jammer to the radar frequency before it receives the radar pulse.

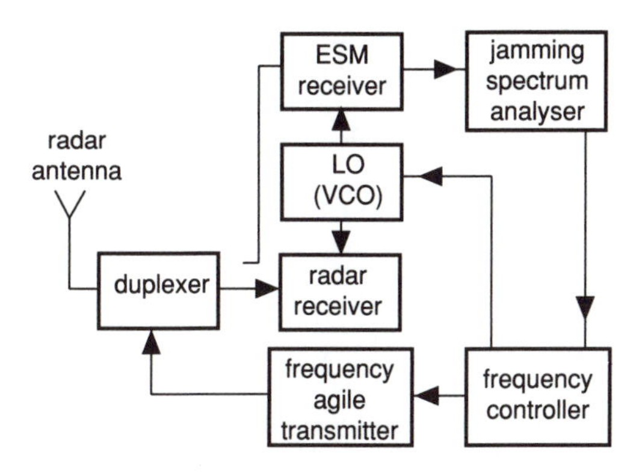

Figure 7.32 *Block diagram of adaptive frequency agility radar*

This means that the radar screen is clean before the position of the jammer. The leading edge of the skin echo from the jammer carrier can be observed no matter how fast the tuning rate of the jammer is. The only way to jam a frequency agility radar is by employing wideband noise jamming. However, this will reduce the power density (watts/MHz) of the jammer if the mean power remains the same. In other words, the development of frequency agility radar forces the jammer to spread the power to a bandwidth at least equal to the operating frequency band of the radar. Therefore, the gain of frequency agility is nearly equal to the ratio of the operating frequency band of the radar to the bandwidth of the narrow-band jammer, e.g. 5–10 MHz. For example, if the operating frequency band of a frequency agility radar equals 300 MHz, the bandwidth of a narrow-band jammer equals 10 MHz; then the gain of frequency agility is approximately equal to 30 (or 14.8 dB).

It seems that there is no way to avoid wideband noise jamming if the bandwidth of the noise is matched to the operating frequency band of the radar. The development of adaptive frequency agility can solve this problem.

7.3.1 *Adaptive frequency agility radar system*

In the early stage of the development of frequency agility radar, the designer makes every effort to vary the transmitting frequency as randomly as possible. Indeed, the transmitting of random variable frequency can reduce the probability of intercept by the ESM receiver. However, from the viewpoint of jamming rejection, random variable frequency is not the optimum way to select the transmission frequency. The best way is to select the transmission frequency on which the jamming is a minimum. This is the principle of adaptive frequency agility. The block diagram of adaptive frequency agility radar is shown in Fig. 7.32.

A wideband ESM receiver and a jamming spectrum analyser detect the existence of jamming signal and analyse its spectrum. The frequency of weakest jamming can be determined. This information is used to control the transmitting frequency so that the frequency agile radar can always select the frequency with minimum jamming.

In this case the designer of the jammer makes every effort to obtain a jamming spectrum with maximum flat. However, this is a futile effort. A jamming signal with flat spectrum at the jammer side will become hollow at the radar side owing to the multipath effect. It has been shown that a 10–20 dB hollow in jamming spectrum can be obtained owing to the multipath effect. This means that the gain of adaptive frequency agility over random frequency agility may be of the same order of magnitude.

The best way to jam adaptive frequency agility radar is by using a standoff electronic tuning narrow-band point jammer. However, the standoff jammer can only shield the target behind it. Meanwhile, the jammer carrier itself will be revealed.

Standoff jamming can be rejected by spatial selectivity as mentioned before. Therefore, the most powerful ECCM radar is a radar which combines adaptive frequency agility and an adaptive sidelobe canceller. Sufficient speed of adaptation is required for both techniques. Open-loop adaptation is the best way to satisfy this requirement.

7.3.2 *Performance of frequency agility radar*

Besides its antijamming capability the frequency agility radar possesses other superior performance:

(i) *Increases the detection range*: If the frequency difference of the adjacent pulses of frequency agility radar is larger than the 'critical frequency', the echoes of these adjacent pulses will be de-correlated. This means that the Swerling case 2 or case 4 can be realised. The detection loss due to slow fluctuation can be avoided. This slow fluctuation of the echo often appears in fixed frequency radar.

Test results showed that, when relatively high detection probability (over 80%) is desired, the detection range of frequency agility radar can be increased 20–30% more for slow fluctuation targets than fixed frequency radar. It is also equivalent to a 2–3 times enlargement of the transmitter power.

(ii) *Increases tracking accuracy*: This is because the use of frequency agility can speed the movement of apparent position around the 'centre of gravity' of a target. This means that it can spread the spectral density of the angular noise to outside the bandwidth of the angular servo system. This can greatly decrease the angular tracking error caused by angular noise. This type of error is the major source of monopulse radar tracking errors for close range and medium range targets.

Test results showed that the glint error of an aircraft target echo in over 2–4 Hz range can be decreased noticeably. For a Ku band radar, its tracking error can be decreased by one-half and the estimated error of the future point can be decreased by one-third. The tracking error for large targets, such as ships, can be decreased by one-half to one-quarter with frequency agility.

(iii) *Suppresses sea clutter and other distributed clutter*: When the difference of the adjacent pulse carrier frequencies is larger than the reciprocal of the pulse width, it is sufficient to de-correlate the echoes from distributed targets, such as sea waves, precipitation and chaff. These de-correlated clutters become noise-like and can be suppressed by video integration. This is because video integration can accumulate correlated target signal but not the de-correlated clutter signal.

Theoretical computations and test results showed that when the number of pulses within the radar beam is about 15–20, the signal-to-clutter ratio can be increased 10–20 dB by the use of frequency agility.

(iv) *Suppresses mutual interference of closed radar*: Whether the nearby radar of the same frequency band is operating in fixed frequency or frequency agility mode, the probability of the same frequency at the same instant is very low. It is approximately equal to the ratio of the radar's bandwidth to the agility bandwidth. Thus frequency agility radars have better electromagnetic compatibility.

(v) *Eliminates multiple roundtrip echoes*: Because there is atmospheric superrefraction, and thus abnormal propagation existing in a great deal of radar (especially coastal surveillance radar), this phenomenon is sometimes called 'atmospheric waveguide'. Owing to this phenomenon the radar receiver can receive very far echoes which are located in the next pulse repetition period. This will cause distant ground clutter or sea clutter embedded in the useful target signal.

Yet, in frequency agility radar, because the carrier frequency of the second transmitted pulse is different from the first, during the second period the receiver is also tuned to the second frequency. Thus the multiple-roundtrip echoes are eliminated.

However, for this reason, frequency agility radar cannot be used directly in high PRF radar with range ambiguity.

(vi) *Reduces the effect of beam splitting owing to the multipath effect*: The multipath effect caused by surface reflection, especially sea surface reflection, will cause beam splitting into many sub-lobes in elevation. The minima in elevation will cause discontinuous tracking of a target. The angular positions of the minima depend on the carrier frequency and the height of the radar. Therefore, varying the radar frequency will change the angular position of the minimum. Although full compensation of these minima requires very large frequency change, the use of frequency agility radar can reduce this multipath effect.

Aside from these, the use of frequency agility can increase range resolution, eliminate blind speed, increase the accuracy of RCS measurement, realise target recognition, and can reduce the effect of refraction of the radome on tracking precision, etc.

7.4 Simultaneous rejection of passive and active interference

In this Section, we will discuss the compatibility of frequency agility and MTI, MTD.

7.4.1 *Introduction*

In a real electronic warfare environment, passive interference (chaff) and active interference (mainly denial ECM) often appear simultaneously. Besides the chaff, ground clutter, sea clutter and sometimes weather clutter always exist in most radars. Therefore, a military radar designer is faced with the problem of how to reject passive interference and active interference simultaneously.

It is well known that the most effective method to reject passive interference is by MTI or MTD techniques, and the most effective method to reject active interference is by frequency agility techniques. However, these two techniques cannot be compatible with each other. The reason is very simple. Since frequency agility will de-correlate the clutter echoes, all clutter rejection methods based on the utilisation of correlation of clutter will be out of order. In other words, frequency agility destroys the coherence between consecutive echoes. This is the main disadvantage of frequency agility technique.

Some methods have been suggested to solve this problem. However, some of them are not valid, and some are not effective. For example, block-to-block frequency agility can be compatible with MTI or MTD as mentioned in the previous Chapter. Block-to-block frequency agility cannot reject electronic tuning narrow-band point jamming. Strictly, block-to-block frequency jumping is not really frequency agility, since the definition of frequency agility is pulse-to-pulse frequency jumping.

Even though it seems that there is no way to solve this problem, some schemes have been proposed. Unfortunately, some of them are classified. We will introduce some unclassified methods in this Chapter.

7.4.2 *MTI system with twin pulse frequency agility*

The earliest scheme to solve this problem is the twin pulse frequency agility radar [26]. This radar transmits a pair of pulses with the same carrier frequency within the radar pulse repetition period. The interval of these two pulses is usually far smaller than the repetition period, and is determined by the longest extended clutter. In the next period, it jumps to a new frequency and again successively transmits two pulses that have the same interval.

The operational principles of this paired pulse MTI system are also different from those of ordinary MTI systems. Firstly, it does not use a coherent oscillator but directly uses echo phase comparison. Here the first pulse echo is delayed a time interval which is equal to the time interval of these two pulses, and then phase compared with the echo of the second pulse. As for the fixed clutter, the phase of two echoes are completely the same, and therefore the output of the phase detector is zero; as for moving targets, although the interval of the two pulses is not large, the phase of their echo already has enough difference, so the output of the phase detector is not equal to zero. Therefore, in this MTI system, we do not need any subtraction circuit, and the phase detector itself is also a cancellation system. Moreover, because the cancellation is not performed in successive periods but in the same period, we also do not need to have a delay line equal to the repetition period, but only a delay line which is equal to the time interval between two pulses.

The paired pulse MTI system possesses the following advantages over the ordinary MTI system:

(i) The blind speed is much higher than that of ordinary MTI. Because the phases of moving target echoes are compared only after going through the time between these two pulses, and this time interval is far smaller than the pulse repetition period, its blind speed is also far higher than the blind speed that is determined by the pulse repetition period. For example, if the time interval between two pulses is equal to 150 μs, its blind speed will be 20 times higher than that of an ordinary MTI system with 3000 μs pulse repetition period.

(ii) In ordinary MTI systems, the clutter cancellation ratio is limited by the antenna scanning modulation. In the paired pulse MTI system, since the phase comparison is carried out in an even shorter time, it can reduce the effect of antenna scanning modulation.

(iii) The paired pulse MTI system does not need a delay line with a delay time equal to the pulse repetition period, but it only needs a delay line with a delay time equal to the time interval between two pulses. This delay line can be easily implemented with a CCD delay line.

Some shortcomings also exist in the twin pulse MTI system. The primary one is the so-called 'phase contamination' problem. When the clutter echo of the first pulse extends to exceed the time interval between two pulses, the clutter echo of the first pulse will mix with the clutter echo of the second pulse. This will cause the clutter of the mixed part to have no way to be eliminated.

There are two ways to solve this problem: one is appropriately by increasing the time interval between the two pulses. The other method is to improve the range resolution, namely to reduce the range resolution cell. This can be realised with pulse compression.

The field test of an experimental system shows that, when operating with wide band pulse compression, the twin pulse MTI system can achieve a 20 dB cancellation ratio. The cancellation ratio for rain clutter of this MTI system without pulse compression is approximately 10 dB. The sub-clutter visibility (SCV) is limited by the pulse compression ratio; when this ratio is equal to 25:1, the optimum SCV is about 14 dB.

7.4.3 *Quad-pulse frequency agility Doppler processor*

Nourse and Crossfield [27] proposed a scheme which transmits a burst of four closely spaced pulses by a magnetron and then processes the echo signal with a Doppler processor which can reject fixed clutter. The block diagram of the transmitter–receiver and timing diagram are shown in Fig. 7.33.

The interpulse spacings were all equal and constant at 1.9 μs, while the sub-pulse lengths were all 0.63 μs (i.e. overall burst lengths of 8.2 μs). All these pulses were generated by a frequency agile magnetron. It is assumed that the frequency agility only carries on from burst to burst and not from sub-pulse to sub-pulse. Since the initial phase of every sub-pulse is different, four COHOs are needed to 'remember' the initial phase of each sub-pulse.

After the received signal goes through the mixer and IF, it goes to the intermediate frequency power distributor. The output of this power distributor is sent to four phase detectors (see Fig. 7.34). Four pairs of in-phase and quadrature-phase outputs can be obtained (the Figure only shows single channels). These outputs are added to four pairs of A/D converters. In order to process them simultaneously, the signals from each channel are delayed by 0, T, $2T$ and $3T$, respectively. Then they are added to the Doppler filter bank.

The filter is conceptually similar to the Kalmus filter, which works on the basis of subtraction of equal and opposite frequency or, in this case, velocity channels, although it was, in fact, arrived at quite independently.

Fig. 7.35 shows the simplest form of the filter, which has the velocity response of Fig. 7.36. In this, the exact square root is taken to obtain the modulus. It may be seen that, except for the 'slow' channel ($V_{0.5}$), there is ample room for clutter

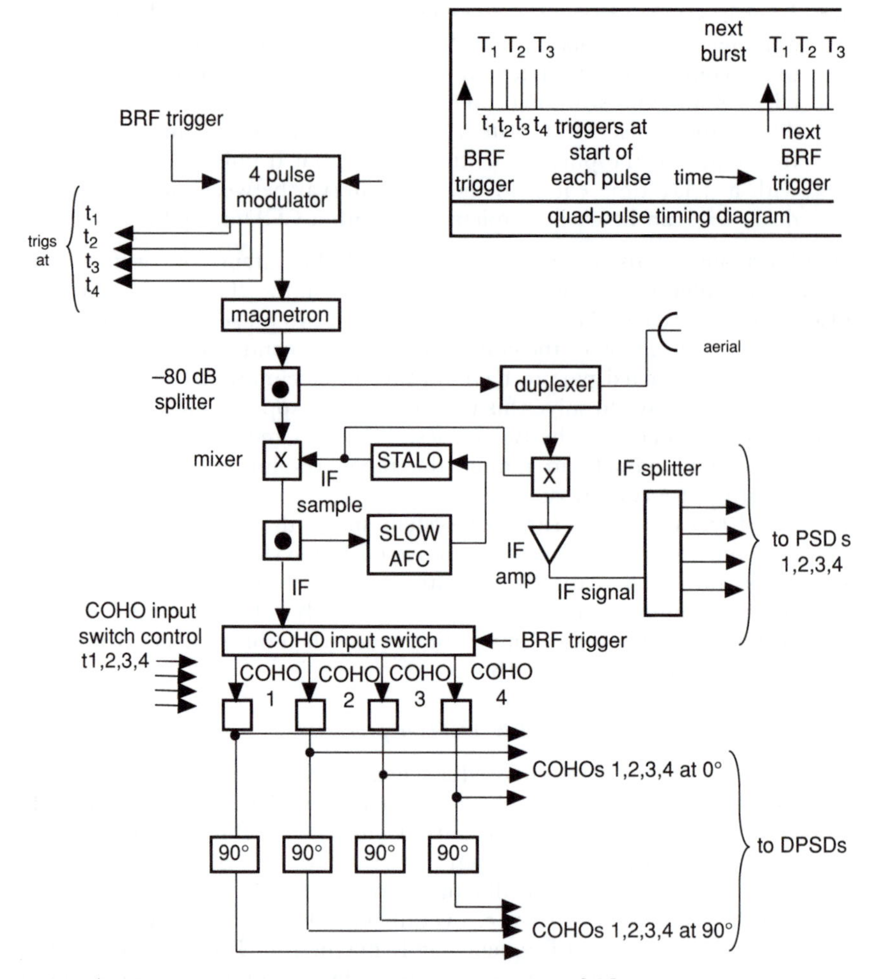

Figure 7.33 *Transmitter–receiver and timing diagram [27]*

spread, and even here there is probably sufficient room. The 'other filters' are necessary to fill in the velocity cover, and are obtained by the use of 45° phase shifts, as shown in Fig. 7.35.

Computer simulated A-scans are shown in Fig. 7.37. Along the *X*-axis are marked 113 range bins; above these are marked, as triangles, various clutter points. Stationary clutter is marked unshaded, and a moving target shaded. The moving target in this instance is in range cell 43, and is moving at 660 km/h. A large clutter point is present at range cell 90. It will be noted that the sidelobes of the target are all of the same sign in all the A-scans, and therefore do not 'cancel'. It will also be noted that the target shows up very clearly after relatively few integrations.

However, there are some particular problems in realising this scheme. If the transmitter employs a rotary tuned magnetron, the typical tune speed is about

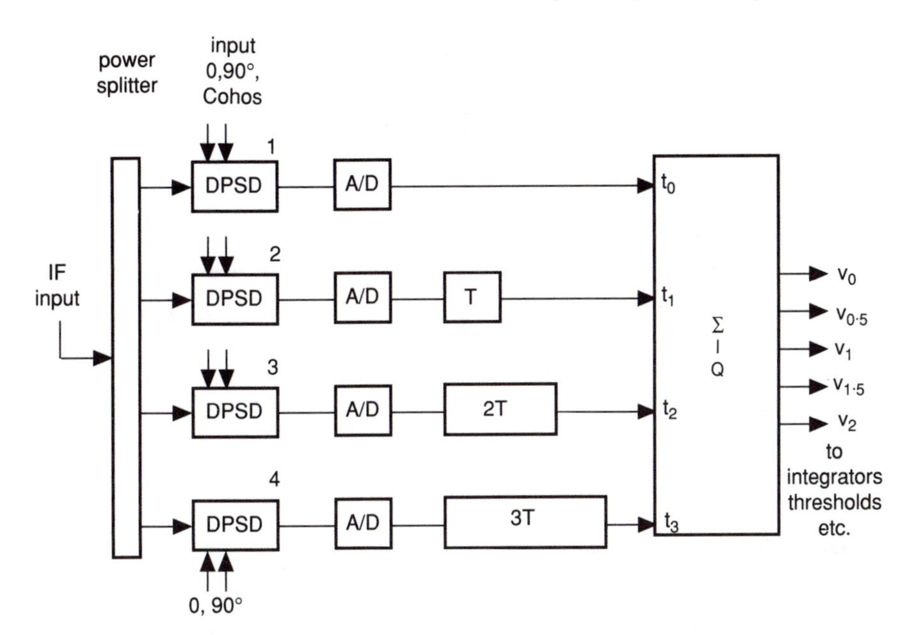

Figure 7.34 *Phase agile receiver [27]*

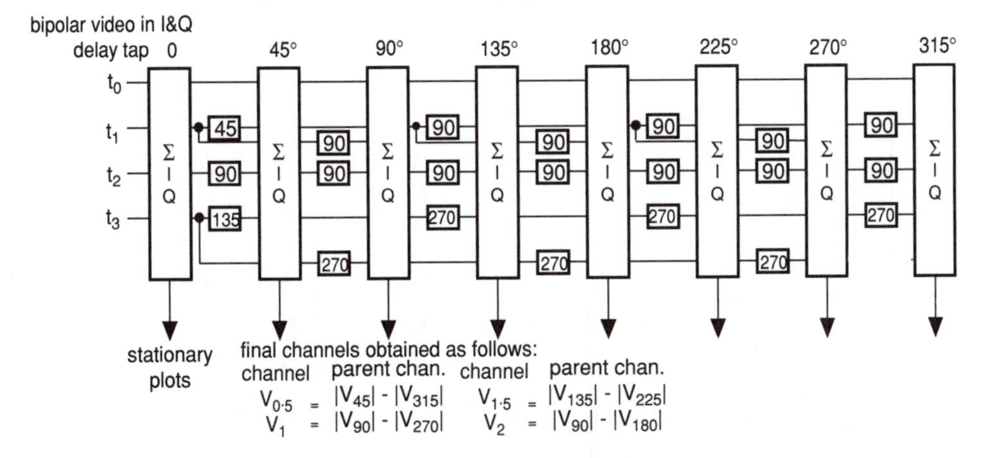

Figure 7.35 *Block diagram of Doppler filter [27]*

several hundred kHz per microsecond. The frequency difference between the first sub-pulse and last sub-pulse of a burst will be several MHz. This frequency difference is sufficient to de-correlate the clutter. Therefore, it is impossible to realise with a frequency-agility magnetron transmitter. It has to be realised with a fully coherent MOPA type transmitter.

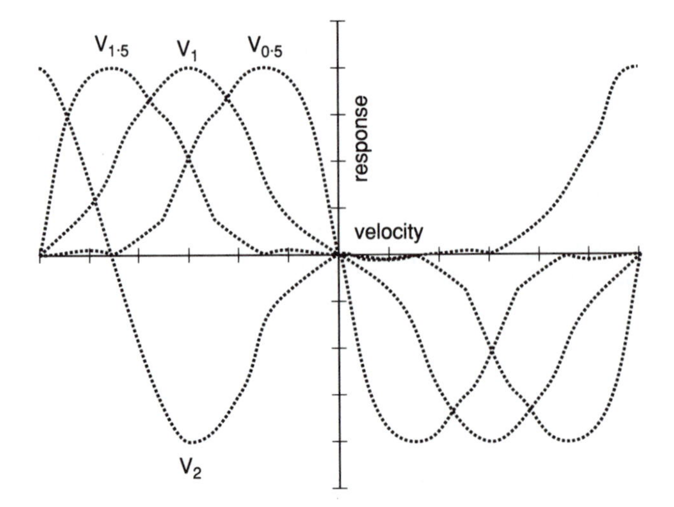

Figure 7.36 *Velocity response of Doppler filter [27]*

Another problem is that, just as in the case of twin pulse frequency agility radar, the transmitter transmits several pulses with the same carrier frequency; even these sub-pulses are within the same repetition period, and it is very easy to be intercepted by an ESM receiver and jammed by an electronically tuned jammer. Strictly speaking, these two radar systems do not belong to frequency agility radar, since they transmit adjacent pulses with the same carrier frequency.

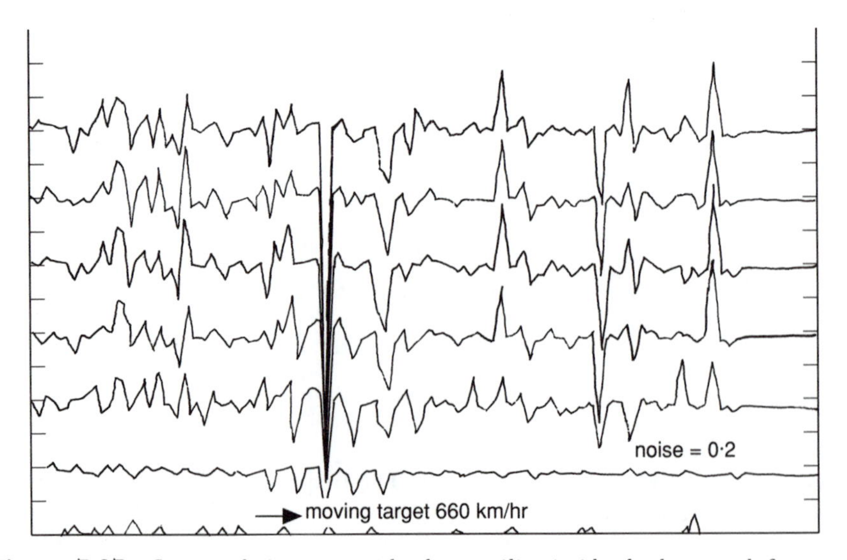

Figure 7.37 *Integrated A-scans; with phase agility inside the burst and frequency agility between bursts [27]*

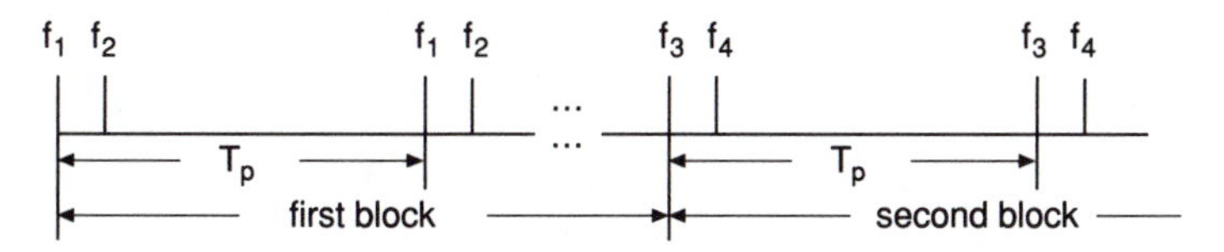

Figure 7.38 *Waveforms of paired pulse frequency agile MTI radar*

7.4.4 *Paired pulse frequency agile MTI system*

To overcome the problem of transmitting two pulses with the same carrier frequency, Taylor and Sinclair [28] proposed a modified paired pulse frequency agile MTI system. It transmits a pair of pulses within a repetition period just as before but with different carrier frequency. In the next period, the same pair of frequencies are transmitted, so that in successive transmission periods there are pulses at the same frequency. In fact, it is a frequency diversity radar with dual channels, each of which operates in block-to-block frequency agility. The waveforms are shown in Fig. 7.38.

The clutter rejection performance of this system is much better than that of the former systems, since block-to-block frequency agility is used, so there are enough pulses to process. When the number of pulses with the same carrier frequency within the block is great enough, MTD can be employed. However, the more of the same frequency pulses there are, the more easily they will be intercepted by the ESM receiver.

7.4.5 *MTI system compatible with random pulse-to-pulse frequency agility*

Petrocchi *et al.* [29] proposed an MTI system which can be compatible with random pulse-to-pulse frequency agility and/or adaptive frequency agility. It can only be realised with a full coherent transmitter. The block diagram is shown in Fig. 7.39.

The automatic frequency selection (AFS) system selects the weakest jamming frequency point according to the information coming from the interference estimator. The control logic controls the frequency synthesiser to output this

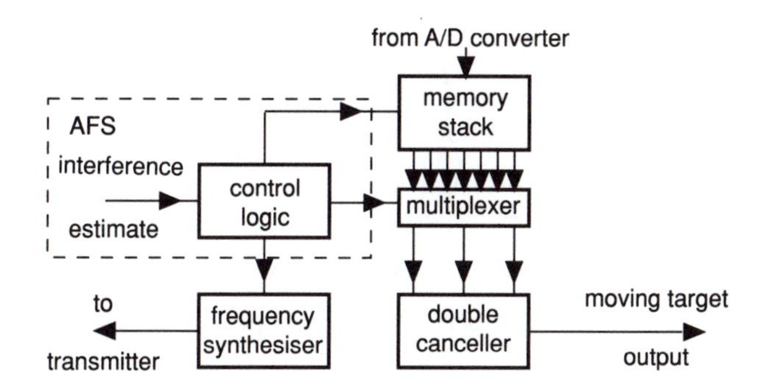

Figure 7.39 *Block diagram of the FA compatible MTI system [29] © IEEE, 1978*

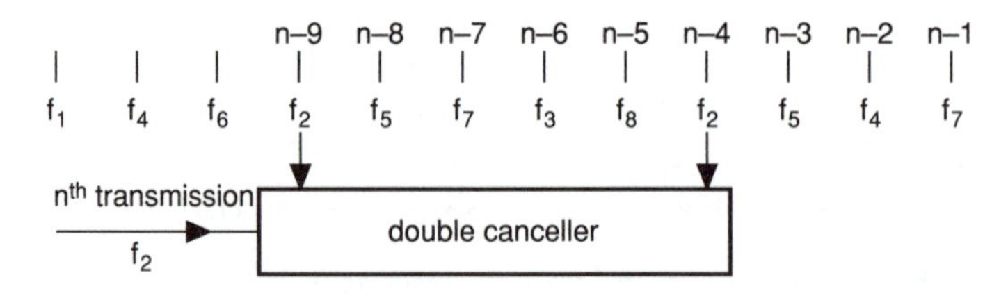

Figure 7.40 *Selection of input signal for the double canceller [29] © IEEE, 1978*

frequency to the transmitter. The usable frequency points are limited, e.g. eight points. The transmitter can also operate in random pulse-to-pulse frequency agility. It selects the frequency from these eight frequencies randomly. The echo signals are stored in a memory stack according to the number of frequency channels and the order of transmission. The multiplexer is controlled by the control logic to select the echo signal from the memory and output to a double canceller. The principle of its operation can be explained by Fig. 7.40.

If the frequency of the last nth input signal to the double canceller is f_2, then the control logic controls the multiplexer to select the nearest two echoes of f_2 from the memory, which are the echoes of $(n-4)$th and $(n-9)$th transmission in this example. So it can ensure that the three signals added to the double canceller are always the echoes of same frequency transmission. The frequencies of transmission are generated with a high stable frequency synthesiser, so the echoes of same frequency transmission are coherent with each other. Therefore, the fixed clutter can be cancelled by the double canceller.

Since the transmission frequency is randomly selected, the intervals between them are also randomly spaced. In this example, the intervals between successive same frequency transmission are 4 and 5 repetition periods. The cancellation performance of the double canceller is different from the ordinary double canceller. The improvement factor IF of the MTI filter is

$$\mathrm{IF}\,(\mathrm{dB}) = CA\,(\mathrm{dB}) + G_n(\mathrm{dB}) \tag{7.64}$$

where CA is the clutter attenuation and G_n is the gain of noise.

$$CA = \frac{\int_{-\infty}^{\infty} S_c(f)\,df}{\int_{-\infty}^{\infty} |H(f)|^2 S_c(f)\,df} \tag{7.65}$$

where $S_c(f)$ is the clutter spectrum and $H(f)$ is the transfer function of the double canceller:

$$H(f) = 1 - 2\,\exp[-j2\pi f K_1 T] + \exp[-j2\pi f(K_1 + K_2)] \tag{7.66}$$

where K_1 and K_2 are the number of repetition periods between the first and second, and second and third, same frequency transmissions. In our example, $K_1 = 4$ and $K_2 = 5$.

If we assume the spectrum of ground clutter has Gaussian shape,

$$S_c(f) = S_0 \exp\left(-\frac{f^2}{2\sigma^2}\right) \tag{7.67}$$

where σ is the standard deviation of the spectral width. Substituting it in eqn. 7.65, we can obtain

$$CA = \frac{1}{6 - 4\exp(1 - K_A^2) - 4\exp(1 - K_B^2) + 2\exp[-(K_A + K_B)^2]} \tag{7.68}$$

where

$$K_A = (2\pi)^{1/2} K_1 T\sigma \tag{7.69}$$

$$K_B = (2\pi)^{1/2} K_2 T\sigma \tag{7.70}$$

The gain of the double canceller is equal to 6; therefore

$$\mathrm{IF} = \frac{3}{3 - 2\exp(1 - K_A^2) - 2\exp(1 - K_B^2) + \exp[-(K_A + K_B)^2]} \tag{7.71}$$

The improvement factor can be calculated with given K_A and K_B. In the expressions for K_A and K_B, T is the pulse repetition period of the radar, and is in the range 200 μs to 1 ms. σ can be assumed smaller than 1.5 m/s; this is equivalent to a wooded hill under 40 km/h wind. The main problem is that K_1 and K_2 in random frequency agility are not constants but random variables. In this case, the IF is not a fixed number, and it can only be indicated by a statistical, P(IF). It represents the probability of achieving some value of IF. It is evident that K_1 and K_2 are functions of the total number of frequency channels. K_1 and K_2 increase with the increasing number of frequency channels. This means that, when the number of frequency channels increases, the average time interval between samples of the same frequency will increase; then the correlation between samples will decrease, and the improvement factor will be worse. Fig. 7.41 shows the P(IF) versus IF for three different numbers of frequency channels, i.e. 3, 4 and 5.

It can be seen from this Figure that, when the number of frequency channels increases from 3 to 5, the improvement factor with same probability of occurrence will decrease by approximately 5 dB. However, if the number of frequency channels is too low, it is disadvantageous for ECCM. Computer simulation shows that the optimum number of frequency channels is about 1/8 the number of pulses within the beamwidth.

In the case of adaptive frequency agility, the probability of each frequency is not equal. The probability of occurrence of some frequency depends on the shape of the jamming spectrum, the multipath effect, the antenna pattern, etc. It is very difficult to calculate it by analytical methods or by computer simulation. In general, when it operates in adaptive frequency agility mode, since the frequency point corresponding to weakest interference may not vary in successive transmission, the probability of operating in the same frequency in successive transmission will be greater than that of the random frequency agility mode. Therefore, it can be expected that the average improvement factor of the former will be better than that of the latter. For example, if three frequencies are selected randomly to illuminate the target, the average improvement factor is about 28 dB. When adaptive frequency agility is used against noise-like jammer, the average improvement factor is about 40 dB. Thus it can be seen that, whether from the viewpoint of rejecting active interference or from the viewpoint of clutter suppression, adaptive frequency agility is always a powerful technique.

The velocity response of stagger PRF MTI system when operated in adaptive frequency agility mode is very similar to that of the fixed frequency mode. Fig. 7.42 shows an example.

7.4.6 *Coherent signal processing compatible with frequency agility*

Buhring [30] proposed a scheme for coherent integration of the echoes from a frequency agility and PRF agility phased array radar. Just like all of the previously mentioned methods, the coherent processing can only be carried on echoes of the same frequency transmission. Since the radar is a phased array radar, it has more flexibility than ordinary radar, especially in the control of dwell time.

It is assumed that the radar has the following features:

- S-band pulse radar with fixed phased array antenna
- $2° \times 2°$ pencil beam scanned over $\pm 45°$ in azimuth and 0° to 30° in elevation
- Minimum unambiguous range 60 km, range resolution 75 m
- Pulse-to-pulse frequency agility over 10% of the band centre frequency, random time intervals between the pulses
- Coherent signal processing with clutter suppression in elevation between 0° and 4°, in directions with weather clutter and for target tracking; constant number of $N = 12$ pulses in a staggered pulse train

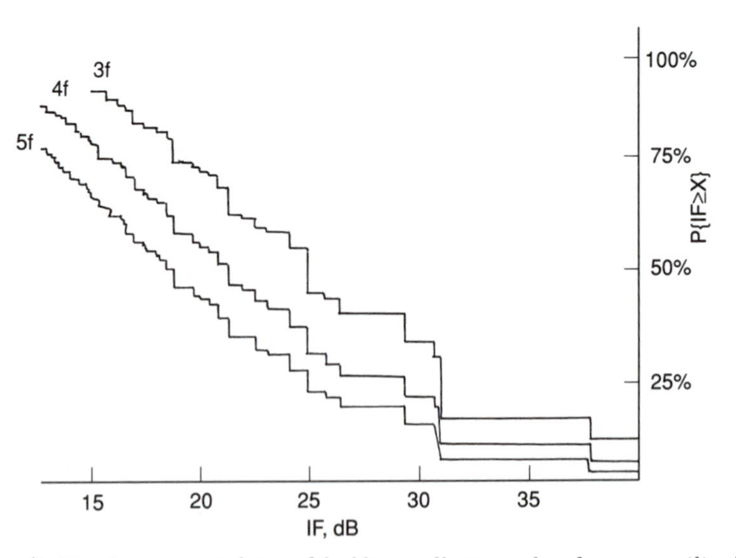

Figure 7.41 *Improvement factor of double canceller in random frequency agility [29] © IEEE, 1978*

- Incoherent integration of echoes obtained on different frequencies in clutter free direction; power management is accomplished by the number of transmitted pulses.

The generator for time and frequency of the pulses was programmed to have the following qualities:

- The time period to the next pulse is chosen with equal probability within a time interval: here 0.4 ms to 1 ms
- The frequency of the next pulse is randomly chosen from a set of frequencies. The probability of a pulse in the channel chosen for coherent processing increases with increasing time distance to the last pulse of this frequency in a manner so that
- The time periods within the pulse train for coherent processing occur with equal probability out of a given time interval. The average dwell time of N pulses in the coherent pulse train is $(N-1)\ (u+v)/2$; here $u=0.4$ ms to $v=8$ ms, so it is equal to 46.2 ms. The standard deviation of the dwell time is $[(N-1)/12]^{1/2}\ (v-u)$, and is equal to 7.3 ms
- The number of frequency channels is chosen to produce nearly equal numbers of pulses in each channel already in a time period equivalent to the dwell time of the coherent pulse train. Here, eight frequency channels were appropriate to the given parameter.

To increase the ratio of coherently to incoherently processed pulses, the pulse generator was programmed to generate two coherent pulse trains in parallel. In

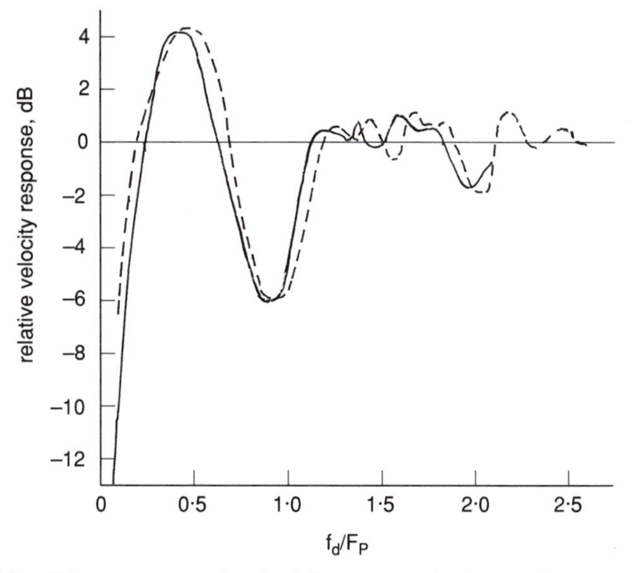

Figure 7.42 *Velocity response for fixed frequency and adaptive frequency agility [29] © IEEE, 1978*

– – – fixed frequency
—— adaptive FA

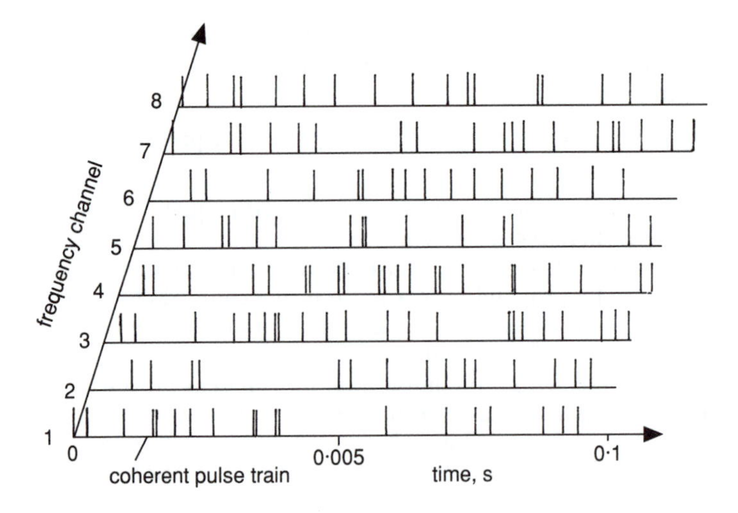

Figure 7.43 *Random frequency agile pulse pattern [30] © IEEE, 1980*

addition, a raster for each pulse train with a grid width of 0.4 ms was introduced to limit the number of required Doppler filters, i.e. to have an unambiguous Doppler region between −1250 Hz and +1250 Hz. The average PRF is 1470 Hz with about 33% of pulse used in coherent processing. Fig. 7.43 shows a sketch of a 100 ms time period of the generator output. The two parallel coherent pulse trains are marked black, beginning in channels 1 and 8.

Since it is assumed that only 13% of all beam directions require coherent processing for ground clutter suppression (82 out of about 650 beam positions), the ratio of pulses for coherent integration is limited to 13% of the total number of pulses.

The pulses required to process coherently are of the same frequency channel but randomly spaced, so the problem is simplified as to how to process randomly spaced pulses coherently. The processing scheme is based on the principle of maximum likelihood ratio detection. The following assumptions are made:

$\boldsymbol{x}=$ vector of N complex sample values, representing the I and Q components of the echoes received from one range bin.

$\boldsymbol{x}=\boldsymbol{n}$ if no target signal is present, $\boldsymbol{n}$ is the vector of the clutter echoes and thermal noise

$\boldsymbol{x}=\boldsymbol{n}+\boldsymbol{s}_g$ if clutter is superposed by a target signal $\boldsymbol{s}$ with the Doppler frequency f_g

$\boldsymbol{s}_g=$ signal vector of a deterministic target model; the vector components are

$$s_{i,g}=\exp(j2\pi f_g\tau_i+j\phi_0) \qquad i=1,2,\ldots,N$$

with τ_i the time distance to the first echo of the sequence and ϕ_0 an arbitrary initial phase

$\boldsymbol{z}_g =$ Doppler filter vector matched to $\boldsymbol{s}_g$
$\boldsymbol{Q} =$ clutter covariance matrix, Hermitian and positive definite; the assumption of an N-dimensional Gaussian process is made. $Q = E\{\boldsymbol{nn}^*\}$ with the expectation $\{\boldsymbol{n}\} = 0$.

The likelihood ratio $\lambda_g(\boldsymbol{x})$ formulates the optimum processing law under the given assumptions [31]

$$\lambda_g(\boldsymbol{x}) = c_0 I_0\{2|\boldsymbol{x}^*\boldsymbol{Q}^{-1}\boldsymbol{z}_g|\} \tag{7.72}$$

c_0 is constant and I_0 is the modified Bessel function.

The extension of eqn. 7.72 to the case of a target signal with unknown Doppler frequency leads to a bank of K Doppler filters. Under the assumption of small target signals a somewhat simplified processing law can be given with

$$\max_g \{|\boldsymbol{x}^*\boldsymbol{Q}^{-1}\boldsymbol{z}_g|\} \geqq \eta \text{ for target present} \tag{7.73}$$

$$g = 1, 2, \ldots, K$$

The maximum output from a bank of Doppler filters is compared with a threshold η to control the false alarm rate. The decision is target present if the output exceeds the threshold.

The clutter fluctuation makes the clutter echo correlation time dependent; therefore, each stagger pattern leads to a specific covariance matrix $\boldsymbol{Q}$. The matrix $\boldsymbol{Q}$ can be decomposed into two triangular matrices $\boldsymbol{Q} = \boldsymbol{DD}^*$. The inverse is $\boldsymbol{Q}^{-1} = (\boldsymbol{D}^*)^{-1}\boldsymbol{D}^{-1}$. So eqn. 7.73 takes the form

$$\max_g\{|\boldsymbol{x}^*(\boldsymbol{D}^*)^{-1}\boldsymbol{D}^{-1}\boldsymbol{z}_g|\} \geqq \eta \tag{7.74}$$

The first triangular matrix $(\boldsymbol{D}^*)^{-1}$ transforms the clutter vector $\boldsymbol{n}^*$ out of $\boldsymbol{x}^*$ into a white noise vector with uncorrelated elements; i.e. after the transformation, the covariance matrix is equal to the identity matrix except for some factors. This transformation is essential for the use of Doppler filters with reduced accuracy. If a target signal is contained in $\boldsymbol{x}^*$, it undergoes a distortion by the first transformation, but the matched filter condition is preserved since the Doppler filter vector $\boldsymbol{z}_g$ undergoes the same transformation with D^{-1}.

The objective is to replace $(\boldsymbol{D}^*)^{-1}$ by a transformation that can be calculated and implemented more easily but without giving up the processing scheme of eqn. 7.74. The approach taken here is a nearby solution. The columns of $(\boldsymbol{D}^*)^{-1}$ contain prediction error filters with increasing length. The actual sample is predicted by a Wiener filter out of the preceding samples and subtracted from the actual sample. The prediction error is determined by the length of the prediction filter and the correlation of the actual sample to the preceding samples. The prediction error filters in $(\boldsymbol{D}^*)^{-1}$ are normalised to achieve equal average error power in each of the filters.

The suboptimum solution considered here uses prediction error filters of constant length M which can be calculated by inversion of submatrices of $\boldsymbol{Q}$ achieved as indicated as follows:

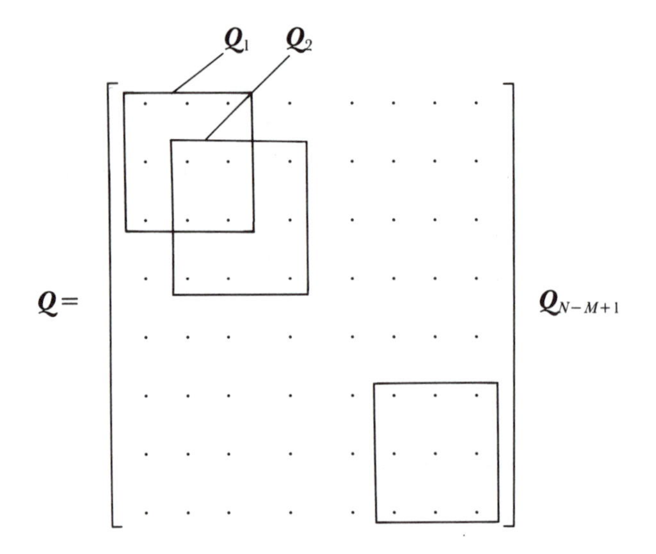

The submatrices can be decomposed into triangle matrices. The inverse of such a triangle matrix contains the prediction error filter of length M in the last row. If the diagonal element in the last row is $\sqrt{w_i}$ and the triangle matrix is normalised to $\sqrt{w_i}$, we achieve for $M=3$

$$\boldsymbol{D}_i^{-1}=\sqrt{w_i}\begin{bmatrix} \cdot & \cdot & 0 \\ \cdot & \cdot & \cdot \\ a_i & b_i & 1 \end{bmatrix}$$

Because of the equivalence of the last row of the inverted triangle matrix to the last row of the inverted matrix these values can also be achieved from $\boldsymbol{Q}_i^{-1}$. The value obtained from the different submatrices of $\boldsymbol{Q}$ filled into the appropriate rows of the $(N\times N)$ matrix $\boldsymbol{P}$, and the $\sqrt{w_i}$ filled into the diagonal matrix $\boldsymbol{W}^{1/2}$, lead to a filter matrix $\boldsymbol{F}$ as an approximation to the triangular matrix $\boldsymbol{D}^{-1}$ (e.g. M=3)

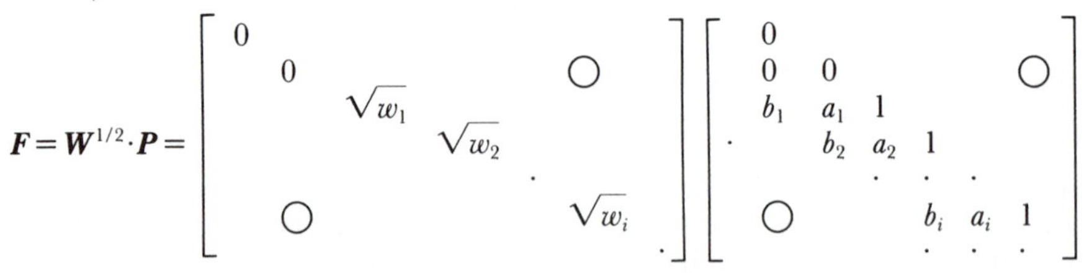

With this approximation, the processing law (eqn. 7.74) becomes

$$\max\{|\boldsymbol{x}^*\boldsymbol{P}^*\boldsymbol{W}\boldsymbol{P}\boldsymbol{z}_g|\}\geqq\eta \text{ for target present} \tag{7.75}$$

Again the prediction error varies because of the different stagger time periods involved. Matrix $\boldsymbol{W}^{1/2}$ normalises this to equal average error power. In the situations where the stagger pattern deviates from an equidistant pulse pattern only slightly or the correlation values are not very sensitive to the time variations involved, the matrix $\boldsymbol{W}$ can be omitted from eqn. 7.75 without causing much loss. Fig. 7.44 shows the block diagram for the implementation of the processing law (eqn. 7.75).

The echo values x_i^* are shifted through a tapped delay-line filter with $(M-1)$ varying filter coefficients representing the matrix $\boldsymbol{P}^*$. The output is multiplied according to $\boldsymbol{W}$. The following tapped delay line has the same structure, but the coefficients are steered by $\boldsymbol{P}$ according to the post-multiplication. $\boldsymbol{P}$ and $\boldsymbol{W}$ are calculated from the pre-assumed clutter model and are dependent on the stagger pattern. The Doppler filters are formed by multiplying $\boldsymbol{z}_g$, and are followed by a CFAR operation with range averaging for each of the Doppler filters and the threshold. Effective clutter suppression in the first tapped delay-line filter allows reduced accuracy in the second tapped delay-line filter and the Doppler filters. This facilitates the implementation of the great number of Doppler filters adapted to each stagger pattern.

Computer simulation was carried out on this scheme of coherent processing. The ground clutter model used in the simulation was assumed to have a fluctuating part with a Gaussian-shaped spectrum and a standard deviation of 0.1 m/s added to a nearly steady state component of the same power and a −40 dB white noise component. The correlation function used was

$$g(\tau)=0.5(1-0.04\tau)+0.5\ \exp\{-(\tau)^2/2\sigma^2\}+10^{-4}\delta(0)$$

with $\sigma=0.08$ s equivalent to the 0.1 m/s standard deviation of the velocity in a 3 GHz radar. For weather clutter, a 1 m/s standard deviation of the velocity was used, i.e. $\sigma=0.008$ s and the correlation function

$$g(\tau)=\exp\{-(\tau)^2/(2\sigma^2)+j2\pi f_d\tau\}+10^{-4}\delta(0)$$

A Doppler shift of $f_d=100$ Hz, equivalent to 5 m/s radial velocity, was assumed. In a situation of mixed ground and weather clutter, the latter was taken to be lower in power than the ground clutter.

The stagger pattern of Fig. 7.45 with a dwell time close to the average dwell time was used in the simulations.

The signal processing gain is given by eqn. 7.76:

$$G=\frac{\boldsymbol{z}_g^*\boldsymbol{P}^*\boldsymbol{W}\boldsymbol{P}\boldsymbol{s}_g\boldsymbol{s}_g^*\boldsymbol{P}^*\boldsymbol{W}\boldsymbol{P}\boldsymbol{z}_g}{\boldsymbol{z}_g^*\boldsymbol{P}^*\boldsymbol{W}\boldsymbol{P}\boldsymbol{Q}\boldsymbol{P}^*\boldsymbol{W}\boldsymbol{P}\boldsymbol{z}_g} \tag{7.76}$$

In Fig. 7.46, the gain is plotted over about half the unambiguous Doppler region for a ground clutter situation. With the prediction error filter of length $M=3$ using the processing scheme of eqn. 7.75, the average gain is less than 2 dB below the gain of optimum processing. Using a conventional two pulse canceller in the scheme of eqn. 7.75, i.e. filling the subdiagonal of the $\boldsymbol{P}$ matrix with $a=-1$ and omitting the matrix $\boldsymbol{W}$, a gain of 4 dB below the optimum was obtained on average. The use of a conventional three pulse canceller resulted in a further reduction of the gain. The use of a Wiener filter with $M=3$ but without $\boldsymbol{W}$ from eqn. 7.75 leads to an average gain of about 4 dB below the optimum.

The gain plotted over the unambiguous Doppler region for weather clutter (Fig. 7.47) shows a good approximation for the Wiener filter with a length of $M=6$ to the optimum processing (1 dB loss on average). When the normalisation by matrix $\boldsymbol{W}$ in eqn. 7.75 is omitted, the loss increases for the same filter to 7 dB on average. The reason for that is the greater sensitivity of the prediction error from the stagger pattern.

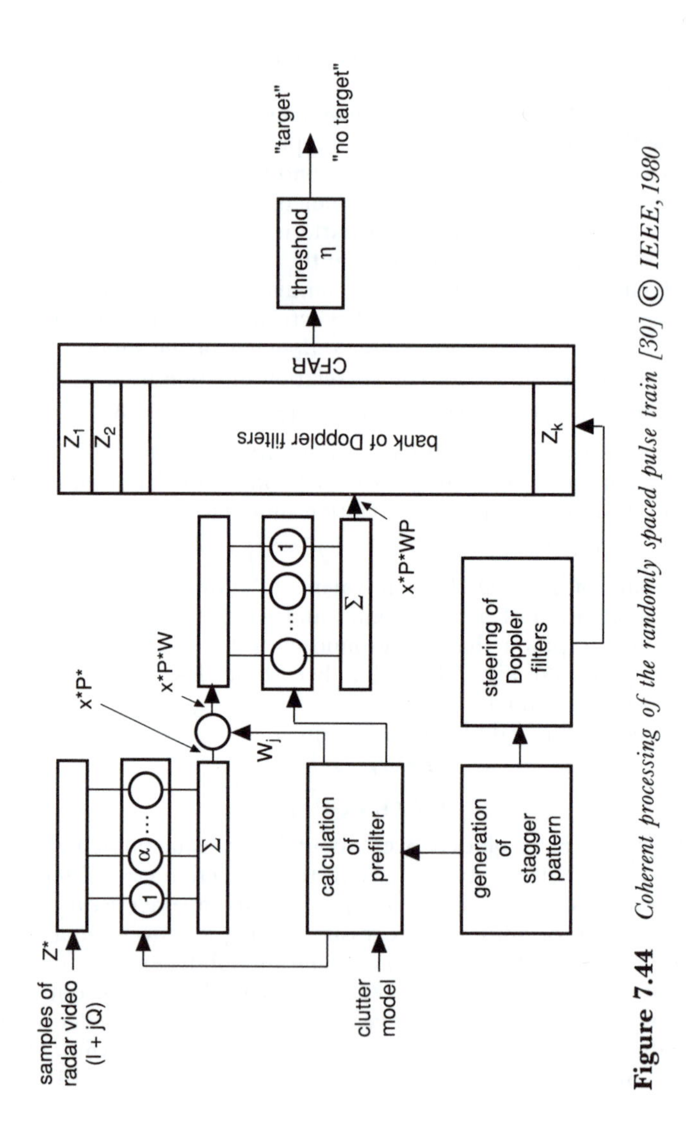

Figure 7.44 *Coherent processing of the randomly spaced pulse train [30] © IEEE, 1980*

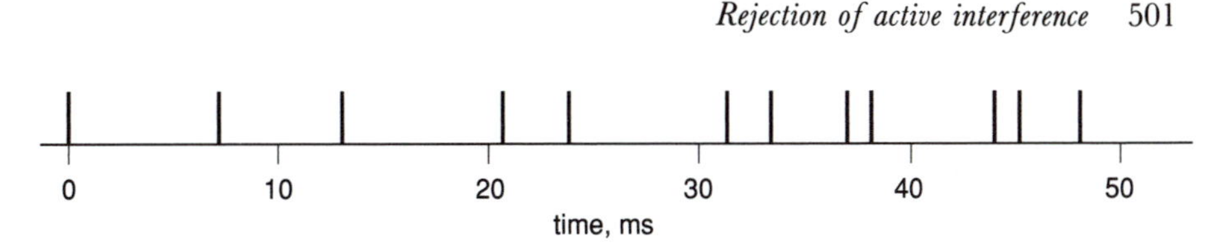

Figure 7.45 *Staggered pulse train used in simulation [30] © IEEE, 1980*

In Fig. 7.48, the gain in a situation of mixed ground and weather clutter is plotted for optimum processing and its approximation by a Wiener filter ($M=6$). This indicates that, for complex realisation of the prefilters and the equivalent clutter model, the clutter suppression within the staggered pulse train is not restricted only to the Doppler shifted region.

Fig. 7.49 shows the variation of output power with the Doppler frequency for constant target signal input power and the signal processor according to eqn. 7.75 in the ground clutter situation. The use of a constant threshold for all Doppler filters would lead to considerable losses. A CFAR operation is needed for each of the Doppler filters in range direction to obtain nearly the same false alarm rate out of each Doppler filter.

In the case of weather clutter, owing to the higher clutter fluctuation, the correlation between the echoes decreases more rapidly. The variation of the output power with constant target signal input power has proved to be very strong in the weather situation. This is undesirable with regard to the required dynamic range in the Doppler filter bank and the CFAR procedure. Therefore it is recommended to be used for weather clutter suppression of the shape of the coherent pulse trains which exhibit a short dwell time.

There are several disadvantages concerned with this scheme of coherent processing:

(i) The ratio of pulses for coherent integration to the total number of pulses is very low. This can be permitted in phased array radar but cannot be

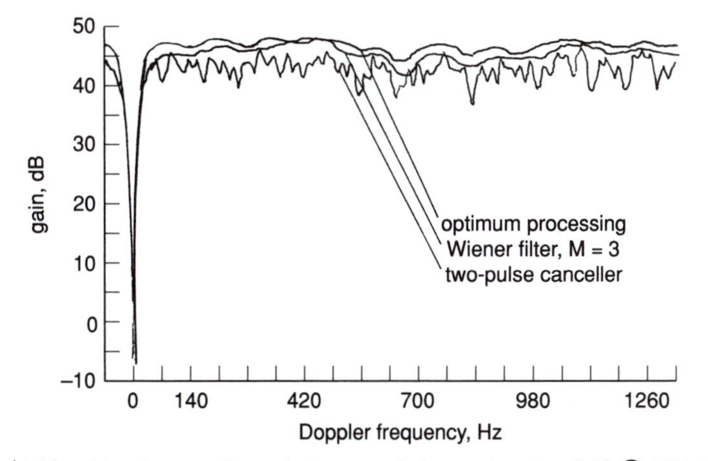

Figure 7.46 *Signal processing gain in ground clutter situation [30] © IEEE, 1980*

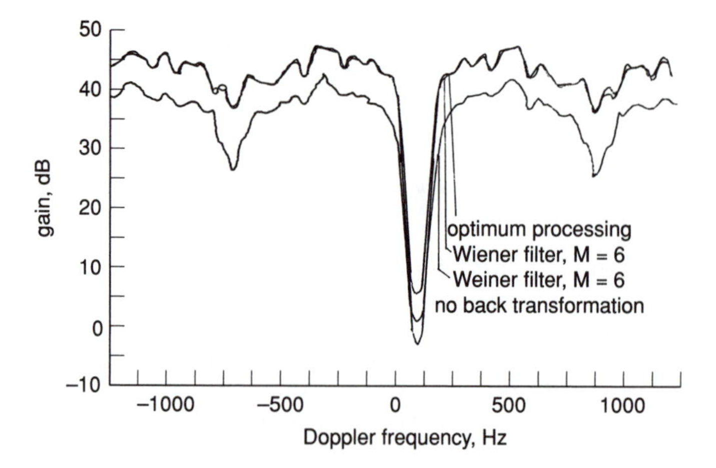

Figure 7.47 *Signal processing gain in weather clutter situation [30] © IEEE, 1980*

permitted in ordinary 2-D search radar. In the latter case, the number of pulses within the beamwidth is very low, unless in the high PRF radar case. Even in high PRF radar, its efficiency is very low if only one channel of eight is used for coherent integration. One way to solve this problem is to integrate the pulses in each channel coherently, and then integrate their output non-coherently. However, it has to be noted that the same target will appear in different Doppler filters in different channels owing to the different Doppler frequency.

(ii) The correlation of clutter echoes decreases with increasing time distance between the echoes, thus leading to a reduction of possible clutter suppression with the pulse train. In this example, the shortest interval between two adjacent pulses within the coherent pulse train is 0.4 ms,

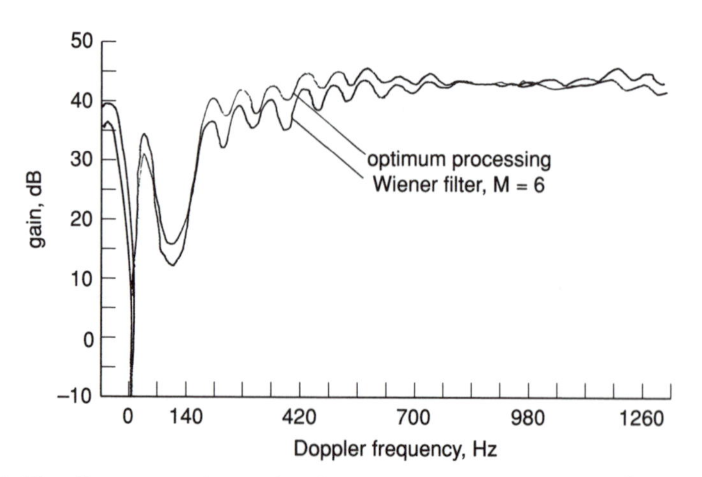

Figure 7.48 *Gain in mixed ground and weather clutter situation [30] © IEEE, 1980*

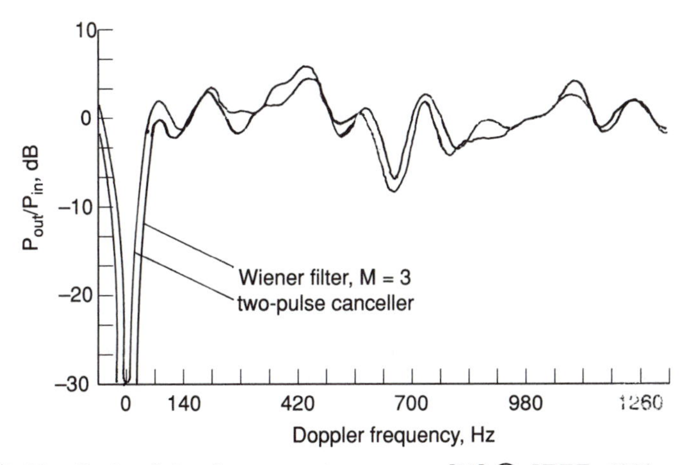

Figure 7.49 *Ratio of signal output to input power [30] © IEEE, 1980*

Table 7.7 *Order of transmission randomly selected from six frequency channels*

	B_1	B_2	B_3	B_4	B_5	B_6	B_7	B_8	B_9	B_{10}	B_{11}	B_{12}
f_1	6	9	16	24	29	33	39	48	53	57	64	70
f_2	3	12	17	21	28	34	38	46	51	55	61	68
f_3	2	10	15	19	25	32	41	44	50	59	62	67
f_4	5	8	14	23	26	31	37	47	49	58	66	72
f_5	1	11	13	22	30	36	40	43	54	56	63	71
f_6	4	7	18	20	27	35	42	45	52	60	65	69

while the longest one is 8 ms. One way to solve this problem is to divide the pulses in blocks of equal length. The length of the block is equal to the number of channels, but the order of transmission is selected randomly. One example of a 12 coherent pulse train from six channels is shown in Table 7.7. It can be seen from Table 7.7 that the average time interval between two adjacent pulses in a coherent pulse train is equal to the number of channels. This can reduce the de-correlation effect of clutter fluctuation considerably.

(iii) In this processing algorithm, the clutter model must be assumed in advance. If the real clutter is different from the assumed model, a worse result for clutter suppression will be obtained.

(iv) The extended dwell time leads to a reduction of the Doppler filter width and to an increase in the number of Doppler filters for a given maximum target velocity. This does not necessarily mean an increase in the number of blind speeds due to the randomly staggered PRF assumed. Unfortunately this goes with an increase of the Doppler filter sidelobes, which in turn may cause signal processing losses where the sidelobes coincide with clutter residue.

The problem of sidelobes is very serious in this algorithm. Computer simulation was performed by the author to study this problem. It is assumed that three clutters, namely ground clutter, weather clutter and chaff clutter, exist simultaneously. The spectrum and the parameters of these clutters are shown in Fig. 7.50.

To simplify the calculation of the coefficients of the filter, a simplified clutter model is assumed. The spectrum is shown in Fig. 7.51.

The result of the signal spectrum from one coherent channel calculated by Buhring's algorithm for a target with a velocity of 120 m/s is shown in Fig. 7.52.

It can be seen that the sidelobes are too high for detection. Sometimes the sidelobe level is higher than the signal and depends on the pattern of spacing. A modified algorithm has been developed. Fig. 7.53 shows the result in the same conditions.

7.5 Frequency synthesiser

In a radar system which has the compatibility of frequency agility and MTI or MTD, another important aspect is the frequency synthesiser; since the frequency agility requires that the radar can change its carrier frequency very fast, while the MTI or MTD requires that the radar can maintain its carrier frequency very stable during the coherent processing interval (CPI). These two requirements are in conflict with each other. It is impossible to realise this requirement with a power oscillator in a non-coherent frequency agility radar.

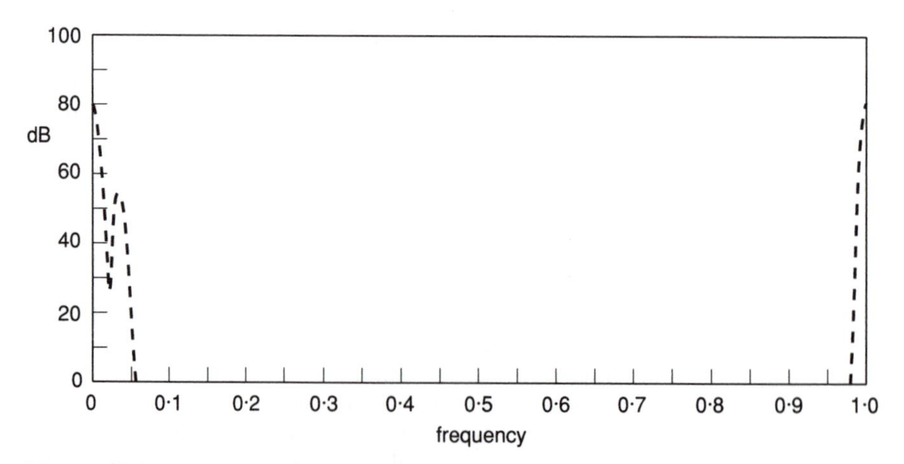

Figure 7.50 *Spectrum of three clutters*

$\sigma_g = 0.0033$ $C_g/N = 60$ dB

$\sigma_w = 0.0033$ $C_w/N = 35$ dB $f_w = 0.03$

$\sigma_c = 0.0033$ $C_c/N = 20$ dB $f_c = 0.04$

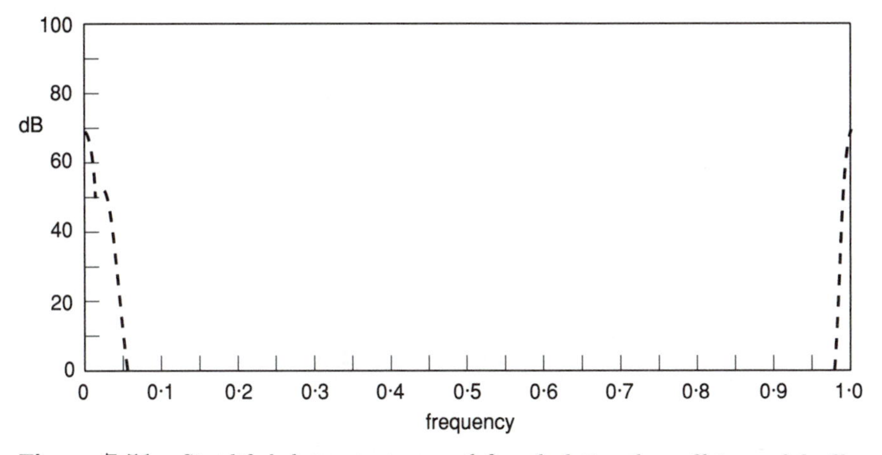

Figure 7.51 *Simplified clutter spectrum used for calculating the coefficients of the filter*

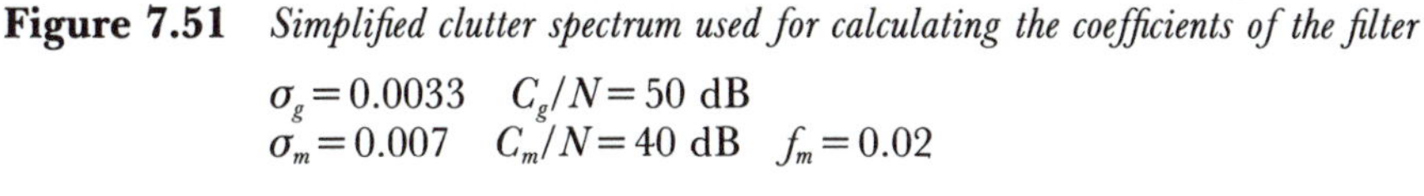

$\sigma_g = 0.0033 \quad C_g/N = 50$ dB
$\sigma_m = 0.007 \quad C_m/N = 40$ dB $\quad f_m = 0.02$

The only way to solve the problem is to develop a highly stable fast switch frequency synthesiser and then to amplify the signal output from the frequency synthesiser with a power amplifier. This is the full coherent radar system.

7.5.1 *Basic requirements of frequency synthesisers*

The basic requirements of frequency synthesisers are as follows:

7.5.1.1 *High stability*

The stability of a frequency synthesiser can be expressed in time domain or in frequency domain. When it is expressed in time domain, it is measured by

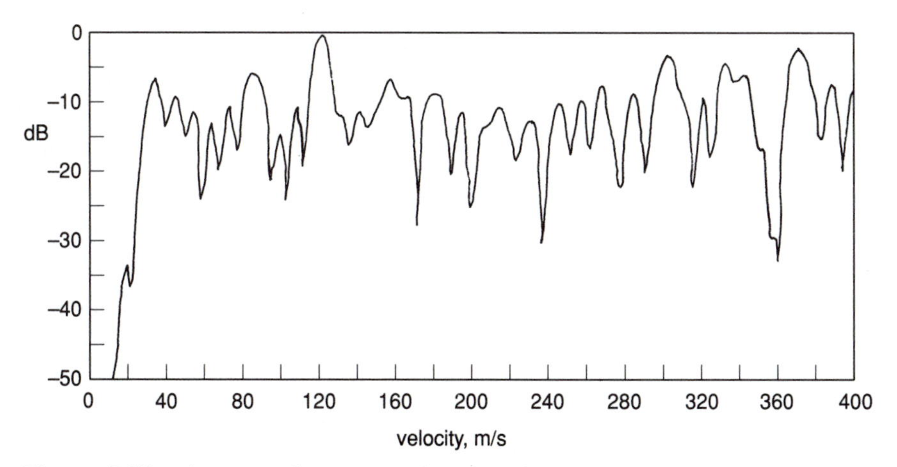

Figure 7.52 *Spectrum of output signal in three clutters*

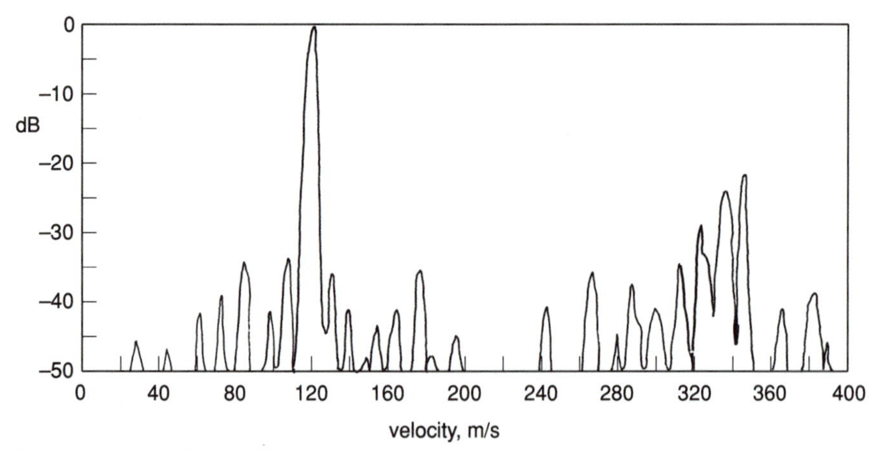

Figure 7.53 *Spectrum of signal from modified algorithm*

relative change of frequency within a definite time period. In radar applications, one is interested in short term frequency stability. This is due to the coherent processing interval being a rather short time period of several radar pulse periods, e.g. several milliseconds to several tens of milliseconds. We often use milliseconds as the time periods for the measurements of frequency stability. The typical value of frequency stability in time domain is of the order of 10^{-9}–10^{-11}/ms, which depends on the carrier frequency of the oscillator and the requirement of improvement factor of the radar.

When the frequency stability is expressed in frequency domain, phase noise is often used.

Consider a frequency source whose instantaneous output signal is

$$e(t) = [E_c + A(t)] \sin[\omega_c t + \theta(t)] \tag{7.77}$$

where E_c and ω_c are the nominal amplitude and angular frequency of the output, respectively. This equation indicates that there is a concentration of noise power surrounding an RF signal. A typical spectral distribution of such a signal is shown in Fig. 7.54. In general, both AM noise and FM noise, $A(t)$ and $\theta(t)$, are present, accounting for the lack of symmetry in the envelope of the spectrum. If it is assumed that

$$A(t) \ll 1 \tag{7.78}$$

and

$$\frac{d\theta(t)}{dt} = \theta(t) \ll 1 \tag{7.79}$$

for any time t, the fractional instantaneous frequency deviation from nominal frequency may be defined as

$$y(t) \equiv \theta(t)/\omega_c \tag{7.80}$$

A definition for the measure of frequency stability is the spectral density, $S_y(f_m)$, of the function $y(t)$, where the spectrum is considered to be one-sided on a per hertz basis. The function $S_y(f_m)$ has the dimensions of 1/Hz.

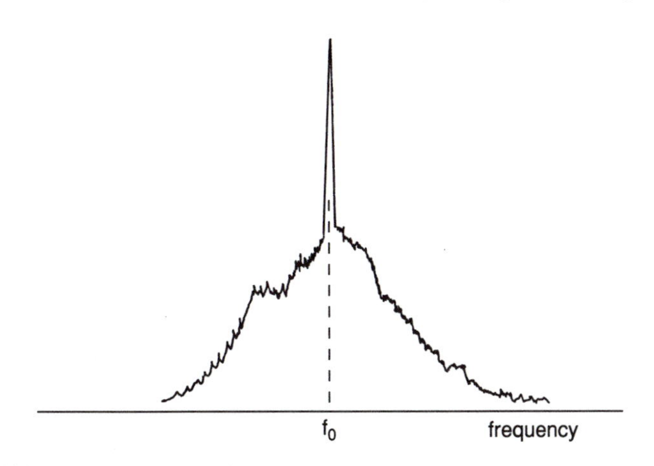

Figure 7.54 *Noise spectrum of an RF signal*

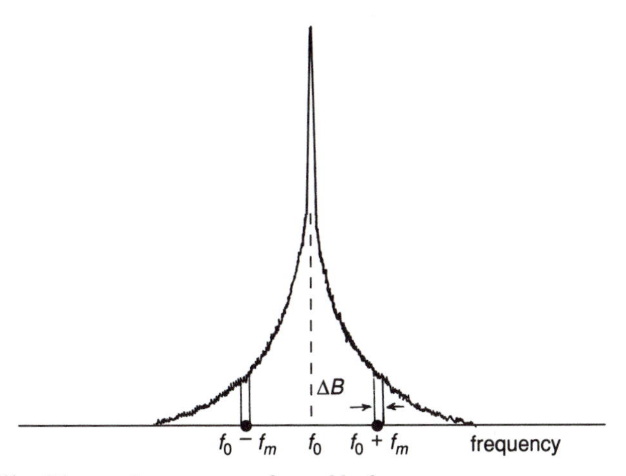

Figure 7.55 *Phase-noise spectrum of a stable-frequency source*

In general, FM noise is the dominating one. The discussion here is limited to considerations of FM noise only.

The FM noise spectrum of an RF signal is shown in Fig. 7.55. Although the noise distribution on each side of the signal is continuous, one can subdivide the spectrum into a large number of strips of width ΔB (the measured passband) located a distance f_m away from the signal, and view the energy in ΔB as being caused by a sinusoidal frequency-modulated signal centred on ΔB with a deviation proportional to the amplitude of the spectrum at f_m.

It has been customary to characterise the noise performance of a signal source as the ratio of the measured power in one noise sideband component, on a per hertz of bandwidth spectral density basis, to the total signal power:

$$\alpha(f_m) \equiv \frac{\text{power density (one sideband, phase only)}}{\text{power (total signal)}} \Bigg/ \text{Hz}$$

In high performance coherent radar systems, the phase noise of a frequency synthesiser is of the order of −90 to −120 dB/Hz at 1 kHz from the carrier frequency.

7.5.1.2 *Fast switching time*
The switching time for a frequency synthesiser used in a frequency agility radar must be less than 5–10% of the repetition period of the radar. The pulse repetition period of a radar depends on its application. It is in the range of 100 μs to 1 ms. Therefore, the switching time of the frequency synthesiser is in the range of 5–50 μs.

There is a transient process in the switching of a frequency synthesiser. Therefore, the switching time of a frequency synthesiser is often measured as the time needed for the phase to settle to within 0.1 rad of the final value. This value must less than 5–50 μs for most radars.

7.5.1.3 *Low spurious and harmonic output*
Since there are often many frequency multipliers, dividers and mixers within the frequency synthesiser, some spurious and harmonics will be present at the output of the frequency synthesiser. The level of these spurious and harmonics are measured as dB less than the carrier (dBc). It must be less than −60–90 dBc for most frequency synthesisers.

7.5.1.4 *Wide working temperature range*
The working temperature range must be the same working temperature range as the radar. It depends on the environment of the radar and is in the range of −10 to +50°C for ground based radar and from −55 to +125°C for airborne based radar.

7.5.1.5 *Rigid and compact*
The short term stability of a crystal oscillator is affected by the mechanical vibration. Special shock absorbers must be mounted around the crystal oscillator.

There are two types of frequency synthesiser: direct frequency synthesiser and indirect frequency synthesiser. We will discuss them in detail.

7.5.2 *Direct frequency synthesiser*

The earliest frequency synthesiser is the direct frequency synthesiser.

The direct frequency synthesiser synthesises the required frequency components with crystal oscillator. RF switches, frequency dividers, frequency multipliers and bandpass filters.

Fig. 7.56 shows a direct frequency synthesiser constructed with frequency dividers, and Fig. 7.57 shows a frequency synthesiser constructed with frequency multipliers. The required frequencies are selected by RF switches. Spurious and harmonics are rejected with bandpass filters.

Frequency mixers are needed when many frequencies are synthesised. Fig. 7.58 shows that 1000 frequencies are synthesised from 10 reference frequencies.

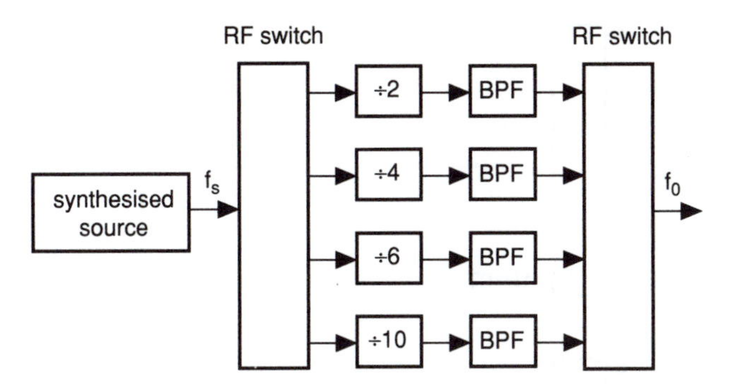

Figure 7.56 *Direct frequency synthesiser constructed with switched frequency dividers and filters [32]. Reproduced by permission of Prentice–Hall*

The main disadvantage of this direct frequency synthesiser is that too many reference frequencies are needed. To solve this problem, a binary-decimal frequency synthesiser was proposed [33]. The block diagram of this frequency synthesiser is shown in Fig. 7.59.

This frequency synthesiser is controlled by a binary-decimal signal. This signal is shown as switches K_1 to K_4 in Fig. 7.59. Only three reference frequencies, i.e. f_1, f_2 and f_i, are needed, where f_i is the input frequency, and f_1, f_2 are two reference frequencies. The relationship between f_1 and f_2 can be expressed as

$$f_2 = f_1 + 10\Delta f \tag{7.81}$$

where Δf is the minimum frequency increment. K_1 to K_4 are frequency selection switches. All the mixers are down converters. If switch K_1 is switched from position '0' to position '1', the frequency of the output of first mixer will be decreased by $10\Delta f$. This frequency change will be reduced by 10 at the output, since the total number of frequency division is equal to $\div 2 \div 2 \div 2 \div 5 \times 4 = \div 10$. This means that the frequency of the output f_o changes $1\Delta f$ when K_1 is switched from '0' to '1'; f_o changes $2\Delta f$ when K_2 is switched from '0' to '1'; f_o changes $4\Delta f$ when K_3 is switched from '0' to '1'; f_o changes $8\Delta f$ when K_4 is switched from '0' to '1'. In other words, the weights of four switches are equal to 1, 2, 4 and 8, respectively. If we choose the input frequency f_i as

$$f_i = (f_1 - 10\Delta f)/1.8 \tag{7.82}$$

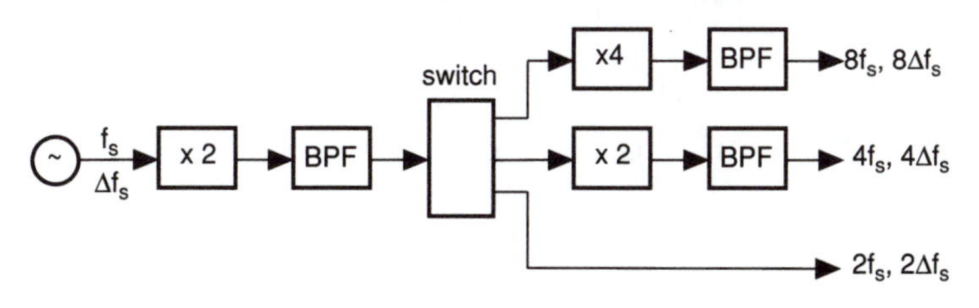

Figure 7.57 *Direct frequency synthesiser constructed with switched frequency multipliers and filters [32]. Reproduced by permission of Prentice–Hall*

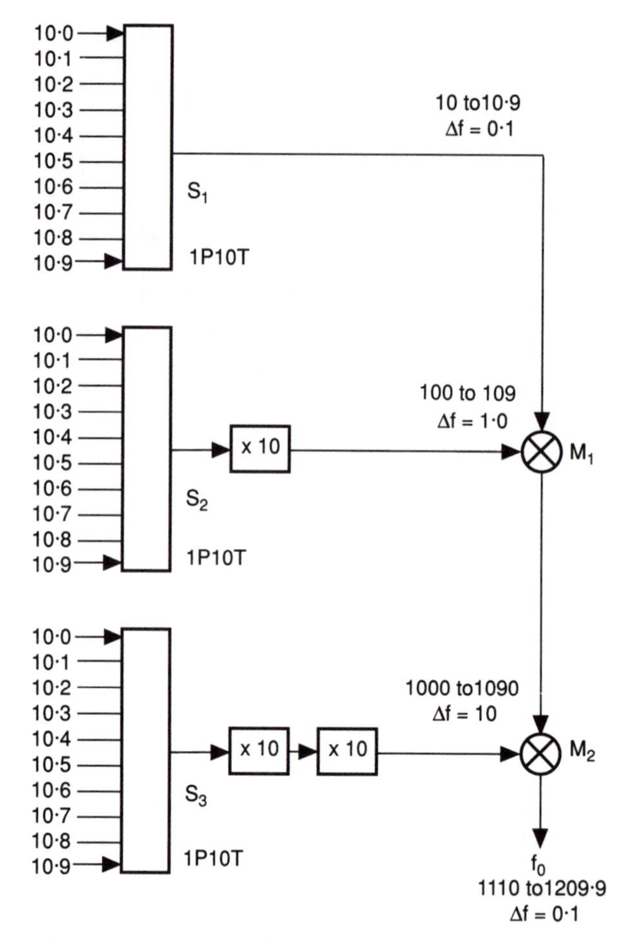

Figure 7.58 *Direct frequency synthesiser with 1000 selectable frequencies [32]. Reproduced by permission of Prentice–Hall*
Note: All frequencies in MHz.

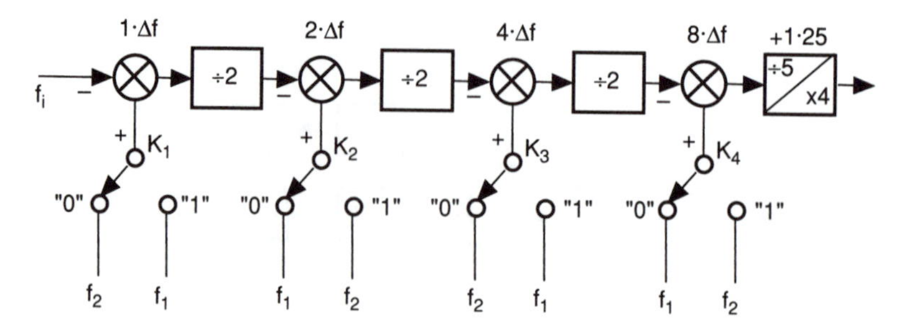

Figure 7.59 *Block diagram of binary-decimal frequency synthesiser [33]*

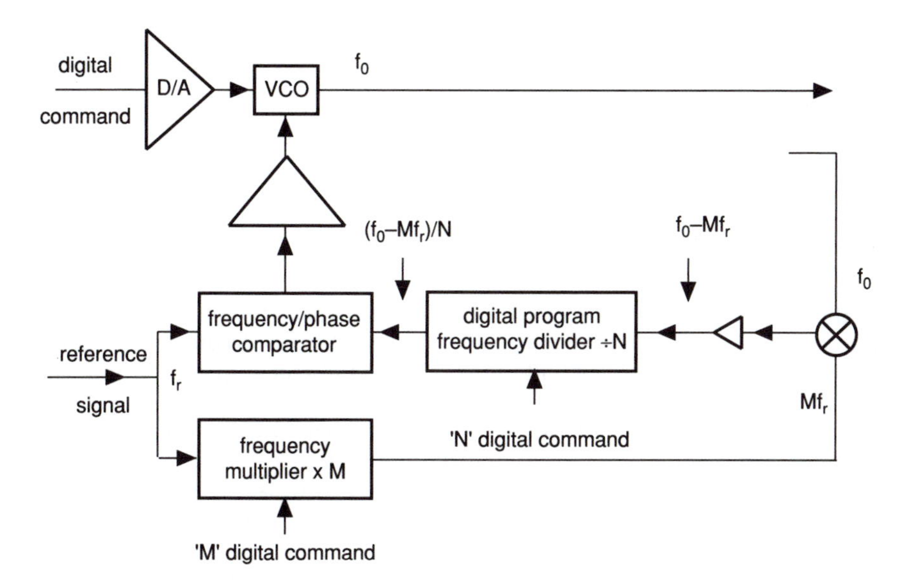

Figure 7.60 *Block diagram of indirect frequency synthesiser implemented with digitally controlled PLL*

then the frequency of the output f_o is equal to the frequency of the input f_i. This means that the frequency will not be changed after passing through this building block. For example, if we hope that $\Delta f = 0.1$ MHz, and we select $f_1 = 64$ MHz, from eqn. 7.81 we obtain $f_2 = 65$ mHz, and from eqn. 7.82 we obtain $f_1 = 35$ MHz. The frequency of the output is also equal to 35 MHz. But any change of f_i will be reduced by 10 at the output. Therefore, it is very convenient to construct a frequency synthesiser if several of these building blocks are cascaded in serial.

7.5.3 *Indirect frequency synthesiser*

Another type of frequency synthesiser is the indirect frequency synthesiser. Phase lock loop (PLL) is used in this type of frequency synthesiser to maintain the coherence between the output signal and the reference signal from the crystal oscillator.

There are two types of PLL: analogue PLL and digital PLL. In principle, PLL belongs to the analogue circuit, since the phase detector, which is used to detect the phase difference between the reference signal and the signal generated by voltage controlled oscillator (VCO), is an analogue one. 'Digital' means that the frequency divider and/or multiplier within the PLL and the VCO are controlled by digital signals. Since it is more convenient to control the frequency synthesiser with a digital signal, the digital PLL is the more commonly used one. Fig. 7.60 shows the block diagram of an indirect frequency synthesiser implemented with digitally controlled PLL.

The output frequency of this frequency synthesiser f_o is equal to

$$f_o = Nf_r + Mf_r \tag{7.83}$$

The digital command controls the frequency of the VCO to the approximate value of desired frequency f_o. At the same time, a pair of appropriate values of M and N, which can satisfy eqn. 7.83, are selected to control the frequency multiplier and frequency divider. Since the output frequency of the frequency divider is close to f_r, the PLL is locked to f_r.

7.5.4 *Comparison between direct and indirect frequency synthesisers*

7.5.4.1 *Phase noise*

In the frequency range close to the carrier frequency, the phase noise of the frequency synthesiser mainly depends on the reference frequency oscillator. Therefore, the phase noise of both the direct and indirect frequency synthesisers are almost the same. When frequency offset increases, the phase noise of these two types of frequency synthesiser becomes different. In the direct frequency synthesiser thermal noise of the amplifier and frequency multiplier becomes the dominant factor. Since short term frequency stability caused by thermal noise is nearly constant, the phase noise will increase with frequency offset.

While in the indirect frequency synthesiser, the phase noise beyond the phase lock bandwidth mainly depends on the VCO. Although it is worse than the crystal oscillator, it is better than the direct frequency synthesiser.

In MTI or MTD systems, the improvement factor is limited by short term frequency stability of the frequency synthesiser. Therefore, the phase noise of the direct frequency synthesiser or indirect frequency synthesiser is of the same order, in the case when they use the same type of crystal oscillator.

7.5.4.2 *Switching time*

The switching time of the direct frequency synthesiser depends directly on the switches. It is very easy to make it as short as several microseconds. The switching time of the indirect frequency synthesiser consists of two parts: the settling time of the VCO and the lock time of the PLL. The former is less than a few tenths of a microsecond and can be neglected. The latter depends on the bandwidth of the PLL. However, the wider the bandwidth, the greater the phase noise will be. Therefore, a compromise must be made. It is impossible to obtain a settling time less than 50 μs for an indirect frequency synthesiser without losing frequency stability.

7.5.4.3 *Spurious and harmonics*

Spurious and harmonics can be very easily generated in direct frequency synthesisers, since there are too many mixers and frequency multipliers. Well designed filters are needed to achieve good spurious suppression, whereas the indirect frequency synthesiser has good spurious suppression capability with the PLL.

7.5.4.4 *Complexity*

The direct frequency synthesiser is more complex than the indirect frequency synthesiser with the same frequency resolution. Many mixers, filters and switches are required in the direct frequency synthesiser. Frequency multipliers are needed to multiply the frequency to microwave band. While in the indirect frequency synthesiser, the VCO can operate in microwave directly. No frequency multiplier is needed, except for short MW.

7.6 References

1 BRICK, D.B., and GALEJIS, J.: 'Radar interference and its reduction', *The Sylvania Technologist*, 1958, **11**, pp. 96–108
2 JOHNSTON, S.L.: 'Radar electronic counter-countermeasures', *IEEE Trans.*, 1978, **AES–14**, pp. 109–117
3 JOHNSTON, S.L.: 'Radar ECM and ECCM' *in* SCANLAN, M.J.H. (Ed.): 'Modern radar techniques', (Collins, 1987) chap 4
4 HANSEN, R.C.: 'Aperture theory' *in* HANSEN, R.C. (Ed.): 'Microwave scanning antennas: Vol. 1, Apertures' (Academic Press NY, 1964.)
5 SHERMAN, J.W.: 'Aperture-antenna analysis' *in* SKOLNIK, M.I. (Ed.): 'Radar handbook' (McGraw-Hill Book Co., NY, 1970)
6 TAYLOR, T.T.: 'Design of circular apertures for narrow beamwidth and low sidelobes', *IRE Trans.*, 1960, **AP–8**, pp. 17–22
7 HANSEN, R.C.: 'Tables of Taylor distributions for circular aperture antennas', *IRE Trans.*, 1960, **AP–8**, pp. 23–26
8 HANSEN, R.C.: 'A one-parameter circular aperture distribution with narrow beamwidth and low sidelobes', *IEEE Trans.*, 1976, **AP–24**, pp. 477–480
9 RUDGE, A.W., and ADATIA, N.A.: 'Offset-parabolic-reflector antennas: A review', *Proc. IEEE*, 1978, **66**, p. 1592
10 FANTE, R., *et al.*: 'A parabolic cylinder antenna with very low sidelobe', *IEEE Trans.*, 1980, **AP–28**, pp. 53–59
11 MARR, J.D.: 'A selected bibliography on adaptive antenna arrays', *IEEE Trans.*, 1986, **AES–22**, pp. 781–798
12 APPLEBAUM, S.P.: 'Adaptive arrays', *IEEE Trans.*, 1976, **AP–24**, pp. 585–598
13 BUCCIARELLI, T., *et al.*: 'The Gram-Schmidt sidelobe canceller'. IEE Radar '82, pp. 486–490
14 BALDWIN, P.J., DENISON, E., and O'CONNOR, S.F.: 'An experimental analogue adaptive array for radar applications', IEEE 1980 Int. Radar Conf., pp. 271–278
15 FIELDING, J.G., *et al.*: 'Adaptive interference cancellation in radar systems'. IEE Radar '77, pp. 212–217
16 FARINA, A., and STUDER, F.A.: 'Evaluation of sidelobe-canceller performance', *IEE Proc.*, 1982, **129**, Pt.F, pp. 52–58
17 GIRAUDON, C.: 'Optimum antenna processing: a modular approach'. Proc. of NATO Advanced Study Institute on signal processing and underwater acoustic, Porto Venere, Italy, 1976
18 WHITE, W.D.: 'Adaptive cascade networks for deep nulling', *IEEE Trans.*, 1978, **AP–26**, pp. 396–402
19 KRETSCHMER, F.K., and LEWIS, B.L.: 'A digital open-loop adaptive processor', *IEEE Trans.*, 1978, **AES–14**, pp. 165–171
20 BRISTOW, T.A.: 'Experimental open-loop canceller for radar', *IEE Proc.*, 1983, **130**, Pt.F, pp. 109–113
21 STONE, D.: 'Open-loop adaptivity for rotating antennas', *IEE Proc.*, 1983, **130**, Pt.F, pp. 114–117
22 LU, Z.L., LIANG, D.N., and LEE, Q.S.: 'Applications of open-loop sidelobe canceller to high-rotation-rate radar', CIE 1986 Int. Conf. on Radar, pp. 624–629
23 GABRIEL, W.F.: 'Spectral analysis and adaptive array supperresolution techniques', *Proc. IEEE*, 1980, **68**, pp. 654–666
24 HUNG, E.K.L.: 'Adaptive sidelobe cancellation by linear prediction'. IEEE 1985 Int. Radar Conf., pp. 286–291
25 MAO, Y.H.: 'Frequency agility radar' (Defence Industry Press, Beijing, 1981) (in Chinese). See also 'Frequency agility radar'. AD–A124003, Dec. 1982 (in English)
26 'An experimental twin pulse MTI S-band radar with pulse compression', RRE TN–743, AD–704, 687, 1969
27 NOURSE, O., and CROSSFIELD, M.: 'Quad-pulse radar-A frequency agile radar with Doppler capability'. IEE Int. Conf. Radar '82, London, pp. 186–190
28 TAYLOR, J.W. Jr., and SINCLAIR, A.L.: 'Dual frequency transmission apparatus for frequency-agile radar system utilizing MTI techniques'. US Patent No. 4, 155 088, 15 May 1979

29 PETROCCHI, G., RAMPAZZO, S., and RODRIGUEZ, G.: 'Anti-clutter and ECCM design criteria for a low coverage radar'. Int. Conf. on Radar, Paris, 1978, pp. 194–200
30 BUHRING, W.: 'Coherent signal processing in a frequency agile jittered pulse radar'. IEEE 1980 Int. Radar Conf., pp. 188–193
31 BRENNAN, L.E., and REED, I.S.: 'Optimum processing of unequally spaced radar pulse trains for clutter rejection', *IEEE Trans.*, 1968, **AES–4**, pp. 474–477
32 RONALD, C.S.: 'Microwave frequency synthesizers', (Prentice–Hall, Inc. NJ, 1987)
33 PAPAIECK, R.J. and COE, R.P.: 'New technique yields superior frequency synthesis at low cost', *EDN*, 20 Oct, pp. 41–44

Chapter 8

Architectures and implementation of radar signal processor

Prof. Y. H. Mao

8.1 Introduction

The performance of radar systems has improved tremendously within the last decade. It is mainly due to the application of digital signal processing and the progress of integrated circuits. The gradual change from small scale integration to very large scale integrated (VLSI) techiques, and the use of microprocessors and special purpose digital signal processing elements have led to dramatic improvements in speed, size, cost and power consumption. Futhermore, it means that the real-time implementation of complex algorithms becomes possible.

The application of digital signal processing in radar systems began from signal detection in video frequency and is now developing into signal processing in intermediate frequency. It seems possible to process the RF signal directly owing to the development of GaAs devices and Josephson junction devices. The percentage of digitised parts in a radar system becomes larger and larger, since it is a cost-effective way to improve the performance of the whole radar system.

'System on silicon' (SOS) or 'System on chip' is very attractive to the radar system designer. For example, the volume of a transistorised radar ranging unit is a rack, and the power consumption is several hundred watts. The same ranging unit made from small scale and medium scale IC occupies only one printed circuit board, and the power consumption is in the range of several tens of watts. Now it is possible to realise the whole radar ranging system in a single chip with LSI technology, the power consumption is less than one watt. Recently, more complex MTD systems can be realised with a single DSP chip plus some high speed RAM and ROM. The volume is no more than one printed board.

Owing to the development of VLSI technology, it is now possible to integrate more than one million transistors into a single chip. More complex systems can be integrated within a single chip, but the main problem is that radar system designers are not familiar with the design of VLSI, especially the layout design. Even though the development of gate arrays simplifies the design procedure, the problem still exists. It can be predicted that there is great potential to employ VLSI technology in radar systems.

Owing to the large bandwidth of radar signals and the requirement of real-time processing, another important specification of the radar signal processor is

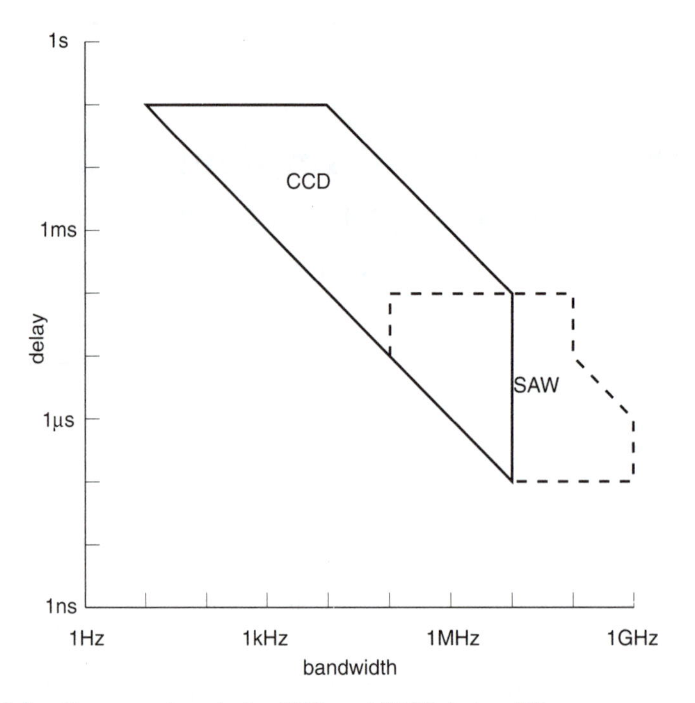

Figure 8.1 *Parameter bounds for CCD and SAW devices [1]*

the speed of processing. Ten years ago, CCD and SAW devices were dominant in many radar signal processing fields owing to their large time–bandwidth product. Fig. 8.1 [1] shows the parameter bounds for CCD and SAW devices. CCD is more attractive owing to the fact that no expensive A/D converter is needed. Mavor and Grant [1] gave a comparison of the performance between CCD and digital circuits which is listed in Table 8.1.

However, the main disadvantage of CCD is the lack of flexibility, which has not been listed in Table 8.1. Dedicated CCD chips have to be designed for different purposes. Low transfer efficiency limits the number of stages. In addition, limited dynamic range and poor linearity prevent its use in some applications. The main uses of CCD in radar signal processing are as analogue delay lines in low-cost MTI, MTD. Tapped delay line can be used in cell average or nonparametric CFAR detector [2]. SAW devices are mainly used as pulse compressors in radar signal processing.

Both CCD and SAW devices can be used in signal transformation, such as Fourier, Hilbert, prime, Hadamard, discrete cosine, slant transformations, in radar signal processing. Fourier transforms are of fundamental importance in signal processing. CCD and SAW devices can realise Fourier transforms by means of analogue chirp-Z transform (CZT) algorithms. Mavor and Grant [1] gave a comparison of parameter bounds for CCD, SAW, and digital techniques against system requirements, which is shown in Fig. 8.2. As there are 10 basic operations in an FFT butterfly computation and $(N/2)\log_2 N$ butterflies per N-point FFT, the computational rate required to perform real-time transforms

Table 8.1 *Comparison of typical performance characteristics of CCD and digital techniques for signal processing [1]*

Parameter	CCD (analogue)	Digital
Information rate	1 kHz–50 MHz	0–200MHz
Delay time	<1 s	Infinite
Time–bandwidth product	<1000 (analogue)	Cost-limited
Linearity (harmonic content)	<1%	Controlled by number of bits N in A/D converter
Dynamic range	~80 dB	2^N:1
Sidelobe level	40–50 dB	2^N:1
Power consumption	~10 μW/bit	~100 μW/bit
Size	Single IC	Multiple IC
Complexity	Medium	Complex
Cost	Low in production	High for large bandwidth

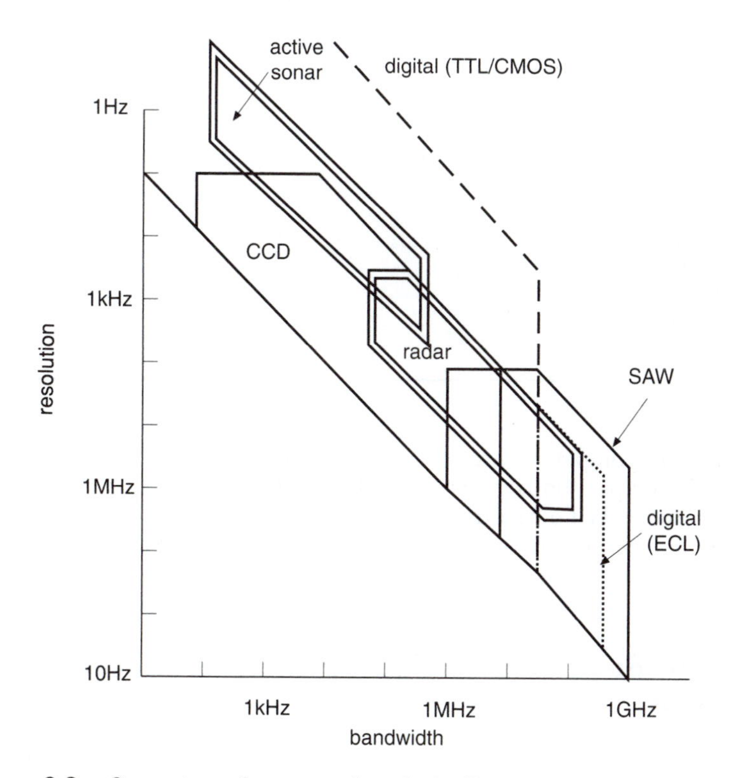

Figure 8.2 *Comparison of parameter bounds for Fourier transform processors based on CCD, SAW and digital techniques against system requirements [1]*

Table 8.2 *Computational speed comparison (in operation/s) [1]*

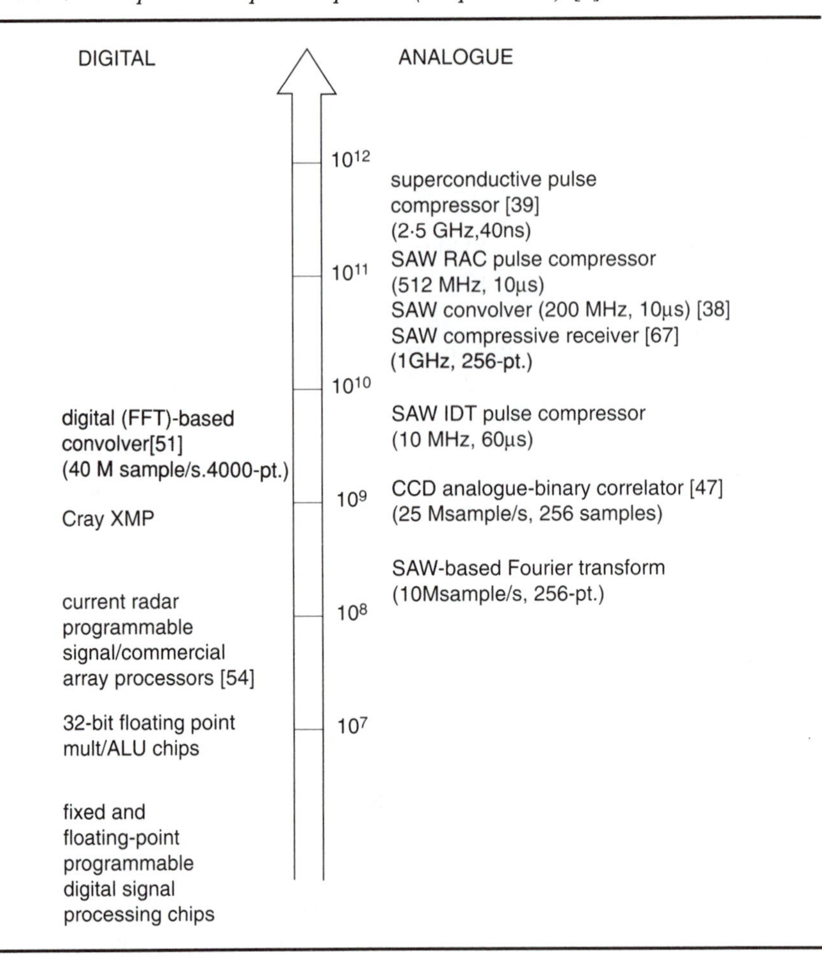

can be directly compared through the equivalent number of operations per second. This comparison is given in Table 8.2 [1].

Although the SAW Fourier transform processor is superior in bandwidth, it is used only in special purpose radar signal processing. We will limit our discussion on digital processors.

8.2 Application of digital signal processing to radar systems

Digital techniques have been used in radar systems for a long time. The radar timing unit and ranging unit were the first digitised units in radar systems. The radar display, servo-system and monitor system were digitised later. We will discuss not the application of digital techniques, but the application of digital signal processing technique in radar systems.

The application of digital signal processing techniques in radar systems can be summarised as follows.

8.2.1 *Radar transmitter*

(i) Adaptive frequency agility (for least jamming or maximal amplitude of echo signal)
(ii) Adaptive waveform generation

8.2.2 *Antenna system*

(i) Synthetic aperture
(ii) Inverse synthetic aperture
(iii) Doppler beamshaping
(iv) Digital beamforming
(v) Adaptive sidelobe cancellation

8.2.3 *Radar receiver*

(i) Matched filters
  (*a*) Digital linear FM (chirp) pulse compression
  (*b*) Phase coded pulse compression
  (*c*) Digital correlator for other waveforms
(ii) Clutter cancellation
  (*a*) Digital MTI
  (*b*) Airborne MTI (AMTI)
  (*c*) Adaptive MTI (ADMTI)
  (*d*) MTD
  (*e*) Adaptive MTI based on spectral estimation
  (*f*) Adaptive MTD

8.2.4 *Digital correction of instability of transmitter*

8.2.5 *Detector*

(i) CFAR detector in time domain
  (*a*) CFAR detector for receiver noise
  (*b*) Parametric CFAR detector for clutter
  (*c*) Non-parametric CFAR detector for clutter
(ii) Non-coherent signal integration
  (*a*) Binary moving window detector
  (*b*) Multi-bit moving window detector
  (*c*) Two-pole filter
(iii) Coherent signal integrator (Doppler filter bank)
(iv) CFAR detector in frequency domain
(v) Sequential detector

8.3 Fundamental requirements of radar signal processor

The fundamental requirements of digital radar signal processors can be summarised as follows.

8.3.1 *Dynamic range*

The required dynamic range of a digital signal processor depends on the system performance specifications. For example, the dynamic range of a digital MTI is determined by the specification of improvement factor. However, the improvement factor of DMTI is different from the improvement factor of the whole radar system. The latter not only depends on the dynamic range but also on the instability of the system, antenna scanning modulation, and spectral spread of the clutter. Once the improvement factor of the system has been determined, the dynamic range of all the parts, from receiver front end to digital signal processor, should not limit this improvement factor specification.

The dynamic range of a digital signal processor is determined by the number of bits of the processor. In general, it is equal to 6 dB/bit. Since the number of bits of a digital processor itself is large enough, such as 16 bits, the dynamic range of a digital processor is mainly determined by the number of bits of the A/D converter. The A/D converter is the 'bottle-neck' of a digital signal processor.

8.3.2 *Number of bits of A/D converter*

Although the dynamic range of a digital signal processor is determined by the A/D converter, it does not mean that the number of bits of the A/D converter is determined by the dynamic range. Some overhead must be considered.

It is well known that the A/D converter will introduce quantisation noise. This quantisation noise will cause, on average, a limit to the improvement factor which is

$$I_{QN} = 20 \log[(2^N - 1)\sqrt{0.75}] \text{ dB} \tag{8.1}$$

This is approximately 1 dB less than 6 dB per bit for $N > 8$. Furthermore, it can be shown that the MTI quantisation loss is a function of the residue standard deviation, σ_R [3]. The relationship between the residue standard deviation and the MTI quantisation loss as determined from simulation for a three pulse canceller is given by Fig. 8.3.

It is seen that, for $\sigma_R < 3$ counts, the quantisation loss increases rapidly. It can be shown that for, $S/C \ll 1$,

$$\sigma_R = \left[\frac{1}{C/N} + \frac{CA}{6}\right]^{1/2} \times 2^N \tag{8.2}$$

where N is the number of bits in the A/D converter, excluding the sign bit. For a specified quantisation loss, the system must be designed to satisfy

$$2^N > \sigma_R / \left[\frac{CA}{6} + \frac{1}{C/N}\right]^{1/2} \tag{8.3}$$

and for $C/N \gg 6/CA$, it is required that

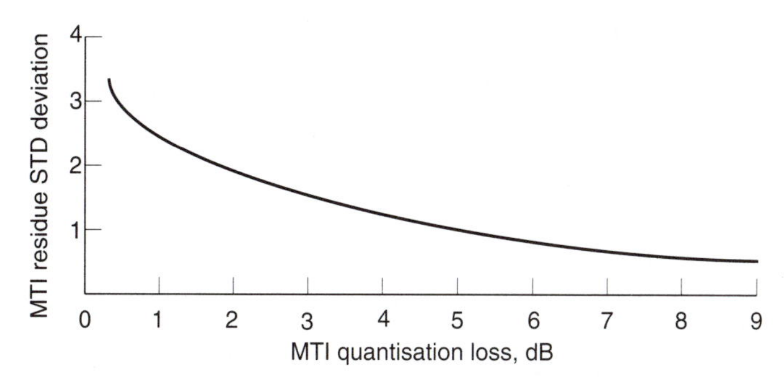

Figure 8.3 *Residue standard deviation versus MTI quantisation loss [3] © IEEE, 1984*

$$N > \frac{10 \log CA}{6} + \frac{10 \log(\sigma_R \sqrt{6})}{3} \qquad (8.4)$$

This means that the number of A/D bits required is one bit per 6 dB of CA plus $10 \log(\sigma_R \sqrt{6})/3$ bit.

The excess number of bits required over the 6 dB/bit rule can be thought of as 'overhead' bits. For a large C/N, the number of overhead bits is a function of MTI quantisation loss, as plotted in Fig. 8.4.

8.3.3 *Sampling rate*

In general, the sampling rate of a digital signal processor can be determined by the Nyquist sampling rate of the signal processed, which is equal to twice the signal bandwidth.

In fact, radar signal processing can be divided into two classes. In the first class, the radar signal is processed in the azimuth direction, such as DMTI, MTD and moving window detector, etc. In this class, the radar signal is sampled at an interval equal to pulse width and then rearranged into azimuth sequences for each range bin. In the second class, the radar signal is processed in the range direction, such as digital pulse compression. In this case, the

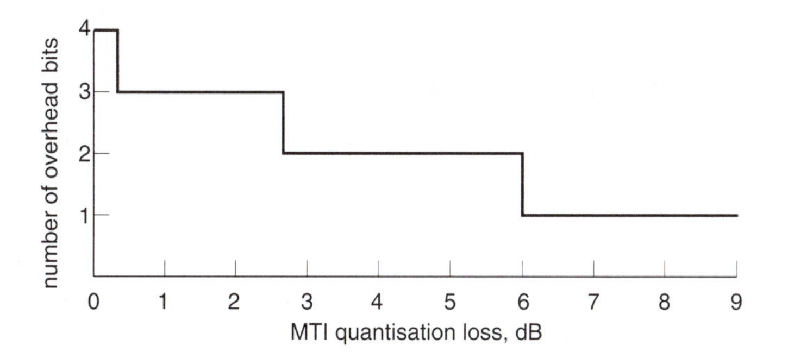

Figure 8.4 *Required number of overhead bits versus MTI quantisation loss [3] © IEEE, 1984*

Nyquist sampling rate must be considered. However, it can be shown that, when quadrature channel is employed, the sampling rate can be one half of the Nyquist rate plus maximum Doppler shift for a linear FM signal.

8.3.4 *Instruction cycle and clock frequency*

The requirement on the instruction cycle depends on the algorithm, the number of data points, and sampling rate of data.

In a general purpose computer, it is measured in 'Million Instructions Per Second' (MIPS).

In the signal processor, the instruction cycle is replaced by 'throughput rate'. Since with most signal processors, including radar signal processors, the data are complex numbers, so the throughput rate is measured in 'Million Complex Operations Per Second' (MCOPS). However, sometimes the 'MCOPS' has another definition in radar signal processors. For example, FFT is the most commonly used algorithm in radar signal processors. The butterfly is a sub-algorithm of the FFT algorithm. In a radar signal processor, which contains the FFT algorithm, the throughput rate is measured in 'Million COmplex radix two butterflies Per Second' (MCOPS). It is well known that one butterfly consists of four real multiplies and six real adds, or one complex multiplication and two complex additions. It means that one (MCOPS) in butterfly operation is equal to 3 MCOPS in general complex number operation.

The instruction cycle of a practical signal processor is determined by the devices and architecture. For example, if DSP chips are used in a radar signal processor, the instruction cycle of the DSP chip is 100 ns. If the required instruction cycle is 10 ns, then 10 DSP chips can operate in parallel. This is because most radar signal processors can be combined by parallel subprocessors. Each processor processes equal amounts of range bin data, since the algorithm is the same for all data.

The clock frequency is related to the instruction cycle but is different from it. In most microprocessors, performing one instruction requires several clock periods. However, in signal processors with Harvard architecture, more than two instructions can be performed within one clock period.

The main clock of the radar signal processor is synchronised with the sampling rate of the A/D converter. In general, it is an integer multiple of the latter, since a polyphase clock is often required in integrated circuits.

8.3.5 *Fixed-point versus floating-point arithmetic*

Most radar signal processors employ fixed-point arithmetic for simplicity. In fixed-point arithmetic, addition, assuming no overflow occurs, is an exact operation, but multiplication is inherently non-exact. The reason for this is that the product of the two n bit numbers is a $2n$ bit number, and if the register length is to be kept constant to n bit, the product should be reduced to n bits. The approximation can be affected by truncation or rounding. This problem is more serious in FFT operation. To avoid overflow in FFT operation, block floating point is often used. It seems that 16 bit fixed-point arithmetic is sufficient for most radar signal processors. Floating-point processors can be used in high performance processors.

8.4 Functional block of radar signal processor

A radar signal processor consists of many functional blocks, each block performs a specified function. They can be summarised as follows.

8.4.1 *A/D converters*

Most radars employ quadrature channels for the baseband video signal, i.e. in-phase (I) and quadrature-phase (Q) channels. Therefore, two A/D converters are needed for two channels. In general, a sample-and-hold (S/H) circuit is needed preceding the A/D converter. The modern A/D converter integrates the sample-and-hold circuit inside the chip.

Recently, a flash type A/D converter has been developed. In this type of A/D converter, a number 2^N of comparators operate in parallel. A very high speed of conversion can be achieved, and no sample-and-hold circuit is required.

Since the output of the phase detector is a bipolar video signal, the A/D converter should handle this bipolar signal.

8.4.2 *Buffer memory and data rearrangement*

Most radar signal processors process the data sequence in azimuth direction as mentioned before. But the output of the A/D converter is arranged in the range direction. Therefore, a buffer memory for data rearranging is needed. This is shown schematically in Fig. 8.5.

8.4.3 *FIR and IIR filters*

FIR and IIR filters are the most commonly used filters for clutter rejection. They can be realised with a multiplier–accumulator chip. Data and coefficients are multiplied in a parallel multiplier and then accumulated to obtain the results.

The functional block of the multiplier–accumulator is shown in Fig. 8.6.

If complex numbers are used, four real multiplies and two real adds are needed for one complex multiplication. Registers and multiplexers are required for intermediate results. Recently, a single chip complex multiplier is available.

The fastest 16 × 16 multiplier was developed by Toshiba with 0.6 μm CMOS technology. The time of multiplication is 7.4 ns, and the power consumption is 400 mW. The fastest commercial available multiplier is the IDT7216. The time

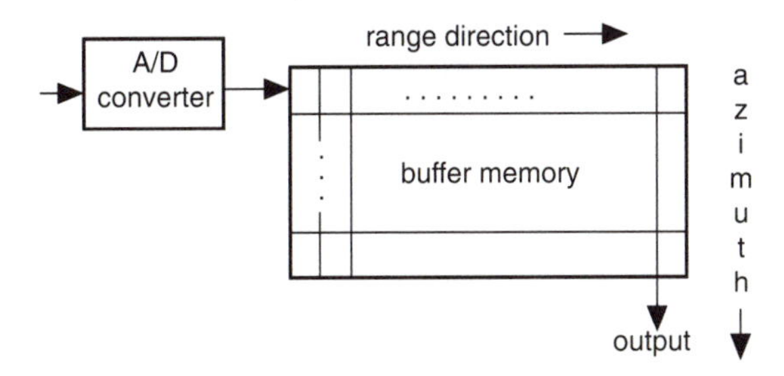

Figure 8.5 *Buffer memory and data rearrangement*

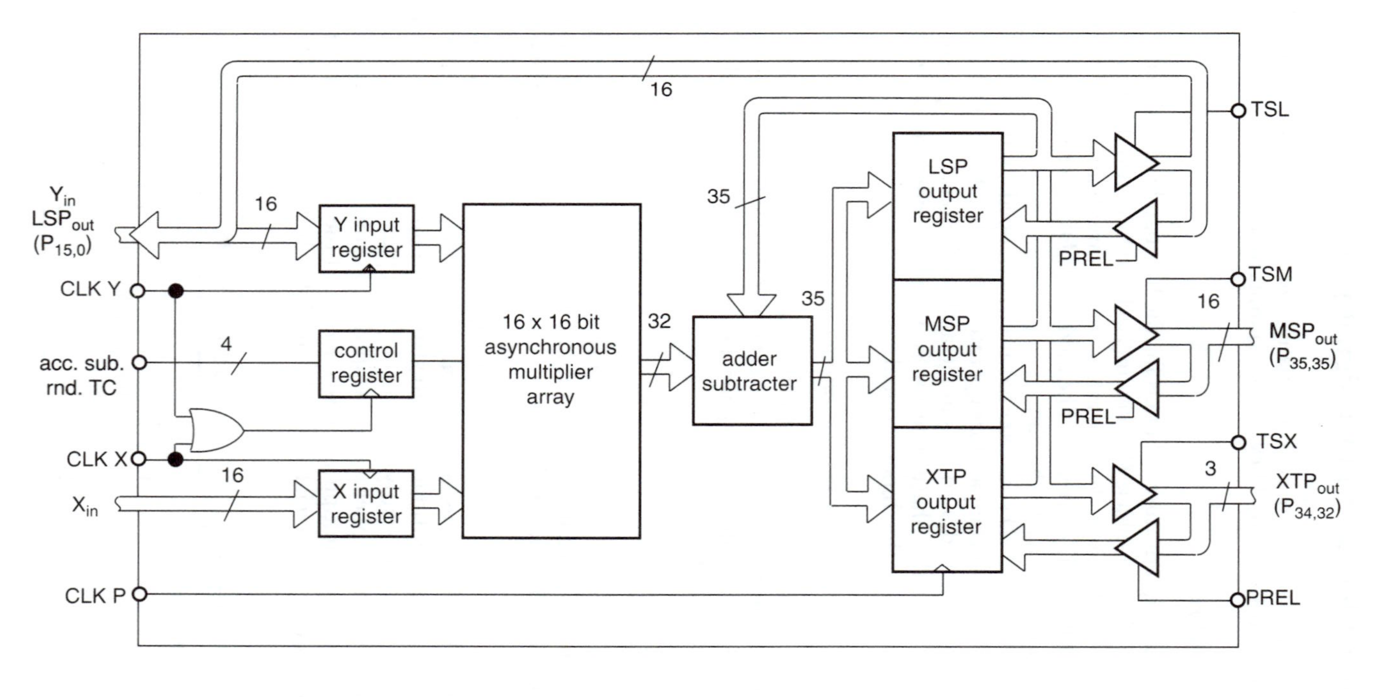

Figure 8.6 *Functional block diagram of multiplier–accumulator*

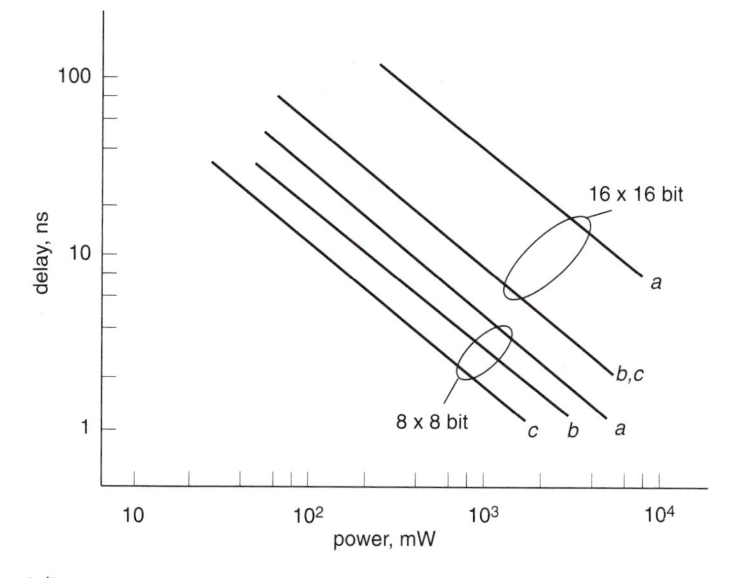

Figure 8.7 *Parallel multiplier speed–power trade-offs for 1 μm design rules [1]*

a Silicon bipolar
b Silicon CMOS
c GaAs

of multiplication is 20 ns. The digital high speed parallel multiplier speed–power trade-offs for 1 μm design rules is shown in Fig. 8.7 [1].

The single chip FIR filter is now also commercially available. For example, the SM5831F developed by Nippon Precision Circuits Ltd. can be configured as 4-, 7- and 8-taps FIR filters. The maximum sampling frequency is 15 MHz. The word length of input data is 9 bits, the coefficients data length is 8 bits, and the word length for internal operation is 14 bits. Higher order FIR filter can be obtained by cascading several chips.

8.4.4 *Fast Fourier transform (FFT)*

The FFT is also an often used functional block in radar signal processors. The signal flow of a typical 8-point FFT is shown in Fig. 8.8.

It consists of several stages or 'passes', which is equal to the power of two of the data points. Each pass consists of several butterflies, one of which is shown in Fig. 8.9.

The parallel multiplier is the heart of a butterfly. One multiplier can serve for several butterflies, if the data rate is relatively low. Additional adders give the results for the next pass. A bit of word growth may result at each stage, hard limiting is not acceptable, and a fixed number of bits are carried at each stage; so a left shift must be performed at each stage. In a 1024-point radix 2 FFT, this implies 10 bits of shift. This could mean the loss of a lot of signal. One way to solve this problem is to shift the data according to the results of overflow detection. Another way is using block floating-point arithmetic as mentioned

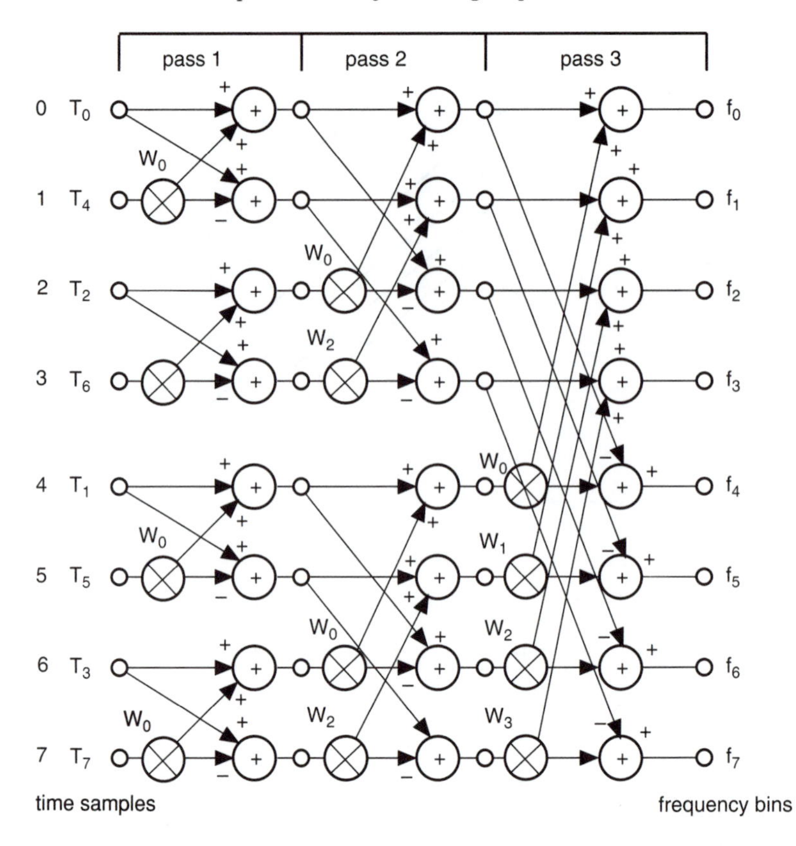

Figure 8.8 *FFT signal flow*

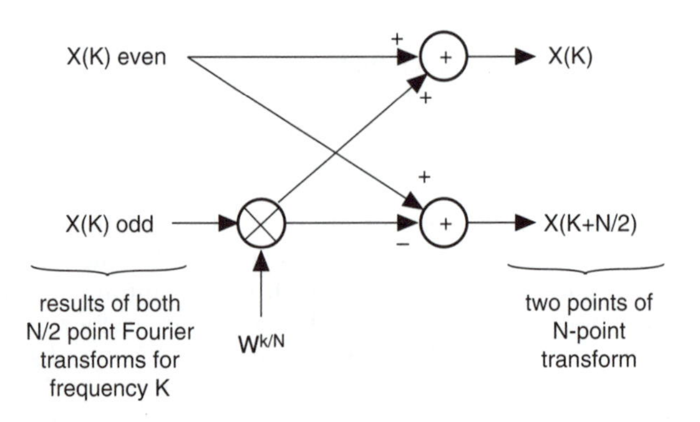

Figure 8.9 *Single butterfly of FFT*

before. The best way is the use of full floating-point arithmetic. Windowing can be performed in the first pass.

Single chip FFT is commercially available from TRW. It performs 32 point FFT within 47 μs.

8.4.5 *Correlator*

The digital correlator is mainly used in phase-coded pulse compression systems. A single chip digital correlator is often used for this purpose. Fig. 8.10 shows a block diagram of a TRW TDC–1023 64-stage digital correlator.

8.4.6 *Fast convolver*

In the radar signal processor, the convolver is used as a matched filter for linear FM pulse compression waveform. In receiver signal processing the echo signal $s_1(t)$ is passed through a matched filter with an impulse response $h(t)$, and the output at time τ, $s_o(\tau)$ is given by the convolution integral

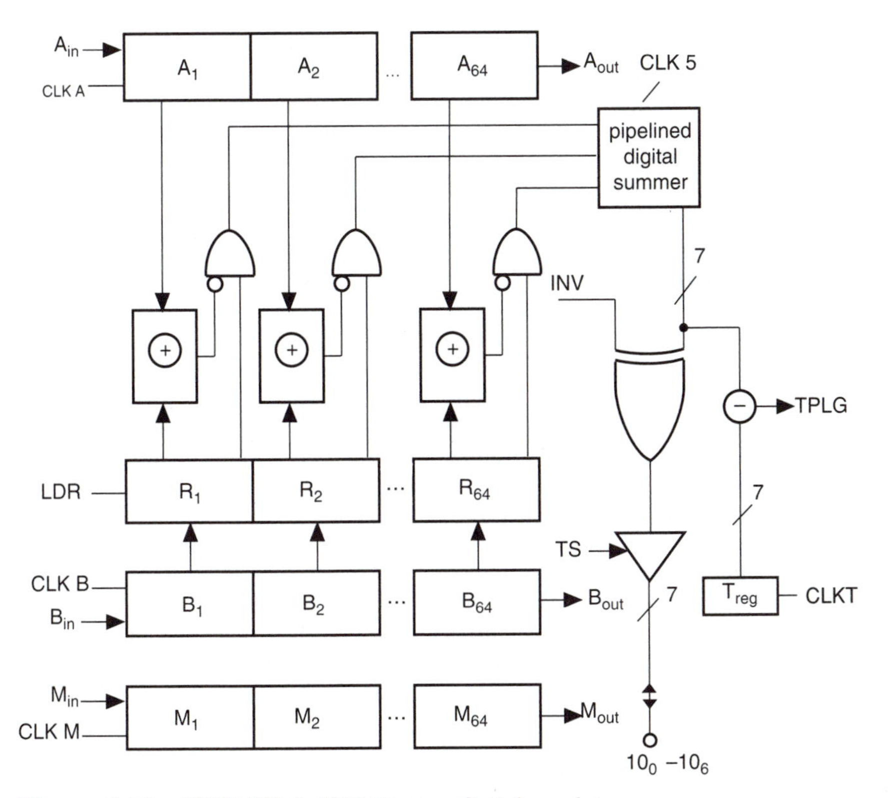

Figure 8.10 *TRW TDC–1023 64-stage digital correlator*

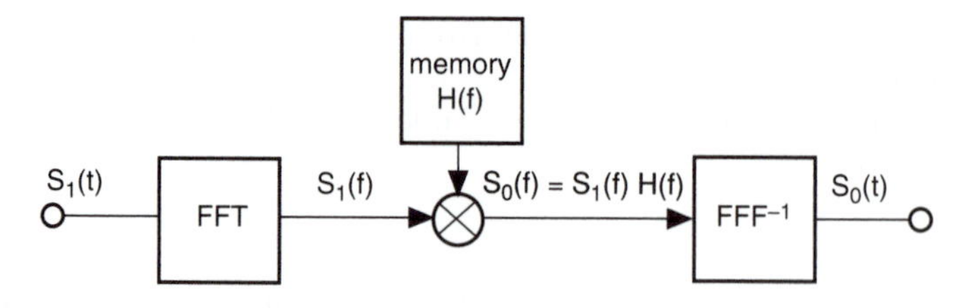

Figure 8.11 *Block diagram of fast convolver*

$$s_o(\tau) = \int_{-\infty}^{\infty} s_1(t)h(\tau - t)\ dt \tag{8.5}$$

Taking the Fourier transform of both sides of eqn. 8.5 yields

$$S_o(f) = S_1(f)H(f) \tag{8.6}$$

where $S_1(f)$ is the spectrum of the signal $s_1(t)$, $H(f)$ is the frequency response of the matched filter, and $S_o(f)$ is the output spectrum. The frequency response of a matched filter is

$$H(f) = S_1^*(f)\ \exp(-j2\pi fT) \tag{8.7}$$

where the asterisk (*) indicates the complex conjugate and T is the duration of the transmitted signal. Using eqn. 8.6, $S_o(f)$ can be obtained in the frequency domain, and by taking the inverse Fourier transform of eqn. 8.6 one obtains the desired output signal $s_o(\tau)$, in the time domain. Since the Fourier transform can be realised with FFT, so this method is called fast convolution. The block diagram of a fast convolver is shown in Fig. 8.11.

8.5 Architecture of radar signal processor

The development of the architecture of radar signal processors is closely related to the development of the devices. For example, the first generation of MTD (MTD–I) developed by Lincoln Lab. was constructed with SSI and MSI integrated circuits. It is a special purpose hardwired signal processor containing approximately 1000 integrated circuits, a core memory and a disc memory. It consists of many diverse logic circuits; this causes increased strain on the maintenance and logistics support organisation. In addition, processor improvements suggested by field experience are difficult to implement in a hardwired processor, often requiring rewiring.

The second generation of MTD (MTD–II) is a parallel microprogrammed processor (PMP) [4]. It has many programmable processing modules in parallel so that a fault in one only partially degrades performance. In addition, since it is completely programmable, algorithm improvements in the field are easily implemented (by a simple replacement of the program memory card). However, most of the integrated circuits are still SSI, MSI and very few LSI (semiconductor memory). The total number of integrated circuits still exceeds 1000. No LSI microprocessors and single chip parallel multipliers were used.

All of the multiplications were programmed by shift and add. Therefore, this not only causes an increase in the number of chips and decrease in speed, but also causes difficulties in programming the microprogram.

Nowadays, most of the radar signal processors are constructed with LSI and VLSI bit-slice or word-slice microprocessors plus single chip parallel multipliers. Disc memory for clutter map is replaced by VLSI semiconductor DRAM. It becomes more compact, faster, more flexible, easier to program, and easier for maintenance.

One of the most important specifications of the radar signal processor is the throughput rate as mentioned before. This is due to the wide bandwidth of radar signal and the real-time processing requirement. The throughput rate of a signal processor at first depends on the devices used. But the architecture of the processor is of the same importance as the devices.

In radar signal processors, pipeline and parallel architecture are widely used to increase the throughput rate of the processor. Most of them have the non-Von Neuman architecture. For example, Harvard architecture is widely used in radar signal processors. In this architecture, separate buses for instruction (or data) and address enhance the speed of operation. Two instructions can be executed in the same cycle. Very long microcode is used to control every section of the pipeline simultaneously. All of these can enhance the throughput rate effectively, even with the same clock frequency.

As an example, we would like to introduce the multiple algorithm programmable processor (MAPP) [5] to explain it in detail. It is designed to provide a high degree of clutter rejection via coherent filtering of pulse train waveforms.

The MAPP is a portion of the total multiple-channel processor. It has been assigned the global function of clutter filtering and/or pulse integration by coherent processing. As shown in Fig. 8.12, it is between the A/D converter and the pulse compression processor. This location allows a reduction in dynamic range in other parts of the processor, since the clutter returns dominate the dynamic range.

The basic requirements on MAPP are as follows.

(i) Implementation of real-time processing of 10 MHz bandwidth 3-pulse MTI with a range extent of 64 miles. Processing to include a type I phase tracking clutter-lock loop with a high C/N lock-on time of less than 6 μs to an equivalent 40 dB clutter improvement factor.

(ii) Implementation of real-time processing of 10 MHz bandwidth pulse Doppler waveform consisting of 16 coherent pulses with an unambiguous range extent of 64 miles. All filter characteristics to be completely independent and controlled by externally loaded complex weights. Sufficient storage for complex weights to allow storage of two independent sets with a capability to switch between sets on any 100 ns real-time boundary.

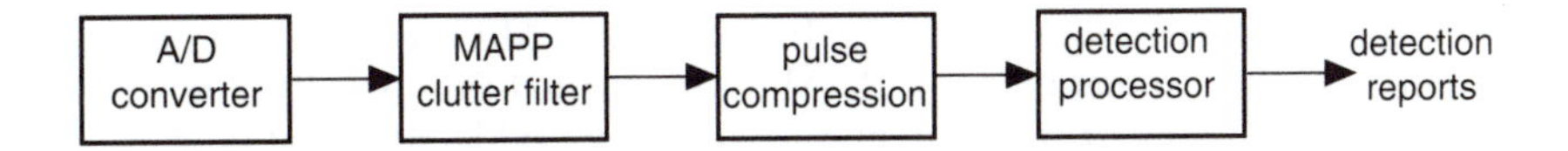

Figure 8.12 *Location of MAPP in the processor [5] © IEEE, 1984*

(iii) Sufficient dynamic range, bit sizing, to achieve 60 dB Doppler and 50 dB clutter improvement in MTI.
(iv) No limitations on minimum range, maximum radar PRF, waveform coding, or other standard radar variables.
(v) Application programming at Assembly language level. No microcode programming to be required.
(vi) Modular hardware to allow implementation of a fractional MAPP for less demanding application.

According to these requirements, the MAPP architecture must satisfy two processing requirements to support two classes of algorithms: contiguous processing and re-entrant processing. Contiguous processing requires a fixed processor configuration. All pre-stored data to be processed flow serially through the MAPP, where each cell receives the same arithmetic manipulation, and then flows out to the rest of the signal processor. No feedback through the MAPP is required. The pulse Doppler algorithm uses this type of processing.

Re-entrant processing requires a changing processor configuration. Packets of data, equal to the processor's pipeline, flow from the memory through the processor, where each cell in the packet receives the same arithmetic manipulation, then back to the memory for later use or back through the processor for another set of operations. The procedure is repeated until all the packets have been processed. The MTI uses this type of processing.

Since the re-entrant process requires feedback around the processor, processing is not done in real time. To compensate for this, many processors must perform the same task in parallel on different blocks of the data. There must be a controller to partition the raw data between the processors, control the parallel processing of the data, collect the results, and distribute in order to the rest of the signal processor.

The MAPP, illustrated in Fig. 8.13, is partitioned into two identical parts identified as even and odd. Splitting MAPP into two halves of eight processors allows dual controllers for improved reliability. A 1750A computer, radar control processor (RCP), provides control commands to the control unit, and the input data interface routes input data to the processors. The output data interface combines the even and odd processed data.

Each processor consists of a processing element (PE) and a buffer memory. Input data are applied directly to the buffer memory where they are sorted efficiently (corner-turned) for later processing by the PE. The memory has two independent 8 k × 24 sections (A and B) allowing for ping-pong operations. That is, while data are read from memory A for processing, data for the next dwell are stored in memory B. This procedure can be repeated for the next dwell by reversing the roles of the memories. After data are processed by the PEs, it is time-multiplexed on the output bus and combined with the other half by the output data interface.

The input data interface also provides a data path from the RCP to the processor's coefficient memory in the PE, where the MTI weights and the pulse Doppler weights are held. It is a RAM and is loaded from the RCP where the tables are maintained in nonvolatile memory. This allows tailoring the coefficents to the operating environment.

The control unit interprets commands as in a conventional computer;

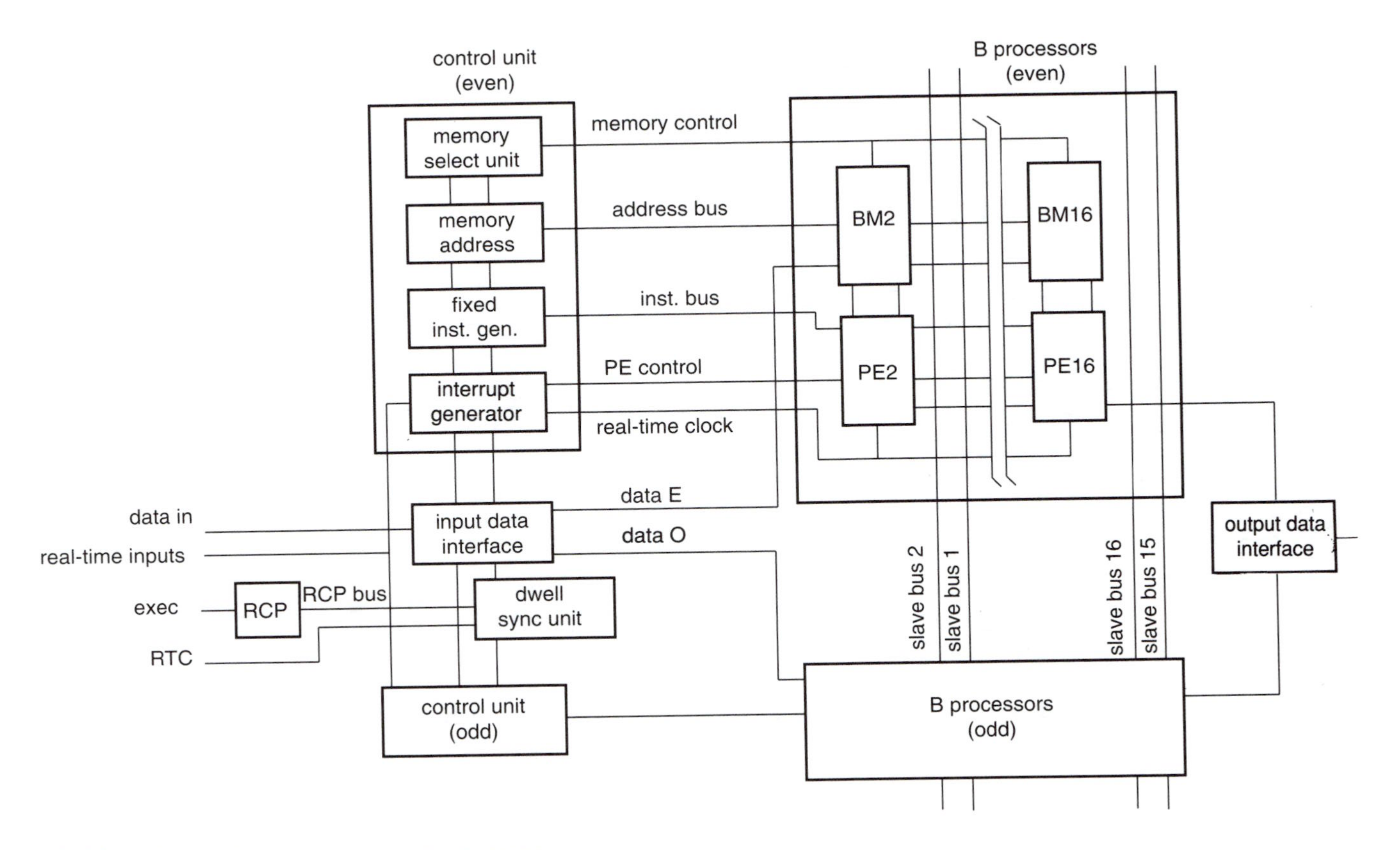

Figure 8.13 *MAPP global architecture [5] © IEEE, 1984*

however, the array of processors is used to simultaneously execute the instructions generated by the control unit. The processing elements are each connected to private memory and are interconnected in parallel by a common instruction bus, coefficient memory bus, interrupt bus and output bus. The memories are interconnected in parallel by a common data-input bus and two memory address buses. The memory I/O control is composed of some bused signals and some discrete signals to allow slicing of data into the processors. With this MAPP structure, a single instruction can operate simultaneously on multiple data streams.

To structure the control, the radar dwell is broken up into a hierarchy of nested time slots as shown in Fig. 8.14. Each dwell is composed of n PRIs where n is 3 to 5 for MTI and up to 16 for pulse Doppler. Each PRI is broken up into m blocks where m is equal to the PRI divided by the number of processors.

A block is classified as an input/output block or a processing block. During an input or output block, the processor ignores the fixed instruction or interrupt, and inputs or outputs data. Each processor has an independent run, input, and output enable for this purpose. A processing block, where the processor accepts the instruction, is composed of a number of epochs that represent the time required to complete a full mode-dependent algorithm. Each epoch computes a result for 8 range cells, so the number of epochs is equal to the number of range cells processed in an input block divided by 8. Each epoch consists of a set of instructions which is repeated until all raw data are processed. Each instruction takes 8 clocks, which is the pipeline length of the PE. If a packet of 8 cells, a quantum, is removed from memory and processed through the PE, the quantum can be re-entered into the processor with a new instruction.

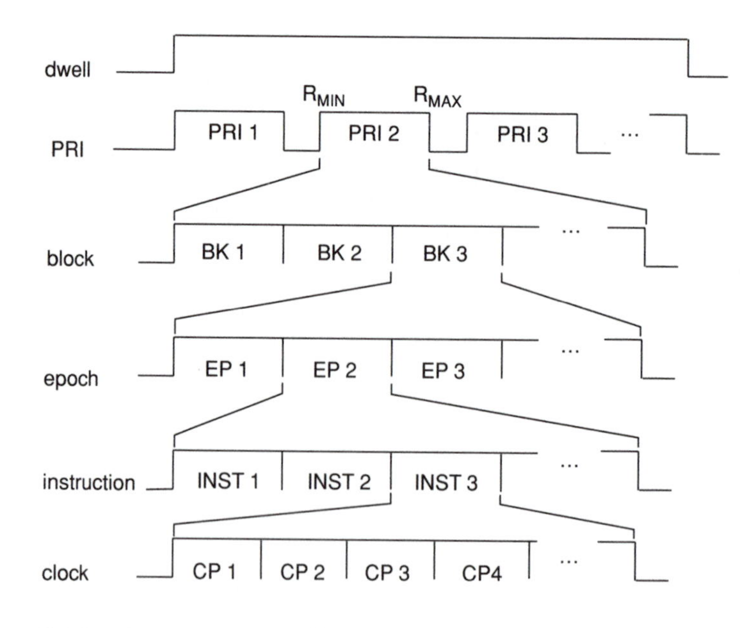

Figure 8.14 *Control timing [5] © IEEE, 1984*

This hierarchy of control translates a dynamic control problem into a set of static ones. Control signals are classified as dwell, PRI, block, epoch, instruction or clock signals, and remain constant over that time slot.

The processing element (PE) is a single module that contains all of the signal path computation logic. A simplified logic diagram of the PE is delineated in Fig. 8.15.

The PE is functionally configured to provide complex multiplication and addition as well as block scaling and a limiter, both of which are programmable. The bit management of the PE is varied down the processing path to ensure an optimum tradeoff between logic complexity and quantisation errors. At the input of the PE the words are sized at 24 bits (12I + 12Q). This size follows through the complex multiplier function. The accumulator is sized for 40 bits (20I + 20Q), allowing for the bit growth associated with the addition of terms in the high performance pulse Doppler algorithm. This expansion of bits ensures no truncation effects. When re-entrant processing is employed, such as in MTI, all outputs of the PE that are recirculated are scaled to the 24 bit complex input.

The logic design of the PE also includes three programmable switch-selectable feedback data paths. Two are 24 bits wide, allowing re-entrant processing with one data path, while the second can return data to memory or the slave bus. The third return bus is 4 bits wide and returns I and Q sign data bits. The input of the PE includes a programmable switch matrix which provides the selection of signal sources, input data or re-entrant paths. A switch matrix is also provided on the output of the PE to route data to the MAPP output register.

The second input to the multiplier is driven from a RAM memory which is initialised with filter coefficients for pulse Doppler processing or trigonometric tables for the rectangular to polar and reverse operations used for the MTI clutter-lock algorithm. This memory is 4 K × 24, allowing storage for forming up to 64 unique filters in pulse Doppler.

The design of the PE provides that all data paths are exactly eight registers in length. Where inherently shorter paths exist, additional registers have been added including the microcoded control path. Forcing the constant register delay has significantly simplified the design of the instruction set and the associated microcode.

In a full MAPP processor, a total of 16 PEs are operated in parallel with range slicing of the input data. Typically a single PE configuration can provide a 4 mile processing window for this case. Fig. 8.16 shows the range capability of MAPP as a function of the number of pulses or filters versus the number of PEs.

One of the advantages of a parallel configuration is that, for applications that do not require the full processing power, less hardware can be configured without disrupting the basic architecture and instruction set.

Since each PE is performing six complex arithmetic functions in a short pipeline at a 10 MHz rate, the total complex computation rate per PE is 60 million complex operations per second (MCOPS). For a 16-PE parallel configuration, a total computation rate of 960 MCOPS is achieved in a single MAPP. The input and output data rates are 240 M bit/s each.

The PE is implemented with commercially available 12 × 12 multipliers and RAM ICs, and two custom designed large scale gate array circuits. The switch

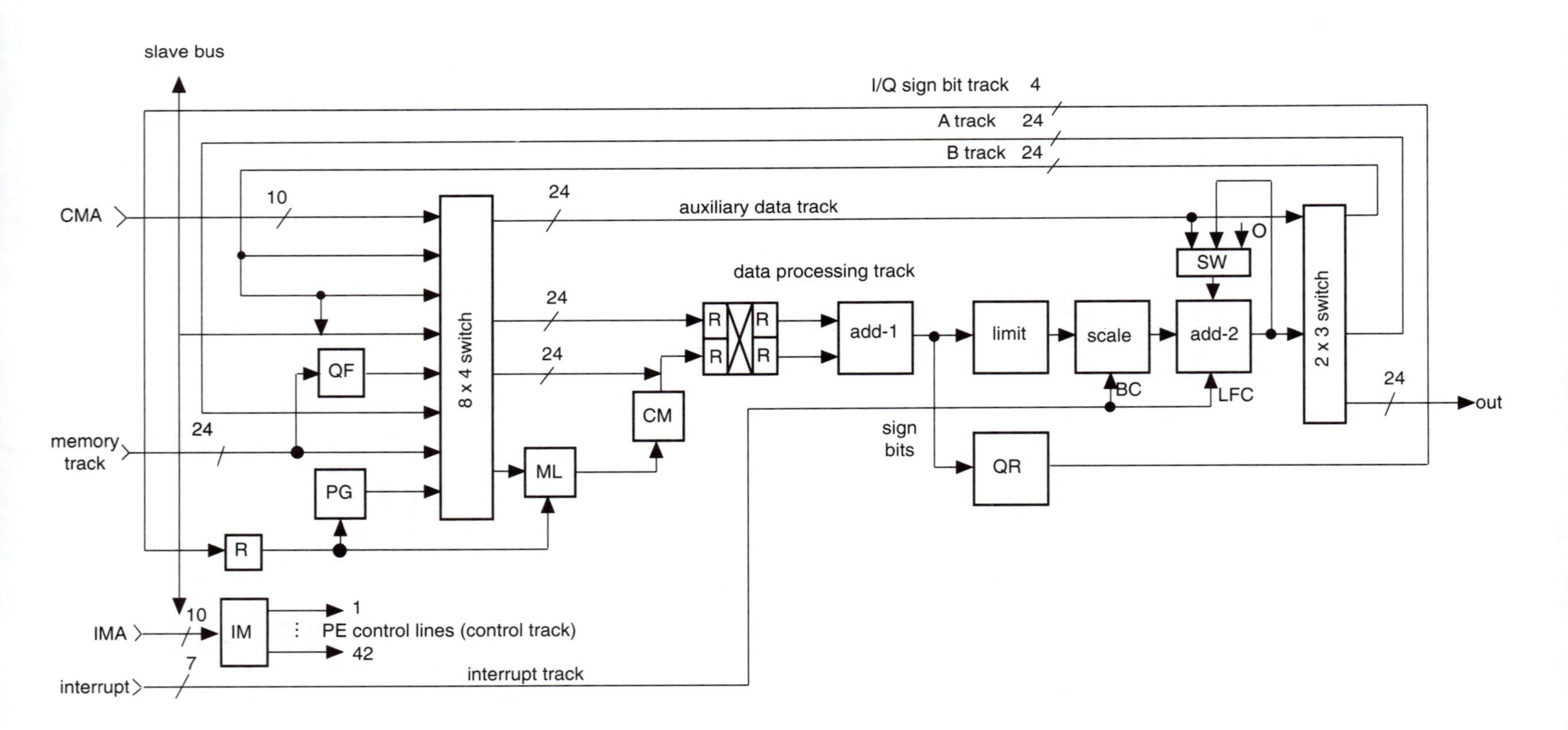

Figure 8.15 *Processing element (PE) [5] © IEEE, 1984*

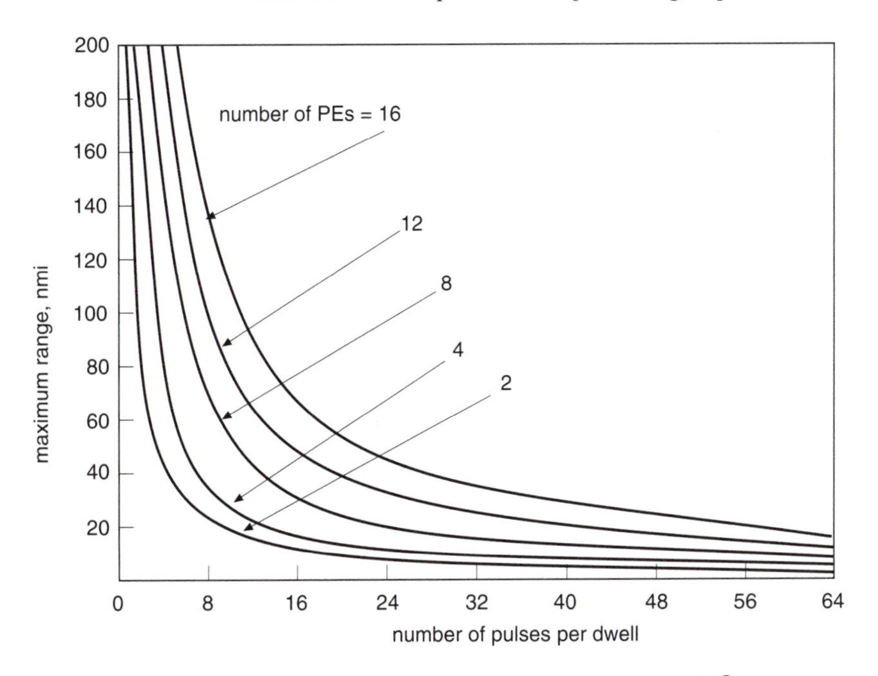

Figure 8.16 *Maximum range constraint versus number of pulses [5] © IEEE, 1984*

matrix is a custom design employing approximately 2500 gates in a 4 bit slice per chip. Six of these chips are required for a PE. The arithmetic processing is a custom gate array employing approximately 6000 active gates on a 10 000 gate chip. It allows full I or Q processing per chip; thus only two chips are required per PE. The two custom designed gate array chips use 2 μm bulk CMOS technology and operate up to 15 MHz clock rate over the full military temperature and supply voltage range.

8.6 Future trends in radar signal processors

The radar signal processor of the future will be primarily digital. It is clear that digital signal processing has the economic momentum and the technical potential to achieve an increasing predominance and ultimately to perform almost all radar processing function. At the same time, it will be a radar system with more intelligence, adaptivity, versatility, reliability, stability, dynamic range and performance than the system of today.

It can be predicted that the future trends of radar signal processor will be a combination of advanced algorithms, architectures and devices.

Many complex algorithms, such as maximum likelihood detection, maximum entropy spectral estimation etc., can be realised in real-time radar signal processors. This is due to the development of the architectures of radar signal processors and custom designed VLSI devices. The basic discipline in a top–down design methodology (as depicted in Fig. 8.17) depends on a fundamental

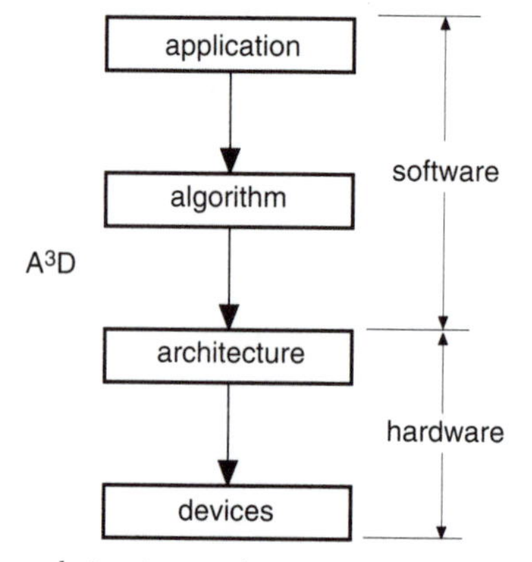

Figure 8.17 *Top–down design integration*

understanding of application, algorithm, architecture and devices. We can call it an integration of A^3D for the sake of simplicity.

The base of this integration is devices. Therefore, we will begin from the development of VLSI and VHSIC devices.

Until now the design rule, or the gate length, of VLSI devices is still scaled down with a factor of 2 per two years. Recently, 1 M bit SRAM has been commercially available, which contains 6.4 million transistors. 16 M bit DRAM has been realised in the laboratory with 0.6 μm CMOS technology.

A very high speed integrated circuits (VHSIC) program was developed by the DoD of USA. The goals of phase I of the VHSIC program is to achieve a 5×10^{11} gate–Hz/cm^2 figure of merit with 1.25 μm technology. It means that the clock frequency is 20 MHz and the number of gates per cm^2 is 25 000. The goal of phase 1 has been achieved in mid-1985. The goal of phase 2 is to realise 1×10^{13} gate–Hz/cm^2 with 0.5 μm technology.

It seems that these goals will be achieved by commercial companies. For example, Hitachi has announced that they have realised a 64 M bit DRAM with a 0.3 μm line width and 5 μm deep channels, which is achieved by using low temperature dry etching. An N channel silicon MOS FET with gate length less than 0.1 μm was made by IBM. A 1/8 frequency divider for 18 GHz signal was developed by NTT with Si ECL logic, which has a gate delay of less than 34.1 ps.

Therefore, it can be said that the development of VLSI technology offers great potential possibilities for the radar signal processor designer. The important problem is to develop VLSI architectures which, on the one hand, should be suitable to realise specified algorithm, and, on the other hand, it should exploit the potential of VLSI technology and also take into account (i) the layout constraint and resultant interconnection costs in terms of area and time, and (ii) the cost of VLSI processors as measured by silicon area and pin count. VLSI architecture design strategies stress modularity, regularity of data and control paths, local communication, and massive parallelism.

A solution to the real-time requirement of signal processing is to use special-purpose array processors, and to maximise the processing concurrency by either pipeline processing or parallel processing or both. As long as communication in VLSI remains restrictive, locally interconnected arrays will be of great importance.

The first such special-purpose VLSI architectures are systolic and wavefront arrays developed by Kung, H.T. [6] and Kung, S.Y. [7].

Array processors derive a massive concurrency for both parallel and pipeline processing schemes (see Fig. 8.18). Parallel processing means that all the processes defined in terms of the data (D_i) and the instruction (i_i) may directly access the **m** processors in parallel and keep all the processors busy (Fig. 8.18*a*). Pipeline processing means that a process is decomposed into many sub-processes, which are pipelined through m processors, i.e. m segments aligned in a chain and each sub-process will be processed in succession (Fig. 8.18*b*).

The concurrency in systolic/wavefront arrays is derived from pipeline processing or parallel processing or both.

A systolic system is a network of processors which rhythmically compute and pass data through the system. Physiologists use the word 'systole' to refer to the rhythmically recurrent contraction of the heart and arteries which pulse blood through the body. In a systolic computing system, the function of a processor is analogous to that of the heart. Every processor regularly pumps data in and out, each time performing some short computation, so that a regular flow of data is kept up in the network [6].

Systolic arrays can be constructed with some basic 'inner product' processing elements (PEs), each performing the operation $Y \leftarrow Y + A \cdot B$. These PEs can be locally connected to perform digital filtering, matrix multiplication and other

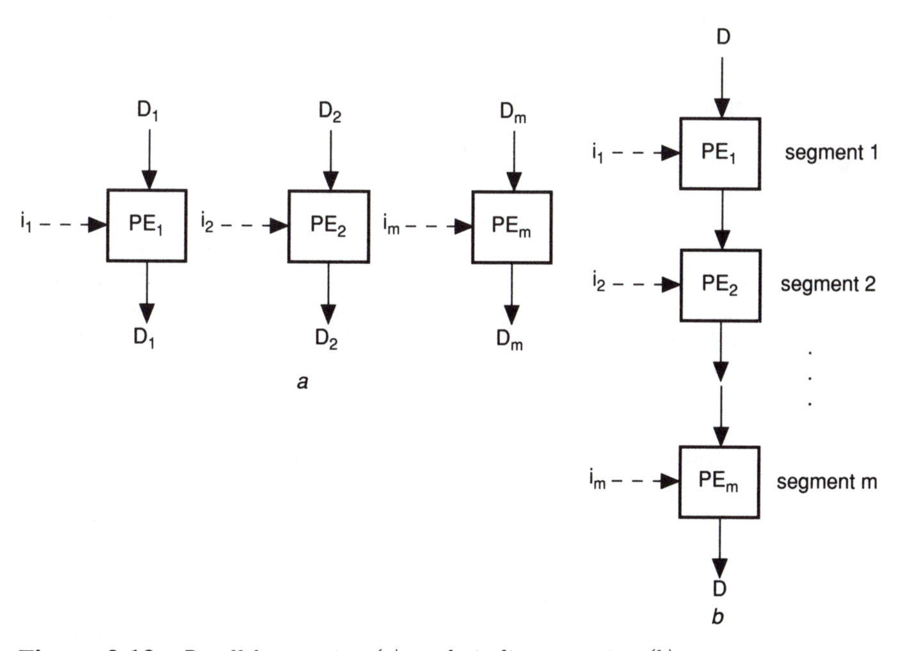

Figure 8.18 *Parallel processing (a), and pipeline processing (b)*

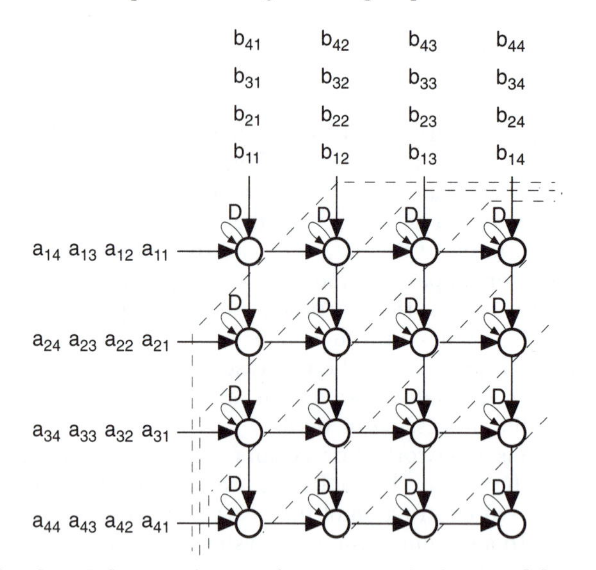

Figure 8.19 *Signal flow graph array for matrix multiplication [6]*

related operations. The principle of operation of systolic arrays can be explained with the example of matrix multiplication.

Matrix multiplication can be represented by a signal flow graph (SFG) as shown in Fig. 8.19.

A straightforward SFG array design is to broadcast the columns A_i and rows B_i instantly along a square array, such as the 4 × 4 array shown in this Figure. Multiply the two data meeting at node (i, j) and add the product to c_{ij}, the data value currently residing in a register in node (i, j). Finally, the new result will update the register via a loop with a delay D and get ready to interact with the new arriving operands. As all the column and row input data continue to arrive at the nodes, all the outer products will be sequentially summed.

Fig. 8.20 shows a systolic array for matrix multiplication. In this example, a two dimensional square array forms a natural topology for the 4 × 4 matrix multiplication problem. All the PEs uniformly consume and produce data in one single time unit. In terms of one 'snapshot' of the activities, the input data are pre-arranged in an orderly sequence. Owing to the systolisation rules, the inputs from different columns of B and rows of A will have to be adjusted by a certain number of delays before arriving at the array. This is why some extra zeroes are introduced here.

In general, the major characteristics of systolic arrays are (i) synchrony; the data are rhymically computed (timed by a global clock) and pumped through the network; (ii) regularity, modularity and spatial locality of interconnections; (iii) temporal locality; (iv) effective pipelinability; (v) expandability, i.e. the size of the array may be indefinitely extended as long as system synchronisation can be maintained.

One problem, however, is that the data movements in a systolic array are controlled by global timing-reference 'beats'. In order to synchronise the activities in a systolic array, extra delays are often used to ensure correct timing.

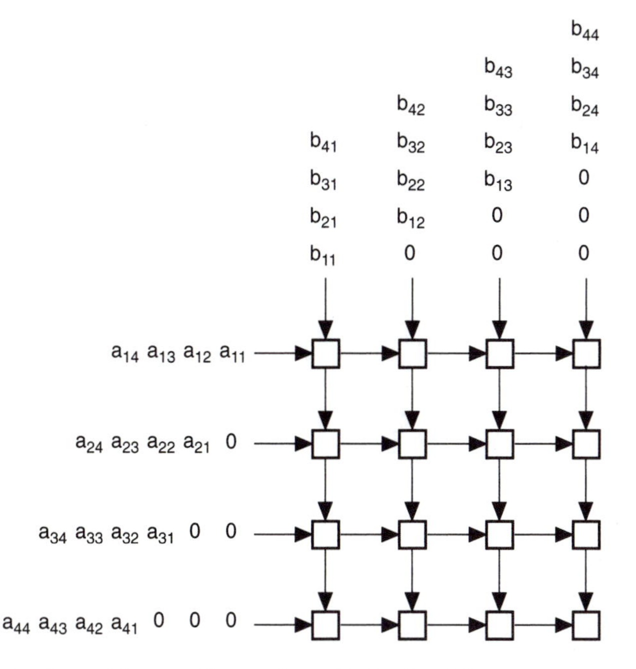

Figure 8.20 *A systolic array for matrix multiplication [6]*

More critically, the burden of having to synchronise the entire computing network will eventually become intolerable for very-large-scale arrays. A simple solution to this problem is to take advantage of the control-flow locality, in addition to the dataflow locality, inherently possessed by most algorithms. This permits a data-driven, self-timed approach to array processing. Conceptually, this approach substitutes the requirement of correct 'timing' by correct 'sequencing'. This concept is used extensively in dataflow computers and wavefront arrays.

A dataflow multiprocessor [8] is an asynchronous, data-driven multiprocessor that runs programs expressed in dataflow graph form. Since the execution of its instruction is 'data driven', i.e. the triggering of instructions depends only on the availability of operands and resources required, unrelated instructions can be executed concurrently without interference. The principal advantages of dataflow multiprocessors are simple representation of concurrent activity, relative independence of individual PEs, greater use of pipelining, and decreased use of centralised control and global memory.

However, for a general purpose dataflow multiprocessor, the interconnection and memory conflict problems remain very critical. Such problems can be greatly alleviated if modularity and locality are incorporated into dataflow multiprocessors. This motivates the concept of the wavefront array processors (WAP).

The derivation of a wavefront process consists of three steps: (i) the algorithms are expressed in terms of a sequence of recursions; (ii) each of the recursions is mapped to a corresponding computational wavefront; (iii) the

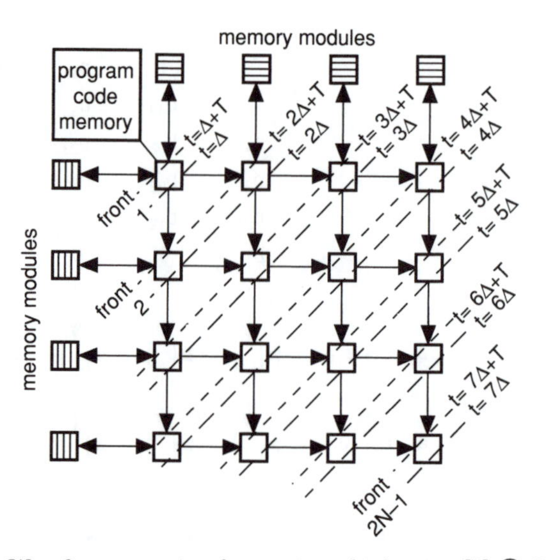

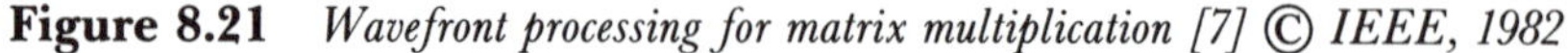

Figure 8.21 *Wavefront processing for matrix multiplication [7] © IEEE, 1982*

—— first wave
– – – second wave
Δ unit time of data transfer
T unit time of arithmetic operation

wavefronts are successively pipelined through the processor array. In fact, it may be stated that a wavefront array is a systolic array in combination with the dataflow principle.

Fig. 8.21 shows the wavefront processing for matrix multiplication.

In this example, the wavefront array consists of $N \times N$ processing elements with regular and local interconnections. The Figure shows the first 4×4 processing elements of the array. The computing network serves as a data wave propagating medium. Hence the hardware will have to support pipelining of the computational wavefronts as fast as resource and data availability allow, which can often be accomplished simply by means of a handshaking protocol. The average time interval T between two separate wavefronts is determined by the availability of the operands and operators. In this case, T is determined by the time needed for the arithmetic operations 'multiply' and 'add'. The speed of wavefront propagation is determined by the time interval Δ, which in this case is equivalent to the data transfer time. In general, the major characteristics of wavefront arays are (i) self-timed, data-driven computation, meaning that no global clock is needed; (ii) regularity, modularity and spatial locality of interconnections; (iii) effective pipelinability.

Although the systolic/wavefront array can be applied to digital filtering, FFT or DFT transformations, etc., they are particularly suitable to matrix computation, such as matrix multiplication, matrix inversion and matrix decomposition.

Some authors make efforts to employ systolic array processor in radar signal processing. Spearman *et al.* [9] showed some applications of systolic arrays to radar signal processing; these are systolic CFAR processor and systolic plot extractor. However, the algorithms for CFAR processing and plot extraction are too simple for systolic/wavefront arrays.

Liu and Xu [10] proposed a multiple systolic array to implement real-time radar target identification. This is based on the least squares sequential classifier model suggested by Therrien [11].

It can be expected that the systolic/wavefront array processors will be employed in radar signal spectral estimation, radar target identification, radar imaging, and other high performance radar signal processing.

8.7 References

1 MAVOR, J. and GRANT, P. M.: 'Operating principles and recent developments in analogue and digital signal processing hardware', *IEE Proc.*, 1987, **134**, Pt. F, pp. 305–334

2 MAO, Y. H., *et al.*: 'A non-parametric CFAR detector implemented with CCD tapped delay line'. IEEE 1985 Int. Radar Conf., pp. 430–434

3 HUGHES, C. J.: 'Performance of digital signal processors for MTI and hard-limited radar detection', Proc. 1984 Int. Symp. on noise and clutter rejection in radars and imaging sensors, Tokyo, pp. 534–539

4 MUEHE, C. E., *et al.*: 'The parallel microprogrammed processor (PMP)'. IEE Int. Conf. Radar '77, pp. 97–100

5 FOGLEBOCH, J. R.: 'A high performance multiple algorithm programmable processor'. IEEE 1984 National Radar Conference, pp. 63–68

6 KUNG, H. T., and LEISERSON, C. E.: 'Systolic array (for VLSI)' Sparse Matrix Symposium, SIAM, pp. 256–282

7 KUNG, S. Y., *et al.*: 'Wavefront array processor: language, architecture, and applications', *IEEE Trans.*, 1982, **C–31**, pp. 1054–1066

8 DENNIS, J. B.: 'Data flow supercomputers', *IEEE Computer*, 1980, pp. 48–56

9 SPEARMAN, R., *et al.*; 'The application of systolic arrays to radar signal processing'. IEEE 1986 National Radar Conference, pp. 65–70.

10 LIU, L. H., and XU, J. G.: 'Multiple systolic array to implement radar target real-time identification'. IEEE Int. Conf. Radar '87, pp. 588–592

11 THERRIEN, C. W.: 'A sequential approach to target discrimination', *IEEE Trans.*, 1978, **AES–14**, pp. 433–440

Chapter 9

Identification of radar target

Prof. Y. H. Mao

9.1 Introduction

The identification of non-co-operative targets by radar has became a subject of the greatest interest over the past decade. This is mainly due to the recent advances in radar components, the availability of fast and efficient signal processing hardware, and the development of automated decision making procedures.

The interest in target identification may be due to several factors, including (i) the desire to know more about a detected target than its mere presence, not only in military radar but also in a variety of civil radar, (ii) the applicability of target identification techniques to numerous research areas, including geophysical and meteorological applications. However, we will limit our discussion here to man-made targets. Also, target recognition in synthetic aperture radar is not included.

Different terms are substituted for 'identification' by different authors, such as 'recognition', 'classification', 'discrimination', etc. Target identification is defined as that process by which a particular target (or class of targets) is distinguished from other targets (or other classes of targets). This definition has intentionally been left general enough to accommodate both the case in which the object is to distinguish between two relatively similar types of targets (e.g. one type of ship relative to another ship) as well as the case in which the goal is to distinguish between one class of targets (the category of ships) and all other possible targets (the category of targets that are not ships).

The target recognition process consists of four steps:

(i) Extracting features from the backscattered electromagnetic waves from the known target
(ii) Establishing the data base of these features for the known target
(iii) Extracting features of the unknown target with a real-time signal processor
(iv) Comparing these features with that in the data base and making decisions.

The most difficult is the first step.

The target identification methodologies may be separated into three broad categories:

(i) Signal generation and feature extraction methods
(ii) Algorithmic procedures
(iii) Theoretical approaches.

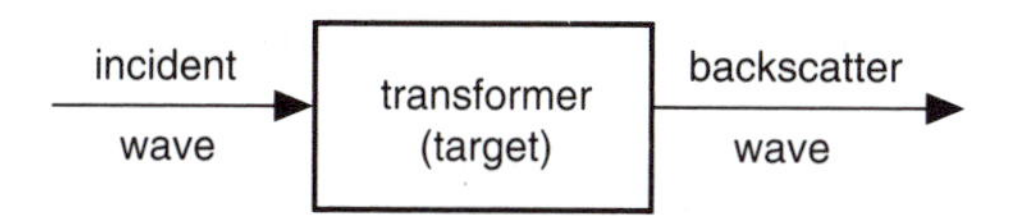

Figure 9.1 *Radar target as a wave transformer*

We will emphasise the first one.

Any target can be seen as a transformer which transforms the incident electromagnetic wave to a somewhat different backscatter wave as shown in Fig. 9.1.

The difference between the incident wave and the backscatter wave depends on the following target features:

(i) Shape of the target
(ii) Aspect of the target
(iii) Movement of the target, including the movement of the moving parts of the target
(iv) Material of the target, including conductivity, dielectric constant, permeability, and even semiconductor nonlinearity in the junction of metal parts.

All these factors will cause the backscatter wave to differ from the incident wave in strength, polarisation and fluctuation with time, etc.

It is somewhat easier to identify a stationary target than a moving target, since the features of a stationary target are time-invariant. Unfortunately, most targets of interest are moving targets.

The fundamental techniques of radar target identification can be listed as follows:

(i) Identification based on the target polarisation properties
(ii) Identification based on pole extraction
(iii) Identification based on frequency domain method
(iv) Identification based on time domain method
(v) Identification based on spatially coherent processing of the received echoes
(vi) Identification based on spectral estimation of coherently received echoes
(vii) Identification based on non-coherent processing of received echoes
(viii) Identification based on increasing the resolution capability of the radar.

Some of these only have a theoretical meaning or represent just a systematic approach. Some have experimental results. However, until now very few techniques can be applied to a real radar system. The reason is that feature extraction is just the first step in target identification. There are also many difficulties in the other steps. In general, radar target identification is very similar to speech recognition or speaker recognition. Unfortunately, the study of target identification is much less advanced than speech recognition.

We will introduce most of the techniques listed above.

9.2 Theoretical approaches to target identification

9.2.1 *Polarisation based target identification* [1]

This is one of the earliest methods used for target identification. It is based on the scattering matrix developed by Sinclair [2].

When excited by a monochromatic wave the polarisation behaviour of a target can be fully described by its 2×2 complex-valued scattering matrix S, which relates the polarisation vector $\boldsymbol{h}_T \triangleq [h_{TA}, h_{TB}]$ of the incident wave to the polarisation vector $\boldsymbol{h}_s \triangleq [h_{SA}, h_{SB}]$ of the backscattered wave, namely

$$\boldsymbol{h}_s = S\boldsymbol{h}_1 \tag{9.1}$$

where

$$S = \begin{bmatrix} S_{AA} & S_{AB} \\ S_{BA} & S_{BB} \end{bmatrix} \tag{9.2}$$

is the scattering matrix, which is suitable for any specified orthogonal polarisation basis A–B used for representing the polarisation vectors. S is a target dependent parameter. In other words, the radar target can be looked on as a polarisation transformer. If we can measure the scattering matrix of a target, this target can be identified.

Unfortunately, the scattering matrix depends not only on the geometrical structure, but also on the target orientation and motion. Therefore many authors tend to find an alternative polarisation parameter, which is relatively independent of the target aspect and motion.

A partial solution to this problem can be obtained by considering the so called polarisation invariants, which are the parameters independent of the pitch angle [3] (see Fig. 9.2).

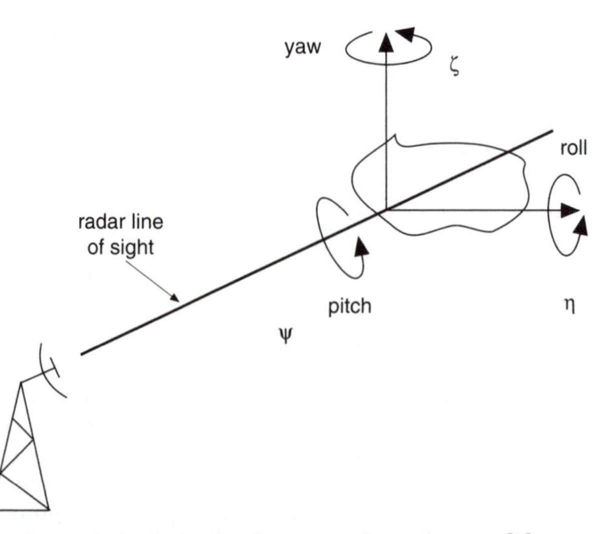

Figure 9.2 *Geometrical relationship between radar and target [3]*

Mathematically they represent invariants of the scattering matrix under unitary transformations, and they can easily be determined once the scattering matrix is known. By exploiting the symmetry of S, it can be shown that

$$S = R(\psi)E^*(\tau)S_D(\nu, \gamma, m)E^{*T}(\tau)R^T(\psi) \tag{9.3}$$

where R is the rotation matrix, E is a unitary matrix and S_D is the diagonalised form of the scattering matrix. Five independent parameters (m, ψ, τ, ν, γ), which uniquely and alternatively represent the above scattering matrix, appear while solving the eigenvalue problem arising when looking for the transmit polarisation which maximises or minimises the backscattered power. The parameters are defined as follows:

m = 'target magnitude'; this is the maximum amplitude of the received signal and is an overall measure of target size or RCS
ψ = 'target orientation angle'; this is a measure of the orientation of the target around the line of sight
τ = 'target helicity angle'; this is a measure of target symmetry with respect to right-hand and left-hand circular polarisation
ν = 'target skip angle'; this can be related to the number of bounce of the reflected signal
γ = 'target polarisability angle'; this is a measure of the target's ability to polarise incident unpolarised radiation.

However, the polarisation invariants still depend on the yaw and roll angles and on target motion. So the answer to the question whether they can be considered good target identification is not yet clear [4].

Another equivalent set of independent target polarisation parameters is obtained through the COPOL nulls [3]: they are two polarisation vectors, $\boldsymbol{x}_1$ and $\boldsymbol{x}_2$, giving rise to a backscattered wave which is polarised orthogonally to the incident wave. The COPOL nulls are generally represented through two couples of independent polarisation parameters which, together with the parameter m ('target magnitude'), still constitute an alternative set of parameters uniquely defining the scattering matrix.

Most of the methods based on polarisation properties of the target require direct measurement of the scattering matrix S. In a monostatic radar the measurement of S typically requires that the orthogonal polarisations A and B are alternately radiated, while two orthogonally polarised channels are simultaneously available on reception.

In low resolution radar, amplitude, phase and polarisation discriminants can rapidly change with target aspect. This behaviour will cause ambiguities in the identification of complex targets. Obviously, ambiguities increase when a reduced set of polarisation discriminants is used. Furthermore, owing to measurement difficulties, polarisation discriminants do not usually account for the absolute phase of the scattering matrix. This is suspected to introduce significant ambiguities in polarisation signature. Polarisation ambiguities can be reduced with high resolution radars, when interference scattering centres are sufficiently spatially resolved.

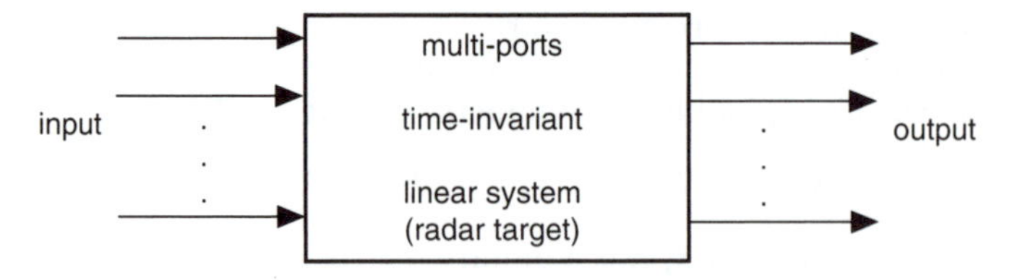

Figure 9.3 *Radar target as a multi-ports time-invariant system*

9.2.2 *Target identification based on poles extraction* [5]

It is assumed in this method that the radar target is a multi-input multi-output time-invariant linear system (see Fig. 9.3) viewed at aspect angles in the range (0–4 π). When two perpendicular-polarisation EM waves illuminate the target at some aspect angle, it corresponds to two input-output pairs. A large number of input–output pairs are provided by observing at different aspect angles.

Therefore the poles of this system may be used to identify the target just as in system identification.

The square of the absolute value of the target transfer function which represents the RCS of the target can be written as

$$\sigma(\omega) = |H(j\omega)|^2 = \sum_{i=1}^{n-1} B_i(j\omega)^{2i} \Big/ \sum_{i=0}^{n} A_i(j\omega)^{2i} \tag{9.4}$$

where $\sigma(\omega)$ is the radar target cross-section which can be measured by conventional radar, and (A_i, B_i) are real constants related to the poles of the transfer function $H(s)$. If a set of m measurements of $\sigma(\omega)$ at different frequencies ω_k $(k=1, 2, \ldots, m)$ is available, it is easy to solve for (A_i, B_i). If $m=2n$, it is suggested to use the method developed originally for network analysis. If $m>2n$, the least square method is usually adopted. Once the n constants A_i, $i=1, 2, \ldots, n-1$, are given, the pole set can be obtained by solving the equation

$$\sum_{i=0}^{n} A_i s^{2i} = 0 \tag{9.5}$$

Eqn. 9.5 has $2n$ solutions which are symmetric about the $j\omega$ axis. The poles located on the right half plane should be omitted. The main advantages of the pole extraction method are that the $\sigma(\omega_k)$ can be obtained with a multi-frequency noncoherent radar and equal sampling intervals are not required.

However, it is evident that the locations of the poles of a target are aspect dependent. The aspect of a target is varying with the movement of the target. Therefore, the assumption of a time-invariant model is invalid for a moving target. Even for a stationary target, the number of frequencies used to illuminate the target must be great enough, so that all of the poles can be detected. If the identification is limited to the same type of targets, e.g. aircraft with similar shape and size, limited frequency range can be adopted. If the identification is carried out for different types of target, such as aircraft and ships, a very wide frequency band is required.

Table 9.1 *Average values of the natural resonance frequencies of aircraft models* ($s^{-1} \times 10^6$) *[6] © IEEE, 1976*

Type	MIG–19	F–4	F–104
f_1	$-8.72+j79.3$	$-16.4+j66.00$	$-19.7\ +j94.0$
f_2	$-6.56+j55.6$	$-10.9+j43.70$	$-7.01+j46.8$
f_3	$-9.35+j66.2$	$-9.9+j130$	$-3.5\ +j131$

9.3 Target identification based on natural resonance

This is a modified method of pole extraction in frequency domain.

It was predicted in theory and partially proved by experiment that a radar target can be considered to be a pole invariant system in the Rayleigh scattering region. This means that the target poles remain invariant even though the target aspect angle changes. For example, Chuang and Moffat [6] measured the dominant natural resonance frequencies of three types of aircraft models. The results are summarised in Table 9.1.

Chen *et al.* [7] suggested a technique to exploit the knowledge of the poles of the target by synthesising a waveform for the incident radar signal such that the reflected signal from a specified target consists of only a single, fully representative, natural resonance mode of the target.

Consequently a particular target can be identified by comparing the received signal with a waveform set, stored in a library, representing the expected returns from each target member of a class of interest. The synthesis of the incident waveforms, constructed as a linear combination of natural modes, has been carried out with reference to simple models of aircraft obtained by using a grid structure [6], and by assuming the impulse response is expressible as a finite sum of dominant natural modes.

The electric field scattered by a target may be divided into an 'early-time' component, representing the forced response when the excitation waveform is traversing the target, and a 'late-time' component that exists after the incident field front has passed over the entire target, i.e. for $t > \tau + 2\tau_t$, where τ is the duration of the incident signal and τ_t is the transit time of the incident wavefront. This corresponds to expressing the target impulse response in the form:

$$h(t, \theta) = \sum_{n=1}^{N} a_n(\theta) \exp(\sigma_n t) \cos[\omega_n t + \psi_n(\theta)] \tag{9.6}$$

where

θ = dependence upon target aspect
$a_n(\theta)$ = aspect-dependent amplitude of the nth natural mode,
$\psi_n(\theta)$ = aspect-dependent phase angle of the nth mode
$\sigma_n + j\omega_n = s_n$ = nth natural frequency.

Mancianti *et al.* [8] suggested synthesising an incident waveform which excites a single mode of resonance. This waveform is constructed as a linear

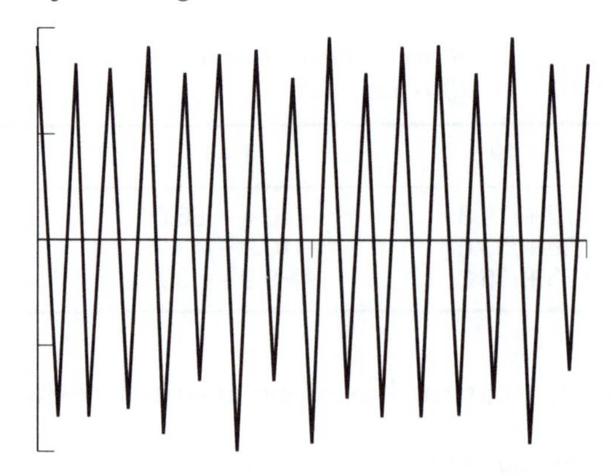

Figure 9.4 *Incident signal to excite the first mode of F–104 aircraft model [8]*

combination of pure sinusoids whose frequencies coincide with the imaginary part of the poles. Thus the single-mode technique concentrates the received energy in a narrow band around the selected resonance mode. Let us assume the incident field $E^i(t)$ as

$$E^i(t) = \sum_{n=1}^{N} b_n \cos \omega_n(t) + c_n \sin \omega_n t \tag{9.7}$$

where b_n and c_n are unknown, orientation-independent, coefficients. By convolving eqns 9.6 and 9.7 and solving for b_n and c_n in order to obtain a single-mode scattered field, the required waveform incident signal is obtained.

For example, by numerical computation the required incident signals to excite each mode of the resonance frequencies listed in Table 9.1 can be obtained. Fig. 9.4 shows the waveform to excite the first natural mode of an F–104 aircraft model. Fig. 9.5 shows the return signals from the three aircraft models, illuminated by the incident signal of Fig. 9.4.

It is important to bear in mind that, even if the amplitude, phase angle and late-time starting point depend on the aspect angle θ, the peculiar feature of the return signal is for it to remain single mode. In fact, the return signal from the matched target in the late time will be a single-mode oscillation, while the returns from other targets are expected to be significantly different.

Therefore, the identification between different targets, belonging to a class of *a priori* known objects, can be realised by comparing the returns received from each target in the class when a waveform for exciting a preselected natural mode of a particular target is transmitted. The comparison is performed by evaluating the maximum of the normalised correlation

$$R(n) = \sum_{i=1}^{l} r(n)c(i-n) \Big/ \left[\sum_{i=1}^{l} c^2(i) \right]^{1/2} \tag{9.8}$$

where $r(n)$ are the Δt spaced samples of the received waveform and $c(n)$ the ones relative to the stored waveforms representing the response of each target to the actual transmitted signal. Note that, even if the amplitude and phase angle of the received signal are aspect-dependent, the use of eqn. 9.8 does not need knowledge of these parameters.

R_{max} is compared with a suitable threshold λ' and then the following decisions are made:

If $R_{max} > \lambda$ 'correct identification' is declared,
If $R_{max} < \lambda$ 'wrong identification' is declared.

A related target identification scheme is shown in Fig. 9.6

To evaluate the algorithm's performance the following probabilities have been computed.

P_{CI} = probability of correct identification
P_{WI} = probability of wrong identification

Some results are shown in Fig. 9.7 where the P_{CI} versus SNR are plotted when the three aircraft models are illuminated by the incident waveform synthesised for exciting each mode of the F–104 model. Analogous results have been proved to be valid also for MIG–19 and F–4 models.

The following general remarks can be made:

- SNRs larger than 24 dB ensure a probability of correct identification practically equal to 1 for $\lambda < 0.9$, whatever mode of the considered aircraft models is excited
- the probability of wrong identification is always less than 10^{-2} for $\lambda = 0.7$ and less than 10^{-3} for $\lambda = 0.85$.

The results obtained by simulation show that the identification sensitivity is practically constant for $\tau > 5\tau_0$ (τ_0 is the period of the natural mode to be

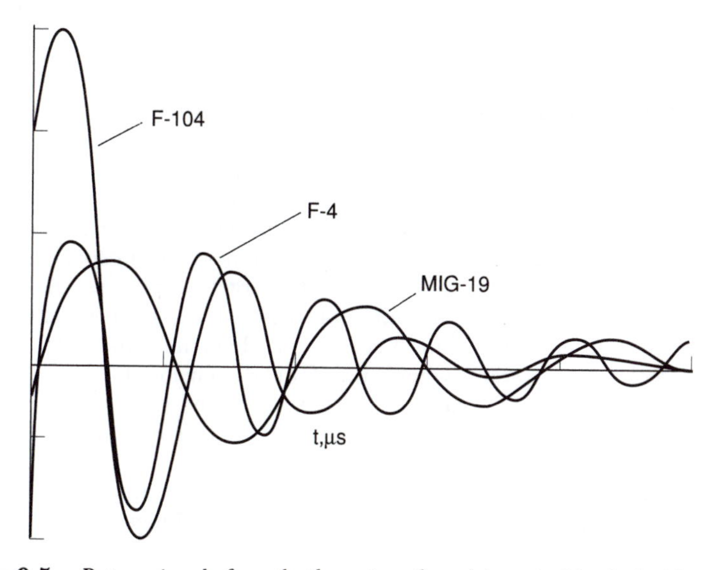

Figure 9.5 *Return signals from the three aircraft models excited by the incident signal of Fig. 9.4 [8]*

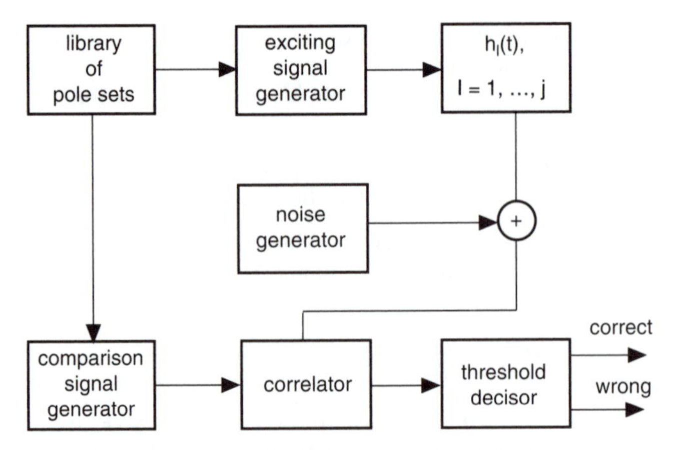

Figure 9.6 *Simulator of target identification procedure [8]*

excited). However, the correlation between the response of a target different from the one to be tested and the comparison waveform rises when τ increases. This occurs because, as the pulse duration is increased, the return signal from the wrong targets will contain mostly the excitation frequency.

There are still many problems in the realisation of this scheme. Firstly, the range of natural resonance frequencies of a target mainly depend on the size of the target. For different types of target, such as the small jet fighter to big bomber, the resonance frequencies may differ several times. So it is very difficult to synthesise a waveform suitable for different types of target. Even for a predetermined type of radar target, the selection of radar wavelength will be restricted by many other factors, e.g. size of the antenna, given frequency band for radar application, etc.

9.4 Target identification with multifrequency radar

This is also a frequency domain method for target identification.

Gjessing [9–11] proposed a 'matched illumination' method to identify a radar target. The basic principle of this method is to transmit a waveform which matches the predetermined target in four dimensions (three in space and one in time). The more *a priori* information about the target we possess, the narrower can the frequency band of our illumination be, or, for a given bandwidth, the shorter is the observation time needed for detection and identification. However, seeking a system where the four dimensional tailoring of the illumination is individual for each dimension will, at best, complicate the issue conceptually. It is also likely that such particular solutions would prove suboptimal.

Accordingly, the problem will be solved in general terms, considering the scattering object (surface) to be characterised by a four dimensional irregularity spectrum (wave-number spectrum).

If we illuminate the object with a set of electromagnetic waves having different but mutually correlated frequencies, i.e. a set of individual phase-locked sine waves (Fourier components) from individual electromagnetic oscillators, then the properties of the backscattered signal are observed; from this we can draw very definite conclusions about the distribution of the scatterers constituting the scattering body along the direction of radiowave propagation.

If we wish to resolve a structure within the scattering object of longitudinal extent Δz, we need an illumination with bandwidth $\Delta f = c/\Delta z$. Fig. 9.8 illustrates this: constructive interference is achieved from an object if that object is illuminated by two waves with frequency difference $\Delta f = c/\Delta z$. With a radially moving target this shows up as maximum-depth modulation in the difference frequency return.

To express this problem mathematically, it is convenient to combine the various factors contributing to the scattered field into one; namely, one which is

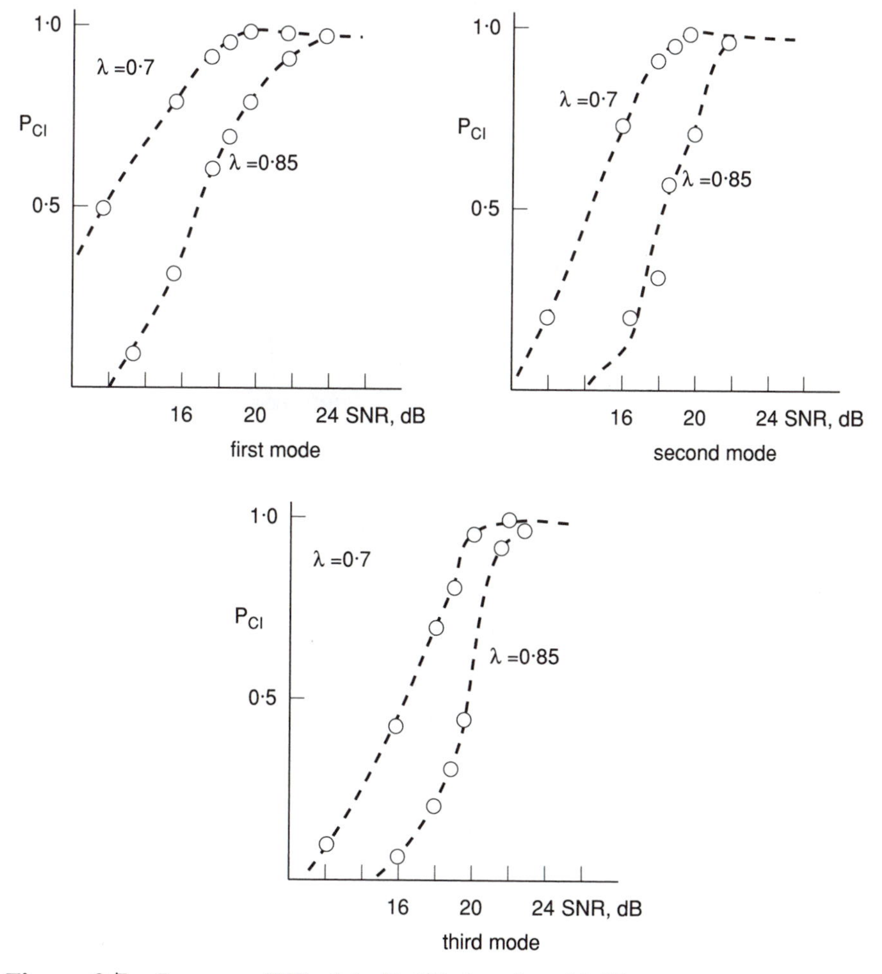

Figure 9.7 *P_{CI} versus SNR of the F–104 aircraft model [8]*

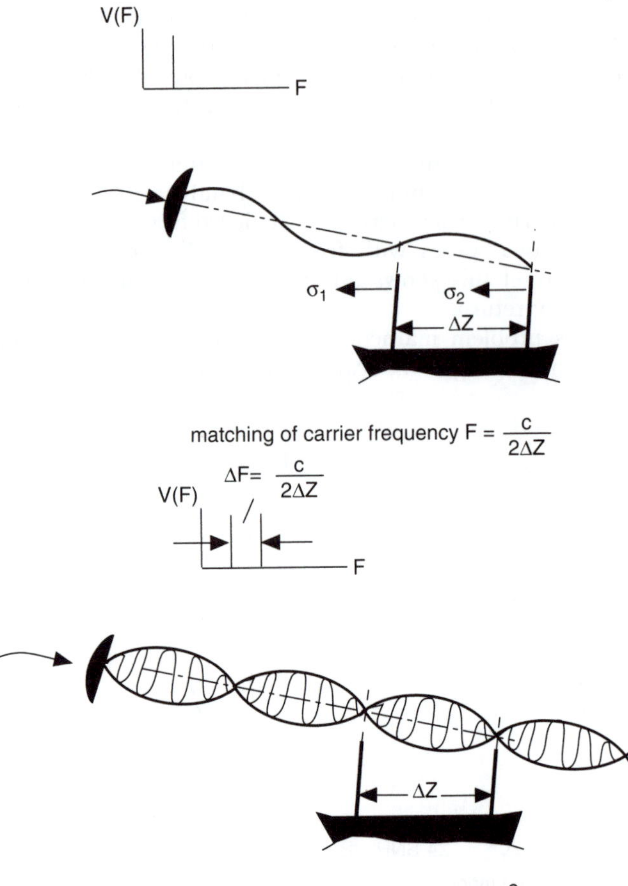

Figure 9.8 *Matching of carrier frequency and beat frequency [11]*

directly related to the scattering cross-section as a function of distance. This function $f(\vec{z})$ is defined as the delay function. This has dimension of field strength per unit length such that the scattering cross-section as a function of distance $\vec{z}$ along the direction of propagation is obtained by squaring the $f(\vec{z})$ function. It is obvious that the field strength of the backscattered wave from an object characterised by the delay function $f(\vec{z})$ is given by

$$V\left(\frac{\omega}{c}\right) \sim E(\vec{K}) \sim \int f(\vec{z}) \exp(-j\vec{K} \cdot \vec{z}) d^3\vec{z} \tag{9.9}$$

Here $\vec{K}$ is the wavenumber difference between the scattered and the incident wave:

$$\vec{K} = \vec{k}_i - \vec{k}_s$$

$$|\vec{K}| = 4\pi/\lambda \sin \theta/2$$

Any scattering object characterised by a distribution in space of the scatterers $(\sigma_E(\vec{r}))$ can be decomposed into its Fourier components $A(\vec{K})$. We will have constructive interference from all these scatterers if we choose a combination of illuminating sine waves $E_T(\vec{K})$ that couple to each of the Fourier components $A(\vec{K})$ of the scattering object. This is illustrated in Fig. 9.9.

As we have already inferred, by introducing the fourth dimension, namely time, the general object is in motion. If we are dealing with translatory motion of a rigid object, then all the Fourier components $\Sigma A(\vec{K})$ move with the same velocity so that each illuminating frequency $\vec{K}$ is subject to a Doppler shift $\omega = \vec{K} \cdot \vec{V}$, where V is the velocity of the object.

The general object, however, is flexible or compressible so that each Fourier component may have a specific and independent motion pattern. This situation is illustrated in Fig. 9.10. However, in general, the power spectra (temporal Fourier transform) of the individual $E(\vec{K}, \omega)$ components may be different and independent of each other.

Our aim now is to derive a simple expression for the correlation (in the frequency domain) between two electromagnetic waves of wavenumbers $E(\vec{K}+$

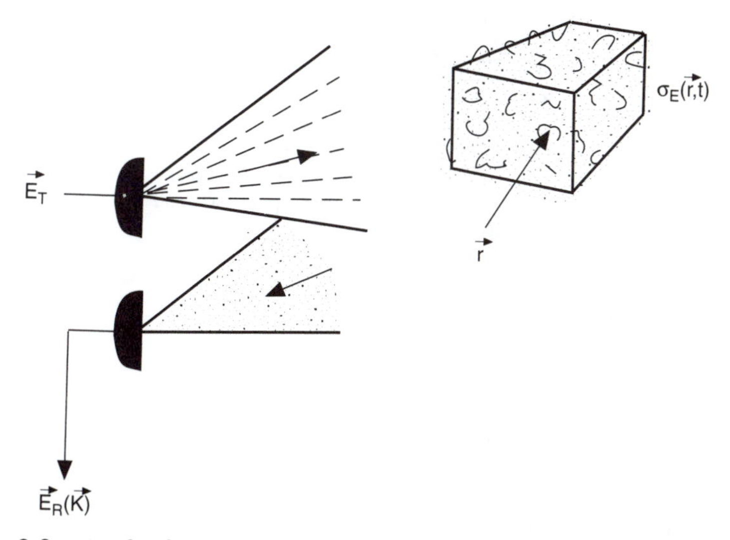

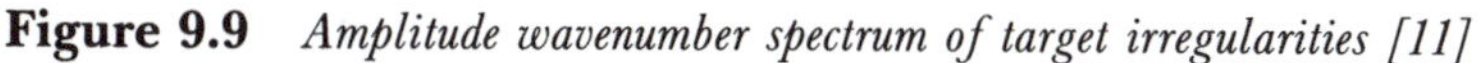
Figure 9.9 *Amplitude wavenumber spectrum of target irregularities [11]*

$$\vec{E}_R(K) \sim \int \sigma_E(\vec{r}, t)\theta^{-j\vec{k}\cdot\vec{r}}\, d^3\,\vec{r}$$

= amplitude wavenumber spectrum of target irregularities = $A(\vec{K})$

Inverse transform gives us

$$\sigma_E(\vec{r}, t) \sim \int A(\vec{K}, \omega)\ \theta^{i\vec{K}\cdot\vec{r}}\ d^3\vec{K}$$

Matched illumination when $E(\vec{K}) = A^*(\vec{K})$

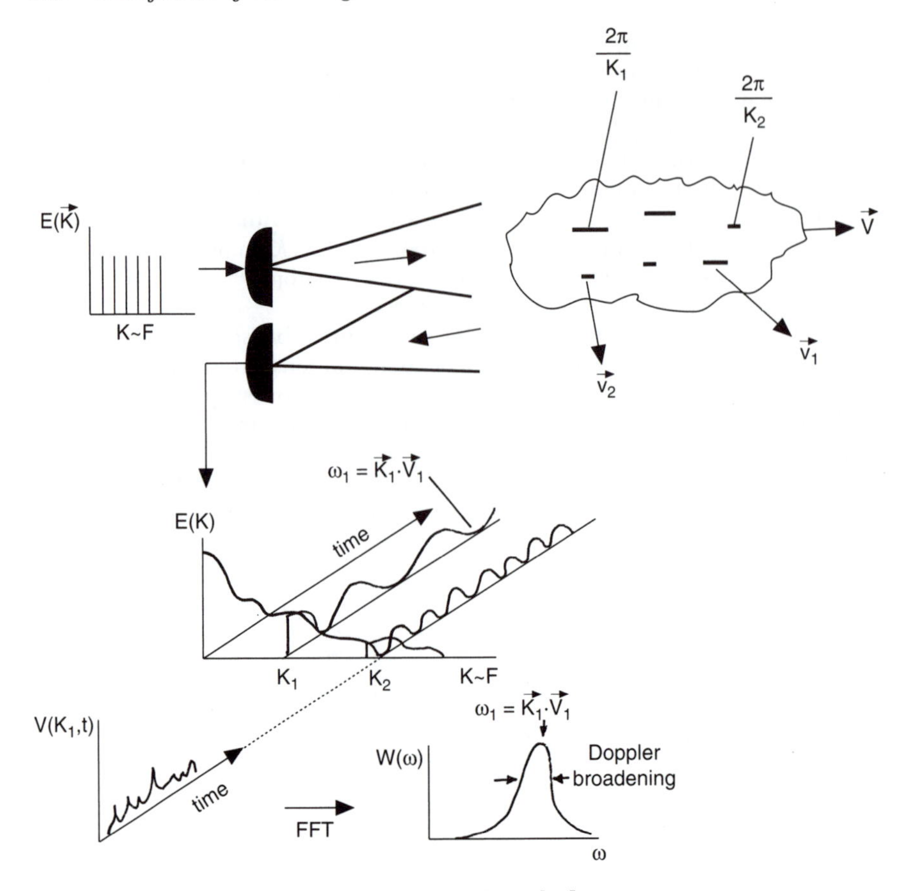

Figure 9.10 *Motion pattern of a flexible object [11]*

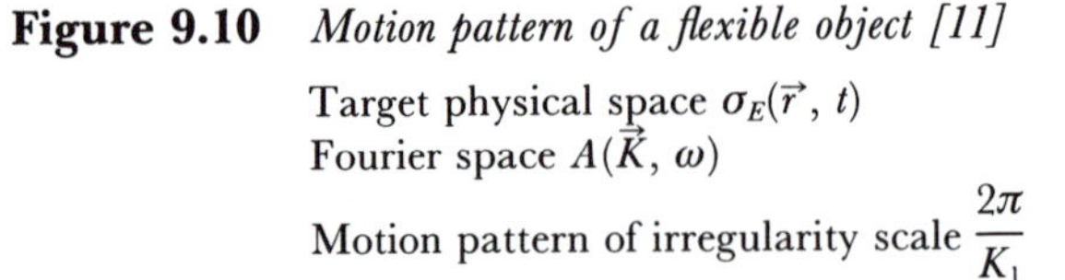

Target physical space $\sigma_E(\vec{r}, t)$
Fourier space $A(\vec{K}, \omega)$
Motion pattern of irregularity scale $\dfrac{2\pi}{K_1}$

$\Delta\vec{K}$) scattered back from the sea surface which is characterised by the delay function $f(\vec{z})$.

This obviously is a four dimensional problem. The normalised correlation in ΔK is then written in the conventional manner:

$$R(\Delta K, t) = \frac{\langle E(K, t)E^*((K + \Delta K), t)\rangle}{\langle |E(K, t)|^2\rangle} \tag{9.10}$$

By substituting for $E(K)$ as obtained from eqn. 9.9 into the general expression of eqn. 9.10, the following result is obtained:

$$R(\Delta f) \sim \int R(\vec{r}) \exp[-j2\pi\Delta f c^{-1}\vec{r}\,]d\vec{r} \tag{9.11}$$

where

$$R(\vec{r}) = \langle |f(\vec{z})|^2 \rangle$$

Eqn. 9.11 states that the envelope of the complex correlation in the frequency domain of waves scattered back from a surface is given by the autocorrelation function $R(\vec{r})$ characterising the surface.

The modulus of the $R(\Delta f)$ correlation is the Fourier transform of the target autocorrelation function. The oscillations give information about range to the object. This brings us to the matched illumination in one dimension, namely in range z.

To examine this theoretical consideration, a large amount of experimental investigation has been performed by Gjessing's organisation. In the case of aircraft identification, the processing scheme is subdivided into four sections in correspondence with the four signature domains as shown in Fig. 9.11.

Signature domain 1: This involves measuring the transverse distribution of the scattering centres and the rate of change of yaw angle.

In this experiment a 6-frequency C-band radar was used. The received power was measured by measuring the AGC voltage of the different frequency channels. Fig. 9.12 shows a plot of the AGC voltage for one of the 6 frequencies as a function of time for two different types of aircraft (F–27 and DC–9–40, respectively).

The processor extracts the amplitude ratio of high and low period amplitude scintillations and also the correlation properties of these over the band of 6 frequencies. On the basis of these a decision diagram can be designed as shown in Fig. 9.13.

Signature domain 2: This domain gives information about the scale selective rattle pattern and is of great significance under certain conditions. The processor first multiplies the complex voltages for the 6 frequencies, thus forming 15 products (beat frequencies). These 'beat voltages' are all simultaneously subjected to Fourier analysis.

Typical Doppler spectra (rattle pattern) are shown in Fig. 9.14 as a selection of 6 out of an ensenble of 15 for two aircraft classes, namely DC–9–40 and British Aerospace 842.

Signature domain 3: Plotting the peak power intensity of the 15 Doppler spectra as a function of beat frequency Δf, domain 3 evolves.

In Fig. 9.15 three values only of Δf are displayed in order to simplify the illustration of the target identification potential of domain 3.

Signature domain 4: This signature domain represents a measure of the rigidity of target. It provides a measure of the degree to which the various scattering centres (scales) constituting the total bulk scattering cross-section move (rattle) in unison.

In evaluating this rigidity performance we shall have to cross-correlate all the 15 beat frequency channels after having performed a frequency shift operation so as to obtain 15 coinciding power spectra, before carrying out the 105 cross-correlation operations (in these there is much redundant information).

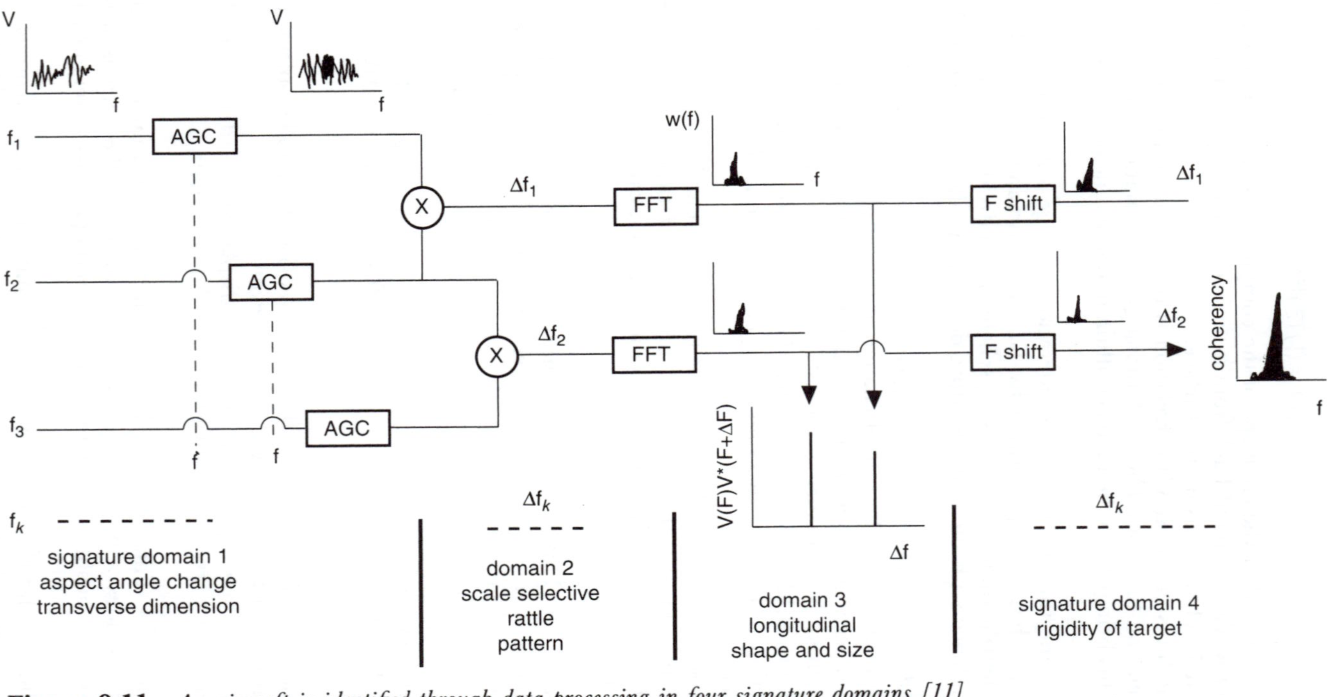

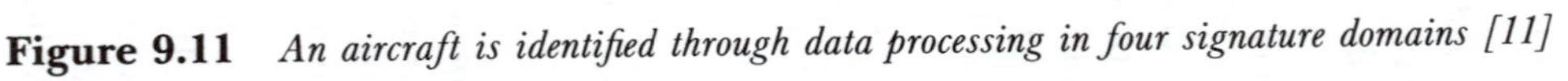

Figure 9.11 *An aircraft is identified through data processing in four signature domains [11]*

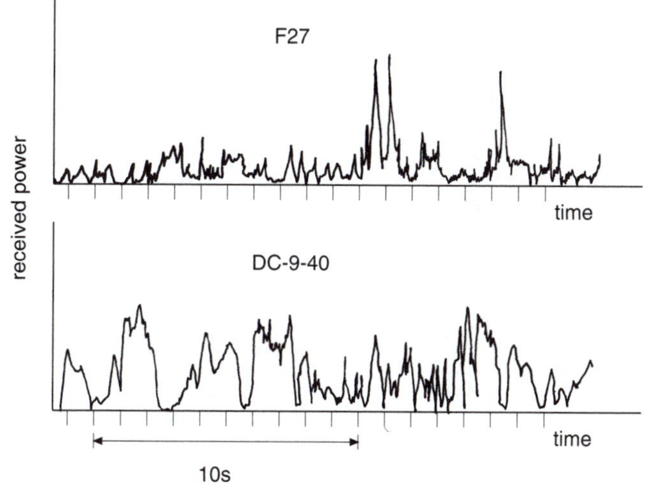

Figure 9.12 *Power versus time recording illustrating the potential of signature domain 1 [11]*

In the current example, for the sake of presenting a comprehensible picture with few words, a simple addition of the 15 frequency shifted spectra are shown for two different aircraft. In the limiting case of random uncorrelated noise (completely flexible object), the 15 summations would lead to a 15-fold enhancement of spectral power density.

In the case of complete correlation (ultimate reigidity), the spectral density will be $\sqrt{15}$ times larger than that corresponding to the flexible object (all scatterers add in phase).

Fig. 9.16 shows that the rigid jet aircraft DC–9–40 has a rigidity factor of 180 whereas the corresponding factor for the propeller plane Super King Air is 54.

Analysis of Doppler spectra associated with the returns from multifrequency radar can also offer a method of suppressing sea clutter and detecting low-flying aircraft and even identifying the type of target [10].

The main disadvantages of this method are:

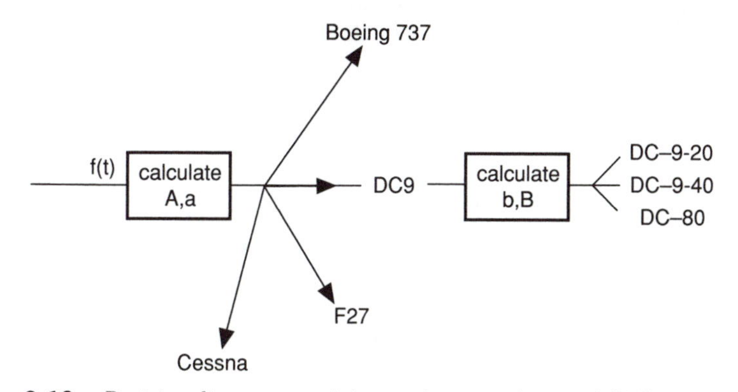

Figure 9.13 *Decision diagram pertaining to signature domain 1 [11]*

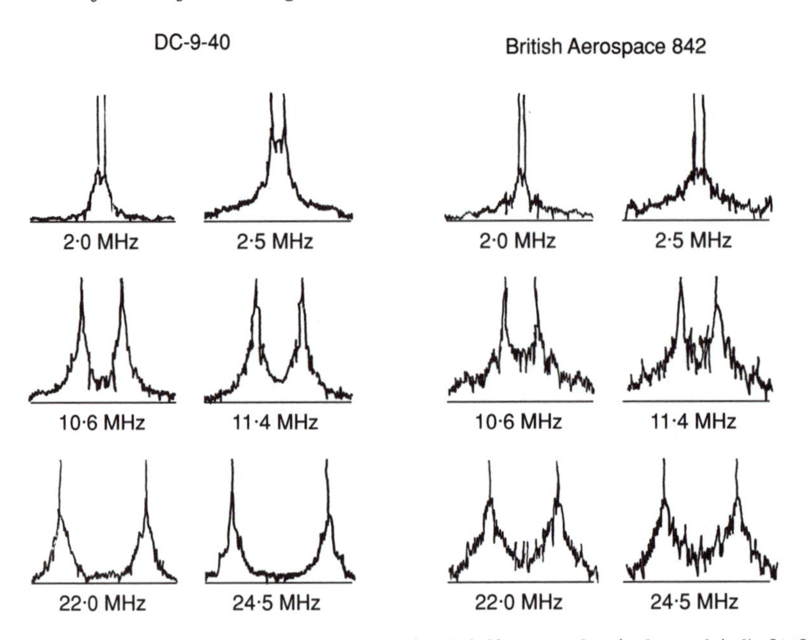

Figure 9.14 *Scale section rattle patterns for 6 different scales (values of Δf) [11]*

(i) It cannot be realised with an ordinary coherent or non-coherent radar
(ii) Very large databases for different types of targets and for different signature domains are needed.
(iii) The algorithm is too complex to be realised in real time.

9.5 Target identification with impulse radar

This technique is based on identification in time domain. It is also assumed that a target can be modelled as a linear time-invariant system with unknown impulse response. In general, the time domain methods involve deconvolution of output signal $v(t)$ from the target, which is related to the input $u(t)$ and target response as

$$v(t) = \int_0^t u(t-s)h(s)ds \tag{9.12}$$

Since $u(t)$ is known *a priori* and $v(t)$ is available from experiments, therefore $h(t)$ can be obtained by solving a linear integration equation. This is equivalent to

$$v_t = \sum_{s=0}^{t} u_{t-s}h_s \tag{9.13}$$

In computationally compact form; h_t is given by the recursive formula

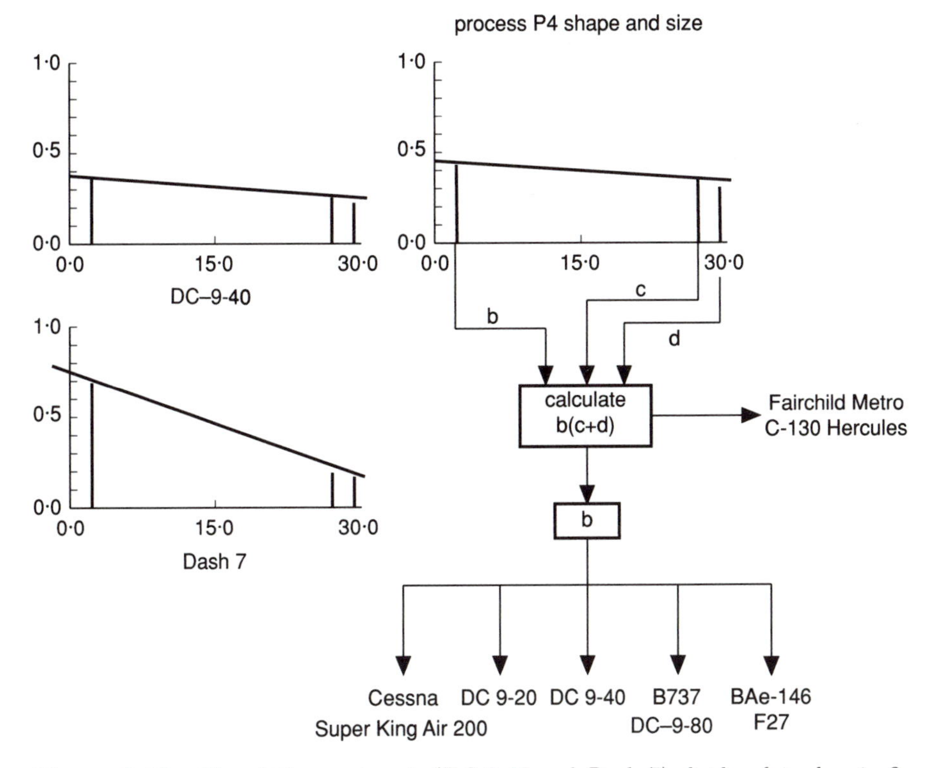

Figure 9.15 *Two different aircraft (DC-9-40 and Dash 7) displayed in domain 3 together with decision diagram for this domain [11]*

$$h_t = \left(v_t - \sum_{s=0}^{t-1} u_{t-s} h_s \right) / u_0 \qquad (9.14)$$

where $u_0 \neq 0$ is assumed.

If the transmission waveform is a delta impulse, then the received signal is the impulse response of the target. Impulse signals, with duration of the order of 1 to 0.1 ns, yield range resolution of the order 15 to 1.5 cm, which is very attractive in target identification.

Hussain [12] suggested a two dimensional decovolution algorithm for target identification with impulse signal.

Consider the three dimensional aircraft shown in Fig. 9.17*a*, which is modelled as a linear system with unique impulse response $h(x, y)$. The impulse response $h(x, y)$ gives the scattering efficiency (or strength) of the aircraft at points in the xy-plane. Thus, it is a function of range x and cross range y of the aircraft. The point scatterers on the surface of the aircraft which yield the backscattered waveform can be identified by two dimensional range resolution. To achieve high resolution in two dimensions x and y, it is assumed that the target in Fig. 9.17*a* is illuminated by an impulse signal $s(x, y)$ with the time variation of a two dimensional rectangular pulse as shown in Fig. 9.17*b*.

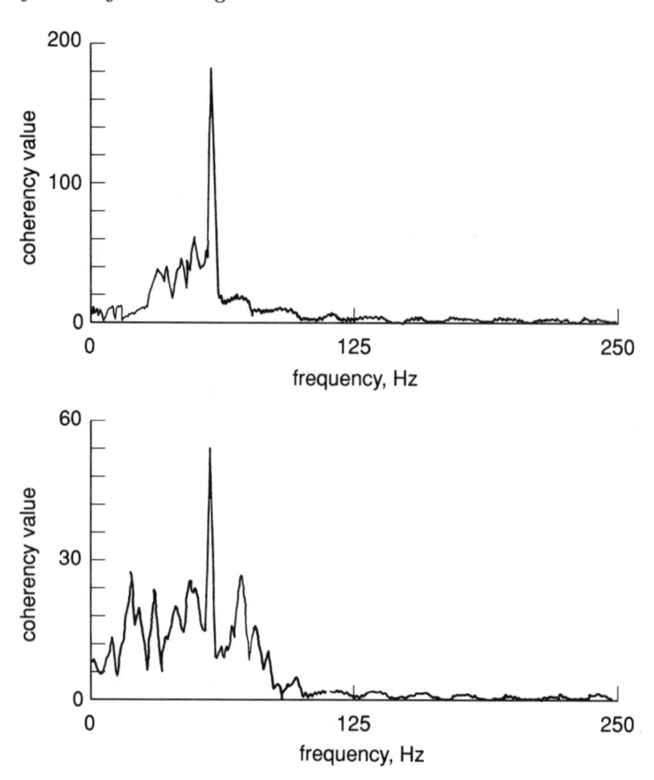

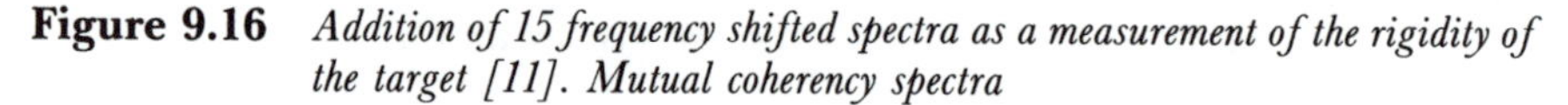

Figure 9.16 *Addition of 15 frequency shifted spectra as a measurement of the rigidity of the target [11]. Mutual coherency spectra*

a JET4.IF6 DC9–40
coherency frequency = 55·10
peak value = 187·90

b PROP8.IF6 Super King Air 200
coherency frequency = 55·10
peak value = 54·24

Since the scattering process from a complex target is modelled as a linear process, the backscattered signal $r(x, y)$ is the convolution of $s(x, y)$ with $h(x, y)$:

$$r(x, y) = s(x, y) * h(x, y) = \int_0^\infty \int_0^\infty s(\lambda, \eta) h(x - \lambda, y - \eta) d\lambda \, d\eta \qquad (9.15)$$

Hence the impulse response $h(x, y)$ of the scattering target can be determined from $s(x, y)$ and $r(x, y)$ by performing a two dimensional deconvolution process in time domain:

$$h(x, y) = r(x, y) \underset{*}{*} s(x) \qquad (9.16)$$

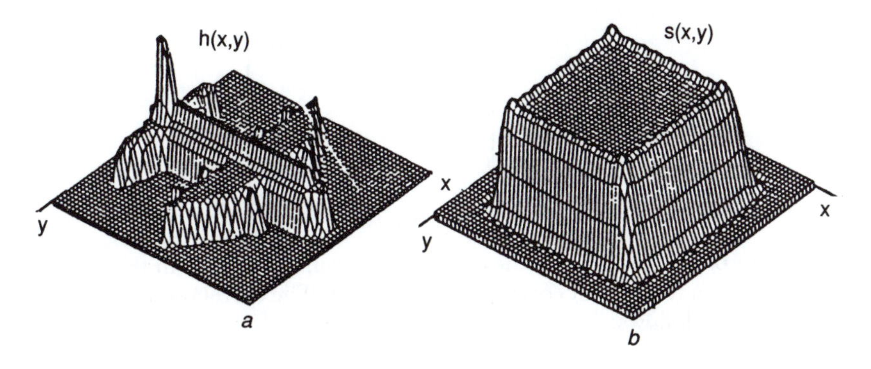

Figure 9.17 *Target modelled as a linear system with impulse response h(x, y) (a); and an impulse signal s(x, y) (b) [12]*

where the double asterist $\ast\!\!\ast$ denotes two dimensional deconvolution. Let the two dimensional sequences $\{s(i, j)\}$ and $\{r(k, l)\}$ represent the signals $s(x, y)$ and $r(x, y)$, respectively,

$$\{s(i,j)\}=\begin{bmatrix} s_{00} & s_{01} & \cdots & s_{0j} \\ s_{10} & s_{11} & \cdots & s_{1j} \\ \cdot & \cdot & \cdots & \cdot \\ \cdot & \cdot & \cdots & \cdot \\ s_{I0} & s_{I1} & \cdots & s_{IJ} \end{bmatrix}, \quad \{r(k,l)\}=\begin{bmatrix} r_{00} & r_{01} & \cdots & r_{0L} \\ r_{10} & r_{11} & \cdots & r_{1L} \\ \cdot & \cdot & \cdots & \cdot \\ \cdot & \cdot & \cdots & \cdot \\ r_{K0} & r_{K1} & \cdots & r_{KL} \end{bmatrix} \tag{9.17}$$

The linear system in Fig. 9.17*a* can now be modelled as a discrete-time system with the impulse response sequence $\{h(m, n)\}$. In discrete-time domain, eqn. 9.15 can be written in terms of summations:

$$\begin{aligned}\{r(k, l)\} &= \{s(i, j)\}^*\{h(m, n)\} \\ &= \sum_{v=0}^{\infty}\sum_{u=0}^{\infty} s(u, v)h(k-u, l-v)\end{aligned} \tag{9.18}$$

From eqn. 9.18, one can determine the sequence $\{h(m, n)\}$ by using the following two dimensional discrete-time deconvolution algorithm:

$$\begin{aligned} h(0, 0) &= r(0, 0)/s(0, 0) \\ h(m, n) &= \frac{1}{s(0, 0)}\,[r(m, n)-w(m, n)], \quad \text{and} \\ w(m, n) &= \sum_{v=0}^{V}\sum_{u=0}^{U} s(u, v)h(m-u, n-v)\end{aligned} \tag{9.19}$$

The deconvolution algorithm in eqn. 9.19 results in the following sequence for the impulse response of the scattering target:

$$\{h(m, n)\} = \begin{bmatrix} h_{00} & h_{01} & \cdots & h_{0N} \\ h_{10} & h_{11} & \cdots & \cdot \\ \cdot & \cdot & \cdots & \cdot \\ \cdot & \cdot & \cdots & \cdot \\ h_{M0} & h_{M1} & \cdots & _{MN} \end{bmatrix} \tag{9.20}$$

A target can be identified from the one whose impulse response in the computer library results in the largest correlation coefficient. The correlation coefficient ρ_q is calculated as follows:

$$\rho_q = C_q / [\tilde{E} E_q]^{1/2}, \quad 0 < \rho_q < 1 \tag{9.21}$$

$$C_q = \sum_{m=0}^{M} \sum_{n=0}^{N} \tilde{h}(m, n) h_q(m, n), \quad q = 1, 2, \ldots, Q \tag{9.22}$$

$$\tilde{E} = \sum_{m=0}^{M} \sum_{n=0}^{N} [\tilde{h}(m, n)|^2, \qquad E = \sum_{m=0}^{M} \sum_{n=0}^{N} [h_q(m, n)]^2 \tag{9.23}$$

In eqns. 9.22 and 9.23, $h(m, n)$ is the calculated impulse response from the backscattered signal, $h_q(m, n)$ is the impulse response of a typical target q stored in the computer library, and Q is the size of the library.

Results of computer simulation are shown in Fig. 9.18*a–d*. In the simulation, a target whose nature is not known *a priori* is illuminated by the signal in Fig. 9.17*b*. The backscattered signal is shown in Fig. 9.18*a*. Three impulse responses of known target, i.e. aircraft, are shown in Fig. 9.18*b–d*. The results of correlating the deconvoluted backscattered signal with all these impulse responses of known targets show that this unknown target is of the type of Fig. 9.18*c*.

In fact, a three dimensional scattering target can be modelled as a linear system whose impulse response is a function of the geometrical shape of the target, scattering efficiency and orientation of the point scatterers on the surface of the target, and target aspect angle. When the target is in motion, the aspect angle will change with time, so that the impulse response will also change with time. Therefore, the target cannot be modelled as a time invariant system. This model is valid only in the case when the target is a stationary target with fixed aspect angle. Therefore, target identification based on the impulse response can be realised in the case of a fixed target. One of the fields of application is to detect underground objects, such as metal pipes, cables, undetonated ordnance etc.

Many types of impulse radars for this purpose have been developed [13–15]. A block diagram of impulse radar is shown in Fig. 9.19.

The transmitter generates a 70 V to 1 kV video pulse with a 0.5–1.0 ns pulse width. This pulse is transmitted through a cross dipole and bowtie antenna. The penetration of the video impulse in soil is about 300 cm. The resolution in depth is about 10 cm to 20 cm. The received signal can be indicated by a display or analysed by a microcomputer. Fig. 9.20 shows the identification process implemented in a microcomputer [16].

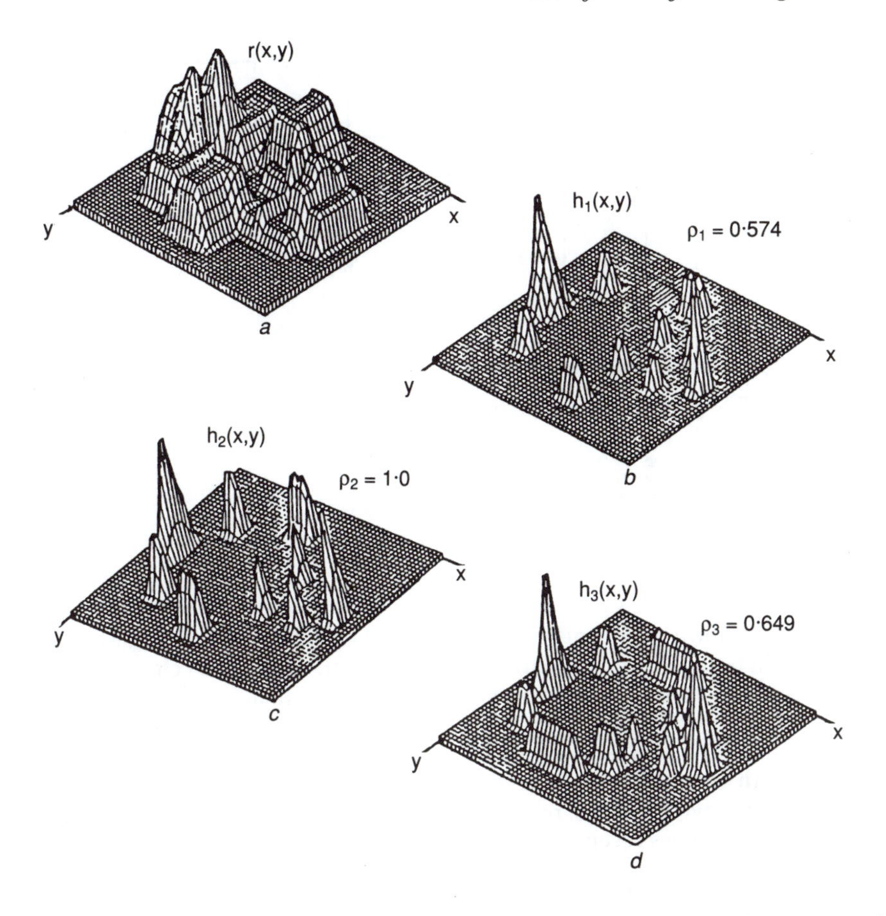

Figure 9.18 *Backscattered signal from an unknown target (a); and impulse responses of known targets stored in a computer library (b–d) [12]*

9.6 Target identification based on spatial processing of coherent echoes

If a high resolution two dimensional target image is desired, one must increase the range and azimuth resolution of the radar system. It is somewhat easier to increase the range resolution by decreasing the pulse width or employing pulse compression. However, it is very difficult to increase the azimuth resolution. Since the azimuth resolution is proportional to the size of the antenna aperture, high azimuth resolution means very large antenna size. For example, 3 dB beamwidth of a rectangular aperture excitation antenna is equal to 0.88 λ/L. If wavelength $\lambda = 0.1$ m, 1 mil beamwidth means 88 m antenna length. This is almost impossible to realise in a real radar system. Furthermore, fixed beamwidth does not mean fixed cross range resolution. The cross range resolution decreases with increasing target range.

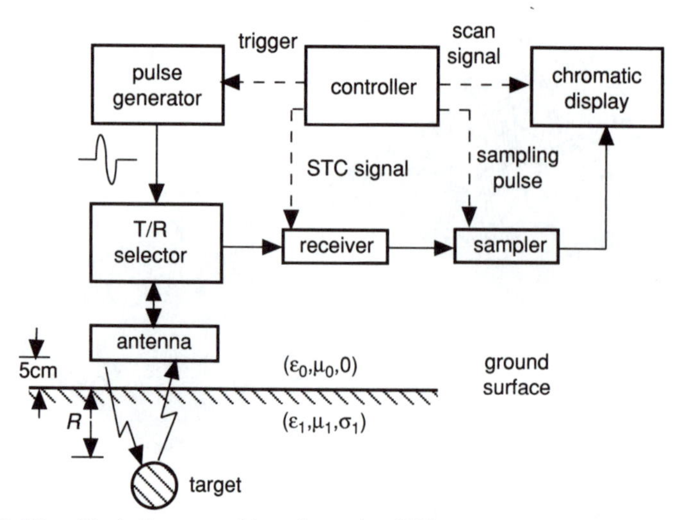

Figure 9.19 *Block diagram of impulse radar [15]*

It is well known that synthetic aperture radar (SAR) can synthesise a very large synthetic aperture by spatially processing the coherent echoes from the ground surface. It utilises the motion of the radar antenna and stores the echoes received from successive positons of the antenna, afterwards combining all these echoes coherently. In this way the effect will be similar to a linear-array antenna whose length is the distance travelled during the reception of the echoes.

Instead of moving a radar relative to a stationary object, it is also possible to process stored echoes received by a stationary radar from a moving target. This processing is called inverse SAR (ISAR) [17, 18]. The difference between a phased array radar, SAR and ISAR is shown in Fig. 9.21.

The ISAR system may be able to distinguish closely spaced targets and even produce a target image in which the different parts of the target are resolved. Such ISAR images may be used to distinguish one type of target from another.

The antenna of ISAR can be a single element antenna or a linear array as in SAR. Although the relative motion between radar and target seems to be the same, the motion of the radar can be controlled in a straight line while the motion of the target is unknown and uncontrollable. Therefore, the most important problem in ISAR is the motion compensation.

Let us assume the simplest condition, namely.that the target is moving along a straight line and the antenna is single element antenna. If two point scatterers separated by ΔA on the basis of target motion are to be resolved, the required integration time can be calculated from the geometry shown in Fig. 9.22.

The discrimination criterion is that the differential range between two scatter points must change by at least half a wavelength $(1/2\lambda)$ during the integration interval T_i. Using this criterion, $\Delta R = \lambda/2$, and $\Delta R \simeq \Delta A \cdot \phi$ for small values of aspect change, ϕ, of interest here. Thus the integration interval becomes

$$T_i = \frac{\lambda R_0}{2\Delta A \cdot v} \tag{9.24}$$

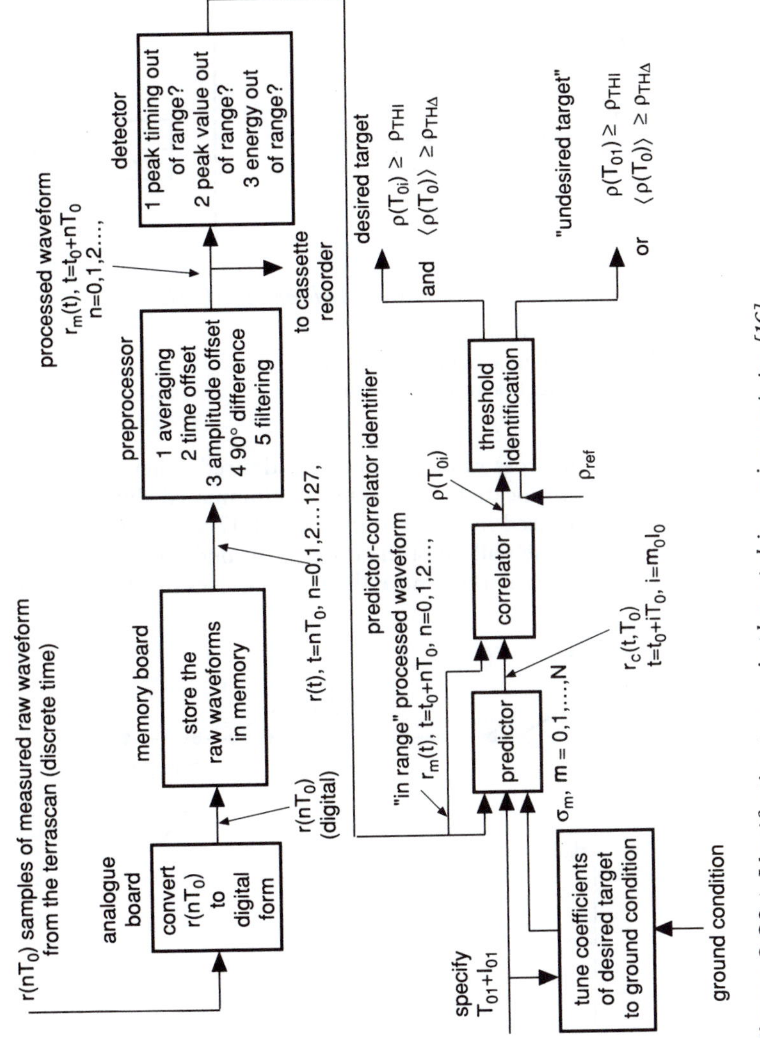

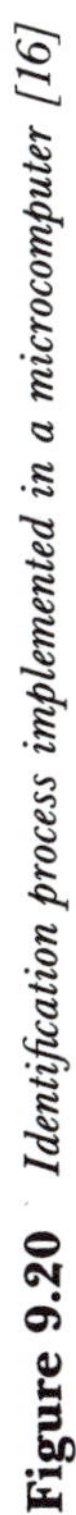
Figure 9.20 *Identification process implemented in a microcomputer [16]*

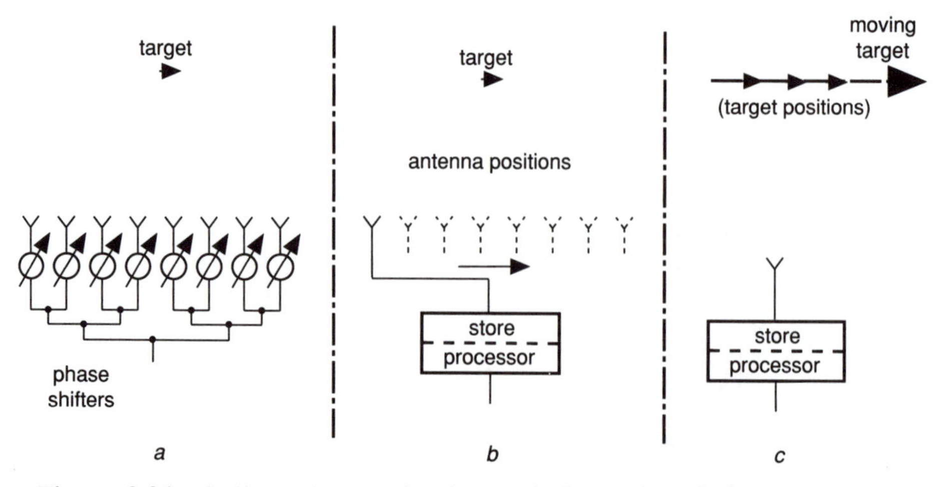

Figure 9.21 *Difference between phased array, SAR and ISAR [17]*

a Linear phased array
b Moving antenna, synthetic aperture
c Moving target, synthetic aperture

which means that if 1.0 m cross range resolution is desired, when a target with the speed of sound and located at 10 km is illuminated by an S-band ($\lambda = 0.1$ m) radar, the required integration time $T_i = 1.47$ s. This indicates that the coherent integration time will be of the order of seconds. If the PRF of the radar is equal to 1000 pulses/s, then more than 1470 pulses must be processed coherently.

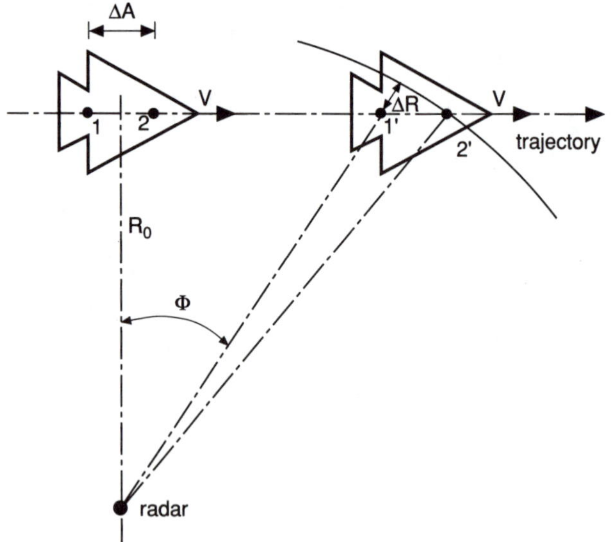

Figure 9.22 *Azimuth resolution obtained from the geometry of differential range change [17]*

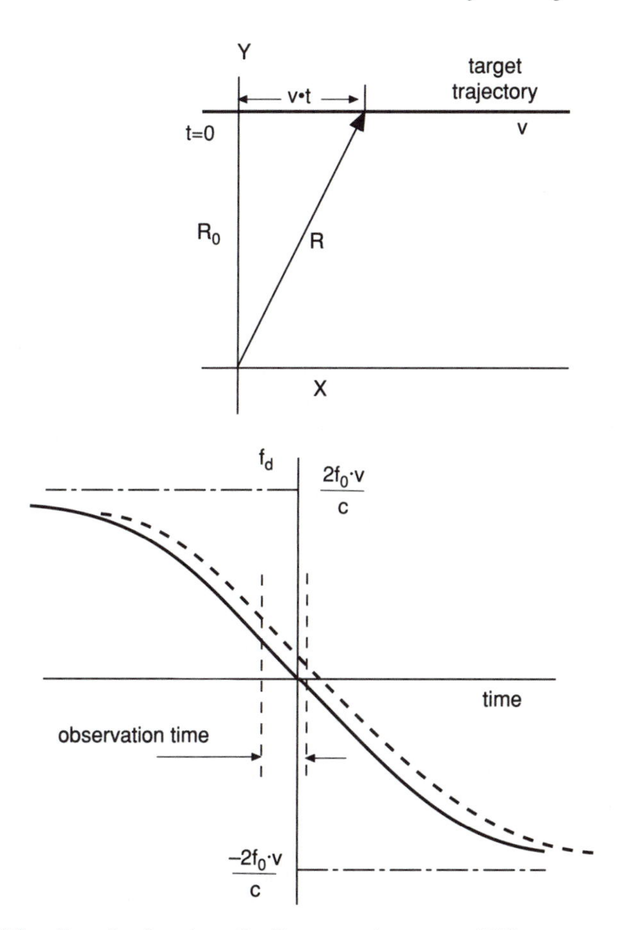

Figure 9.23 *Doppler function of a linear moving target [17]*

The round trip delay function, Δt, is mapped from the time dependent range, $R(t)$, of a moving target by the relationship

$$\Delta t = 2/c \cdot R(t - \Delta t/2) \simeq 2/c \cdot R(t) \tag{9.25}$$

For a target moving with constant velocity on a straight lateral flight path the delay function becomes

$$\Delta t = 2/c \cdot (R_0^2 + v^2 t^2)^{1/2} \tag{9.26}$$

We take the first derivative, which is related to the Doppler shift by $f_d = -f_0 \, \Delta t'$:

$$f_d(t) = -\frac{2f_0}{c} \, (R_0^2 + v^2 t^2)^{-1/2} v^2 t \tag{9.27}$$

This Doppler function is illustrated in Fig. 9.23.

This Doppler function can be expanded into a Taylor series. The expansion of the Doppler function at a particular point, $t = t_1$, becomes

$$f_d(t_1+t) = -\frac{2f_0}{c}\left[\frac{v^2 t_1}{R_1} + \frac{v^2 R_0^2}{R_1^3}\, t\right] \tag{9.28}$$

where R_1 stands for $R_1 = (R_0^2 + v^2 t_1^2)^{1/2}$. Eqn. 9.28 approximates the Doppler function by a linear frequency sweep.

Since we want to discriminate between different scatter points on a moving target, the target spread in Doppler function must be considered. All scatter points belonging to one rigid body will move along the same trajectory with the same speed v, but spread in lateral direction ΔA, so that so far no radial resolution is involved. Therefore the Doppler functions (eqn. 9.27) of particular scatterers will have the same characteristic, but will be shifted in time, $(t - \Delta A/v)$, as indicated by the dashed line in Fig. 9.23.

As is also evident from Fig. 9.23, the target spread has negligible influence on the slope $v^2 R_0^2/R_0^3$ because it can be assumed that the cross-target dimensions, D_A, $D_A \ll R_0$ in $R_1 = [R^2 + v^2(t_1 - D_A)^2]^{1/2}$. Thus the second term of the expansion (eqn. 9.28), representing the sweep rate, will stay the same for all scatter points within the spread of a particular target. In the first term of expansion eqn. 9.28, however, t_1 is affected by the amount $(t_1 - \Delta A/v)$ introducing constant Doppler offsets. The total Doppler spread becomes

$$|D_{fd}| = \left|\frac{2f_0 v}{cR_1} D_A\right| \tag{9.29}$$

From the characterisation of the return signal under the assumed target motion we can deduce the reference signal, which is simply the complex conjugate of that return signal. Since the optimum signal processing involves matched filter reception or correlation processing, it can be implemented by the equation

$$g(\tau) = \int_{T_i} \psi(\tau)\psi_R^*(t-\tau)\,d\tau \tag{9.30}$$

where $\psi(t)$ is the return signal, which can be expressed by

$$\psi(t) = u(t-\tau)\, \exp[j2\pi f_0(t-\Delta t)] \tag{9.31}$$

Applying eqn. 9.30, we take $\tau = 0$ and combine the product of the return signal and reference envelope into $P(t, \tau)$. The optimum processing of a particular reference point then turns out to be

$$g(0) = \int_{T_i} P(t, 0)\, \exp[-j\phi_R(t)]\,dt \tag{9.32}$$

With $\phi(t) = \int 2\pi f_d(t)\,dt$, and taking the integration constant as arbitrary zero, we obtain from eqn. 9.28 the reference phase function

$$\phi_R(t_1+t) = \frac{4\pi f_0}{c}\left[\underbrace{\frac{v^2 t_1}{R_1}\, t}_{\text{steering}} + \underbrace{\frac{v^2 R_0^2}{R_1^3}\,\frac{t^2}{2}}_{\text{focusing}}\right] \tag{9.33}$$

Using phased array terminology, the linear phase variation over the aperture (or equivalent integration interval T_i) causes beam steering towards the

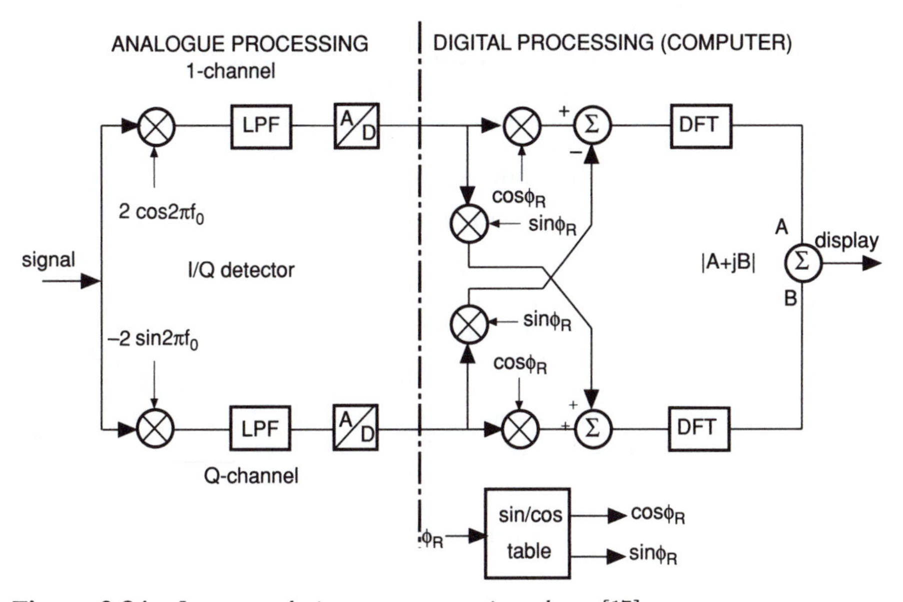

Figure 9.24 *Inverse synthetic aperture processing scheme [17]*

LPF = low pass filter
DFT = discrete Fourier transform

reference point and the quadratic phase variation causes focusing at finite range. As mentioned above, the focusing factor is the same for all scatter points on a particular target (aircraft dimension). To plot the image of the target, we have to vary the steering factor only, to scan overall cross-positions A, by adding the steering term

$$\phi_s(A, t) = \frac{4f_0}{c} \frac{v}{R_1} At \tag{9.34}$$

Thus eqn. 9.32 can be rewritten as follows:

$$g(0, A) = \int_{T_i} \{P(t, 0) \exp[-j\phi_R(t)]\} \exp[-j\phi_S(A, t)]dt \tag{9.35}$$

Eqn. 9.35 can be interpreted as a Fourier transform of the term in brackets, { }, and A becomes the spectral domain variable. The resultant image processor for cross-range dimensions is shown in Fig. 9.24.

The complex conjugate multiplication of the return signal with the reference phase function compensates for the motion of the reference point. Motion compensation of the reference point generally transforms the target motion into plain rotation of the target around the reference point. Therefore, plain target rotation requires no further focusing.

For arbitrary target motion, higher order velocity derivative can no longer be neglected, and search for the proper phase reference function might become very extensive. An accurate reconstruction of the target flight path can be obtained by means of the target phase history. In the case of a radar system with a high range resolution it is possible to use the phase history of a single scatter point on the target as a reference [19].

In the combination of two dimensional target image, high radial range resolution is also required. The radial range resolution, ΔR, is related to the reciprocal of the signal bandwidth, Δf, as follows:

$$\Delta R = c/2\Delta f \tag{9.36}$$

where c is the velocity of light. For reasonable resolution the required bandwidth would be of the order of several hundred MHz to GHz. Although the bandwidth of the transmitted signal is very large, the bandwidth required for the actual information rate is low. Considering, for example, a two dimensional image with 100×100 resolve elements of 10 bit amplitude dynamic range processed in 1 s, the information bit rate would be 10^5 bit/s compared with 10^8 to 10^9 Hz signal bandwidth. As we are interested in a small range window only, not bigger than the target, a processing scheme known as the spectrum analysis technique can be recommended here (see Fig. 9.25*a* and *b*). This technique can reduce the bandwidth by the factor of pulse compression

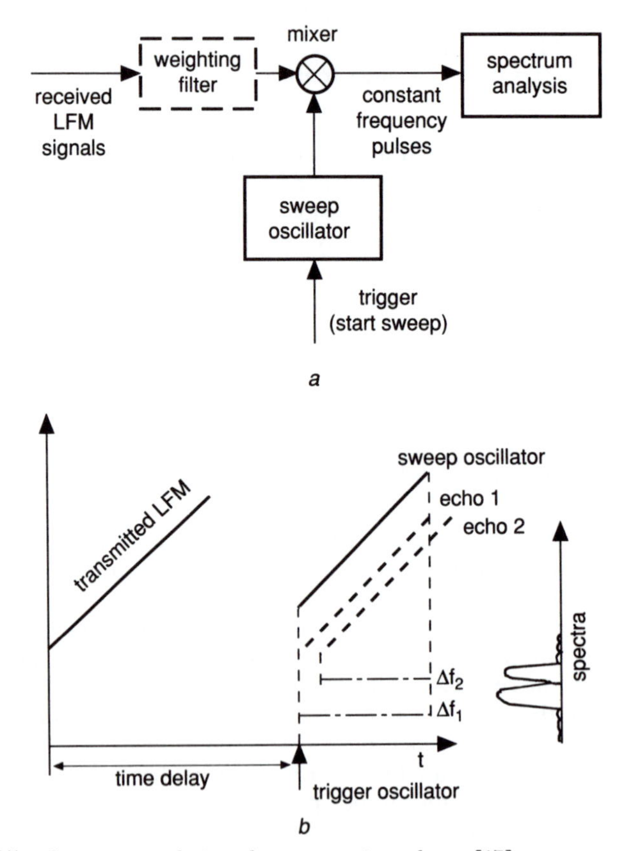

Figure 9.25 *Spectrum analysis pulse compression scheme [17]*

ratio, whilst the range data are converted into parallel frequency channels functioning as range gates.

Finally, the combination of both processing schemes of pulse compression and of synthetic aperture generation provides a grid of range and cross-range resolution on the body of the target. The resultant two dimensional image processor is shown in Fig. 9.26.

After weighting and LFM ('chirp') conversion the return signal enters the filter bank, which functions as parallel range gates. The filter bank is composed of identical low-pass filters, at which frequency offsets are achieved by mixing with different frequencies. For further coherent digital processing (synthetic aperture) each channel is split into its quadrature components (I&Q). For each range channel separately DFT processing is performed, presenting scans of synthetic aperture beams. The applicability of fast Fourier transform (FFT) processing, however, is still limited by high resolution waveform beam pattern interaction. Although the coarse effect has been compensated for by range tracking, the small beam diversions for scanning the target are not taken into account. Alternatively, a more elaborate processing scheme must be used, providing separate motion compensation for each beam.

When an array antenna is used, a somewhat different geometry is involved (Fig. 9.27) [20].

When the aperture is focused at the object plane the lateral extent of the focused spot (cross-range resolution) is approximately

$$\Delta x = \lambda R/2D \tag{9.37}$$

where λ is the wavelength, R is the range and D is the aperture dimension.

In order to derive a synthetic antenna pattern all received echoes must be multiplied by a complex antenna weighting function and coherently summed. The aperture response then becomes

$$S = \sum_{\theta} e(\theta) W(\theta) \exp[-j\phi(\theta)] \tag{9.38}$$

where $e(\theta)$ is the echo obtained from the array element with aspect angle θ, $W(\theta)$ is the amplitude weighting function and $\phi(\theta)$ is the phase weighting function. By means of the phase weighting function the aperture can be focused at any particular point in the object space, which means that the summed echoes must be forced to be in phase for such a point. This can be achieved by compensating for the phase difference along the synthetic antenna, which is equivalent to compensating for the range differences between each element and the point of focus. If the antenna has to be focused at the point (x, y) in Fig. 9.27, eqn. 9.38 must be of the form (neglecting $W(\theta)$)

$$S(x, y) = \sum_{\theta} e(\theta) \exp[-j(2\pi/\lambda) 2r(x, y, \theta)] \tag{9.39}$$

where $r(x, y, \theta)$ is the distance between element θ and the point (x, y). The factor 2 appears owing to the two-way propagation path. From Fig. 9.27 it can be derived that

$$r(x, y, \theta) = R(\theta) [1 + (2/R(\theta)) (y \cos\theta + x \sin\theta) (x^2 + y^2)/R^2(\theta)]^{1/2} \tag{9.40}$$

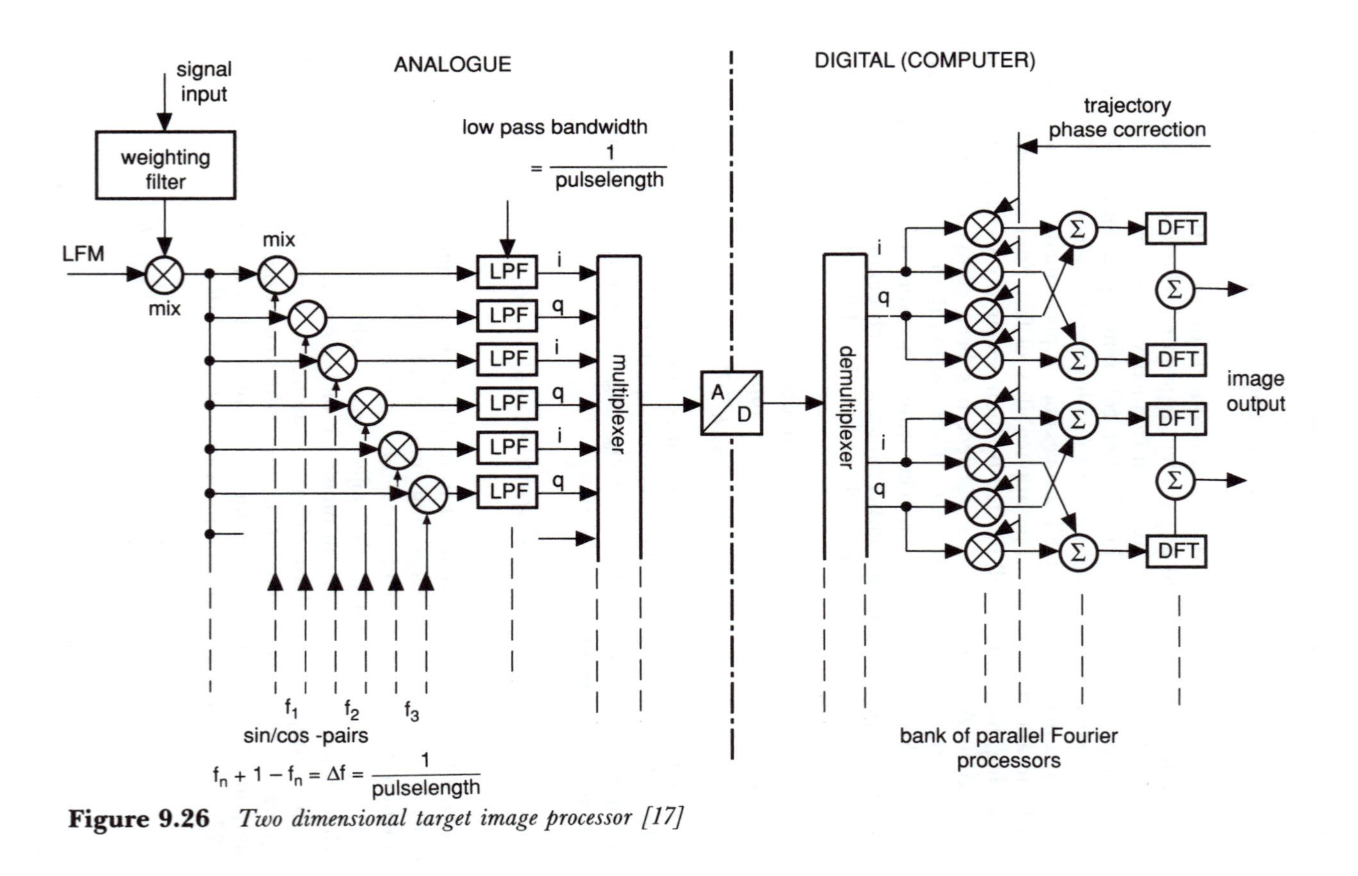

Figure 9.26 *Two dimensional target image processor [17]*

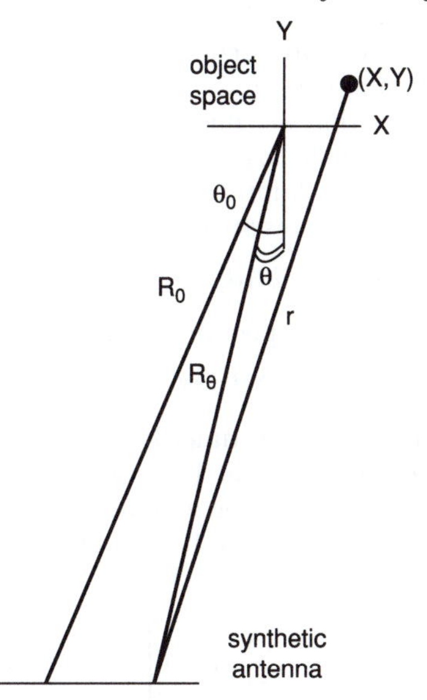

Figure 9.27 *Geometry of ISAR with array antenna [20] © IEEE, 1984*

where $R(\theta)$ is the distance between element θ and the centre of the object space. After substitution of $r(x, y, \theta)$ into eqn. 9.39 and rewriting the square root term in a binomial expansion and assuming $R(\theta) \gg x+y$, the response of the synthetic aperture focused at the point (x, y) becomes

$$S(x, y) = \sum_{\theta} e(\theta) \exp[-j(4\pi/\lambda)R(\theta)]$$

$$\exp[-j(4\pi/\lambda)(y \cos\theta + x \sin\theta)] \qquad (9.41)$$

The first exponential term represents the removal of the motion of the target centre relative to the radar, i.e. motion compensation. The second exponential term represents the focusing operation.

In the case of straight-flight assumption, the phase polynomial for motion compensation can be approximately quadratic:

$$\phi(\theta) = \alpha t_\theta^2 + \beta t_\theta + \gamma \qquad (9.42)$$

where t_θ is the time the echo corresponding to element θ is received. Because the constant γ can be neglectd, only the coefficients α and β are of importance.

In the case of arbitrary motion or for a long observation of straight flight path, the high order terms can no longer be neglected. The echo sequence is roughly pre-corrected by compensating the high order terms using the measured target position data.

The focusing operation can be obtained from eqn. 9.41, which becomes, after motion compensation,

$$S(x, y) = \sum_{\theta} e'(\theta) \exp[-j(4\pi/\lambda)(y \cos\theta + x \sin\theta)] \qquad (9.43)$$

where $e'(\theta)$ denotes the motion compensated echoes.

In general, ISAR imaging is only possible for relatively small apertures, since good motion compensation becomes difficult for long flight paths, because, in that case, Δy will be much larger than the target extent, which is no use for imaging in the y direction. Another consequence of small aperture angle is that the aspect angle can be represented by

$$\theta = \theta_0 + k\Delta\theta, \qquad k = 0, \pm 1, \pm 2, \ldots \tag{9.44}$$

where $\Delta\theta$ is the change in aspect angle between two successive observations:

$$\Delta\theta = v\Delta t \cos\theta_0 / R_0 \tag{9.45}$$

where Δt is the pulse repetition time. Furthermore, the sine and cosine of $k\Delta\theta$ can be approximated by $k\Delta\theta$ and 1, respectively.

Rewriting eqn. 9.43 gives for the magnitude of the one dimensional response, i.e. along the x'-axis tangential to the line of sight to the centre of the motion-compensated target:

$$|S(x')| = \left| \sum_k e'_k \exp\left[-j(4\pi/\lambda)k\Delta\theta x' \right] \right| \tag{9.46}$$

Eqn. 9.46 represents a Fourier transform, indicating that an approximate image can be obtained by Fourier transforming the motion compensated echoes as mentioned before.

Eerland [20] gave some experimental results of ISAR realised with the FUCAS phased array radar system. The system operates around 5500 MHz and has a broadside beamwidth of 4° and a range resolution of 150 m. The PRF of this radar is 1000 Hz. The echo data of a single target was recorded for a reasonable time, e.g. 10–60 s. Therefore, the target was tracked by means of an α–β filter operating on range-monopulse data obtained from incoherent integration of 100 echoes.

Only one dimensional images were obtained owing to the low range resolution. The resulting images of a Boeing 737–200 are shown in Fig. 9.28*a* and a Douglas DC–9–30 in Fig. 9.28*b*, which were obtained by successively processing 2″ of echo data with an overlap of 1″. The cross-range resolution is about 1.2 to 1.0 m for these two Figures, respectively.

Comparison of the images of same aircraft shows similarities but also differences; however, most images give the proper target length, which for the B–737 is 31 m and for DC–9–30 is 36 m. The differences of these images are due to the linear motion compensation used.

Dike *et al.* [21] also gave some computer simulation results of ISAR for aircraft classification. The system simulated was the ALCOR radar collecting data in the coherent wideband mode. The parameters used were a centre frequency of 5659 MHz, bandwidth of 500 MHz, and range samples every 0·176 m. The range resolution is about 0·3 m. Therefore, high resolution two dimensional images can be obtained. The aircraft modelled in these simulations was a Learjet 36A.

Methods were developed for determining the motion of the unknown aircraft, scaling the radar images from Doppler frequency to cross range, and producing overlays of targets corresponding to the radar image projection planes.

The first-order form of image analysis used assumes simple rotation of the target about one axis orthogonal to the radar incidence direction. For a constant rotation rate, the cross range resolution is approximately given by

$$R_{CR} = 1.3(\lambda/2)/\Delta\theta \tag{9.47}$$

where $\Delta\theta$ is the angle through which the target rotates, based on the component of relative rotation vector orthogonal to the radar line of sight (RLOS). The factor 1.3 is used when a Hamming weighting function is applied during processing for sidelobe reduction.

The two dimensional plane in which the target image is produced is defined by the RLOS and the relative angular velocity vector Ω_{tr} between the target and

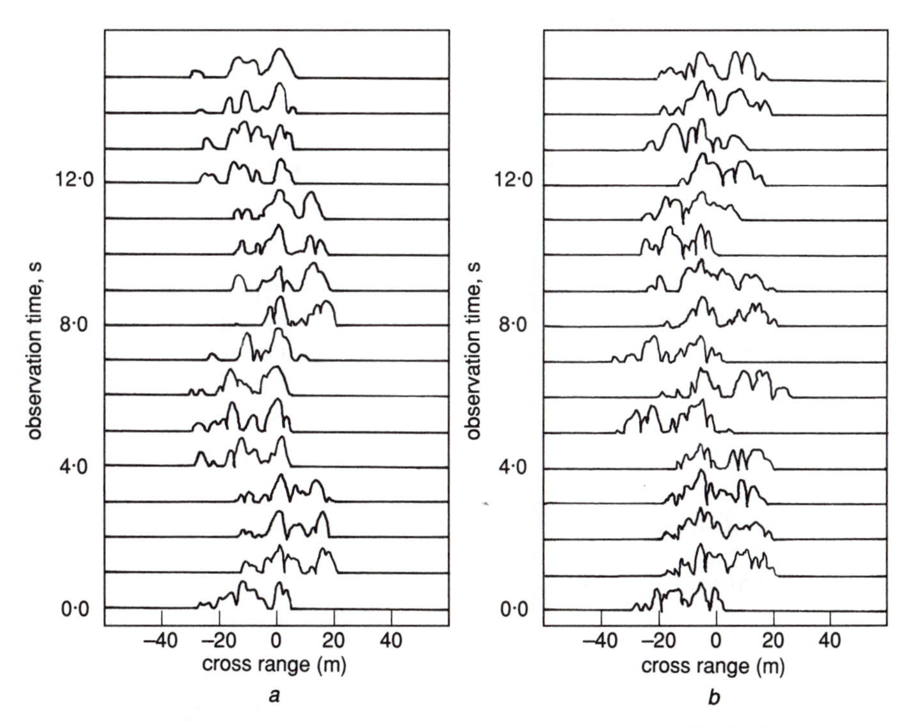

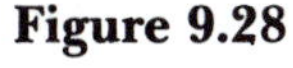
Figure 9.28 *ISAR images of B–737–200 (a); and DC–9–30 (b) [20] © IEEE, 1984*

a B–737–200
$R_0 = 16{\cdot}2$ km
$\theta_0 = 26°$
$V = 202$ m/s
$H = 10{\cdot}3$ km

b DC–9–30
$R_0 = 10{\cdot}4$ km
$\theta_0 = 28°$
$V = 150$ m/s
$H = 1{\cdot}4$ km

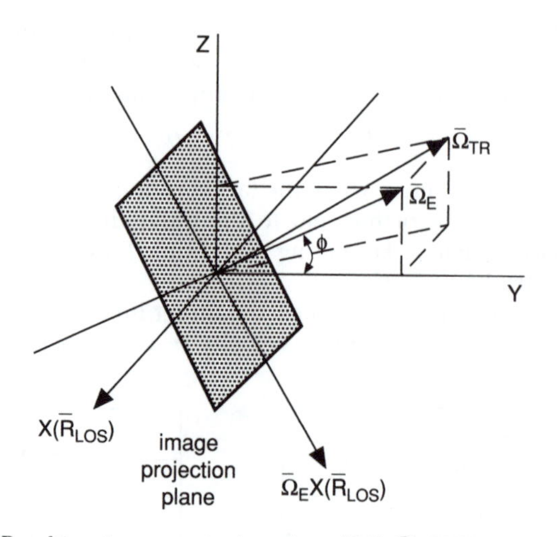

Figure 9.29 *Resulting image projection plane [21] © IEEE, 1980*

the RLOS. The component of Ω_{tr} orthogonal to the RLOS, denoted by Ω_e, defines the normal to the image projection plane. The magnitude of this vector is needed to convert Doppler frequency to cross range and determine the required integration time for a specified cross range resolution. Differences in Doppler frequency Δf_d are then related to the cross range extent ΔR_{CR} between scattering centres by

$$\Delta R_{CR} = \Delta f_d (\lambda/2)/\Omega_e \tag{9.48}$$

An example of the image projection plane for a given Ω_{tr} is illustrated in Fig. 9.29.

The radar imaging simulation program assumes an aircraft modelled as a finite collection of points $\vec{P}_i$. Each point is assigned a radar cross-section σ_i, which is aspect angle independent. The aircraft motion provides updated position, velocity, roll, pitch, yaw and other information on the aircraft. The point scatterer description of the aircraft consisted of scatterers at its nose, engine intake and exhaust, wing pods, and horizontal stabiliser extremities. The time that the image is constructed is the time that the first data sample is taken. Data collection continued until the aircraft had undergone sufficient aspect angle change relative to the radar to give across range resolution of 1.5 ft (appoximately 4.0°).

The algorithms for classification of aircraft depend on first forming two dimensional images scaled from Doppler frequency to cross range. The reference library consists of information describing only the physical shape of the possible target. From the track data, the orientation of the target in the image projection plane is determined. Using this information, the locations and amplitude of scattering centres in the proper target–radar orientation may be determined.

The first algorithm which can be used as a prefilter, is based on metric measurements along four different vectors in the image projection plane. From the motion solution of the aircraft to be identified, the vectors which describe the image projection plane, as well as the wing and velocity unit vectors describing the orientation of the aircraft, can be extracted. The locations of responses within the image projection plane of the aircraft to be identified are obtained by projecting the co-ordinates of the predicted predominant scattering centres of the aircraft we wish to compare with the unknown into the image projection plane, using the slant range and cross range vectors obtained from the motion solution.

The results obtained, when this metric algorithm was used to separate an EA–6B from the remaining 15 aircraft in the library, are shown in Fig. 9.30. The horizontal axis indicates the 10 different image projection planes. The vertical axis is simply the sum total difference (or error) of the projections of responses along the slant range, cross range, wing and velocity vector between the EA–6B and the aircraft plotted. Thus, using this type of 'difference measurement' between aircraft, the EA–6B is most like an F-111A and an MIG 23.

The second algorithm is a comparison, in the same image projection plane, between the predicted locations of the two dimensional responses and the two dimensional image of the unknown target. From the generation of the responses of the reference aircraft, the co-ordinates of their dominant scattering centres are known, so that a two dimensional comparison based on the presence/absence of scattering responses at predicted locations can be made. The results of exercising this algorithm using one of the image projection planes are shown in Fig. 9.31. The vertical axis in each Figure represents the percentage of scattering centres of the RCS response of the reference aircraft that do not have a scattering centre of the RCS responses of the unknown aircraft (EA–6B) within a specified radius. Comparing the EA–6B with itself would result in 0% error.

Berkowitz *et al.* [22, 23] proposed a high resolution radio camera imaging technique based on the concept of spatial sampling suggested by Steinberg [24, 25]. This technique is realised with a large linear array. The signals received by different elements of the array are sampled with very short sampling intervals. The digitised data are recorded and rearranged for off-line processing in the computer. In the data processing, two types of adjustments are required to synthesise very sharp beam focusing on the desired scan point in the radar target: (i) time sampling; and (ii) phase correction. For a specified scan point, the time delay from the transmitter to each array element is calculated and used to pick up the proper sampled data at each element. Such a time delay is also used to compute the required phase correction for the selected data at each array element. Coherent combining of the adjusted data at every element will form the image of the desired target point. Since general distortion of a large array and phase deviation of the local oscillator of a multi-element array are expected, a single reference source in the general direction of the target area is used to calibrate the array. This adaptive beamforming procedure will focus a beam on the reference source, and then the beam is open-loop scanned to the desired target point vicinity. In order to obtain a high resolution two dimensional image, range resolution of such a linear array can also be achieved by transmitting a waveform of very short pulse duration. The procedure and technique presented above is shown in Fig. 9.32.

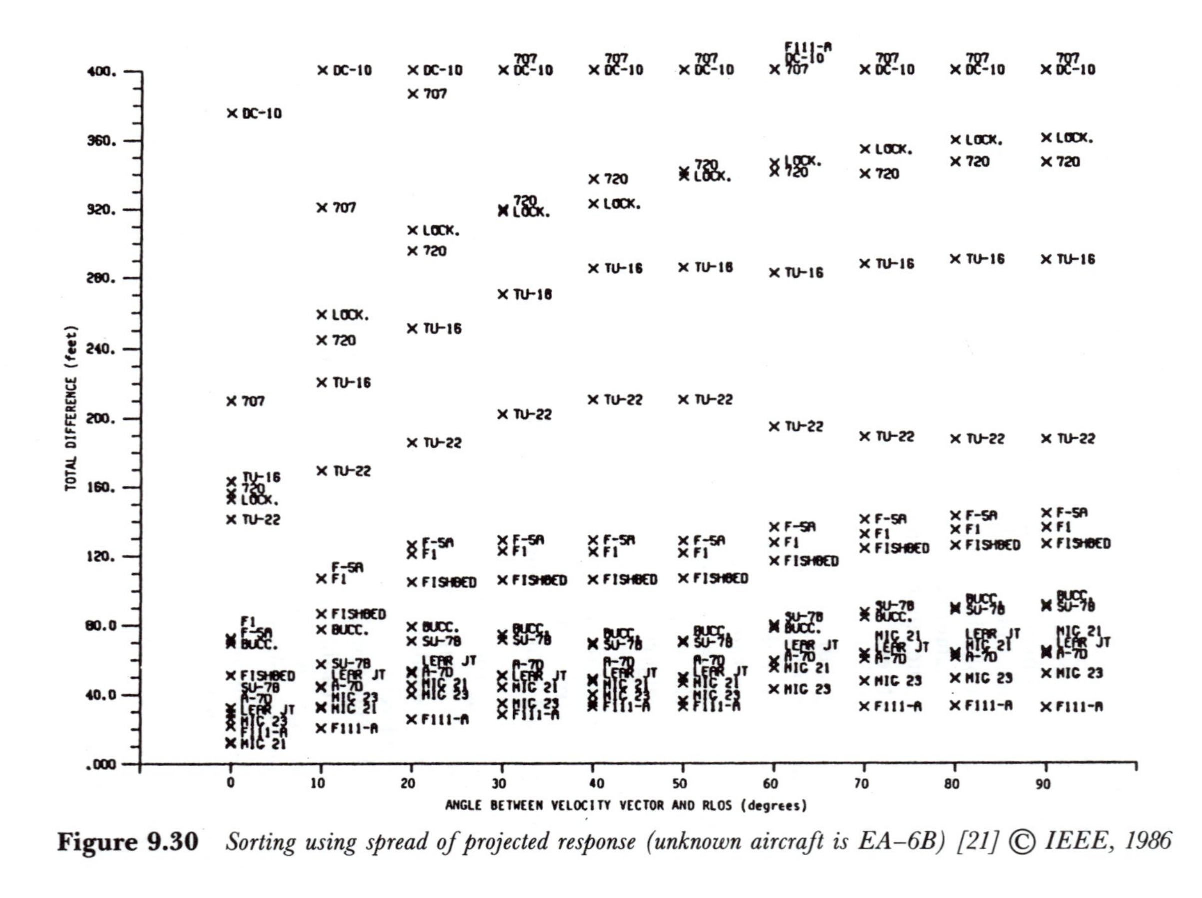

Figure 9.30 *Sorting using spread of projected response (unknown aircraft is EA–6B) [21] © IEEE, 1986*

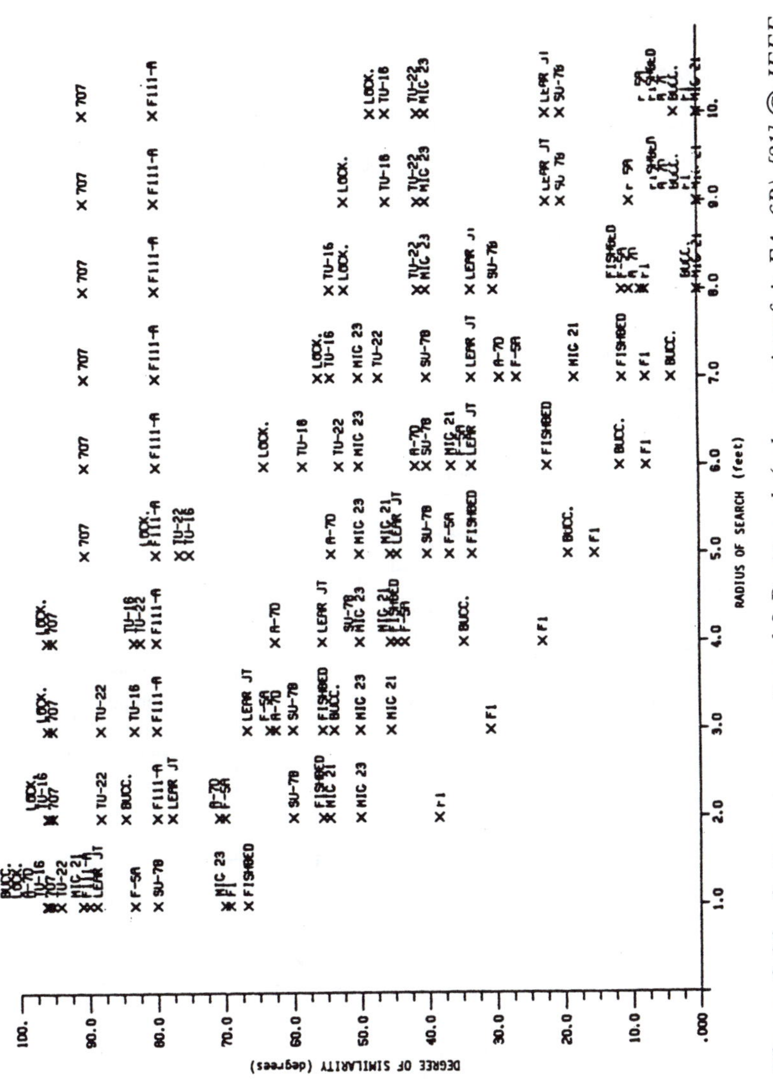

Figure 9.31 *Sorting using nonreciprocal 2-D approach (unknown aircraft is EA-6B) [21] © IEEE, 1986*

The procedure of the fundamental algorithm discussed above is formulated below [22]:

(i) *Illumination signal transmitted:*

$$v_T(t) = m_T(t)\ \exp(j\omega_0 t) \tag{9.49}$$

where m_T is the waveform and ω_0 is the carrier frequency.

(ii) *Received signal at the nth array element:* $v_{Rn}(t)$

(iii) *Frequency conversion:*

$$v_{Vn}(t) = v_{Rn}(t)\ \exp(-j\ \omega_0 t + j\phi_{on}) \tag{9.50}$$

where ϕ_{on} is the phase deviation of the local oscillator at nth element.

(iv) *Signal sampling time at the nth array element:*

$$t_{sn} = T_{In} + (k-1)\Delta t, \quad k = 1, 2, \ldots, k_F \tag{9.51}$$

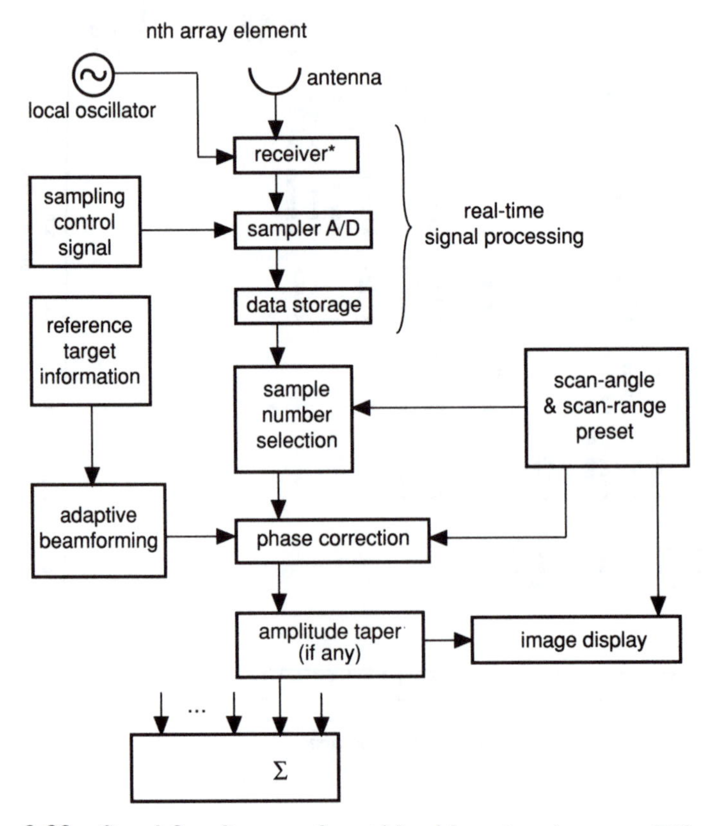

Figure 9.32 *Signal-flow diagram of a wideband large imaging array [22]* *© IEEE, 1984*

* Options: (i) Waveform matched filter, (ii) Pulse compression filter plus spectrum shaping

where Δt is the sample interval, T_{In} is the starting sample time, k_F is the end sample number, and T_{In} and k_F are set by the configuration of array, target and sampling scheme required at the nth array element.

(v) *Sampled data matrix:*

$$[V_{vn}(k)]; \quad k=1, 2, \ldots, k_F; \quad n=1, 2, \ldots, N_e \tag{9.52}$$

where $V_{vn}(k)=v_{Vn}[T_{In}+(k-1)\cdot\Delta t]$, N_e is the total number of array element. For a quadrature sampling scheme, $V_{In}(k)$ and $V_{Qn}(k)$ are obtained.

(vi) *Sample number selection:*
First, the time delay τ'_{jn} for the jth scan point in the image domain and the nth array element is calculated. Then the sample number is computed as follows:

$$k_{jn}=\left[\frac{\tau'_{jn}-T_{In}}{\Delta t}\right]_{\text{round off}}+1 \tag{9.53}$$

With this sample number, k_{in}th data of the sampled data stream at the nth array element are picked up.

(vii) *Sampled number calibration and adaptive beamforming:*
An isolated reference target is used to calibrate the imaging array. With the information on the reference target location, the sample number for such a reference target is computed below:

$$k_{Bn}=\left[\frac{\tau'_{Bn}-T_{In}}{\Delta t}\right]_{\text{round off}}+1, \quad n=1, 2, \ldots, N_e \tag{9.54}$$

where τ'_{Bn} is the time delay for the reference target and the nth array element. Then the strong echoes from the reference target can be associated in the sampled data with the k_{Bn}s and the sample numbers of the recorded data are calibrated. The strong echoes are the beamforming data for the adaptive beamforming process. After picking up the beamforming data, the phase subtraction is performed to make such data in phase. Considering an ideal target model which has no propagation perturbations, the phase being subtracted can be expressed as

$$\phi_{Bn}=-\omega_0\tau'_{Bn}+\phi_{on}+\phi_{oo}+\phi_{Bn}, \quad n=1, 2, \ldots, N_e \tag{9.55}$$

where ϕ_{oo} is a constant phase and ϕ_{Bn} is the beamforming phase error.
The target data of all scan points in the image domain are also phase-subtracted by ϕ_{Bn}. For the jth scan-point we have

$$V_{Dn}(j)=V_{vn}(k_{jn})\exp(-j\phi_{Bn}), \quad n=1, 2, \ldots, N_e \tag{9.56}$$

(viii) *Open-loop scanning:*
For the jth scan point, the required phase correction is performed below:

$$V_{An}(j)=V_{Dn}(j)\exp[j\omega_0(\tau'_{jn}-\tau'_{Bn}), \quad n=1, 2, \ldots, N_e \tag{9.57}$$

Thus the resultant image of the jth scan point is given by

$$\hat{S}(j) = \sum_{n=1}^{N_e} A_n V_{An}(j) \tag{9.58}$$

or

$$|\hat{S}(j)| = \left| \sum_{n=1}^{N_e} A_n V_{An}(j) \right|$$

where A_n is the amplitude taper at the nth element if tapering is desired.

The procedure derived above is valid for both near-field and far-field radar imaging. For a far-field imaging system, the algorithm can be simplified as follows. In the far-field case, the transmitter and reference element are assumed to be located at the origin of the array site. The location of the scan point is specified as (R_m, θ) where θ is the scan angle measured from the array normal and $R_m = m \cdot (c \cdot \Delta t/2)$ is the scan range measured from the origin (m=integer, c=light velocity). In this case, the sample numbers for the scan point (R_m, θ) can be expressed as follows:

$$k_n(R_m, \theta) = m + 1 - \Delta k_n(\theta), \quad n = 1, 2, \ldots, N_e \tag{9.59}$$

where

$$\Delta k_n(\theta) = \left[\frac{X_n \sin\theta}{c\Delta t} + \frac{T_{\mathrm{I}n}}{\Delta t} \right]_{\text{round off}}$$

is the sample number difference at the nth array element for the scan angle θ, and is independent of the scan range R_m. X_n is the location of the nth element.

The required phase correction, which is also independent of the scan range, for open-loop scanning is given below:

$$\phi_{sn} = \frac{\omega_0 X_n}{c} (\sin\theta_B - \sin\theta), \quad n = 1, 2, \ldots, N_e \tag{9.60}$$

From the concepts listed above, it is noticed that the computation of the far-field algorithm is much simpler and faster than that of the near-field algorithm since the sample number difference and the phase correction can be kept constant for different scan range. Four kinds of processing algorithms were developed [23], namely; (i) fundamental algorithms, (ii) range-scan process, (iii) range-scan and simplified sample number selection, and (iv) FFT data processing.

Computer simulation for a V-shaped target (Fig. 9.33) has been performed. System parameters for the simulation are shown in Table 9.2.

The results of simulation for four algorithms are shown in Fig. 9.34*a*, *b*, *c* and *d*, respectively.

This radio camera has also been examined by experiment. A single element movable antenna was used. The transmitter is located at the centre of the array site and the receiving antenna is moved from + sign to − sign (see Fig. 9.35).

The parameters of the experimental equipment are shown in Table 9.3.

The results of experiment for four algorithms are shown in Fig. 9.36*a*, *b*, *c*, and *d*, respectively.

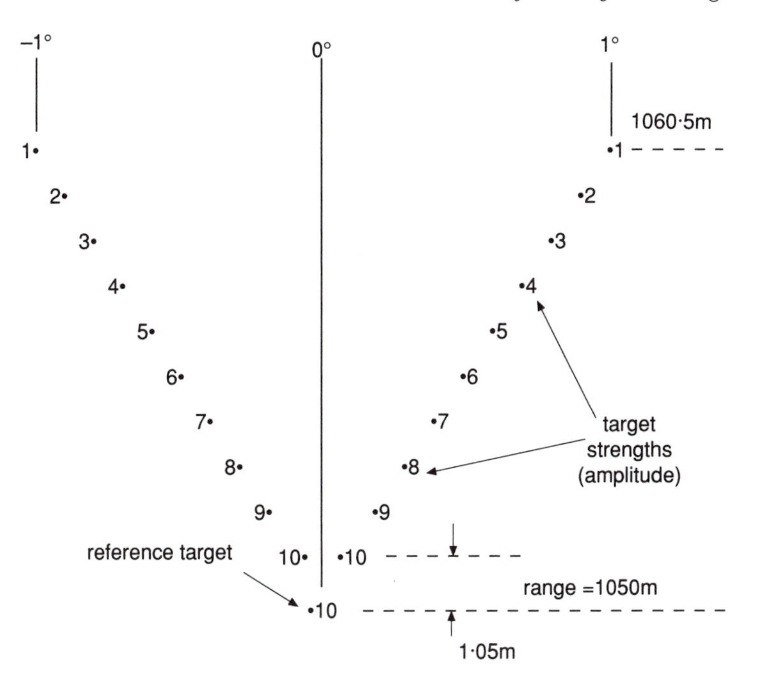

Figure 9.33 *Target configuration of computer simulaton [23] © IEEE, 1984*

In the computer simulation, the following errors are considered: (i) phase deviation of local oscillator; (ii) element location uncertainty; (iii) noise at I and Q channels. Even so, the results of computer simulation are much better than those of the experiment.

Table 9.2 *System parameters for computer simulation [23] © IEEE, 1984*

Parameter	Value
Frequency	9.6 GHz
Pulse duration	7 ns (Gaussian waveform)
Sampling interval	0.875 ns
Starting sample time	6982.5 ns
No. of snapshots	150
No. of array elements	100
Array size	60 m
Reference target	1050 m, 0°
Target area	1047.9 m–1062.6 m −1.2° to 1.2°
No. of scan points	71 × 121
LO phase error	Uniformly random (0,2π)
Element location uncertainty	$N(0, \sigma_{\varepsilon}^2)^*$, $\sigma_{\varepsilon} = 0.03$ m
Noises at I and Q channels	$N(0, \sigma_N^2)$, $\sigma_N = \phi.1$ (SNR = 20 dB)

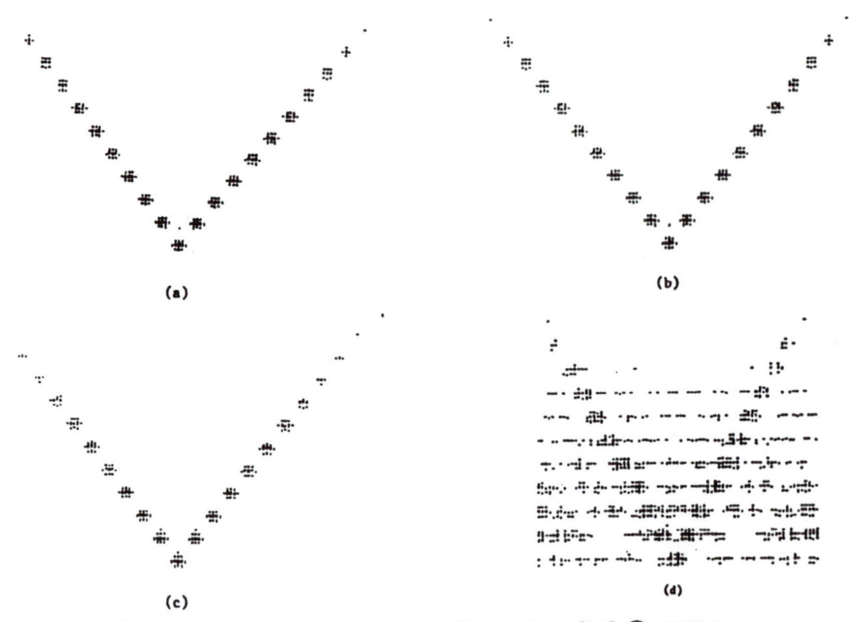

Figure 9.34 *Simulated results with 24.6 dB threshold [23] © IEEE, 1984*

The results for different algorithms are also different. The worst is obtained from the FFT algorithm. However, the computation time of the FFT algorithm is the least. The computation times for different algorithms are shown in Table 9.4.

In any case, this high resolution radio camera is very attractive for target identification. It is very interesting to make some comparisons between the radio camera and ISAR.

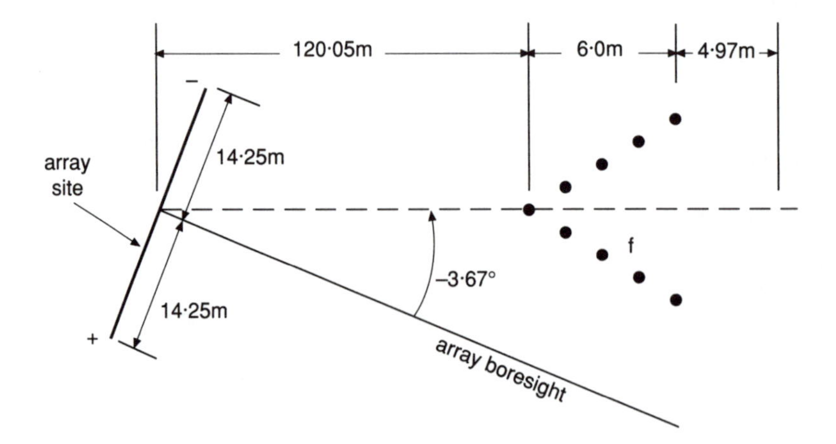

Figure 9.35 *Configuration of the experiment [23] © IEEE, 1984*

Table 9.3 *System parameters of the experimental equipment [23] © IEEE, 1984*

Parameter	Value
No. of array elements	200
Pulse duration	7 ns
Sampling interval	3.5 ns
Starting sample time	787 ns
No. of snapshots	32
No. of scan points	71 × 129
Target area:	
Angular	−5.27° to −2.07°
Range	119.05 m to 129.05 m
Reference target	(−3.67°, 131.02 m)
Frequency	9.6 GHz
Array site	−14.25 m to 14.25 m

Range resolution is improved by reducing the pulse width or increasing the bandwidth of the transmitted waveform. This is the same for the radio camera and ISAR. The main difference between these two techniques is the processing method to improve the cross-range resolution.

In radio camera imaging, coherent signals are obtained by means of spatial sampling from a multi-element linear array and high speed temporal sampling. It does not need relative movement between radar and target. The integration time is very short, and no motion compensation is required. However, a very long linear array is needed. For example, in the case of computer simulation, a

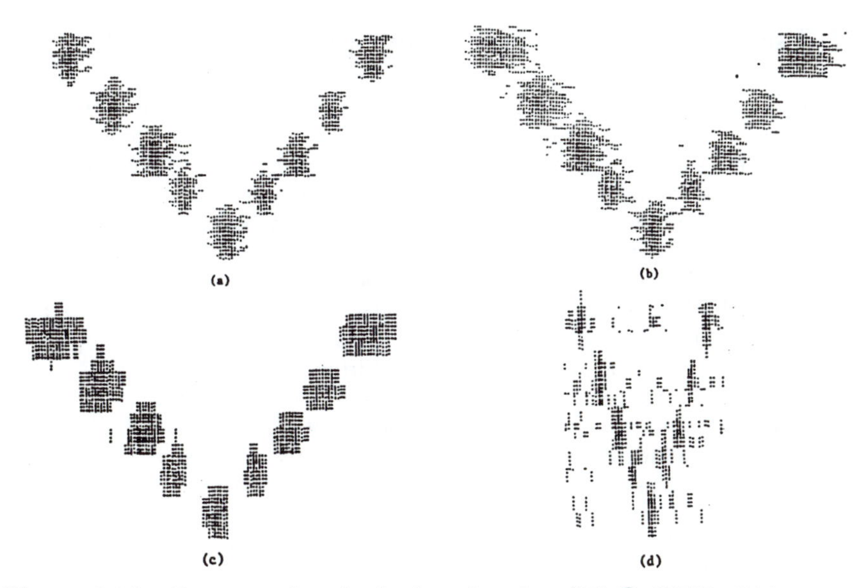

Figure 9.36 *Experimental results for four algorithms [23] © IEEE, 1984*

Table 9.4 *Computation time for different algorithms [23] © IEEE, 1984*

	Computation time (s)	
Algorithm	Simulation**	Experiment**
Fundamental	533.0	1127.2
Range scan	205.79	435.38
Range scan + simplified sample # selection	57.617	118.08
FFT processing	9.68	15.03

** Image scan points: Simulation = 71 × 121; Experiment = 71 × 129.

60 m linear array with 100 elements was used. Even in the experiment, a 28.5 m equivalent linear array with 200 elements was employed.

The radio camera can obtain very sharp images, since no motion compensation is required. Motion compensation becomes more difficult in ISAR than in SAR, since the movement of the target is uncontrollable and unknown. However, a very high speed A/D convertor and high speed signal processor are needed in the radio camera. For example, in computer simulation the sampling rate is as high as 1143 Msamples/s, and a 285 Msamples/s A/D converter was used in the experiment.

Off-line data processing for imaging is needed for both techniques. However, real-time imaging processing is now possible for SAR. This is realised with a high speed floating point array processor. We can expect that real-time imaging for ISAR, even for radio camera, can be realised in the near future.

9.7 Target identification based on spectral estimation

Most of the targets which we are interested in are moving targets or targets with moving parts. The amplitude and phase of radar returns from these moving targets are fluctuating with time. Their spectra depend on the shape, attitude, aspect angle and moving parameters, such as speed, orientation, rotation etc. They are different for different targets. Therefore, it is possible to identify targets from their spectra.

The most well known phenomenon is the Doppler spectrum of the propeller. Sometimes, the Doppler spectrum of compressor blades of a jet engine can also be observed. Fig. 9.37*a* and *b* show the effects of propeller and compressor blades, respectively [26]. They are displaced from the airframe line.

Schlachta [27] suggested classifying different targets by means of signal fluctuation differences; i.e. the fluctuations of amplitude and phase shift that are induced in the statistical behaviour of coherent signals. These signal fluctuation differences may be used in conjunction with a statistical decision test for target classification.

In order to find a statistical model, the empirical density functions of both magnitude and phase of the received complex signal vector elements $z_i = a_i \exp j\phi_i$, $(i = 1, 2, \ldots, N)$, are measured for several flights of the aircraft on flight paths. These measured functions are compared with standardised probability functions.

The histogram of the amplitude samples a_i, which are elements of a signal vector $\boldsymbol{a}$ with length N, can be approximated by a chi-square distribution $a_i \varepsilon \chi^2(\bar{a}, s)$. ($\bar{a}$ is proportional to the radar cross-section; s = degree of freedom.)

The phase shift $\Delta\phi_i \phi_{i+1} - \phi_i$ of consecutive samples follows, with good approximation, a normal distribution $\Delta\phi \varepsilon_N (\overline{\Delta\phi}, \sigma)$, where $\Delta\phi$ is proportional to the radial velocity of the target.

The characteristics of target classes can be expressed by the model parameters σ, a, s. When implementing a technical system for a fluctuation test, useful estimators have to be selected. First, the amplitude estimator was selected as $g_a = \sigma_a / \bar{a}$ (g_a = coefficient of variation) where

$$a = \frac{1}{N} \sum_{1}^{N} a_i$$

and

$$\sigma_a^2 = \frac{1}{N-1} \left(\sum_{1}^{N} a_i^2 - Na^2 \right) \tag{9.61}$$

The variance of the normally distributed variable $\Delta\phi_i$ can be optimally estimated in a similar manner, as shown in eqn. 9.61. Because of ambiguity, the phase shift values are summed by adding the components of the complex phaser with unit length: $\exp j\Delta\phi_i$.

$$r^2 = \frac{1}{N^2} \left[\left(\sum_{i=1}^{N} \cos \Delta\phi_i \right)^2 + \left(\sum_{i=1}^{N} \sin \Delta\phi_i \right)^2 \right]$$

The scattering angle of phase shift is derived by

$$g_p = \arctan \left[\left(\frac{N}{N-1} \, \frac{1-r^2}{r^2} \right)^{0.5} \right] \tag{9.62}$$

The estimators g_a and g_p are applied to stored data from different targets. In the case of aircraft and birds, about 50 to 150 estimates of consecutive scans are

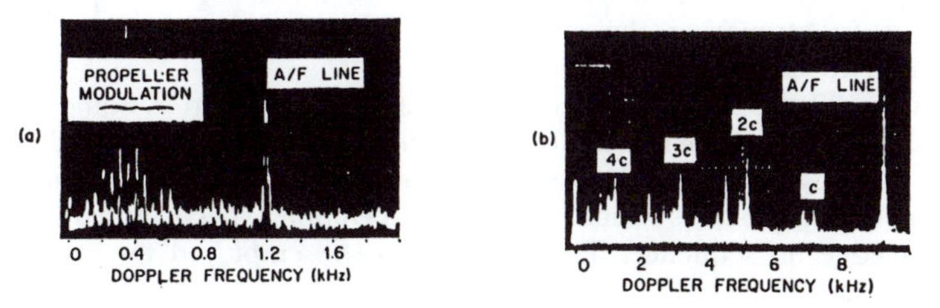

Figure 9.37 *(a) Doppler spectrum of propellers of DC–7 aircraft; (b) Doppler spectrum of compressor blades of jet engines displaced from the airframe Doppler (A/F line) by the rev/s of the compressor, c, and multiples of c [26]. Reproduced with permission of McGraw-Hill*

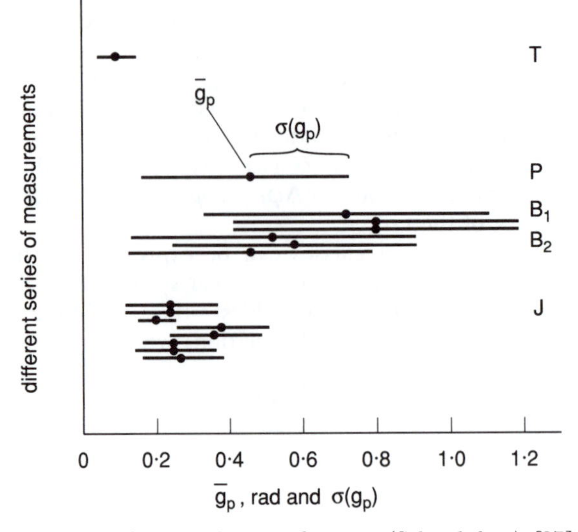

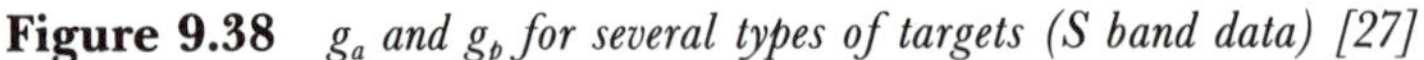
Figure 9.38 *g_a and g_p for several types of targets (S band data) [27]*

T = tower
P = propeller aircraft
B_1 = birds (splitting and converging tracks, changing flight direction, different flocks in radar resolution cell)
B_2 = birds (tracks with predominating flight direction)
J = small jet fighter (different aspects)

averaged during a series of measurements made with a surveillance radar. Each scan contains $N = 15$ samples. Some fixed targets chosen as a reference for system stability are checked with a similar number of scans. Means and standard deviation of g_a and g_p are shown in Figs. 9.38 and 9.39. As can be seen in these Figures, the mean fluctuation is different for several classes of targets. Because of overlapping fluctuation zones, a statistical description of the efficiency of a fluctuation test is required.

Radar echoes can be modelled as an autoregressive time sequence. Therefore, *AR* coefficients are a 'best' possible feature set for radar signal classification, since it preserves all the information contained in the signal sequence. It is equivalent to estimating the spectrum of the radar returns by means of maximum entropy spectral estimation.

Stehwien [28] employed this method to classify radar clutter. Some results of analysis of recorded ground, weather and bird clutter were presented.

Fossi [29] suggested that an autocovariance estimation method can be used for extracting the features which are related to Doppler spectrum parameters of the received echo sequence. Therefore it can be employed for radar signal classification.

It is assumed that the selected feature corresponds to a first-order *AR* representation of the observed sequence for simplicity. In the first-order, zero mean, *AR* process, the estimating parameter is the autocovariance coefficient ρ_1 defined by

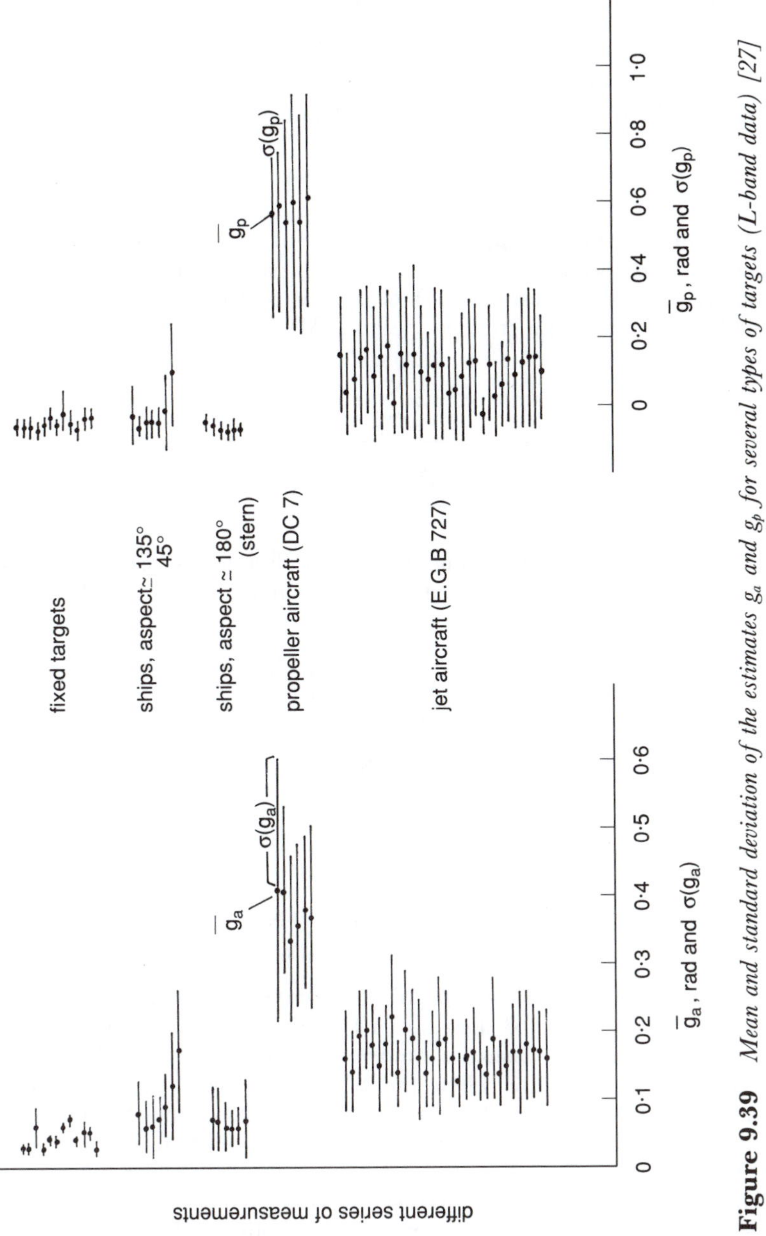

Figure 9.39 *Mean and standard deviation of the estimates g_a and g_p for several types of targets (L-band data) [27]*

$$\rho_1 = \frac{E\{R(1)\}}{E\{R(0)\}} \tag{9.63}$$

where $R(m)$ is the autocorrelation function

$$R(m) = E\{x_i^* x_{i+m}\} \tag{9.64}$$

the coefficient ρ_1 represents the pole of first-order AR model. It also gives an estimate of the mean Doppler frequency f_d of the spectrum given by

$$\bar{f}_d = \frac{1}{2\pi} \arg(\rho_1) \tag{9.65}$$

Furthermore, the module $|\rho_1|$ gives a measure of time series spectral dispersion. This spectral dispersion decreases as $|\rho_1|$ approaches unity.

Among the different estimates $\hat{\rho}_1$ of ρ_1, it is best to select the estimate ρ_{MEM} defined in the maximum entropy method (MEM) and given by

$$\hat{\rho}_{MEM} = \frac{2 \sum_{n=0}^{N-1} x(n) x^*(n-1)}{\sum_{n=0}^{N-1} [|x(n)|^2 + |x(n-1)|^2} \tag{9.66}$$

Since in radar signal processing the FFT is frequently used, an alternative estimate of ρ_1 is directly extracted from frequency samples $\{x_k\}$. The estimated ρ_{spect} is defined by

$$\hat{\rho}_{spect} = \frac{\sum_{i=0}^{N-1} |X_n|^2 \exp\left(j \frac{2\pi}{N} n\right)}{\sum_{n=0}^{N-1} |X_n|^2} \tag{9.67}$$

where

$$X_k = \sum_{i=0}^{N-1} x_i \exp\left(-j \frac{2\pi}{N} ik\right), \quad k = 0, \ldots, N-1 \tag{9.68}$$

The behaviour of the coefficient $\hat{\rho}_{spect}$ for different classes of radar signal has been analysed and the expected distributions on the unit circle have been derived (Fig. 9.40*a*). The inner dashed contour of each class accounts for its extension towards the circle centre when the SNR decreases. A very limited class separability can be obtained at low frequencies. Nevertheless, this ambiguity can be reduced by introducing an MTI filter before the feature extractor (see Fig. 9.40*b*).

Schneider [30] studied the MEM for Doppler spectral analysis of radar echoes from helicopters with rotating blades. Some experimental results were obtained with coherent pulse Doppler radar at VHF and S-band. Pulse repetition frequencies were 625 Hz and 7.1 kHz, respectively. Various types of hovering helicopters were illuminated for several seconds. The MEM spectrum

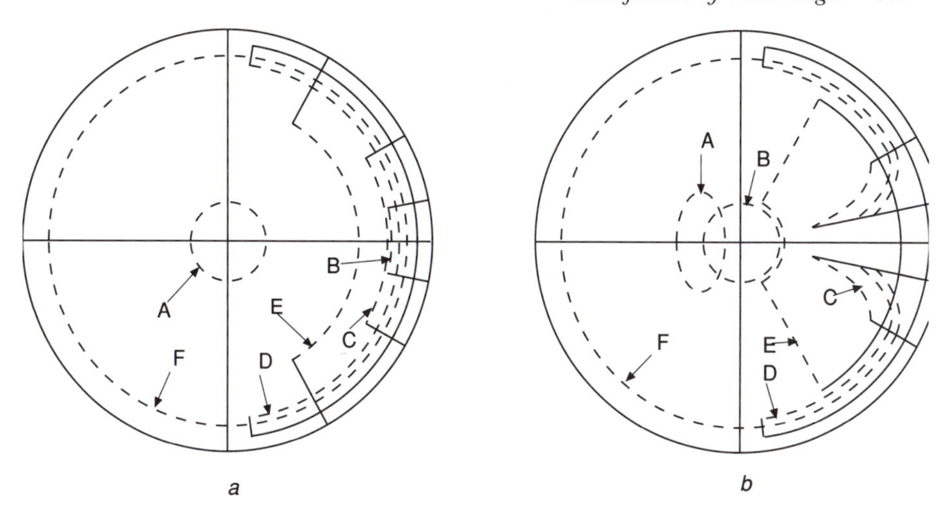

Figure 9.40 *Expected distributions of the autocovariance coefficient without MTI (a); and with MTI (b) filtering for the following class of target [29]* © *IEEE, 1984*

A: Noise; B: Ground clutter; C: Rain clutter; D: Moving helicopter; E: Helicopter hovering; F: Aircraft.

was obtained with Burg's algorithm and compared with the results of a periodogram. The results are shown in Fig. 9.41 (for VHF radar) and 9.42 (for S-band radar).

It appears to be as well suited to determining characteristic parameters as to identifying various types of helicopters.

9.8 Target identification based on non-coherent signal processing

Until now most radars are non-coherent, especially shipborne radars. Therefore, how to identify ships with non-coherent signal processing is one of the important problems of target identification.

Maaloe [31] suggested a method to classify ships with non-coherent marine radar. The characteristic features selected for the classification purpose are:

NP = number of amplitude exceeding a predetermined threshold

VA = variance of the amplitude exceeding the fixed threshold

SM = spatial moment of inertia

PI = product of inertia

TM = template.

The moments are calculated for an area consisting of 21×21 samples placed around the centre of the target, the weight being unity if the amplitude exceeds the predetermined threshold; otherwise it is zero. The spatial moment of inertia, *SM*, is calculated relative to the centre of the target as

$$SM = \sum_{i=1}^{21} \cdot \sum_{j=1}^{21} d_{ij}^2 A_{ij} \tag{9.69}$$

where d_{ij} is the distance from the centre to the point (i, j) and A_{ij} is 1 if the amplitude from the point (i, j) exceeds a predetermined threshold; otherwise it is 0. The product of inertia, *PI*, is calculated as

$$PI = \sum_{i=1}^{21} \cdot \sum_{j=1}^{21} X_{ij} \cdot Y_{ij} \cdot A_{ij} \tag{9.70}$$

where X_{ij} and Y_{ij} are the distances from the two orthogonal axes parallel to the range direction and the azimuth direction, respectively, and passing through the centre of the target. The product of inertia indicates the direction of the target.

The template *TM* is defined as the number of amplitudes backscattered from a target of which 10 out of 15 adjoining amplitudes (3 amplitudes in range and 5 amplitudes in azimuth) exceed a predetermined threshold. The reason for the name template is that the feature depends on the shape of the target.

Experiments were carried on with an X-band horizontal polarisation radar. The azimuth resolution is 0.8° at 3 dB, the pulse length is 250 ns and the PRF is

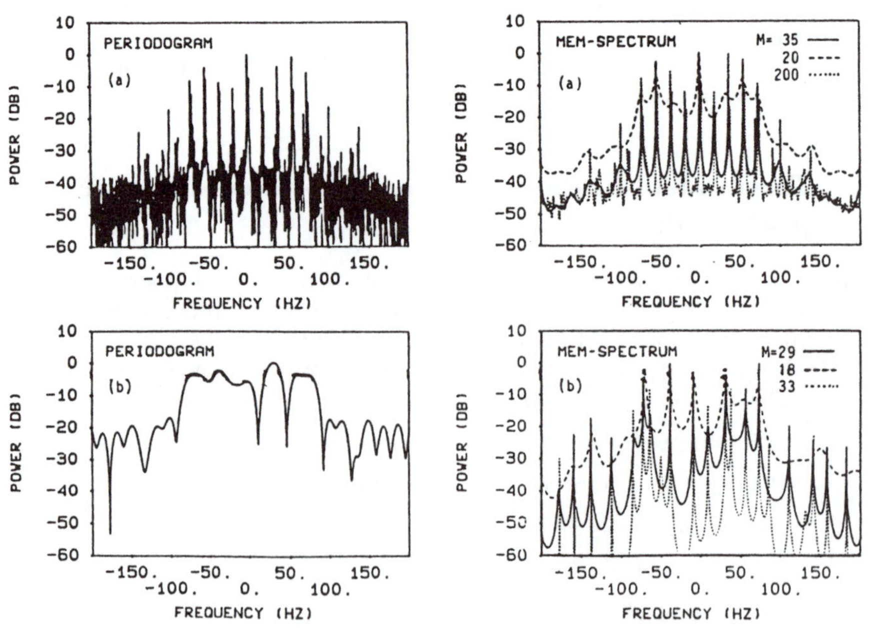

Figure 9.41 *Spectra of real 3-blade helicopter echoes (VHF band) [30]*

Dwell time $T = 6.5$ s (*a*), and $T = 56$ ms (*b*)

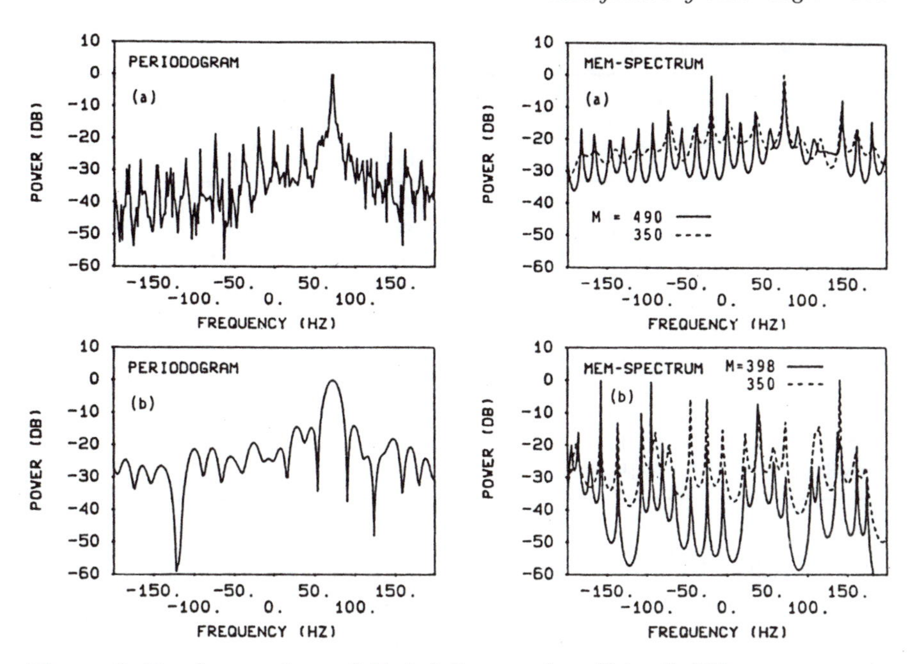

Figure 9.42 *Spectra of same 3-blade helicopter echoes (S-band) [30]*
Dwell time $T=0.57$ s (*a*) and $T=56$ ms (*b*)

1 kHz. The height of the antenna is 15 m above mean sea level. The distance between the radar and the target is in the range interval from 4 nm to 6 nm. At these ranges, the clutter level at sea state 3 is negligible. The identities of the ships are competely unknown.

The stepwise discriminant analysis producing the classification functions for three ships are listed in Table 9.5.

These data were obtained from the radar returns over 130 antenna scans and were used for initial classification. According to these classification functions each measurement (2.64″) of the ships is classified. The results are listed in Table 9.6, showing the number of ships classified into each group and the percentage of correct classifications. Each group is specified by the ship number and the time of measurement.

Table 9.5 *Initial data of the classification functions [31]*

	Ship 1	Ship 2	Ship 3
NP	5.154	5.310	3.597
VA	0.046	0.035	0.020
SM	−3.443	−2.951	−2.168
PI	−3.027	−1.604	−0.168
TM	−3.988	−4.076	−2.924
Constant	−126.071	−108.970	−44.589

Table 9.6 *Experimental results of ship classification [31]*

Group	Percentage of correct	No. of cases classified		
		Ship 1	Ship 2	Ship 3
Ship 1, 0	91.3	73	7	0
Ship 2, 0	92.5	5	74	1
Ship 3, 0	100.0	0	0	80
Ship 1, 5	74.0	37	13	0
Ship 2, 5	94.0	2	47	1
Ship 3, 5	100.0	0	0	50
Ship 1, 10	80.0	40	7	3
Ship 2, 10	100.0	0	50	0
Ship 3, 10	98.0	0	1	49

The classification experiments carried out so far have only dealt with ships sailing in clutter-free environments and following a straight course. It is more difficult to classify the individual ships in various aspect angles and high sea states.

Li *et al.* [32] also proposed an algorithm for ship identification with incoherent radar. The feature extraction is based on spectral analysis, Mellin transformation and coding technique.

Assume that

$$X_i = \{x_{i1}, x_{i2}, \ldots, x_{is}\}, \quad X_i \in V_x$$

is a one dimensional digitised waveform which is the ith sample from radar video returns and V_x is the original signal space defined by X_i:

$$Y_j = \{y_{j1}, y_{j2}, \ldots, y_{jt}\}, \quad Y_j \in V_y$$

is the feature vector, and V_y is the feature vector space.

A feature extractor can be defined as a mapping

$$\begin{aligned} F\colon\ & V_x \rightarrow V_y \\ & X| \rightarrow F(X) \triangleq Y \end{aligned} \tag{9.71}$$

Normally we have

$$\dim(V_y) < \dim(V_x)$$

It is clear that the form of the transformation F will depend on the physical nature of recognition problem. F is represented as follows:

$$F(X) = F_4(F_3(F_2(F_1(X)))) \tag{9.72}$$

where

F_1 = waveform preprocessing

F_2 = maximum entropy spectral transformation

F_3 = Mellin transformation

Table 9.7 *Experimental results of ship identification [32]*

Group	Ship 1	Ship 2	Ship 3	Total	Corr. rate
Ship 1, 1	124	25	46	195	63.6%
Ship 2, 1	7	257	2	266	96.6%
Ship 3, 1	45	49	191	285	67.0%
Ship 1, 2	151	21	8	181	84.0%
Ship 2, 2	5	218	3	226	96.5%
Ship 3, 2	7	42	194	243	79.8%
Ship 1, 3	59	3	2	64	90.1%
Ship 2, 3	0	61	0	61	100%
Ship 3, 3	0	4	62	66	92.5%

Table 9.8 *Some design parameters for NMMW radars [34]*

Wavelength (mm)	1.3 (220 GHz)	0.88 (340 GHz)	0.73 (410 GHz)
Field of view		$10° \times 5°$	
Frame rate		30/s	
Antenna diameter		1 m	
Beamwidth (10^{-3} rad)	1.560	1.056	0.876
Linear resolution at 1 km	1.5 m	1.0 m	0.9 m
No. of beam positions	7200	15700	22900
Beam scan rate (s^{-1})	2.16×10^5	4.71×10^5	6.86×10^5
Dwell time (μs)	4.63	2.12	1.46
Range window/dwell	695 m	318 m	219 m

F_4 = coding transformation, the selection of the final features.

Once the features are extracted, a reference data base should be constructed for every kind of target to be identified. To accomplish the ship identification process, a weighted minimum-distance classifier was suggested as the last step of the recognition system.

Suppose that C reference feature vectors $\boldsymbol{Y} = \{Y_1, Y_2, \ldots, Y_c\}$ are given with Y_i associated with the target class j. A minimum-distance classification scheme with respect to $\boldsymbol{Y}$ is to classify the input feature vector $\boldsymbol{Z}$ as from class j, i.e.

$$Z \in \text{class } j \text{ if } |Z - Y_j| \text{ is the minimum}$$

where the feature vector to be classified $Z = (z_1, z_2, \ldots, z_t)$. $|Z - Y_j|$ is the distance between Z and Y_j and is called weighted minimum distance.

To evaluate the effectiveness of this algorithm, a recognition experiment is conducted for about 600 radar returns (200 per ship) from three ships. The results are shown in Table 9.7.

9.9 Target identification based on millimetre-wave radar

Besides the increase of antenna size, another way to reduce the beamwidth is to employ millimetre or near millimetre waves (NMMW). It is evident that the beamwidth of a 94 GHz MMW radar will be 31 times narrower than that of a 3 GHz radar with same antenna size. However, the detection range of MMW radar is limited by the attenuation of the atmosphere and the power output of the transmitter. Therefore, the application of MMW radar is limited.

Essen *et al.* [33] described a high resolution 94 GHz frequency agility radar used for identifying vehicles. The power output is 4.2 W. The pulse width is 40 ns and 80 ns. The radial resolution, DR, is determined by the total bandwidth of the frequency ramp and described by the equation

$$DR = (c/2) \cdot [1/(DF \cdot n)]$$

where DF is the frequency step and n is the number of steps. The lateral high resolution is accomplished by differential Doppler processing. For a fixed transmit frequency the phase of a backscatter signal from a defined scattering centre is influenced by the movement of the target. For fixed angular velocity the gradient of the phase is proportional to the distance from the centre of rotation. A Fourier transform applied to the complex output of the radial transformation gives the respective lateral position. The lateral resolution, DL, is determined by the ratio between the half wavelength $(\lambda/2)$ and the arc $(\Omega \cdot T)$ corresponding to the observation period T:

$$DL = (\lambda/2) \cdot (\Omega T)^{-1}$$

with the parameters $T = (k \cdot m \cdot n)/PRF$, k = number of averages and m = number of Fourier elements for time series. The parameters were matched to result in the same resolution in range and cross range, namely 37.5 cm. Experimental results showed that it is possible to identify the target by its specific scattering centre distribution and its polarisation characteristics.

9.10 References

1 GIULI, D., GHERARDELLI, M., and FOSSI, M.: 'Using polarization discriminants for target classification and identification'. CIE 1986 Int. Conf. on Radar, Nanjing, China, pp. 889–898
2 SINCLAIR, G.: *Proc. IRE*, 1950, **38**, pp. 148–151
3 HUYNEN, J. R.: 'Phenomenological Theory of Radar Target', *in* USLENGHI, P. I. E. (Ed.): 'Electromagnetic scattering' (Academic Press, NY, 1978)
4 MIERAS, H.: 'Optimal polarization of simple compound targets', *IEEE Trans.*, 1983, **AP–31**, pp. 996–999
5 KE, Y. A., and DALLE MESE, E.: 'Advances on target classification', CIE 1986 Int. Conf. on Radar, Nanjing, China, pp. 226–232
6 CHUANG, C. W., and MOFFAT, D. L.: 'Natural resonances of radar targets via Prony's method and target discrimination', *IEEE Trans.*, 1976, **AES–12** (5)
7 CHEN, K. *et al.*: 'Radar waveform synthesis for single-mode scattering by a thin cylinder and application for target discrimination', *IEEE Trans.*, 1982, **AP–30**, (5)
8 MANCIANTI, M., and VERRAZZANI, L.: 'Incident signal design for target identification by natural resonances'. CIE 1986 Int. Conf. on Radar, Nanjing, China, pp. 239–244
9 GJESSING, D. T.: 'Target recognition by an adaptive radar'. Int. Conf. on Radar, Paris, 1984, pp. 263–268

10 GJESSING, D. T. *et al.*: 'A multifrequency adaptive radar for detection and identification of objects: Results of preliminary experiments on aircraft against a sea-clutter background', *IEEE Trans.*, May 1982, **AP–30**
11 GJESSING, D. T.: 'Matched illumination target adaptive radar for challenging applications'. IEEE Int. Conf. Radar '87, pp. 287–291
12 HUSSAIN, M. G. M., and MASOUD, E. Y.: 'A two-dimensional deconvolution algorithm for target classification and recognition with nonsinusoidal signals'. CIE 1986 Int. Conf. on Radar, Nanjing, China, pp. 233–238
13 EBERLE, A., and YOUNG, J. D.: 'Development and field testing of a new locator for buried plastic and metal utility lines'. Transportation Research Board no. 631, National Academy of Sciences, Washington, DC, Jan. 1977, pp. 47–51
14 CHAN, L., MOFFATT, D. L., and PETERS, L. Jr.: 'A characterization of subsurface radar targets', *Proc. IEEE*, 1979, **67**, pp. 991–1000
15 ARAI, I., and SUZUKI, T.: 'Underground radar systems', *Trans. IECE of Japan*, 1983, **J66–B**, pp. 713–804
16 CHAN, L. C., PETERS, L. Jr., and MOFFATT, D. .L.: 'Improved performance of a subsurface radar target identification system through antenna design', *IEEE Trans.*, 1981, **AP–29**, pp. 307–311
17 PAHLS, J.: 'Radar moving target resolution and imaging'. IEE Int. Conf. Radar '77, London, pp. 285–289
18 PRICKETT, M. J., and CHEN, C. C.: 'Principles of inverse synthetic aperture radar (ISAR) imaging'. IEEE Electronics and Aerospace System Convention (EASCON), Arlington, Sept. 1980
19 CHEN, C. C., and ANDREWS, H. C.; 'Target-motion-induced radar imaging', *IEEE Trans.*, 1980, **AES–16**, p. 2
20 EERLAND, K. K. 'Application of inverse synthetic aperture radar on aircraft'. Int. Conf. on Radar, Paris, 1984, pp. 618–623
21 DIKE, G. *et al.*: 'Inverse SAR and its application to aircraft classification'. IEEE 1980 Int. Radar Conf., pp. 161–167
22 BERKOWITZ, R. S., JUANG, S. T., and YANG, B. N.: 'Wideband waveform and synthesized aperture considerations for a high resolution imaging radar'. IEEE 1984 National Radar Conf., pp. 105–109
23 BERKOWITZ, R. S., JUANG, S. T., and YANG, B. N.: 'Processing techniques for high resolution radio camera imaging'. Int. Conf. on Radar, Paris, 1984, pp. 595–600
24 STEINBERG, B. D.: 'Radar imaging from a distorted array: the radio camera algorithm and experiments', *IEEE Trans.*, 1981, **AP–29**, pp. 740–748
25 STEINBERG, B. D.: 'Microwave Imaging with Large Antenna Arrays' (John Wiley & Sons, NY, 1983)
26 DUNN, J. H., and HOWARD, D. D.: 'Target noise', *in* SKOLNIK, M. I. (Ed.): 'Radar handbook' (McGraw-Hill Book Co., NY, 1970)
27 VON SCHLACHTA, K.: 'A contribution to radar target classification'. IEE Int. Conf. Radar '77, pp. 135–139
28 STEHWIEN, W., and HAYKIN, S.: 'Statistical classification of radar clutter'. IEEE 1986 National Radar Conf., pp. 101–106
29 FOSSI, M., GIULI, D., and PICCILI, L.: 'Autocovariance estimation for classification of radar signals by Doppler spectra'. Proc. 1984 Int. Symp. on Noise and Clutter Rejection in Radars and Imaging Sensors, Tokyo, pp. 257–261
30 SCHNEIDER, H.: 'On the maximum entropy method for Doppler spectral analysis of radar echoes from rotating objects'. IEE Int. Conf. Radar '87, pp. 279–282
31 MAALOE, J.: 'Classification of ships using an incoherent marine radar'. IEE Int. Conf. Radar '82, pp. 274–277
32 LI, Y., LIANG, T. J., and GUO, G. R.: 'Ship identification with an incoherent radar'. IEE Int. Conf. Radar '87, pp. 301–304
33 ESSEN, *et al.*: 'High resolution millimetrewave radar signatures of vehicles'. IEE Int. Conf. Radar '87, pp. 355–359

Chapter 10

Phased array antennas

Prof. R. P. Shenoy

Part 1: Array theory

10.1 Introduction

The present resurgence of activities in radar systems is mainly due to the renewed interest in the field of phased array radars as a result of advances in microwave technology, digital signal processing and integrated circuits. Phased array radars are distinguished from other radars by phased array antennas which are capable of steering the beam electronically in space. This provides greater flexibility and makes the system increasingly versatile by being able to carry out better energy management in the volume of space and optimise the search and track functions. Phased array radars thus represent the best sensor configuration for military applications.

Phased arrays belong to that class of array antennas which provide beam agility by effecting a progressive change of phase between successive radiating elements. The limits of performance of phased array antennas are governed by the characteristics of array antennas and beam scanning effects. Though work on phased array antennas was initiated more than three decades ago, growth in array technology was slowed primarily because of the high cost. The cost per element is still not low enough, but in the last fifteen years increasing demands for radar systems with better performance have resulted in the incorporation of such features as beam agility, multi-target tracking, antenna pattern control, electronic counter-measures and multimoding. Analytical and experimental studies on phased array antennas, including the behaviour of a variety of radiating elements in finite, as well as in large array environments, solutions for large arrays, understanding of edge effects, the phenomenon of array blindness plus advances in signal processing have all led to a better understanding of these systems. There have been equally significant advances in phase shifters, feed networks, microstrip and stripline radiators, and broadbanding of arrays. The final breakthrough that will make the two dimensional phased array system cost-effective will be accomplished by means of printed cicuit technology, utilising PCAAs (printed circuit antenna arrays) based on microstrip patch or printed dipole. These radiators will be fabricated as sub-arrays containing a number of them along with associated phase shifting and driver electronics so that techniques used in digital integrated circuit production can be fruitfully employed. Computer aided design of these arrays as well as computer aided manufacturing will become mandatory. The aim of this Chapter is to draw the attention of radar designers to the newer developments that have taken place in this field in the different areas enumerated above.

Any analysis of phased array antennas can be appreciated better if array antennas, in general, are discussed first. An antenna array can be considered as a matrix of discrete radiators located in an aperture, these being called the radiating elements. Typically these can be monopoles, dipoles, radiating slots on waveguide walls, open-ended waveguides or microstrip radiators, depending on the operating frequency, power handling capability, polarisation, method of feeding etc. The array antenna can be one dimensional, in which case the radiators can be placed in a line (linear arrays) or it can be two dimensional, in which case the placement of the elements can be on a rectangular or on a circular grid. In most cases, the inter-element spacing is kept uniform but nonuniform arrays have also been designed. Generally, the designer has five parameters available to him in the array; these are, the type and the number of radiating elements, inter-element spacing, the amplitude and the phase of the excitation currents/voltages. It is thus possible for antenna designers to combine these in a judicious fashion to optimise the performance of the antenna in its various roles.

Linear antenna arrays have been the subject of considerable analysis because of the insight they provided into understanding beam forming, far field radiation patterns and the relationship between array excitation function and the radiation pattern. In most practical systems, as far as possible the radiating element as well as the inter-element distance are kept fixed and not varied within the array. If an array of $(N+1)$ elements with an inter-element distance d along the array axis is considered, then the field produced at a point P in the far zone will be the superposition of the fields produced by each individual radiating element at that point. If I_m and α_m are the amplitude and phase of the excitation current or voltage of the mth element of the array then from Fig. 10.1, we can write

$$E_A(\theta) = F(\theta) \sum_{m=0}^{N} I_m \exp\{-jmkd \sin\theta + j\alpha_m\} \tag{10.1}$$

Where

$F(\theta)$ = far field function associated with the radiating element; it is the electromagnetic effect created by the radiating element at very large distances as a result of the distribution of unit time varying currents.

I_m = magnitude of the time varying electric current flowing in the mth radiator.

k = propagation constant which is equal to $2\pi/\lambda$.

θ = angle subtended by the point P with the Z-axis of the co-ordinate system.

The summation term of eqn. 10.1 can be represented by

$$S(\theta) = \sum_{m=0}^{N} I_m \exp j\{-mkd \sin\theta + \alpha_m\} \tag{10.2}$$

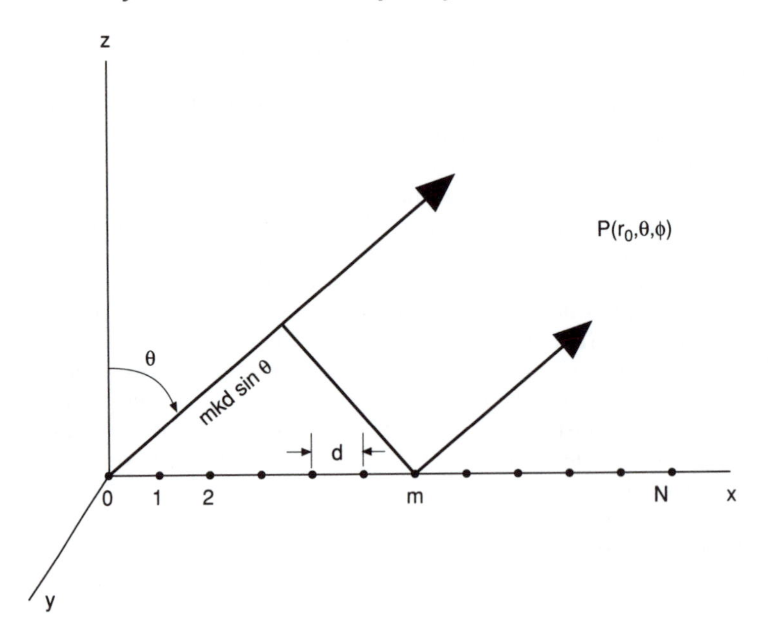

Figure 10.1 *Geometry of a linear array*

By a well known theorem in algebra, the magnitude of $E_A(\theta)$ in eqn. 10.1 can be written as

$$|E_A(\theta)| = |S(\theta)||F(\theta)| \qquad (10.3)$$

and

$$|S(\theta)| = \left|\sum_{m=0}^{N} I_m \exp j\{-mkd \sin \theta + \alpha_m\}\right| \qquad (10.4)$$

$|S|$ is called the array factor and is dependent on the number of radiating elements, their spacing, the magnitude and phase of the excitation function of the individual radiators. The radiation pattern of the linear array is the product of the radiation pattern of the individual radiating element and the magnitude of the array factor. The study of the array therefore 'boils down' to the study of the behaviour of the function $|S|$ in different directions in space. The array factor ideally should have a principal maximum in a specified direction (main lobe) only. This is, however, difficult to achieve and thus there will be secondary maxima (sidelobes) in other directions. The main task of the antenna designer is to ensure that the principal maximum occurs in the desired direction and that the number and levels of the secondary maxima are minimised so that the energy flow (leakage) through the sidelobes is kept to a minimum. The aim is thus to maximise the gain, which is an indicator of the ability of the array to concentrate energy in a given direction.

In order to express the gain functions in terms of the field due to the array at the point P, a spherical co-ordinate system (r, θ, ϕ) is introduced. It should be noted that E_A, F and S can also be expressed in this co-ordinate system. The power gain $G(\theta, \phi)$ can be shown to be equal to

$$G(\theta, \phi) = \frac{\text{Power radiated by the array per unit solid angle in the direction}(\theta, \phi)}{\text{Total power accepted by the antenna}/4\pi}$$

This works out to be

$$G(\theta, \phi) = \frac{4\pi P(\theta, \phi)}{P_{acc}} \tag{10.5}$$

Here $P(\theta, \phi)$ is the power radiated per unit solid angle measured in watts per steradian and P_{acc} is the power accepted by the antenna from the radar transmitter. The power gain is the inherent characteristic of the antenna and takes into consideration the ohmic, dissipative and dielectric losses arising from the conductivity of the metal and the loss tangent in dielectrics, respectively. Transmission as well as the mismatch losses between the transmitter and the antenna are, however, not accounted for. There is another term which is commonly used in antenna literature. It is *directivity* and is defined as

$$D(\theta, \phi) = \frac{\text{Power radiated by the array per unit solid angle in the direction } (\theta, \phi)}{\text{Total power radiated by the antenna}/4\pi}$$

$$= \frac{4\pi |E_A(\theta, \phi)|^2}{\int_0^{2\pi} \int_0^{\pi} |E_A(\theta, \phi)|^2 \sin \theta \, d\theta \, d\phi} \tag{10.6}$$

In this case the power losses in the materials forming the antenna are not accounted for and therefore power gain is always less in magnitude as compared to the directivity. Even though the eqns. 10.5 and 10.6 enable us to compute G and D in any direction in space, the usual practice is to compute these values only in the direction of the main lobe and express them in decibels relative to an ideal isotropic radiator, which is a hypothetical antenna that radiates uniformly in all directions.

10.2 Uniform arrays

A linear array that is made up of radiating elements having equal amplitudes and phase for the excitation currents is called a uniform array. The array factor of a uniform array can be obtained by simply carrying out the following substitutions in eqn. 10.4 (see Fig. 10.1):

$$\begin{matrix} I_m = 1 \\ \alpha_m = 0 \end{matrix}; \quad m = 0, 1, 2, \ldots, N \tag{10.7}$$

Therefore

$$|S| = \left| \sum_{m=0}^{N} \exp j \left\{ -\left(\frac{2\pi}{\lambda}\right) md \sin \theta \right\} \right| \tag{10.8}$$

Eqn. 10.8 is the sum of phasors possessing phase angles which are dependent on θ. However, for a given θ, the phase angles of the phasors are progressive multiples of the basic angle u given by

$$u = \frac{2\pi}{\lambda} d \sin\theta \tag{10.9}$$

Eqn. 10.8 can now be reduced to

$$|S| = \left|\exp j\left(-\frac{Nu}{2}\right)\right| \left|\frac{\sin\left(\frac{N+1}{2}u\right)}{\sin(u/2)}\right| \tag{10.10}$$

The radiation pattern is given by

$$|S| = \left|\left\{\sin\left(\frac{N+1}{2}u\right)\right\} \Big/ \left\{\sin\frac{u}{2}\right\}\right| \tag{10.11}$$

Eqn. 10.11 indicates that the radiation pattern of the array is a periodic function of $\boldsymbol{u}$ with major maxima occurring at $\boldsymbol{u} = 0, \pm 2\pi, \ldots, \pm 2p\pi$, where $\boldsymbol{p}$ is an integer. The occurrence of several main lobes is an undesirable characteristic since energy will be either received or radiated not only from a desired direction but also in an equal measure from other angles. The principal maxima occurring for values of $u = \pm 2\pi, \pm 4\pi, \pm \ldots$, are called grating lobes. It can be seen that the radiation pattern of the antenna, including the sidelobes, repeats between adjacent grating lobes. In a well designed antenna the grating lobes, being undesirable, require to be eliminated. For this purpose, conditions are imposed on the inter-element spacing so that, in the visible space, only one maximum, corresponding to $u = 0$, occurs in the array pattern. Since, $u = .006\,2p\pi$, eqn. 10.9 can now be written as

$$\frac{d}{\lambda}\sin\theta = \pm p \tag{10.12}$$

Again it has to be noted that in the visible region θ is real, and hence the maximum value of $(\sin\theta)$ will be 1. Further, the first grating lobe will appear for $p = 1$ in eqn. 10.12 so that the lowest inter-element spacing at which a grating lobe will appear works out to be

$$\frac{d}{\lambda} = 1 \tag{10.13}$$

Therefore the inter-element spacing should never be allowed to reach the value of one wavelength. Hence, for no grating lobe formation,

$$\frac{d}{\lambda} < 1 \tag{10.14}$$

In the case where the inter-element spacing of the array is less than one wavelength, there is only one principal maximum that is formed in the direction normal to the axis of the array. This is termed the ‘broadside array’. As the inter-element spacing is increased to one wavelength, grating lobes begin to form at the end-fire directions, i.e. at angles $\theta = \pm\pi/2$, while the main beam is formed broadside. Thus, except in some radio astronomy interferometric applications, the inter-element spacing is always kept at a value less than one wavelength, to avoid grating lobe formation.

The nulls of the pattern of the $(N+1)$ element uniform array are obtained from eqn. 10.12, when $\sin\{(N+1)u/2\} = 0$, i.e. for $(N+1)u/2 = (2p+1)\pi$. It can easily be seen that, even in the case of $(d/\lambda) < 1$, between the nulls in the radiation pattern, secondary maxima also occur. These are the peaks of the sidelobes and they are located approximately halfway between the nulls when the number of radiating elements in the array is greater than eight. These locations are given by

$$u = \frac{2\pi}{(N+1)}\left(\frac{2p+1}{2}\right); \quad \text{for } p = 1, 2, \ldots \tag{10.15}$$

Thus the first few sidelobes which are of importance from the design point of view occur at, $u = \pm(3\pi/N+1)$, $\pm(5\pi/N+1)$, . . . , etc. The levels of the first three sidelobes can be obtained from eqn. 10.11 as follows.

The levels of the sidelobes can be computed by taking the derivative of $S(u)$ as expressed in eqn. 10.11 with respect to u and setting it to zero. This leads to a relation $(N+1)\tan(u/2) = \tan\{(N+1)u/2\}$ which has as its first solution $(N+1)u = 2.8606\pi$. This value of u when introduced in eqn. 10.11 provides a sidelobe ratio for the first sidelobe as -13.26 dB. This is independent of the number of elements or the pointing angle of the main beam of the uniform array. Similarly the second and third sidelobe levels can be found and these are -17.9 dB and -21.4 dB, respectively. In general, the sidelobe levels of a uniformly excited array are unacceptably high. The envelope of the sidelobe levels follows the $(1/u)$ law.

The beamwidth of the array is obtained by finding out the half power points of the main beam. This leads to

$$\left\{\sin\left(\frac{N+1}{2}u\right)/\sin\left(\frac{u}{2}\right)\right\} = \frac{N+1}{\sqrt{2}} \tag{10.16}$$

Hansen [1] has shown that the solution of eqn. 10.16 for different values of N leads us to a value of 0.4429 for $(N+1)u/2\pi$ within 1%, so long as the number of elements in the array is greater than 7. If we designate the directions θ_1 and θ_2 as the -3 dB power points on the main beam on either side of the peak, then

$$\sin\theta_1 = \sin(\theta_{B/2}) = 0.4429\,\frac{\lambda}{(N+1)d}$$

$$\sin\theta_2 = \sin(-\theta_{B/2}) = -0.4429\,\frac{\lambda}{(N+1)d} \tag{10.17}$$

where θ_B is the 3 dB beamwidth of the beam. Therefore

$$\sin(\theta_{B/2}) = \frac{0.4429}{(N+1)d}\lambda \tag{10.18}$$

For an array with 20 elements or more, $\sin(\theta_{B/2}) \simeq (\theta_{B/2})$, so that from eqn. 10.18 it can be inferred that

$$\theta_B = \frac{0.8858}{(N+1)d}\lambda \tag{10.19}$$

Eqn. 10.19 brings out clearly that, as the length of the array, i.e. $(N+1)d$ increases, the beamwidth decreases. Also, for a given physical length of the array the beamwidth increases with increase in wavelength or reduction in the operating frequency of the antenna.

10.3 Schelkunoff's theorem

An array with uniformly excited radiating elements has high first sidelobes which is an undesirable feature in present day systems. It would therefore be necessary to taper the excitation currents of the edge elements to bring down the sidelobe levels. The array factor of such antenna arrays may not always be expressed in a closed form solution. Schelkunoff [2] in 1943 showed that a polynomial can be associated with a linear array and that the array factor can be completely analysed in terms of the polynomial characteristics.

The starting point for this is eqn. 10.4 which can be expanded to give

$$|S(\theta)| = |I_N|\,|a_0 + a_1 \exp j(-kd \sin\theta) + \cdots + a_N \exp j(-Nkd \sin\theta)| \tag{10.20}$$

With,

$$\begin{aligned} a_0 &= I_0/I_N, \\ a_1 &= I_1/I_N, \\ &\vdots \\ a_m &= I_m/I_N \\ &\vdots \end{aligned} \tag{10.21}$$

Inserting eqn. 10.9 in eqn. 10.20 and introducing a new variable,

$$\zeta = \exp j(-u) \tag{10.22}$$

Eqn. 10.20 can now be rewritten as

$$|S(\zeta)| = |I_N|\,|a_0 + a_1\zeta + a_2\zeta^2 + \cdots + a_N\zeta^N| \tag{10.23}$$

The second term within the magnitude sign of eqn. 10.23 is an Nth order polynomial corresponding to an $(N+1)$ element array. Denoting this polynomial by $f(\zeta)$ and factorising it, eqn. 10.23 is transformed to

$$\begin{aligned} |S(\zeta)| &= |I_N|\,|f(\zeta)| \\ &= |I_N|\,|\zeta-\zeta_1|\,|\zeta-\zeta_2| \cdots |\zeta-\zeta_{N-1}|\,|\zeta-\zeta_N| \end{aligned} \tag{10.24}$$

Eqns. 10.23 and 10.24 highlight that the Nth order polynomial representing an $(N+1)$ element array can serve as a surrogate of the array factor. The relative excitation currents of the radiating elements are the coefficients of the polynomial and therefore the N roots of the polynomial are functions of the excitation currents.

Eqns. 10.23 and 10.24 provide the means by which the radiation pattern of a linear array can be changed by altering the roots of the polynomial.

Schelkunoff's unit circle representation provides a simple and yet an effective graphical way of carrying out these changes. The variable can be represented by a vector of unity magnitude with its origin coinciding with the origin of the co-ordinate system of the complex plane as in Fig. 10.2. The phase $\{(2\pi d/\lambda)\sin\theta\}$ of this vector is the angle subtended by the vector at the real axis. As θ in the visible space varies from $-\pi/2$ to $\pi/2$, ζ would move along the unit circle from $2\pi d/\lambda$ to $-2\pi d/\lambda$. In effect, it traverses a distance $4\pi d/\lambda$ along the circle in the clockwise direction. Depending on the magnitude of d/λ, this distance can be a portion of the unit circle, the full circumference or several multiples of it. The roots ζ_m $(m=1, 2, \ldots, N)$ of the polynomial are dependent on the ratios of the excitation currents of the radiating elements. The roots of the polynomial can lie anywhere in the complex plane but the array factor will go to zero only when these roots lie on the unit circle. Thus an array with $(N+1)$ radiating elements can at best have N nulls in the radiation pattern when all the roots of the polynomial represented by $f(\zeta)$ lie on the unit circle. If all the roots of the polynomial are off the unit circle, then a null-free radiation pattern can be obtained for the array.

If the array elements have uniform excitation, then

$$a_0=a_1=a_2=\cdots=a_{N-1}=a_N=1 \tag{10.25}$$

Substituting eqn. 10.25 in 10.23 we obtain

$$\begin{aligned} S(\zeta) &= 1+\zeta+\zeta^2+\cdots+\zeta^{N-1}+\zeta^N \\ &= \frac{1-\zeta^{N+1}}{1-\zeta} \end{aligned} \tag{10.26}$$

The generality of eqn. 10.26 is not lost if $I_N=1$ in eqn. 10.23. The roots of this polynomial correspond to

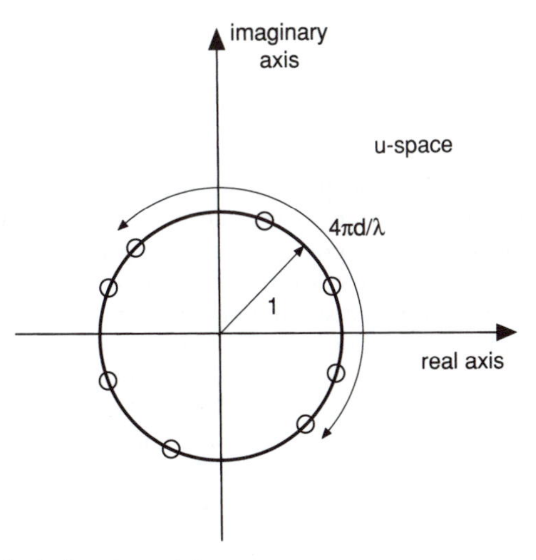

Figure 10.2 *Unit circle representation*

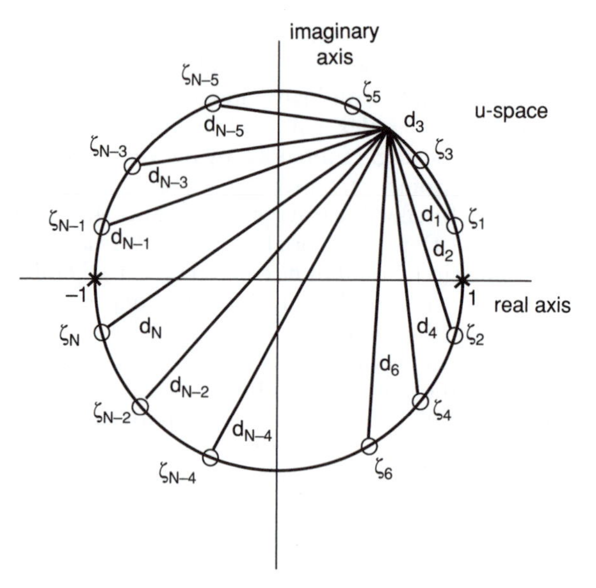

Figure 10.3 *Pattern null representation*

$$\zeta_m = \pm \exp j\{2\pi m/(N+1)\} \tag{10.27}$$

All the roots lie on the unit circle and they are symmetrically placed about the real axis forming complex conjugate pairs. The point $\zeta = 1$ is not a root because the polynomial assumes the peak value of $(N+1)$ at this point. Hence it represents the pointing angle of the main lobe of the array radiation pattern. It is also possible to obtain the magnitude of the radiation pattern from the placement of the roots along the unit circle. Towards this end, the factors of the polynomial, viz. $|\zeta - \zeta_m|$, are equal to d_m, the distance of the point ζ on the unit circle to the mth root as shown in Fig. 10.3. Thus eqn. 10.24 is transformed to

$$|f(\zeta)| = \prod_{m=1}^{N} d_m \tag{10.28}$$

The magnitude of the radiation pattern at any angle θ, corresponding to the point ζ in Fig. 10.3 in the visible space, can be obtained as the product of the distances of the roots from the point ζ. As ζ moves along the unit circle, the magnitudes of $d_1, d_2, \ldots, d_N$ vary, so that the product of these, representing the value of the radiation pattern, also changes. Whenever ζ moves into positions corresponding to the roots of the polynomial one of the distances d_m in the product becomes zero, and thus the radiation pattern also becomes zero. In the case of the uniform array as the variable ζ moves around the circle and away from the point $\zeta = 1$, it would successfully encounter all the roots of the polynomial or nulls in the pattern. There will be a cyclic variation in the value of $|f(\zeta)|$, i.e. build up and decay will alternate in a periodic fashion as ζ moves on that portion of the unit circle away from $\zeta = 1$, which corresponds to the sidelobe region. The sidelobe peaks will occur when ζ is approximately midway between two nulls. Hence, by making ζ move along the unit circle, it is possible to compute the radiation pattern of the array.

As long as the traverse of ζ on the unit circle corresponding to the N roots is less than 2π, the radiation pattern will have only one main lobe in the visible region. In the case of uniform array, therefore

$$\frac{N}{N+1} > \frac{d_{max}}{\lambda} \tag{10.29}$$

If this inequality is met under all conditions then, by shifting the positions of the complex conjugate roots along the unit circle, the radiation pattern can be shaped as per the requirements. For example in Fig. 10.3 the roots ζ_1 and ζ_2 represent the null points of the main lobe, and therefore by shifting ζ_1 and ζ_2 away from the $\zeta = 1$ point, the beamwidth of the lobe is increased. It can be easily checked that if two roots are brought closer, the sidelobe peaks are lowered whereas, if they are moved farther apart, the sidelobe peaks are raised to higher values. While it would be advisable to cluster as many roots as possible around the $\zeta = -1$ point to bring down the level of the sidelobe peaks, a price has to be paid in the broadening of the main lobe. By trial and error it would be possible to arrive at positions for the roots on the unit circle to provide low sidelobe levels without greatly increasing the main lobe width. The process of configuring the positions of the roots on the unit circle for optimum pattern conditions (narrowest main lobe with lowest sidelobe levels) is tedious as the number of roots N increases. Therefore other methods based on study of polynomials will have to be employed.

The directivity of an array denotes the efficiency of the current or voltage distribution functions used to excite the antenna elements of the array. Eqn. 10.6 is used to compute the directivity of a uniform array on the main beam axis as

$$D(\theta_0, \phi_0) = \frac{4\pi |E_A(\theta_0, \phi_0)|^2}{\int_0^{2\pi} \int_0^{\pi} |E_A(\theta, \phi)|^2 \sin\theta \, d\theta \, d\phi} \tag{10.30}$$

The radiation pattern of the array is a product of the radiation field $F(\theta, \phi)$ of the individual radiating element and the array factor $S(\theta)$. Substituting this in eqn. 10.30 we have

$$D(\theta_0, \phi_0) = \frac{F^2(\theta_0, \phi_0) \, |S(\theta_0)|^2}{\int_0^{2\pi} d\phi \int_0^{\pi} F^2(\theta, \phi) \, |S(\theta)|^2 \sin\theta \, d\theta} \tag{10.31}$$

The radiation pattern $F(\theta, \phi)$ of the individual radiating element is considerably broader than the array factor. Further, the array factor has a narrow pencil beam with low enough sidelobes. The main contribution to the integral in the denominator of eqn. 10.31 is in the neighbourhood of the main beam, that is, around the direction θ_0. In such a case, eqn. 10.31 becomes

$$D(\theta_0)=\left[\frac{F^2(\theta_0)}{\dfrac{1}{2\pi}\displaystyle\int_0^{2\pi} F^2(\theta_0)\, d\phi}\right]\left[\frac{|S(\theta_0)|^2}{\dfrac{1}{2}\displaystyle\int_0^{\pi} |S(\theta)|^2 \sin\theta\, d\theta}\right] \tag{10.32}$$

The term in the first bracket of the RHS of eqn. 10.32 represents the directivity of the radiating element in comparison to an isotropic radiator, and is therefore equal to 1. Thus we have for the peak directivity of the array factor of the uniform array comprising elementary radiators, the following expression:

$$D(\theta_0)=\frac{2S(\theta_0)S^*(\theta_0)}{\displaystyle\int_0^{\pi} |S(\theta)||S^*(\theta)| \sin\theta\, d\theta} \tag{10.33}$$

Using the series equivalence of $S(\theta)$ as given in eqn. 10.26, we can write

$$|S(\theta)||S^*(\theta)|=(N+1)+2\sum_{m=0}^{N}(1+N-m)\cos m\zeta \tag{10.34}$$

Substituting eqn. 10.34 in the denominator of eqn. 10.33, we have

$$\int_0^{\pi} |S(\theta)||S^*(\theta)| \sin\theta\, d\theta=2(N+1)+4\sum_{m=1}^{N}\frac{N}{mkd}\sin(mkd)\cos(mkd\cos\theta_0) \tag{10.35}$$

The expression for peak directivity in this case becomes

$$D=\frac{(N+1)^2}{(N+1)+2\displaystyle\sum_{m=1}^{N}\frac{N}{mkd}\sin(mkd)\cos(mkd\cos\theta_0)} \tag{10.36}$$

The second term of the denominator of eqn. 10.36 vanishes when $d/\lambda=p/2$, with $p=1, 2, \ldots,$ because $\sin(mkd)=\sin(mp\pi)=0$. In addition, for other values of d/λ, the other term $\cos(mkd\cos\theta_0)$ can vanish depending on θ_0. Hence, under conditions which make the term $\sin(mkd)\cos(mkd\cos\theta_0)$ vanish, D can be calculated to be equal to $(N+1)$, the total number of elements in the array. Ma [4] has used a graphical method to compute the relative value of D as a function of the inter-element spacing to give a generalised picture.

An alternative approach by Elliott [5] is based on rewriting eqn. 10.3 as

$$S(\theta)=\sum_{m=-N/2}^{m=N/2}(I_m/I_0)\exp j(mu) \tag{10.37}$$

Substituting eqn. 10.37 in eqn. 10.32, we obtain

$$D(\theta_0, \phi_0)=\frac{2\left[\displaystyle\sum_{m=-N/2}^{m=N/2} I_m\right]^2}{\dfrac{1}{kd}\displaystyle\int_{-kd-kd\sin\theta}^{kd-kd\sin\theta} du\left(\sum_{m=-N/2}^{m=N/2} I_m\exp j(mu)\right)\left(\sum_{m=-N/2}^{m=N/2} I_m\exp j(-mu)\right)} \tag{10.38}$$

When $d/\lambda = p/2$ where p is an integer, then eqn. 10.38 reduces to

$$D(\theta_0, \phi_0) = \frac{\left[\sum\limits_{m=-N/2}^{m=N/2} I_m\right]^2}{\sum\limits_{m=-N/2}^{m=N/2} I_m^2} \tag{10.39}$$

From eqn. 10.39 it is evident that the directivity D is a measure of the coherence of radiation, as the numerator of the RHS of eqn. 10.39 is proportional to the square of the total coherent field, whereas the denominator is proportional to the sum of the squares of individual fields due to the radiating elements. If we compare it with eqn. 10.36 we find that the dependence of directivity on scan angle does not figure in eqn. 10.39. This can be explained by the fact that, in the case of a linear array, as the scanning beam moves from broadside to endfire, the disc of main beam radiation folds forwards to form a cone with a broad base. As the beam scans further away from the broadside direction, the two main beams tilt forwards and the wider base of the cone narrows down as the tilt angle increases. Finally, when the -3 dB points of both beams are at the endfire position, the two beams start to fuse to form a single conical beam with directivity in both orthgonal planes. Thus the linear array which had directivity in one plane at broadside ends up with directivity in both orthogonal planes. Even though the beam broadens in one plane owing to the fact that the cone occupied by it in the orthogonal plane shrinks with the tilt, the directivity remains unchanged. In the case of uniform arrays with equi-amplitude excitation, eqn. 10.39 reduces to $D = (N+1)$. The directivity of a linear array with uniform spacing and equi-amplitude and phase excitation, is equal to the number of elements in the array. This is the maximum value of D for a linear array of $(N+1)$ elements.

10.4 Antenna synthesis

The difficulties encountered in finding the excitation currents of the radiating elements to give a radiation pattern with low sidelobes by the unit circle method has given rise to a search for appropriate polynomials which can straightaway meet the requirements or which can be modified for this purpose. The uniform array concept which provides a good understanding of arrays in general will now be replaced by arrays which have non-uniform excitation across the elements.

10.4.1 *Dolph–Chebyshev distribution*

C L Dolph in 1946 [6] showed that Chebyshev polynomials can be used for representing the array factor and that the roots of the polynomial can be placed along the unit circle to yield a radiation pattern with uniform sidelobe peaks of desired values. The radiation pattern is optimum in the sense that the sidelobe peaks specified by the polynomial are the lowest for a given first nulls beamwidth, or conversely the first nulls beamwidth is the narrowest for a specified sidelobe peak value. The Chebyshev polynomial of Nth degree is chosen for the $(N+1)$ element array as the latter has only N (roots) nulls. The Chebyshev polynomial can be written as

$$T_N(x) = (-1)^N \cosh(N \cosh^{-1}|x|); \quad x < -1$$
$$\cos(N \cos^{-1} x); \quad |x| \leq 1$$
$$\cosh(N \cosh^{-1} x); \quad x > 1 \qquad (10.40)$$

Fig. 10.4. shows a typical Chebyshev polynomial which in this case is a fifth degree polynomial. The antenna designer's interest will be in the regions $|x| \leq 1$ and part of $x > 1$. In the region $|x| \leq 1$ the polynomial $T_N(x)$ varies in a cosinusoidal fashion with a peak value of 1. This can therefore be utilised in describing the sidelobe regions. A part of the region $x > 1$ can be used to represent the main beam, because the value of $T_N(x)$ rises steeply in a hyperbolic manner in this region. Depending on the desired sidelobe ratio, the

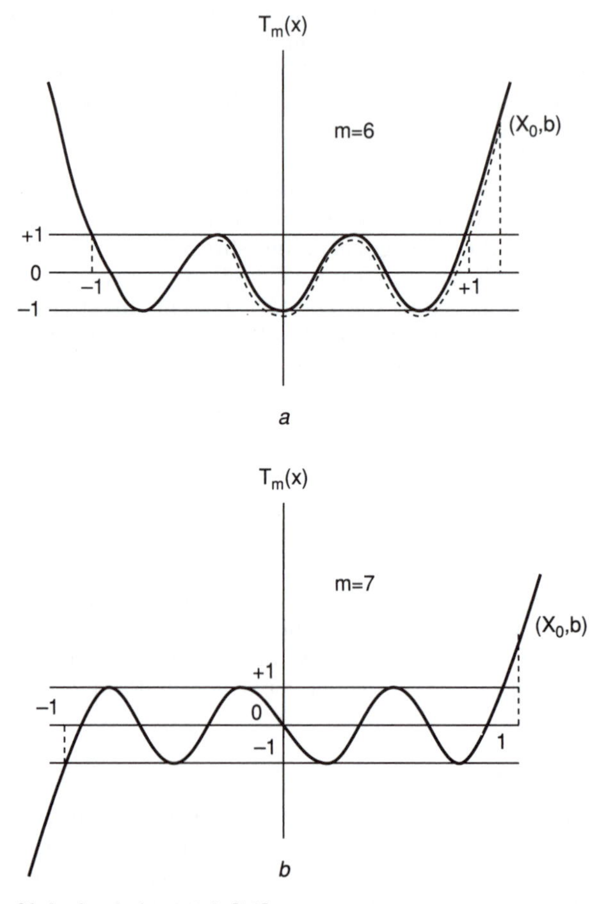

Figure 10.4 *Chebyshev polynomial [25]*

a 6th order
b 7th order

value of $T_N(x)$ in the $x > 1$ region can be selected. If this value is x_0, then $T_N(x_0) = b$, where $\boldsymbol{b}$ is the ratio of the main lobe peak voltage to the sidelobe peak voltage. The sidelobe level is defined as $\text{SLL} = 20 \log_{10} b$. To relate the parameter θ of the visible space to the Chebyshev polynomial, Dolph replaced the variable x by $x_0 \cos(u/2)$. Thus the array factor is now expressed as

$$S(u) = T_N(x_0 \cos u/2) \tag{10.41}$$

It is important that the parameter x should always be higher than -1 because of the nature of the Chebyshev polynomial. When the angle θ in the visible region traverses from $\pi/2$ to 0 and to $-\pi/2$, x moves from $x_0 \cos(\pi d/\lambda)$ to x_0 and then to $x_0 \cos(-\pi d/\lambda)$, respectively. Thus it would be necessary to ensure that, $d/x < 1$ so that $|\cos(\pi d/\lambda)|$ is always below the value $|\cos \pi|$. There is a lower limit for the inter-element spacing arising out of the need to ensure that the traverse of x in the oscillatory region of the Chebyshev polynomial covers the negative values (greater than -1) also. This would result in maintaining the optimal characteristic of the polynomial for the array of $(N+1)$ elements. The lower limit of $\boldsymbol{d}$ in this case is $d > \lambda/2$. Thus the range of values for the inter-element spacing is $\lambda > d > \lambda/2$.

There is a broadening of the main beam with the Chebyshev pattern when it is compared with the pattern of a uniformly excited array of the same length. Elliott [5] has shown that the beam broadening increases with lower sidelobe levels and is approximately in the range 1.32 times to 1.64 times when the sidelobe levels are -40 dB to -60 dB, respectively. The peak directivity of the beam in the case of long arrays with a Chebyshev pattern is 3 dB higher than the sidelobe ratio: i.e., for a Chebyshev array with -40 dB sidelobe, the peak directivity is in the region of 43 dB. It has also been found that there is a directivity compression limit beyond which directivity is bought very dearly in terms of greatly increased array lengths. For a Chebyshev array since all the sidelobes are of equal level, as the array is lengthened, the energy in these additional sidelobes being equal to that in the other sidelobes, the denominator of eqn. 10.33 also increases, resulting in the tapering of the directivity.

The computation of the excitation coefficients of the Dolph–Chebyshev array has been the topic of many papers in the days prior to the era of computational plenty. Bach and Hansen [7] have listed the more important of these where simplified approximate versions have been worked out. It has been found that Chebyshev distribution tends towards large peaks for the excitation of end elements as the directivity compression limit is approached.

The search was therefore on for an 'ideal' source which produces a radiation pattern with equal level sidelobes in the close-in region so that the directivity of the Chebyshev pattern is not disturbed greatly and the peaks of the sidelobes in the far-out region taper quite fast so that there are no large value excitations at the array ends.

10.5 Taylor $\bar{n}$ distribution

Taylor [8] in 1955 developed a distribution which is today called the Taylor $\bar{n}$ distribution that combines the advantages of high directivity with equal level close-in sidelobes and tapered envelope for the far-out sidelobes, as in uniform distribution. The proposed radiation pattern has the sidelobe region represented by a cosine function so that the requirements of equal sidelobes are met. The main beam region is represented by a hyperbolic cosine function so that the pattern function of the continuous line source is required to give

$$S(u_c) = \cos \pi\sqrt{u_c^2 - A^2}; \quad \text{for } u_c \leqslant A$$
$$\cosh \pi\sqrt{A^2 - u_c^2}; \quad \text{for } u_c \geqslant A \tag{10.42}$$

with $u_c = (2L/\lambda)\sin\theta = (Nd/\lambda)\sin\theta$.

At $u=0$, $S(0) = \cosh \pi A$. The value of $S(0)$ is set equal to the sidelobe ratio b so that

$$\cosh \pi A = b \tag{10.43}$$

The problem of equal level sidelobes in the far-out region was resolved by ensuring that these sidelobes conform to the pattern of sidelobes of a uniform array for large values of u_c, so that the $(1/u_c)$ taper for the envelope of the far-out sidelobes takes place. Thus the Taylor distribution divides the sidelobe region into two, namely the close-in and the far-out, and makes the nulls and the sidelobe peaks conform to the mathematical law selected for them. In the close-in region the sidelobe levels are all equal while the positions of the nulls followed the nulls of the first of the equations of eqn. 10.42. In the far-out region the positions of the nulls as well as the envelope of the sidelobe peaks conformed to the nulls and sidelobe peaks of $\{(\sin \pi u_c)/\pi u_c\}$ for large values of u_c. The nulls in the close-in region are formed at

$$u_n = \pm\sqrt{(n-\tfrac{1}{2})^2 + A^2} \tag{10.44}$$

whereas the nulls in the far-out region are formed at

$$u_m = \pm m \tag{10.45}$$

The nulls of the far-out region are perturbed in their positions so that in the transition subregion an integer $\bar{n}$ is selected to make the position of the nth sidelobes of the $\cos(\pi\sqrt{u_c^2 - A^2})$ and that of $\{(\sin \pi u_c)/\pi u_c\}$ functions coincide. The dilation factor σ is now given by

$$\sigma = \frac{\bar{n}}{\sqrt{(\bar{n}-\tfrac{1}{2})^2 + A^2}} \tag{10.46}$$

The modified array factor can now be written as

$$S_T(u_c) = \frac{\sin \pi u_c}{\pi u_c} \frac{\prod_{m=1}^{\bar{n}-1} (1 - u_c^2/u_m^2)}{\prod_{m=1}^{\bar{n}-1} (1 - u_c^2/m^2)} \tag{10.47}$$

From eqns. 10.44–10.47 the roots of $S_T(u_c)$ are now given by

$$u_m = \bar{n}\left[\frac{A^2 + (m-\tfrac{1}{2})^2}{A^2 + (\bar{n}-\tfrac{1}{2})^2}\right]^{1/2} \tag{10.48}$$

The 3 dB beamwidth can be approximated to

$$u_B = \frac{\sigma}{\pi}\sqrt{\{\ln(b+\sqrt{b^2-1})\}^{1/2} + \left\{\ln\left(\frac{b}{\sqrt{2}} + \sqrt{\frac{b^2}{2}-1}\right)\right\}^{1/2}} \tag{10.49}$$

and the aperture distribution can be best expressed in a Fourier series as

$$g(x) = 1 + 2\sum_{m=1}^{\bar{n}-1} F(m, A, \bar{n}) \cos m\pi x \tag{10.50}$$

The aperture variable x is zero at the centre of the line source and ± 1 at both ends. The coefficient $F(m, A, \bar{n})$ is obtained from

$$F(m, A, \bar{n}) = \frac{[(\bar{n}-1)!]^2}{(\bar{n}-1+N)!\,(\bar{n}-1-N)!}\prod_{i=1}^{\bar{n}-1}\left(1-\frac{m^2}{u_i^2}\right) \tag{10.51}$$

The directivity of the Taylor $\bar{n}$ distribution is given by

$$D = \frac{1}{1 + 2\sum_{m=1}^{\bar{n}-1} F^2(m, A, \bar{n})} \tag{10.52}$$

Using eqn. 10.52 it is possible to compute D with ease and find out D_{max} for a given value of $\bar{n}$.

Figure 10.5 shows the radiation pattern of a continuous line source with uniform amplitude distribution. The first sidelobes are too high and therefore

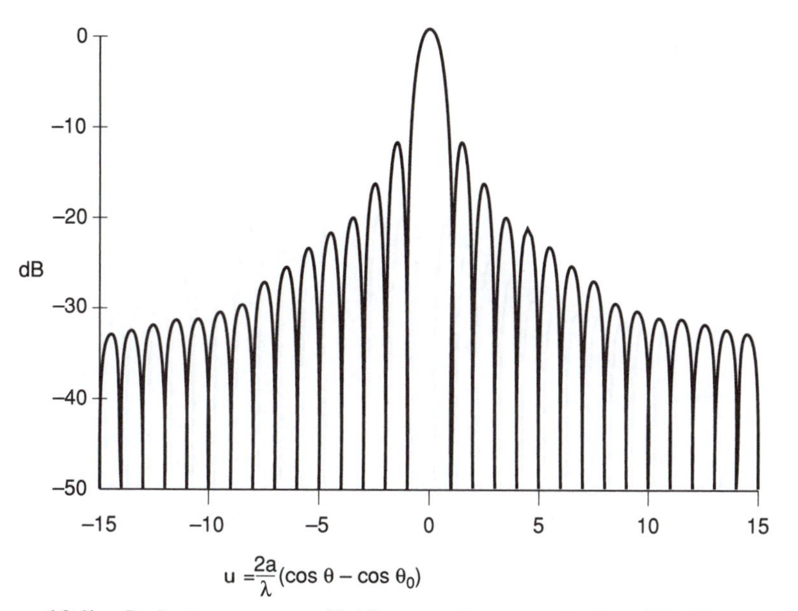

Figure 10.5 *Radiation pattern: Uniform continuous excitation [5]. Reprinted by permission of Prentice-Hall*

the pattern is unacceptable. Fig. 10.6 shows the Taylor distribution for a continuous line source for $\bar{n}=6$ and with -20 dB as the sidelobe level [5]. The radiation pattern highlights the following aspects. Firstly, the sidelobes of the Taylor pattern for $\bar{n}=6$ are not precisely at the design level of -20 dB but are lower. In addition there is a slight droop to this envelope for the close-in sidelobes. If $\bar{n}$ was a larger number the droop would have been less. Also, the increase in beamwidth as a result of the droop would be negligible. The Taylor aperture distribution for the same radiation pattern is shown in Fig. 10.7. It can be seen that it is not monotonic and has edge excitation, which is, however, less than that needed for an equivalent Dolph–Chebyshev array. The discretised Taylor $\bar{n}$ distribution is therefore easier physically to realise than the Dolph–Chebyshev distribution.

$\bar{n}$ is a parameter that influences the beamwidth and hence the directivity. While too large a value of $\bar{n}$ results in a non-monotonic distribution and can even give rise to large peaks at the ends of the line aperture, too small a value of $\bar{n}$ will not permit the transit zones to behave properly. Figs. 10.8 and 10.9 from Hansen [1] illustrate these statements clearly. Fig. 10.8 shows the aperture distribution across the line sources for different values of $\bar{n}$ for three different conditions. These are:

(*a*) SLL $=-25$ dB, $\bar{n}=5$
(*b*) SLL $=-30$ dB, $\bar{n}=7$
(c) SLL $=-40$ dB, $\bar{n}=11$

All the three distributions have a monotonic nature. On the other hand in Fig. 10.9, the aperture distribution for an SLL of -25 dB with $\bar{n}=12$ is distinctly

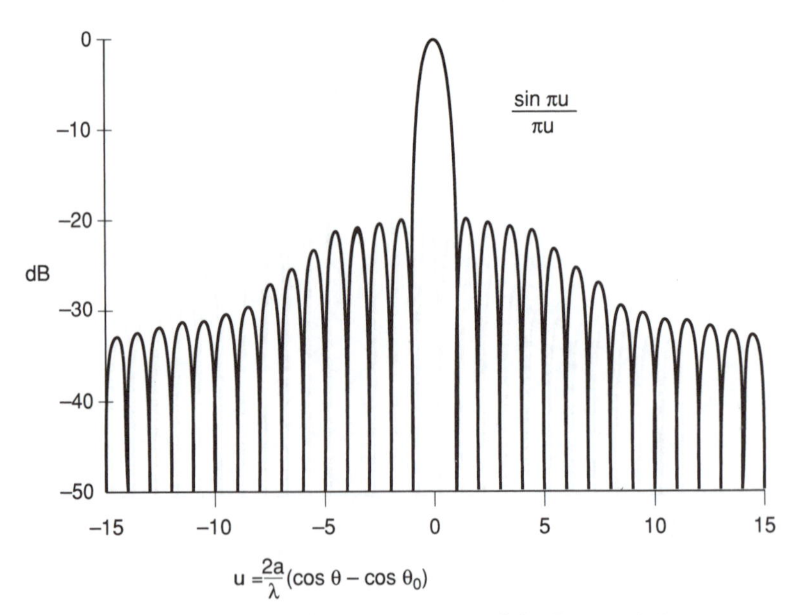

Figure 10.6 *Taylor pattern: Continuous source [5]. Reprinted by permission of Prentice-Hall*

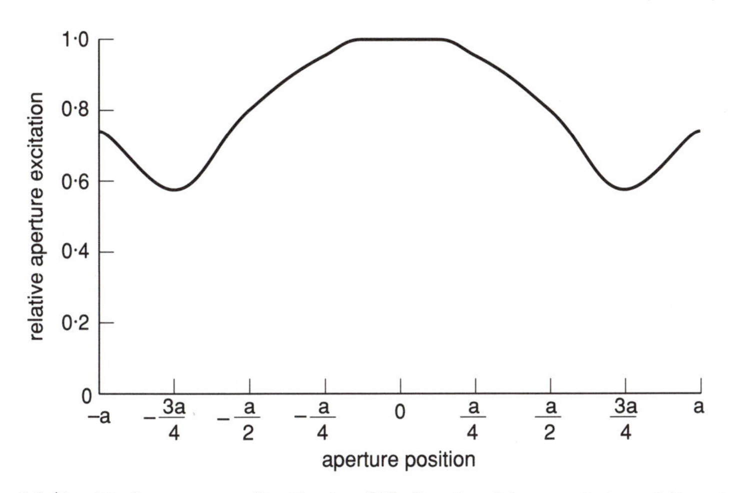

Figure 10.7 *Taylor aperture distribution [5]. Reprinted by permission of Prentice-Hall*

non-monotonic. It not only peaks but also has small oscillations at the edges. This distribution has maximum efficiency as compared with $\bar{n}=5$, though it is only 1.5% higher. In Table 10.1 maximum efficiency values for Taylor $\bar{n}$ distributions are listed for different SLLs.

The Taylor $\bar{n}$ distribution is the most widely used aperture distribution in antenna arrays because of the ease with which it can be adapted to discrete arrays.

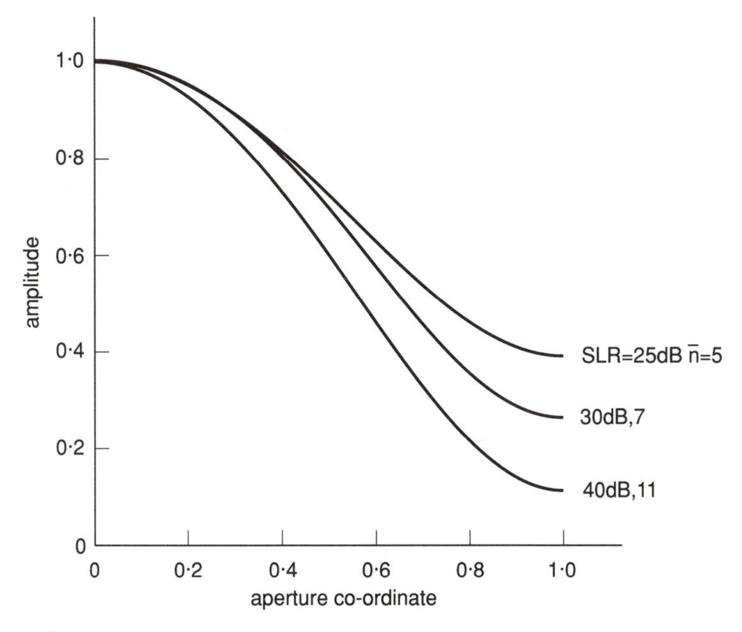

Figure 10.8 *Taylor $\bar{n}$ aperture illuminations [1]*

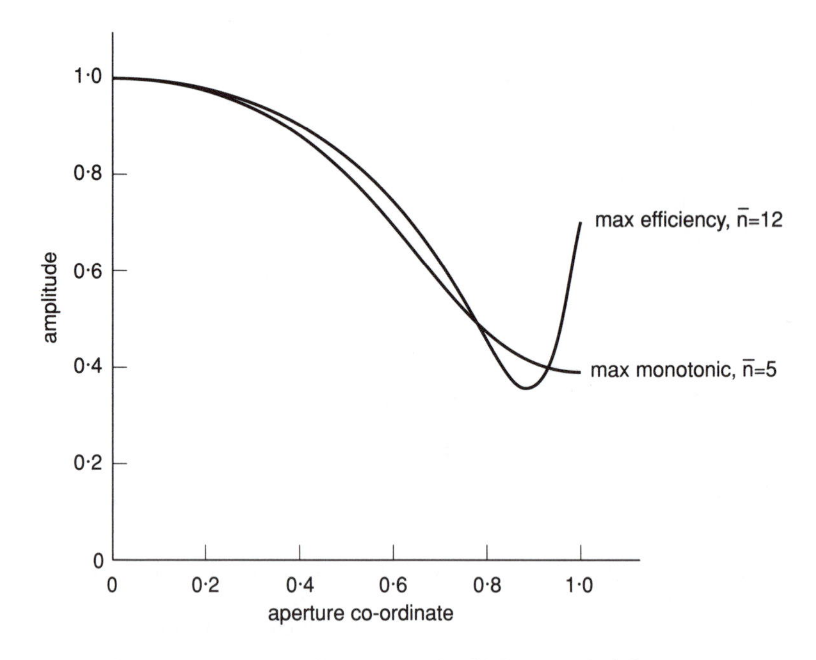

Figure 10.9 *Taylor aperture illumination for SLL = 25 dB [1]*

10.6 Discretisation aspects

In the case of both Dolph–Chebyshev and Taylor $\bar{n}$ distributions which are used to designate the desired radiation pattern, the resulting excitation current distribution across the array is a continuous function. By sampling this at $(N+1)$ points which denote the location of the radiating elements, the excitation currents can be derived. The actual array pattern as a result of the discrete current excitation does not significantly alter the desired radiation pattern. In general this is true so long as the sampling interval is small enough to contain the oscillatory components of the continuous excitation function. In the case of small arrays or in some special cases where the sampling interval is not small enough, discretisation by sampling the resulting continuous line source will not truly provide the desired array pattern.

Table 10.1 *Taylor $\bar{n}$ distribution [1]*

SLL (dB)	$\bar{n}$ for maximum efficiency	maximum efficiency
−20	6	0.9667
−25	12	0.9252
−30	23	0.8787
−35	44	0.8326
−40	81	0.7899

Another way of discretising the excitation function to match the discrete array requirements is to determine the nulls of the desired radiation pattern. These null positions can be known with great precision and they can be conveniently transformed into the corresponding values of ζ. The roots of $f(\zeta)$, the array factor polynomial, are known and these can be utilised to compute the array factor polynomial. This is now expressed as

$$S(\zeta)=\prod_{m=0}^{N}(\zeta-\zeta_m)=\sum_{m=0}^{N}\frac{I_{0m}}{I_{0N}}\zeta^m \tag{10.53}$$

Here I_{0m} is the excitation current for the radiating element $\boldsymbol{m}$ and I_{0N} is the excitation current of the radiating element $\boldsymbol{N}$. Elliott [9] has pointed out that, if the constraints on sidelobe requirements are severe, then the root matched starting pattern given by eqn. 10.53 needs to be processed further. Based on the assumption that the sidelobe levels on the two sides of the main beam of the starting pattern separately approximate to the average sidelobe levels on the two sides of the main beam of the desired pattern, a perturbation method has been suggested [9]. Now, the desired array pattern is expressed as

$$S(\zeta)=\sum_{m=0}^{N}\left(\frac{I_{0m}}{I_{0N}}\right)\zeta^m \tag{10.54}$$

If the departure of the starting pattern from the desired pattern is not too high then the excitation currents for the desired pattern can be considered as small perturbations of the currents for generating the starting pattern. Hence

$$I_m=I_{0m}+\delta I_m \tag{10.55}$$

Since δI_m is small compared to I_m or to I_{0m}, introducing eqn. 10.55 into eqn. 10.54 yields

$$S(\zeta)=S_0(\zeta)+\sum_{m=0}^{N}\left(\frac{\delta I_m}{I_{0N}}\right)\zeta^m \tag{10.56}$$

If ζ_{0r}^p is the position in ζ-space of the peaks of the sidelobes for the starting pattern and if ζ_{00}^p is the position of the main beam peak, then it can be easily seen that

$$\frac{S(\zeta_{0r}^p)-S_0(\zeta_{0r}^p)}{S_0(\zeta_{00}^p)}=\sum_{m=0}^{N}\frac{(\zeta_{0r}^p)^m}{I_{0N}S_0(\zeta_{00}^p)}\,\delta I_m \tag{10.57}$$

As long as the perturbations are small, ζ_{0r}^p is close to the peak position of the rth sidelobe in the desired pattern. It is reasonable to assume that the peak values of the main lobes of the starting and desired pattern are the same. This is expressed as

$$S(\zeta_{00}^p)-S_0(\zeta_{00}^p)=0 \tag{10.58}$$

Introducing eqn. 10.58 in eqn. 10.57, we have

$$\frac{S(\zeta_{0r}^p)-S_0(\zeta_{0r}^p)}{S_0(\zeta_{00}^p)}=\varepsilon_m;\quad m=0,\,1,\,2,\,\ldots,\,N \tag{10.59}$$

ε_m is a pure real number and is known. Introducing eqns. 10.58 and 10.59 in eqn. 10.57 gives us a set of $(N+1)$ simultaneous linear equations in δI_m. Matrix inversion yields the perturbations in the excitation currents of the $(N+1)$ radiating elements of the array. When these are introduced in eqn. 10.53, a new array pattern is computed which is compared with the desired pattern. If it is not satisfactory, the new array pattern becomes the starting pattern and the procedure is repeated all over again until the stipulated criterion for acceptance is reached. The number of iterations in this procedure will depend on the differences between the initial starting pattern and the desired pattern.

10.7 Villeneuve procedure for discretisation

Villeneuve [10] in 1984 developed a technique that provides the excitation of the elements of discrete arrays for a radiation pattern analogous to the Taylor pattern. The starting point is the Dolph–Chebyshev radiation pattern $T_{N-1}(u)$ for an N-element array where N is odd. It is expressed as a product of $(N-1)$ factors containing the roots of the polynomial, so that

$$T_{N-1}(u) = S(u) = \exp j\left\{\frac{N-1}{2}\,u\right\}4^{N-1}\prod_{m=1}^{(N-1)/2}\sin\left(\frac{u-u_m}{2}\right)\sin\left(\frac{u+u_m}{2}\right) \tag{10.60}$$

with $1 \leqslant m \leqslant (N-1)/2$.

As in the case of Taylor distribution the sidelobes in the close-in region will conform to the Chebyshev polynomial, whereas the sidelobes in the far-out will follow the distribution. The nulls in the close-in as well as in the far-out sidelobe regions will now be located, respectively, at

$$u_{0m} = \frac{2\pi m}{N}$$

$$u_m = 2\cos^{-1}\left\{\frac{1}{x_0}\cos(2m-1)\frac{\pi}{2(N-1)}\right\} \tag{10.61}$$

where $m = 1, 2, \ldots, N$.

The radiation pattern polynomial is now transformed to

$$S_V(u) = e^{j((N-1)/2)u}\left[\frac{\sin\dfrac{Nu}{2}}{\sin\dfrac{u}{2}}\right]\left[\frac{\prod_{m=1}^{\bar{n}-1}\sin\left(\dfrac{u-u_m}{2}\right)\sin\left(\dfrac{u+u_m}{2}\right)}{\prod_{m=1}^{\bar{n}-1}\sin\left(\dfrac{u-u_{0m}}{2}\right)\sin\left(\dfrac{u+u_{0m}}{2}\right)}\right] \tag{10.62}$$

$\bar{n}$ is the transition number at which the $\bar{n}$th null due to the $[\{\sin(Nu/2)\}/\{\sin(u/2)\}]$ pattern coincides with the $\bar{n}$th null of the Chebyshev pattern. In order to make this possible, the nulls of the Chebyshev pattern, i.e. the close-in region nulls u_m, are disturbed by a dilation factor from their original position so that $u'_m = \sigma u_m$. The dilation factor is now given by

$$\sigma = \frac{\bar{n}\pi}{N\cos^{-1}\left\{\frac{1}{x_0}\cos(\bar{n}-\frac{1}{2})\frac{\pi}{N-1}\right\}} \tag{10.63}$$

The modified $\bar{n}$ distribution for the desired radiation pattern can now be written as

$$S(u) = e^{j((N-1)/2)u}\left[\frac{\sin\frac{Nu}{2}}{\sin\frac{u}{2}}\right]\left[\frac{\prod_{m=1}^{\bar{n}-1}\sin\left(\frac{u-u'_m}{2}\right)\sin\left(\frac{u+u'_m}{2}\right)}{\prod_{m=1}^{\bar{n}-1}\sin\left(\frac{u-(2\pi m/N)}{2}\right)\sin\left(\frac{u+(2\pi m/N)}{2}\right)}\right] \tag{10.64}$$

Eqn. 10.64 is the Villeneuve $\bar{n}$ distribution. To compute the excitation of the array elements, we go back to eqn. 10.4 with $\alpha_m = 0$, so that

$$S(u) = \sum_{m=-N/2}^{m=N/2} a_m \exp j(mu) \tag{10.65}$$

where the *a*s have been defined in eqn. 10.21. The pattern is sampled at N points defined by $u_p = 2p\pi/N$ and the N excitation coefficients a_m in eqn. 10.65 are evaluated. Since the aperture excitation and the radiation pattern are Fourier transform pairs, from eqn. 10.64 we can easily derive

$$a_m = \frac{1}{N}\left[S(0) + 2\sum_{p=1}^{\bar{n}-1} S\left(\frac{2p\pi}{N}\right)\cos\left(\frac{2mp\pi}{N}\right)\right] \tag{10.66}$$

with

$$S(0) = N\frac{\prod_{q=1}^{\bar{n}-1}\sin^2\left(\frac{u'_q}{2}\right)}{\prod_{q=1}^{\bar{n}-1}\sin^2\left(\frac{u_q}{2}\right)} \tag{10.67}$$

and

$$S\left(\frac{2p\pi}{N}\right) = \frac{N(-1)^p\prod_{q=1}^{\bar{n}-1}\sin\left(\frac{p\pi}{N}-\frac{u'_q}{2}\right)\sin\left(\frac{p\pi}{N}+\frac{u'_q}{2}\right)}{\sin\left(\frac{p\pi}{N}\right)\sin\left(\frac{2p\pi}{N}\right)\prod_{\substack{q=1\\q\neq p}}^{\bar{n}-1}\sin\left(\frac{(p-q)\pi}{N}\right)\sin\left(\frac{(p+q)\pi}{N}\right)} \tag{10.68}$$

Villeneuve has proved by numerical examples that the excitation of a discrete array with the Villeneuve $\bar{n}$ radiation pattern can be computed with very high accuracy without recourse to an intermediate continuous distribution. Hansen [11] has computed the directivity of Villeneuve distribution for small arrays and has showed that modest low sidelobe levels can be obtained at modest efficiencies.

10.8 Generalised Villeneuve arrays

The Villeneuve array combines the high efficiency equal sidelobe level feature of the Dolph–Chebyshev pattern with the $(1/u)$ taper for the sidelobe envelope of the uniform distribution. This is the Taylor equivalent for discrete radiating element arrays. McNamara [12] has proposed a generalised Villeneuve $\bar{n}$ distribution which provides additional control on the taper of the far-out region sidelobe envelope. In this case, the Chebyshev nulls in the close-in region, as well as the nulls in the far-out region, are disturbed by the dilation parameter. In addition, the nulls in the far-out region are disturbed by a parameter called taper parameter from their original positions derived from the $\{\sin(Nu/2)/\sin(u/2)\}$ pattern. The new roots are now given by

$$\begin{aligned} u'_m &= \sigma u_m; & m \leqslant \bar{n} \\ u'_m &= u_m + (\nu+1)(u_{0m} - u_m); & m \geqslant \bar{n} \end{aligned} \tag{10.69}$$

with

$$\sigma = \frac{u_{\bar{n}} + (\nu+1)(u_{0\bar{n}} - u_{\bar{n}})}{u_{\bar{n}}} \tag{10.70}$$

Here ν is the taper parameter which decides the slope of the sidelobe envelope in the far-out sidelobe region. If $\nu = 1$, then $\sigma = 1$, and the array pattern reduces to a regular Dolph–Chebyshev pattern. If $\nu = 0$, then the array is the Villeneuve $\bar{n}$ array described in the previous Section. When $\nu < 0$, the slope of the taper for the envelope of the sidelobe levels in the far-out region are shallower than $(1/u)$ and if $\nu > 0$, the slope of the taper is steeper. In this case $\bar{n}$ has to be governed by a minimum value based on the consideration that under no circumstances can the main lobe beamwidth between the first nulls of the Villeneuve $\bar{n}$ array be less than the Chebyshev array. Therefore

$$u'_1 \geqslant u_1 \tag{10.71}$$

If the inequality is introduced in the second of eqns. 10.69, then $u'_m \geqslant u_{0m}$ irrespective of the value of ν. Consequently,

$$\bar{n} \geqslant \left(\frac{N}{\pi}\right) u_{\bar{n}} \tag{10.72}$$

Therefore for each array size and for each desired SLL the minimum value of $\bar{n}$ has to be determined so that inequality 10.72 is not violated. If a value of $\bar{n}$ less than $\bar{n}_{min}$ is used, then there is a strong likelihood of the close-in sidelobe peaks having irregular increases over the Chebyshev values.

In general, the taper parameter ν reduces the efficiency of the array and results in higher beam broadening if the value of $\bar{n}$ is brought closer to the minimum value indicated in inequality 10.72. By and large, the directivity of the generalised Villeneuve $\bar{n}$ array is less than that of the Chebyshev array and it peaks over a narrow range of values of $\bar{n}$ for each ν. However, the Chebyshev array reaches the directivity compression limit sooner than the Villeneuve array. Finally by choosing $\bar{n}$ higher than $\bar{n}_{min}$ is no guarantee that edge brightening of the array does not take place. Thus, a proper choice of $\bar{n}$ needs to be made so that efficiency, monotonicity and pedestal heights for edge excitation can be traded off against the desired sidelobe level for discrete arrays.

10.9 Scanning arrays

A linear array with progressive phase shift between successive radiating elements provided by microwave phase shifters forms the main beam in space in a direction other than broadside. By varying the progressive phase shift in a programmed manner, the main beam can be moved from broadside towards endfire. This is the principle of electronic scanning and hence needs to be examined from the point of view of the behaviour of the main beam and sidelobes as well as the appearance of grating lobes, even though at broadside the latter aspects are fully under control The process is initiated with the uniform array having progressive phase shifts between the elements so that the main beam is formed in the desired direction θ_0. In such a case the progressive phase shift introduced between the radiating elements nullifies the phase shift introduced in the direction θ_0 due to the path difference from each radiating element to the far field point. Thus u in equation 10.9 is transformed to $(2\pi d/\lambda)(\sin\theta - \sin\theta_0)$ and eqn. 10.12 changes to

$$(\sin\theta - \sin\theta_0) = \pm p\,\frac{\lambda}{d} \tag{10.73}$$

Now θ_0 can lie anywhere in real space, that is, $-\pi/2 \leqslant \theta_0 \leqslant \pi/2$. In such a case $\boldsymbol{u}$ will vary from $(2\pi d/\lambda)(1+\sin\theta)$ to $(2\pi d/\lambda)(1-\sin\theta)$ so that the traverse of the parameter u on the unit circle is still $4\pi d/\lambda$ but the starting and the end points have shifted. Schelkunoff's unit circle remains unaffected.

It is now necessary to investigate the likely restrictions on the spacing between consecutive radiating elements for avoidance of grating lobe formation when the main beam is being pointed in directions other than the broadside direction. Now eqn. 10.73 can be rewritten for grating lobe formation as

$$\sin\theta = \sin\theta_0 \pm \frac{p\lambda}{d} \tag{10.74}$$

where $p = 1, 2, \ldots, N-1$. The first grating lobe will occur for $p = 1$. Let the angle of the grating lobe be designated as θ_g. In that case eqn. 10.74 becomes

$$\sin\theta_g = \sin\theta_0 \pm \frac{\lambda}{d} \tag{10.75}$$

Since the purpose of array design is to avoid grating lobe formation, $\sin\theta_g$ has to be greater than 1 for all θ_0 so that θ_g is outside the real space. Eqn. 10.75 now leads to two equations

$$\begin{aligned} |\sin\theta_0| + \lambda/d &> 1 \\ |\sin\theta_0| - \lambda/d &< -1 \end{aligned} \tag{10.76}$$

The condition set by first of the inequality 10.76 is that, if $\lambda/d > 1$, then even when $|\sin\theta_0| = 0$, the equation will be satisfied and no grating lobes will be formed. This condition is the same as that arrived at in inequality 10.14. The condition that is set by the second of the inequality 10.76 is more demanding, and thus we obtain for the scanning arrays the criterion for avoiding the appearance of grating lobe in visible space with beam scan as

$$d/\lambda < \frac{1}{1+|\sin\theta_0|} \tag{10.77}$$

If the beam is to be scanned close to the endfire direction where $\theta_0 \approx (\pi/2)$, an inter-element spacing less than half a wavelength would be adequate. However, for the limit value $d/\lambda = \frac{1}{2}$ of the inter-element spacing, the grating lobes approach the visible space with half of the grating lobe visible when the scanning beam approaches the hemispherical limit. Thus the inter-element spacing plays a vital role in deciding the limits of scan in an electronically scanned array.

10.10 Planar arrays

Planar arrays have found application in modern surveillance radar systems with electronic scanning in the elevation plane and mechanical rotation in azimuthal plane. Of the many planar configurations, the rectangular configuration has found maximum usage in mobile radar systems, while hexagonal, oval and other boundaries have been tried out for two dimensional phased array antennas. A number of grid configurations have been suggested for placement of the radiating elements such as the rectangular/square grid, the triangular/hexagonal grid etc. Excitation of such arrays also offers two distinct possibilities, i.e. separable excitation and non-separable excitation. Of the many combinations of array boundaries, lattice configurations and array excitation options, only a few have found wider application than all others.

The planar array with rectangular/square lattice and with separable excitation for the rows and the columns is a natural extension of the work carried out for linear arrays. Fig. 10.10 indicates a planar array with a rectangular grid and a rectangular boundary with N columns and N rows along the X and Y directions respectively. The column elements are separated d_x apart while the row elements have a grid spacing of d_y. The mnth element of the array will be at (md_x, nd_y) assuming that one edge of the array is the origin of the co-ordinate system. The array factor can now be expressed as

$$S(\theta,\phi) = \sum_{m=1}^{N}\sum_{n=1}^{N}\left(\frac{I_{mn}}{I_{00}}\right)\exp j\{k\sin\theta(md_x\cos\phi + nd_y\sin\phi)\} \tag{10.78}$$

where I_{mn} is the excitation current of the mnth element, k is the propagation constant $(=2\pi/\lambda)$ and I_{00} is the excitation current of the centre element. Variables $\boldsymbol{u}$ and $\boldsymbol{v}$ are now defined as

$$\begin{aligned} u &= kd\sin\theta\cos\phi \\ v &= kd\sin\theta\sin\phi \end{aligned} \tag{10.79}$$

so that eqn. 10.78 becomes

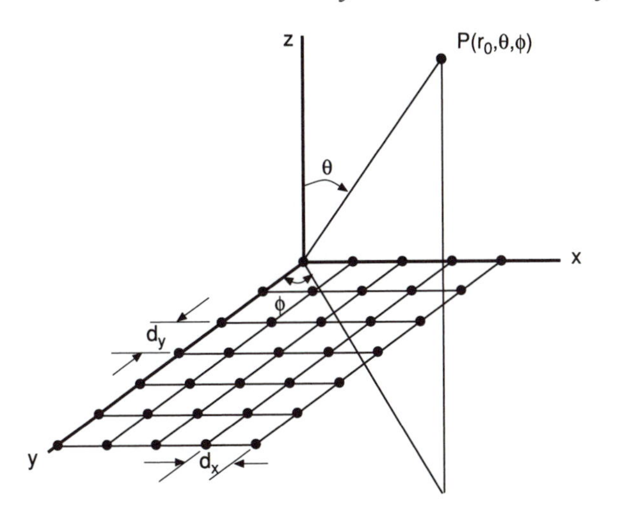

Figure 10.10 *Planar array: Rectangular grid*

$$S(u, v) = \sum_{m=1}^{N} \sum_{n=1}^{N} \frac{I_{mn}}{I_{00}} \exp j(mu + nv) \tag{10.80}$$

The array factor is a doubly periodic function of $\boldsymbol{u}$ and $\boldsymbol{v}$ in the u, v plane. For further analysis of the array characteristics it is necessary to define the excitation function of the array. In the first instance, a separable excitation distribution is considered so that the array factor is a product of the excitation distribution of two linear arrays, namely the row array and the column array. If the same excitation distribution is used for all rows of the array but with different excitation levels, it is possible to write

$$\frac{I_{mn}}{I_{m0}} = \frac{I_{0n}}{I_{00}} \tag{10.81}$$

Eqn. 10.80 is now transformed to

$$S(u, v) = \sum_{m=1}^{N} I_m \exp j(mu) \sum_{n=1}^{N} I_n \exp j(nv) \tag{10.82}$$

where,

$$\frac{I_{m0}}{I_{00}} = I_m, \qquad \frac{I_{0n}}{I_{00}} = I_n$$

Eqn. 10.82 shows that the array factor in this case is the product of the array factors of two linear arrays—one laid out along the X axis and the other along the Y axis. All the results that had been earlier developed for linear arrays about the visible space, the efficiency of the excitation functions and the excitation functions that will produce low sidelobes are applicable in this case also.

10.10.1 *Grating lobes*

In order to relate the formation of grating lobes to a unit circle in the u', v' plane of the array, a new set of variables $\boldsymbol{u}'$ and $\boldsymbol{v}'$ are defined as

$$u' = \frac{u}{kd_x} = \sin\theta\cos\phi$$
$$v' = \frac{u}{kd_y} = \sin\theta\sin\phi \qquad (10.83)$$

Substituting eqn. 10.83 into eqn. 10.82,

$$S(u, v) = S_x(\theta, \phi)S_y(\theta, \phi) = \sum_{m=1}^{N} I_m\, e^{j(2\pi d_x/\lambda)mu'} \sum_{n=1}^{N} I_n\, e^{j(2\pi d_y/\lambda)nv'} \qquad (10.84)$$

Eqn. 10.84 makes explicit the product relationship of a planar array with its constituent orthogonal linear arrays with separable excitations. In the principal planes of the array, i.e. in the Y, Z and X, Z planes, the behaviour of the array with respect to grating lobe formation and beam broadening is identical to that of the respective linear arrays. At other scan angles (outside the principal planes) the situation is more complicated. It is therefore necessary to establish conditions for single beam formation out of the two linear arrays, the requirements for keeping the grating lobes out of the visible space, and the shape of the beam as it scans the space.

If we consider both the column array and the row array from the main beam at the point P in space (r, θ_0, ϕ_0) but with different progressive phase shifts α_x and α_y, respectively, along the column and row array, then

$$kd_x\cos\theta_x - \alpha_x = 0 = kd_x\sin\theta_0\cos\phi_0 - \alpha_x$$
$$kd_y\cos\theta_y - \alpha_y = 0 = kd_y\sin\theta_0\sin\phi_0 - \alpha_y \qquad (10.85)$$

The planar array in this case radiates only in the $z>0$ hemisphere as it does in actual planar phased array antennas. From eqn. 10.85, the following equations can be derived easily:

$$\tan\phi_0 = \frac{\alpha_y d_x}{\alpha_x d_y}$$
$$\sin^2\theta_0 = \frac{\alpha_x^2}{(kd_x)^2} + \frac{\alpha_y^2}{(kd_y)^2} \leqslant 1 \qquad (10.86)$$

Eqns. 10.86 define the conditions for single beam formation in the case of separable excitations. The equality relation in the second of eqns. 10.86 indicates that it is better to avoid the $\theta_0 = \pi/2$ condition, which is the endfire beam position in the case of both arrays and which thus forms the start for disappearance of the single beam.

An easy way of depicting the formation of grating lobes in a planar array with rectangular lattice is the graphical method. For this purpose we define a new set of relationships from eqn. 10.83 as

$$\tan\phi = \frac{v'}{u'}$$
$$(u')^2 + (v')^2 \leq 1 \tag{10.87}$$

The second of eqns. 10.87 defines the boundary of the visible space as that being inside the unit circle in the (u', v') space, and on the unit circle for $\theta = \pi/2$. The main beam of the array is at the centre of the (u', v') space as seen in Fig. 10.11. The space outside the unit circle in Fig. 10.11 is the invisible space in which θ will be imaginary. The tiling of the u', v' space will be similar to that of the x, y space of the physical array. The lattice structure of the u', v' space will be rectangular for a rectangular lattice configuration of the array [13]. When the main beam is scanned in physical space, the point corresponding to the main beam in the (u', v') space also shifts to a new position, and all the grating lobes present in the (u', v') space move accordingly as shown in Fig. 10.11. The points in the (u', v') space where the grating lobes will be formed can be obtained from eqn. 10.77 which sets the limit to the inter-element distance as a function of the scanning angle. Using this relation, it can be easily shown that so long as

$$(\lambda/d_x) > 2$$
$$(\lambda/d_y) > 2 \tag{10.88}$$

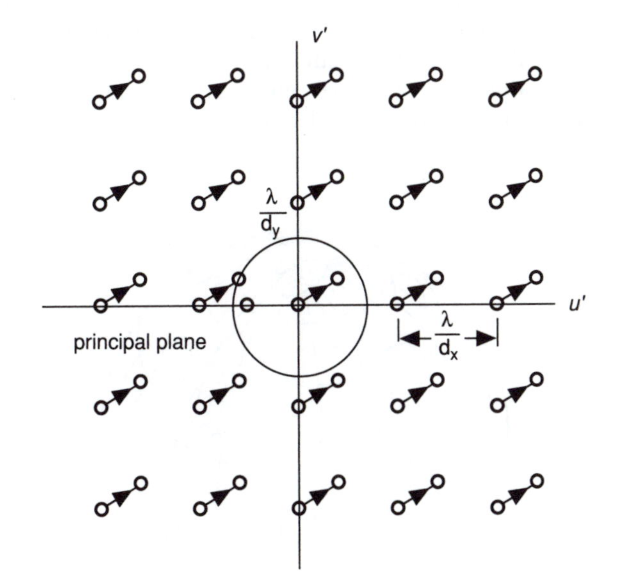

Figure 10.11 *Grating lobes in u–v space [1]*

the grating lobes do not appear. The positions of the inverse lattice have been marked in Fig. 10.11 and also their shift along with the main lobe. As long as the distance between the circumference of the unit circle and the inverse lattice points of the grating lobes is greater than one, there is no change of appearance of grating lobes as the beam is being scanned in the $z>0$ hemisphere.

10.10.2 *Beamwidth and directivity*

The array factor of the planar array from eqn. 10.62 is a product of the array factor of two linear arrays. As the progressive phase shift across the row as well as column arrays tilt the beams away from broadside, by virtue of eqns. 10.86 these two beams intersect to produce a single main pencil beam in the direction (θ_0, ϕ_0). Only the main beam in the $z>0$ hemisphere is relevant to the discussion here as the other beam in the lower hemisphere is not generated owing to the geometry of planar array construction. Assuming a rectangular array with the same excitation function it can be stated that in the zenith direction the main beam cross-section is elliptical with the major axis in the direction of the longer side of the array. As the beam is scanned in one of the principal planes $(\phi=\pi/2)$ from broadside towards $\theta=\pi/2$, the axis of the beam lying in that plane is elongated during the downward traverse of the beam. The shape of the ellipse will constantly change with beam scan. The elongated axis of the elliptical cross-section of the main beam will now lie along the longitude passing through the $\phi=\pi/2$ plane. In a similar manner if we consider the beam being scanned in the $\phi=0$ plane from broadside towards $\theta=\pi/2$, the other axis of the ellipse undergoes the elongation process and thus the elliptical cross-section of the beam is changed continuously. Fig. 10.12 shows the typical main beam cross-section as it is scanned away from broadside. For a given angle θ as the beam scans from $\phi=0$ to $\phi=\pi/2$, the change in the cross-section of the ellipse and the interchange of major and minor axes of the ellipse take place smoothly. However, the orientation of the axes of the ellipse along the lattitude and longitude lines cannot be assumed except when the beam is scanned in the two principal planes of the array. In

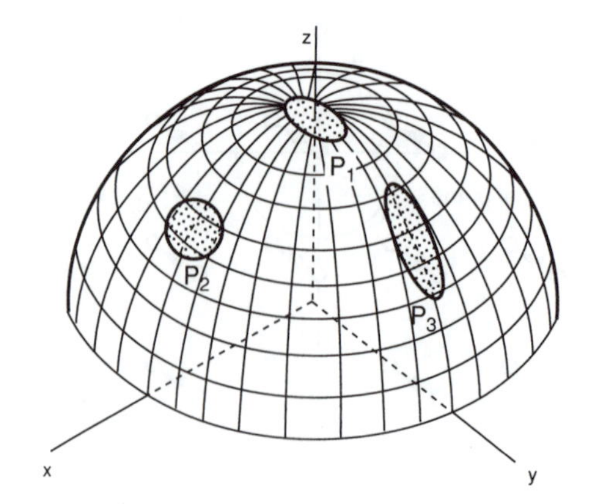

Figure 10.12 *Beam shape with beam scan [25]*

this case, the beamwidth of the array is given by the area beamwidth which is a measure of the area inside the -3 dB contour of the pencil beam cross-section [5]. It has been shown that B, the area beamwidth of the planar array is independent of ϕ in the following way:

$$B = \theta_{xB}\theta_{yB}\sec\theta_S \tag{10.89}$$

where

θ_{xB}, θ_{yB} = 3 dB beamwidths in the broadside direction of the x-directed and y-directed linear arrays, respectively

θ_S = pointing angle of the beam as it is being scanned

This holds good for pencil beams so long as they are not scanned closer to $\theta = \pi/2$ by several beamwidths. The aerial beamwidth relation above also leads to a definition of the peak directivity of a planar array as

$$D = D_x D_y \cos\theta_S \tag{10.90}$$

where

D_x, D_y = peak directivities of the x-directed and y-directed linear arrays respectively

The directivity in the case of a planar array is dependent on the scan angle of the beam away from the zenith.

10.10.3 *Sidelobe region*

In the two principal planes namely $\phi = 0$ and $\phi = \pi/2$, the nulls as well as the sidelobe levels of the radiation pattern of the planar array are fully governed by the aperture excitation functions of the Y-directed and X-directed linear arrays. For sidelobes occurring in other planes, the levels of the peaks will be lower than those obtained in the two principal planes. Since the planar array pattern is a product of the two linear array radiation patterns, the voltage sidelobe level ratio expressed in dB will be a sum of the two individual SLLs (sidelobe levels). For example if the X-directed linear array has been designed for -30 dB SLL and the Y-directed linear array has been designed for -35 dB SLL, the sidelobe levels of the planar array in all planes other than the two principal planes will be -65 dB. The reduction in the sidelobe levels in all other planes leads to the broadening of the main beam and loss of directivity. Though there is over-achievement as far as sidelobe level requirements are concerned, this undercuts the important aspect of directivity. Thus there is a necessity to explore other avenues of search for optimum non-separable radiation patterns.

10.11 Non-separable radiation pattern

The non-separable radiation pattern for a square array is based on the general technique of transformation of a linear array excitation function to a planar array excitation function with quadrant symmetry so that a pattern with non-degraded ring sidelobes is obtained. The only restriction for the applicability of this technique is the need to have equal numbers of elements for the row as well as for the columns. If the inter-element distance along the rows or columns of the array is kept equal, then the array is a square array. If the inter-element distances in the row are not the same as in the column, then a rectangular array is obtained.

The non-separable pattern response function is symmetrical and real, so that the excitation coefficients of the array are, respectively, real and symmetrical in both the X and Y dimensions. This results in the necessity for cosine terms only in the expansion of both the array excitation and array pattern functions. As a result, both the number of independent points to define the pattern polynomial and the number of non-zero coefficients of the excitation function are reduced. Tseng and Cheng [15] represented the array factor of $(2N\times 2N)$ element array as

$$S(u, v)=4\sum_{m=1}^{N}\sum_{n=1}^{N} I_{mn}\cos\left(\frac{2m-1}{2}u\right)\cos\left(\frac{2n-1}{2}v\right) \tag{10.91}$$

where I_{mn} = excitation current in the mnth element

$$u=\frac{2\pi d_x}{\lambda}(\sin\theta\cos\phi-\sin\theta_0\cos\phi_0)$$

$$v=\frac{2\pi d_y}{\lambda}(\sin\theta\sin\phi-\sin\theta_0\sin\phi_0) \tag{10.92}$$

The beam is being formed in the direction (θ_0, ϕ_0). A new set of variables ν and μ are now defined

$$\mu=\frac{\pi d_x}{\lambda}(\sin\theta\cos\phi-\sin\theta_0\cos\phi_0)$$

$$\nu=\frac{\pi d_y}{\lambda}(\sin\theta\cos\phi-\sin\theta_0\cos\phi_0) \tag{10.93}$$

Eqn. 10.91 becomes

$$S(\mu, \nu)=4\sum_{m=1}^{N}\sum_{n=1}^{N} I_{mn}\cos(2m-1)\,\mu\cos(2n-1)\nu \tag{10.94}$$

In order to ensure that the pattern in any cross-section (ϕ-cut) is the same, any polynomial representing the pattern should have the variable

$$w=w_0\cos\mu\cos\nu \tag{10.95}$$

so that

$$S(\mu, \nu)=4P_{2N-1}(w_0\cos\mu\cos\nu) \tag{10.96}$$

where $P_{2N-1}(\ )$ is an odd polynomial of $(2N-1)$th degree, which represents the desired radiation pattern of the array. The parameter w_0 controls the sidelobe level in that, $P_{2N-1}(w_0)=b$, the sidelobe level ratio. The odd polynomial $P_{2N-1}(w_0\cos\mu\cos\nu)$ can be expressed as

$$P_{2N-1}(w_0\cos\mu\ \cos\nu)=\sum_{S=1}^{N} a_{2S-1}w_0^{2S-1}\cos^{2S-1}\mu\cos^{2S-1}\nu \tag{10.97}$$

Eqns. 10.94, 10.96 and 10.97 can be linked to each other by using the relation [5]

$$\cos^{2S-1}\mu = \frac{1}{2^{2S-2}} \sum_{p} \binom{2S-1}{S-p} \cos(2p-1)\mu \tag{10.98}$$

Introducing eqn. 10.98 into eqn. 10.97 and equating it to the RHS of eqn. 10.94, the following expression for the excitation coefficients of the array in terms of the coefficients of the polynomial is obtained:

$$I_{mn} = \sum_{S=(m,\,n)}^{N} \frac{a_{2S-1}}{2^{2S-1}} \binom{2S-1}{S-m} \binom{2S-1}{S-1} \left(\frac{w_0}{2}\right)^{2S-1}$$

$$w_0 = \cosh\left[\frac{1}{2N-1} \cosh^{-1} \mathrm{SLL}\right] \tag{10.99}$$

in which $(m, n) = m$ if $m \geqslant n$; and $(m, n) = n$ if $m < n$. Tseng and Cheng used a Chebyshev polynomial for their radiation pattern. In such a case,

$$T_{2N-1}(w_0 \cos\mu \cos\nu) = \sum_{S=1}^{N} (-1)^{N-S} \frac{2^{2S-2}(2N-1)}{N+S-1} \times \binom{N+S-1}{2S-1} w_0^{2S-1} \cos^{2S-1}\mu \cos^{2S-1}\nu \tag{10.100}$$

Comparing eqn. 10.100 with eqn. 10.97, a_{2S-1} can be obtained. This is now introduced in eqn. 10.99 to yield

$$I_{mn} = \sum_{S=(m,\,n)}^{N} (-1)^{N-S} \frac{2N-1}{2(N+S-1)} \binom{N+S-1}{2S-1} \binom{2S-1}{S-m} \binom{2S-1}{S-1} \left(\frac{w_0}{2}\right)^{2S-1} \tag{10.101}$$

Eqn. 10.101 enables one to obtain the excitation current of the mnth element in the array in terms of the desired sidelobe level ratio on the assumption that a Chebyshev polynomial of $(2N-1)$th degree represented the desired radiation pattern. This is the well known Tseng and Cheng non-separable excitation distribution first elaborated by them in 1968.

Tseng and Cheng have also presented an alternative method for determining the excitation current I_{mn} which is based on sampling rather than the coefficient matching carried out above. From eqn. 10.101 it is immediately apparent that I_{mn}, the excitation current, is independent of the scan angle (θ, ϕ) of the main beam and also of the inter-element spacings. This is indeed a great relief because it is now possible for the main beam to scan the space by merely varying the relative phases of the excitation currents I_{mn}. Another welcome feature is that, of the $4N^2$ elements in the array, the excitation currents of only $N(N+1)/2$ radiating elements need be known because $I_{mn} = I_{nm}$. Further, there is symmetry about the X and Y axes as well as about the $x=y$ lines.

The beamwidth of the planar array for the non-separable distribution is given by

$$(\theta_{3dB})_0 = 2\ \sin^{-1}\left[\frac{\lambda}{\pi d\ \cos\ \theta_0}\ \cos^{-1}(w/w_0)\right] \tag{10.102}$$

with

$$T_{2N-1}(w_1) = \frac{b}{\sqrt{2}} \tag{10.103}$$

Eqn. 10.103 holds good in the two principal planes of the planar array when the main beam is scanning the space in the hemisphere $z > 0$. Since the beam broadening is maximum in the plane which is at 45° to the principal planes, the 3 dB beamwidth in that plane is obtained as

$$(\theta_{3dB})_{\pi/4} = 2\ \sin^{-1}\left[\frac{\sqrt{2}\lambda}{\pi d\ \cos\ \theta_0}\ \cos^{-1}(w_1/w_0)\right] \tag{10.104}$$

To compare the efficiency of the non-separable type of distribution of the planar array excitation, the 3 dB beamwidth of the separable distribution in the plane which is at 45° to the principal planes is obtained as

$$(\theta_{3dB})^{SD}_{\pi/4} = 2\ \sin^{-1}\left[\frac{\sqrt{2}\lambda}{\pi d\ \cos\ \theta_0}\ \cos^{-1}(w_2/w_0)\right] \tag{10.105}$$

where,

$$w_2 = \cosh\left[\frac{1}{2N-1}\ \cosh^{-1}\left(\frac{b}{\sqrt{2}}\right)^{1/2}\right] \tag{10.106}$$

In the plane which is at 45° to the principal planes, the worst beam broadening takes place for both separable and non-separable types of radiation patterns. However, the beam broadening is worse for the separable type of radiation patterns. This is highlighted in Fig. 10.13 where the ring-like nature of the individual sidelobes as well as the fact that peak sidelobe levels are equal can be clearly noted.

From eqns. 10.102, 10.104 and 10.105 it is clear that, for a given main beam pointing direction, the beamwidth decreases as the inter-element spacing d/λ increases. This is as it should be, because an increase in d/λ increases the total antenna aperture. *Prima facie* it would appear that we stand to gain by increasing d/λ. While this would be an advantage in non-scanning types of arrays, in the case of scanning arrays there is a further limitation due to the likelihood of grating lobes being present. It is also necessary to maintain $d/\lambda < (d_{max}/\lambda)$, where d_{max} is the maximum permissible inter-element spacing for a specified scan extent θ_m of the beam. This condition is expressed as

$$\frac{d_{max}}{\lambda} = \frac{1 - \dfrac{1}{\pi}\ \cos^{-1}(1/w_0)}{(1 + \sin\ \theta_m)} \tag{10.107}$$

where w_0 is expressed in terms of SLL and the number of radiation elements through the second of eqns. 10.99. Table 10.2 gives the variation in the inter-

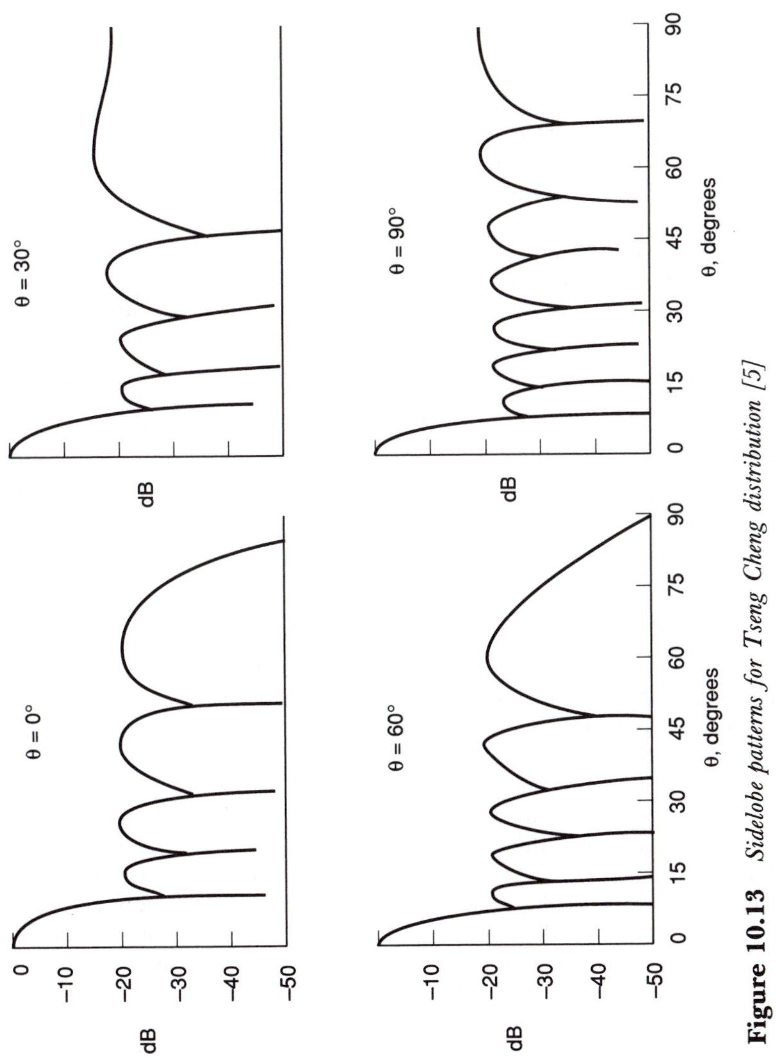

Figure 10.13 *Sidelobe patterns for Tseng Cheng distribution [5]*

Table 10.2 *Variation of inter-element spacing (d/λ)*

Sidelobe levels (SLL)	Inter-element spacing for different scan limits		
	30°	45°	60°
30 dB	0.58	0.51	0.47
40 dB	0.64	0.56	0.54

element spacing for different sidelobe levels in the case of a square planar array of 1600 elements for different scan limits.

For more stringent sidelobe level conditions, the inter-element spacing is higher, thus calling for a larger aperture for the arrays. As the limits of scanning increase, it is found that the inter-element spacing decreases.

In carrying out the steps leading to the determination of the excitation coefficients I_{mn} for Tseng and Cheng's non-separable distribution, Elliott's [5] procedure of using any odd polynomial of $(2N-1)$th degree, represented by $P_{2N-1}(w)$ had been followed. This makes it possible to consider polynomials other than Chebyshev if the design requirement calls for control of the heights of individual ring sidelobes in the Chebyshev radiation pattern. If no radical departure in the sidelobe structure is sought it is only rational to assume that the desired polynomial $P_{2N-1}(w)$ differs from the Chebyshev polynomial $T_{2N-1}(w)$ slightly and this can be represented as [16]

$$P_{2N-1}(w) - T_{2N-1}(w) = \sum_{S=1}^{N} \delta_{2S-1} w^{2S-1} \tag{10.108}$$

Eqn. 10.108 signifies that each of the coefficients of the Chebyshev polynomial is perturbed by $\boldsymbol{\delta}$ and that this perturbation is not large enough to affect the main beam gain so that the equality condition $P_{2N-1}(w_0) = T_{2N-1}(w_0)$ can be imposed. Further, it is also considered that the $(2N-1)$ peaks of the sidelobes of the Chebyshev polynomial are broad enough to maintain them at approximately the same (μ, ν) values for the new polynomial. In that case, if w_m^0 $(m=1, 2, \ldots, 2N-1)$ are the values of w at which these peaks occur in the Chebyshev polynomial, then $P_{2N-1}(w_m^0) - T_{2N-1}(w_m^0)$ would come to be the difference in the levels of the mth sidelobe between the two patterns. Therefore in eqn. 10.108, the LHS for the $2N$ values of $w_0, w_1, \ldots, w_{2N-1}$ are known and the δs would thus be determined. The individual coefficients of the polynomial can now be computed to obtain $P_{2N-1}(w)$. It has been stated [5] that in most cases two or three iterations are sufficient to arrive at the polynomial.

10.11.1 *Non-separable distribution for rectangular arrays*

Tseng and Cheng's method for determining the excitation coefficients for a non-separable array pattern is applicable for arrays which have equal numbers of elements in the rows and columns and

$$w_0 = \cosh[(\cosh^{-1} b)/(2N-1)] \tag{10.109}$$

For representing the radiation pattern of the rectangular array, two polynomials, one of the order $(2M-1)$ in $\boldsymbol{u}$ and the other of the order $(2N-1)$ in $\boldsymbol{v}$, respectively are now chosen. They are as follows [17]:

$$S_{B1}(u, v) = T_{2M-1}[w_0(\cos u)f(v)] \qquad (10.110)$$

$$S_{B2}(u, v) = T_{2N-1}[x_0(\cos v)g(u)] \qquad (10.111)$$

Eqns. 10.110 and 10.111 are of the same form as eqn. 10.100 in the known variables $(\cos u)$ and $(\cos v)$, respectively. Along the principal axes eqns. 10.110 and 10.111 should reduce to the same function of a single variable. This would lead to the condition,

$$f(0) = g(0) = 1 \qquad (10.112)$$

The design eqns. 10.110 and 10.111 have to be forced to approximate to equations of order $(2M-1)$ in $\boldsymbol{u}$ and $(2N-1)$ in $\boldsymbol{v}$ to be valid at least along one grid line. If eqn. 10.110 is to be forced to approximate to a polynomial of order $(2N-1)$ in $\boldsymbol{v}$, let it be along a grid line of constant $\boldsymbol{u}$, say $\boldsymbol{u} = u_c$. Both eqns. 10.110 and 10.111 will have to be equal at that grid line, so that

$$T_{2M-1}\,(w_0 \cos u_c \cdot f(v)) = T_{2N-1}(x_0 \cos v \cdot g(u_c)) \qquad (10.113)$$

Since eqn. 10.113 is true for all values of $\boldsymbol{v}$, it is true for $\boldsymbol{v} = 0$ in which case eqn. 10.113 can be rewritten with the help of eqn. 10.110 as

$$x_0 g(u_c) = \cos\left[\frac{M-1}{N-1} \cos^{-1}(w_0 \cos u_c)\right] \qquad (10.114)$$

$x_0 g(u_c)$ can now be determined from eqn. 10.114. This is then substituted in eqn. 10.113 to give

$$w_0 f(v) = \cos\left[\frac{N-1}{M-1} \cos^{-1}\{x_0 g(u_c) \cos v\}\right] / (\cos u_c) \qquad (10.115)$$

By substituting eqn. 10.115 in eqn. 10.110, a polynomial of order $(2M-1)$ in $\boldsymbol{u}$ is obtained. Similarly the polynomial of order $(2N-1)$ in $\boldsymbol{v}$ along the grid line $\boldsymbol{u} = u_c$ is also determined. This polynomial completely matches the desired pattern in all respects along the grid line. A similar exercise can be carried out on eqn. 10.111 to obtain

$$x_0 g(u) = \cos\left[\frac{M-1}{N-1} \cos^{-1}\{w_0 f(v_c) \cos u\}\right] / (\cos v_c) \qquad (10.116)$$

By substituting eqn. 10.116 in eqn. 10.111, a polynomial of order $(2N-1)$ in $\boldsymbol{v}$ and a polynomial of order $(2M-1)$ in $\boldsymbol{u}$ along the grid line $\boldsymbol{v} = v_c$ are determined. The design polynomials are Chebyshev functions along the grid lines in the u, v space. For example, the polynomial represented by eqn. 10.110 is a Chebyshev function of order $(2M-1)$ along $\boldsymbol{v}$ = constant grid lines and of order $(2N-1)$ in u along one grid line $\boldsymbol{u} = u_c$. The realisable array patterns based on the excitation coefficients of the radiating elements are equal to the desired array patterns along all the grid lines $\boldsymbol{v}$ = constant in one dimension and along the selected grid line $\boldsymbol{u} = u_c$ in the other dimension. Similar remarks hold good for eqn. 10.111 of the polynomial of order $(2N-1)$ in the v dimension.

Evaluation of numerous designs using both design polynomials of eqns. 10.110 and 10.111 has been carried out [17]. Using different values of u_c and v_c, computations for all combinations of odd and even numbers of elements in rows and columns have been carried out for a wide range of aspect ratio 1:1 to 1:6. The desired array pattern was aimed at −40 dB SLL and the computed response was within 1 dB of the design level. Approximately equal performance was also obtained for all combinations of even and odd numbers of elements in the rows and columns. Only in certain array designs where both the number of elements in row as well as in column was odd, the computed sidelobe was 3 dB over the design level. By changing the u_c or v_c values, it is possible to reduce this deviation also.

10.12 Iterative procedures for array pattern synthesis

There are a number of synthesis techniques which require iterations to arrive at an optimal pattern with respect to a specific performance parameter such as nulls in fixed directions, maximum envelope of sidelobes etc. Unlike the separable and non-separable synthesis techniques which allow much lesser degrees of freedom than the available $4N^2$ degrees of freedom, the iterative techniques permit a larger number of degrees of freedom. However, these techniques require extensive computational capabilities. Nevertheless, these are of interest especially in the case of adaptive arrays where null placement is an important requirement. A number of papers have been published in this area since 1953 and an excellent review paper has been written by Cheng [18].

In this case we go back to eqn. 10.1 which relates the far field of the array to the excitation coefficients in the most general form. We consider that the array has N elements:

$$E_A(\theta, \phi) = \sum_{m=1}^{N} I_m\{\exp j(-\alpha_m)\}\{\exp j(-mkd \sin \theta)\} \qquad (10.117)$$

If

$$[J] = \begin{bmatrix} I_1 \, e^{-j\alpha_1} \\ I_2 \, e^{-j\alpha_2} \\ \vdots \\ I_N \, e^{-j\alpha N} \end{bmatrix} \qquad (10.118)$$

$$[F_0] = \begin{bmatrix} e^{-jkd \sin \theta} \\ e^{-j2kd \sin \theta} \\ \vdots \\ e^{-jNkd \sin \theta} \end{bmatrix} \qquad (10.119)$$

where [ ] represents a matrix. In eqn. 10.118, the complex excitation currents/voltages of the radiating element are represented by a column matrix $[J]$. $[F_0]$ is a column matrix representing the phases of the fields at the observation point due to the path-length differences of the radiating elements. The directivity of the array as expressed in eqn. 10.6 can be written as

$$D(\theta, \phi) = \frac{[J^{\dagger}][A][J]}{[J^{\dagger}][B][J]} \tag{10.120}$$

$[^{\dagger}]$ represents the conjugate transpose of the matrix. $[A]$ and $[B]$ are $(N \times N)$ square matrices defined as

$$[A] = [F_0][F_0^{\dagger}] \tag{10.121}$$

$$[B] = \begin{bmatrix} b_{11} & b_{12} & b_{13} & \cdots & b_{1N} \\ b_{21} & b_{22} & b_{23} & \cdots & b_{2N} \\ \vdots & \vdots & \vdots & & \vdots \\ b_{N1} & b_{N2} & b_{N3} & \cdots & b_{NN} \end{bmatrix} \tag{10.122}$$

with

$$b_{mn} = \frac{1}{4\pi} \int_0^{2\pi} d\phi \int_0^{\pi} F_m(\theta, \phi)\, e^{-jmkd \sin\theta}\, d\theta \tag{10.123}$$

with $F_m(\theta, \phi)$ as the far field power pattern of the element.

In eqn. 10.120 both the matrices $[A]$ and $[B]$ are Hermitian and in addition $[B]$ is positive definite and non-singular. The largest eigenvalue $\lambda_m = (D_m)$ of the associated equation, the eigenvector corresponding to λ_m and the maximum directivity can now be expressed as

$$[A][J] = D_{max}[B][J]$$
$$[J_{max}] = [B^{-1}][F_0]$$
$$D_{max} = [F_0^{\dagger}][B^{-1}][F_0] \tag{10.124}$$

Eqns. 10.124 solve the problem of optimisation by adjustment of the excitation coefficients of the array for $[J_{max}]$ so that D_{max} is reached. It is to be noted that, for an array of N elements, the optimisation procedure involves the determination of $2N$ parameters. Further, the adjustment of excitation amplitudes and phases of the currents or voltages at the radiation elements for maximisation of directivity also changes the radiation pattern. Therefore the antenna designer may well like to impose certain other constraints such as the directions in which a certain number of nulls have to be created with reference to the direction of maximum directivity or the maximum envelope for the sidelobe peaks etc.

The imposition of constraints or auxiliary conditions effectively reduces the number of independent parameters available to the designer to a value less than $2N$. For example, the constraint that is most often sought to be imposed is that of achieving maximum directivity in the direction of the target and at the same time minimisation of interference by creating nulls in the direction of M jammers $(M < N)$. Looking back at eqn. 10.117 it can be easily seen that the radiation pattern is the inner product of the space vector $[F]$ and the excitation vector $[J]$, so that

$$E_A(\theta, \phi) = \langle [F_0], [J] \rangle = [F_0^{\dagger}][J] \tag{10.125}$$

In the direction of the desired target $E_A(\theta, \phi)$ is maximum $(=1)$ whereas in the direction of jammers it is zero. In effect, the excitation vector $[J]$ is required to be orthogonal to $\boldsymbol{M}$ independent constraint vectors $[F_c]$ while it has to maximise directivity in the direction (θ_0, ϕ_0) of the desired target. The N-dimensional Hilbert space is thus divided into two mutually orthogonal spaces: an M-dimensional subspace containing the constraint vectors $[F_c]$ and an $(N-M)$ dimensional space where the excitation vectors $[J]$ must lie.

The construction of an optimum excitation vector $[J]$ can be carried out by the method of alternate orthogonal projections (AOP). This was first spelt out by Youla [19] and later it has been applied to pattern synthesis by Prasad [20]. Recently it has been expanded in scope by Ng [21] for the general case of N-element arrays with multiple null constraints. The optimum excitation vector $[J]$ is constructed in an appropriately selected vector subspace from its known projection in the linear vector space containing the constraint vectors. The recursive algorithm that is evolved for the construction of the optimum excitation vector uses only the operations of the projections on to an appropriate set of vector subspaces. The N-dimensional Hilbert space $\mathcal{H}$ has $[J]$, $[F_d]$, $[F_{ci}]$ $(i-1, \ldots, m)$ vectors as elements of this space with

$$[F_d^+][J]=1$$
$$[F_{ci}^+][J]=0 \tag{10.126}$$

$\mathfrak{R}_1$ and $\mathfrak{R}_2$ are two linear vector subspaces in $\mathcal{H}$ with $\mathfrak{R}_1^\perp$ and $\mathfrak{R}_2^\perp$ as the orthogonal complements of $\mathfrak{R}_1$ and $\mathfrak{R}_2$, respectively. $[P_1]$ and $[Q_1]$ are the linear projection vectors projecting $[J]$ in $\mathfrak{R}_1$ and $\mathfrak{R}_1^\perp$, respectively. Similarly $[P_2]$ and $[Q_2]$ are linear projection vectors that project $[J]$ in $\mathfrak{R}_2$ and $\mathfrak{R}_2^\perp$, respectively. The projection vectors $[P_1]$ and $[Q_2]$ can be expressed in terms of $[F_d]$ and $[F_{ci}]$ as

$$[P_1]=[F_d]([F_d^+][F_d])^{-1}[F_d^+]$$
$$[Q_1]=[I]-[P_1]$$
$$[Q_2]=[F_{ci}]([F_{ci}^+][F_{ci}])^{-1}[F_{ci}^+]$$
$$[P_2]=[I]-[Q_2] \tag{10.127}$$

where $[I]$ is the identity matrix. Using Youla's orthogonal projection theorem the recursive algorithm for array pattern synthesis is given as

$$[J_{k+1}]=[F_d]+[Q_1][P_2][J_k] \tag{10.128}$$

where $[Q_1]$ and $[P_2]$ have been identified in eqn. 10.127 and the subscripts (k), $(k+1)$ signify the iteration numbers. $[J_{k+1}]$ converges in the norm to $[J]$, the required optimum excitation vector. The error norm converges to zero rapidly, unless the desired signal vector $[F_d]$ is close to the interference directions. Both Prasad and Ng have shown that the projection operators for the algorithm can be easily computed from $[F_d]$, the desired signal direction vector, and the given set of null constraint direction vectors $[F_{ci}]$.

10.13 Taylor circular $\bar{n}$ distribution

The Taylor circular pattern is another method by which the disadvantages of the separable distributions can be taken care of. A planar aperture with circular boundary is assumed in this case with a linear excitation current density so as to ensure the constancy of the peaks as well as the number of sidelobes for the radiation pattern in every ϕ cut. The Taylor circular distribution (analogous to the Taylor linear distribution) will be most suitable for circularly shaped planar arrays. The start in this case is uniform excitation of a circular aperture which leads to the array pattern,

$$S_p(u_c) = \frac{J_1(\pi u_c)}{\pi u_c} \tag{10.129}$$

where

$S_p(u_c)$ = pattern due to the uniform excitation

$J_1(\pi u)$ = Bessel function of the first order and argument (πu)

$u_c = (2\boldsymbol{a}/\lambda)\sin\theta$, where $\boldsymbol{a}$ is the radius of the aperture

$S_p(u_c)$ as a function of u_c has a main pencil beam plus a family of sidelobes that decay in height. The sidelobe occurs at higher values of $u_c\{1 \leqslant (2a/\lambda)\}$. The pencil beam in this case is surrounded by ring sidelobes with close-in sidelobes having peaks which are undesirable. To produce $\bar{n}$ roughly equal close-in sidelobes at desired level, $\bar{n}$ zeros on each side of the main beam are modified by removing the previous zeros and replacing them by new zeros. The array factor is now given by

$$S_{CT}(u_c) = \frac{J_1(\pi u_c)}{\pi u_c} \frac{\prod_{n=1}^{\bar{n}-1} (1 - u^2/u_n^2)}{\prod_{n=1}^{\bar{n}-1} (1 - u^2/\nu_n^2)} \tag{10.130}$$

where ν_ns are the zeros of the Bessel function $J_1(\pi u_c)$ and u_{cn} are the new zeros given by

$$u_{cn} = \pm \sigma \sqrt{A^2 + (n - \tfrac{1}{2})^2} \tag{10.131}$$

with

$$-20 \log_{10} \cosh A = SLL$$

The dilation factor σ is expressed as

$$\sigma = \frac{\nu_{\bar{n}}}{\sqrt{A^2 + (\bar{n} - \tfrac{1}{2})^2}} \tag{10.132}$$

The aperture distribution in this case becomes

$$g(p) = \frac{2}{\pi^2} \sum_{m=0}^{\bar{n}-1} \frac{S(\nu_m)}{J_0^2(\pi \nu_m)} J_0(\nu_m p) \tag{10.133}$$

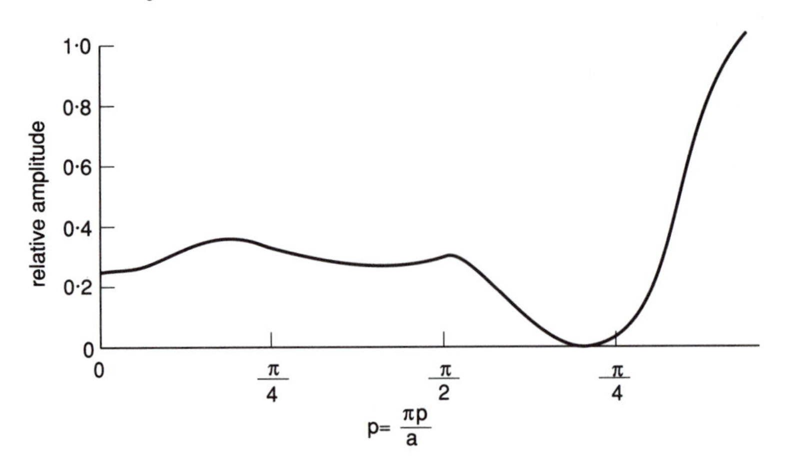

Figure 10.14 *Aperture distribution for Taylor circular illumination [5]. Reprinted by permission of Prentice-Hall*

where, $p=\pi a/\rho$, ρ being the running variable for the radius of the aperture.

$$S(\nu_0)=1$$

$$S(\nu_m)=-J_0(\pi\nu_m)\frac{\prod_{n=0}^{\bar{n}-1}(1-\nu_m^2/u_n^2)}{\prod_{n=0}^{\bar{n}-1}(1-\nu_m^2/\nu_n^2)} \tag{10.134}$$

Aperture efficiency is computed to be equal to

$$\eta=\frac{1}{1+\sum_{n=1}^{\bar{n}-1}\frac{S^2(\nu_n)}{J_0^2(\pi\nu_n)}} \tag{10.135}$$

Fig. 10.14 indicates the type of aperture distribution that is needed if the first five close-in sidelobes are modified to approximately -15 dB SLL as well. Fig. 10.15 shows the array pattern. Table 10.3 indicates the beam broadening factor of the Taylor circular distribution with sidelobe level and $\bar{n}$ factor [1]. u_3 is the 3 dB beamwidth.

The selection of $\bar{n}$ affects the aperture distribution as in the case of the Taylor linear function. Too large a value of $\bar{n}$ will produce a non-monotonic distribution with a large peak at the ends of the aperture, while too small a value will not allow the transition-zone zeros to behave properly. There is thus an appropriate value of $\bar{n}$ for each desired SLL from the point of efficiency to produce maximum directivity. Table 10.4 lists some typical values taken from Reference 22. The values of $\bar{n}$ that give an excitation efficiency which has maximum directivity increase for a given SLL are listed.

10.14 Discretisation in planar arrays

The discretisation of planar arrays by coefficient matching the desired radiation pattern at $2N \times 2N$ points, where $2N$ is the number of elements in any row or column, provides the means to compute the excitation coefficients for radiating elements of the array. These excitation coefficient values are then used to compute the desired pattern in the u, v plane. In the case of sampling of continuous distributions even with rectangular grid and circular or oval array boundaries, the pattern realised as a result of the excitation coefficients leads to satisfactory results in the sidelobe region as long as the number of elements is large. As the number of elements gets fewer, conventional sampling leads to more degradation for an array with rectangular grid and circular boundary.

On lines similar to the improved discretising technique for linear arrays, Elliott [5] has developed a procedure in which a desired radiation pattern can be achieved by an iterative process. The starting pattern in this case is computed by assuming an array of m concentric ring sources spaced at short distances from the centre to the periphery of the circular array. The excitation ring currents are so adjusted that all the $(m-1)$ nulls of the starting pattern coincide with that of the desired pattern. The starting pattern ring currents are then used to compute the starting pattern. If it deviates from the desired pattern in the sidelobe region, each of the ring current sources are perturbed and the resulting m simultaneous linear equations of the current perturbation factor are solved to obtain the perturbation to be effected at each of the m sources. This procedure is continued until the realised pattern and the desired pattern are satisfactorily matched within the accuracy limits.

The continuous ring source, however, has to be replaced in practice by discrete radiators having the same excitation. As proved by Elliott [5], unless the number of radiators replacing each ring source is large, in addition to the ϕ-dependent component of the excitation current, a second component which has

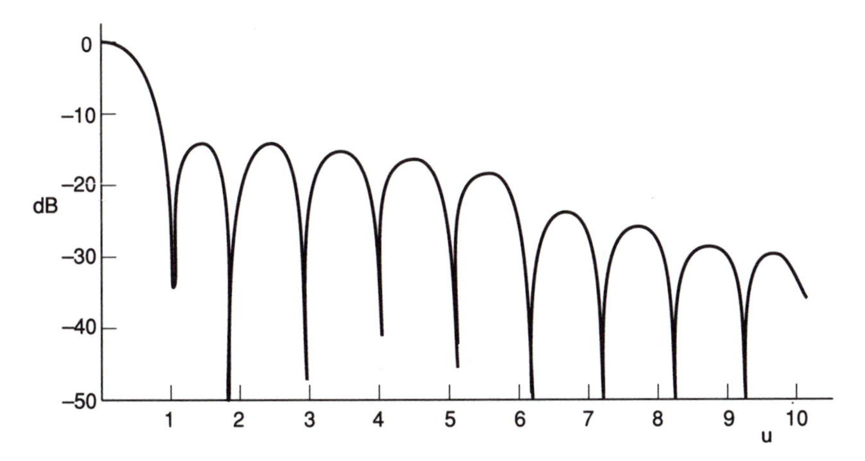

Figure 10.15 *Radiation pattern: Taylor circular distribution [5]. Reprinted by permission of Prentice-Hall*

Table 10.3 *Taylor circular distribution characteristics (Hansen [1])*

SLL (dB)	3 dB beamwidth	Value of σ factor for $\bar{n}$					
		3	4	5	6	7	8
25	0.0489	1.1792	1.1525	1.1296	1.1118	1.0979	1.0870
30	0.5284	1.1455	1.1338	1.1180	1.1039	1.0923	1.10827
35	0.5653	—	1.1134	1.1050	1.0951	1.0859	1.0779
40	0.6000	—	1.0916	1.0910	1.0854	1.0789	1.0726

Table 10.4 *Excitation efficiency and $\bar{n}$ [22]*

SLL (dB)	Value of $\bar{n}$ and corresponding efficiencies that are optimal					
	3	4	5	6	8	10
25	0.09151	0.9324	0.9404	0.9379	0.9064	0.8526
30	0.8377	0.8482	0.8623	0.8735	0.8838	0.8804

period of $2\pi/N_m$ in ϕ and an amplitude proportional to J_{Nm} is generated with all its spatial ϕ harmonics ($m=1, 2, \ldots$).

The Taylor circular aperture distribution can be sampled to obtain excitation of a planar array with an elliptical boundary. For this purpose the elliptical contour is transformed into a circle by the use of the scaling factor such that $\mu'=\mu$ and $\nu'=(a/b)\nu$, where (μ, ν) refer to the co-ordinates in the plane of the actual array with elliptical boundary and (μ', ν') refer to the co-ordinates in the virtual array which is circular. Thus the *mn*th element of the circular array has the radial co-ordinate $\rho_{mn}[(\mu'_{mn})^2+(\nu'_{mn})^2]^{1/2}$. The Taylor pattern is modified to have a beamwidth in the *YZ* plane (see Fig. 10.10), which is a/b times that of the beamwidth in the *XZ* plane. The ring sidelobes have an elliptical contour, and so long as the inter-element spacing is less than 0.75λ and the array is large, the pattern degradation is virtually eliminated.

For planar arrays with rectangular or triangular lattice and arbitrary boundaries, the technique of collapsed distribution is likely to give better results [23]. The concept of collapsed distribution is based on the fact that, when planar rotation is introduced in the array so that the rows and columns of the array are no longer aligned with the new X' and Y' axes, then the radiation pattern of the linear array resulting from the projections of all the elements of the array on to any of the new axes will be the same as the pattern of the actual planar array in that plane, provided the excitation I_{mn} of the elements is unchanged in both cases. If this technique is now applied to a planar array then from eqn. 10.78 we obtain the array factor as

$$S(\theta, 0)=\frac{1}{I_{00}}\sum_{n=1}^{N(m)} I_{mn}\sum_{m=1}^{M} \exp j\{mkd_x \sin\theta\} \tag{10.136}$$

$$S(\theta, \pi/2)=\frac{1}{I_{00}}\sum_{n=1}^{M(n)} I_{mn}\sum_{n=1}^{N} \exp j\{nkd_y \sin\theta\} \tag{10.137}$$

The collapsed distributions $\Sigma_{m=1}^{M(n)} I_{mn}=a_n$ and $\Sigma_{n=1}^{N(m)} I_{mn}=b_m$ can be selected in such a way that they give exactly the desired radiation patterns in the XY and YZ planes (see Fig 10.10). In this manner, the pattern degradation in the principal planes is avoided in sampling continuous planar distributions. The sidelobe topography in the two principal planes will, however, be different. The question now 'boils down' to one of finding out the spread of the collapsed distributions, i.e. the individual excitation values I_{mn} from the known values $\Sigma_{m=1}^{M(n)} I_{mn}=a_n$, $\Sigma_{n=1}^{N(m)} I_{mn}=b_m$. There is no unique answer, but a starting distribution that has been suggested [23] is as folows:

$$I_{mn}^{0}=\frac{1}{2}\left[\frac{a_n}{M(n)}+\frac{b_m}{N(m)}\right] \tag{10.138}$$

It can be seen from eqn. 10.138 that each row (column) is uniformly excited but the result is usually a distribution that is not too far from what will ultimately be achieved. An improvement over this method is the conjugate gradient method.

10.15 Pattern synthesis by simple arrays

Iterative procedures have the advantage that they can either lead to an optimum radiation pattern with respect to a given characteristic or to realisable patterns that match the desired pattern very closely. However, there has been, and will always be, an urge to find simpler procedures which enable computations to be carried out easily by an engineer. One such procedure evolved by Laxpati [24] utilises a small planar array as the building block. Synthesis in this case involves the designation of pattern nulls. The procedure is non-iterative and employs a two-dimensional convolution process. A large array can now be built up as a combination of the simpler arrays. For example, a simpler array of 4 elements can be arranged either in the rhomboid which has a triangular lattice, or in diamond lattice which leads to an equilateral triangular arrangement, and in rectangular lattice, which can be realised with a rectangular arrangement. If $S_{e,p}(\theta, \phi)$ represents the array factor of the canonical 4-element array called the basic array (BA), the array factor of the larger array out of N canonical arrays, now designated as $(BA)_N$, is given by

$$S(\theta, \phi)=\prod_{p=1}^{N} S_{e,p}(\theta, \phi) \tag{10.139}$$

Since the array factor is the Fourier transform of the aperture excitation function, the convolving of the excitation functions of N canonical arrays will result in eqn. 10.139. From eqn. 10.139 it is easily seen that the nulls of the function $S(\theta, \phi)$ are given by the union of the nulls of the N radiation patterns of the individual canonical arrays. Also if m of these N canonical arrays have a common null, then this repeated null will become a multiple null of the pattern $S(\theta, \phi)$. Thus the synthesis of a large array such as $(BA)_N$ with desired nulls in the radiation pattern $S(\theta, \phi)$ is reduced to the synthesis of a 4-element canonical array with prescribed nulls. A 4-element canonical array can be represented by a third order polynomial for its radiation pattern which means it can have at most three nulls. Hence, in the synthesis of the radiation pattern of a large array such as the $(BA)_N$, N such arrays are needed. The designer has only $3N$ nulls available to him out of a possible N^2 nulls for the $(N+1) \times (N+1)$ array. The null synthesis procedure for large square arrays $[(N+1) \times (N+1)]$ will be as follows:

(*a*) Choose an appropriate canonical array depending on the array lattice and array boundary. For an equilateral triangle lattice, the choice of a diamond shaped 4-element canonical array would be right. A rhomboid shaped 4-element canonical array can be the choice if the triangular lattice of the array has arbitrary dimensions (see Fig. 10.16*a–c*)
(*b*) Distribute the $3N$ nulls equally to the N canonical arrays.
(*c*) Synthesise each of the N canonical arrays so that the excitation coefficients of the radiating elements are determined. This is a relatively simple matter in this case.
(*d*) Determine the aperture excitation of the large array by performing the convolution product as indicated below.

Referring to Fig. 10.16*a* which shows a canonical rhomboid array, the locations of the elements are indicated by 1, 2, 3 and 4 at distances d_1, d_2, d_3 and d_4. The excitation coefficients of the array for producing the desired pattern with three nulls is I_p (for element p) where $p = 1, 2, 3$ and 4. The aperture illumination in this case is given by

$$A_{C1}(x, y) = \sum_{p=1}^{4} I_p \delta(\rho - d_p) \tag{10.140}$$

where $\delta(\rho - d_p)$ is the two dimensional delta function locating a point d_p in the (x, y) plane and ρ is a two dimensional radial vector. When this array is convolved with another canonical array C_2, this leads to a 9-element array $(BA)_2$, the aperture illuminaton of which is obtained from

$$[\text{APERTURE ILLUMINATION}]_{BA_2}$$
$$= [\text{APERTURE ILLUMINATION}]_{C_1} * [\text{APERTURE ILLUMINATION}]_{C_2}$$

i.e.

$$A_{BA_2}(x, y) = \sum_{p=1}^{4} \sum_{q=1}^{4} I_{1p} I_{2q} \delta(\rho - d_p - d_q) \tag{10.141}$$

The convolution is indicated by *.

In eqn. 10.141 there are only nine distinct terms corresponding to the nine different delta functions, and the coefficients of these are the excitations of the

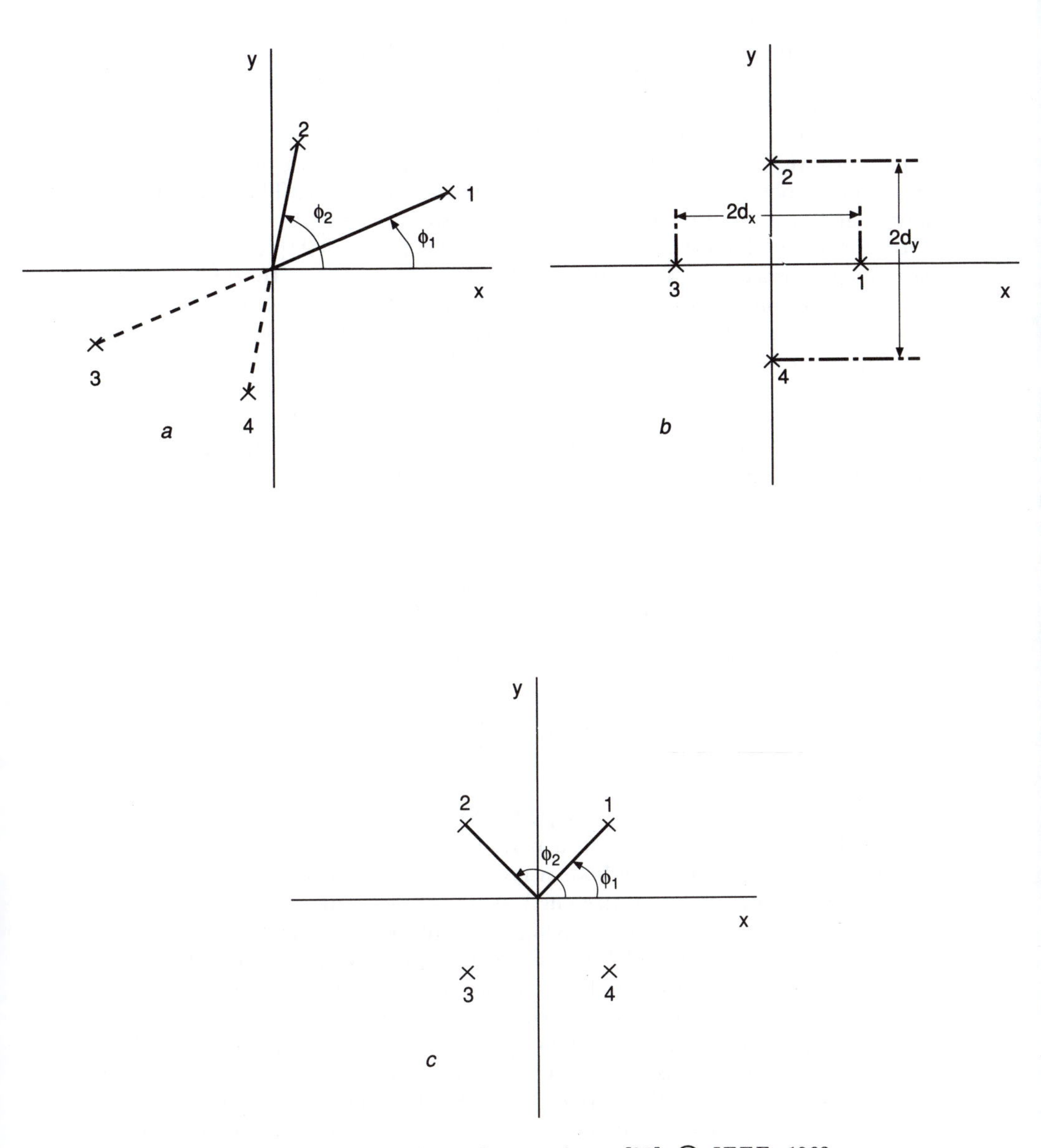

Figure 10.16 *Some typical canonical array shapes [25].* © *IEEE, 1982*

a Rhomboid canonical array
b Diamond canonical array
c Rectangular canonical array

radiating elements of the array generated by this convolution process. The inter-element distances and the orientation of the array are also preserved in this process. The design procedure as indicated above does not call for any symmetry in the excitation functions or in the radiation pattern that is desired. If quadrant symmetry is called for then $I_1 = I_3$ and $I_2 = I_4$ for the 4-element canonical array. In this case there is only one degree of freedom available to the designer; i.e. he has to have either I_1 or I_2 as the reference excitation current or voltage and, depending on the pattern null, adjust the other accordingly. There is no difference between this design procedure and that of the separable excitation method in the number of degrees of freedom that is in the control of nulls. For a $[(N+1)\times(N+1)]$ element planar array, in both cases N nulls will be at the disposal of the designer out of the N^2 available. This procedure is simple and easy to implement in the case of square arrays.

In the case of rectangular arrays of $[(M+1)\times(N+1)]$ elements where $M \neq N$, the procedure suggested by Laxpati is different. In this case, for example, one convolves the basic canonical array $(M+1)$ times or $(N+1)$ times depending on whether N or M is greater, respectively. If we consider that $M > N$, then the canonical array is convolved $(N+1)$ times and a large planar array $(BA)_N$ is designed. The $(N+1)\times(N+1)$ planar array is the largest subarray of the $(M+1)\times(N+1)$ desired array. The $(BA)_N$ array is now convolved with a linear array of $[(M-N)+1]$ elements. In general, for obtaining a rectangular array of $(M+1)\times(N+1)$ elements,

$(BA)_{NM} =$
- Convolve the largest subarray $(BA)_N$ with a linear array of $[(M-N)+1]$ elements for $M > N$
- Convolve the largest subarray $(BA)_M$ with a linear array of $[(N-M)+1]$ elements for $N > M$.

The array factor for the final rectangular array is given by

$$S(\theta, \phi) = S_l(\theta, \phi) \prod_{p=1}^{\mathrm{nin}[M_1]} S_{e,p}(\theta, \phi) \qquad (10.142)$$

where $S_l(\theta, \phi)$ is the array factor of $[(M-N)+1]$ linear arrays. Certain aspects of convolving a linear array with a planar array need to be understood for a better appreciation of this method. A rectangular array can be considered as a square array which is being stretched along a particular direction or along a specified lattice direction by convolving with a linear array. Therefore the linear array will have its axis parallel to the direction along which the square array is being stretched. Secondly, the linear array, being in the plane of the square array, will have a radiation pattern which is ϕ-dependent (azimuth-dependent) with rotational symmetry in the planes perpendicular to the x, y plane in which both the square as well as the linear arrays lie. Therefore the $(M-N)$ nulls of the radiation pattern of the $[(M+1)\times(N+1)]$ planar array assigned to the linear array will have to be projected on to the x, y plane for the realisation by the linear array. Laxpati suggests that the nulls of the linear array may be effectively used to control the radiation pattern of the $(BA)_{MN}$ planar array in the $\phi = 0$ or $\phi = \pi/2$ planes. Patterns in other ϕ planes can be better controlled by the planar subarray which is being convolved with the linear array. Thus

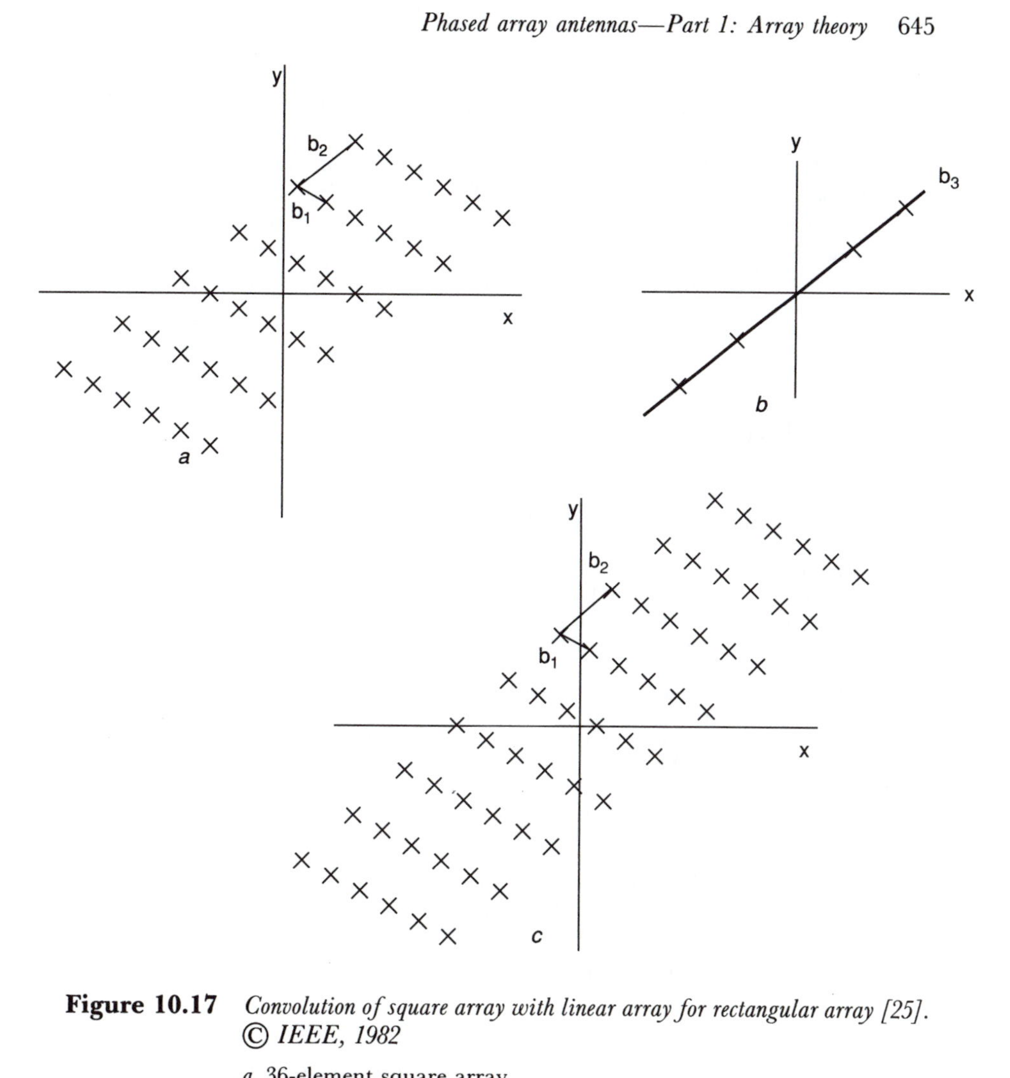

Figure 10.17 *Convolution of square array with linear array for rectangular array [25]. © IEEE, 1982*

a 36-element square array
b 4-element linear array
c 54-element rectangular array

independent control of the nulls in two orthogonal planes, i.e. azimuth and elevation, is now possible. Fig. 10.17 shows the 4-element linear array which is convolved with the thirty 6-element planar arrays to produce the 54-element rectangular array. For arriving at synthesis of arrays with various boundaries such as the hexagon, diamond, squashed hexagon etc, a simple canonical array is convolved with arrays of similar configuration first to evolve a larger subarray. This in turn is convolved with a linear array whose axis lies in the direction in which elongation of the final array is called for, to obtain the desired final array configuration. It can also be modified to carry out such radiation patterns as Taylor in specified planes. The aperture illumination efficiency of these designs has been evaluated and found to compare favourably with designs based on separable aperture excitation. The greatest advantage is its simplicity and ease of computation.

The methods discussed under pattern synthesis for planar arrays has focused on equispaced arrays, and it has been a fertile field of inquiry for the past half century. The advent of fast and powerful computers has now made possible the utilisation of sophisticated techniques to generate patterns with arbitrary sidelobes at reasonable cost and time. It is also possible to synthesise shaped beam patterns with a specified ripple and a choice of excitations.

10.16 References

1 HANSEN, R. C.: Chaps. 9 and 10 *in* RUDGE, A. W., *et al.* (Eds.): 'The handbook of antenna design', (Peter Peregrinus, London, 1986)
2 SCHELKUNOFF, S. A.: 'A mathematical theory of linear arrays', *BSTJ*, 1943, **22**, pp. 80–107
3 RUBINE, E., and BOLOMEY, J. C. H.: 'Antennas: 1. Introduction generale': (Masson, Paris, 2nd edn., 1986) chap. V
4 MA, M. T.: 'Theory and applications of antenna arrays' (John Wiley, NY, 1974) Chaps. 1 and 2
5 ELLIOTT, R. S.: 'Antenna theory and design' (Prentice–Hall of India (P) Ltd., New Delhi, 1985) Chaps. 4–6
6 DOLPH, C. L.: 'A current distribution for broadside arrays which optimises the relationship between beamwidth and side lobe level', *Proc. IRE,* 1946, **34**, pp. 335–348
7 BACH, H., and HANSEN, J. E.: 'Uniformly spaced arrays' *in* COLLIN, R. E., and ZUCKER, L. F. J. (Eds.); 'Antenna Theory. Part I' (McGraw Hill, NY, 1969) Chap. V
8 TAYLOR, T. T.: 'Design of line source antennas for narrow beam width and low side lobes', *IEEE Trans.*, 1955, **AP–7**, pp. 16–28
9 ELLIOTT, R. S.: 'On discretising continuous aperture distributions', *IEEE Trans.*, 1977, **AP–25**, pp. 617–627
10 VILLENEUVE, A. T.: 'Taylor patterns for discrete arrays', *IEEE Trans.*, 1984, **AP–32**, pp. 1089–1093
11 HANSEN, R. C.: 'Aperture efficiency of Villeneuve arrays', *IEEE Trans.*, 1985, **AP-33**, pp. 666–668
12 McNAMARA, D. A.: 'Generalised Villeneuve distribution', *IEE Proc.*, 1989, **136** Part II, pp. 245–249
13 VON AULOCK, W. H. 'Properties of phased arrays', *Proc. IRE,* 1960, **48**, pp. 1715–1727
14 ELLIOTT, R. S.: 'The theory of antenna arrays' *in* HANSEN, R. C. (Ed.): 'Microwave scanning antennas: Vol. II' (Academic Press, NY, 1966)
15 TSENG, F. I., and CHENG, D. K.: 'Optimum scannable planar arrays with an invariant side lobe level', *Proc. IEEE*, 1968, **56**, pp. 1771–1778
16 ELLIOTT, R. S.: 'Synthesis of rectangular planar arrays for sum patterns with ring side lobes of arbitrary topography', *Radio Science*, 1977, **12**, pp. 653–657
17 AUTREY, S. W.: 'Approximate synthesis of non-separable design responses for rectangular arrays', *IEEE Trans.*, 1987, **AP-35**, pp. 907–912
18 CHENG, D. K.: 'Optimisation techniques for antenna arrays', *Proc. IEEE*, 1971, **59**, pp. 1664–1674
19 YOULA, D. C.: 'Generalised image restoration by the method of alternate orthogonal projections', *IEEE Trans.*, 1978, **CAS–25**, pp. 694–702
20 PRASAD, S.: 'Generalised array pattern synthesis by the method of alternating orthogonal projections', *IEEE Trans.*, 1980, **AP-28**, pp. 328–332.

21 NG, T. S.: 'Array pattern synthesis by the method of alternating orthogonal projections: The general case', *IEE Proc.,* 1985, **132**, Part H, pp. 451–454
22 RUDDUCK, R. C., *et al*,: 'Directive gain of circular Taylor patterns', *Radio Science*, 1971, **6**, pp. 1117–1121
23 ELLIOTT, R. S.: 'Antenna pattern synthesis. Part II: Planar Arrays', *IEEE Antennas & Propagation Society Newsletter*, April 1986, pp. 5–10
24 LAXPATI, S. R.: 'Planar array synthesis with prescribed pattern nulls', *IEEE Trans*., 1982, **AP–30**, pp. 1176–1183
25 OLINER, A. A., and MALECH, R. G.: 'Mutual coupling in infinite scanning arrays', *in* HANSEN, R. C. (Ed.): 'Microwave scanning antennas. Vol. 2' (Academic Press: Peninsula Publishing, 1985)

Part 2: Mutual Coupling Effects

10.17 Introduction

The theory discussed in Part 1 is based on an ideal array in which the radiating elements are electrically uncoupled from each other. The currents and the fields induced in each of the radiators of the array are functions of the applied excitation only and are not modified by the parasitic fields. However, in actual practice for electronically scanning arrays, owing to the mutual coupling effects of the neighbouring radiating elements, the parasitically induced currents and fields alter the radiation and the reflection (impedance) characteristics with beam scan. In extreme cases, mutual coupling produces 'blindness' and no power is radiated from or received by the array. The blindness effect has been found in waveguide arrays as well as in some dipole and microstrip radiator arrays. In this Chapter our focus is on planar arrays a sketch of which is shown in Fig. 10.18.

To appreciate the effect of the variation in the input impedance characteristics on the radiation pattern in an array environment, a dipole phased array, in which a constant driving voltage is applied to each radiating element, is considered. This is the forced excitation model of the array, and in this the current at the terminals of the radiating element is a function of the current distribution along the element. In the array environment the current distribution is a superposition of the currents induced on the element by all the elements of the array, and it can be easily related to the well established concept of mutual impedance (admittance) based on the induced EMF method. The input

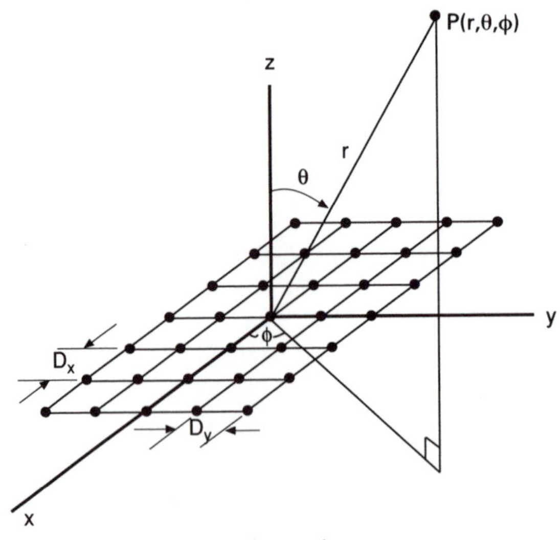

Figure 10.18 *Planar array: rectangular grid*

impedance of the element in the array environment is now a function of the self impedance of the element as well as of the mutual impedances due to the voltages (currents) of the other array elements. The input impedance of the element in the array environment is better known as the active impedance or more appropriately as the scan impedance [1]. For the main beam to scan the volume of space, inter-element phase shifts have to be progressively applied across the array and this in turn results in the variation of the phase and amplitude relationship of the currents induced by the other elements at the designated radiator. In effect, this changes the current distribution along the element, the current at the terminals of the radiator and finally the scan impedance. The variation of the current at the terminals of the radiator affects the scan element pattern with scan angle. If the array environment is uniform, then the current variations with scan of any one element only need be known. The array pattern being the product of the array factor and the element pattern, it will also reflect the changes that occur in the element pattern. The scan element pattern can be measured with one antenna element excited and all other elements present in the array being match terminated.

In actual practice, the simple feed networks employed in phased array antennas are more likely to be the constant power types where an impedance mismatch of the element results in lesser voltage being applied to it. This enables us to model the array as a free excitation type with each area of the feed network being equivalent to a voltage source that is matched to the radiation resistance of the element at broadside. In this case, the voltage incident at the input port of the element and the voltage reflected from the input port back to the feed are linked through the scattering matrix which provides a measure of the active reflection coefficient at the antenna input port. The reflection coefficient can be expressed as a function of the coupling coefficients of the other array elements to the designated element. The scattering matrix appoach accurately represents most phased arrays. The variation of the radiation pattern of the array with beam scan is now linked to the active reflection coefficient at the antenna input port due to the variations of the coupling coefficients. In the array environment, the feed system no longer stays matched to the array. The power transmission coefficient from the feed to the array is now written as $(1-|\Gamma(\theta, \phi)|^2]$ where Γ is the reflection coefficient. When $\Gamma(\theta, \phi)=1$, no power is transmitted to the array and at that angle scan blindness occurs. The array gain is now the product of the isolated element pattern, the array factor of uncoupled elements and the power transmission coefficient from the feed network to the array. The isolated element pattern is taken in the array environment but with all the other elements in the open-circuited condition. It can be easily shown [2] that the scan element gain and the gain of the antenna element in the array with all the other elements being open-circuited is given by

$$G_a(\theta, \phi)=\frac{R_i G_e(\theta, \phi)}{R_a(\theta, \phi)}\,[1-|\Gamma(\theta, \phi)|^2| \qquad (10.143)$$

where

$G_a(\theta, \phi)$ = scan element gain pattern
$R_a(\theta, \phi)$ = scan element radiation resistance at (θ, ϕ)

R_i = input radiation resistance of the element
$G_e(\theta, \phi)$ = open-circuited element pattern

This expression establishes that the pattern null in the scan element gain and the complete reflection condition are identical and correlate with the blindness of the array.

The blindness phenomenon has been examined in terms of leaky waves which come into existence to cause the null [3]. In the case of a single excited element, when it radiates a space wave it also excites leaky waves on the neighbouring portions of the passively terminated array. The radiated portion of the leaky wave interferes with the space wave radiating in that direction to produce a null in the element pattern. Similarly the same phenomenon takes place on the other side of the array so that a symmetrical pattern is generated. In the case when all the elements of the array are excited, and the array operates in the customary mode, with beam scanning, the field distribution in space becomes asymmetric. This asymmetry in space is reflected as an asymmetry in the aperture field and is accounted for by the addition of the first higher order mode in the feed system even though the mode is below cut-off. The presence of this additional mode produces field cancellation in the aperture at some beam scan angles, as a consequence of which the active conductance becomes zero and all the power is reflected back in to the feed system. The forced field existing on the array surface resembles the field of a surface wave on a modulated structure. It is neither the field of an actual surface wave nor an independently excited wave. It has meaning only at exactly the scan angle corresponding to the total reflection of power back into the feed. The occurrence of blindness in arrays in the early days of phased arrays led to the analysis of infinite arrays, studies of the behaviour of the array element in a simulator and fabrication of a small array of the elements prior to the launching of the full scale effort. It is presently realised that the blindness condition can be reduced or eliminated by making the array lattice dimension sufficiently below the cut-off condition of the next higher mode of the feed guide or of the unit cell throughout the operating frequency range [4].

10.18 Element-by-element method

There are two principal methods by which mutual coupling effects in phased arrays are analysed. These are the element-by-element approach and the infinite array approach. In the element-by-element approach it is recognised that the aggregate mutual coupling effects at each radiator due to the other radiators in the array are not uniform and hence will have to be determined on an element-by-element basis. On the other hand in the case of infinite arrays it is assumed that the aggregate mutual coupling effects at each radiator due to the other radiators in the array is uniform. In this case by determining the aggregate mutual coupling effects for one typical radiator the behaviour of the array is known. The element-by-element method gives an accurate solution but it is computationally intensive even for moderately large arrays. The infinite array approach provides an approximate solution for large arrays and it is computationally simpler.

In the element-by-element approach, initial investigations were mostly with dipole arrays which formed the basis for design of phased arrays with single mode radiators. For example, in the case of narrow slots, Babinet's principle made it possible to consider them as complementary to narrow flat strip dipoles which in turn can be replaced without much error by thin cylindrical dipoles. The equations governing the dipole array can be written as

$$\begin{aligned} V_1 &= I_1 Z_{11} + I_2 Z_{22} + \cdots + I_N Z_{1N} \\ V_n &= I_1 Z_{n1} + I_2 Z_{n2} + \cdots + I_N Z_{nN} \\ &\;\cdot\cdot \qquad\qquad \cdots \quad \cdot \\ &\;\cdot\cdot \qquad\qquad \cdots \quad \cdot \\ V_N &= I_1 Z_{N1} + I_2 Z_{N2} + \cdots + I_N Z_{NN} \end{aligned} \tag{10.144}$$

Here I_n is the current at the terminals of the nth radiating element, V_n is the voltage applied and Z_{nn} and Z_{mn} are the self and the mutual impedances, respectively. For single mode radiating elements a set of N equations have to be solved. Now, based on the induced EMF method, the mutual impedance between two antennas is defined as the open circuit voltage induced in the terminals of one antenna when the other is excited by a unit current. Logically, the mutual impedances should be determined in the array environment with all other elements being present. Owing to difficulties encountered in the determination of mutual impedance in the array environment, a first order approximation was made by considering mutual impedance values between pairs of elements outside the array environment, to build up the impedance matrix of the array. Implicit in this step was the assumption that these radiators are canonical minimum scattering (CMS) antennas, that is, they are electromagnetically 'invisible' in the open-circuited condition. A second assumption was that the current flow on the radiators was sinusoidal. The measured scan element pattern and beamwidth of the radiator was found to have reasonable agreement with the computed values based on these assumptions in the case of thin dipole arrays and other single mode radiating elements. However, it clearly falls short of the requirements in the case of arrays with multimode radiating elements or with elements which have both dimensions comparable to the wavelength of operation.

From eqns 10.144 it is possible to rearrange the terms to derive the scan impedance of any radiating element as

$$Z_n^a = V_n / I_n = Z_{nn} + (I_m / I_n) Z_{nm} \tag{10.145}$$

The scan impedance of the nth radiator is the sum of its self impedance and a mutual coupling term which is a weighted sum of its mutual impedances with all the other radiators in the array. The weighting factor is (I_m/I_n), the relative excitations of the radiators.

Ideally, the imaginary part of Z_{nn} should be made to cancel the imaginary part of the second term on the RHS of eqn. 10.145 so that Z_n^a is pure real. For this purpose, so long as the radiator is a single mode antenna element, its self impedance Z_{nn} can be very accurately determined both by theoretical analysis and by measurement. The value of Z_{nn} is primarily a function of the dimensions of the radiator and therefore the designer has the freedom to adjust the dimensions at resonance. The second term on the RHS is a summation of the product of ratios of currents and mutual impedances. The current ratios are determined by the desired radiation pattern, whereas Z_{nm} is a function of the radiator dimensions and the inter-element distance. In the case of such single mode elements as monopoles, dipoles, waveguide-fed broad-wall slots and the patch radiator, very accurate analytical methods based on an integral formulation of Maxwell's equations and applying the method of moments have been developed over the years for determination of mutual coupling. These methods are now preferred to the more expensive and less accurate experimental methods [5]. Thus the designer arrives at the final value of the scan impedance of the nth element Z_n^a ($n=0, 1, 2, \ldots$) iteratively so that it is pure real. The scan impedance will not be identical for all the elements, even for moderately large arrays because the aggregate mutual coupling converges slowly and also the array is nonuniformly excited. Finally, since it is the excitations of the peripheral radiating elements that influence the sidelobe levels most, determination of the magnitude and phase of these elements become a critical aspect of design. The computation of mutual coupling will have to be carried out for each of the N elements of the array and for each beam position.

10.19 Infinite array method

As the array size increases in a regular manner, an element in the centre of the array will eventually be affected very little by the increase in size. For example when the array size has gone beyond 1000 elements, for many types of radiators the mutual coupling effect becomes negligible and the element can be considered to be in an infinite array environment. The scan impedance as well as the radiation pattern of any radiating element in such large arrays can be very well approximated by the behaviour of a central radiator. Thus, for large arrays, a uniformly excited infinite array approach is more practical for computation of mutual coupling effects. In such a case, the radiation and the reflection characteristics of all the elements in the array are assumed to be identical and the array pattern can now be computed as the product of the element pattern and the array factor. Any large array is now considered as a finite portion of an infinite array, and all the inner radiating elements have a similar array environment to the central radiator [6].

For an infinite array, the periodicity in the placement of radiators in a cellular fashion enables the designer to apply Floquet's theorem which permits us to describe the fields in the open region in terms of a complete orthogonal set of modes. The modal description repeats itself periodically except for a multiplicative exponential factor so that the fields at the radiating element of the array differ from that of the adjacent elements by only a phase shift. Based on this concept Wheeler [7] developed the unit cell or periodic cell around each radiating element and analysed the radiation resistance of the radiating element (the dipole in this case) in an infinite array and phased for broad side radiation. The cells are alike, contiguous and each cell can be extended indefinitely in the open region above the array in a direction normal to the array face by a unit cell waveguide with two electric walls (zero electric field) and two magnetic walls (zero magnetic field). When the array is radiating or receiving at broadside, each unit-cell waveguide points vertically upwards and samples a portion of the outgoing or incoming TEM wave with its electric field perpendicular to the electric walls and magnetic field perpendicular to the magnetic walls (see Fig. 10.19).

Oliner and Malech [8] modified this concept in such a way as to include the admittance variation of the radiator with scan angle. A straightforward extension of the periodic cell in the case of scanning arrays would have necessitated tilting of each unit-cell waveguide in the direction of the arbitrary scan angle, so that the properties of the electric and magnetic walls are retained. This would also require a provision to be made for separate and additional unit-cell waveguides for each of the grating lobes that may appear in the scan volume. Instead, these authors retained the unit-cell and the unit-cell waveguide normal to the plane of the array as in Wheeler's method. However, they postulated that the unit-cell waveguide would accommodate any arbitrary scan angle of the beam and also the grating lobe that may appear in scan directions away from broadside direction. The walls of the unit-cell waveguide, however, will no longer be electric and magnetic walls but will be different. The opposite walls are identical to each other except that the support fields differ by a Floquet shift. The modes in the unit-cell waveguide are no longer TEM but will be a TM mode or a TE mode or a combination of both, depending on the plane of scan. For any arbitrary scan angle the combined contributions of TM and TE mode are required. An alternative to TM and TE mode set is the LSM and LSE mode set which are characterised by a vanishing magnetic or electric field component, respectively, in one of the transverse directions. Each of the LSM and LSE modes are orthogonal to each other and the set is complete. The main lobe is represented in the unit-cell waveguide by the lowest propagating mode of each kind (TM and TE). When the beam is scanning, in case a grating lobe

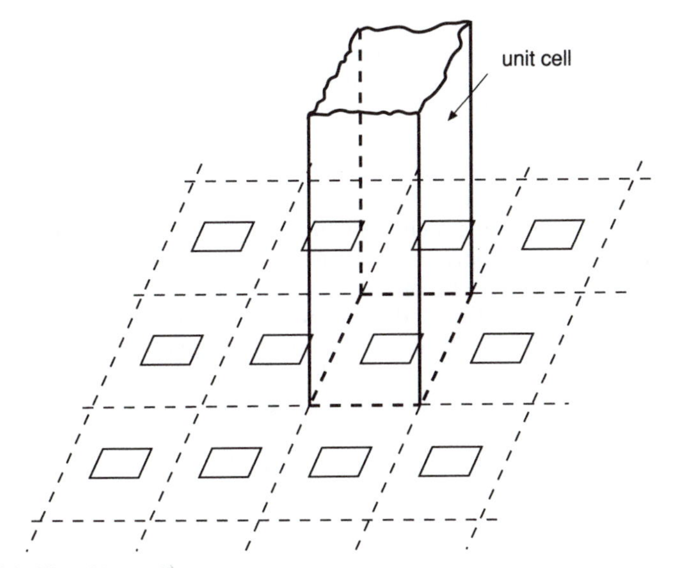

Figure 10.19 *Unit cell concept*

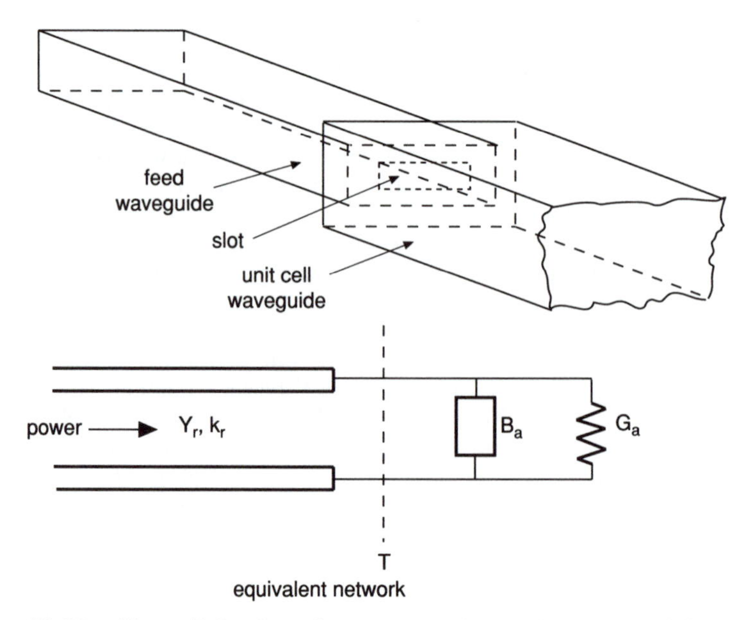

Figure 10.20 *Unit cell for slot radiator array and equivalent network [8]*

appears, the next higher order TE, TM mode pair propagating within the unit-cell waveguide will have to be taken note of. Hence, with each main lobe a pair of TE and TM modes are associated, and in the case of well designed arrays, only the lowest of the mode pair will be propagating in the unit-cell waveguide. Depending on the plane and the direction of scan, the relative proportion of the modes within the mode pair propagating in the unit-cell wave guide will be varying, thereby giving rise to different scanning behaviour. Since the total field within the unit-cell waveguide is a superposition of the two modes, the scanning behaviour of the radiator is translated into changes in scan impedance and it can vary significantly with beam scan.

The infinite array effectively divides the total space into the radiation half-space in which the unit-cell waveguides are situated, and the feed network half-space. Though the analysis is general enough to cover all types of feed networks, for the purposes of providing a clear picture, each radiator is assumed to be fed independently by an appropriate transmission line. For narrow slot radiators, a rectangular waveguide with its broad wall parallel to the length of the slot and excited in the dominant TE mode can be selected as the feed network. The radiator is now at the junction of two different waveguides, its impedance being across both of them (see Fig. 10.20). The susceptance of the radiator can be considered as related to the sum of the energy stored on both sides of the radiator. The conductance, on the other hand, is related to the radiation pattern of the radiator and thus to the power radiated in the radiation half-space. Therefore for efficient transfer of power from the feed network to the radiator the characteristic impedance of the transmission line has to be matched to the conductance of the radiator through a compensating network. Thus, unless this is inserted between the feed and the radiator, the efficiency of the array will vary during scan, which will result in uneven radiation of power by the main beam of the array with beam scan.

The active conductance G_a of the slot radiator in an infinite array normalised to waveguide characteristic impedance Y_r can now be expressed as

$$\frac{G_a}{Y_r}=\frac{ab\lambda_g}{2\lambda D_x D_y}\left[\frac{1-(a_s/a)^2}{\cos(\pi a_s/2a)}\right]^2\left[\frac{\cos\,(k_{x0}a_s/2)}{1\text{-}(k_{x0}a_s/\pi)^2}\,\frac{\sin(k_{y0}b_s/2)}{(k_{y0}b_s/2)}\right]^2 H_{00} \quad (10.146)$$

In eqn. 10.146, a and b refer to the feed guide dimensions, a_s and b_s refer to the slot dimensions, D_x and D_y, the array lattice dimensions, λ is the wavelength in free space, λ_g is the wavelength in the guide, k_{x0} and k_{y0} are related to the wave number by the following expressions:

$$\begin{aligned} k_{x0}&=k\,\sin\,\theta_0\,\cos\,\phi_0 \\ k_{y0}&=k\,\sin\,\theta_0\,\sin\,\phi_0 \end{aligned} \quad (10.147)$$

and H_{00} is given by

$$H_{00}=[1-\sin^2\,\theta_0\,\cos^2\,\phi_0]/(\cos\,\theta_0) \quad (10.148)$$

H_{00} is the scanning factor which provides the link between the variation of the conductance of the slot and the scan direction. When the beam is scanning in the H-plane (parallel to the plane of the slot array), ($\cos\,\phi_0$) in eqn. 10.148 will be unity, so that $H_{00}=[\cos\,\theta_0]$. In the E-plane scan, H_{00} will be equal to $[\cos\,\theta_0]^{-1}$ as $\phi_0=\pi/2$. As the scan angle is widened or as the lattice dimensions are increased or a combination of both is tried, one or more grating lobes will appear in the scan volume. In such a case eqn. 10.146 will have an additional term for each grating lobe appearance. The electric field distribution on a resonant slot being very much similar to the current distribution in a strip dipole in free space, an equivalent array of dipoles can be invoked to model the slot array. In such a case, the scan impedance of the slot is simply related to that of the dipole by a scale factor. The computations that have been carried out earlier for an infinite dipole array can be used for understanding the variation of conductance with beam scan for slot arrays. Fig. 10.21 brings out the variation of the scan resistance of half wave dipoles in free space in the infinite array environment with scan angle [8]. The inter-element spacing for this array is half wavelength. The scan resistance goes to infinity when the beam is being scanned in the H-plane and as the beam nears the scan angle of 90° away from broadside. When the beam is scanning in the E-plane, the scan resistance goes to zero as the scanning angle approaches 90° away from broadside. The variation of scan resistance with scan angle for the slot array will be similar except that the scan planes are interchanged.

In the computation of scan susceptance of the radiator, the power stored in the higher order modes on both sides of the array has to be taken into account. On the feed guide side of the array the computation of the susceptance requires knowledge of the placement of the radiator in the guide. For example, the slot can be off-centred on the broad wall or it can be an inclined slot on the narrow wall of the guide. For the susceptance contributed by the higher order modes in the unit-cell waveguide no simple and concise expression is available. The expression given below is due to Hansen [2] who has cast the results of an earlier analysis in unit-cell form and symbols. The active reactance is thus given by

$$\frac{B_a}{Y_r} = -\frac{ab\lambda_g}{2D_xD_y\lambda^3}\left[\frac{1-(\pi a_s/2a)^2}{\cos(\pi a_s/2a)}\right]^2 \sum_n \sum_m \left[\frac{(1-k_{xn}a_s)^2}{\cos k_{xn}a_s}\right]^2 \left[\frac{\sin k_{ym}b_s/2}{k_{ym}b_s/2}\right]^2 H_{nm} \tag{10.149}$$

In eqn. 10.149, H_{nm}, $(k_{xn}a_s)$ and $(k_{ym}b_s)$ stand for the generalised scanning factor and the distribution factors, respectively. The distribution factors take into account the effect of the field distribution along and across the slot for the *mn*th mode. They can be expressed as

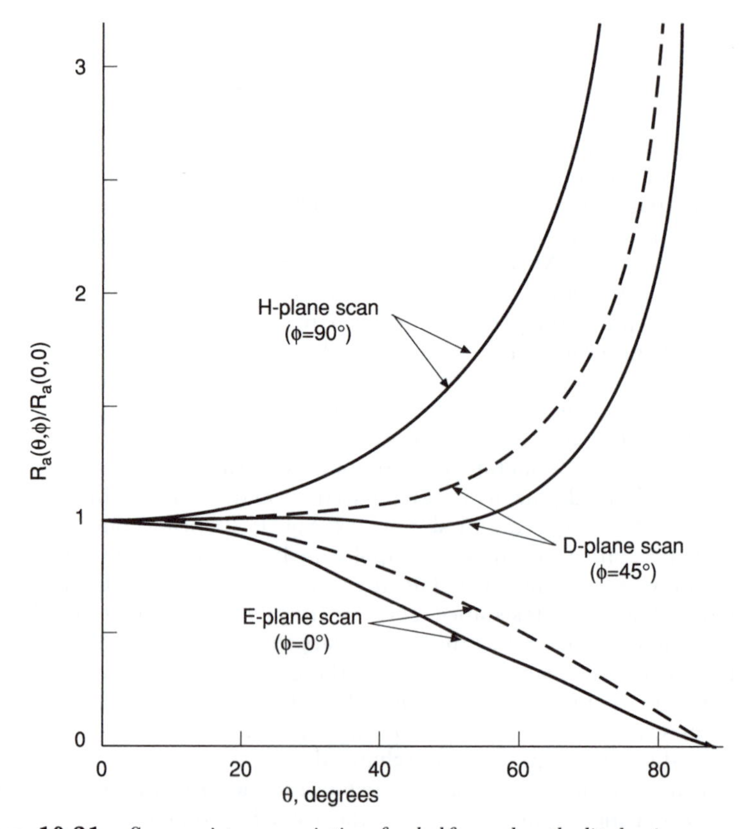

Figure 10.21 *Scan resistance variation for half-wavelength dipoles in array configuration [8]*

—— half-wave dipoles
--- infinitesimal dipoles

$$H_{nm}=\frac{\left(\sin\theta_0\cos\phi_0+\dfrac{n}{D_x}\right)^2-1}{\sqrt{\left(\sin\theta_0\cos\phi_0+\dfrac{n}{D_x}\right)^2+\left(\sin\theta_0\sin\phi_0+\dfrac{m}{D_y}\right)^2-1}} \tag{10.150}$$

and

$$k_{xn}=k\left(\sin\theta_0\cos\phi_0+\frac{n}{D_x}\right); \qquad k_{ym}=k\left(\sin\theta_0\sin\phi_0+\frac{m}{D_y}\right) \tag{10.151}$$

The reactance term contains an infinite sum over all the non-propagating modes. This modal sum is known to be slowly convergent. Figs. 10.22*a*, *b* and *c* indicate the variation in susceptance of the half-wave strip dipole in free space in the array environment when the beam scan takes place in the *E*-plane ($\phi=90°$), *H*-plane ($\phi=0°$) and the *D*-plane ($\phi=45°$), respectively, for different inter-element spacings and for different dipole widths [8]. It can be seen that the effect of the width of the strip dipoles on the reactance values is significant and that as the width increases the reactance becomes more negative.

The effect of the finiteness of an array on the scan element pattern has been analysed by Hansen [1] for dipole arrays by computing the radiation pattern of a central element with an infinite number of elements, and comparing it with the scan element pattern of a central element for 101×101 and 41×41 planar dipole arrays. For the purposes of computation, the large but finite array can be considered as a finite portion of an infinite array so that all the inner radiating elements have the same array environment. The edge effects are reflected by the finite number of the terms used for summing. This has a distinct advantage in that the N^3 (N being the number of elements of the array) time dependence of computation due to the element-by-element method is eliminated. The computations have been carried out for inter-element spacings of a quarter, a half and seven-tenths of the wavelength with and without a reflector screen. For example, for inter-element spacing of $0.5\ \lambda$, the scan element *H*-plane patterns of the central dipole radiator in the 41×41 and the 101×101 array environment are superimposed on the *E*-plane pattern of an infinite dipole array in Figs. 10.23 and 10.24. The *H*-plane pattern of an infinite dipole array clearly follows the *E*-plane pattern and the oscillations on the finite array pattern are found to be the same as the number of elements in a row or in a column of the finite array. The oscillations are due to the edge elements having an environment which is not the same as that of the inner elements. It can also be noted that as the angle of scan moves away from the broadside, the amplitude of the oscillations increases and they tend to have an amplitude which is independent of the size of the array. In the case of arrays with an inter-element spacing of $0.7\ \lambda$, the *H*-plane patterns in Figs. 10.25–10.27 bring out clearly that the notch in the pattern due to the grating lobe is broader and slightly above the -15 dB level for the 41×41 array. This array is a coarse approximation to the infinite array. Fig. 10.25 brings out the results of a 101×101 array in which the notch due to the grating lobe is sharper and deeper (-17 dB or so). The graph in Fig. 10.26 is the *H*-plane pattern of a 41×201 array. The notch in this graph is very sharp and as deep as -19 dB. These graphs clearly demonstrate that as far as

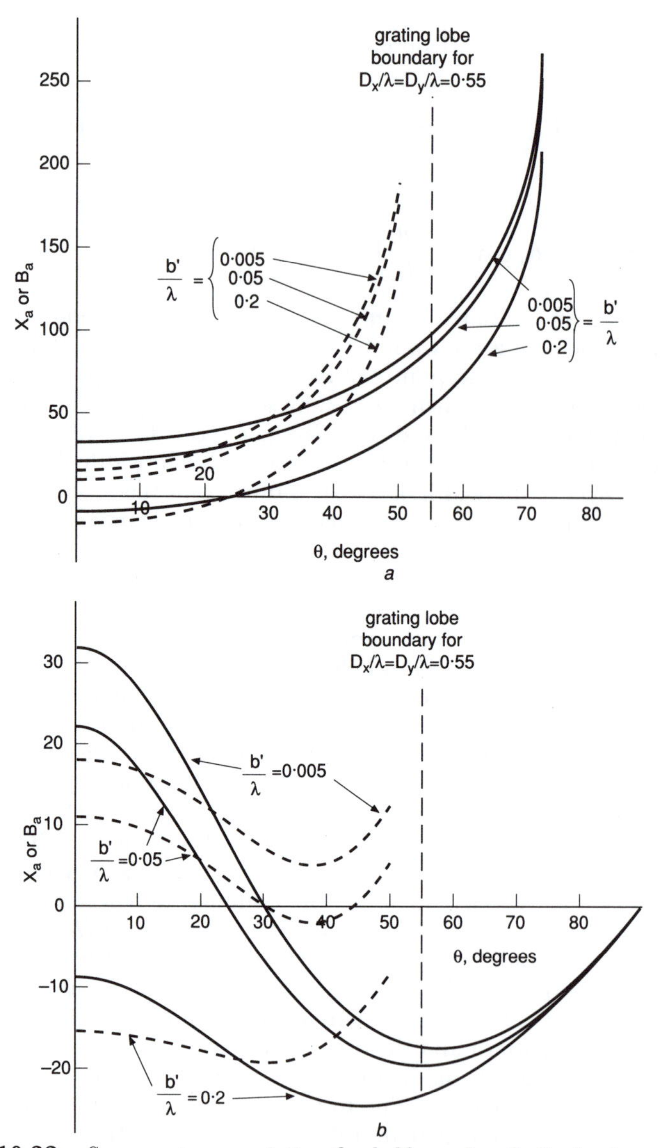

Figure 10.22 *Scan reactance variation for half-wavelength dipoles in array configuration [8]*

a H-plane scan, $\phi = 0°$
b E-plane scan, $\phi = 90°$
c Diagonal plane scan, $\phi = 45°$

—— $\dfrac{D_x}{\lambda} = \dfrac{D_y}{\lambda} = 0.50$

- - - $\dfrac{D_x}{\lambda} = \dfrac{D_y}{\lambda} = 0.55$

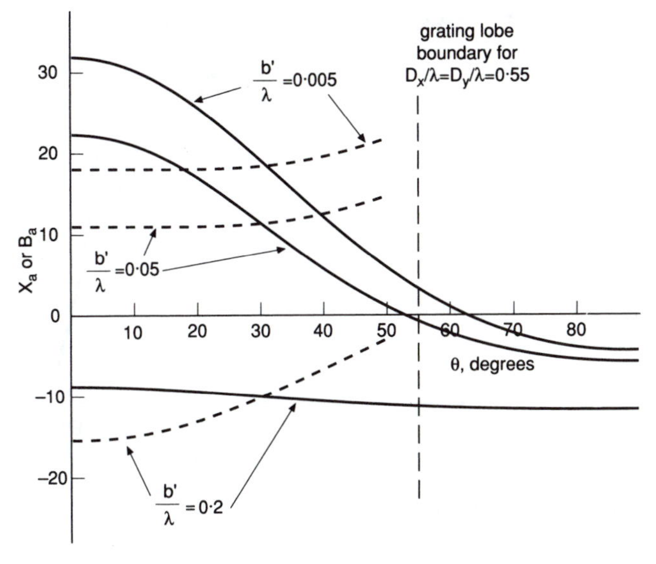

c

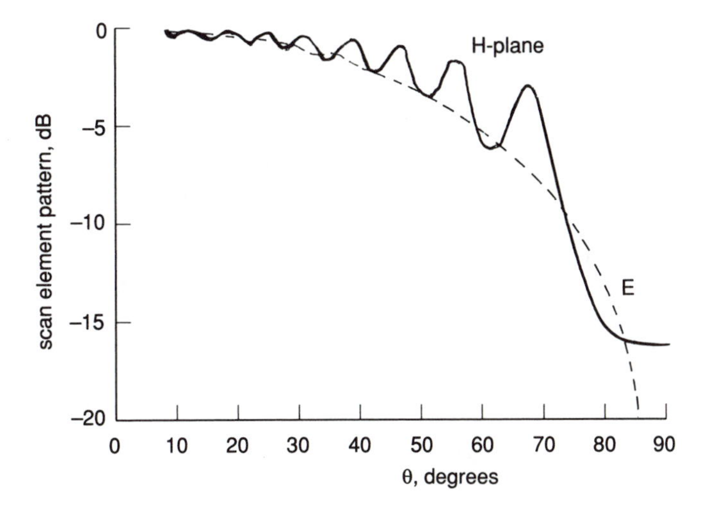

Figure 10.23 *Scan element pattern for 41 × 41 dipole array [1]*

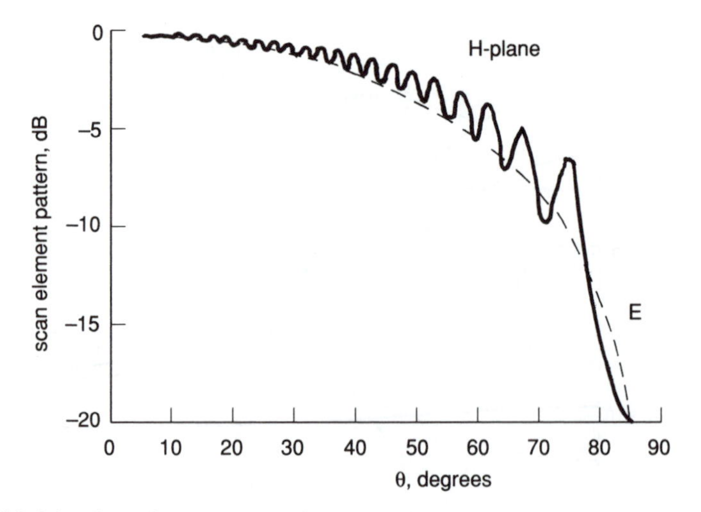

Figure 10.24 *Scan element pattern for 101 × 101 diple array [1]*

the *H*-plane pattern is concerned, the number of elements in the transverse direction is less critical and that 41 elements or more are needed in this direction to produce a sufficiently deep null signifying the appearance of a grating lobe. On the other hand, a comparison of Figs. 10.24 and 10.26 brings out that, by increasing the number of elements in the longitudinal direction from 41 to 201, the notch is deeper and sharper.

It is quite clear from Figs. 10.21 and 10.22 that the scan impedance of a radiator in the array environment varies significantly with beam scan. However none of these figures exhibit scan blindness before the appearance of the grating lobe. The appearance of scan blindness would have been preceded by a steep

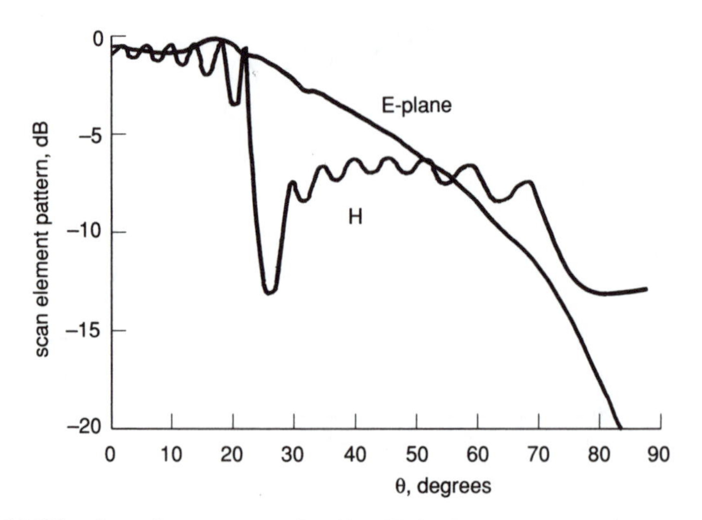

Figure 10.25 *Scan element pattern for 41 × 41 dipole array with 0.7λ square grid [1]*

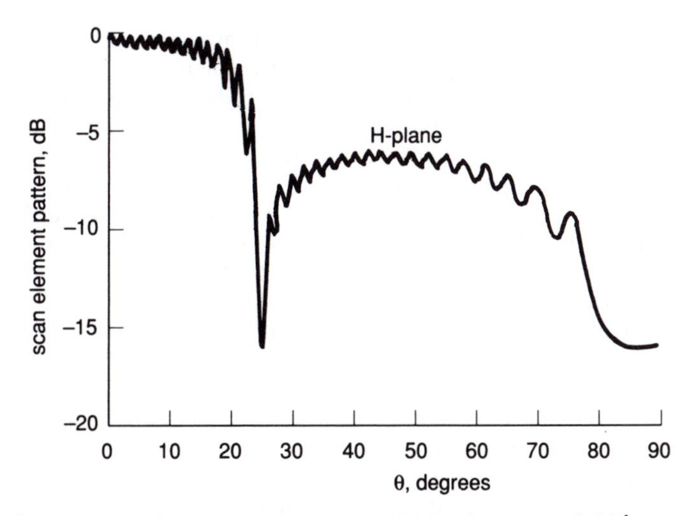

Figure 10.26 *Scan element pattern for 101 × 101 dipole array with 0.7λ square grid [1]*

fall in the scan resistance values near a specific scan angle in any of the scan planes, the value in the limit reaching zero. This would correspond to a deep notch in the radiation pattern of the radiator at that angle. The analyses so far discussed in this Chapter are based on the assumption that the electric field distribution along a slot radiator or the current distribution on a dipole is sinusoidal and scan invariant. Further, the radiation conductance was obtained from energy considerations without accounting for the feed network. Since both the narrow resonant slot and the thin half wave dipole are single mode radiators, the assumption of sinusoidal field or current along the radiator and

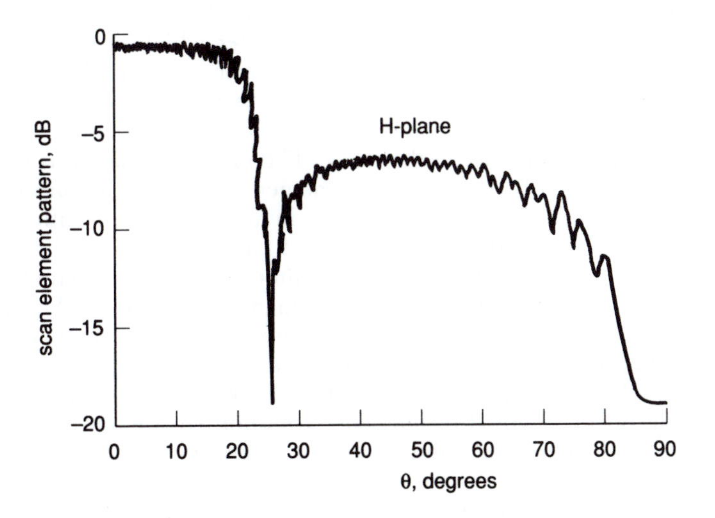

Figure 10.27 *Scan element pattern for 41 × 201 dipole array with 0.7λ square grid [1]*

its scan invariance is not a large deviation from reality. Hence the results obtained from analysis provide a representative picture of the phenomenon. However, in the case of open-ended waveguide radiators which are multimode elements, the measured radiation pattern of a central element in an array exhibits deep notches at scan angles less than those corresponding to a grating lobe formation in the visible space (see Fig. 10.28). It shows the normalised power pattern of a central element of an array of open-ended rectangular horns having a deep notch well inside the grating lobe circle indicated by the dot/dash line. Further, the Figure also contains the results computed using a single mode grating lobe series for an infinite array. A third curve shown in the Figure refers to the full modal solution which predicts scan blindness at a scan angle well before the onset of the grating lobe. The full modal solution discards the earlier 'one pair of TE/TM modes per grating lobe' assumption and includes waveguide higher order modes on either side of the radiator for computing the scan impedance. It is therefore evident that any analysis of mutual coupling between radiating elements of an array has to be based on expressing the fields in the radiation as well as in the feed region by a complete set of orthogonal modes.

10.20 Integral equation formulation

The starting point in the case of integral equation formulation for the analysis of mutual coupling is Maxwell's equations. In the case of wire radiators, the boundary condition that has to be satisfied is that the tangential electric field at the surface of the radiator is zero. For the transmitting array, it is assumed that the excitation source of the radiator is a slice generator or a delta function at the point of excitation. The total field at the surface of the radiator is the sum of the field due to the source and the fields induced on it by the other radiators on the array as well as by its own radiated field. These fields are expressed in the form of an integal equation in which the current along the radiator appears in the integrand. The current is the unknown parameter which needs to be determined. There will be one such equation for each radiator in the array. If, however, the array is assumed to be an infinite array, then one needs to work with only one such equation since the infinite array provides a uniform environment for all radiators in the array. The integral equation is converted into a matrix equation by the method of moments in which the current is now expressed as the sum of basis or expansion functions each with a weight coefficient which is to be determined. These basis functions, which are simple and well known, can either be entire domain functions or sub-domain functions. If M such functions are used to describe the unknown parameter, then a system of M linear equations with M unknowns is formed. These M equations can now be expressed in matrix form and solved by digital computers [10] with fast and computationally efficient procedures that have been evolved over the years. The solutions due to the method of moments are approximate and the degree of accuracy is a function of validity of the expansion functions and their complexity.

In the case of aperture radiators, the planar array is considered to be a ground plane with perforations corresponding to the aperture radiators. This

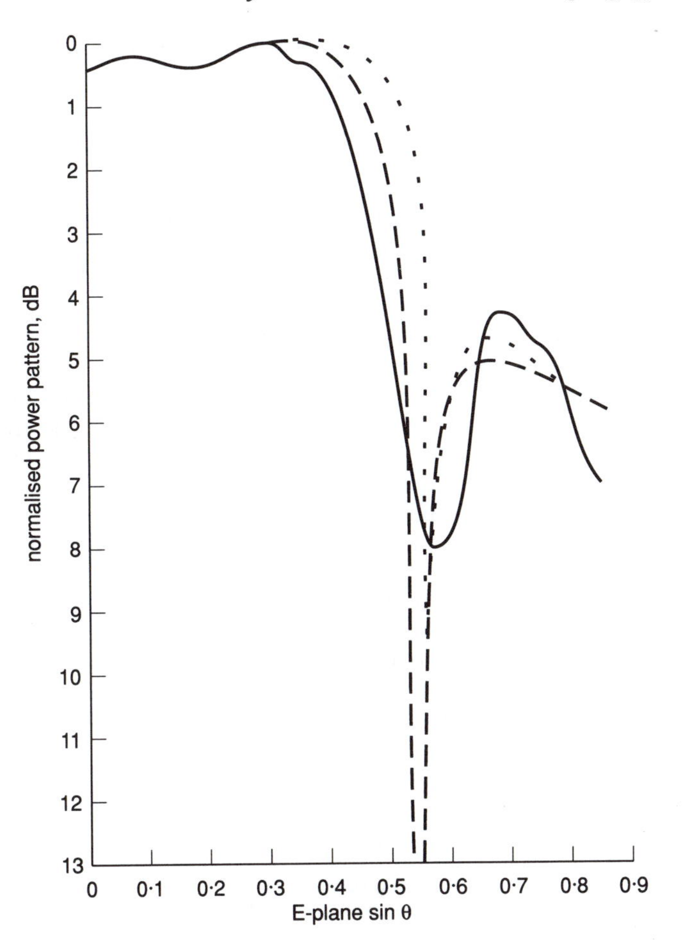

Figure 10.28 *Scan element pattern of infinite waveguide radiator array [41]. Reprinted by permission of © Wiley-Interscience, 1972*

— experiment (13 × 13 array)
–– modal analysis
--- 'grating lobe series' one-mode solution

Rectangular grid $a/\lambda = 0{\cdot}5898$
$c/\lambda = 0{\cdot}5898$
$b/\lambda = 0{\cdot}6439$
$d/\lambda = 0{\cdot}6439$

array ground plane, which coincides with the xy-plane of a Cartesian co-ordinate system, divides the space into two half-spaces; the feed half-space ($z < 0$) containing the impressed sources and the feed network and the radiation half-space ($z > 0$) into which the aperture radiators radiate the fields. By covering the apertures with a sheet of an electric conductor, the array ground plane is without perforations and the isolation between the half-spaces is total. The original aperture electric field is now replaced by equivalent magnetic currents of equal magnitude but of opposite sign on either side of the array ground plane at positions corresponding to the aperture radiators as shown in Fig. 10.29. By this, the first boundary condition, namely the continuity of the tangential electric field across the aperture, is satisfied. In order to satisfy the second boundary condition, namely the continuity of the tangential magnetic field at the aperture, the components of the field on either side of the array ground plane are identified and then equated to each other. On the feed half-space ($z<0$) side there are two components to the aperture magnetic field. These are the tangential magnetic field at the aperture due to the impressed source plus the image, and the tangential magnetic field due to the equivalent magnetic current. On the radiation half-space ($z>0$) side the aperture magnetic field arising out of the equivalent magnetic current and the image is computed as follows. Firstly, the magnetic field component in the radiation half-space is related to the equivalent magnetic current at the array ground plane corresponding to the positions of the aperture radiators, and this is expressed in the form of an integral equation with the free-space Green's function as the kernel. Next, the field point is made to lie on the array ground plane so that the integral equation is transformed into an equation of the field components tangential to the array ground plane at positions corresponding to those where the apertures had been situated. This is summed over all the apertures contained in the array plane and then equated to the resultant tangential magnetic field from the feed half-space side. The tangential fields are

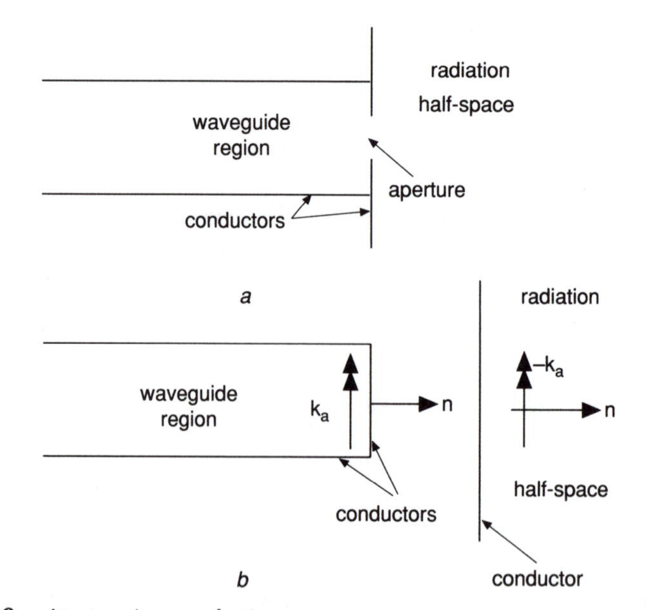

Figure 10.29 *Aperture in a conducting screen*

a Original problem
b Equivalent problem

now expressed in terms of the TE and TM modes of the waveguide, and the method of moments with Galerkin's procedure is now employed with these mode functions as expansion (basis) and as weight functions. As before, the solutions obtained are approximate and the degree of accuracy will be dependent on the degree of computational complexity.

The integral form of Maxwell's equations used in these formulations expresses the electric and magnetic fields in a given volume of space and arising from sources confined to a finite region through a source–field relationship more familiar as the Green's function. These are dealt in great detail in textbooks on electromagnetics and on antennas. While there are a variety of Green's functions to choose from, in the case of mutual coupling problems, the Green's function corresponding to infinite medium is made use of. Two kinds of integral equations are used: the first one, namely the Fredholm equation of the first kind, has the unknown variable appearing only under the integral, and the second one, i.e., the Fredholm equation of the second kind, in which the unknown also appears outside the integral. The expressions for the electric field $\vec{E}(R)$ and the magnetic field $\vec{H}(R)$ in the volume of space can be written as [12]

$$\vec{E}(r) = -\frac{1}{4\pi} \int_v \left\{ j\omega\mu \vec{J}\, G(r, r') + \vec{K}\, \nabla' G(r, r') + \frac{\nabla ! \vec{J}}{j\omega\varepsilon} \nabla' G(r, r') \right\} dV' \tag{10.152}$$

$$\vec{H}(r) = \frac{1}{4\pi} \int_v \left\{ -j\omega\varepsilon \vec{K}\, G(r, r') + \vec{J} \times \nabla G(r, r') + \frac{\nabla ! \vec{K}}{j\omega\mu} \nabla' G(r, r') \right\} dV' \tag{10.153}$$

Hence, $G(r, r')$ is the Green's function which is given by

$$G(r, r') = [e^{-jk|r-r'|}]/[|r-r'|] \tag{10.154}$$

$\vec{J}$ and $\vec{K}$ are the electric current and magnetic current densities, k, μ and ε are respectively the propagation constant, permeability and the permittivity of the free space, r' and r are the distances, respectively, of the source and the observation point from the origin. An alternative form of these integral equations for the electric and magnetic fields which is also used extensively in analysis is as follows [12]:

$$\vec{E}(r) = \frac{1}{4\pi j\omega\varepsilon} \int_v \{\vec{J} \cdot (k^2\bar{I} + \nabla\nabla) G(r, r') - j\omega\varepsilon \vec{K} \times \nabla G(r, r')\} dV' \tag{10.155}$$

$$\vec{H}(r) = \frac{1}{4\pi j\omega\mu} \int_v \{\vec{K} \cdot (k^2\bar{I} + \nabla\nabla) G(r, r') + j\omega\mu \vec{J} \times \nabla G(r, r')\} dV' \tag{10.156}$$

Here $\bar{I}$ is the unity dyad. The term ∇' represents operation with respect to the primed co-ordinates. Eqns. 10.152, 10.153, 10.155 and 10.156 are for fields due to volumetric distribution of sources. For current sources constrained on a surface S the integral representations are modified by replacing the volume densities by surface densities and the volume integral by the corresponding surface integral. The electric and the magnetic current densities in eqns. 10.152, 10.153, 10.155 and 10.156 can be considered as currents induced on the surface of a scatterer by an incident plane wave. In such a case, $\vec{E}(r)$ and $\vec{H}(r)$ in these equations are the scattered fields, $\vec{E}^s$ and $\vec{H}^s$. Using the following boundary conditions for a conducting scatterer:

$$\hat{n}\times(\vec{E}^i+\vec{E}^s)=0$$
$$\hat{n}\times(\vec{H}^i+\vec{H}^s)=\vec{J} \qquad (10.157)$$

Here $\vec{E}^i$ and $\vec{H}^i$ are the electric and magnetic fields incident on the body. With these changes, eqns. 10.152, 10.153, 10.155 and 10.157 can be transformed to become the well known electric field integral equation (EFIE) and the magnetic field integral equation (MFIE). There are other integral equation formulations derived from such methods as the Rayleigh–Ritz variational method and the reaction concept. These are also used in the analysis of mutual coupling effects. Of this the reaction concept is of particular interest because it enables one to relate the impedance matrix for mutual coupling computations elegantly. The reaction concept is due to Rumsey [13] and it expresses reaction as a measure of coupling between two sources a and b. In isotropic media, the reciprocity theorem enables us to state that the reaction of source a on b is equal to the reaction of source b on source a. Thus the integral relationship can be written as

$$\iiint_v 4(\vec{E}_a\cdot\vec{J}_b)-(\vec{H}_a\cdot\vec{K}_b)-(\vec{E}_b\cdot\vec{J}_a)+(\vec{H}_b\cdot\vec{K}_a)\}dV=0 \qquad (10.158)$$

Here $\vec{E}_a$, $\vec{E}_b$ are the electric fields, $\vec{H}_a$ and $\vec{H}_b$ are the magnetic fields, $\vec{J}_a$ and $\vec{J}_b$ are the electric current densities and $\vec{K}_a$ and $\vec{K}_b$ are the magnetic current densities due to the sources a and b, respectively. In the case of metallic bodies, $\vec{K}_a$ and $\vec{K}_b$ are equal to zero so that eqn. 10.158 becomes

$$\iiint_v (\vec{E}_a\cdot\vec{J}_b-\vec{E}_b\cdot\vec{J}_a)dV=0 \qquad (10.159)$$

Eqn. 10.159 is useful for developing a basic formula for mutual impedance between antenna elements. The unknown in the integral equation formulation for mutual coupling is the surface electric current $\vec{J}$ or the corresponding tangential magnetic field at the antenna element. The current or the field is approximated by a set of basis or expansion functions and the MOM with Galerkin's procedure is now employed to solve for the current.

It is now proposed to illustrate the application of the MOM technique to the integral equation formulation of Maxwell's equations for computing the mutual coupling in single-mode as well as multi-mode radiator planar electronic scanning arrays.

10.21 Phased arrays with dipole radiators

The dipole belongs to the linear element class of radiators and has been the subject of studies for over 50 years with respect to current distribution, mutual coupling and mathematical representation. The dipole consists of a cylindrical radiator of length L with diameter a which is small compared with the wavelength. It is excited at the centre and since it is made of high conductor material, the current is confined to the surface of the radiator. The current distribution is modelled as an infinitely thin sheet of current forming a radius a. Therefore, the only significant component of current on the dipole is the axial current $I(z)$, the z-axis being the axis of the dipole. Consequently, the vector potential A will have only the z-component A_z which can be expressed in terms of the impressed voltage V_i as,

$$\left(\frac{\partial^2}{\partial z^2}+k_0^2\right)A_z(z)=-j\frac{k_0^2}{\omega}V_i\delta(z) \tag{10.160}$$

Eqn. 10.160 can be solved for A_z, so that we can write

$$A_z(z)=-j\sqrt{\omega\varepsilon_0}\left\{C_1\cos k_0z+C_2\sin k_0z+\frac{V_i}{2}\sin k_0|z|\right\} \tag{10.161}$$

Here k_0 is the free space propagation constant, ω is the frequency of operation and ε_0 is the permittivity of the medium. C_1 and C_2 are constants to be determined. The vector potential can also be expressed by an integral equation that relates the current flowing in the radiator to the vector potential at the surface of the radiator as follows:

$$A_z(z)=\frac{1}{4\pi\mu_0}\int_c\int_{-L/2}^{L/2}G(z,z')I(z')\,dz' \tag{10.162}$$

with

$$G(z,z')=\frac{e^{-jk_0R}}{R} \tag{10.163}$$

where R is the distance between the observation point (x, y, z) on the surface of the dipole and the source point (x', y', z'). The RHS of eqn. 10.162 represents the integration around the circumference of the dipole. The only unknown parameter in eqn. 10.162 is the axial current $I(z)$.

In the infinite array environment, the current flow in each of the radiating elements will contribute to the build up of the magnetic vector potential at the surface of each radiator in the array. Therefore, the resultant vector potential at the surface of any radiator of the infinite array is the sum of the vector potentials generated at the surface by each of the radiating elements of the array. Fig. 10.30 shows the geometrical configuration of the dipole array. From considerations of elementary geometry, the distance parameter R from the designated radiator to the mnth radiator can be expressed as

$$R^2=(z-z'-2nd)^2+r_m^2+a^2-2ar_m\cos\phi' \tag{10.164}$$

with

$$r_m^2=(mb)^2+a^2+2|m|ab\cos\phi \tag{10.165}$$

Here b is the distance between the columns of the array and $2d$ is the distance between the rows. The vector potential at the surface of the designated element can now be written as [14]

$$A_z^a(z)=\frac{1}{4\pi\mu_0}\int_{-L/2}^{L/2}\int_0^{2\pi}\int_0^{2\pi}\sum_{m,n=-\infty}^{\infty}I_m(z')G(z,z')\,dz'd\phi'd\phi \tag{10.166}$$

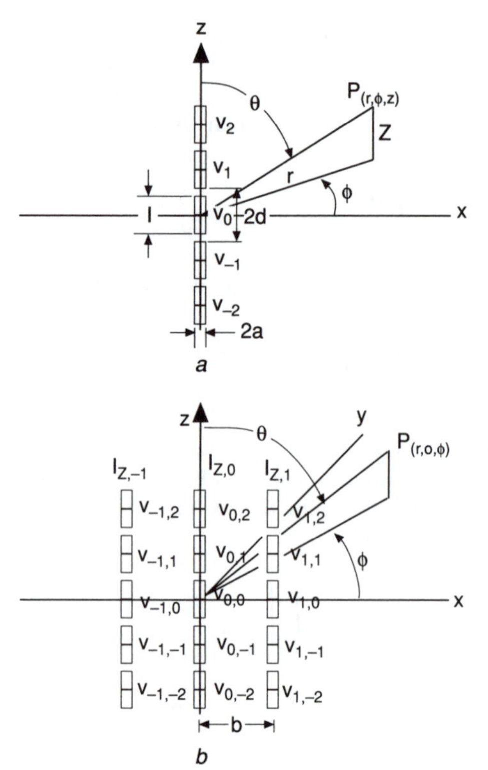

Figure 10.30 *Infinite dipole array [14].* © *IEEE, 1968*

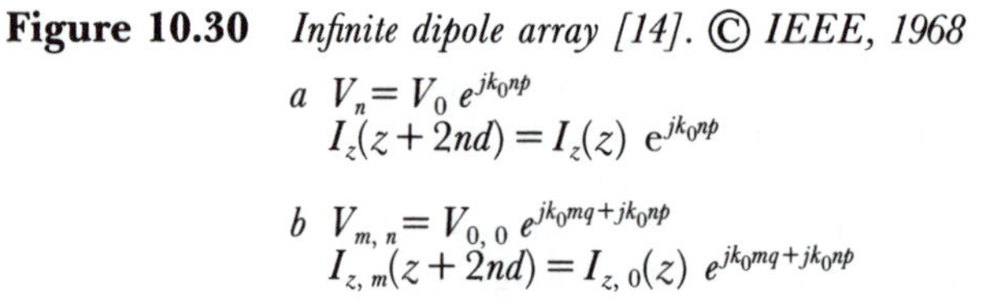

a $V_n = V_0\, e^{jk_0 np}$
$I_z(z+2nd) = I_z(z)\, e^{jk_0 np}$

b $V_{m,n} = V_{0,0}\, e^{jk_0 mq + jk_0 np}$
$I_{z,m}(z+2nd) = I_{z,0}(z)\, e^{jk_0 mq + jk_0 np}$

The driving voltages for the elements in the array are progressively phased along the rows and columns. From Floquet's theorem, the currents in these radiators have the same progressive phase shift as the voltages. Thus

$$V_{mn} = V_{00}\, e^{jk_0(mq+np)} \tag{10.167}$$

$$I_{mn} = I_{00}(z) e^{jk_0(mq+np)} \tag{10.168}$$

where p and q are constants. Substituting eqn. 10.168 in eqn. 10.166, we have

$$A_z^a(z) = \frac{1}{4\pi\mu_0} \int_{-L/2}^{L/2} \int_0^{2\pi} \int_0^{2\pi} dz'\, d\phi'\, d\phi\, \frac{I(z')}{4\pi^2} \times \sum\sum_{m,\,n=-\infty}^{\infty} \frac{\exp[jk_0(mq+np) - jk_0\sqrt{(z-z'-2nd)^2 + r_m^2 + a^2 - 2ar_m \cos\phi'}]}{\sqrt{(z-z'-2nd)^2 + r_m^2 + a^2 - 2ar_m \cos\phi'}} \tag{10.169}$$

Since the dipole elements are thin, the vector potential along the designated element satisfies eqn. 10.161 so that, by combining it with eqn. 10.169, we obtain

$$\int_{-L/2}^{L/2}\int_0^{2\pi}\int_0^{2\pi} dz'\, d\phi'\, d\phi$$

$$\times \sum_{m,\,n=-\infty}^{\infty}\sum \frac{\exp[\,jk_0(mq+np)-jk_0\sqrt{(z-z'-2nd)^2+r_m^2+a^2-2ar_m\cos\phi}\,]}{\sqrt{(z-z'-2nd)^2+r_m^2+a^2-2ar_m\cos\phi}}$$

$$=-j\,\frac{4\pi}{z_0}\left\{C_1\cos k_0z+C_2\sin k_0z+\frac{V_i}{2}\sin k_0|z|\right\} \tag{10.170}$$

Eqn. 10.170 can be rewritten as

$$\int_{-L/2}^{L/2}\int_0^{2\pi}\int_0^{2\pi} dz'\, d\phi'\, d\phi\, \frac{I_z(z')}{4\pi^2}$$

$$\times \sum_{m=-\infty}^{\infty} e^{jk_0mq} \sum_{n=-\infty}^{\infty} e^{jk_0p(2nd)/2d}\, \frac{e^{-jk_0\sqrt{(z-z'-2nd)^2+r_m^2+a^2-2ar_m\cos\phi}}}{\sqrt{(z-z'-2nd)^2+r_m^2+a^2-2ar_m\cos\phi}}$$

$$=-j\,\frac{4\pi}{z_0}\left\{C_1\cos k_0z+C_2\sin k_0z+\frac{V_i}{2}\sin k_0|z|\right\} \tag{10.171}$$

The last term for summation on the LHS of eqn. 10.171 is in a form to which Poisson's formula can be applied. It can be expressed as

$$f(2nd)=\frac{1}{2d}\sum_{n=-\infty}^{\infty} F\left(\frac{n\pi}{d}\right) \tag{10.172}$$

$$F(\omega)=\int_{-\infty}^{\infty} e^{j\omega t} f(t)\, dt \tag{10.173}$$

Also,

$$\int_{-\infty}^{\infty} e^{j\omega t}\, \frac{\exp[-jk_0\sqrt{C^2+(\xi-t)^2}]}{\sqrt{C^2+(\xi-t)^2}}\, dt=e^{j\omega t}\, 2K_0(C\sqrt{\omega^2-k_0^2}) \tag{10.174}$$

where $K_0(x)$ is the modified Bessel function of the second kind. If $k_0>\omega$, then replace $\sqrt{\omega^2-k_0^2}$ by $j\sqrt{k_0^2-\omega^2}$. Using eqns. 10.172–10.174 in eqn. 10.171, we obtain

$$\int_{-L/2}^{L/2}\int_0^{2\pi} d\phi \sum_{m=-\infty}^{\infty} \frac{I_z(z')}{2\pi^2}\, dz' \sum_{n=-\infty}^{\infty}\int_0^{2\pi} \exp\left[\,jk_0\left(\frac{n\pi}{k_0d}+\frac{p}{2d}\right)(z-z')\right]$$

$$\times\; K_0\left[k_0\sqrt{r_m^2+a^2-2ar_m\cos\phi'}\;\sqrt{\left(\frac{n\pi}{k_0d}+\frac{p}{2d}\right)^2-1}\,\right] d\phi$$

$$=-j\,\frac{4\pi}{z_0}\left\{C_1\cos k_0z+C_2\sin k_0z+\frac{V_i}{2}\sin k_0|z|\right\} \tag{10.175}$$

Now we use Graf's formula,

$$K_0(\sqrt{u^2+v^2-2uv\cos\alpha}) = \sum_{m=-\infty}^{\infty} K_m(u)K_m(v)\cos m\alpha \qquad (10.176)$$

and introduce eqn. 10.176 in eqn. 10.175 and then carry out the integration once with ϕ' as variable and next with ϕ as variable. The result is

$$\int_{-L/2}^{L/2} I(z,z')K(z,z')\,dz' = -j\frac{4\pi}{z_0}\left\{C_1\cos k_0z + C_2\sin k_0z + \frac{V_i}{2}\sin k_0|z|\right\} \qquad (10.177)$$

where,

$$K(z,z') = \frac{1}{2\pi^2}\sum_{m=-\infty}^{\infty}\exp\left[jk_0\left(\frac{n\pi}{k_0d}+\frac{p}{2d}\right)(z-z')\right]D_n \qquad (10.178)$$

with,

$$D_n = K_0(k_0\gamma_n a)I_0(k_0\gamma_n a) + 2\sum_{m=1}^{\infty}K_0(k_0\gamma_n mb)I_0^2(k_0\gamma_n a)\cos k_0mq \qquad (10.179)$$

and,

$$\gamma_n = \sqrt{\left(\frac{n\pi}{k_0d}+\frac{p}{2d}\right)^2 - 1} \qquad (10.180)$$

Eqn. 10.177 is the integral equation that has to be solved with $I(z')$ as the unknown. It cannot be solved analytically but the use of the MOM technique provides us with an approximate solution with high accuracy. From the mathematical point of view, there are certain rules to be followed in the choice of expansion and weighting functions [15, 16]. Firstly, the expansion functions should be in the domain of the integro-differential operator, that is, they have to satisfy the differentiability criterion and the boundary conditions. The expansion functions $\{x_i\}$ should be so chosen that the set $\{Ax_i\}$ should be a complete set for the range of the operator.

There are two principal categories of expansion or basis functions, namely, the entire domain functions and the subdomain functions. The entire domain functions are defined over the complete length of the radiator and can be based on Fourier, McClaurin, Chebyshev, Hermite and Legendre polynomial series. Subdomain representation involves terms for the expansion functions valid over only a small part of the entire dipole length. Hence, a number of subdomain functions will have to be used for representing the current flow on the surface of the dipole radiator. Pulse expansion functions with point matching, piecewise linear expansion functions and piecewise sinusoidal expansion functions in Galerkin's procedure and three-term sinusoidal functions with point matched weights have been used in the case of dipole arrays [17]. These numerical models for the current flow not only approximate the current distribution in an acceptable way, but also match the boundary conditions along the dipole radiator length. In this case the boundary conditions specify that for the centre-fed dipole of length L, the currents $I(L/2) = I(-L/2) = 0$. It has also been pointed out that, in the case of subdomain expansion functions, with samples of the order of six to twenty per wavelength, accuracies of about 10% or better in antenna input impedance are possible if the adjacent segment lengths are not changed too abruptly. Further, it is advisable to use shorter but equal segments near the feed point at the centre of the radiator and on either side of the feed region. For the cylindrical model of the radiator, the point matching procedure is straightforward and it yields a good approximate solution. However, care has to be taken to ensure that the matching points on the radiator are away from regions of zero produced by the basis functions.

While numerical modelling of the currents or fields at the radiators and mathematical computation should form the mainstay of phased array antenna design, validation by experimental data, at least in the beginning, is the most satisfactory way of checking the correctness of the numerical models. In the case of the dipole radiator, input impedance is the recommended parameter for validation as it is most sensitive to the feed region details.

There have been several higher order solutions for the current flow along the linear radiator developed by Prof. R. W. P. King and his colleagues [18, 19], which are more realistic in the computation of mutual coupling effects between columns or rows of dipoles with parallel axes. The five-term solution for the current flow has been found to be most comprehensive in this respect and has been used for computation of mutual coupling effects in a large array of dipole radiators [14]. The current model assumed in this case is given by

$$I_z(z) = \sum_{i=1}^{5} Q_i J_i(z) \tag{10.181}$$

with

$$\begin{aligned}
J_1(z) &= \sin k_0|z| - \sin k_0|L/2| \\
J_2(z) &= \cos k_0|z| - \cos k_0|L/2| \\
J_3(z) &= \cos \frac{k_0}{2} z - \cos \frac{k_0}{2} (L/2) \\
J_4(z) &= \sin k_0 z - \frac{2z}{L} \sin k_0(L/2) \\
J_5(z) &= \sin \frac{k_0}{2} z - \frac{2z}{L} \sin \frac{k_0}{2} (L/2)
\end{aligned} \tag{10.182}$$

Upon substitution of eqns. 10.182 in eqn. 10.177 and point matching the resulting equations at seven points along the length of the radiator, namely, at $z = 0$, $\pm L/6$, $\pm L/12$, $\pm L/2$, a set of 7×7 simultaneous equations are obtained. These are easily solved for Q_i, C_1 and C_2. As shown in Fig. 10.31, the five-term model for the current flow gives satisfactory results for the scan admittance of the infinite dipole array. The scan element conductance diminishes sharply around $\phi = 62°$ with the polar angle fixed at 90°. As the array is scanned in the E-plane ($\phi = 90°$), the scan conductance increases sharply at $\theta = 80.4°$ and also at 48.2°. The array blindness at 80.4° is due to the appearance of the grating lobe, and the one at 48.2° is due to mutual coupling effects. Fig. 10.31 also brings out that there is a difference in the admittance values based on the sinusoidal Fourier series current model and the five-term current model. It is evident from the Figure that the five-term current model is nearer to the experimental data.

10.22 Waveguide backed slot radiator array

The waveguide fed longitudinal slot array has slots which are formed on the broad wall of the waveguides and are offset from the centre of the broad wall as shown in Fig. 10.32. The slots are of half-wave resonant length and the excitation of the slots is a function of the offset from the centre. A single waveguide called the main line guide and placed behind the array transverse to all the waveguides housing the radiators feeds the slot radiator array as shown in Fig. 10.33. The method of feeding is by means of either centre-inclined or centre-offset slots on the main line guide in a sequential manner. The power coupled to each guide containing the radiating slots is a function of the tilt and

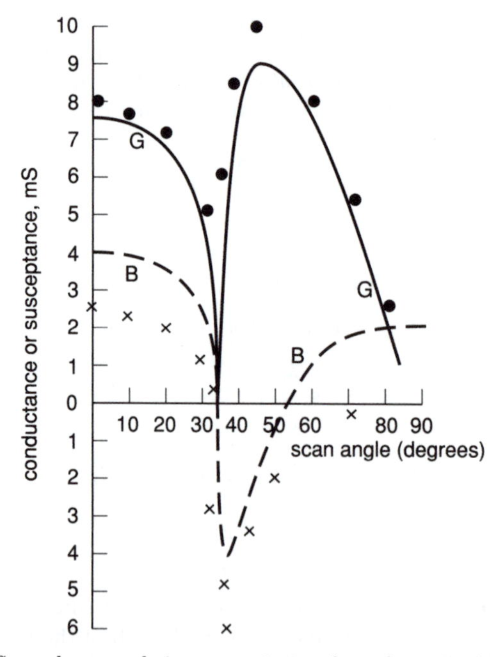

Figure 10.31 *Scan element admittance variation for infinite dipole array [14]* *© IEEE, 1968*

● ● ● sinusoidal
× × × theory
—— 5-term
--- method

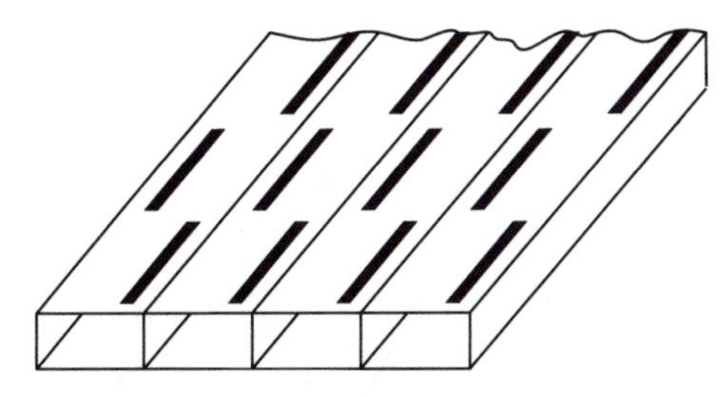

Figure 10.32 *Planar array of waveguide-fed longitudinal slot radiators [20]. Reprinted by permission of Prentice Hall*

the length of the coupling slot. The popularity of the waveguide fed longitudinal slot array is due to the fact that it can be flush mounted and it is sturdy because of the box-like construction.

Primarily, the radiating element is narrow band, as seen from Fig. 10.34, the data for which were collected through carefully conducted experiments [20]. As the slot is offset from the centre of the broad wall, the resonant length of the slot and the admittance also vary as shown in Fig. 10.35. The normalised conductance at resonance of the slot on the broad wall of a rectangular guide excited by the TE_{10} mode is represented by [21]

$$g(x) = C \sin^2 \frac{\pi x}{a} \tag{10.183}$$

where x is the offset from the centre line, a is the broad wall dimension, b is the narrow wall dimension of the guide and C is a constant that is a function of the transverse dimensions, the propagation constant of the electromagnetic wave in the waveguide and the free-space wavelength of operation. Also the admittance of the slot seen as a shunt element on an equivalent lossless transmission line of characteristic admittance G_0 has been represented by

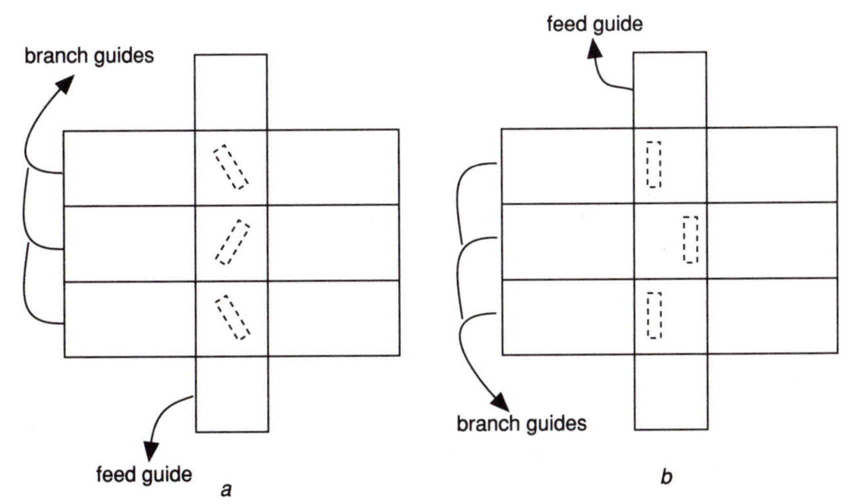

Figure 10.33 *Feed mechanism to slot array [26]. © IEEE, 1991*

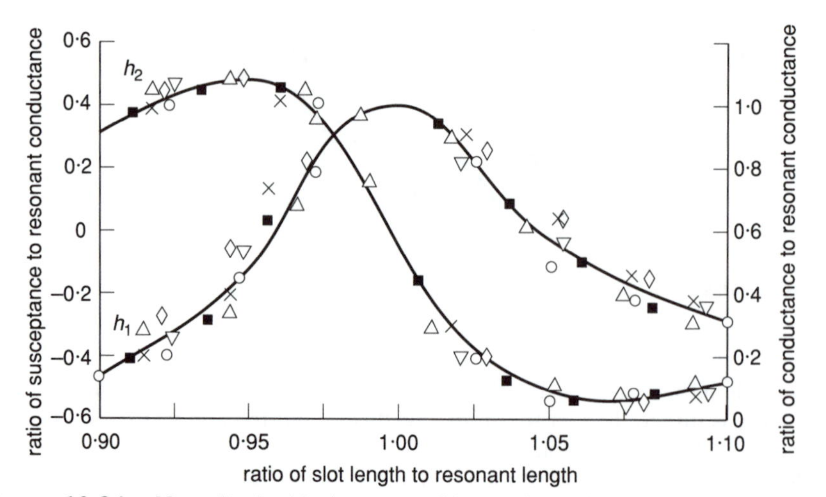

Figure 10.34 *Normalised self-admittance of longitudinal shunt slot [47]. Reproduced by permission of McGraw-Hill*

$a = 0{\cdot}9000$ inches
$b = 0{\cdot}4000$ inches
$t = 0{\cdot}0500$ inches
$w = 0{\cdot}0625$ inches
$f = 9{\cdot}375$ GHz

OFFSETS (inches)

○ $x = 0{\cdot}029$
■ $x = 0{\cdot}054$
△ $x = 0{\cdot}074$
× $x = 0{\cdot}104$
▽ $x = 0{\cdot}127$
◇ $x = 0{\cdot}153$

$$\frac{Y(x, l)}{G_0} = [h_1(y) + jh_2(y)]g(x) \tag{10.184}$$

with $y = l/l_r$, l_r being the resonant length and l the length of the slot. $h_1(y)$ and $h_2(y)$ symbolise the complex admittance of the slot. Eqns. 10.183 and 10.184 have been derived based on the experimental data in the early days of phased array design. While the accuracy achieved in the experiment was adequate at that time, it is now considered that this procedure does not provide the necessary accuracy in the value of the slot self impedance, it also puts an increasingly unnecessary restriction on the design of high performance low sidelobe level arrays [21, 22]. The earlier assumption of the slot being a shunt impedance across a lossless transmission line is no longer valid especially for increasing offset of the slot radiator from the centre line. The electric field in the slot aperture in such cases is not found to be symmetrical because it is a resultant of the dominant waveguide mode TE_{10} and also of the higher order modes such as the TE_{20}. Hence the mutual coupling effects have to be related to the fields scattered in and outside the waveguide by the radiators.

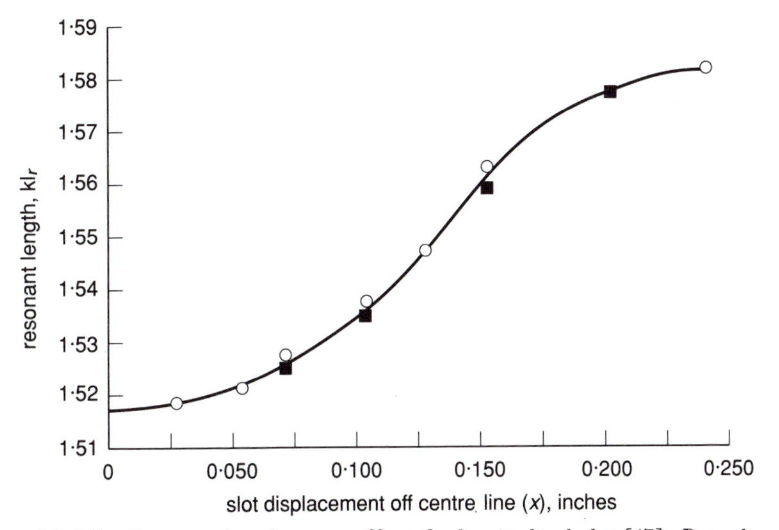

Figure 10.35 *Resonant length versus off-set for longitudinal slot [47]. Reproduced by permission of McGraw-Hill*

Guide width, $a = 0{\cdot}900$ inch
Guide height, $b = 0{\cdot}400$ inch
Wall thickness, $t = 0{\cdot}050$ inch
Slot width, $w = 0{\cdot}0625$ inch

■ experimental points determined by admittance measurements
○ experimental points from radiation pattern measurements

Accordingly, a design procedure that is more general than the earlier procedures and based on the fundamental integral equations that describe the array without prior assumptions about the nature of the slot aperture fields for internal mutual coupling will be discussed. The analysis described below is an abbreviated form of the design described in Reference 23.

The planar array antenna consists of a number of waveguides, say N, stacked in the x-direction as shown in Fig. 10.36. Each waveguide has M longitudinal slots cut in the broad wall at a distance $\lambda_g/2$ apart on their centres. The longitudinal slots are aligned with the z-direction and they are off-centered with respect to the centre line of the broad wall in such a way that alternate slots are on the same side of the centre line. Uniform spacing between the radiating elements in the x- and z-directions is assumed though the inter-element spacings in these two directions may not be identical. For the purposes of analysis it is assumed that each waveguide is excited separately in accordance with the overall excitation scheme chosen for the array. The waveguide dimensions are so selected that the dominant mode propagating down the guide and providing excitation to the slots on the broad wall of the waveguide is the TE_{10} mode with the propagation constant β_{10}. Each of these waveguides is short-circuited at a distance $\lambda_g/4$ beyond the centre of the last slot. The residual energy left in the TE_{10} mode after the last slot is thus reflected back in the negative z-direction to the slot radiators. The short circuit can therefore be

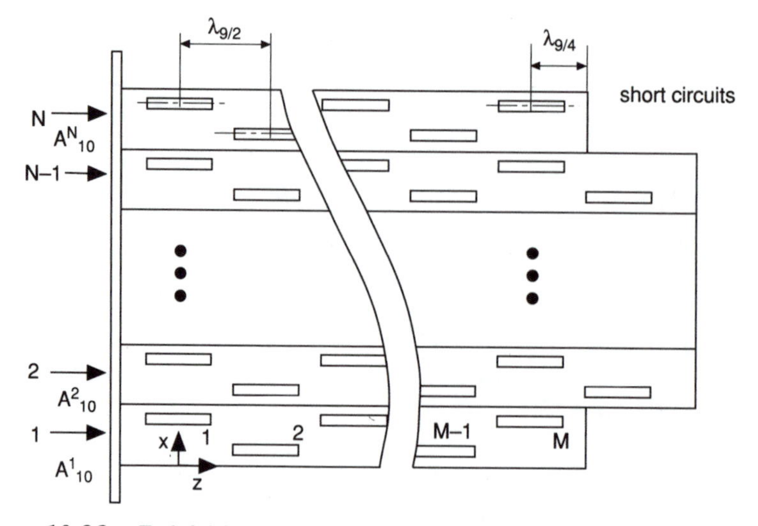

Figure 10.36 *End-fed longitudinal slot array [23]*

accounted for by considering that internally each waveguide is mirrored about the plane defined by the short circuit and this is equivalent to the waveguide being fed asymmetrically from both ends. Each waveguide will now have twice as many ($2M$) slots as the original number and the field in the image slot apertures is the mirror image of the field in the corresponding physically realised slot. The number of unknowns in this case remains unchanged. As far as the radiation half-space is concerned, only the real slot apertures come into play. Since the slots are narrow, the aperture electric field in each slot is entirely transverse, that is, there is only the E_x field component present. This has been borne out by experimental data [21] which show that the maximum longitudinal electric field in the slot is one three-thousandth of the maximum electric field in the transverse direction. The magnetic field in the aperture of the slot will be the longitudinal field H_z. The finite thickness t of the wall of the waveguide is taken into account by considering that each radiator has two apertures interconnected by a short section waveguide. The inner aperture which is looking into the waveguide will be designated as aperture 1 and all fields in the waveguide associated with this aperture will be designated by the subscript 1. Similarly the outer aperture of the slot radiator looking into the radiation half-space outside of the waveguide array will be aperture 2 and the fields associated with this aperture will have the subscript 2. The radiators are distinguished from each other by the co-ordinates of the centre of the slot. For example, the co-ordinates (x_{sm}, z_{sm}) represent the centre of the mth slot radiator on the broad wall of the sth waveguide of the array. Since we have to deal with E_x and H_z fields only, the subscripts x and z are dropped for these fields in this analysis to reduce the number of symbols to be used with the field components.

The analysis is based on computing the H field across the aperture 1 of any radiator, say the mth radiator, on the sth waveguide. In this case the total H field is made up of the impressed field, fields due to scattering by other radiators of the sth guide at the mth slot, the fields scattered by the image slots and self scattered fields by the mth slot itself. Similarly the H field at aperture 2 of the mth slot is the superposition of the field radiated by the slot plus the fields due to all other $\{(M \times N) - 1\}$ slots of the array at the mth slot. The H field of aperture 1 interacts with the H field of aperture 2 through the waveguide modes that exist in the waveguide cavity of length t formed between the two apertures.

10.22.1 *Computation of the fields at aperture* 1

Fig. 10.37 indicates the cross-sectional view of the mth slot in any of the sth branch waveguides. P_m is any point in aperture 1 of the mth slot of the sth waveguide with the local co-ordinates (u_m, v_m) as shown in Fig. 10.38. u_m and v_m are the distances measured from the centre of the aperture with the co-ordinates (x_m, z_m). Thus the tangential magnetic field at aperture 1 of the mth slot in the waveguide can be expressed as

$$H^{tot}_{1m}(P_m) = H^{inc}_{1m}(P_m) + H^{sscat}_{1m}(P_m) + \sum_{\substack{p=1 \\ p \neq m}}^{M} H^{mscat}_{1p}(P_m) + \sum_{p=1}^{M} H^{im}_{1p}(P_m) \tag{10.185}$$

where the superscripts *tot* stands for the total field, *inc* for the impressed field, *sscat* for the self-scattered field, *mscat* for the mutual scattered field and *im* for the image slot fields. The subscript $1m$ denotes aperture 1 of the mth slot in any of the waveguides.

The excitation of the waveguide being TE_{10} mode, and by taking the centre of the first slot in the waveguide as reference, it is possible to express the field at the mth slot of the sth branch guide as

$$H^{inc}_{1m}(P_m) = 2jA^{s}_{10} \cos\left[\frac{\pi}{a}(x_m + u_m)\right] \cos[(\beta_{10}(z_m + v_m)] \tag{10.186}$$

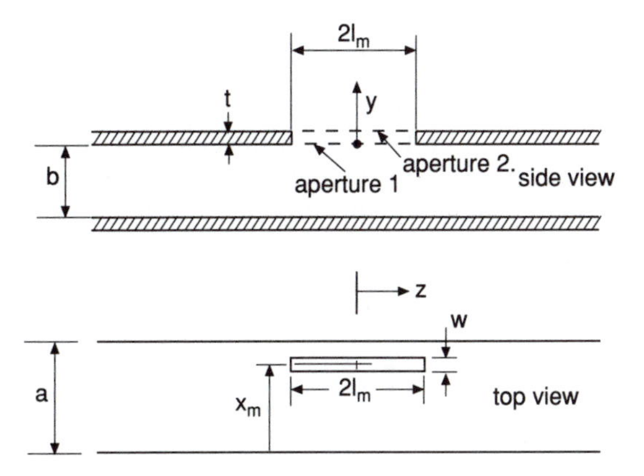

Figure 10.37 *Cross-sectional view of branch guide [23]*

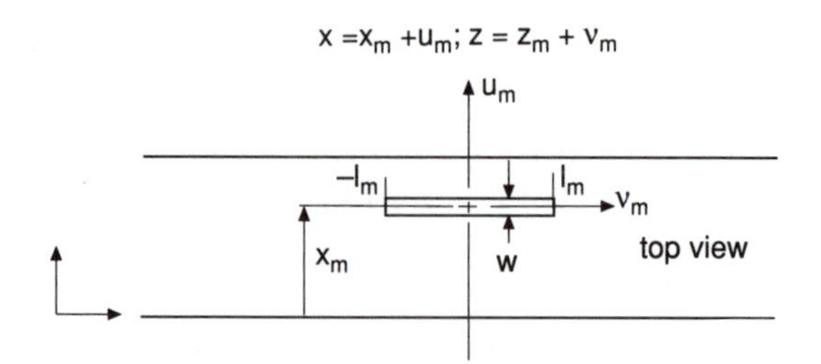

Figure 10.38 *Local co-ordinate system for mth slot [23]*

where A^s_{10} is the complex amplitude of the TE_{10} wave inside the *s*th waveguide, *a* and *b* are the broad wall and narrow wall inner dimensions of the waveguide and β_{10} is the propagation constant which is expressed as, $\beta_{10}=\sqrt{(\pi/a)^2-k^2}$, with, $k=2\pi\sqrt{\varepsilon_r}/\lambda$, and ε_r is the permittivity of the medium.

For computing the scattered fields inside the waveguide, Stevenson's expression [24] for the longitudinal component of the magnetic field anywhere in the guide for an arbitrary transverse electric field $E_x(x, z)$ in the upper broad wall slot aperture is used. It is given by

$$H_z(x, y, z)=\frac{1}{2j\omega\mu_0 ab}\sum_{p=0}^{\infty}\sum_{q=0}^{\infty}\frac{\varepsilon_p\varepsilon_q}{\gamma_{pq}}\cos\left(\frac{p\pi x}{a}\right)\cos\left(\frac{q\pi y}{b}\right)$$

$$\times\iint_{\text{Aperture}} E_x(x', z')\cos\left(\frac{p\pi x'}{a}\right)\left(\frac{\partial}{\partial z'}z+k^2\right)$$

$$\times e^{-\gamma_{pq}|z-z'|}\,dx'\,dz' \qquad (10.187)$$

Here, $\gamma_{pq}=\sqrt{(p\pi/a)^2+(q\pi/b)^2-k^2}$ is the propagation constant of the TE_{pq} mode and (x, y) are measured from the waveguide side wall and lower wall respectively. ε_p and ε_q are Neumann numbers which are equal to 1 for $p=0$ or $q=0$ and 2 for all other values of *p* and/or *q*.

From eqn. 10.187, the expression for the last three terms on the RHS of eqn. 10.185 will be arrived at. For this purpose, the magnetic field is expressed in the local co-ordinates (u'_m, v'_m) and (u_n, v_n) of the slots, the first set of co-ordinates representing the position of the source and the second set representing the observation point. In that case, eqn. 10.187 becomes

$$H^{sscat}_{1sn}(u_n, v_n)=-\frac{1}{2j\omega\mu_0 ab}\sum_{p=0}^{\infty}\sum_{q=0}^{\infty}\frac{\varepsilon_p\varepsilon_q}{\gamma_{pq}}\cos\left[\frac{p\pi}{a}(x_n+u_n)\right]$$

$$\times\iint_{\text{Aperture}} E(u'_n, v'_n)\cos\left[\frac{p\pi}{a}(x_n+u'_n)\right]\left(\frac{\partial^2}{\partial v'^2_n}+k^2\right)e^{-\gamma_{pq}|z_n+v_n-z_n-v'_n|}\,dx'_n\,dz'_n$$

$$(10.188)$$

In the case of self scatter, $x_m = x_n$, $z_m = z_n$, because the source and the observation point are on the same aperture. Care has to be taken to evaluate the integral of eqn. 10.188, as the first derivative of the exponential term is discontinuous as a result of which the second derivative has a singularity. Bearing this in mind, eqn. 10.188 can be expressed for the self scatter term in eqn. 10.185 as

$$H_{1sn}^{sscat}(P_n) = \int_{-l_n}^{l_n} \int_{-w/2}^{w/2} G_1^{sscat}\ (u_n, v_n, u'_n, v'_n) E_1(u'_n, v'_n)\ du'_n\ dv'_n \quad (10.189)$$

where $G_1^{sscat}(\ )$ is the Green's function and is given by

$$G_1^{sscat}(u_n, v_n, u'_n, v'_n) = \frac{1}{2j\omega\mu_0 ab} \sum_{p=0}^{\infty} \sum_{q=0}^{\infty} \frac{\varepsilon_p \varepsilon_q}{\gamma_{pq}} \cos\left[\frac{p\pi}{a}(x_n + u_n)\right] \times \cos\left[\frac{p\pi}{a}(x_n + u'_n)\right]\left(\frac{\partial^2}{\partial v'^2_n} + k^2\right) e^{-\gamma_{pq}|v_n - v'_n|} \quad (10.190)$$

In eqn. 10.190, (u_n, v_n) and (u'_n, v'_n) are on the same slot aperture and $2l_n$ is the length of the nth slot and W is its width.

In the case of mutual scatter, since the slots do not overlap there is no possibility of the source and the observation point coinciding. Therefore, in eqn. 10.188, the differentiation can be carried out before integration. In this case the third term on the RHS of eqn. 10.185 becomes

$$H_{1sn}^{mscat}\ (P_n) = \int_{-l_m}^{l_m} \int_{-w/2}^{w/2} G_1^{mscat}\ (u_n, v_n, u_m, v_m) E_1(u_m, v_m)\ du_m\ dv_m \quad (10.191)$$

where

$$G_1^{mscat}\ (u_n, v_n, u_m, v_m) = \frac{1}{2j\omega\mu_0 ab} \sum_{p=0}^{\infty} \sum_{q=0}^{\infty} \frac{\varepsilon_p \varepsilon_q}{\gamma_{pq}} (k^2 + \gamma_{pq}^2) \cos\left[\frac{p\pi}{a}(x_n + u_n)\right] \times \cos\left[\frac{p\pi}{a}(x_m + u_m)\right] e^{-\gamma_{pq}|z_n + v_n - z_m - v_m|} \quad (10.192)$$

In the case of image slots, the field in any image slot is the mirror image of a corresponding real slot. This is expressed as

$$E_1(x, z)\,|_{image} = E_1(x, -z)\,|_{real} \quad (10.193)$$

Thus,

$$H_{1sn}^{im}(P_n) = -\int_{-l_m}^{l_m} \int_{-w/2}^{w/2} G_1^{im}\ (u_n, v_n, u_m, v_m) E_1(u_m, v_m)\ du_m\ dv_m \quad (10.194)$$

where,

$$G_1^{im}(u_n, v_n, u_m, v_m) = -G_1^{mscat}\ (u_1, v_n, u_m, v_m) \quad (10.195)$$

Thus based on eqns. 10.189, 10.191 and 10.194, eqn. 10.185 can now be written as

$$H_{1sn}^{tot} = H_{1sn}^{inc}(P_n) - \int_{-l_n}^{l_n} \int_{-w/2}^{w/2} G_1^{sscat}(u_n, v_n, u_n', v_n') E_1(u_n', v_n')\, du_n'\, dv_n'$$

$$- \sum_{\substack{m=1 \\ m \neq n}}^{M} \int_{-l_m}^{l_m} \int_{-w/2}^{w/2} G_1^{mscat}(u_1, v_n, u_m, v_m) E_1(u_m, v_m)\, du_m\, dv_m$$

$$- \sum_{m=1}^{M} \int_{-l_m}^{l_m} \int_{-w/2}^{w/2} G_1^{im}(u_n, v_n, u_m, v_m) E_1(u_m, v_m)\, du_m\, dv_m \qquad (10.196)$$

10.22.2 *Computation of the fields at aperture 2*

Aperture 2 of the slot radiator interfaces with the radiation half-space external to the array. The total field at aperture 2 is the superposition of the radiated fields of all the slots in the array. The continuity of the magnetic field at aperture 2 is governed by the equation

$$H_{2sn}^{tot}(Q_n) = H_{2sn}^{rad}(Q_n) + \sum_{s=1}^{N} \sum_{p=1}^{M}{}' H_{2sp}^{ext}(Q_n) \qquad (10.197)$$

where, $H_{2sn}^{rad}(Q_n)$ is the field at the point Q_n on the aperture 2 of the nth slot on the sth waveguide which is responsible for the radiated field of this element and $H_{2sp}^{ext}(Q_n)$ is the field produced at the point Q_n on aperture 2 of the nth radiator on the sth branch guide due to the radiation pattern of the pth slot on the sth waveguide. The summation symbol Σ' in this case includes all the $(M \times N)$ slots of the array except the nth slot on the sth branch guide.

In the case of these slots, the electric field in the aperture is entirely transverse, and hence the expression for the field radiated by any of the slots is given by

$$H_z^{rad}(x, y, z) = \frac{1}{2\pi j\omega\mu_0} \iint_{Aperture} E_x(x', z') \left(\frac{\partial^2}{\partial z'^2} + k_0^2 \right) \frac{e^{-jk_0 R}}{R}\, dx'\, dz' \qquad (10.198)$$

where $R = \sqrt{(x-x')^2 + (y-y')^2 + (z-z')^2}$ is the distance between the source point in the aperture to any arbitrary observation point in the radiation half-space. For the point Q_n on aperture 2 of the nth slot radiator, eqn. 10.198 becomes

$$H_{2sn}^{rad} = \int_{-l_n}^{l_n} \int_{-w/2}^{w/2} G_2^{rad}(u_n, v_n, u_n', v_n') E_2(u_n', v_n')\, du_n'\, dv_n' \qquad (10.199)$$

with,

$$G_2^{rad} = \frac{1}{2\pi j\omega\mu_0} \left(\frac{\partial^2}{\partial v_n'^2} + k_0^2 \right) \frac{e^{-jk_0 R_1}}{R_1} \qquad (10.200)$$

where $R_1=\sqrt{(u_n-u'_n)^2+(v_n-v'_n)^2}$. For the point Q_n on the aperture, the tangential field due to the mth slot is given by

$$H_{2sn}^{ext}(Q_n)=\int_{-l_m}^{l_m}\int_{-w/2}^{w/2} G_2^{ext}(u_n, v_n, u_m, v_m)E_2(u_m, v_m)\, du_m\, dv_m \qquad (10.201)$$

with

$$G_2^{ext}(u_n, v_n, u_m, v_m)=\frac{1}{2\pi j\omega\mu_0}\left(\frac{\partial^2}{\partial v'^2_n}+k_0^2\right)\frac{e^{-jk_0R_2}}{R_2}$$

and, (10.202)

$$R_2=\sqrt{(x_n+u_n-x_m-u_m)^2+(z_n+v_n-z_m-v_m)^2}.$$

Thus, based on eqns 10.199 and 10.201, eqn. 10.197 can be written as

$$H_2^{tot}(Q_n)=\int_{-l_n}^{l_n}\int_{-w/2}^{w/2} G_2^{rad}(u_n, v_n, u'_n, v'_n)E_2(u'_n, v'_n)\, du'_n\, dv'_n$$
$$+\sum_{s=1}^{N}\sum_{m=1}^{N}{}' \int_{-l_n}^{l_n}\int_{-w/2}^{w/2} G_2^{ext}(u_n, v_n, u_m, v_m)E_2(u_m, v_m)\, du_m\, dv_m \qquad (10.203)$$

10.22.3 *Computation of waveguide cavity fields*

The total fields at apertures 1 and 2 of the nth slot on the sth waveguide are linked to each other by the fields that propagate in the short section waveguide of length t (thickness of waveguide walls) between the two apertures. This waveguide has a width $2l_n$ and height w and only the TE_{mn} modes propagate in this waveguide section. The magnetic field at aperture 1 will be the linear superposition of the fields created by the electric fields at apertures 1 and 2, and in a similar fashion the magnetic field at aperture 2 is the linear superposition of the magnetic fields due to electric fields at apertures 1 and 2. Thus

$$H_1^{slot}(P_n)=\int_{-l_n}^{l_n}\int_{-w/2}^{w/2} G_1^{slot}(u_n, v_n, u'_n, v'_n)E_1(u'_n, v'_n)\, du'_n\, dv'_n$$
$$+\int_{-l_n}^{l_n}\int_{-w/2}^{w/2} G_2^{slot}(u_n, v_n, u'_n, v'_n)E_2(u'_n, v'_n)\, du'_n\, dv'_n \qquad (10.204)$$

$$H_2^{slot}(Q_n)=-\int_{-l_n}^{l_n}\int_{-w/2}^{w/2} G_2^{slot}(u_n, v_n, u'_n, v'_n)E_1(u'_n, v'_n)\, du'_n\, dv'_n$$
$$-\int_{-l_n}^{l_n}\int_{-w/2}^{w/2} G_1^{slot}(u_n, v_n, u'_n, v'_n)E_2(u'_n, v'_n)\, du'_n\, dv'_n \qquad (10.205)$$

where (u_n, v_n) refer to the observation point and (u'_n, v'_n) to the source co-ordinates on either aperture 1 or aperture 2 of the nth slot. G_1^{slot} and G_2^{slot} are given by

$$G_1^{slot}(u_n, v_n, u'_n, v'_n) = \frac{1}{j\omega\mu_0 wl} \sum_{p=1}^{\infty} \sum_{q=1}^{\infty} \varepsilon_q \frac{k_0^2 - (p\pi/2l)^2}{\gamma_{pq}^{slot}[\tanh(\gamma_{pq}^{slot} t)]} \sin\left[\frac{p\pi}{2l}(v+l)\right]$$
$$\times \cos\left[\frac{q\pi}{w}(u+w/2)\right] \sin\left[\frac{p\pi}{2l}(v'+l)\right] \cos\left[\frac{q\pi}{2w}(u'+w/2)\right] \quad (10.206)$$

and,

$$G_2^{slot}(u_n, v_n, u'_n, v'_n) = \frac{1}{j\omega\mu_0 lw} \sum_{p=1}^{\infty} \sum_{q=1}^{\infty} \varepsilon_q \frac{k_0^2 - (p\pi/2l)^2}{\gamma_{pq}^{slot}[\sinh(\gamma_{pq}^{slot} t)]} \sin\left[\frac{p\pi}{2l}(v+l)\right]$$
$$\times \cos\left[\frac{q\pi}{w}(u+w/2)\right] \sin\left[\frac{p\pi}{2l}(v'+l)\right] \cos\left[\frac{q\pi}{w}(u'+w/2)\right] \quad (10.207)$$

Introducing eqns. 10.204 and 10.205 in eqns. 10.196 and 10.203, the following set of coupled integral equations for the nth slot on the sth waveguide are obtained:

$$H_{1sn}^{inc}(P_n) = \iint_{S_n} [G_1^{slot}(P_n, P'_n) + G_1^{sscat}(P_n, P'_n) + G_1^{im}(P_n, P'_n)] E_1(P'_n)\, dS'_n$$
$$+ \sum_{\substack{m=1 \\ m \neq n}}^{M}{}' \iint_{S_m} [G_1^{mscat}(P_n, P_m) + G_1^{im}(P_n, P_m)] E_1(P_m)\, dS_m$$
$$+ \iint_{S_n} G_2^{slot}(P_n, P'_n) E_2(P'_n)\, dS'_n \quad (10.208)$$

and

$$0 = \iint_{S_n} G_2^{slot}(P_n, P'_n) E_1(P'_n)\, dS'_n$$
$$+ \iint_{S_n} [G_1^{slot}(P_n, P'_n) + G_2^{rad}(P_n, P'_n)] E_2(P'_n)\, dS'_n$$
$$+ \sum_{S=1}^{N} \sum_{\substack{m=1 \\ m \neq n}}^{M}{}' \iint_{S_m} G_2^{ext}(P_n, P_m) E_2(P_m)\, dS_m \quad (10.209)$$

Here P'_n is the point (u'_n, v'_n) on the nth slot, P_m is the point (u_m, v_m) on the mth slot, S_m and S_n are the surface areas of the apertures of the mth and nth slot, respectively. Eqns. 10.208 and 10.209 have to be satisfied simultaneously for every slot in the array, that is, for $s=1$ to N and $m=1$ to M.

10.22.4 *Matricisation of equations*

In eqns. 10.208 and 10.209, the electric fields across apertures 1 and 2 are the unknowns which have to be determined. The method of moments converts the integral equations into a set of algebraic equations; that is to a matrix equation which can be solved by standard matrix inversion algorithms. The unknown aperture fields are expressed in terms of known but unspecified expansion or basis functions appropriately weighted by unknown weight coefficients as follows:

$$E_1(P_n) = \sum_{j=1}^{k_n} a_n(j) f_n(j, P_n) \tag{10.210}$$

and,

$$E_2(P_n) = \sum_{j=1}^{k_n} b_n(j) f_n(j, P_n) \tag{10.211}$$

where a_n and b_n are the unknown coefficients and f_ns are simple and well known algebraic or trigonometric functions. The basis functions used in this case for apertures 1 and 2 are the same and the number of terms for summation is also kept the same. This is not a restriction, however, for solving the equations. Eqns. 10.208 and 10.209 can be treated as operator equations and are subjected to Galerkin's procedure with the expansion as well as weighting functions given by $f_n(j, P_n)$. Using eqns. 10.210 and 10.211 in eqns. 10.208 and 10.209 and based on Galerkin's procedure, we obtain

$$\iint_{S_n} f_n(j, P_n) H_{1sn}^{inc}(P_n)$$

$$= \sum_{j=1}^{k_n} a_n(j) \iint_{S_n} f_n(j, P_n) \left[\iint_{S_n} G_1^{self}(P_n, P'_n) f_n(j, P'_n)\, dS'_n \right] dS_n$$

$$+ \sum_{m=1}^{M} \sum_{j=1}^{k_n} a_m(j) \iint_{S_n} f_n(j, P_n) \left[\iint_{S_m} G_1^{mut}(P_n, P_m) f_m(j, P_m)\, dS_m \right] dS_n$$

$$+ \sum_{j=1}^{k_n} b_n(j) \iint_{S_n} f_n(j, P_n) \left[\iint_{S_n} G_2^{slot}(P_n, P'_n) f_n(j, P'_n)\, dS'_n \right] dS_n \tag{10.212}$$

and

$$0=\sum_{j=1}^{k_n} a_n(j)\iint_{S_n} f_n(j,P_n)\left[\iint_{S_n} G_2^{slot}(P_n,P_n')f_n(j,P_n')\,dS_n'\right]dS_n$$
$$+\sum_{j=1}^{k_n} b_n(j)\iint_{S_n} f_n(j,P_n)\left[\iint_{S_n} G_2^{self}(P_n,P_n')f_n(j,P_n')\,dS_n'\right]dS_n$$
$$+\sum_{m=1}^{M}{}'\sum_{j=1}^{k_n} b_n(j)\iint_{S_n} f_n(j,P_n)\left[\iint_{S_m} G_2^{ext}(P_n,P_m)f_n(j,P_m)\,dS_m\right]dS_n \quad (10.213)$$

with

$$G_1^{self}(P_n,P_n')=G_1^{slot}(P_n,P_n')+G_1^{sscat}(P_n,P_n')+G_1^{im}(P_n,P_n') \quad (10.214)$$
$$G_1^{mut}(P_n,P_m)=G_1^{mscat}(P_n,P_m)+G_1^{im}(P_n,P_m) \quad (10.215)$$
$$G_2^{self}(P_n,P_n')=G_1^{slot}(P_n,P_n')+G_2^{rad}(P_n,P_n') \quad (10.216)$$

Eqns. 10.212 and 10.213 are a system of equations that can be solved for the coefficients $a_n(j,P_n)$ and $b_n(j,P_n)$ for a specified basis function $f_n(j,P_n)$ and the number k_n of these functions. Once these coefficients are determined, with the aid of eqns. 10.210 and 10.211 the electric fields across the two apertures for each slot in the $(M\times N)$ array will be known. The common form for the matrix equation in mutual coupling analysis is to cast it in the form

$$[Y]\,[B]=[F] \quad (10.217)$$

where $[Y]$ is a $2(M\times N)$ by $2(M\times N)$ admittance matrix, $[B]$ is a $2N$ column matrix of the weighting coefficients of the electric field across the apertures and $[F]$ is a $2N$ column matrix of the impressed fields. Eqn. 10.217 can be expanded to

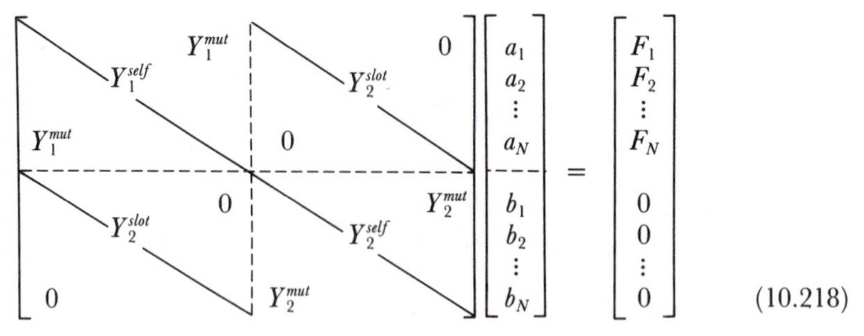

(10.218)

Each quadrant of the $[Y]$ matrix is partitioned into $(M\times N)$ by $(M\times N)$ submatrices. The order of each submatrix is dependent on the number of basis functions necessary for an accurate estimation of the electric field of the slot aperture. It can be easily deduced that a generic term in these quadrant admittance matrices is given by

$$Y_{b,\,mn}^{(i,\,j)}=\int_{-l_m}^{l_m}\int_{-w/2}^{w/2} f_m(i,\,u_m,\,v_m)$$

$$\times\left[\int_{-l_n}^{l_n}\int_{-w/2}^{w/2} G_b(u_n,\,v_n,\,u_m,\,v_m)f(j,\,u_n,\,v_n)\,du_n\,dv_n\right]du_m\,dv_m \qquad (10.219)$$

where, the subscript $b=1$ or 2, and G stands for any of the Green's functions defined in eqns. 10.202, 10.207, 10.214–10.216.

10.22.5 *Computational aspects*

The transverse electric field in any resonant slot is not constant either in the longitudinal or in the transverse direction. The field variations in these two directions are independent, and hence the basis function $f_n(i,\,P_n)$ can be written as the product of the transverse and longitudinal basis functions, so that

$$\sum_{i=1}^{k_n} a(i)f_n(i,\,P_n)=f_n^{tr}(u_n)\sum_{i=1}^{k_n} a(i)f_n^{long}(i,\,v_n) \qquad (10.220)$$

where

$$f^{tr}(u_n)=\frac{w/2}{\sqrt{(w/2)^2-u_n^2}}, \qquad -w/2<u_n<w/2 \qquad (10.221)$$

and,

$$f_n^{long}(i,\,v_n)=\begin{cases}\sin k_0(v_{n,\,i+1}-v_{n,\,i})/\sin k_0\Delta_n; & v_{n,\,i}\leqslant v_n\leqslant v_{n,\,i+1}\\ \sin k_0(v_n-v_{n,\,i-1})/\sin k_0\Delta_n; & v_{n,\,i-1}\leqslant v_n\leqslant v_{n,\,i}\\ 0; & \text{elsewhere}\end{cases} \qquad (10.222)$$

with $\Delta_n=$ subdomain extent $=v_{n,\,i+1}-v_{n,\,i}=v_{n,\,i}-v_{n,\,i-1}$ and $v_{n,\,i}=-l_n+i\Delta_n$; $(i=0,\,1,\,2,\,\ldots,\,k_n)$.

It is to be noted that, though $f_n^{tr}(u_n)$, the transverse basis function chosen in eqn. 10.221 is an accurate description of the field variation in a slot cut in a waveguide of zero wall thickness, it is appropriate for all thicknesses typically encountered in practice owing to the fact that it automatically satisfies the condition that the field should vary as $r^{-1/2}$, where r is the distance from the slot edge. Eqn. 10.222 indicates that the longitudinal basis functions are sinusoidal and are subdomain types. It has been suggested [23] for ease of computation that, even though the slots may vary in length on any guide, the number of subdomain functions be kept the same irrespective of the slot length, viz. K_0. This would mean that the extent of a subdomain from slot to slot can vary. These basis functions are point matched at the centre line of the slot in the x-direction since this is the usual line at which admittances are measured or computed. In that case one of the co-ordinates of the point of observation, viz. u_n, decreases to zero so that the Green's function $G_b(u_n,\,v_n,\,u_m,\,v_m)$ now becomes $G(0,\,v_n,\,u_m,\,v_m)$ in eqn. 10.219 and in subsequent computations. It is now proposed to discuss the computational aspects of the admittance matrix $[Y]$. Firstly, each element of the matrix is a submatrix of order $(K_0 \times K_0)$. Secondly, the $[Y]$ matrix is of order $2(M \times N)$ by $2(M \times N)$ and can be considered as having four quadrant matrices of order $(M \times N)$ by $(M \times N)$. Thus, at first sight, the computational effort needed appears to be enormous. Therefore there is a need to examine each quadrant matrix from the point of view of ease of computation. We start with the left upper quadrant submatrix $[Y_1^{self}]$, the upper triangular and lower triangular submatrix $[Y_1^{mut}]$. Each of these needs to be analysed and it can be seen from eqn. 10.214 that the diagonal submatrix $[Y_1^{self}]$ is the sum of three other submatrices. Thus

$$[Y_1^{self}] = [Y_1^{im}] + [Y_1^{sscat}] + [Y_1^{slot}] \tag{10.223}$$

The elements of all the submatrices are integrals of the type indicated in eqn. 10.219. In the case of $[Y_1^{im}]$, by virtue of eqn. 10.195 and by virtue of the fact that the source and the observer do not overlap, the integral of eqn. 10.219 can be solved in a straightforward manner. However, in the case of $[Y_1^{sscat}]$ each of its elements has, by virtue of eqns. 10.189, 10.190 and 10.210, an integral of the form

$$\int_{-l}^{l} f_n(j, v_n') \left(\frac{\partial^2}{\partial v_n'^2} + k_0^2 \right) e^{-\gamma_{mn}|v_n - v_n'|}$$

which has a singularity for the second derivative of the exponential at $v_n = v_n'$. This has been evaluated [23] near the singularity to give

$$\int_{-l}^{l} f_n(j, v_n') \left(\frac{\partial^2}{\partial v_n'^2} + k_0^2 \right) e^{-\gamma_{mn}|v_n - v_n'|}\, dv_n'$$

$$= -2\gamma_{mn} f(j, v_n) + (k_0^2 + \gamma_{mn}^2) \int_{-l}^{l} f(j, v_n')\, e^{-\gamma|v_n - v_n'|}\, dv_n' \tag{10.224}$$

After substituting eqn. 10.224 into eqn. 10.219 and performing the integration, the final result provides us with the expression for the element $Y_1^{sscat}(i, j)$ in the form of a double infinite series. The double summation can be performed on a computer, but the summation has been found to be unacceptably slow especially for $i = j$ and $|i - j| = 1$. However, since the value of $y_1^{sscat}(i, j)$ is dependent only on the absolute distance $|i - j|$, and as subdomain cells of the same length have been used, the matrix $[Y_1^{sscat}]$ is Toeplitz-symmetric. Thus, only one row or column need be computed.

The last submatrix of eqn. 10.223 that has to be examined is the matrix $[Y_1^{slot}]$. It arises owing to the 'coupling' between the internal (aperture 1) and the external aperture (aperture 2) through the cavity formed between them. Hence the two submatrices, namely $[Y_1^{slot}]$ and $[Y_2^{slot}]$, will be evaluated together because the same procedure applies to both. In the expression for the element of the matrix $[Y_2^{slot}]$, the subscripts of the subdomain length Δ and the slot length $2l$ have been deleted so that these representations are valid for any slot in the array. The element of this matrix is given by

$$y_1^{slot}(i, j) = \int_{-l}^{l} \int_{-l}^{l} \int_{-w/2}^{w/2} G_1^{slot}(0, v_n, u'_n, v'_n) f^{tr}(u'_n) f^{long}(i, v_n) f^{long}(j, v'_n)\, du'_n\, dv'_n\, dv_n \tag{10.225}$$

The integration can be easily carried out to yield,

$$y_1^{slot} = \frac{8\pi k_0^2}{j\omega\mu_0 l \sin^2(k_0\Delta)} \sum_{p=1}^{\infty} \sum_{q=0}^{\infty} \frac{\varepsilon_q}{\gamma_{pq}^{slot} \sinh(\gamma_{pq}^{slot}\, t)} \cos^2(q\pi/2)$$
$$\times \frac{1}{k_0^2 - (p\pi/2l)} J_0(q\pi/2) \sin\left[\frac{p\pi}{2l}(v_i + l)\right] \sin\left[\frac{p\pi}{2l}(v_j + l)\right]$$
$$\times \sin^2\left[\frac{\Delta}{2}(k_0 + p\pi/2l)\right] \sin^2\left[\frac{\Delta}{2}\left(k_0 - \frac{p\pi}{2l}\right)\right] \tag{10.226}$$

The integration is similar to that carried out on eqn. 10.223. Since the source and the observer do not overlap, in this case also, the computation of the elements of the matrix $[Y_1^{mut}]$ is straightforward. In the above equation, the product term can be written as

$$\sin\left[\frac{p\pi}{2l}(v_i + l)\right] \sin\left[\frac{p\pi}{2l}(v_j + l)\right] = \frac{1}{2}\left\{\cos\frac{p\pi}{2l}(v_i - v_j) - \cos\frac{p\pi}{2l}(v_i + v_j)\right\} \tag{10.227}$$

Thus the matrices $[Y_1^{slot}]$ or $[Y_2^{slot}]$ can be considered as the sum of two submatrices. The first one is dependent on the absolute distance $|i-j|$ and is Toeplitz symmetric. The second one is dependent on $|i+j|$ and is reverse Toeplitz symmetric. In both cases only one row or column need be computed. We now examine the upper and lower triangular matrices $[Y_1^{mut}]$ to complete the analysis on the computability of the left upper quadrant matrix of the admittance matrix $[Y]$. The element of the submatrix $[Y_1^{mut}]$ is the sum of the integral equation 10.219 with the Green's function G_b ( ) being obtained from eqns. 10.192 and 10.195. The integration of each of these integrals is similar to the one carried out for the elements of $[Y_1^{im}]$ of eqn. 10.223. Since the source and the observation point do not overlap in this case also, computation of the upper and lower triangular matrix $[Y_1^{mut}]$ is possible.

In the upper right and lower left quadrant matrices of eqn. 10.218, only the diagonal matrix $[Y_2^{slot}]$ is to be evaluated. Since this matrix is similar to the matrix $[Y_1^{slot}]$, the procedure outlined in the previous paragraph can be followed in this case also. Thus, we are now left with the examination of the lower right quadrant of admittance matrix $[Y]$ for computability. The lower right quadrant matrix is made up of three submatrices, namely, the diagonal submatrix $[Y_2^{self}]$, the upper and lower triangular submatrices $[Y_2^{mut}]$. The diagonal submatrix is itself made up of two other submatrices $[Y_1^{slot}]$ and $[Y_2^{rad}]$. The elements of the submatrix $[Y_2^{rad}]$ is obtained from eqns. 10.219 and 10.200. The Green's function of eqn. 10.200 covers the specific case of the source and observation point being on the same aperture, whereas eqn. 10.202 is more general as it represents the case where the observation point and the source are on different apertures. It is also the Green's function for the matrix $[Y_2^{mut}]$. Therefore, we shall examine both the matrices for their computability. The elements of matrices $[Y_2^{mut}]$ and $[Y_2^{rad}]$ are given by

$$y_{2,\,mn}^{mut}(i, j) = \int_{-l_m}^{l_m} \int_{-l_n}^{l_n} \int_{-w/2}^{w/2}$$

$$\times G_2^{ext}(0, v_n, u_m, v_m) f_n^{long}(i, v_n) f_n^{long}(j, v_m) f^{tr}(u_m)\, du_m\, dv_m\, dv_n \quad (10.228)$$

$$y_{2,\,mn}^{rad}(i, j) = \int_{-l_n}^{l_n} \int_{-l_n}^{l_n} \int_{-w/2}^{w/2}$$

$$\times G_2^{rad}(0, v_n, u'_m, v'_m) f_n^{long}(i, v_n) f_n^{long}(j, v'_n) f^{tr}(u'_n)\, du'_n\, dv'_n\, dv_n \quad (10.229)$$

We shall examine eqn. 10.228 first since it is the general case. It is a triple integral with v_n, u_m and v_m as the variables for integration. Since the Green's function $G_2^{ext}(0, v_n, u_m, v_m)$ as expressed in eqn. 10.202 contains the factor $\left(\frac{\partial^2}{\partial v_n'^2} + k_0^2\right)$, the first integration of the triple integral will be with respect to the variable v_n [25]. This leads us to the expression

$$y_{2,\,mn}^{mut}(i, j) = \frac{k_0}{2\pi j\omega\mu_0 \sin(k_0\Delta_n)} \int_{-w/2}^{w/2} \int_{v_{m,\,i-1}}^{v_{m,\,i+1}} f_m^{long}(i, v_m) f^{tr}(u_m)$$

$$\times \left[\frac{e^{-jk_0R_{j+1}}}{R_{j+1}} + \frac{e^{-jk_0R_{j-1}}}{R_{j-1}} - 2\cos k_0\Delta_n \frac{e^{-jk_0R_j}}{R_j}\right] du_m\, dv_m \quad (10.230)$$

with

$$R_j = \sqrt{(x_m - x_n - u_m)^2 + (z_m + v_m - z_n - v_{n,\,j})^2} \quad (10.231)$$

$v_{n,j}$ in the above equation represents the local co-ordinate of the jth cell or subdomain of the nth slot aperture. R_{j-1} and R_{j+1} can be expressed in terms of $v_{n,j-1}$ and $v_{n,j+1}$, respectively, in the same manner. By virtue of the fact that $f^{tr}(u_m)$ as represented by eqn. 10.221 gives rise to singularities at the slot edges, a co-ordinate transformation is effected by putting $(w/2)\sin\theta = u_m$. Eqn. 10.230 is then transformed to

$$y_{2,\,mn}^{mut}(i, j) = \frac{k_0 w}{4\pi j\omega\mu_0 \sin(k_0\Delta_n)\sin(k_0\Delta_m)} \left[\int_{-\pi/2}^{\pi/2} \int_{v_{m,\,i-1}}^{v_m} \sin\{k_0(v_m - v_{m,\,i-1})\}\right.$$

$$\times \left\{\frac{e^{-jk_0R_{j+1}}}{R_{j+1}} + \frac{e^{-jk_0R_{j-1}}}{R_{j-1}} - 2\cos k_0\Delta_m \frac{e^{-jk_0R_j}}{R_j}\right\} dv_m\, d\theta$$

$$+ \int_{-\pi/2}^{\pi/2} \int_{v_{m,\,i}}^{v_{m,\,i+1}} \sin k_0\{v_{m,\,i+1} - v_m\}$$

$$\left.\times \left\{\frac{e^{-jk_0R_{j+1}}}{R_{j+1}} + \frac{e^{-jk_0R_{j-1}}}{R_{j-1}} - 2\cos k_0\Delta_m \frac{e^{-jk_0R_j}}{R_j}\right\} dv_m\, d\theta\right] \quad (10.232)$$

Eqn. 10.232 has no singularities and is therefore computable.

As far as eqn. 10.229 is concerned we follow the same procedure as in the case of eqn. 10.228 and arrive at the following equation:

$$y_{nn}^{rad}=\frac{k_0 w}{2\pi j\omega\mu_0 \sin^2(k_0\Delta_n)}\left\{\int_0^{\pi/2}\int_{v_{n,i-1}}^{v_{n,i}} \sin k_0(v_n - v_{n,i-1})\right.$$

$$\times\left[\frac{e^{-jk_0R_{j+1}}}{R_{j+1}}+\frac{e^{-jk_0R_{j-1}}}{R_{j-1}}-2\cos k_0\Delta_n\frac{e^{-jkR_j}}{R_j}\right] dv_n\, d\theta$$

$$+\int_{-\pi/2}^{\pi/2}\int_{v_{n,i}}^{v_{n,i+1}} \sin k_0(v_{n,i+1}-v_n)$$

$$\left.\times\left[\frac{e^{-jk_0R_{j+1}}}{R_{j+1}}+\frac{e^{-jk_0R_{j-1}}}{R_{j-1}}-2\cos k_0\Delta_n\frac{e^{-jk_0R_j}}{R_j}\right] dv_n\, d\theta\right\} \tag{10.233}$$

with,

$$R_j=\sqrt{(w/2)^2\sin^2\theta+(v_n-v_{n,j})^2} \tag{10.234}$$

Similarly, expressions for R_{j-1} and R_{j+1} can be obtained by replacing $v_{n,j}$ in eqn. 10.234 by the corresponding term. In eqn. 10.233, the integrand has a singularity whenever the source and the observer cells or subdomains on the slot aperture coincide. In that case $i=j$ and R_j goes to zero. In a similar fashion, for the case $|i-j|=1$, singularities are generated in the integrand because R_{j+1} and R_{j-1} go to zero. All the terms involving R_{j-1}, R_j and R_{j+1} are similar in form and therefore only one of them need be assessed for computability. Thus the two integrals with R_j in the denominator of the integrand are

$$I_{1a}=\int_0^{\pi/2}\int_{v_{n,i-1}}^{v_{n,i}} \sin k_0(v_n-v_{n,i-1})\frac{e^{-jk_0R_j}}{R_j}\,dv_n\,d\theta$$

$$I_{1b}=\int_0^{\pi/2}\int_{v_{n,i}}^{v_{n,i+1}} \sin k_0(v_{n,i+1}-v_n)\frac{e^{-jk_0R_j}}{R_j}\,dv_n\,d\theta \tag{10.235}$$

To remove the singularity, the variables in eqn. 10.235 are transformed first, by substituting $v_n - v_{n,i}=-wy/2$, and $v_{n,i+1}-v_n=wy/2$. The resulting two equations are then combined to give [23]

$$I_1=I_{1a}+I_{1b}=2\int_0^{\pi/2}\int_0^{2\Delta n/w} \sin k_0\left(\Delta_n-\frac{w}{2}y\right)\frac{e^{-jk_0w\sqrt{\sin^2\theta+y^2/2}}}{\sqrt{\sin^2\theta+y^2/2}}\,dy\,d\theta \tag{10.236}$$

Eqn. 10.236 can be written as the sum of three specific integrals as follows:

$$I_1=\int_0^{\varepsilon}\int_0^{\sqrt{\varepsilon^2-\theta^2}} D(y,\theta)\,dy\,d\theta+\int_0^{\varepsilon}\int_{\sqrt{\theta^2-w^2}}^{2\Delta n/w} D(y,\theta)\,dy\,d\theta$$

$$+\int_{\varepsilon}^{\pi/2}\int_0^{2\Delta n/w} D(y,\theta)\,dy\,d\theta \tag{10.237}$$

and

$$D(y, \theta) = \sin k_0\left(\Delta_n - \frac{w}{2}y\right) \frac{e^{-jk_0 w\sqrt{\sin^2\theta + y^2/2}}}{\sqrt{\sin^2\theta + y^2/2}} \tag{10.238}$$

In eqn. 10.237, only the first integral contains the singularity. Therefore only this integral need be assessed further for computability. Since the singularity occurs at the lower end point (0, 0) of the domain of integration, a circle of small radius is drawn around it. For this purpose another transformation of the variables is carried out by substituting $r \cos \phi = \sin \theta$, and $r \sin \phi = y$, r being the radius of the circle pertaining to the quadrant containing the singularity. The projection of the radius along the y-axis is taken to be y, where ε is small. For such small values, $r \cos \phi = \sqrt{\varepsilon^2 - \sin^2 \theta}$, so that the projection of the radius on the θ-axis would be $\sqrt{\varepsilon^2 - \theta^2}$. With this transformation, the first integral of eqn. 10.237 becomes

$$2\int_0^{\pi/2}\int_0^{\varepsilon} \sin\left[k_0\left(\Delta_n - \frac{w}{2} r \cos\phi\right)\right] e^{-jk_0 rw/2}\, dr\, d\phi$$

It is now possible to compute the integral in eqn. 10.237 so that the elements of the matrix $[Y_2^{rad}]$ are determined for all three specific cases, $i=j$, $i-j=-1$ and $i-j=1$. The elements of the admittance matrix $[Y]$ can therefore be computed.

In eqn. (10.218) we now turn our attention to the fields due to the forcing function represented by the matrix $[F]$. The elements of this matrix are given by

$$F_m(i) = \int_{-l_m}^{l_m} H_z^{inc}(0, v_m) f_m(i, v_m)\, dv_m \tag{10.239}$$

The impressed field at the nth slot radiator on the sth waveguide is H_z^{inc} and it is obtained from eqn. 10.186. Making the appropriate substitution and carrying out the integration we obtain

$$F_m(i) = -4jA_{10}^s \frac{k_0}{k_0^2 - \beta_{10}^2} \frac{\cos k_0\Delta_m - \cos\beta_{10}\Delta_m}{\sin k_0\Delta_m} \cos\left(\frac{\pi x_m}{a}\right) \cos[\beta_{10}(x + v_{m,i})] \tag{10.240}$$

The resulting system of eqn. 10.218 can now be solved to obtain the unknown coefficients $a(j)$ and $b(j)$. These are later substituted in eqns. 10.210 and 10.211 to obtain the aperture electric fields.

The aforesaid analysis has brought out that, for a given slot length and offset from the centre line of the broad wall of the guide, and for a given wall guide thickness, the field across each slot of the guide can be determined for a given impressed field. In actual arrays, each of the N branch waveguides housing M slots is fed by means of resonant coupling slots on the main feed guide as shown in Fig. 10.33. In the main feed guide, the impressed source propagates the dominant TE_{10} mode to excite the resonant coupling slots. In a recent study [25], it has been shown that it is necessary to include the coupling due to the TE_{20} mode in computing the field at the resonant aperture of the feed guide. Depending on the offset, the inclination and the length of the resonant slot, the power available in each branch guide is determined. For determining the actual field across each slot aperture of the branch guide, the input reflection coefficient has to be computed. The backscattered fields from each of the M slots towards the source are to be considered and of these only the dominant mode will contribute as the higher order modes are attenuated before they reach the coupling slot of the feed. The expression for this backscattered field is obtained from eqn. 10.187 with $p=1$ and $q=0$. The contribution of all the slots including the image slots is now summed and referred to the co-ordinates of the coupling slot to obtain the energy reflected back. The ratio of the total field (which is the sum of the impressed and reflected fields) to the reflected field at the coupling slot is the input reflection coefficient. The final expression for the input reflection coefficient is obtained as

$$\Gamma_1 = 1 + \frac{2j(\pi/a)^2 k_0 w \psi(1)}{A^s_{10}\omega\mu_0\beta_{10}ab(k_0^2-\beta_{10}^2)} \sum_{p=1}^{M} (-1)^p \cos\left(\frac{\pi x_p}{a}\right) \frac{\cos\beta_{10}\Delta_p - \cos k_0\Delta_p}{\sin k_0\,\Delta_p}$$

$$\times \sum_{j=1}^{K_0} a_p(j)\cos\beta_{10}v_{p,j} \qquad (10.241)$$

The symbol $\psi(x)$ is defined later in eqn. 10.245.

The far field electric vector potential of the array can now be obtained as the superposition of the far field vector potential of each of the slots of the array. The earlier assumption of a pure transverse electric field across the slot enables us to write,

$$F_\theta(\theta,\phi) = -\frac{4\sin\theta}{\mu_0}\sum_{m=1}^{MN}\int_{-l_m}^{l_m}\int_{-w/2}^{w/2} E_{2m}(u'_m, v'_m)e^{jk_0 d_m}\,du'_m\,dv'_m \quad (10.242)$$

with,

$$d_m = (x_m+u_m)\sin\theta\cos\phi + (z_m+v_m)\cos\theta \qquad (10.243)$$

The expression for the electric field E_{2m} is available from the eqn. 10.211. From eqn. 10.242 it is quite clear that it is not possible to express the electric vector potential of the array as the product of an array factor and the element electric vector potential. The design becomes computationally intensive and a simplification has been suggested [23]. For this purpose, the electric vector potential of the array is obtained after carrying out the integration in eqn. 10.242:

$$F_\theta(\theta,\phi) = -\frac{8\psi(\theta,\phi)}{\mu_0 k_0\sin\theta}\sum_{m=1}^{MN} e^{jk_0[x_m\sin\theta\cos\phi + z_m\cos\theta]}$$

$$\times\left[1-\cos\theta\,\frac{\sin(k_0\Delta_m\cos\theta)}{\sin k_0\Delta_m}\right]\sum_{j=1}^{K_0} b_m(j)\,e^{jk_0 v_{m,j}\cos\theta} \quad (10.244)$$

with,

$$\psi(\theta,\phi) = \int_{-w/2}^{w/2} f^{tr}(u_m)\,e^{jk_0 u_m\sin\theta\cos\phi}\,du_m \qquad (10.245)$$

For typical cell or subdomain and slot lengths, $k_0\Delta_m \sim \dfrac{\pi}{K_0+1} \ll 1$, so that

$$\frac{\sin(k_0\Delta_m \cos\theta)}{\sin(k_0\Delta_m)} \cong \cos\theta \tag{10.246}$$

Considering the fact that the fine structure of the array is derived from the first exponential factor in eqn. 10.244, it is suggested that little error in radiation pattern is likely to occur if the excitation coefficient of the mth slot is defined as

$$C_m = \sum_{j=1}^{K_0} b_m(j) \tag{10.247}$$

The array factor is then obtained as

$$S(\theta, \phi) = \sum_{m=1}^{MN} C_m\, e^{jk_0(x_m \sin\theta\cos\phi + z_m \cos\theta)} \tag{10.248}$$

If higher accuracies are desired then it is better to revert to eqn. 10.244.

10.22.6 *Design procedure*

Once the desired radiation pattern has been selected and the fraction of the power that would be coupled into each of the branch guides is computed, the next step is to decide on the offset and the dimensions of each of the M slots corresponding to the desired radiation pattern. For this purpose, the initial values of the dimensions and offsets of the radiators are chosen based on the earlier analysis or experimental data. With these offsets and dimensions the fields at the aperture of each slot in the branch guides are computed on the basis of the present analysis. The input reflection coefficients of each of the branch guides is also computed on the basis of eqn. 10.241. In all likelihood these first sets of values for the aperture fields and the input reflection coefficients will yield a pattern that is different from the desired radiation pattern. It is now assumed that any changes from the initial set of values in the dimensions or offsets of the slot radiators to reach the design goal will not alter the form of the electric field along the slot. Therefore the alterations will have the effect of multiplying the aperture field values by a complex number which is different for each of the M slots. This complex multiplier is the product of two factors. The first one is determined by the ratio of the aperture field at a given slot to the aperture field at the first slot (slot nearest to the resonant coupled feed slot). The second factor is the input reflection coefficient, which governs the absolute value of the field at each of the slots.

Once the two factors of the multiplier are computed, the radiation pattern can now be obtained based on eqn. 10.248, but with the following modification:

$$S(\theta, \phi) = \sum_{m=1}^{MN} C_m B_m\, e^{jk_0(x_m \sin\theta\cos\phi + z_m \cos\theta)} \tag{10.249}$$

The symbol B_m introduced for the terms under the summation sign represents the multiplier factor at each of the slots of the array. At this point the direction along which the aperture fields must be modified to achieve the desired radiation pattern is clear. Therefore the next set of new slot lengths and offsets are determined and the procedure is repeated until the desired radiation pattern is realised. It has been stated [23] that only two or three iterations are needed to arrive at the final values.

Two case studies have been cited by Gulick and Elliott [23]. The first case study was to validate the new procedure. This involved the design of an array in which the slots can be represented as shunt elements and for which the earlier procedure had been found to give very accurate values for the slot characteristics. The second case study illustrates the inaccuracies of the earlier procedure and how these are reduced by the new procedure. In this case, the aperture electric field of a self-resonant offset slot in a reduced height wave guide was computed for which it was known that the earlier procedure could not satisfactorily account for the phase variation and the skewness of the field along the slot.

10.23 Rectangular waveguide radiator array

Open ended waveguides in the form of sectoral, pyramidal and conical horns have been used as radiating elements at microwave frequencies. Unlike the dipole or the waveguide backed longitudinal slot, the physical dimensions of the waveguide aperture radiators are comparable to the wavelength as a result of which they are capable of multimode radiation. Therefore, in an electronic scanning array of such radiators, mutual coupling strongly influences the array radiation pattern, scan impedance and the polarisation characteristics. Hence the study of mutual coupling effects in an array environment assumes significance. In the case of an array with circular waveguide radiators a first order analysis of the effects of mutual coupling through the dominant TE_{11} mode has been investigated by several authors [27, 28]. Direct evaluation of the coupling as well as an asymptotic formula to facilitate quick computation were the results of these studies. The values obtained by the two methods differed significantly in the case of an H-plane scan. The integral equation formulation of the mutual coupling effects of the circular waveguide array has been attempted in Reference 29 to successfully derive expressions for all mode couplings occurring in a finite array. A new formula based on asymptotic expansion of Green's function in terms of the distance between the radiator apertures was derived for the cross coupling between modes in separate waveguide apertures. This formula gives satisfactory results for waveguide separations greater than one wavelength. Further, good agreement was found with the experimental results of the earlier quoted studies [27]. The packing of circular apertures in a planar array is not as efficient as the stacking of rectangular waveguide radiator elements. Higher aperture efficiency can be realised by arranging the antenna elements closer, as in an array with rectangular/square cross-section aperture radiators. The rectangular aperture also provides the designer with higher degrees of freedom owing to the two beams being different in the two orthogonal planes. The infinite array of rectangular waveguide radiators was the first configuration that was subjected to analysis by experimentation in a limited-number array [9]. It was found that deep nulls occur in the radiation pattern with beam scan before the onset of the grating lobe. This phenomenon cannot be accounted for by using a single mode solution. Another interesting aspect of this work is the existence of cross polarised modes at the array face. The behaviour of small finite planar arrays of rectangular waveguides has been analysed by different methods. In one, an expression was obtained [30] for the mutual admittance between two identical apertures by relating it to the power radiation pattern of the element in the form of a Fourier transform function. The near field coupling between two collinear open ended waveguides has been determined [31] by formulating the problem as a set of simultaneous integral equations and solving them by expanding the aperture fields in a Fourier series. Instead of the Fourier series approach the method of moments and a single mode approximation to the aperture field has also been attempted [32]. Improvements to this first order analysis by taking into account the higher order modes has been suggested. Experimental results indicate good agreement with theory. In yet another approach [33], the problem of mutual coupling between rectangular waveguide apertures in a finite array has been cast in the form of an integral equation on the basis of the tangential electric and magnetic fields being continuous across the radiating apertures. The integral equation is solved by the method of moments for the equivalent magnetic current in the aperture region and a single expansion function was used to approximate the electric field in each aperture. In a slightly different approach [34], the method of moments with piecewise sinusoidal functions for the basis and the weighting functions has been used for a finite array in an infinite ground plane. The analysis is found to be in good agreement with the available data and is particularly efficient in E- and H-plane scans for the dominant TE_{10} mode. The integral equation formulation coupled with the method of moments and equating the aperture fields as combinations of TE and TM modes [35] expresses the mutual coupling effects between the waveguide aperture radiators in terms of coupling between individual modes. The integral equation based on the Green's function approach resulted in quadruple integrals which were reduced to double integrals to decrease the computational load. The method has been extended [36] to cover different size rectangular waveguide radiating elements. Excellent agreement between theory and experiment has also been found.

In this Chapter, the analysis of the mutual coupling effects of arrays using open ended waveguide radiators will be based on integral equation formulation and the use of the method of moments to define the aperture fields at a designated radiator in the presence of the other elements of the array. In the case of infinite arrays the spectral domain method-of-moments approach appears to be computationally simpler. For the finite array, the analysis will be based on expressing the radiated field as an integral with Green's function kernel and matching the tangential magnetic fields at the radiator aperture. The fields are expressed in terms of the TE and TM mode functions. These form the basis and weight functions in the Galerkin approach for the method of moments. The basic block for analysing the mutual coupling effects of the array is the mutual admittance between any two waveguide aperture radiators. The analysis will show that substantial mutual coupling power is contained in the higher order mode interactions, and therefore a first order analysis based purely on dominant TE_{10} mode is not adequate.

In the analysis of both infinite and finite arrays the starting point is the matching of the tangential fields at the waveguide radiator aperture. The waveguide radiators are assumed to be flush with the ground plane, which in this case is the entire x, y plane of a Cartesian co-ordinate system having periodically spaced rectangular apertures of dimensions $a \times b$ corresponding to the broad wall and the narrow wall of a guide (see Fig. 10.39). The apertures are spaced at distances D_x and D_y along the x- and y-directions, respectively on the x, y plane. The array plane, being at $z=0$, divides the field space into two half-spaces, namely the region $z<0$, which is the feed space containing the feed networks in which the fields due to the impressed source propagate in the dominant TE_{10} mode of the guide, and the region $z>0$, which is the radiation half-space. Since the task of deducing the complete current distribution for this structure is a forbidding one, the analysis is based on the estimation of the wave fields, which can be determined easily. The equivalence principle is first applied to separate the fields in the two half-spaces. For this purpose, the aperture radiators at the array plane are replaced by a perfect electric conductor (PEC) sheet to create a smooth and continuous conducting plane at $z=0$. The two half-spaces are thus completely isolated and the field in each region can be individually determined. In the feed half-space, as a result of the replacement of the aperture by PEC, an equivalent magnetic current corresponding to the aperture electric field is substituted at the array plane. It can be expressed as

$$\vec{K}_a = \hat{z} \times \vec{E}_a \tag{10.250}$$

Here $-\hat{z}$ is the unit normal to the array plane in the feed half-space, $\vec{E}_a$ is the aperture electric field before the placement of the PEC and $\vec{K}_a$ is the surface magnetic current. To ensure continuity of the tangential component of the electric field, namely $\vec{E}_a$, at the position, a surface magnetic current of equal magnitude but of opposite sign, that is $-\vec{K}_a$, has to be placed at the same position on the array plane but on the radiation half-space side (see Fig. 10.29). The electromagnetic fields also have to satisfy the second boundary condition,

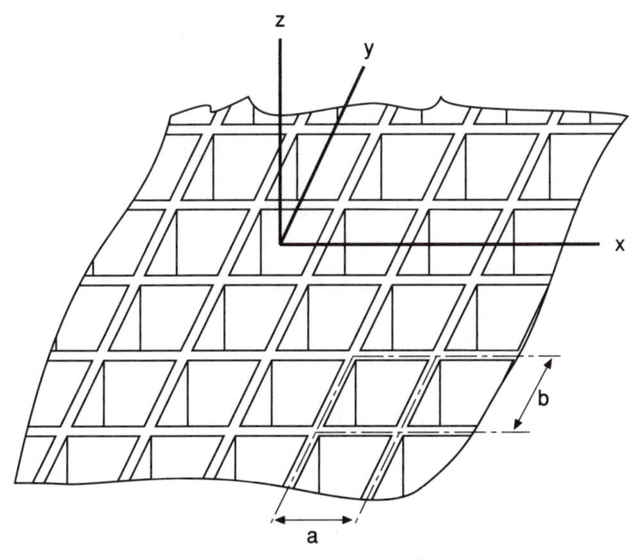

Figure 10.39 *Infinite array of rectangular waveguide apertures*

namely the continuity of the tangential magnetic fields at the aperture. From the feed half-space side, the magnetic field at the aperture has two components. The first one is the magnetic field due to the equivalent magnetic current $\vec{K}_a$ and its image. The image contribution arises because of the aperture in the array being replaced by a PEC and thereby providing a smooth and conducting plane. The second component is the incident field produced by the source and its image which is equivalent to twice the value of the field due to the feed network. The fields at the aperture from the radiation half-space will be the field due to the equivalent magnetic current $-\vec{K}_a$ and its image. The boundary condition requires that the aperture magnetic fields from both the half spaces be equal so that the continuity of the tangential magnetic field is maintained. Thus

$$2\vec{H}^t_{inc} + \vec{H}^t_{wg}(2K_a) = \vec{H}^t_{rad}(2K_a) \tag{10.251}$$

where,
$\vec{H}^t_{inc}$ = field due to the impressed source
$\vec{H}^t_{wg}$ = field due to the equivalent magnetic current on the feed half-space
$\vec{H}^t_{rad}$ = field due to the equivalent magnetic current from the radiation half-space.
Since $\vec{H}^t_{rad}(-2K_a) = -\vec{H}^t_{rad}(2K_a)$, eqn. 10.251 becomes

$$2\vec{H}^t_{inc} + \vec{H}^t_{wg}(2K_a) = -\vec{H}^t_{rad}(2K_a) \tag{10.252}$$

Eqn. 10.252 is the basic operator equation for determining the equivalent magnetic current K_a which is related to the tangential component of the electric field in the aperture. One can also consider $\vec{H}^t_{wg}(2K_a)$ in eqn. 10.252 as the field in the direction of the feed network as a result of the mismatch between the source and the aperture radiator due to coupling through free space of other radiators of the array.

10.23.1 *Infinite array*

The analysis of the infinite array based on eqn. 10.252 will be taken first. The tangential magnetic field H^t_{rad} at the aperture will now be expressed in the form of an integral equation with the free space Green's function while the other two terms on the LHS of eqn. 10.252 will be related to the TE_{mn} and TM_{mn} waveguide modes. The electric and the magnetic fields in the radiation half-space are expressed in terms of the magnetic vector potential $\vec{A}(\vec{r})$ as follows:

$$\vec{E}(\vec{r}) = -j\omega\mu_0\left\{\vec{A}(\vec{r}) + \frac{1}{k_0^2}\nabla(\nabla\cdot\vec{A}(\vec{r})\right\}$$

$$\vec{H}(\vec{r}) = \nabla\times\vec{A}(\vec{r}) \tag{10.253}$$

$$\vec{A}(\vec{r}) = \iint_{sources} G(r, r')\vec{J}(r)\, ds' \tag{10.254}$$

where r is the distance parameter, $\vec{J}(r)$ is the electric current and $G(r, r')$ is the Green's function for free space and is expressed as

$$\frac{e^{-jk|\vec{r}-\vec{r}'|}}{|\vec{r}-\vec{r}'|}$$

It is now proposed to apply the spectral domain technique which is a powerful analytical tool that permits considerable simplification of the computation of the method of moments. The spectral domain method was developed by Prof. R. Mittra and his colleagues [37, 38] in the 1970s and has been extensively used in the study of the characteristics of frequency selective surfaces to which category the infinite array of open-ended waveguide apertures belong. The spectral domain method permits considerable simplification of the MOM calculations for planar surfaces because it eliminates the singularity of the spatial Green's function. In addition, the spatial convolution of the surface current and the Green's function is transformed to a simple algebraic multiplication when the computations are carried out in the spectral domain. In the present instance the analysis of mutual coupling effects in the infinite waveguide radiator arrays will be largely based on the spectral domain procedure described in Reference 39. We begin with the plane wave spectrum or expansion [40] for the field vectors $\vec{F}$ in the radiation half-space and which is obtained as follows:

$$\vec{F}(x, y, z) = \int_{-\infty}^{\infty}\int \vec{F}(k_x, k_y)e^{j(k_x x+k_y y)}\, e^{jk_z z}\, dk_x\, dk_y \tag{10.255}$$

$$k_z^2 = k^2 - k_x^2 - k_y^2 \tag{10.256}$$

$F\,(k_x, k_y)_{z=0^+}$ is the value of the field quantity at the array plane on the side of the radiation half-space. (x, y, z) is any point in the radiation half-space. In the visible region, k^2 is greater than $k_x^2+k_y^2$ and the plane waves are uniform propagating waves. When k^2 is less than $k_x^2+k_y^2$, the positive imaginary root is taken so that the plane waves are nonuniform and evanescent. For the infinite periodic structure such as the present infinite open ended waveguide radiator array Floquet's theorem is applicable for the fields at $z=0^+$ in the radiation half-space. Floquet's theorem is essentially an extension of the Fourier series theorem and relates the fields at any one lattice of the array to those of any other cell by a linear phase shift corresponding to the relative phase of the incident plane wave field over the two cells. Since the infinite array of waveguide radiators has spatial periodicities D_x and D_y in the x- and y- directions respectively, one can write for the field vector at $z=0^+$,

$$\vec{F}(x+D_x, y+D_y) = \vec{F}(x, y)e^{j(k_x D_x+kyDy)}\, e^{j2\pi}\, e^{j2\pi} \tag{10.257}$$

If we now designate

$$\left.\begin{aligned} k_{xm} &= k_x + \frac{2m\pi}{D_x} \\ k_{yn} &= k_y + \frac{2n\pi}{D_y} \\ k_{zmn} &= k^2 - k_{xm}^2 - k_{yn}^2 \end{aligned}\right\} \tag{10.258}$$

Eqns. 10.257 and 10.258 imply that k_x and k_y are now limited to changes in discrete steps. In the light of this, the Fourier integral in eqn. 10.255 reduces to the following Fourier series:

$$\vec{F}(x, y, z) = \sum_m \sum_n \vec{F}(k_{xm}, k_{yn})_{z=0^+} \, e^{j(k_{xm}x + k_{yn}y)} \, e^{jk_{zmn}z} \tag{10.259}$$

Inserting the expression from eqn. 10.259 into the two eqns. 10.253, one obtains

$$\begin{bmatrix} E_x(k_{xm}, k_{yn}) \\ E_y(k_{xm}, k_{yn}) \end{bmatrix}_{z=0^+} = \frac{1}{j\omega\varepsilon} \begin{bmatrix} k^2 - k_{xm}^2 & -k_{xm}k_{yn} \\ -k_{xm}k_{yn} & k^2 - k_{yn}^2 \end{bmatrix} \begin{bmatrix} A_x(k_{xm}, k_{yn}) \\ A_y(k_{xm}, k_{yn}) \end{bmatrix}_{z=0^+} \tag{10.260}$$

$$\begin{bmatrix} H_x(k_{xm}, k_{yn}) \\ Hy(k_{xm}, k_{yn}) \end{bmatrix}_{z=0^+} = -\sqrt{k^2 - k_{xm}^2 - k_{yn}^2} \begin{bmatrix} 0 & 1 \\ -1 & 0 \end{bmatrix} \begin{bmatrix} A_x(k_{xm}, k_{yn}) \\ A_y(k_{xm}, k_{yn}) \end{bmatrix}_{z=0^+} \tag{10.261}$$

Since the aperture electric field and the surface magnetic current are related by $\hat{z} \times \vec{E} = \vec{K}_a$, eqns. 10.260 and 10.261 yield the following expression for the tangential magnetic field at the aperture:

$$H(k_{xm}, k_{yn})_{z=0^+} = -\frac{1}{2\omega\mu_0} \begin{bmatrix} \dfrac{k^2 - k_{xm}^2}{\sqrt{k^2 - k_{xm}^2 - k_{yn}^2}} & \dfrac{-k_{xm}k_{yn}}{\sqrt{k^2 - k_{xm}^2 - k_{yn}^2}} \\ \dfrac{-k_{xm}k_{yn}}{\sqrt{k^2 - k_{xm}^2 - k_{yn}^2}} & \dfrac{k^2 - k_{yn}^2}{\sqrt{k^2 - k_{xm}^2 - k_{yn}^2}} \end{bmatrix} [2K_a(k_{xm}, k_{yn})]_{z=0^+} \tag{10.262}$$

Now, it is seen that at the array plane ($z = 0^+$) eqn. 10.259 is changed to

$$\vec{F}(x, y)_{z=0^+} = \sum_{min}\sum \vec{F}(k_{xm}, k_{yn})_{z=0^+} \, e^{j(k_{xm}x + k_{yn}y)} \tag{10.263}$$

From eqn. 10.263 it is seen that the field quantities $\vec{E}$, $\vec{H}$, or $\vec{A}$ on the side of the radiation half-space can be expressed at the waveguide radiator aperture as an infinite sum in the spectral domain. In the present instance our interest is in the $\vec{H}^{\,t}_{rad}$ component of eqn. 10.252. Therefore, introducing eqn. 10.263 into eqn. 10.262 one obtains at the waveguide radiator aperture,

$$\vec{H}^{\,t}_{rad}(x, y) =$$
$$-\frac{1}{2\omega\mu_0} \sum_{min}\sum \begin{bmatrix} \dfrac{k^2 - k_{xm}^2}{\sqrt{k^2 - k_{xm}^2 - k_{yn}^2}} & \dfrac{-k_{xm}k_{yn}}{\sqrt{k^2 - k_{xm}^2 - k_{yn}^2}} \\ \dfrac{-k_{xm}k_{yn}}{\sqrt{k^2 - k_{xm}^2 - k_{yn}^2}} & \dfrac{k^2 - k_{yn}^2}{\sqrt{k^2 - k_{xm}^2 - k_{yn}^2}} \end{bmatrix} 2K_a(k_{xm}, k_{yn}) e^{j(k_{xm}x + k_{yn}y)} \tag{10.264}$$

Eqn. 10.264 is the tangential magnetic field at any of the apertures of the array due to the equivalent magnetic currents of all the aperture radiators of the array. Introducing eqn. 10.264 in the LHS of eqn. 10.252 it is now possible to write

$$2\vec{H}^{t}_{inc}+\vec{H}^{t}_{wg}(2K_a)=-\frac{1}{2\omega\mu_0}\sum_{min}\sum \begin{bmatrix} \dfrac{k^2-k_{xm}^2}{\sqrt{k^2-k_{xm}^2-k_{yn}^2}} & \dfrac{-k_{xm}k_{yn}}{\sqrt{k^2-k_{xm}^2-k_{yn}^2}} \\ \dfrac{-k_{xm}k_{yn}}{\sqrt{k^2-k_{xm}^2-k_{yn}^2}} & \dfrac{k^2-k_{yn}^2}{\sqrt{k^2-k_{xm}^2-k_{yn}^2}} \end{bmatrix} 2\vec{K}_a(k_{xm},\,k_{yn})]e^{j[k_{xm}x+k_{yn}y]} \tag{10.265}$$

with

$$\vec{K}_a(k_{xm},\,k_{yn})=\frac{1}{D_xD_y}\int_{-a/2}^{a/2}\int_{-b/2}^{b/2}\vec{K}_a(x,y)\;e^{-j(k_{xm}x+k_{yn}y)}\;dk_x\;dk_y \tag{10.266}$$

Now, the magnetic fields on both sides of the aperture, namely, $\vec{H}^{t}_{wg}$ and $-\vec{H}^{t}_{rad}$ can be expressed in terms of the functions related to the TE and TM modes of the rectangular guide. However, the magnetic field $\vec{H}^{t}_{inc}$ can be expressed only in terms of the TE_{mn} mode which propagates on the feed network of the array. These mode functions are also the basis and the testing functions for Galerkin's procedure. These mode functions are normalised for unit power transfer in each mode and can be expressed as

$$\vec{\psi}^{TM}_{pq}(x,y)=C_{pq}\left[\hat{x}\frac{p\pi}{a}\cos\frac{p\pi(x+a/2)}{a}\sin\frac{q\pi(y+b/2)}{b}\right.$$
$$\left.+\hat{y}\frac{q\pi}{b}\sin\frac{p\pi(x+a/2)}{a}\cos\frac{q\pi(y+b/2)}{b}\right] \tag{10.267}$$

$$\vec{\psi}^{TE}_{pq}(x,y)=C_{pq}\left[\hat{x}\frac{q\pi}{b}\cos\frac{p\pi(x+a/2)}{a}\sin\frac{q\pi(y+b/2)}{b}\right.$$
$$\left.-\hat{y}\frac{p\pi}{a}\sin\frac{p\pi(x+a/2)}{a}\cos\frac{q\pi(y+b/2)}{b}\right] \tag{10.268}$$

$$\vec{E}^{TM}_{a}=e^{\pm jk_{zpq}z}\vec{\psi}^{TM}_{pq} \tag{10.269}$$

$$\vec{H}^{TM}_{a}=e^{\mp YTM}_{pq}\vec{\phi}^{TM}_{pq} \tag{10.270}$$

$$Y^{TM}_{pq}=\frac{1}{z_0}\left(\frac{k}{k_{zpq}}\right) \tag{10.271}$$

$$\vec{\phi}^{TM}_{pq}=\hat{z}\times\vec{\phi}^{TM}_{pq} \tag{10.272}$$

In a similar manner we can define

$$\vec{E}^{TE}_{pq}=e^{\pm k_{zpq}z}\vec{\psi}^{TE}_{pq} \tag{10.273}$$

$$\vec{H}^{TE}_{pq}=\mp Y^{TE}_{pq}\vec{\phi}^{TE}_{pq} \tag{10.274}$$

$$Y_{pq}^{TE} = \frac{1}{z_0}\left(\frac{k_{zpq}}{k}\right) \tag{10.275}$$

$$\vec{\phi}_{pq}^{TE} = \hat{z} \times \vec{\psi}_{pq}^{TE} \tag{10.276}$$

$$C_{pq} = \sqrt{\frac{2\varepsilon_p \varepsilon_q}{ab\tau_{pq}^2}} \tag{10.277}$$

$$\tau_{pq}^2 = \left[\left(\frac{p\pi}{a}\right)^2 + \left(\frac{q\pi}{b}\right)^2\right] \tag{10.278}$$

and ε_p, ε_q are Neumann numbers defined in Section 10.22.1. In the above equations, the superscript refers to the mode category and the subscript refers to the order of the mode. In eqns. 10.269, 10.270, 10.273 and 10.274 the upper sign refers to the waves propagating in the negative $(-z)$ direction and the lower sign to the waves in the positive $(+z)$ direction. In order to utilise eqns. 10.267–10.278 to solve eqn. 10.265 we first express the two terms on the LHS of this equation in terms of mode functions as

$$\vec{H}_{wg}^{t} = -\sum_{p,q}\left[a_{pq} Y_{pq}^{TE} \vec{\phi}_{pq}^{TE}(x, y) + b_{pq} Y_{pq}^{TM} \vec{\phi}_{pq}^{TM}(x, y\text{-}\right] \tag{10.279}$$

$$\vec{E}_a \times \hat{z} = -K_a(k_{xpq}, k_{ypq}) = \sum_{p,q} [E_{pq}^{TE} \vec{\phi}_{pq}^{TE}(k_{xp}, k_{yq}) + E_{pq}^{TM} \vec{\phi}_{pq}^{TM}(k_{xp}, k_{yq})] \tag{10.280}$$

$$\vec{H}_{inc}^{t} = Y_{pq}^{TE} \vec{\phi}_{\pi 6}^{\mathrm{TE}} \quad \text{or} \quad Y_{pq}^{TM} \vec{\phi}_{pq}^{TM} \quad \text{as the case may be} \tag{10.281}$$

In these equations a_{pq}, b_{pq}, E_{pq}^{TE} and E_{pq}^{TM} are the unknown coefficients which have to be determined. Since the feed network propagates only the TE_{pq} mode, the requirement of the continuity of aperture fields gives rise to the following equations:

$$\begin{aligned} a_{pq} + \delta_{p-p_i,\, q-q_i} &= E_{pq}^{TE} \\ b_{pq} &= E_{pq}^{TM} \end{aligned} \tag{10.282}$$

Introducing eqns. 10.279–10.282 in eqn. 10.265 and rearranging the terms leads to

$$\begin{aligned} 2Y_{mn}^{TE} \vec{\phi}_{mn}^{TE} = \sum_p \sum_q & \\ \times \Bigg[E_{pq}^{TE} \Bigg\{ & Y_{pq}^{TE} \vec{\phi}_{pq}^{TE}(x, y) - 2 \sum_{m,n} G_p(k_{xm}, k_{yn}) \vec{\phi}_{pq}^{TE}(k_{xm}, k_{yn}) e^{j(k_{xm}x + k_{yn}y)} \Bigg\} \\ + E_{pq}^{TM} \Bigg\{ & Y_{pq}^{TM} \vec{\phi}_{pq}^{TM}(x, y) - 2 \sum_{m,n} G_p(k_{xm}, k_{yn}) \vec{\phi}_{pq}^{TM}(k_{xm}, k_{yn}) e^{j(k_{xm}x + k_{yn}y)} \Bigg\} \Bigg] \end{aligned} \tag{10.283}$$

with

$$G_p(k_{xm}, k_{yn}) = -\frac{1}{2\omega\mu_0}\begin{bmatrix} \dfrac{k^2 - k_{xm}^2}{\sqrt{k^2 - k_{xm}^2 - k_{yn}^2}} & \dfrac{-k_{xm}k_{yn}}{\sqrt{k^2 - k_{xm}^2 - k_{yn}^2}} \\ \dfrac{-k_{xm}k_{yn}}{\sqrt{k^2 - k_{xm}^2 - k_{yn}^2}} & \dfrac{k^2 - k_{yn}^2}{\sqrt{k^2 - k_{xm}^2 - k_{yn}^2}} \end{bmatrix} \tag{10.284}$$

Eqn. 10.283 represents the fact that the tangential magnetic field at the aperture of any radiator is continuous on either side of the array plane and has three main components, namely the magnetic field due to the mode propagating in the waveguide from the feed network, the magnetic field reflected from the aperture towards the feed system and the magnetic field generated due to the coupling of aperture fields of other radiators of the array. The last two components of the tangential magnetic field at the aperture contain the higher order waveguide modes also. Galerkin's procedure is now applied by taking the inner product of the weighting function and the individual terms of eqn. 10.283. This enables us to find the amount of power in mode rs at a given aperture due to the mode pq of all the apertures of the array by convolving in the spatial domain. Therefore dotting both sides by the test function $\vec{\phi}_{rs}^{TE*}$ and integrating over the aperture radiator, the following equations are obtained:

$$\begin{aligned} 2Y_{rs}^{TE}D_{rs}\delta_{p'-r,\,q'-s} = \sum_{pq} E_{pq}^{TE}\Big\{ & Y_{rs}^{TE}D_{rs}\delta_{p-r,\,q-s} \\ & - (2D_xD_y)\sum_{m,n} \vec{\phi}_{rs}^{TE*}(k_{xm}, k_{yn}) \cdot G_p(k_{xm}, k_{yn})\vec{\phi}_{pq}^{TE}(k_{xm}, k_{yn})\Big\} \\ & - E_{pq}^{TM}\Big\{(2D_xD_y)\sum_{m,n} \vec{\phi}_{rs}^{TE*}(k_{xm}, k_{yn}) \cdot G_p(k_{xm}, k_{yn})\vec{\phi}_{pq}^{TM}(k_{xm}, k_{yn})\Big\} \end{aligned} \tag{10.285}$$

and

$$D_{rs} = 1 \tag{10.286}$$

Using $\vec{\phi}_{rs}^{TM*}$ as the next set of test functions, we obtain the following equation

$$\begin{aligned} 2Y_{rs}^{TM}D_{rs}\delta_{p'-r,\,q'-s} = \sum_{p,q} E_{pq}^{TE}\Big\{ & (2D_xD_y) \\ & \sum_{m,n} \vec{\phi}_{rs}^{TM*}(k_{xm}, k_{yn}) \cdot G_p(k_{xm}, k_{yn})\vec{\phi}_{pq}^{TE}(k_{xm}, k_{yn})\Big\} \\ & - E_{pq}^{TM}\Big\{ Y_{pq}^{TM}D_{rs}\delta_{p-r,\,q-s} \\ & - (2D_xD_y)\sum_{m,n} \vec{\phi}_{rs}^{TM*}(k_{xm}, k_{yn}) \cdot G_p(k_{xm}, k_{yn})\vec{\phi}_{pq}^{TE}(k_{xm}, k_{yn})\Big\} \end{aligned} \tag{10.287}$$

In eqns. 10.285 and 10.287 to obtain E_{pq}^{TE} and E_{pq}^{TM}, the unknown modal coefficients, it is necessary to compute $\vec{\phi}_{pq}^{TE}$, $\vec{\phi}_{pq}^{TM}$, $\vec{\phi}_{pq}^{TE*}$, and $\vec{\phi}_{pq}^{TM*}$. The expressions for $\vec{\phi}_{pq}^{TE}(k_{xm}, k_{yn})$ and $\phi_{pq}^{TM}(k_{xm}, k_{yn})$ work out to be

$$\begin{aligned}\vec{\phi}_{pq}^{TE}(k_{xm}, k_{yn}) = \frac{1}{C_{m,n}^{1} D_x D_y}\Bigg\{ & \hat{x}\frac{p\pi b}{4j}\,[j^{p}\,\mathrm{sinc}(k_{xm}a - p\pi)/2 - j^{-p}\,\mathrm{sinc}(k_{xm}a + p\pi)/2] \\ & \times [j^{q}\,\mathrm{sinc}(k_{yn}b - q\pi)/2 + j^{-q}\,\mathrm{sinc}(k_{yn}b + q\pi)/2] \\ & + \hat{y}\frac{q\pi a}{4j}\,[j^{q}\,\mathrm{sinc}(k_{yn}b - q\pi)/2 - j^{-q}\,\mathrm{sinc}(k_{yn}b + q\pi)/2] \\ & \times [j^{p}\,\mathrm{sinc}(k_{xm}a - p\pi)/2 + j^{-p}\,\mathrm{sinc}(k_{xm}a + p\pi)/2]\end{aligned} \quad (10.288)$$

and

$$\begin{aligned}\vec{\phi}_{pq}^{TM}(k_{xm}, k_{yn}) = \frac{1}{CD_x D_y}\Bigg\{ & -\hat{x}\frac{q\pi a}{4\pi j}\,[j^{p}\,\mathrm{sinc}(k_{xm}a - p\pi)/2 - j^{-p}\,\mathrm{sinc}(k_{xm}a + p\pi)/2] \\ & \times [j^{q}\,\mathrm{sinc}(k_{yn}b - q\pi)/2 + j^{-q}\,\mathrm{sinc}(k_{yn}b + q\pi)/2] \\ & + \hat{y}\frac{p\pi b}{4j}\,[j^{q}\,\mathrm{sinc}(k_{yn}b - q\pi/2) - j^{-q}\,\mathrm{sinc}(k_{yn}b + q\pi)/2] \\ & \times [j^{p}\,\mathrm{sinc}(k_{xm}a - p\pi)/2 + j^{-p}\,\mathrm{sinc}(k_{xm}a + p\pi)/2]\end{aligned} \quad (10.289)$$

where

$$\mathrm{sinc}\ x = \frac{\sin x}{x} \quad (10.290)$$

Eqns. 10.288 and 10.289 provide us with the values of the modal functions $\vec{\phi}^{TE}$ and $\vec{\phi}^{TM}$ in the spectral domain. These values are then introduced in eqns. 10.285 and 10.287 along with the values of the admittances Y^{TE} and Y^{TM} of the TE and TM modes in the guide, the Green's function $G_p(k_{xm}, k_{yn})$ in the spectral domain and D_{rs} from eqns. 10.271, 10.273, 10.284 and 10.286, respectively. Each of eqns. 10.285 and 10.287 are reduced to a set of linear equations with the E_{pq}^{TE} and E_{pq}^{TM} $(p, q = 1, 2, 3, \ldots)$ as the unknowns. The determination of E_{pq}^{TE} and E_{pq}^{TM} leads to determination of the field across the waveguide aperture radiator and then the radiation pattern of the array. Each term of eqns. 10.285 and 10.287 has the dimensions of current. The terms of interest to us are of the form $[E_{pq}^{TX}\Sigma_{m,n}\,\vec{\phi}_{rs}^{TX}\cdot G_p\vec{\phi}_{pq}^{TX}]$ where TX stands for either TE or TM. Since E_{pq}^{TX} has the dimension of volts, the summation term has the dimension of admittance, and therefore it represents the admittance due to the coupling of mode *rs* in an aperture to the mode *pq* in all the apertures of the array. Thus the admittance matrix $[Y]$ can be built up by considering each mode in turn of the designated aperture radiator and summing it with respect to all the modes existing in the waveguide aperture radiators of the array.

It is seen from the above discussion that the higher TE and TM modes at the aperture of any waveguide radiator of the array contribute to the imaginary part of the Poynting vector of the radiator aperture, and thus give an indication of the energy stored in the higher order modes. Even though the higher order modes do not propagate, they contribute to the far field pattern of the array.

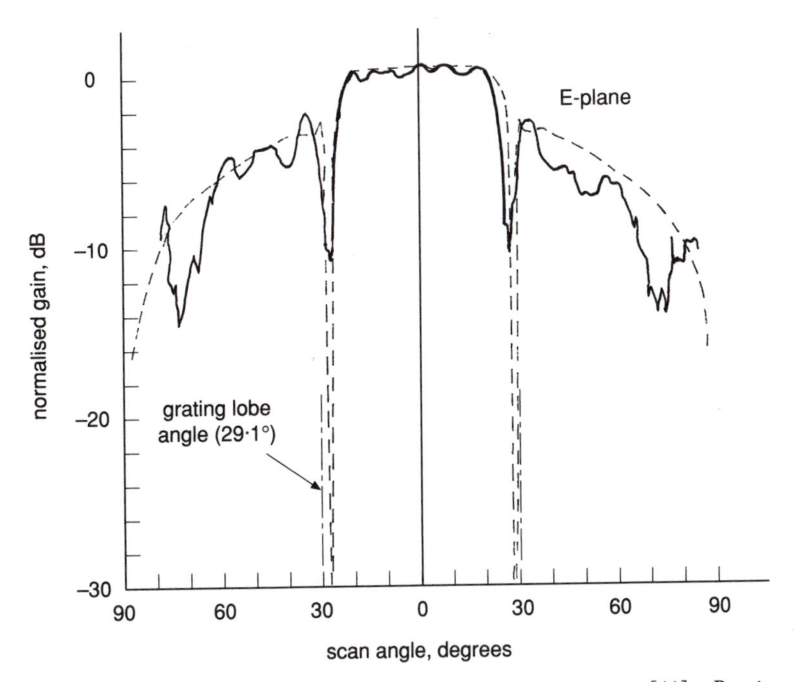

Figure 10.40 *Scan element gain of waveguide aperture array [41]. Reprinted by permission of © Wiley-Interscience, 1972*

Square grid
$b = d = 0{\cdot}6729\ \lambda$
$a = c = 0{\cdot}6305\ \lambda$

—— experiment: 19 × 19 array (Amitay and Wu)
--- multimode theory

Experimental data support the theory of existence of, and coupling among, higher order modes at the aperture of the waveguide radiator in an array environment. Farrell and Kuhn [9] have shown that the first ten higher order modes along with the dominant TE_{10} mode are adequate from the point of view of computing the co-polarised and cross-polarised patterns of the array. These findings show that beyond the first 17 modes the energy contained in the other higher modes is less than 0.5% for all grid sizes. Amitay and his colleagues [41] have shown that the principal contributions are made by the TE_{11}, TM_{11}, TE_{12} and TM_{12} modes. Fig. 10.28, which shows the *E*-plane scan for a square grid array of square waveguides very clearly, brings out that the modal theory can predict the scan blindness at a scan angle before the onset of the grating lobe. This is a significant advance over the single mode analysis which can only predict the grating lobe appearances. Fig. 10.40, also taken from Reference 41, depicts the experimental and calculated radiation patterns of a 19 × 19 array with a square grid of side 0.6729λ of rectangular waveguide aperture radiators. In this case, the multimodal analysis is able to predict the scan blindness and the angle at which it occurs. The theoretical prediction is very close to the experimental results. These two examples serve to emphasise that in an array environment mutual coupling effects between aperture fields of the radiators generate higher order modes even though the feed network does not favour their propagation. The multimodal analysis permits us to consider the effect of mutual coupling as a sum of the coupling between individual modes as well as between pairs of radiators. Fortunately, the number of higher order modes to be considered is limited.

10.23.2 *Finite array*

In the case of finite arrays, the uniformity of the environment for all the radiators cannot be assumed. Consequently mutual coupling effects will have to be determined on an element-by-element basis at each radiator. Owing to the absence of infinite periodicity in the array plane the application of Floquet's theorem to transform the Fourier integral based plane wave expansion of eqn. 10.255 to a Fourier series as in eqn. 10.259 is not valid in this case. Thus in eqn. 10.252, which expresses the continuity of the tangential magnetic field at the waveguide radiating aperture, only the field component $\vec{H}'(2K_a)$ will not be the same. It has to be altered to meet the finite array environment.

We begin with the plane wave expansion of the magnetic field vector at $z=0^+$ at the array face on the radiation half-space side. For this purpose let $z=0$ in eqn. 10.255 so that

$$\vec{H}(x, y)_{z=0^+} = \iint_{-\infty}^{\infty} \vec{H}(k_x, k_y)_{z=0^+} e^{j(k_x x + k_y y)} dk_x dk_y \tag{10.291}$$

Applying the spectral domain technique to eqns. 10.253 and substituting $\hat{z} \times \vec{E} = \vec{K}_a$, one obtains

$$\vec{H}(k_x, k_y) = G_A(k_x, k_y)\,[2\vec{K}_a(k_x, k_y)]_{z=0^+} \tag{10.292}$$

with

$$G_A(k_x, k_y) = -\frac{1}{2\omega\mu_0}\begin{bmatrix} \dfrac{k^2-k_x^2}{\sqrt{k^2-k_x^2-k_y^2}} & \dfrac{-k_x k_y}{\sqrt{k^2-k_x^2-k_y^2}} \\ \dfrac{-k_x k_y}{\sqrt{k^2-k_x^2-k_y^2}} & \dfrac{k^2-k_y^2}{\sqrt{k^2-k_x^2-k_y^2}} \end{bmatrix} \tag{10.293}$$

Introducing eqn. 10.292 in eqn. 10.291, we have

$$\vec{H}(x, y)_{z=0^+} = \iint_{-\infty}^{\infty} G_A(k_x, k_y)\,[2\vec{K}_a(k_x, k_y)]_{z=0^+}\, e^{j(k_x x + k_y y)} dk_x dk_y \tag{10.294}$$

The tangential magnetic field produced at an aperture by the equivalent magnetic current $2\vec{K}_a(k_x, k_y)$ is given in eqn. 10.294 on the assumption that both the source and the observer are on the array plane. The observer (field point), corresponding to a designated aperture radiator, has the co-ordinates (x, y) and the source corresponds to the *mn*th waveguide aperture radiator with the co-ordinates (x_m, y_n). Eqn. 10.294 can be used to compute the field produced at the designated aperture by all other aperture radiators of the array:

$$\vec{H}^{\,t}_{rad}(x, y) = \sum_{m,n} \iint_{-\infty}^{\infty} G_A(k_x, k_y)\,[2\vec{K}_a(k_x, k_y)]\; e^{j[k_x(x-x_m)+k_y(y-y_n)]} dk_x dk_y \qquad (10.295)$$

Eqn. 10.295 expresses the tangential magnetic field induced at a designated aperture radiator as a result of the aperture electric field (equivalent magnetic current) of another radiator at a distance given by $(x-x_m, y-y_n)$ away from it. In the finite array environment at each aperture radiator it would be necessary to add the contribution of each of the *mn*th radiators of the array, that is to say, compute eqn. 10.295 for all values of (m, n) starting from $m = 1, 2, 3, \ldots, M$, and $n = 1, 2, 3, \ldots, N$. If eqn. 10.295 is compared with eqn. 10.264, the difference between the computatons of $\vec{H}^{\,t}$ for the finite and the infinite array becomes apparent. While in the case of infinite periodic structures the variables k_x and k_y assume discrete values known as Floquet's harmonics, in the case of the finite array these variables are continuous and therefore the integral of eqn. 10.264 remains.

Since the aperture fields can be considered to be made of TE and TM modes, the method of moments can be applied with the modal functions as the basis and weighting functions and the equivalent magnetic current in eqn. 10.295 as the unknown to be determined. Therefore we have

$$\vec{H}^{\,t}_{rad}(x, y) = 2 \sum_{p,q} \sum_{m,n}^{MN} \iint_{-\infty}^{\infty} G_p(k_x, k_y) E^{TE}_{pq}\, \vec{\phi}^{\,TE}_{pq}(k_x, k_y)\; e^{j[k_x(x-x_m)+k_y(y-y_n)]} dk_x dk_y$$
$$+ \iint_{-\infty}^{\infty} G_p(k_x, k_y) E^{TM}_{pq}\, \vec{\phi}^{\,TM}_{pq}(k_x, k_y)\; e^{j[k_x(x-x_m)+k_y(y-y_n)]} dk_x dk_y \qquad (10.296)$$

The method of moments procedure is completed by taking the inner product on both sides of eqn. 10.296 with the weighting functions. For this purpose, the dot product of each of the terms of eqn. 10.296 with the complex conjugate of the TE and TM mode functions has to be carried out first. The infinite integral is now transformed to a sum of $(M \times N)$ finite integals, the domain of integration for each being the surface of the radiator aperture. This is due to the fact that the electric field over the array plane is zero everywhere except at the aperture radiator surfaces. Two sets of equations are obtained, one with TM mode functions and the other with TE mode functions. These are:

$$\langle \vec{\phi}^{\,TE}_{rs}, \vec{H}^{\,t}_{rad} \rangle = (2D_x D_y) \sum_{p,q} \sum_{m,n} \left\{ \iint_{-\infty}^{\infty} \vec{\phi}^{\,TE*}_{rs}(k_x, k_y) \cdot G_p(k_x, k_y) \vec{\phi}^{\,TE}_{pq} E^{TE}_{pq} \right.$$
$$\times e^{j[k_x(x-x_m)+k_y(y-y_n)]}\, dk_x dk_y$$
$$+ \iint_{-\infty}^{\infty} \vec{\phi}^{\,TE*}_{rs}(k_x, k_y) \cdot G_p(k_x, k_y) E^{TE}_{pq} \vec{\phi}^{\,TM}_{pq}(k_x, k_y)$$
$$\left. \times e^{j[k_x(x-x_m)+k_y(y-y_n)]}\, dk_x\, dk_y \right\} \qquad (10.297)$$

and,

$$\langle \vec{\phi}^{TM}_{rs}, \vec{H}^{t}_{rad}\rangle = (2D_x D_y) \sum_{p,q} \sum_{m,n} \left\{ \iint_{-\infty}^{\infty} \vec{\phi}^{TM*}_{rs}(k_x, k_y) \cdot G_p(k_x, k_y) E^{TE}_{pq} \vec{\phi}^{TE}_{pq}(k_x, k_y) \right.$$
$$\times e^{j[k_x(x-x_m)+k_y(y-y_n)]}\, dk_x dk_y$$
$$+ \iint_{-\infty}^{\infty} \vec{\phi}^{TM*}_{rs}(k_x, k_y) \cdot G_p(k_x, k_y) E^{TM}_{pq} \vec{\phi}^{TM}_{pq}(k_x, k_y)$$
$$\left. \times e^{j[k_x(x-x_m)+k_y(y-y_n)]}\, dk_x dk_y \right\} \tag{10.298}$$

Each of the terms of eqns. 10.297 and 10.298 have the dimension of current. In the terms containing the integral on the RHS of these equations, E^{TE} and E^{TM} have the dimension of voltage. Since E^{TE} and E^{TM} are constants, they can be outside the integral sign. The resulting integrals then have the dimension of admittance. Hence, the elements of the admittance matrix due to coupling between the *rs*th TE mode at the designated aperture and the *pq*th TE mode of the (m, n)th aperture radiator of the array can be obtained from eqn. 10.297 as

$$J^{mn}_{pq,\,rs} = (2D_x D_y) \iint_{-\infty}^{\infty} \vec{\phi}^{TE*}_{rs}(k_x, k_y) \cdot G_p(k_x, k_y) E^{TE}_{pq} \vec{\phi}^{TE}_{pq}(k_x, k_y)$$
$$\times e^{j[k_x(x-x_m)+k_y(y-y_n)]}\, dk_x dk_y \tag{10.299}$$

In a similar manner, the admittance matrix elements for the TE→TM, TM→TM and TM→TE mode coupling can be derived from eqns. 10.297 and 10.298. The determination of the admittance matrix comprising such elements is likely to be computationally intensive and therefore it would be necessary to examine the integrand of eqn. 10.299 to reduce the complexity. From eqns. 10.288 and 10.289 it is evident that the mode functions $\vec{\phi}^{TE}_{pq}(k_x, k_y)$ are sinc functions and are well behaved everywhere in the domain of integration. However, eqn. 10.299 has, in the denominator of the Green's function $G_p(k_x, k_y)$, a term of the type $\sqrt{k^2-\xi^2-\eta^2}$ which can give rise to branch point singularities. The analytical behaviour of such square root functions have been investigated in the literature and the branch points and the branch cuts have been determined so that the contour of integration can be suitably modified [42]. The function theoretic manipulation involving the integral of eqn. 10.299 has been discussed in Reference 39. By two changes in the variables and by suitably altering the path of integration, the branch point singularity can be eliminated in the integrand.

An alternative method that has been tried out to simplify the computation is the fast Fourier transform [43]. In both approaches the main difficulty is to ensure that the integral converges especially in the region $k \leq |\xi|$, and $|\eta| < \infty$. This has been attempted by subjecting the variables to one more transformation from rectangular to spherical co-ordinates. Each term in the resulting expression is treated as a separate integral and evaluated by choosing the contours suitably [44] and truncating the rapidly oscillating and decaying imaginary part to reduce the computational load for numerical integration.

The method used in this Chapter for reduction of the computational load in the expression for the admittance is based on Lewin's procedure of change of variables and the range of integration [36, 45]. The first step is to express the tangential magnetic field induced at the aperture radiator due to other radiators of the array, that is the term $\vec{H}^{t}_{rad}$ in eqn. 10.252 in terms of the spatial Green's function and the aperture electric field. From eqns. 10.253 and 10.254 it is possible to derive the following equation for the tangential magnetic field $\vec{H}^{t}_{rad}$:

$$\vec{H}^{t}_{rad} = -\frac{j}{2\pi\omega\mu_0}(k^2 + \nabla_t\nabla_t)\cdot\hat{z}\times\iint_{-\infty}^{\infty} ds'\vec{E}(\vec{r})G(r, r') \qquad (10.300)$$

where

$$G(r, r') = \frac{e^{-jk|r-r'|}}{|r-r'|} \qquad (10.301)$$

and

$$|r-r'| = \sqrt{(x-x')^2 + (y-y')^2 + z^2} \qquad (10.302)$$

Here ∇_t is the transverse component (transverse to the direction of propagation, z) of the del operator. In eqns. 10.300–10.302 the primed co-ordinates refer to all the apertures whose electric field contributes to the tangential magnetic field at the designated aperture radiator (observer). Since all the aperture radiators are at the array plane, $z = 0^+$, eqn. 10.300 becomes an integral equation in the field components tangential to the array plane. The method of moments is now applied with the TE and TM mode functions forming the expansion and weighting functions in the Galerkin's procedure. Substituting eqns. 10.279, 10.281 and 10.300 in the tangential magnetic field continuity equation 10.252 we obtain

$$2Y^{TE}_{mn}\vec{\phi}^{\,TE}_{mn} = \sum_p \sum_q \times\left[E^{TE}_{pq}\left\{Y^{TE}_{pq}\vec{\phi}^{\,TE}_{pq} + \frac{j}{2\pi\omega\mu_0}(k^2+\nabla_t\nabla_t)\cdot\iint_{-\infty}^{\infty} ds'\vec{\phi}^{\,TE}_{pq}(x', y')G(r, r')\right\}\right.$$
$$\left. + E^{TM}_{pq}\left\{Y^{TM}_{pq}\vec{\phi}^{\,TM}_{pq} + \frac{j}{2\pi\omega\mu_0}(k^2+\nabla_t\nabla_t)\cdot\iint_{-\infty}^{\infty} ds'\phi^{TM}_{pq}(x', y')G(r, r')\right\}\right] \qquad (10.303)$$

The inner product with the weighting function is now formed for all the terms of eqn. 10.303. The resulting equations are

$$2Y^{TE}_{rs}D_{rs}\delta_{m'-r,\,n'-s} = \sum_p \sum_q \left[E^{TE}_{pq}\left\{Y^{TE}_{rs}D_{rs}\delta_{p-r,\,q-s}\right.\right.$$
$$\left. + \frac{j}{2\pi\omega\mu_0}\iint_{-\infty}^{\infty} ds\vec{\phi}^{\,TE*}_{rs}(x, y)\cdot(k^2+\nabla_t\nabla_t)\cdot\iint_{-\infty}^{\infty} ds'\vec{\phi}^{\,TE}_{pq}(x', y')G(r, r')\right\}$$
$$+ E^{TM}_{pq}\left\{\frac{j}{2\pi\omega\mu_0}\iint_{-\infty}^{\infty}\vec{\phi}^{\,TE*}_{rs}(x, y)\cdot(k^2+\nabla_t\nabla_t)\right.$$
$$\left.\left.\times\iint_{-\infty}^{\infty} ds'\phi^{TM}_{pq}(x', y')G(r, r')\right\}\right] \qquad (10.304)$$

and

$$2Y_{rs}^{TM} D_{rs} \delta_{m'-r,\, n'-s} = \sum_{p} \sum_{q} \Bigg[E_{pq}^{TM} \Bigg\{ D_{rs} \delta_{p'-r,\, q'-s} + \frac{j}{2\pi\omega\mu_0} \iint_{-\infty}^{\infty} ds \vec{\phi}_{rs}^{TM*}(x, y) \cdot (k^2 + \nabla_t \nabla_t) \cdot \iint_{-\infty}^{\infty} ds' \vec{\phi}_{pq}^{TM}(x', y') G(r, r') \Bigg\}$$

$$+ E_{pq}^{TE} \Bigg\{ \frac{j}{2\pi\omega\mu_0} \iint_{-\infty}^{\infty} ds \vec{\phi}_{rs}^{TM*}(x, y) \cdot (k^2 + \nabla_t \nabla_t) \times \iint_{-\infty}^{\infty} ds' \vec{\phi}_{pq}^{TE}(x', y') G(r, r') \Bigg\} \Bigg] \tag{10.305}$$

Eqns. 10.304 and 10.305 are similar to eqns. 10.285 and 10.287 except that the summation over (m, n), that is $(\Sigma_{m,n})$ has been replaced by integrals. The quadruple integrals in eqns. 10.304 and 10.305 represent the admittance at the designated aperture due to the mutual coupling of the aperture electric fields of other radiators in the array.

The element of the admittance matrix representing the mutual coupling between the TE_{rs} mode of the designated aperture radiator and the TE_{pq} mode of the aperture radiator at a distance $|r-r'|$ is given by

$$y_{rs,\, pq} = \frac{j}{2\pi\omega\mu_0} \iint_{-\infty}^{\infty} ds \vec{\phi}_{rs}^{TE*}(x, y) \cdot (k^2 + \nabla_t \nabla_t) \cdot \iint_{-\infty}^{\infty} ds' \vec{\phi}_{pq}^{TE}(x', y') G(r, r') \tag{10.306}$$

The domain of integration in eqn. 10.306 is the $z=0$ conducting plane containing the array. Since the tangential electric fields vanish over this plane except at the positions where the waveguide apertures are situated, the infinite integral of eqn. 10.306 reduces to a sum of integrals over all the waveguide radiator apertures of the finite array. Therefore the mutual admittance between TE_{rs} mode and TE_{pq} mode of the ith and jth waveguide aperture radiators can now be written as

$$y_{rs,\, pq}^{i,j} = \frac{j}{2\pi\omega\mu_0} \iint_{A_i} ds \vec{\phi}_{rs}^{TE*}(x, y) \cdot (k^2 + \nabla_t \nabla_t) \cdot \iint_{A_j} ds' \vec{\phi}_{pq}^{TE}(x', y') G(r, r') \tag{10.307}$$

where A_i and A_j are the surfaces corresponding to the ith and jth waveguide aperture radiators. In view of the symmetry of the Green's function with respect to its variables, $\nabla G(r, r) = -\nabla' G(r, r')$. This relation is introduced in eqn. 10.307 and one of the del operators is moved inside the integral sign. After differentiation of the Green's function, integration by parts with respect to the primed variable is carried out first and of the unprimed variable next. This leads to

$$y^{i,j}_{rs,pq} = \frac{jk^2}{2\pi\omega\mu_0}\left[\iint_{A_i} ds\, \vec{\phi}^{\,TE*}_{rs}(x, y) \cdot \iint_{A_i} ds'\, \vec{\phi}^{\,TE}_{pq}(x', y')G(r, r')\right.$$

$$\left. + \iint_{A_i} ds\, \vec{\nabla}_t \cdot \vec{\phi}^{\,TE*}_{rs}(x, y) \iint_{A_j} ds'\, \vec{\nabla}'_t \cdot \vec{\phi}^{\,TE}_{pq}(x', y')G(r, r')\right] \qquad (10.308)$$

In the case of TE modes, there is a longitudinal magnetic field component which has also been taken into account. From Maxwell's equations it can be inferred as

$$\nabla \cdot (\vec{H}^{\,t}) = jk_{zpq}\vec{H}^{\,l} \qquad (10.309)$$

Here $\vec{H}^{\,t}$ is the transverse magnetic field of the TE mode and it is tangential to the radiator aperture. $\vec{H}^{\,l}$ is the longitudinal magnetic field and propagates in the z-direction with a velocity k_{zpq}. Using eqn. 10.309, eqn. 10.308 can be rewritten as

$$y^{ij}_{rs,pq} = \frac{jk^2}{2\pi\omega\mu_0}\left[\iint_{A_i} ds\vec{\phi}^{\,TE*}_{crs}(x, y) \cdot \iint_{A_j} ds'\vec{\phi}^{\,TE}_{cpq}(x', y')G(r, r')\right] \qquad (10.310)$$

with

$$\vec{\phi}^{\,TE}_{cpq} = \vec{\phi}^{\,TE}_{pq} + j\hat{z}\vec{\phi}^{\,TE}_{zpq} \qquad (10.311)$$

$\vec{\phi}^{\,TE}_{cpq}$ is the composite modal function containing the transverse as well as the longitudinal components of the magnetic field of the pqth TE mode. $\vec{\phi}^{\,TE}_{crs}$ is the corresponding composite modal function of the TE mode. In the case of the TM modes eqn. 10.308 becomes

$$y^{i,j}_{rs,pq} = \frac{jk^2}{2\pi\omega\mu_0}\left[\iint_{A_i} ds\vec{\phi}^{\,TM*}_{rs}(x, y) \cdot \iint_{A_j} ds'\vec{\phi}^{\,TM}_{pq}(x', y')G(r, r')\right] \qquad (10.312)$$

The term within the square brackets in eqn. 10.310 is a sum of three quadruple integrals because each of the modal functions $\vec{\phi}^{\,TE}_{cpq}$ and $\vec{\phi}^{\,TE}_{crs}$ have three components in the x, y and z directions. It can be seen that the modal functions ϕ are regular trigonometric functions and are well behaved everywhere within the domain of integration. However, the Green's function requires special attention since it has a square root term $\sqrt{(x-x')^2+(y-y')^2}$ in the denominator. Therefore the term within the square brackets of eqn. 10.310 is now expressed as three separate integrals as shown below:

$$I_C = C_x I_x + C_y I_y + C_z I_z \qquad (10.313)$$

where

$$C_x = \frac{2pr}{a^2}\sqrt{\frac{\varepsilon_p\varepsilon_q}{ab\tau^2_{pq}}}\sqrt{\frac{\varepsilon_r\varepsilon_s}{ab\tau^2_{rs}}}$$

$$C_y = \frac{2qs}{b^2}\sqrt{\frac{\varepsilon_p\varepsilon_q}{ab\tau^2_{pq}}}\sqrt{\frac{\varepsilon_r\varepsilon_s}{ab\tau^2_{rs}}}$$

$$C_z = \frac{2\tau^2_{pq}\tau^2_{rs}}{\pi^2 k^2}\sqrt{\frac{\varepsilon_p\varepsilon_q}{ab\tau^2_{pq}}}\sqrt{\frac{\varepsilon_r\varepsilon_s}{ab\tau^2_{rs}}} \tag{10.314}$$

$$I_x = \iint_{A_i} ds \iint_{A_j} ds' G(r, r')\left\{\sin\frac{r\pi}{a}(x+a/2)\cos\frac{s\pi}{b}(y+b/2)\right\}$$
$$\times\left\{\sin\frac{p\pi}{a}(x'+a/2)\cos\frac{q\pi}{b}(y'+b/2)\right\} \tag{10.315}$$

$$I_y = \iint_{A_i} ds \iint_{A_j} ds' G(r, r')\left\{\cos\frac{r\pi}{a}(x+a/2)\sin\frac{s\pi}{b}(y+b/2)\right\}$$
$$\times\left\{\cos\frac{p\pi}{a}(x'+a/2)\sin\frac{q\pi}{b}(y'+b/2)\right\} \tag{10.316}$$

$$I_z = \iint_{A_i} ds \iint_{A_j} ds' G(r, r')\left\{\cos\frac{r\pi}{a}(x+a/2)\cos\frac{s\pi}{b}(y+b/2)\right\}$$
$$\times\left\{\cos\frac{p\pi}{a}(x'+a/2)\cos\frac{q\pi}{b}(y'+b/2)\right\} \tag{10.317}$$

In each of eqns. 10.315–10.317 a quadruple integral appears to integrate over the domain of the source and of the observer apertures, respectively. Lewin's procedure enables us to bring the order of integration down to a double integral.

Each of the quadruple integrals in eqns. 10.315–10.317 is similar and therefore the procedure followed for any of them in reducing the order of integration is applicable to others. Eqn. 10.315 will now be taken up to apply Lewin's method. The first step is to change the order of integration of the quadruple integral. Instead of integrating over the source (x, y) and the observer (x', y') we consider integration over (x, x') and (y, y') domains. Eqn. 10.315 is then rearranged to read

$$I_x = \int_0^b dy \int_0^b dy' \cos\frac{s\pi}{b}(y)\cos\frac{q\pi}{b}y'$$
$$\times\int_0^a dx\int_0^a dx' G(x-x', y-y')\sin\frac{r\pi}{a}x\sin\frac{p\pi x'}{a} \tag{10.318}$$

A transformation in the variables is now introduced such that, $\sigma = x - x'$, and $\nu = x + x' - a$. Also, the product of the two sine functions is changed by a trigonometric identity to a sum or difference of two cosine functions. Further the product $dx\,dx'$ changes to $\frac{1}{2}d\sigma\,d\nu$. Thus

$$I_x=\frac{1}{4}\int_0^b dy\int_0^b dy' \cos\frac{s\pi}{b}y \cos\frac{q\pi}{b}y' \int d\sigma \int d\nu G(\sigma, y-y')$$
$$\times\left[\cos\left\{\frac{\pi}{2a}(r+p)\sigma+\frac{\pi}{2}(r-p)+\frac{\pi}{2a}(r-p)\nu\right\}\right.$$
$$\left.-\cos\left\{\frac{\pi}{2a}(r-p)\sigma+\frac{\pi}{2}(r+p)+\frac{\pi}{2a}(r+p)\nu\right\}\right] \tag{10.319}$$

In eqn. 10.319 the end points of integration for σ and ν have not been specified as they have to be derived from Fig. 10.41. It can be easily seen that when

(*a*) $x=x'=0$, then, $\sigma=0$, and $\nu=-a$.
(*b*) $x=a$ and $x'=0$, then $\sigma=a$, and $\nu=0$.
(*c*) $x=0$ and $x'=a$, then $\sigma=-a$, and $\nu=0$.
(*d*) $x=x'=a$, then $\sigma=0$, and $\nu=a$.

The range of integration is now defined. A second change of variables is now introduced to express the variable ν in terms of the variable σ. Consider now the interval of integration OA (see Fig. 10.41). The variable ν can be represented by $\nu=\sigma-a$ in this leg. Similarly it can be shown that, for the intervals of integration, AB, BC and CO, the variable ν can be represented by $\nu=a-\sigma$, $\nu=a+\sigma$ and $\nu=-(a+\sigma)$, respectively. In addition, in eqn. 10.319 another trigonometrical identity is used to express each of the cosine functions as a sum of the product of two cosine and two sine functions. The integration with respect

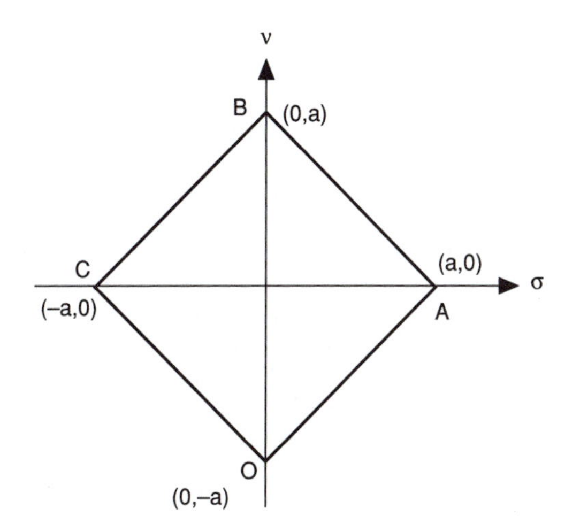

Figure 10.41 *Domain and range of integration for σ and ν*

to the variable is now carried out over the entire range of integration in terms of σ over each of the intervals specified. The terms are then grouped so that

$$
\begin{aligned}
I_x = \int_0^a d\sigma \Bigg[& \frac{a}{\pi(r-p)} \cos\left\{\frac{\pi}{2a}(r+p)\sigma + \frac{\pi}{2}(r-p)\right\} \sin\left\{\frac{\pi}{2a}(r-p)\sigma - \frac{\pi}{2}(r-p)\right\} \\
& - \frac{a}{\pi(r+p)} \cos\left\{\frac{\pi}{2a}(r-p)\sigma + \frac{\pi}{2}(r+p)\right\} \sin\left\{\frac{\pi}{2a}(r+p)\sigma - \frac{\pi}{2}(r+p)\right\}\Bigg] \\
& \times \int_0^b dy \int_0^b dy' [G(\sigma, y-y') + \cos r\pi \cos p\pi \; G(-\sigma, y-y')] \\
& \times \sin\frac{s\pi}{b} y \sin\frac{q\pi}{b} y'
\end{aligned}
\tag{10.320}
$$

The double integral with variables y and y' is subjected to the same procedure as the earlier two variables x and x'. In this case we define $\zeta = y - y'$ and $\xi = y + y' - b$. The range of integration is derived in a similar manner and the values of ξ in terms of the different intervals of integration are also determined. Ultimately eqn. 10.320 is changed to

$$
\begin{aligned}
I_x = \int_0^a d\sigma \Bigg[& \frac{a}{\pi(r-p)} \cos\left\{\frac{\pi}{2a}(r+p)\sigma + \frac{\pi}{2}(r-p)\right\} \sin\left\{\frac{\pi}{2a}(r-p)\sigma - \frac{\pi}{2}(r-p)\right\} \\
& - \frac{a}{\pi(r+p)} \cos\left\{\frac{\pi}{2a}(r-p)\sigma - \frac{\pi}{2}(r+p)\right\} \sin\left\{\frac{\pi}{2a}(r+p)\sigma - \frac{\pi}{2}(r+p)\right\}\Bigg] \\
& \times \int_0^b d\zeta \left[\frac{b}{\pi(s-q)} \cos\left\{\frac{\pi}{2a}(s+q)\zeta + \frac{\pi}{2}(s-q)\right\}\right. \\
& \times \sin\left\{\frac{\pi}{2a}(s-q)\zeta - \frac{\pi}{2}(s-q)\right\} + \frac{b}{\pi(s+q)} \cos\left\{\frac{\pi}{2a}(s-q)\zeta\right. \\
& \left.\left. - \frac{\pi}{2}(s+q)\right\} \sin\left\{\frac{\pi}{2a}(s+q)\zeta - \frac{\pi}{2}(s+q)\right\}\right] \\
& \times [G(\sigma, \zeta) + \cos r\pi \cos p\pi \; G(-\sigma, \zeta) + \cos s\pi \cos q\pi \; G(\sigma, -\zeta) \\
& + G(-\sigma, -\zeta) \cos r\pi \cos p\pi \cos s\pi \cos q\pi]
\end{aligned}
\tag{10.321}
$$

The quadruple integral of eqn. 10.319 has been reduced to a double integral with simple trigonometric functions in the integrand. This procedure is then applied to the other two integrals given in eqns. 10.316 and 10.317. The determination of I_x, I_y and I_z leads us to the completion of calculating the element $y^{ij}_{rs,pq}$ of the mutual admittance matrix. A generalised expression for the integrand of eqn. 10.321 has been provided in Reference 35 to cover the mutual coupling between the modes TE→TM, TM→TM and TM→TE.

Eqn. 10.321 needs to be be processed further in the case of mode coupling in the same aperture. By converting the variables (σ, ζ) to polar co-ordinates the singularity of the integral can be removed. The procedure is similar to that followed in Section 10.22.5 for eqn. 10.237.

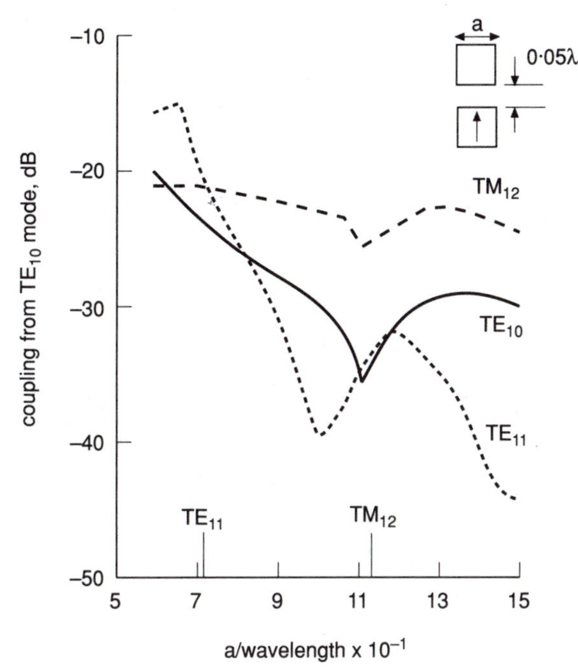

Figure 10.42 *Mode power coupling between square waveguides [46]*

It is very clear from the analysis that the mutual coupling between the fundamental TE_{10} modes between the waveguide apertures of the array cannot explain the practical results since a significant amount of coupled power is shared by the mutual coupling among the higher order modes. This aspect is effectively brought out in Fig. 10.42 which shows the mode power coupling between two square waveguide apertures with a wall separation of only 0.05λ [46]. The aperture sizes are varied from 0.6λ to 1.5λ. The modes shown are TM_{12}, TE_{10} and TE_{11}. Of these, the TM_{12} mode appears to be the most dominant. Because of the asymmetry of the mode field, radiation by this mode from the aperture distorts the co-polar pattern of the dominant TE_{10} mode. The TM_{12} mode is also rich in cross-polarisation and this contributes to the raising of the cross-polarised radiated power. Fig. 10.43, also from Reference 46, highlights the fact that the TE_{10} mode of one waveguide aperture radiator excites four other higher order modes, namely TE_{11}, TM_{11}, TE_{12} and TM_{12} modes, in another waveguide aperture of the same array when the wall to wall distance is 0.05λ. The variation of the power in these modes with aperture size variation has also been brought out. It is evident that, for $a/\lambda < 1$, the TE_{11} mode is excited most in the other waveguide in comparison to the other higher order modes. For a/λ between 1 and 1.4, all the five modes TE_{10}, TE_{11}, TM_{11}, TE_{12} and TM_{12} are excited to about the -30 dB level. An increase in the inter-aperture distance results in the reduction of the coupled power by about 3 dB for the $TE_{10} \rightarrow TE_{11}$ mode coupling. For other mode couplings the power decreases slightly. It has been found that effectively the inter-mode coupled power is limited to about seven modes over the dominant TE mode. If these

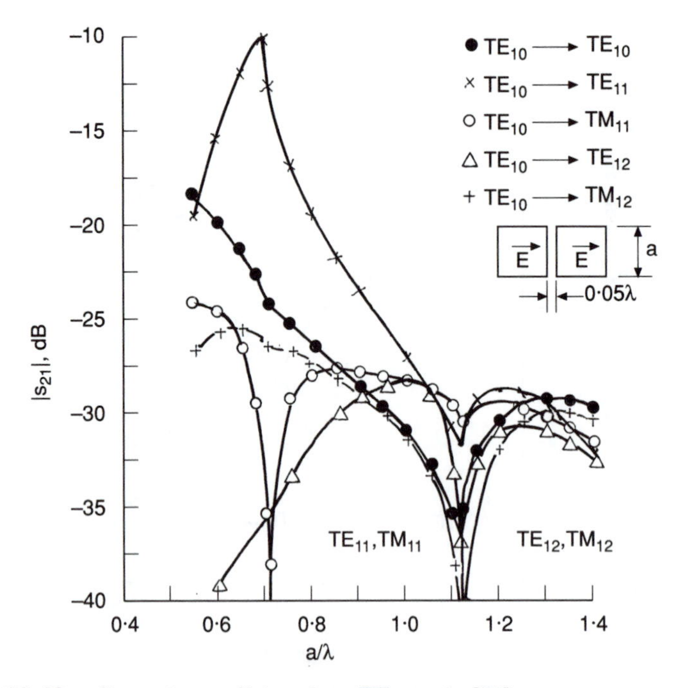

Figure 10.43 *Conversion coefficient from TE_{10} mode [35]*

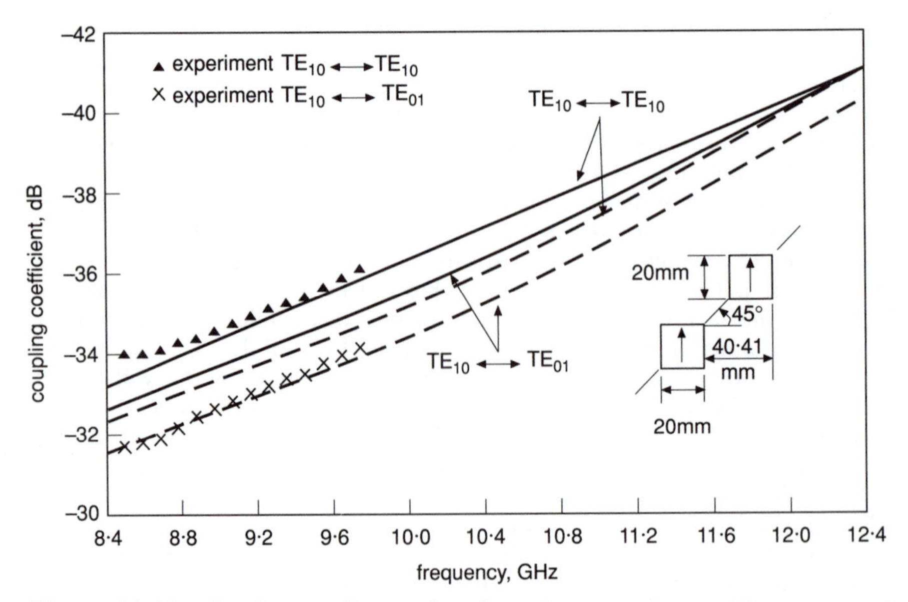

Figure 10.44 *Coupling coefficients for identical waveguides at 45° to principal polarisation [36].* © *IEEE, 1990*

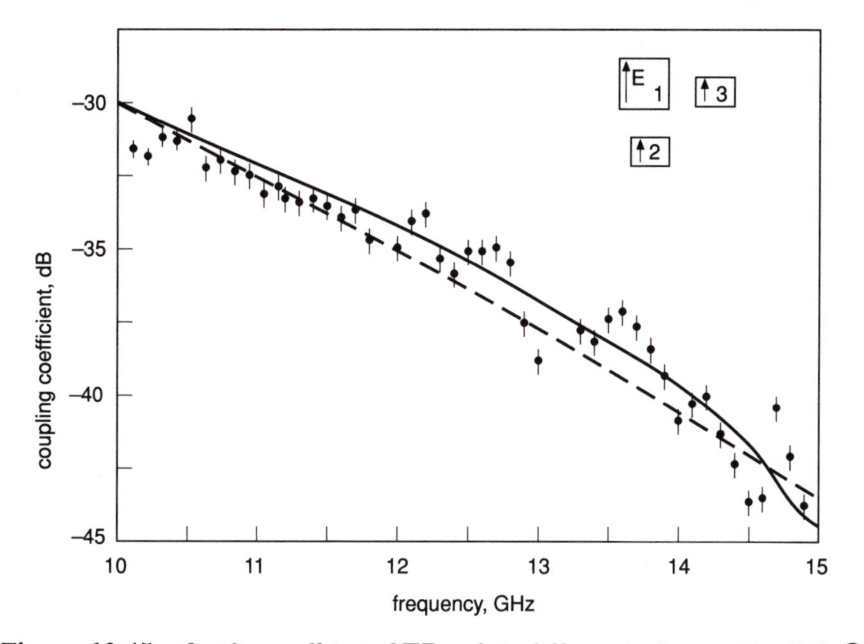

Figure 10.45 *Coupling coefficient of TE mode in different sized waveguides [36].* © *IEEE, 1990*

● ● ● experiment
—— theory, 7 modes
--- theory, single mode

square aperture radiators had been realised by flaring a square waveguide in the manner of a pyramidal horn, the higher order modes generated at the aperture due to mode coupling from other aperture radiators of the array would have travelled down the tapered horn until they reach the throat of the horn or at that cross-section where it is cut off to a particular higher order mode. The higher order modes would have been reflected back towards the radiating aperture where the phase of the returning waves not only alters the aperture field but also causes further mutual coupling. Fortunately it has been found that, after three iterations, this effect can be neglected. The experimental results are reported to have good agreement with the analytical findings. A comparison of the effectiveness of the multi-modal analysis with the first order approximation [36] (only $TE_{10} \rightarrow TE_{10}$ coupling) is presented in Fig. 10.44. The results of the two analyses are compared with the experimental results for two square waveguides of side 20 mm and operating in the 8.2–12.4 GHz region. The waveguides are so arranged that the coupling in the 45° plane is considered.

The multi-modal analysis in the case of different size waveguide apertures has also been validated in experiments [36]. The main finding of the analysis, namely that only the first seven modes need be considered to account for the coupling power has been verified by experiment. Fig. 10.45 shows the analytical and experimental results for the $TE_{10} \rightarrow TE_{10}$ mode coupling in a 3-element array of different size waveguide apertures. Fig. 10.45 refers to the results for two apertures of this array, one of them being a square waveguide of side 22.8 mm and the other a rectangular aperture radiator of dimensions 15.7 × 7.7 mm placed at a distance 30 mm away from the centre of the square guide. It is evident from the Figure that there is good agreement between the experiment and analysis.

Over the years, analysis of mutual coupling effects in electronically scanning arrays has progressed to a point where the electromagnetic effects are understood and theory is accurate enough to predict the behaviour of the array.

10.24 References

1 HANSEN, R. C.: 'Evaluation of the large array method', Proc. IEE, vol. 137, Part H, April 1990, pp 94–98.
2 RUDGE, A. W., MILNE, K., OLVER, A. D., and KNIGHT, P. (Eds.): 'Handbook of antenna design' (Peter Peregrinus, London 1986) Chap. 10
3 OLINER, A. A.: 'Surface wave effects and blindness in phased array antennas', Proc. Phased Array Ant. Symp. 1970 (Artech House, Norwood, USA, 1972) pp. 107–112
4 KNITTEL, G. H.: 'Phased array antennas', *in* BROOKNER, E. (Ed.): 'Radar technology' (Artech House, Norwood, USA, 1979) Chap. 21
5 ELLIOTT, R. S.: 'Basic considerations in the design of arrays', Proc. 15th Euro. Microwave Conf. 1985, pp. 561–565
6 KO, W. L., and MITTRA, R.: 'Scattering by a truncated periodic array', *IEEE Trans.*, 1988, **AP-36**, pp. 496–503
7 DIAMOND, B. L.: 'Small arrays: Their analysis and their use for the design of array elements', *in* OLINER, A. A., and KNITTEL, G. H. (Eds.) 'Phased array antennas', (Artech House, Norwood, USA, 1972) pp. 127–131
8 OLINER, A. A., and MALECH, R. G.: 'Mutual coupling in infinite scanning arrays', *in* HANSEN, R. C. (Ed.) 'Microwave Scanning Antennas'. vol. II (Academic Press, NY, 1966) Chap. 3
9 FARRELL, G. F., and KUHN, D. H.: 'Mutual coupling in infinite planar arrays of rectangular waveguides', *IEEE Trans.*, 1968, **AP-16**, pp. 405–414
10 HARRINGTON, R. F.: 'Field computation by moment methods', (McMillan, NY, 1968)
11 HARRINGTON, R. F., and MAUTZ, J. R.: 'Electromagnetic transmission through an aperture in a conducting plane', *in* HANSEN, R. C. (Ed.): 'Moment methods in antennas and scattering', (Artech House, Boston, USA, 1990) pp. 358–364
12 POGGIO, A. J., and MILLER, E. K.: 'Techniques for low frequency problems', *in* LO, Y. T., and LEE, S. W. (Eds.): 'Antenna Handbook', (Van Nostrand Reinhold, NY, 1988)
13 RUMSEY, V. H.: 'Reaction concept in electromagnetic theory', *Phys. Rev.*, 1954, **94**, pp. 1483–1491
14 CHANG, V. W. H.: 'Infinite phased dipole array', *Proc. IEEE*, 1968, **56**, pp. 1892–1900.
15 SARKAR, T. K.: 'A note of the choice of weighting functions in the method of moments', *IEEE Trans.*, 1985, **AP-33**, pp. 436–441
16 SARKAR, T. K., DJORDJEVIC, A. R., and ARVAS, E.: 'On the choice of expansion and weighting functions in the numerical solution of operator equations', *IEEE Trans.*, 1985, **AP-23**, pp. 988–996
17 MILLER, E. J., and DEADRICK, F. J.: 'Some computational aspects of thin wire modelling', *in* MITTRA, R. (Ed.): 'Numerical and asymptotic techniques in electromagnetics' (Springer-Verlag, NY, 1975) pp. 90–127
18 KING, R. W. P., MACK, R. B., and SANDLER, S. S.: 'Arrays of cylindrical dipoles' (Cambridge University Press, 1968) pp. 282–283
19 CHANG, V. W. H., and KING, R. W. P.: 'Theoretical study of dipole array of *N* parallel elements', *Radio Science*, 1968, **3**, pp. 411–424

20 ELLIOTT, R. S.: 'Antenna theory and design' (Prentice Hall India, 1981) Chap. 8
21 STERN, G. J., and ELLIOTT, R. S.: 'Resonant length of longitudinal slots and validity of circuit representations: Theory and experiment', *IEEE Trans.*, 1985, **AP-33**, 1264–1271
22 SANGSTER, A. J., and McCORMICK, A. H. I.: 'Theoretical design/synthesis of slotted waveguide arrays', *Proc. IEE*, 1989, **136**, Part H, pp. 39–46
23 GULICK, J. J., and ELLIOTT, R. S.: 'The design of linear and planar arrays of waveguide-fed longitudinal slots', *Electromagnetics*, 1990, **10**, pp. 327–347
24 STEVENSON, A. F.: Theory of slots in rectangular guides', *J. Appl. Physics*, 1948, **19**, pp. 34–38
25 STUTZMAN, M., and THIELE, G. A.: 'Antenna theory and design', (Wiley, NY, 1981) Chap. 7
26 RENGARAJAN, S. R.: 'Higher order mode coupling effects in the feeding waveguide of a planar slotted array', *IEEE Trans.*, 1991, **AP-39**, pp. 1219–1223
27 BAILEY, M. C., and BOSTIAN, C. W.: 'Mutual coupling in a finite planar array of circular apertures', *IEEE Trans.*, 1974, **AP-22**, pp. 178–184
28 STEYSKAL, H.: 'Mutual coupling analysis of a finite planar waveguide array', *IEEE Trans.*, 1974, **AP-22**, pp. 594–597
29 BIRD, T. S.: 'Mode coupling in a planar circular waveguide array', *Proc. IEE, Microwave, Optics & Acoustics*, 1979, **3**, pp. 172–180
30 BORGIOTTI, G. V.: 'A novel expression for the mutual admittance of planar radiating elements', *IEEE Trans.*, 1968, **AP-16**, pp. 329–333
31 MAILLOUX, R. J.: 'Radiation and near field coupling between two collinear and open ended waveguides', *IEEE Trans.*, 1969, **AP-17**, pp. 49–55
32 MAILLOUX, R. J.: 'First order solutions for mutual coupling between waveguides which propagate two orthogonal modes', *IEEE Trans.*, 1969, **AP-17**, pp. 740–746
33 FENN, A. J., THIELE, G. A., and MUNK, B. A.: 'Moment method analysis of finite rectangular waveguide phased arrays', *IEEE Trans.*, 1982, **AP-30**, pp. 554–564
34 LUZWICK, J., and HARRINGTON, R. F.: 'Mutual coupling analysis in a finite planar rectangular waveguide antenna array', *Electromagnetics*, 1982, **2**, pp. 25–42
35 BIRD, T. S.: 'Mutual coupling in finite coplanar rectangular waveguide arrays', *Electron. Lett.*, 1987, **AP-23**, pp. 1199–1207
36 BIRD, T. S.: 'Analysis of mutual coupling in finite arrays of different sized rectangular waveguides', *IEEE Trans.*, 1990, **AP-38**, pp. 166–172
37 ITOH, T., and MITTRA, R.: 'Spectral domain approach for calculating the dispersion characteristics of microstrip lines', *IEEE Trans.*, 1973, **MTT–21**, pp. 496–499
38 RAHMAT SAMII, Y., ITOH, T., and MITTRA, R.: 'A spectral domain analysis for solving microwave discontinuity problems', *IEEE Trans.*, 1974, **MTT–22**, pp. 372–378
39 SCOTT, C.: 'The spectral domain method in electromagnetics', (Artech House, Mass., USA, 1989) Chaps. 1 and 4
40 HANSEN, J. E. (Ed.): 'Spherical near-field antenna measurements', (Peter Peregrinus, London, 1988) Chap. 7
41 AMITAY, N., GALINDO, V., and WU, C. P.: 'Theory and analysis of phased array antennas', (Wiley InterScience, NY, 1972) Chap. 5
42 FELSEN, L. B., and MARCUVITZ, N.: 'Radiation and scattering of waves', (Prentice Hall, NJ, 1973) pp. 444–473
43 MOHAMMADIAN, A. H., MARTIN, N. M., and GRIFFIN, D. W.: 'A theoretical and experimental study of mutual coupling in microstrip antenna arrays', *IEEE Trans.*, 1989, **AP-37**, pp. 1217–1223
44 KITCHENER, D., RAGHAVAN, K., and PARINI, C. G.: 'Mutual coupling in a finite array of rectangular apertures', *Electron. Lett.*, 1987, **23**, pp. 1169–1170
45 LEWIN, L: 'Advanced theory of waveguides', (Iliffe & Sons, London, 1951) chap. 6
46 KITCHENER, D., and PARINI, C. G.: 'Mutual coupling in finite arrays of rectangular apertures', Proc. International Conference on Antennas and Propagation, 1989, ICAP-89, pp. 243–248
47 JASIK, H.: 'Antenna engineering handbook', Figs. 9.5, 9.7, 9.9 and 9.10 (McGraw-Hill, 1961)

Part 3: Active aperture arrays

10.25 Introduction

Active aperture phased array has come into prominence in the last five years mainly because it is the only technology which holds out the promise of realising affordable electronic scanning radar sensors. The search for an economical method to incorporate electronic scanning has been initiated in the late 1970s because of the low rate of induction of phased array radars even though the technology was more than one decade old. While it has come to be accepted that the cost of phased array antennas cannot be lowered to a level where they would be as widely used as the mechanically scanning reflector antennas, it should be possible to improve the precision in design and production of phased arrays to reduce costs.

It is a fact that in most surveillance radar sensors, be they airborne, ground based or shipborne, a minimum amount of electronic control has already been used for beam forming, shaping and scanning by employing planar phased array antennas. Even with such limited application, substantial progress has been made in the understanding of mutual coupling aspects and in compensatory microwave circuit design, phase shifting networks and integrated radiators. Such phased array antennas are currently being assembled by integrating individual radiating elements, phase shifting and matching networks and further testing on a unit-by-unit basis. This has proved expensive and alternative approaches which make full use of production techniques perfected in large volume production of integrated circuits are being implemented. The active aperture array concept, in which the individual radiating element is easily combined with a module which transmits RF power of adequate level in phase synchronism with other such modules in the array and receives RF energy in the receive mode, thus appears to be the most promising solution.

The idea of active aperture phased array dates back to 1964 when the AN/FPS–85 development was initiated for a radar operating in the UHF band with 4660 active elements in the phased array [1]. The active array development appears to have been motivated by the need for realisation of a large power-aperture product, since each of these active elements generated 10 kW of peak RF power. Interestingly, work on solid state active arrays was started in the same year for developing MERA (Molecular Electronics for Radar Application) an experimental X-band system [2]. The primary objective was to improve on the conventional approach of interconnecting many RF functions so that more microwave circuit functions can be accommodated in a limited space. The microwave integrated circuit packages also provided other advantages such as higher reliability and increased performance due to reduced interconnections. MERA was completed in 1969 and was followed by RASSR (Reliable Advanced Solid State Radar) to demonstrate that a practical active aperture phased array radar system can be built, and to bring out that reliability improvement due to solid state devices is actually available in such systems [3]. However, the technology of RASSR was not followed through in production for other radars built after 1975, because a direct power source and a low noise amplifier at microwave frequencies were not available in the desired configuration.

The advent of the field effect transistor effectively closed this gap so that it would be possible to have solid state T–R (transmitter–receiver) modules with low noise figures in the receive path and higher output powers for transmission, manufactured in sufficiently large numbers [4, 5]. Since then, there have been significant advances in the technology of active and passive components used in T–R modules, the emphasis being reduction of size and volume for both types and better efficiency for power devices. Currently opinion exists that, for radars operating in the lower microwave frequencies (up to 6 GHz), hybrid T–R modules may be a better answer firstly because they are less expensive and secondly because they can deliver power levels beyond the reach of a single chip. This picture is, however, changing towards higher functional integration, and currently the linear gallium arsenide integrated circuit, more familiarly known as the monolithic microwave integrated circuit or MMIC, will replace other circuit configurations because it combines the field effect transistor and the microwave integrated circuit technologies as a single chip. As a result of this, the parasitics are reduced and reliability is improved, leading to enhanced performance. The MMIC also has potentially reduced cost, and reduced manufacturing time scales for high volumes. However, the MMICs place an additional burden on the microwave designer in that he must commit to a final circuit very early because the small size chip prevents tuning or changing circuits during testing to peak the performance. Further, for realising the benefit of cost reduction, the antenna configuration has to be compatible with the transmission line configuration of the T–R module to which the radiator will have to be joined. Since microstrip lines are used extensively in the T–R module for connecting the variable network elements, the only compatible configuration would be the PCA or the printed circuit antenna. PCA would therefore replace the waveguide excited slot or aperture or the coaxial transmission-line compatible rounded dipole as the antenna element in active aperture phased arrays. In the last two decades considerable information and understanding of the behaviour of these elements in the finite as well as in the infinite array environment had been gathered by experiment and confirmed by theoretical studies. These will be of use in resolving problems likely to arise in printed circuit array antennas.

In an active aperture array, therefore, the RF power sources as well as the RF portion of the receiver chain are distributed across the array. Phased array radars in future can be built by going to active aperture so that the thousands of T–R modules needed in the sensor can be economically manufactured by techniques of lithography. Printed circuit antenna arrays (PCAA), MMIC techniques for T–R modules, VLSI chips and fibre optic techniques are expected to play a large part in the development of future radar systems. Active aperture radar is therefore: ‘A favoured design that provides greater benefits with the forthcoming increases in signal and computer control capabilities. The active array offers advantages in areas of power management, beam steering, target detection and system performance. Active arrays may also give advantages in weight and volume' [6]. Hence there should be no doubt about the inevitableness of these advances in radar systems of the future, especially as the present efforts to reduce the cost of fabrication bears fruit in the next few years.

It is interesting to note that similar conclusions can be arrived at, if one views the progress that has taken place in technology. Already the active aperture 'technology has matured to allow its application in production radar systems in the 1990s. The evolution of the monolithic microwave integrated circuit (MMIC) technology for active aperture in AESA (Active Electronically Scanning Array) application has been the final step in system efficiency and cost reduction, making it the preferred technology for all future multimode radar developments' [7]. The main arguments in support of the conclusions are as follows.

Over the years, the power output of the T–R modules has gone up and the efficiencies have also shown a significant upward trend. For example in X-band, solid state devices have progressed from 1 W peak power and junction efficiency of 15% to approximately 10 W peak and junction efficiencies of 40% in less than two decades.

There has been considerable progress in the last 15 years in MMIC realisation from a single function module to multifunction modules containing power amplifier, low noise amplifier, phase shifting network, limiters, switches etc. The progress, especially in the last five years, has been very significant as evidenced from a comparison of typical MMIC multifunction modules of 1986 and of today. The present module has a lower component count (59 versus 348), fewer subassemblies (2 versus 5) and fewer interconnections (200 versus 561).

There is a major programme of development aimed at ensuring not only the millimetre and microwave monolithic circuits but also the processes by which such circuit can be designed rapidly, repeatedly and productionised at lower costs. In addition, there are other programmes for expanding the performance, reliability and cost envelopes of the components used in these modules.

A key component in implementing such arrays with minimum performance degradation is the power supply for T–R modules. The effort is to develop compact, highly reliable power supplies to give low voltage, high current low noise power with densities of 50 W/in^3 as against the present capability of 10 W/in^3.

There are two important radar system developments, one for airborne and the other for ground based applications, that have been initiated for demonstrating the key performance factors, cost and reliability–availability–maintainability (RAM) features of active phased array radars [8]. The first one is the Ultra Reliable Radar (URR) programme using an 813 mm diameter solid state phased array (SSPA) containing 1980 T–R modules. This radar will generate different beam shapes such as pencil and wide pencil, cosecant squared, wide fan beam, thinned pencil and dual beam modes by controlling the state of each T–R module. The second one is an active aperture radar (AAR) employing MMIC, VHSIC, optical and bistatic technologies to meet the requirements of the late 1990s. AAR has significantly expanded multifunction capabilities and is expected to achieve 50% savings in acquisition cost, a 10:1 saving in life cycle cost, a 7:1 increase in availability, a 50–60% saving in prime power and a 50% reduction in weight and volume. This radar will have

6000 T–R modules and can be used both as a rotating or as a fixed array operating in conventional, monostatic and bistatic modes. A 1000 module AAR is planned to verify many of the key performance, cost and RAM features.

In this Chapter it is therefore planned to go into the present status of the T–R modules, bringing out the progress in GaAs MMIC single chip and multiple chip amplifiers, microwave digital phase shifters, multiple chip and single chip T–R modules and their manufacturability. In addition the current status on issues of broadbanding printed circuit antennas, the feed configurations that are compatible with the manufacturing techniques and the packaging aspects will also be looked into.

10.26 Basic configuration

The active aperture concept of RF power sources being distributed across the array eliminates the need for a feed network for distributing a large amount of RF power from a distant single high power source to the antenna aperture. On the other hand, because of the space combining aspect of RF power in these systems, the feed network deals with relatively low level signal distribution to and from the aperture. Thus each T–R module which is placed in the immediate vicinity of the radiating element of the active array will need microwave, bias, IF and digital lines to provide and process the data. One basic configuration of an active aperture array is shown in Fig. 10.46. For the sake of clarity three separate networks for feed and control are shown away from the active array. The RF distribution in this case generates the necessary microwave signals and transmits through the distribution network these signals to the power amplifiers in the transmit path of the T–R module so that phase and frequency synchronisation among the modules is realised for space combining to provide the effective radiated power (EP) at the target. The microwave distribution network also provides the local oscillator (LO) signals necessary for converting the signals received through the receive channel of the T–R module into IF signals for combining, mixing and signal processing. It is also possible to conceive the combining of the received signal at the video level because the active aperture phased array lends itself easily to digital beam forming owing to the distributed nature of the transmit and receive channels. More than any other configuration, the active aperture system provides great scope for control of transmit waveform, radiation characteristics of the antenna, signal processing and target management.

In the transmit mode of operation, the use of MMIC power amplifiers with each T–R module eliminates all high potentials and the necessary safety mechanisms that need to be built into radar transmitters for safety of operating personnel. Secondly, the distributed nature of the RF sources across the array provides for graceful degradation and thus removes additional constraints that are imposed today on reliable operation of dual or standby transmitters. Since low level RF signals are transmitted to each of the T–R modules, it is possible to incorporate with greater ease desired digital modulation for the transmitted waveforms. Computer control of the transmitter waveform can be carried out without the attendant problems faced in high power transmitters. Further, the advantage of the distributed nature of the RF sources is in controlling the

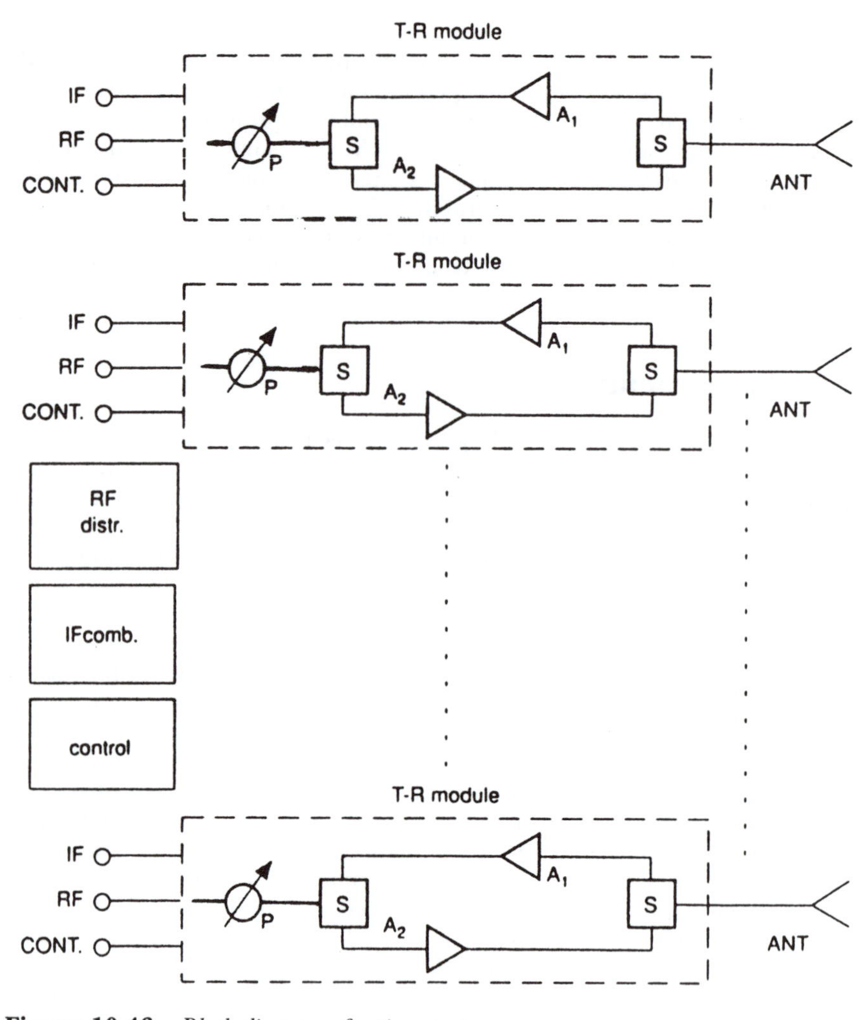

Figure 10.46 *Block diagram of active aperture array*

A_1 = LNA
A_2 = PA
S = switch
P = phasor

illumination function across the array so that, in the transmit mode, wide pencil, pencil and shaped beams can be generated on command from a common processor to alter the amplitude of the output signal either by gain control or by attenuator and the phase of the signal through the microwave digital phase shifter. In the receive mode, since flexibility exists in the system to carry out combining the received signal through the array at the video level, it is possible to have a radiation pattern different from the transmit pattern and even to have multiple beams without loss of gain in the system.

These flexibilities, available for changing the transmit waveform and having separate transmit and receive radiation patterns for the array, open up greater possibilities for optimised target management. Target management can be defined as more than simple detection but a varying and relevant response by the sensor towards a target during its traverse in the coverage volume of the radar. The response can have several measures such as variable and non-uniform RF illumination of the coverage volume, adaptive radiation characteristics of the antenna, repertoire of transmitter waveforms relevant to responses sought from the radar and a mix of scan, track-while-scan and track modes.

The distributed nature of power generation as well as the receiver chain enables us to rewrite the classical radar equation in terms of T–R module and antenna characteristics. In what follows, it is assumed that a completely filled array is utilised and the illumination function of the array antenna remains the same in transmission as well as in reception.

If,

G_e = gain of the radiating element,
P_t = average output power of the T–R module
L_t = losses in transmission including propagation
L_r = loss in the receiving path
N = the number of T–R modules which is equal to the number of radiating elements
η = efficiency of the excitation function of the array as compared with uniform illumination.

Then, the power illuminating a point target at a distance R from the radar is given by

$$P_{inc} = \frac{P_t L_t G_e N^2 \eta}{4\pi R^2} \tag{10.322}$$

Assuming a radar cross-section of σ for the point target, the power reflected by the point target is given by

$$P_{rf} = \frac{P_t L_t G_e N^2 \eta \sigma}{4\pi R^2} \tag{10.323}$$

The power received at the radar receiver is given by

$$\begin{aligned} P_{rec} &= \frac{P_t L_t G_e N^2 \eta \sigma}{(4\pi R^2)^2} \frac{\eta N G_e \lambda^2 L_r}{4\pi} \\ &= \frac{P_t L_t L_r G_e^2 \lambda^2 N^3 \eta^2 \sigma}{(4\pi)^3 R^4} \end{aligned} \tag{10.324}$$

Now $P_{min} = F_n KTB\,(S/N)_{OUT}$

Here F_n is the noise figure of the receiver, K is Boltzmann's constant, T is the absolute temperature in degrees Kelvin, B is the bandwidth and $(S/N)_{OUT}$ is the signal to noise ratio at the output of the receiver.

$$R^4_{max} = \frac{P_t L_t L_r G_e^2 N^3 \eta^2 \lambda^2 \sigma}{(4\pi)^3 F_n KTB\,(S/N)_{OUT}} \tag{10.325}$$

The dependence of the maximum detection range for a given radar cross-section of the target on the T–R module is brought out clearly in eqn. 10.325. The three parameters that affect the range are:

(*a*) N, number of T–R modules
(*b*) P_t, average power of the T–R module
(*c*) F_n, noise figure of the low noise amplifier in the T–R module.

Prima facie it would appear from eqn. 10.325 that the most effective way of improving the detection range would be to increase the parameter N or the number of T–R modules in a filled active aperture array. This is because, unlike in a passive aperture system, the number of T–R modules affects the RF power available for transmission as well as the gain of the antenna. Constraints of physical aperture and beamwidth considerations can restrict the increase in the number of T–R modules. If, however, the second parameter of the T–R module is taken into consideration then N^3P_t can be considered as the power-aperture product so that, by a judicious mix of increase of N and P_t, the detection range can be improved. The developments that have taken place in improving the efficiency of the output amplifiers in T–R modules bear testimony to the importance given to this parameter. The noise figure of the receive path in the T–R module, which is the third parameter, has also received considerable attention in the last four years. Mention also needs to be made of the fact that, since the maximum detection range of the active aperture radar system is also directly dependent on the number of T–R modules, for proliferation of active aperture systems, reducing the cost of T–R modules in production is an important consideration. The T–R module is therefore a key subsystem for active aperture radars.

10.27 Transmitter–receiver modules

The possibility of using solid state devices in radar transmitters was initiated in the 1960s because of the obvious advantages of higher reliability, graceful degradation, on-line replacement, low voltage operation, elimination of modulators etc. The use of low voltage power supplies also results in substantial improvement in the transmitter stability because of lower modulation sidebands. Enhanced values for the MTI improvement factor come about mainly as a result of better control on the ripple in the power supplies which would be below 50 V. Solid state transmitters with capability of 70 dB improvement factor have been fabricated without much effort. This approach proved that distributing RF power sources across the array to feed a group of radiating elements and providing a return path for the received signals can be engineered in field systems with higher reliability. The use of T–R modules behind the radiating aperture thus became a viable architecture. However, the technology of fabrication in this case was microwave integrated circuits (MIC) or hybrid MIC which is not compatible with array lattice configurations.

Simultaneously with this effort of building solid state power amplifiers, attention was also focused in the 1960s on GaAs as a semiconductor material owing to its superior mobility, electron velocity and microwave properties. Over 10 years of R&D have been spent on this semiconductor as a material for discrete field effect transistors (FETs). GaAs is now refined to a point where it is good enough for most analogue applications, with 4 in diameter wafers already available in the market. It was realised right from the beginning that the T–R module architecture for phased arrays needs reduction in cost and size of these subsystems. This could only be guaranteed by replacing MIC with monolithic microwave integrated circuit technology. In MMIC all components, both passive and active, are incorporated into a single semiconductor die permitting complete operation by the application of DC and/or microwave signals. MMIC offers several advantages, the more important of them being:

(*a*) It enables us to compress many separate functions into a single substrate and thus substantially reduce the severity of problems due to bonding and interconnects and improved reliability.
(*b*) The size and weight of the circuits are usually much smaller, allowing either subsystem size reduction or incorporation of more functions within the same volume.
(*c*) All circuit functions can be integrated so that highly reproducible performance can be obtained in batch production.
(*d*) Simpler packaging schemes can be considered owing to reduced circuit sizes.
(*e*) More than anything else, it is possible to achieve economical high volume production. Eventually the T–R module size is limited by the power output and the number of RF, DC and control line inter-connections.

It is thus evident that T–R module development is emerging as a business rather than a purely R& D effort with specific goals of making them available in the 1–30 GHz region [9–13]. These super components have been the subject of many studies and investigations [14–19] from the point of view of design, fabrication, production yield, process tolerance and cost. It is now certain that gallium arsenide monolithic microwave integrated circuit technology will be the technology vehicle through which the minimisation of cost and enhancement of production yield and process tolerance is achieved. It has almost reached the status of a universal panacea, something that will solve all the problems that have been earlier confronted in the effort to achieve affordable phased arrays. In a very short period of time MMICs have progressed from a single function circuit to more complex multi-function modules, the emphasis being reduction of size and volume, higher rating for power devices and better efficiency for power amplifiers with gradual decrease in the cost of production. A comprehensive and detailed account of the progress made in T–R modules, as well as towards the realisation of active aperture arrays, is provided by Pengelly [20] and Shenoy [21, 22], respectively. Ideally, the T–R module would contain amplifiers for transmit and receive states as well as for electronic phase shifting for both states. Thus it would carry out four RF functions, namely power amplification, phase shift/control, low noise amplification and switching. Because of the fact that the active aperture uses individual T–R module at the array radiating elements, this configuration can be substantially more efficient on the basis of overall prime power versus detection capability. For the same effective radiated power, the active aperture configuration can be smaller and lighter than the conventional system. A generic form of the T–R module is shown in Fig. 10.47.

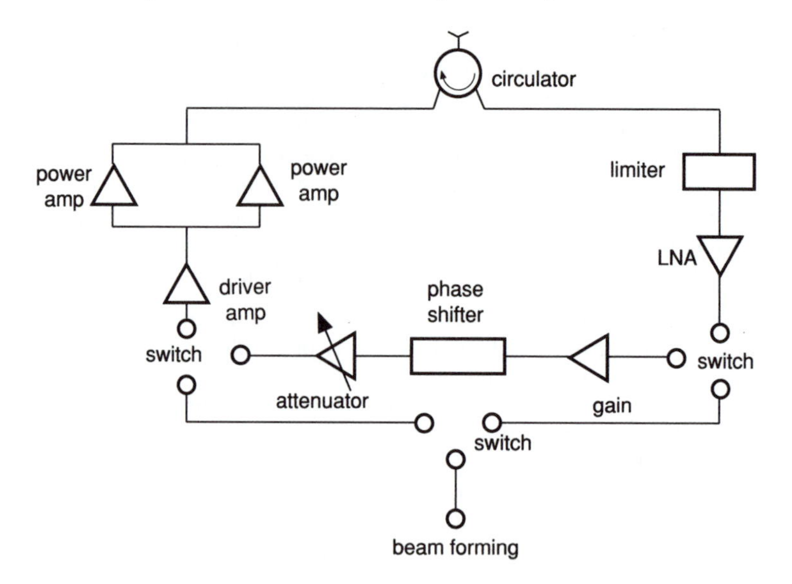

Figure 10.47 *T–R module block diagram [148]*

Basically from the systems point of view the placement of the T–R module in the vicinity of the radiating element imposes constraints in respect of its circuit configuration. Some of the more important ones are as follows.

(*a*) The input microwave signal at the transmit channel of the T–R module would be in the region of a few nanoWatts. If the module has to provide output powers in tens of watts, then multistage power amplification will be necessary.

(*b*) The power dissipation at the T–R module would be primarily due to the power amplifier stages. To keep this at a low value, high efficiency modes of amplifier operation such as class B would have to be used. Since class B amplification is inherently non-linear, over a large number of modules the behaviour of the amplifiers over the design bandwidth and the temperature range, as well as for variations in primary power supply noise, will not be identical except with tight manufacturing tolerances for the processes. In addition, for taking advantage of adaptivity, a means for varying the power by about 20 dB will have to be built in.

(*c*) Procedures for testing and calibration of the T–R modules will have to be accurate, and yet simple and economical.

(*d*) Variation of the output of the power amplifier will have to be achieved either through a resistive attenuator or by variable power or gain amplification without much loss of efficiency.

(*e*) The receive channel of the T–R module should have enough gain to offset the effect of likely post-LNA losses on the overall noise figure. In addition, since very little spatial filtering of the signal entering through the radiating element takes place, the LNA stages must have a high third-order intercept point. Since this can result in high losses of the order of watts in

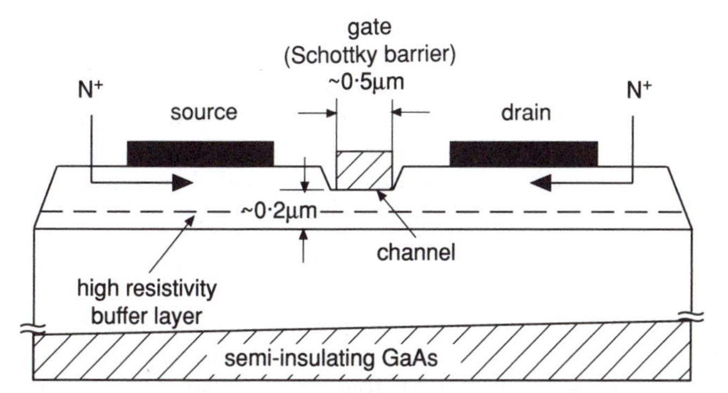

Figure 10.48 *GaAs MESFET for microwave applications [151]*

the LNA, trade-off studies with respect to noise figure, intercept point and power dissipation need to be carried out first.

From the point of view of cost and criticality, the key circuit functions are those of power amplification, low noise amplification and switching/control. These will determine the affordability of phased arrays.

10.27.1 *The device*

The active device most commonly used in GaAs MMIC is the field effect transistor using Schottky barrier gates, i.e. MESFET (see Fig. 10.48). Three metal electrodes, namely gate, source and drain, are connected to a very thin ≈0.2 μm semiconductor active channel layer. The active channel is created beneath the gate by ion implantation of donor atoms into semi-insulating material or by growing epitaxial *N*-type GaAs on chromium doped semi-insulating GaAs substrate. Contacts to both source and drain is through ohmic metallisation. The gate is a Schottky barrier metal contact about 0.5–1.0 μm length so that when a positive drain-source voltage is applied the current will flow, yielding characteristics as shown in Fig. 10.49. Thus, for small biases the source–drain terminals behave like a linear resistor while for higher biases current saturation occurs. Since these devices are depletion mode FETs, on reverse biasing the gate, the active channel height as well as the source–drain current is reduced. The source–drain current flow can therefore be regulated by the bias applied to the gate electrode. Small voltages that are applied to the gate control large source–drain currents so that the FET can be considered as an amplifying device. On the whole, the GaAs FET has better noise performance, higher power gain per stage, excellent thermal stability due to negative temperature coefficient of the drain current and is insensitive to amplifier class operation. However, in device design and in the selection of the technological processes, care has to be taken to minimise parasitic resistances and capacitances which limit the device performance. For example, if low noise performance is desired, the source and drain parasitic series resistances need to be reduced. For power amplifier applications additional features such as high channel current, high gate–drain and gate–source breakdown voltages, efficient heat removal, efficient DC/RF conversion and good power combining efficiency have to be considered. High RF output power means high channel currents and large gate widths for power MESFETs. Presently the general trend in power density (output power per unit gate width) over the frequency band between 1 and 30 GHz indicates that it decreases with increase in frequency of operation. Most of the data show that the power density falls between 0.3 to 0.5 W/mm of gate width. Since the power density of the the device is limited, it would be necessary to connect several gate fingers in parallel to achieve higher output power. It would also be necessary to combine the output of multiple chips of large gate widths with appropriate matching networks for realising the power output needed in active aperture phased array radar. The maximum device size that can be used at a given frequency is governed by:

(*a*) The maximum transverse length, which should be below one-sixth of the wavelength at the operating frequency to avoid unequal phase and amplitudes to each of the fingers so that degradation of power density is reduced.
(*b*) Unit gate finger width so that excessive gate attenuation is reduced.
(*c*) Circuit losses which increase with device size due to corresponding reduction in impedance. The lower values of impedance also degrade the bandwidth because matching becomes more difficult.

For applications requiring instantaneous bandwidths of 30–40%, smaller gate width devices with low enough Qs are preferred because they can be easily matched with conventional reactive matching circuits. Therefore, the accepted design technique for power amplifiers is to combine several of such chips with combining circuitry so that higher output powers with good input/output VSWR are achieved. The emphasis in this case is on maximising the total RF output power realised per wafer and not necessarily by individual chips. Better power-added-efficiency and higher percentage of active area are preferred for maximising the sum output power per wafer. A large peripheral power device is

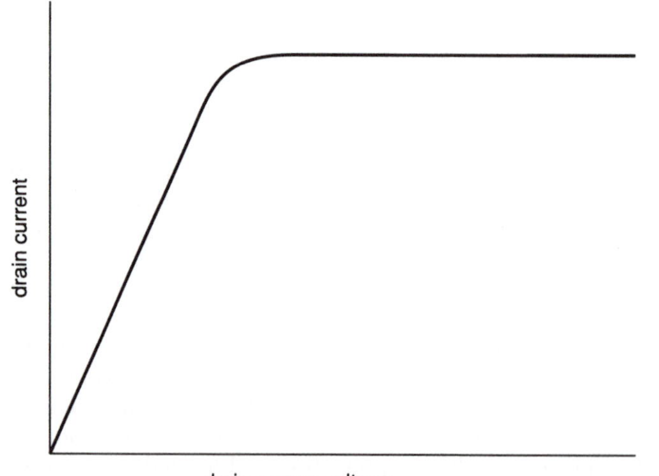

Figure 10.49 *DC current/voltage characteristics of a MESFET [150] © Artech House, 1989*

thus subdivided into several small devices which are identical, partially matched and eventually combined together. The impedance levels of the smaller FETs being much higher than the aggregate FET, uniform phasing across the aggregate FET is achieved owing to identical phasing at each of the smaller FETs so that maximum power is delivered. However, in these cases the overall matching network is larger and more complex. The larger circuitry may as well give rise to odd mode or push–pull oscillations which will generate second, and other higher-order, even harmonics.

The requirement of high breakdown voltages between gate and drain and between gate and source arise from the fact that in class A power FET operation the gate–drain sinusoidal voltage peaks can be of the order of 16 V when RF signals are applied to the gate. Breakdown voltages will thus have to be somewhat higher than this. An equally important consideration is thermal management of the device. This is directly linked to the heat that is produced within the device during the process of conversion from DC to RF, and is specified in terms of the power added efficiency (PAE) which is defined as

$$\mathrm{PAE} = \frac{P_{out} - P_{in}}{P_{dc}} \tag{10.326}$$

where P_{out} is the RF output power of the amplifier, P_{in} is the RF input power to the amplifier and P_{dc} is the power supplied to the device. The power dissipated can now be obtained from eqn. 10.326 as

$$\begin{aligned} P_{dis} &= P_{dc} - P_{out} + P_{in} \\ &= \frac{P_{out}[(1 - PAE) - (1 - PAE)/G]}{PAE} \end{aligned} \tag{10.327}$$

where G, the gain of the RF device, is >1 (for $G \geqslant 10$ dB)

$$P_{dis} = P_{out}(PAE^{-1} - 1)$$

For a power-added efficiency of 50%, the heat that is dissipated will be the same as the RF power. Thus the aim of combining smaller peripheral devices with a matching network is not only for maximising the power output but also the power-added-efficiency. Fig. 10.50 shows the data for power-added efficiency in graphic form for MESFET amplifiers up to 1989 [23]. From Fig. 10.50 it is clear that the power-added efficiency of MESFET amplifiers is relatively high compared to other solid state devices. A minimum PAE of 20% can be obtained at frequencies up to 40 GHz. Higher power-added efficiencies of 60% to 70% at the lower end of the microwave region (up to 6 GHz) have been realised by optimising the MESFET devices and circuits by controlling the device load impedance at the fundamental and harmonics. Besides improving the PAE, thermal management also requires efficient heat removal from the devices. It is generally accepted that channel temperatures in FETs should not rise above 175°C even if the device were to be operated at ambients of 125°C. A rise of 50°C in the channel temperature above ambient would indicate that the thermal resistance between the FET channel (top side) and the heat sink (rear-side) should be no greater than 42°C per watt. Since GaAs has poor thermal conductivity (one-third as compared to silicon) thermal resistance can be lowered by reducing the thickness of the GaAs substrate. From considerations of mechanical handling in production, substrate thicknesses below 100 μm are not practical, as a result of which selective thinning of the substrate under the active area as well as spreading the heat source over a wider area have to be attempted. Thus MMIC amplifiers required to produce high microwave power use physically larger FETs than their discrete counterparts.

10.27.2 *Power amplifier*

There are numerous reports on MMIC power amplifiers that have been developed for applications in phased array radars. Initially, the designs were mostly class A even though these did not lead to PAEs higher than 40%. In the last few years with improvements in FET chip design tuned-B, class-F operation has been realised with PAEs higher than 50%. This approach, which has been very successful at lower frequencies, can be incorporated in GaAs MMIC amplifiers to obtain higher PAEs. High efficiency operation [24] is achieved by controlling the harmonic loads presented to the transistor. This is, in essence, waveshaping of the voltage and current waveforms at the FET to reduce the overall power dissipated in the transistor. The FET in this case is operated close to the B-mode of operation with a drain current waveform of a half rectified sine wave. The amplifier design provides a power match at the fundamental, by tuning the output circuit to pass only the fundamental frequency. The second harmonic is presented with a short circuit to suppress second harmonic components in the voltage waveform. The short circuit can be achieved through a series resonance. The third harmonic is open-circuited to provide the proper component of third harmonic voltage under optimum RF drive conditions. The open-circuit impedance at third harmonic suggests a low

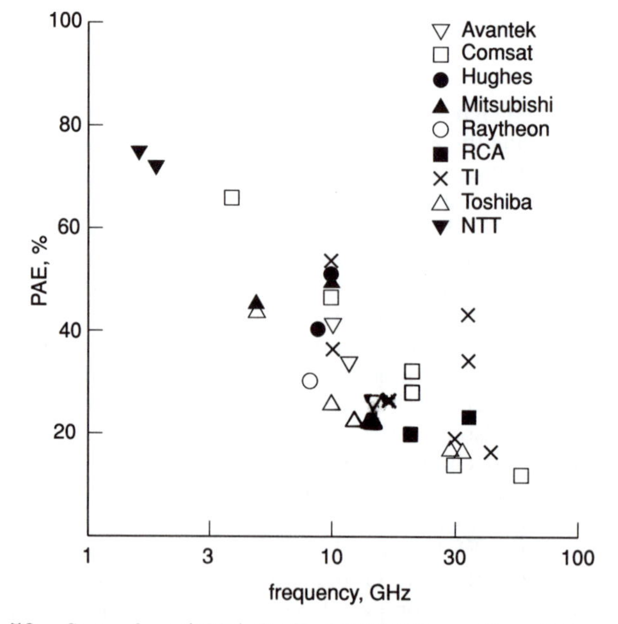

Figure 10.50 *State-of-art (1989) PAE: MESFET amplifiers [23]*

pass topology for the fundamental matching circuit. The increase in efficiency comes about as a result of the reduction of the time average of the voltage–current product at the FET. The amplifiers are usually fabricated in microstrip line configuration using semi-insulating GaAs substrates. The active channels are formed by any of the technologies of ion implantation, vapour phase epitaxy, molecular beam epitaxy etc., and the gates are defined either by optical or by electron beam lithography. Lumped as well as distributed circuit elements have been used for impedance matching with silicon nitride as the dielectric medium for parallel plate capacitors. The most important characteristic of this technology is that both active and passive circuit elements can be formed at the same time on a GaAs chip using compatible processing techniques. Performance characteristics of monolithic circuits vary owing to processing tolerances, layout considerations, yield optimisation and circuit complexity. The system designer has to understand these limitations when selecting these components. For example, module amplitude errors of ± 1 dB and $\pm 5°$ in phase can limit the sidelobe levels of the antenna to -40 dB. However, the actual tolerances achieved in batch processing are far tighter than the tolerances indicated in the aforesaid example.

The developmental activities in this area in the last decade have been considerable. From a modest beginning with circuits designed for S- to X-bands, the range of applicability has been extended to as high as 100 GHz. The coverage in this volume will, however, be limited to important developments in 1–35 GHz region. Single chip power amplifier designs are currently available that can deliver 2.5 W at L-band with 40 dB gain and 27% efficiency, 6 W at S-band with 12 dB gain and 22% efficiency and 6 W at X-band with 15 dB gain at 20% efficiency [25]. It has also been found that, at S-band, 12 W output power can be obtained by combining two 6 W chips (actually by selecting chips capable of giving excess of 6.5 W) in parallel. Nishiki and Nojima [26] have built a 1.7 GHz harmonic reaction amplifier with 2.7 W output power, 9 dB gain and 75% PAE. They have also demonstrated an output of 5 W at 2.0 GHz with 70% PAE and 9 dB gain. Geller and Goettle [27] have recently developed a 1 W GaAs MESFET amplifier with gain of 13 dB, PAE of 65% and operating over the frequency region 3.7 to 4.2 GHz. The amplifier is operated in class AB mode terminating in low reactance for second harmonics. It is highly miniaturised, having dimensions of 91 mm × 51 mm. Quasi-monolithic form has been used in this case so that the passive circuits are fabricated monolithically on a GaAs substrate separately from the device. Bahl *et al.* [28–30] have reported several developments of MESFET amplifiers in the C-band. A single chip MMIC power amplifier with 10 W output, 36% PAE and 0.625 W/mm power density has been reported by them. They have used a 4 mm gate periphery FET that is optimised for maximum power and efficiency at C-band by reducing the source series resistance and thermal resistance. The gate fingers are 250 μm wide and there are 16 gate fingers in each FET cell. A substrate thickness of 75 μm was chosen as a compromise between reducing the thermal resistance and being able to design and handle power MMICs. The 10 W power amplifier chip has two 8 mm gate periphery FETs in single-ended amplifier configuration with their outputs combined in parallel on the chip. Each of the single-ended amplifiers used a pair of reactively combined standard 4 mm gate FETs for a net 8 mm of FET

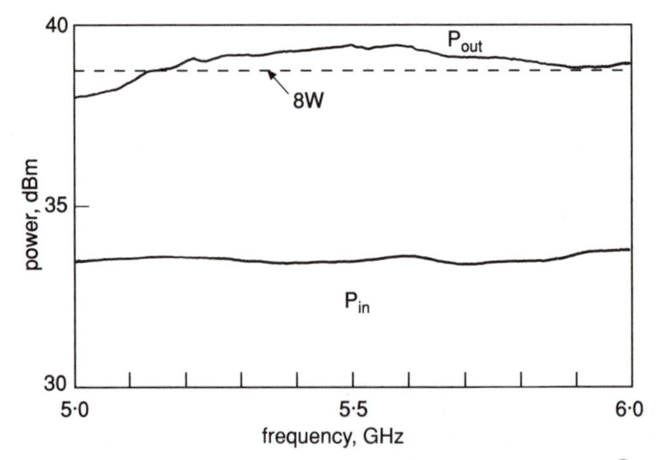

Figure 10.51 *Bandwidth and gain of 10 W C-band amplifier [30].* © *IEEE 1989*

gate periphery. An average IC yield well over 70% has been obtained from the wafers. Fig. 10.51 shows the power output as a function of frequency at 1 dB gain compression. The gain flatness is ±0.5 dB over the 5.2–6 GHz frequency range. A second development by this group is a 12 W power amplifier using four 3.7 W monolithic chips over 5–6 GHz bandwidth with 34% PAE, and 7 dB gain. At saturation, the output was 14.5 W with 3.6 dB gain. The 3.5 W chip uses two 4 mm gate periphery FETs with a power density of 0.45 W/mm. Each amplifier was designed to provide conjugate impedance match at the input. The output impedance matching circuit was optimised for maximum power output. For power-combining the output of the four amplifiers, a four-way divider/combiner was designed on a 635 μm thick alumina substrate. In a third development, they have achieved high efficiency class-B, low cost amplification with 70% PAE, 8 dB gain and 1.7 W power output in the 5–6 GHz region. A dynamic range of about 10 dB with over 40% PAE and almost constant gain has been realised. Since the individual GaAs FETs do not provide sufficient power, they have combined two 2.5 mm gate periphery FETs with a power density of 0.7 W/mm, with the combiner circuits being external to the chips. This is a single-ended amplifier having reactive termination to short-circuit the second and third harmonics. The 2.5 mm FET is matched to 50 Ω input and output, the matching networks being a combination of lumped and distributed elements. The power FETs are fabricated using the refractory metal multi-functional self-aligned gate (MSAG) MMIC process.

In the X-band region Avsarala *et al.* [31] have developed a 7.2 mm GaAs power FET which can give an output of 35.3 dBm with 7 dB gain and 40% PAE at 10.2 GHz. Compact single and two stage balanced amplifier modules have also been tested. The single stage amplifier module had an output power of 37.7 dBm, 34.8% PAE and 6.7 dB gain while the two stage amplifier had 15.1 dB gain and 33.1 % PAE for the same output power. This module has been targeted for use in X-band phased array T–R modules; Khatibzadeh [32] has developed a high efficiency harmonic tuned class-B power amplifier with an output circuit designed to provide optimum load for the fundamental and short circuit at the second harmonic. The characteristics at 10 GHz are: 0.45 W output, 60% PAE, and 7 dB gain with the second harmonic level being lower than −40 dB. Monolithic power amplifier developments at 35 GHz have been reported [33] using FETs on AlGaAs heterobuffer with a very heavily doped active layer. An average output power density of 0.71 W/mm with 5.2 dB gain and 34% PAE has been realised. Six 400 μm FET amplifiers were monolithically integrated to minimise the parasitics. The resulting device gave 0.6 W output power with 2.8 dB gain at 34 GHz.

In these developments no mention has been made of the behaviour of these amplifiers when they are backed off from their rated output. The necessity for this arises when the element excitation across the array is required to have a taper with the same T–R module being used throughout the array. Even though in class-B operation, the overall efficiency degradation over a wide range of output powers is less compared to class-A operation, it is nevertheless not acceptable in phased array applications. A technique has been developed and analysed by Geller *et al.* [34] in which they combine the modification of the drain and gate bias voltages of the FET power amplifier as a function of output power back off together with a reduction in the RF input drive. This concurrent modification of the DC bias and RF drive allows maximum modulation of the permitted device drain voltage and current at any given drive level, and thus maintains the device in a state of 'extended saturation' over a large output power range. Using this technique an experimental 1 W C-band power amplifier capable of 65% PAE at its rated output power was built. It was found to maintain a minimum efficiency of 55% for a 10 dB output back off range whereas under fixed bias conditions the PAE would have dropped to 18%. Another amplifier which is a 3-stage 2 W power amplifier was designed and tested for 51% PAE at the rated output. The efficiency dropped to 41% under 10 dB power back-off for the variable bias techniques, whereas for fixed bias it was around 10%. However, the phase variations over the gain control range are not known.

In the quest for maximisation of RF output power for a given size and weight one of the more promising solid state technologies is that of heterojunction bipolar transistor(HBT). While FET is still an excellent device, in power amplifier applications it has the disadvantages of lower power density, higher $1/f$ noise and lower breakdown voltage under nonlinearity in operation. HBTs on the other hand score over FETs in these areas because of their inherently high power density capabilities. Coupled with the vertical structure of the device, this leads to greater compactness for monolithic circuit applications. HBTs also have other advantages, namely high operating voltage capability and low reactive impedance at input and output (broad band applications). Bayraktaroglu [35] has indicated that HBT unit cells for power amplifiers in the 1–10 GHz have been built. For example in the 3 GHz frequency range, single-stage class-B amplifier produces 1 W power output at 61% PAE and 12.3 dB associated gain. The amplifier is turned on with the input RF signal and dissipates no DC power when idle even though it is biased at all times. Bayraktaroglu predicts that by the end of 1991, 1 W monolithic power amplifiers with HBT devices will match the present performance of discrete devices in these frequency ranges. This would be a great advantage because HBTs are essentially GaAs devices, and hence are compatible with MMIC technology.

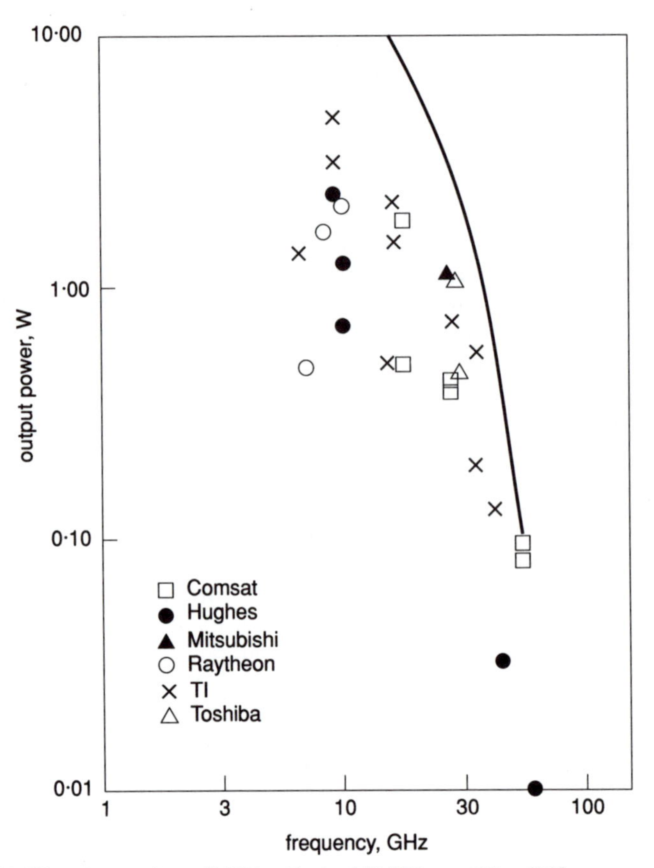

Figure 10.52 *State-of-art (1989): GaAs MMIC amplifier [23]*

Fig. 10.52 provides the status of GaAs MMIC amplifier development in the world as at 1989 [23]. The main factor that has to be noted is that the power output of the GaAs MMIC amplifier is nearly one order less than that of the discrete power MESFET devices. It is, however, evident from Fig. 10.53 that the power density in the case of the GaAs MMIC amplifier chip is not different from that for the discrete power MESFET device [23]. Because of the fact that power devices with much smaller gate widths are used in MMIC amplifiers as compared to MESFET devices, the power output of the former is lower.

10.27.3 *Phase shifters*

An equally critical and important constituent in the T–R module is the phase shifter. Silicon PIN diodes have been widely used for switching in phase shifters owing to their high performance. However, there is a shift in activity in this area towards the GaAs FET which not only offers total integrability with low noise amplifiers and power amplifiers, but also provides bidirectional switching with relatively low DC power requirements. Fast switching speeds (less than a fraction of a nanosecond), high power handling capability (greater than several

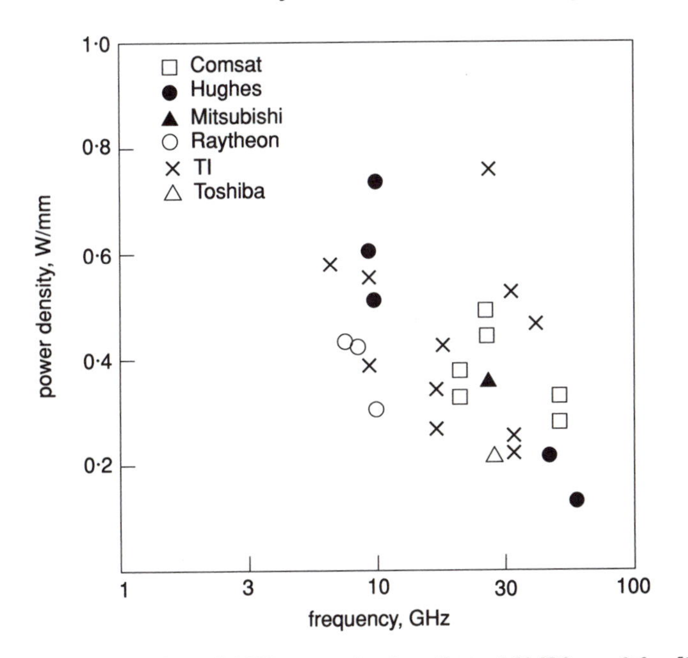

Figure 10.53 *State-of-art (1989) power density: GaAs MMIC amplifier [23]*

watts), wide bandwidths, low power consumption, small size, low cost and light weight have enhanced the potential of FETs as microwave and millimetre-wave control devices. The key parameter for design is the on-state resistance of the MESFET, which can be kept at low values by having heavily doped and thicker layers to minimise the gate periphery. The performance requirements for the phasor network are stringent because of the need to obtain consistent gain and phase characteristics of the module. Phase shifting has been realised from one or more combination of any of the four phasor circuit techniques, namely, loaded line, reflection, switched line and high-pass/low-pass type. For small phase bits up to 45° phase shift, the loaded line phase shifter has been generally preferred because it provides a constant phase shift over a wide frequency range with a low VSWR. The reflection type phase shifter requires a hybrid coupler and appropriate reflection planes to achieve the desired phase shift. In the switched line phase shifter, phase delay is obtained by switching between two transmission paths of different lengths. For small as well as large phase shifts, the reflection type as well as the switched line type phase shifters are frequently employed. These three types, however, use distributed line lengths needing relatively large substrate area, which would enhance the cost of the phase shifter. The high-pass/low-pass switched combination phase shifter, on the other hand, dispenses with the distributed elements and therefore can be made compact as well as lower cost. Phase shifts are realised by switching the signal path containing either a high pass or low pass circuit. Therefore the design of this phase shifter is different because the MESFET can be used not only for switching but also for realisation of high-pass/low-pass networks. The low resistance of the ON state is used to short out unwanted elements of the network

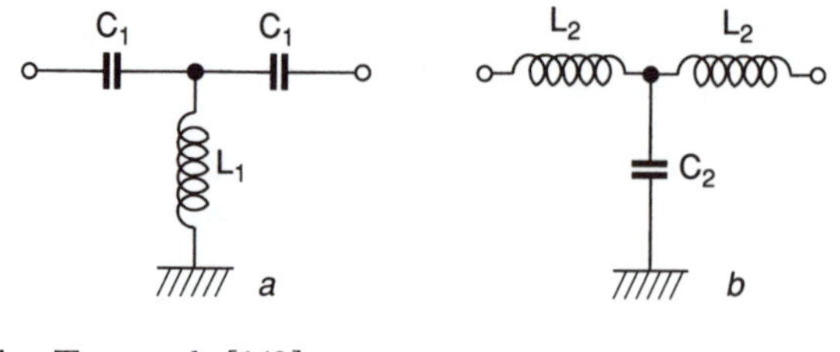

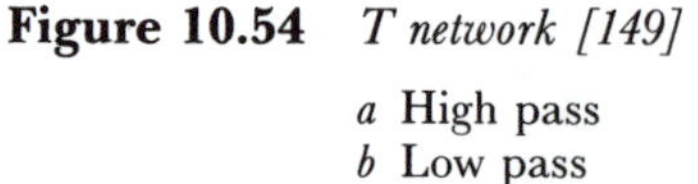
Figure 10.54 *T network [149]*

a High pass
b Low pass

while the series connected capacitor and resistor in the OFF state are used as part of the network. The phase shifter has a wide bandwidth and constant phase shift over the frequency band. For phase bits up to 90°, T networks, such as that shown in Fig. 10.54, are used for high-pass/low-pass with integrated switches, and for the 180° phase bit, a typical network as shown in Fig. 10.55 can be implemented. This network has significantly lower process sensitivity than the low-pass/high-pass circuit in which the switch capacitance is integrated. It also requires less chip area. Though the 22.5° phase bit in the case of the low-pass/high-pass circuit is significantly larger than the series loaded-line design, the higher processing yield more than outweighs the decrease in unit area yield caused by the larger chip size. In general, for low insertion loss in the high pass state the reactance of the series capacitors must be low while the reactance of the shunt inductor is kept high. For low insertion loss in the low pass state the reactance of the series inductor is low while the reactance of the shunt capacitor is maintained high. Careful modelling, simulation and layout are necessary for achieving high performance in phasor networks. A schematic diagram of a typical switched high-pass/low-pass phase shifter network is shown in Fig. 10.56.

There have been several developments in the 1–35 GHz region, of GaAs MMIC phase shifters over the last decade. Some of the work reported in the literature is reproduced to provide an estimate of the current status. In the L-band, Maloney *et al.* [36] have developed a 5-bit phase shifter for operation at

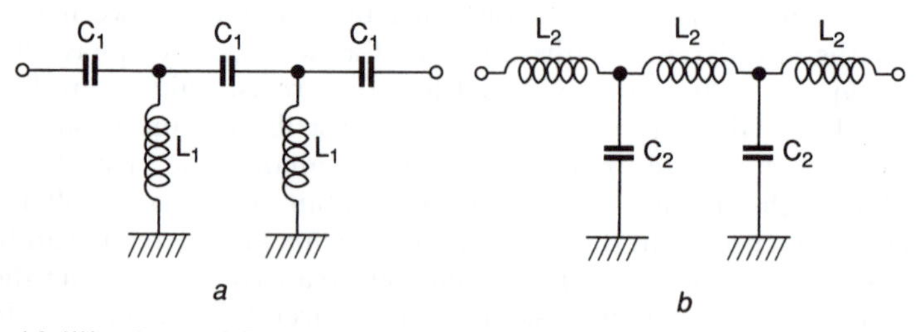

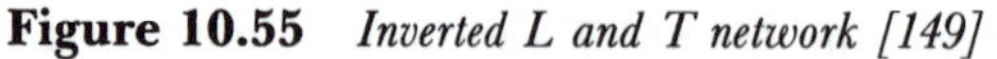
Figure 10.55 *Inverted L and T network [149]*

a High pass
b Low pass

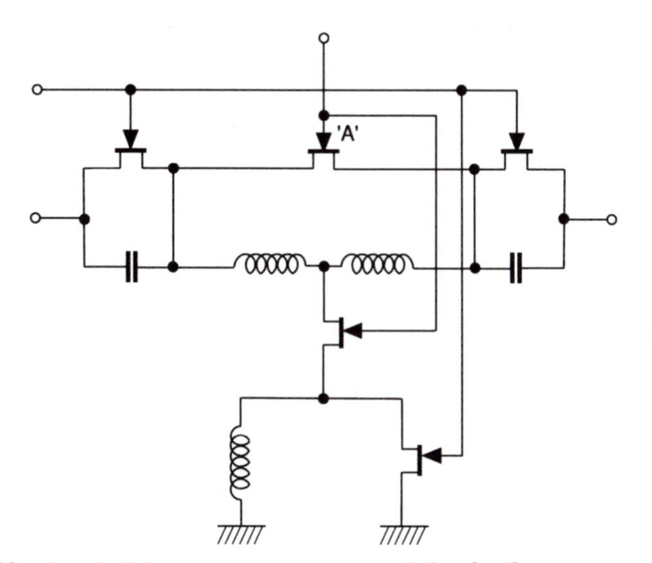

Figure 10.56 *Switched, high-pass/low-pass phase shifter [149]*

L-band. They have used the loaded line configuration for low phase bits (11.25°, 22.5° and 45°) and high-pass/low-pass elements for the larger phase bits (90° and 180°). The circuit consists of a series of FET switches with a T high-pass section and a π low-pass section in the two transmission paths. The phase shifter has an insertion loss of 7.5 ± 1 dB, a return loss higher than 10 dB and an RMS phase error of ±5° at the centre of the band. The dimensions of the phase shifter are 4.95 mm + 3.45 mm. Andricos *et al.* [37] report the development of a C-band (5–6 GHz) 6-bit phase shifter with reflection type phase shifter for large bits (90° and 180°) and loaded line phase shifter for smaller phase bits (11.25°, 22.5° and 45°). An analogue bit is included to adjust the phase shift between 0° and 11°. In order to reduce the overall chip length, transmission lines are folded with sufficient spacing to avoid inter-line coupling. The phase shifter contains 12 FETs, with an insertion loss of 6 dB, return loss higher than 10 dB and an RMS phase error of ±2° at the centre of the band. Excellent phase accuracy with good chip-to-chip phase tracking over the temperature range 25°C to 80°C has also been reported. The phase shifter is 9.43 mm in length and 4.2 mm in width. A 4-bit phase shifter operating in the X-band (8.5–11.5 GHz) has been developed by Ayasli *et al.* [38] with loaded line configuration for small phase bits (22.5° and 45°) and switched line for higher phase bits (90° and 180°). It uses ten FETs and has an insertion loss of 5.1 ± 0.6 dB with a return loss greater than 10 dB and an RMS error of ±11.4° at the centre of the band. In order to obtain good switching characteristics from the FETs, the total capacitance in parallel with the output resistance is resonated. For the basic SPST switch, high impedance was obtained with negative bias voltage for the shunt FET and a low impedance state was achieved with zero gate bias. Moye *et al.* [39] have combined, on a single chip, a 6-bit phase shifter with digital drivers for operation in the 7.2–10.2 GHz band. The integration of the drivers with the chip eliminates six of the twelve control lines normally used. Of the six phase bits, the 180°, 90° and 45° phase bits are of the switched line type consisting of π or T-type high-pass/low-pass phase shift networks while the 22.5°, 11.25° and 6.125° phase bits are of the embedded-FET type with the devices becoming part of the phase shifting network. The phase shifter performance shows an RMS phase error of less than 6°, mean insertion loss of 9 ± 0.5 dB and an overall mean return loss of 20 dB. The power dissipation of the digital driver is around 13.6 mW and the propagation delay through the driver is 180 ps. The phase shifter chip is 2.9 mm × 2.9 mm × 0.1 mm. Schindler and Miller [40] have reported the development of a 3-bit phase shifter using the high-pass/low-pass phase shift network covering the frequency range 18–40 GHz. It uses 16 FETs and has an insertion loss of 9 ± 3 dB with return loss greater than 10 dB. The RMS phase error is $\pm 10°$. In this case the capacitance between the gate and source of the device is used in the realisation of the high-pass/low-pass transmission paths. Broad banding is achieved by having additional FETs to effect third-order high-pass and low-pass filters in the transmission paths.

There is an alternative concept in phase shifter design which leads to compact size and is also less sensitive to process variations. This is the vector modulator in which the microwave signal is fed to a quadrature power splitter with associated network to generate two orthogonal signal vectors in any selected quadrant. The amplitude of each signal is now adjusted so that the resultant signal has a phase between 0° and 90° in any selected quadrant (see Fig. 10.57). Fazal Ali *et al.* [41, 42] have reported single chip C-and X-band MMIC 5-bit phase shifters based on this concept. The phase shifters use an off-chip balun to generate two signals 180° out of phase. High-pass/low-pass filter networks perform quadrant switching of these signals into the quadrature power splitter. The quadrature power splitter is implemented with the same lumped element high-pass/low-pass filter network as that used for quadrant switching at the input. A differential phase shift of 90° between the signal vectors is realised and the amplitudes of the signals are adjusted with a variable attenuator consisting of a set of parallel-connected dual gate FETs biased as common source amplifiers.

The chip operates over the frequency range 5.5–8.5 GHz with RMS phase error of less than 5° and insertion loss of less than 13 ± 1 dB. The X-band phase shifter operates in the 7–12 GHz range and has an RMS phase error of around 4° and an insertion loss of 10 ± 1 dB. Both chips have a high yield so that a compact and cost effective single chip phase shifter is possible.

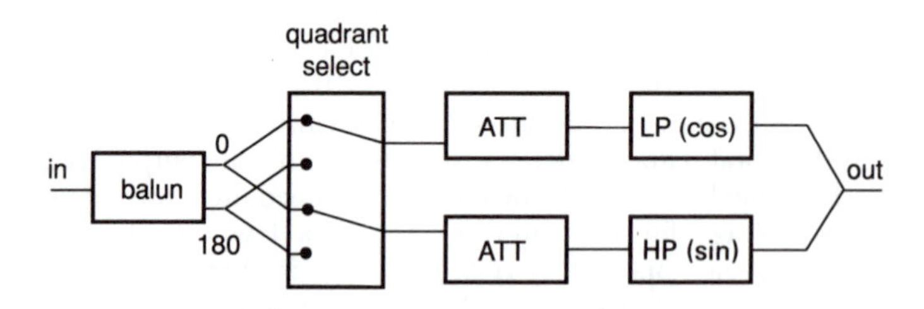

Figure 10.57 *Block diagram of vector modulator [42]. © IEEE, 1990*

The technological progress in power amplifiers and phase shifters, which are the two most important circuit functions in a T–R module, has been significant. On the other hand, the circuit function representing low noise amplification did not receive as much benefit of monolithic technology as the other two until recently. Advances in AlGaAs high electron mobility transistor (HEMT) processing technology have opened up new possibilities of cost effective monolithic low noise amplifiers. To design a low noise amplifier with good input and output return losses over a given bandwidth, design trade-offs between noise figure, input match, gain and bandwidth have to be initially worked out. Dixit *et al.* [43] and Upton *et al.* [44] have designed LNAs using HEMT devices in the MMIC configuration for possible incorporation in a T–R module. Dixit *et al.* realised designs with single ended or balanced topology and multi-section Chebyshev noise matching circuits for optimum broadband performance. For obtaining decade bandwidth distributed topology was used. Two-stage amplification with shunt feedback inter-stage matching was used for broadband stability and improved gain response. The transistor noise parameters and the layout parasitics were also integrated in the optimisation process. The LNA in the 2–7 GHz region achieved less than 2.5 dB noise figure and about 9 dB gain. Another design in the 5–11 GHz region gave less than 2.5 dB noise figure and a minimum of 10 dB gain. The wideband LNA design covering the 2–20 GHz frequency region achieved a noise figure around 3.5 dB and a gain of 9.5 dB. One of the LNAs designed by Upton *et al.* covered the 7–11 GHz region. The amplifier had 1.2 dB noise figure and 15 dB gain. Series feedback between the source of the transistor and the ground was used for improving the input match while maintaining good noise figure performance. Distributed microstrip line elements have been used for matching circuits. A wideband distributed low noise amplifier has also been designed by them for the 2–18 GHz region. The measured performance indicates less than 3.5 dB noise figure and 11.8 dB gain. The availability of such low noise monolithic amplifiers in the near future makes possible their incorporation in the multi-chip T–R modules for active aperture systems.

10.27.4 *Multiple-chip and single-chip T–R modules*

Simultaneously with the efforts in the development of the MMIC power amplifier and phase shifter, efforts have also been launched for the development of multi-chip and single-chip T–R modules. The thrust in this case is towards low cost T–R modules through higher levels of functional integration. While GaAs MMIC technology has the potential to realise the technical goals, the complexity of the module suggests that the partial solution to the cost problem is high yield and a higher level of integration that would reduce the number of chips and thus result in a low chip test and module assembly cost. Therefore the developmental activities would be viewed in ascending order of functional integration that is realised.

Aumann and Willwerth [45] have described a T–R module operating in the 1.25–1.35 GHz band, using a 12-bit voltage controlled attenuator and a phase shifter operating at intermediate frequency to obtain the required precision for phase and amplitude characteristics. However, this T–R module has an RF output power of 100 mW in the transmit channel and a noise figure of 2.5 dB and 60 dB dynamic range for the receiver channel. This has been built more for

Table 10.5 *Typical performance characteristics of T–R modules [25]*

Parameter	L-band	S-band	X-band
Transmit channel			
Peak power output	11 W	10 W	2 W
Gain	36 dB	33 dB	30 dB
Efficiency	30%	17%	15%
Gain error	0.1 dB	0.4 dB	0.2 dB
Phase error	6°	5°	6°
Receive channel			
Noise figure	3.0 dB	3.6 dB	4 dB
Gain	30 dB	24 dB	23 dB
Gain error	0.8 dB	0.5 dB	0.5 dB
Phase error	6°	4°	7°
Physical attributes			
Volume	4.0 in^3	2.4 in^3	0.7 in^3
Weight	4.0 oz	2.4 oz	0.7 oz

proving the concept in the UHF to L-band that higher precision in phase and amplitude is best obtained and in a compact package by having them operational at intermediate frequencies. The authors consider that it is entirely feasible to realise the module in hybrid MIC. Borkowski [25] reports that T–R modules with the performance characteristics shown in Table 10.5 can be manufactured for use in active aperture systems.

The physical size of some of the additional but necessary module components such as circulators, DC/RF connectors and logic circuitry need to be reduced by the use of advance material technologies involving multiple layer/DC interconnects and less expensive module housing materials before they can be used in an active aperture system. Green *et al.* [46] have reported the development of an S-Band T–R module with eight GaAs devices (five of them are MMICs) and having a performance of 2 W RF output with a 20% bandwidth. The receive channel has a noise figure slightly higher than 4 dB and a gain greater than 20 dB. The module has an accurate 4-bit phase shifter. The physical dimensions of the module are 40 mm × 117.5 mm × 10 mm excluding the heat sink, and it weighs 100 g also excluding the heat sink. The MMIC circuits used in the module have been fabricated on 200 μm thick GaAs. The power FETs have been fabricated on 100 μm thick GaAs to obtain better thermal properties. The development of a low noise MMIC receive module operating at 4–6 GHz has been reported by Fazal Ali *et al.* [47]. It contains two stages of MIC low noise amplifiers, two stages of single ended MMIC amplifiers, two monolithic active baluns interfacing with a balanced phase shifter and a single ended amplifier, a 2-bit monolithic phase shifter with on-chip active isolator and the associated support circuitry including logic decode, level shift and voltage regulation. The module has a minimum of 50 dB gain, a maximum noise figure of 2.2 dB and is realised in a housing of dimensions, 2.2 in × 1.1 in × 0.6 in. Across the 5.0–5.25 GHz region, the module had 52 dB of gain and a maximum phase error of 4°, and an amplitude deviation of

±0.3 dB. For the phase shifter and in the two stages of single ended MMIC amplification, lumped element circuit designs enabled a high yield of about 80% with good amplitude (less than 1 dB) and phase tracking (less than 5°) from sample to sample. Priolo *et al.* [48] have reported the development of a high density MMIC T–R module for use in an array of dual polarised tapered notch printed antenna elements. The module features a low insertion loss MMIC shunt reflective switch, three distributed power amplifier stages each with a gain of 5.4 ± 0.4 dB, a dual gate variable gain amplifier with 10 dB gain adjustment, a 5-bit phase shifter based on the vector modulator concept, a built-in-test MMIC and thick film hybrid digital interface along with a voltage regulator. In the receive mode there are two stages of low noise amplification. The performance characteristics of the module are: output power 30 dBm (maximum), nominal transmit and receive gains of 30 dB and 25 dB, respectively. The phase shift over the 10 dB gain variation is less than 6.5° and the amplitude variation due to changing phase states is +3 and −2 dB. The isolation between the transmit and receive states is not less than 6 dB. The receive channel has a dynamic range of 60 dB for 10 MHz bandwidth. A three-chip set for the receive path of a T–R module for operation in the 5–6 GHz frequency range has been announced by Willems *et al.* [49]. A block diagram of the receive path consisting of a variable gain LNA chip, a 6-bit phase shifter chip with integrated SPDT switch and a transmitter buffer amplifier chip is shown in Fig. 10.58. The variable gain LNA chip has a two-stage low-noise amplifier, a variable attenuator, a step attenuator and a buffer amplifier. The maximum noise figure is 2.5 dB and the maximum gain is 28 dB. The yield for this chip was 45%. The 6-bit phase shifter chip contains loaded line elements up to the 45° phase bit and a reflection type phasor network for the higher bits. The functional yield of this chip was about 40%. The transmit/receive buffer amplifier chip provides gain in the forward or in the reverse direction by means of two single pole double throw switches at each end of the chip. The gain of the amplifier is about 6 dB. The yield for this chip was 50% or greater. Andricos [152] has reported the development of a T-R module chip set consisting of nine individual monolithic substrates combined in four packages and the power output amplifier, for use in an active aperture radar. The T-R module operates

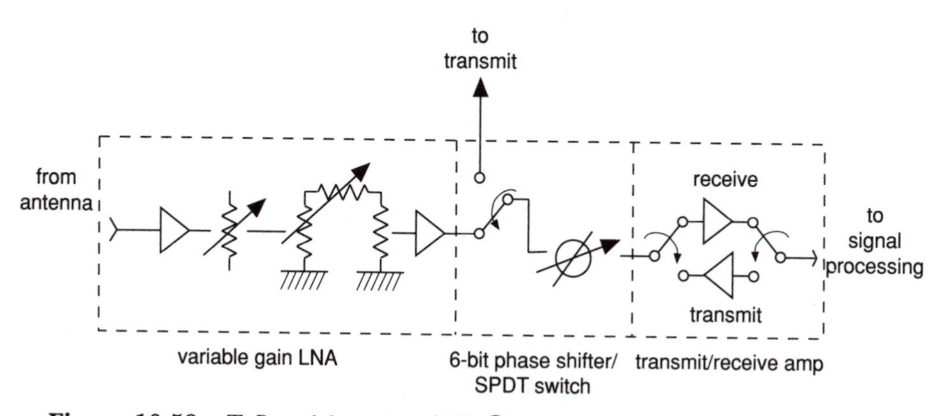

Figure 10.58 *T–R module receiver [49].* © *IEEE, 1990*

in the 5.2 to 5.8 GHz band with 10 to 12 W output power, an overall efficiency of 20% and a gain flatness of ±1 dB over the band. It has a low noise amplifier with 5 to 6 dB noise figure and a six bit programmable phase shifter. The phase shifter has 3° RMS phase tracking and 0.25 dB RMS gain tracking. It is also claimed that with no corrections the transmit mode gain tracking is within +1.5–2 dB, while phase tracking is within 15° over the temperature range of −55°C to +85°C. With correction, they are reduced to 1 dB peak amplitude and 7° peak phase. The chip set has a 12 to 15 W power drainage at 25% duty cycle. It has been stated that with internal power conditioning all DC sensitivities are reduced to negligible values and that a +10% voltage change on all bias voltages produced no measurable gain or phase changes. The T-R module occupies a volume given by $4\frac{7}{8}$ in × $1\frac{3}{8}$ in × 1 in, and weights only 8 ozs. Yau *et al.* [50] have designed a low cost transceiver in the X-band using monolithic transceiver and power MMIC chips. The transceiver consisting of a voltage controlled oscillator, a two-stage gain/buffer amplifier, a 3 dB coupler and a double balanced mixer is a single MMIC chip with dimensions, 120 mil × 150 mil. It provides 10 dB gain with approximately 13 dBm of output power.

The power MMIC chip has two stages of amplification to provide about 14 dB of gain and a total output power of about 23 dBm. A two-stage LNA, each stage having a noise figure of less than 3 dB with gain of 20 dB, has been provided at the IF port to reduce the second stage noise contribution. The entire transceiver has been designed to be compatible with a planar microstrip antenna. Wisseman [51] *et al.* have shown, in over 12 single chip T–R modules developed for X-band airborne radar purposes, that an output power of 0.5 W with 12.3% efficiency in the transmit mode and a noise figure of 5.3 dB with a gain of 13.8 dB in the receive mode has been achieved. The phase shifter in this case has four phase bits, and the fully integrated T–R module chip has the dimensions of 13.0 mm × 4.5 mm × 0.15 mm. It contains 25 FETs, 24 resistors, 43 capacitors and 48 through-the-substrate vias. A single chip T–R module capable of operation in the 2–20 GHz frequency band has been reported by Schindler *et al.* [52]. It includes four stages of power amplification, a four-stage low noise amplifier chain, and two T–R switches using the standard series–shunt FET configuration. The power as well as the low noise amplifiers use distributed element topology. The physical size of the chip is 3.6 mm × 4.9 mm and has 44 FETs, 69 resistors, 44 capacitors, 60 spiral inductors, 60 via holes and 13 bias terminals. In the receive mode the gain is not less than 16 dB and the noise figure is in the region of 8.5–11 dB. The output power in the transmit mode is in the region of 23–25 dBm at 2 dB gain compression. The RF yield for the module is about 18%, which is considered good in view of the complexity of the chip.

It is evident from the efforts that are being made that the T–R module development and technological advancement will generally be along the following lines:

(*a*) T–R module development in the L- to Ka-band based on GaAs MMIC will continue with the aim of producing them in large numbers and at economical prices for phased array and ECM applications.

(*b*) Even though some system requirements may need higher output powers, most T–R modules will be designed for RF power outputs up to 30 W with a 10% to 20% duty cycle.

(*c*) While system designers of the future would prefer wider bandwidths for radars operating in an ECM environment, bandwidths in excess of 30% for the T–R modules are not likely to be available.

(*d*) Further efforts in the power amplifier area will be towards enhancing the power added efficiency as well as reduction in the number of stages so that module complexity and prime power consumption are minimised.

(*e*) In low noise amplifiers the trend is towards LNA MMICs based on HEMT devices.

(*f*) The trend in phase shifter development will be towards configurations with high-pass/low-pass filter networks for phase switching.

(*g*) Multi-chip T–R modules will be the trend in this area. They can be fitted well within the 0.5–0.6 wavelength spacing normally used as lattice dimensions for phased array antennas.

Investigations have been conducted on T–R modules to study the effect of aging so that the likely change in performance of the active aperture system as a result of the changes in the characteristics of T–R modules may be appreciated [53]. The T–R modules had been fabricated by regular production processes.

These were meant for X-band operation with seven RF circuit assemblies, a control and logic circuit assembly, a dual orthogonal feed radiating element and an aluminium housing. The effect of aging, i.e. variations in the phase and amplitude characteristics of the modules, were checked after several tests under various conditions to simulate aging. The performance checks were made by placing a near-field probe in front of each T–R module and the output power was measured and recorded. It was found that a general degradation of phase accuracy occurs with age. In addition, the standard deviation of error in most cases was higher after aging, thereby indicating that the sidelobe levels may not be easily constrained. Using these values, radiation patterns were computed which show that, after aging, well defined nulls no longer exist between sidelobes and some of the sidelobes merge. They were attributed to the fabrication techniques used in hybrid MIC production and to the use of epoxy cement to seal the modules. The present improvements in production technology and better techniques for hermetic sealing are expected to eliminate these deleterious effects of the outside environment in future T–R modules.

10.28 Manufacturability and cost

As GaAs MMIC technology matures and these chips become readily available there will be increasing demands on bettering the performance in production to bring down the cost. It can be readily seen that, since the typical processing sequence in the case of T–R modules uses several semiconductor technologies, high volume applications benefit most from cost considerations. To meet the low cost objective, the process must start at the design stages, where the emphasis would have to be to avoid physically large and low yield circuits as well as extremely complex circuit designs which are likely to result in low yield.

Specifically, some of the design considerations for high yield low cost T–R modules can be spelt out as follows [25]:

(i) Power amplifiers should generate high levels of power output with high efficiencies. Hence;
(*a*) Total gate periphery of the device is at a premium. The load impedance presented to the final device needs to be carefully chosen to maximise power output and efficiency.
(*b*) Since losses in the output circuit of the final stage significantly affect power output and efficiency, off-chip matching may be necessary to minimise losses.
(*c*) Since GaAs is a poor thermal conductor, heat sinking of the chip is mandatory, the trade-off being amplifier size and weight.
(*d*) For efficient multiple stage designs the final stage of the amplifier should reach saturation before the preceding stages.
(ii) The use of MESFETs as switching elements in the T–R module imply that they should have high isolation and low insertion loss. Therefore;
(*a*) The FET design should keep the off/on resistance ratio as high as possible. The trade-off in this case is between short gate length (hence lower processing yield) and insertion loss (channel length largely determines the on resistance as well as insertion loss).
(*b*) The value of the parasitic drain–source capacitance affects the off-state isolation. This capacitance largely depends on the source–drain spacing of the FET.
(iii) Low noise amplifiers require high gain, low noise figure and linearity. Hence;
(*a*) Multiple-stage linear designs require proper device sizing of successive stages to maintain low intermodulation distortion products. The trade-off is higher dissipated power for better linearity.
(*b*) Circuit losses at the input before the first stage degrade the noise figure. Since thin transmission lines on GaAs substrate tend to be lossy, off-chip matching may be more suitable.
(*c*) A low noise condition requires biasing close to the pinch-off voltage of FET. Both gain and noise figure are highly dependent on the pinch-off voltage when the biasing is close to pinch-off. Since pinch-off voltage varies significantly for devices on the same substrate, the bias condition has to be chosen carefully. Thus gain and noise figure are to be traded off against repeatable performance.
(iv) Phase shifters require consistent phase and amplitude response in all the states. They are also required to be compact and less sensitive to process variations. Thus phase shifter designs would generally be high-pass/low-pass filter network switched phase bit configurations with lumped elements for the networks.

All these constraints will have to be borne in mind and various trade-off studies have to be carried out before the T–R module design can be firmed up. In the normal circumstances of microwave circuit design, several months would be spent in the laboratory breadboarding components intended to fulfil the estimated requirements. At the level of circuit integration for the GaAs MMIC modules, it is not possible to breadboard in the same medium that is used for final fabrication. In addition, the finished circuits in MMIC process technologies have no provisions for making any circuit adjustments to peak the device performance. Both these aspects lead to the conclusion that the circuit designs must be inherently correct as fabricated. A variety of high frequency software solutions for providing assurance of circuit designs, in spite of the transistor and circuit element tolerances, have been suggested. The CAE design therefore starts from the system simulation function, that is, top–down design. This is mainly because components such as amplifiers, mixers, oscillators, filters, switches and phase shifters have specifications that are driven by the subsystem and system requirements. The component requirements that have been generated out of the system simulation now drive the device specification. For the CAE system to succeed, the software must allow for the exchange of information on the three hierarchical levels, namely, system simulation, circuit simulation and layout, and device simulation and characterisation, among various aspects of design.

The start of the design cycle being system simulation, the software should provide the designer with the capability to model the radar system and permit the observation of swept frequency and power system response simulations, spectrum response simulations, mixer spurious signal inter-modulation products, parameter trade-offs and linear control-loop simulations. In all cases of microwave systems, S-parameter data and component models to simulate linearity and non-linearity in both the frequency and time domain have to be used. The circuit simulation and layout layer generally begins with linear circuit design and analysis for design optimisation. In this case, frequency-domain linear simulator software can be used to analyse extremely complex devices (e.g. MMICs, amplifier networks, microstrip and stripline circuits etc.) and to provide such output as noise, gain, insertion loss, phase shift, group delay and S-parameters. The program should also enable the designer to tune, optimise and predict the yield of microwave/RF devices and circuits. To ensure that the circuits come as close as possible to design goals, a library of a wide variety of verified element models as well as optimisers and S-parameters from manufacturers of microwave bipolar and GaAs FET, HBT and HEMT devices must be available. For situations where synthesis of passive networks needs to be carried out, programs must be available which describe the source and output termination in terms of *RLC* combinations or as S-parameters. Such standard synthesis packages as Chebyshev, elliptic and Butterworth are now contained in the software for matching or for filter design to meet a specification. There is provision to select lumped, distributed or mixed elements to design the circuit. The synthesis tool, in addition, should be able to identify many useful network topologies that are unlikely to be found by any other means. It should also independently optimise transformation approximations and analyse the network response. The linear analysis in the frequency domain has to be combined with the non-linear analysis in time domain to analyse steady-state performance of microwave networks by harmonic balance techniques. These are fast, efficient methods for designing networks and they permit the evaluation of gain compression, conversion loss and distortion characteristics as a function of frequency and power. Provision should also exist to use time domain analysis to simulate the response of non-linear microwave circuits to arbitrary waveforms such as square, impulse and modulated signals. The final design must extend beyond ordinary optimisation of electrical performance. Since the most critical factor is cost, yield must be quantitively considered in the light of statistically based variations in the fabrication process. Hence the design process would include yield optimisation through design centering. For this purpose the designer starts with the optimised design and the nominal circuit parameters. Using circuit parameter tolerances with a design centering feature, the designer arrives at a design with the best yield/performance characteristics.

The cost reduction of the T–R module will also depend on the elimination of peaking after assembly of the module. This will be a function of the modelling and computer aided design techniques of planar circuits [54–56]. The presently available software solutions have been found very CPU intensive when applied to situations where small component size and close inter-component separations is the rule. Therefore the MMICs need modelling accuracies beyond those provided by these CAD packages. The need to fabricate complete circuit functions on small chip areas has brought the inter-component distances to 100 microns for a chip of say 200 micron thickness. Under these circumstances, inter-component electromagnetic coupling would significantly affect the performance. The layout of the MMIC chip for reducing the 'real estate' in GaAs, component placement, aspect ratio and orientation will result in changes in strip width, transmission line bends and other discontinuities that can cause significant performance changes. Apart from layout aspects, fabrication of multilayer multi-dielectric structures with non-negligible metal thickness add further deviations to the performance. Incorporation of coupling effects in MMIC circuit realisation is therefore one of the most important requirements for efficient design. In addition, as the frequency of operation increases to the millimetre wave region, the phenomenon of radiation, surface waves and package modes also comes into existence. The finite metal thickness of conductors with thickness/separation values in the range 0.1–0.3 affect the performance of tightly coupled structures with different potentials. Of the various approaches available for MMIC design, the standard cell apears to be the most attractive as much of the time and cost associated with design can be reduced by establishing a library of pre-designed and characterised standard cells. These cells can be used to build complex subsystem designs [57]. According to the experience of most MMIC designers, almost 80% of the needs could be met by an appropriate cell library which typically will include generic gain blocks, special purpose gain blocks, mixers, switches, phase shifters, oscillators, attenuators, isolators, baluns etc. The design automation system available today is mostly workstation based because this is a less expensive option and also provides localised control of software and hardware. These workstation-based design automation systems, in addition, provide an excellent environment for networking across software and hardware platforms.

From the point of view of manufacturing, monolithic circuit elements can be viewed as consuming 'real estate' on a GaAs substrate and the processing complexity of each step has an impact on the yield for individual elements. For field effect transistors which are the active devices in the GaAs MMIC, dopant materials have to be introduced on the top side surface of the GaAs substrate by ion implantation to form the active transistor layer. The perimeter of the active channel region is then delineated by any of several patterning techniques – E-beam, photolithography etc. – on the GaAs substrate. Deposited dielectric films and metal layers form the passive components and also interconnect all the elements of the circuit. In this case, even if typical processing yields may exceed 95% per millimetre of FET active channel periphery and greater than 99% per picofarad for capacitors, the net yield for a complex circuit may be low. For cost effective active aperture radar systems to be a reality, the cost of production of the T–R module has to be brought down. The manufacture of the module will involve fabrication of the MMICs and passive circuits, piece-parts placement, automated testing and tuning [18, 58]. The technology of the MMICs will greatly influence the subsequent processes to be followed for the T–R module and therefore MMIC producibility aspects have to be analysed separately. Regardless of the MMIC technology, the majority of the substrate area is taken up by passive circuits commonly referred to as thin film networks. These will contain transmission line components, general transmission line interconnects, bias circuitry and other surface-mounted active and passive components. Thus, passive circuits are generally associated with input and output matching circuits used with power amplification stages, and below 18 GHz they occupy a sizable portion of the area of the standard 50 × 50 mm substrate.

This area limitation gives rise to an appreciable circuit unit cost since even a small production run of T–R modules needs a large number of substrates. For example, only three typical 5-bit phase shifters can be accommodated on the present GaAs substrate whilst ten can be fitted on a 75 × 75 mm substrate and 18 on a 100 mm square substrate. As the processing costs of the largest substrate are the same as the costs of the smallest, true cost savings can be realised only by eliminating some of the producibility problems such as breakage and process nonuniformity. In addition, these circuits have to be designed so that a minimum number of post-fabrication operations will be required to reach the final product. Since the off-chip circuit has to be connected to other components/subsystems of the T–R modules, accurate placement with respect to each other has to be carried out before they are connected. Automated component placement presents a unique challenge because the components are significantly smaller than those used in nonmicrowave MCMs and automated component placement systems for MMIC are not yet commercially available.

Automated component interconnection usually follows component placement and attachment. It is particularly critical in microwave applications because it is normally implemented by automated wire-bonding machines. Wire loop height and repeatability from assembly to assembly are keys to the maintenance of controllable wire inductance and to controlled electrical performance. Computer controlled rather than operator controlled wire bonding is to be employed to meet the reliability requirements. In the case of automated component placement, the present challenge lies in the definition of the wire bonding parameters such as force, ultrasonic power and heat, reference delineation, loop configuration, sequencing, lighting and magnification levels etc., for a variety of components. The components are thus mounted separately on a number of carrier plates, such as multilayer ceramic or copper-polyimide substrates, and soldered by a high temperature solder such as gold-tin and gold-germanium. The assembled carrier plates are tested and then reflow soldered to the housing by means of low-temperature solders such as indium based or tin based alloys. Thse are then tested for their output performance. Fig. 10.59 shows the performance of several thousand X-band T–R modules fabricated at Texas Instruments. As can be seen, the output performance of the entire production is very highly grouped, mostly because of computer-assisted

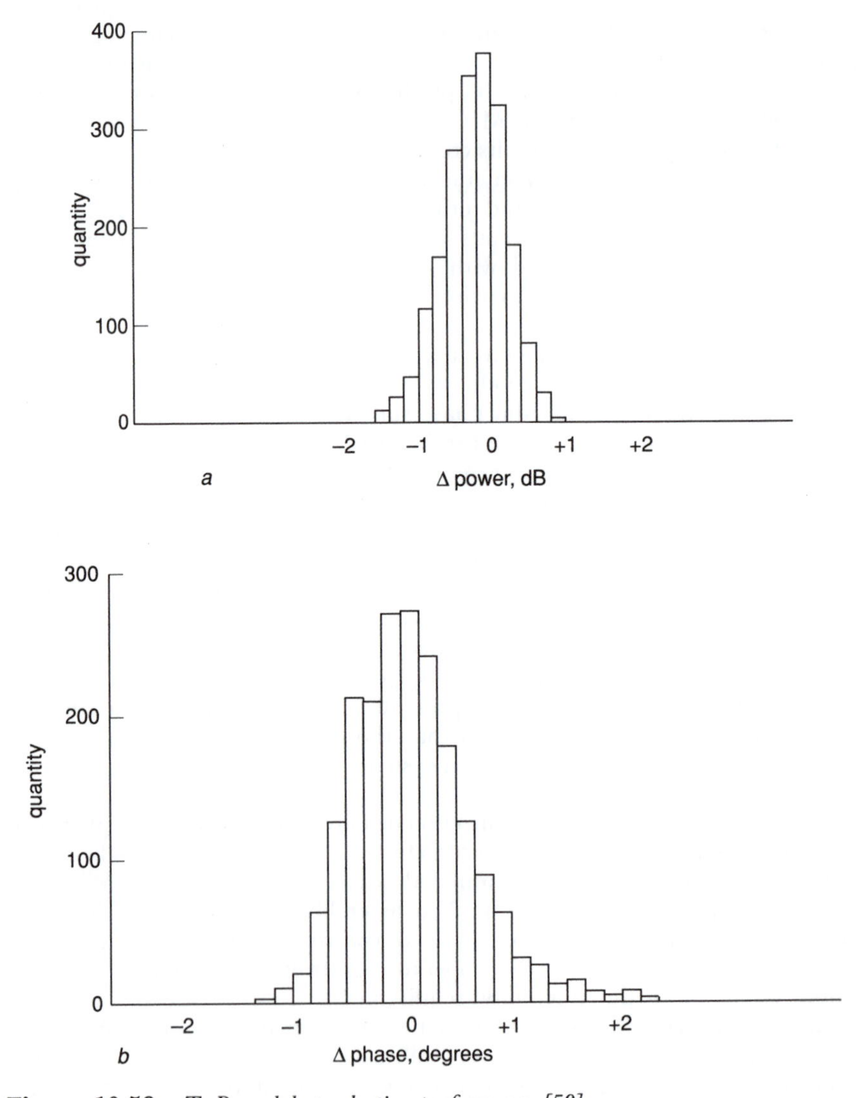

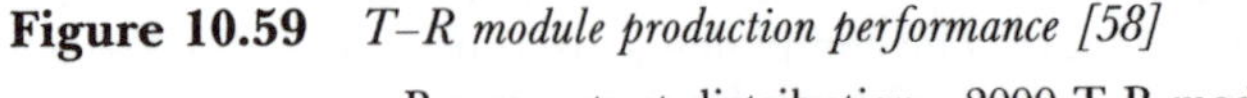

Figure 10.59 *T–R module production performance [58]*

a Power output distribution—2000 T-R modules
b Phase matching for T-R modules

manufacturing (CAM). The phase performance data shown alongside also confirm the repeatability in performance that can be obtained with CAM for T–R modules.

10.28.1 *MMICS and process control*

As the MMICs used in the T–R modules are the largest contributors to the cost of the module, the manufacture of these high-performance and complex chips in a repeatable manner at affordable prices, places new demands on manufacturing techniques perfected for digital integrated circuits. The key to affordability is manufacturing control and to achieve this, statistical process control (SPC) is the main control tool required [63]. At every chip site on a processed GaAs wafer there are small variations in materials and in technological process parameters that lead to variations in the microwave response of the chips. Consequently, a number of yield models have been formulated and used by process engineers [15, 64–66]. In all of these yield models, the objective is to predict the yield of a new design, and to work towards the stabilisation of the process so that acceptable yield is achieved and SPC methods are put in for monitoring and control.

Statistical design or design centering can be approached for a new MMIC either by constructing an archive of data from several thousand trials or by computer modelling which exploits the physics-based model. Collection of experimental data to build an empirical equivalent circuit model is a time-consuming and costly process as it has to be repeated for each new design. On the other hand, physics-based device modelling for active devices and 3-D modelling for passive structures are still computationally intensive. However, although it is a protracted process in the initial stages, these techniques are more fruitful because they are generalised enough to cover new designs without requiring fabrication and testing of a large number. Once the data on new designs are available, statistical parameter models can be derived and the system can be analysed fully by Monte Carlo simulation of the statistical variation of the constituent components. From these data it is now possible to study the tolerance and sensitivity of the components and process parameters with reference to the yield within the specification window. This step has to be iterated as many times as necessary to come close to the desired yield and performance specifications. Finally, the nominal values of the parameters are determined for SPC purposes. The results of various yield value programs of GaAs MMIC chips are now being published by system houses engaged in the manufacture of these MMICs. It is evident from these publications that extensive statistical process control methods, which are in force in digital integrated circuit fabrication, are being fully utilised for MMICs and that, in most cases for active devices, empirical equivalent circuit models have been used.

Technology transportability of MMIC to several foundries is an important consideration both from the point of view of evolving process tolerant designs and that of cost effectiveness in manufacture. At present, the interaction of the system designer with a GaAs foundry is necessary for this purpose. The interaction can occur in two ways: In the first a circuit specification is provided and the foundries would design and fabricate the MMIC. This can be done only by those foundries which have the design capability. Alternatively, the MMIC can be designed in accordance with the active device models and layout rules of the foundry. In the first method there is no technology transfer involved and the design and process details would remain proprietary of the foundry. The second method leads to a better understanding of the process and device modelling as well as the circuit layout rules in a wider class of engineers and scientists. This in turn will accelerate the development of process tolerant designs. The experience of system designers in channelling designs through processes in several foundries has led to insights about the impediments faced by the GaAs industry in lowering the cost [67–69].

Most GaAs foundries offer the general-purpose depletion-mode process for active devices with gate lengths varying from 0.4 to 1 μm. Generally they guarantee yields by setting the specifications to be met by parameters that are measured on process control monitors (PCMs) in various locations on the GaAs wafer. These parameters include the sheet resistance of metallisation layers and resistor layers, the capacitance density of capacitors and the basic DC parameters of the active devices such as the maximum channel current, transconductance, pinch-off voltage, gate-source and gate-drain breakdown voltages, the gate length and the cut-off frequency of the FETs. Usually, at least 12 PCMs are measured on each wafer and, if each specification is met on at least three-quarters of the PCMs, the wafer is considered to be electrically good. It has been found that the specification parameters listed above are not reliable indicators of performance because in actual practice, parameters such as FET output resistance, input resistance, gate-drain capacitance and metallisation resistivity must also be specified. Difficulties have also been experienced in the foundries meeting their own specifications due to lack of strict process control. In FETs, for example, side gating and back gating have ruined the functioning of the circuits operating with positive supply voltages of 8 V or greater and also when both positive and negative supply voltages are present. Back gating and side gating affect n-resistors as well as FETs and this makes biasing difficult. Another problem that often occurs is low-frequency substrate oscillation which destroys the phase noise characteristics of the oscillator chips and can render amplifier chips useless for most applications. The large signal characteristics of FETs have been left largely uncovered as a result of which reduced power efficiency due to excess drain-source current, lower than expected output and higher than expected optimum load impedance have been experienced. Further, it has been reported that the foundries have not yet fully stabilised their processes and consequently the yield statistics for each process parameter vary from quarter to quarter and year to year. Also, layout rules differ widely from foundry to foundry and consequently chips fabricated at one foundry can be more than four times as large as chips with identical circuit topology fabricated at a different foundry. Zaghloul *et al.* [59] have carried out a statistical analysis on the MMIC circuits fabricated for an active phased array for controlling the phases and amplitude of the radiating elements. The transmit portion of the T–R module in this case had a 5-bit digital phase shifter, buffer amplifier, 5-bit digital attenuator, driver amplifier and high power amplifier with a module controller. The amplitude and phase spreads in their case were found to be ± 1.5 dB and $+10°$, respectively with 33% of the modules having a spread of $+2°$ and ± 0.25 dB in phase and amplitude characteristics. Such tight grouping has been attributed to repeatability of bonding, parts placement and tuning that have been carried out with computer controlled manufacturing methods. Vorhaus [60] states that, during the production of more than 2000 power amplifier modules at the rate of 100 modules per week, it was possible to ensure excellent module-to-module uniformity by extensive qualification of the dyes for production of the power FETs and careful control of the assembly process on such critical factors as die attach and component bonding. The power amplifier module had a nominal rating of 5 W output at 30% PAE for the frequency band 9.2–10.2 GHz. Performance data show that 69% of the modules were within 0.18 dB of the population mean in power output and 66% of the modules were within 1.9% of the overall population average PAE of 33.2%.

Geen *et al.* [61] have reported that they have been able to obtain yields of around 50%. This improvement is attributed to a new production process that has cut out a large number of stages originally used to produce an MMIC circuit. The number of masks needed in the process have been reduced from 14 to 8, largely by changing the way in which the multi-layer process has been organised. The passivation layers are shared with dielectric deposition for capacitors, and each mask is designed to provide the maximum number of elements at each stage. They report average RF yields between 40% and 57% and very close circuit performance matching both within a wafer and from wafer to wafer. Parisot and Magarshak [62] have based their designs of GaAs MMICs on the standard cell approach for individual components. According to them, these data are state-of-art for a given process, and since the accuracy of the models and repeatability of the process is high, their final design yield is impressive. Three types of circuits have been productionised by them; these are: a two-stage 0.5–9 GHz distributed amplifier, a 2–18 GHz distributed amplifier and a 4.5–6.2 GHz 1 W amplifier. Process yields of better than 50% have been achieved even after taking into consideration the dispersion of the process.

Currently opinion is gaining ground that extending the present CAM techniques to cover several thousand modules a month (which would be more than one order higher than the current output) requires more than building a large factory. It has to be an operator-less factory where all work handling is accomplished by means of robotics and assembly is computer controlled.

10.29 Feed and control network

The active aperture concept eliminates the need for a feed network for distributing large amount of RF power from a distant single source to the antenna aperture. Instead, because of the space combining aspect of RF power in these systems, the feed network deals with relatively low level signal distribution to and from the aperture. The feed and control network must, however, allow the passage of a variety of signals as follows:

(*a*) RF signals of high stability in amplitude and phase for combining in space in the transmit mode and for coherent processing in the high performance Doppler processing mode.

(*b*) IF return signals with a dynamic range of at least 70 dB, if down conversion is used at each array element.

(*c*) Digital signals at base band to control accurately phase and amplitude as well as to provide a timing reference at each active array element. The requirements of excitation current for RMS sidelobe levels lower than 40 dB with pencil-beam electronically-steered arrays indicate that phase and amplitude control in the receive mode may have to be tight.

Thus each T–R module would need several microwave/RF lines, bias lines and digital lines to provide/process the data. This presents an extremely complex signal distribution network with topology and interference being quite severe, as thousands of T–R modules are involved. Signal interfaces at RF/IF and base band will be needed for each module and these may be provided from a centralised control or fully/partially distributed control configurations. The most promising solution is the use of an optical fibre distribution network interconnecting the monolithically integrated optical components with the T–R modules, consisting of GaAs MMIC chips, so that the conventional coaxial or waveguide distribution network is replaced by fibre optic links and the size and weight of the feed system is reduced by at least one order of magnitude. The merging of the microwave/millimetre wave techniques with light wave technology facilitates the introduction of such novel signal processing methods as wide band true time delaying for beam forming. Other desirable features of optical networks are high speed, good electrical isolation, elimination of grounding and immunity to electromagnetic pulses (EMP).

The three functions required to be accomplished by the optical feed network are: phase and frequency synchronisation of the array, control signal distribution for beam forming and steering, and transmission of communication and data signals from a central processing unit to the T–R modules. These three tasks may be combined on the same FO link or they may be transmitted as three different signals through different fibres. The basic building blocks of the feed network are: optical sources with facilities for modulation (microwave-to-optical converter), fibre optic links and optical detector (optical-to-microwave down converter)/receiver. These can be discrete components or form integrated opto-electronics having compatibility in production processes with the GaAs MMICs that form the T–R modules. Hence GaAs based integrated opto-electronic circuits (IOC) appear to be the best bet for use in active aperture phased array systems.

10.29.1 *Optical sources*

Of the several optical sources that are available the semiconductor injection laser is most compatible with fibre optics and integrated optics. This type of laser requires, at a minimum, a *p–n* junction material of extremely high optical quality, two parallel mirrors perpendicular to the junction which defines the length of the laser and two electrical contacts [70]. Since it is required to function with low powers, the electrical current through the device and the optical field propagating in the device have to be confined in both vertical and lateral dimensions to about 1 μm. This is accomplished in the vertical dimension by a heterostructure of *p*-type GaAs sandwiched between layers of *n*-type AlGaAs and *p*-type AlGaAs, which have different bandgaps and indices of refraction. The early versions of laser diodes emitted in the wavelength range of 0.8–0.9 μm. The more recent semiconductor laser diode, emitting in the wavelength range of 1.1–1.6 μm, is fabricated from quaternary semiconductors, e.g. InGaAsP lattice matched to InP. These lasers are of great interest in optical fibre transmission systems because monomode fibres have the lowest attenuation in this wavelength band. For example, the attenuation is 0.5 dB per km at $\lambda = 1.3$ μm, and 0.2 dB per km at $\lambda = 1.55$ μm. In addition, the dispersion characteristics of single mode fibres are minimum at about $\lambda = 1.29$ μm.

The genesis of activities relating to semiconductor sources for optical communication dates back to the 1960s when it was found that GaAs has a wide band gap and a higher electron mobility than silicon. When multiple semiconductor layers of aluminium and other dopants with wider band-gap and lower refractive index were epitaxially grown on the GaAs substrate, they provided better carrier and optical confinement for laser action. This is the well known double heterostructure laser; a thin active layer of GaAs sandwiched between *p*-type and *n*-type $Ga_{1-x}Al_xAs$ cladding layers. There is a difference between the refractive indices of GaAs and the ternary $Ga_{1-x}Al_xAs$ with different molar fractions x, and therefore the two cladding layers give rise to a three-layered dielectric waveguide. Most of the energy is concentrated in the higher index ternary layer for the fundamental mode of propagation. When a positive bias is applied, electrons are injected from the *n*-type GaAlAs material and holes are injected from the *p*-type GaAlAs into the active GaAs region. The electrons are prevented from entering into the *p*-layer and the holes are prevented from entering into the *n*-layer by a potential barrier due to the energy gap. The holes and electrons are thus confined to the active GaAs region. This double confinement of injected carriers as well as the optical mode energy to the same region results in a low-threshold continuously-operated room-temperature semiconductor laser. It can be considered to be a current-sensitive device, the output of which at low current values consists of spontaneous emission in all directions. As the bias current is raised to high current densities in excess of the threshold, the light output from the laser increases and also changes from spontaneous emission to stimulated emission. Fig. 10.60 shows a light intensity against current characteristic of a double heterostructure laser. The extrapolation of the steep incline of the curve to zero optical power gives the threshold current I_t. The external differential quantum efficiency, defined as the number of photons emitted per radiative electron-hole pair recombination above threshold, is calculated from the steep portion of Fig. 10.60. Experimentally it has been observed that the threshold current increases exponentially with temperature. The output of the laser is predominantly TE modes and, in the waveguide formed by the active GaAs with the *p*-type and the *n*-type GaAlAs layers, they are totally internally reflected at the walls to form a standing wave pattern. Only the fundamental TE mode is most often observed at the active GaAs layer when its thickness is of the order of 0.3 μm. The double heterostructure laser is incapable of confining the current and radiation in the transverse direction, as a consequence of which it supports more than one transverse mode and gives rise to mode hopping. It has relatively broad gain and multimode emission in the form of a series of peaks separated from each other by a few angstroms. The output intensity is observed to change abruptly in a random fashion even though the total intensity of the light output appears to have stabilised. In general, the behaviour of the multimode laser spectra is not completely understood [71], but smooth mode changes and sharply defined mode hops have been observed with increase in the device driving current. The output power in each longitudinal mode of the laser varies with time owing to electron and photon density fluctuations, and under transient conditions even for a single mode laser there are periods when the device operates in many modes. The transient spectrum, which is broad in the transient state, narrows down to one prominent mode. The power fluctuations of the multimode laser can be analytically described and solved to yield fluctuations in intensity of individual modes in the output intensity. The fluctuations in the output intensity give rise to a relative intensity noise (RIN) spectrum in the power output from the laser. This is the dominant source of noise and is defined as

$$\mathrm{RIN} = \overline{\Delta P^2}/P_0 \tag{10.328}$$

where $\overline{\Delta P^2}$ is the mean square intensity fluctuation in 1 Hz bandwidth centred at the frequency of interest and P_0 is the average optical intensity. This is AM noise and, as shown in Fig. 10.61, its amplitude spectrum has a broad and pronounced resonance peak at the relaxation oscillation frequency of the laser. Since the noise intensity of the individual mode is anticorrelated below the relaxation oscillation frequency of the laser, the RIN of all the modes is typically 40 dB lower than that of an individual mode. In addition, as the driving current

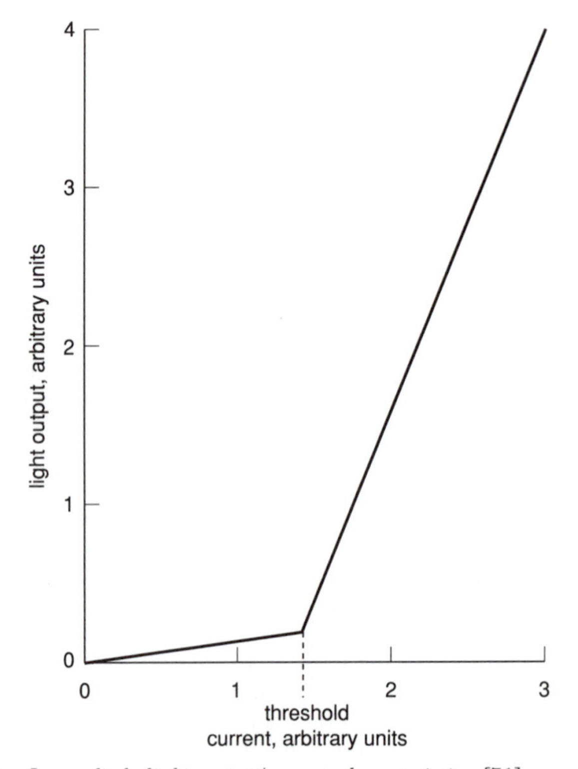

Figure 10.60 *Laser diode light-output/current characteristics [71]*

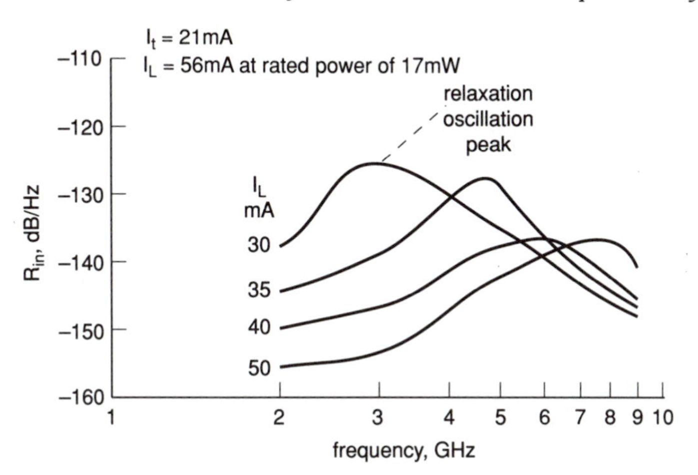

Figure 10.61 *Variation of RIN with bias current [72]. © Artech House, 1990*

of the laser is increased to above the threshold value, more lasing modes appear and the RIN of all the modes is expected to decrease signficantly as the current is increased.

The problem of mode hopping is overcome by providing optical and carrier confinement in the transverse direction. In the buried hetero-structure configuration, the active GaAs layer is surrounded by a lower index GaAlAs region so that the electromagnetic structure is that of a rectangular waveguide. The transverse dimensions and the refractive index discontinuities are chosen so that only the lowest order transverse mode can propagate. Commercially available GaAs/GaAlAs buried hetero-structure lasers operate at a peak wavelength of 0.84 μm at room temperature. The threshold current is typically 25 mA and the beam width in the plane of propagation ($\theta_\parallel$) and in the perpendicular plane ($\theta_\perp$) are about 35° and 25°, respectively. The 3 dB bandwidth of the laser is typically about 10 GHz. An InP/InGaAsP planar buried hetero-structure laser is made up of a mesa structure with an undoped GaAsP active layer, a *p*-type InP cladding layer and a *p*-type cap layer. A semi-insulating burying layer is grown on both sides of the mesa to reduce parasitic capacitance. It operates at a peak wavelength of 1.31 μm at room temperature and has a low threshold current, typically about 17 mA. It has a high external differential quantum efficiency of 42% and the beam widths $\theta_\parallel$ and $\theta_\perp$ are 22° and 32°, respectively. The measured bandwidth of the laser is about 16 GHz at room temperature. The disadvantage of this laser is its increased thermal resistance. In both of these lasers, radiation is generated by a Fabry-Perot resonator cavity whose end walls are mirror facets formed by two parallel cleavage planes of the semiconductor crystal. The mirror facets result in the disruption of the wafer surface and, as a consequence, pose problems in the fabrication of optical integrated circuits in which the laser diodes are to be monolithically integrated within the semiconductor wafer. This is overcome in the distributed feedback (DFB) laser in which a Bragg type diffraction grating is incorporated into the multilayer structure along its length. The grating is produced by corrugating the surface between the two semiconductor layers that make up the laser diode. The corrugation provides feedback for light in a very narrow spectral range and therefore the lasing occurs with a pure single frequency. A typical InP/InGaAsP distributed feedback laser operates at a peak wavelength of 1.3 μm at room temperature with a bandwidth of 13.9 GHz. In a recent development in DFB lasers, Tsuji *et al.* [74] have reported the performance of a distributed feedback laser with a multiquantum well structure for the active region to obtain an improved performance. The spectrum width of this laser under 10 Giga bits per second modulation is around 0.5 nm which is one-third of that of conventional DFB. The threshold current is as low as 15 mA and the output power is 5 mW at 85°C. The lowest reported RIN is −160 dB/Hz and the spectral width is typically as small as 0.24–0.25 nm under modulation [73, 75]. This is about one tenth of that of the Fabry-Perot laser diodes discussed earlier. In addition, these lasers have a wavelength stability against temperature of about 0.9 Å/°C. In this case, the laser shifts to shorter wavelengths with ageing. However, the overall reliability of the laser is considered to be good. The DFB laser is favoured in active aperture radar applications. For the single mode laser, theory predicts and experiments confirm that the low frequency RIN increases for driving currents below the threshold value. The RIN reaches a maximum with the driving current close to the threshold value and decreases thereafter with increase in the current. For the double heterostructure direct feedback (DH–DFB) laser the RIN is observed to be typically 30 dB lower than the conventional laser for identical modulation depths, frequency and driving currents.

There are two methods for modulating a laser diode: direct and indirect. In direct modulation the driving current is generally maintained close to the threshold and the modulation current is superimposed on the steady-state value. A typical measured normalised-amplitude response of a laser diode as a function of the modulating frequency and for several bias currents, clearly shows that the measured response is flat at low frequencies but has a response peak in the range of 1–4 GHz before it drops down to the 40 dB per decade attenuation region of the response [72]. The frequency corresponding to the resonant peak is known as the relaxation frequency. It can be expressed as

$$\omega_r = \sqrt{\frac{AP_L}{\tau_p}} \tag{10.329}$$

Here P_L is the internal photon density, A is the optical gain coefficient and τ_p is the photon life time. The gain coefficient can be increased to five times its normal value by cooling the laser to liquid nitrogen temperature. The photon density P can be increased by biasing the laser at higher currents which also results in a high optical output power density which can cause catastrophic damage at about 1 MW/cm^2, for a mirror reflectivity of 0.3. This sets the upper limit on the maximum photon density and also on the maximum modulation bandwidth. A third way of enhancing the modulation bandwidth is to reduce the photon life-time by decreasing the length of the cavity. Thermal effects due to heating will, however, limit the maximum attainable modulation bandwidth. Goldberg *et al.* [76, 77] reported that direct modulation is limited to frequencies below approximately 20 GHz and is accompanied by a large harmonic content due to the nonlinearity of the laser diode modulation response. Direct modulation is the simplest way of modulating the laser output and therefore it is widely used. In active aperture systems the fibre optic link will be based on direct modulation for ease of implementation. The output of the directly modulated laser has an almost linear relationship with the bias current above the threshold value. The optical spectrum of the directly modulated laser is broad and gives rise to excessive pulse dispersion over long hauls. However this is not a limitation for phased array applications, since the fibre network is expected to be less than 1 km.

The output of a laser diode is coupled to a single mode fibre by any of several methods such as hemispherical microlens, spherical lens, high index lens or by butt-coupling. The measured coupling loss between the laser and the single mode fibre is found to vary from 2.5 to 4.0 dB for the lens couplings and it is greater than 10 dB for butt-coupling.

External modulation of the laser, which is the second option, provides high spectral purity but is not recommended for active aperture systems at present owing to complexity, bulk and higher insertion losses.

10.29.2 *Optical detectors*

Semiconductor optical detectors are two-terminal devices that convert optical inputs into electrical carriers. Optical detectors are also characterised by fast response time and low power consumption. Over the last few years, several types of detectors have been developed, for example the InGaAs PIN diode, germanium avalanche photo diode and the InGaAs avalanche photo diode. Although all the three types of detectors are suitable for active aperture radar applications, the InGaAs PIN diode represents the most widely used optical detector in fibre communications. One of the main reasons for its popularity is that the absorption of optical energy is extremely efficient. For example, at a wavelength of 830 nm, more than 90% of the incident light is absorbed in a GaAs detector 2 μm thick. PIN photo diodes have better temperature stability and lower operating voltages than other types of detectors. The current output by the PIN photo diode is of the order of nanoAmperes for a microWatt of incident light and therefore it needs to be amplified to a useful voltage level for further signal process. It therefore requires several stages of amplification to raise the level of the detected signals to digital levels. Since the detector and the first stage amplifier determine the overall performance of the receiver, the electrical compatability (bias voltage, resistance, capacitance etc) of the detector and the amplifier must be carefully analysed. The detector and amplifier have also to be designed as a unit so that their monolithic integration can provide a significant improvement in bandwidth.

The PIN photo diode consists of *p*-type and *n*-type semiconductor regions separated by an intrinsic layer. In normal operation, the intrinsic region is fully depleted of carriers on application of a large reverse-biased voltage. If the light is then incident on the diode (depletion region), free electron-hole pairs called photo carriers are generated because the photon with energy equal to or greater than the band gap energy excites the electron in the valence band to the conduction band. As the carriers traverse the depletion region, a displacement current is induced with one electron flowing for every carrier. This current flow is the photo current. The detector being basically a square law device, the output photo current I_D of the detector is a linear function of the incident optical power P_0 so that

$$I_D = RP_0 \tag{10.330}$$

where R is the responsivity which is a measure of the photo current that is generated when optically illuminated and is expressed in A/W, i.e. as a ratio of the photo diode active area to the width of the depletion region. Typical values for responsivity are between 0.7 and 0.8 A/W.

The PIN photo diode detector has a mesa structure and its fabrication is initiated by growing InGaAs epitaxially on the InP substrate. It has to be lattice matched to InP to obtain defect-free thermodynamically stable epitaxial growth. A mole fraction of 0.53, i.e. $In_{0.53}Ga_{0.47}As$, yields a nearly perfect lattice match with the substrate and a band gap of 0.72 eV. The use of InP as the semi-insulating substrate helps in providing greater isolation between devices and also lowers the dielectric losses. Further, InP is transparent to optical illumination in the range 1.1–1.7 μm. It also makes it possible to monolithically integrate the PIN photo diode detector with a post-detection MESFET. For purposes of high speed and high quantum efficiency, a heterostructure consisting of an $n+$ InP substrate, an n-type InP buffer layer, an undoped $n-$InGaAs depletion layer and a $p+$ InGaAs contact layer in mesa configuration is used. An aperture in the substrate metallisation is used to illuminate the diode. A silicon nitride anti-reflection coating over the aperture minimises the reflections. Since the receiver noise is proportional to the leakage current, the diameter of the mesa is kept small, typically 40 μm, to keep the diode capacitance low, and also to minimise the number of active area defects. A typical value of the leakage current is 5 nA at about -10 V bias. At higher voltages, the leakage current increase exponentially.

In general, the limiting value of the minimum optical power required to obtain a given signal-to-noise ratio, assuming 100% modulation, is

$$(P_0)_{min} = \frac{2h\nu B}{\eta_d}(S/N) \tag{10.331}$$

Here h is Planck's constant, ν is the light frequency, B is the bandwidth and η_d is the quantum efficiency. The frequency response of the detector is limited by the RC-time constant of the PIN diode and the transit time effects [70]. Since present PIN photo detectors produce a relatively low current of 0.85 μA which has to be amplified in several stages to obtain a reasonable voltage level, the packing of the fibre optical receiver becomes an important issue.

To find a way of solving it in a cost effective manner, it is to be noted that the dominant source of noise in the fibre optic detector receiver using a PIN photo detector is preamplifier noise. Therefore, to maximise sensitivity, MESFETs are used to minimise the noise of the preamplifier [78]. Secondly, to achieve a high photonic-to-electronic conversion efficiency, precision optical couplings have to be employed without in any way increasing the cost unduly. Thus the packaging has to repeatedly and efficiently couple the opto-electronic transducers to their photonic inputs and microwave outputs. One proposed method is the on-wafer integration of all necessary components using GaAs MMIC technology. Monolithic realisation would improve the performance of the photonic/microwave receiver and also achieve the goal of low cost. Using GaAs

MMIC technology, miniaturised fibre-pigtailed drop-in microwave-to-photonic and photonic-to-microwave modules are being fabricated for easy and repeatable integration with the rest of the radar sensor. A typical photonic-to-microwave module will have an InGaAs PIN photo detector, a 54° etched silicon turning mirror for backside illumination of the detector and an 8 μm core single-mode optical fibre. The detector's dark current is less than 0.5 μA and the optical fibre-to-detector coupling is greater than 80%. The module detects 1300 nm wavelength light with a responsivity of greater than $0.85A/W$. The module has the dimensions $3.8 \times 1.8 \times 1.1$ mm [79].

10.29.3 *Optical fibre*

The transmission of optical signals is effected by an optical fibre which can be multimode step index, multimode graded index or single mode. The most important characteristics are the attenuation and the dispersion, both of which are distance dependent. The factors that contribute to attenuation in an optical fibre are the absorption and scattering of light on its transmission through the fibre. At lower wavelengths, that is below 1500 nm, Rayleigh scattering occurs owing to the microscopic variations in the material density and in the composition. The power that is scattered and also the attenuation are inversely proportional to the fourth power of the wavelength. At wavelengths higher than 1.5 μm, the absorption loss is due to infrared absorption caused by resonances associated with the lattice vibrations of the atoms. The region of minimum attenuation lies between Rayleigh scattering and the infrared absorption regions. From the data available on ultra-low-loss germano-silicate single mode optical fibre [80], it is observed that the attenuation can be brought down to less than 1.0 dB/km over the wavelength range 1.1–1.7 μm. Typical attenuation of single mode dispersion shifted fibres is about 0.35 and 0.28 dB/km, respectively, at 1.35 μm and 1.55 μm wavelengths. In fibre optic networks with multiports where power combining and dividing has to be carried out by fibre splicing, a loss of about 0.2 dB per splice needs to be considered as part of the attenuation in transmission.

The dispersion of a single mode fibre is caused by both material and waveguide dispersion. Material dispersion arises as a result of variation of the refractive index of the core material with wavelength and causes the speed of light to be different for each wavelength. Thus a pulsed signal after propagating through the fibre broadens at the output, the broadening being dependent on the material dispersion, wavelength and the length of the fibre. Waveguide dispersion is caused by variation in the modal propagation constant along the radial direction of the fibre. It causes the speed of light in the fibre to attain a value which is between velocities in the core and in the cladding material. This type of dispersion also contributes to pulse broadening. For a single mode silica core optical fibre the material dispersion and the waveguide dispersion have opposite signs. The zero dispersion wavelength is about 1.27 μm for this fibre. The attenuation graph also takes a dip at 1.3 μm and at 1.55 μm. Beyond 1.55 μm wavelength the material dispersion dominates. Commercially available single mode fibres have a bandwidth–distance product of 125 GHz km. Therefore for the short distances used in active aperture systems neither the attenuation nor the dispersion is significant, and hence for practical purposes they are taken to be zero. A parameter of equal importance as the dispersion is the maximum optical power above which degradation takes place owing to the nonlinear phenomenon. Actually the measured limits are around 10–20 mW [81]. A great advantage of the optical fibre is its low sensitivity to temperature, which is one order weaker than for the coaxial cable or microstrip lines.

10.29.4 *Fibre optic link*

It is clear from the earlier paragraphs that the optical fibre link to be used in the active aperture array will be a short haul direct modulated link. It is now necessary to arrive at the power budget for such a network so that the microwave signals available at the T–R module will satisfy the system requirements. The analysis presented here is based on the line of reasoning given in Reference 72.

A fibre optic link in its essential form consists of a semiconductor laser at one end of the optical fibre of a predetermined length and a photodiode at the other end. The bias current I_L flowing into the laser diode is converted into optical power at a certain efficiency η_L of the laser. This is the differential quantum efficiency of the source and is determined by the slope of the output optical power versus the bias current curve of the emitter. A fraction of the laser output power is lost owing to imperfect coupling η_{LF} between the laser and the optical fibre. The optical signal travelling in the fibre suffers attenuation which is denoted by η_F. In addition, there is a splicing loss η_c to be considered, as the optical fibre network can be a multiport network with connectors for power combining and division. Once the optical signal reaches the other end of the fibre it again suffers a loss owing to imperfect coupling η_{FD} (coupling efficiency) between the fibre and the detector. The optical power now incident on the active area of the photodiode detector is converted into current I_D with a conversion efficiency η_D (responsivity) which is determined by the slope of the bias current versus the optical power of the detector. The bias current I_L of the laser emitter is related to the bias current of the photodiode detector by the expression

$$I_L = \eta I_D \tag{10.332}$$

where,

$$\eta = \eta_L \eta_{LF} \eta_F \eta_C \eta_{FD} \eta_D \tag{10.333}$$

Eqn. 10.332 can be expressed in terms of the electrical power transfer coefficient or function ζ^2 as

$$P_R = \zeta^2 P_i \tag{10.334}$$

Here P_i is the electrical power supplied to the laser diode and P_R is the power delivered by the detector. ζ can be expressed as

$$\zeta^2 = \frac{\eta_L^2}{R_L} \zeta_F^2 \eta_D^2 R_D \tag{10.335}$$

where,

R_L = incremental drain impedance of the laser about the driving point
R_D = load impedance of the detector
ζ_F = transfer efficiency of the optical fibre and is equal to $(\eta_{LF} \eta_F \eta_C \eta_{FD})$

The interest of the system designer is to determine the noise figure of the link so that the likely reduction in the RF signal can be computed at the T–R module. For this purpose it has to be noted that the relative intensity noise of the laser defined in eqn. 10.328 is the total noise at the output of the laser. It is made up of the noise component associated with carrier injection and recombination plus the thermal noise N_{in} from the RF signal source. Thus the equivalent of the noise power generated in the laser is given by

$$N_L = (RIN)[I_b - I_t]R_L B - N_{in} \tag{10.336}$$

where,

$$N_{in} = KTB \tag{10.337}$$

Here

K, T and B have been defined earlier
I_b = bias current of the laser
I_t = threshold current of the laser

Now the noise power N_D at the output of the detector is due to the shot noise associated with the average photocurrent I_D of the detector so that

$$N_D = 2eI_D BR_D \tag{10.338}$$

where e is the electronic charge. With the aid of eqns. 10.332, 10.333, 10.335 and 10.336 eqn. 10.338 becomes

$$N_D = 2e(I_b - I_t)\zeta B\sqrt{R_L R_D} \tag{10.339}$$

The total electrical noise power at the output of the link, that is, N_{out}, can now be determined as

$$N_{out} = \zeta^2 N_L + \zeta^2 N_{in} + N_D \tag{10.340}$$

Eqn. 10.340 can now be expanded to give the expression

$$N_{out} = \zeta^2[(RIN)(I_b - I_t)^2 R_L B - KTB] + \zeta^2 KTB + 2e(I_b - I_t)\zeta B\sqrt{R_L R_D} \tag{10.341}$$

The noise figure F_{L1} of the link is expressed as

$$F_{L1} = \frac{(S_{in}/N_{in})}{(S_{out}/N_{out})} \tag{10.342}$$

Using eqns. 10.335, 10.337 and 10.341 in eqn. 10.342 the noise figure of the link becomes,

$$F_{L1} = \frac{(RIN)(I_b - I_t)^2 R_L}{KT} + \frac{2e(I_b - I_t)\sqrt{R_L R_D}}{\zeta KT} \tag{10.343}$$

Eqn. 10.343 brings out clearly that the noise figure of the directly modulated fibre optic link is a function of the characteristics of the optoelectronic components, namely, the laser diode, the photodetector and the optical fibre. In Fig. 10.62 the noise figure F_{L1} is plotted as a function of the link transfer function ζ, for different values of RIN for a typical link. It can be seen that the noise figure attributable to the optoelectronic components is large, and thus RIN is the dominant source of noise of the link. The noise figure is directly proportional to the RIN and relatively insensitive to the link transfer function ζ in the shaded portion of the graph, this being the normal operational region of the link.

It is possible to improve the overall link noise figure if amplifiers with noise figures F_1 and F_2, both being lower than the link noise figure given in eqn. 10.343, are used before and after the link. In that case,

$$F_{L2}=F_1+\frac{F_{LI}-1}{G}+\frac{F_2-1}{G} \tag{10.344}$$

where G is the gain of the input amplifier. It would appear from eqn. 10.344 that the overall link noise figure F_{L2} can be minimised if RIN is kept low and the gains ζ^2 and G are maximised. The RIN is reduced by selecting a laser such as the DH–DFB laser. The gain ζ^2 is maximised by lowering the coupling and fibre losses. However, care has to be taken before increasing the gain G of the input amplifier as it can lead to laser burn-out and negative peak clipping.

The graph connecting the laser power output to the driving current as shown in Fig. 10.63 is assumed to be directly modulated at the point I_b by a sinusiodal microwave signal. The modulation index m is defined as

$$m=\frac{\Delta I}{I_b-I_t} \tag{10.345}$$

where ΔI is the variation in the microwave signal current above the driving point. From Fig. 10.63 the maximum operating current I_{PK} can be expressed as,

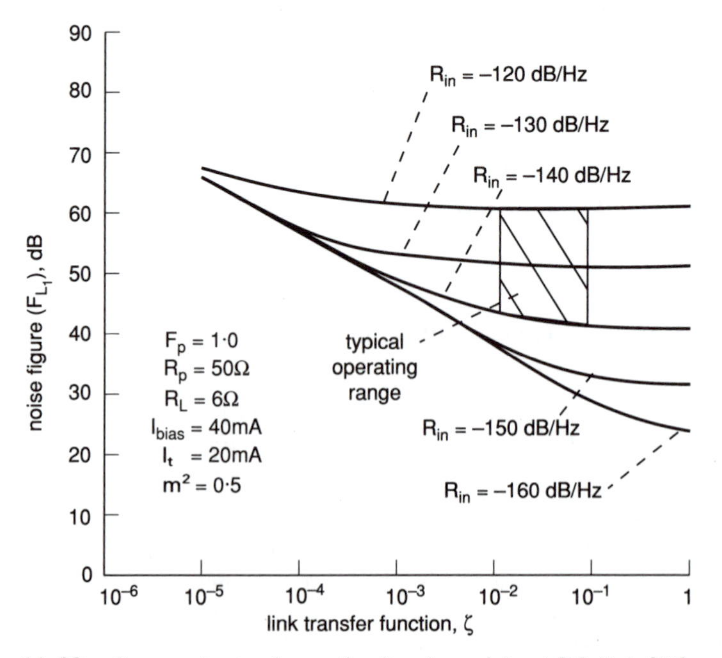

Figure 10.62 *Computed noise figure of a directly modulated FO link [82]*

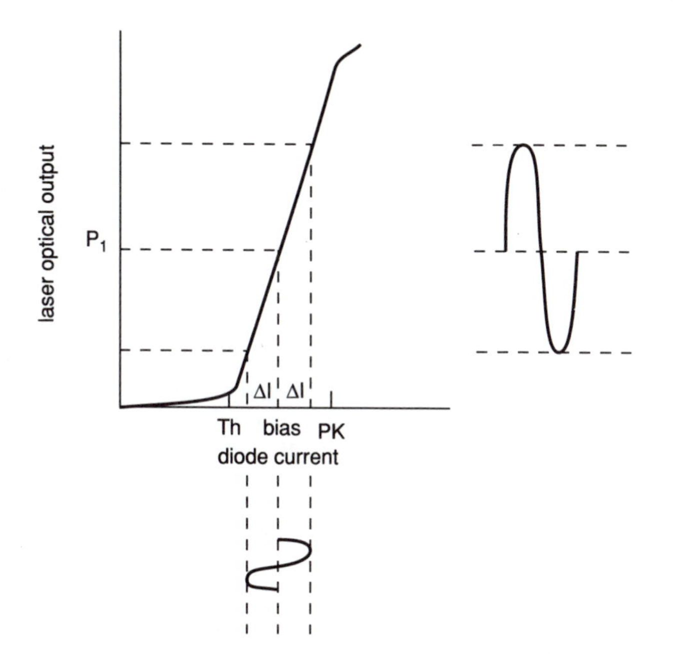

Figure 10.63 *Laser diode optical output versus bias current [72] © Artech House, 1990*

$$I_{PK} = I_t + 2(I_b - I_t) \tag{10.346}$$

Substituting eqn. 10.346 into eqn. 10.345

$$\Delta I = \frac{m(I_{PK} - I_t)}{2} \tag{10.347}$$

The peak microwave signal P_0 that can be dissipated in the incremental drive impedance R_L of the diode without waveform clipping is now given as

$$P_o = \frac{(\Delta I)^2 R_L}{2} \tag{10.348}$$

Substituting eqn. 10.347 into eqn. 10.348 gives

$$P_o = \frac{m^2(I_{PK} - I_t)^2 R_L}{8} \tag{10.349}$$

The gain G of the input amplifier for an input P_i is now given as

$$\begin{aligned} G &= P_o/P_i \\ &= \frac{m^2(I_{PK} - I_t)^2 R_L}{8P_i} \end{aligned} \tag{10.350}$$

Substituting eqn. 10.350 into eqn 10.344, the following expression is obtained for the overall noise figure F_{L2}:

$$F_{L2} = F_1 - \left[\frac{8P_i}{m(I_{PK} - I_t)^2 R_L}\right] + \left[\frac{2P_i(RIN)}{m^2 KT}\right] + \left[\frac{8e\sqrt{R_D/R_L}P_i}{m^2 \zeta KT(I_{PK} - I_t)}\right] + \left[\frac{8P_i(F_2 - 1)}{m^2 \zeta^2 (I_{PK} - I_t)^2 R_L}\right] \quad (10.351)$$

The directly modulated fibre optic link performance according to eqn. 10.351 is shown for a typical link in Fig. 10.64. For large input signal levels the link noise figure is degraded from the noise figure of the amplifier. On the other hand for low input signal levels the link noise figure is given by the noise figure of the first amplifier, namely F_1. By using such lasers as the buried heterostructure direct feedback type, the RIN of which is lower, the link noise figure is reduced. For high performance Doppler radars the near carrier noise of the microwave source has to be −110 dB below the carrier level, which means that the signal-to-noise ratio for the RF signal out of the fibre optic link has to be at least 145 dB if not better. From Fig. 10.64 it is quite clear that, if a GaAlAs/GaAs buried heterostructure laser is used, then the link will not meet the S/N requirement of the microwave signal for the high performance Doppler radar, because the RIN of the laser at 0.84 μm wavelength is about −120 dB/Hz. On the other hand, if a InGaAsP–InP distributed feedback laser operating at 1.3 μm wavelength is used, since it has an RIN of −145 dB/Hz, a noise figure of 35 dB is permitted

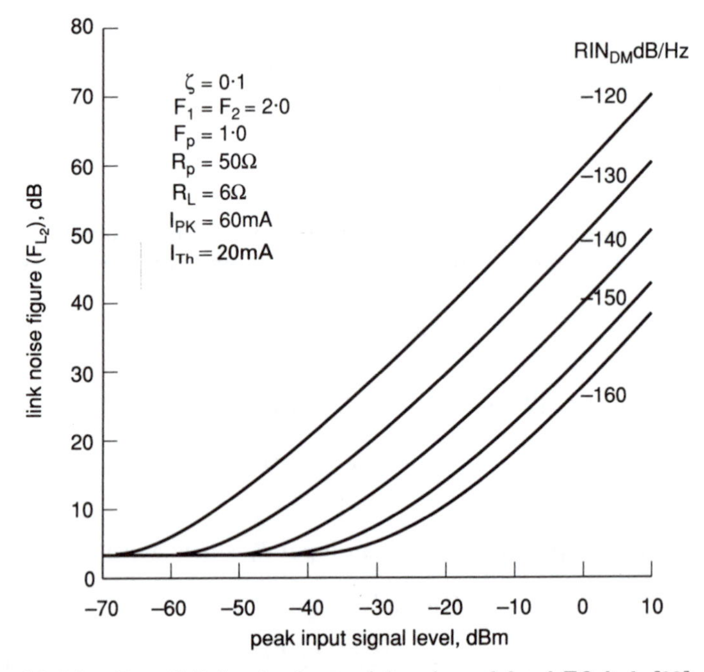

Figure 10.64 *Overall link noise figure of directly modulated FO link [82]*

for the link to meet the microwave source quietness requirement of Doppler radars. Fig. 10.64 also indicates that the peak input RF signal power required is around −8 to −12 dBm which, after amplification to the maximum signal level that can be dissipated by the laser and direct modulation, propagation through the FO link and then demodulation by the photodetector, will deliver nanowatts of RF signal to the T–R module if it is a single unbranched link. A large amount of gain and more than one stage of gain would be needed to amplify the microwave signal to the desired level. At microwave frequencies the realisation of such large amounts of phase-stable gain becomes more difficult. These difficulties are being overcome by injection locking of microwave oscillators.

The nonlinear characteristic of the laser diode current at higher values with optical power intensity gives rise to higher harmonics and intermodulation distortion (IMD) in directly modulated links. If f_1 and f_2 are two frequencies which are close to each other, and if the composite signal of these frequencies is used to intensity modulate a laser, which, being a nonlinear device, will give rise to harmonics at $2f_1$, $2f_2$, $3f_1$,... etc, second-order intermodulation products at $f_1 - f_2$, $f_2 + f_1$, third-order intermodulation products at $2f_1 - f_2$, $2f_2 - f_1$, . . . and so on. In the frequency range 5–15 GHz, the second harmonic to fundamental power ratio is typically about −10 dB, but it can be as much as −5 dB for a modulation index of 30% and a laser driving current of three times the value of the threshold current [81]. This is not of consequence in the present case because the bandwidth of the modulating signal is less than an octave, and so the higher harmonics and second order intermodulation products would be outside the link pass band. But in the case of the third order intermodulation products some of the frequency components are within the link pass band and they would affect the link linearity. The dynamic range of the link is therefore reduced by the presence of the third order intermodulation products to the spurious free dynamic range (SFDR). The SFDR is defined as the level of IMD suppression achieved when that level equals the link signal-to-noise ratio.

10.29.5 *Phase and frequency synchronisation*

It is quite evident that the microwave signal power available at the T–R module from the fibre optic link will not be adequate to drive the microwave circuits in the module. Therefore, there is a requirement for amplification or regeneration of these microwave signals with high AM noise compression and low AM–PM noise conversion. Berceli *et al.* [83] have carried out investigations on the optimum method of amplifying microwave signals after transmission through fibre optic links with phase and frequency synchronisation at T–R modules. Two different configurations have been studied at the receiving end. The first contains two microwave amplifiers to be used in tandem after the PIN photodetector. The second configuration utilises an injection locked oscillator after the PIN photodetector (see Fig. 10.65). Figs. 10.66 and 10.67 show the typical AM configuration of a single tuned amplifier and that of an injection locked oscillator, respectively, at the band centre as a function of input power. In the case of a single tuned amplifier the AM compression is low for low input powers. AM compression can be enhanced with increased input power, but this also reduces the gain of the amplifier. Hence two stages of amplifiers are needed. In the case of the injection locked oscillator, when the gain is high, i.e.

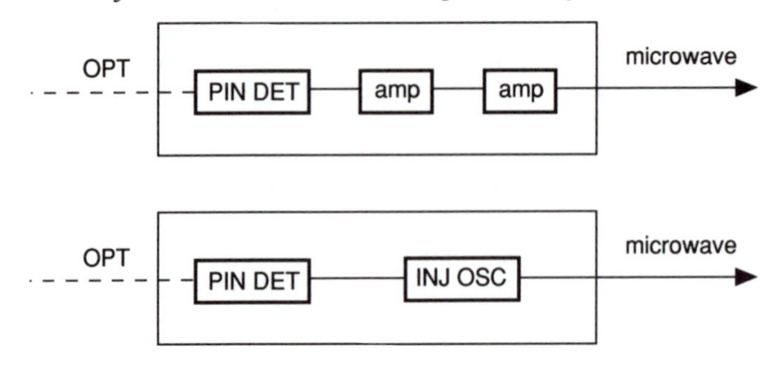

Figure 10.65 *EO–microwave frequency translation [83]*

the input power is low, the AM compression is very high. As the gain is reduced (i.e. the input power is increased), the AM compression is also lowered. Hence the injection locked oscillator is a better option than the tuned amplifier in reducing the AM noise, since the former simultaneously exhibits a high AM compression and a high gain. Another aspect that has been investigated is the AM–PM conversion, which has to be kept low for high performance. Figs. 10.68 and 10.69 indicate the AM–PM conversion of the amplifier and the injection locked oscillator, respectively, as a function of the relative frequency deviation. The magnitude of AM–PM conversion in this case decreases the increase of the input power level. Furthermore, it is also possible to make an optimum adjustment to effect zero AM–PM conversion so that AM noise is not converted to phase noise. In addition, in the case of the injection locked oscillator, phase noise can be significantly reduced by having a high Q resonator coupled with the injection locked oscillator. Therefore, for high performance Doppler radar systems based on active aperture phased array antennas, the use of an injection locked oscillator is the preferred mode of phase and frequency synchronisation of T–R modules using the fibre optic distribution network.

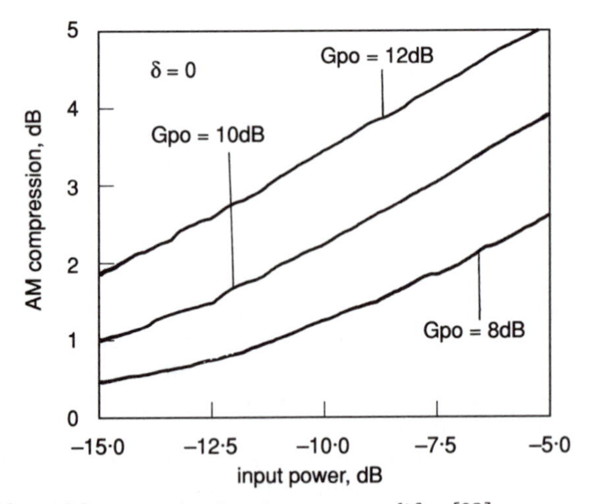

Figure 10.66 *AM compression in microwave amplifier [83]*

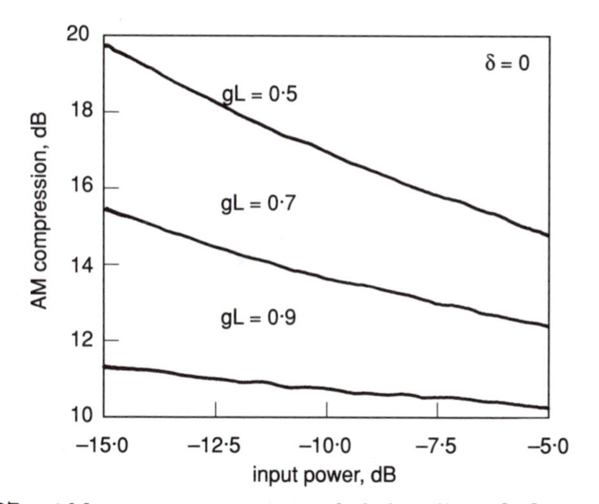

Figure 10.67 *AM compression in injection locked oscillator [83]*

In their recent experiments Berceli and his colleagues [84] have explored the possibility of devising an efficient microwave injection locked oscillator which will not only amplify the signals but also provide significant amplitude and phase noise suppression. A block diagram of the scheme is shown in Fig. 10.70. It shows two FET amplifiers which are operated as wideband amplifiers, and a dielectric resonator that acts as a narrow band feedback element to determine the free running oscillator frequency. The microwave signal from the FO link provides the reference for injection locking. This signal V_i is fed through matching circuits to a low noise almost linear amplifier. The second stage amplifier is a high gain FET in a nonlinear mode to increase the oscillator

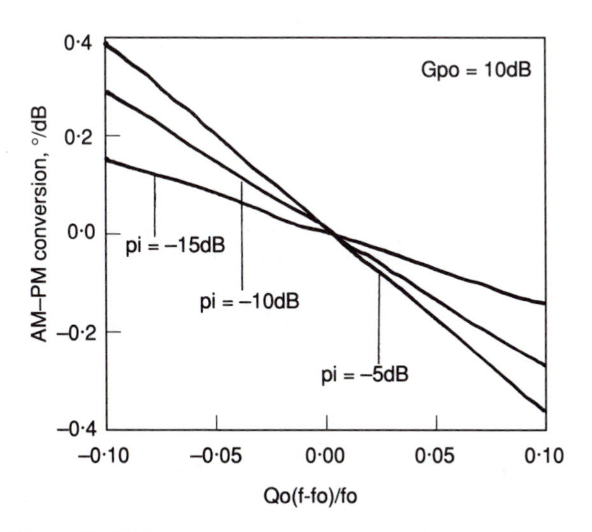

Figure 10.68 *AM–PM conversion: Microwave amplifier [83]*

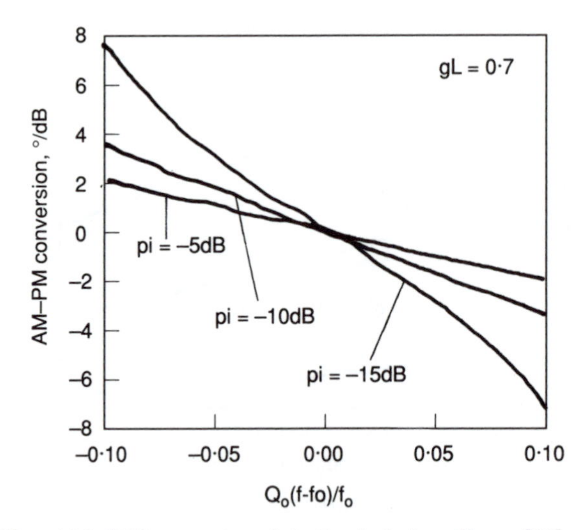

Figure 10.69 *AM–PM conversion: Injection locked oscillator [83]*

output as well as to compress the amplitude of oscillation. The two stage injection locked oscillator provides significant AM compression because of the enhanced gain. Since the bandwidth of both amplifiers is wider than the locking bandwidth, there is no frequency dependence over the locking range. The oscillation frequency is controlled by the resonant frequency of the dielectric resonator, and hence frequency and phase stability over a wide temperature is ensured. In addition, because of the two stage design, the reference signal can be lower than that for a single stage design. The nonlinear circuit developed by the authors shows that the locking bandwidth B_L is expressed as

$$B_L = \frac{f_c}{Q_e b_c} \sqrt{\frac{p_i}{p_o}} \tag{10.352}$$

Here Q_e is the external Q of the resonator, b_c is the feedback coupling factor, p_i is the input reference power and p_o is the output power of the injection locked

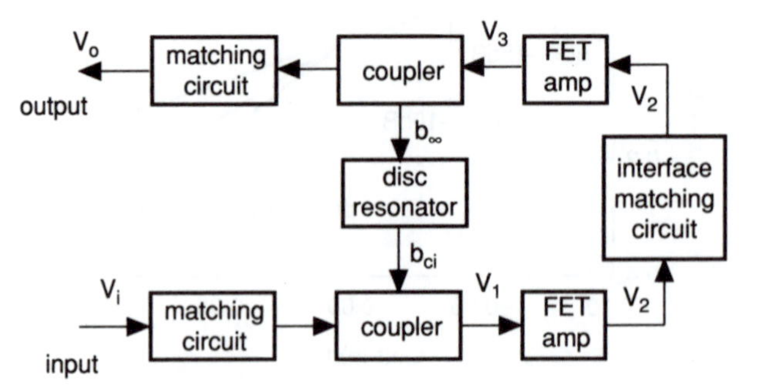

Figure 10.70 *Two-stage injection locked oscillator [84]. © IEEE, 1991*

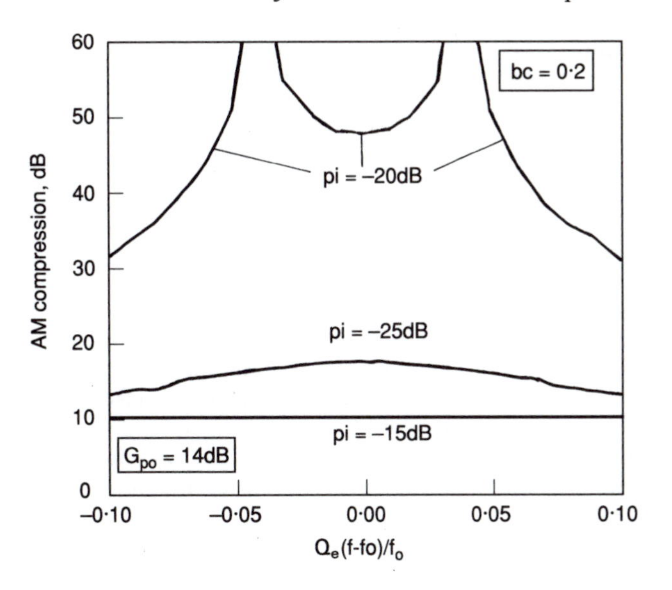

Figure 10.71 *AM–PM conversion for different input powers [84].* © *IEEE, 1991*

oscillator. As the locking bandwidth is inversely proportional to the feedback coupling, an increased locking bandwidth can be obtained by properly choosing the value of b_c. Similarly the advantage of increased gain capability with respect to the locking band is obvious.

The experimental results obtained are shown in Figs. 10.71 and 10.72. Fig. 10.71 shows the AM compression versus frequency deviation for different

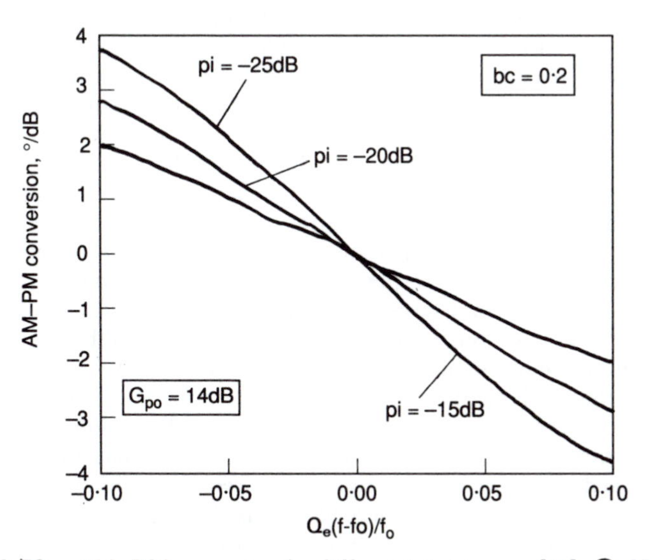

Figure 10.72 *AM–PM conversion for different input powers [84].* © *IEEE, 1991*

normalised input powers with overall feedback coupling factor of 0.2. For an input power of −20 dBm a very high degree of AM compression was obtained over a wide band. In general, maximum AM compression does not occur when the power output of the injection locked oscillator is maximum. It was also noted that, as the input power is reduced, both the AM compression and the gain increase. This ensures that the AM noise originating in the laser and in the fibre optic link can be substantially suppressed. Fig. 10.72 shows the AM–PM conversion response for different input powers as a function of frequency deviation. It highlights that the magnitude of the AM–PM conversion increases with decrease in the input level. Since the AM noise has already been compressed significantly the increase in AM–PM conversion does not result in noticeable noise conversion. Fig. 10.72 also shows that an optimum adjustment can be found where the AM noise is not converted into phase noise. However, it is necessary to note that, since the phase of the injection locked oscillator is a function of the frequency difference between the free running frequency and the reference, owing to manufacturing tolerances the oscillators have to be adjusted individually.

In the case of direct modulation, the modulating frequency should preferably be below half the relaxation frequency, to keep the noise floor as low as possible. Therefore, the upper limit of the modulating frequency would be around 10 GHz. Hence, for active aperture radars operating in the X-band or above, a low-phase noise signal for high Doppler performance systems will have to be obtained by subharmonic injection locking. Studies of the nonlinear characteristics of solid state active devices and the effect of oscillator circuit topology on subharmonic injection locking, have led to the conclusion [85] that the nonlinear active device generates the nth harmonic response of the injected signal and locks the free-running oscillator in a similar way to the injection locking with the fundamental. The locking bandwidth is proportional to the square root of the ratio of the power in the nth harmonic to the power output by the device. The minimum phase noise degradation of the nth order subharmonically locked oscillator is n or (20 ln) dB. The phase noise degradation can be minimised by increasing the input power before it reaches the 3 dB compression point. Experimentation has proved that a large locking range, minimal noise degradation with minimum injection power level can be obtained by proper design of the circuit topology. Since the nonlinear characteristic of the laser is optimum at frequencies close to the large-signal relaxation oscillation frequency, a laser which has a relaxation frequency that is very near the subharmonic frequency is selected and used for direct modulation of the laser. It is also necessary that the response peak at the relaxation frequency of the laser is pronounced so that large signal modulation takes place. Under these conditions, the laser output is rich in harmonics of the subharmonic injection frequency and these harmonics are likely to be at a much higher level than those achieved by the fundamental modulation at the same harmonic frequency.

Although the frequency synchronisation of the T-R modules can be carried out by injection locking, experiments have revealed that the phase of the injection locked oscillator is a function of the frequency difference between the free-running frequency and the reference. Thus the phase of the oscillators will vary due to both manufacturing tolerances of the oscillators and the temperature gradient across the array. The phase difference would result in inaccurate phase shifting of the individual radiating elements and thus disturb the radiation pattern of the array. Daryoush *et al.* [86] have been successful in correcting the asynchronism in phase by a technique of phase correction which is a combination of injection locking and phase-locked loop to ensure both frequency and phase coherence of the local oscillators of the active aperture array. This has been termed the ILPLL (injection locked phase-locked loop) and the incorporation of the ILPLL microwave oscillator into the T-R module now becomes necessary. This would also lead to changes in the configuration of the T-R module (see Fig. 10.80).

10.29.6 *Control network*

The requirement for phase and frequency synchronisation of microwave signals is most stringent on the FO link, and once these are met, the use of the link for control and communication signals in the active aperture array would be relatively easy. While the modulation frequency, bandwidth and noise figure requirements for the link can be more easily met, there are two major aspects of the requirements that need to be looked into before design. The first is the control architecture, i.e. whether control should be centralised or distributed or a hybrid scheme needs to be evolved. The next is whether the control should be direct or indirect. If the control is indirect, the optical link transmits the control signals but the computations are carried out digitally and the control is implemented via the microwave elements. If the control is direct, the optical signals are used for phase shifting, and also for beam forming.

The major considerations in determining the control architecture are the update rate and the amount of hardware that is required. The update rate is dependent on the number of control bits required per element, the number of elements and the maximum data transmission rate for the array. In centralised control, all computations leading to the values for the phase command for each element of the array for beam positioning are carried out in a CPU and transmitted serially to each phase shifter. For simple logic connections since 2 MHz per second is the practical limit, as the size of the array becomes larger, the execution time increases directly as the number of elements. Waldron *et al.* [87] have shown that, for a typical phased array, the execution time for this type of architecture is the highest of all the three options available. An improvement on this is the add-at-element architecture where the elements are located on a rectangular grid and each element of the array is identified by a column and a row number. The CPU centrally computes the phase angles pertaining to a given beam direction in terms of bearing and elevation angles and transmits these values along each row and each column of the array. If $m\alpha$ is the phase command for the mth row and $n\beta$ is the phase command for the nth column, then for the (m, n)th element of the array, the total phase change $(m\alpha+n\beta)$ is effected at the phase shifter by simple addition of $m\alpha$ to $n\beta$. Even though the execution time in this case increases linearly with the number of elements, for medium and small arrays the add-at-element scheme appears to be advantageous if the array can be described by a rectangular or square lattice and if no compensation is to be added at each T–R module. These compensations may arise as a result of performance variations of the T–R modules owing to frequency, temperature, supply voltage, aging etc. In the distributed control architecture the CPU transmits to each radiating element along a common serial bus, the proposed direction of the beam and also the desired changes in the gain of the T–R module. Only, for phase steering, 16 bits each for the two direction cosines are more than sufficient since it gives a beam granularity finer than 1% of the beamwidth. Allowing 10 bits for selection of a gain command and several bits for instruction codes and synchronisation, the information to specify phase and gain can be sent in less than 30 μs by the CPU for any number of T–R modules. The distributed control scheme requires minimal control connections since the actual phase and gain commands as well as the compensations are generated by the controller at each T–R module on receipt of a serial command input. The time limitation in this case is the time to select the desired correction term by transmitting a memory address to all modules in parallel. Waldron *et al.* have tried out two types of distributed controllers, namely, one using a microprocessor and the other a fully custom design. The customised design has been found to be more advantageous because of the lower chip count, higher speed and lower power consumption. The customised controller consists of three chips, the heart of the system being a beam steering controller dedicated to communications interface, T–R module control, beam steering calculations, self-test and interface to a calibration circuit. The control signals for each of these architectures can be transmitted to the T–R module over an optical link.

In the indirect control method the function of the fibre optic link is to transmit the beam control signals from the CPU to the T–R module where they are demodulated, decoded and transmitted to the module controller to activate the conventional phase shifting and gain control circuits. In the case of phase shifter control one method envisages the multiplexing of the command for a certain number of phase shifters, and transmitting them serially from one end over the FO link. At the receiving end the signals are demultiplexed and converted to parallel data for the phase shifter. In this connection it has been reported that a hybrid GaAs optical controller has been tried out in a high speed FO link with a PIN photodetector and MESFET demultiplexer, and that the serial data has been transmitted to as many as 16 phase shifters [88]. An alternative method that is being tried out uses an optical beam former spatial filter to generate appropriate phase commands for each phase shifter. These commands are then transmitted on dedicated FO links, one per phase shifter, to the T–R modules where they are detected by a light sensitive MESFET [89–91] (see Fig. 10.73). The MESFET is biased at or near pinch-off where it appears most sensitive to light. The end of the optical fibre is positioned over the active area of the FET for optimum light responsivity. The drain–source voltage of the FET varies with the light intensity. An inverting operational amplifier is used to scale the FET detector output voltage either to provide phase commands to the phase shifter or to the variable gain of the power MMIC amplifier for controlling the output to the antenna element. Fig. 10.74 indicates the variation in the gain of the amplifier with change in optical intensity. It is important to note that a change of nearly 15 dB in gain is achieved with only 250 μW of optical power. Similarly Fig. 10.75 indicates the variation in phase shift with optical power. Recent experiments with an X-band 6-bit phase shifter have demonstrated that optical control of the phasor is possible. The indirect control method appears to be well suited for use with the distributed control approach indicated earlier.

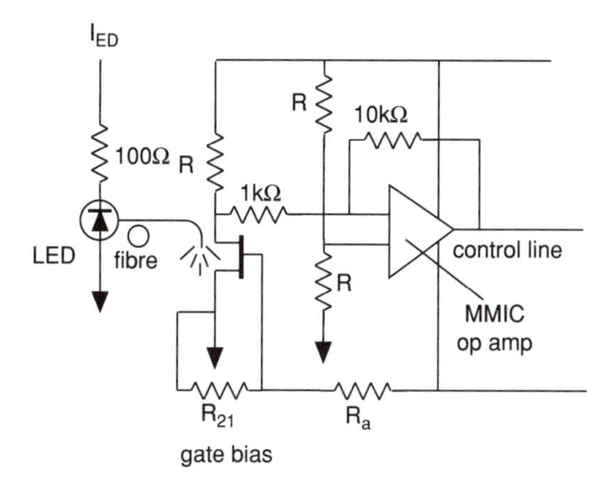

Figure 10.73 *Optical control of GaAs MMIC amplifier [91]. © IEEE, 1988*

The direct control method on the other hand utilises the fibre optic network to distribute control signals and emphasises compatibility with parallel optical processing methods. To control the phase and amplifier gain in the T–R module through optical methods, the gain and phase have to be a function of the optical intensity, which means

$$G = G_n(\alpha_n)$$
$$F = F_n(\psi_n) \qquad (10.353)$$

where α_n is the optical intensity to the amplifer and ψ_n is the optical intensity to the phase shifter of the nth T–R module. Thus the aim is to optically generate the required control functions $G_n(\alpha_n)$ and $F_n(\psi_n)$ that control the gain and phase of the T–R module. In this method the concept of using progressive phase shift between radiating elements for beam forming is still retained. Only, the way of

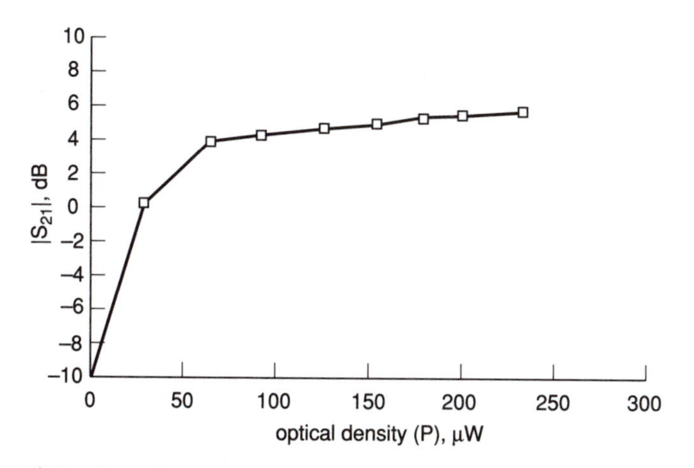

Figure 10.74 *Optical gain control of GaAs MMIC amplifier [90]*

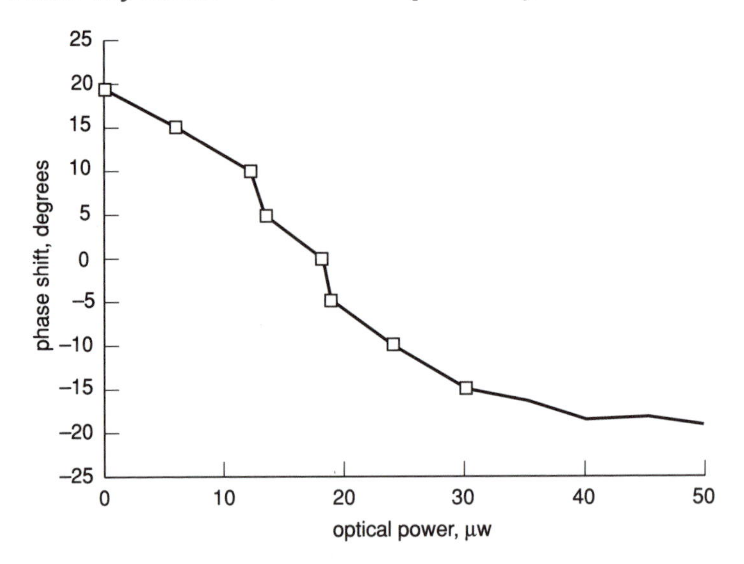

Figure 10.75 *Phase shift as a function of optical intensity [90]*

generating the control functions is by utilising optically controlled microwave devices that are incorporated in the T–R module. A microwave lateral PIN diode has been tried out in the attenuator circuit of the power amplifier chain. In the phase shifter circuit the lateral diode terminating in a 50 Ω transmission line serves as an optically controlled microwave reflection-type phase shifter [89]. The optical port in the device represents an additional terminal through which the device impedance is controlled by injection of additional carriers.

A second approach is to do away with the concept of progressive phase shift between radiating elements and reintroduce true time delayers between radiating elements for beam formation. This approach is broad band but could not be elegantly implemented earlier with heavy and bulky microwave transmission lines. Kam *et al.* [92] and Lee *et al.* [93] have considered the options of using analogue and digital real time delays in the optical domain for beam formation and steering. In the case of Kam, the analogue time delay mechanism at optical frequencies is based on a piezo-electric crystal. On application of a voltage, the circumference of the optical crystal is elongated and the optical fibre wrapped around it is lengthened, resulting in a delay. Lee suggests the use of switchable fibre optic delay lines to produce the required time delays, the switching being effected by PIN photodiodes. Ng *et al.* [94] have demonstrated true time delay steering using switchable FO time shifters in a microwave phased array antenna. Seeds *et al.* [95, 96] have described a coherent beam forming technique using two optical frequencies with an offset that is equal to the desired microwave frequency of the array. If $E_1 = E_{1M}$ (sin $\omega_1 t$) and $E_2 = E_{2M}$ sin$(\omega_2 t + \phi)$ are two optical signals which are combined to illuminate a photodetector, the current at the output of the detector will be proportional to $E_1 E_2$ {cos$[(\omega_1 - \omega_2)t + \phi]$} owing to the square law detection process. Since $(\omega_2 - \omega_1)$ is equal to the microwave frequency of the phased array, a phase shift at the optical frequency is directly translated to the microwave output. Therefore the two optical frequencies are differentially phase modulated with the required base band data and combined in an optical power combiner/splitter. The output of the splitter is fed individually to the T–R modules. The two optical frequencies are combined finally on a single polarisation axis and then transmitted to the array face.

It is now clear that the most important task for the fibre optic network relates to the phase and frequency synchronisation of the array for the purposes of power combining in space and for coherent detection. The other two tasks, namely, the distribution of control signals for beam forming and beam steering and the transmission of communication and data signals from CPU to T–R modules, are relatively easier from the point of view of bandwidth and spectral purity. All these three tasks can be implemented on one fibre optic link especially up to 12 GHz, but from the point of view of design, three separate links would be preferable so that optimisation of each can be carried out independently. By using fibre optics, the reference and control signal distribution network is considerably simplified. It is also ensures light weight, low power consumption and compactness.

10.29.7 *Power supply requirements aspects*

One of the advantages claimed for active aperture phased arrays is that of graceful degradation. The random failure of individual T–R modules makes only undetectably minor changes to the radiation pattern of a large array. The losses in the transmitted signal in such cases reduce accuracy in estimation of target parameters in a very minor way but do not represent loss of total desired radar function. However, very little has been mentioned about the effect of loss of a percentage of radiating elements due to power supply failures. In view of the graceful degradation characteristic of the active aperture phased arrays, the important reliability parameter in this case is the mean time between critical failures (MTBCF) rather than the MTBF. In an array of N T–R modules we define the array to have reached the critical failure stage when q of the T–R modules are out of action as a result of which either the power gain of the antenna has fallen below an acceptable value or the sidelobe levels have gone beyond an acceptable level. On the other hand the MTBF in this case is the average time taken for any single T–R module to be out of service from this ensemble of N T–R modules. If the entire system is being supplied by only one power supply unit with an MTBF of T, the MTBCF is the same as the MTBF because the failure of the power supply unit results in total failure of the array. In reality also, the use of a centralised power supply for all the T–R modules is not recommended because it has to generate very high currents at low voltages and distribute the DC supply to all modules across the antenna aperture. By and large, it is therefore advisable to have independent power supplies for individual T–R modules or parallel the output of these individual power supplies for each sub-array with an over-capacity. The percentage of over-capacity is maintained at a value slightly higher than the percentage of T–R module failures that determine the critical failure of the array. The power requirements of a single T–R module can be conveniently obtained from compact and highly reliable units as the present state of the art in power supplies indicates that power densities in the region of 50 W/in^3 are possible.

10.30 Printed circuit antennas and arrays

10.30.1 *Antenna element*

Printed circuit antennas (PCA) are not new. They were first proposed by Deschamps in the USA in 1953, but there was not much interest at that time since not many practical applications could be found. Four decades later, PCA has become a growth area and the volume of literature covering experimental as well as theoretical aspects testifies to the vigour and tempo of these activities [97, 98]. A comprehensive account of the development of the PCA as well as the printed circuit antenna array (PCAA) up to the advent of the 1980s is available [99, 100]. Of the many configurations that are possible in printed circuit antennas, the patch radiator and the printed circuit dipole have found maximum favour with research workers and system designers. In microstrip radiators, even though the freedom of design is restricted to the plane of the radiator, a multiplicity of shapes have been attempted. Some of these are square, rectangular, circular, elliptical, annular, triangular, a strip or a thin rectangle. The characteristics of all these principal shapes are generally similar, with the fundamental modes giving rise to broadside radiation pattern. Bandwidth and physical area vary with the shape. The generic microstrip patch radiator is an area of metallisation supported above a ground plane and fed by a transmission line against the ground plane at an appropriate point or points (see Fig. 10.76). The metallised area above the ground can assume any of the principal shapes enumerated earlier. The printed circuit dipole is a strip dipole that is supported above a ground plane and fed at the centre by an appropriate transmission line against the ground plane. In the case of the printed circuit dipole the distance between the ground plane and the radiator is higher than in the case of the patch radiator. These patch radiators can have coplanar feed, in which case the input line is fed against the upper metallic plane. It can also be fed from a coaxial line, in which case the metallic patch on the upper conductor is fed by the inner line of the coaxial feed. For the PCD, both the outer and inner line of the coaxial transmission line end up on the two halves of the dipole.

Of these two microstrip radiator configurations, the patch radiator is an interesting and useful radiating element that offers the promise of eventual production by monolithic integrated circuit techniques. Hence, much of the future active aperture array technology is likely to be built around this element [101]. The microstrip patch has typically about one-third to one-half of the free space wavelength for its longest dimension while the dielectric thickness is normally in the range of 0.03–0.05 λ. A commonly used dielectric for such radiators is Teflon or polytetrafluoroethylene often set in a reinforcing glass fibre matrix. Besides its radiation characteristics, the popularity of the patch radiator is attributed to its low profile, conformability to nonplanar surfaces, and ease of manufacture in large volume with photolithographic techniques. These advantages, however, must be weighted against the disadvantage of the patch being a high Q radiator, with Q factors sometimes exceeding 100 for the thinner elements. Therefore the bandwidth is narrow, around 1–3% in most cases. It is to be noted that the bandwidth referred to is the impedance bandwidth, i.e. the variation of the input impedance of the radiator with frequency. The radiation pattern of the patch radiator, on the other hand, remains stable over a wider frequency range of operation. Hence the concern of designers has been to minimise the impedance variation or lower the Q of the radiator to increase the bandwidth of operation.

There are several theories for microstrip elements with varying degrees of accuracy and complexity. Of these, the transmission line model and the cavity model provide the best physical insight into the microstrip radiator. The transmission line model is the simplest in its concept because it views the radiator as a transmission line propagating in the quasi-TEM mode and terminated at its two ends by two radiating apertures, very much like radiating slots. Though the results, especially of the input impedance, are good enough for many applications, they are not accurate enough for CAD purposes. Moreover the transmission line model requires modification to account for the

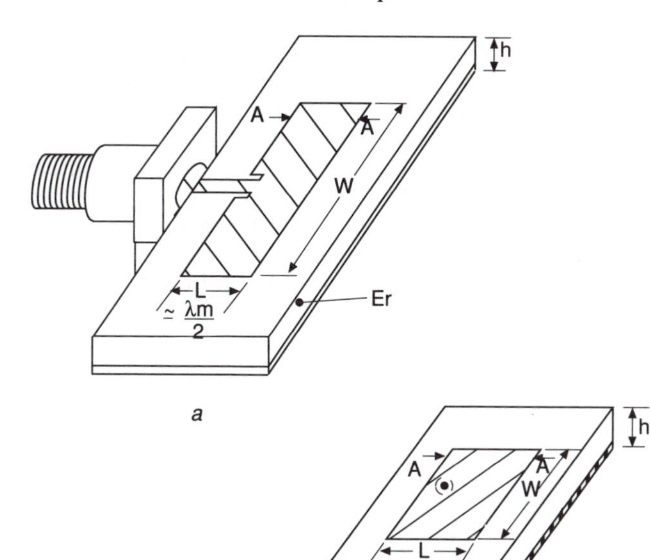

A
A
W
h
L
≈ λm/2

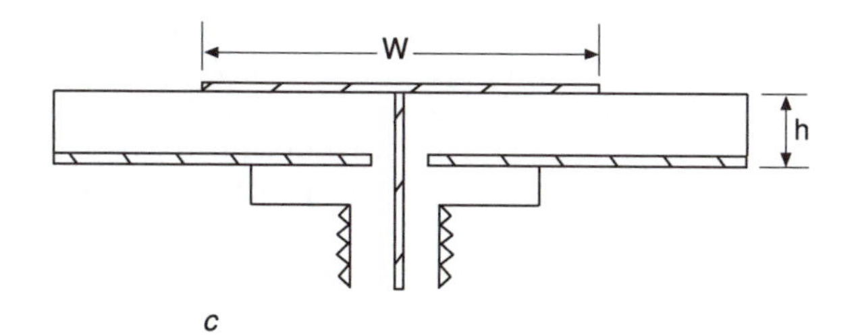

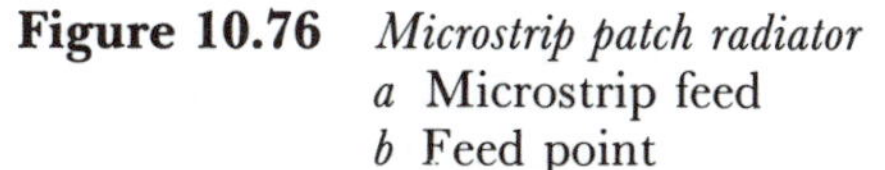

Figure 10.76 *Microstrip patch radiator*
a Microstrip feed
b Feed point
c Coaxial probe feed

fringing fields so that the resonant frequency can be accurately ascertained from its dimensions. Lastly it is applicable only for rectangular or square patch geometries. The cavity model provides a deeper insight into the phenomenon of radiation because it considers the volume of space under the patch and between the ground plane as a leaky cavity and brings out the type of excitation for the different modes that can provide the radiation field. Fig. 10.77 provides the possible radiation patterns from a rectangular patch for different modes that can be excited. Based on this model, the rectangular microstrip patch dimensions for a given frequency of operation can be expressed as

$$f_{res} = \frac{c}{2\sqrt{\varepsilon_r}} \sqrt{\left(\frac{m}{L}\right)^2 + \left(\frac{n}{W}\right)^2} \tag{10.354}$$

where ε_r is the dielectric constant, L and W are the length and width of the patch, c is the velocity of light, and m and n are integers with values starting from zero. The fundamental mode in this case will be either $m=1$, $n=0$ or $m=0$, $n=1$, depending on the mode of excitation. Eqn. 10.354 is the zeroth order of approximation, since it does not take into account the inhomogeneous structure and the fringing fields of the actual microstrip configuration. It can still form the basis for computation of the radiator dimension provided the relative dielectric permittivity ε_r is replaced by an effective value $\varepsilon_{eff}(a)$ and the dimensions of the cavity are lengthened at each end by $\Delta L(W)$ and by $\Delta W(L)$ to account for the radiator. In that case

$$f_{res} = \frac{c}{2\sqrt{\varepsilon_{eff}(a)}} \sqrt{\left(\frac{m}{L+2\Delta L(\omega)}\right)^2 + \left(\frac{n}{W+2\Delta W(L)}\right)^2} \tag{10.355}$$

$$\varepsilon_{eff}(a) = \begin{cases} \varepsilon_{eff}(L) & \text{if } m=0 \\ \varepsilon_{eff}(W) & \text{if } n=0 \\ \dfrac{\varepsilon_{eff}(L)\,\varepsilon_{eff}(W)}{\varepsilon_r} & \text{otherwise} \end{cases} \tag{10.356}$$

The effective dielectric constant is a function of the dimensions of the radiator and the dielectric constant of the substrate. The accuracy of this parameter has been greatly improved over the years to also include frequency dispersion effects [103]. Similarly the displacement $\Delta W(L)$ or $\Delta L(W)$ of the microstrip patch radiator has been refined over the years [104] and it provides for an accuracy better than 2.5% in the range $0.1 < W/h$ or $L/h < 100$ and for dielectric constants less than 50, with h being the thickness of the substrate. When these values of ε_{eff}, $\Delta W(L)$ and $\Delta L(W)$ are used in eqn. 10.355, the best accuracy is found for (h/λ_d) less than 0.05 and the percentage difference between the measured and computed values is better than 5% for higher values of (h/λ_d) [105], with λ_d being the wavelength in the substrate. A major portion of the percentage difference is due to the tolerances and uncertainties in fabrication. Because the microstrip radiator is a very narrow band, even small dimensional variations result in sizable departures in the magnitudes of the resistance and reactance of the radiator from the values at resonance. This would necessitate tighter tolerances which would contribute to higher costs. Therefore there is a need for innovation in manufacturing to overcome these hurdles.

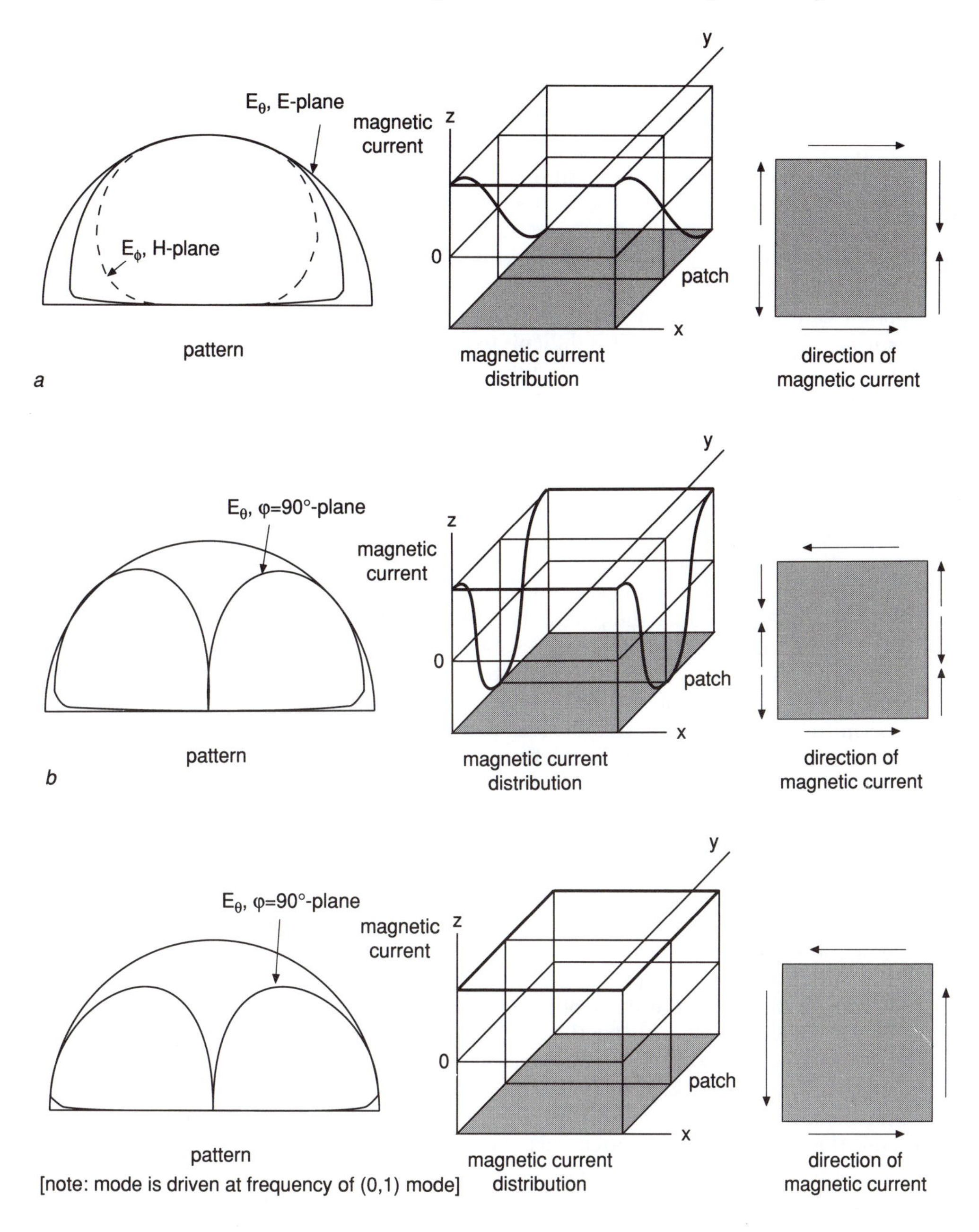

Figure 10.77 *Radiation patterns and magnetic current distributions [102]*
a TM_{01} mode
b TM_{02} mode
c Resonant frequency of TM_{01} mode

The cavity model also provides greater insight into the problems of feeding the patch radiator. It treats both the coaxial probe and stripline feed as having identical source distribution. At frequencies near each of the resonant modes, it computes the location of the feed as well as the resonant resistance of the radiator. It thus provides a satisfactory explanation for the variation of the resonant resistance with the position of the feed and also a method for matching the antenna to the feed. The feed reactance values obtained from the cavity model are not quantitatively accurate, though it predicts precisely the trends in feed inductance as a function of feed location. The optimum feed location in the case of patch radiators is, however, not unique. For some modes a suitable feed point can be found at the edge for the rectangular patch. In some other cases such as the circular patch it is difficult to find an optimal feed point on the edge. One is required to feed at a point interior to the patch. The simple impedance matching techniques described above do not extract the largest possible bandwidth from the radiators. For the microstrip patch operating in the lowest mode, a typical bandwidth is from less than 1% to several percent for thin substrates satisfying the criteria $h/\lambda < 0.07$ in the case of dielectrics having a permittivity of 2.3; and $h/\lambda < 0.023$ for substrates with permittivity in the region of 10.0. Considerable efforts have been made in the last ten years to improve the impedance bandwidth by such means as altering the shape of the patch, addition of active components, impedance matching networks and increasing the volume/area under the patch. The last option, namely, increasing the area or volume under the patch is promising. The straightforward way would be to increase either the permittivity of the dielectric substrate or increase its thickness. Analysis of the behaviour of the microstrip patch with varying permittivity and thickness indicates that, as long as (h/λ) is small, a linear dependence of bandwidth on thickness can be assumed [106]. It is also to be noted that, as the substrate thickness increases, the radiation resistance and the efficiency of the patch decrease but the directivity of the radiation pattern shows an increase. It is further seen that, as the permittivity of the substrate increases, for the same thickness, the bandwidth, directivity and the efficiency of the radiator decrease while the radiation resistance increases. The analysis clearly leads to the conclusion that the radiated power is shared by the space wave which is linked to the desired radiation pattern and by the surface wave modes, the dissipation of power in the latter being undesirable. With increasing substrate thickness, the share of the radiated power in the surface wave modes, $(h/\lambda)^3$, increases, whereas for the space wave component of the radiated power, the radiation is $(h/\lambda)^2$. Since the surface wave power contributes to the sidelobe region, it has to be kept as low as possible. Pozar [107] and Schaubert [108] have shown that, as the substrate thickness increases, the radiation resistance and the efficiency of the patch decrease but the bandwidth of the radiator and the directivity of the radiation pattern show an increase. It has also been observed that, as the permittivity of the substrate increases, for the same thickness, the bandwidth and the directivity of the radiator decrease and the radiation resistance increases. Further, as the substrate thickness increases, there is an inductive shift which in some cases is so large that the resonant frequency of the patch radiator could not be defined [109]. All of these experiments lead to the conclusion that the inductive shift is caused primarily by the excitation of the higher order modes. This is clearly seen from the leaky cavity model in which, as the cavity dimension in the direction normal to the patch surface increases, there will be higher order modes with field dependence on the normal. These higher order modes lead to significant differences in the radiated power and consequently in the patch quality factor. It is therefore evident that, from the point of view of microstrip patch radiator performance, low permittivity thick substrates are needed to meet the bandwidth requirements. On the other hand, in the case of T-R modules, the microstrip line configuration used has high permittivity and a thin substrate. There is therefore a need to match these two for the integrability of the PCAA (printed circuit antenna array) with the T-R module in the active aperture radar sensor.

10.30.2 *Feed configuration*

It is now necessary to discuss the methods of excitation of the microstrip patch radiator so that the technique which will satisfy the requirements of wider bandwidth and integrability with the MMIC (T-R modules) can be determined. Basically there are four methods of feeding the microstrip patch radiators and these are, microstrip line edge-feed, coaxial line probe-feed, microstrip line proximity-feed and the aperture-coupled feed. The microstrip line edge-feed and the coaxial line probe-feed are very similar in operation and were the first to be used for the excitation of the patch radiator. These have been shown in Figs. 10.76*a* and *c*. The microstrip line edge-feed is in the same plane as the patch radiator and is connected to the radiator at an edge. Although the feed and the radiator can be fabricated in one step, thereby bringing down the cost, the asymmetry of the feed gives rise to higher order modes which generate cross polarisation components. In addition, as the microstrip feed line is in the same plane as the radiator, the spurious radiation from the feed network is higher and unavoidable. In the case of the probe feed, the inner conductor of the coax is connected to the underside of the patch and the outer conductor is joined to the ground plane. As the feed line is normal to the plane of the radiator, the impedance match between the patch and the line can be realised by varying the position of the feed point. Any asymmetry of the feed point, however, gives rise to higher order modes and cross-polarisation. Also, there is only one degree of freedom available to the designer for feed matching and this is not always adequate. Despite the fact that the feed is below the ground plane of the patch, feed radiation exists due to the bare inner conductor. In an active aperture radar, as there can be more than 1000 such connections to be made between the feed line and the patch radiator, costs are likely to be high. There are also significant difficulties associated with the accurate analysis of probe-fed microstrip-patch antennas. These include the need to enforce current continuity from the probe feed to the patch radiator. Since the current on the patch exhibits a singularity at the feed point, the representation of the current in terms of simple basis functions is virtually impossible. Recently, for substrates with thickness greater than 0.05 λ, a two-step approach has been developed [110, 111] to rigorously model the probe-feed and remove *a priori* the current singularity at the probe tips. The microstrip line proximity-fed patch radiator has two substrates; the patch on the top substrate and the feed line on the lower substrate. In this case additional capacitance is added by electromagnetically coupling the radiator to the feed line [112]. This method is an advance over the previous two because it provides independent control for radiator and the microstrip line characteristics. It also leads to two degrees of freedom for input impedance matching, namely independent choice of the ratio of patch width to microstrip line width and the open stub length of the feed. In addition, the spurious radiation is considerably reduced. Chen *et al.* and Katehi *et al.* [113, 114] have shown that bandwidths greater than 10% can be obtained with an appropriate configuration of the electromagnetically-coupled patch (EMCP) antenna. However, there are some disadvantages which have to be borne in mind. First, the fabrication is complicated because of the need for accurate alignment of the substrates to achieve the correct excitation of the radiator. Secondly, integrabilty of the radiator with the T-R module is not easy. The aperture-coupled feed for the patch radiator is the fourth method of excitation that is possible (see Fig. 10.78). In this case, it is more practical to have two substrates, one on top of the other, with a common ground plane. The microstrip feed line is etched on the lower substrate and then coupled through a nonresonant aperture in the common ground plane to the microstrip-patch radiator on the upper substrate. It has many advantages, such as easy integrability with active devices, simplicity of heat removal and lower cost due to the possibility of etching the slot and the patch in one single operation. There are four degrees of freedom available to the designer: the size of the coupling slot; the position of the slot; the feed-substrate characteristics and the feed-line width. The coupling of energy is through the magnetic field and, by positioning the aperture at the centre of the radiator, maximum coupling is achieved and at the same time the cross-polarisation component is reduced to a negligible value. This method of feeding the patch radiator has been considered superior and is currently the most popular candidate for the active aperture radar, even though the complexity in fabrication is higher. The main reasons for its popularity are the impedance bandwidth improvement and better integrability with MMICs. Bandwidths as high as 20% have been reported for a single-patch radiator in this configuration by lengthening the coupling slot so that it is near resonance [115]. However, precautions must be taken before enlarging the slot dimensions to ensure that the back radiation is not unduly enhanced. The aperture-coupled feed system leads to purer linear polarisation and therefore, in applications calling for circular polarisation, two feed points are to be employed to excite two orthogonal modes in the patch with 90° phase shift between the excitations.

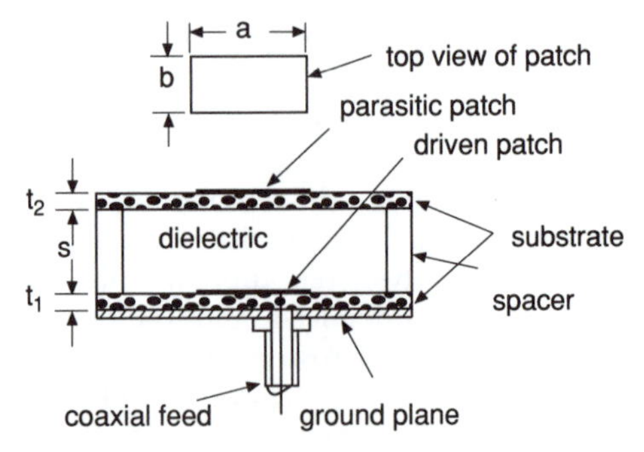

Figure 10.78 *Electromagnetically coupled double patch radiator [123].* © *IEEE, 1990*

Alternatively, it has been suggested that an appropriately oriented and positioned single-coupling aperture is used to generate circular polarisation [116].

10.30.3 *Analysis of aperture-coupled patch radiator*

In view of the inadequacy of the earlier analytical methods to account for the surface wave effects, full wave analysis using exact Green's function for the dielectric substrate and the method of moments, has been carried out. For example, Sullivan and Schaubert [117] have analysed the aperture-coupled rectangular microstrip-patch radiator using a set of expansion modes on the feed line whereas Pozar [118] made use of the reciprocity theorem to handle the currents on the feed line. In Reference 117 the dielectric substrate is assumed to be infinite in extent but finite in thickness. The electric surface currents on the patch radiator and on the feed line as well as the electric field in the slot are assumed to be in the direction of propagation. The equivalence principle is then invoked and the aperture in the ground plane is closed by a perfect electric conductor, so that the feed-line region (lower region) is isolated from the radiator region (upper region). On both sides of the closed aperture, equivalent magnetic surface currents are introduced to ensure the continuity of the electric field across the original aperture. Coupled integral equations are now formulated by using Green's functions for the grounded dielectric substrate along with the boundary conditions that the tangential electric fields at the patch radiator in the upper region and the tangential electric fields at the feed line in the lower region are zero and the magnetic field at the coupling aperture is continuous. The analysis includes all coupling effects, the radiation and the surface wave effects of the two substrates. The current on the feed line is expressed in terms of travelling waves incident and reflected, and they are represented by full-domain sinusoidal expansion functions plus seven piece-wise sinusoidal basis functions in the vicinity of the open-circuit stub on the feed line and the coupling aperture. As far as the current on the patch radiator is concerned, it is represented by five piece-wise sinusoidal basis functions. The method of moments solution based on the Galerkin procedure is formulated for the three coupled integral equations. It has been reported that, even after employing time-saving measures, each data point required about 30 minutes of central processing unit time on a VAX 11/750 running under the VMS operating system in a multiple user environment. The combined results of the analysis lead to the following:

(*a*) The aperture-coupled microstrip-patch radiator can be tuned by the open-circuited stub of the feed line, which is approximately a quarter of a wavelength. For better accuracy, a length extension to the stub has to be considered to account for the fringing fields at the end of the open stub. In most practical cases, the length extension is approximately equal to 0.4 h_f where h_f is the thickness of the feed substrate. The coupling aperture and the patch radiator appear as a series load along the open-circuited feedline.

(*b*) The length of the coupling aperture has a small effect on the resonant frequency of the radiator. As the length of the aperture is reduced, the coupling between the antenna and the feed line decreases. On the other hand, if the length of the coupling aperture is increased, the resonant frequency of the patch decreases but the input impedance at resonance increases. Thus, the open-circuited stub of the feed line can be used to cancel out the reactance of the patch radiator and the length of the coupling aperture helps in obtaining the desired resistive part of the impedance for match.

(*c*) For small off-sets of the aperture from the centre of the patch radiator in the direction of the resonant dimension of the radiator, the variation of the input impedance is relatively insensitive. However, for large off-sets, it decreases substantially as it moves away from the centre towards the edge. The patch radiator appears as a stub with slightly capacitive input impedance in series with the smaller coupling aperture which is inductive.

(*d*) When the aperture off-set is in the direction orthogonal to the resonant dimension, there is little change in the coupling factor so long as the entire slot is covered by the patch. However, experimental data show that the coupling factor increases as the slot edge coincides with the patch edge and then decreases as the slot emerges from the patch cover.

(*e*) The variation of the dielectric constant of the substrate used for the feed line affects the coupling factor. It increases with the increase in the dielectric constant but the resonance frequency remains unchanged. The increase in coupling factor is attributed to the fact that the slot appears electrically longer as the dielectric constant increases.

(*f*) The increases in the thickness of the feed substrate reduces the coupling factor without changing the resonant frequency. This is because, with increase in the thickness of the substrate, the distance between the feed line and the coupling slot increases thereby reducing the coupling factor.

(*g*) Apertures which are in the range of 0.01–0.03 λ yield a front-to-back ratio of about 20 dB for the radiation. Therefore, increase of the slot length beyond this range of values brings in undesirable radiation from the slot.

The effect of a dielectric superstrate on the performance of the aperture-coupled patch radiator has been analysed by Huang and Hsu [119] by making use of the reciprocity theorem to handle the feed line. Two sets of Green's functions in the spectral domain are used in the analysis. The grounded double-layer dielectric slab Green's function models the half-space region containing the substrate-superstrate configuration, whereas the single-layer dielectric slab Green's function models the half-space region containing the microstrip line. Using these Green's functions in the electric field integral equations and then applying the method of moments, the electric field in the coupling aperture and the current on the patch radiator can be obtained. From their analysis it is observed that the resonant frequency of the patch radiator decreases monotonically as the thickness of the superstrate increases. The rate of decrease is larger for thinner superstrates and it becomes less as the superstrate becomes thicker. With increase in the thickness of the superstrate, the resonant input resistance decreases and correspondingly the bandwidth increases. The effect of the permittivity of the superstrate on the input resonant resistance is similar to that on the thickness parameter. The effect of the superstrate on the radiation efficiency has also been analysed. For a given thickness of the substrate there is an appropriate superstrate thickness which yields the maximum radiation efficiency. This figure can be further improved if the substrate is made thinner and the superstrate thicker. Although the superstrate makes matching poorer, input match can be obtained by proper adjustment of the patch radiator dimensions and the length of the coupling aperture.

10.30.4 *Bandwidth enhancement*

It is now clear that, except for its narrow-band performance, the aperture-coupled microstrip-patch radiator is suitable as the radiator for active aperture radar systems. The basic limitation that has to be overcome is the small electrical volume occupied by the patch element. Broadly, the efforts to improve the bandwidth can be categorised as either designing an impedance matching network or making use of parasitic elements to increase the electrical volume. Of these, the second technique, namely increasing the volume under the radiator, will be of interest. The simplest way to increase the volume is to use a thicker and/or higher permittivity substrate. Although the use of a thicker substrate does improve the impedance bandwidth of the patch radiator, it inevitably leads to unacceptable spurious feed radiation, surface wave generation and feed inductance. Another technique is that of using parasitically coupled elements to produce a staggered twin peak resonance. If the parasitic patch were to be coplanar with the active radiating element, it would be disadvantageous in array applications. The best way to make use of parasitic elements is by stacking them above the feed patch with a dielectric in between. Double stack patches with probe feed have been reported to have a bandwidth of 18–23% [120, 121]. Experiments have been carried out by Lee *et al.* [122, 123] to study the bandwidth enhancement as well as the radiation pattern characteristics. In their experiments they used air to separate the parasitic patch and the feed patch (see Fig. 10.79). The major findings are that, depending on the spacing s between the probe-fed patch and the radiating (parasitic) patch above it, the characteristics of the antenna can be divided into three regions: Region 1 is associated with relatively large bandwidth, region 2 has an abnormal radiation pattern and region 3 is associated with narrow bandwidth and high gain. Our interest is therefore in stacked patch radiators with region 1 spacing which is near to or less than 0.20 of the wavelength. The best performance obtained is a bandwidth of 19.6% at X-band and a gain of 7.0 dB for a double-stack patch of dimensions 1.5×1.5 cm with an air separation distance of 0.203 cm at the operating frequency of 10.4 GHz. The feed-patch substrate had a thickness of 0.0762 cm and the radiating patch was of thickness 0.0254 cm. Both the substrates had a dielectric constant of 2.17. This configuration of two patch radiators on two different substrates and with air separation is difficult to use in large planar arrays and therefore the results of this study have to be considered as an affirmation of achieving higher impedance bandwidth without deterioration of the radiation pattern. In addition, the updated version of the T-R module shown in Fig. 10.80 is a sealed unit, so it would be preferable to have the RF coupling between the T-R module and the antenna by electromagnetic coupling and not through a feed probe. Fig. 10.81 shows a blown up version of an improved double-stack patch radiator, the incorporation of which in the active aperture radar system would be easier.

Analytical studies of mutlilayered microstrip-patch antennas are more complicated than the studies carried out for the aperture-coupled single

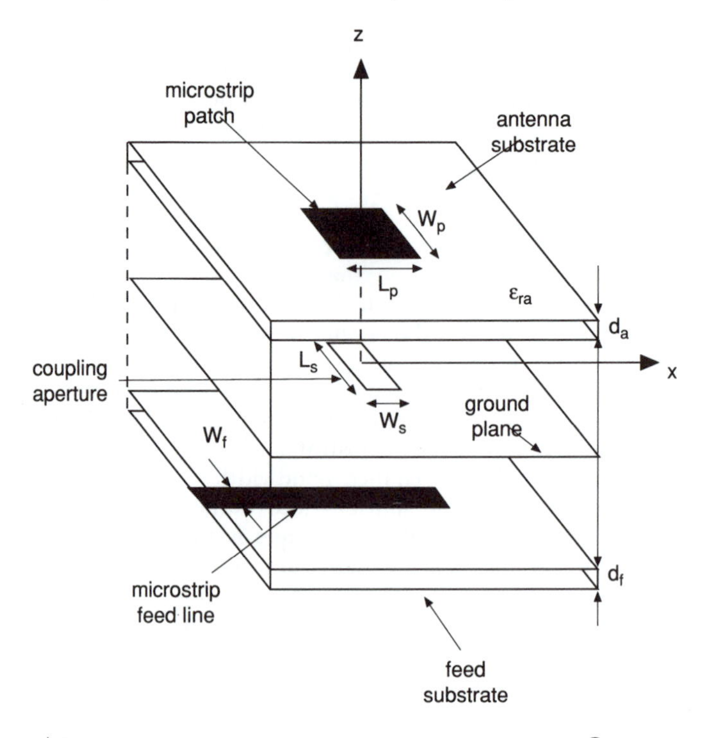

Figure 10.79 *Aperture coupled microstrip patch radiator [107].* © *IEEE, 1989*

microstrip-patch radiator, due to the presence of a second dielectric layer and a second patch radiator. Barlatey *et al.* [124] used entire domain functions for the method of moments solution of the mixed potential integral equation (MPIE) formulation. A complete and rigorous treatment of the Green's function has been provided so that numerical characterisation of surface waves is possible. The MPIE has been solved with the Galerkin procedure and, by introducing auxiliary variables, the overall computation time has been brought down. Theoretical results for the input impedance of a double stack have been computed and found to be in good agreement with measurements. A typical configuration is a double-stack patch with each patch radiator of dimensions 8 × 8 mm, printed on two identical substrates of dielectric 2.33 and thickness 1.57 mm. The bandwidth achieved was 15% for a VSWR of two or less and with the centre frequency in the band 8.5–11 GHz. Hall and Mosig [125] have extended Barlatey's study by using a more accurate mathematical representation of the currents on the probe-feed. Triangular basis functions have been used for the currents on the probe and special attachment basis functions used for the currents at the probe-patch junction. They have also used rectangular subdomain basis functions for the patch radiator current so that arbitrarily shaped patches can be analysed. They have claimed that the results of their analysis are an improvement over the results obtained in Reference 124.

Damiano *et al.* [127], Meck *et al.* [128], Hassani *et al.* [129] and Croq and Papiernik [130] have all carried out full-wave analysis of stacked-patch

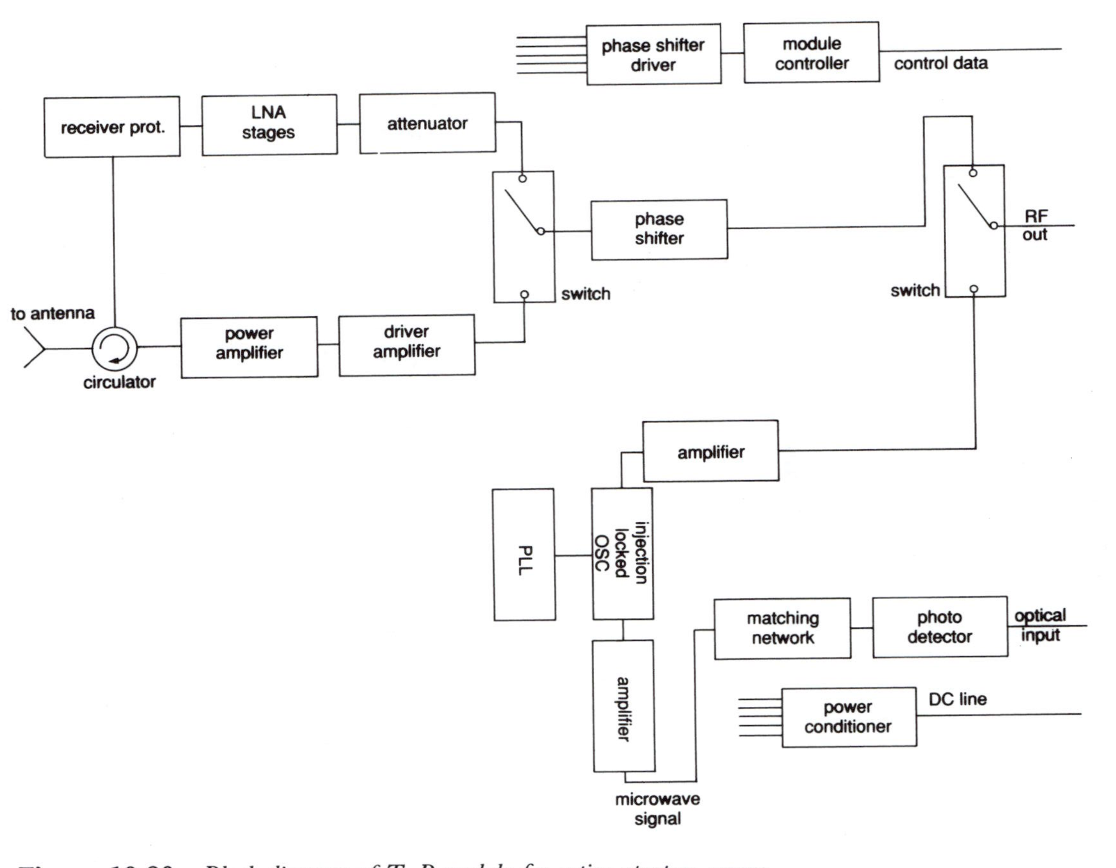

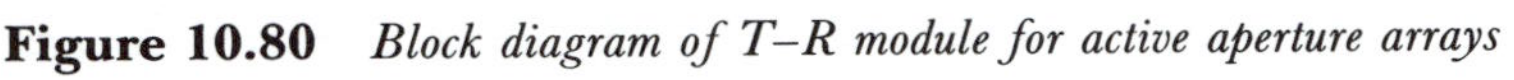

Figure 10.80 *Block diagram of T–R module for active aperture arrays*

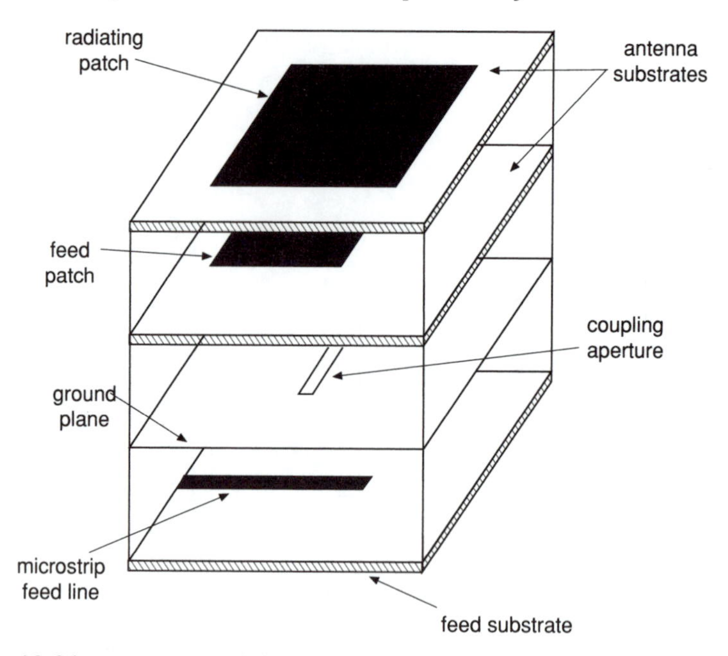

Figure 10.81 *Aperture coupled double patch radiator*

antennas based on the reaction integral theorem and then expressed the electric field integral equation in the spectral domain and solved for the patch radiator currents by using the method of moments in the Galerkin procedure. In Reference 127, the patch currents are expanded in terms of a basis of eigenfunctions related to the geometry of the patch. These are entire domain functions and the stability and convergence of the moment method is dependent on them. Various geometric shapes of the feed-patch radiator and of the parasitic patch have been studied. It was found that circular patches offer better electrical characteristics in comparison with a square configuration. In Reference 128, the probe-feed is represented by an additive inductive load. The longitudinal as well as the lateral currents on the patch radiators have been expanded by entire domain basis functions. An engineering model of the double-stack patch radiator was fabricated with Nomex honeycomb material (dielectric constant 1.2) covered with Kevlar. Input impedance and pattern measurements were carried out and it has been found that there is 'qualitatively good agreement' between analysis and experiment. In Reference 129, the current on the probe is represented by a delta function at the junction between the probe and the patch. Piece-wise sinusoidal expansion functions have been used to represent the currents flowing in the longitudinal direction. Uniform current flow is assumed for the lateral currents. It was found that a double-peak response is obtained with the amplitude at the upper resonance being larger than that at the lower resonance. The computed resonance frequencies were within 3% of the measured values. The input impedance at the upper resonance frequency is sensitive to conductor and dielectric losses. Croq and Papiernik [130] have analysed the behaviour of the aperture-coupled double-stack rectangular-patch radiator by expanding the patch currents in terms of five entire domain sinusoidal basis functions. Also, the electric field in the coupling aperture has been described by a single piece-wise sinusoidal basis function. A bandwidth of 25% with a VSWR less than 1.6 has been achieved and the computed result was within 4% of the measured value. The directivity of the antenna was found to be between 7.6 and 9.0 dB over the bandwidth, and the cross-polarised component was very low; lower than 40 dB near broadside. The worst front-to-back ratio of the antenna pattern is in the range of −18 to −21 dB which is consistent with the measured values of between −17 to −21 dB. A trade-off has to be made between the bandwidth and back radiation.

Das and Pozar [131–133] have analysed the aperture-coupled double-stack patch as a multilayered printed antenna and used the multiport scattering approach with the spectral domain technique to solve for the various currents and fields at the coupling slot, the feed patch and the parasitic patch. A set of exact spectral domain Green's functions for a general purpose multilayered medium that includes material (conductor and dielectric) losses and losses caused by leakage of power to surface wave modes are first derived. The multiport scattering analysis provides a methodology to express the magnetic field at the slot in terms of the field scattered by the current induced on the strip line and the field induced at the slot by the incident wave. The input impedance of the slot can now be computed. For computational purposes, five entire domain sinusoidal functions each were assumed for expansion of the longitudinal and lateral components of the patch current and one piece-wise sinusoidal function was used for the slot. Nontravelling wave modes were included on the feed line along with the travelling wave modes, but the results were not significantly different from those with travelling wave modes alone. A comparison of the computed results was made with measurements carried out in a waveguide simulator for a patch radiator. The computed values of the impedance compare reasonably well with the measurements. The impedance locus has a loop in the operating frequency range of 3–4 GHz due to the overlapping of the two modes corresponding to the feed and the parasitic patch. This was repeated in both computation and measurement but with a slight frequency shift for the two sets of values. The analysis and the computation clearly show that the available bandwidth, which is a function of the relative levels of excitation as well as of the separation between the resonant frequencies of the two modes, is determined by the relative dimensions of the two patches and their separation distance. The bandwidth of the aperture-coupled double-stack patch antenna can be optimised by suitably tuning out the reactance and matching it to the characteristic impedance of the transmission line. A bandwidth of 12% was obtained in contrast to the 3% bandwidth obtained without the top parasitic patch and its substrate. A better bandwidth can be obtained by using a top patch larger than the feed patch, however, if the dimensions are increased beyond a limit, the two patch radiators radiate independently, thereby indicating the onset of dual frequency operation and cessation of broadband operation of the antenna. If the top parasitic patch is smaller than the feed patch, the former is more isolated from the latter and therefore is not strongly excited. Consequently, the weakly-excited resonant mode of the top patch fails to enhance the total bandwidth of the antenna. This is true only for cases with thin substrates and may not be valid for electrically thick substrates due to the absence of a significant fringing field of the bottom patch for coupling to the top patch.

It is very clear from this analysis that the aperture-coupled double-stack microstrip-patch radiator is suitable as a radiating element for the active aperture array radar.

10.30.5 *Behaviour in an array environment*

In addition to the bandwidth, the suitability of a radiating element for an electronic scanning array is also governed by the absence of scan-blindness within the beam scan limits and by the level of spurious radiation generated as a consequence of excitation of the elements of the array. Scan-blindness refers to a condition in which, for certain angles, no real power can be transmitted or received by the array. This has been experimentally observed and theoretically derived for a number of different types of arrays. Printed circuit array antennas also show scan-blindness effects because of the presence of the dielectric substrate. As the beam scans near the angle at which scan-blindness occurs, the propagation constant of a Floquet mode (corresponding to the infinite array) approaches the propagation constant of a TM surface-wave mode of the (unloaded) dielectric substrate. A 'forced surface wave' is excited so that no real power enters or leaves the array surface. In this situation an electromagnetic wave will be propagating up and down the dielectric substrate and the field above the surface of the array is evanescent. Scan-blindness of this type can be predicted by comparing the propagation constants of the surface waves of the dielectric substrate and the various Floquet modes. It is possible to completely eliminate scan-blindness by decreasing the element spacing. For electrically thin substrates or for substrates with a lower dielectric constant, scan-blindness will occur closer to the end-fire direction. For printed circuit antennas, scan-blindness can also occur due to leaky waves [134]. This is particularly noticeable when the patch radiators are operating at frequencies away from resonance. The scan-blindness due to leaky waves can occur when the propagation constant of the leaky wave in the unloaded dielectric substrate is near to the propagation constant of one of the Floquet modes.

Analysis of an infinite array of aperture-coupled single-layered microstrip patches has been carried out by Pozar [126]. Fig. 10.82 indicates the variation of the magnitude of the reflection coefficient against scan angle for an array using aperture-coupled patch radiators. The characteristics of the array are as follows:

Antenna substrate:	Permittivity, 2.55
	Array lattice, square, 0.5 λ per side
Patch radiator:	Form, square
	Dimension, 0.29 λ per side
	Substrate thickness, 0.02 λ
Ground plane:	Slot length, 0.10 λ
	Slot width, 0.010 λ
Feed substrate:	Permittivity, 12.8
	Thickness, 0.02 λ

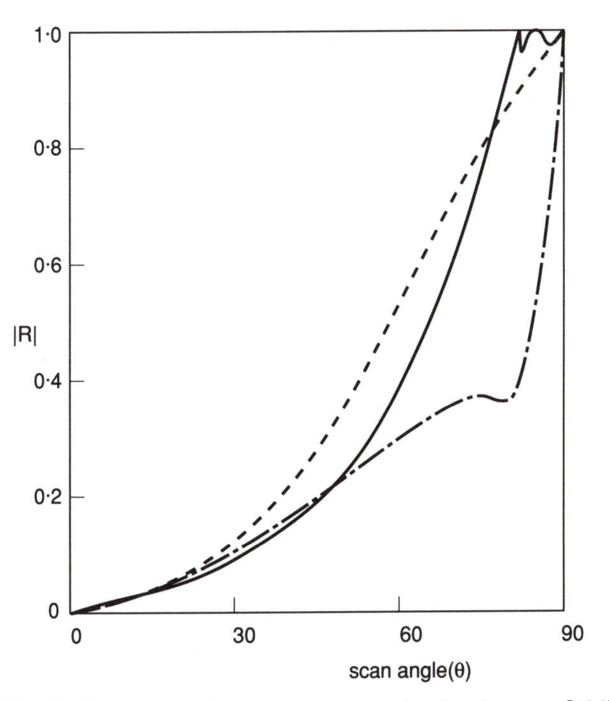

Figure 10.82 *Reflection coefficient versus scan angle: Patch array [126]. © IEEE, 1989*

—— E-plane
--- H-plane
—·— D-plane

It has been reported that, as the array is scanned off-broadside, impedance mismatch occurs in all the scan planes. There are two distinct angles at which scan-blindness can occur in this array. The scan-blindness at 82.9° off-broadside in the E-plane is associated with the thinner feed substrate. The blindness associated with the antenna substrate is at an angle 85.6° off-broadside. As long as the thickness of the feed substrate (GaAs with dielectric constant around 12.8) is maintained at 0.02 λ or less, scan-blindness due to the feed can be made to approach the end-fire direction so that the feed substrate is not a factor for scan-blindness. Generally, the substrate used in GaAs MMICs has a thickness of 150 μm which works out to be less than 0.02 λ for frequencies up to 40 GHz. Therefore, in the present application, scan-blindness due to the feed substrate is not present.

Pan and Wolff [135] analysed the behaviour of a finite array of slot-coupled microstrip-patch radiators by combining the equivalence principle and the reciprocity theorem. In their method, the coupling slots are closed and replaced by equivalent surface magnetic current densities. To avoid the modelling of the nonuniform current distribution on the feed line, the discontinuity due to the slot on the feed line is replaced by voltage reflection coefficients on the feed lines after invoking the reciprocity theorem. The integral equations for the unknown surface current densities on the patches and the slots in the array are obtained by enforcing the boundary conditions on the patches and slots. Spectral domain Green's functions are used to take advantage of the simplicity in the resulting computation. In order to consider the effect of excitation source, an N-port equivalent network is used to represent the N-element array. The N-element impedance matrix is evaluated directly from the network voltage and current matrices of order $(N)^2$. The theory has been verified by computing the mutual impedances and scattering coefficients for a four-element and an eight-element linear array and comparing it with measurements.

Pozar and Kaufman [136] have investigated the effects of excitation amplitude and phase accuracies, mutual coupling effects, diffraction effects, positioning accuracies, errors due to imperfect matching and feed-network isolation in the practical realisation of low sidelobe-level single-layered patch-radiator arrays. The effects due to the narrow bandedness of the radiator for low sidelobe realisation are discussed in Section 10.33. The mutual coupling and the surface wave effects have been found to have negligible influence on the sidelobe levels of such arrays.

Das and Pozar [133] have computed the E-plane scan behaviour of an infinite array of the slot-coupled double-stack patch antenna with a unit cell of 4.11×4.11 cm at a frequency of 3.75 GHz. It can be seen that the scan-blindness angle is close to the horizon and the behaviour of the double-stack array is similar to that of a normal single-layered microstrip-patch array. The authors have indicated that the occurance of a scan-blindness angle limits the bandwidth optimisation. They also state that the thicker the substrate, the closer the blindness angle is to the broadside.

Lubin *et al.* [137] have analysed the behaviour of a double-stack patch radiator in the environment of an electronically scanning array. Their findings which are as follows, point the way forward for the design of the array with the double-stack microstrip-patch radiator.

(*a*) The lower resonance peak in the frequency response of the double-stack patch corresponds approximately to that of the lower patch with substrates in place and the upper patch removed. The higher frequency peak corresponds approximately to the resonance of the upper patch with a conducting ground. The lower resonance peak decreases in amplitude, whereas the amplitude of the higher resonance peak builds up with the increase in size of the upper patch.

(*b*) The change in ratio between the sizes of the feed patch and the radiating patch has a stronger effect on the peak amplitude of the resonance than on the resonant frequency.

(*c*) To achieve an adequate bandwidth, it is easier to adjust the separation of the resonant frequencies by varying the ratio of the dielectric constants of the substrates.

(*d*) In the array environment, for the same element spacing, a lower average dielectric (average of the upper and lower substrates) causes the leaky wave resonance to recede from the passband, so that the bandwidth increases. However, this also leads to a more rapid variation of the impedance with scan angle and thus increases the mismatch in the usable frequency band.

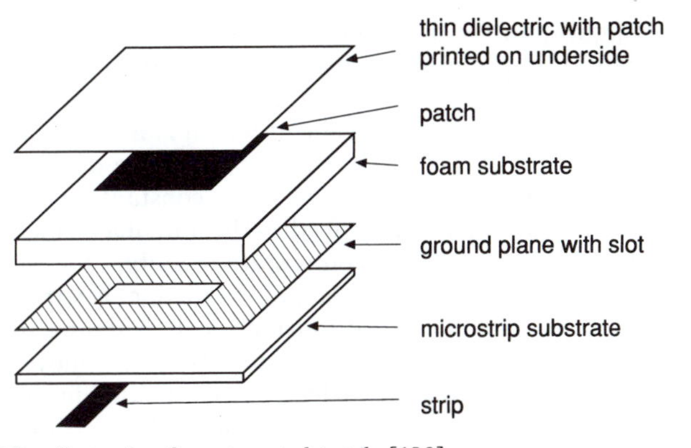

Figure 10.83 *Strip-slot-foam-inverted patch [138]*

However, there are some practical challenges which have to be overcome before the full potential of the printed circuit antenna array can be exploited. Foremost among them are tolerance control and reduction of uncertainities in high volume production. Variations in the characteristics of the substrate, mask stretching and ageing and etching tolerances would put a limit on the maximum size of the array antenna that can be handled as an integral unit. There is also a need to find electrically equivalent, mechanically acceptable and inexpensive dielectric material for the substrate. Two alternatives have been suggested to keep the costs down. The first is the use of low cost polypropylene substrates which would need thick metal backing to prevent board distortions. The second alternative is to use the concept of strip-slot-foam inverted radiator as shown in Fig. 10.83 [138].

10.31 Antenna and T–R module coupling

In the design of active aperture arrays, the mechanical structure, thermal management, module replaceability and size constraints are interdependent factors unlike the situation in conventional phased arrays. Generally speaking, active array designs can be classified into two configurations, which are referred to as 'tile' or 'brick' configurations. The tile concept places the GaAs circuits on the same plane as the patch radiator. Hence the available aperture space is shared between the radiating elements and the circuits of the T–R modules. Fig. 10.84 is a schematic of such an arrangement. The patch radiators in this case can be edge-fed by microstrip lines. Schaubert *et al.* [139] have pointed out a number of problems of using this approach. Firstly a single layer substrate will not have for all frequency bands enough surface area to accommodate radiating elements, phase shifters and feed networks without crowding and without generating deleterious coupling between active circuitry and radiating elements. Since the antenna element spacings are in the region of half wavelength, to avoid grating lobes, even at C-band, the routing of the feed network and the bias lines becomes a serious problem. Hence closer spacing of antenna elements is ruled out, thereby limiting the maximum scan range of the array. This configuration can also give rise to severe spurious feed radiation. The brick configuration provides a way out of this situation by removing the active circuits, phase shifters etc. away from the same plane as the radiating elements. Fig. 10.85 shows one arrangement which is in one sense a two-layer design. Here a grounded layer of GaAs holds the active devices and the feed network and a cover layer of a low dielectric constant material holds the radiating elements. Coupling from the feed to the antenna elements is made by holes or by proximity coupling. This configuration doubles the usable substrate area and also provides substrates matched to the distinct electrical functions of radiation and circuitry. The dielectric constant of the antenna substrate being lower than the GaAs used for the feed substrate, the movement of scan blindness angles towards broadside is effectively eliminated in this case. In this configuration spurious signal radiation from the feed substrate has not been fully eliminated. On the other hand, spurious radiation from the feed may be more harmful because of the strong coupling to the radiating element directly above. Fig. 10.86 shows that the gains in bandwidth and maximum scan range are not as high as expected for the two-layer substrate. For 10% bandwidth and a GaAs layer thickness of 0.02 λ, scan blindness occurs around 68° compared with 63° for a single layer GaAs geometry. Fig. 10.87 shows an improvement. Here the GaAs substrate containing the active devices and feed networks is on one side of the ground plane and a low dielectric constant substrate is bonded to the other side of the ground plane for placement of the radiating elements. Coupling takes place through apertures in the ground plane. This type of design has the advantage of using the best substrate for a given function: the GaAs substrate for the phase shifters plus active devices and a low dielectric constant substrate for the antenna elements.

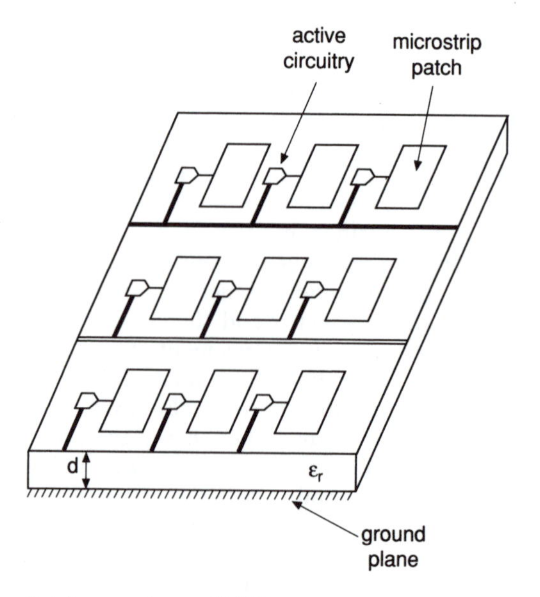

Figure 10.84 *Patch array layout [141]*

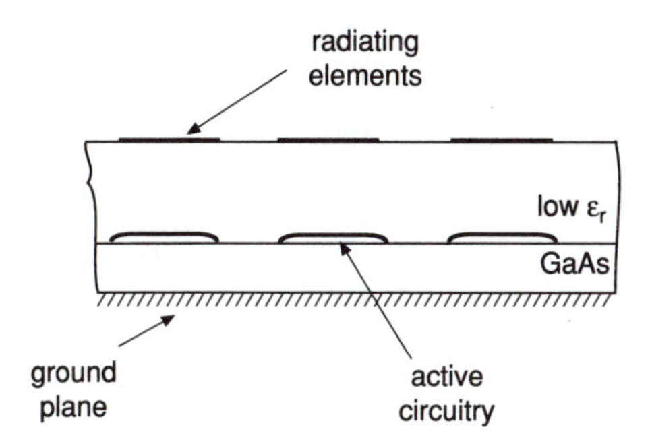

Figure 10.85 *Two-layer feed system [141]*

Another version using two substrates has been described by Pozar *et al.* [140] where the feed substrate is now oriented perpendicular to the array substrate as shown in Fig. 10.88. Coupling is again through an aperture in the ground plane beneath the microstrip patch and the feed line employs proximity coupling on one side of the aperture. This geometry combines the best features of the earlier configurations that had been discussed. From the geometry of Fig. 10.88, it can be seen that the patch radiator is printed on the low level dielectric constant

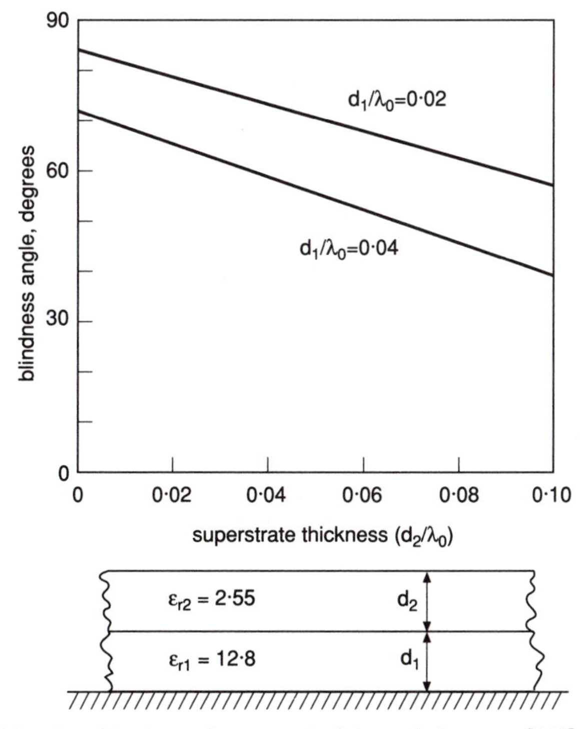

Figure 10.86 *Scan blindness phenomena in 2-layer feed system [141]*

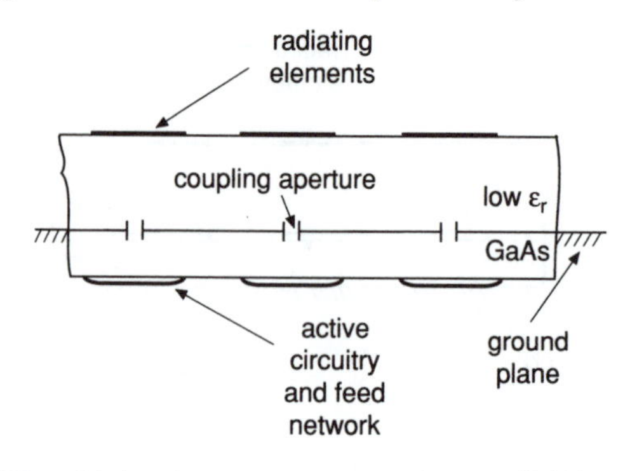

Figure 10.87 *Two-sided active aperture array geometry [141]*

substrate that is mounted, vertically. The feed substrate which is perpendicular to the antenna substrate is positioned so that it bisects the coupling aperture located behind the patch antenna and the microstrip feed line passes in close proximity to the aperture. It terminates in an open circuit stub which is used for tuning. The ground planes of the two substrates are electrically connected at the junction except in the vicinity of the coupling aperture where the ground plane has been partially removed underneath the feed line. This prevents the aperture from being shorted out by the feed substrate ground plane and also increases the coupling from the feed line to the aperture by causing the field lines to bend from the microstrip line over to the ground plane of the antenna substrate. The aperture is centred behind the patch to obtain maximum coupling. The *E*-plane of the patch is parallel to the feed line and therefore the longer dimension of the aperture should be at right angles to this direction. The size of the aperture

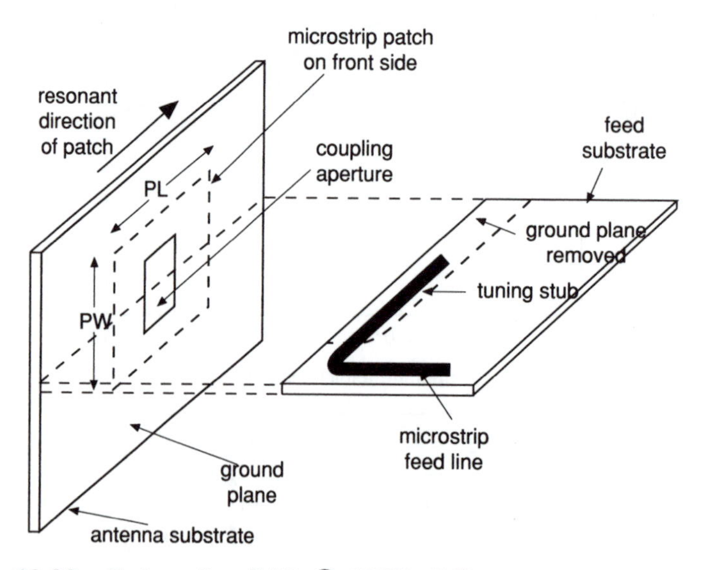

Figure 10.88 *Feed coupling [140]. © IEEE, 1987*

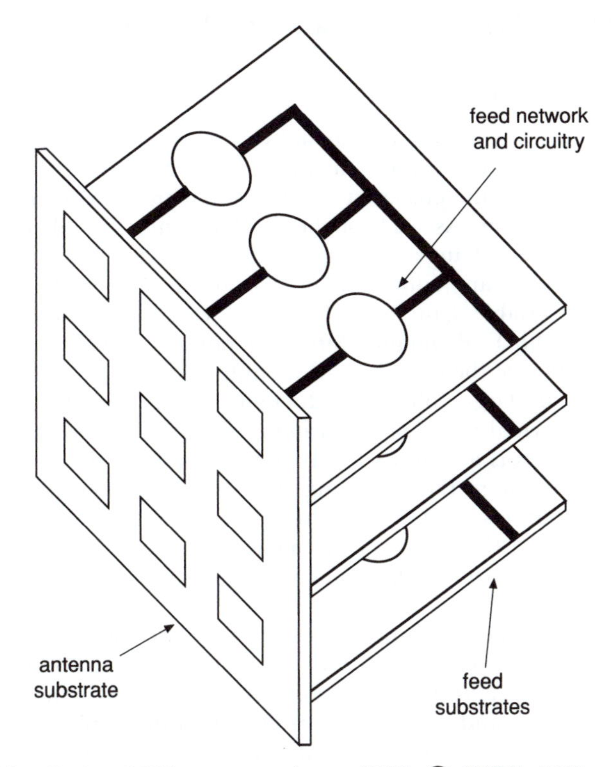

Figure 10.89 *Feed and RF source attachment [140].* © *IEEE, 1987*

controls the amount of coupling between the feed line and the patch and also the back radiation. Experimentally it has been verified that a front-to-back ratio of 20 dB or more can be achieved if the aperture is appropriately designed. Active aperture phased arrays can be fabricated by using stacked feed substrates as shown in Fig. 10.89. This configuration lends itself easily for reflection as well as for the phased-lens type of arrays, and has the following features [142]:

(*a*) It accommodates more functions
(*b*) It is possible to seal hermetically either at individual T–R module level or in groups to ensure better performance
(*c*) In this case replaceability is easier and more practical
(*d*) Since the system is amenable to automatic assembly techniques, cost can be brought down
(*e*) There is enough space between rows to accommodate the adequate mechanical structures needed to provide structural strength
(*f*) There is more surface area available for heat removal, and hence thermal management becomes easier
(*g*) The depth behind the array is an additional dimension available to the system designer to avoid crowding of MMIC circuits, and enables better control of radiating and thermal efficiencies.

10.32 Array illumination: Low sidelobe levels

10.32.1 *VGA and VPA efforts*

For low sidelobe level performance an active aperture phased array may require tapering of the power across the array aperture. Furthermore, for efficient beam forming and steering, the phases at the individual antenna elements have to be adjusted electronically. As each of the T–R modules would have a multibit digital phase shifter, change of phase of the microwave field at the radiating element will not pose any difficulty, whereas for the change of amplitude of the field either a variable gain amplifier or a variable attenuator has to be incorporated in the T–R module. Since the power amplifier cost represents a substantial portion of the overall T–R module cost (29%), substantial savings can be effected by large volume production of T–R modules that are nearly identical. The normal mode of operation of the final power amplifier stage in these T–R modules is a saturated or nonlinear mode so that small-signal gain variations that occur in the preceding stages of the power amplifier may not show up as variations in the final output. However, these small-signal gain variations can affect the efficiency, harmonic power and insertion phase of the output, so that gain variation becomes an important parameter of the amplifier module design.

Snow *et al.* [143] have investigated the performance of variable gain MMIC power amplifiers in the C, X and Ku bands, so that the effect of small signal and large gain variations, and their impact on phase variations, can be known. Dual gate FETs are used, and digitally controlled programmability is introduced to provide the amplifier with precise amplitude quantisation over a wide range of gain values with minimum insertion phase variations so that the likely complexity in calibration is minimised. Maximisation of power and efficiency with reduced per chip cost is accomplished by employing small signal inter-stage design augmented by large signal load-pull analysis and off-chip output matching on a high dielectric constant substrate ($\varepsilon_r = 37$). Three variable gain amplifiers covering C/X-band, X-Band and X/Ku band have been designed and fabricated. For the CXVGA, small signal gain is 5.5 ± 0.5 dB in the frequency range 5.5–10.5 GHz with 64 gain/attenuation states. Across this band linearity is better than 2.5% and incidental phase variation with state is less than 10° over a 20 dB dynamic range of gain. In the case of XKuVGA the incident phase shift is less than 10° from 6 GHz to 18 GHz, over a 20 dB dynamic range. In the two stage XVPA covering the frequency range 7–11.5 GHz, incident phase variation was less than 10° over a 20 dB dynamic range. Over the first 10 dB of dynamic range this variable power amplifier showed less degradation in efficiency than the conventional approach of operating a saturated power amplifier in the linear region. Culbertson *et al.* [144] have designed a 3 W X-band two-stage variable gain amplifier also using dual gate FETs. Both amplifier stages are controlled in this case and the incident phase variation is less than ±6° for 20 dB of dynamic range. Both these developments have established confidence in the reproducibility of fairly complex variable power gain amplifier with minimum degradation in power, efficiency and phase variation. For sidelobe levels of the order of −40 dB below the main peak, these incident phase variations will not be acceptable, and therefore the fifth phase bit in the 5-bit phase shifter of the T–R module can be analogue bit to neutralise the phase change accompanying the gain variations. Therefore the option of using amplitude taper for the aperture illumination of the array is very much alive.

10.32.2 *Errors and sidelobe level*

The realisation of low sidelobe levels in an antenna has been related to the number of elements and the RMS error in phase, amplitude or position of the element by the expression [6]

$$g_{sl_1}=\frac{\sigma_e^2}{\eta_a N} \tag{10.357}$$

where N is the number of elements and η_a is the aperture efficiency. This relation is valid for elements above 50. Since in most cases the element spacing is of the order of $\lambda/2$, we can write

$$N=\frac{4A}{\lambda^2}=\frac{G}{\pi\eta_a} \tag{10.358}$$

where A is the physical aperture and G is the gain of the antenna. From eqns. 10.357 and 10.358 we can therefore write

$$g_{sl_1}=\frac{\pi\sigma_e^2}{G} \tag{10.359}$$

Thus eqn. 10.359 indicates the contribution of sidelobes to the radiation pattern of the antenna owing to random inaccuracies in phase, the amplitude of the field radiated by the antenna or random inaccuracies in the placement of the element in the array.

The peak sidelobes in the principal planes of the phased arrays designed for low sidelobe radiation pattern without random errors has been shown to follow the relationship

$$g_{sl_2}=\frac{1}{K_{ss}N} \tag{10.360}$$

where

$$K_{ss}\simeq 1000 \text{ for broadside radiation}$$
$$\simeq 300 \text{ for scanning to } 60^\circ$$

Another contributory factor to sidelobe levels is the quantisation of the phase shifts in a phased array due to the employment of multibit phase shifters. As the phase shifters change phase state, there is a random distribution of errors from the true value across the aperture. This will be

$$\sigma_\phi=\frac{2\pi}{2^m\sqrt{12}} \tag{10.361}$$

which can be related to the sidelobe level through eqn. 10.359 by replacing σ_e by σ_ϕ. Thus there are four error sources, namely the amplitude and the phase excitation errors, the position error and the random distribution of phase state error. In the case of the active aperture array, one more source of error, namely the incident phase variation due to gain variation, has to be included as part of the phase excitation error of the T–R module. All these errors give rise to a 'noise'-like addition or a random component to the designed sidelobe levels. This is assumed to be Gaussian and has a uniform variance over angle. Thus sidelobes in the visible space are seen to comprise two components; a design level component that is a function of the illumination/excitation function chosen, and hence can be mathematically described. In addition there is a residue that is over the design level sidelobe, which is random. Sidelobe predictions can now be made with respect to the three parameters of interest, i.e. specified sidelobe level which is based on the system requirements, designed sidelobe level which is defined as the sidelobe level that can be computed for a perfect array based on the illumination function, and the residual sidelobe level which is contributed by the imperfections of the array in phase and amplitude of array excitation and element position. These predictions are based on a Rician probability function

$$p(E/R) = (E/R^2)\ \exp\{-(E^2+P^2)/2R^2\} I_0\left(\frac{EP}{R^2}\right) \tag{10.362}$$

where

E = ensemble of design plus random variable
R = RMS random component
P = peak design component
I_0 = Bessel function of zero order, imaginary argument

Statistical predictability provides a measure of assessing antenna vulnerability to sidelobe interference, because from eqn. 10.362, it is possible to obtain the percentile of 'pop ups' above a specified level in the all-angle space. The designer can now decide on the proper choice between over-design and tight tolerances in his measured response to the desired specifications. The choice is either to evolve a design that is close to specifications and maintain the tight tolerances or have an alternative design that has substantially lower sidelobes and relax the tolerances. The range for such a trade-off is quite broad for the same probability of pop-ups above the specified level of sidelobes, and the decision is therefore governed by the relative risks and costs of enlarging the array to reducing the efficiency of illumination in comparison to the risks and costs of maintaining tight tolerances.

10.33 Low sidelobe level printed circuit antenna array

10.33.1 *Design and practical realisation*

It has already been pointed out that, owing to the narrow bandwidth of the single layered microstrip patch radiator, relatively small changes in its dimensions shift the resonance frequency and consequently cause rapid changes in the driving point impedance of the radiator. In a similar fashion, the variations in the permittivity of the substrate will affect the resonance frequency and the driving point impedance. Consequently, this causes changes in the amplitude and phase of the radiated field from its desired value. If these changes were uniform for all radiators then the progressive phase shift between elements remains unaltered from the desired values. However, owing to manufacturing tolerances these errors in the phase of the radiated field across the array will not be the same. Invariably the phase errors contribute to an increase in the sidelobe levels of the radiating pattern and set a lower limit to the sidelobe levels. The computations of Pozar [136], for a microstrip radiator of dimensions 1.902 cm × 1.85 cm on a substrate of permittivity 2.2 and thickness 0.16 cm and operating at 5 GHz, shows that, for a 1% change in resonance frequency, there can be a phase change of about 12.6° and an amplitude change of 0.2 dB in the radiated field (see Figs. 10.90 and 10.91). Considering that a linear broadside array of 100 microstrip elements with Chebyshev illumination cannot accept a phase error of more than 1° for an average sidelobe level of −50 dB, the manufacturing tolerances in the dimensions as well as in the permittivity of the substrate will be the crucial factors. From Fig. 10.90 and from eqn. 10.354 it can be seen that dimensional tolerances of the order of 0.1% or less have to be achieved. This is not an easy task.

Another important design consideration that would have a significant influence on the realisation of the array is the energy reflected into the feed network from the radiator. This arises owing to imperfect match between the feed network and the radiator. The mismatch gives rise to reflection of the electromagnetic wave from the radiator down the feed network and possibly travelling to other radiators of the array. The maximum phase error will be caused when the reflected field at the radiator is 90° out of phase with the excitation and is expressed as

$$\Delta\phi(\max) = \tan^{-1}|\Gamma||I| \tag{10.363}$$

where $|\Gamma|$ is the magnitude of the reflection coefficient of the element and $|I|$ is the feed network isolation. In practical microstrip radiator arrays element losses and network isolation has been stated to be of the order of 20 dB, so that the

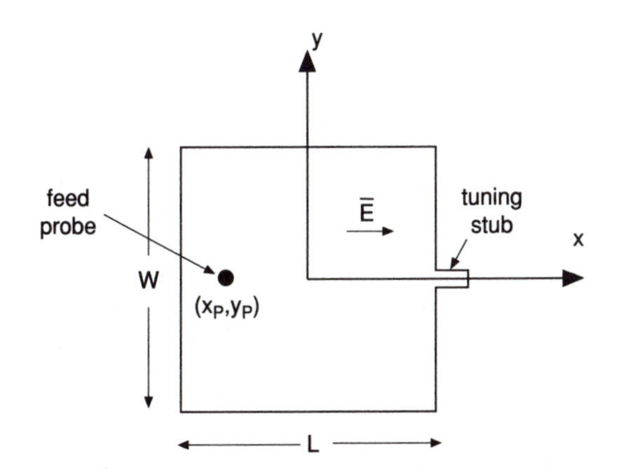

Figure 10.90 *Geometry of rectangular patch [136].* © *IEEE, 1990*

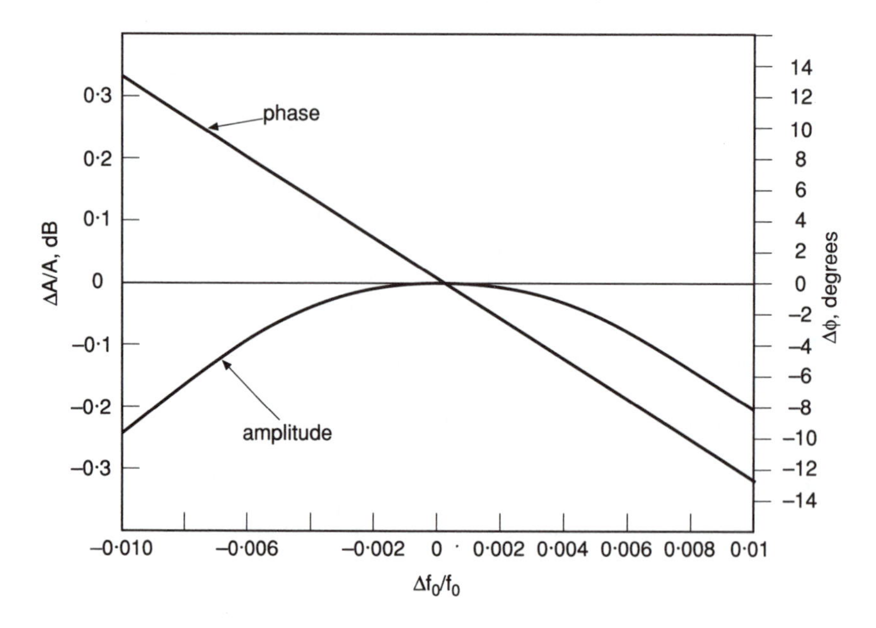

Figure 10.91 *Normalised amplitude and phase errors in radiated field [136]*

L = 1·902 cm
W = 1·85 cm
ε_r = 2·22
d = 0·16 cm
X_p = 0·22 cm
Y_p = 0·0
f = 5·0 GHz

maximum phase error is about 5.7° [136]. This phase error also needs to be reduced for low sidelobe level arrays.

A third design consideration in low sidelobe level microstrip patch radiator arrays is the level of cross-polarisation. Normally, the cross-polarisation level of a properly designed patch element is about 20 dB below the co-polarisation. Hence in an array of identical patch radiators the cross-polarisation level would be given by the product of the array factor and the cross-polarisation pattern of the single radiator and this would be significantly lower for large arrays. However, it has been found that the cross-polarisation level of the patch radiator array is influenced by the geometric form of the radiator as well as by the imprecision in the positioning of the feed. Fig. 10.92, taken from Reference 136, shows the calculated cross-polarisation level of a probe-fed rectangular microstrip patch radiator versus the probe positioning error for different widths. The patch length in this case is 1.9 cm, the permittivity of the dielectric 2.2, the substrate thickness 0.16 cm and the operating frequency 5 GHz. It can be seen that the cross-polarisation component of a square patch is highest and is greater than −20 dB. It is reduced when the width of the patch is less than its length, and therefore, from the aspect of minimising the cross-polarisation component, the rectangular form is preferred. These effects, due to the narrow bandwidth of the radiator, are not significant enough in the case of the double stack patch radiator.

10.33.2 *Stepped amplitude distribution*

A major issue in active aperture arrays is to find a way of achieving low sidelobe levels in the transmit patterns since it is difficult to achieve as smooth an illumination function as one can have in a conventional array. An option which has been investigated by Lee [145] is based on a few finite groupings of the T–R modules depending on their power output, and configuring the aperture in such a way that the step sizes and arrangement of the power modules are optimised for low sidelobes. The method is based on aperture synthesis techniques which control the far field pattern to achieve low sidelobe levels with only four or five groups of T–R modules. The design effort in this case is not concentrated on how close a desirable illumination taper can be approximated by a minimum number of power levels of finite steps but to apportion the aperture properly into different power zones and adjust the levels to suppress the peak sidelobe. The aperture of the array is first approximated by an ellipse to simplify the formulation. This is permissible because the power levels at the periphery of the

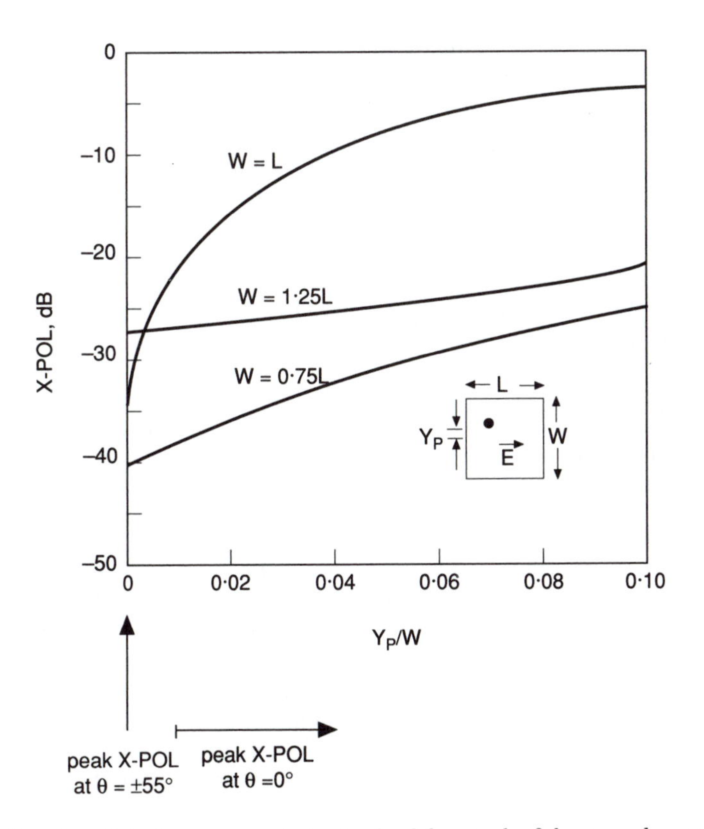

Figure 10.92 *Calculated cross-polarisation level for a probe-fed rectangular patch [136]*

array are low owing to power tapering. For sidelobe levels up to −35 dB five groups of T–R modules, that is, five stepped power zones, are used for the elliptical aperture of the array. The stepped power zones are chosen to have a common aspect ratio a_i/b_i, where a_i is the semi-major and b_i is the semi-minor axis of the ellipse. To optimise for low sidelobes the initial starting point is a specific Taylor distribution that yields a peak sidelobe level about 5 dB below the desired sidelobe level. Then the aperture is divided in such a way that the envelope of the stepped function follows the selected Taylor aperture distribution. The overall far-field pattern of the antenna is simply the total sum of the m contributing fields given by

$$F(\theta, \phi) = [f(\theta, \phi)(\hat{a}_\theta \cos\phi - \hat{a}_\phi \sin\phi \cos\theta)] \tag{10.364}$$

with

$$f(\theta, \phi) = \sum_{i=1}^{m} 2\pi a_i b_i E_i J_1(u_i)/u_i \tag{10.365}$$

and

$$u_i = k_0 a_i \sin\theta \sqrt{\cos^2\phi + (b_i/a_i)^2 \sin^2\phi} \tag{10.366}$$

where a_i, b_i refer to the semi-major and semi-minor axes of the ellipse, and E_i is the corresponding voltage amplitude of the ith level of the aperture. The far-field pattern of the elliptic aperture differs from that of an equivalent circular aperture only by a transformation factor defined by the square root in eqn. (10.366) and the pattern exhibits an elliptical symmetry around the azimuthal angle. To find an optimum set of b_i and E_i for a given aspect ratio, one starts with the initial stepped power profiles and computes the far-field pattern to determine the sidelobe error. A gradient search method is employed to optimise the parameters b_i and E_i by varying these parameters one at a time to compute the variations of the error. The objective is to find the global minimum in the n-dimensional parametric space. Each time the parameter is changed, the far-field pattern is recomputed and the gradient of the error found, and a new set of parameters which could reduce the error is established. The process is repeated until the error is minimised to a preset value. Although the final solution is not unique in the rigorous sense, this process converges rapidly to a fairly stable solution with no oscillations, especially in the case of large arrays with (1000+) elements. An example with five stepped power aperture zones resulted in a −36 dB peak sidelobe level in all planes for an aspect ratio of 1.5 with very few iterations. The partition of the aperture is fairly reasonable and the power steps were around 3 dB, which is a convenient figure to handle. The idealised elliptical power zones into which the antenna aperture has been divided and the assumption of continuous and uniform apertures superimposed on each other requires that the zig zag boundary that actually results owing to the triangular form of the array-lattice/grid be corrected. It has been found that by checking the percentage area of the module at the boundary in each power zone and allocating that module to the inner or outer contour depending on where the maximum area lies, the low sidelobe level structure of the radiation pattern is preserved in the resulting configuration. The solution is thus not sensitive to the lattice structure of the array when the number of elements are

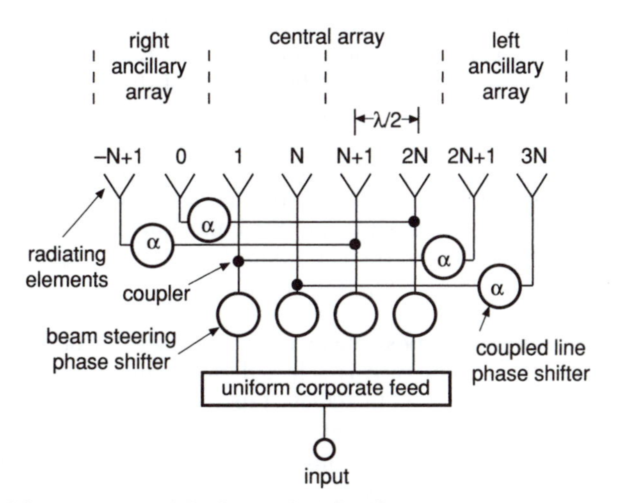

Figure 10.93 *Principle of Dufort method [146]*

large. Even after taking into account the random errors in a realistic manner an average sidelobe level of −35 dB can be achieved with five groupings of the T–R modules. The greatest advantage of Lee's method is in the simplicity of the feed network as no complex power dividing network is employed to control the aperture taper. For example, to achieve −35 dB sidelobe level a 14 dB amplitude taper with Taylor distribution has been worked out by Lee to be accomplished in five steps of 0 dB, −2.62 dB, −3.06 dB, −3.19 dB and −5.11 dB. The disadvantages of this method are as follows. There is no analytical way to link the number of groupings of the T–R modules to the desired sidelobe levels. Once the required taper has been decided upon, step sizes of the power cannot always be matched in numbers to the distribution in manufacturing. Finally for ultra-low sidelobe level requirements, the number of groupings of the T–R modules will go well beyond 5, especially if sidelobe levels of −50 dB are expected. The cost savings in such cases may not be significant at all.

10.33.3 *Lossless constrained synthesis*

Dufort [146] has advocated the use of a network of hybrids and phase shifters to control the taper across the aperture and the beam scanning with identical T–R modules. The concept can be understood better by considering a linear array of $4N$ elements of which $2N$ elements in the middle are from the central array, while N elements on the right become the right ancillary array and N elements on the left become the left ancillary array. Fig. 10.93 depicts the principle of beam formation and beam shift for an 8-element linear array. The elements $-N+1$ to 0 form the right ancillary array, the elements 1 to $2N$ form the central array and the elements $2N+1$ to $3N$ form the left ancillary array. The radiating elements of the central array are connected to the uniform corporate feed through beam steering phase shifters designated as ϕ in Fig. 10.93. The radiating elements of the left ancillary array are connected to the radiating elements in the right half of the central array through a phase shifter designated as α and a transmission line coupler. Similarly the radiating elements of the right ancillary array are connected to the left half of the central array through the phase shifter α and the transmission coupler. If ϕ is the progressive phase shift between consecutive elements of the central array and if α is the phase shift between an element of the central array and a corresponding element of the right or left ancillary array, then the continuity of the phase gradient between the end elements of the central array and the immediate neighbouring elements of the ancillary arrays leads to the condition

$$e^{j(2N\phi-\alpha)}=1 \tag{10.367}$$

Continuous beam steering is possible by choosing any value of ϕ between 0 and 2π. The value of α in each case is selected by means of eqn. 10.367. Since identical T–R modules are used to achieve tapered distribution across the array the sum of the power fed to the central array element and its corresponding ancillary array element has to be maintained equal to any other similar pair of elements in the array. In such a case,

$$a^2(n)+a^2(2N+n)=\text{constant} \tag{10.368}$$

where $a(n)$ is the amplitude of excitation at the array element n and $a(2N+n)$ are the amplitude of the excitation at the array element $(2N+n)$. A generalised amplitude taper

$$a(n)=\cos\,\phi(n) \tag{10.369}$$

can be considered so that

$$a(2N+n)=\sin\,\phi(n) \tag{10.370}$$

Here even though $\phi(n)$ has been arbitrarily chosen, eqn. 10.368 is satisified by this choice. It is now clear that tapered illumination of the array is possible by identical T–R modules with this method. It is also possible to steer the beam in the same way as in any electronic scanning array.

The selection of $\phi(n)$ for low sidelobe design is based on optimisation of signal-to-noise ratio or noise minimisation in the presence of noise angular spectrum for a signal at broadside. The radiation pattern is a consequence of this optimisation process. The array factor is very low in regions of high noise and has high gain in the broadside direction. This is, in fact, the steady state solution that an adaptive array would seek. In the case of a linear array of M elements with inter-element spacing of $\lambda/2$ and excitation $a(n)$ for the nth element in the array, the field $E(\psi)$ produced by it is expressed as

$$E(\psi)=\sum_{0}^{M-1} a(n)\ e^{jn\psi} \tag{10.371}$$

where,

$$\psi=\frac{2\pi d}{\lambda}\sin\theta=\pi\sin\theta \tag{10.372}$$

The array is pointed to free space where the noise angular spectrum $p(\psi)$ is present and whose Fourier coefficients are H_l.

Thus

$$p(\psi)=\sum_{-\infty}^{\infty}\frac{H_l}{2\pi}e^{jl\psi} \tag{10.373}$$

and

$$H_l=\int_{-\pi}^{\pi}p(\psi)e^{-jl\psi}\,d\psi \tag{10.374}$$

One of the element excitations, namely, $a(0)$ is taken as reference and set equal to 1. The excitations of the other elements are unconstrained both in phase and amplitude and they are selected in such a way that the received noise power N_r is a minimum. Thus

$$N_r=\int p(\psi)g(\psi)\,d\psi \tag{10.375}$$

and,

$$g(\psi)=|E(\psi)|^2 \tag{10.376}$$

The optimisation should reduce the final array factor $g(\psi)$ in angular regions where $p(\psi)$ is large. One can therefore, expect that $p(\psi)g(\psi)$ will be approximately constant and $g(\psi)$ will be the reciprocal of the imposed noise angular spectrum. Substituting eqns. 10.371 and 10.373 into eqn. 10.375, the expression for the received noise power N_r becomes

$$N_r=\sum_{0}^{M-1}\sum_{0}^{M-1}a^*(n)a(m)H_{n-m} \tag{10.377}$$

The noise is minimised when $\partial N_r/\partial a^*(n)=0$; with $n=1, 2, \ldots, N-1$, and with $a(0)=1$. In that case,

$$\sum_{1}^{M-1}a(m)H_{n-m}=-H_n,\; n=1, 2, \ldots, N-1 \tag{10.378}$$

For the lossless tapered illumination sought by Dufort, the field $E(\psi)$ produced by a linear array of $4N$ elements placed at distance d apart would be given by

$$E(\psi)=\sum_{1}^{2N}b(n)\cos(n-1/2)\psi \tag{10.379}$$

where $a(N+n)=b(n)$. It is also assumed that $a(n)$ is real and symmetrical about the array centre, so that $a(2N+1-n)=a(n)$. $b(n)$ has to satisfy the energy constraints expressed in eqns. (10.369) and (10.370). These can now be written as

$$b(n)=\cos\phi(n) \tag{10.380}$$

$$b(2N+1-n)=\sin\phi(n) \tag{10.381}$$

The noise angular spectrum chosen is

$$p(\psi) = 1 \text{ for } 0 \leqslant \psi < \psi_1 \tag{10.382a}$$

$$p(\psi) = P_1 + P_2\delta(\psi - \psi_2), \begin{cases} \psi_1 < \psi < \pi \\ \psi_1 < \psi_2 < \pi \end{cases} \tag{10.382b}$$

If $P_2 = 0$, the noise angular spectrum is flat between $\pm\pi$ except for a notch at $\psi = 0$, which is very deep if P_1 is very large. The noise spectrum will tend to suppress the factor in the region ψ_1 to π and allow the array factor to pop up in the notch where the signal is. The purpose of the delta functions with P_2 is to have a roving noise source which can be used to suppress any residual sidelobe. The noise power N_r can now be written in terms of $\phi(n)$ as

$$N_r = \sum_1^N \sum_1^N H(n, m) * \cos\phi(n) * \cos\phi(m)$$

$$+ 2 \sum_1^N \sum_1^N H(n, 2N+1-m) * \cos\phi(n) * \sin\phi(n)$$

$$+ \sum_1^N \sum_1^N H(2N+1-n, 2N+1-m) * \sin\phi(n) * \sin\phi(m) \tag{10.383}$$

with

$$H(n, m) = (1 - P_1)\left[\frac{\sin(n-m)\psi_1}{(n-m)} + \frac{\sin(n+m-1)\psi_1}{(n+m-1)}\right]$$

$$+ \pi P_1 \delta_{n,m} + 2P_2 \cos(m - 1/2)\psi_2 \,.\, \cos(n - 1/2)\psi_2 \tag{10.384}$$

The signal power is given by

$$|E(0)|^2 = |\sum_1^{2N} b(n)|^2 = \left\{\sum_1^N (\cos\phi(n) + \sin\phi(n))\right\}^2 \tag{10.385}$$

For SNR to be a minimum, the gradient, $\partial(SNR)/\partial\phi(l) = 0$, for $l = 1, 2, \ldots, N$.

As pointed out by Dufort, these are highly nonlinear equations which are difficult to solve. A recursive solution has been suggested for which one proceeds towards the peak of SNR in small steps along the gradient until it vanishes. The numerical implementation as well as the critical choice of the step size for convergence has been worked out. For $N = 10$, $P_1 = 10^5$, $P_2 = 0$ and $\psi_1 = 1.7\pi/2N$, Dufort has found the first sidelobe of the resulting radiation pattern at 24.5 dB down from the main lobe with an efficiency of −0.83 dB. The remaining sidelobes are below 38 dB from the main lobe. By letting $P_2 = 46$ dB the first sidelobes are reduced to −27.5 dB with the efficiency increasing to −0.76 dB. However, other sidelobes, especially the second and third, are comparable to the first in their peak values. In the second example a radiation pattern with −23 dB for the first sidelobe level was arrived at with an efficiency of −0.45 dB. A cosine distribution with −23 dB level sidelobes would have an efficiency of −0.91 dB. Thus the lossless constrained synthesis can provide high efficiency low sidelobe level radiation patterns with sidelobes as low as −27.5 dB.

In this method it can be seen that an additional phase shifter has to be catered for and the insertion loss of such a component has to be taken into account. Furthermore, to realise the optimum values of $\phi(n)$ in hardware, a variable analogue phase shifter will have to be used. Dufort's method for realisation of low sidelobe levels using identical T–R modules not only calls for a complex network but also introduces an additional phase shifter after the T–R module which should preferably be of analogue type. The feed network can be considerably simplified if continuous beam steering is replaced by discrete beam pointing positions without in any way leaving gaps in the search volume. The 3 dB beamwidth of an aperture of $4N$ elements with inter-element spacing d is roughly $0.866\lambda/4Nd$, where λ is the wavelength, N is large and the array is uniformly excited. If the inter-element spacing is $\lambda/2$, then the 3 dB beamwidth becomes $0.866/2N$. The peak-to-null spacing of the main beam of $4N$ elements with uniform illumination is $1/2N$. Hence, if the discrete beam positions are maintained at 0, $1/2N$, $1/N$, $3/2N$, . . . , $m/2N$ where $m=1, 2, \ldots, 4N-1$, then the beams covering the entire search space will overlap most likely at −4 dB points. However, for low sidelobe levels of −40 dB, the main beam broadening will be about 1.5 times, and thus by spacing the beams every $1/2N$, the resulting overlap for consecutive beam positions will be higher than −2 dB points. Thus, from the point of view of surveillance of a given volume of space, the discrete beam positions at every $1/2N$ is adequate. To shift the beam in angular space by $m/2N$, the progressive phase shift to be introduced between adjacent elements will be $m\pi$ where $m=1, 2, \ldots$ etc. This means that the value of α, the inter-element phase shift between the right ancillary array and the appropriate member of the left half of the central array as computed from eqn. 10.367, will be either 0 or π. It is now possible to evolve a distribution network replacing the phase shifter α after the power amplifier in Fig. 10.93 by the network shown in Fig. 10.94. The network outputs $\cos\theta_1$ and $\sin\phi_1$ thus provide the desired amplitude tapering. With the phase shift set at $+\phi_1$ on the left and $-\phi_1$ on the right, the reversal of the sign does not change the power split but merely reverses the phase of the right output. Thus the pair of ϕ_1 phase shifters performs the power splitting function as well as the function of the earlier α phase shifter. It is evident from Fig. 10.94 that there is no need to have a separate phase shifting network to be included for the beam steering phase shift ϕ_{nm}. It can be incorporated into the power splitting phase shifters, thus simplifying the network. Thus the network of Fig. 10.94 achieves low sidelobes, eliminates separate phase shifters for power splitting, phase gradient continuity and beam steering, and uses identical T–R modules which can be operated at maximum power-added efficiency. For the receive network, a duplexer/switch is to be installed at each radiating element so that there are no restrictions on the receive beam forming network. Lossless constrained synthesis can provide satisfactory results for sidelobe levels up to −27 dB from the peak. For lower sidelobe levels, it may not be possible to find ϕ_1 values with corresponding $\cos\phi_1$ and $\sin\phi_1$ amplitude tapers which will be optimum. This would result in loss of efficiency plus beam broadening. Instead it has been suggested that the lossless distribution which satisfies the energy constraints be the starting point for the final distribution. This distribution is now attenuated to an efficient Taylor $\bar{n}$ distribution which produces the desired sidelobes. In the limited examples quoted by Dufort for −35 dB sidelobe level the attenuation loss is of the order of 0.6 dB and the overall efficiency is less by 1.5 dB. Dufort has also extended the concept to planar arrays. In this case the array is partitioned into quartets of elements and the connectivity of each element of the quartet is arranged in such a way that beam steering is possible and that a measure of amplitude tapering is also provided. If the array has separable excitation along the two principal axes the typical power distribution of the quartet would be given by $\cos\phi_1 \sin\phi_2$, $\sin\phi_1 \sin\phi_2$, $\cos\phi_1 \cos\phi_2$ and $\sin\phi_1 \cos\phi_2$. Fig. 10.95 shows the building block that produces this distribution. This network is not unique and others can be constructed by various combinations of quadrature and magic tees. The major limitation is that the feed network is quite complex and that further development work needs to be carried out to simplify the topology of the criss crossing feed lines and also to minimise the losses.

10.33.4 *Phase-only synthesis*

Deford and Gandhi [147] approached the problem of low sidelobe level radiation patterns in active aperture systems by phase-only antenna pattern synthesis. The problem of minimising the peak sidelobe level for a given beamwidth by phase-only synthesis is considered in terms of an objective function

$$F(\vec{\psi}) = \max_{\hat{r}\in S} |E(\hat{r}, \vec{\psi})|/|E(\hat{r}_o, \vec{\psi})| \tag{10.386}$$

where $\vec{\psi}$ is the vector of unknown phases, $\hat{r}$ is the field point, S is the set of all points in the sidelobe region of the array pattern, $\hat{r}_0$ is the location of the main beam and E is the radiation pattern. The value of $\vec{\psi}$ that produces the global

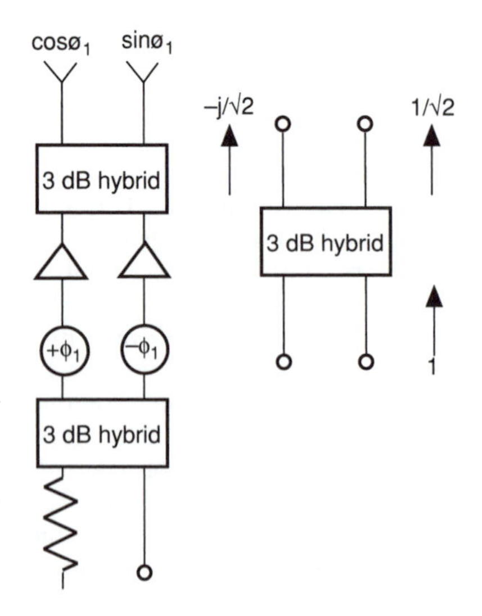

Figure 10.94 *Simplified feed-linear array [146]*

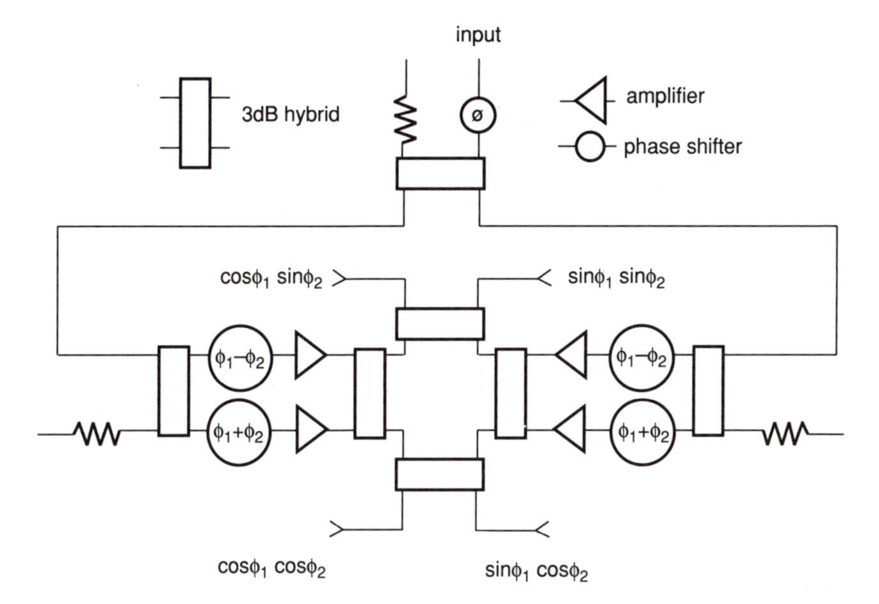

Figure 10.95 *Quartet building block [146].* © *IEEE, 1988*

minimum of F for equal excitation amplitude can easily be seen to be the pattern that has the minimum possible sidelobe level. As far as $F(\vec{\psi})$ is concerned, in the sidelobe region there are $N-1$ sidelobe peaks if $\vec{\psi}$ is a vector of N elements. When the phase of the excitation of any of the radiating elements of the array is altered, $\vec{\psi}$ is altered and $F(\vec{\psi})$ will also change. It is difficult to predict the position of the highest sidelobe level or its value and whether this peak sidelobe level value is really the maximum for any $\vec{\psi}$ vector. It is also not possible to state that $F(\vec{\psi})$ is convex, i.e. there is only one minimum which is the global minimum. On the other hand there is more than one local minimum which cannot be easily distinguished from the global minimum. However the vast majority of the minima are clustered fairly close to values of the vector $\vec{\psi}$. The global minimum is determined by taking the lowest of the set of local minima. For a large planar array the determination of a single minimum can itself need a large amount of computer time, thereby making this a very time consuming and expensive exercise. Another aspect of $F(\vec{\psi})$ is that, though it is continuous, it is piecewise differentiable. This arises as a consequence of the fact that, for every value of $\vec{\psi}$, there is a peak sidelobe value but the position of this peak value is discontinuous on the $(N-1)$ dimensional surfaces. As one changes the phase values of the radiating elements to reduce the sidelobe level of the mth peak which has been found to be the maximum, there is every likelihood that the sidelobe level of the mth peak decreases but an rth peak pops up with a higher level and r may not be near m, thereby resulting in discontinuity across the $(N-1)$ dimensional surfaces. These surfaces on which the first derivative of F is undefined appear as ridges because of their appearance in one dimension. $F(\vec{\psi})$ is at least piecewise differentiable in the ridge bound regions and therefore it is possible to apply descent algorithms. The existence of ridges implies that it is possible to increase F by taking a step in the steepest descent direction when the step crosses such a surface, so long as the step size is small enough. The net effect of this discontinuity in the location of the maximum of $(N-1)$ sidelobe peaks is to slow down the convergence of the descent methods than if F was differentiable.

Deford and Gandhi have used the steepest descent method (SDM) as the numerical tool to minimise the functional F and define the step vector as

$$\vec{S} = -\alpha \vec{\nabla} F = \mathrm{Re}\left[\frac{E(\hat{r}_m)\vec{\nabla} E(\hat{r}_m)}{|E(\hat{r}_m)|} - \frac{E(\hat{r}_0)\vec{\nabla} E(\hat{r}_0)}{|E(\hat{r}_0)|}\right] \quad (10.387)$$

where the gradient is taken with respect to $\vec{\psi}$, α is a positive constant, and $\hat{r}_m$ is the location of the maximum field in the sidelobe region; E is the complex conjugate of E. The phase vector $\vec{\psi}$ is updated at each iteration by adding to it the step vector $\vec{S}$. The gradient of F becomes undefined when $\hat{r}_m$ is multiple valued, i.e. $\hat{r}_m = \hat{Z}_1, \hat{Z}_2, \ldots, \hat{Z}_p, p > 1$. By forcing $\hat{r}_m$ to take one value $\hat{Z}_q$ within the set of possible P values, this is circumvented. The initial conditions can be any set of phase values for the vector $\vec{\psi}$ except the equiphase condition, since F is stationary for this value of $\vec{\psi}$. Even though the starting values of the phases of the excitation at the radiating elements have a great bearing on the location of the peaks and nulls in the sidelobe region, the sidelobe level of the final pattern is only a weak function of the initial phases. It has been found that the maximum variation in peak sidelobe levels between runs using different initial conditions for a given array is less than 1 dB. The computation of linear and planar arrays has revealed that a large majority of the lobes in the sidelobe region have the same amplitude as one approaches the minimum, and this can be used as an indication of the convergence of the solution; i.e. the more nearly equal the sidelobe amplitudes are, the closer is the pattern to the local optimum pattern. Also it has been found that, for minimum peak sidelobe arrays of any size, the phases for elements near the centre of the array are always quite uniform. Thus the initial phase values can be slight perturbations of the equiphase condition. The sidelobe levels of phase-only arrays have an approximately linear dependence on the logarithm of the number of elements in the case of linear arrays, or on the logarithm of the number of elements along one of the orthogonal axes in the case of planar arrays with separable excitations, i.e. $\log_2[N/10]$ dependence. The trade-off between beamwidth and sidelobe level that exists for amplitude taper arrays is also valid for phase-only arrays with the difference that, for phase-only arrays, the range over which these two parameters may be varied is somewhat more limited. The beamwidth broadening has been found to vary from 5% to 20% as compared with uniform excitation in the case of linear large arrays. The relative efficiency (which is the ratio of the peak power density in the main beam to the peak power density for a uniformly excited array with the same radiated power per element) is understandably higher for phase-only synthesis. As seen from Fig. 10.96 the efficiency falls off approximately linearly with $(\log N)$, but it is higher than the values for Taylor and Hamming distributions. Deford and Gandhi indicate that the number of operations by the numerical procedure per element for planar arrays is N^3 dependent and hence it is computationally intensive. There is no analytical approach yet to yield a pattern corresponding to a specific local minimum of F.

A different approach for the analytical solution was taken by Deford and Gandhi. This is based on statistical methods in which the pattern is treated as a random process related to the random variable that describes the phase of the excitation of the antenna elements. The excitations can be described by a Fourier transform and the expected pattern is treated as the expected functional of the transform. Array synthesis now boils down to optimisation of the sidelobe level of the expected pattern by adjusting the parameters that determine the form of the probability density functions describing the excitation of the elements. The analysis thus assumes that phase of the excitation of the mth element, i.e. ϕ_m, is a random variable that is uniformly distributed in the interval $[-r_m, r_m]$. The far field pattern which is represented by the random variable E can be written as

$$E = \sum_{m=1}^{N} e^{j\phi_m} \cos(m-0.5)\psi \tag{10.388}$$

where

$$\psi = \frac{2\pi d}{\lambda} \sin \theta$$

The associated power pattern is given by

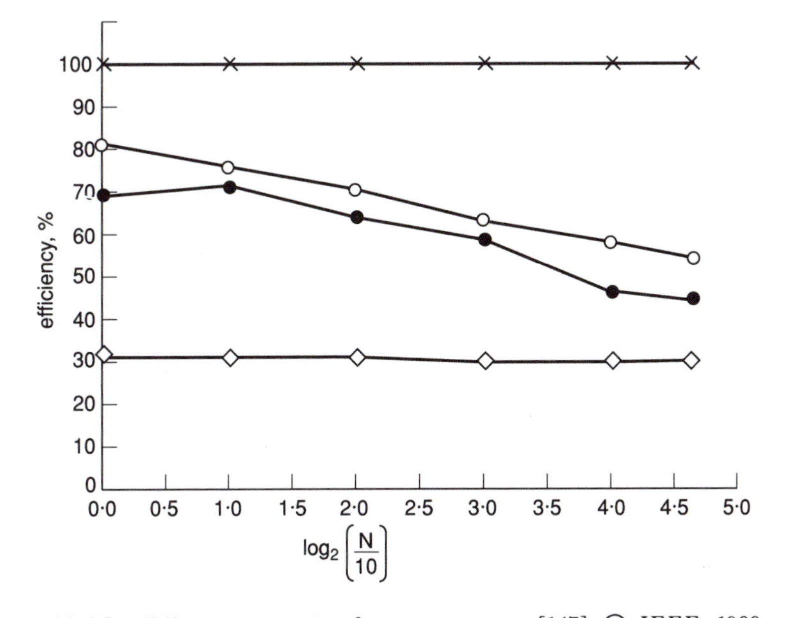

Figure 10.96 *Efficiency versus size for array antennas [147]. © IEEE, 1988*

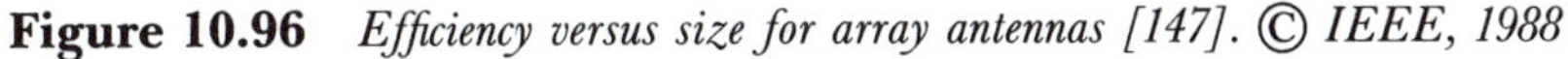

○ phase-only synthesis
◇ Hamming
● Taylor
× uniform

$$\mathcal{P}=\sum_{m=1}^{N}\sum_{l=1}^{N}\cos(\phi_m-\phi_l)\cos(m-0.5)\psi\cos(l-0.5)\psi \qquad (10.389)$$

To determine the value of r_m for each element the sidelobe of the expected value of the power pattern, i.e. $\mathcal{E}\{\mathcal{P}\}$, should be a minimum as a function of r_m, with $m=1, 2, \ldots, N$. The expected value is expressed as

$$\mathcal{E}\{\mathcal{P}\}=\left[\sum_{m=1}^{N}\frac{\sin r_m}{r_m}\cos(m-0.5)\psi\right]^2 + \sum_{m=1}^{N}\left(1-\frac{\sin^2 r_m}{r_m^2}\right)\cos^2(m-0.5)\psi \qquad (10.390)$$

An examination of eqn. 10.390 reveals that the first term on the RHS is the power pattern of the array with all the elements driven in phase, but with amplitudes given by $\sin r_m/r_m$. Further, for a given set of $\sin r_m/r_m$ values, the first term has a magnitude that is dependent on N^2 while the second term exhibits only a linear dependence on N. Hence for large N, only the first term dominates and the factor $\sin r_m/r_m$ in the first term can represent any given distribution. The phase envelope of the element excitations is now defined by the values of for $m=1, 2, \ldots, N$. It can be determined from

$$\frac{\sin r_m}{r_m}=W(m),\ m=1, 2, \ldots, N \qquad (10.391)$$

for the values of r_m. With the values of r_m known, the probability density function of each random variable ϕ_m is determined, and the phase of the mth element is obtained by sampling the random variable ϕ_m. The relative efficiency obtained by this method is lower than that obtained by the numerical search process. Both the numerical search technique and the analytical method can be applied to arrays of arbitrary geometry.

10.34 References

1 BROOKNER, E.: 'Radar Technology', (Artech House, Boston, USA, 1979), p. 34
2 BILLAM, E.R.: 'New technologies in phased array radar'. Proc. Military Microwaves, MM–85, pp. 57–66
3 HARTWELL, T.E.: 'RASSR array evaluation'. IEEE NAECON–77 Record, 1977, pp. 1164–1172
4 EDWARDS, W.J., REES, D.S., and KAMMERLY, R.T.: 'Advanced microwave technologies'. IEEE EASCON Record, 1982, pp. 419–423
5 TSAI, W.C., GRAY, R., and GRAZIANO, A.: 'The design and performance of supercomponents: High density RF modules', *Microwave J.*, Nov. 1983
6 MORCHIN, W. C.: Airborne early warning radar', (Artech House, Boston, USA, 1990)

7 LOCKERD, R.M., and CRAIN, G.E.: 'Airborne active element array radars come of age', *Microwave J.*, Jan. 1990, pp. 101–117

8 SPARKS, R.A.: 'Recent microwave applications to radar in the US'. Proc. 19th Euro. Microwave Conference, 1989, pp. 96–106

9 MAGARSHAK, J.: 'European MMIC Activities'. IEEE MTT-S Digest, 1988, pp. 645–648

10 KIM, B. *et al.*,: 'GaAs Power MESFET with 41 Percent Power Added Efficiency at 35 GHz', *IEEE Electron Devices Letters*, 1988, **EDL–9**, pp. 57–58

11 CHI, J. *et al.*,: 'High performance monolithic power amplifier using all RIE etch process'. IEEE GaAs IC Symposium Digest, 1987, pp. 159–161

12 IGI, S. *et al.*,: 'High power and high efficiency GaAs FETs in C-band'. 1988 IEEE MTT–S Digest, p. 834

13 YAMADA, Y. *et al.*,: 'X- and Ku-band high power GaAs FETs'. IEEE MTT–S Digest, 1988, pp. 847–850

14 NASTER, R.J., and LANG, M.R.: 'Affordable MMIC design for phased arrays', *Microwave J.*, March 1987, pp. 145–150

15 LADBROOKE, P.: 'Statistical design for yield of MMICs'. Proc. Military Microwaves, MM–88, 1988, pp. 386–390

16 CHILTON, R.H.: 'MMIC TR modules and applications', *Microwave J.*, Sept. 1987, pp. 131–146

17 PITZALIS, O.: 'Microwave to mm wave CAE: Concept to production', *Microwave J.*, State of the Art Reference Supplement, Sept 1989, pp. 15–49

18 PAVIO, A.M., PAVIO, J.S., and CHAPMAN Jr, J.E.: 'The high volume production of advanced radar modules', *Microwave J.*, Jan. 1989, pp. 139–148

19 MCQUIDDY, D.N., GASSNER, R.L., HULL, P., MASON, J.S., and BEDINGER, J.M.: 'Transmit/receive module technology for X-band active array radar', *Proc. IEEE*, 1991, **79**, pp. 308–341

20 PENGELLY, R.S.: 'Monolithic microwave integrated circuits'. Parts I and II, *IEEE MTT-S News Letter*, Winter & Spring 1989, pp. 39–48, 24–33

21 SHENOY, R.P.: 'Active aperture phased arrays', *Alta Frequenza*, 1989, **LVIII**, pp. 137–149.

22 SHENOY, R.P.: 'Recent advances in radar technology', *Indian J. Radio & Space Physics*, 1990, **19**, pp. 359–371

23 SHIH, Y. and KUNO, Y.: 'Solid state sources from 1 to 100 GHz', *Microwave J.*, State of Art Reference Supplement, Sept. 1989, pp. 145–161

24 KOPP, W.S., and PRITCHETT, S.D.: 'High efficiency power amplification for microwave and millimeter frequencies'. IEEE MTT–S Digest, 1989, pp. 857–858

25 BORKOWSKI, M.T., and LAIGHTON, D.G.: 'Decreasing cost of GaAs MMIC modules is opening up new areas of application', *Electronic Progress*, 1989, **XXIX**, pp. 32–41

26 NISHIKI, N., and NOJIMA, T.: 'Harmonic reaction amplifier–A novel high efficiency and high power microwave amplifier'. IEEE MTT–S Digest, 1987, pp. 963–966.

27 GELLER, B.D., and GOETTLE, P.E.: 'Quasi-monolithic 4-GHz power amplifiers with 65 percent power-added-efficiency'. IEEE MTT–S Digest, 1988, pp. 835–838.

28 BAHL, I.J., GRIFFIN, E.L., and ANDRICOS, C.: 'A 14 W C-band power amplifier employing MMIC chips', *Microwave J.*, May 1989, pp. 295–303

29 BAHL, I.J., GRIFFIN, E.L., GEISSBURGER, A.E., ANDRICOS, C., and BURKIEWA, T.F.: 'Class-B power MMIC amplifiers with 70% power-added-efficiency', *IEEE Trans.* 1989, **MTT–37,** pp. 1315–1320

30 BAHL, I.J., WANG, R., GEISSBURGER, A.E., GRIFFIN, C.L., and ANDRICOS, C.: 'C-band, 10W MMIC class-A amplifier manufactured using the refractory SAG process', *IEEE Trans.*, 1989, **MTT–37**, pp. 2154–2158

31 AVSARALA, M., DAY, D.S., CHAN, S., GREGORY, P., and BASSET, J.R.: 'High efficiency small size 6W class AB X-band power amplifier module using a novel MBE GaAs FET'. IEEE MTT–S Digest, 1988, pp. 843–846

32 KHATIBZADEH M., and TSERNG, H.Q.: 'Harmonic tuning of power FETs at X-band'. IEEE MTT–S Digest, 1990, pp. 989–992

33 KIM, B., CAMILLERI, N., SHIH, H.D., TSERNG, H.Q., and WURTELA, M.:

'35 GHz GaAs power MESFETs and monolithic amplifiers', *IEEE Trans.*, 1989, **MTT–37**, pp. 1327–1332
34 GELLER, B.D., ASSEL, F.T., GUPTA, R.K., and CLINE, P.K.: 'A technique for the maintenance of FET power amplifier efficiency under back off'. IEEE MTT–S Digest, 1989, pp. 949–952
35 BAYRAKTAROGLU, B.: 'Heterojunction bipolar transistors respond to high power needs of military microwave circuits'. Proc. Military Microwaves, 1990, MM–90, pp. 27–31
36 MALONEY, P.R., MEZGER, M.A., and SASONOFF, J.P.: 'L-band GaAs transceiver components'. IEEE GaAs Symposium Digest, 1985, pp. 121–124
37 ANDRICOS, C., BAHL, I.J., and GRIFFIN, E.L.: 'C-band 6-bit monolithic phase shifter', *IEEE Trans.* 1985, **MTT–33**, pp. 1591–1596
38 AYASLI, Y., PLATZKER, A., VORHAUS, J., and REYNOLDS, L.D.: 'A monolithic single chip X-band four-bit phase shifter', *IEEE Trans.*, 1982, **MTT–30**, pp. 358–363
39 MOYE, C. SAKAMOTO, G. and BRAND, M: 'A compact broadband six bit MMIC phasor with integrated digital drivers'. IEEE MTT–S Digest, 1990, pp. 457–460
40 SCHINDLER, M.J., and MILLER, M.E.: 'A 3/4-bit, K/Ka-band MMIC phase shifter'. IEEE Microwave and mm-Wave Monolithic Symposium Digest, 1988, pp. 95–98
41 ALI, F., MITCHELL, S., MURPHY, A., ADAR, A., HO, P., and PODELL, A.: 'Production results of X-band GaAs MMIC 5-bit phase shifters'. IEEE MTT–S Digest, 1990, pp. 633–636.
42 ALI, F., MITCHELL, S., MOGRI, J., and PODELL, A.: 'A single chip C-band GaAs monolithic 5-bit phase shifter with on chip digital decoder'. IEEE MTT–S Digest, 1990, pp. 1235–1237.
43 DIXIT, R., NELSON, B., JONES, W., and CARILLO, J.: 'A family of 2–20 GHz broadband low noise AlGaAs HEMT MMIC amplifiers'. IEEE Microwave & mm-Wave Monolithic Circuits Symposium Digest, 1989, pp. 15–19
44 UPTON, M.A.G., SNOW, K.H., GOLDSTACK, D.I. *et al.*,: 'Monolithic HEMT LNAs for radar EW and Comm'. IEEE Microwave and mm-Wave Monolithic Circuit Symposium Digest, 1989, p. 105–109
45 AUMANN, H.M., and WILWERTH, F.G.: 'Intermediate frequency transmit/receive modules for low side-lobe phased array application'. Proc. IEEE National Radar Conference, 1988, pp. 33–37
46 GREEN, C.R., LANE, A.A., TOMBS, P.N., SHUKLA, R., SUFFOLK, J., SPARROW, J.A., and COOPER, P.D.: 'A 2-watt GaAs Tx/Rx module with integrated control circuitry for S-band phased array radars'. IEEE MTT–S Digest, 1987, pp. 933–936
47 ALI, F., MITCHELL, S., MOGHE, S., HO, P., and PODELL, A.: 'A C-band low noise MMIC phased array receive module'. IEEE MTT–S Digest, 1988, pp. 951–953.
48 PRIOLO, M., ST. ONGE, G., COUGHLIN, W., BUGEAU, J., and MEHARRY, D.: 'Transmit receive modules for 6 to 16 GHz multifunction arrays'. IEEE MTT–S Digest, 1990, pp. 1227–1230
49 WILLEMS, D., BAHL, I., POLLMAN, M. *et al.*,: 'Multifunction chip set for T/R module receive path'. IEEE MTT–S Digest, 1990, pp. 113–116
50 YAU, W., HERMAN, M.I., MENDOLIA, G., WEN, C.P., and CHEN, J.C.: 'An X-band low cost GaAs monolithic transceiver'. IEEE MTT–S Digest, 1990, pp. 827–830.
51 WISSEMAN, W.R., WITKOWSKI, L.C., BREHM, G.E., COATS, R.P., HESTON, D.D., HUDGENS, R.D., LEHMANN, R.E., MACKSEY, H.M., and TSERNG, H.C.: 'X-band GaAs single chip T/R radar module', *Microwave J.*, Sept. 1987, pp. 167–172
52 SCHINDLER, M.J., CHU, S.L., KAZIOR, T.E., BERTRAND, A.M., and SIMON, K.M.: 'A single chip 2–20 GHz T/R module'. IEEE MTT-S Digest, 1990, pp. 117–120
53 SLIVA, R.J.: 'Analysis shows effects of aging on T/R modules', *Microwaves and RF*,

1988, pp. 79–86
54 PUCEL, R.A.: MMIC modelling and CAD—where do we go from here?'. Proc. 16th Euro. Microwave Conf., 1986, pp. 61–70
55 HANSEN, R.H., ARNOLD, R.G., and EDDISON, J.G.: 'A comprehensive CAD approach to the design of MMICs up to mm-wave frequencies', *IEEE Trans.*, **MTT-36**, 1988, pp. 208–219
56 RIZZOLI, V., and NERI, A.: State-of-the-art and present trends in non-linear microwave CAD techniques', *IEEE Trans.* **MTT-36**, 1988, pp. 341–365
57 GOYAL, R. (Ed.): 'Monolithic microwave integrated circuits: technology and design' (Artech House, Mass, USA, 1989)
58 PAVIO, A.M.: 'Deliver 100,000 modules a month? Of course we use CAM!'. Proc. 18th Euro. Microwave Conf., 1988, pp. 38–48
59 ZAGHLOUL, A.I., POTUKUCHI, J.R., MOTT, R.C., and KROLL, R.T.: 'Analysis and measurement of tolerances in MMIC active phased arrays'. Proc. 19th Euro. Microwave Conf., 1989, pp. 966–971
60 VORHAUS, J.L.: 'High volume manufacturing of a high efficiency X-band power module for phased array applications', 1989, *IEEE MTT-S Digest*, pp. 769–772
61 GEEN, M.W., CONLON, R.F.B., BUCK, B.J., CHARLTON, R.W., and JENKINS, A.P.: 'GaAs MMIC yield evaluation'. Proc. 17th Euro. Microwave Conf., 1987, pp. 448–453
62 PARISOT, M. and MAGARSHACK, J.: 'Production of high performance reproducible and reliable GaAs MMICs'. Proc. Military Microwves Conf., MM-88, pp. 391–396.
63 BREHM, G.E.: 'Statistical process control for GaAs MMIC wafer fabrication'. Proc. Military Microwaves, MM-90, 1990, pp. 375–380
64 BHARJ, S.S.: 'MMIC manufacturability'. Proc. Military Microwaves, MM-88, 1988, pp. 378–385
65 MATSUNO, C.T. and SHARMA A.K.: 'Forecasting MMIC yields and costs', *Microwave J.*, 1988, pp. 105–128
66 NORRIS, G.B., and BARRATT, C.A.: 'GaAs MMIC yield modelling'. Proc. IEEE GaAs IC Symp., 1990, pp. 317–320
67 STONEHAM, E.B. and O'SULLIVAN, P.A.J.: 'Working with nine different foundries'. Proc. IEEE GaAs IC Symp., 1990, pp. 11–14
68 PERKINS, W.H., and JANSEN, N.: 'MIMIC transportability', *IEEE MTT-S Digest*, 1990, pp. 109–112
69 KEMMERLY, R.T., and FAYETTE, D.F.: 'Affordable MMICs for Airforce systems', *Microwave J.*, 1991, pp. 172–175
70 HUTCHESON, L.D. (Ed.): 'Integrated optical circuits and components' (Marcel Dekker Inc., Switzerland, 1987) Chapter 7
71 ADAMS, M.J., STEVENTON, A.G., DEVLIN, W.J., and HENNING, I.D.: 'Semiconductor lasers for long wavelength optical fibre communication systems' (Peter Peregrinus Ltd, London, 1987) Chapter 3
72 Simons, R.: 'Optical control of microwave devices' (Artech House, Boston, 1990) Chapter 5
73 JOINDOT, I., and BOISROBERT, C.: 'Laser RIN calibration by extra noise injection', *Electron. Lett.*, **25**, 1989, pp. 1052–1053
74 TSUJI, S., and CHINONE, N.: 'Semiconductor optical devices for lightwave systems', *Hitachi Review*, **40**, 1991, (2), pp. 181–184
75 YANO, M., and KOTAKI, Y.: 'Extremely low-noise facet-reflectivity-controlled InGaAsP distributed-feed back lasers', *IEEE J. Lightwave Technol.*, **LT-4**, 1986, pp. 1454–1459
76 GOLDBERG, L., ESMAN, R.D., and WILLIAMS, K.J.: 'Optical techniques for microwave generation, transmission and control', *IEEE MTT-S Digest*, 1990, pp. 769–772
77 GOLDBERG, L., ESMAN, R.D., and WILLIAMS, K.J.: 'Generation and control of microwave signals by optical techniques', *IEE Proc. J*, **139**, 1992, pp. 288–294
78 WOJTUNIK, H.J.: 'PIN diodes provide low-cost detectors for fibre lasers', *Laser Focus World*, 1992, pp. 115–120
79 ACKERMAM, E., WANUGA, S., CANDULA, K., SCOTTI, R., MCDONALD,

W., and GATES, J.: 'A 3 to 6 GHz microwave/photonic tranceiver for phased array interconnects', *Microwave J.*, 1992, pp. 60–71

80 LI, T.: 'Structures, parameters and transmission properties of optical fibers', *Proc. IEEE*, **68**, 1980, pp. 1175–1180

81 LIVNE, A.: 'System design considerations in fibre-optic networks for active array satellite antenna'. Proc. 19th Euro. Microwave Conf., 1989, pp. 951–957

82 KEN, H.W., GEE, C.M., and BLAUVALT. H.: 'High speed optical modulation techniques' *Optical Technology for Microwave Applications II*, **545**, 1985, pp. 2–9

83 BERCELI, T., HERCZFELD, P.R., DARYOUSH, A.S. and ROSEN, A.: 'Performance optimisation of optically controlled phased array antenna T/R modules'. Proc. 19th Euro. Microwave Conf., 1989, pp. 945–950

84 BERCELI, T., JEMISON, W.D., HERZCFELD, P.R., DARYOUSH, A.S., and PAOLELLA, A.: 'A double-stage injection locked oscillator for optically fed phased array antennas', *IEEE Trans.*, **39**, 1991, pp. 201–207

85 ZHANG, X., ZHOU, X., ALINER, B., and DARYOUSH, A.S.: 'A study of subharmonic injection locking for local oscillators', *IEEE Microwave and Guided Wave Letters*, **2**, 1992, pp. 97–99

86 DARYOUSH, A.S., FRANSISCO, M., SAEDI, R., POLIFKO, D., and KUNATH, R.: 'Phase control of optically injection locked oscillators for phased arrays', *IEEE MTT-S Digest*, 1990, pp. 1247–1250

87 WALDRON, T.P., CHIN, S.K., and NASTER, R.J.: 'Distributed beam steering control of phased array radars', *Microwave J.*, 1986, pp. 133–146

88 JEMISON, W.D., BERCELI, T., HERCZFELD, P.R., KASEMSET, D., and JACOMB-HOOD, A.W.: 'Optical control of digital phase shifter', *IEEE MTT-S Digest*, 1990, pp. 233–236

89 HERCZFELD, P.R., PAOLELLA, A., DARYOUSH, A.S., JEMISON, W., and ROSEN, A.: 'Optical control of MMIC based T-R modules', *Microwave J.*, 1988, pp. 303–321

90 HERCZFELD, P.R., PAOLELLA, A., DARYOUSH, A., JEMISON, W., and ROSEN, A.: 'Optical phase and gain control of a GaAs transmit-receive module'. *Proc. 18th Euro. Microwave Conf.*, 1988, pp. 831–836

91 PAOLELLA, A., and HERCZFELD, P.R.: 'Optical control of a GaAs MMIC transmit/receive module', *IEEE MTT-S Digest*, 1988, pp. 959–962

92 KAM, M., WILCOX, J., and HERCZFELD, P.B.: 'Design for steering accuracy in antenna arrays using shared optical phase shifters', *IEEE Trans.* **AP-37**, 1989, pp. 1102–1108

93 LEE, J.J., YEN, H.W., NEWBERG, I., and WECHSBERG, W.: 'A dual band array controlled by fibre optic delay lines', *IEEE AP-S International Symposium Digest*, 1989, pp, 690–693

94 NG, W., WALSTON, A., TANGONAN, G., LEE, J.J., and NEWBERG, I.: 'Optical steering of dual band microwave phased array antenna using semiconductor laser switching', *Electron. Lett.*, **26**, 1990, pp. 791–792

95 SEEDS, A.J., BLANCHFLOWER, I.D., GOMES, N.J., KING, G., and FLYNN, S.J.: 'New developments in optical control techniques for phased array radar', *IEEE MTT-S Digest*, 1988, pp. 905–908

96 SEEDS, A.J., MCMILLAN, W.I., PESCOD, C.R., WALE, M.J., and BIRKMAYER, W.S.: 'Optical control of phased array antennas: a European perspective', *IEEE MTT-S Digest*, 1990, pp. 1343–1346

97 JAMES, J.R., and HALL, P.G. (Eds.): 'Handbook of microstrip antennas' (Peter Peregrinus Ltd, London, UK, 1989)

98 GUPTA, K.C., and BENALLA, A.: 'Microstrip antenna design' (Artech House, Boston, USA, 1988)

99 CARVER, K.R., and MINK, W.J.: 'Microstrip antenna technology', *IEEE Trans.*, **AP-29**, 1981, pp. 1–24

100 MAILLOUX, R.J., MCILVENNA, J.F., and KEMWEIS, N.P.: 'Microstrip array technology', *IEEE Trans.*, **AP-29**, 1981, pp. 25–37

101 MAILLOUX, R.J.: 'Phased array theory and technology', *Proc. IEEE*, **70**, 1982, pp. 246–291

102 LO, Y.Y., and LEE, S.W.,: 'Antenna handbook: theory application and design'

(Van Nostrand Reinhold, NY, 1988) pp. 10–22

103 HAMMERSTAD, E., and JENSEN, O.: 'Accurate models for microstrip computer aided design', *IEEE MTT-S International Microwave Symposium Digest*, 1980, pp. 407–409

104 KIRSCHNING, M., JANSEN, R.H., and KOSTER, N.H.L.: 'Accurate models for open end effect of microstrip lines', *Electron. Lett.*, 1981, **17**, (3), pp. 123–125

105 DEARNLEY, R.W., and BAREL, A.R.F.: 'A comparison of models to determine the resonant frequencies of a rectangular microstrip antenna', *IEEE Trans.*, **AP-37**, 1989, pp. 114–118

106 PERLMUTTER, P., SHTRIKMAN, S., and TRAVES, D.: 'Electric surface current model for the analysis of microstrip antenna with application to rectangular elements', *IEEE Trans.*, **AP-33**, 1985, pp. 301–311

107 POZAR, D.M.: 'Considerations for millimetre wave printed antennas', *IEEE Trans.*, **AP-31**, 1983, pp. 740–747

108 SCHAUBERT, D.H., POZAR, D.M., and ADRIAN, A.: 'Effect of microstrip antenna substrate thickness and permittivity: comparison of theories with experiment', *IEEE Trans.*, **AP-37**, 1989, pp. 1217–1223

109 SCHAUBERT, D.H., POZAR, D.M., and ADRIAN, A.: 'Effect of microstrip antenna substrate thickness and permittivity: comparison of theories with experiment', *IEEE Trans.*, **AP-37**, 1989, pp. 677–682

110 LIU, C. C., HESSEL, A., and SHMOYS, J.: 'Performance of probe-fed microstrip patch element phased arrays', *IEEE Trans.*, **AP-36**, 1988, pp. 1501–1509

111 LIU, C.C., HESSEL, A., and SHMOYS, J.: 'Performance of probe-fed microstrip patch element phased arrays', *IEEE Trans.*, **AP-36**, 1988, pp. 1501–1509

112 OLTMAN, H.G., and HUEBNER, D.A.: 'Electromagnetically coupled dipoles', *IEEE Trans.*, **AP-29**, 1981, pp. 151–157

113 CHEN, C.H., TULINTSEFF, A., and SORBELLO, R.M.: 'Broad band two-layer microstrip antennas', *IEEE AP-S International Symposium Digest*, 1984, pp. 251–254

114 KATEHI, P.B., ALEXOPOULOUS, N.G., and HSIA, I.Y.: 'A bandwidth enhancement method for microstrip antennas', *IEEE Trans.*, **AP-35**, 1987, pp. 5–12

115 SULLIVAN, P.L., and SCHAUBERT, D.H.: 'Analysis of an aperture coupled microstrip antenna', *IEEE Trans.*, **AP-34**, 1986, pp. 977–984

116 AKSUN, M.I., CHUANG, S., and LO, Y.T.: 'On slot-coupled microstrip antennas and their application to CP operation—theory and experiment', *IEEE Trans.*, **AP-38**, 1990, pp. 1224–1230

117 POZAR, D.M.: 'A reciprocity method of analysis for printed slot and slot coupled microstrip antennas', *IEEE Trans.*, **AP-34**, 1986, pp. 1439–1446

118 KINZEL, J.A., EDWARD, B.J., and REES, D. 'V-band space based phased arrays', *Microwave J.*, 1987, pp. 89–102

119 HUANG, C., and HSU, P.: 'Superstrate effects on slot coupled microstrip antennas', *IEEE Trans. on Magnetics*, **27**, 1991, pp. 3868–3871

120 TSAO, C.H., HWANG, Y.M., KILBURN, F., and DIETRICH, F.: 'Aperture coupled patch antennas with wide bandwidth and dual polarisation capabilities', *IEEE Antennas & Propagat. Symposium Digest*, 1988, pp. 936–939

121 ITTIPIBON, A., CLARKE, B., and CUHACI, M.: 'Slot coupled stacked microstrip antennas', *IEEE Antennas Propagat. Symposium Digest*, 1990, pp. 1108–1111

122 LEE, R.Q., LEE, K.F., and BOBINCHAK, J.: 'Electromagnetically coupled rectangular patch antenna', *JINA* (*Journeés Internationales De Nice Sur Les Antennas*), 1988, pp. 240–244

123 LEE, R.Q., and LEE, K.F.: 'Experimental study of the two layer electromagnetically coupled rectangular patch antenna', *IEEE Trans.*, **AP-38**, 1990, pp. 1298–1302

124 BARLATEY, L., MOSIG, J.R., and SPHICOPOULOUS, T.: 'Analysis of stacked microstrip patches with a mixed potential integral equation', *IEEE Trans.*, **AP-38**,

1990, pp. 608–615

125 HALL, R.C., MOSIG, J.R.: 'The analysis of dual layer coaxially fed microstrip antenna', *Proc. HINA*, France, 1990, pp. 274–277

126 POZAR, D.M.: 'Analysis of infinite phased array of aperture coupled microstrip patches', *IEEE Trans.*, **AP-37**, 1989, pp. 418–444

127 DAMIANO, J.P., BENNEGUEOUCHE, J., and PAPIERNIK, A.: 'Study of multilayer microstrip antenna with radiating elements of various geometry', *IEE Proc. H*, 1990, **137**, pp. 163–170

128 MECK, U., SAURER, B.: 'Double patch microstrip antenna for broad band application', *Proc. JINA*, 1988, pp. 250–254

129 HASSANI, H.R., and MIRSHEKAR-SYAHKAL, D.: 'Full-wave analysis of stacked rectangular microstrip antennas'. *Proc. Int. Conf. on Antennas Propag.*, 1989, ICAP-89, pp. 369–373

130 CROQ, F., and PAPIERNIK, A.: 'Stack slot-coupled printed antenna', *IEEE Microwave and Guided Wave Letters*, **1**, 1991, pp. 288–290

131 DAS, N.K., and POZAR, D.M.: 'Full-wave spectral domain computation of material, radiation and guided wave losses in infinite multi-layered printed transmission lines', *IEEE Trans.*, **MTT-39**, 1991, pp. 54–63

132 DAS, N.K., and POZAR, M.: 'Multiport scattering analysis of general multi-layered printed circuit antennas fed by multiple feed ports: part I-theory', *IEEE Trans.*, **AP-40**, 1992, pp. 469–481

133 DAS, N.K., and POZAR, D.M.: 'Multiport scattering analysis of general multi layered printed circuit antennas fed by multiple feed ports: part II-applications', *IEEE Trans.*, **AP-40**, 1992, pp. 482–491

134 LIU, C.C., SHMOYS, J., and HESSEL, A.: 'E-plane performance trade-offs in two-dimensional microstrip—patch element phased arrays', *IEEE Trans.*, **AP-30**, 1982, pp. 1201–1206

135 PAN, S.G., and WOLFF, I.: 'Computation of mutual coupling between slot-coupled microstrip patches in a finite array', *IEEE Trans.*, **AP-40**, 1992, pp. 1047–53

136 POZAR, D., and KAUFMAN, B.: 'Design considerations for low sidelobe level microstrip arrays', *IEEE Trans.*, **AP-38**, 1990, pp. 1176–1185

137 LUBIN, Y., and HESSEL, A.: 'Wide-band wide angle microstrip stacked-patch-element phased arrays', *IEEE Trans.*, **AP-39**, 1991, pp. 1062–1070

138 HALL, P.S.: 'Review of practical issues in microstrip antenna design' *Proc. JINA* 1990, pp. 268–273

139 POZAR, D.M., and SCHAUBERT, D.H.: 'Scan-blindness in infinite phased array of printed dipoles', *IEEE Trans.*, **AP-32**, 1984, pp. 602–610

140 POZAR, D.M., and JACKSON, R.W.: 'An aperture coupled microstrip antenna with a proximity feed on a perpendicular substrate', *IEEE Trans.*, **AP-35**, 1987, pp. 728–731

141 POZAR, D.M., and SCHAUBERT, D.H.: 'Comparison of architectures for monolithic phased array antennas', *Microwave J.*, 1986, pp. 93–104

142 KINZEL, J.A., EDWARD, B.J., and REES, D.: 'V-band space-based phased arrays', *Microwave J.*, 1987, pp. 89–102

143 SNOW, K.H., KOMIAK, J.J., and BATES, D.A.: 'Wide band variable gain amplifier in GaAs MMIC', *IEEE MTT-S Digest*, 1988, pp. 183–187

144 CULBERTSON, R.B., and ZIMMERMAN, D.C.: 'A 3-Watt X-band monolithic variable gain amplifier', *IEEE MTT-S Digest*, 1988, p. 174

145 LEE, J.J.: 'Side lobes control of solid state array antennas', *IEEE Trans.*, **AP-36**, 1988, pp. 339–344

146 DUFORT, E.C.: 'Low side lobe electronically scanned antenna using identical transmit/receive module', *IEEE Trans.*, **AP-36**, 1988, pp. 349–356

147 DEFORD, J.F., and GANDHI, O.P.: 'Phase-only synthesis of minimum peak side lobe patterns for linear and planar arrays', *IEEE Trans.*, **AP-36**, 1988, pp. 191–201

148 CLARIDGE, P.A., *et al.*: 'Design of phased array antenna using solid state T-R modules', Proc. 19th European Microwave Conf., 1989, p. 1132
149 LANE, A.A., and MYERS, P.A.: 'GaAs MMIC phase shifters', *Proc. Military Microwaves*, *MM*-88, p. 215
150 GOYAL, R. (Ed.): 'Monolithic microwave integrated circuits: technology and design' (Artech House, Norwood, MA, 1989) p. 209
151 BAR-LEV, A.: 'Semiconductors and electronic devices' (Prentice Hall, 1984, 2nd edn.) p. 410
152 ANDRICOS, C.: 'C-band T-R module for high performance phased array radar'. IEEE National Radar Conf. Proc., 1991, pp. 67–70

Chapter 11

Bistatic radars

M. R. B. Dunsmore
Defence Research Agency, Great Malvern, Worcestershire, WR14 3PS, UK

11.1 Background

11.1.1 *Description of bistatic radars*

A bistatic radar is, by definition, one in which the transmitter and the receiver are deployed at two separate locations; either or both of these locations can be changing with time. These sites are separated by a baseline whose length is usually comparable with the ranges to the targets of interest, as illustrated in Fig. 11.1. As in monostatic radars, the location of a target can be determined by a time delay measurement and an antenna bearing, provided that the length and orientation of the baseline are known. In the bistatic case, the time is that for propagation from transmitter to target to receiver and its value defines a prolate spheroid with the transmitter and receiver sites as foci. The location of

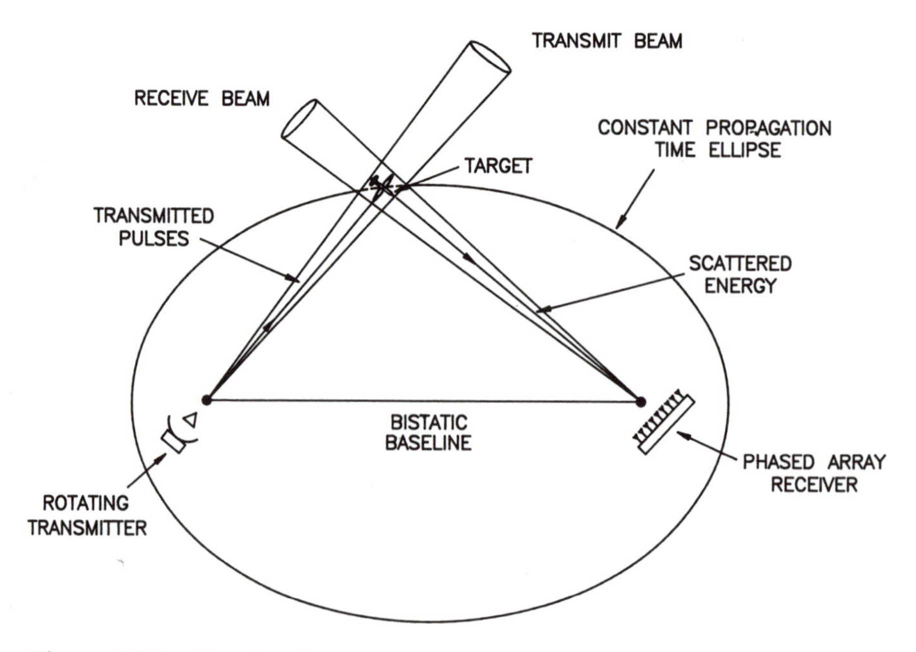

Figure 11.1 *Bistatic radar geometry*

the target on the surface of this spheroid is then determined by the azimuth and elevation angles of the transmitter antenna.

For a bistatic radar to have a data rate comparable with that provided by a monostatic radar, the receiver antenna has to be able to follow the directions from which energy might be scattered towards it by any targets encountered by the transmitter's pulses. This 'pulse chasing' requirement is one of the most salient features of a bistatic radar. Since the resultant antenna scanning rates can exceed a million degrees per second, for pulses travelling near to the baseline, the development of effective bistatic radars has had to await the development of phased array antennas. The orientation of the receiver antenna is a dependent variable and does not contribute to the accuracy of target location unless this antenna has a monopulse capability.

Some bistatic radars have only been called bistatic in the narrow sense of using separate sites as a method of protecting the receiver from the transmitter's pulses – two examples of this are the British 'Chain Home' radar system, deployed shortly before the Second World War, and the Russian ABM related radars, according to the US publication 'Soviet Military Power'. The use of separate antennas is convenient for phased array radars since it avoids requiring duplexing circuits for the array elements; an example of this is the US AN/FPS–85 UHF ABM radar.

The first published use of the term 'bistatic radar' appears to have been by Siegel in 1955. Unfortunately, whoever coined the name obviously was not a student of classical Greek, as it should really have been called a distatic radar, to match the use of the term 'monostatic radar'. It is probably too late now to correct the name; in any case, strict etymological standards could lead, for example, to systems with two mobile platforms being called dikinetic. In this review, we will consider only those bistatic radars where the combination of baseline length, target ranges and antenna beamwidths will normally require the receiver antenna beam(s) to be pointed in direction(s) different from those of the transmitter beam.

11.1.2 *History of bistatic radars*

The first demonstration of bistatic radar operation can be attributed to Heinrich Hertz, at his laboratory at the Technische Hochschule in Karlsruhe in 1888, although he did not describe his experiments as such. He first succeeded in transmitting electromagnetic waves, with wavelengths of 50 to 60 cm, from Leyden jars fitted with a spark gap and a loop of wire and detecting them with sensitive wire coils fitted with another spark gap. He then showed that these waves from his spark-gap transmitter could be reflected by metal mirrors, back to his detector. Hertz's work provided the experimental proof of the existence of the electromagnetic waves predicted by James Clerk Maxwell in 1864.

A working device for the detection of ships, based on these experiments, was tested by Hulsmeyer [1904] in Germany in the early 1900s. However, nothing appears to have been done to exploit these early demonstrations, even though for many years before the development of radar systems there were reports of radio waves in communication systems being reflected by objects.

During the early 1930s, most of the major countries (United Kingdom,

United States, France, Italy, the Netherlands, Russia and Japan) were attempting to develop radars, as a means of countering the threat posed by military aircraft, especially bombers. Prior to the development of the duplexer by the US Naval Research Laboratory in 1936, these radars were predominantly what we would now describe as bistatic CW interference systems. They detected targets as they approached the transmitter–receiver baseline by observing the beat frequency between their Doppler-shifted reflections and the direct signal from transmitter to receiver. This is the same as the flickering effect that can be observed on a TV set, especially when receiving a weak signal, if an aircraft passes close to the baseline between the TV transmitter and the receiver.

UK radar development followed from the secret memorandum by R. A. Watson-Watt to the UK Air Ministry on the 12th February 1935, entitled 'Detection and location of aircraft by radio methods'. This led to a bistatic CW radar demonstration by A. F. Wilkins on the 26th February 1935. The transmitter was the BBC Empire short-wave radio station at Daventry which operated at 6 MHz and radiated 10 kW of power in a southerly direction, with an azimuth beamwidth of 60° and a main vertical lobe at an elevation of 10°. A Handley Page 'Heyford' bomber flew up and down the beam, and a receiver, seven miles away at the village of Weedon in Northamptonshire just off the A5, was able to observe fluctuations in its received signal owing to interference between the direct signal and the signal reflected from the aircraft.

The success of the Daventry experiment led to the development of the CH (Chain Home) radar. This was the first radar to form part of a complete air defence system and the first to be used under wartime conditions. The first 20 early warning CH stations started their continuous watch on Good Friday 1939, providing cover from Ventnor on the Isle of Wight to the Firth of Tay. By the end of the Second World War there were approximately 50 stations around the coasts of Britain. Each radar site was bistatic in the narrow sense of having the transmitter and the receiver separated by hundreds of metres, as a means of protecting the receiver from the transmitter's energy. The CH radars used a low PRF of 25 Hz to minimise the effects of range ambiguous echoes from the ionosphere and each CH station had to be synchronised to a preset point on the 50 Hz waveform of the UK National Grid in order to avoid mutual interference effects (known as 'running rabbits'). Since the transmitters radiated wide beams at known times, a receiver could operate with a more distant transmitter when its own local transmitter was being attacked or jammed. Thus the UK system was using true bistatic operation as an ECCM technique even before the term 'radar' had been coined [Wardrop, 1992].

The first trials of an airborne bistatic radar appear to have been those in the UK, in the Autumn of 1936, when a small team led by Dr E. G. Bowen installed a receiver on a Heyford bomber and operated it with a ground based transmitter. The system used a wavelength of 6.7 m and the receiver had a half-wave dipole antenna connected between the undercarriage wheels. The transmitter radiated 3–4 μs pulses with a peak power of 30–40 kW from a single dipole antenna on the roof of the Red Tower at Bawdsey Manor, near Felixstowe. Because of the bistatic geometry, the range measurements were only correct when the Heyford was between the ground transmitter and the target aircraft. It became known as the RDF1.5, since the ground air warning system was initially known as RDF1 and the projected airborne system was called RDF2* [Bowen, 1987].

The earliest example of bistatic radar operation with a hostile, non-co-operative transmitter was achieved by the Germans during the Second World War. As illustrated in Fig. 11.2, they deployed a Klein Heidelberg radar at Oostvoorne in Holland in the summer of 1944 and used it as a bistatic receiver to detect aircraft over the southern part of the North Sea, with the British Chain Home radars, operating in the 20–30 MHz band, as non-co-operative illuminators [Hoffman, 1968]. Similar systems were deployed by the German Luftwaffe on the island of Rømø, off the West coast of Denmark [Price, 1978], and on the English Channel [zur Heiden and Klippel, 1979]. They measured the time difference of arrival between the direct signal from the transmitter and the reflected echo from the target; this difference, when combined with the transmitter location and the bearing to the target, defined the location of the target.

The Japanese Army developed a bistatic CW radar and had deployed it by 1940 as their Type A radar (the Type B was a pulsed system). A chain of these radars formed part of the Air Defence System which guarded the Japanese

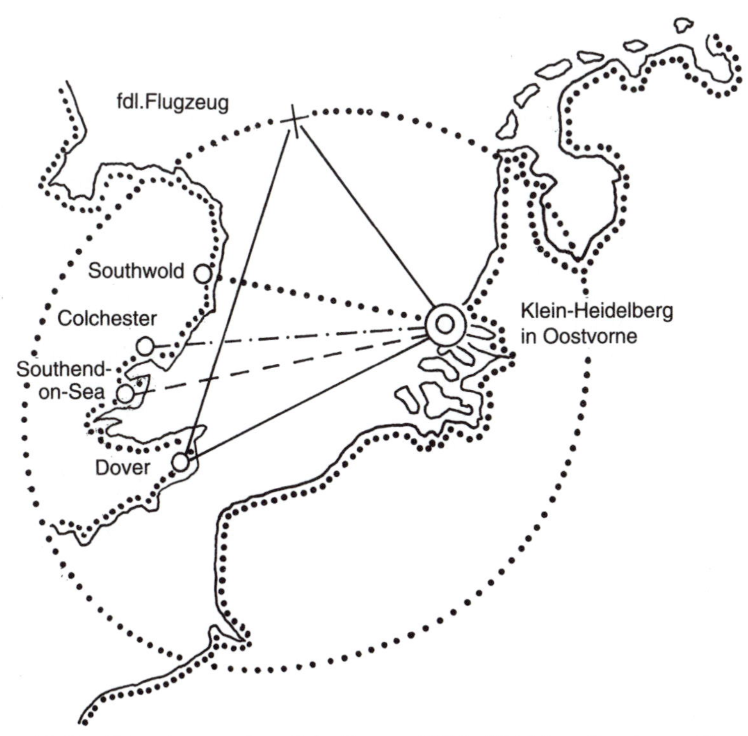

Figure 11.2 *Bistatic operation of the German 'Klein Heidelberg' receiver with a British 'Chain Home' transmitter during the Second World War*

* Radio direction finding (RDF) was the first term used in the UK to describe what later became known as a radar system.

mainland. The Type A sets detected the forward scattered signals from aircraft as they crossed the baseline between a transmitter and a receiver. The lengths of these baselines normally lay in the range from about 40 to 150 miles; the longest Type A baseline lay between Taiwan and Shanghai, a distance of over 400 miles [Nakajima, 1988].

11.1.3 *Deployment options for bistatic radars*

Target detection broadside to the baseline is likely to be one of the preferred modes of operation, since it is especially suited to covert bistatic receivers using the signals from existing air defence surveillance radars. Such radars usually have PRFs that are too low to permit unambiguous and unaliased Doppler measurements on targets in the presence of volume clutter. Higher PRFs could be used, but conventional monostatic operation would be seriously impeded by range ambiguities causing the small echoes from long range targets to be aliased onto the large returns from local ground clutter. Bistatic radars do not suffer from strong clutter returns as do monostatic radars, since the clutter cannot be simultaneously at short ranges from both the transmitter and the receiver. Suitably deployed ground-based systems can avoid ground clutter altogether by using terrain blockage to ensure that no areas of ground will be simultaneously within the lines of sight from both the transmitter and the receiver. These radars can resolve range and velocity ambiguities by geometry, or by PRF agility, or by both.

The detection of targets on or near to the baseline is similar to the operation of some of the first radar systems. The forward scatter RCS of a target is normally larger than the backscatter, or monostatic, RCS and its value is independent of most RCS reduction techniques (this is discussed in more detail in Sections 11.11.1.3 and 11.13.2). However, there are problems in using this enhanced RCS. The target echo will arrive at the receiver at the same time as the direct signal from the transmitter, as well as all the ground clutter echoes from along the baseline. Even if the target can be detected, its position on the baseline and its velocity cannot be determined.

The detection of targets near to an extension of the baseline is potentially more useful. A target's bistatic RCS tends towards its monostatic RCS (they become the same when the target is exactly on the extended baseline) and the target range and velocity measurement techniques are virtually the same as in the monostatic case. This approach allows the operation of transmitters in safe areas, for the benefit of receivers in, or near, hostile territory; typical examples are the US Sanctuary system [Fleming and Willis, 1980] and the Bistatic Alerting and Cueing (BAC) system [Glaser, 1986]. It is also suitable for airborne interceptors using signals from Airborne Early Warning (AEW) aircraft in principle, although conventional AI radars operate at higher frequencies than AEW radars and therefore would not be suitable without modifications.

11.1.4 *Comparison with monostatic radars*

With conventional monostatic radars, the contours for constant target round-trip delay time, for zero target Doppler shift and for constant target detection range, are all spheres centred on the location of the radar. The directions for maximum observed Doppler shift are radial lines.

With bistatic radars, and considering first the two dimensional case where the target locations are restricted to lie on a plane through the baseline, the contours for constant target round-trip delay time and thus for zero target Doppler shift become ellipses with the transmitter and receiver sites as foci. The contours of constant detection range, or, equivalently, of constant received power, become ovals (the ovals of Cassini). The directions for maximum Doppler shift become confocal hyperbolae orthogonal to the ellipses. Unlike the monostatic case, a target travelling along a contour for maximum Doppler shift does not give a direct indication of its true velocity, for the value of the observed Doppler shift will depend on the bistatic angle at the target, namely on the directions to the transmitter and the receiver. (This is discussed in more detail in Section 11.5.6.)

In the three dimensional case, the ellipses become prolate spheroids while the hyperbolae become hyperboloids of two sheets.

11.2 Countermeasures immunity of bistatic radars

In any potential future conflict that involved the intrusion of hostile aircraft, surveillance and tracking radars could need to resist a wide range of countermeasures, especially barrage and deception jamming, chaff, anti-radiation missiles and Stealth* aircraft. In a bistatic radar system, the vulnerability of the passive, covert receiver to all forms of direct attack is removed, or at least markedly reduced. Higher PRFs can be employed to provide the unambiguous Doppler measurement capability needed to discriminate against chaff, while the bistatic geometry should reduce the effectiveness of RCS reduction techniques based on body shaping. The transmitter is, of course, still vulnerable to direct physical attack, but a bistatic receiver can operate with the target illuminations provided by any of several monostatic radars, or by its own dedicated illuminators, which may be fixed or mobile.

11.2.1 *Target detection in chaff*

Chaff is perceived as being a serious threat to the operational effectiveness of conventional surveillance radars and it has been one of the principal reasons for the development of bistatic radars. Not surprisingly, details of its probable RCS densities and its dispersion and drift characteristics are not readily obtainable in the open literature, though the general principles of chaff deployment are well known. Chaff directed against tracking radars is normally deployed in small clouds, to create decoy targets that can break the lock of trackers. Chaff directed against surveillance radars is likely to be spread from aircraft in fairly continuous streams (or corridors), with the aim of preventing the determination of the exact location of any following aircraft. The initial dispersion of the chaff filaments in such corridors is likely to be similar to the spread of the condensation trails left by the jet engines of a high flying aircraft and arises from the turbulence in its wake. However, chaff filaments are heavier than water droplets and they fall more quickly through the atmosphere. If all the filaments

* To be precise, the term Stealth should refer to limiting the susceptibility to all detection techniques but it will often be used here when only one technique is of interest, namely the reduction of the radar cross-section.

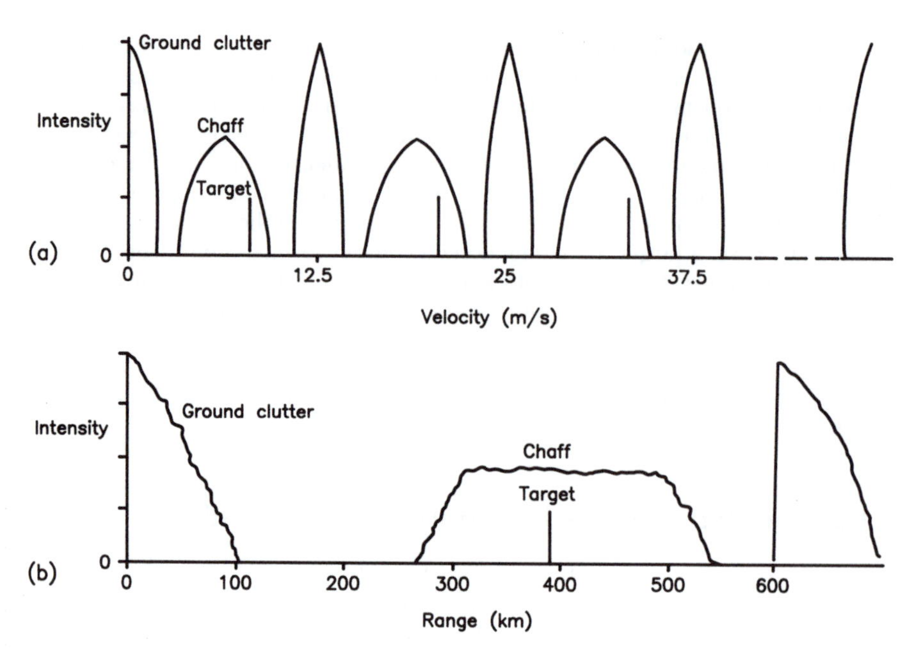

Figure 11.3 *Low PRF (250 Hz) operation of a 3 GHz radar: unambiguous range but highly ambiguous velocity*

were straight and had a uniform distribution of weight along their length, then they might all tend to fall at a uniform rate, probably with a horizontal orientation. This is unlikely to occur in practice. Some filaments will fall slowly with a horizontal orientation, some will spiral down like sycamore seeds, some will fall more quickly with a vertical orientation and some will fall more quickly still owing to the individual filaments not separating correctly (an effect usually known as 'birds-nesting'). The effect of these varying orientations and fall rates is to spread the cloud in height and, probably, make the polarisation of its RCS height dependent as well. Since wind speed varies with altitude, sometimes quite markedly, the filaments can be dispersed over tens of kilometres along the wind directions. The effect on a surveillance radar, especially if it has a limited ability to resolve targets in elevation, is to create the appearance of chaff corridors which are tens of kilometres wide and which contain a spread of velocities that matches the velocities of the wind. Radars intended to detect targets in moving chaff clouds or corridors have to use waveforms that can accommodate clutter rejection notches up to 50 m/s (100 kt) wide.

Bistatic radars permit the use of the higher PRFs that are required for Doppler processing to achieve reliable discrimination of targets that are concealed by chaff – this effect is illustrated in Fig. 11.3*b*. In a conventional low PRF surveillance radar, the Doppler spread of turbulent chaff or rain can be nearly as wide as the frequency range over which the radar can achieve unambiguous Doppler measurements. Target returns are then aliased onto chaff returns even though the two may be travelling at very different velocities – as illustrated in Fig. 11.3*a*. Radars with higher PRFs can achieve unambiguous

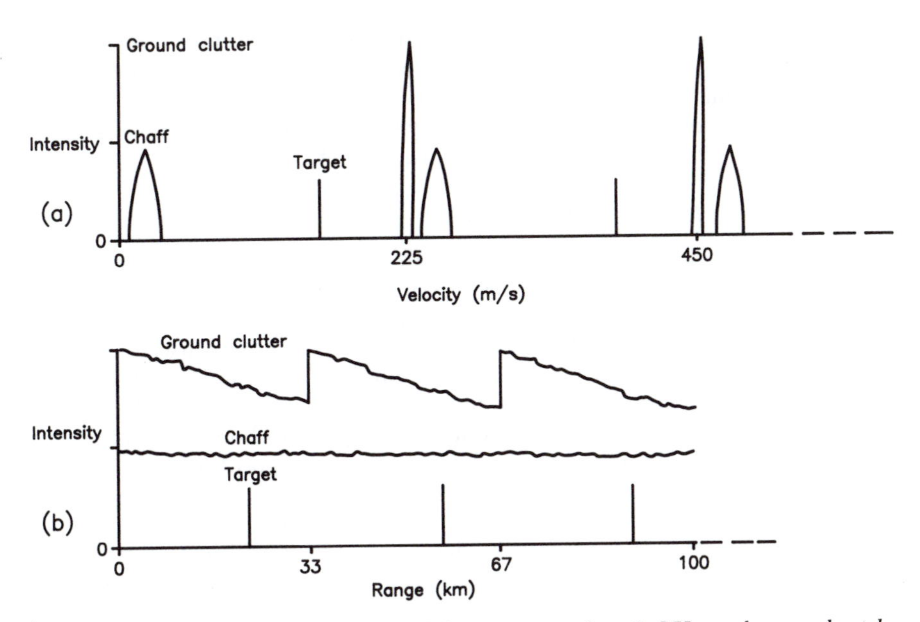

Figure 11.4 *Medium PRF (4.5 kHz) operation of a 3 GHz radar: moderately ambiguous in both range and velocity*

Doppler measurements, as shown in Fig. 11.4*a*, but the signal processor then has to cope with the simultaneous reception of aliased targets and clutter at ambiguous ranges. In the case of surveillance radars looking out at low angles to achieve horizon coverage, not only is the range measurement then highly ambiguous, but the large ground clutter returns at short ranges are superimposed on the aliased target returns, as shown in Fig. 11.4*b*, much as the chaff was superimposed on targets in the velocity domain, when using low PRFs.

The same type of aliasing problem occurs at the higher PRFs with other forms of extended clutter, such as rain, snow and clouds. However, their RCS densities are usually smaller than for ground clutter and they are not concentrated at short ranges. Even with a coherent monostatic radar, it can be very difficult to achieve sufficient sub-clutter visibility to detect all the available targets.

With well-sited ground-based bistatic radars, terrain blockage can screen the receiver from energy reflected from ground illuminated by the transmitter and the problem of ground clutter vanishes. For other bistatic radar deployments, the clutter returns are still smaller than for monostatic radars, since the clutter cannot be at short ranges from both the transmitter and the receiver. The bistatic geometry can also assist in reducing the effects of aliasing with other forms of distributed clutter, since returns are only received from the overlap region between the transmitter and receiver beams, rather than potentially from all ranges as happens with monostatic radars. The geometry aids in the resolution of range ambiguities when using high PRFs, since any ambiguities would correspond to target locations outside the volume defined by the beams' intersection. However, the use of PRF agility may be more expedient as this will allow the use of techniques and hardware already developed for certain monostatic radars. The geometry can then be used to eliminate spurious target detections through the antenna's sidelobes. The transmitter and receiver antenna orientations will still be required by the signal processor for geographic correction, immediately prior to the final display of target data.

11.2.2 *Target detection in active ECM*

Bistatic receivers are obviously not vulnerable to jamming signals that are not incident upon them but are directed towards their transmitters. Even if the receiver's position is revealed to a jammer, it will remain less vulnerable to active ECM than a monostatic receiver. The receiver may use a range of waveforms, with different RFs and modulations, coming from several different transmitters. The receiver antenna can also implement sidelobe blanking or null steering. Such techniques are equally applicable to both monostatic and bistatic radars, so bistatic radars can benefit from the algorithms developed for use in phased array radars.

Responsive deception jammers on aircraft were developed for use against conventional monostatic radars. The deception signals will normally only be emitted during periods when the radar's main beam is illuminating the jammer and they will be delayed by several tens of microseconds after the arrival of each transmitter pulse at the jammer. When these signals reach the receiving antenna, its beam(s) will usually not be aligned towards the jammer (the reason for this is made clear by Section 11.9.1 on antenna scanning for pulse chasing) and the jamming signals will only be able to enter via the antenna's sidelobes; the sidelobe gain will, typically, be 30 to 50 dB less than the main beam gain.

The maximum power available from an airborne jammer is constrained by the size, weight, prime power and cooling provided by the platform on which it is carried. This power could be increased by increasing the efficiency of its microwave power amplifier, but the technology of these is now mature and significant improvements are unlikely. The principal ways of enhancing the effectiveness of the jammer are therefore to concentrate its available power in the directions of its intended target systems and in the parts of the frequency spectrum that they are using. These techniques play into the hands of a bistatic radar as its location and its choice of illuminators are both unlikely to be revealed to the jammers. The use of adaptive pattern control would provide higher antenna gains in the directions of the perceived targets but would reduce the probability of jamming reaching the bistatic receiver solely by chance. The use of such directional beams will have the advantage of reducing EMC problems, especially that of preventing jammers on adjacent aircraft from triggering each other. It may also reduce a jamming aircraft's vulnerability to SAMs with a home-on-jam mode, or to ARMs, when these have to approach the aircraft from directions away from the jammer's main beam.

11.2.3 *Detection of low-RCS targets*

The designers of military systems that are intended to be operated in, or near, hostile territory, now seek to limit their susceptibility to all detection techniques, whether by active or passive forms of radar, infra-red, acoustic, magnetic or other sensors. This concept of achieving minimum detectability in order to enhance survivability is commonly referred to as Stealth. Of the detection methods available, radar techniques are likely to provide the longest ranges against targets that have deliberately minimised all their signatures and emissions. It is not feasible to minimise all the aircraft's signatures simultaneously.

An aircraft cannot be made completely invisible to radar but the initial detection range can be reduced, albeit only in proportion to the fourth root of the aircraft's radar cross section. The longer this initial detection is delayed, the shorter the time for the opposing forces to react. The reduction in radar cross-section confers several operational advantages. The aircraft can now afford to operate in the middle air space where its fuel consumption is minimised and both the airframe and the crew avoid the stresses of low-level flight. The jamming power or clutter level needed to conceal the aircraft are both markedly reduced since they are both directly proportional to the RCS. Thus, for example, the chaff clouds required to achieve range-gate pull-off would be reduced in size by several orders of magnitude. Even if the aircraft is eventually tracked by a terminal threat radar, the RCS may be insufficient to operate the missile's fuse correctly.

The radar energy that is incident upon a target must either be absorbed or re-radiated; of the re-radiated energy, only that which reaches a radar receiver presents any threat to the target. The RCS of an aircraft can be reduced by applying RAM, but this will add weight and increase the aerodynamic drag. RAM is therefore only used in certain critical areas, such as on jet engine inlets and exhausts and on wing edges and wing roots. The body then has to be constructed and shaped so as to minimise diffuse scattering and to control the angular distribution of specular reflections so that the re-radiated energy will be of minimum use to radars.

Two main examples of using body shaping have so far appeared in the open technical literature; both are typified by US aircraft, the B–2 Stealth bomber and the F–117A Stealth fighter.* The outer skin of the F–117A appears to be made entirely in the form of flat surfaces and the normals to these surfaces appear to be constrained to lie in a small number of directions. These flat surfaces will give very large RCSs but these RCSs will only be seen over very small solid angles around the normals. Once the RCS, at any particular aspect angle, is large enough for the aircraft to be detected reliably, increasing this RCS confers little further benefit to the radar. For an aircraft like the F–117A, the most significant monostatic threat radars are likely to be those close to its direction of flight and the shaping of the aircraft's skin should ensure that none of the normals to its flat surfaces falls within a 'cone-of-silence' about this direction.

Bistatic radars and multistatic radars should achieve better detection performance against such aircraft than will monostatic radars, since they can be deployed to benefit from the energy reflected away from the cone-of-silence. Of course, a similar advantage may accrue to monostatic radars which observe the aircraft at aspect angles outside the cone-of-silence. There is also no guarantee that energy redirected from the cone-of-silence aspect angles will go anywhere useful to bistatic (or multistatic) radars. If the Stealth aircraft do indeed choose

* Although the F–117A is normally called the Stealth fighter, it appears to have been conceived as a strike platform for low intensity warfare and its main operational role in the Gulf War was for covert strike by night.

to operate in the middle air space, then the terrain blockage problems with bistatic radars, discussed in Section 11.4.3, will be markedly reduced.

Active Stealth techniques will also be countered by bistatic radars, since the correct nulling signal cannot be derived for an unknown receiver location. Actually, active Stealth is probably not feasible when the radar wavelength is much less than the typical target dimensions, owing to the rapid variations of RCS with target aspect angle.

The forward scatter RCS, that is observed when a target crosses the baseline between the transmitter and receiver, can be much larger than the monostatic or back-scatter RCS and is not reduced by body shaping or RAM coating. However, as noted in Section 11.1.3, its practical use is limited.

11.2.4 *Covert operation of the bistatic receiver*

The operation of bistatic radar receivers should be completely covert, as they do not need to radiate electromagnetic energy that can be detected by hostile ESM systems. The system design must avoid any unintentional radiation, for example, owing to local oscillator leakage through the down-converter and into the antenna. Communication channels from the receiver will have to use cables or low-probability-of-intercept (LPI) microwave links. The mobile receivers must also use passive techniques to determine their positions, such as pre-surveyed sites or GPS–NavStar receivers. Since they are small and mobile, they can avoid being located by other means.

11.2.5 *Protection of transmitters from ARMs and direct attack*

The transmitter site in a bistatic system advertises its presence and so is vulnerable to ARMs and other forms of direct attack. Even if a transmitter is put off the air, either temporarily, while an ARM is reduced, or even permanently, owing to its destruction, the bistatic receiver can change to operating with another transmitter. Alternatively, there are several techniques for protecting transmitters, including:

(i) the use of LPI waveforms
(ii) the use of several netted transmitters, with switching between them
(iii) the use of decoy transmitters
(iv) the use of small, mobile transmitters.

Bistatic radars are particularly suitable for LPI operation, using waveforms with very high duty cycles or even pseudo-noise modulated CW. However, such waveforms usually have ‘thumb tack’ ambiguity functions and are thus intolerant to Doppler shifts.

Small, mobile transmitters are, of course, likely to provide lower peak powers than large fixed transmitters, since the high peak power tubes are physically larger, operate at higher voltages and may require screening against X-rays. Unless the average powers of the fixed and mobile transmitters can be made comparable by the use of high duty cycle waveforms in the mobile transmitter, the maximum bistatic detection ranges would be reduced and the system might need to operate with shorter baselines.

11.2.6 *Effectiveness of ESM, RWRs and decoys*

The locations of non-emitting bistatic receivers cannot be determined by ESM techniques. Even if their locations are determined by some other means, their vulnerability to jamming is still less than that of a comparable monostatic system since ESM techniques cannot determine the antenna patterns used by the receiver. In contrast, nearly all monostatic radars advertise the patterns that they use for both transmission and reception.

Radar warning receivers will be less effective in warning of a detection risk, since the signal that would need to be incident on a target to permit its detection by a distant monostatic radar could be much greater than that needed to permit its detection by a covert, adjacent bistatic receiver.

The RCSs from trihedrals decrease rapidly with bistatic angle, as described in Section 11.11.1. Corner reflectors (trihedrals) therefore cannot reflect large echoes towards bistatic receivers, so small targets cannot easily be given artificially large RCSs to act as decoys.

11.3 Design and operational advantages

11.3.1 *Design advantages*

11.3.1.1 *Low dynamic range requirements*

There are design and operational advantages to add to the ECCM advantages, arising from the separation of the transmitter and the receiver. The dynamic range of the reflected signals, particularly clutter, collected by a bistatic receiver is much lower than in the monostatic case. This is because no reflector can be simultaneously close to both the transmitter and the receiver. Consequently, the receivers do not require duplexers or protection circuits and there is no dead zone near to the receiver, owing to the recovery time of the protection circuits. Since the transmitter waveforms are not constrained by the need for receiver protection and receiver look-through periods, they can be in the form of long coherent bursts, or even CW, as indicated in Section 11.2.4. These higher duty cycles will allow lower peak transmitter powers to be used.

The use of higher duty cycle waveforms will be particularly significant for active phased array antennas, where each radiating element will have its own solid-state RF amplifier. The use of these active phased array antennas may present some problems, especially in a military environment, that would not arise when using a single transmitter, usually a tube, and a corporate feed network to the radiating elements. The face of the active phased array antenna will have a raised temperature and will present an enhanced IR signature. It may be feasible to limit the consequences, by using coatings that limit the IR emissions that occur within the atmospheric IR transmission windows. Placing the active elements on the array face also makes them more vulnerable to electromagnetic pulses and to RF weapons.

11.3.1.2 *System reliability and safety*

If a bistatic receiver is operated as an adjunct to a monostatic radar, the overall system reliability/availability will be greater than for the monostatic radar on its own. The bistatic radar can provide target data even when the receiver on the monostatic radar is out of action, whether this is a consequence of system malfunctions or simply owing to routine maintenance. Such reliability could be of major importance for air traffic control safety.

The phased array antenna required in a bistatic receiver can often provide greater reliability than a reflector antenna with a single downconverter. This arises from the phased array antenna using many separate receiving elements (or, in some cases, sub-arrays), each with its own downconverter. The complete array will exhibit graceful degradation if any components or subsystems fail, though some of this graceful degradation may be offset by the extra reliability problems introduced by the 'pulse chasing' controller.

Bistatic operation can avoid some of the safety problems that arise from the operation of a conventional monostatic radar, since the transmitter can be deployed in areas where its emissions do not present any radiation hazard, whether to personnel or to other electronic systems.

11.3.2 *Operational advantages*

11.3.2.1 *Supplementary roles for ESM and jammer location*

Since bistatic receivers are designed to make passive use of signals radiated by other radars, it will naturally be efficient to make use of their ability to receive and process any other incident signals that lie within their operating bandwidths, as will be discussed in Section 11.7.2. These bandwidths will only be about 10–12%, or possibly up to 15% with poorer antenna patterns, since bistatic receivers have to use phased array antennas. The available signals are therefore likely to come from other surveillance radars and from jammers that are attempting to suppress them.

As Milne [1977] indicates for multistatic radars, triangulation can be used to locate sources of repetitive pulses; passive detection techniques, employing correlation and triangulation, can be used to locate noise-like emitters. The principal difficulties with such passive systems lie with the need for co-operation between receiving sites (in particular, the need for wideband data links to enable correlation or interferometric methods to be used) and the problems of co-ordinate conversion arising from the hyperbolic contours of constant time difference between the same signals when received at two different sites.

The location of emitters may be achieved by simple triangulation techniques, but, with only two receivers, the number of 'ghosts' would increase rapidly with the number of emitters of each type. The addition of further receivers, or collateral data on target locations, can help to eliminate these ghosts. Correlation detection techniques, involving the integration of the cross-products of signals received at two or more sites, can provide an alternative method of resolving triangulation ambiguities with noise-like sources, provided they are not radiating directionally decorrelated beams. However, these techniques can involve the use of wideband data links to carry the signals between sites. The sensitivity will depend on the ratio of the noise from the emitter in the common volume to the noise from other emitters in the individual beams; it will also depend upon the product of the RF bandwidth and the observation time. The optimum value of this time will be limited by the correlation time of the noise and by the time the emitter is in the common volume of the receiver beams; this could typically be one second for high speed airborne targets.

11.3.2.2 *Extending the life of conventional radars*
Old air defence radars, and most air traffic control radars, will have little resistance to modern ECM techniques. However, if they are adapted to incorporate the synchronisation capabilities described in the next Section, they will be able to provide co-operative illumination of targets for bistatic receivers. This will extend the useful life of such conventional radars and it will add to the range of radars that have to be countered by approaching jammers. Their lower value will allow them to continue to be used at times when the ARM threat might have forced a high value radar temporarily to switch off. These older radars, with reflector antennas, fed from travelling wave tubes or other high power tubes, would also be more tolerant of ARM strikes than would a modern phased array radar, with active devices on, or near, its array face.

Even if these older radars are not coherent, the geometrically dependent advantages of bistatic operation will still be available – for example, the avoidance of directional jamming of the receiver, the reduction of ground clutter and, possibly, the limiting of the effectiveness of RCS reduction techniques.

11.4 Design challenges and operational problems

The penalty that has to be paid for the ECM and other advantages of bistatic radars lies in the increased technical complexity of the radar. With recent advances in phased array antennas, microwave signal generators and satellite navigation, this penalty is now heavily outweighed by the advantages of bistatic operation in most military scenarios.

11.4.1 *Synchronisation aspects*

The receiver has to be able to reproduce the timings and frequencies of the transmitter waveforms before it can measure accurately the time delays and Doppler shifts of target echoes. It also needs knowledge of the transmitter's antenna orientation and, in the case of a mobile platform such as an aircraft, its location and motion. In principle, the receiver could be provided directly with a sample of the transmitter's waveform, either by line-of-sight transmission if one of the platforms were airborne, or by land line or troposcatter if both sites were ground-based. In practice, more reliable and accurate operation can be achieved by sending a description of the transmitter's frequency and modulation down a narrow band data link, along with the orientation, location and motion data, and then recreating the transmitter reference signal at the receiver site. If the transmitter uses prescribed RF and PRF sequences, then the description of the transmitter waveforms requires very low data rates and the bandwidth of the data link is determined by the time synchronisation procedure and by the data rates needed to follow changes in the antenna's orientation.

11.4.1.1 *Transmitter frequencies*
The transmitter frequency and the local oscillator frequency in the receiver both have to be sufficiently accurate to allow the RF bandwidth to be converted down to the receiver's IF bandwidth with an error less than about 10% of the frequency resolution of the radar. For a system using incoherent detection, this

implies a frequency accuracy of about one part in 10^4 (for example, to 0.3 MHz in 3 GHz). Errors greater than this would have to be compensated by increasing the receiver's bandwidth, with a consequent reduction in the radar's detection range.

If the radar is using coherent detection and Doppler processing, then the radar's frequency resolution is the Doppler channel bandwidth and the transmitter and receiver frequencies have to be stable to about one part in 10^9 (for example, to 3 Hz in 3 GHz); this stability has to be maintained during the radar's coherent processing period of, typically, 100 ms. The choices for the frequency standards lie between caesium atomic standards (stable to one part in 10^{13}), rubidium standards (stable to one part in 10^{11}) or crystal oscillators (typically, stable to one part in 10^7). The levels of frequency stability and coherence that are required for Doppler analysis mean that the RF reference signals, at both the transmitter and receiver, have to be generated by frequency synthesisers, locked to rubidium atomic standards that have long-term stabilities of the order of 1 part in 10^{11} per month.

11.4.1.2 *Pulse timing*

The pulse timing presents greater difficulties; errors in the time reference are cumulative, since it is derived by counting pulses from the atomic standard. Even with a rubidium standard, errors of $\pm 5\ \mu s$ per day are typical. This is comparable with the accuracy that can be achieved by using the reception of direct breakthrough signals from the transmitter, due to troposcatter or diffraction. This error has to be removed by regularly resetting the timing clocks, at both transmitter and receiver, using an external reference signal, such as those from Loran C, GPS–NavStar or Glonass (the Russian equivalent of the US NavStar), with appropriate allowances for the geographic separation of the two sites. This external reference signal requires some distinct modulation that repeats at a sufficiently low rate for the data link to be able to co-ordinate the resetting of the clocks at the two sites.

The range cell size is usually the same as the radar's (compressed) pulse length; for a surveillance radar this would typically lie in the range 0.1 to 1.0 μs. A slowly varying timing error equivalent to several range cells will be tolerable but a pulse-to-pulse timing stability equivalent to less than 10% of a range cell width, over the duration of each coherent pulse burst, will be required by the Doppler processor. A timing error of 1 μs per day would require a long term frequency stability of one part in 10^9. A pulse-to-pulse stability of 10 ns, with 1 kHz PRF, would require a short term stability of one part in 10^5, over each coherent processing period of 100 ms.

11.4.1.3 *Antenna orientations*

The transmitter antenna's azimuth must be supplied to the receiver with sufficient accuracy for its antenna to achieve pulse chasing without appreciable loss of scattered energy. Typical surveillance radars have rotating reflector antennas with shaft encoders which define the antenna azimuths with a resolution of 12 bits. If they complete each rotation in about 15 s, sending the full 12 bit word for each of the resulting 4096 discrete azimuths would imply a data rate of 3277 bit/s. For a typical PRF of 275 Hz, the number of discrete azimuths is nearly the same as the number of pulses per revolution of the antenna. The data rates for transferring the transmitter's antenna azimuth can be reduced by sending only the least significant bit (LSB), or a few of the LSBs, and summing the number of the transitions in the LSB(s) at the receiver. This technique can give steadily increasing errors if the antenna azimuth does not increase continuously but exhibits occasional small decreases owing, for example, to wind effects on a reflector antenna. These errors can be limited by incorporating resynchronisation facilities that allow the estimate of the azimuth angle to be reset at the end of each rotation.

11.4.1.4 *Platform locations and motions*

The locations of both the transmitter and receiver sites (or platforms) need to be known in three dimensions before a measured echo delay time can define a prolate spheroid on which the target must lie. Similarly, the velocities of moving platforms have to be resolved in three dimensions before a measured echo Doppler shift can be converted into an apparent target velocity. A GPS–NavStar receiver and an inertial navigation system (INS) will be able to complement each other in providing accurate locations and velocities of moving platforms. GPS will be able to provide position updates to the INS at about 1 s intervals, while the INS will be able to provide the GPS receiver with the velocity information it requires to correct for the effects of platform motion. GPS–NavStar will also complement maps when ground-based systems are deployed to sites that have not been surveyed. Indeed, the location accuracies achieved with GPS-NavStar can exceed those achieved with maps. NavStar will be able to provide latitude, longitude and height.

11.4.2 *Multibeam electronically scanned antenna*

A phased array antenna is required to provide the set of electronically controlled beams that can follow the directions from which echoes may appear, as they are reflected from targets; with medium PRF waveforms, several such pulses may be in the antenna's field of view simultaneously and each will require its own independent beam to chase it. An equal number of signal processing channels is required to process all the available target data. If the deployment geometry and the antenna beamwidths do not allow range ambiguities to be resolved by geometrical techniques (as discussed in Section 11.7.1), then these signal processing channels will have to resolve the range and velocity ambiguities that are a consequence of using medium PRF waveforms.

The need for 'pulse chasing' requires a phased array design that does not introduce amplitude or phase modulation of the successive echoes from each resolution cell, since these echoes would normally be Fourier transformed to provide Doppler information. Scanning or switching of the antenna beams should not present any problems to the signal processors unless the phase control settings change between successive echoes from a target – for example, owing to non-repeatability of the antenna or to targets moving with significant components of tangential motion.

One choice of antenna design is a phased array that provides a set of contiguous beams covering the whole sector of interest. The alternative is a phased array that creates scanning beams that adapt to match the size of the resolution cells occupied by the uncompressed radar pulses; the widths of such beams are derived in Section 11.9.1.

11.4.3 *Terrain blockage effects*

Terrain blockage effects can remove ground clutter problems for bistatic radars whose transmitter and receiver antennas are sited so that they cannot observe common areas of ground. However, both the bistatic transmitter and receiver require a clear line of sight to the target and this limits the effectiveness of bistatic radars against low-level targets. The extent of this problem is assessed in Section 11.5.5 for a theoretical smooth Earth. The problem will be exacerbated in the real world if the sites are not properly assessed to allow for terrain blockage, before the transmitter and the receiver are deployed. Mobile radars could have repertoires of pre-surveyed sites; alternatively, they could carry their own digital land maps in their system control computers. Nowadays, precise deployment can be achieved at the sites using GPS–NavStar receivers.

A further consequence of this terrain blockage is that it can prevent direct line-of-sight communications between the transmitter and receiver sites, so forcing them to resort to tropospheric scatter, diffraction from crests, satellites, radio relays or land lines. All of these can introduce indefinite time delays that depend on the route followed by the data; hence the need for the synchronisation techniques described in Section 11.4.1.

11.5 Basic formulas of bistatic radar performance

11.5.1 *Approximations and assumptions*

The aim of this review is to explain the major concepts and design principles of bistatic radars without involving complicated equations. The following assumptions have therefore been made:

(i) The transmitter and the receiver are ground-based and stationary.
(ii) The transmitter, receiver and target (scatterer) can be treated as lying in the same plane, with linear propagation between them.
(iii) Since propagation is two dimensional within this plane, the orientation of each antenna can be defined by a single (azimuth) angle.
(iv) The antenna azimuths will be defined in the mathematician's convention, with angles measured anticlockwise from the x-axis.
(v) Multipath effects are neglected.

The consequences of using mobile (airborne) platforms are discussed in Section 11.12. The second assumption requires a transformation to a 4/3 Earth model; even this is not exactly true in practice, owing to atmospheric effects, as discussed in Section 11.4.3, but these effects are small and would only be significant in the final stages of the design or when actually operating the radar. A bistatic radar will probably be as susceptible to multipath effects as a monostatic one, though with different effects occurring on the transmitter–target and on the target–receiver paths.

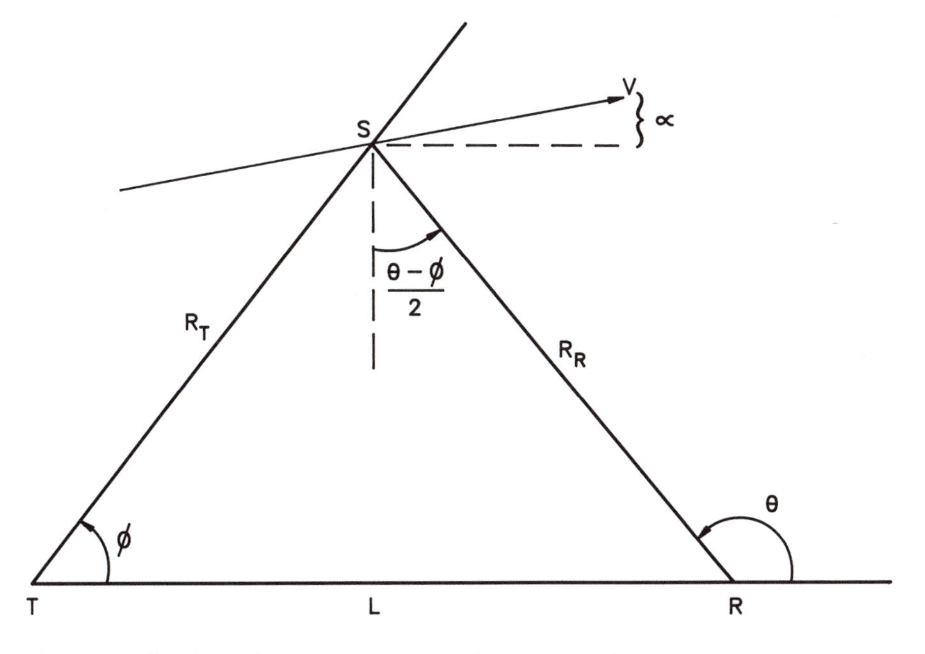

Figure 11.5 *Simplified co-ordinate system for bistatic radar analysis*

11.5.2 *Co-ordinate system and notation*

The simplified co-ordinate system is illustrated in Fig. 11.5 and the following notation has been adopted:

Angles (rad)
ϕ = transmitter antenna azimuth
θ = receiver antenna azimuth
β = bistatic angle subtended at the target
$= \theta - \phi$
α = target heading

Lengths (m)
L = baseline length
R = path lengths, in bistatic calculations
λ = radar wavelength
r = ranges normalised as $2R/L$

Times (s)
t = time since a pulse left the transmitter
τ = pulse length
T = pulse repetition interval = 1/PRF

Velocities (m/s)
V = target velocity
c = velocity of light

Subscripts
M = monostatic case
B = bistatic case
T = transmitter
R = receiver
S = scatterer (target)

11.5.3 *Detection ranges*

The bistatic version of the radar equation can readily be derived (for example, Skolnik [1980]) as

$$P_R = \frac{P_T G_T G_R \lambda^2 \sigma_B}{(4\pi)^3 R_T^2 R_R^2 L_{PT} L_{PR} L_S} \tag{11.1}$$

where P_R = received signal power, W
P_T = transmitter power, W
G_T = transmitter antenna gain in the direction of the target
G_R = receiver antenna gain in the direction of the target
σ_B = bistatic radar cross-section, m^2
R_T = transmitter-to-target path length, m
R_R = receiver-to-target path length, m
L_{PT} = propagation loss over the transmitter-to-target path length
L_{PR} = propagation loss over the receiver-to-target path length
L_S = system losses
λ = radar wavelength, m

This equation can be converted into the corresponding monostatic form by substituting

$$\begin{aligned} G^2 &= G_T G_R \\ R^4 &= R_T^2 R_R^2 \\ \sigma_M &= \sigma_B \\ L_P^2 &= L_{PT} L_{PR} \end{aligned} \tag{11.2}$$

Treating all the terms on the right-hand side of the bistatic equation, apart from the radar cross-section and the path lengths, as constants, and substituting

$$K = \frac{P_T G_T G_R \lambda^2}{(4\pi)^3 L_{PT} L_{PR} L_S} \tag{11.3}$$

reduces the bistatic radar equation to

$$R_T^2 R_R^2 = \frac{K \sigma_B}{P_R} \tag{11.4}$$

This equation can then be written in normalised form with

$$\begin{aligned} r_T &= \frac{2R_T}{L} \\ r_R &= \frac{2R_R}{L} \end{aligned} \tag{11.5}$$

and

$$C = \frac{2}{L}\left(\frac{K\sigma_B}{P_R}\right)^{1/4}$$

as

$$r_T r_R = C^2 \tag{11.6}$$

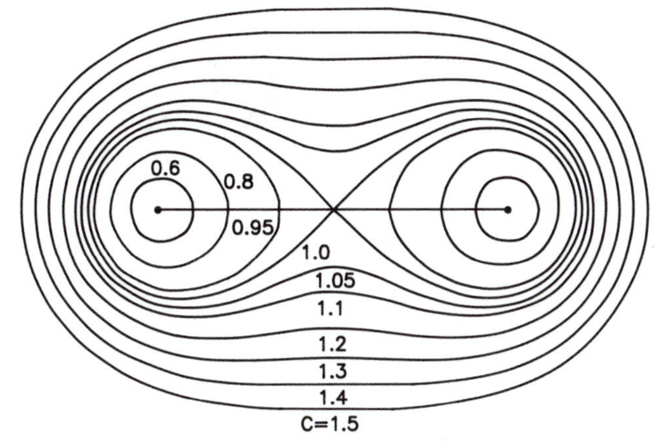

Figure 11.6 *Ovals of Cassini: contours of constant target detectability*

which represents the ovals of Cassini. Examples of the curves defined by this equation are shown in Fig. 11.6. These curves are contours of constant sensitivity or, equivalently, of constant target detection capability. In the limit when $C \gg 1$, these contours tend towards the circles that would be obtained for a monostatic radar. However, when $C \leqslant 1$ the situation is very different from the monostatic case and the curves split into two separate contours round the transmitter and the receiver. This means that a target could be detected in the vicinity of the bistatic receiver even when it was well outside the detection range of an equivalent monostatic radar located at the transmitter site.

11.5.4 *Iso-delay contours*

The iso-delay contours for bistatic radars have constant range sums to the transmitter and receiver sites and are consequently ellipses (prolate spheroids in the three dimensional case) with these sites as foci. At a time t after a pulse of length τ was emitted from the transmitter, the area from which scattered energy could reach the receiver lies between the contours for time t and for time $t+\tau$, as illustrated in Fig. 11.7; of course, only the part of this area that lies within the transmitter antenna's beamwidth would actually provide echoes.

The internal bisector of the bistatic angle β is known, from analytic geometry, to be perpendicular to the ellipse while its external bisector is tangential to the ellipse. When $\beta<\pi$, the separation δ of the elliptical contours, measured along the perpendicular, is then

$$\delta = \frac{c\tau}{2} \sec\left(\frac{\beta}{2}\right) \tag{11.7}$$

11.5.5 *Line-of-sight constraints*

Except for the small number of radars that use ground wave propagation or rely on reflection from the ionosphere, the volume within which a monostatic radar is able to detect targets is restricted by the need to have a line of sight from the

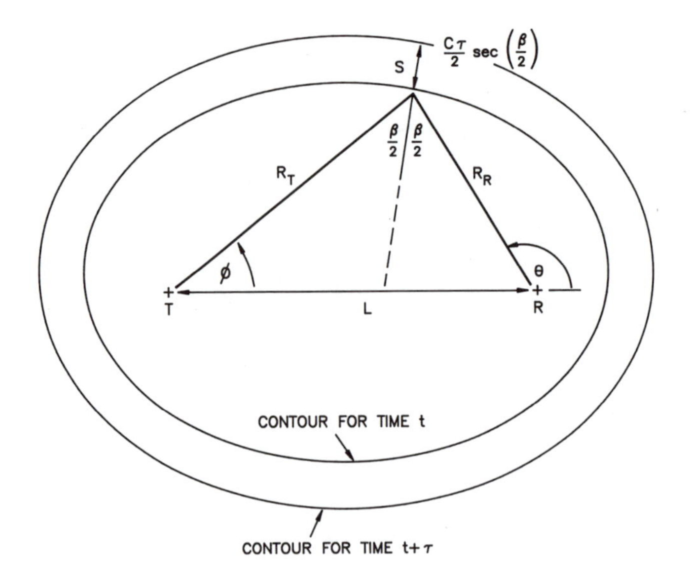

Figure 11.7 *Iso-delay contours for a bistatic radar*

radar to the target. Bistatic radars are even more seriously affected since, to be detected, a target must be simultaneously visible from both sites.

With the simple 4/3 Earth propagation model, the maximum range R_M, at which a target at a height H_S can be seen from a monostatic radar with an antenna at a height H_T is given by

$$R_M = 130(\sqrt{H_S} + \sqrt{H_R}) \tag{11.8}$$

where all the dimensions are in kilometres; the target and radar heights are both assumed to be much less than the Earth's radius.

The surface area A_M over which a target at a given height can be seen by a monostatic radar is simply a circle centred on the radar and of radius R_M, as defined by eqn. 11.8, i.e.

$$A_M = \pi R_M^2 \tag{11.9}$$

The corresponding surface area A_B, over which the same target could be observed by a bistatic radar is defined approximately by the overlap between two circles, centred on the transmitter and the receiver, as illustrated in Fig. 11.8. The area of this overlap region can be shown to be [Holmes, 1988].

$$A_B = \frac{\pi}{2}\left(R_T^2 + R_R^2\right) - R_T^2(\sin^{-1}\mu + \sqrt{1-\mu^2}) - R_R^2(\sin^{-1}\nu + \sqrt{1-\nu^2}) \tag{11.10}$$

where

$$\mu = \frac{1}{2R_T L}\,(L^2 + R_T^2 - R_R^2)$$

$$\nu = \frac{1}{2R_R L}\,(L^2 - R_T^2 + R_R^2)$$

with L = baseline length
R_T = maximum slant range of target visibility from the transmitter
R_R = maximum slant range of target visibility from the receiver

Eqn. 11.10 applies for the general case where $R_T + R_R \geqslant L$. There are four special cases:

(i) $L=0,\ R_T=R_R=R$	monostatic case	$A_B=\pi R^2$
(ii) $R_T+R_R \leqslant L$	no overlap	$A_B=0$
(iii) $R_T-R_R \geqslant L$	receiver centred case	$A_B=\pi R_R^2$
(iv) $R_R-R_T \geqslant L$	transmitter centred case	$A_B=\pi R_T^2$

The ratio of bistatic to monostatic target visibility has been plotted in Fig. 11.9 as a function of baseline length, for a range of target heights H_S. The transmitter and receiver antennas have both been assumed to be at zero height above ground level. This Figure can be used to assess the trade-off between minimum target altitude and maximum baseline length, for any required target detection range. It indicates that differences in coverage between bistatic and monostatic radars decrease with increasing target altitude and with decreasing baseline length.

Fig. 11.9 illustrates the extent of the disadvantage of bistatic radars, relative to monostatic radars, in detecting low-flying targets. However, military aircraft only operate at low altitudes when this is essential for their survival. Whenever possible, they operate at medium to high altitudes to achieve the fuel economy which will extend their range and/or increase their payload. The line-of-sight

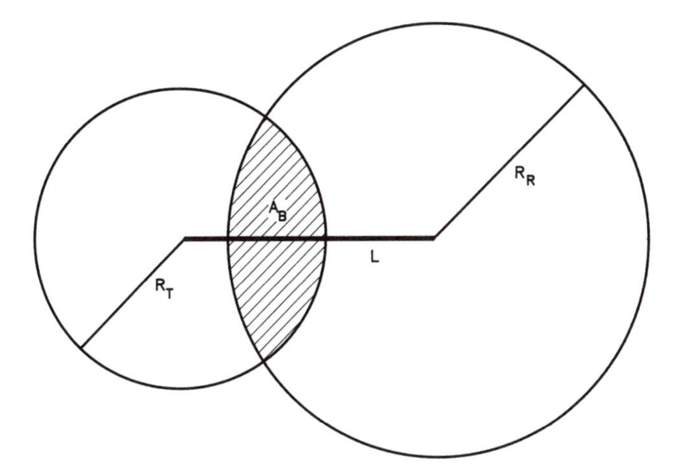

Figure 11.8 *Area of target visibility by a bistatic radar*

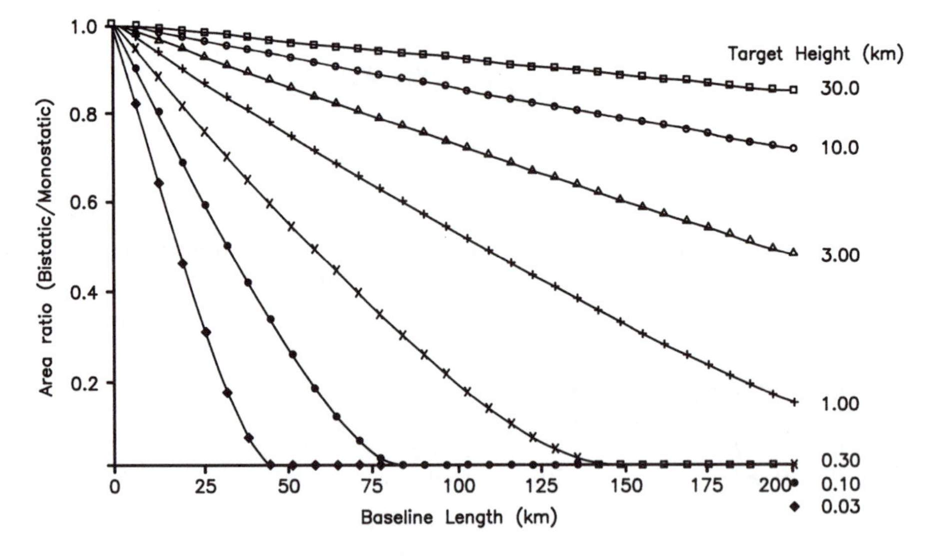

Figure 11.9 *Ratio of bistatic to monostatic target visibility as a function of baseline length and target height*

problem may become less significant with the advent of stealthy aircraft, if they take advantage of their low RCSs and operate at higher altitudes. The removal of the line-of-sight problem may only be of academic interest if the stealthy aircraft still cannot be detected, but that is a problem to be reviewed in Section 11.11.1.

The combined effects of the contours of constant target detection capability (described in Section 11.5.3) and the line-of-sight constraints on system performance are illustrated in Fig. 11.10, for a bistatic system deployed in the UK. This Figure assesses operation with the Byson radar transmitter at the Defence Research Agency at Great Malvern and the BEARS bistatic receiver at the GEC–Marconi Research Centre at Great Baddow (the reason for choosing these systems and locations will become clear when the UK Argus bistatic radar programme is described in Section 11.13).

The line-of-sight contours (thin, light lines) are those for the Byson radar at the DRA, when using its two lowest beams (0–4° elevation cover), while the BEARS antenna at Great Baddow uses its fixed elevation cover with its 90° azimuth scan range. The detection contour (the oval shown as a broad, dark line) would apply for a constant target RCS of 10 m^2, independent of bistatic angle, with both the transmitter and receiver antennas having full 360° coverage, with constant gain.

In practice, a full 360° coverage by the receiver would require a minimum of three phased array antennas, each covering a 120° sector, though better performance would be achieved with four antennas, each covering a 90° sector. The gain of each phased array antenna would also have a secant dependence on the angle of its boresight. The Byson antenna does not operate with cover to its West since this would be screened by the Malvern Hills.

11.5.6 *Doppler frequencies*

The Doppler shift f_D is determined by the time rate of change of the ratio of the signal path length to the radar wavelength λ; thus the Doppler shift for the target S in Figure (11.5) will be

$$f_D = -\frac{1}{\lambda}\frac{d}{dt}(R_T + R_R) \tag{11.11}$$

where R_T and R_R are, respectively, the ranges from the transmitter and the receiver to the target.

If the target S moves with velocity $V (V \ll c)$ in the plane containing the transmitter and the receiver, at an azimuth angle α relative to the baseline, then

$$f_D = -\frac{V}{\lambda}(\cos(\theta - \alpha) + \cos(\phi - \alpha))$$

$$= -\frac{2V}{\lambda}\cos\left(\alpha - \frac{\theta + \phi}{2}\right)\cos\left(\frac{\theta - \phi}{2}\right)$$

This expression for the Doppler shift consists of three parts, namely

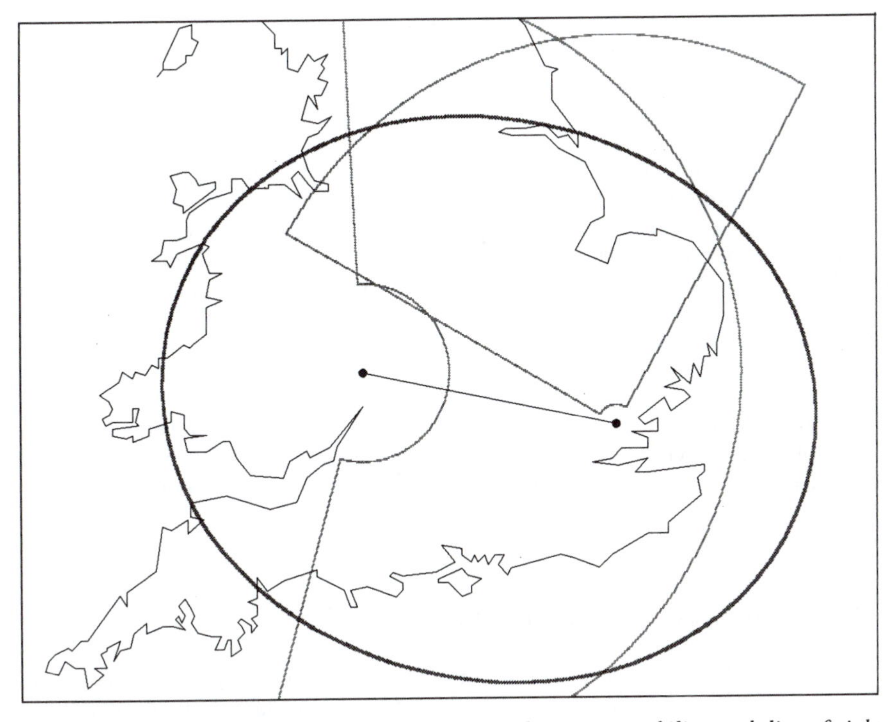

Figure 11.10 *Combined effects of the target detection capability and line-of-sight constraints for a bistatic radar deployed in the UK*

$\frac{2V}{\lambda}$ the maximum Doppler frequency that can be attained

$\cos\left(\alpha - \frac{\theta+\phi}{2}\right)$ which shows the dependence on the target's trajectory

$\cos\left(\frac{\theta-\phi}{2}\right)$ which shows the dependence on the bistatic angle

The second term will be zero when

$$\alpha = \frac{\theta+\phi}{2} \pm \frac{\pi}{2} \tag{11.12}$$

namely when the target moves along the external bisector of the bistatic angle, i.e. when it moves along the elliptical, iso-delay contour defined by $R_T + R_R =$ constant (see Section 11.5.4). This result also follows directly from integration of eqn. 11.11. The second term will have extrema of ± 1 when

$$\alpha = \frac{\theta+\phi}{2} \text{ or } \frac{\theta+\phi}{2} \pm \pi \tag{11.13}$$

namely when the target moves along the internal bisector of the bistatic angle, i.e. along the hyperbolae defined by

$$R_T + R_R = \text{constant}$$

that are orthogonal to the iso-delay ellipses.

The third term will have extrema of ± 1 when

$$\theta = \phi \tag{11.14}$$

This indicates that the full Doppler shift can only be attained by targets on the extended baseline or for targets at ranges much greater than the length of the baseline. Thus a target moving along a hyperbolic path does not normally produce the maximum Doppler shift. This contrasts with the monostatic case where a target on a radial trajectory always produces the maximum Doppler shift for its velocity.

The third term will be zero when

$$\theta = \phi + \pi \tag{11.15}$$

This corresponds to targets lying on the baseline. The second term will also be zero for targets lying on the baseline, since this is the limiting case of the elliptical iso-delay contour, when it degenerates into a line defined by

$$R_T + R_R = L$$

11.6 Atmospheric propagation effects

This Section briefly reviews those atmospheric propagation effects that could have significant influence on the operation of bistatic radars; it does not attempt to provide a comprehensive survey of all the possible characteristics of electromagnetic wave propagation in the atmosphere.

Atmospheric propagation effects can be more significant for bistatic radars than for monostatic radars, owing to the need for the receiver antenna to follow the directions from which target echoes may appear. For a bistatic radar receiver, an error in its pointing direction of more than one antenna beamwidth, in either azimuth or elevation, would lead to any target echoes being missed during such pulse chasing. For a multistatic radar, when three or more receivers are being used to give accurate locations of targets in three dimensions (as described in Section 11.15.5) precise allowance has to be made for the curvatures and delays of the paths.

During its transit from the transmitter antenna to the target and then back to the receiver antenna, a radar signal will pass through regions of the atmosphere with varying refractive indices n. Consequently, it will follow curved paths that can, in theory, be derived from Fermat's principle (namely that the path followed by the signal will be the one which exhibits the minimum propagation time). The curvature of the path means that the range and bearing of a target, as seen by the radar, are not usually those that would follow from simple geometry. Also, the Doppler shift observed by the radar depends on the velocity of the target with respect to the path of the radar signal and on the local wavelength of the radar signal; this wavelength is a function of the refractive index of the medium through which the target is travelling. The medium is unlikely to produce significant variations in the Doppler shift unless, for example, the target is travelling so rapidly that it is surrounded by a sheath of ionised gases.

The errors in the location of targets would be unlikely to be sufficiently large to cause significant problems for a monostatic surveillance radar. The transmit and receive paths for a monostatic radar will be identical, even under conditions of anomalous propagation, except when operating at VHF, or below, with a space surveillance radar which has propagation through the ionosphere. Electromagnetic waves at frequencies below the VHF band are reflected by the ionosphere; those above VHF do not have any significant interaction with the electrons in the ionosphere. Even in the VHF case, the differences in the transmit and receive angles are typically much less than an antenna bandwidth. There is consequently no difficulty in ensuring that a monostatic receiver collects all the energy that targets reflect back in the direction of its transmitter.

If both the transmitter and the receiver antennas have narrow beams, then it may not always be feasible for a bistatic receiver to chase the reflections from its transmitter's pulses with adequate accuracy. This will be especially true for a ground-based system under anomalous propagation (ducting) conditions, since the paths of the two beams cannot then be predicted accurately. The implications on the antenna design are that the system would need a receive antenna with a relatively wide beam in elevation. Correction for serious bearing errors probably could not be achieved for simple predictive pulse chasing (as described in Section 11.9.1), but could require a monopulse receiver antenna with error correction feedback to the pulse chasing computer. Anomalous propagation (anaprop) and other atmospheric propagation effects are significant for ground-based radars and even more significant for shipborne radars. Their relevance to airborne radars depends on the altitudes of these radars and their targets – tropospheric propagation effects are not readily discernable above about 10 km.

The resultant errors in pulse chasing or target location are of two types, namely (i) systematic errors that can, in principle, be corrected by calibration of the atmosphere and (ii) random errors, arising from atmospheric irregularities whose statistics will be known, though not their actual values. The smaller the scale of an irregularity and the faster its fluctuations, the more difficult will the compensation of the resultant errors become. In practice, accurate compensation for systematic atmospheric effects could require regular calibration by facilities covering the detection volume of the radar. The development of GPS–NavStar will aid the characterisation of the ionosphere. GPS operates on two L-band frequencies so that the variable delay introduced by the ionosphere can be eliminated. The calculation could be modified to provide the total electron content of the ionosphere and, hence, the delay corrections for other frequencies.

11.6.1 *Atmospheric regions*

There are three distinct regions of the atmosphere that are of interest in assessing radar propagation, namely the troposphere, the stratosphere and the ionosphere. In each region, the velocity of propagation of radar signals is c/n, where c is the velocity of light *in vacuo* and n is the local refractive index. The resultant error in the radar range is then

$$\begin{aligned} \delta R &= c\int_L \frac{1}{V}\,dl - R \\ &= \int_L n\,dl - R \end{aligned} \tag{11.16}$$

where integration takes place along the curved path L and R is the true range to the target. Since n is very close to unity, for many purposes it is convenient to use the refractivity

$$N = (n-1)\times 10^6 \tag{11.17}$$

11.6.1.1 *Troposphere*

The troposphere is the lowest region of the Earth's atmosphere, extending upwards from the Earth's surface, in which the temperature decreases with height, except within local layers with temperature inversion. It extends to an altitude of about 9 km at the poles and 17 km at the equator. The velocity of propagation in the troposphere is independent of frequency, except at millimetre wave frequencies close to absorption bands. The shape of a complex electromagnetic wave, with a spectrum of almost any width, remains constant during propagation as all the components of its spectrum move at the same velocity.

The refractive index is greater than unity and normally decreases monotonically with height

$$n > 1 \qquad \frac{dn}{dh} < 0 \tag{11.18}$$

and this causes radar waves to bend back towards the Earth, so creating some radar coverage beyond the geometrical horizon, in a manner that is frequently represented by the 4/3 apparent Earth radius. This approximation assumes a linear variation of n with h, whereas the true variation is closer to exponential. It works well for ground-based radars operating at low elevation angles but is not adequate for bistatic radars that are airborne or are operating at large elevation angles.

The total content of tropospheric gases is very stable, with 90% of them being present below 20 km, and changes in N of only about 10% occur in the lower 2 km.

11.6.1.2 *Stratosphere*

The stratosphere is the region between the end of any significant concentration of atmospheric gases, at a height of about 20 km, and the start of the ionised plasma of the ionosphere, at a height of about 50 km. The refractive index is virtually constant at unity within this region. There is no dispersion: radar waves travel in straight lines and pulses retain their shapes.

11.6.1.3 *Ionosphere*

The presence of electrons in the ionosphere is primarily caused by the ionisation of atmospheric gases exposed to the Sun's energy, though there are further contributions from cosmic rays, radioactive decay of substances in the Earth's crust and charge separation within clouds. The effective part of the Sun's energy lies mainly in the X-ray and ultraviolet parts of the spectrum and virtually all of it is absorbed at heights above 60 km. These electrons form a continuous but non-uniformly ionised sheath around the Earth, at heights between approximately 50 and 600 km, which can have at least four peaks in their concentration, designated as the D, E, F1 and F2 layers. Of these, the F2 layer is the most intensely ionised and it is the only one which is always present.

At the upper boundary of the ionosphere (1100–1300 km), the concentration of electrons is low owing to the low density of gases; in the lower layers, it decreases as the solar radiation which causes the ionisation is absorbed. The concentration depends on the time of day (maximum near midday), on the season of the year (maximum during the summer) and on solar activity (maximum at the peak of the eleven year cycle of sunspot activity). The lower boundary of the ionosphere can be considered approximately constant at 80 km during the day and at 100 km at night. The altitude of the maximum electron concentration, the peak of the F2 layer, is at about 300 km and the maximum electron concentration is typically $1_{10}12$ electron/m^3 on a summer day.

11.6.2 *Wave propagation in a vertically stratified model of the atmosphere*

When estimating the systematic propagation errors for monostatic radars, it is convenient to assume a vertically stratified model of the atmosphere, in which its parameters change only with altitude and remain constant in spherical layers. With this assumption of horizontal homogeneity, which is seldom realised under normal atmospheric conditions, no azimuth errors are predicted for monostatic radars. This is not the case for bistatic radars, although the errors are slight. As the beam from the transmitter curves away from a linear path, the correct bearing from the receiver for pulse chasing will also change.

In this vertically stratified model, Fermat's principle reduces to Snell's law. From this, it follows that, as the wave propagates through the layers of the atmosphere [Bean and Dutton, 1966],

$$nR \cos \theta = \text{constant} \tag{11.19}$$

where R is the distance from the centre of the Earth to the bottom of the layer, n is the refractive index of the layer and θ is the angle of entry of the wave into the next layer.

In the limit, as the layer is reduced to an infinitesimal thickness dh, the length dl of the ray within the layer becomes

$$dl = \frac{dh}{\sin \theta} \tag{11.20}$$

and if the refractive index within a layer is $n + dn$, while that of the preceding layer is n, then

$$d\theta = -\frac{\cot \theta}{n} dn \tag{11.21}$$

The total angular refraction, or bending, τ of the ray between two points can then be derived as

$$\tau_{1,2} = -\int_{n_1, \theta_1}^{n_2, \theta_2} \cot \theta \frac{dn}{n} \tag{11.22}$$

The curvature $(1/\rho)$ of a radar beam, propagating at an angle θ to the local horizontal in an atmosphere whose refractive index n is a function only of the height, is given by [Bean and Dutton, 1966]

$$\frac{1}{\rho} = -\frac{\cos \theta}{n} \frac{dn}{dh} \tag{11.23}$$

This expression is usually approximated to

$$\frac{1}{\rho} \approx -\frac{dn}{dh} \tag{11.24}$$

since the refractive index in the atmosphere is very close to unity and only rays which are close to the horizontal are of particular interest in long range air defence radars. These assumptions are not always true; θ could be any angle from 0 to $\pi/2$ for a space surveillance radar or a space based radar. The refractive index could be markedly less than unity for HF propagation in the ionosphere. Eqn. 11.23 shows that the curvature of a radar beam depends on the gradient of the refractive index and it is taken to be positive when the refractive index decreases with height, that is the beam curves downwards. This curvilinear propagation can be transformed to a hypothetical Earth having a radius kR, such that the curvature of a ray propagating in its atmosphere is then zero (the beam would become a straight line). The multiplier k is then given by

$$\frac{1}{kR} = \frac{1}{R} + \frac{\cos \theta}{n} \frac{dn}{dh} \tag{11.25}$$

The typical lapse rate of the atmospheric refractive index is $-3.9_{10}-5$ km^{-1} at sea level, where $n \approx 1.000319$. For elevation angles that are close to zero, this equation then yields the familiar $k = 4/3$ approximation, so increasing the

Earth's apparent radius from 6366 km to 8488 km. Eqn. 11.25 can also be used to predict the lapse rate for which the Earth's effective radius becomes infinite. This allows initially horizontal rays to follow the curvature of the Earth – an effect commonly known as ducting or trapping. This occurs when

$$\frac{dn}{dh} = -\frac{1}{R}\frac{n}{\cos\theta} \tag{11.26}$$

This gives a lapse rate of $-1.57_{10}-7\ \mathrm{m}^{-1}$ in the refractive index or -157 N/km in the refractivity.

The factor k can be derived from either meteorological observations or by propagation measurements. In general, k is derived from propagation measurements and represents a spatial average which could only otherwise be obtained from many simultaneous meteorological soundings along the propagation path. The distribution of k values derived by this means displays less variability than that derived from single point meteorological measurements. The variability decreases with increasing path length.

When the atmosphere is sufficiently sub-refractive (large positive values of the gradient of the refractive index ≡ low k factor values), the ray paths can be bent to such an extent that the Earth appears to obstruct the direct paths between the transmitter and the target or the receiver and the target. Measurements of the minimum effective value of k that are exceeded for approximately 99.9% of the time in a Continental temperate climate have been made for a range of path lengths by Boithias and Battesti [1967]. These indicate minimum k values of 0.57 for a 20 km path, but k increases to 0.92 for a 100 km path and to 0.99 for a 200 km path.

11.6.3 *Tropospheric propagation*

11.6.3.1 *Effective refractive index*

At radio frequencies, the tropospheric refractivity may be approximated as (CCIR V[1982])

$$N = (n-1)\times 10^6 = 77.6\,\frac{p}{T} + 3.73(10^5)\,\frac{e}{T^2} \tag{11.27}$$

where T is the absolute temperature in deg K and p and e are, respectively, the atmospheric and the water vapour pressures in millibars.* This expression for the refractivity of air is independent of frequency in the range 100 MHz to 30 GHz; for frequencies above 100 GHz, the error is less than 0.5%. The relationship between the water vapour concentration $\rho(\mathrm{kg/m^3})$ and the water vapour pressure e (mb) is given by

$$\rho = 0.217\,\frac{e}{T} \tag{11.28}$$

For an ICAO standard atmosphere, which represents standard conditions in the temperate zone, the temperature T (in deg K) and the pressure (in mb) are given by

* The bar is a remnant of the CGS system of units, where 1 bar $= 10^6$ dyn cm^{-2}. One millibar is equal to 100 Pa in the SI system of units. The sea level pressure of 1013 mb for the ICAO standard atmosphere corresponds to 760 mmHg.

$$T = 288(1 - 0.02257h)$$
$$p = 1013(1 - 0.02257h)^{5.26} \tag{11.29}$$

where h is the height in kilometres and the equations are valid up to an altitude of 11 km (i.e. up to the tropopause). For the ICAO standard atmosphere at sea level, with $T = 288$ °K, $p = 1013$ mb and $e = 10.2$ mb (corresponding to 60% relative humidity), the water vapour contribution to N is only 14% and, as an approximation for dry air, eqns. 11.27 and 11.29 give

$$N(h) = 319(1 - 0.02257h)^{4.26} \tag{11.30}$$

The reference atmosphere for refraction, recommended by CCIR V [1982], is

$$N(h) = N_A \exp(-b_A h) = 315 \exp(-0.136h) \tag{11.31}$$

where $N(h)$ is the refractivity at a height h, measured in kilometres above sea level, and N_A is the value of the refractivity at the surface of the Earth. The accuracy of eqn. 11.31 can be improved by measuring the local refractivity, at height h_s above sea level, and then using the equation to convert N_A to the sea level value N_0.

Eqn. 11.27 is often separated into wet and dry terms

$$D = 77.6 \frac{p}{T} \quad \text{and} \quad W = 3.73 * 10^5 \frac{e}{T^2} \tag{11.32}$$

Both of these terms decrease with height, but at different rates, so leading to a bi-exponential model

$$N(h) = D \exp\left(-\frac{h}{h_D}\right) + W \exp\left(-\frac{h}{h_W}\right) \tag{11.33}$$

Typical values for the scale heights are $h_D = 9$ km and $h_W = 2.5$ km. The dry coefficient D is fairly constant at about 273, while the wet coefficient W causes most of the variation in $N(h)$.

11.6.3.2 *Tropospheric propagation anomalies*

Microwave propagation in the troposphere frequently exhibits irregular behaviour, with its vertical refractivity profile rarely following the standard lapse rate of -39 N/km that leads to the 4/3 Earth approximation. Instead, the troposphere is generally composed of horizontally stratified layers of differing refractivity gradients. The most significant propagation anomalies occur when the refractivity gradient is such that local radius of curvature of the beam is less than the local distance from the centre of the Earth. A trapping or ducting layer is then formed and the radar signal propagates much as in a waveguide. For a horizontally directed wave at ground level, a ducted mode occurs for a layer gradient $\leqslant -157$ N/km (based on an average Earth radius of 6371 km).

Ducting can be caused by temperature and/or humidity inversion. Temperature inversion can occur at night, near the ground at heights below about 300 m, when the surface temperature falls faster than that of the air. Ducting extends over smaller areas, and is less pronounced for propagation over land, since hills and variations in the ground cover prevent conditions being suitable. However, ducting may occur for ranges of several hundred kilometres over the sea.

According to Anderson *et al.* [1984], there are two mechanisms that can produce such trapping layers in the maritime environment. In the first case, a trapping layer is created by an advection process in which a mass of warm dry air blows from the land over a cooler sea, thus causing evaporation which increases the humidity near the surface and so decreases the temperature. This produces either a surface or an elevated duct which may affect frequencies as low as 100 MHz. Such advection ducts are important where there are enclosed seas with adjacent hot, dry land, as in the Mediterranean Sea, the Red Sea and the Arabian Gulf. They also occur in coastal regions such as those of the North Sea. In the second case, a very persistent surface trapping layer is produced by water evaporation at the air–sea interface. This evaporation duct is a nearly permanent propagation mechanism created by a rapid decrease in moisture content in the atmosphere immediately above the ocean's surface. Air adjacent to the surface is saturated with water vapour and the vapour pressure decreases nearly logarithmically with increasing height until an ambient value of water vapour content is reached, which is dependent upon the general meteorological conditions. Typical evaporation duct heights are between a few metres and approximately 30 m, with a world-average value of 13.6 m. The annual mean evaporation duct heights for waters around the UK are about 5–6 m, while those in the Mediterranean are about 11–14 m. Because these ducts are shallow, strong trapping is observed infrequently for frequencies below 2 GHz.

For low-altitude, over-water applications, the evaporation duct has been shown to be a reliable propagation phenomenon that can significantly increase beyond-the-horizon propagation for frequencies above 2 GHz. Measurements have shown that the extent of the signal enhancement (above what would have been achieved with diffraction alone) increases with increasing frequency. For example, analysis of measurements by Anderson *et al.* [1984] in the Aegean Sea showed that the median received signal power on a 35 km path was 2, 15, 27 and 30 dB above diffraction for frequencies of 1, 3, 9.6 and 18 GHz, respectively.

Although propagation within an atmospheric duct is similar to that in a waveguide, one or both of its boundaries will be diffuse and the actual planes of reflection will depend on the angle of incidence of the wave into the duct. Normally, the angle between the radar beam and the duct direction will not be greater than about 1° if efficient coupling is to be achieved. A duct supports only certain modes of propagation and will not readily support propagation below a critical wavelength. Kerr [1951] has shown, for a simplified model of wave propagation in a surface duct of height d, that propagation can only occur for wavelengths that are shorter than the critical wavelength

$$\lambda_{max} = 2.3\left(-\frac{\Delta n}{\Delta h}\, d^3\right)^{1/2} \qquad (11.34)$$

where λ_{max}, Δh and d are in consistent units.

Atmospheric ducts are generally of the order of 10–20 m in height and never more than 150–200 m. Extended range propagation is more likely to be experienced at the higher microwave frequencies than at the lower ones. The cut-off wavelength does not mark a sharp boundary between regions of propagation and no propagation; the ducts act like leaky waveguides and radars whose wavelength is several times the cut-off wavelength may still be affected by them.

11.7 Waveform design

As with monostatic radars, the bistatic radar is constrained in its choice of transmitter waveforms. In the simplest versions of bistatic radars, the PRF is limited by the need to ensure that, at any one time, only one range resolution cell lies within the overlap volume of the transmitter and receiver beams and only one pulse lies within the receiver antenna's scan sector. However, more complicated versions can accommodate higher PRFs by using multiple PRFs to resolve range ambiguities and by having several receiver beams, each with its own signal processor, to cover the required sector.

The maximum operating bandwidth of a bistatic radar's antenna is constrained by the bandwidths of the individual components from which it is constructed and by the interactions between the individual elements in its array, such as mutual coupling effects and grating lobes. The maximum instantaneous bandwidth is limited by the scanning technique.

If the processing and display of data are carried out at the receiving site, then it is clearly desirable for the receiver to have a direct influence over the transmitter's waveforms, so that they can be adapted to suit the observed target scenarios. This would be similar to the ability of a monostatic radar with a phased array antenna to adapt its scan sequences to give such facilities as sequential detection and target tracking interleaved with scanning. In practice, this is only likely to be feasible with dedicated illuminators.

11.7.1 *PRF constraints*

The PRF has to be sufficiently high to allow most of the interesting targets to be separated from chaff and rain, but without producing an inconveniently small unambiguous range. This normally implies the use of medium PRF waveforms, with a moderate number of both range and velocity ambiguities. These ambiguities then have to be resolved, either by geometrical techniques (an option that is denied to monostatic radars) or by operating with a sequence of coherent bursts of pulses with appropriate PRFs (in the same manner as is used with monostatic radars, especally those used for airborne interception (AI) and airborne early warning (AEW)).

There is a maximum PRF, determined by the deployment geometry and the antenna beamwidths, beyond which range ambiguities can no longer be resolved simply by geometry, and PRF agile operation is required. There is a similar trade-off between the maximum PRF and the maximum width of the receiver antenna's scan sector; this arises from the need to have only one pulse within the sector at any one time. The relationship between the PRF and the scan sector also determines the number of independently scanned beams that are required for the phased array antenna to be able to chase pulses across its full scan range; one independent signal processing channel will, of course, be required for each of the beams.

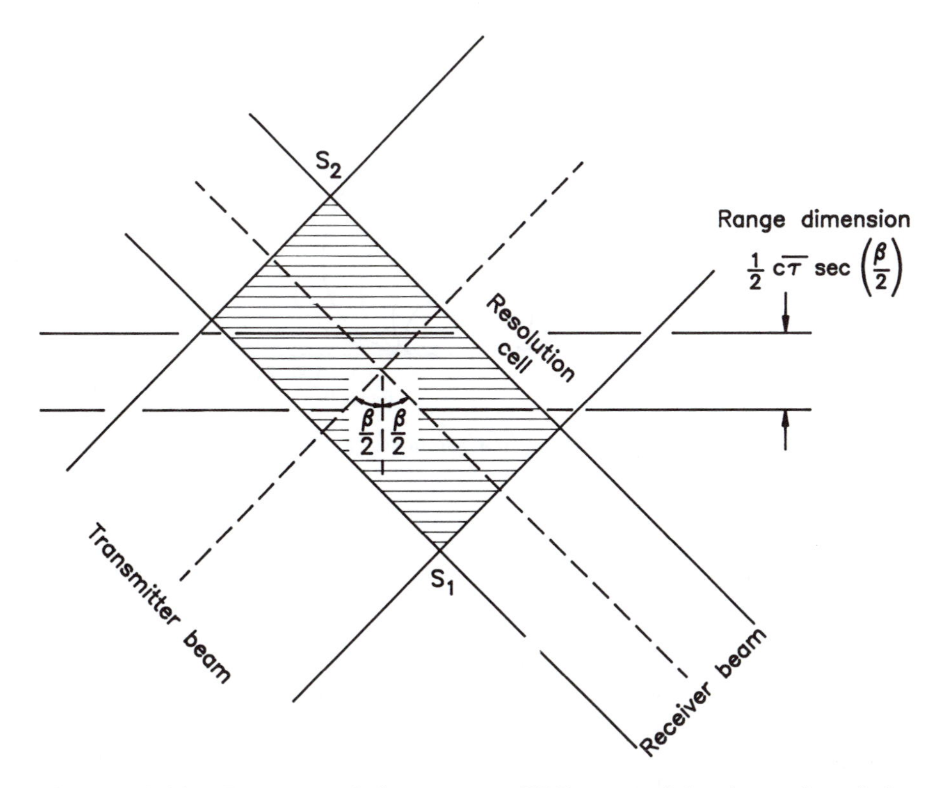

Figure 11.11 *Evaluation of the maximum PRF permitted by the overlap of the transmitter and receiver beams*

The use of signal processors with an ability to process PRF agile waveforms is desirable, even though it is not always essential. This ability allows the receiver to make use of a wider range of transmitter waveforms. The technology for such processors has already been developed, for use in conventional monostatic AI and AEW radars that employ medium to high PRF waveforms.

11.7.1.1 *Maximum PRF permitted by beam overlap*

A bistatic radar system is only sensitive to targets that lie in its beam crossover region and the size of this region determines the accuracy with which a target can be located on the prolate spheroid defined by the propagation time of its echoes. This is illustrated in Fig. 11.11, for the two dimensional ellipse. At least one of these beams will normally be narrow to provide acceptable spatial resolution in the 'azimuth' co-ordinate; the transmitter beam will usually be preferred since the receiver beam cannot provide azimuth resolution unless it has a monopulse capability. Another reason for preferring the transmitter to have the narrower beam is that its susceptibility to location by ESM and to attack by ARMs is thereby reduced.

The range resolution is determined by the distance between successive elliptical contours, separated in bistatic delay by the radar's (compressed) pulse

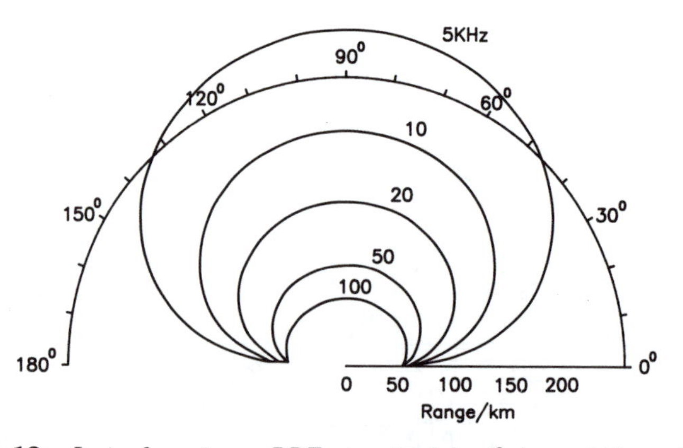

Figure 11.12 *Loci of maximum PRFs permitted by 1° beamwidths and a 100 km baseline*

width. The size of the radar's resolution cell is then the portion of the strip between two successive contours that also lies within the beam crossover region. If the bistatic geometry ensures that only one resolution cell ever lies within the beam overlap region, then the range ambiguities that occur with medium to high PRF bistatic radars can be resolved by geometrical techniques.

If the transmitter PRF is too high in relation to the size of the beam overlap region, then more than one resolution cell can lie within it. The resultant range ambiguities can only be resolved by PRF agility. Even with radars that can operate with PRF agility, a target in the presence of rain or chaff will still have clutter echoes from more than one resolution cell aliased onto it and this will reduce its signal-to-clutter ratio.

Referring to Fig. 11.11, it can be seen that the minimum pulse repetition interval T occurs when targets at S_1 and S_2 are illuminated by successive transmitter pulses. The value of T is given approximately by

$$T = \frac{1}{c}\,[TS_2 + S_2R - TS_1 - S_1R] \qquad (11.35)$$

where the notation is the same as that used in Section 11.5.

This equation has been evaluated by Jackson [1986], by assuming that the limits of the beam overlap are defined by the beams' conventional 3 dB beamwidths. The maximum PRF that avoids range ambiguity is given by

$$PRF = \frac{c\left(1 + \left(\dfrac{B_T + B_R}{2\sin\beta}\right)^2\right)\tan\left(\dfrac{\beta}{2}\right)}{R_TB_T + R_RB_R} \qquad (11.36)$$

where B_T and B_R are the beamwidths of the transmitter and receiver antennas, respectively. This PRF depends on the target locations and Fig. 11.12 shows how it varies for a 100 km baseline and 1° beamwidths. The loci of constant PRF are symmetrical about the baseline; in this example, they are also symmetrical about the perpendicular bisector of the baseline, but this is only because both antennas have been allocated the same beamwidth.

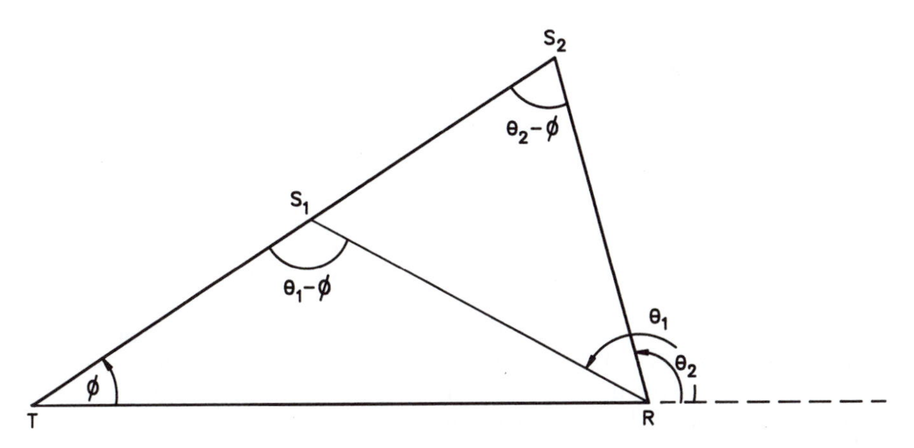

Figure 11.13 *Evaluation of the maximum PRF permitted by the width of the receiver antenna's scan sector*

11.7.1.2 *Maximum PRF permitted by the receiver antenna's scan sector*

The maximum transmitter PRF is constrained by the need to have one, and only one, pulse within the receiver antenna's scan sector at any one time. Conversely, the maximum azimuth sector width over which a beam on the receiver antenna can chase transmitter pulses is constrained by the need to chase every pulse that crosses the sector. If the transmitter's PRF is too high for the receiver's sector width, the receiver will not be able to collect all the energy that may be scattered towards it by all the targets that may lie on the line of flight of the pulses. If only every Nth pulse can be chased, then each target's integrated signal-to-noise ratio will be reduced by N (assuming coherent processing); unless the integration time is increased to compensate, the receiver's target detection capability will be reduced. There will be an even more serious effect on the receiver's Doppler processing. If the effective sampling ratio N is fixed, then the unambiguous Doppler range will be reduced by N; if it is not fixed, the Doppler processor cannot give valid results.

The trade-off between the transmitter's PRF and the antenna's sector width can be evaluated from Fig. 11.13, using simple trigonometry. If energy scattered from transmitter pulse n by target S_2 reaches the receiver site R at the same time as energy scattered from transmitter pulse $(n+1)$ by target S_1, then the minimum pulse repetition interval T is given by

$$\begin{aligned} T = \frac{1}{c}\,&[TS_2 + S_2R - TS_1 - S_1R] \\ &+ (\text{beam reset time}) \\ &+ (\text{uncompressed pulse duration}) \\ &+ (\text{insertion delay of the filter}) \end{aligned} \tag{11.37}$$

The beam reset time, namely the time taken to return the antenna's beam to the start of its scan, will normally be a few orders of magnitude less than the sector scan time. The pulse duration term in the equation is a consequence of the need to have the receiver matched to the transmitter pulse and this cannot be achieved until a complete uncompressed pulse has had time to travel through the filter; this will be a minimum of one pulse duration after the start of the scan. When the bistatic receiver is operating with a conventional monostatic radar, the ratio of the pulse duration to the pulse spacing will be equal to the duty cycle of the transmitter and this will normally only be a few percent. These two terms are therefore sufficiently small, in most situations, to be neglected in this simplified analysis of beam scanning. This low duty cycle approximation will not be valid for some LPI radars where the uncompressed pulse duration could become as long as the pulse repetition interval.

The insertion delay correction will be required when the pulse is compressed with an analogue filter. The insertion delay will be the further delay between the trailing edge of the uncompressed pulse entering the filter and the trailing edge of the compressed pulse leaving the filter.

From the law of sines it follows that

$$\frac{TS_1}{\sin\theta_1}=\frac{S_1R}{\sin\phi}=\frac{L}{\sin(\theta_1-\phi)} \tag{11.38a}$$

and

$$\frac{TS_2}{\sin\theta_2}=\frac{S_2R}{\sin\phi}=\frac{L}{\sin(\theta_2-\phi)} \tag{11.38b}$$

Combining eqns. 11.37 and 11.38 gives

$$\frac{cT}{L}=\frac{\cos\left(\dfrac{\theta_1-\theta_2}{2}-\phi\right)-\cos\left(\dfrac{\theta_1-\theta_2}{2}+\phi\right)}{\cos\left(\dfrac{\theta_1-\theta_2}{2}\right)-\cos\left(\dfrac{\theta_1+\theta_2}{2}-\phi\right)} \tag{11.39}$$

If we now put $\theta_1=\theta+\Delta\theta$ and $\theta_2=\theta-\Delta\theta$, where θ is the azimuth at the centre of the scanned sector and $2\Delta\theta$ is the sector width, eqn. 11.39 gives the PRF

$$PRF=\frac{c}{L}\,\frac{\cos(\Delta\theta)-\cos(\theta-\phi)}{2\sin(\Delta\theta)\sin\phi} \tag{11.40}$$

The dependence of the maximum PRF on the transmitter and receiver azimuths is illustrated in Fig. 11.14 for a 40° scan sector and a 100 km baseline.

Alternatively, the receiver antenna's sector width $2\Delta\theta$ can be derived from the transmitter's PRF by solving

$$\cos(\Delta\theta)-\frac{2L\sin\phi}{cT}\sin(\Delta\theta)-\cos(\theta-\phi)=0 \tag{11.41}$$

to give the sector's half-width as

$$\theta=2\arctan\left[\frac{-\dfrac{2L}{cT}\sin\phi+\left(\left(\dfrac{2L}{cT}\sin\phi\right)^2+\sin^2(\theta-\phi)\right)^{1/2}}{1+\cos(\theta-\phi)}\right] \tag{11.42}$$

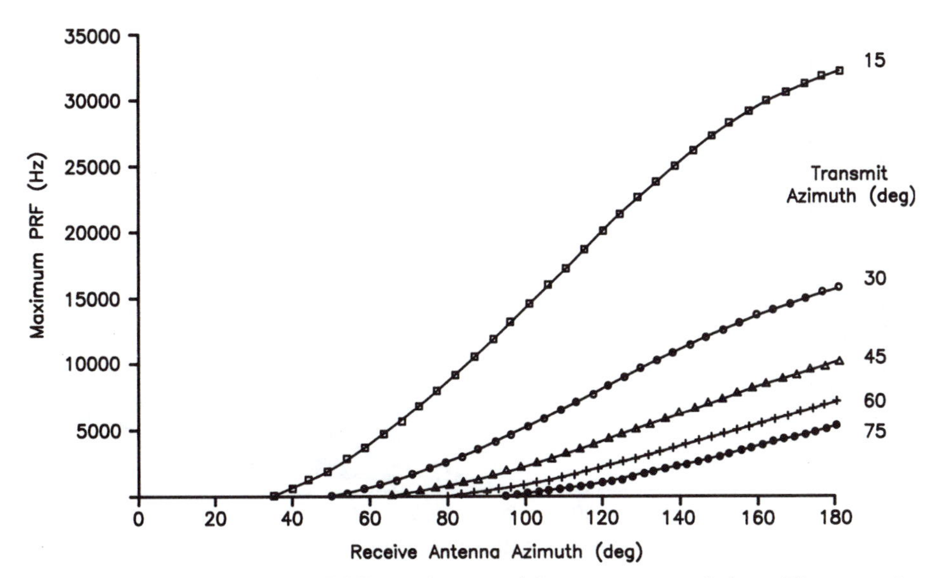

Figure 11.14 *Maximum PRF as a function of the antenna azimuth for a 40° sector and a 100 km baseline*

The positive sign is required for the square root term, since $\Delta\theta$ must be less than $\pi/2$ and $\tan(\Delta\theta/2)$ must therefore be positive.

The dependence of the receiver antenna scan sector on the receiver azimuth and the transmitter PRF is illustrated in Fig. 11.15 for a 75° transmitter azimuth and a 100 km baseline.

The minimum pulse repetition interval T can also be seen as defining the separation between two ellipsoids, one passing through S_1 and the other through S_2. As the transmitter beam scans, the positions of S_1 and S_2 will change as they move along their respective ellipsoids. The area that can be covered by a single signal processor is then that part of the area between the two ellipsoids which also lies within the fields of view of both the transmitter and the receiver antennas.

The receiver antenna's scan sector may also be limited by the number of range cells that can be handled by each signal processor. If T_{proc}, the product of the number of range cells and the compressed pulse length, is less than the pulse repetition interval T discussed above, then the separation between the ellipsoids will need to be reduced to T_{proc}. If it is not, there will be blanks in the radar's coverage. This reduced separation will limit the antenna scan angle that can be accommodated by a single processor, though it will allow the use of a larger PRF.

11.7.2 *RF bandwidth constraints with phased array antennas*

Radars that use Doppler effects to aid target detection have to radiate coherent bursts of pulses, with constant PRF and RF over the duration of each burst; this makes them more vulnerable to set-on jammers and to ARMs. Burst-to-burst RF agility over wide bandwidths can help to restore the radar's ECCM,

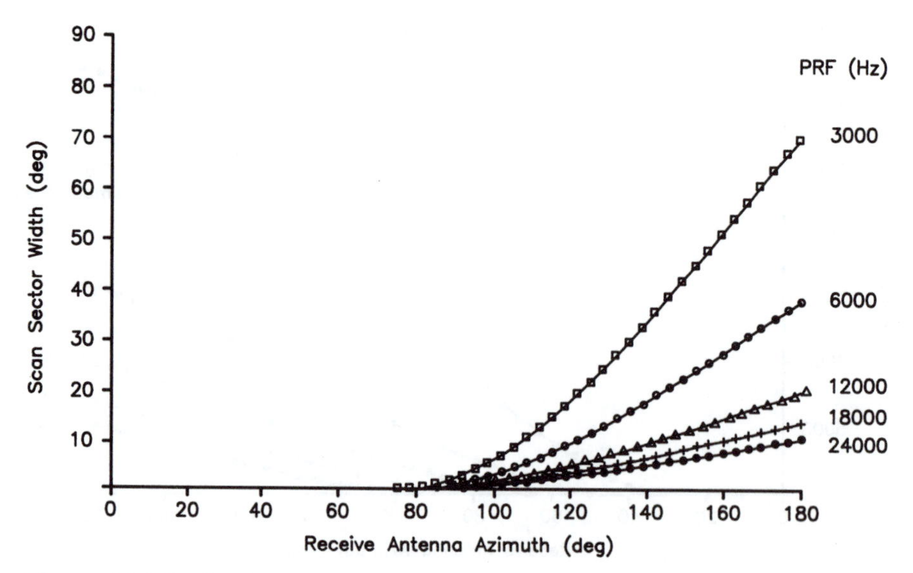

Figure 11.15 *Receiver scan sector versus receiver azimuth and PRF for a 75° transmitter azimuth and a 100 km baseline*

especially for bistatic receiver sites whose locations are not known to their opposing jammers. It also reduces the risk that targets will produce RCS nulls on a sufficient number of the PRFs in a sequence for the ambiguity resolution algorithms to be defeated. The maximum RF bandwidth of a phased array antenna is constrained by the use of phase steering (instead of time delay steering), by mutual coupling effects between its adjacent elements and by the bandwidths of its components.

11.7.2.1 *Maximum instantaneous bandwidth*

The maximum instantaneous bandwidth defines the minimum (compressed) pulse width that can be processed by the antenna. Of the various types of electronically scanned phased array antennas, only those arrays whose beams are steered by true time delay elements, including the arrays with 'optical' beam forming systems, have beam patterns which are largely independent of frequency and can therefore be used with very short (compressed) pulse lengths.

For a simple phase scanned array as shown in Fig. 11.16, the phase difference $\delta\phi$ between two adjacent radiating elements, required to steer the beam to a direction θ away from the normal to the array, is given by

$$\delta\phi = kd \sin \theta \tag{11.43}$$

where the wavenumber $k = 2\pi/\lambda$ and d is the distance between the elements. If the operating frequency is changed, without altering the phase shifter settings, it follows from eqn. 11.43 that the beam is steered and the rate of change of θ with operating frequency f is

$$d\theta = -\frac{\tan\theta}{f}\,df \tag{11.44}$$

If a phased array antenna is excited by a signal with a non-zero spectrum width, the result is a 'spreading' of the beam on both sides of the desired direction, with a resultant loss of gain. This spreading will cause some degradation in the angle measurement accuracy that can be obtained from overlapping-beam amplitude comparison systems or from monopulse systems. The particular values of spectrum width and beam spreading to be used in assessing the effects are to some extent a matter of opinion. Garnham [1977] recommends a conservative estimate in which the spectrum width for good fidelity is measured at the -10 dB level and the beam is scanned through less than $\pm 0.2\,\beta$ by the frequencies at the edge of this spectrum. The half-power beamwidth β of a phased array antenna is approximately

$$\beta \approx \frac{c}{fD}\sec\theta \tag{11.45}$$

where D is the overall length of the array and the exact value of β depends on the illumination taper of the array.

For an unmodulated pulse of length τ seconds, the spectral width at the -10 dB level is approximately $1.5/\tau$ Hz, so from eqns. 11.44 and 11.45 with

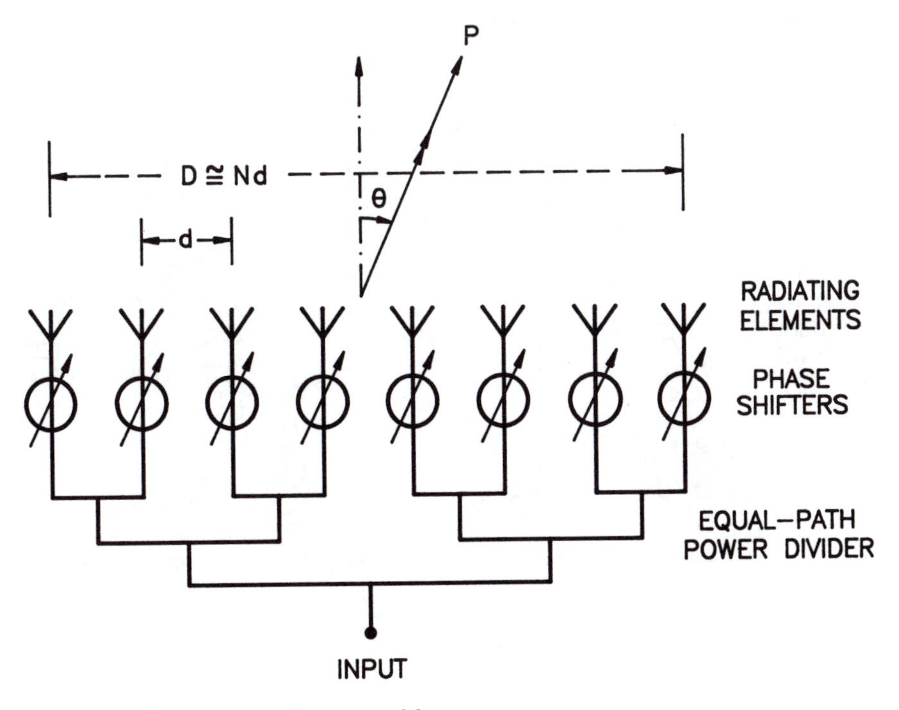

Figure 11.16 *Simple phase-scanned linear array*

$d\theta = 0.4\,\beta$, the criterion for tolerable beam distortion at a scan angle θ becomes approximately

$$c\tau \geqslant 4D \sin\theta \tag{11.46}$$

For modulated pulses, the resultant compressed pulse length should be used in eqn. 11.46. This equation is also consistent with the requirement

$$c\tau \gg D \sin\theta \tag{11.47}$$

which ensures that the energy from all parts of the array has time to contribute to the true antenna pattern. The need for this aperture fill time is illustrated in Fig. 11.17.

For an array with a uniform aperture weighting, the length of the array can be expressed, in terms of the beamwidth B, as

$$D = \frac{c}{fB} \tag{11.48}$$

where f is the frequency in Hz. The criterion for tolerable beam distortion then becomes

$$\tau \geqslant \frac{4 \sin\theta}{fB} \tag{11.49}$$

For a 4° beamwidth antenna operating on boresight at 3 GHz, $\tau \geqslant 25$ ns.

11.7.2.2 *Maximum operating bandwidth*

The maximum operating bandwidth is determined by the bandwidths of the individual components from which the array is constructed. It is wider than the

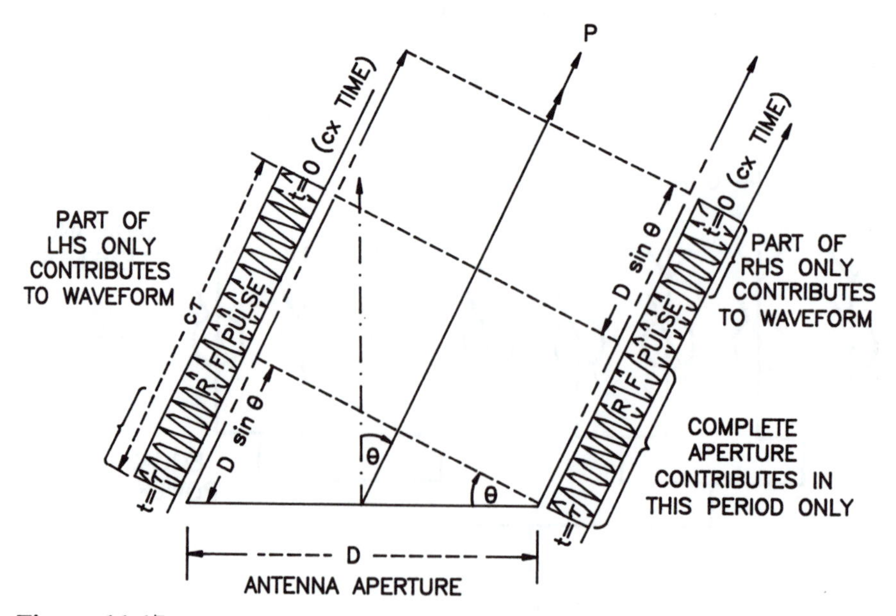

Figure 11.17 *Aperture fill-time effects in a phase-scanned array excited with a short RF pulse*

instantaneous bandwidth but the phase shifters have to be adjusted to maintain the beam orientation despite changes in frequency over this bandwidth. Mutual coupling effects between adjacent radiating elements will probably be the main constraint on the operating bandwidth of a phased array antenna. However, these are now well understood for a wide range of radiators and array configurations, though stringent performance requirements may demand dual tuning and reduced lattice spacings. Such antennas can currently be designed to achieve operating bandwidths ≤10% without any serious design problems or compromises. Raising the bandwidth to 12–15% would involve some design problems, including increased reflection coefficients and sidelobe levels. Bandwidths of 15–20% might be achieved with state-of-the-art receivers, but would be improbable for the high power sources and phase shifters at the transmitter.

11.8 Illuminator options

11.8.1 *Co-operative illuminators*

For an existing monostatic radar to act as a co-operative illuminator for a bistatic receiver, it will require the following modifications and additions:

(i) The transmitter frequencies and the timings of the PRF sequences will have to be locked to an atomic clock, so that they can both be reproduced at the receiver with sufficient precision for the accurate measurement of the time delays and Doppler shifts of target echoes.
(ii) The transmitter site will require a data link to the receiver site; this will relay complete details of the waveforms that the illuminating radar is about to radiate, together with the current orientation of the transmitter antenna. If the transmitter is mobile, the data link will also need to relay the platform's location and velocity in three dimensions. The use of a data link between transmitter and receiver may be unattractive, both for the security of the radar and to avoid the costs that might be incurred in modifying the radar to provide access to these data.

The radar can, of course, act as its own data link by modulating data onto some of its transmitter pulses, when its antenna scans past the direction of the receiver site, as described by Coleman and Alter [1985]. The very large ERP of a radar transmitter will then provide communication with considerable resistance to jamming.

11.8.1.1 *Existing monostatic surveillance radars*

Long-range ground-based surveillance radars do not have parameters which are particularly well matched to the needs of bistatic systems and there is no reason to believe that the monostatic performance of such surveillance radars would readily be compromised in order to enhance their performance as bistatic illuminators. The locations of surveillance radars will be compromised as soon as they transmit, and there is consequently a risk that many would be disabled or forced to restrict their transmissions soon after the start of any hostilities. There are further technical and operational problems:

(i) The transmitter will need to be modified to support PRFs that will be some 10 to 20 times greater than those required for its normal surveillance modes, if the bistatic system is to demonstrate its ability to detect targets in dense volume clutter. These PRFs may exceed the duty cycle limits of the transmitter.

(ii) There may be a loss of monostatic cover by the illuminating radar owing to range ambiguities and the fold-over of close-in ground clutter, introduced by the use of the medium PRF required by the bistatic receivers. However, this may be an acceptable loss if the monostatic cover is already seriously compromised by the effects of dense, turbulent chaff.

(iii) The monostatic radar receiver (and the bistatic receiver, if its position is revealed) will be susceptible to intelligent jamming during the long coherent bursts of pulses that will be required for Doppler processing via an FFT. It will also be easier for ARMs to lock on to such long bursts.

(iv) When decoys are deployed around a radar transmitter, their cover pulses have to be longer than the radar pulses so that they can cover the radar pulse for ARMs approaching from any possible direction. With medium PRF operation, the duty cycle problems for such decoy transmitters may be even greater than those for the radar transmitter.

(v) The elevation coverages of the transmitter and receiver antennas may not be compatible. Also, the bistatic receiver may not have the bandwidth to use all the energy radiated by a transmitter, whether the transmitter's bandwidth is used as frequency agility for ECCM or for frequency scanning of its antenna.

11.8.1.2 *Airborne radars*

AEW and AI radars are usually capable of pulse Doppler operation. Hence, they will have PRFs that are sufficiently high to allow bistatic target detection in chaff. The complications associated with airborne operation are reviewed later, in Section 11.12. In addition, the PRFs may be too high for optimum bistatic operation, for reasons described in Section 11.7.1, although AI radars normally operate at higher RFs than do AEW radars.

11.8.2 *Non-co-operative illuminators*

This Section reviews some of the options available for bistatic radar operation with non-co-operative illuminators and the salient constraints on the use of such illuminators. Some possible future techniques are also discussed in Section 11.17.1.

11.8.2.1 *Other radars*

The bistatic receiver will not normally have any information on the waveforms and antenna orientations of non-co-operative radars, unless these can be derived from observation of the radar's operation. Unfortunately, future radars can be expected to use random RF and, possibly, PRF agility to defeat responsive jamming. This PRF agility cannot be completely random as their PRF sequences will be constrained by the need to resolve range and velocity ambiguities. The RF sequences will probably be derived by random number generators but the range of the agility will be constrained by electromagnetic compatibility (EMC) considerations and by the bandwidths of the microwave components used in the radar. In addition, the scanning of their phased array antennas will incorporate sequential detection and target updating, so making the scanning patterns dependent on the target environment. Even surveillance radars with mechanical azimuth scanning will probably incorporate some phase steering to allow 'look-back' for sequential detection procedures.

11.8.2.2 *Sources other than radars*

The use of illuminators that do not offer waveforms like those from radars has been considered by Griffiths and Long [1986], especially in relation to the use of TV waveforms. So far, they have reported mainly negative experimental results when using a UHF PAL TV signal, but these appear to be due to equipment limitations. They note that the ambiguity function of the waveform is of key importance, as this will determine the range resolution, the spacing of range ambiguities, the range sidelobe level and the achievable Doppler resolution. They conclude that such illuminators should have the following properties:

(i) Adequate transmitter power for the required coverage – this power will be that in the portion of the modulation spectrum which is to be used for bistatic radar purposes.
(ii) A radiation pattern that is either omni-directional (floodlight coverage) or a single pencil beam.
(iii) A modulation bandwidth that is consistent with the required range resolution – the ambiguity function should approximate to a 'thumb tack' (i.e. high resolution with an absence of ambiguities and sidelobes in both the range and Doppler co-ordinates)

The receiver must also be able to recreate the signal – when this signal has unpredictable modulations, direct reception of it is essential.

Apart from a few military systems, communication systems usually have frequency bands that are fixed, or, at least, predictable. Their antenna patterns also tend to be fixed, though they may floodlight a wide area, as occurs with communication satellites. However, their signals will vary in an apparently random manner that may be superimposed on a fixed modulation pattern, such as time division multiplexing or for frame and line rates on TV channels. Such fixed modulation patterns will create unwanted peaks in the signals' ambiguity functions.

11.8.3 *Dedicated illuminators*

These problems with existing illuminators can be avoided by designing bistatic systems that operate with their own dedicated illuminators, preferably mobile ones. Such a bistatic radar system would be much better suited than a monostatic one to operating in a low-probability-of-intercept manner, by using, for example, pseudo-noise (PN) modulated continuous transmissions. Owing to the separation of the transmitter and the receiver, the receiver does not require blanking, so there is not a dead period immediately after the transmitter pulse. Also, there will not be any large, short range echoes from ground clutter and the dynamic range of the received target echoes is smaller. This will reduce the problem in LPI systems of having small and/or long range targets obscured by large and/or short range targets. It should therefore be feasible to move from medium PRF modulations up to continuous signals with pseudo-noise phase modulations. These should have very low probabilities of intercept by conventional ESM systems that have been designed to detect pulsed radars.

For an LPI system to achieve equivalent performance to a pulsed system, each pulse would have to be replaced with a PN modulation pattern which had an element length equal to the required compressed pulse length and an overall duration equal to that of the PRI in the pulsed version. The simplest PN patterns are maximal length sequences which are generated with shift registers, with appropriate feedback connections, and have lengths of $2^N - 1$, where N is the number of stages in the register. One approach would therefore be the use of such sequences which were truncated to match the length of the PRI. However, the high levels of range sidelobes resulting from the pulse compression of a truncated sequence would probably not be acceptable, and PN patterns might have to be devised to match the durations of the PRIs.

The pulse chasing mode of antenna scanning cannot be implemented as readily with LPI modulations as when using the conventional low duty cycle pulses. The antenna beam has to dwell in each direction for a sufficient time to allow the reception of a complete PN pattern. This can most readily be achieved by using a fan of contiguous beams from a phased array antenna and providing a correlator/pulse compression filter (PCF) for each of the beam outputs. However, there will be transients and energy losses if a PCF is switched between beams at times when it is still collecting the full, uncompressed echo pulse from a target. This implies having a PCF for each beam, unless a switching arrangement can be devised to allow them to share a smaller number of PCFs.

The simultaneous, or alternating, use of several illuminators would enhance ECM resistance and the use of orthogonal PSK modulations should reduce mutual interference. It should also provide greater target location accuracy, since a target's bistatic range can be measured more precisely than can its bearing.

11.8.4 *Antenna scan rates*

When the bistatic receiver operates with a conventional, low PRF, scanning radar, the signal processor will be constrained to operating on the same number of hits per target as the monostatic radar – typically about 5 to 10. This may be enough for MTI operation but is not really enough for any significant Doppler processing, especially since the amplitude and phase of the set of echoes will be modulated by the antenna pattern as it scans across each target. If the radar uses pulse-to-pulse RF agility, then any sort of Doppler processing will be prevented.

For a bistatic system to be able to achieve useful Doppler resolution, the transmitter's PRF has to be increased, typically to about 1 to 10 kHz. The transmitter antenna's scan rate then has to be sufficiently slow to ensure that the set of echoes, as required by the FFT processor, can be collected from every target, even those at the maximum operating range; typically, a 32-pulse FFT would be required. If a sequence of several different PRFs is used, to resolve range and velocity ambiguities, then the scanning of the antenna would have to be slowed down even further. Such transmitter operation is unlikely for ground-based radars, except for dedicated illuminators and, possibly, monostatic radars whose conventional operation was prevented by ECM.

Suitable target illuminations might be provided by AI and AEW radars, since they are designed to operate with medium to high PRF agile modes. The antenna scan sector achieved by each receiver beam will be restricted in the manner described in Section 11.7.1, if the radars use PRFs of several tens of kHz.

11.9 Receiver antenna requirements

With a monostatic radar, the antenna can normally be expected to collect all the transmitter energy that is scattered back towards it, since the transmitter–target and target–receiver paths are identical and the transmit and receive functions have the same antenna patterns. The assumption of identical paths will not be exactly true when the propagation medium is gyrotropic, for example, with space surveillance radars operating at VHF or below, as their signals pass through the ionosphere; nevertheless, the differences between the transmit and receive directions will not be significant. Also, some radars, like those mentioned in Section 11.1.1, use separate, but adjacent, antennas as a method of protecting the receiver from the transmitter pulses. Other radars, like the GEC 'Martello', use one antenna array to develop different antenna patterns for the transmit and receive functions.

For a bistatic radar to have a data rate comparable with that of an equivalent monostatic radar, the receiver antenna has to be able to generate beams that follow the directions from which energy may be scattered towards it by any targets encountered by the transmitter pulses. These antenna beams also have to adapt so as to collect as much as possible of the energy that is scattered back towards the receiver by every target; the phased array antenna options for achieving this are described in Section 11.9.2. The collection of all the energy from each pulse is particularly important in the case of pulse compression waveforms, since, if these are truncated before entry to the pulse compression filter, the range sidelobes will increase while the peak amplitude of the main lobe will fall. Conversely, the beams should not be too wide as this will allow the antenna to collect more energy from interfering sources while reducing its gain for the wanted signals. The remainder of this Section considers antenna scanning in the simplified case where the transmitter, receiver and target can be treated as lying in the same plane, with linear propagation between them. The approximations and assumptions are the same as those used in Section 11.5, when deriving the basic formulae of bistatic radar performance.

11.9.1 *Antenna scanning for pulse chasing*

11.9.1.1 *Scan angles*

The scan angles and scan rates for pulse chasing can be analysed using the geometry illustrated in Fig. 11.18. From the law of sines,

$$\frac{R_T}{\sin\theta}=\frac{R_R}{\sin\phi}=\frac{L}{\sin(\theta-\phi)} \tag{11.50}$$

Assuming that a receiver antenna beam is being scanned so that its boresight always points in the direction from which the centre of any target echoes would be received, then, if a pulse is received at a time t after it left the transmitter antenna,

$$ct = R_T + R_R \tag{11.51}$$

From the rules of proportion

$$\frac{R_T + R_R - L}{\sin\theta + \sin\phi - \sin(\theta - \phi)} = \frac{R_T + R_R + L}{\sin\theta + \sin\phi + \sin(\theta - \phi)} \tag{11.52}$$

It follows from eqns. 11.51 and 11.52 that

$$\frac{ct - L}{ct + L} = \frac{\sin\left(\frac{\theta + \phi}{2}\right)\cos\left(\frac{\theta - \phi}{2}\right) - \sin\left(\frac{\theta - \phi}{2}\right)\cos\left(\frac{\theta - \phi}{2}\right)}{\sin\left(\frac{\theta + \phi}{2}\right)\cos\left(\frac{\theta - \phi}{2}\right) + \sin\left(\frac{\theta - \phi}{2}\right)\cos\left(\frac{\theta - \phi}{2}\right)} \tag{11.53}$$

whence

$$\tan\left(\frac{\theta}{2}\right) = \frac{ct + L}{ct - L}\tan\left(\frac{\phi}{2}\right) \tag{11.54}$$

This expression is only valid at times when $ct \geqslant L$, i.e. after the pulse has had time to reach the receiver. The receiver antenna azimuth has been plotted in Fig. 11.19 as a function of the transmitter antenna azimuth and the pulse delay time. There are some interesting special cases:

if $ct \gg L$	then $\theta \to \phi$	i.e. operation approximates to the monostatic case
if $\phi = 0$ and $ct = L$	then $\theta = \pi$	i.e. operation along the baseline (forward scatter)

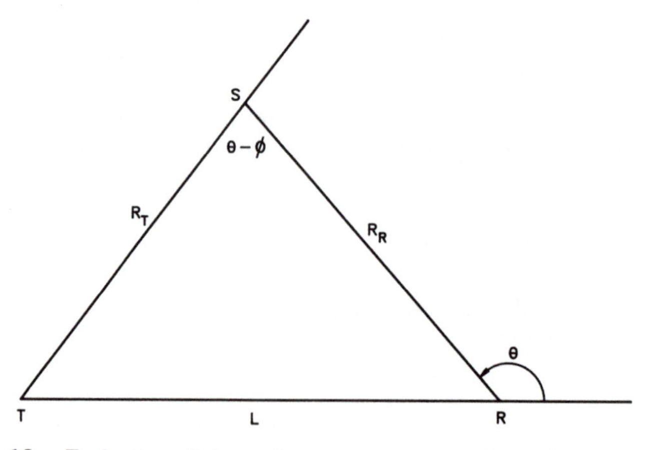

Figure 11.18 *Evaluation of the receiver antenna scan angles and scan rates for pulse chasing*

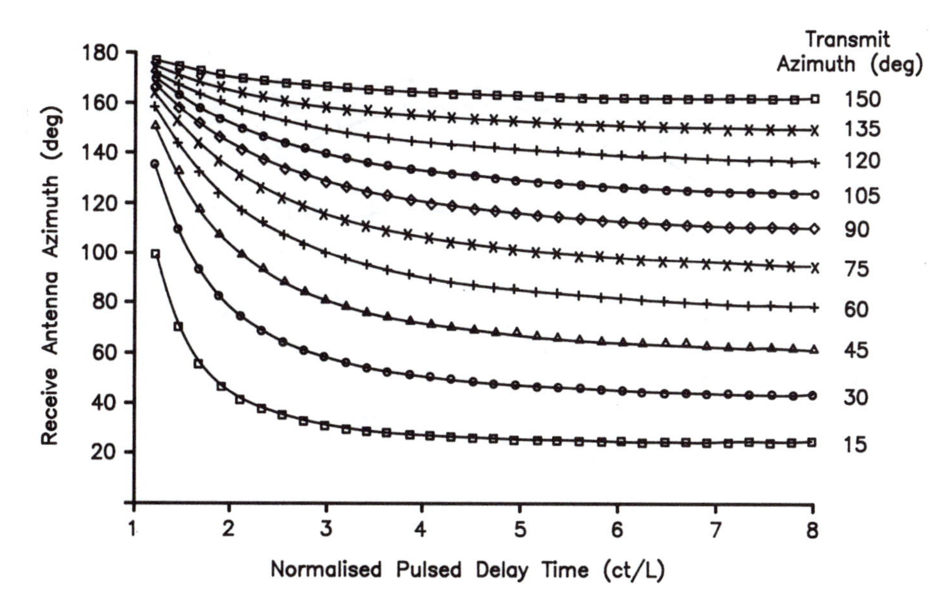

Figure 11.19 *Receiver antenna azimuth angle versus pulse delay time and transmitter antenna azimuth*

if $\phi=0$ and $ct>L$	then $\theta=0$	i.e. operation along the extended baseline
if $\phi=\pi$ and $ct>L$	then $\theta=\pi$	

11.9.1.2 *Scan rates*

The antenna scanning rate $d\theta/dt$ could be obtained from differentiation of eqn. 11.54, but a more convenient expression for differentiation is

$$\frac{ct}{L}=\frac{\sin\theta+\sin\phi}{\sin(\theta-\phi)} \tag{11.55}$$

whence

$$\frac{d\theta}{dt}=-\frac{c}{L}\,\frac{1-\cos(\theta-\phi)}{\sin\phi} \tag{11.56}$$

This expression shows that, for any fixed transmitter antenna orientation defined by ϕ, the scan rate $d\theta/dt$ will depend on the bistatic angle subtended at the target, namely $(\theta-\phi)$; the scan rate is also inversely proportional to the length L of the baseline. The loci of constant scan rates are illustrated in Fig. 11.20 for a 100 km baseline.

When $ct\gg L$, then, as before, $\theta\rightarrow\phi$ and the operation of the radar tends towards the monostatic case and the scan rate drops towards zero. For smaller values of the round-trip distances $c\tau$, some care is needed in interpreting eqn. 11.56. For a non-zero value of ϕ, the scan rate approaches a maximum as θ approaches π; this occurs when the resolution cells lie close to the baseline, so

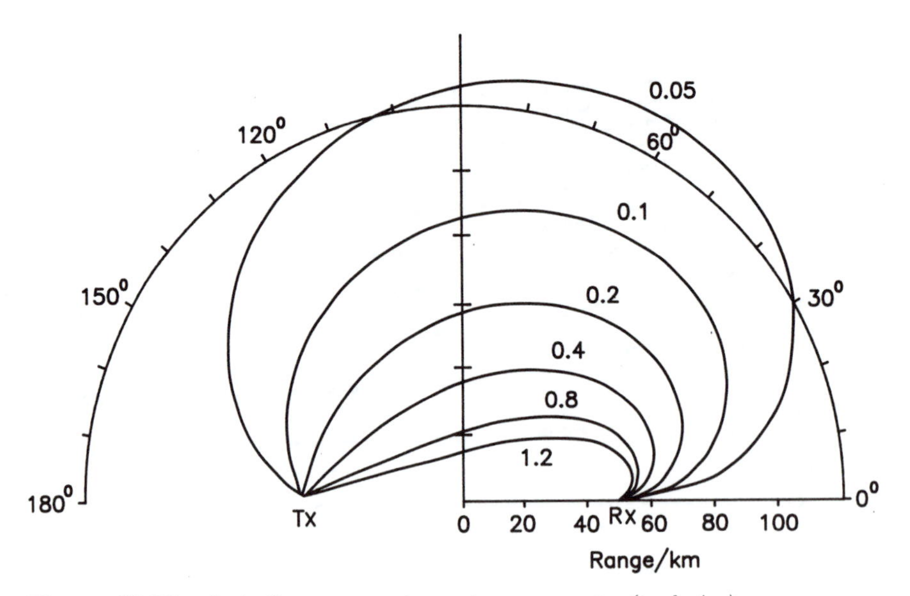

Figure 11.20 *Loci of constant receiver antenna scan rates (in deg/μs)*

that the receiver antenna is pointing close to the direction towards the transmitter and $c\tau \rightarrow L$. This limiting value is

$$\left[\frac{d\theta}{dt}\right]_{\max} = -\frac{c}{L}\frac{1+\cos\phi}{\sin\phi} \tag{11.57}$$

This maximum scan rate becomes infinite when $\phi = 0$, corresponding to the instant when the transmitter pulse passes through the receiver antenna after being transmitted along the baseline. It can be noted from eqn. 11.56 that the maximum scan rates do not occur when the receiver beam is perpendicular to the direction of propagation of the transmitter pulses (namely at a bistatic angle of $\pi/2$) as might have been expected from a superficial examination of Fig. 11.18. This is because the antenna has to follow the directions from which target echoes may reach the receiver rather than pointing at the current position of the pulse.

11.9.1.3 *Beamwidth*

The geometry for assessing the minimum receiver beamwidth is shown in Fig. 11.21. Jackson [1986] has described how the beamwidth must, at any given time, embrace all the directions from which reflected energy might arrive. The transmitter beamwidth and the pulse duration both make distinct contributions to the required receiver beamwidth.

For vanishingly short pulses, simultaneous echoes could be received from any targets on the appropriate range ring and inside the transmitter beamwidth. As was pointed out in Section 11.5.4, the transmitter and receiver beams intersect the elliptical range rings at the same angle, and it follows that the receiver beamwidth B_{R1}, to allow for the transmitter beamwidth B_T, is

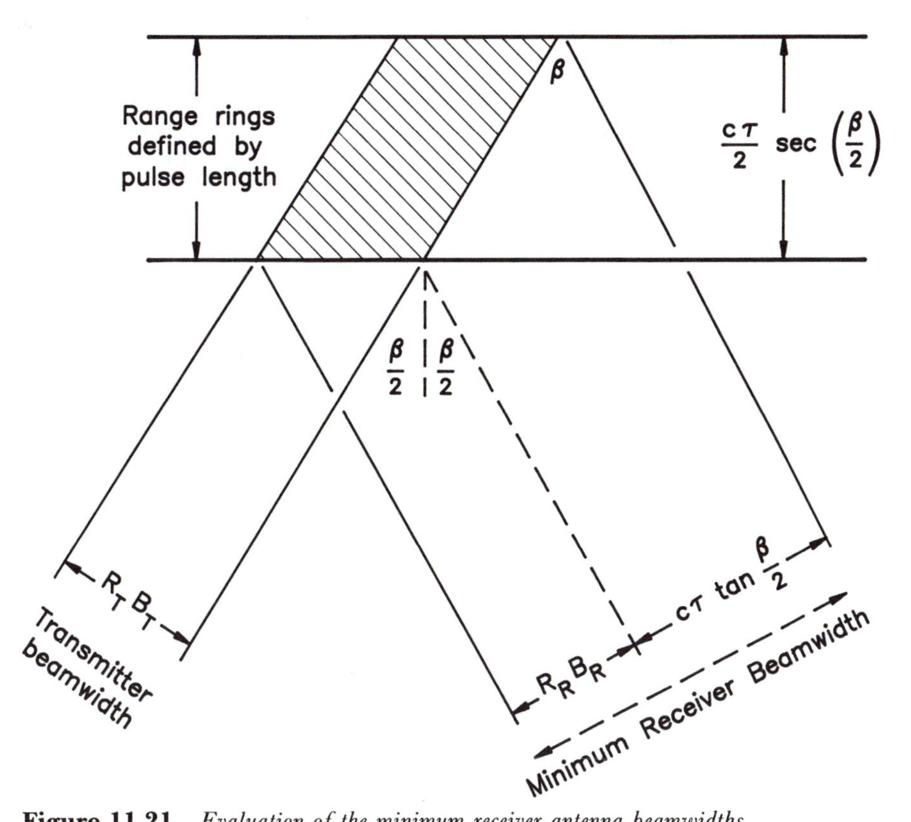

Figure 11.21 *Evaluation of the minimum receiver antenna beamwidths*

$$B_{R1} = \frac{B_T R_T}{R_R} \tag{11.58}$$

The pulse length contribution to the receiver beamwidth B_{R2} is the angle subtended at the receiver by the proportion of the transmitter sightline which lies between range rings that are separated by δ, the spatial equivalent of the (uncompressed) pulse duration. Using the value for δ given by eqn. 11.7, the pulse length contribution becomes

$$B_{R2} = \frac{c\tau}{R_R} \tan\left(\frac{\beta}{2}\right) \tag{11.59}$$

Using eqn. 11.50, the total receiver beamwidth B_R can then be expressed as

$$B_R = B\,\frac{\sin\theta}{\sin\phi} + \frac{2c\tau \sin^2\left(\frac{\beta}{2}\right)}{L \sin\phi} \tag{11.60}$$

The pulse length contribution to the beamwidth is small when the antenna azimuths are nearly equal and $\beta \rightarrow 0$, i.e. at long range or on the extended baseline.

11.9.1.4 *Elevation patterns*

The receiver antenna will need to accommodate to, or compensate for, deviations in the path of the pulses owing to the atmospheric propagation effects discussed in Section 11.6. The predictable deviations could, in principle, be accommodated by vertical scanning of the receive antenna, possibly assisted by a monopulse elevation tracking capability with feedback to the pulse chasing computer. However, it is simpler to use an antenna with a broad (possibly tapered) elevation cover. It also greatly simplifies the beam forming problems in the phased array antenna. Such an antenna cannot provide any height resolution and will be slightly more susceptible to interference owing to a lack of spatial resolution. Height information could be obtained if the transmitter radiates pulses whose RFs or modulations vary in a well-defined manner with their elevation angles.

With phased array antennas, beams with wide elevation cover bend outwards by progressively greater amounts as the azimuth and elevation angles move away from boresight – an effect that is normally known as 'coning'.

11.9.2 *Phased array antenna options*

The spread of angles within which target echoes may reach the receiver antenna at any given time will depend on the antenna pattern at the transmitter, the bistatic geometry and the length of the uncompressed transmitter pulse. As described in Section 11.7.1, there may be several such cones of angles simultaneously, depending on the PRF of the transmitter. Ideally, the receiver should be able to match a receiver beam to each cone and then adapt it as each corresponding pulse is chased. The phased array antenna has two main ways in which it can collect all the available echo energy without truncating any of the pulses:

(i) The antenna can provide an independent beam to chase each of the transmitter pulses, but the widths of its beams have to be matched to the transmitter beamwidth and pulse length, as derived in Section 11.9.1.3. Variations in gain and phase shift during pulse chasing also have to be minimised because of their effects on pulse compression.
(ii) The antenna can point a beam in each direction long enough to collect all the pulse energy returned from within its beamwidth. To collect all the available energy the antenna has to generate a fan of contiguous beams covering the sector of interest and to switch the signal processor(s) between these beams at times that can be derived from eqn. 11.54. This is not a straightforward exercise, since parts of a pulse can be in more than one beam simultaneously, for example, as the pulse crosses from one beam to the next.

The design and operation of phased array antennas have been described by many authors, but the review of beamforming options by Wallington [1985] is well suited to the assessment of antennas for bistatic radar receivers.

RF beamformers precede the receivers in an antenna and their losses will affect the system's sensitivity unless each element in the array is preceded by an RF amplifier. The need for such networks to be substantially lossless constrains any multiple beam set to have orthogonal beams, as the formation of non-orthogonal beams requires the sharing of energy between these beams.

Multiple individually scanned or adaptive beams cannot be produced with an RF network as orthogonality of beams cannot be guaranteed. RF beamformers are, however, ideally suited to single fixed beam applications and such networks exist in many systems, providing a wide variety of beamshapes. A typical example is the stripline cosec-squared feed used to combine the outputs from the vertical sub-arrays in the antennas used in RSRE's bistatic radar programme (see Section 11.14). The use of RF beamformers to generate single fixed beams leaves the succeeding beamforming stages with the much simpler task of providing scanning in a single dimension.

IF networks can be used to form non-orthogonal beamsets, since the receiver gain prior to the beamforming network means that it need not be lossless. Although this allows considerable freedom in defining beamshapes, a minimum loss is incurred and this implies gains of 40–50 dB in the individual receivers. Such IF beamforming networks require identical, stable receivers for each array element (or for each sub-array if an RF beamformer is used); these receivers need a complicated local oscillator distribution network, plus calibration facilities to maintain adequate phase and gain tracking.

IF beamformers are well suited to the production of large numbers of fixed beams, fully populating wide antenna fields of view; the best performance so far has been achieved by the resistive beamforming network. This network operates by sampling four quadrature components of each (sub-) array element's signal and performing vector summation over the complete array.

RF phase shifters in the receivers can scan the complete set of beams, formed at IF. IF phase shifters can produce independent steering of beams; although this technique is complex and suffers from problems with stability and spurious coupling, it can compensate for beam squint and other phase errors and it produces very agile beams.

Digital beamformers are considered further in Section 11.17, on possible future developments.

11.10 Signal processing considerations

The signal processing requirements for a bistatic receiver are likely to be very similar to those for a monostatic receiver operating with the same transmitter waveform, although the bistatic receiver will require synchronisation to the transmitter waveform if it is to achieve coherent processing and it is likely to observe markedly different clutter Doppler spectra. A coherent bistatic receiver is likely to include most, or all, of the sub-systems shown in Fig. 11.22 (this is actually the block diagram of the UK Argus bistatic receiver, which is described in Section 11.14.1). The order of the sub-systems may vary slightly in other applications; for example, in the US Sanctuary radar (described in Section 11.15.1), some Doppler processing precedes the pulse compression. If the radar uses digital pulse compression, rather than analogue pulse compression, the in-phase and quadrature analogue-to-digital converter would need to precede the pulse compressor.

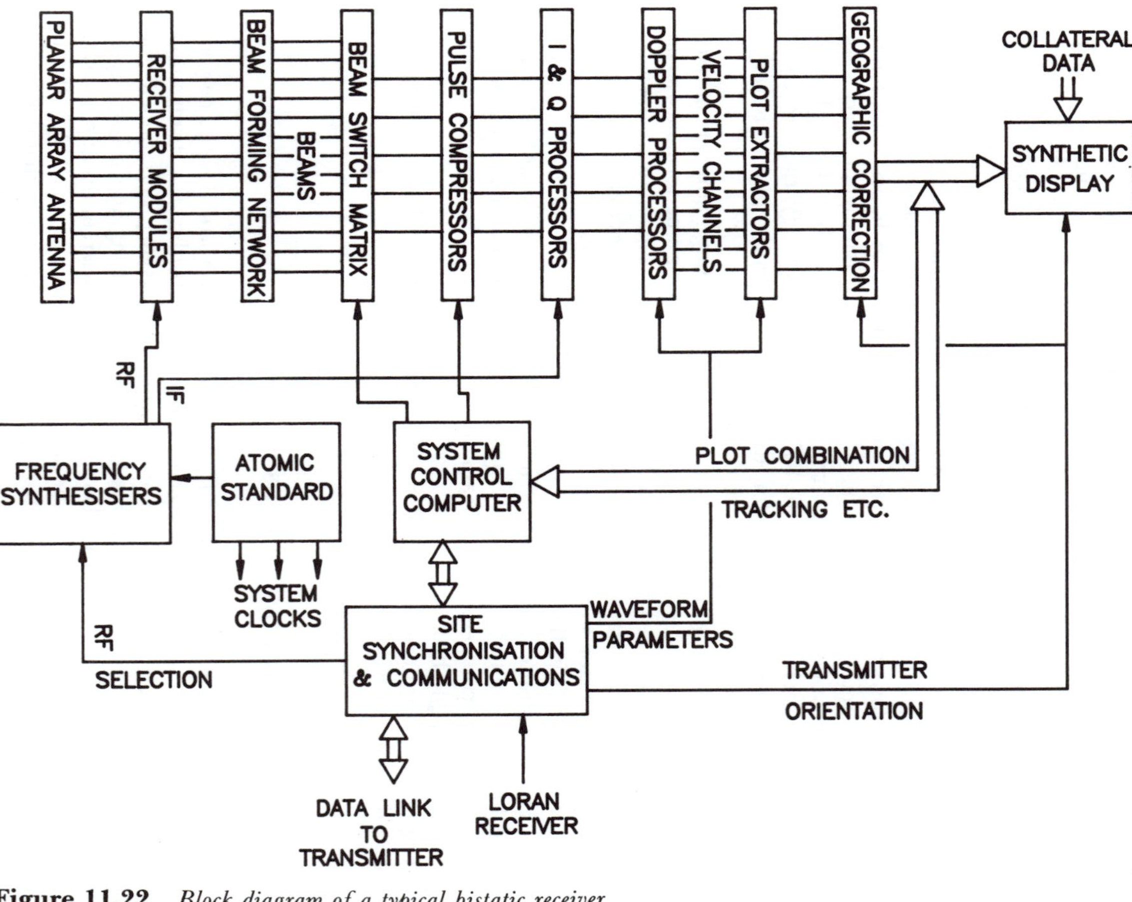

Figure 11.22 *Block diagram of a typical bistatic receiver*

11.10.1 *Pulse-compression/matched-filtering*

Like monostatic radars, bistatic radars benefit from the use of pulse compression modulations to achieve greater pulse energy without sacrificing range resolution. High range resolution reduces the radar's resolution cell size and hence increases the target to clutter ratio; it can also be required in some multistatic deployments where several range measurements are being used to provide accurate target location in three dimensions. The bistatic radar needs to ensure that the complete receiver from the antenna to the output from the pulse compressor is matched to the received waveforms. If they are not, the amplitude of the mainlobe of the compressed signal will be reduced while the range sidelobes will be increased. The first effect is usually tolerable but the second is not, for it will lead to false target detections, especially in a processor that has to resolve range and velocity ambiguities.

The capability of bistatic radars to make use of pulse compression techniques is greater than in the monostatic case, since the separation of the transmitter and the receiver avoids the need for receiver protection and reduces the dynamic range of the echo signals at the receiver. This can allow the radar to use much higher duty cycles, possibly to the extent of using continuous transmissions, with a probable reduction in the vulnerability of the transmitter to detection by ESM. Analogue devices can be used, but the problem with them lies in the need to have matched pairs of pulse compressors for each waveform that is used. If only limited waveform flexibility is required, then pulse expansion and compression can be achieved simply and reliably with SAW devices. However, the expander and compressor will no longer be collocated, as they would with a monostatic radar, and close temperature control will be required at the two sites to maintain accurate matching. This requires a degree of equipment co-ordination that would be inconvenient even for a bistatic receiver intended for dedicated use with co-operative radars.

The use of digital pulse compression techniques avoids the need for separate pulse compressors for each of the waveforms the bistatic receiver may use. Instead, the pulse compressor simply changes to a different set of weights. In the bistatic case, these could be held in store, ready for use with any of the transmitters operating in the vicinity of the receiver. This form of pulse compression also offers the capability of capturing one of the initial pulses from a burst radiated by any radar and then processing it to derive the weights required to compress the remaining pulses. The complex conjugate of the FFT of the initial pulse, received directly from the transmitter, would give the matched filter response. The complex FFT of each echo would then be multiplied in the frequency domain by the matched filter response, followed by an inverse FFT to yield the compressed pulse. In practice, the complete set of returns for all the range cells in a PRI would be processed as an array, rather than as individual signals.

Pulse chasing by a continuously scanning phased array antenna is not always well suited to pulse compression since, in practice, it could cause phase and amplitude modulation of the received signal as the beam followed the pulse. The gain of the scanning antenna will have the usual secant dependence on the scan angle of boresight, possibly complicated by an attempt to match its beamwidth to the pulse size (as described in Section 11.9.1.3). A multibeam phased array antenna will truncate the pulse compression waveforms unless each beam has a pulse compression filter connected to it for as long as it is able to collect reflected target energy. In the extreme case of modulated CW this would imply one filter per beam, but with lower duty cycles, a smaller number of filters could be switched between beams.

The sampling and analogue-to-digital conversion (by the I&Q processors) of the signals from the pulse compression filters has to occur at intervals equal to the compressed pulse length and at times that ensure that they act as matched filters; for example, sampling should not occur while the input to an I&Q processor is switching between beams.

11.10.2 *Doppler processing*

The Doppler processing is normally based on the use of the Fast Fourier Transform (FFT) and an FFT processor will require 2^N samples for each resolution cell, where N is an integer. The illuminating radar has to operate at a fixed RF and PRF while an array of 2^N samples is acquired for each resolution cell; the time to fill this array is sometimes known as the coherent processing interval. Since the FFT process implicitly assumes that the signal from each resolution cell is periodic, with its period equal to the time taken to acquire the set of samples, the set of samples has to be weighted so that the amplitudes taper towards zero at each end of the set.

The Doppler data are resolved into frequency channels that are PRF/(FFT length) wide. From a knowledge of the sidelobe patterns generated by the weighting function, it is possible to interpolate between frequency channels, so improving the Doppler resolution by a factor, typically, of about 8. This reduces the errors in the subsequent velocity ambiguity resolution and aids target tracking.

11.10.3 *Plot extraction*

The operations carried out in the plot extractor of a bistatic radar would normally include some, though not necessarily all, of the following functions [Pell and Crossfield, 1984]:

CFAR processing
Range masking
Doppler masking
Range ambiguity resolution
Doppler ambiguity resolution
Second threshold (moving window integration)
Multiple beam data combination
Plot message generation

11.10.3.1 *CFAR processing*

Modern monostatic radars commonly convert their target data to log modulus form before the derivation of the background average. The resultant average is then a geometric one, rather than an arithmetic one, and this confers several advantages. For an average over N samples, the contribution of each sample to the geometric average is the Nth root of its value while, in the arithmetic average, it is its value divided by N. This means that a geometric average is much less affected by the occurrence of a few isolated large signals amongst a large set of other small signals. It therefore reduces the influence of target signals on the background averages of neighbouring range cells. It also suppresses the effect on the background average of interfering pulses from other radars or from responsive jammers.

Subtraction of a logarithmic background average from a logarithmic target signal produces the logarithm of their ratio. Detection thresholds, therefore, do not have to be adjusted to compensate for any fluctuations in the overall gain of the receiver or the antenna. For the same set of data, the geometric average is always less than or equal to the arithmetic average (equality occurs when the data are all equal). Thus the apparent signals after the subtraction of an arithmetic background average will be smaller than with a geometric average. The setting of the detection thresholds will involve the usual trade-offs between probability of detection and false alarm rate in the local jamming and target environment.

11.10.3.2 *Range and Doppler masking*

A typical target will be extended in both range and Doppler. The range spread can arise from such effects as:

- The range sidelobes on pulse compression
- The target crossing between range cells
- The target spreading over several range cells.

The last of these range spreading effects would only occur for very large targets or for high resolution radars.

The Doppler spread can arise from such effects as:

- The weighting function that is applied to the set of echoes before the FFT
- The sideband noise on the transmitter and/or the local oscillators in the receiver
- Modulation generated by the target itself (from propellers or jet engines).

Each target therefore appears as a distributed peak in range–Doppler space. This distributed peak has to be processed to yield one plot in range–Doppler space, normally at either the peak or the centroid of the distribution, and the remainder of the distribution is masked. If it were not, multiple false targets would be generated for each true target by the ambiguity resolution processes.

11.10.3.3 *Range and Doppler ambiguity resolution*

Since bistatic radars will usually employ medium to high PRFs to reduce aliasing effects in the frequency domain, they will have to employ burst-to-burst PRF agility to resolve the range and velocity ambiguities. These PRF sequences should be devised to incorporate some redundancy so that the loss of data from one PRF (or possibly even several PRFs), due to jamming or RCS nulls, will not defeat the ambiguity resolving circuits in the signal processor. The PRF sequences will also have to allow for the expected range and Doppler spreads of the expected targets lest the signal processor should generate spurious correlations.

11.11 Bistatic RCS characteristics of targets and clutter

11.11.1 *Bistatic RCS of targets*

The detection capabilities of radars cannot be assessed without RCS data for all their potential targets, defined as functions of the target's orientation relative to the radar. Calibrated radars could be used to observe an aircraft in flight, but this approach cannot provide complete and accurate records, although fading rates and statistical distributions of RCS can be obtained for typical scenarios. The range of available aspect angles is limited by each aircraft's flight envelopes; also, aircraft are unable to fly consistently along straight lines but suffer from pitch, yaw and roll in amounts determined by the extent of local atmospheric turbulence. Unless the complete aircraft can be mounted and rotated at an RCS measurement range, the RCS data have to be estimated by computer simulation or measured at scaled frequencies on accurate scale models of the aircraft. RCS measurements on both models and full-size aircraft have usually been made for military purposes and so are not available in the open literature. This is particularly true for bistatic RCSs.

11.11.1.1 *Predictions of bistatic radar cross-sections*

Accurate mathematical equations exist for both the monostatic and bistatic RCSs of a range of simple structures, such as spheres, cones, plates, dihedrals and trihedrals; in fact, spheres and trihedrals are used to calibrate both RCS measurement ranges (whether full-size or scaled) and to test computer simulations. Estimates of the RCSs of aircraft and missiles with relatively smooth shapes can also be derived from combinations of such simple structures. In the high frequency limit ($a > 3\lambda$), the bistatic RCS of a perfectly conducting sphere of radius a is

$$\sigma_B = \pi a^2 \tag{11.61}$$

for all bistatic angles, except those close to π [Ruck *et al.*, 1970*a*].

More complex structures have RCSs that are due to the combination of many signals that, for example, may be reflected by trihedrals, specularly reflected by smooth flat surfaces or diffracted by edges. These signals combine vectorially to create a distribution of power over the complete solid angle of 4π steradians, that is dependent on the aspect and polarisation of the transmitter with respect to the target. For targets with highly conducting surfaces, the integral of the scattered power over 4π steradians will be almost as large as the total incident power intercepted by the target. This would not be true for other target surfaces, for example Stealth aircraft, where some of the RCS reduction could be achieved by absorptive surface coatings. As the targets become larger and/or the frequencies become higher, the number of wavelengths between the various scattering or reflecting surfaces increases and the reflected power therefore varies more rapidly as a function of the aspect angles to both the transmitter and the receiver.

Estimates of bistatic radar cross-sections of simple targets can be obtained from the theorem of Crispin, Goodrich and Siegel [1959] which states that: 'For perfectly conducting bodies which are sufficiently smooth, in the limit of vanishing wavelength, the bistatic cross section is equal to the monostatic cross section at the bisector of the bistatic angle between the directions to the transmitter and the receiver'. They used the physical optics approximation to show that the theorem is approximately true only if the bistatic angle is considerably less than 180°. The theorem is clearly not valid as the bistatic angle approaches 180° since it does not predict the forward scatter effects described in Section 11.11.1.3. The assumption of smooth, perfectly conducting bodies does not apply to most current aircraft. Indeed, the US F–117A Stealth fighter appears to go to the opposite extreme, with a skin that is completely composed of flat surfaces. As aircraft are made more stealthy, by the elimination of specular reflections from edges, flat surfaces and trihedrals, they may well have shapes that come closer to the smooth body assumption, though a surface with a RAM coating is clearly not perfectly conducting. However, the theorem does imply that the range of bistatic RCSs is approximately the same as the range of monostatic RCSs.

The theorem of Crispin *et al.* does not apply when the radar wavelength is long compared with the target dimensions. Kell [1965] has shown, for small bistatic angles β, that the bistatic RCS, σ_B, is very closely approximated by the monostatic RCS, σ_M, measured on the bisector of the bistatic angle at a frequency lower than the true frequency by a factor $\cos(\beta/2)$.

$$\sigma_B(f) = \sigma_M\left(f\cos\left(\frac{\beta}{2}\right)\right) \tag{11.62}$$

This generalised monostatic-bistatic equivalence theorem is not applicable to bodies in which multiple scattering is important, nor to convex bodies in which the contributions from creeping waves are relatively important. Bickel [1966] concluded that the monostatic equivalence theorem can be made exact for non-intersecting scattering centres for which the scattering can be described by physical optics but that the theorem cannot be extended to depolarising bodies. The usefulness of Kell's theorem is limited, since it only applies for small bistatic angles and requires monostatic RCS data to be measured as a function of frequency.

In practice, the bistatic RCS is likely to be rather smaller than the monostatic RCS owing to the loss of corner reflector effects and the shadowing of resonating structures. Flat surfaces will still be able to give specular reflections, but these will occur when the bisector of the bistatic angle is perpendicular to the plane of the surface. A similar effect will occur for dihedrals and edges. The effects of looking down tubes will be lost, or at least reduced, at large bistatic angles; this will be especially significant for aircraft where reflections due to the jet engines can be lost.

11.11.1.2 *Measurements of bistatic radar cross-sections*

Plots of the monostatic RCSs of aircraft and missiles, as functions of azimuth and elevation angles, tend to be dominated by the peaks due to the specular reflections from the fuselage and from the dihedrals at the wing roots, at azimuth angles of 90° and 270°; there will usually be smaller, narrower peaks due to specular reflections from the leading edges of wings. These plots usually also have lesser peaks due to multiple reflections within the engine intakes and exhausts, at azimuths close to 0° and 180°. The observed RCSs at other angles will have substantial contributions from various other scattering centres, especially trihedrals, and these will provide fluctuating lower RCS levels between the peaks.

The plots of the bistatic RCSs also exhibit similar specular reflections owing to the fuselage and wing roots, though these tend to become slightly smaller as the bistatic angle increases, partly because of reductions in the projected cross-section area of the target. The engine intakes and exhausts make a smaller contribution to the RCS, due to shadowing effects as the bistatic angle increases. The contributions from trihedrals decrease rapidly with bistatic angle. The angular range over which the monostatic RCS of a trihedral is within 3 dB of its maximum value is a cone about its axis of symmetry. According to Peters [1962], square, circular and triangular trihedrals have, respectively, 23, 32 and 40° cones. The contributions of trihedrals to the bistatic RCS fall off more rapidly than this; experimental measurements by Peters [1959] of the bistatic RCS of a triangular trihedral showed that the enhanced reflection effect had almost disappeared for bistatic angles greater than about 10°.

With monostatic tracking radars, phase interference from two or more dominant scatterers causes the apparent phase centre of the radar reflections from a complex target to wander with changes in the target's aspect angle. This effect, known as 'glint', can cause loss of track. The apparent RCS of a glinting target is anti-correlated with the amplitude of the glint angle. Fawcette [1980] has observed that operation at bistatic angles greater than 15 to 20° can drastically reduce glint. However, the RCS was also reduced by about 5 to 10 dB. Glint is, of course, much more important in missile radars than in surveillance systems. Even here, the improvement in the bistatic case may not be very significant. The electrical centre will be within the physical bounds of the target for the majority of the time, while the tracking circuits will smooth out many of the glint spikes, so reducing the effect of glint even further.

11.11.1.3 *Forward scatter radar cross-sections*

When the bistatic angle at a target, measured between the directions to the transmitter and the receiver, is equal, or close, to π then the RCS is enhanced by forward scattering. This forward scatter RCS can be many times the backscatter RCS (i.e. the RCS observed by a monostatic radar) and will not be reduced by the normal RCS reduction techniques of body shaping and coating with radar absorbent material.

Siegel [1955] has used the physical optics approximation to show that, if the wavelength λ of the radar signal is small compared with the target dimensions, then the forward scatter RCS, σ_F is

$$\sigma_F = 4\pi \frac{A^2}{\lambda^2} \tag{11.63}$$

where σ_F, A and λ must all be in consistent units, and A is the area within the contour separating the illuminated and shadow regions of the target.

In the lower frequency limit, where $k_0a < 0.4$ or $\lambda > 16a$, the pattern of the forward scatter RCS of a sphere has the form [Ruck *et al.*, 1970*a*]

$$\sigma(\pi - \alpha) = 4\pi a^2 (k_0 a)^2 \left(\frac{J_1(k_0 a \sin \alpha)}{k_0 a \sin \alpha} \right)^2 \tag{11.64}$$

As α tends towards zero, the quantity inside the brackets reduces to $\frac{1}{2}$ and

$$\sigma(\pi) = 4\pi \frac{A^2}{\lambda^2} \quad \text{with } A = \pi a^2 \tag{11.65}$$

For example, if a sphere has a backscatter RCS of 0.25 m^2 at 3 GHz, then its forward scatter RCS will be 78.5 m^2, an enhancement of 25 dB. However, the backscatter and forward scatter RCSs are equal for normal incidence on a conducting plate.

In the high frequency limit, the forward scatter RCS for large convex-shaped conducting bodies remains comparatively constant with frequency. The half-power width θ of the lobe of forward scattered energy is then approximately

$$\theta = \frac{\lambda}{D} \tag{11.66}$$

where D is the length of the characteristic dimension of the area A, in the direction along which the lobe width is being measured. According to Babinet's principle, the radiated diffraction pattern of the forward scattered energy is actually the same as the pattern that would be produced by a plane electromagnetic wave incident upon a plane, perfectly conducting screen, with a hole in it of the same cross-sectional area as that of the contour separating the illuminated and shadow regions of the object.

11.11.1.4 *Comparison of monostatic and bistatic radar cross-sections*

The only simultaneous monostatic and bistatic radar cross-section measurements of which the author is aware are those commissioned by RSRE as part of its Argus technology demonstrator programme. The monostatic measurements were made by the Byson radar at RSRE while the bistatic measurements were made by the Argus receiver at the GEC–Marconi Research Centre (for details of these systems, see Section 11.14.1). The first phase of this programme established the synchronisation and other techniques and then made some preliminary measurements. Fig. 11.26 shows a 30 s sequence of these measurements that compared the RCSs for a large civil airliner flying over the southern part of the UK at a height of approximately 2750 m [Matthewson *et al.*, 1990]. These showed a bistatic spread of RCS values that was about 20 dB less than the monostatic spread and had slightly less rapid fluctuations in RCS.

11.11.2 *Bistatic RCS of clutter*

11.11.2.1 *Chaff clutter*

Mack and Reiffen [1964] have analysed the average bistatic RCS of a cloud of lossless resonant half-wave dipoles, with the assumption that the dipoles are uniformly distributed over all orientations. If the transmitting and receiving polarisations are both perpendicular to the plane of the transmitting and receiving propagation vectors, then the average RCS is independent of the bistatic angle and has the same value as in the monostatic case, i.e. 0.153 λ^2. If the polarisations are both in the plane of incidence, the multiplying term falls from 0.153 at a bistatic angle of 0° (the monostatic case) to 0.038 at an angle of 90° – a reduction of 6 dB. For crossed transmitting and receiving polarisations, the term falls to 0.05 at 0° and to 0.043 at 90° – reductions of 4.8 and 5.5 dB, respectively.

These reductions in the apparent RCSs of chaff clouds may not be achieved in practice, since long-range ground-based surveillance radars tend to use horizontal polarisation, and undistorted, metallised glass fibre chaff, with a uniform thickness, could tend to fall with a predominantly horizontal orientation.

As chaff is wind driven, its velocity spectrum should be similar to that of rain, as discussed in Section 11.11.2.2.

11.11.2.2 *Precipitation clutter*

Bistatic scattering from various forms of precipitation can be a source of interference in communications links as well as being a source of clutter for bistatic radars. The problem with communication links occurs when two antennas are operating in overlapping frequency bands and have intersecting beams but are only weakly coupled, if at all, on a direct path via their sidelobes. If a CW signal is transmitted, the bistatic signal is usually Rayleigh distributed and has fluctuations with periods in the millisecond range. It is believed that the received signal will resemble Gaussian noise if the transmitted signal has a modulation bandwidth $\gg 1/T$, where T is the time delay spread of the scattering paths. The short term spectrum of this noise will display intense frequency selectivity on a scale of $1/T$, with the structure itself changing randomly on the millisecond scale [McEwan, 1989].

Falling raindrops, subject only to the effects of gravity and the surface tension of the water, assume an oblate spheroidal shape with the symmetry axis close to the vertical [CCIR, 1982]. If the radar wavelength is large compared with the diameter D of a scattering particle ($\lambda > 8D$), the particle's E-plane and H-plane bistatic RCSs are given by (Ruck *et al.*, 1970*b*]

$$\sigma_E = \frac{\pi^5 D^6}{\lambda^4} |K|^2 \cos^2\beta$$
$$\sigma_H = \frac{\pi^5 D^6}{\lambda^4} |K|^2 \qquad (11.67)$$

The format of Ruck's equations has been altered to match that normally used for the raindrop RCS. Thus, in the low frequency limit, the E-plane cross-section depends on the bistatic angle β, with maxima in the forward and backward directions and a null at $\beta = \pi/2$. The H-plane cross-section is independent of β and is the same as for the monostatic case. This differs from the low frequency bistatic RCS of a perfectly conducting sphere, described in Section 11.11.1.1. A further difference from the perfectly conducting sphere is that the forward scatter and backward scatter RCSs are identical.

The term $|K|$ is given by

$$|K| = \frac{\varepsilon - 1}{\varepsilon + 2}$$

where ε is the dielectric constant of the scattering particle. Over the frequency range 3–10 GHz, and at temperatures between 0 and 20 °C, $|K|$ for water is 0.93 ± 0.004. $|K|$ for ice is about 0.197 at all temperatures and at any frequency in the centimetre band. The RCS density of rain can be derived by summing the contributions of the individual scatterers as defined by eqn. 11.67. This RCS density depends on the sum of the sixth power of the diameter of each of the particles in a unit volume. The monostatic value of this sum is known to radar meteorologists as Z, the radar reflectivity factor, and is usually expressed in $\mathrm{mm}^6/\mathrm{m}^3$. The E-plane and H-plane bistatic values are then

$$Z_E = Z\cos^2\beta = \sum_i D_i^6\cos^2\beta$$
$$Z_H = Z = \sum_i D_i^6 \qquad (11.68)$$

The reflectivity is dominated by the largest drops, owing to the sixth power dependence on the drop size. Thunderstorm rain therefore has a larger Z than stratiform rain, while orographic rain has a much smaller Z. Raindrops normally have diameters in the range 0.2–5.0 mm, so that Rayleigh scattering will occur for wavelengths down to about $\lambda \geqslant 30$ mm (frequencies $\leqslant 10$ GHz). Experimental measurements have shown that Z ($\mathrm{mm}^6/\mathrm{m}^3$) can be related to the rainfall rate R (mm/h) by various empirical formulae of the form

$$Z = \alpha R^\beta \qquad (11.69)$$

There is considerable variation among the reported results for α and β (see, for example, Battan [1973], Table 7.1). This is because of the variability of raindrop distributions with time and location and because of the difficulties in obtaining accurate quantitative measurements. The following results are considered to be fairly typical:

Stratiform rain	$Z = 200\,R^{1.6}$	(11.70*a*)
Orographic rain	$Z = 31\,R^{1.71}$	(11.70*b*)
Thunderstorm rain	$Z = 486\,R^{1.37}$	(11.70*c*)
Snow	$Z = 2000\,R^2$	(11.70*d*)

The accuracy of these expressions is best for uniform drop sizes and is normally not better than ±50%.

The apparent large value of Z for snow arises from its low fall rate, which is only about 10–20% of that of raindrops. The concentration of snowflakes must therefore be larger than that of raindrops to yield the same equivalent fall rate R. Frozen snow tends to give poorer scattering than rain since $|K|$ for ice is only about 20% of that for water. However, a snowflake on the point of melting develops a surface layer of water, with a resultant scattering that is larger than when it is fully melted. The effect of this large scattering on radar displays is known as the 'bright band'.

Non-precipitating clouds containing only liquid water have little effect at frequencies below about 100 GHz. They usually have particle sizes in the range 10–45 μm radius, so their RCS density is much less than that for rain; being so small, they are virtually spherical and do not cause any measurable cross-polarisation effects [McEwan, 1989].

From eqns. 11.67 and 11.68, it follows that the effective RCS density σ_H, for rain in the H-plane case in $\mathrm{m}^2/\mathrm{m}^3$, is

$$\sigma_H = 3.5*10^{-14} f^4 \alpha R^\beta \tag{11.71}$$

where f = radar frequency, GHz
R = rainfall rate, mm/hr

11.11.2.3 *Surface clutter*

There is a significant body of data about bistatic surface clutter, since, under the title of forward scatter clutter, it is of great importance in the assessment of multipath effects. There is an even larger body of data about monostatic ground clutter but there does not appear to be any reliable relationship between bistatic and monostatic ground clutter RCSs. The difficulty in deriving a relationship can be illustrated by an extreme example of the problem. No bistatic or monostatic RCS would be observed for signals incident upon a highly absorbent surface; the same absence of monostatic RCS would be observed for non-normal incidence on a flat, smooth, conducting land or sea but a large bistatic RCS would occur in the direction of specular reflection. The data on both forms of clutter have been reviewed by several authors, notably by Long [1983] and by Ruck *et al.* [1970*c*], and only a few salient aspects have been mentioned in this Section.

Unfortunately, there is great variability in the reported data, the most important reason for this being the variability of the surfaces being measured and in the inability of the various experimenters adequately to describe or quantify these surfaces. Further variations are introduced by the measurement systems, since the scattering coefficient σ_0 is derived from measurements of received powers which will depend on the antenna patterns and on the deployment of these antennas. Long, when discussing sea clutter, has observed that: ‘the wide variability of the data makes analysis difficult . . . usually, even the most rudimentary description of the sea surface conditions (wave height, wave direction, wind speed, wind direction, or a qualitative description) is not reported, not only because these measurements are difficult but also because no one knows precisely what characteristics of the sea surface are important to the radar problem . . . it is virtually hopeless to obtain meaningful results by comparing data at one wavelength from one observer with those at some other wavelength from another observer’.

In a simple theory (see, for example, Long [1983]), the magnitude of the reflection coefficient from the earth is the product of three terms, namely

(i) the magnitude ρ of the reflection coefficient of a plane, smooth earth
(ii) the divergence factor D that describes the reduction in reflection caused by the Earth’s curvature
(iii) a factor R that depends on the surface roughness.

The reflection coefficient ρ can be derived exactly in some simple cases. For example, if a plane electromagnetic wave is incident from free space upon a smooth, flat, dielectric surface, then the reflection coefficients for horizontal and vertical polarisation are given by Fresnel’s equations.

Ground clutter need not be a significant problem for ground-based bistatic radars. The amplitudes of the clutter echoes are normally smaller than for monostatic radars since the clutter cannot be simultaneously at short ranges from both the transmitter and the receiver. Suitably deployed ground-based systems can sometimes avoid ground clutter altogether by using terrain blockage to ensure that no area of ground is simultaneously within the lines of sight from both the transmitter and the receiver. It is this reduction in ground clutter effects that allows bistatic radars to operate with medium to high PRFs for extended Doppler measurement capabilities.

In the airborne case, the ground clutter is still unlikely to be at short ranges from both the transmitter and the receiver, but other complications can arise. The locations and relative motions of the aircraft will alter the distribution of ranges and velocities where targets are obscured by clutter while the amount of clutter detected through the antenna sidelobes can be increased. These problems are considered further in the following Section.

11.12 Airborne bistatic operation

The first trials of an airborne bistatic radar appear to have been those in the UK, in the Autumn of 1936, when a small team led by Dr E. G. Bowen installed a receiver on a Handley Page Heyford bomber and operated it with a ground based transmitter on the roof of the Red Tower at Bawdsey Manor. The system used a wavelength of 6.7 m and the receiver had a half-wave dipole antenna connected between the undercarriage wheels. Because of the bistatic geometry, the range measurements were only correct when the Heyford was between the ground transmitter and the target aircraft. With the Heyford circling at about two to three thousand feet above Bawdsey, aircraft were regularly detected at ranges of 8 to 10 miles, with a maximum range of around 12 miles. A transmitter was added to the Heyford in March 1937 [Bowen, 1987].

The semi-active seekers on some types of anti-aircraft missiles are examples of airborne bistatic radar operation and such missiles have been in use since the 1950s. These seekers rely on the target being illuminated by the transmissions from a radar that is normally on, or near, the launch platform. Their receivers then process the target echoes using a transmitter reference signal that is collected by a separate rear-facing antenna. This technique is used in the British Sea Dart naval surface-to-air missile (SAM) and the Bloodhound ground-launched SAM; it is also used in the American Hawk SAM and the Sparrow air-to-air missile (AAM).

11.12.1 *Doppler effects of platform motion*

The motion of one or both platforms can have deleterious effects on the performance of bistatic radars. Three types of platform motion have to be considered, namely

(i) uniform motion
(ii) erratic motion
(iii) vibration.

Uniform motion of the platforms relative to each other will alter the Doppler frequencies measured by the receiver. The resultant Doppler frequency f_D can be derived from a vector equation as

$$f_D = \frac{(\boldsymbol{V}_T - \boldsymbol{V}_S) \cdot \boldsymbol{U}_T + (\boldsymbol{V}_R - \boldsymbol{V}_S) \cdot \boldsymbol{U}_R}{\lambda} \tag{11.72}$$

where $\boldsymbol{V}_T$ = transmitter platform's velocity vector
$\boldsymbol{V}_R$ = receiver platform's velocity vector
$\boldsymbol{V}_S$ = scatterer's (target's) velocity vector
$\boldsymbol{U}_T$ = unit vector from the transmitter to the scatterer
$\boldsymbol{U}_R$ = unit vector from the receiver to the scatterer

The Doppler spectra of chaff and rain clutter will also be moved and broadened. These changes need not present any significant problems to a well designed CFAR processor which will automatically adjust its detection thresholds in both range and Doppler to maintain the required false alarm rates. Similarly, the conversion from Doppler frequencies to relative target velocities will be straightforward, provided the platform's locations and velocity vectors are known at the receiver with sufficient accuracy.

Erratic motion of the platforms, due to wind shear, buffeting or rapid manoeuvres, is unlikely to have any serious effects on the performance of the Doppler processing. The periods of such motions are typically in the range from tenths of a second up to several seconds. They are thus at least two orders of magnitude longer than either the PRIs of typical low PRF surveillance radars or the coherent pulse bursts of medium to high PRF AI or AEW radars. The effect of any erratic motion on the platform's velocity vector will have to be included in the conversion from Doppler frequency to target relative velocity. The effects of any erratic platform motion will also have to be compensated in the commands to the receiver's phased array antenna if accurate pulse chasing is to be maintained.

Vibration can have two effects. It can perturb critical analogue sub-systems in the signal processor, such as the atomic clocks and the frequency synthesisers, so leading to errors in the reference signals to the Doppler processor. It can also produce pulse-to-pulse phase changes in the received signals, so broadening Doppler spectra and degrading the ability of the Doppler processor to distinguish targets from clutter. The vibration on a large aircraft in cruising flight is likely to be acceptable, but it could cause problems on smaller aircraft, especially when manoeuvring, and it probably would not be acceptable on most helicopters.

11.12.2 *Synchronisation*

The receiver aircraft requires the location and velocity of the transmitter aircraft in three dimensions. The transmitter aircraft therefore requires a navigation system that provides a continuous, highly accurate measure of these parameters. This can most effectively be provided by an inertial navigation system that is updated by GPS–NavStar. The receiver aircraft will also require the true azimuth and elevation angles of the transmitter (referred to the ground) when each pulse leaves the transmitter.

For efficient operation with a dedicated or, at least, co-operative illuminator, there will need to be a continuously available, narrow band data link from the transmitter to the receiver. In principle, it would be feasible to modulate details of future transmitter parameters onto a few transmitter pulses during each scan of the transmitter's antenna. However, this approach could not accommodate sudden changes in the transmitter aircraft's trajectory. A more attractive option would be a dedicated data link, preferably a low-probability-of-intercept one.

This could be achieved with pseudo-noise modulations, with frequencies in the atmospheric absorption bands or with high gain tracking antennas.

If the receiver is not deployed on a fighter that will use the target information solely for its own purposes, then a further data link will be required to return target data, so that they can be added to the recognised air picture. The data bandwidth can be very small if the echoes are processed on the receiver aircraft, so as to reduce them to target plots.

11.12.3 *Ground clutter effects*

The relative motions of the platforms will alter the distribution of ranges and velocities where targets are obscured by clutter and the broadening of the clutter spectra may reduce the overall target detection capability. The ground clutter spectrum will not be centred on the zero Doppler filter of the Doppler processor, as it would be with a wholly ground-based deployment, and the broadening of its spectrum will be significant if one, or both, of the platforms moves at high speed and carries an antenna with a relatively wide beamwidth. The rain clutter spectrum will also be moved and broadened. Sidelobe detection of clutter can present more serious problems to a bistatic radar, since clutter returns can be obtained via the transmitter antenna's mainbeam and the receiver antenna's sidelobes, and vice versa. Thus, clutter echoes can experience only one-way sidelobe attenuation where two-way attenuation would have occurred with monostatic operation. Since the Doppler frequencies of the bistatic clutter echoes depend on range and geometry, the ground clutter spectrum can be manipulated by controlling the relative motions of the platforms with respect to their targets.

The contours of constant Doppler shift for (stationary) ground clutter can be derived from the general Doppler shift equation 11.72 with $\boldsymbol{V}_S=0$; then

$$f_D=\frac{1}{\lambda}\,(\boldsymbol{V}_T\cdot\boldsymbol{U}_T+\boldsymbol{V}_R\cdot\boldsymbol{U}_R) \tag{11.73}$$

If we assume a flat Earth and neglect the aircraft heights, this equation can be simplified to

$$f_D\lambda = V_T\cos(\phi-\alpha_T)+V_R\cos(\theta-\alpha_R) \tag{11.74}$$

where the transmitter aircraft has a speed V_T on a bearing of α_T (measured in the mathematician's convention), while the receiver aircraft has a speed of V_R on a bearing of α_R. The bearings to the patch of ground clutter are θ_T and θ_R from the transmitter and receiver aircraft, respectively. These assumptions are not very realistic but this simple model can provide an approximate representation of the contours which depends only on the various angles and not on distances.

The value of the Doppler shift where the velocity vectors cross will be

$$f_D\lambda=\pm V_T\pm V_R$$

with the positive sign applying for platform motion towards the crossing point. This point will be at infinity when the platforms are on parallel headings. If ϕ is allowed to vary over the range 0 to 2π, the values of θ which define a constant Doppler contour are derived from

$$\theta = \alpha_R \pm \cos^{-1}\left(\frac{1}{V_R}\left[f_D\lambda - V_T\cos(\phi - \alpha_T)\right]\right) \qquad (11.75)$$

If V_R is zero, θ will have to be varied and values of ϕ calculated to avoid division by zero. If both platforms are stationary, the contours will not exist as all the ground clutter will have zero Doppler shifts.

The amplitudes of the echoes from clutter lying within the receiver's mainlobe will require effective rejection techniques to avoid high false alarm probabilities. Since all the aircraft trajectories and antenna pointing directions can be obtained or predicted before each coherent processing interval, the CFAR unit can be instructed to delete all the returns from within the ranges occupied by mainbeam ground clutter. This approach was adopted in the US 'Sanctuary' test bed – the proportion of the range–velocity space that had to be erased was dependent on the geometry, but generally lay in the range ≈3–8% [Fleming and Willis, 1980].

11.13 Other deployment options

11.13.1 *Multistatic systems*

The basic concept of a multistatic radar system would involve a set of transmitter and receiver sites, spaced so that each receiver can operate with two or more transmitters, and vice versa. The simplest implementation would be a conventional radar which had retained its monostatic capability while being augmented with a bistatic receiver. Multistatic radar systems will have all the advantages of bistatic systems, as described earlier, plus [Milne, 1977]:

(i) Multistatic radars can achieve greater target location accuracy, since they have resolution cells that are defined by two or more range measurements. With monostatic and bistatic radars, target location is derived from a range measurement and from the orientation of the antenna(s). The radar's resolution cell therefore has a large cross-section, equal to that of its antenna beam(s) at the target's location, but usually a much smaller depth defined by the radar's (compressed) pulse length.

(ii) Moving targets cannot readily present zero Doppler shifts to two receiver sites simultaneously, especially if the locations of these sites are not known. In theory, a target could move along a semi-ballistic contour representing the intersection between two prolate spheroids, but this is clearly impractical and could be negated by having three or more receivers.

(iii) Targets can be viewed from several aspects simultaneously; this will reduce the risk of all sites observing RCS nulls and will provide additional protection against jamming. It may also act as a counter to Stealth.

(iv) Pulsed emitters can be located by triangulation.

(v) Passive detection techniques, employing correlation and triangulation, can be used to locate noise-like interfering sources.
(vi) The multiplication of transmitter and receiver combinations allows individual sites to be closed down for re-location without serious compromise to the overall system performance.
(vii) Similarly, individual transmitter sites can be turned off while ARMs are seduced by decoys. Alternatively, cooperative operation of transmitter combinations can be used, for example alternating transmissions; simultaneous transmissions might cause an ARM to home on the centroid of the signals it received.

The price that has to be paid for all these advantages lies in the need for communication and synchronisation between sites, and in the added cost and complexity. In addition, the line-of-sight problem with bistatic radars is further aggravated when three or more sites require a line of sight. However, networking existing radars for multistatic operation may greatly enhance their overall effectiveness at moderate cost.

11.13.2 *Forward scatter operation*

The enhanced forward scatter RCS, described in Section 11.1, has been proposed as a counter to Stealth techniques since it will not be reduced by the normal RCS reduction techniques of body shaping and coating with radar absorbent material. However, the enhanced forward scatter RCS seems unlikely to be of much practical use, for the following reasons:

(i) The observed Doppler shift of a target will fall to zero as it crosses the baseline.
(ii) The radar loses its ability to measure range as the target crosses the baseline. The radar may detect the presence of the target, but will not be able to measure where it lies on the baseline.
(iii) Owing to the loss of range measurement capability, all the ground clutter returns from along the baseline will arrive at the receiver at the same time as the target's echo.
(iv) When the receiver looks directly along the baseline, it will intercept the full incident transmitter power. Apart from any other detectability considerations, the power ratio between the transmitter breakthrough and the target's forward scatter may exceed the dynamic range of the receiver (this would typically be about 70–90 dB).
(v) The forward scatter RCS will occupy a small solid angle about the direction of the baseline, so the enhanced RCS will not be observed unless the target passes close to the baseline.
(vi) Even if a target does pass very close to the baseline, the duration of the enhanced RCS may be very short, especially if the component of the target's velocity across the baseline is large.

Forward scatter effects will only be of interest for the detection of targets which are reasonably likely to pass through, or very close to, the baseline(s) between a bistatic (or multistatic) transmitter and its receiver(s). They are consequently likely to be only of limited use in space-based radars where targets will have freedom of movement over three dimensions. These effects could, nevertheless, be useful in cases where the targets' freedom of movement is constrained, as with aircraft that are flying low to evade other radars. Such a fence system could be particularly useful against aircraft that were designed for low radar detectability.

11.13.3 *Space-based bistatic radars*

Bistatic radars have already been used in planetary exploration. A system using an orbiting CW transmitter and an Earth-based receiver was used for lunar surface exploration by the USA on Lunar Orbiters 1 and 3, on Explorer 35 and on Apollo missions 14 and 15. A similar system with a satellite transmitter and an Earth-based receiver was used by the USSR on their Venera 9 and 10 satellites to estimate the mean-squared slope and dielectric constant of the surface of Venus. The USA has also made bistatic measurements of the thickness of Saturn's rings and of the electrical properties of the surface of Mars (see, for example, Glaser [1986] for a summary).

Since the positions and velocities of satellites can be predicted with great precision, they are ideal candidates as the platforms for the transmitter and/or receiver in a bistatic SAR. Tomiyasu [1978] proposed such a SAR using two satellites, and noted that a direct communication link would be required between them to allow measurement of the phase of the received signal relative to the phase of the transmitted signal.

Lee and Coffey [1985] have proposed the use of a bistatic illuminator in geostationary orbit for future tactical air surveillance; its signals would then be used by ground or airborne receivers deployed over the area of interest. They have estimated that it should be technically feasible to develop such a radar by the mid-1990s, by which time there would have been sufficient progress on large power supplies, deployable antennas, high power amplifiers and, of course, space transportation. Although the use of such a remote orbit makes target detection more difficult, it has the advantage of improving satellite survivability, requires fewer satellites for global coverage and experiences less clutter spreading than for lower orbits.

11.14 UK bistatic radar programmes

The UK interest in bistatic radars preceded even the use of the term radar. As described in Section 11.1, the first airborne bistatic radar trials took place in the Autumn of 1936, using a transmitter at Bawdsey Manor, on the East coast of the UK, and a receiver in a Heyford bomber. Once the first phase of the Chain Home system was deployed in 1939, the receiver at one site could operate in a bistatic mode with the transmissions from a neighbouring site, since they were all synchronised to the UK's 50 Hz National Grid. The UK interest in such bistatic operation faded once the invention of the duplexer allowed the transmitter and receiver to share the same antenna.

11.14.1 *The RSRE programmes – Argus, BEARS and RABIES*

The main UK bistatic radar activities have been run by the Royal Signals and Radar Establishment (RSRE).* The resurgence in the UK interest in bistatic operation arose from radar techniques studies carried out by RSRE, the GEC–Marconi Research Centre and what was then Plessey Research and Technology, during the period 1976 to 1980. These studies considered the problems that would be faced by air defence radars operating in a hostile environment and their conclusions included the recommendation that the potential advantages of bistatic radars merited further research and experiments.

In the longer term, bistatic radar designers may well be faced with situations as complex as operation on airborne platforms, while using the transmissions from non-co-operative illuminators. However, RSRE chose to develop first the simplest form of bistatic radar for which there is an established need; even then, they developed and tested critical sub-systems before attempting the complete radar. By building a ground-based system, they avoided not only the need to transfer location and motion data between platforms but also the effects of ground clutter and platform induced Doppler and, of course, the problems inherent in flying aircraft. The illuminating radar throughout nearly all of these developments has been the Byson research radar at RSRE, but the aim was to be able to use the UKADGE radars (such as the GEC Martello and the Plessey AR320), as well as developing independent mobile illuminators.

The research programme over the first half of the 1980s involved studies by most of the UK's radar related firms into both ground based and airborne bistatic radars. These studies addressed all the technical problems, including synchronisation, electronically scanned antennas and bistatic radar cross-sections. In parallel, RSRE, assisted by MRC and Plessey, carried out some initial trials on synchronisation, geographic correction and the bistatic radar cross sections of targets. Relatively little of the details of this work was published, though there were several papers that reviewed the advantages of bistatic and multistatic radars.

The conclusions of these studies and the success of the initial experiments led, via the RABIES and BEARS programmes, to the funding of the Argus Technology Demonstrator Programme by the UK MoD, over the period 1986–1992. During this programme, a complete bistatic receiver was successfully constructed and tested. The bistatic radar provided by Argus and Byson was then used to acquire data on targets of opportunity, principally civilian aircraft in the major UK airlanes. The first open publication of some of the work on Argus was a set of three papers in the plenary session that opened the Radar '87 Conference in London [Dunsmore, 1987; Soame and Gould, 1987; Bovey and Horne, 1987]. Further papers at the same Conference described the Byson research radar [Lappage *et al.*, 1987] and the development of calibration techniques using a metal sphere towed behind an airship [Reader *et al.*, 1987].

The Argus bistatic radar receiver has a block diagram similar to that shown

* Since this text was completed in July 1990, the Royal Signals and Radar Establishment has become part of the UK Defence Research Agency. Although, this text has been updated immediately prior to publication, the title RSRE has been retained since that was the name of the Establishment during the period when the bulk of the bistatic radar work was carried out.

in Fig. 11.22 and it comprises five major sub-systems:

(i) *Phased array antenna*: The phased array provides a set of contiguous beams covering a 90° sector in azimuth, with each beam having the same elevation pattern. Each elevation sub-array of the antenna incorporates an RF beamforming network to generate a fixed elevation pattern, and its output is connected to a low-noise, image-rejection receiver that feeds an IF beamforming network (BFN). The resultant set of beams from this network is connected to a beam switch matrix, whose switching times are controlled to simulate several independently scanning beams that chase all the transmitter pulses that pass through the antenna's field of view. Each of these simulated beams is connected to:

(ii) *Signal processor*: This provides a data analysis channel for each of the independent antenna beams. Each channel incorporates sets of SAW IF pulse compression filters, followed by a complex-sample analogue-to-digital converter (also known as an I&Q processor). The subsequent stages include such digital processing as fast Fourier transformation, log modulus extraction, background averaging and threshold detection, range and velocity ambiguity resolution and compensation for geographic errors.

(iii) *Site synchronisation and communication system (SS&CS)*: This uses rubidium atomic clocks and Loran C receivers at each site, together with a duplex data link between them, to allow the Argus receiver to recreate the timing of the transmitter's pulses. The data link also transfers the transmitter's waveform parameters and the orientation of the transmitter's antenna.

(iv) *System control computer*: This is responsible for the overall management of the receiver, under the operator's direction. The control functions include such tasks as the generation of the beam switching times for pulse chasing, the allocation of signal processing channels to antenna beams, the combination of the outputs from all the signal processing channels and the recording of target data.

(v) *Synthetic display*: This combines the plots from all the processing channels with geographic information and other collateral data.

As preludes to the construction of the full Argus bistatic receiver, RSRE had sponsored the development of:

BEARS: Bistatic Experimental Array Receiving System: This is a phased array antenna whose azimuth width and number of beams are approximately one third of those originally proposed for a full size Argus antenna. The BEARS antenna is illustrated in Fig. 11.23, without its protective radome and with one of its elevation sub-arrays protruding from the array face. Its analogue beam-forming techniques are described by Wallington [1985]. It was developed jointly by RSRE and GEC–Marconi Radar Systems Ltd.

RABIES: Range Ambiguous BIstatic Experimental System: This was the single channel prototype of the six channel Argus signal processor, combined with the site synchronisation and communication system. It was developed under contract by what is now Roke Manor Research Ltd.

The BEARS antenna achieved the microwave performance required for bistatic operation while RABIES was successfully tested against both synthetic and genuine targets in the presence of both chaff and jamming.

Figure 11.23 *BEARS phased array antenna, without its radome*

For the field trials of RABIES, the complete system was installed in a converted AA4 Mk. VII radar cabin, which retained only its single beam reflector antenna and turning gear from its original equipment. RABIES was first deployed to an RAF station in the Cotswolds, south of RSRE, where it observed targets in the clear, then to another RAF station on the East Coast of the UK where it observed multiple targets in the presence of chaff and active jamming. In each case the target illuminations were provided by the Byson research radar at RSRE; this radar uses a simple reflector antenna, shown in Fig. 11.24, to minimise the extent to which the antenna can introduce unquantified modulations and distortions on its radiated signals. This can provide coherent sequences of 4.6 or 4.5 μs pulses with RF agility over a 12% range in E/F band, PRF agility at rates up to 4.5 kHz and 2 or 10 MHz non-linear FM for pulse compression; more details are given by Lappage *et al.* [1987]. Although Byson is a surveillance radar that normally scans continuously at rates up to 5 rev/min, it was operated in a tracking mode during most of the RABIES trials to provide continuous illumination of the targets. The target bearings were relayed to the RABIES cabin so that its antenna could also follow the targets. This tracking technique maximised the amount of target data available for subsequent off-line analysis; operating Byson in a scanning mode would have provided target data for, typically, only about 20 ms during a 15 s scan period.

The RABIES targets included dedicated trials Canberras,* as well as various targets of opportunity, mainly commercial aircraft in the major UK airlanes. In addition, the trials examined direct breakthrough from transmitter to receiver owing to troposcatter, forward scattering from aircraft crossing the baseline and various forms of volume clutter. Some co- and cross-polar measurements were also made, since the Byson radar could provide both vertical and circular polarisations, while the AA4 MkVII antenna and the BEARS antenna were respectively vertically and horizontally polarised.

The RABIES processor was operated mainly as a medium PRF radar, although it also has a low PRF capability. It relies on range ambiguity resolution to distinguish separate targets within the antenna's field of view and it derives the unambiguous velocity estimates as aids to subsequent target tracking processes. The range and velocity ambiguities were resolved by using PRF agility, with up to six PRFs in each sequence. These sequences of PRFs were designed by Aldous [1987], using a random search routine. They allow target observations achieved on any two PRFs in a sequence to provide unambiguous Doppler resolution anywhere in the specified velocity range; they also permit unambiguous range resolution, even for targets whose echoes spread over as many as ±3 contiguous range cells. Each coherent burst of pulses, at a constant RF and PRF, had to be long enough for even targets at the

Figure 11.24 *Byson radar at RSRE*

* The English Electric 'Canberra' is an old design of twin-jet bomber, which was reproduced in the USA as the B-57. It has a smoothly curved fuselage that gives it a low RCS for its size.

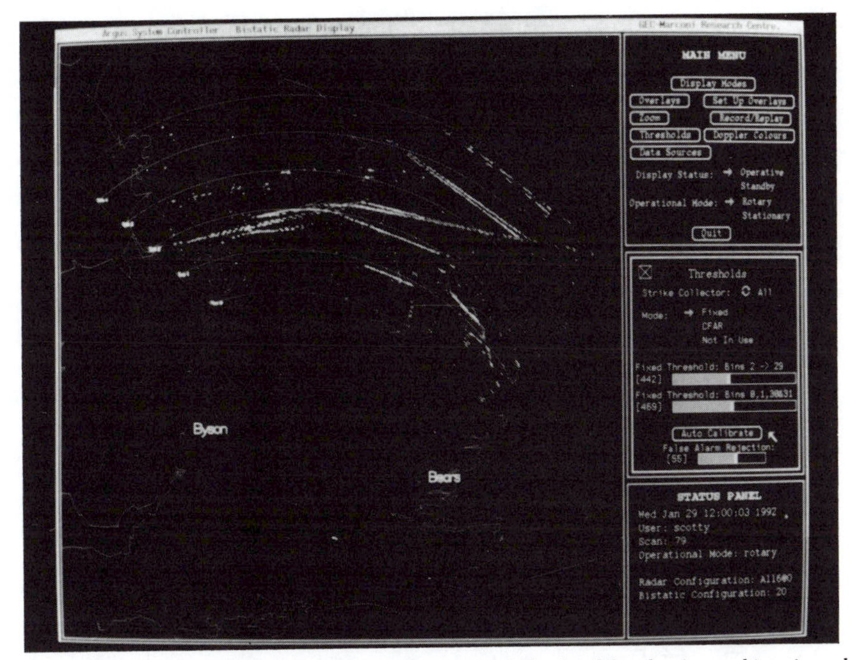

Figure 11.25 *Aircraft detection over the south of England by the Argus bistatic radar operating in its 'pulse chasing' mode*

maximum design range to provide sufficient echoes to fill whichever length of FFT was currently in use.

The complete Argus bistatic receiver incorporates an enhanced version of BEARS and six enhanced versions of the RABIES signal processor. These have been combined with a system controller and a synthetic display, based on a Sun work station, which provides a range of operator interfaces and display facilities that are comparable with those in a conventional radar. Throughout its trials, Argus has been operated from the GEC–Marconi Research Centre at Great Baddow, using RSRE's Byson radar as the target illuminator. Apart from the tests of its target detection in chaff laid by the Royal Air Force, these trials have used targets of opportunity, mainly civilian aircraft in the major UK airlanes. The system has operated reliably in its pulse chasing mode, without any evidence of glitches as targets have passed between successive antenna beams and between signal processors. It has successfully provided area coverage comparable with that from a monostatic radar, detecting targets at ranges of hundreds of kilometres from the two sites. A typical example of the synthetic display showing aircraft detection over the North of England is given in Fig. 11.25. Most of the aircraft are in the 'Blue 1' airlane which runs East from Manchester and then turns South-east towards Holland.

Operator reports during the various bistatic radar trials indicated that the monostatic and bistatic radar echoes had markedly different fading characteristics. An initial direct comparison of the echoes used HDDR recording systems operating simultaneously at the RABIES + BEARS version of the bistatic

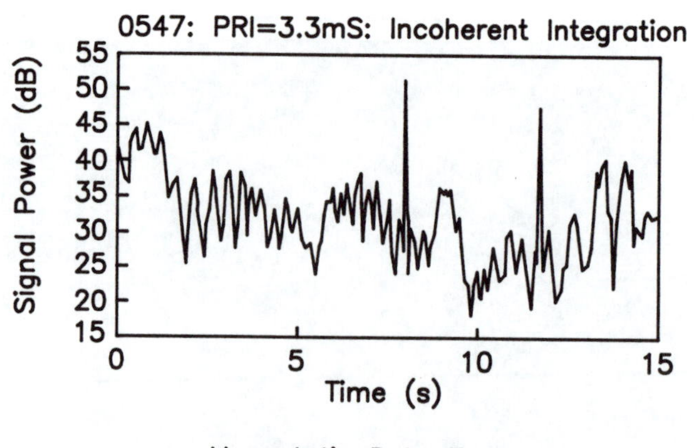

Monostatic Data Excerpt

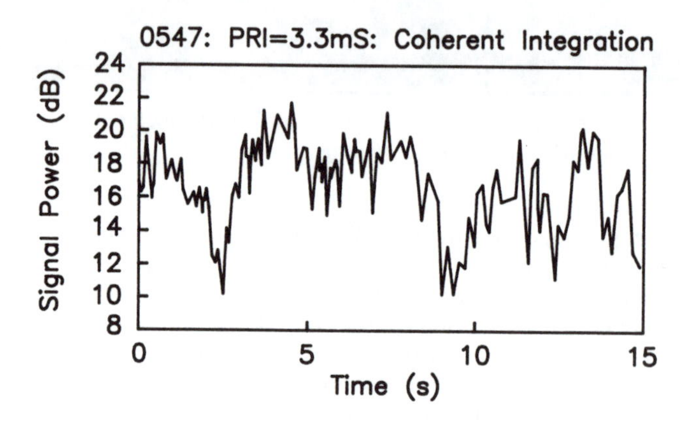

Bistatic Data Excerpt

Figure 11.26 *Comparison of bistatic and monostatic aircraft RCSs*

receiver and the Byson monostatic radar. A typical comparison of the different monostatic and bistatic RCS fluctuations for an aircraft is given in Fig. 11.26 [Matthewson *et al.*, 1990].

11.14.2 *Independent bistatic receivers for co-operative radar transmitters*

Schoenenberger and Forrest [1982] have addressed the problem of providing reference signals at independent bistatic receivers, using radars which co-operate in the sense of providing known waveforms but which do not provide any direct data links to the receivers. They noted that the required reference data are the transmitter antenna's azimuth angle, the transmitter's pulse train timing and its RF carrier. If these data had to be sent via a data link, its bandwidth could be very large. For example, it would be several tens of MHz to provide timing accuracies of the order of tens of nanoseconds – even more if the RF carrier had to be sent to provide a coherent reference for MTI processing.

In practice, most of these data can be derived from direct reception of the transmitter signal as it scans over the receiver site; signals scattered from a fixed target at a known location could also be used. Estimation of the transmitter antenna's azimuth relies on its scan rate being constant, at least for the duration of several complete rotations – this is something that would not usually happen with phased array antennas. However, the rotation rate stability for conventional reflector antennas is very good, even under adverse weather conditions. Provided the bistatic receiver can measure accurately the times when the antenna's boresight is aligned along the baseline, then any subsequent beam direction can be inferred from the elapsed time, typically to an accuracy of about 1°.

Continuous, reliable pulse train timing cannot be achieved solely from direct breakthrough. Even if the transmitter antenna's pattern had general sidelobe levels that could be detected for most of the time, it would have some nulls and it could also be corrupted by multipath effects. However, the waveform characteristics of most illuminators of interest will be known and will be reasonably constant, since their timings are likely to have been derived from a temperature controlled crystal oscillator. The PRF can therefore be derived from a 'flywheel' generator which is regularly resynchronised. Even if such resynchronisation only occurred once every 10 s, a timing accuracy of ±1 μs would only require a voltage-controlled crystal oscillator (VCXO) with a stability of $1_{10}-7$. However, for transmitters with random PRF agility, a direct link would be required.

Unless Doppler processing is required, the transmitter frequency only needs to be known to an accuracy less than the IF bandwidth. This would readily be achieved – even random frequency agility at the transmitter would not be a severe problem, owing to the availability of fast frequency measuring techniques.

Using these principles, Schoenenberger and Forrest developed a totally independent bistatic receiver at University College, London. It operated with the 600 MHz civil ATC radar at London Heathrow Airport, some 25 km from their receiver. Their receiver used a single omnidirectional antenna. This obviously gave a relatively poor detection range and little protection from other interfering signals (including multipath) but it avoided the need for pulse chasing. The target location accuracy therefore depended on the accuracy with which the transmitter antenna's orientation and the bistatic delay time could be measured. The azimuth resolution was determined by the transmitter's beamwidth. Their rotation generator estimated the transmitter antenna to be in one of 2048 discrete positions within the full 360° horizontal rotation. Errors in determining the centre of the mainbeam and in the alignment of the baseline, plus wind effects, degraded the overall azimuth accuracy to ±1.6°. This was deemed to be adequate as the transmitter had a 3 dB beamwidth of 2.1°.

The transmitter's known PRF and stagger pattern were recreated by counting the signals from a VCXO. Its initial synchronisation was performed during the mainbeam dwell period, with its 25 direct path pulses. This 'flywheel' generator provided a PRF signal accurate to ±1 μs over the complete 6 s rotation period – which was adequate as the radar's pulse length was 4 μs.

To complete the bistatic receiver, the target and clutter echoes were presented on a raster scan synthetic display with numerical correction of the display distortion that arose from the use of a bistatic geometry. The observed range performance was in good agreement with predictions and aircraft targets of about 10–20 m^2 RCS were reliably detected out to 75 km and tracked out to ranges in excess of 130 km.

11.14.3 *Television-based bistatic radar*

Griffiths and Long [1986] have addressed the problem of developing a bistatic radar which employed as its illuminator any convenient radar like transmission, or 'illuminator of opportunity'. This was to have the advantage of being a completely covert radar system which did not advertise the presence of its transmitter or receiver sites. Their experiments used a bistatic receiver at University College, London, to detect aircraft in the vicinity of London Heathrow Airport, employing the 625 line PAL TV signals from the Crystal Palace transmitter. These signals included the use of TV pictures designed to make the transmissions more closely resemble a pulsed signal. The horizontal plane geometry of Crystal Palace, UCL and Heathrow is shown in Fig. 11.27.

The Crystal Palace station transmits with an ERP of 1 MW, using horizontally polarised radiation with a uniform azimuth distribution. However, this power was spread over four TV channels, only one of which could be used at any one time, so reducing the effective ERP by 6 dB. In addition, the vertical radiation pattern had its peak at 1° below the horizontal, with a −15 dB null at about 2.5° above the horizontal and a −6 dB lobe at about 5° above the horizontal. The effective ERP in the lobe was therefore ≈62.5 kW. Although this transmitter power is of the same order as some radar systems, it is produced by a low gain antenna. Normal monostatic radar systems have large aperture, high gain antennas for both the transmission and reception functions. The oval of Cassini shown in Fig. 11.27 corresponds to $R_T R_R = 2_{10}8$ m^2 and was derived from this power, with the assumptions of a 20 m^2 bistatic RCS for a large aircraft, the wavelength for the 567.25 MHz BBC 2 signal, a 10 dB detection threshold, a 15 dB receiver antenna gain and a 2.5 MHz receiver bandwidth. Although the TV channel was 8 MHz wide, most of the power was concentrated into this smaller bandwidth centred on the vision carrier.

Griffiths and Long concluded that the separation of targets from the direct signal and clutter in their quasi-CW system require extensive signal processing under all but the most favourable conditions. They achieved successful target detection only on the few occasions when the target's aspect gave a high value of bistatic RCS, the receiver antenna's boresight coincided with the target and the data were successfully captured by their 8 bit digitiser and store. Their difficulties in target detection were attributed mainly to the 48 dB dynamic range of their MTI canceller system – this was unable to cope with clutter-to-signal ratios which may have been as much as 40 dB more than this value.

They also concluded that the ambiguity function of the illuminator waveform was of key importance since it would determine the achievable resolution and ambiguities in range and Doppler. For a transmitter signal with complex modulations, such as those for TV, the power in the part of the spectrum used for radar purposes had to be commensurate with the required coverage, and the bandwidth of this part of the spectrum should also be commensurate with the required range resolution. Ideally, the ambiguity function should be a 'thumb tack'. In most of these respects, the TV waveform was not ideal as its autocorrelation function showed broad peaks at 64 μs intervals, corresponding to the line sync pulses. For any such quasi-CW illuminator, the range sidelobe level would be high and the radar best suited to Doppler rather than range measurements.

11.15 US bistatic radar programmes

In the United States, the first radar detection was achieved in 1922 by Taylor and Young of the US Naval Research Laboratory (NRL), who detected a wooden ship using what would now be described as a bistatic CW radar. Hyndland of NRL accidentally detected an aircraft at a range of 3.2 km while working with direction finding equipment at 33 MHz. Taylor, Young and Hyndland again used a bistatic radar to detect an aircraft at a range of 80 km in

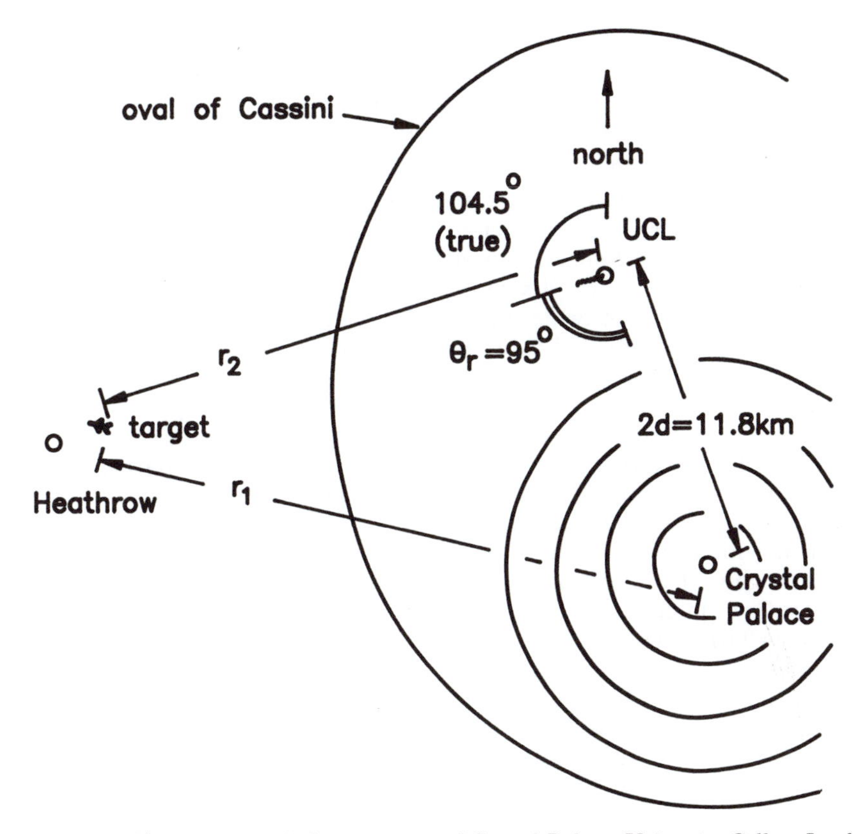

Figure 11.27 *Horizontal plane geometry of Crystal Palace, University College London and London Heathrow Airport, used in the television based bistatic radar*

1932 [Skolnik, 1985]. After the invention of the duplexer by NRL in 1936, US attention switched to monostatic radars and there was little further interest in bistatic radars for the next forty years.

During the 1970s and the early 1980s, the US Defense Advanced Research Projects Agency (DARPA) and the US Department of Defense sponsored several research and development programmes aimed at assessing the suitability of bistatic radars for a variety of mainly military applications. Although papers on bistatic radar techniques continue to appear, they no longer seem to give the same attention to the performance of actual systems. The characteristics of five of the most significant US bistatic radar programmes are summarised in this Section. Some further details of other US bistatic and multistatic systems can be obtained from the review article by Glaser [1986] and in the book by Willis [1991].

11.15.1 *Sanctuary*

In the Sanctuary experimental bistatic radar system, a ground based tracking receiver was operated with a co-operative airborne illuminator, in order to assess the performance and the reduction in vulnerability of tactical systems, especially to ARM attack. The objectives of the programme were stated by Bailey *et al.* [1977] to be:

(i) the gathering of data on performance capability limitations
(ii) the demonstration of surveillance and tracking
(iii) the gathering of data for engineering and trade-off studies for future military bistatic radar systems.

This paper by Bailey *et al.* appears to be one of the first unclassified papers on the Sanctuary system – its only reference is to a secret paper presented to the July 1976 Tri-Service Radar Symposium.

The Sanctuary radar was developed by the Technology Service Corporation (TSC) for the Defense Advanced Research Projects Agency (DARPA) and the US Army Missile Command (MiCom). The configuration of the experimental test-bed system is shown in Fig. 11.28. The following details of the system were obtained mainly from the paper by Fleming and Willis [1980].

The airborne illuminator was mounted in an A–3 aircraft and radiated a modulated 1.6 kW CW waveform at 1.385 GHz. Its 'floodlight' staring antenna gave 14.5 dB gain with a 36° azimuth beamwidth.

The ground-based Sanctuary receive antenna was a multibeam phased array antenna, with a beamwidth of ≈6° in both azimuth and elevation, and a boresight gain of 27.2 dB on each beam. These receive beams were used in pairs, stacked in elevation to cover 12°. The peak receiver antenna sidelobes were at ≈0 dBi. With its 3.65 m square aperture, it was claimed in *Defense Electronics* [1981] that, at the time of its construction, it was the largest stripline circuit ever built. It used 504 dipole elements and was capable of scanning through ±45° in azimuth, under the control of 5 bit diode phase shifters. The array provided two sum beams for detection and two difference beams for tracking, with the two sum beams being located at ±3.2° about the array normal. The array also carried two low gain antennas that were used for direct path cancellation and sidelobe blanking. There was an adaptive canceller, similar to SLC, to cancel the signal received by the direct path.

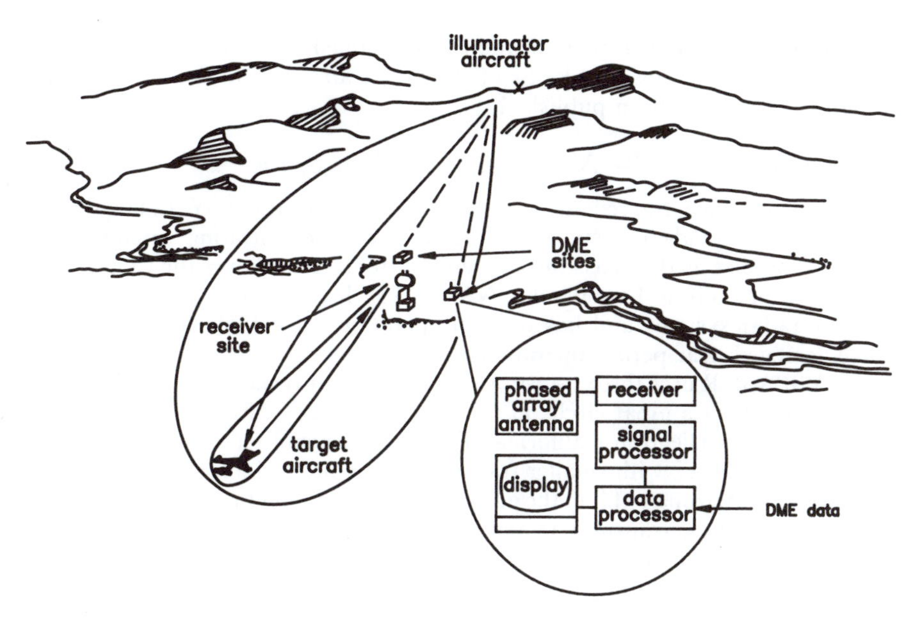

Figure 11.28 *US Sanctuary test-bed system*

The transmitted waveform could be either CW or pulsed; CW was considered preferable in order to maximise the average power at minimum cost. It used pseudo-noise coding of 1024 bits (each 0.67 μs long) which was repeated 128 times during each coherent processing interval of 87 ms. The scan rate of the receiver antenna gave a dwell time on target of 87 ms, to match the duration of this processing interval. This waveform could be pulse compressed to give −56 dB sidelobes. The frequency reference was provided by stable crystal oscillators – these were found to be quite adequate, with about one order of magnitude to spare in stability. The receiver's signal processor used 11 500 integrated circuits, consumed 10 kW and gave 48 dB of coherent gain in the 1024 range cells.

In this type of bistatic radar, only one-way sidelobe rejection of clutter was obtained and the strength of the clutter returns in the main beam was such that provision had to be made for blanking it to avoid a very high false alarm rate. Since the range–Doppler location of the main beam clutter is deterministic, main beam filter blanking of clutter could be employed. The equivalent 'range' resolution was 100 m and the clutter penalty was that about 3–8% of the range–velocity space had to be excised from the radar's coverage.

The flight trials of the system were conducted successfully during July–September 1980 at the Pacific Missile Test Center (PMTC) at Point Magu in California. The illuminator aircraft operated at 29 000 feet and the targets were A6 and A7 jet aircraft, operating at speeds of 300 and 450 kt. Each target aircraft was fitted with a C-band tracking beacon. The false alarm rate was stated to be very low and operationally acceptable for air defence; the maximum detection range was greater than 100 km over a wide spread of bistatic angles. However, no indication was obtained of the differences between the bistatic and monostatic RCSs of the target aircraft. No details of further progress have since been published.

11.15.2 *NAVSPASUR—Naval Space Surveillance System*

The Naval Space Surveillance System is a multistatic CW radar that is intended to provide detection and tracking data on orbiting satellites and ballistic missiles as they pass through an electronic 'fence' over the Continental USA, at ranges out to 15 000 miles. NAVSPASUR is one of the oldest of the US bistatic radar systems and has been in operation since 1960; it is claimed to have achieved near-perfect operability.

NAVSPASUR uses three transmitter sites and four receiver sites interspersed alternately along a great circle, at a latitude of 33.5 °N, from Fort Stewart in Georgia to San Diego in California. The main transmitter is located at Lake Kickapoo in Texas, with smaller transmitters at Jordan Lake in Alabama and Gila River in Arizona to fill in the gaps in the illumination provided by the main transmitter. All the transmitters and receivers have stationary, vertical, planar, fan beams which are aligned along the great circle. Targets passing through the beams from the transmitters scatter echo signals that are intercepted by the receivers, using multiple phased array interferometers to measure their angles of arrival. Each transmitter operates with two receivers, one on each side of it. The signals are then processed at the receiver sites and the resultant data sent to NAVSPASUR headquarters at Dahlgren, Virginia, where they are filtered, correlated and recorded. Calibration of the system apparently was achieved by the observation of satellites with known ephemerides.

The transmitters (at the Lake Kickapoo site) originally used 50 kW travelling wave tubes for each sub-array of 144 dipoles, but these were later replaced by 300 W solid state modules at each of the 2556 dipoles in the full linear array. At the time of its completion, the modernised NAVSPASUR was the highest average power solid state transmitter ever produced. It produces an average effective radiated power in excess of 98 dBW, including the antenna gain.

The system stability has been improved to $5_{10}-12$ from a caesium atomic standard instead of $5_{10}-10$ from a crystal, while the spectral purity has been improved to a 0.2 Hz bandwidth and −70 dBc harmonics instead of a 5 Hz bandwidth and −40 dBc harmonics. This has extended the system's target detection performance (one of the aims of the improvement programme was to double the detection range) and has increased its ability to discriminate between targets [Francoeur, 1989].

11.15.3 *BAC—Bistatic Alerting and Cueing*

The main features of the Bistatic Alerting and Cueing (BAC) concept were the detection of air and ground targets using existing radars as non-co-operative illuminators, as illustrated in Fig. 11.29. The pay-off from such a concept was to be a real-time, passive surveillance capability at each low-cost, independent BAC receiver. An additional application was to be its use to enhance the survivability of existing monostatic surveillance radars. A BAC unit could be collocated with, or integrated with, a monostatic radar which could then remain silent until alerted and cued to the target direction.

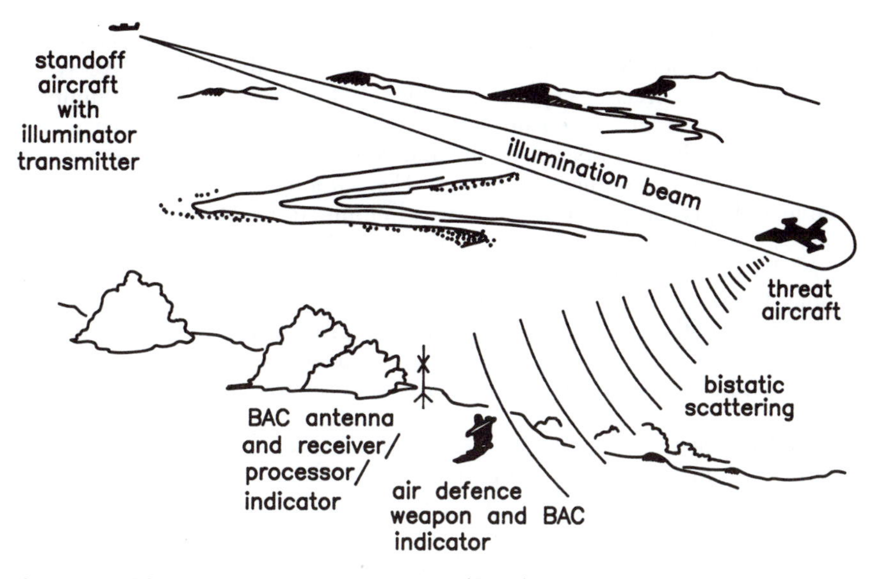

Figure 11.29 *Bistatic alerting and cueing (BAC) concept*

The BAC system was developed by TSC for DARPA and the US Army Harry Diamond Laboratories, with partial funding by the US Air Force Wright Aeronautical Laboratory and Westinghouse. It was field tested at the Naval Weapons Center, China Lake and Dugway Proving Grounds. The system was designed to detect airborne targets using radiation from the E-3A 'Sentry' Airborne Warning and Control System (AWACS) and to detect ground targets using radiation from the Joint Surveillance Target Attack Radar System (JSTARS), carried by a C–18 aircraft.

Synchronisation is one of the most critical problems in operating a bistatic system with a non-co-operative illuminator – this problem was successfully solved during the BAC programme [Thompson, 1989]. The programme was also intended to investigate the effects of clutter and ECM and to supply an inexpensive receiver for the bistatic detection of ground and airborne targets.

The first stage in the BAC system's operation was an ≈30 s illuminator search mode, during which its antenna would step across a set of eight contiguous, staring beams, covering a 120° sector. All the illuminators whose RFs matched those in its memory were identified and tagged with the appropriate beam number and the 'scan-by' time. Since the BAC surveillance volume is only a small proportion of the illuminator's search volume, the BAC receiver only needs to synchronise to the illuminator for the short periods, defined by the 'scan-by' time, when the illuminator's beam passes over the ground-based BAC site. The designation of the illuminator's RF from the illuminator search mode is not effective if the frequency changes during every coherent dwell (a characteristic of modern radars, as an ECCM and to improve detection performance). In this case, the system uses the clutter echoes from the first two pulses in a coherent dwell to determine the current RF, assuming that these arise from the clutter lying in the illuminator's main beam.

Although the illuminators used by the BAC system did not co-operate directly with it, their RFs and PRFs were known. This allowed the BAC system to incorporate a precise, crystal-controlled LO for each unique illuminator frequency – the dual conversion architecture of the receiver allowed the RF band to be split into sub-bands by the first down conversion and then to have a separate LO for each of the resulting first IFs. Although this approach provided a stable and spectrally pure LO that was immune to ECM and multipath, the effects of long-term drift could be significant. They appeared as a bias in the observed Doppler shifts of targets – an error of only $1_{10}-5$ would produce a Doppler bias of $\approx$30 kHz. However, the IF filters were wide enough to pass all these signals without attenuation. After FFT processing, the CFAR circuitry recognised the fixed clutter returns and used these as the zero Doppler reference. The CFAR system also demonstrated coherent clutter suppression of >50 dB.

After RF synchronisation, PRF synchronisation was achieved by range gate counting on pulses on the direct path from the illuminator's main beam, with receiver blanking being used to reject pulses that did not lie in the expected range of PRIs. Thereafter, range gate counting was used to measure the excess delays between pulses received directly from the illuminator and those reflected from targets. Using the notation of Section 11.5.2 and Fig. 11.5, this defined a distance $D=R_T+R_R-L$ for each target. Solution of the bistatic triangle then required two further items of data; of the four options considered by Thompson, the most convenient would use the range to the transmitter and the angle subtended at the receiver by the transmitter and the target.

Although the programme demonstrated successful synchronisation during its $2\frac{1}{2}$ years of field testing, no details appear to have been published of its achieving pulse chasing to increase the efficiency of the bistatic receiver. Also, the synchronisation techniques only apply to two specific illuminators of known RFs and PRFs.

11.15.4 *BTT/TBIRD—Bistatic Technology Transition*

The Bistatic Technology Transition/Tactical Bistatic Radar Demonstrator (BTT/TBIRD) was an experimental airborne bistatic synthetic aperture radar (SAR) system which was developed jointly by Xontech and Goodyear Aerospace for DARPA and the US Air Force Wright Aeronautical Laboratory. The objectives of this programme were to demonstrate clutter tuning, forward looking SAR operation and sideways looking MTI operation. The main technical challenges arose from the use of airborne platforms, namely motion compensation, control of the platforms and antenna pointing. Real time processing and the effects of clutter spreading were also investigated.

Time synchronisation between the systems on the transmitter and receiver aircraft was provided by rubidium atomic clocks. As was described in Section 11.12, compensation for platform motion requires a knowledge of the locations and velocity vectors of both platforms. With BTT/TBIRD, this was achieved by using inertial navigation equipment on the illuminating aircraft and telemetering its relative position and velocity to the receiver, in real time. The growth of position and velocity errors in the inertial navigation systems was limited by using frequent position updates from distance measuring equipment.

11.15.5 *Kiernan Re-Entry Measurement Site*

The Kiernan Re-Entry Measurement Site (KREMS), on Kwajalein Atoll in the Pacific, is equipped with a multistatic measurement system, developed by the Lincoln Laboratory of the Massachusetts Institute of Technology. This installation, illustrated in Fig. 11.30, uses the high power 'Tradex' L-band and 'Altair' UHF radars on Roi-Namur Island to collect bistatic signature data and to perform high accuracy coherent tracking of re-entry bodies. These radars illuminate the targets and collect the monostatic echoes. The L-band and UHF bistatic echoes are collected at a remote unmanned receiver station located at a distance of about 50 km to the south-east on Gellinam Island. A second, unmanned L-band receiver is deployed at about 40 km to the south on Illegini Island [Salah and Morriello, 1980; and *Aviation Week & Space Technology*, 1980].

The GTE Sylvania Altair transmits at 415 MHz, from a 24 dBi gain antenna, for multistatic operation. This radar's 17 MW peak output and its wide beamwidth permit the acquisition, tracking and target-cluster evaluation at long ranges. The RCA Tradex transmits at 1.32 GHz, from a 36 dBi gain antenna, for multistatic operation. It offers a variety of waveforms that are suitable for high resolution measurements. However, for the multistatic operation, it provided pulses either as a continuous train or in 32 pulse bursts. Both remote sites collect L-band data for distance measuring and signature analysis. The UHF data are only needed for signature studies and are received at Gellinam.

The multistatic system can be calibrated by operating it as an interferometer and using the known positions of radio stars as references. Other steps taken to

Figure 11.30 *Kiernan re-entry measurement site*

ensure high measurement accuracy include calibration techniques for measuring the atmospheric effects on the delays through the microwave links between the sites. The time differentials used to calculate the target trajectories are accurate to $\leqslant 0.5$ ns.

The KREMS system provides three-dimensional position and Doppler data, with accuracies of better than 4 m and 0.1 m/s (at the 1σ level), respectively, throughout re-entry. However, the geometric precision of the tri-static system deteriorates over the last few hundred metres prior to impact, since all three range vectors fall into a common plane and the target's height can no longer be resolved adequately.

11.16 USSR bistatic radar activities

The USSR* technical literature gives little guidance on any of their bistatic radar development programmes, other than those used in space exploration, though it does include many theoretical papers. This is not surprising in view of the military significance of the ability of bistatic radars to counter most of the effects of ECM and, possibly, Stealth. The Soviet Union will certainly be aware of the Western interest in bistatic radars and they are unlikely to have neglected their development. It is worth noting that Soviet military radar developments have long had to comply with the requirements of the concept of Radio Electronic Combat (Radio Electronniya Bor'ba), which is defined in military manuals as: 'The measures adopted to hinder the enemy's use of his electronic equipment by jamming, artillery fire, air strikes or physical attack while, at the same time, ensuring the reliable operation of Soviet radio electronic equipment while the enemy is trying to suppress it'. It is improbable that the Soviet Union is not investigating military bistatic radars – whether they have actually developed and deployed them is another matter.

At the beginning of 1934, Soviet military engineers and leading Soviet scientists came to accept the significance of 'Radio Detection' (Radio Obnaruzhenie) for air defence. They embarked at once on an intensive research and development programme that converged on the use of the pulsed method [Erickson, 1988]. However, the earliest Soviet radars were bistatic. In his book 'Out of the past of radar', Lobanov [1969] has described how a group led by B. K. Shembel at the Leningrad Electrophysics Institute (LEPI) demonstrated the detection of aircraft at distances of 5 to 6 km in October 1935, using a CW Doppler radar. This used two separate 2 m diameter parabolic transmitting and receiving antennas, mounted side by side, and radiated 8 W generated by a split-anode magnetron at a wavelength of 25 cm (1200 MHz).

The 'Rapid' early warning set was tested from the roof of the LEPI building in 1934 – this became the prototype for the subsequent 'Reven' system. At the end of 1938, a limited production order was issued for the Reven and, after field trials in 1939, it was accepted into service under the designation RUS–1 (Radio Ulavlivatel Samoletnoy). This CW bistatic system operated at 75 MHz with a

* This text was completed in July 1990 and since then the Republics of the USSR have separated into the Commonwealth of Independent States (CIS). The title of this section still refers to the USSR since that was the name of the state when the radars were developed. Although much more information has become available in the West, nothing yet appears to have been published that affects the material in this chapter.

35 km baseline between each transmitter and receiver; 45 of them had been deployed in the Caucasus and the Far East by the time of the German invasion (Operation Barbarossa) in June 1941. The RUS–2 entered service in 1940; like the RUS–1 it operated at 75 MHz. It still had some of the characteristics of a bistatic system, with the transmitter and receiver on separate vehicles, each with its own Uda–Yagi antenna, but work proceeded on a single antenna version. With its 50 kW transmitter it was capable of detecting targets with a height of 500 m at ranges of 10–30 km.

The Committee for Radio, attached to the State Defence Committee, was established in July 1943. The term 'radar' was then brought into official Soviet usage in place of the earlier 'radio location' (Radio Lokatsiya).

The increase in the effective scattering cross-section of a target as it approaches the baseline between the transmitter and receiver of a bistatic radar (i.e. at bistatic angles $\beta \approx \pi$) is a well known phenomenon in the radar literature of the Soviet Union. The advantages and the feasibility, in principle, of detecting targets using this forward scattering have been noted by Zakhar'yev and Lemanskiy [1972] and by Aver'yanov [1978]. Shneyderman [1972] and Galkin *et al.* [1974] have also indicated that forward scattering can be useful when reliable detection is required even in the absence of any information on the shape and orientation of the target or when its surface is coated with RAM.

More recently, the Soviet Union has used bistatic radars in planetary exploration. A system with an orbiting satellite transmitter and an Earth-based receiver was used on the Venera 9 and 10 satellites to estimate the small scale roughness, the large scale slopes and the dielectric constant of the surface of Venus. The system gave maximum resolutions on the surface of $\approx$10 km, compared with $\approx$150 km for the US Pioneer satellites. Comparison between the Venera and Pioneer results showed good agreement of the average parameters in the overlapping regions.

Some insight as to current USSR intentions may be inferred from a study of other published documents, such as a comparison of the ABM Treaty, signed by President Nixon and General Secretary Brezhnev in Moscow on the 26th May 1972, and the description of the Soviet ABM system given in the 1986 and later editions of the US Department of Defense publication 'Soviet Military Power'. These describe the Soviet ABM system as having four classes of sensor, namely launch detection satellites, over-the-horizon radars (OTHRs), early warning radars and battle management radars.

There are three OTHR systems located at Kiev, Komsomolsk and Nikolayev. The coverages of the OTHRs at Kiev and Komsomolsk are shown as crossing in the middle of the Continental United States, while that at Nikolayev looks eastwards towards China (see Fig. 11.31). All three OTHR systems are likely to be bistatic in the narrow sense of each having transmitters and receivers that are well separated, in the same way as the US OTH–B system with its transmitter at Moscow, Maine, and its receiver about 161 km away at Columbia Falls, Maine. However, this separation will be small in comparison with the ranges to the targets and the resultant bistatic angles will be negligible. On the other hand, the lines of sight from the Kiev and Komsomolsk radars will meet at bistatic angles of about 75°. This would provide two monostatic echoes and two distinct bistatic echoes from each target, provided the monostatic and bistatic echoes could be separated at each

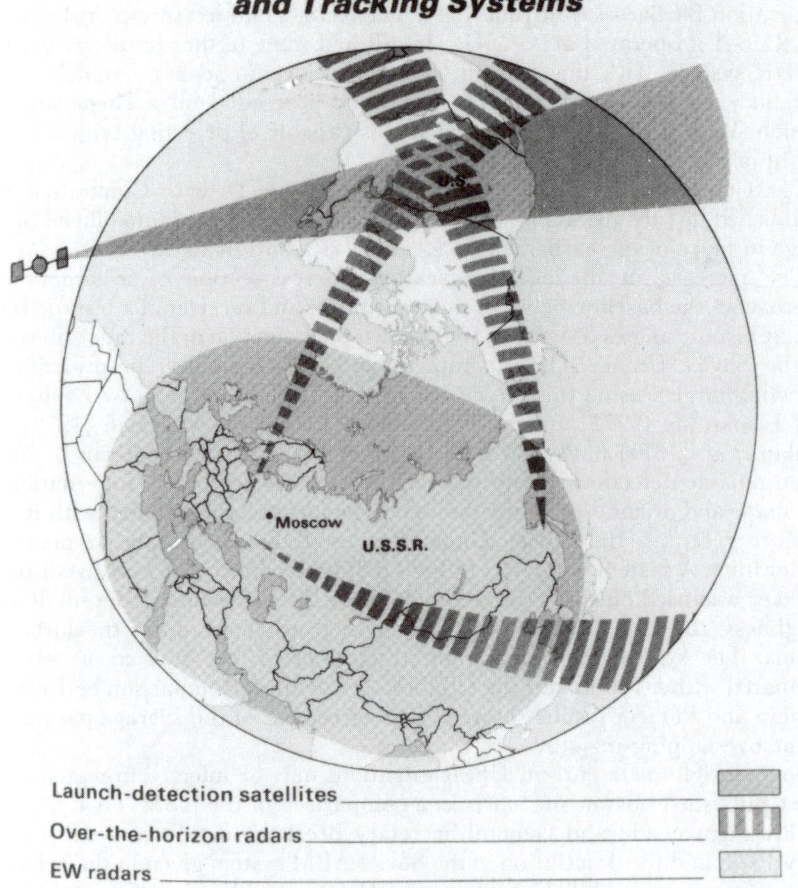

Figure 11.31 *Coverage of the Russian bistatic missile detection and tracking systems*

receiver. This could be achieved by frequency diversity or by phase coding the waveforms – the Soviet OTHRs are believed to use PSK modulation [Judd, 1988]. Even if the two OTHRs work on the same frequency, the bistatic echoes will be different, owing to the non-reciprocal reflection from the ionosphere.

The early warning part of the ABM system is described as having 11 of the older Hen House Ballistic Missile Early Warning (BMEW) radars deployed at six locations on the periphery of the Soviet Union. These are being supplemented by a network of eight large phased array radars (LPARs), also on the periphery of the Soviet Union, plus a further LPAR that is well inside USSR territory, near Krasnoyarsk in Siberia. A typical Hen House and the LPARs at Krasnoyarsk and at Pechora are illustrated in Fig. 11.32*a*, *c* and *d*, respectively. The overall coverage predicted for these early warning radars is shown in Fig. 11.33. It can be seen that there is a considerable degree of overlap in the

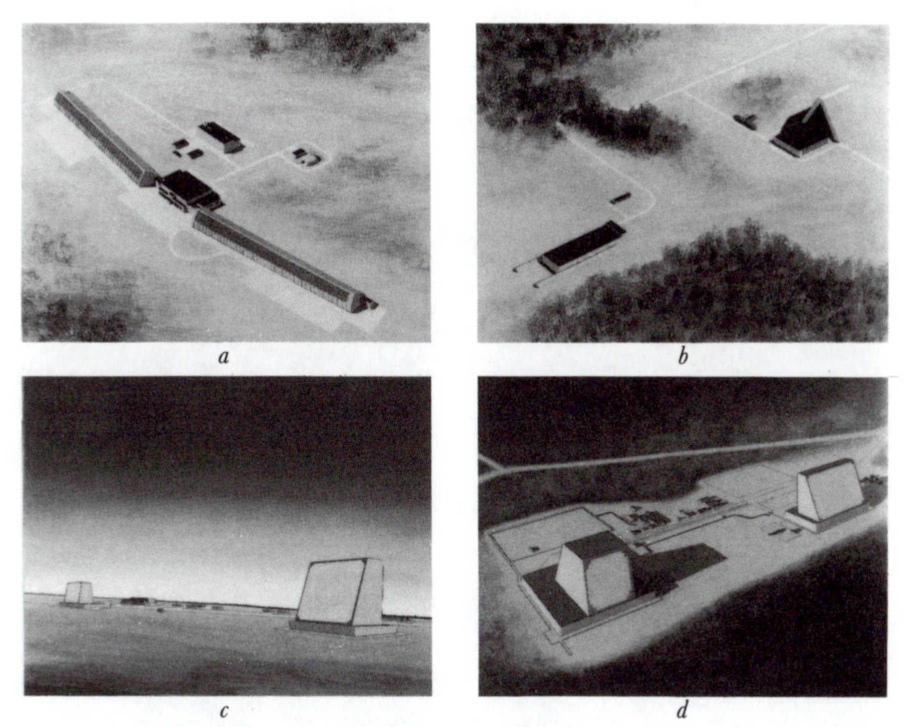

Figure 11.32 *Artist's impressions of some of the main Russian ABM early warning radars: (a) Hen House, (b) Dog House, (c) Krasnoyarsk LPAR and (d) Pechora LPAR*

coverage of some radars, particularly of that at Pechora with those near Krasnoyarsk and near Murmansk. There is further overlap between the coverages shown for the LPARs and those for the Hen Houses. It is therefore probable that the Soviet Union will at least have investigated the advantages of their operation in a bistatic mode. However, the Krasnoyarsk LPAR may not be able to continue to contribute to any bistatic operation as the former USSR Foreign Minister, Eduard Schevardnadze, agreed that its location was a breach of the ABM Treaty and stated that it would be demolished.

The battle management role is currently filled by the Dog House radar (see Fig. 11.32*b*) and the Cat House radar, both south of Moscow, together with the new Pill Box radar complex at Pushkino (shown under construction in Fig. 11.34). Pill Box will provide 360° coverage against incoming missiles and is expected to become operational in the early 1990s.

Article III of the Treaty allows each country to deploy one ABM system within a radius of 150 km centred on its national capital. The deployment of radars within this system is restricted to six complexes, each of which is to be circular, with a diameter of no more than 3 km. None of the ABM-related phased array radars that were being developed at the time by the USA would have required such a large site, so it may be assumed that this size was included in the Treaty at the behest of the Soviet Union. However, a baseline of only

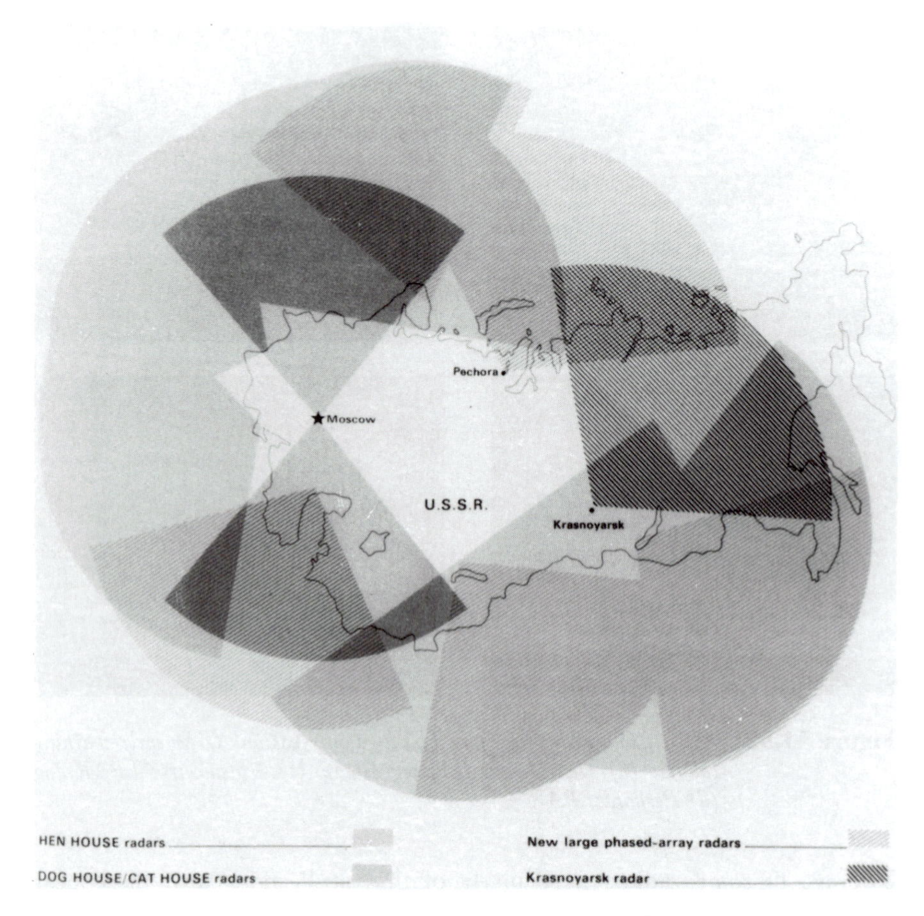

Figure 11.33 *Predicted overall coverage of the Russian ABM early warning, target tracking and battle management radars*

3 km is so short in comparison with the typical target ranges (hundreds of kilometres) that any radars deployed on these sites would only be bistatic in the narrow sense of having the receiver well separated from the transmitter so as to protect it from the transmitter's pulses. Fig. 11.34, from the 1988 edition of 'Soviet Military Power', shows the latest Moscow ABM radar at Pushkino to be a four-faced truncated pyramid, with one rectangular aperture and one much larger circular aperture on each face. This indicates that the modern Soviet radars for ABM terminal defence can now use transmitter and receiver antenna apertures that are adjacent, though they do not share the same aperture. The use of separate antennas avoids the need for duplexers and reduces the need for receiver protection; it will thereby improve the overall performance by reducing the system losses on both transmission and reception. It will also permit the use of different antenna patterns for transmission and reception.

Figure 11.34 *Pill Box radar complex under construction at Pushkino*

With up to six well separated radar complexes there is an opportunity to develop longer baseline bistatic, or even multistatic, systems, and this would be suspected if the radar complexes were developed close to the limits of the prescribed deployment area. Multistatic systems could use multiple range measurements to give very accurate positional data on a small number of re-entry bodies (REBs) but would probably be confused by large numbers of targets. At very short ranges, the 3 km diameters of the six radar complexes would allow radars within each complex to locate targets accurately. This multistatic range measurement approach would be similar to that used by the USA at their Kiernan Re-entry Measurement Site on Kwajalein Atoll.

Article VI of the Treaty allows both parties to have an unspecified number of radars for early warning of strategic ballistic missile attack, provided these are at locations along the periphery of their national territory and oriented outwards. In considering the design options for these radars we can note that the re-entry bodies they have to detect will typically be narrow cones with a length of about 2 m. The radars' deployment around the periphery of USSR territory, with outward orientations, means that these cones will normally approach the radars with an aspect that is close to nose-on. For radars whose wavelength is much shorter than the length of such cones, the RCS is at a minimum for the monostatic nose-on aspect and is extremely low. The RCS of such cones does have large values over a narrow range of aspect angles that are close to perpendicular to the surface of the cone. There are two main ways of increasing the apparent RCS. One is to operate the peripheral radars with wavelengths that are comparable with the length of the RVs – this would require radar frequencies of about 150 MHz and would not have a significant effect on the nose-on RCS of an REB. The other would be to operate these radars at higher frequencies and equip them to operate in both monostatic and bistatic modes, so as to avoid the minimum RCS condition. The overlap in coverage, shown in Fig. 11.33, implies that useful improvements in detection performance could arise from such bistatic operation.

11.17 Future developments

The principal impetus for the development of bistatic radars over the past couple of decades has come from military customers who appreciated the operational advantages that might be gained from the use of radar receivers that were covert and substantially immune to electronic countermeasures; there was the further prospect of an improved ability to detect stealthy aircraft. This interest has been encouraged by advances in the technologies of phased array antennas, signal processing and synchronisation that have reduced their probable costs. It is now becoming possible to achieve bistatic operation through the use of systems that have been deployed for other purposes – in particular, phased array antennas and GPS-NavStar global positioning systems are becoming more common on advanced military aircraft and ships.

The future developments of bistatic radars will depend on the roles that these radars are expected to play. Military roles usually involve the supply of relatively small numbers of systems but place a premium on resistance to countermeasures and on robustness in a wide range of environments, with cost being of lesser importance. Civilian applications require low costs for a mass market and they count on achieving this partly through economies of scale and partly through adopting less challenging specifications. In between the military and the civilian roles, there is the application of bistatic radars to counter-terrorism and law enforcement. For example, very light aircraft are very difficult to detect when used by terrorists; small bistatic radar receivers could be operated by counter-insurgency forces to detect these aircraft at short ranges – an approach similar to the US Bistatic Alerting and Cueing system described in Section 11.15.3.

Some of the military advantages of bistatic radars are likely to become less significant if the threat of direct conflict between Superpowers continues to recede and more attention will then turn to their civilian roles. The anti-stealth potential may aid air traffic control by locating and tracking small air vehicles that are not fitted with SSR transponders, such as private light aircraft, gliders and balloons. These might represent threats to airliners – they might also be used by smugglers, or by terrorists in low intensity conflicts.

11.17.1 *Use of non-co-operative illuminators*

The use of non-co-operative/non-dedicated radar illuminators reduces the cost of bistatic operation and reduces the overall system vulnerability to direct attack. Many current and proposed bistatic radar systems have been developed to make passive use of target illumination provided by monostatic radars deployed for air defence, airborne early warning or airborne interception. In the future, bistatic receivers can be expected to reduce their dependence on any one type of illuminator, especially ones which have not included the optimisation of bistatic radar operation among their design requirements. Operation with any available radars in the correct frequency band is one option and this can be aided by the development of synchronisation schemes that rely on direct breakthrough or on reflections from targets with known locations. Unfortunately, their freedom in using non-co-operative illuminators will be constrained; the receiver cannot predict the illuminator waveforms and antenna orientations unless they follow some predictable sequences. RF and PRF agility could tend to defeat any processing that required a sequence of coherent pulses, while monostatic radars with phased array antennas can be expected to use scanning sequences that adapt to their observed target environment rather than following any regular pattern.

A bistatic radar intended to operate with non-co-operative illuminators will require:

(i) *Synchronisation*: The illuminator timing can be achieved by the use of direct reception of the illuminator signal or its reflection from known targets in known locations.

(ii) *Adaptive pulse compression*: Samples of the illuminator pulse will need to be captured and processed to derive the weights required for digital pulse compression.

(iii) *Multibeam antennas with a signal processor for each beam*: This seems to be the only way of operating with a radar whose illuminator antenna's orientation is not known – for example, owing to the antenna being a phased array whose beam positions are being influenced by the sequential detection, and subsequent tracking, of targets.

If provided with an illuminator trigger signal and a digital pulse compressor, each beam and signal processor pair can detect and process any echoes from targets positioned in the beam. The range resolution will be as good as the accuracy of the synchronisation and the pulse compression, while the azimuth resolution will depend on the widths of the individual antenna beams, unless a beam interpolation capability is included. Without data on the non-co-operative illuminator antenna's orientation, the range information may be ambiguous unless the illuminator is using PRF agility, with a suitable set of PRFs that can be identified by the bistatic radar and used by its signal processing channels.

The calculations involved in deriving the locations of targets can be illustrated from Fig. 11.18, making the same assumptions and using the same symbols as those used in Section 11.5. The bistatic receiver will be able to measure the differential time delay $\delta\tau$ between the times of arrival of the direct signal from the sidelobes of the non-co-operative illuminator and the echo signal from a target in its main beam. This will define an equivalent distance D, such that

$$\begin{aligned} D &= R_T + R_R - L \\ &= \frac{\delta\tau}{c} \end{aligned} \tag{11.76}$$

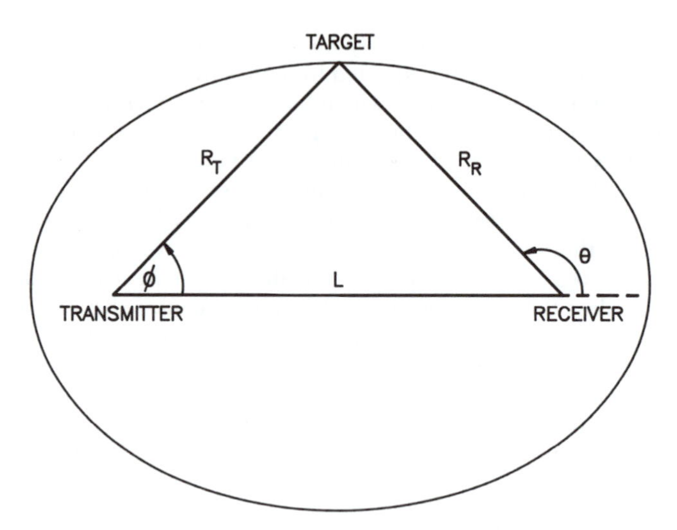

Figure 11.35 *Evaluation of target locations for a non-co-operative bistatic radar deployment*

The receiver will also be able to measure the angle θ between the directions to the illuminator and the target. To define the dimensions of the bistatic triangle in Fig. 11.35, the length L of the baseline to the illuminator also has to be known. The range to the target can then be derived, from an application of the cosine rule, as

$$R_R = \frac{0.5D(D+2L)}{D+L(1+\cos\theta)} \tag{11.77}$$

There are five ways in which the length of the baseline to the illuminating radar can be obtained:

(i) For a ground based radar, its fixed geographic location should be known.
(ii) For a satellite borne radar, the ephemeris data will indicate the current location of the satellite.
(iii) For an airborne radar:
  (*a*) the aircraft's inertial navigation system and/or its GPS–NavStar receiver will define its continuously changing location and this can be transferred to the bistatic receiver via a data link or
  (*b*) the location of one of the aircraft can be measured by another radar or
  (*c*) the location of one of the targets observed by the bistatic radar can be measured by another radar and eqn. 11.77 can then be solved to give L as a function of R_R.

When the non-co-operative illuminator uses a reflector antenna with a constant scan rate, it should be possible to estimate the angle at the transmitter between the directions to the target and to the receiver (angle ϕ in Fig. 11.18). The length of the baseline will not then be required. If the time interval between successive passes of the illuminator antenna's main beam over the receiver antenna is T_0, while the target echo reaches the receiver at a time interval T_S after the main beam, then the angle ϕ is given (in radians) by

$$\phi = 2\pi \frac{T_S}{T_0} \qquad (11.78)$$

This equation is only strictly correct when the receiver remains at a constant bearing from the illuminator. The correction will be straightforward if only the receiver is moving and can measure the changing bearing of the illuminator.

The distance from the receiver to the target can be derived from eqns. 11.76 and 11.50 as

$$D = R_R\left(\frac{\sin \phi}{\sin \theta} + 1 - \frac{\sin(\theta - \phi)}{\sin \theta}\right)$$

whence

$$R_R = \frac{D \sin \theta}{\sin \theta + \sin \phi - \sin(\theta - \phi)} \qquad (11.79)$$

11.17.2 *Digital pulse compression*

Bistatic radars use pulse compression, in the same way as monostatic radars, to achieve high energy pulses without loss of range resolution. Indeed, their ability to make use of such techniques is greater than in the monostatic case, since the separation of the transmitter and the receiver avoids the need for receiver protection and reduces the dynamic range of the echo signals at the receiver. This could allow the radar to use much higher duty cycles, possibly to the extent of using continuous transmissions, with a consequent reduction in the vulnerability of the transmitter to detection by ESM systems. The problem with using analogue pulse compressors in bistatic radars lies in the need to have matched pairs of filters for each waveform used, with the expander being at the transmitter and the compressor at the separate receiver. This requires a degree of equipment co-ordination which would be inconvenient even for a bistatic receiver intended for dedicated use with a specific transmitter. It would make the use of non-co-operative transmitters virtually impossible.

Future bistatic receivers can be expected to employ digital pulse compressors, as their computing costs are reduced and their speed and bandwidth increased. The versatility of digital pulse compression could allow the bistatic receiver to adapt more readily to using illuminators that did not provide waveforms dedicated to its needs. The use of digital pulse compresson avoids the need for separate compressors for each of the waveforms the bistatic receiver may use. Instead, the pulse compressor simply changes to different sets of weights; these can be held in store, ready for use with any of the transmitters, operating in the correct frequency band, in the vicinity of the receiver.

This form of pulse compression also offers the capability of capturing one of the initial pulses radiated by a radar and processing it to derive the weights required to compress the remaining pulses. Clearly, it will only work for radars with pulse compression modulations that do not vary on a pulse-to-pulse basis.

Coherent radars are most suitable as their waveforms will normally provide the sets of pulses required for discrete Fourier transforms. Since there will be a delay while an illuminator pulse is captured and processed, there may be a need to investigate the problem of processing truncated sequences – for example, ones whose length is not 2^N samples.

11.17.3 *Adaptive beamforming and null steering*

Adaptive beamforming will be an important requirement for the phased array antennas in future bistatic receivers. Even though the covert receivers in bistatic radars are expected to evade most jamming, there will still be a residual risk from widebeam jamming, from directional jamming that is inadvertently aligned on the receiver and from other radars in the same frequency band. Various adaptive techniques already exist for beamforming carried out at RF, IF or baseband. With the increasing performance and availability of digital processing, the trend is likely to be towards digital beamforming; a review of the techniques is given by Wardrop [1985]. Digital beamforming would have the two advantages of being able to deploy scanning beams:

(i) whose widths adapted to match the angle subtended by the illuminator pulse as it propagated across the receiver antenna's field of view
(ii) whose sidelobes had nulls in the directions of known interfering emitters.

The digital processing would also be able to compensate for any gain and phase shift in a scanning beam, so avoiding any corruption of the performance of the pulse compressors.

The function of an adaptive beamformer in a radar is to suppress the gain of the antenna in the directions of jamming and other interfering signals, while maintaining its gain in all other directions. The simplest type of adaptive processing is the sidelobe canceller (SLC), in which the signals from a number of low gain auxiliary antennas are combined with the signals from thenprimary antenna. The differences in gain between the primary and auxiliary antennas are sufficient to allow signals to be suppressed in the sidelobes of the primary antenna without any risk of significant distortion of its mainlobe. SLC techniques are suitable where the antenna only requires a small number of nulls ($\leq$ the number of auxiliary antennas).

For multiple beam systems, like the BEARS antenna described in Section 11.14.1, it is possible to carry out the adaptive processing in 'beam space', namely processing the multiple beam outputs rather than the outputs from individual array elements or from sub-arrays. In this case, the outputs from a number of beams, presumably those directed towards jammers, would be selected as the inputs to the adaptive network. The performance of such a system should be much better than that of a conventional sidelobe canceller, since the gains and patterns of the auxiliary beams will now be very close to those of the primary beam. The increase in the gains of the auxiliary beams means that the weights used in the adaptive combiner will be lower, so reducing the noise added to the adapted beam. The narrowing of the auxiliary beams means that the pattern of the adapted beam should only be distorted in the directions of the jammers [Ward, 1990].

11.18 Acknowledgments

The preparation of this paper has been assisted by many discussions on bistatic radars with staff at the Royal Signals and Radar Establishment (now part of the Defence Research Agency), the GEC–Marconi Research Centre at Great Baddow and Plessey Research at Roke Manor (now Roke Manor Research).

The RSRE bistatic radar programme was started by Dr. C. Pell and much of RSRE's contribution to the signal processing techniques was made by Dr. R. G. Aldous, while Mr. J. E. Summers supervised the measurements of RCS and the feasibility studies for airborne operation. The first phase of the trials described in this report (the RABIES Evaluation Studies) was managed by R. Lappage and all the dedicated flight trials aircraft were organised by Fl. Lt. M. A. Parnell. The second (Argus) phase was co-ordinated by Mr. S. E. Trigg, while Mr. Lappage was in charge of the Byson radar.

BEARS was constructed and tested by the team at the Marconi Research Centre led by P. L. Williams, while their work on pulse chasing was led by Mr. T. A. Soame. The MRC digital activities were led by Mr. P. Matthewson and Mr. D. Gould. RABIES was constructed and tested by the team at Plessey Research (Roke Manor) Ltd. led initially by C. K. Bovey and later by Dr. G. A. Bent.

I am also grateful to Dr. C. S. Griffin, formerly of SD–Scicon, for his detailed comments on this Chapter.

11.19 References

ANDERSON, K.D., RICHTER, J.H., and HITNEY, H.V. (1984): 'Tropospheric propagation assessment' in 'Characteristics of the lower atmosphere influencing radio-wave propagation' Agard Conference Proceedings 346 (Neuilly-Sur-Seince, France)

AVER'YANOV, V.Ya (1978): 'Raznesennyye radiolokatsionnye stantsii i sistemi' ('Widely spaced radar stations and systems') (Nauka i Tekhnika Press, Minsk)

BAILEY, J.S., GRAY, G.A., and WILLIS, N.J. (1977): 'Sanctuary signal processing requirements'. 11th Asilomar Conf. on Circuits, Systems and Computers, Pacific Grove, CA, (IEEE 1978), pp. 310–315

BATTAN, L.J. (1973): 'Radar observation of the atmosphere' (University of Chicago Press)

BEAN, B.R., and DUTTON, E.J. (1966): 'Radio Meteorology'. National Bureau of Standards Monograph 92

BICKEL, S.H. (1966): 'Polarization Extension of the Monostatic-Bistatic Equivalence Theorem'. Mitre Corp. Report TM–4242

BOITHIAS, L., and BATTESTI, J. (1967): 'Protection against fading on line-of-sight radio-relay systems (in French)', *Ann. des Télécomm.*, **22**, 9–10, pp. 230–242

BOVEY, C.K., and HORNE, C.P. (1987): 'Synchronisation aspects for bistatic radars'. Radar '87, London, IEE Conference Publication 281, pp. 22–25

BOWEN, E.G. (1987): 'Radar days'. (Adam Hilger)

COCKBURN, Sir R. (1988): 'The radio war', *in* BURNS, R. W. (Ed.): 'Radar Development to 1945', (Peter Peregrinus, London) chap. 24

COLEMAN, J.O., and ALTER, J.J. (1985): 'Communication through a surveillance Radar'. IEEE International Radar Conf, pp. 308–311

CRISPIN, J.W., GOODRICH, R.F., and SIEGEL, K.M. (1959): 'A theoretical method for the calculation of the radar cross-sections of aircraft and missiles'. University of Michigan, Report 2591–1–M, AF 19(604)–1949, AFCRC–TN–59–774

DAVIES, K. (1965): 'Ionospheric propagation'. National Bureau of Standards Monograph 80

DUNSMORE, M.R.B. (1987): 'Bistatic radars for air defence'. Radar '87 Conference, London, IEE Conference Publication 281, pp. 7–11

EASTON, R.L., and FLEMING, J.J. (1960): 'The Navy space surveillance system', *Proc. IRE*, **48**, pp. 663–669

ERICKSON, J. (1988): 'The air defence problem and the Soviet radar programme 1934/35–1945' *in* BURNS, R. W. (Ed.): 'Radar developments to 1945' (Peter Peregrinus, London) chap. 16

FAWCETTE, J. (1980): 'Vulnerable radars seek a safe sanctuary', *Microwave System News*, **10**, pp. 45–50

FLEMING, F.L., and WILLIS, N.J. (1980): 'Sanctuary radar'. Military Microwaves '80 Conference Proceedings, London, (Microwave Exhibitions & Publishers), pp. 103–108 (The details quoted from the Conference Publication have been augmented by notes made by Dr. J. Clarke during the author's oral presentation)

FRANCOEUR, A.R. (1989): 'Naval space surveillance system (NAVSPASUR) solid state transmitter modernization'. Proc IEEE National Radar Conf., Dallas, pp. 147–152

GALKIN, V.I. *et al.* (Eds.) (1974): 'Radiotekhnicheskiye sistemi v raketnoy tekhnike' ('Radio systems in missile technology') (Voyenizdat Press, Moscow)

GARNHAM, R.H. (1977): 'Minimum pulse length allowable in various types of electronically scanned array antenna'. RRE Memo 3023

GLASER, J.I. (1985): 'Bistatic RCS of complex objects near forward scatter'. *IEEE Trans.*, **AES–21**, pp. 70–79

GLASER, J.I. (1986): 'Fifty years of bistatic and multistatic radar', *IEE Proc.* **133**, Pt. F, pp. 596–603

GRIFFITHS, H.D., and LONG, N.R.W. (1986): 'Television-based bistatic radar', *IEE Proc.*, **133**, Pt. F, pp. 649–657

HERTZ, H. (1962): 'Electric waves', (Dover)

HIATT, R.E., SIEGEL, K.M., and WEIL, H. (1960): 'Forward scattering by coated objects illuminated by short wavelength radar', *Proc. IRE*, **48**, pp. 1630–1635

HOFFMAN, K.O. (1968): 'Die Geschichte der Luft nach Richlentruppe. Band II–Die Weltkrieg' (Neckargemund–Kurt Vowinckel Verlag)

HOLMES, D.G. (1988): Private communication

HORNE, C.P., and BOVEY, C.K. (1987): 'Synchronisation for bistatic air defence radars'. Radar '87, London, IEE Conference Publication 281, pp. 22–25

HULSMEYER, C. (1904): 'Hertzian wave projecting and receiving apparatus adapted to indicate or give warning of the presence of a metallic body, such as a ship or train, in the line of projection of such waves'. British Patent 13170, 2 Sept.

JACKSON, M.C. (1986): 'The geometry of bistatic radar systems', *IEE Proc.* **133**, Pt. F, pp. 604–612

JUDD, F.C. (1983): 'Over-the-horizon radar systems – Part 2', *Practical Wireless (GB)*, **59**, 9, pp. 44–47

KELL, R.E. (1965): 'On the derivation of bistatic RCS from monostatic measurements', *Proc. IEEE*, **53**, pp. 983–988

KERR, D.E. (Ed.) (1951): 'Propagation of short radio waves'. MIT Radiation Lab. Series, Vol. 13, (McGraw-Hill)

LAPPAGE, R., CLARKE, J., PALMER, G.W.R. and HUIZING, A.G. (1987): 'The Byson research radar'. Radar '87, London, IEE Conference Publication 281, pp. 453–457

LEE, P.K. and COFFEY, T.F. (1985): 'Space-based bistatic radar: Opportunities for future tactical air surveillance', Proc. IEEE International Radar Conf, pp. 322–328

LOBANOV, M.M. (1969): 'Iz proshlovo radiolokatsii' ('Out of the past of radar') (Military Publisher of the Ministry of Defence, Moscow)

LONG, M.W. (1975): 'Radar reflectivity of land and sea', (Artech House)

MACK, C.L., and REIFFEN, B. (1964): 'RF characteristics of thin dipoles', *Proc. IEEE*, **52**, pp. 533–542

MATTHEWSON, P., WARDROP, B., and GOULD, D.M. (1990): 'An adaptive bistatic radar demonstrator'. Military Microwaves '90 Conference Proceedings, London (Microwave Exhibitions & Publishers), pp. 471–476

McEWAN, N.J. (1989): 'Nature of precipitation, cloud and atmospheric gases and their effects on propagation mechanisms from VHF to millimetre waves' *in* HALL, M.P. M., and BARCLAY, L.W. (Eds.): 'Radiowave Propagation' (Peter Peregrinus, London) chap. 10

MILNE, K. (1977): 'The principles and concepts of multistatic radars'. Radar '77 Conference, London, IEE Conference Publication 155, pp. 46–52

NAKAJIMA, S. (1988): 'The history of Japanese radar development to 1945' *in* BURNS, R.W. (Ed.): 'Radar development to 1945', (Peter Peregrinus, London) chap. 18

PELL, C., and CROSSFIELD, M.D. (1984): 'Signal and data processing architectures for multistatic surveillance radars'. Proceedings Radar '84 Conference, Paris, pp. 436–443

PETERS, L. Jr. (1962): 'Passive, bistatic radar enhancement devices', *Proc. IEE*, **109C**

PETERS, L. Jr. (1959): 'Theory of corner reflectors as bistatic radar enhancement devices'. Antenna Laboratory, Ohio State University, Report 768–5, Contract AF 33(615)–5078

PRICE, A. (1978): 'Instruments of darkness – the history of electronic warfare' (Macdonald and Jane's Publishers, London), pp. 216–217

READER, K.N., CLARKE, J., and LAPPAGE, R. (1987): 'A new calibration method for pulse-Doppler and MTI radars'. Radar '87, London, IEE Conference Publication 281, pp. 458–461

RUCK, G.T., BARRICK, D.E., STUART, W.D., and KRICHBAUM, C.K. (1970): 'Radar cross section handbook' (Plenum Press), (*a*) Section 3.2 'Perfectly conducting spheres', (*b*) Section 3.3 'Dielectric spheres – lossy and lossless', (*c*) Section 9.1 'Rough surfaces – analytical models'

SALAH, J.W., and MORRIELLO, J.E. (1980): 'Development of a multistatic measurement system'. Proc. IEEE International Radar Conf., pp. 88–93

SCHOENENBERGER, J.G., and FORREST, J.R. (1982): 'Principles of independent receivers for use with cooperative radar transmitters', *Radio & Electronic Engr.*, **52**, 2, pp. 93–101

SHNEYDERMAN, Ya A. (1972): *Zarubezh Radioelektron*, (7), p. 102

SIEGEL, K.M. *et al.* (1955): 'Bistatic radar cross sections of surfaces of revolution'. *J. Applied Physics*, **26**, 3, pp. 297–305

SKOLNIK, M.I. (1980): 'Introduction to radar systems' (McGraw-Hill), chap. 36

SKOLNIK, M.I. (1985): 'Fifty years of radar'. *Proc. IEEE*, **73**, 2, pp. 182–197

SOAME, T.A., and GOULD, D.M. (1987): 'A description of an experimental bistatic radar system'. Radar '87, London, IEE Conference Publication 281, pp. 12–16

THOMPSON, E.C. (1989): 'Bistatic radar noncooperative illuminator synchronization techniques'. Proc. IEEE National Radar Conf, Texas, pp. 29–34

TOMIYASU, K. (1978): 'Bistatic synthetic aperture radar using two satellites'. IEEE EASCON Record, pp. 106–110

WALLINGTON, J.R. (1985): 'The role of analogue beamforming in radar', *GEC J. Research*, (3), pp. 25–33

WARD, C. (1990): Private communication

WARDROP, B. (1985): 'The role of digital processing in radar beamforming', *GEC J. Research*, **3**, pp. 34–45

WARDROP, B., and MOLYNEUX-BERRY, M.R. (1992): 'Active–passive bistatic surveillance for long-range air defence'. Microwave J. (USA), **35**, (6), pp. 98, 101–104, 108, 110–111

WATSON-WATT, R.A. (1935): 'Detection and location of aircraft by radar methods' *in* BURNS, R.W. (Ed.): 'Radar development to 1945', (Peter Peregrinus, London), Appendix to chap. 20

WILLIS, N.J. (1991): 'Bistatic radar' (Artech House, USA)

ZAKHAR'YEV, L.N., and LEMANSKIY, A.A. (1972): 'Rasseyaniye voln "chernymi" telami' ('Scattering of waves by "black" objects') (Sovetskoye Radio Press, Moscow)

ZUR HEIDEN, D., and KLIPPEL, L. (1979): 'Passive ground-based system for target detection' *in* 'Briefings on multistatic radar systems for air defence'. Joint Meeting at NATO HQ, Brussels, 5 Dec.

Aviation Week & Space Technology (1980): 'Multistatic mode raises radar accuracy', 14 July, pp. 62–69

CCIR V (Comité Consultatif International des Radiocommunications) (1982): 'Propagation in non-ionized media'. XVth Plenary Assembly, Geneva 1982, 'Radio meteorological data'. Report 563–2, pp. 96–123

Defense Electronics (1981): 'Largest microstrip antenna tested', **13**, 2, pp. 90–92

TRE (1941): 'Camoflaging of an aircraft at centimetre wavelengths' File 4/18, ref HGB/MH, s/no. 5273, 27 August.

US Department of Defense (1986): 'Soviet military power'

US Department of Defense (1988): 'Soviet military power – An assessment of the threat'

Chapter 12

Space-based radar

Prof. Ing. G. Galati and Ing. M. Abbati

Part 1: Synthetic aperture radar (SAR) and space-based SAR missions

12.1 Introduction

The coming of the space era allowed us to develop satellite-borne radar systems able to overcome some of the limitations of ground-based (and also airborne) radars. In fact, the performance of a ground-based radar with regard to low altitude coverage is seriously affected by environmental factors such as earth curvature or geographical factors, drawbacks not completely eliminated using either OTH techniques in the HF band or the ducting effect. HF solutions in particular do not allow steady operation, because of the variable behaviour of the ionosphere, and call for high power and very large antennas.

The former satellite-borne radars were employed for remote sensing tasks, to observe the Earth surface for scientific purposes and to analyse its resources, with many resultant applications (for example, in geology and oceanography). The need for good ground resolution, in order to get satisfactory images of the interested area, accelerated the use of synthetic aperture radar (SAR) techniques. At present, the only known operational space-based radars are SAR, altimeters and scatterometers, employed from 1978 up to now.

Therefore, it is important to recall the fundamental stages in the SAR evolution, showing the remarkable results obtained, and also the intrinsic limitations, which lead to finding new and more effective solutions, especially for surveillance functions.

12.2 Working principle of SAR

It is possible to distinguish between unfocused and focused SAR [1]. In the *unfocused SAR* case, the working principle is very simple: the various radar echoes arriving at the physical antenna in its subsequent positions during the platform motion are coherently integrated, without any phase weighting. In order to get significant integration gain, a maximum length for the synthetic antenna is assumed, so that the maximum two-way path difference between the central point and one of the extremes of the array, with respect to the same point on the Earth, is at the most $\lambda/4$ (a quarter of wavelength).

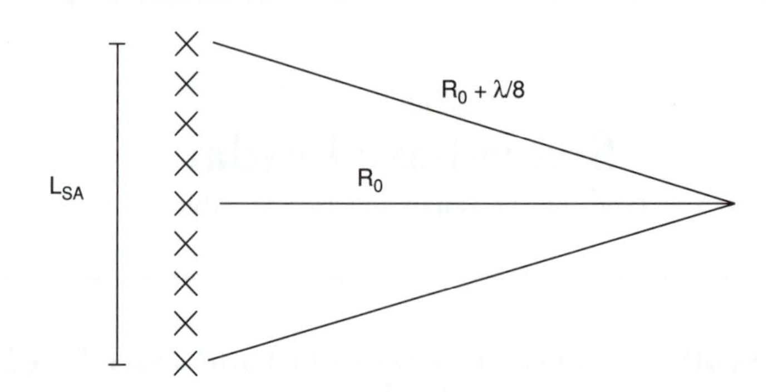

Figure 12.1 *Geometry of unfocused SAR*

With this assumption, the echoes arriving on the different antenna positions from the same point are 'nearly in phase'; therefore, they can be added without significant integration losses.

From Fig. 12.1, we have:

$$(R_0+\lambda/8)^2=R_0^2+(L_{sa}/2)^2 \tag{12.1}$$

where

R_0 = distance between the central point of the synthetic antenna and the on-Earth point
L_{sa} = synthetic antenna length

Then

$$L_{sa}^2=\lambda(R_0+\lambda/16) \tag{12.2}$$

and, since $R_0 \gg \lambda$:

$$L_{sa} \simeq \sqrt{\lambda R_0} \tag{12.3}$$

Remembering that the pattern of a synthetic array has to be referred to two-way operation, and therefore the phase differences at the aperture are twice those of a real array, the synthetic antenna beamwidth in the azimuthal plane is given by

$$\theta_{sa}=\frac{\lambda}{2L_{sa}} \tag{12.4}$$

and the azimuthal resolution is

$$d_A=\theta_A R_0=\frac{\lambda R_0}{2L_{sa}} \tag{12.5}$$

Replacing the expr. 12.3 in expr. 12.5, we have

$$d_A=\frac{\sqrt{\lambda R_0}}{2} \tag{12.6}$$

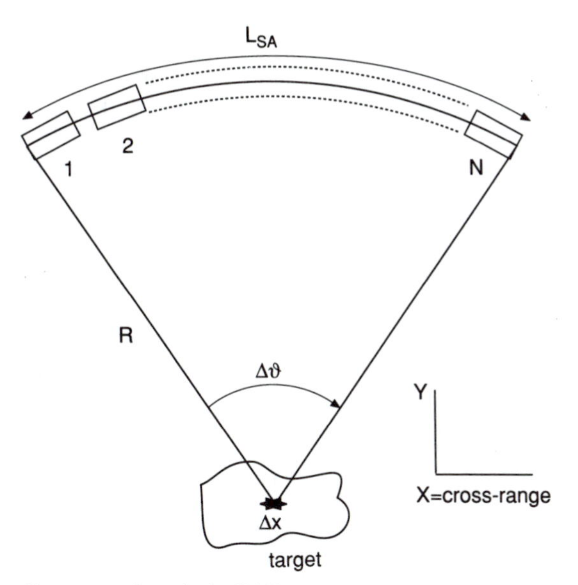

Figure 12.2 *Geometry of spotlight SAR*

showing that, for the unfocused SAR, the azimuthal resolution does not depend on the antenna dimensions.

Although the unfocused SAR can be implemented without any particular computations, it does not permit very good angular resolution values, because, as shown before, the synthetic antenna length is limited by expr. 12.3.

If phase correction of the received pulses before integration is accepted, longer synthetic antennas, and better resolutions, can be obtained at the expense of an increase of the system complexity.

In the *spotlight-SAR* case, the physical radar antenna, while moving, is always pointed at the same object, for a span corresponding to the angle $\Delta\theta$. The synthetic antenna length is given by (Fig. 12.2)

$$L_{sa} = R\Delta\theta \tag{12.7}$$

and the (two-way) synthesised beamwidth is again given by expr. 12.4. The cross-range resolution is obtained from exprs. 12.4 and 12.7:

$$\Delta x = R\theta_{sa} = R\frac{\lambda}{2R\Delta\theta} = \frac{\lambda}{2\Delta\theta} \tag{12.8}$$

It depends on the observation angle $\Delta\theta$ and the wavelength λ, but not on the distance R. Of course, the same result is obtained when the antenna is fixed and the target rotates with respect to it: this is the inverse SAR (ISAR) operating principle.

In the traditional 'strip-map' SAR operation, linear motion of the physical antenna (not a circular one any more) is assumed. Taking into account that in the strip-map mode the observation angle $\Delta\theta$ is, at most, equal to the real antenna beamwidth, and using the expr. 12.8 for the cross-range resolution, from the strip-map SAR geometry (Fig. 12.3), we have

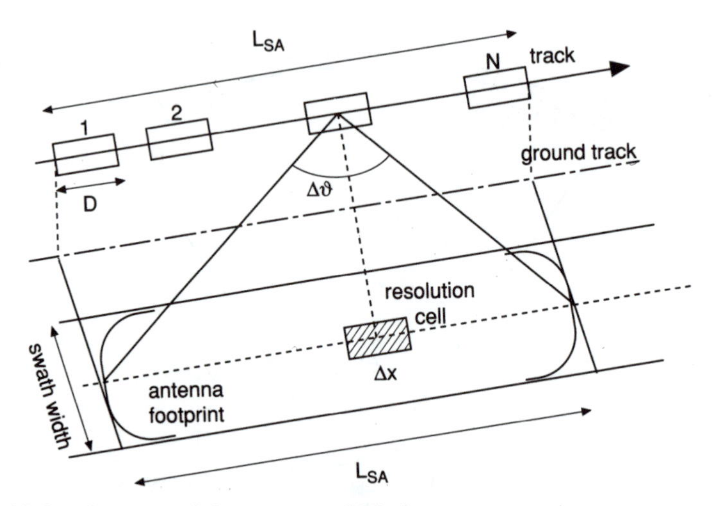

Figure 12.3 *Geometry of the strip-map SAR (perspective view)*

$$d_A = \Delta x = \frac{\lambda}{2\Delta\theta} = \frac{\lambda}{2\lambda/D} = \frac{D}{2} \tag{12.9}$$

Therefore, the strip-map SAR resolution depends on the physical antenna dimension D only.

The spotlight-SAR technique is preferred to the traditional one when the resolution requirement is of paramount importance. By means of azimuthal

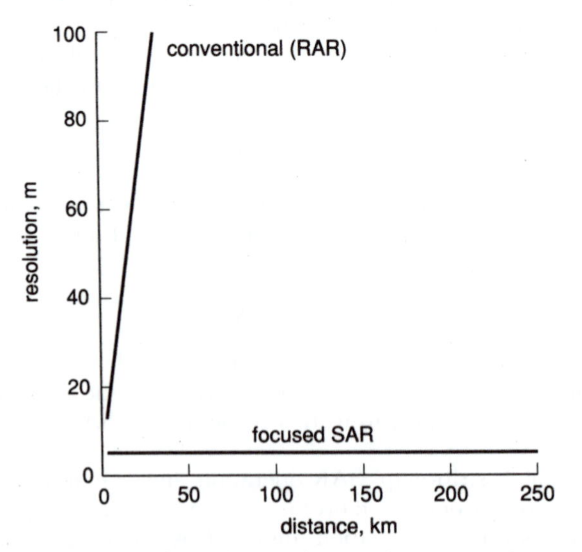

Figure 12.4 *Comparison of angular resolution of conventional antenna (RAR) and focused SAR ($D = 10$ m, $\lambda = 3$ cm)*

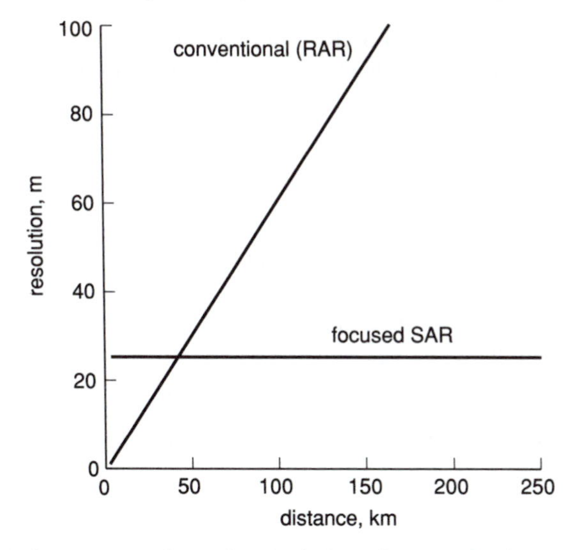

Figure 12.5 *Comparison of angular resolution of conventional antenna (RAR) and focused SAR ($D = 50$ m, $\lambda = 3$ cm)*

beam scanning, the SAR, during satellite motion, illuminates the same fixed spot on the Earth surface, allowing an increase of coherent processing time of the echoes from the spot, and so improving the azimuthal resolution. The price to be paid is, of course, reduced (discontinuous) coverage, as compared with the full (continuous) coverage of the swath in the strip-map mode.

It is interesting to compare the azimuthal resolution values of focused SAR and conventional (real) antenna, firstly with an antenna diameter $D = 10$ m and a wavelength $\lambda = 3$ cm (Fig. 12.4), and secondly with $D = 50$ m and $\lambda = 3$ cm (Fig. 12.5). Obviously, for reasonable D, the best performance with regard to the azimuthal resolution is given by the focused SAR, but on increasing the antenna dimension D the situation changes, and for very large D, a conventional antenna, e.g. a real aperture radar (RAR), gives the best results when focused.

The azimuthal resolution analysed above refers to 'one-look' processing, i.e. to the exploitation of the full synthetic aperture to get the best achievable spatial resolution. However, in this way, only one sample of the target reflectivity is obtained, resulting in a 'speckled' image [1, 2]. The speckle can be reduced by generating a number n_L of independent reflectivity samples (looks) and adding them after modulus extraction (incoherent integration). The independent samples are obtained by dividing the synthetic aperture into n_L apertures of length L_{sa}/n_L each, and therefore by increasing the azimuthal resolution cell by a figure equal to n_L.

Comparing the RAR azimuthal resolution at a distance R_0 (if R_0 is smaller than the near-field distance D^2/λ, the RAR antenna should be focused at the range R_0):

$$d_{RAR} = R_0 \frac{\lambda}{D}$$

with the unfocused and focused SAR resolution (eqns. 12.6 and 12.9), degraded by the number of looks, the RAR resolution is better than the SAR resolution at near-field distances; it results:

$$d_{RAR} < d_{foc.SAR} \text{ for } D > \sqrt{\frac{2}{n_L} \lambda R_0} \tag{12.10}$$

$$d_{RAR} < d_{unfoc.SAR} \text{ for } D > \frac{2}{n_L} \sqrt{\lambda R_0} \tag{12.11}$$

As an example, in an airborne application ($\lambda = 0.031$ m, $R_0 = 10^4$ m, $n_L = 2$) from eqn. 12.10 we have

$$d_{RAR} < d_{foc.SAR} \text{ for } D > 17.6 \text{ m}$$

Such a value of D is compatible with a distributed array conformal to an aircraft [34].

12.3 Main space-based SAR achievements and missions

The main space-based SAR realisations and applications are explained in this Section, with reference to Earth missions [for planetary missions it is necessary to refer to other documents, such as 'Space-based radar handbook' by L. J. Cantafio, (Ed.)].

The first space-based SAR mission started in June 1978, with the launch of the SEASAT satellite [2], operating on a 800 km height orbit for more than three months and collecting images over an area of about 126×10^6 km^2 and, in particular, including North America, Western Europe, North Atlantic, North Pacific and the North Pole. The antenna consisted of a 10.7×2.16 m^2 L-band ($\lambda = 0.235$ m) array, without any possibility of beam steering; the antenna dimensions were related to the range and Doppler frequency ambiguity level requirements. The antenna beam was pointed with an off-nadir angle $\alpha = 20.5°$ perpendicular to the velocity vector of the satellite (Fig. 12.6); a swath s of about

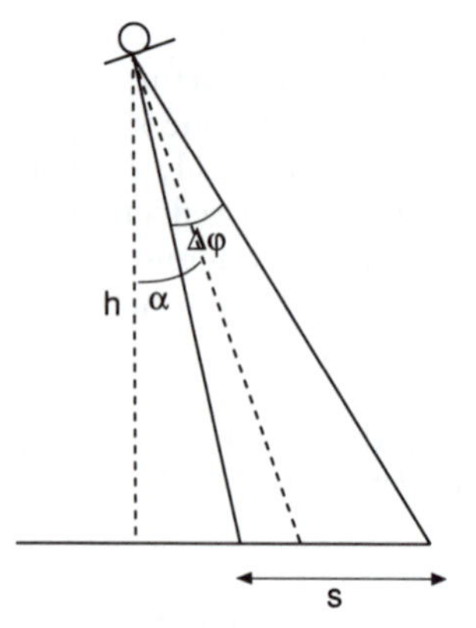

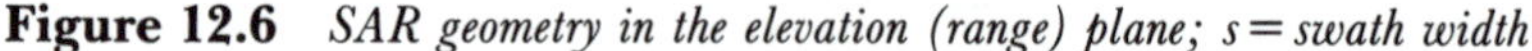

Figure 12.6 *SAR geometry in the elevation (range) plane; s = swath width*

100 km was obtained, resulting from the values of the across-track dimension of the antenna and of the off-nadir angle. The azimuthal resolution was 25 m, with 4-looks processing. A range resolution of 25 m was obtained by means of pulse-compression techniques, using a solid-state low-peak-power transmitter.

The Space Shuttle missions permitted further development of imaging radar. In November 1981 the Space Shuttle *Columbia* carried the SIR–A (shuttle imaging radar) in a low (about 250 km) orbit [2]. This mission lasted only three days, but an area of about 10×10^6 km^2 was covered, including large regions of North and South America, Africa, Asia and Australia.

Although the L-band wavelength and the antenna dimensions of 9.4×2.16 m^2 were not very much different from the ones of SEASAT SAR, the off-nadir angle was changed from 20° to 47°, without antenna beam steering capabilities. Owing to the differences in the SIR–A geometry with respect to the SEASAT, the swath width was about 50 km, instead of the 100 km of SEASAT. Moreover, 6-look processing allowed a better radiometric resolution, but with a worse azimuthal resolution (38 m instead of 25 m for SEASAT). Some technical data about SIR–A mission are shown in Fig. 12.7.

The data resulting from these two SAR missions were used in several applications. For example, the SAR data, suitably processed, were used to build geological maps of very large regions, emphasising the various mountain reliefs. The SEASAT and SIR–A mission helped to develop oceanography, so as to point out, on maps, waves, streams and, for example, stains due to polluting materials. Moreover, SEASAT in particular collected a number of images of large glacial zones of the North Pole, with the possibility of analysing movements of glacial masses.

Finally, many applications concerned agriculture, where SAR, together with optical techniques, permitted the observation of the growth of different kinds of crops in some regions.

With regard to the ensuing Shuttle missions, in October 1984 *Challenger* put SIR–B into an orbit of height variable between 200 and 400 km [3]. In seven days images of a 6.5×10^6 km^2 area were collected.

In comparison with its predecessors, SIR–B presented a substantial difference, namely the possibility of mechanical beam steering so as to change the incidence angle from 15° to 60°, and to extract further informations (Fig. 12.8) by stereoscopic techniques. As a consequence, the range resolution could be varied from 17 to 58 m, while the 10.7×2.16 m^2 antenna assured an azimuthal resolution of 25 m (4-look processing), as SEASAT.

In Fig. 12.9 an outline of SIR–B is shown, while in Fig. 12.10 some technical data about the mission are shown. Even in the SIR–B case, the data obtained were used in a large number of applications especially in geology, sub-surface radar penetration, vegetation science and oceanography.

Future developments include the realisation of SIR–C/SAR–X and EOS (Earth Observation Satellite) missions, that will conclude the cycle begun by the SEASAT. With regard to the SIR–C mission, three flights are forecast for the 1990s [4], starting in 1994. (Note that space program plans change frequently so the dates quoted here are merely illustrative.)

A substantial difference between SIR–C and the previous missions is that SIR–C is able to work both in frequency and in polarisation diversity. For this purpose, the antenna is formed from two planar microstrip arrays, one for

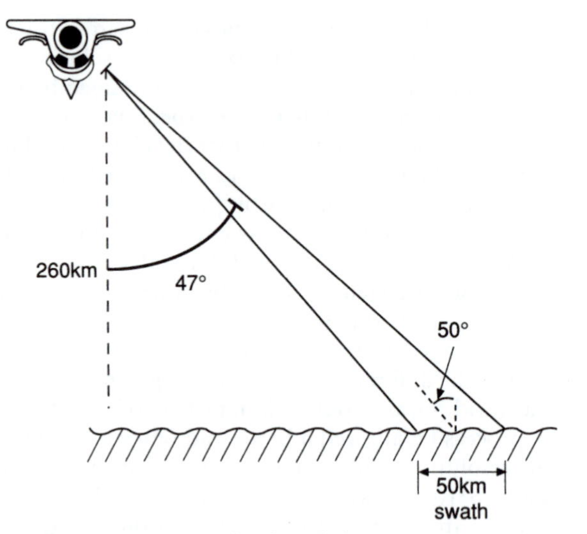

Figure 12.7 *SIR–A: Geometry and technical data*

System characteristics
- frequency: 1·278 GHz
- wavelength: 23·5 cm (L-band)
- polarisation: HH
- look angle: 47°

Image characteristics
- swath width: 50 km
- azimuth resolution: 40 m
- range resolution: 40 m
- number of looks: 6

Instrument components
- transmitter/receiver
- antenna
- optical recorder

Instrument characteristics
- peak power: 1 kW
- antenna dimensions: 9·4 × 2·16 m
- bandwidth: 6 MHz

Data collection
- optical data: 8 h
- total coverage: 10 million km^2
- processing: optical

Orbit characteristics
- altitude: 260 km
- inclination: 38°

Mission
- launch date: 12 November 1981
- mission duration: 2·4 days
- vehicle: Columbia
- flight: STS-2

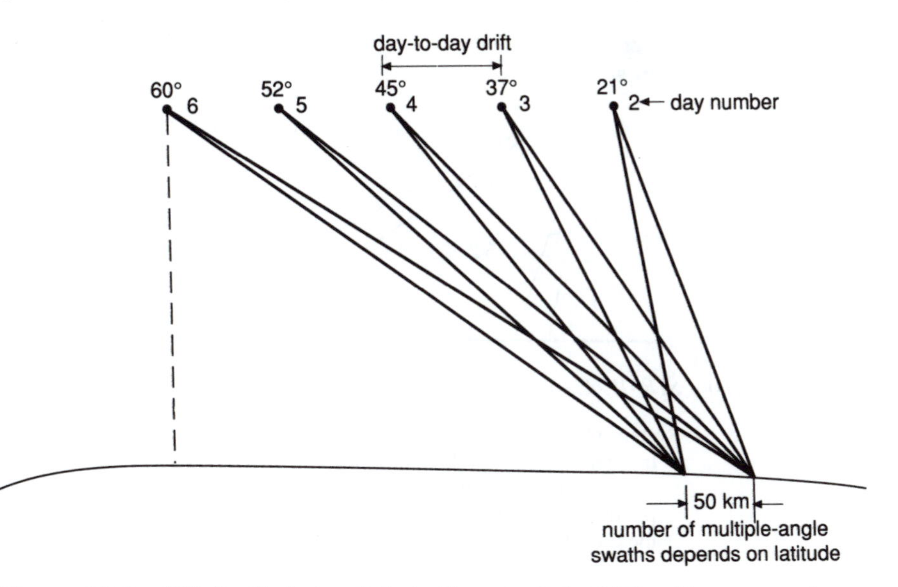

Figure 12.8 *SIR–B: Variation of the incidence angle (multiple orbit stereoscopy)*

L-band and the other for C-band, each of which can be fed to transmit and receive in horizontal or vertical polarisation. In particular, the L-band array is composed of two rows of nine panels, every panel including nine rows of six

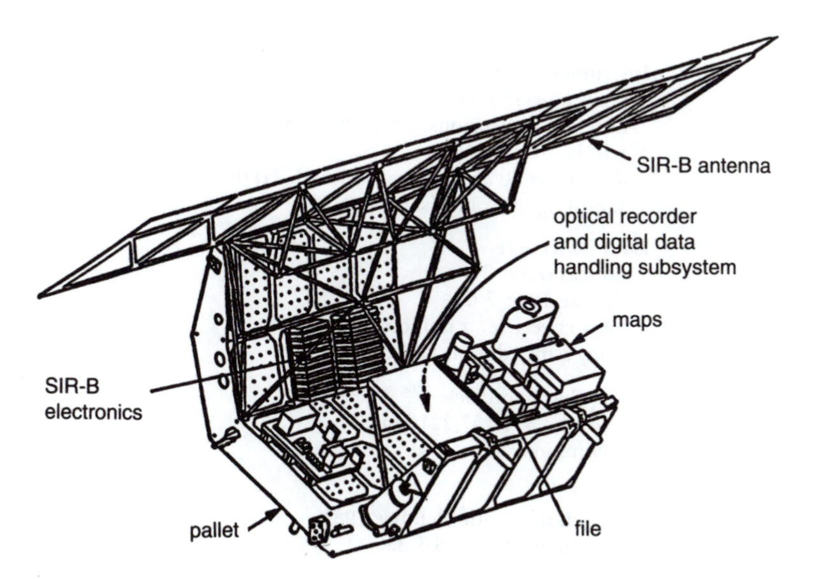

Figure 12.9 *SIR–B on the pallet*

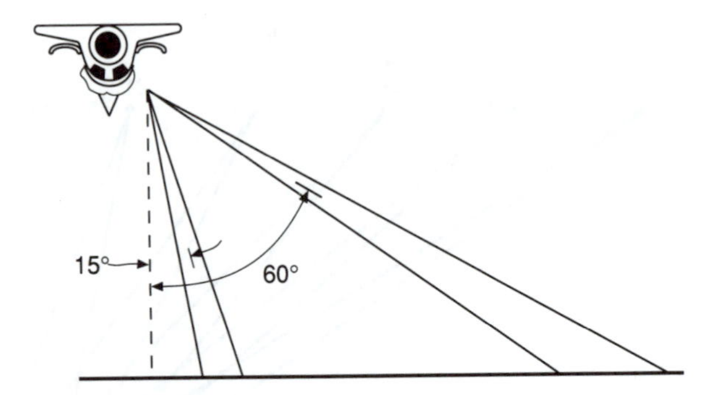

Figure 12.10 *SIR–B: Geometry and technical data*

System characteristics
- frequency: 1·282 GHz
- wavelength: 23·4 cm (L-band)
- polarisation: HH
- look angles: 15–60°

Image characteristics
- swath width: 20–40 km
- azimuth resolution: 25 m
- range resolution: 58–17 m
- number of looks: 4

Instrument components
- transmitter/receiver
- antenna (mechanically tiltable and folding)
- digital data handling sub-system (DDHS)
- high data rate recorder (HDRR)
- optical recorder (OR)

Instrument characteristics
- peak power: 1 kW
- antenna dimensions: 10·8 × 2·16 m
- bandwidth: 12 MHz
- data rate: 30·35 Mbits/s (through HDRR)
- sampling: 3–6 bits/sample

Data collection
- optical data: 8 h
- digital data: 7 h
- total coverage: 5 million km^2
- processing: digital and optical

Orbit characteristics
- altitude: 354,257,224 km
- inclination: 57°

Mission
- launch date: 5th October 1984
- mission duration: 8·3 days
- vehicle: Challenger
- flight: 41-G

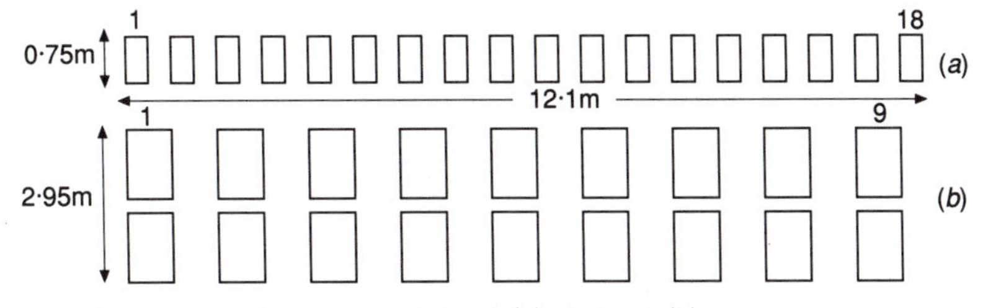

Figure 12.11 *SIR–C antennas: C-band (a), L-band (b)*

patches each, for an overall aperture dimension of 12.1×2.95 m^2 (Figs. 12.11 and 12.12).

The C-band array is formed by a single row of 18 panels, each row of which includes 18 rows of 18 patches, for an overall aperture dimension of 12.1×0.75 m^2 (Figs. 12.11 and 12.12). Such antenna dimensions assure the same beamwidth in the elevation plane for both L-band and C-band, but the beamwidth can be varied by a phase weighting; the selection of a pre-programmed phase function permits widening the elevation beamwidth from the nominal value of 4.7° to 32° [4].

Another fundamental difference from the previous missions is in the electronic steering of the antenna main beam, which is limited to ±23° in the elevation plane and to ±2° and ±1° (for L- and C-band, respectively) in the azimuthal plane. It is possible in such a way to change the off-nadir angle from 15° to 55°, thus obtaining a range resolution and a swath width variable between 10 and 60 m and between 15 and 60 km (Fig. 12.13), respectively. An azimuthal resolution of 30 m is obtained by a 4-look processing.

In Fig. 12.14 the main characteristics of SIR–C are summarised, and in Fig. 12.15 a block diagram of the sensor is shown. The SIR–C mission will permit many experiments in various areas, such as geology, hydrology, vegetation analysis, glaciology and oceanography.

Better performance will be possible by the integration of SIR–C with the German–Italian sensor SAR–X, using an X-band, vertically polarised planar array of 12×0.33 m^2, with the possibility of mechanically steering the antenna beam by means of the antenna rotary joint of the SIR–C [5]. In Fig. 12.16 some technical data on SAR–X are reported.

As already mentioned, the SAR missions started with the SEASAT SAR and have the EOS mission as their destination. The EOS (Earth Observation Satellite) mission is forecast for the end of the 1990s, for a duration that will probably exceed ten years [6]. EOS–SAR will be a three-frequency multipolarisation radar system, with two antennas, one horizontally polarised and the other vertically polarised, for each operating frequency (L-, C-, X-band) (Fig. 12.17).

From Fig. 12.18, where some system parameters are reported, it can be noted that EOS–SAR includes and enhances all the capabilities of the previous SAR missions. Nevertheless, EOS represents a fundamental innovation with respect to its predecessors, namely the possibility of on-board SAR image generation, without quality losses, using the sensor data to reduce the downlink data volume to the ground station. This innovation will allow direct image

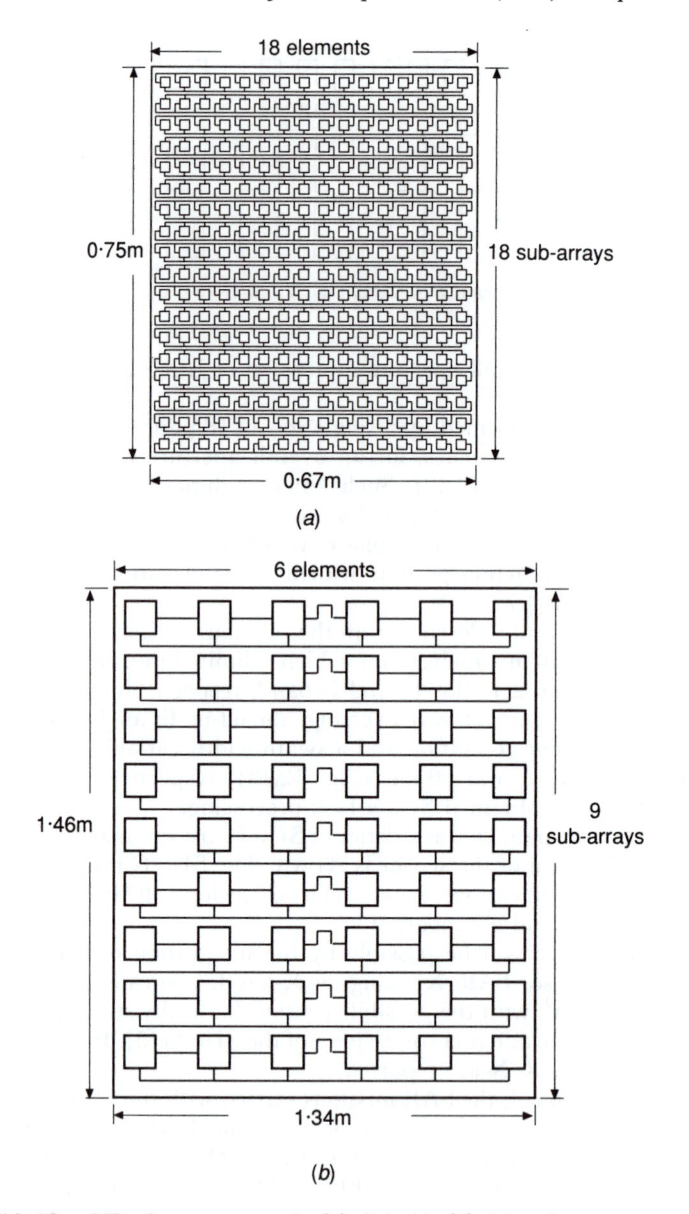

Figure 12.12 *SIR–C antenna panel: (a) C-band, (b) L-band*

distribution to the users, with a consequent increase of interactivity between system and users.

Besides the American experiments, in the European environment space-based SAR activities are increasing. The European Space Agency (ESA) mission ERS–1 (Earth Resources Satellite) started in 1991. This satellite carries an active microwave instrument (AMI) working in the C-band, with both SAR

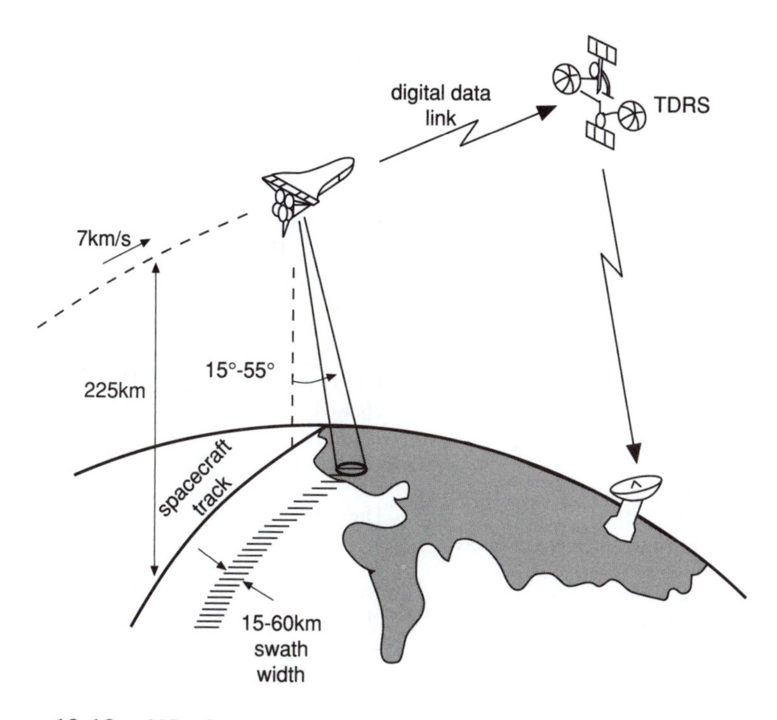

Figure 12.13 *SIR–C geometry*

and scatterometer functions (Fig. 12.19). In the 'image-mode' the AMI activates the SAR antenna for ten minutes each orbit, working at a frequency of 5.3 GHz, with vertical polarisation, an incidence angle of 23° and a swath 100 km wide.

Besides typical full-swath SAR operation, ERS–1 may accomplish a 'wave-mode' operation, where data are collected from 5 km^2 spots every 200–300 km. This kind of spatial sampling allows data volume reduction, so as to store on board the data received from the various spots, without the necessity for immediately sending them to ground. In such a way, the presence of a ground station in the field of view of the satellite is not mandatory, and, as a consequence, it is possible to obtain images from zones, such as the oceans, not provided with ground stations.

As stated before, ERS–1 AMI can work as a scatterometer, as well as an SAR, in order to investigate wind behaviour. The scatterometer mode of operation is based on the simple concept that a superficial wind causes change of reflectivity of the underlying sea surface. Therefore, if three squinted antennas are used, from the three different measurements of the radar cross-section of the same area, it will be possible to extract wind velocity and direction.

It is important to emphasise that ERS–1 also includes a K_u-band (13.5 GHz) radar altimeter, developed in Italy, in order to perform precise height measurements on oceans and glacial zones.

Parameter	L-Band	C-Band
Orbital altitude	225 km	225 km
Orbital inclination	57°	57°
Transmitter peak power	4300 W	2250 W
Modulation	Linear FM	Linear FM
Pulse width	8.4,16.9,33.8 μs	8.4,16.9,33.8 μs
RF centre frequency	1248.75 MHz	5298.75 MHz
Bandwidth	10.20 MHz	10.20 MHz
RCVR noise temperature	575 K	655 K
RCVR gain	56 to 101 dB	56 to 101 dB
Antenna size	12.08 × 2.95 m	12.08 × 0.75 m
Polarisation	HH,HV,VV,VH	HH,HV,VV,VH
Antenna gain	36.4 dB	42.8 dB
Data processing	Digital	Digital
Off-nadir angle	15° to 55°	15° to 55°
Electronic beam steering	23°	23°
Data recording	Digital	Digital
Number of high-rate record channels	4	4
Bit rate/channel	45 Mbit/s	45 Mbit/s
Range resolution	10 to 60 m	10 to 60 m
Azimuth resolution	30 m	30 m
Number of azimuth looks	4	4
Swath width	15 to 90 km	15 to 90 km

Figure 12.14 *SIR-C technical data*

Together with the European ERS–1, further SAR evolution in the 1990s is characterised by the Japanese JERS–1 and the Canadian RADARSAT.

With regard to the JERS–1 mission, begun in 1992, a working frequency of 1.2 GHz (L-band) and horizontal polarisation are used with a $12 \times 2.2\ m^2$ antenna, placed at an off-nadir angle of about 44°. In this way, an azimuthal resolution of about 20 m and a swath width of 75 km are obtained.

The RADARSAT mission is forecast for the middle 1990s and will use a C-band SAR (5.3 GHz) and vertical polarisation both in transmission and reception. The Sun-synchronous polar orbit will have an altitude of 1000 km. Unlike JERS–1, in the RADARSAT mission it will be possible to steer the beam in elevation so as to change the off-nadir angle from 20° to 45°, and to make range resolution and swath width variable (wide swath operation). The main purpose of RADARSAT is the monitoring of the Arctic regions to protect shipping and petroleum exploration platforms from ice floes. The satellite will also carry a scatterometer and a very high resolution radiometer.

For an immediate comparison, Fig. 12.20 gives illustrative data on the satellite and Shuttle SAR missions described here.

Many of these missions seem to converge on the Polar Platform project; this is a small part of a bigger project whose purpose is the realisation of an orbiting Space Station, on which a large number of sensors, such as SAR, will be integrated in order to enhance knowledge about the Earth (atmosphere, ground and sea surfaces).

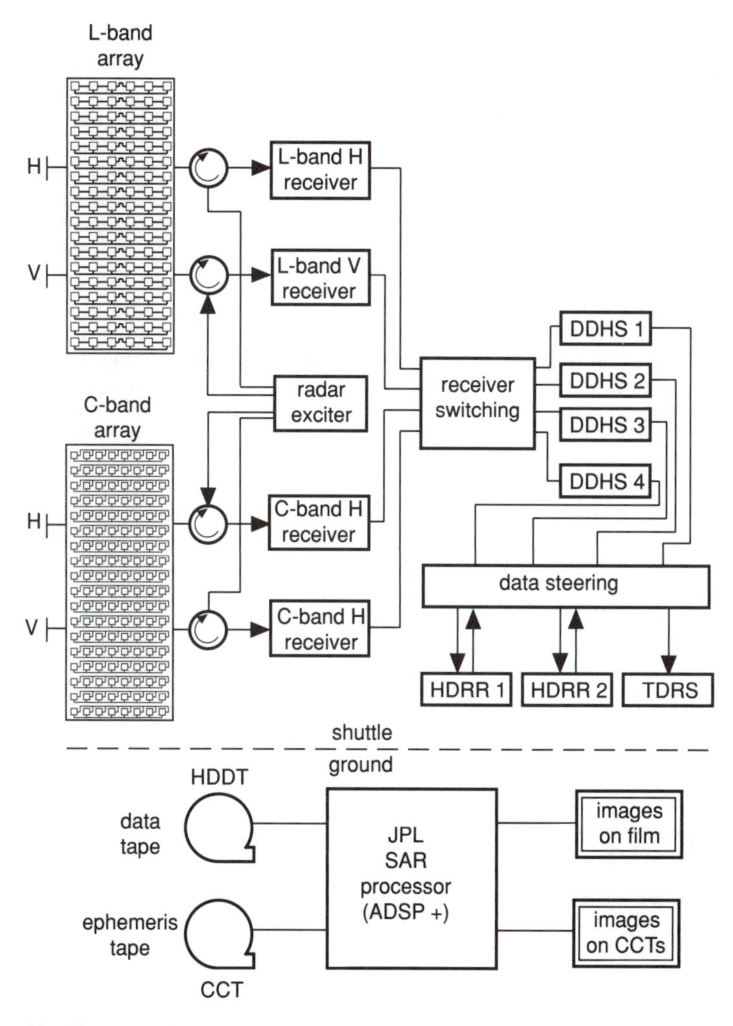

Figure 12.15 *SIR–C: Block diagram*

12.4 SAR coverage rate and wide swath

Besides geometric and radiometric resolution, an important requirement for space-based SAR is the area coverage rate (ACR), i.e. the area surveyed in unit time. For a conventional strip-map SAR, the ACR is simply the product of the platform speed V_s and the swath width s:

$$ACR = sV_s \qquad (12.12)$$

A typical ACR value for a low-Earth-orbit satellite SAR such as SEASAT is 750 km^2/s.

The ACR is directly connected both to the repeat time, i.e. the time between two surveys of the same point on the Earth surface, and to the amount of area

Antenna peak gain (1-way)	43.2 dB
Peak power	3300 W
PRF range	1240–1860 Hz
Pulse length	40 μs
Chirp bandwidth (fine/coarse)	19/9.5 MHz
Gain control range	+20 dB
Transmit path loss	1.8 dB
Receive path loss	3.5 dB
System noise figure	3.4 dB
ADC bit number (I/Q)	4/4 or 6/6
Range compression weighting function	Hamming, factor 0.75
Azimuth compression weighting function (per look)	Hamming, factor 0.75
Number of azimuth looks	4, non-overlapping
Look Doppler bandwidth	275 Hz
Range resolution	<10/20 m
Azimuth resolution	<30 m
Radiometric resolution	<2.5 dB($\sigma = -18$ dB)
Peak sidelobe ratio (PSLR)	>20 dB (range and azimuth)
Instantaneous dynamic-range	>20 dB

Figure 12.16 *SAR–X technical data*

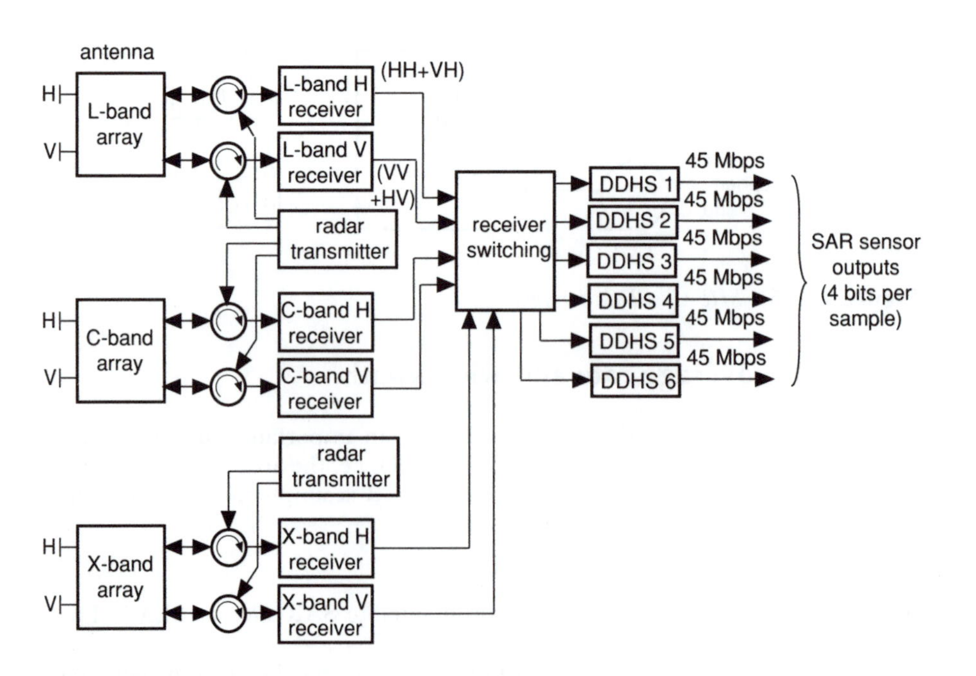

Figure 12.17 *EOS: Block diagram of SAR sensor*

Nominal altitude	824 km
Orbital inclination	98.7°
Mission duration	10 or more years
Orbital time	101 min
Off-nadir angle	15 to 45 degrees
Incidence angle	15 to 60 degrees
Radar frequency	1.253 GHz (L-band) 5.303 GHz (C-band) 9.6 GHz (X-band)
Pulse length	50 μs
Radar chirp bandwidth	19 MHz (linear FM)
Range resolution	7 m (slant), <30 m (ground)
Antenna dimensions	2.92 × 20.2 m (L-band) 0.75 × 20.2 m (C-band) 0.40 × 20.2 m (X-band)
Peak power	4.9 kW(L-band) 4.6 kW(C-band) 4.4 kW (X-band)
PRF	750 to 1100 Hz
Polarisation	HH,VV,VH,HV (L-,C-band) HH,VV(X-band)

Figure 12.18 *EOS–SAR technical data [6]*

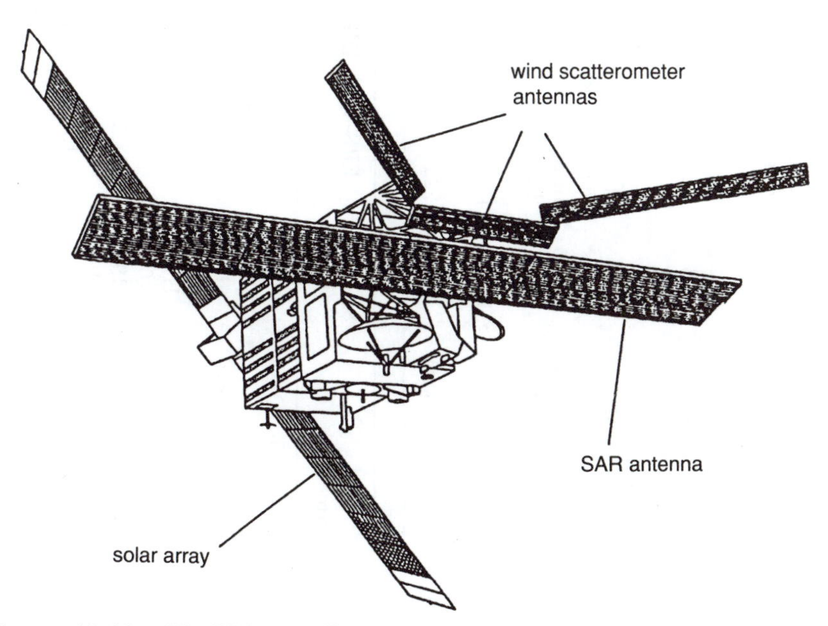

Figure 12.19 *The ERS–1 satellite*

Parameters	SIR-A	SIR-B	SIR-C	XSAR
Flight Year	1981	1984	1994	1994
Resolution (m)	40	25	15–60	15–60
Swath width (km)	50	30	30–100	10–45
Frequency (GHz)	1.28	1.28	1.28(L)&5.3(C)	9.6
Polarisation	HH	HH	Quad.	VV
Look angle	47°	15–60°	15–60°	15–60°
Bandwidth (MHz)	6	12	10&20	10&20
Bit rate (Mbps)	N/A	34	46/channel	46
Antenna length (m)	9.4	10.7	12.1	12.1
Antenna width (m)	2.16	2.16	2.8(L) 0.8(C)	0.4
Peak power (W)	1000	1000	3600(C) 5500(L)	3000
Orbit altitude (km)	259	225	250	250

Parameters	SEASAT (US)	ERS–1 (ESA)	JERS–1 (Japan)	RADARSAT (Canada)	EOS/SAR (USA)	SAR of the Polar platform (ESA)
Year of launch	1978	1991	1992	1994/5	end of century	end of century
Resolution (m)	25	30 or 100	20	28	10–20	30–100
Swath width (km)	100	100	75	130	50–300	400–500
Frequency (GHz)	1.28	5.3	1.2	5.3	1.2–5.3–9.6	5.3
Polarisation	HH	VV	HH	VV	Quad.	VV
Look angle	20°	23°	44°	20–45°	10–55°	23° (imaging)
Bandwidth (MHz)	19	13.5	15	11–17	10–20	—
Data rate (Mbps)	110	165	60	110	300	100
Antenna length (m)	10.74	10	12	14	18	13
Antenna width (m)	2.16	1	2.2	TBD	4	1.5
Peak power (W)	1000	4800	744	1200	TBD	1500
Orbit altitude (km)	800	785	568	1000	800	800

Figure 12.20 *Past and future SAR missions*

that can be surveyed. For example, consider a low Earth orbit SAR with a swath of 80 km; if a complete coverage of the Earth is required, the repeat time at the Equator is 35 days; conversely with a three days' repeat time only 9% of the Equator is covered.

Some space-based SAR applications, mainly related to the monitoring of varying phenomena, require larger ACR values (i.e. shorter repeat times and/or larger coverage areas) than the ones specified before. To show the electrical and geometrical radar parameters affecting the ACR, it is necessary to refer to the pulse repetition frequency (PRF) limitations in order to avoid range and Doppler ambiguities [1].

In a simplified flat-Earth geometry (see Fig. 12.6) we have

$$PRF_{min} = \frac{2V_s}{D} \leq PRF \leq \frac{c}{2s \cdot \sin\alpha} = PRF_{max} \tag{12.13}$$

where D is the azimuthal antenna dimension, s the swath width and α the off-nadir angle. From eqn. 12.13 we have

$$V_s = \frac{PRF_{\min} D}{2}; \quad s = \frac{c}{2 \sin\alpha \ PRF_{max}} \tag{12.14}$$

that, substituted in eqn. 12.12, gives

$$ACR = \frac{c}{2}\frac{D}{2}\frac{PRF_{\min}}{PRF_{max} \sin\alpha} \tag{12.15}$$

Typical values of the term $PRF_{min}/(PRF_{max} \sin\alpha)$ are close to unity, and typical antenna lengths are about 10 m, resulting in ACR values (from eqn. 12.15) of about 750 m^2/s. Eqn. 12.15 shows that, for a fixed value of the PRF ratio and a fixed geometry, it is possible to increase the ACR only by increasing D.

As an example, let us consider a requirement for SAR surveying capability of any point on the Earth with a 1-day repeat time. The resulting ACR is of the order of 6 000 km^2/s, and, using eqn. 12.15 with $PRF_{min}/(PRF_{max} \sin\alpha)$ equal to 1, we obtain, for the antenna azimuthal dimension, values of about 80 m; the corresponding swath width for $V_s = 7.5$ km/s, resulting from eqn. 12.12, is of the order of 800 km.

Some techniques designed to get the same result, i.e. swath widening, without requiring for very large antennas and then being compatible with the capabilities of SAR satellites, can be analysed. Perhaps the SCANSAR technique is the most promising [7]. Basically, a SCANSAR is a SAR with the possibility of a beam scanning in the elevation plane. In the beginning the antenna beam is steered in order to illuminate the nearest subswath for the time necessary to get an image of that subswath. Then the operation is repeated, steering the beam on the adjoining subswath. The cycle finishes when the sub-satellite point reaches the end of the first subswath, this being necessary to steer the antenna beam back to the initial elevation angle, in order to accomplish continuous coverage without any gaps in the interested area (Fig. 12.21).

It is obvious that, if N_S is the number of different positions of the beam on the elevation plane, the swath width is made N_S times larger, but, on the other hand, since the length of the synthetic antenna is divided into N_S subswaths, the azimuthal resolution is N_S times worse than that of a traditional SAR.

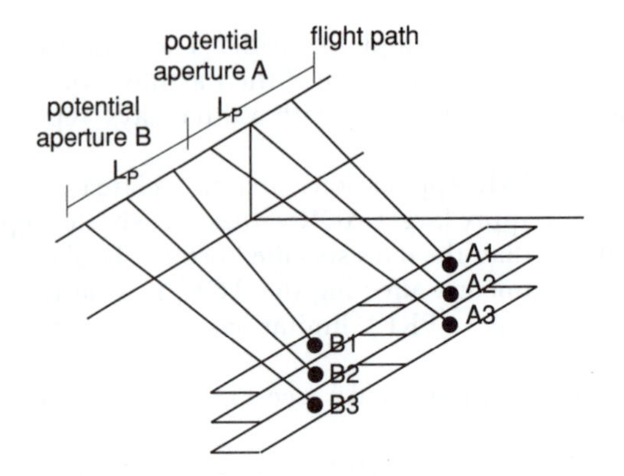

Figure 12.21 *SCANSAR: Working principle*

Therefore, both SCANSAR and large-antenna techniques affect the azimuthal resolution.

Besides SCANSAR, a further group of techniques allows the swath to be widened using 'small' antennas in the elevation plane, able to illuminate larger regions because of their wide beams, and causing a range ambiguity phenomenon. This kind of ambiguity can be eliminated at the receiver level by separating the pulses coming from adjoining subswaths, either working with a multi-carrier system, or with digital beam forming, or using different transmitted waveforms.

These solutions, although allowing the use of antennas of affordable dimensions, cause an increase of the system complexity owing to the multiple channels needed in order to accomplish the various diversity techniques. In conclusion, high-resolution SAR seems to be confined to limited ACR applications, while large ACR values call for large azimuth resolution cells and large antennas. In such conditions SAR approaches the RAR concept. This trend is confirmed by the SAR equation, as shown in Part 2.

Part 2: Space-based radar surveillance

12.5 Operational requirements

The significant results obtained with SAR techniques suggest the use of satellite-borne radar sensors for purposes other than imaging. Unlike ground-based and airborne radars, a space-based radar has the remarkable capability of observing any point on the Earth surface, without the field-of-view limitations typical of a ground-based radar. The possible applications of space-based radars for purposes other than remote sensing are briefly discussed in the following.

In the past years, the race to armaments and the development of more and more sophisticated weapons, especially nuclear, pushed many governments to ratify Treaties in order to control the situation, such as the Antarctic Treaty, the Partial Test Ban Treaty etc. [8]. A periodic check on the preservation of these Treaties is an important matter, and, for this purpose, a space-based radar with good ground resolution and with the capability of observing large areas in reasonable times was considered as a suitable solution.

Remaining in the military environment, it is necessary to consider the development of projects financed by some governments to guarantee from space the defence of the national borders against possible aircraft or missile attacks. In particular, the USA studied the possibility of implementing an operational space-based radar for the CONUS (Continental United States) surveillance and for assistance to the naval forces, tasks only partly performed by present ground-based radars [9]. One should also emphasise the economic efforts, supported by the Canadian government, to design a space-based radar system to protect the whole national territory, including the large areas surrounding it.

Coming to civilian applications, space-based radar represents a tool for solving some typical problems of air traffic control (ATC), and being able to accomplish both primary (without any co-operation from the aircraft) and secondary (using aircraft on-board transponders) surveillance [10, 11, 12].

Modern primary radars are able to cover areas at a maximum range of 370–420 km, because of limits mainly due to earth curvature. This long-range radar is called en-route-radar (ERR) and is different from medium-range terminal area radar (TAR), which has a smaller revisit time in order to deal with the manoeuvres of aircraft near the airport. Space-based radar could be useful for air-route surveillance at zones farthest from the airports, e.g. deserts, oceans or large forests, overcoming the limitations of HF radio links in zones not provided with radar coverage.

Besides primary surveillance, it is necessary to accomplish secondary surveillance, in order to know the identity and the elevation of the aircraft. A secondary surveillance radar (SSR) uses airborne transponders to detect and locate aircraft; the identity and altitude of the aircraft are transmitted downlink to the SSR ground station. The most recent version of SSR, the 'Mode S', is capable of interrogating each aircraft separately, using its unique address, and exchanging digital data with it (uplink and downlink). Although different from primary radar, secondary radar [13] also suffers from the typical limitations of the environment (earth curvature, geography, etc.). Therefore, future space-based radar should include the mode-S secondary radar function [14].

It is important to emphasise the necessity for high-level integration of the functions, previously described, in a single system, in order to balance the high costs necessary for the realisation of these systems.

Up to the present, no space-based radar system for civilian surveillance is operating.

With regard to the military surveillance functions, it is known that the Soviet RORSAT, (Radar Ocean Reconnaissance Satellites) co-operating with EOSAT (Electronic Ocean Reconnaissance Satellites), have naval detection and identification functions. Typical RORSAT orbits have a height of between 200 and 425 km and are inclined at 65° to the equatorial plane.

The COSMOS 954 fall (1979) enabled us to know some characteristic of such satellites. In the Western world, especially in the USA (Lincoln Laboratory of MIT), in Canada (CRC, Ottawa, SPAR, Canadian Astronautics) and elsewhere, various studies have been funded [33, 35].

12.6 Surveillance capability of space-based radar

A fundamental surveillance requirement of space-based radars (SBR) is to detect targets that cross a certain fence surrounding the area to be covered by the radar (Figs. 12.22 and 12.24).

This is a classical problem of 'cumulative probability of detection'. The target is lost if it is not detected when it is crossing the fence; the resulting number of

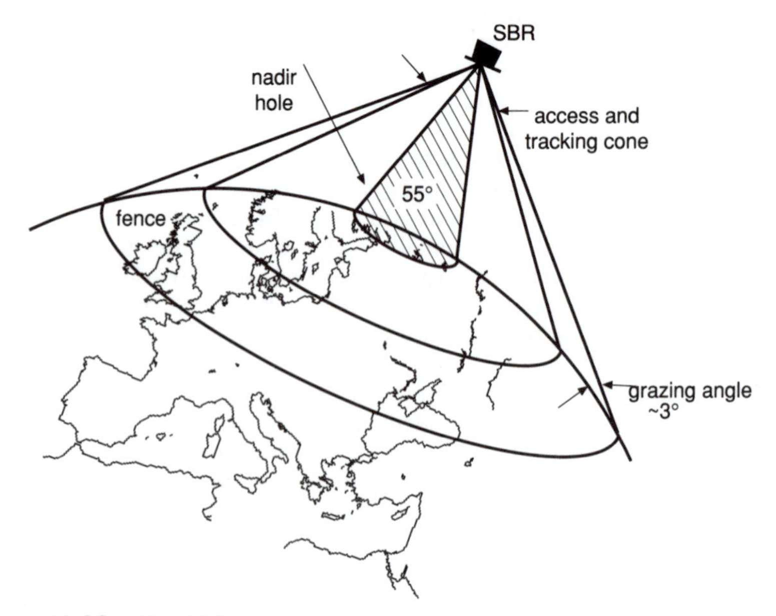

Figure 12.22 *The SBR concept*

opportunities n is simply the ratio between the time needed for the target to cross the fence, T_f, and the scan time, or data renewal interval, T_R. Of course,

$$T_f = \frac{w}{v}$$

where w is the width of the fence and v is the aircraft speed perpendicular to the fence. The probability of loss of the target, P_L, is related to the single-scan detection probability P_D and to the number of opportunities n as follows:

$$P_D = 1 - P_L^{\,1/n}$$

As an example, assuming $P_L = 10^{-4}$ we have the following P_D–n pairs:

n	1	2	3	4	5	6	8	10
P_D	0.9999	0.990	0.953	0.90	0.84	0.78	0.68	0.60

The selection of the P_D–n pair, for a given target fluctuation model, allows the specified value of P_L to be obtained with minimum transmitted energy (or with other optimisation criteria). Here, reference is made, where needed, to the case $n = 4$, $P_D = 0.9$.

The main parameter of a space-based surveillance system is the area coverage rate (ACR) introduced in Part 1. In the applications considered, the ACR values result from (i) a search contribution on the fence, (ii) a search contribution on spots (air traffic sources) and (iii) a tracking contribution within the area of interest [15]. In practice, the ACR values are mainly

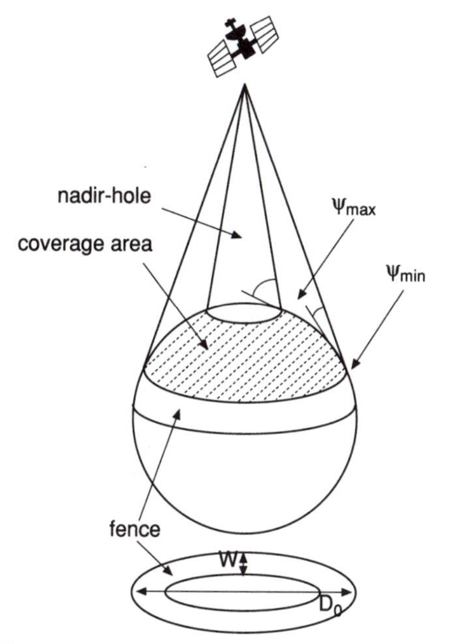

Figure 12.23 *SBR geometry*

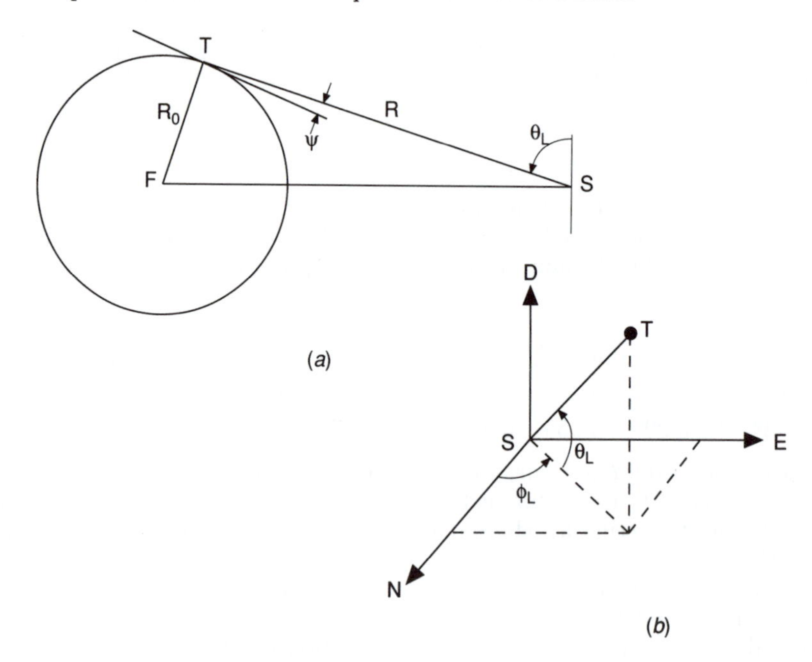

Figure 12.24 *(a) Satellite-Earth geometry*
(b) Co-ordinate system north-east-down (NED)

determined by contribution (i), since, in many environments, contributions (ii) and (iii) are at least one order of magnitude smaller than contribution (i) [15].

If T_R is the data renewal interval (or revisit time) and if n is the number of opportunities for detecting a target crossing the fence with a speed v (orthogonal to the fence), then the minimum fence width w must be:

$$w = nvT_R$$

Moreover, if D_0 is the diameter of the surveillance zone, the fence area (Figs. 12.22 and 12.24) is given by

$$A = \pi D_0 w = \pi D_0 nvT_R$$

So, the fence area covered in time unit, or area coverage rate (ACR), is

$$ACR = A/T_R = \pi D_0 nv \tag{12.16}$$

depending on the surveillance requirements, that is on D_0 and the nv product.

Let us now consider a conventional strip-map SAR, whose antenna pattern intercepts on the Earth surface an area whose dimension in the range direction is the swath width s, i.e. the width of the 'strip' surveyed by the radar. Obviously the swath width is related to the area coverage rate and the SAR velocity V_s as follows:

$$ACR = sV_s$$

and comparing this with eqn. 12.16, the swath s can be obtained:

$$s = \pi D_0 nv / V_s \tag{12.17}$$

which, for civilian aircraft ($v = 250$ m/s), is about 45% of D_0, if $n = 4$ and $V_s = 7000$ m/s.

Therefore, if an SAR is used for surveillance purposes, it should have to guarantee a swath of thousands of kilometres, e.g. 2300 km for a European coverage ($D_0 \simeq 5100$ km), and so the performance of the actual SAR missions, where the swath width does not exceed 100 km, are not satisfactory.

The Doppler bandwidth $[-V_s/D, V_s/D]$, where D is the antenna length, which gives the SAR its good azimuthal resolution, becomes a drawback in a surveillance task, from the main beam clutter widening owing to satellite motion; in particular, the value V_s/D is related to the minimum detectable velocity, or MDV, of a target: if the target has a lower speed than the MDV, it cannot be detected because its Doppler frequency is superimposed on the main beam clutter spectrum. On narrowing the clutter bandwidth by increasing D, the MDV becomes lower and the moving-target detection performance improves.

From eqn. 12.13, the increase of the swath s, due to the surveillance requirements previously discussed, causes a decrease of PRF_{max}, i.e. a narrower useful band, from the clutter bandwidth, to detect possible moving targets, without Doppler ambiguity. Therefore, together PRF_{max} reduction, it is necessary to decrease even PRF_{min}, increasing the dimension D of the antenna, in order to keep the useful detection bandwidth unchanged.

Even the surveillance radar equation (see Reference 22 p. 64) leads to the main result of increasing the antenna dimension in order to maximise the ACR. In fact, this equation is

$$R^4_{max} = \frac{P_{av} A_e \sigma T_R}{4\pi k_B T_S (SNR)_{min} L_S \Omega_R} \tag{12.18}$$

where

- R_{max} = maximum range
- P_{av} = average power
- A_e = antenna effective area
- σ = radar cross-section of the target
- T_R = revisit time required for surveillance (data renewal interval)
- k_B = Boltzmann constant
- T_s = system noise temperature
- $(SNR)_{min}$ = minimum signal-to-noise ratio required to obtain the assigned P_D and P_{FA}
- L_S = system losses
- Ω_R = scan solid angle

At the maximum range from the radar, the ratio Ω_R/T_R is related to the area coverage rate (ACR) as follows:

$$ACR = \frac{R^2_{max} \Omega_R}{T_R} \tag{12.19}$$

which, replaced in eqn. 12.18, gives the SBR surveillance equation

$$R^2_{max} = \frac{P_{av} A_e \sigma}{4\pi k_B T_S (SNR)_{min} L_S ACR} \tag{12.20}$$

Thus a large ACR calls for a high average power and/or large antenna dimensions.

With regard to clutter, if d_R and d_A are the dimensions of the resolution cell in range and azimuth, and σ_0 is the clutter superficial reflectivity, the signal-to-main-beam clutter ratio (neglecting ambiguities) is

$$S/C=\frac{\sigma}{\sigma_0 d_A d_R} \tag{12.21}$$

and its maximisation calls for a high resolution waveform and, again, a large antenna.

12.7 Space-based radar architectures

From the previous discussion it emerges that the basic element of a space-based radar is a large antenna that allows the fundamental surveillance requirements to be satisfied, especially for the ACR, besides providing fairly good angular resolution and minimum main-beam clutter.

A first system choice considers the array structure, whether regular or random. If a regular structure is chosen, the resulting distances between the various elements of the array are much greater than a wavelength in order to get dimensions of some tens or hundreds of metres with an acceptable number of elements. Therefore, grating lobes having the same amplitude as the main lobe are produced [16].

In order to avoid this problem, a random array structure may be preferred, although it gives rise to a pattern with high sidelobe levels with respect to a conventional array [17].

Another possible choice is between a monostatic structure, where the same antenna transmits and receives, and a bistatic or multistatic structure, where a single antenna transmits and one or more receives [18].

A bistatic structure has some advantages over the monostatic one, including a better resistance to jamming, and a potential for better performance in detecting low radar cross-section targets, by using the forward scattered energy [18]. Among the shortcomings of a bistatic structure are an increase of system complexity, and energy losses due to the non-coincidence of the transmission and reception footprints. This last drawback could be overcome by synthesising more reception beams, at the expense of a further increase in system complexity. Very often the previous problems call for a monostatic solution or, at most, a quasi-bistatic solution, with a bistatic angle very close to zero.

In the following we consider the choice of coherent or independent operation of the array elements. It can be shown that coherent operation is more advantageous for tracking, because, at parity of the received signal-to-noise ratio, it is possible to track a large number of targets with the assigned revisit time [16].

Once the array structure has been fixed, it is necessary to consider other problems involved in the system design.

Firstly, it is necessary to set a suitable operating frequency. The capability of maximising the target radar cross-section, if the target dimensions are comparable with the wavelength, together with the robustness with respect to atmospheric factors such as rain, lead to a low operating frequency, e.g. VHF. This choice causes some drawbacks such as the necessity of using large antennas, impairment of the ionospheric propagation and narrow bandwidth [9]. On the other hand, higher frequency operation allows the use of smaller antennas, but the higher the working frequency, the more serious are the disturbing atmospheric effects (attenuation). According to most authors, the optimum frequency range includes the L- and S-band.

Besides the operating frequency, it is necessary to set the fundamental orbital parameters of the satellite, or the constellation of satellites, that will carry the radar sensors. With regard to the orbit, a circular orbit rather than an elliptical one is to be preferred. In fact, an elliptical orbit permits a good performance for the coverage of particular regions on the Earth when the satellite is at the apogee and moves slowly. However, the apogee is the point of the orbit with the poorest performance, both for signal-to-noise ratio and signal-to-clutter ratio, because, being at the maximum range from the Earth, on one hand the useful received power (proportional to $1/R^4$) is minimised, and, on the other hand, the intercepted footprint on the Earth surface (proportional to R^2), and the clutter power is maximised [19].

With regard to the satellite altitude, low-altitude operation permits the use of lower powers, and reduces the amount of energy for the insertion of the satellite in its orbit, but it calls for a larger number of satellites on the same orbit for continuous coverage of a given region on the Earth.

Conversely, high-altitude operation calls for a reduced number of satellites for continuous coverage, but this advantage is partially balanced by the burden of inserting the satellites in their orbit. Moreover, the increase of the satellite–Earth range on one hand reduces the received power (for equal transmitted power), but, on the other hand, it allows the satellites to move at lower speeds, so as to narrow the clutter spectrum, just proportional to the satellite orbital speed [20, 21], the latter being fixed once the satellite altitude is set.

If the antenna beam is steered to a given fixed point T on the Earth surface, and the SBR position is indicated by S while the centre of the Earth is indicated by F (Fig. 12.24), a convenient co-ordinates system is the NED one, [20, 23]. This system, centred on the SBR, has the axis D (down) towards the point F, the axis N towards the North and the axis E (East) perpendicular to D and N in the east direction.

In the NED system the angle between the plane STF and the N axis is the azimuthal observation angle ϕ_L and the angle between the NE plane and the antenna steering direction ST is the elevation observation angle θ_L. If V_s is the satellite speed, the radial velocity V of the target is

$$V = V_s \cos\phi_L \cos\theta_L \tag{12.22}$$

In Fig. 12.24 the satellite geometry and the relationship between θ_L and the grazing angle ψ are shown [20, 23].

The platform motion (Fig. 12.25) causes a shift of the average Doppler frequency of the clutter and a widening of its spectrum. The former effect is due to the component of the platform velocity parallel to the antenna pointing direction, and the latter is explained as follows.

Since the antenna beam footprint on the Earth surface includes a large number of points having different positions (and then different radial velocities)

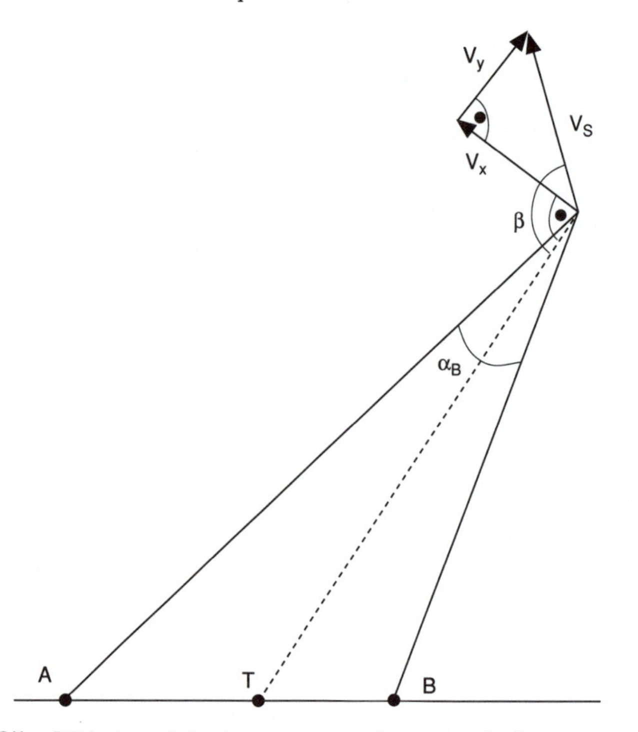

Figure 12.25 *Widening of the clutter spectrum due to the platform velocity*

with respect to the radar, velocity dispersion and then clutter spectrum widening is expected. Consider the elevation plane containing the satellite velocity vector and the target; from Fig. 12.25 it results that the Doppler frequencies f_A, f_T and f_B of the points A, T and B on the Earth's surface are related as follows:

$$f_A - f_T = \frac{2}{\lambda} V_x \sin\frac{\alpha_B}{2}$$

and

$$f_B - f_T = -\frac{2}{\lambda} V_x \sin\frac{\alpha_B}{2}$$

with $V_x = V_s \sin\beta$.

Therefore, the main beam clutter bandwidth due to the satellite motion is approximately equal to:

$$\Delta f_D = f_A - f_B = \frac{2}{\lambda} \alpha_B V_S \sin\beta \tag{12.23}$$

where

α_B = antenna beamwidth in the plane considered;

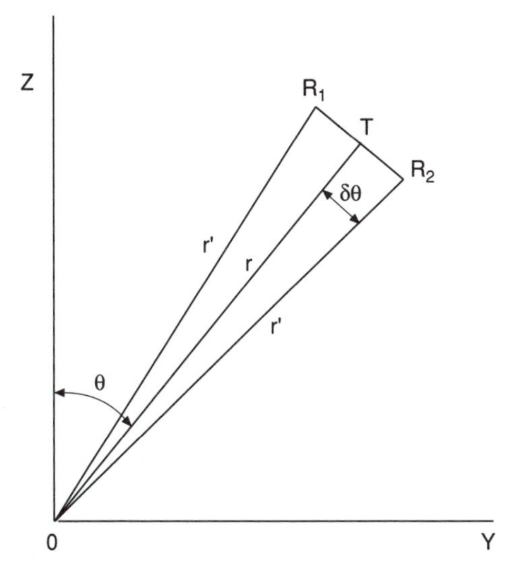

Figure 12.26 *DPCA technique: Position of phase-centres in transmission (T) and in reception (R)*

β = angle between the velocity vector and the antenna pointing direction (radar–target line)

The clutter spectrum widening due to the satellite motion seriously affects the space-based radar performance for moving targets detection.

As explained before, the clutter bandwidth is related to a minimum detectable velocity (MDV). The narrower the antenna beam, the lower the velocity spread within the footprint, and hence the MDV. Therefore the use of large antennas or of a distributed array of large dimensions, provided that the sidelobe levels are low enough to neglect the power received by them, is suitable with regard to the performance against clutter [20].

Finally, it is necessary to emphasise the effects of the azimuthal observation angle ϕ_L and of the grazing angle ψ on the clutter bandwidth. As shown in References 20 and 23, if ϕ_L is close to 0° the clutter spectrum widens only on the

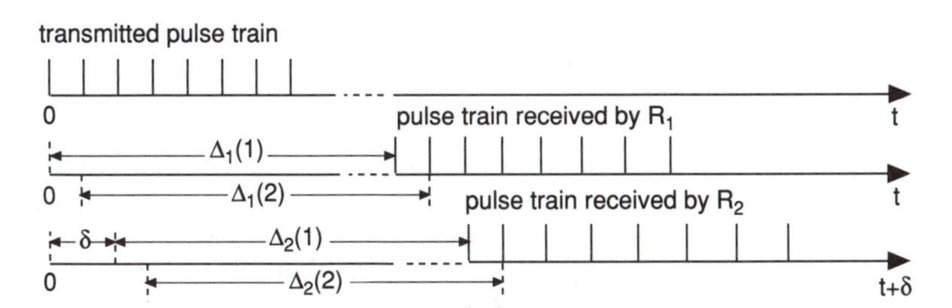

Figure 12.27 *Transmitted and received pulse trains*

negative speed side, but as ϕ_L approaches the value 90° the spectrum becomes symmetrical. Therefore the situation $\phi_L = 0$ could be advantageous for detecting only approaching targets, because of the limited clutter spectrum widening for positive values of the target speed.

With regard to the grazing angle ψ, the performance of signal-to-clutter ratio improves slightly as ψ changes from a minimum value of 3°, related to propagation and atmospheric attenuation problems, to a maximum value of 50°–55°, which is not generally exceeded because of the increasing ground reflectivity and the reduced target radial speed components (Fig. 12.24). Such an improvement is fundamentally due to reduction of the footprint area as the grazing angle increases [20, 23].

It is worth mentioning a particular technique used to improve clutter suppression, namely is the displaced phase centre antenna (DPCA) technique [24–26]. A phase centre for the transmitting array, and two or more, phase centres for the receiving array, are accomplished by beamforming techniques (Fig. 12.26), so that, by choosing suitable values for the PRF and for the distance between the two phase centres, two pulses transmitted at different times arrive at the two receiving phase centres with the same delay. This technique allows cancellation of the platform speed effects between two pulses, so as to obtain an improvement by subtraction of the two pulses (Fig. 12.27). This means that the clutter bandwidth is reduced; equivalent clutter spectra four times narrower are generally obtained.

It is important to note that a substantial portion of a radar system includes the signal and data processing circuits. In a space-based radar it would be possible to accomplish all the processing on-board, in order to reduce the amount of digital data to be sent to ground. This advantage must be balanced by the increased system complexity, owing to the processing payloads [27, 28].

To complete this view of the main problems in a space-based radar system design, it is necessary to remember the problems connected with the choice of the most suitable radar waveform [29]; moreover, the space-based radar design would have to consider all the interaction phenomena with the atmosphere (troposphere and ionosphere), such as those affecting the polarisation of the transmitted waveform, e.g. Faraday rotation and squint rotation [30, 31], as well as prime power generation and distribution [32].

Finally it is possible to summarise the fundamental advantages of a large phased array, possibly a distributed random array, apart from permitting a large ACR: flexibility and modularity typical of an array structure; ability to allow graceful degradation with respect to the 'monolithic' structures; the possibility of electronically steering the antenna beam and of implementing adaptivity techniques, in order to lower the sidelobe levels in fixed points for alleviating clutter and jamming problems, which are particularly serious because of the large field of view of a space-based radar; good resolution due to the large array dimensions.

On the other hand it is necessary to emphasise some of the main disadvantages of a distributed array structure: high costs owing to the complexity and its management; high sidelobe levels because of the random nature of the array; necessity of accomplishing array self-cohering techniques in order to compensate for all the movements of the array elements with suitable phase shifts on the various array elements.

The technological developments, especially in integrated electronics techniques, will lower the array element costs, and the prospect of integrating a large number of functions in the same surveillance and tracking system is a driving factor for the realisation of new space-based radar systems.

12.8 Space-based radar configuration

12.8.1 *Constellation of SBR satellites*

As explained before, the main reason for using a space-based radar rather than a ground-based or an airborne one is the increased capability for large areas coverage.

The use of a *constellation of satellites* is mandatory to accomplish the continuous coverage of a given region from an orbit much lower than the geostationary one, since a single satellite, because of its motion with respect to the Earth, would guarantee coverage of the interested region for a small fraction of its orbital period. In fact, the fence area has to be dynamically shared between the satellites of the constellation in order to guarantee uninterrupted surveillance with the specified value of the revisit time and with values of the grazing angle

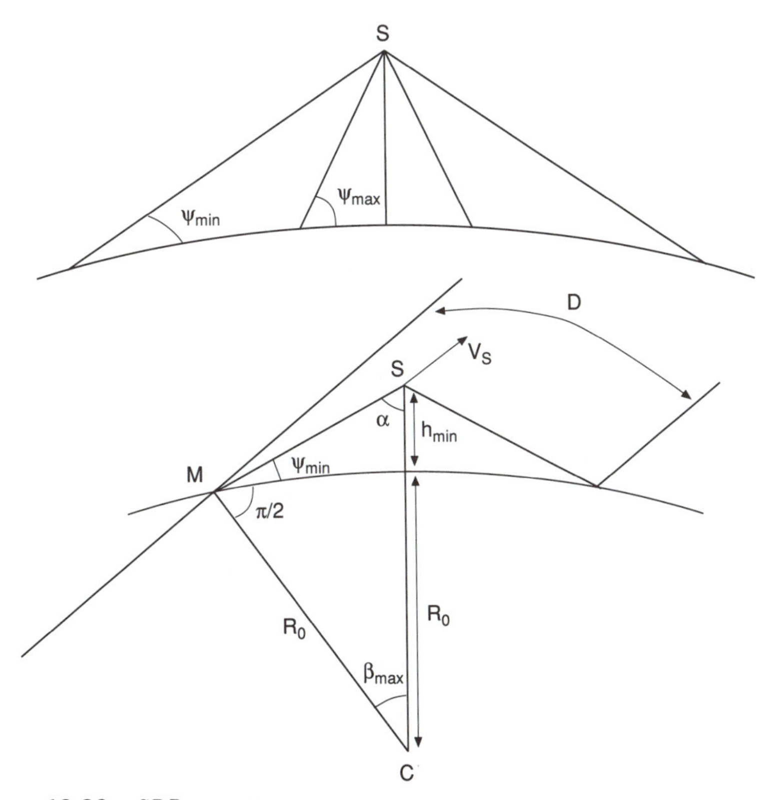

Figure 12.28 *SBR geometry*

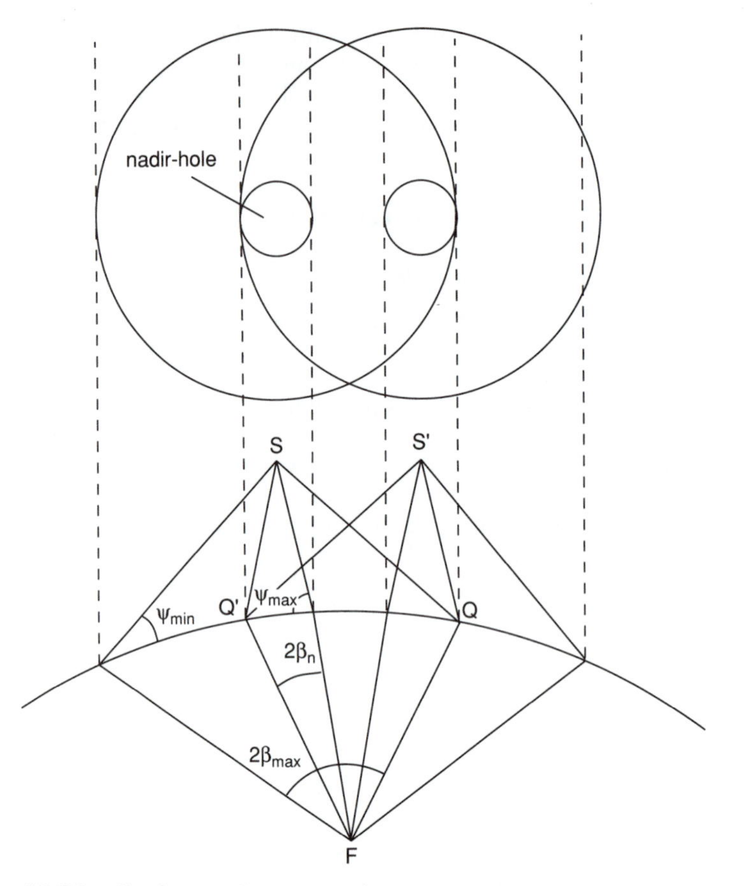

Figure 12.29 *Overlapping between contiguous satellites, k = 1*

(i.e. the angle between the direction of propagation and the tangent plane to the Earth surface) within a specified interval related to propagation and backscatter characteristics.

For the sake of simplicity, the ensuing evaluations of the required area coverage rate are performed by considering a simplified geometry in which the sub-satellite point (i.e. the point of the Earth surface at the nadir of the satellite) is at the centre of the surveillance area. A circular orbit is assumed for the reasons explained in the previous Sections.

The choice of constellation geometry leads to the establishment of the number of satellites N_s in the constellation, and of the common height of the satellites. The main criterion for selecting the optimum satellite altitude H_{op} is to minimise, according to the power budget (see eqn. 12.20), the quantity

$$N_s P_{av} A_e$$

where

N_s = number of satellites in the constellation, required for continuous coverage of the surveillance area

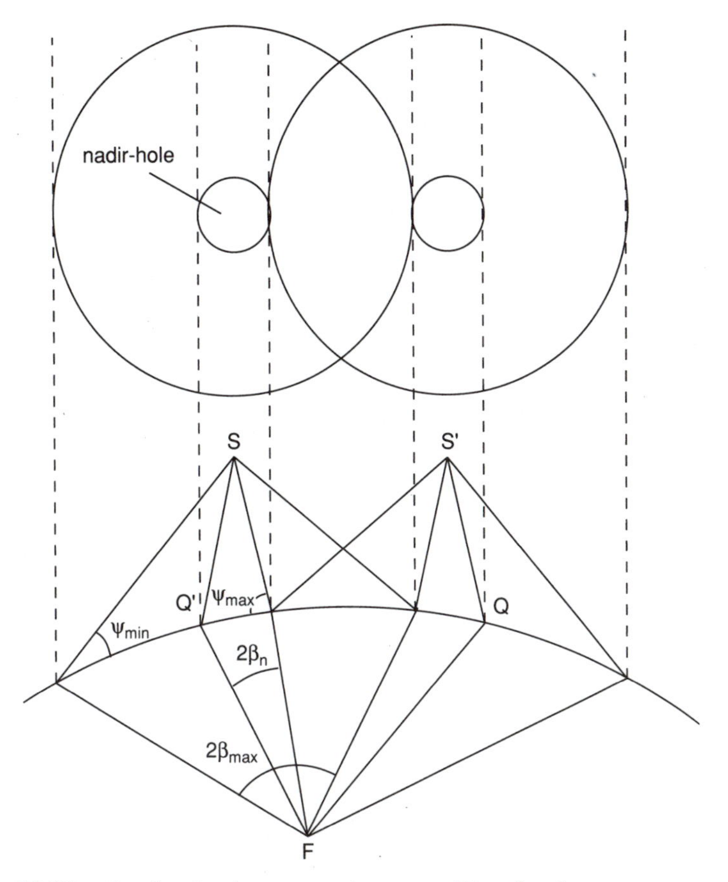

Figure 12.30 *Overlapping between contiguous satellites, k = 0*

P_{av} = average transmitted power of the radar
A_e = effective area of the radar antenna

Two methods can be followed to evaluate the number of satellites, according to whether the 'nadir hole' (i.e. the shade cone whose axis is the connecting line between the sub-satellite point and the satellite, and whose half-width angle is related to the maximum grazing angle ψ_{max} (Fig. 12.28) by which the satellite can observe the Earth surface) is considered or not. The upper value ψ_{max} for the grazing angle ψ, i.e. the lower value of the nadir-hole dimension, is due to the high clutter reflectivity and to the reduced Doppler frequencies of moving targets, when the grazing angle increases.

Besides ψ_{max}, it is necessary to note a minimum value ψ_{min} for the grazing angle. This value depends on the atmospheric propagation effects and is set at 3°.

The method that takes the nadir hole into account is chosen here to calculate the number of satellites N_s. In Figs. 12.29 and 12.30 the satellites' geometry is

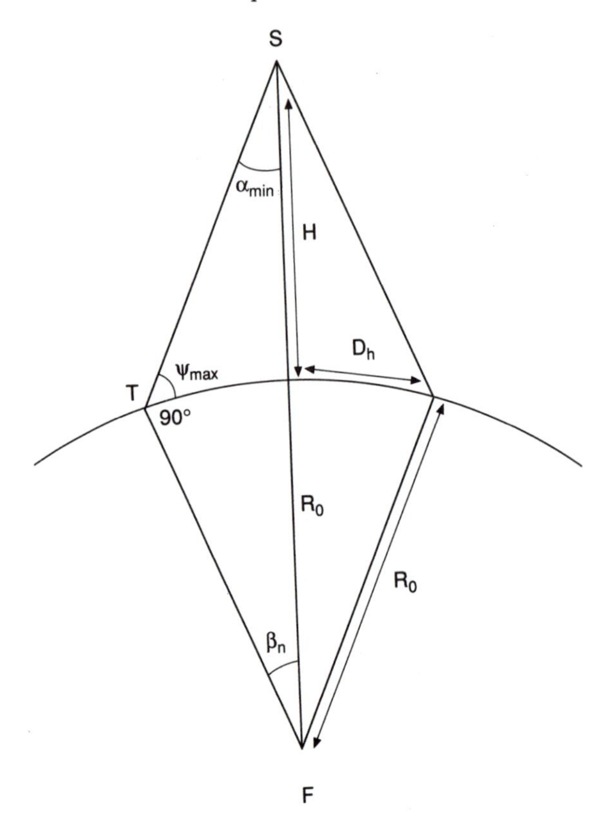

Figure 12.31 *Nadir-hole geometry*

shown, and the overlapping angle β_s between two contiguous satellites is defined as

$$\beta_s = 2k\beta_n \qquad (12.24)$$

where k is the overlapping factor referred to the nadir hole (if $k=1$ the nadir hole of a satellite is entirely covered by the contiguous one (Fig. 12.29); if $k=0$ the nadir hole of a satellite is uncovered (Fig. 12.30)), and β_n is the angle at the centre of the Earth corresponding to the maximum grazing angle ψ_{max} and is given by (Fig. 12.31)

$$\beta_n = \pi/2 - \psi_{max} - \alpha_{min} \qquad (12.25)$$

where α_{min} is the minimum off-nadir angle, which is easily computed from the STF triangle of Fig. 12.31 as

$$\sin(\alpha_{\min}) = \frac{R_0}{R_0 + H} \cos(\psi_{max}) \qquad (12.25a)$$

where: R_0 = Earth radius
H = satellite altitude.

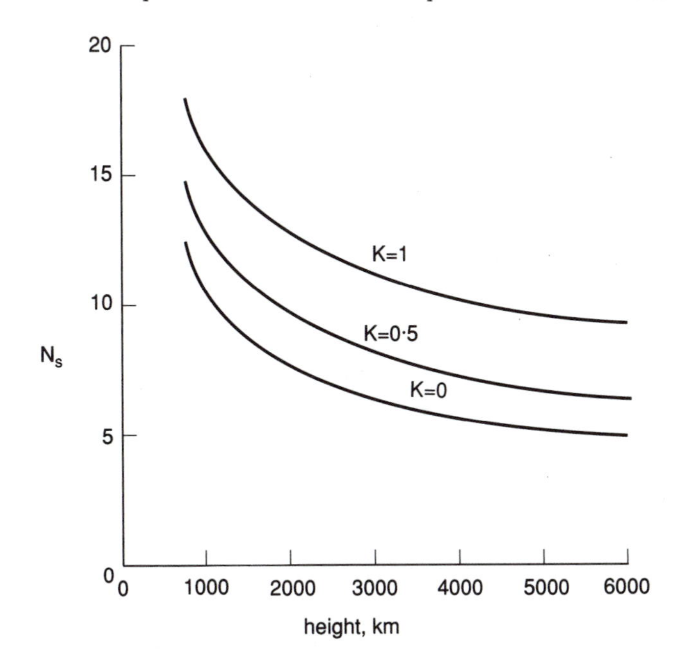

Figure 12.32 *Number of satellites N_s versus height*

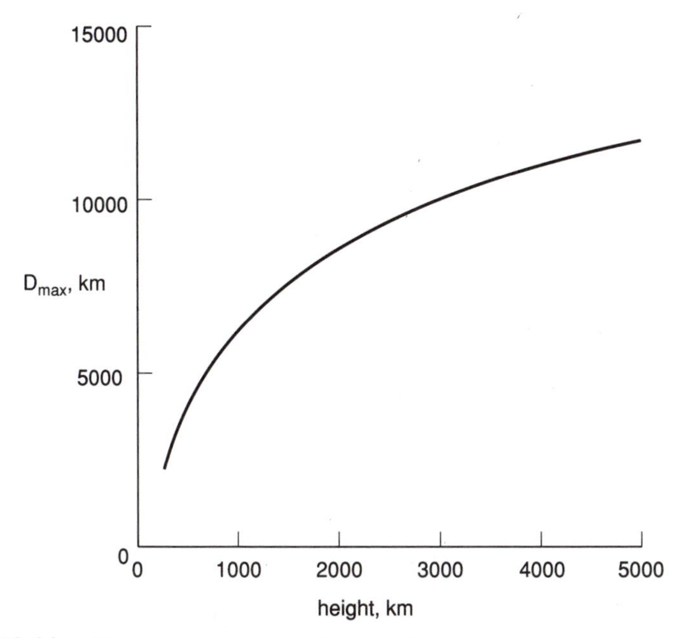

Figure 12.33 *Maximum diameter of the surveillance area versus height*

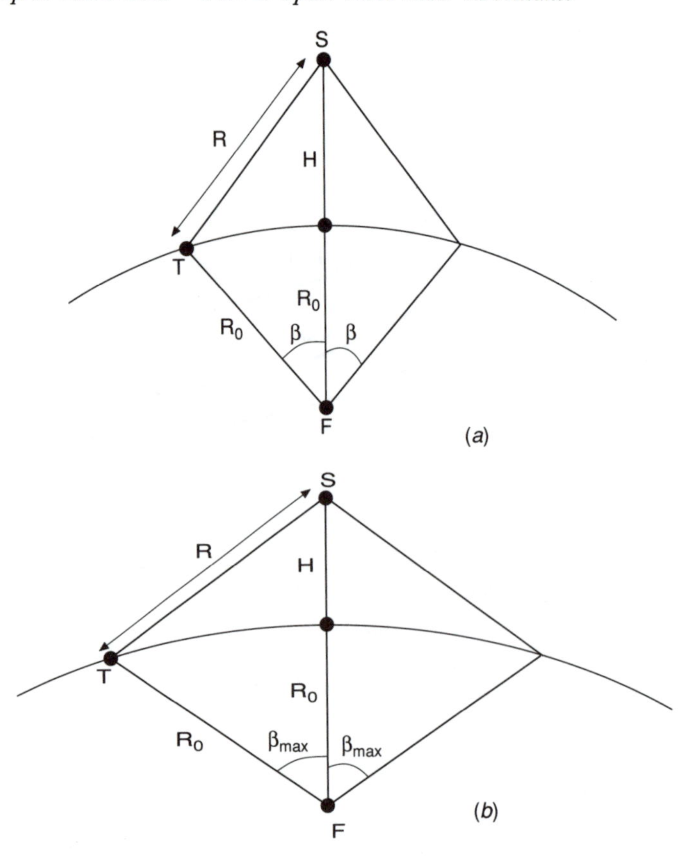

Figure 12.34 *Distance R in the average (a) and in the worst (b) case*

From Fig. 12.29, or Fig. 12.30, one can obtain the number of satellites

$$N_s = \frac{2\pi}{\angle \mathrm{SFS'}} \tag{12.26}$$

where β_s is defined by eqn. 12.24 and

$$\angle \mathrm{SFS'} = \beta_{max} + (\beta_n - \beta_s) = \beta_{max} + \beta_n(1-2k) \tag{12.26a}$$

The resulting expression for the number of satellites is

$$N_s = \frac{2\pi}{\beta_{max} + \beta_n \cdot (1-2k)} \tag{12.27}$$

In Fig.12.32 the N_s values are plotted versus the satellites' height for different values of k (diameter of the surveillance region $D = 5300$ km, $\psi_{min} = 3°$, $\psi_{max} = 55°$). In Fig. 12.33 the maximum diameter D_{max} of the surveillance area ($D_{max} = 2\beta_{max}R_0$, where β_{max} is defined in Fig. 12.28) is shown versus the height: the method considered allows continuous cover of any region with a diameter not greater than D_{max}.

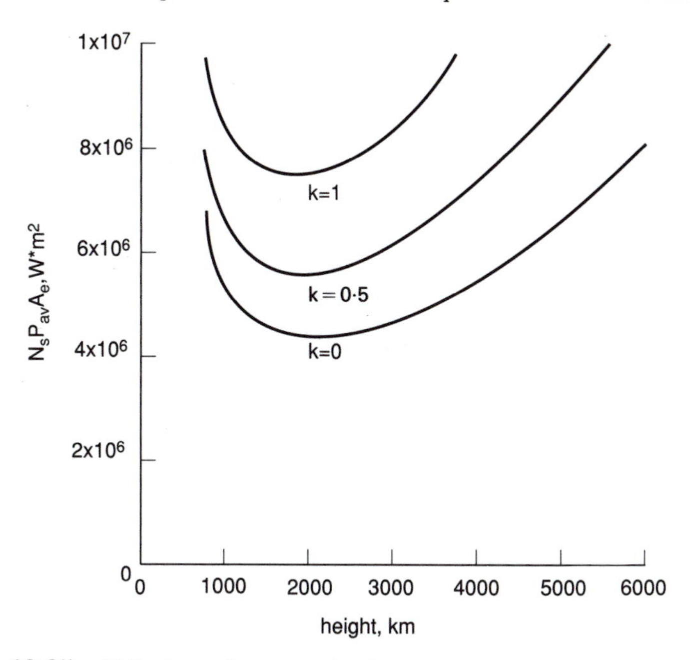

Figure 12.35 *$N_sP_{av}A_e$ product versus height*

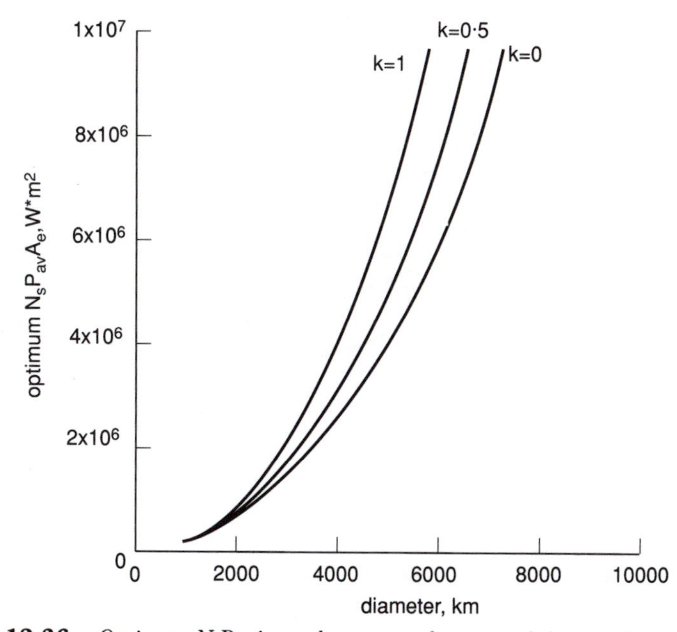

Figure 12.36 *Optimum $N_sP_{av}A_e$ product versus diameter of the surveillance area*

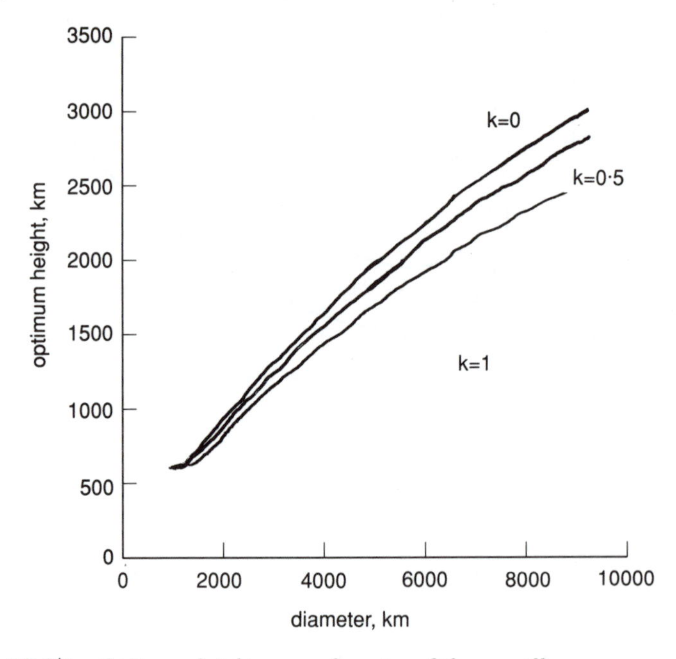

Figure 12.37 *Optimum height versus diameter of the surveillance area*

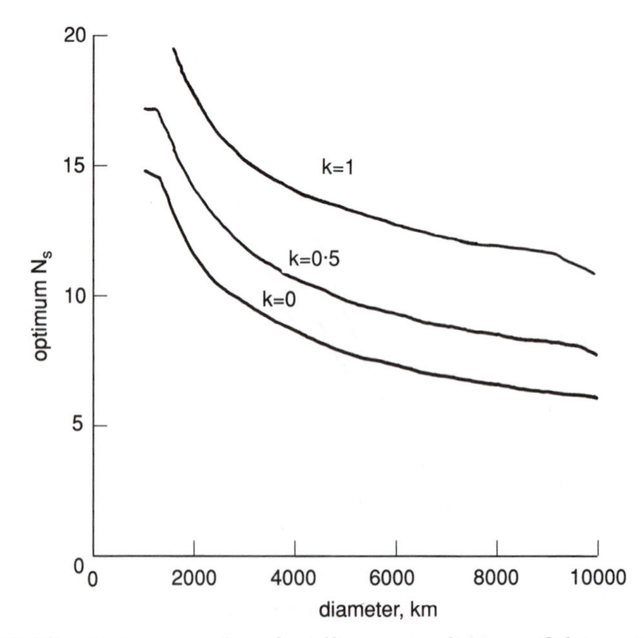

Figure 12.38 *Optimum number of satellites versus diameter of the surveillance area*

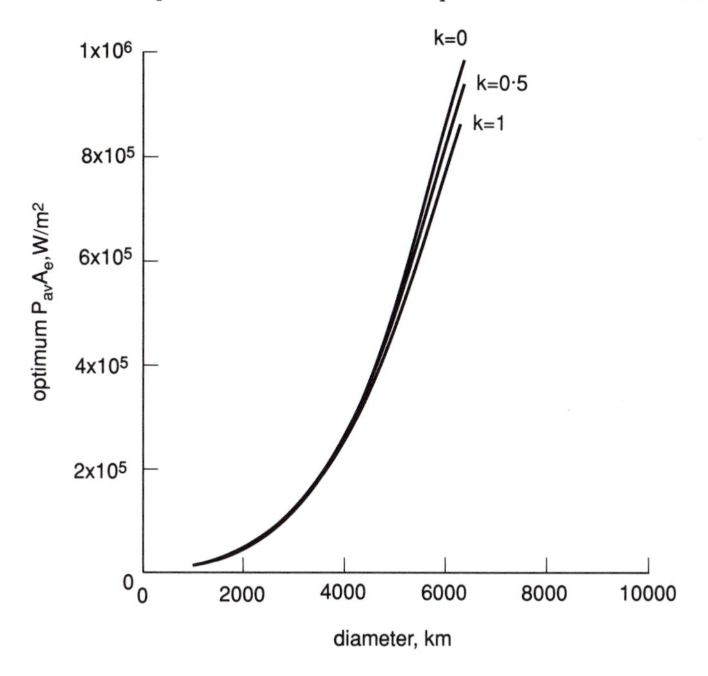

Figure 12.39 *Optimum $P_{av}A_e$ product versus diameter of the surveillance area*

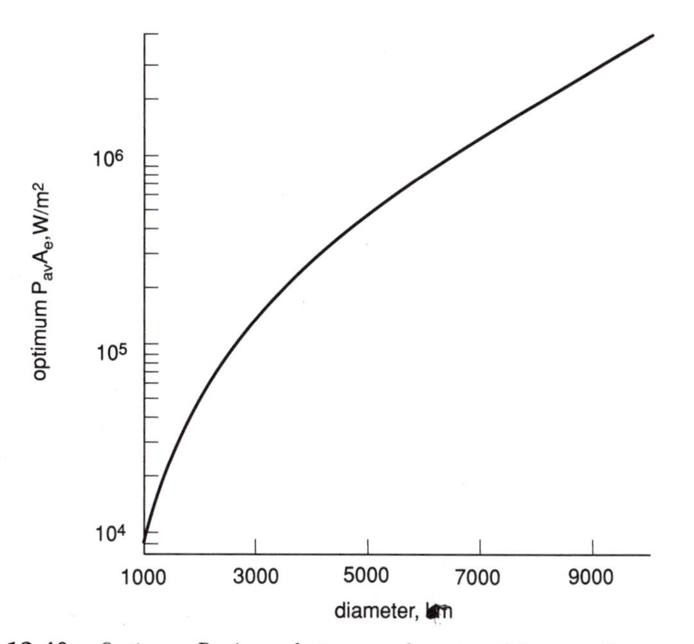

Figure 12.40 *Optimum $P_{av}A_e$ product versus diameter of the surveillance area*

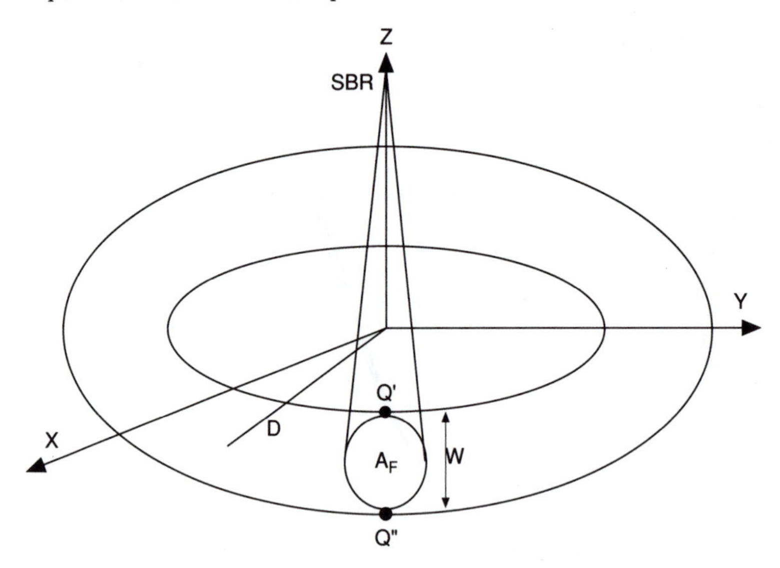

Figure 12.41 *Geometry for determination of the antenna beamwidth*

12.8.2 *Power-aperture budget*

Once the number N_s of satellites has been evaluated, the product $P_{av}A_e$ has to be determined and, for this purpose, the SBR surveillance equation is used:

$$P_{av}A_e = \frac{(ACR)\ R^2 \cdot (SNR)_{min} 4\pi k_B T_S L_S}{\sigma} \tag{12.28}$$

where R is the target–radar distance and the other symbols have been defined in eqn. 12.18; the system losses L_S include the term $1/\cos\alpha$ as beam steering losses, where α is the off-nadir angle.

The distance R can be determined in an average-case hypothesis assuming that the sub-satellite point is at the centre of the surveillance area, and that the satellite observes the edge of such an area (Fig. 12.34*a*, $\beta = D/(2R_0)$). Note that the distance R considered here is only approximately equal to the average case, because it would be necessary to average all the values of R^2, with respect to the off-nadir and azimuth angles, inside the surveillance area excluding the nadir hole. The resulting expression can be obtained from the STF triangle of Fig. 12.34*a*, and is

$$R^2 = R_0^{\,2} + (R_0 + H)^2 - 2R_0(R_0 + H)\cos\beta \tag{12.29}$$

On the other hand, the distance R could be calculated in a worst-case hypothesis, simply by supposing that the satellites observe the Earth surface with the minimum grazing angle ψ_{min} (Fig. 12.34*b*), and using expr. 12.29 with $\beta = \beta_{max}$.

In the following the first hypothesis about R, i.e. the average case, will be used. In the above frame, it is possible to evaluate the $N_s P_{av} A_e$ product and to obtain its minimum value versus the satellite height. The results are shown in Fig. 12.35, for different values of the parameter k.

It is necessary to emphasise that, although the ACR is evaluated with a diameter of the surveillance area, $D = 5300$ km, it would be possible to cover any region whose diameter is not greater than the maximum one of Fig. 12.33.

Other significant results have been obtained by varying the diameter of the surveillance area and considering the optimum configuration, i.e. the one with the minimum value of the product $N_s P_{av} A_e$ as explained before. In Figs. 12.36–12.39 the curves of the optimum $N_s P_{av} A_e$ product (optimum height, optimum number of satellites and optimum $P_{av} A_e$ product) are shown versus the diameter D of the surveillance area.

To conclude this Section, it is possible to state that the curves obtained using the above method with k between 0.5 and 1, and with the average-case hypothesis for the distance R, are representative of a large set of possible operating conditions. Since the $P_{av} A_e$ product is very important for primary radar design, and since this product is only slightly affected by the k factor and

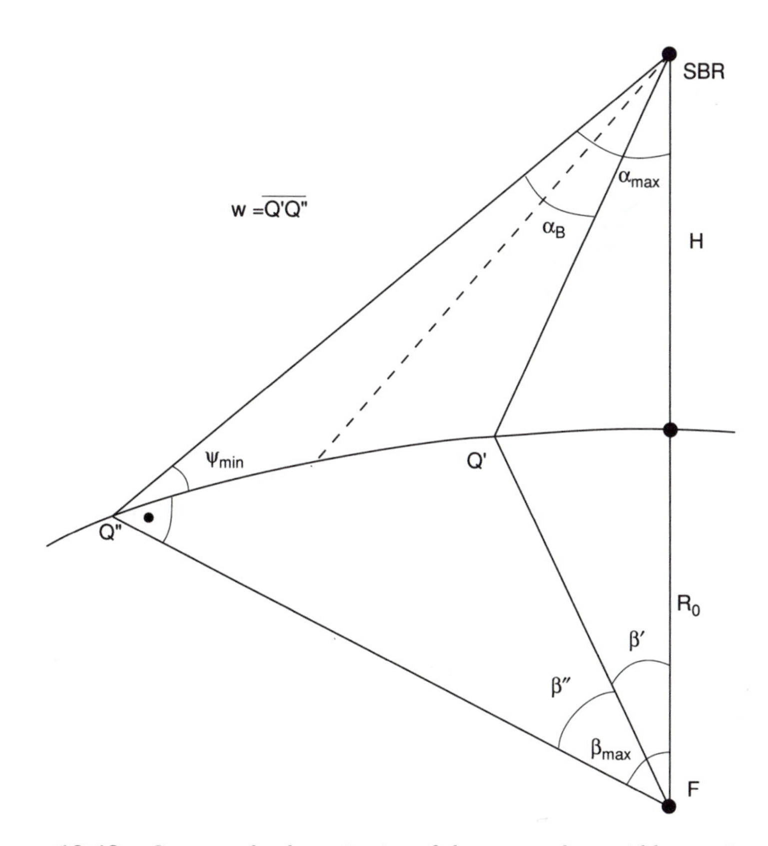

Figure 12.42 *Geometry for determination of the antenna beamwidth, cut view*

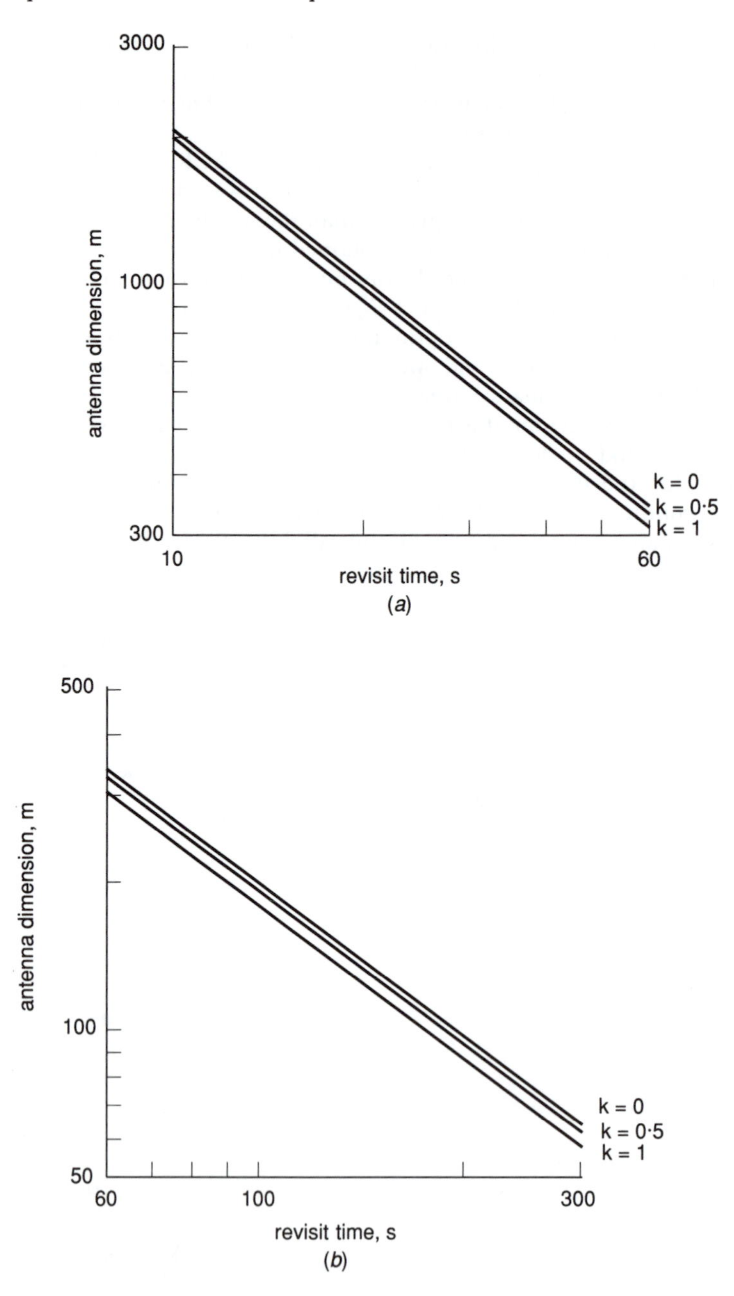

Figure 12.43 *Antenna dimension versus revisit time*

by the chosen computing method, the curve of Fig. 12.39 with $k=0.5$ is a landmark for the basic SBR design, and for this reason such a curve is also shown in Fig. 12.40 using a logarithmic scale for the y-axis.

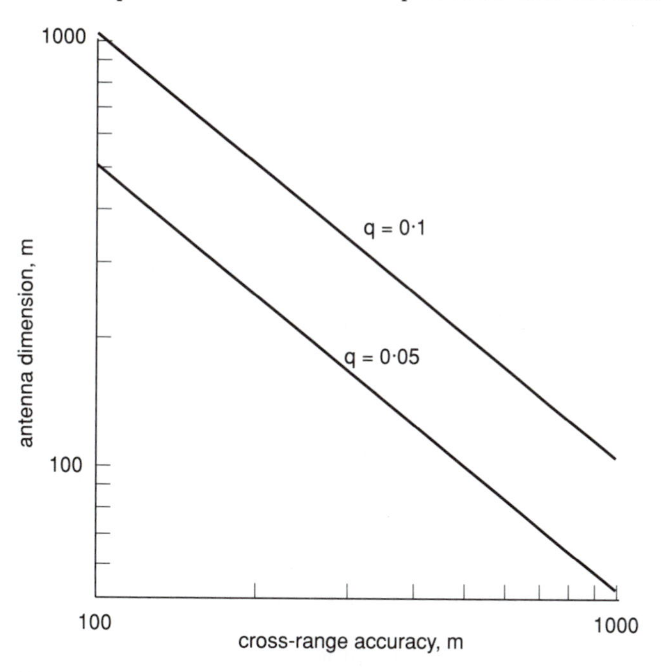

Figure 12.44 *Antenna dimension versus cross-range accuracy*

12.8.3 *Antenna dimensions*

Once the number of satellites and their optimum height have been set, the beamwidth and the dimensions of the radar antenna placed on each satellite can be determined.

Let us suppose that a region with a given (large) diameter, surrounded by a fence, has to be covered, i.e. the targets' acquisition has to be performed once they cross the fence, and that the targets are tracked inside the surveillance area. The available time is shared between the surveillance and the tracking, and most of the surveillance time has to be spent in surveillance on the fence, as explained before.

Therefore, the revisit time T_R can be defined as the time required for the surveillance of the fence surrounding the interested region. From the discussion in Section 12.6, this is related to the fence width w by the following equation:

$$w = nvT_R$$

Thus, a short revisit time, i.e. a fast surveillance system, calls for a narrow fence. Note that the power–aperture product (eqn. 12.28) is related to ACR, and ACR $= \pi Dnv$, independent of T_R; therefore the power–aperture product does not depend on w. Once the fence width has been obtained from the selected value of the revisit time, it is possible to determine the antenna beamwidth α_B in the elevation plane, assuming that the diameter of the antenna footprint is equal to the fence width (Fig. 12.41), and that the sub-satellite point is at the centre of the surveillance region.

From the geometry of Fig. 12.42, approximating the arc Q′Q″ with the line segment Q′Q″, one gets the following expression:

$$\alpha_B = \cot^{-1}\left(\frac{\cos \alpha_{max}}{\sin \alpha_{max}} + \frac{R_0 \sin \beta'}{w \sin \alpha_{max} \sin \psi_{min}}\right)$$

where β' is the angle at the centre of the Earth corresponding to the off-nadir angle $\alpha_{max} - \alpha_B$.

If a ‘square’ antenna structure is assumed (equal beamwidths in the azimuth/elevation planes), from the α_B value it is possible to obtain the (unweighted) antenna dimension L as

$$L = 0.89 \frac{\lambda}{\alpha_B}$$

In Fig. 12.43 the antenna dimension L is plotted versus the revisit time for L-band operation ($\lambda = 0.3$ m). The operating conditions of Fig. 12.43 are the same as before, i.e. optimum height and D = 5300 km, $\psi_{min} = 3°$; $\psi_{max} = 55°$.

From Fig. 12.43 it is confirmed that the faster the system, i.e. the shorter the revisit time, the larger must be the antenna. For revisit times shorter than 5 or 6 min, a distributed random array structure is required to obtain such small beamwidths.

The azimuthal accuracy and resolution is another factor calling for very large antennas. For example, the antenna azimuthal dimension L can be plotted versus the cross-range RMS error σ_A related to the azimuthal error σ_θ:

$$\sigma_A = R\sigma_\theta$$

$$\sigma_\theta = 0.89 \frac{\lambda}{L} q$$

where q depends on the integrated signal-to-noise ratio and on the kind of processing (e.g., monopulse) used. In Fig. 12.44 typical results are shown for the same operating conditions as before and for $q = 0.10$ and $= 0.05$, and for a distance of about 3800 km. With regard to the MDV parameter previously described, it is possible to show that a typical value required of 50 m/s could be accomplished with an antenna dimension equal to, or greater than, about 40 m. Since the surveillance and accuracy requirements (T_R, σ_θ) call for larger antennas (diameter ~100–200 m), the new MDV value will be much lower than the previous one, so as to improve the capability of the SBR in the detection of moving targets.

12.8.4 *Example of SBR architecture and parameters*

In this Section a comparison between a ‘monolithic’ (traditional) array, whose dimensions are 20×10 m^2 (effective area = 100 m^2), and a distributed random array whose diameter, in principle, can exceed some hundred of metres, is given.

In the former case, the selected values for the antenna dimensions cause a large revisit time (greater than 5 min as one can observe from Fig. 12.43) and a poor cross-range accuracy (greater than 1 km, as can be observed from Fig. 12.44) owing to the large distances involved.

Consider now the latter case of a distributed array with an effective area of 100 m^2: a value of 60 s for the revisit time T_R, which is typical of a fast surveillance system, can be set. With this assumption, the $P_{av}A_e$ product and the antenna dimension L can be evaluated. For a diameter of the surveillance region equal to 5300 km, we have $P_{av}A_e \simeq 5 \times 10^5$ W m^2 (from Fig. 12.40) and $L \simeq 300$ m (from Fig. 12.43).

Once the antenna dimensions have been defined, the azimuthal accuracy and the MDV can be determined. Finally, the antenna effective area A_e can be defined; for example, consider 100 array modules of $A_{e1} = 1$ m^2 each (i.e. the same effective area as in the monolithic case), from which, since $P_{av}A_e$ equals 5×10^5 W m^2, an overall average power P_{av} of about 5 kW is obtained. Such average power levels, and the related prime power levels, are of the same order of magnitude as those given in Reference 32.

In the distributed array case, instead of starting the system design from the T_R requirement, it would be possible to start from the MDV requirement, from which the parameters can be obtained.

With regard to the radar pulse repetition frequency PRF and the dwell time, it can be stated that the former parameter is chosen between a minimum value related to the MDV requirement and a maximum value related to the range ambiguities, while, if the number of pulses in the dwell time obtained from the system input is less than that desired, the latter parameter can be increased by synthesising more receiving beams by means of digital beamforming techniques.

12.9 Acknowledgment

The valuable suggestions kindly contributed by Dr. Albert W. Bridgewater on space-based radar are gratefully acknowledged.

12.10 Bibliography

1 FITCH, J.P.: 'Synthetic aperture radar' (Springer-Verlag, 1988)
2 ELACHI, C., BICKNELL, T., JORDAN, R.L., and WU, C.: 'Spaceborne synthetic-aperture imaging radars: applications, techniques and technology', *Proc. IEEE*, Oct. 1982, pp. 1174–1209
3 CIMINO, J.B., HOLT, B., and RICHARDSON, A.H.: 'The Shuttle Imaging Radar B (SIR-B) Experiment Report'. NASA, Jet Propulsion Laboratory, Pasadena, California, Mar. 1988
4 'Shuttle imaging radar: C Science Plan'. NASA Jet Propulsion Laboratory, Pasadena, California, Sept. 1986
5 MILLER, D.: 'X-SAR specification, design and performance modeling', *IEEE Trans. Geoscience and Remote Sensing*, 1989, (2), pp. 170–176
6 LIU, K.Y., and ARENS, W.E.: 'Spacecraft on-board SAR image generation for EOS-type missions', *IEEE Trans. Geoscience and Remote Sensing*, 1989, (2), pp. 184–191
7 MOORE, R.K., CLAASSEN, J.P., and LIN, Y.H.: 'Scanning spaceborne synthetic aperture radar with integrated radiometer', *IEEE Trans. Aerospace and Electronic Systems*, 1981 (3), pp. 410–420
8 CANTAFIO, L.J.: 'Space-based radar for the United Nations International Satellite Monitoring Agency', *Microwave J.* Dec. 1984, pp. 115–121
9 BROOKNER, E., and MAHONEY, T.F.: 'Derivation of a satellite radar architecture for air surveillance'. IEEE EASCON 1983, pp. 465–475

10 CLINE, J.K., and TORRETTA, D.C.: 'Aircraft track initiation with space-based radar'. IEEE EASCON 1983, pp. 123–129
11 CAIME, W.J.: 'Space-based radar application to air traffic control', IEEE International Radar Conference, Washington, 1985, pp. 312–321
12 VAISNYS, A.: 'System and technology considerations for space-based air traffic surveillance'. American Institute of Aeronautics and Astronautics, Space System Technology Conference, S. Diego, June 1986, pp. 31–36
13 STEVENS, M.C.: 'Secondary surveillance radar', (Artech House, Boston and London, 1988)
14 VAISNYS, A.: 'Surveillance satellite system study result: Vol. 1', Jet Propulsion Laboratory/California Institute of Technology, Pasadena, Ca. USA, Jan. 1985
15 DI VITO, A., GALATI, G., and PISTOIA, D.: 'Design options for space-based radar systems', International Radar Conference, Paris, 1989, pp. 586–592
16 HEIMILLER, R.C., BELYEA, J.E., and TOMLINSON, P.G.: 'Distributed array radar', *IEEE Trans. AES*, Nov. 1983, pp. 831–839
17 STEINBERG, B.D.: 'Principles of aperture and array system design', (J. Wiley & Sons, 1976)
18 DUNSMORE, M.R.B.: 'Bistatic radars', *Alta Frequenza*, 1989, **LVIII**, pp. 53–80
19 FAUBERT, D., and KERR, M.P.: 'The performance of space-based radars orbiting on elliptical orbits'. Defence Research Establishment, Report 995, Ottawa, 1988, pp. 1–23
20 BIRD, J.S., and BRIDGEWATER, A.W.: 'Performance of space based radar in the presence of earth clutter', *IEE Proc.* Part F, Aug. 1984, pp. 491–500
21 TSANDOULAS, G.N.: 'Space-based radar', *Science*, 1987, **237**, pp. 257–262
22 SKOLNIK, M.I.: 'Introduction to radar systems', (McGraw Hill, 1980, 2nd edn.)
23 ROOK, B.J., BIRD, J.S., and BRIDGEWATER, A.W.: 'Detection of near-earth targets by space-based radar: development and use of computer simulations', Communications Research Centre, Report 1389, Ottawa, 1985, pp. 1–35
24 KELLY, E.J., and TSANDOULAS, G.N.: 'A displaced phase center antenna concept for space-based radar applications'. IEEE EASCON, Sep. 1983, pp. 141–148
25 TAM, W., and FAUBERT, D.: 'Displaced phase center antenna clutter suppression in space-based radar application'. IEE/IEEE Radar Conference, London, Oct. 1987, pp. 385–389
26 TAM, W., and FAUBERT, D.: 'Improvement in the detection performance of a space-based radar using a displaced phase center antenna'. International Symposium on Antennas and Propagation, Boston, Jun. 1987, pp. 1–4
27 ARMSTRONG, C.V.W., and FATHI, E.T.: 'A fault-tolerant multimicroprocessor-based computer system for space-based signal processing', *IEEE Micro*, Dec. 1984, pp. 54–65
28 TESHIROGI, T., and CHUJO, W.: 'A multiple-access link in an inter-satellite data relay system using an on-board multibeam antenna', *IEEE Proc.* 1985, pp. 786–790
29 HARDIN, R.H.: 'System optimization techniques for radar surveillance from space,' *IEEE AES Magazine*, Apr. 1989, p. 914
30 BROOKNER, E., and HALL, W.M., and WESTLAKE, R.H.: 'Faraday loss for L-band radar and communications systems', *IEEE Trans. AES*, Jul. 1985, pp. 459–468
31 GOLDSTEIN, G.B.: 'The combined effect of Faraday and squint rotation for an electronically scanned, satellite-borne radar'. Technical Information Service, AIAA, A82–18097
32 MOODY, M.H., and MASKELL, C.A.: 'Electrical power distribution on space based radar satellites'. *IEEE AES Magazine*, Nov. 1989, pp. 10–20
33 COGHLAN, J. 'Countering the Cruise Missiles', *Defence Electronics*, Sept. 1987, pp. 77–86
34 STEINBERG, B.D.: 'Microwave imaging with large antenna arrays' (J. Wiley & Sons, 1983)
35 CHAN, K.W., and MARTIN, R.: 'Rotating reflector systems with limited electronic scan', Proc. Second Intl. Conf. on Electromagnetics in Aerospace Applications, Sept, 1991, Politecnico di Torino, Italy, pp. 263–266

Chapter 13

Evolution and future trends of radars

PROF. M. H. CARPENTIER

13.1 Emerging technologies

Most of the evolution of the radar systems in the future will be the immediate consequence of the evolution of technologies. While in the 1950s and even in the 1960s, the evolution of radar was mainly the product of finding new concepts such as the 'ideal receiver', matched filter, pulse compression, correlation, monopulse, electronic scanning, Doppler filtering, it now appears that the likelihood of the emergence of any new principle is very small.

On the other hand, the evolution of technology will allow better utilisation of known principles, or even for possible new applications of well known principles.

13.1.1 *Silicon components*

Firstly radar will take advantage of the evolution of silicon technology, i.e. evolution in the field of discrete components (Impatt silicon diodes used at millimetric wavelengths), but essentially in the field of digital VLSI.

Reduction in the geometry of silicon CMOS and/or bipolar VLSI will proceed regularly from now. At present, a geometry of 1.2 μm is common and permits, for instance, 0.5 G flops in a volume of a few tens of litres (format of 24 bits). Further reduction down to 0.8–0.9 μm could be expected within 3–4 years, allowing for more power in digital programmable signal processing, and probably followed by 0.5 μm technology in silicon VLSI to be available at the end of the 1990s. Then more than 10 G flops will be available within a few tens of litres, with a format of 32 bits, permitting new possibilities of digital computation for signal processing.

Even if we could reasonably expect that the pace of the evolution of silicon VLSI technology will decrease, geometrics of around 0.2–0.3 μm, possibly using X ray microlithography with synchrotron-like equipments, seem likely by 2010, allowing a digital signal processing power of about a few hundred G flops to be possible in a reasonable size (with format of 48 bits), and giving birth to the implementation of completely multistatic radar systems.

13.1.2 *III–V compounds*

The other essential evolution of technology in radar systems is the utilisation of so-called III–V components (AsGa, AsAlGa, AsGaInP, As Al Ga In P).

The first advantage of the III–V (gallium arsenide) components initially was the mobility of the electrical carriers, four times that of silicon. In fact, this advantage decreases when the geometry is reduced, but several other advantages appear from using such compounds, namely:

- The possibility of the Gunn effect, being used in oscillator devices
- The possibility of realising solid state laser devices (basically in the near infrared region, between 0.8 and 1.6 μm) and relevant optoelectronic devices
- The possibility of using the same material as a substrate ('semi-insulating material'), opening up the possibility of realising in monolithic form all the functions needed in microwaves [so-called microwave monolithic integrated circuits (MMIC)]
- The possibility of realising [with the assistance of a new process for epitaxy, such as molecular beam epitaxy (MBE) or metal organic chemical vapour deposit (MOCVD)] very thin successive layers of various compounds to obtain either new structures of components (such as two-dimensional-electron-gas FET, realised in 1981 by Thomson–CSF as a TEGFET, also later called HEM FET for high electron mobility FET) or artificial crystals with better performance.

III.V components will offer new possibilities in:

- Generation of relatively high power levels in centimetric/millimetric wavelengths (X-, K- and W-band)
- Generation of power at micrometre wavelengths
- Low-noise microwave amplifiers
- Realisation of controlled phase shifters and controlled delay shifters
- Realisation of high performance analogue/digital converters, strictly necessary before using silicon high performance programmable digital signal processing devices.

III–V analogue/digital converters will, at the beginning of the 1990s, probably provide the possibility of encoding at a speed of 1 GHz with an accuracy of 8 bits (or at a speed of 0.1 GHz with an accuracy of 15 bits); and progress will, slowly but surely, improve that level of performance, possibly up to a speed of 0.5 GHz with an accuracy of 15 bits by the end of the 1990s (1 GHz, 13 bits). That type of performance will complete the silicon signal processing possibilities, giving the necessary input to the very high performance digital programmable signal processing.

The other essential field of application of III–V compounds, and particularly MMIC's, will be the realisation, now beginning at a slow rate but expected to grow at an increasing rate in the 1990s, of the so-called active phased-array antennas.

That type of antenna is composed of transmitter–receiver modules approximately every half-wavelength, each T–R module having two functions, one of transmission and one of reception:

- Transmission consisting, from a very low level frequency/phase reference, of a phase (delay) shifter (digitally controlled) followed by a power amplifier immediately feeding an elementary antenna

- Reception (after microwave duplexing) consisting of a microwave (low noise) amplifier, followed by a protecting limiter and a phase (delay) shifter

Many advantages will ensue from active phased (delay) array antennas:

- Reduction of losses to nearly zero on transmission (recovering 1.5 to 3.5 dB), and similarly on reception
- Reliability improved by redundancy and by voltage reduction
- Possibility of transmission of more sophisticated signals (amplitude modulation), which are difficult to reproduce for spoofing
- Possibility of more flexible installation, opening up the possibility of better utilisation of carriers and, as a result, of better angular resolution
- Possibility of sampling, about every half-wavelength, the electromagnetic field along the receiving antenna, permitting better angular resolution.
- Voltage reduction.

13.1.3 *Microwave tubes*

Even if, in general, the progress in microwave tubes will not be as spectacular as in solid state, three main evolutions in microwave tubes will have significant effect on radar systems:

- Improvement of efficiency of 'O' type tubes (TWT or klystrons) by better recovery of accelerated electrons (multistage electron collectors) and by reduction of peak power (increasing the duty cycle)
- Standardisation of controlling grid in 'O' type tubes
- Development of fast-wave tubes (gyrotrons and equivalent).

Present fast-wave tubes, which could be considered to lie between the lasers and conventional travelling-wave tubes, are at present mainly used in scientific experiments (for nuclear fusion), where they are expected to deliver CW powers of some hundreds kilowatts at 100–150 GHz.

The present gyrotrons are only used as oscillators, but nothing prevents this type of technology from providing amplifying tubes, and their properties facilitate the production of a high level of (CW) power at very high frequency (millimetric).

13.1.4 *Miscellaneous*

Progress made in capacitors (possibility of storing more joules per litre) is relevant to the progress in TWTs equipped with a controlling grid, and helps to reduce the difficulties (specifically with regard to the size) encountered when using the 'hard-tube' type modulator, which come from the size of the tank capacitor.

Progress has been made in generating low-level very-high-frequency/phase-stable signals, allowing for the possibility of very accurate Doppler measurements.

II–VI compounds are also following a slow but steady pace of improvements, opening up more possibility of their use in the 3–11 μm infrared zone for sensing, detection, and possibly for transmission.

13.2 High resolution in range and in angle

The evolution of components, for transmission and reception, will open up the possibility of radars transmitting very wide instantaneous bandwidths (a few hundreds of megahertz). Two main advantages are expected from the very good associated range resolution (around 1 m):

- Some identification of the target: an account of the number of echoes in a small given zone, rough evaluation of the size of the target, classification of the target, etc.
- Some possibility of the association of Doppler returns with range: presently many targets deliver several Doppler responses for the same target (coming from various mobile parts of the target) giving saturation in signal processing and/or in data processing. Very good range resolution will permit the identification of the Doppler response by allowing for an association with a particular part of the target, thus facilitating the relevant data handling.

For a long time it has been stated that there was no limit to angular resolution, provided

- Signal-to-noise ratio is high enough
- Relevant computation is achieved

In fact, the signal-to-noise ratio is very high in two very interesting cases: when the 'signal' is a signal transmitted by a jammer (by nature) or when the signal is the skin return from a target, but long time after detection (this is very often the case when one really needs an accurate angular measurement which takes place a long time after detection, at a time when the target is much closer to the radar, and providing a fairly 'high signal').

If the signal-to-noise ratio is very high it is possible, from analysis of the varation of the (electromagnetic) field along the antenna, to know more about the 'targets' which are producing that field.

Measurements realised up to now on classical antenna structures have been very simple: addition of all vector fields ('sum' signal), addition of all vector fields for the right part subtracted from the equivalent sum for the left part ('difference' signal of monopulse radars), slightly more sophisticated measurement ('deviation' channel – ('voie écart') by S. Drabowitch, where the sum of the central vector fields is subtracted from the sum of lateral vector fields). In fact these were simple and not very clever since it was, in practice, assumed that the (electromagnetic) field was zero outside the antenna (which is the only impossible assumption).

If complete sampling of the electromagnetic field is made all along the antenna (every half-wavelength), because an active phase-array structure is used – not too much perturbed by noise – realistic complementary assumptions could be made to derive realistic information on an angular target situation. It could be assumed that useful targets are point targets in limited number and/or it could be assumed that targets (jammers) are transmitting independent signals during most of the time. These are independent of the useful signal (skin return of 'useful' targets) and/or that the electromagnetic field outside the antenna aperture is more likely to be accorded 'maximum entropy' than to be zero.

All these assumptions, which could be combined, and which depend on the situation, permit one to obtain reasonable angular information on the targets; much more than can be expected from the so-called 'angular resolution power', provided a complicated computation can be achieved, which will soon be possible owing to the rapid increase in the power of digital programmable signal processing.

13.3 Use of new frequencies

The evolution of technology is beginning to fill the gap between centimetric wavelengths and the infrared.

Even if atmospheric attenuation is troublesome when using millimetric wavelengths, they still have the advantage of facilitating angular measurements and atmospheric attenuation is not significant outside the atmosphere or at short ranges. 90 GHz radar systems are being used more and more, thanks to semi-conductor evolution, and 140 GHz is beginning to follow. Consideration is also being given to the use of powerful gyrotrons in the above frequency bands in the future.

An important advantage of using higher frequencies becomes clear when considering the use of micrometric (infra-red) wavelengths in radars (LIDARS). The use of these very short wavelengths will permit a very accurate radial velocity measurement (provided 'coherent' micrometric radar is used). This allows, for instance, the detection of the motion of the ailerons of an aircraft, indicating that the aircraft *will* turn, when, at present, radars are only able to verify that the aircraft has turned, a few seconds later.

On the other hand, there would be advantages in using low frequencies between HF and UHF in radar systems, since the radar cross-section of targets is very difficult to reduce at those frequencies (stealth techniques are not very efficient at metric wavelengths). Unfortunately, lack of angular resolution and reduced frequency availability generally prevents us presently using those frequencies.

13.4 Advanced system concepts

The situation could change when more progress in signal processing has been achieved (0.1 to 1 T Flops, 48 bits), because it will then be possible to use a fully multistatic arrangement of transmitter–receiver antennas, spread all over a territory of thousands of square kilometres:

- With the capability of processing in a coherent manner all the signals received by all the antennas, which need a high dynamic range together with complex computation.

Obvious advantages will come from that architecture such as

- Possibility of good angular resolution, owing to the size of the complete antenna arrangement, even if the frequency is low

- Reduced vulnerability of such an architecture to physical destruction (or failure)
- Possibility of transmitting a very sophisticated modulated signal, preventing copying by an active spoofer

Bistatic arrangements will be used before that 'final solution', in which one transmitter will be associated with two receivers, one at least being far from it (the transmitter, which is easier to be detected, could be further from the 'enemy' target).

In any case the evolution of the radar systems is clearly proceeding towards a situation in which

- Electronic scanning (passive or active) will be able to provide users with an enormous number of different antenna patterns
- Transmission (from one microwave tube or many solid-state amplifiers) will be able to have many different modes (especially in the case of an active array)
- Signal processing will be fully programmable, providing the user with an enormous quantity of potential processing

The total number of possible configurations will be 'more than infinite'. In an actual situation, some configurations would be optimum, others would be excellent, others fairly good, etc.

How will the computer operate, which will manage the radar system? In general, if it tries to find the optimum configuration, that will take a very long time. It is clear that generally its task will be to find a fairly good solution (possibly an excellent one) but in a limited (and short) time rather than finding a better solution, but taking too long to do so.

This gives an indication of how the methodologies of 'artificial intelligence' will find a place in the management of future radar systems (i.e. expert systems).

Index